Physiology

PHYSIOLOGY

Third Edition

By 11 Authors

Edited by Ewald E. Selkurt, Ph.D.

Professor and Chairman, Department of Physiology,
Indiana University School of Medicine,
Indianapolis

Little, Brown and Company

Boston

In memory of Dr. Robert Winslow Bullard,
colleague, collaborator, and friend

Library of Congress catalog card No. 73-152434

Third Edition

Fifth Printing

ISBN 0-316-78049 (C)
0-316-78047 (P)

Printed in the United States of America

Contributing Authors

THE CONTRIBUTORS to *Physiology,* with the exception of Drs. R. W. Bullard and P. C. Johnson, are all members of the Department of Physiology, Indiana University School of Medicine, Indianapolis.

EWALD E. SELKURT, Ph.D., Professor and Chairman; Editor

WILLIAM McD. ARMSTRONG, Ph.D., Professor

ROBERT W. BULLARD,* Ph.D., Professor

JULIUS J. FRIEDMAN, Ph.D., Professor

KALMAN GREENSPAN,† Ph.D., Professor

PAUL C. JOHNSON,‡ Ph.D., Professor

LEON K. KNOEBEL, Ph.D., Professor

THOMAS C. LLOYD, JR., M.D., Professor

WARD W. MOORE, Ph.D., Professor

SIDNEY OCHS, Ph.D., Professor

CARL F. ROTHE, Ph.D., Professor

---

*Department of Anatomy and Physiology, Indiana University, Bloomington. Dr. Bullard was killed in an accident on Mt. McKinley in June, 1971, while engaged in research into high-altitude physiology.

†Joint Appointment in Department of Medicine.

‡Professor and Chairman, Department of Physiology, University of Arizona College of Medicine, Tucson.

Preface

THE THIRD EDITION of this text has adhered to the format of a compact, yet comprehensive handling of facts essential to a good foundation in mammalian physiology particularly suited to the needs of professional courses in medicine, dentistry, and the allied health professions. Development of so-called core curricula has made this style of text particularly useful.

Accordingly, established facts are emphasized, and complementary subject matter taught by other basic science disciplines – for example, anatomy and biochemistry – has been limited to a review of pertinent essentials. Although basic principles are stressed, sufficient clinical application has been included to provide a foundation for subsequent development of the subject matter according to clinical department. We are indebted to our clinical colleagues for many helpful suggestions in this direction.

On the other hand, it is believed that the text is firmly grounded on basic principles established by the disciplines of biology, chemistry, physics, mathematics, and biophysics. For this reason, it has met the needs of advanced undergraduate college courses and graduate courses, particularly in the biological sciences. It is a useful foundation text for emerging disciplines in biomedical engineering.

We are indebted to our secretary, Mrs. Juanita Ward, and to Mr. James Glore and his staff of the Department of Medical Illustrations of Indiana University School of Medicine for invaluable services.

E. E. S.

Indianapolis

Contents

Physiology

1

The Cell Membrane and Biological Transport

William McD. Armstrong

CELLS are the smallest and simplest organized units in which integrated physiological function can be recognized. The various tissues of which the body is composed are made up of different kinds of cells. Thus the physiological activity of the body as a whole and of its various parts depends, in large measure, on the proper functioning of its component cells.

Living cells are complex and highly organized. In addition to a fluid phase (cytoplasm) and a nucleus, the interior of a typical cell contains a number of *inclusions* or *organelles* (nucleolus, mitochondria, lysosomes, etc.). Modern research has shown that these subcellular particles are themselves highly organized structures which play a vital role in the overall activity of the cell. For present purposes, however, we shall be concerned mainly with the exchanges of material that take place between the cell as a whole and its external environment.

Cytoplasm or intracellular fluid differs markedly in composition from the external medium (blood or interstitial fluid) bathing the cell. In intact, normally functioning cells some of these differences in composition remain virtually constant over long periods of time. At the same time, there is a continuous exchange of material between the cell interior and the external environment, with a variety of substances entering the cell from the outside, and other substances leaving it. Evidently cells must possess a mechanism or mechanisms by which the entry and exit of dissolved solutes are regulated. Two alternative kinds of mechanism have been postulated. One invokes the idea that the cytoplasm as a whole is capable of selectively retaining certain substances in the cell interior while effectively excluding others. The other assigns the control of the transfer of materials between the cell and its environment to a special region at or near the periphery of the cell, the *cell membrane* or *plasma membrane.* At present, although the idea of functional selectivity by the cytoplasm as a whole is vigorously championed by some workers (Ling, 1962), most physiologists accept the special role of the membrane in the regulation of cellular composition as the most satisfactory interpretation of the available experimental evidence. Therefore, in the present chapter, cellular transport mechanisms will be discussed from this point of view.

STRUCTURE OF THE CELL MEMBRANE

The idea that the cell membrane is a discrete entity, structurally as well as functionally distinguishable from the bulk of cellular material, originated mainly from two lines of evidence: permeability studies and electrical studies.

PERMEABILITY STUDIES

When systematic investigation of the penetration of cells by a variety of dissolved substances began toward the end of the last century, it was quickly realized that living cells are remarkably selective toward substances dissolved in the extracellular fluid. Some substances were found to enter cells readily, while others reached the cell interior very slowly if at all. A large number of studies using many different substances, both nonelectrolytes and electrolytes, resulted in two important generalizations which served as a basis for subsequent theories of membrane structure. These generalizations were arrived at from a comparison of the permeability of cells to polar and nonpolar molecules, respectively.

First, it was observed that charged ions or highly polar molecules, unless they are very small, penetrate cells with difficulty if at all. Nonpolar molecules, on the other hand, penetrate most cells with comparative ease. These findings are conveniently summarized in terms of the *partition coefficient* or *distribution coefficient* of the dissolved substance. This is defined by the *partition* or *distribution law,* which states that the distribution of a solute between two immiscible or partly immiscible solvents is given by the equation.

$$K = S_1/S_2 \tag{1}$$

where K is the partition coefficient and S_1 and S_2 are the equilibrium concentrations of solute in the two liquid phases. The ease with which a nonpolar compound penetrates the cell membrane has been found to be correlated with its partition coefficient between typical fat solvents (e.g., ether, chloroform, benzene) and water, a quantity which has been termed the *lipid solubility* of the compound in question. High lipid solubility is associated with high permeability, and vice versa.

The factor of molecular size also plays a part in determining the permeability of cells to nonpolar molecules. In general, however, it is less important than lipid solubility, though the relative importance of these two factors may vary from cell to cell. This point is clearly brought out by comparing the rates of penetration of a homologous series of organic compounds such as the straight-chain monohydric alcohols. Here it is found, with the lower members of the series at least, that permeability increases rapidly with molecular size. The explanation of this is as follows. The straight-chain monohydric alcohols consist of a single polar group (–OH) attached to a hydrocarbon residue. The overall properties of the molecule are the result of a kind of "balance" between the polar (hydrophilic) properties of the hydroxyl group and the nonpolar (lipophilic) properties of the hydrocarbon moiety. Increasing the size of the latter "tips the balance" in the direction of increasingly lipophilic properties on the part of the molecule as a whole, with a consequent increase in lipid solubility and permeability.*

The role of lipid solubility in cell penetration is further underlined by the behavior of weak electrolytes, i.e., electrolytes which undergo *reversible dissociation* of the type

$$HA \rightleftharpoons H^+ + A^- \tag{2}$$

within the physiological pH range. It is found that whereas weak electrolytes (e.g., acetic acid) penetrate cells readily in the *undissociated* state, they fail to do so when

*As one goes up the series in the direction of increasing lipid solubility, the apparent permeability may reach a maximum beyond which it begins to decrease. This may in part reflect the influence of increasing molecular size, but it is more probably due to the fact that the water solubility decreases rapidly as the size of the hydrocarbon moiety is increased, so that eventually the compound in question can be presented to the cell in minute concentrations only.

dissociated. This is in keeping with the fact that the lipid solvent/water partition coefficients of these compounds are much greater in the undissociated than in the dissociated state.*

ELECTRICAL MEASUREMENTS

The obvious importance of lipid solubility in determining the permeability characteristics of organic molecules prompted the suggestion that the cell membrane is predominantly lipid in nature. This conclusion was reinforced by electrical studies. The electrical impedance of a crushed or injured cell, in which the membrane has been wholly or partially destroyed, is low, being comparable to that of the extracellular fluid or to that of a salt solution of similar electrolyte composition. In contrast, the impedance of intact cells is many times higher, suggesting the presence of poorly conducting lipid material in the membrane. The development in recent years of the microelectrode technique (see p. 18) has made possible the direct measurement of the electrical impedance across the membrane of several types of cells. Such measurement has confirmed the earlier idea that the resistance to current flow across intact cells resides principally in the region of the membrane.

MEMBRANE STRUCTURE

There is thus a very substantial body of evidence to support the idea that the cell membrane consists primarily of a layer of lipid. On this basis the permeability of dissolved solutes should be governed mainly by the relative ease with which they can be taken up by and pass through this lipid layer. This, as we have seen, offers a satisfactory explanation, as a first approximation at least, for the known permeability properties of many compounds. There are, however, certain notable exceptions. Reference has already been made to the fact that ions and highly polar molecules, if sufficiently small, readily move across the membranes of most cells. Prominent examples among substances normally occurring in the body are urea and the monovalent ions K^+ and Cl^-. These substances have a negligibly small solubility in lipid solvents and so do not fit easily into the concept of the membrane as a lipid layer. These facts gave rise to a second theory of membrane structure, the so-called *pore theory.* According to pore theory the membrane contains aqueous pores or channels. Molecules or ions which are small enough to pass through the pores can thus enter or leave the cell freely. Larger molecules or ions cannot. In other words, the membrane acts as a molecular sieve, allowing some substances to pass through it while preventing others from doing so.

This theory at once raises the question of the possible role of proteins in membrane structure. The existence of pores or channels in a predominantly protein structure is more easily visualized than the existence of such pores in an oil-like layer of lipid. Proteins are ubiquitous throughout the cell, and it would be rather surprising not to find them in the membrane. In any event, direct experimental evidence supports the idea that cell membranes contain proteins. First, chemical analysis of membrane fractions obtained from cells has revealed that they contain large amounts of lipids (mainly cholesterol and phospholipids) and proteins (Stein, 1967). By weight, the amount of protein is about one to two times the amount of lipid though this proportion is variable for different kinds of membranes; e.g., mito-

---

*A much fuller discussion of the role of lipid solubility, molecular size, and dissociation in the permeability of nonelectrolytes and weak electrolytes is given in Davson (1964), Chapter 8.

chondrial membranes are relatively rich in protein.* Second, cells have been found to exhibit a low surface or interfacial tension when in contact with aqueous solutions. Since a lipid membrane would be expected to have a high surface tension under these conditions, these observations suggest the presence in the membrane of highly surface-active material such as protein. Finally, it may be noted that chemical agents which interact specifically with proteins have been found to destroy the selective permeability of the membrane.

Given the presence of both lipid and protein in cell membranes, one may ask how these substances are organized to produce a structure possessing the permeability and electrical properties outlined above. An important contribution to this question was the work of Gorter and Grendel (Davson, 1964) which utilized the surface film technique. If lipid molecules such as long-chain fatty acids and their esters, cholesterol, or phospholipids containing one or more polar groups, together with a relatively large hydrocarbon residue, are spread at the phase boundary between a polar and a nonpolar substance (such as water and air or water and oil), they orient themselves so that their polar groups are in contact with the polar phase while their nonpolar portions are oriented toward the nonpolar phase. Under suitable conditions such molecules can be spread as a layer one molecule thick *(monolayer)* at an air/water or oil/water interface. The orientation of a lipid monolayer at an air/water interface is shown in schematic form in Figure 1-1A. The circles represent the polar groups of the lipid, and the straight lines represent the hydrocarbon portion of the molecules. The area occupied by the individual molecules will depend on a number of factors, including the size and position of the polar groups, the structure of the hydrocarbon residues, and the degree of lateral adhesion between them (Davson, 1964).

Gorter and Grendel extracted the lipids from red blood cell ghosts (see p. 12) and spread them as a monolayer. From the area of this monolayer they calculated that the lipid content of the red blood cell membrane was just enough to cover the surface of the cell if it were arranged as a bimolecular leaflet (Fig. 1-1B). They suggested that a bimolecular lipid layer of this type forms an integral part of the red blood cell membrane.

Such an arrangement satisfies several of the properties of cell membranes already mentioned. For example, it accounts satisfactorily for the correlation observed between ease of penetration and lipid solubility. If the principal permeability barrier at the surface of the cell is a layer of lipid of the type illustrated in Figure 1-1B, it is evident that high lipid solubility should greatly facilitate the entry of a substance into the cell. In addition, the model illustrated represents a stable configuration, on a physicochemical basis, for polar-nonpolar molecules such as lipids to assume at an interface between two polar phases. It allows the hydrophilic groups of the lipid molecules to "anchor" themselves in the aqueous phase on either side of the bimolecular layer while at the same time permitting mutual interactions to occur between the hydrocarbon residues. These cohesive forces between the hydrocarbon portions of the lipid molecules help to give overall stability to the structure.

On the other hand, the model shown in Figure 1-1B ignores the structural role of protein in the membrane and would be expected, contrary to what is found with living cells, to exhibit a relatively high surface tension against an aqueous solution. The model shown schematically in cross section in Figure 1-1C meets these objections. In this model, originally proposed by J. F. Danielli (Davson, 1964), the bimolecular leaflet of lipid is sandwiched between two layers of adsorbed protein, adsorption being accomplished by a considerable degree of unfolding of the poly-

---

*Other macromolecules, e.g., polysaccharides and nucleic acids, occur in relatively small amounts only in cell membranes. However, mucopolysaccharides play a major role in certain properties of the cell surface, e.g., immunological responses (Stein, 1967).

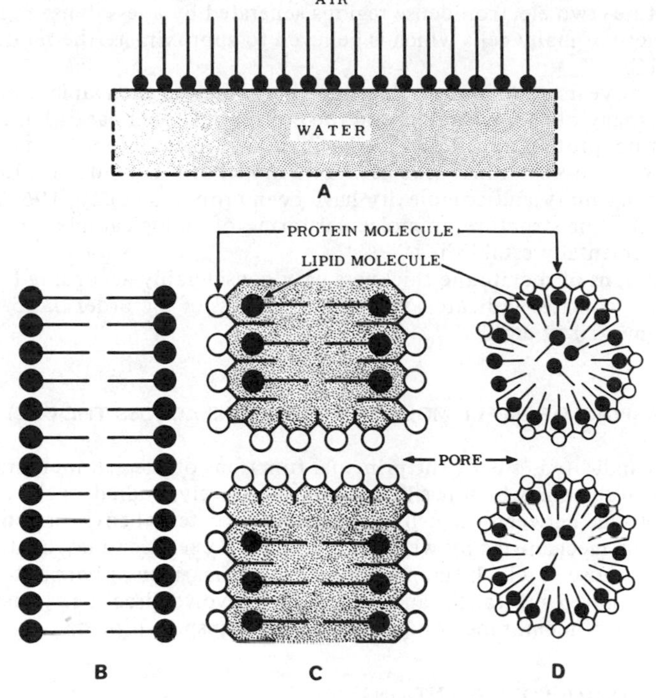

FIGURE 1-1. A. Orientation of monolayer of polar-nonpolar molecules at an air/water interface. B. Suggested orientation of bilayer of lipid molecules in cell membrane. C. Model of cell membrane proposed by J. F. Danielli. D. "Micellar" model of cell membrane.

peptide chains of the protein molecules. The adsorbed protein may be considered to form a fairly open mesh which does not offer much resistance to the passage of small molecules or ions. Thus, over most of the surface of the cell, the lipid layer can still be regarded as the principal permeability barrier to many substances.

Figure 1-1C also shows a possible structure for aqueous pores in the membrane. These are thought to arise from discontinuities in the lipid matrix and to possess a lining of protein which gives them their hydrophilic character. The permeability of cells to small ions and polar molecules is readily explained by the existence of such pores in the membrane.*

Figure 1-1C is clearly an oversimplification. Living membranes are of course much more complex and labile than is suggested by this representation. They are also dynamic structures in which a variety of metabolic and other activities take place continuously. Nevertheless, the Danielli model has been a very useful and successful tool in membrane physiology and is still accepted by many physiologists as essentially correct in broad outline, although opinions differ regarding finer

*Recent work, however, has led to the concept that ion transport across cell membranes is frequently mediated by interactions within the membrane which result in the transfer of ions as part of a hydrophobic (lipophilic) complex (see the discussion of ion transport in this chapter, and Pressman and Haynes, 1969).

details. It is of interest to note that electron microscopy has revealed the existence of a structure (two electron-dense regions separated by a less dense region) at or near the periphery of many cells which is believed to approximate the model shown in Figure 1-1C.

In recent years a number of plausible alternatives to the Danielli model have been proposed, many of them built around the concept of spherical or lamellar lipid/protein or lipoprotein micelles as the basic units of membrane structure. Figure 1-1D shows a cross-sectional schematic of a model of this kind. A number of models of varying ingenuity and complexity have been proposed (Lucy, 1968; Hendler, 1971). However, the fine structure, in molecular terms, of biological membranes has not yet been experimentally established.

Estimations of membrane thickness vary considerably according to the technique used. Most estimates indicate a probable thickness of the order of 75 to 150 angstroms (1 angstrom = 10^{-8} cm).

THE TRANSPORT OF SMALL MOLECULES AND IONS ACROSS THE CELL MEMBRANE

As already indicated, one of the principal functions of the cell membrane, and without question the one which has been most extensively studied, is the control of solute exchange between the cell interior and the external environment. Broadly speaking, the mechanisms by which this control is exercised may be divided into two classes, those in which the membrane acts primarily as a barrier to the free diffusion of dissolved substances and those which involve direct participation of the membrane or of membrane components in the transport process.

DIFFUSION IN AQUEOUS SOLUTIONS

Diffusion (Davson, 1964) is the tendency for a solute to distribute itself spontaneously and uniformly throughout the whole space available to it. If a solute is present initially in a higher concentration in one part of a solution than in another, it is found that in time this concentration difference disappears, the solute concentration becoming uniform throughout the solution. As long as a concentration difference exists between one part of a solution and another, the spontaneous tendency for net solute movement from the region of higher to that of lower concentration persists. When the concentration difference or concentration gradient disappears, there is no further tendency for net solute movement to occur between different regions of the solution, which then is said to have reached a state of equilibrium with respect to the solute.

The rate of solute diffusion (V) between two regions of a solution which contain unequal concentrations of solute is obviously equal to $-dS/dt$, the amount of solute passing from the region of higher to that of lower solute concentration in unit time. At constant temperature, this quantity is given by Fick's equation:

$$V = -\frac{dS}{dt} = DA\,(S_1 - S_2) \tag{3}$$

In this equation S_1 and S_2 are the solute concentrations in the regions of higher and lower concentration respectively, A is the cross-sectional area of the boundary between these regions (i.e., the area across which diffusion is taking place), and D is the *diffusion coefficient* or *diffusivity* of the solute. It is apparent from equation 3 that D is numerically equal to the rate of diffusion across unit area (1 cm^2) when the concentration difference ($S_1 - S_2$) across the boundary layer is unity.

Thus D is a measure of the inherent ability of the solute molecules to move through the solution. It is also clear from equation 3 that the net rate of diffusion at constant temperature depends on the concentration gradient ($S_1 - S_2$) and on the magnitude of D. D has been found to depend on temperature, becoming larger as the temperature is increased. It also varies with the molecular weight of the dissolved substance, becoming smaller as the molecular weight increases. D likewise depends to some extent on molecular shape as well as on molecular size. With large molecules in particular, the extent and strength of their interactions with other molecules, and hence the overall resistance to their diffusion in aqueous solution, are governed by shape as well as size. In these circumstances, the relationship between D and molecular weight can be rather complex.

WATER MOVEMENT – OSMOSIS AND OSMOTIC PRESSURE

The foregoing discussion of solute diffusion relates to free diffusion in aqueous solutions. In other words, it is concerned with systems in which no restraints to diffusion exist other than those imposed by the inherent resistance, within the solution itself, to the free movement of solute. Situations of particular interest to the physiologist arise in systems in which free diffusion is restricted in various ways, since in many cases the passage of solutes across the cell membrane can be described adequately by assuming that the membrane acts as a simple restraint to free diffusion (see p. 14).

A special case of restricted diffusion arises in the presence of a membrane which permits water to move across it but is completely impermeable to dissolved substances *(semipermeable membrane).* Consider, for example, a vessel divided into two compartments by such a membrane. If the compartment on one side of the membrane contains pure water and the other compartment contains an aqueous solution, water will flow spontaneously from the side containing pure water to the side containing the solution. This spontaneous net flow of water is called *osmosis* or *osmotic flow.* Its origin may be explained as follows. All natural processes tend to proceed spontaneously in the direction of equilibrium. In the system under consideration there are, initially, two concentration gradients across the membrane: a solute concentration gradient from the side containing the solution to the side containing pure water, and (because of the diluting effect of the solute on the water in the side containing the solution) a concentration gradient for water in the opposite direction. In the absence of external restraints both water and solute would diffuse freely in the direction of their respective concentration gradients until mixing was complete. Because of the restraint imposed on the system by the semipermeable membrane, the solute cannot diffuse into the side containing pure water. The only net movement of material which can take place across the membrane is a flow of water into the side containing the solution. If, instead of pure water and a solution, two solutions containing unequal concentrations of solutes were used, a similar osmotic flow of water would take place. In this case osmosis would occur from the more dilute to the more concentrated solution.

In systems such as these it is clear that, unless some additional restraint is imposed, water will continue to move in the direction of its concentration gradient as long as the gradient exists, i.e., until the concentrations of water on both sides of the membrane become equal. In the case of two solutions this would be achieved when the solute concentrations on both sides of the membrane became equal. In a system containing pure water and an aqueous solution, osmotic equilibrium cannot, in principle at least, be realized in this way at any finite solute concentration. However, it is evident that if a force or pressure equal and opposite to the force generated by the concentration gradient for water across the membrane could be applied to the side containing the solution, osmotic flow could be prevented. The pressure required to prevent osmotic flow of water into a given solution is called the *osmotic pressure* of that solution.

There are many ways in which osmotic flow can be prevented by the application of an external force. One of the simplest, which also permits the osmotic pressure to be determined directly, is to utilize the force of gravity. A schematic illustration of a device, called an *osmometer,* which makes use of this principle is shown in Figure 1-2. In this apparatus the solution whose osmotic pressure is to be measured

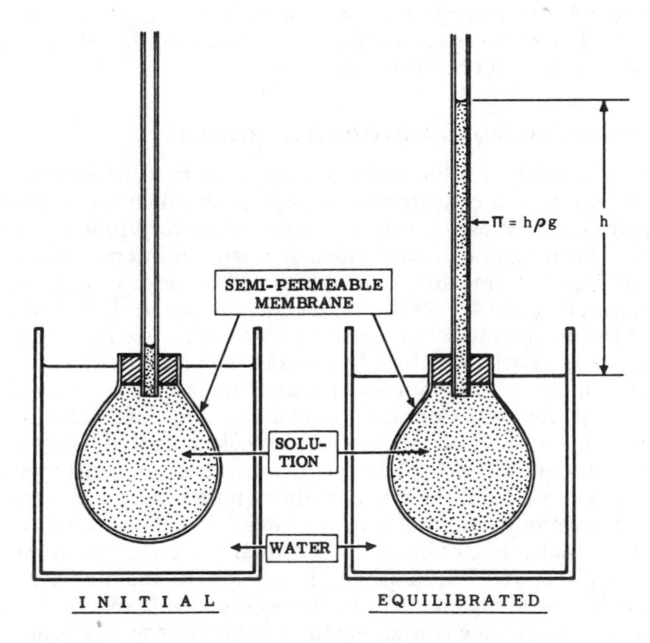

FIGURE 1-2. A simple osmometer demonstrating the osmotic pressure developed when a solute is present on one side only of a semipermeable membrane.

is placed inside a thin semipermeable bag which is then immersed in a vessel containing water. A fine capillary tube open at the top is inserted into the bag. Initially the apparatus is adjusted so that the liquid levels in the capillary and in the outer vessel, apart from the slight rise due to capillary action, are the same. As water enters the bag because of osmosis, the level of the liquid in the capillary rises until the hydrostatic pressure developed is just sufficient to balance the osmotic driving force across the wall of the bag. If the volume of the column of solution in the capillary is very small compared to the total volume enclosed in the bag (so that the total amount of water which enters the bag and its diluting effect on the solution within the bag are negligible), the osmotic pressure of the original solution is given by

$$\pi = h \rho g \tag{4}$$

where π is the osmotic pressure, h is the height of the column of solution in the capillary necessary to balance the osmotic driving force, ρ is the density of the solution, and g is the acceleration due to gravity. For moderately dilute solutions

ρ does not differ significantly from the density of water; hence the height in centimeters of the column in the capillary will give the osmotic pressure directly in centimeters of water. Since 1034 cm water is equivalent to 76 cm mercury or 1 standard atmosphere, π is readily obtained in either of these units.

OSMOTIC PRESSURE AND SOLUTE CONCENTRATION: UNITS OF OSMOTIC CONCENTRATION

It is clear from the preceding discussion that the magnitude of the osmotic pressure in any given solution depends only on the difference between the concentration of water in that solution and its concentration in the pure liquid. This dependence, however, is of little practical use in physiology. Useful and meaningful relationships can be obtained if osmotic pressure is expressed in terms of solute concentration. Fortunately, despite the fact that the solute, aside from its diluting effect on the concentration of water in the solution, is not a fundamental factor in the generation of osmotic pressure, a simple relationship between osmotic pressure and solute concentration does exist. This is because, in moderately dilute solutions, there is a simple complementary relationship between the concentration of water and that of solute.

Consider a solution containing n_1 moles of water and n_2 moles of solute per unit volume. The *molar fractions* of water and solute, respectively, are $X_1 = n_1/(n_1 + n_2)$ and $X_2 = n_2/(n_1 + n_2)$. Hence, $(X_1 + X_2) = 1$, and $X_2 = (1 - X_1)$. Evidently it is the factor $(1 - X_1)$ which determines the osmotic pressure of a solution in contact with pure water. Therefore, the osmotic pressure is directly proportional to X_2 or solute concentration. For a dilute solution the quantitative relationship between osmotic pressure and solute concentration is given by the van't Hoff equation

$$\pi = CRT \tag{5}$$

where C is the solute concentration, R is the gas constant, and T is the absolute temperature.

It is apparent from equation 5 that if the concentration of solute in a given solution is known, its osmotic pressure can readily be calculated. Conversely, given the osmotic pressure, the solute concentration can be obtained from this equation. It is important to realize that there is a difference between concentrations as they relate to osmotic activity and ordinary chemical concentrations. Since, as has already been pointed out, the solute has no intrinsic effect on osmotic pressure, the osmotic pressure of a solution is independent of the nature of the solute particles and depends only on their number per unit volume. In other words, in a given volume of solution, equal numbers of dissolved particles will contribute equally to osmotic pressure whether the particles are large molecules, small molecules, or ions. For this reason the osmotic activities of solutions containing equal *chemical* concentrations of different solutes will not necessarily be identical. Consider, for example, two solutions, one containing 0.1 M sucrose and the other containing 0.1 M NaCl. Although the concentrations of these two solutions are equal in chemical terms (moles per liter), their osmotic pressures are not the same because NaCl exists in solution as Na^+ and Cl^- ions. Consequently the osmotic pressure of a 0.1 M solution of NaCl is approximately twice that of a 0.1 M sucrose solution. Thus it is apparent that, with electrolyte solutions, if one wishes to relate the osmotic pressure or effective osmotic concentration of solute to its chemical concentration, he must multiply the term C in equation 5 by a factor G (the osmotic coefficient), where G is the number of ions produced by 1 molecule of electrolyte.

The situation is further complicated by the fact that this simple relationship between the number of ions formed by an electrolyte and its osmotic activity holds only for very dilute solutions. Because of the attractive forces between ions of opposite charge and between individual ions and water molecules, the value of G for a given electrolyte varies with concentration. Also, the concentration dependence of G is different for different electrolytes. At physiological concentrations the divergence of G from its limiting value in dilute solutions is sufficiently great to affect appreciably the accuracy of results calculated on the basis of equation 5. Further, the physiologist is frequently confronted with solutions, such as blood or urine, which contain complex mixtures of solutes, both electrolytes and nonelectrolytes, in widely different concentrations. In such a situation the necessity for a practical unit of osmotic concentration which is independent of the variation of G with concentration for different individual solutes will readily be appreciated. The *osmole* (Osm) is such a unit. A solution having an osmotic pressure of 22.4 standard atmospheres is said to have an effective osmotic concentration of one osmole per liter or to be an *osmolar* solution, regardless of its chemical composition. For any individual substance the osmole is defined as the weight in grams which gives rise to an osmotic pressure of 22.4 standard atmospheres when dissolved in one liter of solution. It will be apparent that for osmotic purposes the concentration of any given solution can be expressed directly in terms of *osmoles per liter* or *osmolarity.** In physiological work the milliosmole (mOsm) is usually employed as the unit of osmotic concentration (1 Osm = 1000 mOsm). In the case of dilute solutions one may write for nonelectrolytes (e.g., sucrose, urea), *milliosmoles* = *millimoles* and for electrolytes, *milliosmoles* = G × *millimoles,* where G has its ideal value, i.e., the number of ions formed by each molecule of the electrolyte. For solutions containing a single solute, these "ideal" relationships may be used for approximate purposes at physiological concentrations.

THE DETERMINATION OF OSMOTIC CONCENTRATIONS

In principle, the osmolarities of physiological solutions can be determined by direct measurement of their osmotic pressures, utilizing the technique illustrated in Figure 1-2 and converting these pressures to Osm or mOsm. Because of technical complications, however, this method is unsuitable for routine physiological or clinical investigations, in which rapid determination of the osmolarities of large numbers of samples is usually required. An alternative method for the determination of osmolarities which is at once simpler in practice, more rapid, and more accurate than all but the most elaborate instruments for the direct measurement of osmotic pressure is based on the relative freezing points of solutions.

Osmotic pressure is one of the so-called *colligative* properties of dilute solutions; i.e., its magnitude is related to the concentration (number per unit volume) of dissolved particles and is not affected (in very dilute solutions at least) by such factors as their size, shape, or chemical composition. Other colligative properties of dilute

---

*From a physicochemical point of view, *osmolality* (number of osmoles per kilogram of water) is preferable to *osmolarity* (number of osmoles per liter of solution), the osmole being defined as the weight in grams of solute which, when dissolved in 1 kg water, yields a solution having an osmotic pressure of 22.4 standard atmospheres. However, in the present discussion, osmolarities rather than osmolalities will be employed for two reasons. First, at normal physiological concentrations, the error introduced by using osmolarities instead of osmolalities is sufficiently small to be neglected. Second, in dealing with physiological fluids such as blood plasma and urine it is often more practicable to refer one's results to a volume of solution rather than to an amount of water as a standard. It is assumed that the solutions in question are not subject to wide variations in temperature.

solutions are vapor pressure lowering, depression of the freezing point, and elevation of the boiling point. These four properties of solutions are very simply related so that if one of them is known for a given set of circumstances, the others may readily be calculated.

In the case of freezing point depression it can be shown that, in very dilute solutions, the amount by which a nondissociating solute like sucrose lowers the freezing point of pure water is 1.86°C per mole. This factor, the *cryoscopic constant,* is the same for all nondissociating solutes providing they can be considered to behave ideally. Under similar conditions one would expect an electrolyte like NaCl, which gives rise to 2 ions per molecule, to lower the freezing point of water by 2×1.86 or 3.72°C per mole, and so on. In practice it is found that the factor 1.86°C per mole is subject to the same kind of concentration dependence as the osmotic coefficient G. Therefore, for practical purposes an osmolar (or osmolal) solution may be defined as one which lowers the freezing point of water by 1.86°C regardless of the precise conditions used or of its exact chemical composition. Since, as already pointed out (p. 10), such a solution has an osmotic pressure of 22.4 atm, the following relations between osmolarity, freezing point lowering, and osmotic pressure are at once apparent:

$$\text{milliosmoles} = \frac{\Delta T_f \times 1000}{1.86}$$

and

$$\text{osmotic pressure (atm)} = \frac{\Delta T_f \times 22.4}{1.86}$$

where ΔT_f is the difference between the freezing point of the solution and that of pure water (0°C).

Nowadays a number of instruments (osmometers) are available with which the freezing points of solutions can be rapidly and accurately measured and which give a direct readout in mOsm of the results obtained. Thus freezing point lowering (cryoscopy) is at present the usual method of choice for the determination of osmolarities in physiological and clinical investigations.

OSMOTIC BEHAVIOR OF RED CELLS: HEMOLYSIS

Red blood cells have been a frequent choice in osmotic and permeability studies for a number of reasons, including the following: the ready availability in quantity of these cells (1 mm^3 of human blood contains about 5 million red cells), their ease of handling, and the fact that, since they are free cells whose whole surface is exposed to the external medium, the volume changes which they undergo in response to changes in their osmotic environment and the rates of entry of various substances into them are easily determined. Although there are, undoubtedly, differences in detail in the osmotic behavior of different cells, the red cell may be considered a fairly typical representative, in this respect, of animal cells in general.

The total osmotic concentration of mammalian blood plasma is about 300 mOsm. Present estimates indicate that in most cases, including red cells, the intracellular osmolarity in vivo is essentially the same as that of plasma. This intracellular osmolarity is principally due to the presence, inside the cell, of K^+, Na^+, and Cl^- ions, together with smaller amounts of HCO_3^-, phosphates, amino acids, etc. In red cells as in many other cells the sum of Na^+ and K^+ ions greatly exceeds the total amount of small anions present. To preserve electroneutrality it is necessary to postulate the existence within the cell of large nondiffusible or "fixed" anions. These are partly the ionized anionic groups (e.g., $-COO^-$ groups) of cellular proteins and also include smaller anionic molecules (e.g., adenosine triphosphate – ATP) which are "nondiffusible" in the sense that they do not readily move across the cell membrane.

When red cells are placed in a solution containing 300 mOsm per liter of a nonpenetrating or virtually nonpenetrating solute such as NaCl,* their volume does not change, since there is no osmotic gradient across the membrane. If they are placed in a NaCl solution of greater concentration than this, they shrink because of osmotic loss of water. In such shrunken cells the surface is often roughened or *crenated.* Conversely, if the cells are suspended in a NaCl solution containing less than 300 mOsm per liter, they swell on account of osmotic entry of water from the external solution. Swelling in red blood cells is accompanied by a change in shape. Normally the cells are biconcave discs with an average major diameter of about 7.2 μ. When water enters them as the result of an osmotic gradient, the cells become progressively spherical. In this way they can increase their volume by a maximum of about 67 percent without any appreciable change in surface area. According to present ideas, the red blood cell membrane does not possess any appreciable tensile strength. Thus, if the cells cannot, by becoming spherical, neutralize the osmotic gradient imposed on them, they cannot further increase their volume without damage to the membrane. Entry of water beyond the point at which the cells become fully spherical results in damage to the membrane and escape of the characteristic red protein *hemoglobin* from the cell interior. This phenomenon is called *hemolysis.* The residual cells left following hemolysis are called *ghosts* or *stroma.* The cell volume beyond which hemolysis occurs is the *critical hemolytic volume.* Within the limits imposed by hemolysis, swelling and shrinking of red cells are reversible.

OSMOLARITY AND TONICITY

Any two solutions which have the same osmotic pressure are *isosmotic.* Solutions which have a greater or smaller osmotic pressure than a given reference solution are *hyperosmotic* and *hypo-osmotic,* respectively, compared to the reference solution. A solution containing 300 mOsm of NaCl (approximately 0.15 M) is *isosmotic* with the contents of the red cell. In addition, since such a solution does not cause any volume change in the cells, it is said to be *isotonic.* Solutions containing NaCl in excess of 300 mOsm per liter induce a net loss of water from the cells and are *hypertonic.* Conversely, solutions containing less than 300 mOsm NaCl per liter cause the cells to take up water and are *hypotonic* with respect to the cell contents. It is evident that similar considerations apply to other nonpenetrating solutes.

With penetrating solutes such as urea, the situation is quite different. A solution containing 300 mOsm urea (approximately 0.3 M) is *isosmotic* with the fluid contents of the red cell. Nevertheless, if red cells are placed in such a solution, they rapidly hemolyze because urea readily penetrates the cell membrane. Since the cell is virtually impermeable, over short periods of time, to the osmotically active substances it initially contains, the entry of urea cannot be balanced by an equivalent exit of osmotically active material and must therefore result in an increased intracellular osmolarity. This sets up an inwardly directed osmotic gradient for water as a result of which water flows into the cell. Urea is normally present only in very small amounts in red cells; it is therefore clear that, unless the volume of

---

*Although it is now known that Na^+ and K^+ ions can exchange fairly freely across the red cell membrane (p. 22), net transmembrane movements of salt are slow. For this reason NaCl may be regarded, for practical purposes, as a nonpenetrating solute in osmotic experiments of relatively short duration.

external fluid is exceedingly small, the cells cannot compensate for the osmotic gradient resulting from the entry of urea and will in consequence hemolyze. Since water moves across the cell membrane very rapidly, the osmotic response to the entry of urea will be correspondingly rapid. Thus, there will be a virtually simultaneous movement of water and urea into the cells, and the time required for hemolysis to occur will depend on the rate at which urea moves across the membrane. An equivalent situation will arise if any other penetrating solute is substituted for urea. Thus, the time taken for hemolysis to occur in an isosmotic solution of a given solute is an index of the rapidity with which that solute penetrates the cell, and hemolysis has been widely used to determine relative rates of penetration. In practice, since all the red cells in a suspension do not hemolyze simultaneously, the time necessary for a definite amount of hemolysis to occur is usually employed.

It will be apparent that the above considerations concerning hemolysis by a penetrating solute are not limited to the case in which the initial external solute concentration is osmotically equal to that of the cell contents but can be extended to other initial concentrations of the penetrating solute. Hence it can be concluded that a solution containing a penetrating solute only, or a mixture of penetrating solutes, whether or not it is isosmotic with the cell interior, cannot be isotonic. In fact, from the definition of tonicity given above, *all* solutions of penetrating solutes must be regarded as *hypotonic,* whatever their actual osmolarity may be.

NONOSMOTIC HEMOLYSIS

In addition to hemolysis resulting from an osmotic gradient across the membrane or from the presence of a penetrating solute in the external solution (osmotic or hypotonic hemolysis), hemolytic injury to red cells may stem from a number of causes, among them excessive heat or mechanical agitation and a variety of "hemolytic agents," that is, substances which impair or destroy the selective permeability of the membrane by specific interaction with one or more of its components. Hemolytic agents of this kind include fat solvents such as chloroform and ether, surface-active agents (detergents, bile salts, saponin, etc.), heavy-metal ions, and certain snake venoms.

The resistance of red cells to osmotic hemolysis or its reciprocal (fragility) is an important clinical parameter. In certain diseases (e.g., pernicious anemia) this resistance may be significantly greater than normal. In others, such as hemolytic jaundice, it may be lower than in normal cells (Harris, 1960).

MECHANISMS OF BIOLOGICAL TRANSPORT

Apart from special mechanisms, such as *pinocytosis,* which involve engulfment by the cell of microscopic droplets of extracellular fluid and which are thought to be of importance in the transfer of large molecules such as proteins across the membrane, the satisfactory classification of biological transport processes has proved difficult. Much of the difficulty has centered upon the precise definition of "active" transport (Curran and Schultz, 1968). The following classification rests on currently accepted views of membrane function and should be sufficiently rigorous for all except the specialist in this field.

In the absence of exact knowledge concerning the molecular mechanisms involved, the classification of membrane transport processes has thus far been based primarily on two considerations: (1) the energy requirements for *net* solute movement across the membrane under a given set of conditions and (2) the kinetic pattern of the transport process. These will now be discussed in turn.

ENERGY REQUIREMENTS FOR TRANSPORT – "PASSIVE" AND "ACTIVE" TRANSPORT PROCESSES

From an energetic or thermodynamic point of view membrane transport processes are classified according to the direction, relative to the physicochemical driving force or energy gradient, of *net* movement of the substance being transported. If this movement is in the direction of the energy gradient (downhill movement), the process is described as *passive transport.* Conversely, if net movement occurs against the energy gradient (uphill movement), the term *active transport* is used to describe it (Ussing, 1960; Wilbrandt and Rosenberg, 1961). It should be stressed that it is the direction of *net* transport, relative to the driving force, that determines the energy characteristics of the transport process. Frequently, a given solute is found to cross the membrane simultaneously in both directions. In this case there are two *unidirectional fluxes,* an *influx* and an *outflux,* to be considered. The *net flow* or net flux of solute is of course the difference between them.

Clearly, active transport, as we have defined it, requires energy to bring about net movement of the transported substance against the physicochemical or "passive" driving force across the membrane. In the case of an uncharged species this force may usually be equated to its transmembrane concentration gradient. The transport of ions involves additional factors such as the contribution of electrical forces to the net energy gradient across the membrane. The energy for active transport is derived from the metabolic energy of the cell through coupling mechanisms (the exact nature of which is as yet unknown) as shown by the fact that active transport processes are usually sensitive to metabolic inhibitors. This, however, has led to some confusion in the definition of active and passive transport processes since many "downhill" processes occur which show a dependence on metabolism. In this chapter active transport will be defined as *net or accumulative transfer against an energy gradient which is dependent on metabolism.* Transport processes in the direction of the energy gradient, whether or not they depend on metabolism, will be described as downhill or *equilibrating* processes regardless of the mechanisms involved. This classification approximates closely to that suggested by Wilbrandt and Rosenberg (1961).

TRANSPORT KINETICS

Diffusion-Type Transport Processes (D-Kinetics)

In many cases the movement of dissolved substances across cell membranes can be described adequately by treating the membrane as a simple external restraint to the free diffusion of solute in an aqueous system consisting of two phases (the external medium and the intracellular fluid) in which it exists at different concentrations. The restraint imposed by the membrane on solute movement may be total or partial; that is to say, the movement of a solute from the external fluid to the cell interior or vice versa may be totally prevented or hindered to varying degrees. Membranes which possess the ability to prevent some solutes from passing through them while allowing others to pass with more or less ease are called *selectively permeable* or *permselective* membranes.

Where the rate of movement of a solute across the cell membrane is governed by diffusion forces, this rate obeys an equation which is formally analogous to Fick's diffusion equation (p. 6). If S_o is the concentration of a substrate in the external fluid and S_i is its concentration inside the cell, the net rate of diffusion, V, is given by the equation

$$V = PA\,(S_o - S_i) \tag{6}$$

In this equation A is the area of the cell membrane and P is a permeability coefficient which is analogous to the diffusion coefficient D (p. 6). Because of uncertainty about the actual area of the membrane, it is usually impossible to determine exactly the absolute value of P. However, minimal estimates of A can be made from which maximal values of P can be computed with reasonable certainty. These calculations show clearly that the value of P for a given substance is usually much smaller than its diffusion coefficient in free solution, indicating a considerable resistance to diffusion on the part of the membrane.* By comparing the rates of entry into the cell of different substances under similar conditions, it is possible to obtain comparative estimates of P. It was on the basis of such studies that the marked correlation between P (or the resistance to diffusion offered by the membrane) and lipid solubility in the case of nonpolar molecules, and between P and molecular size in the case of polar molecules, was demonstrated by early workers in the field of permeability.

Equation 6 yields information about several important kinetic characteristics of diffusion-type transport. For example, it indicates that the rate and direction of the net transmembrane movement of any uncharged substance which results from diffusion forces depends on its concentration gradient across the membrane. If $S_o > S_i$ the rate of inward movement will exceed the rate of outward movement, and net entry will occur. The converse will be true if $S_o < S_i$. If $S_o = S_i$ the rates of diffusion in both directions will be equal and there will be no net transfer of material across the membrane in either direction. The end point of a diffusion process of this type is therefore the attainment of equal concentrations of substrate on both sides of the cell membrane. In other words, diffusion across the cell membrane is an *equilibrating* process. This has the important consequence that it is unnecessary to postulate the existence of any source of energy other than that arising from the concentration gradient itself to drive the process of diffusion. Consequently, transport processes of this type are often referred to as passive processes (Wilbrandt and Rosenberg, 1961).

Equation 6 also predicts that, in diffusive transport, the net rate of transmembrane movement, V, increases linearly as the concentration gradient across the membrane increases. Figure 1-3A shows a graph of this linear relationship between the net rate of inward diffusion and $(S_o - S_i)$ for an inwardly directed diffusion gradient. For an outwardly directed gradient, a similar relationship would exist between the net rate of outward diffusion and $(S_i - S_o)$. If, for an inwardly directed diffusion process, the rate of penetration is measured under conditions in which $S_i = 0$ (initial rate of entry), a plot of V against S_o yields a curve similar to Figure 1-3A.

Carrier-Mediated Transport (E-Kinetics)

With many substances of physiological importance, e.g., sugars, amino acids, and ions, the experimental curve relating the rate of entry to the outside concentration is frequently of the type shown in Figure 1-3B. At first, as S_o is increased, the rate of entry also increases. Ultimately, however, an external substrate concentration is reached at which the rate of entry becomes maximal and is not further increased in the presence of higher external substrate concentrations. Kinetic behavior of this type is referred to as *saturation kinetics* and is frequently exhibited by systems in

---

*In a few cases, e.g., Na^+ and K^+ ions, it has been possible, using special techniques, to measure the rates of diffusion of exogenous substances within the cell. In the absence of binding to high-molecular-weight cytoplasmic constituents, the diffusion coefficients so obtained were close to the values observed in free solution.

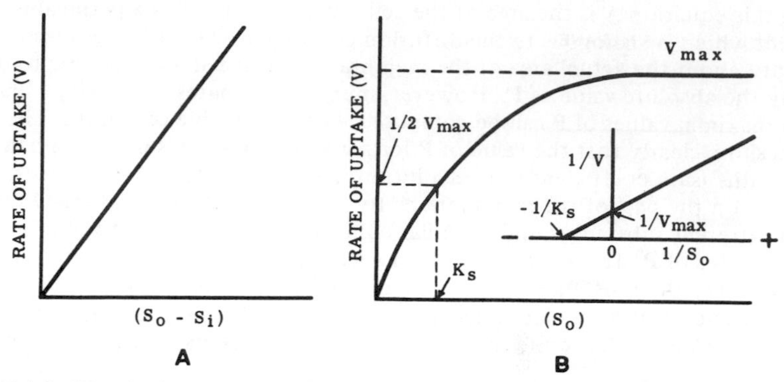

FIGURE 1-3. Kinetic characteristics of transport processes. A. Diffusion-type transport. B. Carrier-mediated transport. (From H. Stern and D. L. Nanney. *The Biology of Cells.* New York: Wiley, 1965. P. 326.)

which the kinetic process involves the reversible combination of the substrate with a receptor site, e.g., enzyme reactions. In the case of membrane processes it is postulated that the transported solute or *substrate* enters the cell by combining reversibly with a specific membrane component or *carrier* at the outer surface of the membrane. The carrier-substrate complex so formed moves across to the inner surface of the membrane, where it dissociates, releasing free substrate to the cell interior. The carrier then moves back across the membrane to the outer boundary, where it combines with a second molecule of substrate and the cycle begins again. In this way a relatively small number of carrier molecules operating cyclically can transport large amounts of substrate. On this basis the occurrence of a maximum in the rate of transport with increasing substrate concentration is readily explained as being due to saturation of the available carrier sites when the external substrate concentration reaches a sufficiently high level.

A model for a simple carrier process of this type is diagrammed in Figure 1-4A. Applying the law of mass action to the reversible combination of substrate with carrier, one may write

$$S + C \rightleftharpoons CS \qquad \text{and} \qquad \frac{[C] \times [S]}{[CS]} = K_S \tag{7, 8}$$

where K_S is the *dissociation constant* for the complex CS. The brackets denote concentration. K_S may be calculated from the equation

$$\frac{V}{V_{max}} = \frac{[S]}{[S] + K_S} \tag{9}$$

where V is the rate of transport at a substrate concentration (S) and V_{max} is the maximal or saturation rate of transport. (This equation is identical with the well-known Michaelis-Menten equation of enzyme kinetics.)

As shown in Figure 1-3B, K_S is numerically equal to the substrate concentration when $V = ½ V_{max}$. If the association-dissociation steps on both sides of the membrane (equation 7) are *rapid* compared to the movement or *translocation* of the carrier substrate complex across the membrane, then, as indicated by equation 8, $1/K_S$ gives a measure of the *affinity* of the carrier for the transported substrate. Note that this affinity is *reciprocally* related to the magnitude of K_S.

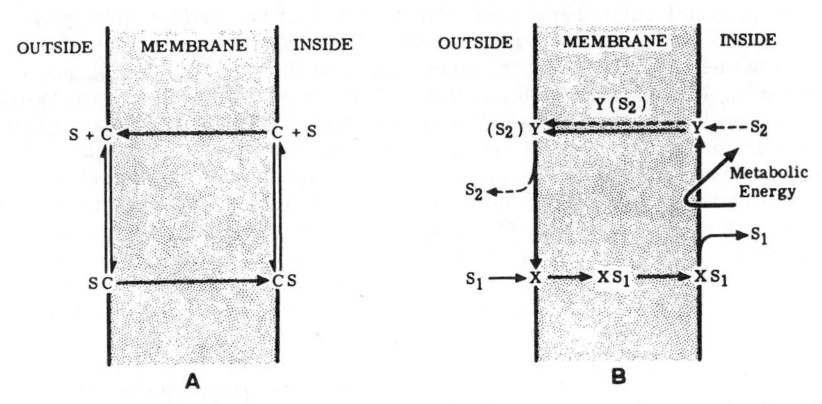

FIGURE 1-4. Diagram of two types of carrier-mediated transport. A. A symmetrical carrier of the equilibrating type. B. A carrier capable of accumulative transport.

Both K_S and V_{max} are more conveniently determined from a plot of $1/V$ against $1/[S]$ as shown in the insert in Figure 1-3B. With this plot, a straight line is obtained which, when extrapolated to cut the axes of the graph, gives an intercept equal to $1/V_{max}$ on the $1/V$ axis and an intercept equal to $-1/K_S$ on the $1/[S]$ axis.

A symmetrical carrier of the type illustrated in Figure 1-4A, which has the same affinity for substrate on both sides of the membrane, can lead only to equalization of the substrate concentration across the membrane. As in the case of simple diffusion, the direction of *net* transport will depend upon the direction of the substrate concentration gradient. When $[S]_{inside} = [S]_{outside}$, the same amount of substrate will be carried across the membrane in each direction and no net transport of S will occur. Equilibrating processes of this kind do not, in the strict thermodynamic sense, require an external source of energy. However, the production of carrier molecules by the cell may be dependent on metabolism. Consequently, equilibrating carrier processes may show some sensitivity to metabolic inhibitors. This has occasionally led to their being described erroneously as "active" transport processes. The entry of sugars and amino acids into human red cells occurs by way of equilibrating carrier processes.

The transport of sugars and amino acids by cells such as those of kidney tubule epithelia and intestinal mucosa also shows saturation kinetics. Unlike red cells, however, these cells accumulate sugars and amino acids against a concentration gradient. The accumulation process depends on metabolic energy and is inhibited by metabolic poisons such as cyanide and dinitrophenol.

A model of a carrier system capable of accumulative or active transport is shown in Figure 1-4B. One important difference between this model and the one illustrated in Figure 1-4A lies in the fact that, in the model shown in Figure 1-4B, the carrier is asymmetrical, i.e., it undergoes a change at the inner surface of the membrane from a form X with a relatively high affinity for the substrate to a form Y which has a relatively low substrate affinity. The reverse takes place at the outer surface of the membrane, where the X form of the carrier is regenerated.

Certain other features of the carrier model shown in Figure 1-4B may be noted. Since the transport process effected by the carrier is an "uphill" or active process, it requires a continuous supply of energy. In the reversible transformation of the carrier between the X and Y forms (which might be, for example, a configurational or other intramolecular change) one step at least must be energy requiring. As indicated in Figure 1-4B, it is rather generally believed that the energy-requiring step is the one which occurs at the inner surface of the membrane. This provides a mechanism by which metabolic energy can be continuously fed into the system.

Again, as indicated by the solid arrows in Figure 1-4B, an accumulating carrier process or "pump" can serve to transport a single substance S_1 across the cell membrane, the carrier returning across the membrane in the free or uncombined form during the second half of the cycle.* Alternatively, as indicated by the dotted arrows, such a mechanism can act as a so-called exchange pump in which the Y form of the carrier combines with a second substrate S_2, which is transported out of the cell in an amount equivalent to the amount of S_1 transported inward. The essential requirement for an exchange pump is that there be a reciprocal relationship between the affinities of two forms of the carrier for S_1 and S_2. The transport of sodium and potassium ions by red cells is a well-known example of an exchange pump (p. 29).

Biological transport processes which show saturation kinetics also exhibit a number of other characteristics which are easily explained on the basis of carrier theory. For example, among a series of closely related compounds such as sugars and amino acids, some may be transported readily while others may be transported very slowly or not at all. This *substrate specificity* may be regarded as arising from the possession of a highly specific structure by the combining site on the carrier molecule which requires an equally specific structure on the part of the substrate and the carrier-substrate complex if transport is to be easily effected. A consequence of substrate specificity is *competition,* in which two or more structurally similar substrates compete for the available sites. In this situation the relative amount of each substrate which passes through the membrane will depend on its relative affinity for the carrier and on its relative concentration in the vicinity of the carrier sites. The existence of substrate specificity and competition has been amply demonstrated in many cell species.

THE TRANSPORT OF IONS

The Energy Gradient for Ion Transport: The Resting Potential

If a microelectrode (tip diameter 1 μ or less) is inserted into a cell (Fig. 1-5), a *potential difference* is observed between it and a reference electrode immersed in the external medium. This potential difference, which in resting cells is always such

---

*This statement is of course an approximation. Since there are finite though different affinities between carrier and substrate on both sides of the membrane, there will always be some "reverse" flux of S from the low-affinity to the high-affinity side providing there is a finite concentration of S in the compartment bounded by the former. Using primes to denote the kinetic parameters pertaining to the low-affinity side (e.g., V' is the unidirectional flux of solute from this side to the high-affinity side), one can describe the kinetics of the system by the following equation:

$$V_{net} = V - V' = V_{max}\left[\frac{[S]}{[S] + K_s} - \frac{[S']}{[S'] + K'_s}\right] \quad (10)$$

This equation shows that when $V = V'$ (i.e., $V_{net} = 0$), $[S']/[S] = K'_s/K_s$. The ratio K'_s/K_s is thus the *maximum accumulation ratio* of the system and can be used to predict the extent to which a metabolically linked carrier of the type under discussion can accumulate a transported solute on one side of the membrane.

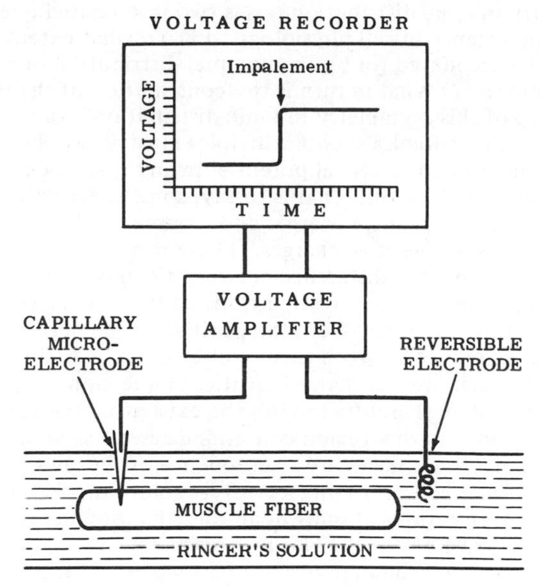

FIGURE 1-5. Diagram of the measurement of a resting membrane potential. An open tip capillary microelectrode filled with KCl solution is inserted into the cell, and the potential difference between it and a reversible half-cell (e.g., calomel or silver/silver chloride) immersed in the bathing medium is recorded. At the moment of impalement there is a sharp deflection in the voltage trace.

that the intracellular electrode is *negative* compared to the external reference electrode, is the *resting potential* (E_R) or *membrane potential* of the cell. By convention, the potential of the external electrode is taken as zero at all times. Typical values for the resting potential are about 90 mv for skeletal muscle and about 70 mv for nerve.

The existence of a potential difference across a membrane implies that a charged particle such as an ion moving through the membrane will be subjected to an electrical force which may, according to the direction of movement and sign of charge of the ion, either assist or hinder its passage. In the case of the cell membrane, the potential difference is oriented in such a way as to assist the movement of cations into the cell and oppose their outward movement from the cell interior. The opposite is true for anions. Thus, it is clear that the net energy gradient for ions across the cell membrane cannot be specified in terms of concentration differences alone but must also take into account the effect of electrical forces.

The Origin of Membrane Potentials: Resting Potentials of Muscle and Nerve

Living cells are characterized by an unequal distribution of ions between the cell interior and the external environment. The most important quantitative inequalities relate to Na^+, K^+, and Cl^-. The interior of a typical mammalian cell contains about 20 to 30 times as much K^+ as Na^+. In blood plasma or interstitial fluid this ratio is reversed. Chloride is also unequally distributed between the cell and its environment, being present in lower concentration in the cell interior than in the external fluid.

This asymmetry in ionic distribution raises two interrelated questions which are of fundamental importance in cell physiology: (1) To what extent can the resting potential of cells be accounted for by the unequal distribution of Na^+, K^+, and Cl^- across the membrane? (2) What in turn is the contribution of the resting potential to the maintenance of this asymmetry in ionic distribution? In discussions of the origin of cell membrane potentials, two principles must always be borne in mind. First, the development of an electrical potential in any system depends on separation of charge within that system. That is to say, somewhere within the system there must be an excess of positive charges and, consequently, somewhere else there must be an excess of negative charges. These regions must be spatially separate, and further, if a potential difference between them is to be maintained for any length of time, there must be a restraint on the flow of electrical charge, i.e., current should not flow too freely from the positive to the negative region. Otherwise the potential difference between them will be quickly dissipated. In the case of aqueous solutions ions are the charged entities in question. A potential difference in an aqueous system therefore implies the existence of a region containing an excess of cations and another region containing an excess of anions.

On the other hand, it is a necessary consequence of the laws of thermodynamics that in any *finite* volume of an electrolyte solution, whatever inequalities may exist in the concentrations of individual ions, the total numbers of positive and negative charges must be equal. This is the principle of *electroneutrality.* If cytoplasm and extracellular fluid are regarded as two essentially electroneutral solutions separated by a membrane, the question of the origin of membrane potentials can be stated very simply as follows: How can charge separation be achieved in such a system without violating the condition of electroneutrality?

Granted an asymmetrical distribution of individual ionic species across a membrane, and regardless of how this asymmetry came about, two kinds of electrical potentials may be expected: *equilibrium potentials* and *diffusion potentials.* Both have been used as a basis for interpreting cell membrane potentials, but diffusion potentials offer a more satisfactory and complete explanation of the observed behavior of cellular systems.

In the simplest case, an equilibrium potential arises when a membrane is permeable to one ionic species only. For example, if a cell membrane is permeable to K^+ only and there is an unequal distribution of K^+ ions across it, the following situation may be visualized. Because of their unequal distribution across the membrane, K^+ ions will tend to diffuse from the inside of the cell where their concentration is high to the outside. This diffusion of K^+, unaccompanied by any compensatory movement of anions, would, if unchecked, result in a net transfer of positive charge to the outside of the cell in violation of the electroneutrality principle. Therefore, the tendency for K^+ to diffuse across the membrane is immediately countered by the development of a potential difference of a size and orientation sufficient to prevent any further movement of K^+. The end result, in the system under discussion, will be a slight charge separation, in the *immediate region of the membrane* only. This will involve only a very minute quantity of K^+ ions, will cause the outside of the membrane to become positively charged with respect to the inside, but will not significantly affect the overall electroneutrality of the solutions on each side of it. The magnitude of the membrane potential generated in this situation is given by the Nernst equation

$$E_M = E_K = \frac{2.303\ RT}{zF} \log \frac{[K_i^+]}{[K_o^+]} \tag{11}$$

where R is the gas constant, T the *absolute* temperature, F the Faraday (96,500 coulombs), and $[K_i^+]$ and $[K_o^+]$ denote the K^+ concentrations (strictly speaking the *activities,* but in most physiological investigations concentrations may be substituted without undue error) inside and outside the cell respectively; z is the valence of the ionic species involved, and for K^+ and other *monovalent* ions, $z = 1$.

This equation leads to certain theoretical predictions about the behavior of the membrane potential which can be tested by actual measurements on the systems involved. First, according to equation 11, it is obvious that at any given temperature E_M will depend only on the ratio $[K_i^+]/[K_o^+]$. Further, equation 11 predicts that the relationship between E_M, expressed in *millivolts* (mv), and log $(K_i^+]/[K_o^+]$ should be a straight line with a slope (2.303 RT/F) of 58, at 18°C, or 61, at 37°C, for a 10-fold change in the ratio $[K_i^+]/[K_o^+]$.* In other words, equation 11 states that a membrane which obeys it behaves as an ideal K^+ electrode. In muscle and nerve there is evidence that the fiber membrane, under certain conditions, approximates closely the behavior of a K^+ electrode. Figure 1-6, taken from an investigation of the resting potential in frog sartorius muscle (Hodgkin and Horowicz, 1959), illustrates this point. In the experiment shown in this figure $[K_o^+]$ was altered under conditions in which $[K_i^+]$ remained virtually constant (140 mM). In this situation E_M, according to equation 11, should be a linear function of log $[K_o^+]$. The straight line AB in Figure 1-6 is the theoretical plot obtained using equation 11. The circles are the experimental values of E_R under the same conditions. It is clear from

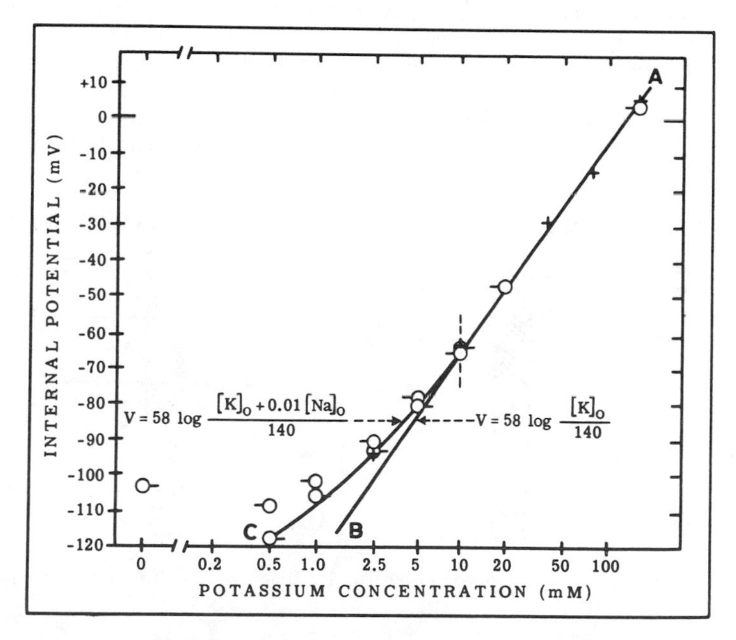

FIGURE 1-6. Effect of external potassium on the membrane potential of a single fiber taken from the frog sartorius muscle. (From Hodgkin and Horowicz, 1959.)

---

*In this chapter the experimentally observed potential difference across a cell membrane will be called the resting potential (E_R). The theoretical membrane potential calculated from a specific equation or set of equations will be denoted by E_M. E_K, E_{Na}, etc., will be used to indicate the equilibrium (Nernst) potentials of individual ions. For simplicity a value of 60 is assigned to the factor 2.303 RT/F in subsequent equations of the Nernst and similar types.

Figure 1-6 that when the external K^+ concentration is greater than 10 mM there is excellent agreement between the observed values of E_R and the calculated values of E_M; thus, when the external K^+ concentration is relatively high, the resting potential of frog muscle is satisfactorily accounted for by equation 11. When the external K^+ concentration is less than 10 mM the observed potential deviates rather markedly from the values predicted by equation 11. This is discussed later (p. 23).

Apart from the fact that the observed behavior of the resting potential in skeletal muscle deviates noticeably from the predictions of equation 11 for $[K_o^+]$ values below 10 mM, which is, of course, the situation of greatest relevance to normal physiological conditions (an essentially similar type of behavior is seen with nerve fibers), attempts to explain E_R as a Nernst equilibrium potential encounter other serious difficulties. In addition to the restriction that the membrane should be permeable to just one ionic species, the Nernst equation is valid only when there is no flow of current across the membrane. There is, at this time, overwhelming evidence against both of these assumptions. Following the introduction of radioactive tracers into biological research, it soon became clear that cell membranes are permeable to Na^+ and Cl^-, though, in general, their permeabilities to these ions are considerably less than their permeabilities to K^+. Also, in many instances current flow across cell membranes can be demonstrated. These and other findings have led to an alternative formulation of the membrane potential in terms of diffusion potentials.

Diffusion potentials arise where a boundary such as a membrane separates two solutions containing different concentrations (activities) of individual ionic species and where the individual ionic permeabilities of the membrane also differ. A simple system illustrating the development of a diffusion potential is illustrated in Figure 1-7A. In this illustration, a membrane permeable to both K^+ and Cl^- separates a KCl solution from water. Because of the concentration gradient, both K^+ and Cl^-

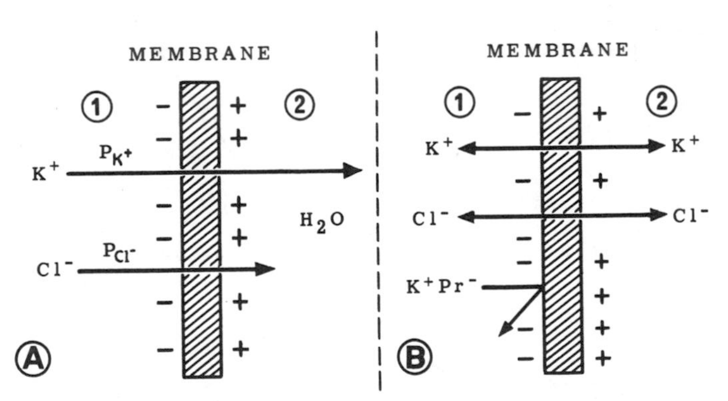

FIGURE 1-7. A. Origin of a diffusion potential. B. Diagram of a Gibbs-Donnan equilibrium.

diffuse across the membrane. Since the membrane is assumed to be more permeable to K^+ than to Cl^- (as indicated by the lengths of the arrows labeled P_{K^+} and P_{Cl^-} respectively), K^+ initially diffuses more rapidly than Cl^- from side 1 to side 2. This quickly generates a potential difference across the membrane such that side 2 is positive with respect to side 1. This potential has a retarding effect on K^+ ions and an accelerating effect on Cl^- ions, and, once a very small amount of K^+ has leaked across the membrane, the potential stabilizes at a value which permits K^+ and Cl^- ions to diffuse from side 1 to side 2 at the same rate.

Several important differences exist between diffusion potentials and equilibrium potentials. Diffusion potentials permit a flow of current across the membrane. Also, in diffusion systems such as that illustrated in Figure 1-7A, unless the volume of solution on each side of the membrane is infinite, the membrane potential will dissipate with time as the salt concentration on side 2 approaches that on side 1. By contrast, an equilibrium potential does not decay with time and, in fact, represents an extreme case of a diffusion potential in which all permeability factors except one are zero. Thus in fluid systems of finite volume such as the cells of the body together with the extracellular and vascular fluid spaces, the maintenance of steady state diffusion potentials requires the operation of additional mechanisms to conserve the ionic concentration differences which give rise to these potentials. The mechanisms are discussed below (pp. 24–31).

To return to a consideration of the resting potential as a diffusion potential: It is evident that, since cell membranes show a wide spectrum of permeabilities to different ions, and since both cytoplasm and extracellular fluid contain a large number of diffusible ionic species, all of these could, in principle, contribute to the potential difference observed across the membrane. Fortunately, there is ample evidence to show that, quantitatively, K^+, Na^+, and Cl^- are the only ions that need to be considered. On this basis, and with the assumption that the potential gradient across the membrane is linear (the *constant field* assumption), the following equation for the membrane potential can be derived (Goldman, 1944; Hodgkin and Katz, 1949):

$$E_M = 60 \log \frac{P_K [K_i^+] + P_{Na} [Na_i^+] + P_{Cl} [Cl_o^-]}{P_K [K_o^+] + P_{Na} [Na_o^+] + P_{Cl} [Cl_i^-]} \quad (12)$$

In this equation, usually called the *Goldman* or *constant field* equation, the expressions in brackets denote concentrations as before (equation 11), and P_K, P_{Na}, and P_{Cl} are the permeability coefficients for K^+, Na^+, and Cl^-. According to equation 12, the contribution of each ion to the overall membrane potential is determined by its concentration ratio across the membrane and by the magnitude of its permeability coefficient.

Equation 12 has proved to be exceptionally useful in studying the relationship of the resting potential to transmembrane ionic concentration differences and in determining the relative permeabilities of K^+, Na^+, and Cl^- in many cell types. As an example of its use let us return to the data shown in Figure 1-6. In this experiment Cl^- was replaced by sulfate (a virtually impermeant anion) so that only K^+ and Na^+ need be considered. As already mentioned, when $[K_o^+]$ was greater than 10 mM, the resting potential followed the predictions of equation 11 for a potassium equilibrium potential. In terms of equation 12 this is explained by postulating that in skeletal muscle $P_K \gg P_{Na}$ so that when $[K_o^+] > 10$ mM the contribution of Na^+ to the resting potential becomes negligible, and equation 12 reduces to equation 11. Thus, for these conditions, there appears to be little to choose between the equilibrium approach and the diffusion approach (in fact, the former seems to be a great deal simpler!). However, equation 11 offers no explanation for the fact that when $[K_o^+]$ is reduced below 10 mM the resting potential ceases to be a linear function of $\log [K_i^+]/[K_o^+]$ and becomes progressively less than the value predicted by this equation. In terms of the Goldman equation this discrepancy can be explained simply as being due to an increasing effect of Na^+ on the resting potential as $[K_o^+]$ is progressively lowered. As already mentioned, Na^+ is usually present in much lower concentrations inside the cell than in the external medium. Since the membrane is permeable to Na^+, one would expect a continuous net diffusion of Na^+ ions into the cell (which must be continuously removed from it under steady state conditions,

p. 28). This inward "leak" of Na^+ would, on the basis of equation 12, be expected to affect the resting potential. For the conditions shown in Figure 1-6 one would, in fact, predict that the membrane potential would obey an equation of the type

$$E_M = 60 \log \frac{[K_i^+] + a\,[Na_i^+]}{[K_o^+] + a\,[Na_o^+]} \qquad (13)$$

where $a = P_{Na}/P_K$. The curve AC in Figure 1-6 is drawn to fit such an equation, a being taken as 0.01 – that is, assuming a K^+ permeability 100 times greater than the Na^+ permeability. It is clear that, when $[K_o^+]$ is less than 10 mM, this curve fits the observed values much more closely than the line AB which is drawn on the basis of equation 11.

Thus, equation 12 can be used to determine the relative permeabilities of the membrane to any two ions providing the conditions are such that E_R depends essentially on these two ions only. Similar methods to those illustrated by Figure 1-6 have been used to determine P_{Na}/P_{Cl} in skeletal muscle (Hodgkin and Horowicz, 1959) and hence the ratio $P_K : P_{Na} : P_{Cl}$ in this and other tissues. Finally, it should be reemphasized that in any situation in which the permeability of one ionic species greatly exceeds that of all others, equation 12 reduces to the appropriate form of the Nernst equation. In other words, under these conditions, one would expect E_R to approximate closely the equilibrium potential of the highly permeant ion. The application of the Nernst equation to the resting potentials of cells is therefore frequently justified in practice even though one is not dealing with an equilibrium situation of the kind for which it is theoretically valid.

The Regulation of Intracellular Ionic Composition

We have seen that the analysis of resting potentials in terms of ionic diffusion potentials gives a reasonably satisfactory answer to the first of the two questions posed on page 20; i.e., one can say, on this basis, that the resting potentials of living cells can in large measure be accounted for by the asymmetry in ionic distribution between the cytoplasm and the external medium. The second question, the role of the resting potential in the maintenance of this asymmetry, will now be examined in the broader context of the overall regulation of the electrolyte content of cells.

As already pointed out, the maintenance of steady state diffusion potentials across a membrane requires that mechanisms be brought into play which prevent the dissipation of the ionic gradients giving rise to these potentials. In the case of living cells this raises the question to what extent the gradients can be maintained by purely physicochemical or "passive" forces and to what extent active, metabolism-linked transport processes must be invoked to explain the normal steady state distribution or "intracellular homeostasis" with respect to inorganic ions.

The Distribution of K^+ and Cl^-: The Gibbs-Donnan Equilibrium

A situation in which an unequal distribution of ions across a membrane can arise as a result of purely physicochemical forces and which is of particular relevance to the behavior of cell membranes as well as to a number of other physiological functions (Chaps. 22, 23, and 24) is that shown diagrammatically in Figure 1-7B. Two solutions, one containing KCl only and the other containing KCl together with KPr (the potassium salt of a large anion Pr^-), are separated by a membrane which is permeable to water, K^+, and Cl^-, but not to Pr^-. If the membrane were permeable to Pr^- as well as to K^+ and Cl^-, all three ions would diffuse freely until they were equally distributed throughout the whole solution on both sides of the membrane. By making the membrane impermeable to one ionic species (Pr^-) one imposes a restriction on free diffusion which gives rise to a number of important consequences. Of particular interest in the present context is the fact that the impermeability of the membrane to Pr^- has a marked effect on the equilibrium distribution of the diffusible ions K^+ and Cl^- across the membrane. This distribution is given by the Gibbs-Donnan relationship, which states that *at equilibrium the products of the concentrations of the diffusible ions on each side of the membrane are equal.* For the conditions illustrated in Figure 1-7B, this may be written:

$$[K_1^+] \times [Cl_1^-] = [K_2^+] \times [Cl_2^-] \qquad (14)$$

or

$$[K_1^+] \;/\; [K_2^+] = [Cl_2^-] \;/\; [Cl_1^-] \qquad (14a)$$

Certain consequences of this relationship will be immediately apparent. Clearly, since the solution on each side of the membrane must be, as a whole, electrically neutral, $[K_2^+]$ must be equal to $[Cl_2^-]$. Also, since, in the solution containing the nondiffusible ion Pr^-, the condition of overall electroneutrality requires that there be sufficient K^+ ions present to balance the negative charges on the Pr^- ions as well as those on the Cl^- ions, $[K_1^+]$ must be greater than $[Cl_1^-]$. Again, equation 14 requires that the sum of $[K_1^+]$ and $[Cl_1^-]$ be greater than the sum of $[K_2^+]$ and $[Cl_2^-]$.* Thus for the situation shown in Figure 1-7B, $[K_1^+] > [K_2^+]$ and $[Cl_1^-] < [Cl_2^-]$.

These considerations lead to the important generalization that in a Gibbs-Donnan system at equilibrium there is a concentration gradient across the membrane for each species of diffusible ion present. This gradient (the Gibbs-Donnan ratio) is identical in magnitude for each diffusible ionic species but is opposite in direction for cations and anions. Thus, the equilibrium condition cannot be specified in terms of chemical concentrations alone since the individual concentration gradients would, if unopposed, result in an equal distribution of diffusible ions across the membrane. In the Gibbs-Donnan situation the concentration gradients for diffusible ions which exist at equilibrium must therefore be neutralized by an equal and opposite driving force. This force is an equilibrium potential, generated by the unequal distribution of diffusible ions across the membrane and, for the system illustrated in Figure 1-7B, given by the Nernst equation in the following form:

$$E_M = 60 \log \frac{[K_1^+]}{[K_2^+]} = 60 \log \frac{[Cl_2^-]}{[Cl_1^-]} \qquad (15)$$

Note that the side of the membrane in contact with the solution containing the nondiffusible anion is negative with respect to the opposite side.

The forces giving rise to this potential are similar to those already discussed in the case of potassium equilibrium potentials (p. 20). The main point to note is that the Gibbs-Donnan equilibrium represents a state of balance between two opposing tendencies: the tendency for each diffusible ion to achieve an equal chemical concentration throughout the system, and the tendency for the system as a whole to

---

*This follows from the fact that if $ab = c^2$, where a and b are unequal quantities, then $(a + b) > 2c$.

A well-known geometrical illustration of this relationship is the fact that the sum of any two adjacent sides of a rectangle is greater than the sum of any two sides of a square of equal area (Davson, 1964).

remain electrically neutral.* This has major implications for the distribution of K^+ and Cl^- ions between many types of cell and the external environment.

Often the observed ratios of intracellular to extracellular concentrations of K^+ and Cl^- are close to those predicted by the Gibbs-Donnan relationship. This observation, together with the fact that at the intracellular pH values which prevail in vivo the cell interior is known to contain relatively large quantities of nondiffusible anions, raises the question of the extent to which the Gibbs-Donnan equilibrium can account for the distribution of K^+ and Cl^- across the cell membrane without the intervention of other forces. The matter will be examined in detail for two types of cell, skeletal muscle fibers and red blood cells. These may be considered to represent two extreme situations with respect to the maintenance of intracellular K^+ concentrations. Other cells of the body show intermediate types of behavior. Some, like cardiac muscle and nerve fibers, approximate skeletal muscle in behavior. Others, e.g., smooth muscle, epithelia, liver cells, diverge rather widely from the pattern seen in skeletal muscle.

The essence of the situation is as follows: The chemical force tending to cause net movement of an ion into or out of a cell is its intracellular/extracellular concentration ratio. It is frequently possible, by direct chemical analysis, to obtain fairly accurate estimates of this ratio for Na^+, K^+, and Cl^-. The electrical force tending to drive ions across the membrane in one direction or the other is the resting potential. This too can often be measured accurately. The two forces can be in the same or in opposite directions, but in either case their *algebraic* sum (taking into account their orientation with respect to the membrane) is the net force or *electrochemical potential gradient* which tends to displace the distribution of a given ion from the steady state condition and which must be overcome if a steady state is to be conserved. The analysis of ionic distribution thus resolves itself into the determination of the magnitude and orientation of the electrochemical gradients involved and the identification and measurement of the restoring forces which maintain the steady state distribution of ions across the cell membrane.

The electrochemical potential gradient for a given ion can be determined as follows: By inserting its observed intracellular/extracellular concentration ratio into the Nernst equation one can calculate its equilibrium potential. This is the potential required to balance exactly the driving force arising from its concentration gradient across the membrane. The electrochemical potential gradient is the difference between this equilibrium potential and the measured resting potential.† Obviously, if E_R is equal in size and opposite in direction to the calculated equilibrium potential for any ion, that ion is in electrochemical equilibrium across the membrane and there is no net force acting upon it.

In Table 1-1 this type of analysis is applied to frog skeletal muscle and human red blood cells. The Na^+, K^+, and Cl^- concentrations listed are average figures from the published literature (Harris, 1960). The data shown for skeletal muscle make it clear that, within the limits of experimental error, $E_R = E_{Cl}$. Therefore one may

---

*It is important to realize that the term *equilibrium* in a Gibbs-Donnan system applies only to the diffusible ions present and not to the system as a whole. Obviously, in the case shown in Figure 1-7B, there cannot be an equilibrium in the case of Pr^-, which is confined to one side of the membrane. Also, the presence of Pr^- on one side only of the membrane, together with the presence on the same side of an *excess* of total diffusible ions, gives rise to an osmotic gradient which, unless it is opposed, will cause water to move into the side containing the nondiffusible ion. This osmotic pressure difference is called the *colloid osmotic pressure* and is of importance in such physiological functions as transcapillary fluid movement (Chap. 12).

†In effect, this procedure is equivalent to converting the observed chemical (concentration) driving force into an equivalent electrical force, which can then be compared with the measured electrical force across the membrane.

TABLE 1-1. Ionic Concentrations, Measured Resting Potentials, and Calculated Equilibrium Potentials in Frog Skeletal Muscle and Human Red Blood Cells

	Na^+	K^+	Cl^-	E_R	E_{Na}	E_K	E_{Cl}
Frog Skeletal Muscle							
Fibers	13	140	~ 3	−90	+56	−105	−86
Plasma	110	2.5	90				
Human Red Blood Cells							
Cells	19	136	78	− (7 to 14)	+55	−86	−9
Plasma	155	5	112				

Note: Potentials in millivolts (minus sign indicates inside negative). Concentrations for cells and fibers in milliequivalents per liter cell water. Plasma concentrations in milliequivalents per liter.

conclude that in this tissue Cl^- ions are in electrochemical equilibrium ("passively" distributed) across the fiber membrane and that no direct energy expenditure is required to maintain a steady state with respect to Cl^- in the living muscle fiber.* In other words, the inwardly directed concentration gradient for Cl^- is balanced by an equal and opposite outwardly directed electrical gradient.

With regard to potassium, it is seen that in skeletal muscle fibers, under normal physiological conditions, E_K is significantly greater than E_R, though both are oriented in the same direction. This means that an E_R value some 10 to 15 mv greater than that observed would be required to balance the outwardly directed concentration gradient for this ion. In other words, K^+ is present within the fibers in excess of the amount corresponding to electrochemical equilibrium, although the excess is relatively small. Because of this net outwardly directed electrochemical potential gradient, K^+ must leak continuously out of the fibers and the loss must continuously be made good. Thus, there is a necessity in skeletal muscle for an inwardly directed energy-requiring potassium pump which can transport K^+ against an electrochemical gradient. The nature of this pump will be discussed later in connection with Na^+ transport (p. 28).

As for the distribution of K^+ and Cl^- between red blood cells and plasma, it is seen from Table 1-1 that the situation is qualitatively similar to but quantitatively different from that in muscle. In recent years several workers have succeeded in penetrating red blood cells with microelectrodes and have recorded potentials in the range 7 to 14 mv (inside negative). Once again, these figures are in essential agreement with the calculated value of E_{Cl} (Table 1-1), indicating an equilibrium distribution of Cl^- across the red blood cell membrane.

By contrast, the calculated value of E_K (Table 1-1) shows that the concentration of K^+ ions inside the red blood cell is far in excess of the amount required for electrochemical equilibrium. An inwardly directed active potassium pump must therefore be a major component of the potassium transporting machinery in these cells. Such a pump, showing saturation kinetics and depending on a continuous

*Inasmuch as metabolism is necessary to the life of cells, there is, of course, an *indirect* dependence of all membrane properties on metabolism.

direct supply of metabolic energy, has, in fact, been shown to exist and has been studied in considerable detail (Hoffman, 1966).

The Distribution of Sodium: The Sodium Pump

Inspection of Table 1-1 shows clearly that the situation with respect to intracellular Na^+ is not only quantitatively but qualitatively different from that of K^+ and Cl^-. In both muscle and red blood cells, E_{Na} is different from E_R both in magnitude and in orientation; i.e., E_{Na} requires that the inside surface of the membrane be *positive* with respect to the outside whereas in reality the opposite is true.* This means that the intracellular Na^+ concentration in muscle and red blood cells is much *lower* than that required for electrochemical equilibrium and that both the chemical and electrical gradients tend to "push" Na^+ into these cells. Further, the same appears to be true for virtually all cells, not only those of man and other animals, but also higher plant cells and microorganisms. Since cells are permeable to Na^+ and since many cell species exist in an environment which is rich in Na^+ ions, there is an almost universal requirement for a mechanism which removes Na^+ from the cell interior as fast as it enters. Removal of Na^+ is usually accomplished against an electrochemical potential gradient. Consequently, the mechanism involved must be an active process. It is generally referred to as the *sodium pump.*

The precise nature of the sodium pump in a number of cell species has been the subject of much discussion. In general, the movement of Na^+ into or out of cells is found to be associated with a converse movement of K^+. For example, when the metabolism of isolated muscles or red cells is blocked by low temperature or metabolic inhibitors, the cells accumulate Na^+ and lose K^+ in approximately equivalent amounts. When normal metabolic activity is restored, these changes are reversed, Na^+ ions being extruded and K^+ ions being taken up by the cells. Since the active nature of the Na^+ extrusion process is clearly established on energetic grounds alone for virtually all animal cells, much interest has centered upon the nature of the associated inward movement of K^+.

There is little doubt that the underlying reason for the occurrence of Na^+-K^+ exchange in association with the operation of the Na^+ pump is the necessity for maintaining overall electroneutrality in the cell and its surroundings. The movement, in any finite quantity, of a charged species such as Na^+ across the membrane would, if unaccompanied by compensatory movements of other charged particles, quickly generate a potential gradient which would effectively prevent further movement of the charged species. In the case of the Na^+ pump, electroneutrality is very simply achieved by a counterflow of K^+ ions. Thus each Na^+ ion leaving the cell is balanced by a K^+ ion entering it. The exchange is effected without altering the overall osmotic equilibrium of the cell and also serves to replace K^+ lost by outward diffusion.

Two mechanisms have been proposed by which K^+ ions could move into the cell in exchange for Na^+ ions. The first postulates the existence of a *linked* or *coupled pump* in the cell membrane; that is, for each Na^+ ion pumped out of the cell a K^+ ion is pumped in. It has been suggested that this exchange pump operates by way of an asymmetrical carrier such as that shown in Figure 1-4B. In this case the X form of the carrier molecule as illustrated in the figure is supposed to have a high affinity for K^+ and a low affinity for Na^+. In the Y form of the carrier

*This "inversion" of sign of E_{Na} is of fundamental importance in relation to the phenomenon of "overshoot" in the action potentials of muscle and nerve (Chaps. 2, 3B, and 14).

the relative affinities for K^+ and Na^+ are reversed.* It should be noted that a coupled pump of this kind does not give rise to a potential difference across the membrane and thus would not be expected to influence E_R. The reason is that in such a system there is a close correspondence between the rate of outward movement of Na^+ and the rate of inward K^+ movement. Pumps of this type are said to be *non-electrogenic.*

A second mechanism proposed for Na^+-K^+ exchange is one in which the outward movement of Na^+ is the only active process. According to this concept K^+, on account of its much greater permeability, is taken up in preference to Na^+ by simple diffusion from the external medium to balance the excess negative charges inside the cell that would otherwise result from the outward movement of Na^+. Since there is no reason to suppose that the rate of outward Na^+ movement will correspond exactly to the rate of inward diffusion of K^+, such a mechanism can generate a diffusion potential which can appear as a component of the resting potential. Mechanisms of this kind, therefore, are frequently called *electrogenic* pumps.

In red blood cells convincing evidence for the existence of a coupled Na^+-K^+ pump has been obtained (Hoffman, 1966; Glynn, 1968). Part of this evidence may be summarized as follows: In the presence of glucose, K^+ influx in the red blood cell shows two components, a component which follows diffusion kinetics and a saturable component. In the absence of glucose or in the presence of glycolytic inhibitors, only the diffusion component of K^+ influx is observed. The saturable component of K^+ influx has a K_s with respect to $[K_o^+]$ of 2 mM.

Na^+ efflux likewise shows a diffusion component and a metabolism-dependent component which is saturable with respect to $[K_o^+]$; that is, as $[K_o^+]$ is increased, Na^+ efflux reaches a maximal value. The K_s in terms of $[K_o^+]$ for the saturable component of Na^+ efflux is virtually the same as the K_s for K^+ influx (2mM). These facts point to a very tight obligatory coupling of Na^+ efflux to K^+ influx in the red cells, although recent evidence indicates that the coupling ratio is not one to one but is probably 3 Na^+ out to 2 K^+ in (Glynn, 1968).

In nerve and muscle the active extrusion of Na^+ alone could, in principle, account for the maintenance of normal steady state Na^+ and K^+ concentrations within the cell. However, because certain metabolic poisons (e.g., cyanide, azide, and dinitrophenol) markedly inhibit K^+ influx in nerve under conditions in which K^+ efflux is not much affected, and because Na^+ efflux in muscle and nerve has been shown to depend strongly on external K^+, it has been suggested that part at least of the influx of K^+ in these tissues is an active process linked to Na^+ efflux. Recent studies with denervated skeletal muscles from the frog and the rat, previously allowed to accumulate Na^+ by immersion in a cold K^+-free solution, indicate that, during the early stages of extrusion of this accumulated Na^+, E_R is considerably higher than E_K. This suggests a relatively large contribution from an electrogenic Na^+ pump to the overall resting potential under these conditions. In muscles with an intact nerve supply evidence for a more tightly coupled Na^+-K^+ exchange has been obtained (Dockry et al., 1966).

---

*In the last three or four years this interpretation of Na-K exchange across the cell membrane has received dramatic and unexpected support from the discovery of a series of naturally occurring antibiotics which have remarkable effects on ion transport in isolated mitochondria and which can facilitate the transfer of ions across artificial lipid bilayers. In the latter systems they have been shown to function as cyclic "carriers." The possibility that these substances can be regarded as artificial models of naturally occurring membrane carriers has opened up new and exciting perspectives in the molecular biology of membrane transport. A discussion of this topic is outside the scope of the present chapter. The reader is referred to Pressman and Haynes (1969) for a fuller account.

The Sodium Pump and Metabolism: The Role of Adenosine Triphosphate (ATP)

The dependence of the Na^+ pump on a supply of metabolic energy has been amply demonstrated in numerous investigations. In all cases studied it has been found that interference with the normal metabolic activity of the cell, whether this proceeds by way of oxidative pathways or through glycolysis, inhibits the active transport of ions. The nature of the precise coupling of metabolic energy to the task of driving Na^+, and in some cases K^+, ions across the cell membrane in opposition to an electrochemical gradient is, as yet, an unsolved problem in membrane physiology. However, recent studies have indicated that the phosphate bond energy of ATP, which is known to be the immediate energy source for the majority of cellular functions, plays a vital role in the active transport of Na^+ by several kinds of cell.

Na^+ efflux and K^+ influx in squid nerve are inhibited when cyanide or dinitrophenol is incorporated in the bathing solution. Removal of the inhibitors from the bathing solution results in restoration of the normal fluxes. The changes in Na^+ efflux are paralleled by changes in the ATP and arginine phosphate content of the nerves. In the presence of the inhibitors these compounds disappear from the nerve fibers. When the inhibitors are removed, ATP and arginine phosphate are resynthesized. Further, it has been shown that Na^+ efflux can be restored to normal by microinjection of ATP or arginine phosphate into fibers poisoned with cyanide or dinitrophenol.

Red blood cell ghosts prepared by careful hypotonic hemolysis (p. 11) can be "reconstituted" by immediate reimmersion in isotonic saline solution. The membranes of the reconstituted cells recover their normal permeability properties for some hours following reconstitution. During reconstitution certain substances, including ATP, to which the membrane is normally impermeable can be incorporated in the cells. Cells to which ATP has been administered in this way show pump activity for Na^+ and K^+. In the absence of ATP the reconstituted red blood cell ghosts do not pump these ions.

The discovery of a Na^+ and K^+ specific ATPase (which splits ATP to form adenosine diphosphate and inorganic phosphate and uses the magnesium salt of ATP as a substrate) in the membranes of a number of cell species has shed further light on the stimulation of active ion transport by ATP (Skou, 1965; Hoffman, 1966; Glynn, 1968). This enzyme, which was first discovered in the membrane of crab nerve but which has now been detected in the membranes of many kinds of cell, including red blood cells, shows some remarkable analogies with the Na^+ and K^+ transporting system in these cells. The most extensive studies to date in this area have been made with red blood cells, and it seems appropriate to conclude this section with a brief description of some of the more striking results obtained (Hoffman, 1966; Glynn, 1968).

First, the enzyme found in the membrane fraction of red blood cells behaves in many ways like the Na^+-K^+ exchange pump in these cells. Both the enzyme and the pump require Na^+ and K^+ together for operation, neither ion alone being effective. NH_4^+ ions can substitute for K^+ but not for Na^+ in both systems. Both systems are inhibited by the cardiac glycoside ouabain. Finally, the concentrations at which Na^+, K^+, NH_4^+, and ouabain exert their half maximal effects are the same for both the pump and the enzyme.

Second, two kinds of red blood cell are found in sheep. One is a normal (HK) cell containing a high concentration of K^+ and a relatively low concentration of Na^+. The other (LK) has a relatively high Na^+ and a relatively low K^+ content. The difference between the two kinds of cell is apparently due to a single gene mutation. It has been found that the failure of LK sheep cells to accumulate K^+ to the normal extent is due to the fact that in them the Na^+-K^+ pump is only about one-fourth as active as it is in HK cells. Both types of cell possess an identical Na^+ and K^+ requiring ATPase, but in LK cells the ATPase activity is only one-fourth as great as it is in HK cells.

Third, a recent series of experiments by I. M. Glynn and his associates (Garrahan and Glynn, 1967; Glynn and Lew, 1970) has provided a fascinating additional demonstration of the tight coupling between ATP metabolism and cation transport in red blood cells. When the amount of energy available from ATP hydrolysis is compared with the energy required to move 3 Na^+ out of the cell and 2 K^+ into it under physiological conditions, it is found that the energy "surplus" available to the pump is only about 4 kcal per mole ATP hydrolyzed. This suggested that, under extremely "unfavorable" conditions (i.e., a very high outward concentration gradient for K^+ and a similarly high inward gradient for Na^+), it might be possible to make the sodium pump "run backward" and synthesize ATP. Garrahan and Glynn (1967) achieved such conditions by suspending reconstituted ghosts (made rich in K^+ and inorganic phosphate during reconstitution) in high-Na^+ media. They were then able to demonstrate ATP synthesis during the entry of Na^+ into the cells. This synthesis was inhibited by cardiac glycosides. Very recently Glynn and Lew (1970) have shown that the ouabain-sensitive K^+ efflux from starved intact human red blood cells incubated in high-Na^+, K^+-free media (associated with a ouabain-sensitive Na^+ entry and believed to be due to a reversal of the pump mechanism which normally carries Na^+ out of and K^+ into the cells) is accompanied by ATP synthesis. Further, the stoichiometric ratio involved (1 mole ATP synthesized/2 to 3 equivalents of K^+ leaving the cell) is of the order expected from an ATP-driven Na^+-K^+ pump of the type believed to exist in the red blood cell membrane.

The Sodium Pump and the Regulation of Cell Volume

Unlike plant cells, which possess a tough outer wall capable of withstanding large pressure differences, animal cells, as illustrated by the above discussion of red blood cell hemolysis (p. 11), have little mechanical ability to resist volume changes due to osmotic gradients. If all the diffusible ions present in the body were distributed across cell membranes in accordance with the Donnan relationship (as Cl^- frequently is and K^+ sometimes is approximately), the resulting colloid osmotic pressure would cause cell swelling and damage on account of osmotic entry of water. By keeping Na^+ "off balance" relative to its electrochemical equilibrium, the Na^+ pump successfully counteracts this tendency for osmotic swelling to occur. Colloid osmotic swelling of cells has been demonstrated under conditions in which the Na^+ pump is inhibited, and it has been found that where the inhibition of the pump is reversible swelling can also be reversed (Tosteson, 1964).

Ion Transport in Epithelial Cells

So far this chapter has been concerned with transport phenomena in "symmetrical" cells, i.e., cells in which the membrane (aside from small localized special areas such as the end-plate zones in skeletal muscle fibers) is homogeneous all around the cell periphery. Such cells may be said to discriminate between two spatial regions only, the cell interior and the external milieu. In other words, once a molecule or ion enters the cell there is no reason why it should emerge in any one specific direction rather than another. Thus the cell *as a whole* has no vectorial properties in relation to membrane transport. A schematic summary of ion transport in a symmetrical cell is shown in Figure 1-8A.

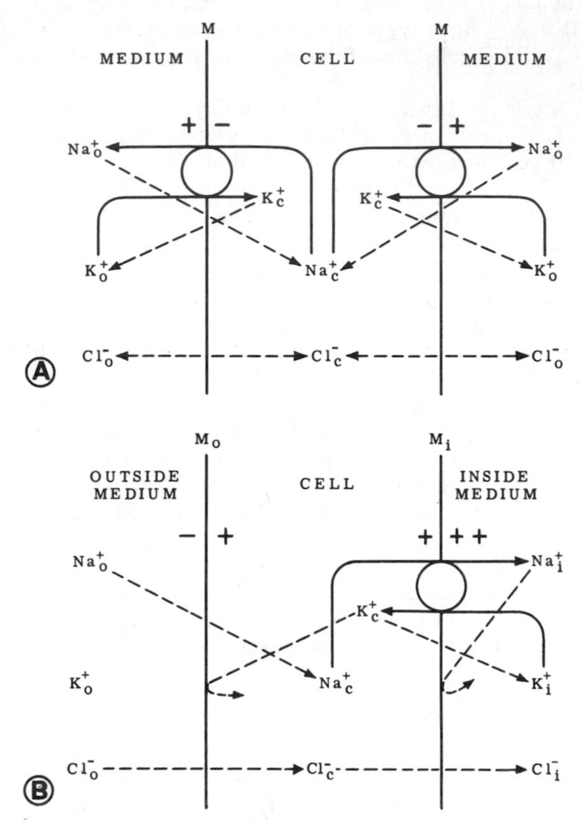

FIGURE 1-8. A. Symmetrical cell – regulation of ionic content. M (membrane): Na^+ leaks in, pumped out; K^+ leaks out, pumped in (possibly as coupled Na^+-K^+ pump); Cl^- in equilibrium across membrane. B. Asymmetrical cell (frog skin epithelium) – transcellular transport. M_o, M_i (outer and inner membranes): $M_o : P_{Na} \gg P_K$ – no pump; $M_i : P_K \gg P_{Na}$ – Na^+ pump. Na^+ leaks in through M_o, pumped out through M_i. A potential is generated which can drag a permeable anion (e.g., Cl^-) across the cell.

Several groups of cells in the body, however, show a completely different behavior because the membrane bounding one part of these cells has different properties from the membrane surrounding the remainder. Such cells can effect net *transcellular* movement of ions and other solutes from the solution bathing one side of them to the solution bathing the other, often against an energy gradient.

These are *epithelial cells.* They usually occur in sheets or layers in which individual cells are joined together by discrete junctional areas. Examples of highly functional epithelia (from the point of view of membrane transport) in the human body are the gastric and intestinal mucosa, the tubular epithelium of the kidney, and the cells lining the ducts of exocrine glands. Many of these will be discussed in detail in the appropriate parts of this book. However, they are sufficiently alike as a group and sufficiently unlike all other cells of the body to merit separate consideration here. Since a number of epithelial systems will be discussed at length later, it is proposed at this point to emphasize one preparation, the isolated surviving amphibian skin. This has served for years as an experimental model for the investigation of epithelial function in general, and many fundamental features of epithelial ion transport were first discovered through its use (Ussing, 1960).

It has been known for over a century that isolated frog skin when mounted between two physiological solutions maintains an electrical potential difference between its outer and inner surfaces. This potential difference is normally oriented so that the inside surface is positive with respect to the outside. In addition, the skin has a remarkable capacity for net Na^+ transport from the outer to the inner solution.

Net Na^+ transfer can be effected against both electrical and chemical potential gradients and therefore constitutes a clear and unequivocal example of active transport according to Rosenberg's definition (Wilbrandt and Rosenberg, 1961). As might be expected, the Na^+ transport process is extremely sensitive to a number of metabolic inhibitors (Ussing, 1960).

Not surprisingly, the view that these two processes, the maintenance of an electrical potential difference and active Na^+ transport, are somehow interrelated gained wide acceptance among physiologists. The clarification of the mechanisms underlying the interrelationship is due largely to the work of H. H. Ussing and his collaborators (Ussing, 1960). In particular, their introduction of the so-called short circuit technique provided an experimental tool which proved to be extremely useful not only in the study of isolated amphibian skin but in the investigation of the physiological behavior of a number of other isolated epithelial tissues (Schoffeniels, 1967). The principle of this technique is illustrated in Figure 1-9. The skin (S) is mounted between two identical Ringer solutions. The potential difference across it is measured by a pair of reversible electrodes (A, A' in Fig. 1-9) placed as close as possible to its outer and inner surfaces without actually touching them. The potential is recorded by the millivoltmeter (P). By means of a source of electromotive force (D) and a potential divider (W), a variable current can be passed through the skin using a second pair of electrodes (B, B'). This current can be applied in such a way as to oppose the spontaneous skin potential and can gradually be increased

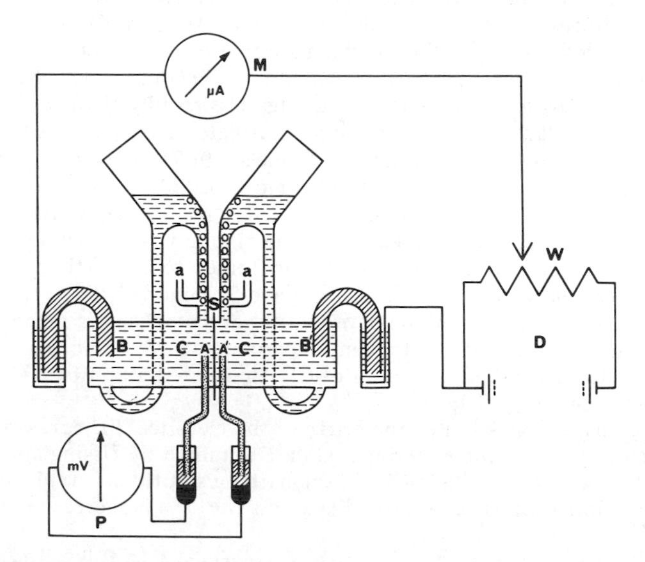

FIGURE 1-9. Apparatus for the determination of open circuit potential difference, short circuit current, and isotopic fluxes across isolated frog skin. See text for explanation of apparatus. (From Ussing and Zerahn, 1951.)

until the potential becomes zero. Under these conditions the current flowing in the external circuit and measured by the microammeter (M) is the *short circuit current.*

There are, therefore, two fundamental electrical parameters to be interpreted in the analysis of this system, the short circuit current and the spontaneous or open circuit potential across the skin. The latter is of course the potential difference (PD) observed when no external current is drawn from the battery (D). Let us consider each of them in turn.

First, what is the meaning of the short circuit current? It is evident that under short circuit conditions all external electrical, chemical, osmotic, etc. gradients which could cause a passive net flow of ions across the skin have been eliminated. In this situation the short circuit current must represent the algebraic sum of all *active net ionic fluxes* across the skin. Under normal circumstances the short circuit current is found to correspond to a net transfer of positive charge from the solution bathing the outside surface to the solution bathing the inside.

How can one identify which ions contribute to this current flow? The amounts of ions involved are so minute, compared to the total ionic content of the bathing solutions, that direct chemical estimates would be extremely difficult. The use of radioactive tracers permits the unidirectional fluxes of ions from outside the skin to inside (influx) and from inside to outside (efflux) to be determined accurately. Net flux is obviously the difference between these. The tracer method is particularly powerful when two different isotopes can be used simultaneously as in the case of Na^+. In this case one can label the solution on one side of the skin with ^{24}Na and the solution on the other side with ^{22}Na and determine influx and efflux simultaneously in the same preparation.

The application of tracer methods to the isolated frog skin has shown that, under normal conditions, short circuit current and net Na^+ flux from outside to inside are equal. Thus, one may conclude that in this preparation Na^+ is the *only* ion which shows *net* movement across the skin under short circuit conditions and is therefore the only ion which normally is actively transported across the tissue. Under open circuit conditions the situation is different. The electrical PD across the skin constitutes a driving force for *all* ions which are free to move. Thus, in this situation Cl^-, to which the skin is permeable, will also move (in the net sense) from the outside bathing solution to the inner one though less rapidly than Na^+.* The capacity for net transcellular Na^+ transport found in isolated frog skin has been shown to be a property of epithelia in general (Schoffeniels, 1967). Frequently, however, the situation under short circuit conditions is more complex than that exhibited by this tissue. In addition to active Na^+ transport there may be active transport of other ions. For example, in mammalian small intestine in vivo the evidence is convincing that besides an active transport of Na^+ from lumen to blood there is an active transport of Cl^- in the same direction (Curran and Solomon, 1957). In vitro studies using the short circuit technique have not yet detected such a mechanism in isolated mammalian intestine, but recent experiments have demonstrated the presence of an active lumen-to-blood transfer of Cl^- in isolated amphibian small intestine (Quay and Armstrong, 1969).

As regards the question of the origin of the electrical PD across the skin, let us first consider some of the experimental data obtained by Ussing and his collaborators (Ussing, 1960). When Cl^- in the bathing solutions is replaced by an impermeant anion such as sulfate, the PD across the skin is increased. (In Cl^- media

*This is of considerable importance to the frog! The sodium-transporting mechanism of the skin enables the intact animal to absorb salt from dilute solutions such as pond water and thus plays a vital role in its overall electrolyte regulation.

this potential is frequently in the range 50 to 70 mv. Replacement of Cl^- with sulfate increases it by some 20 to 30 mv.) Under these conditions the potential across the skin shows a linear dependence on the logarithm of the Na^+ concentration in the medium bathing its outer surface, and the slope of this line is close to the value predicted by the Nernst equation.*

On the other hand, the K^+ content of the outside bathing medium has very little effect on the skin potential. The opposite is found when the K^+ and Na^+ concentrations of the medium bathing the inner surface are changed. Here the skin potential shows a linear dependence on log $[K_i^+]$ which is reminiscent of the situation found with skeletal muscle (Fig. 1-7), but Na^+ has virtually no effect. In addition, specific Na^+ pump inhibitors such as the cardiac glycoside ouabain when applied to the inner surface of the skin abolish the PD across it but have little effect when applied to the outer surface.

Ussing (1960) has combined these and other observations in the model of transcellular transport illustrated in Figure 1-8B. Before discussing this in detail, however, a word of caution is in order. The model illustrated assigns the task of net Na^+ transport across the whole skin to a single layer of cells (one of which is shown schematically in Figure 1-8B). At one time it was thought that a single epithelial cell layer (the *stratum germinativum*) did in fact control net Na^+ transport in this tissue. Recent electron microscope and other studies (Schoffeniels, 1967) have shown that frog skin epithelium is a highly complex tissue, so that the "single cell layer" model is an oversimplification. It is nevertheless valuable because it illustrates the principles involved in a clear and simple fashion and also because it can, with suitable modifications, be applied to other epithelia (e.g., intestine, kidney tubules) in which transcellular Na^+ transport can with confidence be thought of as being effected by a single layer of cells.

In the model shown in Figure 1-8B the outer membrane of the epithelial cell is regarded as being virtually a Na^+ electrode ($P_{Na} \gg P_K$) and the inner membrane is considered a virtual K^+ electrode ($P_K \gg P_{Na}$). In the absence of a penetrating anion, Na^+ diffuses into the cell from the outside medium, causing a diffusion potential which is given by the equation

$$E_o = 60 \log \frac{[Na_o^+]}{[Na_c^+]} \qquad (16)$$

Similarly, K^+ diffuses outward across the inner membrane of the cell, giving rise to a second diffusion potential:

$$E_i = 60 \log \frac{[K_c^+]}{[K_i^+]} \qquad (17)$$

(in these equations o, c, and i refer to the outside bathing medium, cell interior, and inside medium, respectively). In the model illustrated the Na^+ pump in the inner membrane is assumed to be a coupled (nonelectrogenic) pump.† Therefore the total PD across the skin is given by

$$E_T = E_o + E_i \qquad (18)$$

---

*Note that in this case the PD increases with increasing external Na^+ as one would expect since the intracellular Na^+ in this tissue (as in virtually all tissues) is low and raising the external Na^+ concentration will raise the ratio $[Na_o^+]/[Na_c^+]$ (see equation 16).

†Recent studies indicate that this is not always the case in frog skin. At least it is now known that a tight one-to-one coupling of Na^+ and K^+ cannot always be assumed.

In the presence of an ion such as Cl^- to which both outer and inner membranes are permeable, both E_o and E_i will be reduced by the effect of Cl^- permeability. The appropriate equation for the total skin potential will then be the sum of two Goldman equations which take into account the effect of the Cl^- permeabilities of both membranes:

$$E_T{}' = E_o{}' + E_i{}' = 60 \left[\log \frac{[Na_o{}^+] + \beta [Cl_c{}^-]}{[Na_c{}^+] + \beta [Cl_o{}^-]} + \log \frac{[K_c{}^+] + \gamma [Cl_i{}^-]}{[K_i{}^+] + \gamma [Cl_c{}^-]} \right] \quad (19)$$

In this equation the coefficients β and γ represent $P_{Cl}{}^o/P_{Na}{}^o$ and $P_{Cl}{}^i/P_K{}^i$ respectively. In both the presence and the absence of a penetrating anion the model predicts that the total skin potential should be the sum of two potentials in series. In agreement with this prediction, two potential jumps (of increasing positivity) were recorded when a microelectrode was slowly advanced through the skin from the outside to the inside (Schoffeniels, 1967).

As in the case of other cells the Na^+ pump plays a vital role in maintaining the diffusion gradients for Na^+ and K^+ across the two membranes which are necessary to sustain a steady potential across the skin. Inhibition of the pump allows Na^+ to accumulate inside the cell and K^+ to diffuse out of it, dissipating these gradients and causing the potential to drop to zero.

Co-Transport

The movement of Cl^- ions across the isolated open-circuited frog skin is an example of the phenomenon of co-transport. Co-transport takes place where the energy gradient created by the active transport of an ion (in the present case the electrical gradient generated by active Na^+ transport) is used to drive another ion or molecule in the same direction across a membrane. Co-transport has been much studied in recent years, particularly in the small intestine, where some of its most striking manifestations have been found (Armstrong and Nunn, 1970). In intestinal epithelia net transcellular Na^+ transport from lumen to blood is thought by many to provide the energy gradient necessary for the active transport of sugars and amino acids as well as for passive transfer of water and of other ions. Some aspects of intestinal co-transport are discussed in Chapter 26. At the present time the mechanisms by which energy coupling between intestinal Na^+ transport and intestinal transport of sugars and amino acids is achieved are not known. However, this field is one of the most active ones in current membrane research.

REFERENCES

Armstrong, W. McD., and A. S. Nunn, Jr. (Eds.). *Intestinal Trasnport of Electrolytes, Sugars, and Amino Acids.* Springfield, Ill.: Thomas, 1970.

Christensen, H. N. *Biological Transport.* New York: Benjamin, 1962.

Curran, P. F., and S. G. Schultz. Transport Across Membranes: General Principles. In W. O. Fenn and H. Rahn (Eds.), *Handbook of Physiology.* Section 6: Alimentary Canal, Vol. III. Washington, D.C.: American Physiological Society, 1968. Chap. 65.

Curran, P. F., and A. K. Solomon. Ion and water fluxes in the ileum of rats. *J. Gen. Physiol.* 41:143–168, 1957.

Davson, H. *A Textbook of General Physiology,* 3d ed. Boston: Little, Brown, 1964.

Dockry, M., R. P. Kernan, and A. Tangney. Active transport of sodium and potassium in mammalian skeletal muscle and its modification by nerve and by cholinergic and adrenergic agents. *J. Physiol.* 186:187–200, 1966.

Garrahan, P. J., and I. M. Glynn. The incorporation of inorganic phosphate into adenosine triphosphate by reversal of the sodium pump. *J. Physiol.* 192:237–256, 1967.

Glynn, I. M. Membrane ATP-ase and cation transport. *Brit. Med. Bull.* 24:165–169, 1968.

Glynn, I. M., and V. L. Lew. Synthesis of adenosine triphosphate at the expense of downhill cation movements in intact human red cells. *J. Physiol.* 207:393–402, 1970.

Goldman, D. E. Potential, impedance and rectification in membranes. *J. Gen. Physiol.* 27:37–60, 1944.

Harris, E. J. *Transport and Accumulation in Biological Systems,* 2d ed. New York: Academic, 1960.

Hendler, R. W. Biological membrane ultrastructure. *Physiol. Rev.* 51:66–97, 1971.

Hodgkin, A. L., and P. Horowicz. The influence of potassium and chloride ions on the membrane potential of single muscle fibers. *J. Physiol.* 148:127–160, 1959.

Hodgkin, A. L., and B. Katz. The effect of sodium ions on the electrical activity of the giant axon of the squid. *J. Physiol.* 108:37–77, 1949.

Hoffman, J. F. The red cell membrane and the transport of sodium and potassium. *Amer. J. Med.* 41:666–680, 1966.

Ling, G. N. *A Physical Theory of the Living State.* Boston: Blaisdell, 1962.

Lucy, J. A. Ultrastructure of membranes: Bimolecular organization. *Brit. Med. Bull.* 24:127–134, 1968.

Pressman, B. H., and D. H. Haynes. Ionophorous Agents as Mobile Ion Carriers. In D. C. Tosteson (Ed.), *The Molecular Basis of Membrane Function.* Englewood Cliffs, N.J.: Prentice-Hall, 1969.

Quay, J. F., and W. McD. Armstrong. Sodium and chloride transport by isolated bullfrog small intestine. *Amer. J. Physiol.* 217:694–702, 1969.

Schoffeniels, E. *Cellular Aspects of Membrane Permeability.* New York: Pergamon, 1967.

Skou, J. C. Enzymatic basis for active transport of Na^+ and K^+ across cell membrane. *Physiol. Rev.* 45:596–617, 1965.

Stein, W. D. *The Movement of Molecules Across Cell Membranes.* New York: Academic, 1967.

Tosteson, D. C. Regulation of Cell Volume by Sodium and Potassium Transport. In J. F. Hoffman (Ed.), *The Cellular Functions of Membrane Transport.* Englewood Cliffs, N.J.: Prentice-Hall, 1964.

Ussing, H. H. The Alkali Metal Ions in Isolated Systems and Tissues. In O. Eichler and A. Farah (Eds.), *Handbuch der Experimentellen Pharmakologie,* Vol. 13. Berlin: Springer, 1960. Pp. 1–195.

Ussing, H. H., and K. Zerahn. Active transport of sodium as the source of electric current in the short-circuited isolated frog skin. *Acta Physiol. Scand.* 23:110–127, 1951.

Wilbrandt, W., and T. Rosenberg. The concept of carrier transport and its corollaries in pharmacology. *Pharmacol. Rev.* 13:109–183, 1961.

2

General Properties of Nerve

Sidney Ochs

ONE of the most fundamental properties of cells is *excitability,* defined as the capacity of cells to respond to changes in the external environment. Excitation is demonstrated in a unicellular organism such as the ameba. When this cell is subjected to either mechanical or chemical stimulation of part of its surface, the response is the formation of pseudopods and movement of the ameba. Much evidence suggests that these pseudopodal reactions are excited at the surface membrane. Special physicochemical reactions of the membrane, such as have been discussed in Chapter 1, in turn control excitability, as will be discussed in Chapter 3A.

The primitive changes due to excitability which are found in the single cell become more highly developed in the multicellular organism. In the course of evolutionary specialization (individuation), cells arise whose chief functions are concerned with excitation and conduction. The *neuron* is such a cell. Other cells, the *effectors,* become specialized for contraction and movement of the organism – e.g., in higher forms, the muscles (Chap. 3B).

PHYLOGENETIC ASPECTS OF NERVE CELL FUNCTION

In the coelenterates (Fig. 2-1) one sees clearly the development of special sensory cells which respond selectively to environmental stimulation. Special motor (effector) cells are found which can change the shape of the body wall and the position of the tentacles, thereby enabling the animal to move about, capture food, and perform its other functions. Also developed are intervening neurons which act to control the spread of incoming excitation from specialized receptor cells to the muscular effector cells. These intervening cells are a primitive nervous system which has been referred to as a *nerve net.* Study of the diffuse spread of excitatory activity in the nerve-end system shows the operation of the simplest nervous system. In the higher organisms, the nerve net type of neural control is preserved in some of the visceral organs. The rhythmic movements of the hairlike cilia lining the inside surface of the trachea are controlled by such a nerve net. The cilia act to carry mucus out of the lung. As another example, the nerve net plexus in the intestinal tract produces the waves of peristalsis which propel food along the canal.

The fundamental plan of the higher life forms made its appearance very early in evolution. In the linearly oriented organism represented by the segmented worm, ganglia arise within each segment. These ganglia contain neurons which receive

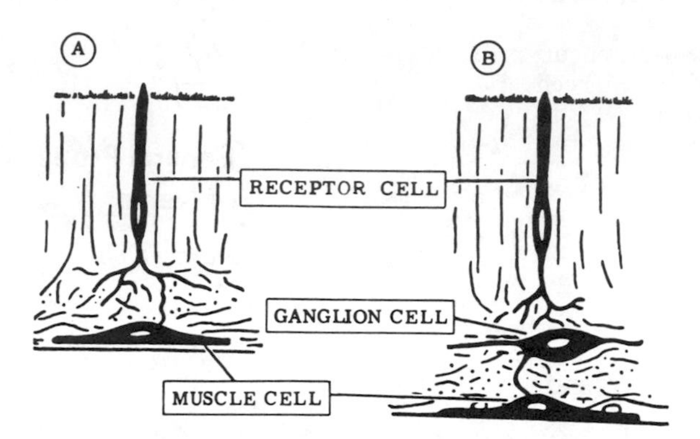

FIGURE 2-1. Example of primitive nervous systems. In A, a specialized receptor cell of a sea anemone makes contact with a muscle cell. In a more complex arrangement in the wall of the organism shown in B, a ganglion cell is interposed between sensory and motor elements. (From P. Bard, Receptor Organs and Discharges in Sensory Nerves. Modified after Parker, Prentiss, and Bayliss. In P. Bard [Ed.]. *Medical Physiology,* 11th ed. St. Louis: Mosby, 1961.)

sensory information from sensory fibers located in the ectoderm, the fiber endings responding selectively to chemical, radiant, and mechanical stimuli. Larger ganglia develop at the head end of the organism. The neurons of the ganglia make contact with one another at junctions which are called *synapses* (Gr. *synapsis,* "contact"). Sensory cells and interneurons within the ganglia eventually excite neurons which are motor in function. These *motoneurons* have elongated axonic portions which in turn synapse on and excite muscle fibers of the segmental musculature.

An appropriate stimulus causes a *reflex* to take place. In a reflex, sensory neurons (afferents), central neurons, and motoneurons (efferents) are involved. The central neurons are arranged in certain patterns, and upon sensory stimulation an appropriate motor nerve discharge is elicited, thus producing a reflex response (see Chap. 5). The fibers in the nerve cords passing between the ganglia and connecting the central nervous neurons with one another constitute the means whereby the segments of the organism are brought into coordinated activity. It would not be possible for the animal to move unless each segment acted at the right time with respect to other segments, thus bringing each appendage into position to exert its force in the movement of the whole organism. This conjoining of activity of the various parts of the body was pointed out by Sherrington as the essential role of the nervous system: to integrate the various cells of the body so that the organism behaves as a coordinated entity.

As an animal progresses forward in its environment, the head end is brought first into contact with the source of sensory stimuli. The gradual phylogenetic development of more neurons in the headward parts under the influence of greater activity there was designated in 1917 by Ariens Kappers as *neurobiotaxis.* The massive and complicated brain that is encountered in the higher forms is the result of the tremendous augmentation of the numbers of sensory and motor cells, seen only as a simple ganglion in the anterior part of primitive nervous systems.

Whether in the center or at the periphery, neurons are chiefly engaged in the process of excitation and conduction. These two functions of cell activity will be treated separately for simplicity of exposition, but first the structure of the nerve will be considered.

ANATOMY OF THE NERVE AND ITS PROCESSES

Several representative neurons found in the vertebrate are shown in Figure 2-2. The sensory neuron with its cell body in the dorsal root ganglion has one long fiber passing outward to innervate the skin, muscle, or other parts of the body. At its termination in the periphery the afferent fiber ending is often combined with other special

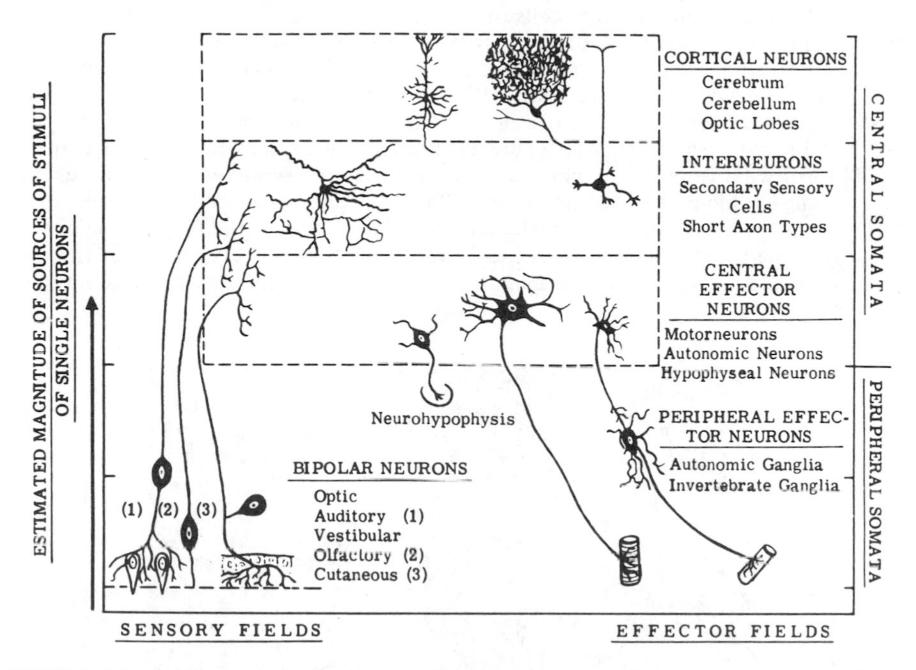

FIGURE 2-2. Varieties of neurons. Sensory neurons in the higher animal forms are typically bipolar cells, shown to the left. Both the fiber portion receiving sensory input and the portion synapsing centrally are long. Effector neurons often have a long terminal axon, except for those of the autonomic nervous system. Shown in the dashed box are cells with very great arborizations of dendrites. (From D. Bodian. *Cold Spring Harbor Symposia on Quantitative Biology* 17:3, 1952).

associated cells to form a sensory receptor organ. Impulses pass in the usual direction from the receptor ending to the cell body, then along another fiber process of the sensory cell into the central nervous system (CNS), where synapses are made on cells within the CNS. A motoneuron such as is typically found in the ventral portion of the gray matter of the spinal cord, the ventral horn of the spinal cord, may have a large, wide, ramifying system of treelike branches known as the *dendrites* (Gr. *dendron,* "tree"). These arise from the cell body of the neuron, which is also referred to as the *soma* or *perikaryon.* Many synapses take place on the dendrites as well as on the soma. At the synapse, as can be observed in electron microscope studies, there is no merging of the membrane of one cell with another. Across the cleft, one cell excites another by the release of transmitter substance (see Chaps. 3A and 5).

Most neurons have all their portions within the CNS. The cerebral cortex, the thin layer of gray matter over the surface of the brain, contains as one of its characteristic cells the pyramidal cell (Fig. 2-2). This cell takes its name from the pyramidal form of the cell body. The apex of the pyramid is pointed radially outward to the surface, from which a large dendrite takes origin – the *apical dendrite.* The apical dendrite passes upward for some distance before dividing into smaller and smaller branches and, by thus multiplying, makes available a relatively large area of dendritic surface for receiving synapses terminating on the branches. This allows the neuron to be influenced by many other cells of the cortex or from other regions in the brain. Its axon passes to other parts of the cortex or to other regions of the CNS. Another type of cortical neuron is the stellate (L. *stellatus,* "starry") cell. It has a short axon and performs an integrative function with a relatively smaller region. Regions of the CNS where cell bodies and dendrites and their connections predominate are gray in color compared with the regions where axon tracts are dense, which appear whiter.

Most of what is known of nerve excitability has been obtained from studies made on the long axonic part of the nerve cell. The axon plus associated structures is referred to as the nerve fiber. In particular, the sciatic nerve of the frog has been a favorite tissue for examination. As shown in Figure 2-3, the nerve trunk of the frog sciatic nerve (as is typical of mammalian nerves also) is composed of a large number

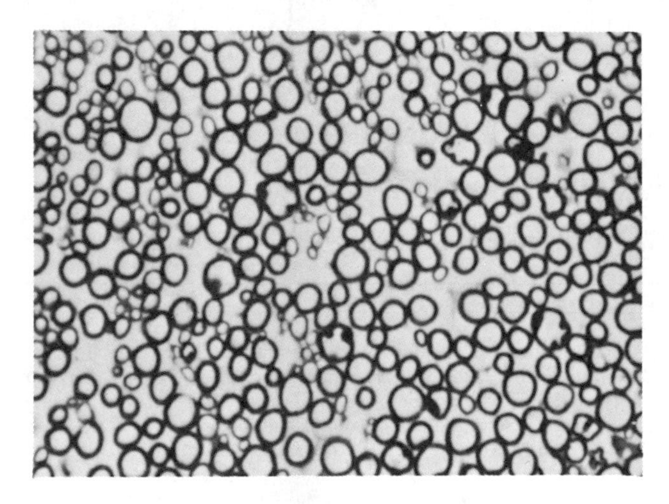

FIGURE 2-3. Cross section of myelinated nerve fibers. The dense bands around the fibers are myelin sheaths. The smallest groups of fibers are the nonmyelinated fibers, which typically are found in bundles, and are not visible here. ×425.

of fiber axons which are enclosed within a fibrous sheath (the epineurium). The diameters of individual nerve fibers range from a fraction of 1 μ up to approximately 20 μ. The fibers above 1 μ in diameter have a myelin sheath around them and are therefore known as *myelinated nerve fibers.* Myelin has a large content of lipid material. Sections of nerve stained with dyes having an affinity for myelin (e.g., osmium tetroxide) show the fiber as a ring of myelin (Fig. 2-3). Fibers below 1 μ in diameter are usually not observed with light microscopy and are referred to as *nonmyelinated.* Recent electron microscope studies of the myelin sheath of the myelinated fibers have revealed that the myelin is composed of layers which are regularly arranged in the form of lamellae. The myelin layers are laid down by the *Schwann cell* at intervals along the length of the fibers. The axon is not covered with myelin at the *nodes* between Schwann cells. As will be discussed later, special properties of conduction are conferred on the nerve fiber by the myelination. Electron microscope studies of growing nerves during embryogenesis have revealed that myelin growth occurs by a jelly-roll wrapping process around the axon to produce the typical lamellated appearance (Fig. 2-4). The nerve fibers in the CNS are similarly myelinated by their analogous satellite cells – the *glial cells.* To be compared

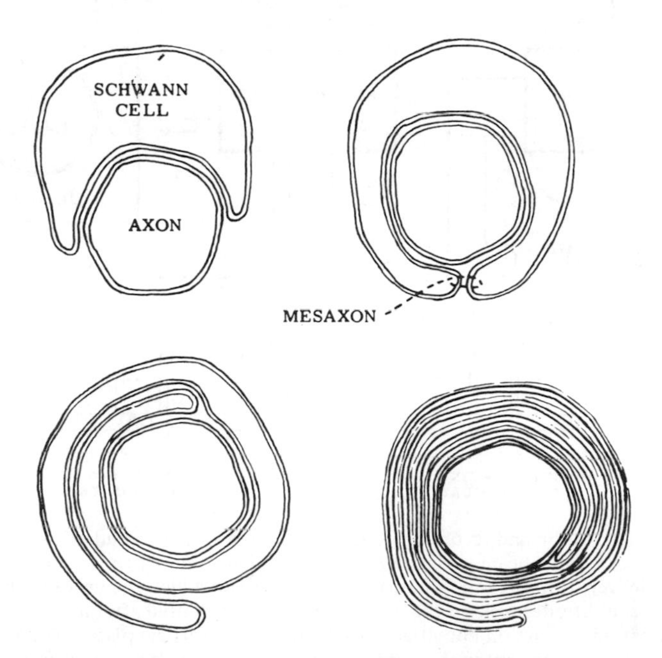

FIGURE 2-4. Embryological development of the myelin sheath. Early in development the Schwann cell becomes approximated to the axon. The membranes of the Schwann cell meet to form the mesaxon. One surface pushes forward and in so doing wraps a double membrane around the nerve. Myelin is thus laid down in a jelly-roll fashion. (From B. B. Geren. *Exp. Cell Res.* 7:559, 1954.)

with the layered myelinated nerve structure are the thinner nonmyelinated fibers. These have been shown in electron micrographs to have one layer of myelin around them. The giant fibers of squid and cuttlefish are a special type of nonmyelinated nerve fiber which under the electron microscope are seen to have several irregular myelin-like layers around them supplied by satellite cells analogous to the Schwann cell. These nerves attain diameters up to 1 mm. The use of such large-diameter fibers has supplied the basis for a modern theory of nerve excitation, to be considered later in this chapter.

EXCITATION AND RECORDING OF THE ACTION POTENTIAL

The frog sciatic nerve has long been favored for the study of excitation and conduction because it can survive for many hours after it has been removed from the animal.

The isolated sciatic nerve is placed in a moist air chamber where electrical contact can be made to it through electrodes at various points along its length. As indicated in Figure 2-5, two stimulating electrodes are used to convey a brief pulse of current through a portion of the nerve at one end. When a current of adequate strength is used, the fibers become excited and a *nerve impulse* passes along the nerve. If the

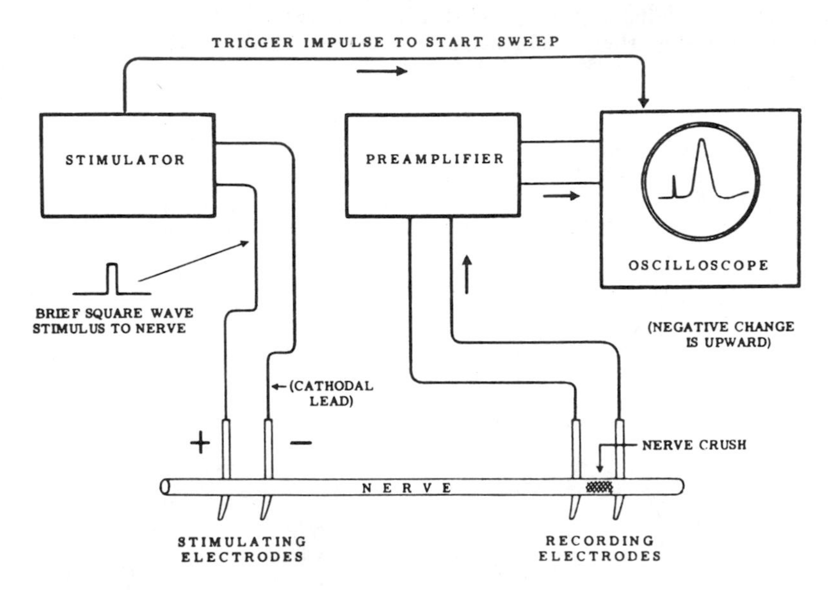

FIGURE 2-5. Excitation and recording of action potentials. The stimulator starts the horizontal sweep of the oscilloscope beam with a trigger pulse, and at some set time thereafter a brief pulse of current is delivered through the stimulating electrodes on which the nerve is placed. Farther along the nerve, electrodes pick up the propagated action potential after a conduction delay. After amplification, the action potential is displayed on the vertical plates of the oscilloscope. The combined time movement in the horizontal direction and voltage changes in the vertical direction give an X-Y graphic display.

sciatic nerve were still innervating a muscle and the nerve fibers so excited were motor nerves, a brief contraction (twitch) of the muscle innervated by those fibers would result. In earlier investigations nerve properties were studied by the muscle response although muscle contraction (see Chap. 3B) made the analysis difficult. In spite of this some simple observations can be made using the muscle twitch as an index of effective nerve excitation. Stimulation below a strength effective to excite is called a *subthreshold stimulus.* A strength of stimulation which causes a just discernible motor response is referred to as a *threshold stimulation.* A further increase of stimulation causes more fibers to respond, and the twitch increases in size. When all the nerve fibers are excited, this is a *maximal response.* An additional increase in strength, a *supramaximal* stimulation, does not further increase the twitch size.

More accurate studies of excitability can be made by recording the electrical changes which constitute the nerve impulse. Two recording electrodes along the sciatic nerve are connected through an amplifier and then to an oscilloscope, as shown in Figure 2-5. The nerve impulse is a small brief negative voltage change (giving rise to the term *action potential* or *spike* for the nerve impulse) which is picked up by these electrodes. The response must be considerably amplified before it can be displayed upon the screen of an oscilloscope.

Since much recent work in neurophysiology involves use of the oscilloscope, the underlying principles of this instrument will be mentioned. The oscilloscope is a vacuum tube device by which a small beam of electrons is focused on the inner face of the tube and visualized by the fluorescence it produces. Voltages are applied to horizontal and vertical plates arranged along the path of the beam so that the electron beam can be displaced horizontally and vertically on the face of the screen. A special sweep circuit is used to apply appropriate voltages to the horizontal plates to make the beam move uniformly horizontally from the left to the right side of the tube screen. The rate of this movement controlled by the sweep generator constitutes the time axis. After a horizontal sweep is completed the beam is shut off and it returns to the starting position until a trigger pulse starts the movement. If, while the beam is moving in the horizontal X direction, voltages are applied to the vertical Y plates, the beam will give rise to a time display of the impressed voltage.

CHARACTERISTICS OF THE ACTION POTENTIAL

The brief electrical signal seen on the oscilloscope trace when the stimulating pulse has passed through the stimulating electrodes and the nerve is an *artifact* caused in large part by an adventitious capacitative coupling to the recording electrodes. The artifact serves as a convenient indication of the time when a stimulus has been applied to the nerve. After delivery of the stimulus, a *latency* is seen as the oscilloscope beam moves along for a time with no further voltage change. The latency depends on the speed at which the nerve impulse is propagated along the nerve from the stimulating electrodes to the recording electrodes, and on the distance it travels to the first recording electrode to make it negative with respect to the other electrode. The latency is usually taken from the cathode of the stimulating electrode, which is the electrode closest to the first recording electrode (Fig. 2-5). There is, however, some variation in the site of origin of excitation under the cathode lead of the stimulating electrode. (The actual point at which the excitation starts may, because of spread of currents in the epineural sheath, be initiated a millimeter or more away.)

At the usual distances of several centimeters between stimulating and recording electrodes the conduction velocity is computed to be approximately 40 meters per second for the fastest action potential of the frog sciatic nerve. As the action passes the first electrode, it becomes negative. The action potential continues on past the first electrode until it reaches an intermediate position where both recording electrodes are relatively equally negative, so that no difference in potential exists between them. The beam, as shown in A of Figure 2-6, returns to a position of isoelectricity, i.e., back to the baseline level. Then, as the action potential passes under the second recording electrode, the latter becomes relatively more negative with respect to the first electrode. The beam of the oscilloscope is directed in the reverse (downward) direction and later returns to the baseline, thus constituting the *diphasic action potential.*

To convert the diphasic action potential to the more useful *monophasic action potential,* the nerve is crushed between the two recording electrodes. The nerve impulse can then pass by only the first electrode, the crush preventing passage of a conducted action potential to the second electrode. The monophasic action potential is ordinarily arranged so that the negative deflection is recorded in the upward direction on the face of the oscilloscope.

The action potential discharge is said to be *all-or-none.* A propagated all-or-none action potential passing along a fiber is analogous to a burning dynamite fuse which

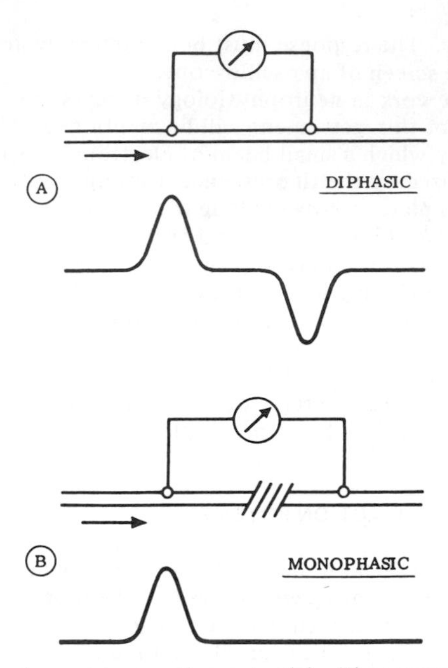

FIGURE 2-6. Diphasic and monophasic action potentials. The arrows show the direction of an action potential as it moves along a nerve in A past the first and then the second recording electrode. As it passes each electrode, that electrode becomes relatively negative with respect to the other, accounting for the deflection first one way, then the other, giving rise to the diphasic action potential. In B a monophasic action potential is seen because the crush of the nerve prevents the action potential from passing to the second recording electrode.

obtains its energy from the amount and condition of the powder all along its length. If, for example, a region of the fuse is encountered which is damp or deficient in powder, the burning may be slowed and sputter and failure may occur. But if this area is passed, burning resumes in the next region. So in nerve conduction the action potential does not decrement down the length of the fibers but develops its fullest possible amplitude in all the regions encountered in its propagation.

While the form of the monophasic action potential is usually a simple smooth wave of increasing negativity with return to baseline within approximately 0.5 to 1.0 msec, some smaller voltages may appear afterward. A *negative afterpotential* is seen as a delay in the return of the action potential, merging with another smaller and much longer lasting *positive afterpotential.* Uncertainty exists regarding the significance of the afterpotentials recorded from multifibered nerve trunks. Afterpotentials recorded from giant single fibers and from small nonmyelinated nerve trunks can be directly related to ionic events and metabolism (see below).

The fibers of a mixed nerve such as the frog sciatic nerve have different diameters. Figure 2-3 indicates the variations of size. The action potential excited by use of the lower stimulus strengths comes from the most excitable fibers and, as has been found by Erlanger and Gasser (1937), from the group of fibers with the largest diameters. These fibers are classified as the *A fiber group,* which is divided into subgroups of successively smaller sizes. Greek letters are used to designate the various subgroups of the A group: *alpha, beta, gamma,* and *delta.* The alpha fibers, ranging in diameter from about 15 to 18 μ in the frog sciatic nerve, are the most excitable subgroup of fibers of the A group. With just threshold stimulus strengths only the most excitable fibers of the alpha group are excited, and the response is small in size. As the stimulation strength is increased, more fibers of the alpha group are added to the responding population until the action potential response reaches a maximum. Further increase in stimulation strength does not increase the size of the alpha group response. With greater strengths a hump appears on the descending portion of the alpha action potential. This addition of potential is due to the excitation of the beta group of fibers (Fig. 2-7). Still greater stimulation strength similarly

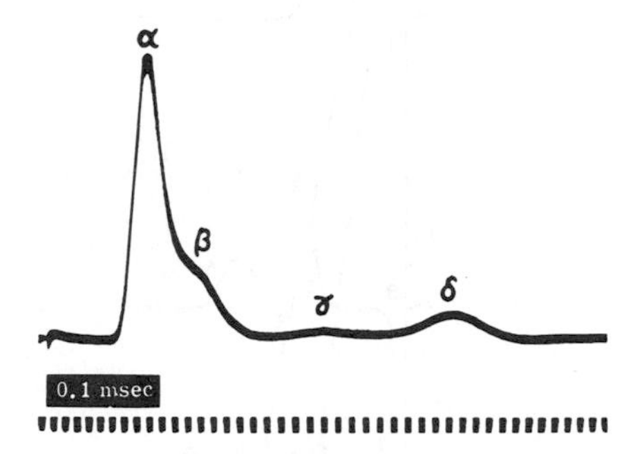

FIGURE 2-7. Compound action potential. The A fibers are subdivided into subgroups labeled with Greek letters having successively higher stimulation thresholds and slower conduction velocities. (Note: γ seen only in motor nerves.) (From H. S. Gasser. *Association for Research in Nervous and Mental Disease.* Proceedings [1942] 23:48, 1943.)

excites the gamma group. Successively later appearing responses are due to the fact that the smaller fiber groups have not only a higher threshold but also a lower conduction velocity. The gamma group does not represent a separate group in skin nerves, as shown in Figure 2-7, but it is present in motor nerves (Chap. 5).

The action potentials of these different nerve groups may be better separated by increasing the distance between the stimulating and recording electrodes. Just as in a race the faster and slower runners are bunched together at the start but later spread out as the race proceeds, so in nerves the action potentials traveling in the various fiber groups will show a better spread if the stimulating-recording distance is increased, as seen for the alpha and beta groups in Figure 2-8.

If a second stimulus is initiated too soon after a first effective stimulus, it does not excite an action potential. The nerve is said to show absolute refractoriness; i.e., it is inexcitable no matter how strong the stimulus. In the alpha group of A fibers of the frog sciatic nerve, absolute refractoriness lasts only a little longer than 1 msec and is followed by a period of *relative refractoriness,* in which some response is possible if the strength of stimulus is increased. As the fibers recover their full excitability, the amplitude of response increases along an S-shaped curve until the action potential equals that of the control response. The duration of the relative refractory period for the alpha group of the frog sciatic nerve is approximately 3 to 5 msec.

STIMULATING CURRENTS AND EXCITABILITY

Electrical conduction in electrolyte solutions takes place by the movement of ions. The major ions of intercellular (interstitial) fluids in mammalian tissue are Na^+ and

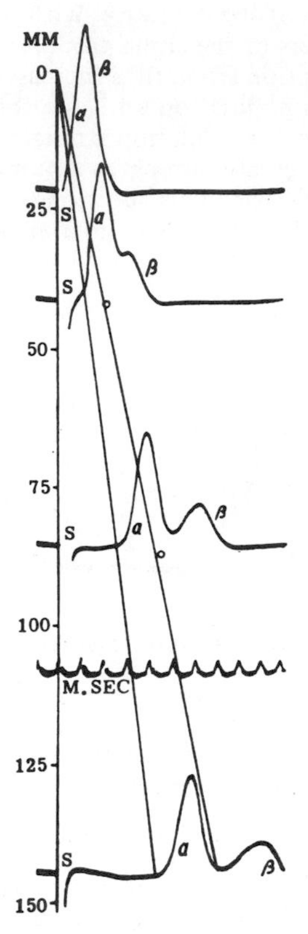

FIGURE 2-8. Separation of alpha (α) and beta (β) groups. If the distance between stimulated and recording sites along a nerve is successively increased, then the α and β groups become easier to distinguish. The velocities of the two groups are indicated by the differing slopes of the two lines drawn through the beginning (the foot) of their action potentials. At a farther distance the foot of the β group is clearly separated from that of the α group. (From Erlanger and Gasser, 1937).

Cl^-. The positively charged Na^+ moves toward the cathode and the negative Cl^- moves toward the anode when a current is impressed on tissues. Nerves and body tissues in general can be considered as electrolyte solutions of various ionic compositions. Ions moving between electrodes because of an imposed difference of potential constitute a current flow; the number of ions flowing is directly related to the quantity of current. Lines drawn between the electrodes signify the direction, and the number of lines the density, of the current.

As a general rule, nerve cells are excited in the region close to the cathode. If lines of current are drawn from the anode through and into the individual fibers of the nerve trunk and then back to the cathodal electrode of the stimulating pair of electrodes, one must take into account two major paths of current flow (Fig. 2-9A). One path of current flow passes through the ionic solution between the nerve fibers, in the intercellular compartment. This intercellular current is ineffective as regards excitation. Another path of current flow is from the anode inside the fibers for some distance before leaving the fibers at the cathode. It is passage of current outward across the membrane that causes excitation to take place. The lines of current are pictured in Figure 2-9 and show only the form of current flow with no particular designation of the species of ion along the indicated current line. The actual ions may differ at different regions along the paths of these current flows. By convention, the direction of current flow is from anode to cathode.

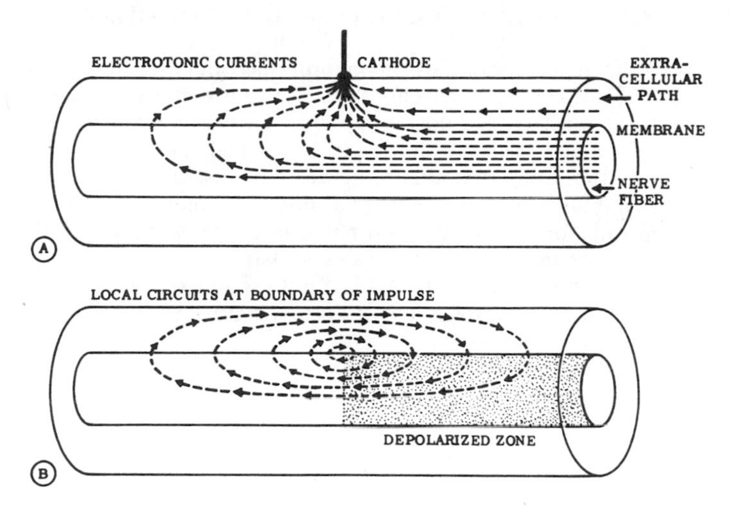

FIGURE 2-9. Paths of applied current and local circuits resulting from an action potential. In A the lines of current from an applied source are shown passing from the nerve to the cathode. Part of the current passes between the fibers, and part inside. The latter current is the effective portion, and excitation occurs in the region of the cathode as current passes outward across the nerve membranes and depolarizes the membranes. In B a similar direction of outward current in the region in front of an active part of the fibers depolarizes and can excite that region of nerve.

The effective part of the current is that which flows within the fibers, and it is a small portion of the total current applied, for two reasons. First, the membranes of the fibers (and other excitable cells such as muscle) offer a high resistance to the passage of electrical current. This relatively high resistance can be pictured by suitable resistance elements in an electrical model of the nerve (Fig. 2-10). Second,

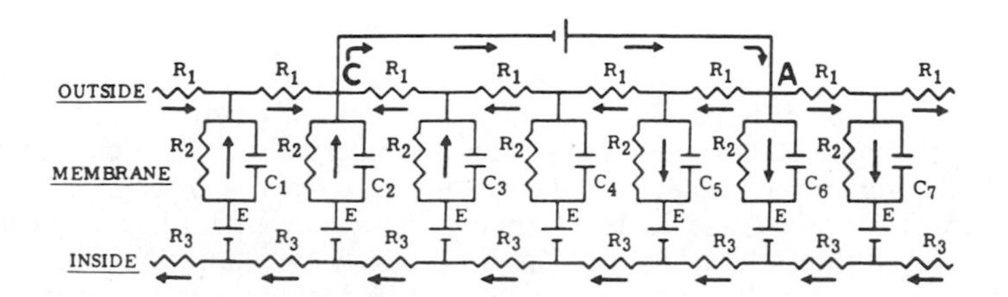

FIGURE 2-10. Electrical model of the membrane. The arrows in this model show the direction of current through an electrical network representing the nerve membrane. The model is composed of resistance and capacitance elements with a battery representing resting membrane potential.

there is a high resistance of the axoplasm due to the relatively small diameter of the axons. Resistance is inversely related to the cross-sectional area of the fiber. And resistance increases with the length of the fiber. Excitation occurs in the membrane of the axon because the current flowing outward across the membrane reduces the resting membrane potential; i.e., it *depolarizes* the fiber. When depolarization reaches a critical level, an action potential is excited. In a later section of this chapter the ionic events related to excitation and the production of an action potential will be discussed.

Because of the high resistance of the membranes (and also their electrical capacitance), application of electrical currents gives rise to distributed potentials along the nerve to which the term *electrotonus* is applied. This term refers not only to distributed voltages along the nerve resulting from applied currents but also to potentials produced by the action potential of nerve. The distributed voltages have an effect on the membranes of nerve fibers and muscle to change their excitability. Electrotonic spread along the single giant nerve fibers showing the relation to excitability changes and conduction will be discussed later.

The relationship between the duration of a pulse of exciting current and the strength adequate to excite is the *strength-duration curve* (Fig. 2-11). The curve is

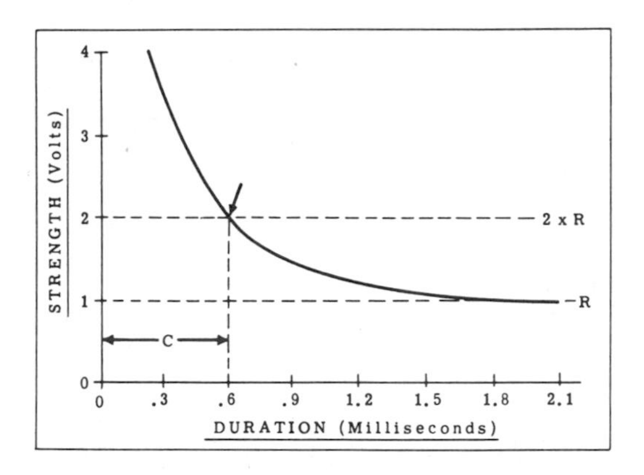

FIGURE 2-11. Strength-duration curve. The curve is derived from the strength required to attain a given level of response when various durations of stimulation are used. The longer pulse durations approach the rheobase (R). Chronaxie (C) is the duration of the stimulus required for excitation at a strength of stimulation twice rheobase (arrow).

determined by the level of strength of the stimulating current required to reach the same threshold when different durations of the stimulating current are used. As the duration of the pulse is shortened, the strength required rises, first slowly, then more rapidly. With the longer stimulus durations the curve reaches an asymptote called the *rheobase* (Gr. *rheo,* "time"). The term *rheobase* is often defined as the strength required to reach threshold when using an indefinitely long duration pulse. The term *chronaxie* is defined as the duration of stimulus required at a strength twice the rheobase. The more excitable tissues in general have shorter chronaxies. Muscle, for example, has a longer chronaxie than nerve.

Stimulation of certain points over the surface of the body via electrodes causes excitation of muscle responses. These *motor points* are places where a motor nerve enters the muscle. The strength-duration curve over the motor point shows the short chronaxie and a strength-duration curve typical of nerve excitation.

IONIC BASIS OF MEMBRANE POTENTIALS

For many years the presence of an electrical potential across the membrane of nerve and muscle was known from indirect evidence. Crushing of the nerve or placing KCl on one part of it causes a negative potential which appears in recordings made with a direct current (DC) amplifier. One electrode is placed on the injured part of the nerve or muscle, the other at least several millimeters away on intact tissue. The *injury potential* or *demarcation potential* is due to a relatively large current passing from the intact part of the fiber to the injured portion. The voltage is not simply a static electrical potential due to the special arrangement of molecules present in the membrane. In monomolecular layer films such a static voltage difference is found. The fact that current can be continuously drawn from the nerve is evidence that some type of "battery" must exist across the nerve cell membrane. The source of potential is the asymmetry of ions across the membrane. The ionic considerations developed in Chapter 1 apply to both resting and injury potentials. The Nernst equation discussed in Chapter 1 shows an *electrical driving force* acting in equilibrium with the *chemical driving force,* i.e., the ratio of K^+ across the membrane. Alteration of the voltage across the membrane will eventually result at equilibrium in a new ratio of K^+, and, conversely, changing the K^+ ratio will result at equilibrium in a new transmembrane voltage. An important point to note is that only a small number of ions need move across the membrane to produce the voltages recorded, of the order of several picamoles (pM = 10^{-12}M).

With the giant nerve axons obtained from squid and cuttlefish, in which diameters may be as high as 0.5 mm or more, the transmembrane potential found by inserting an internal electrode was approximately 50 to 60 mv, with the inside negative. Analysis of the K^+ concentration of axoplasm squeezed out from the giant axon and the K^+ in the bathing solution outside showed that the potential could be accounted for by the Nernst formulation.

Later the microelectrodes developed by Ling and Gerard in 1949 were used in extended measurements of a variety of cells. The tips of the microelectrodes measure 0.5 μ or less and can pass through the membrane with little damage. By this means nerve fibers and cell bodies as well as muscle cells from various species have been found to have a resting membrane potential of the order of 50 to 90 mv.

The results of changing external K^+ have in general confirmed the Nernst relationship, except that at the lower concentrations of K^+ a deviation from the expected voltage is found due to a small leakage of Na^+ into the cell. In muscle fibers there is also a rather large contribution of Cl^- to the resting membrane potential. Because Cl^- is low in concentration inside the muscle fiber and high outside, and has the reverse charge, it acts as a "battery" with the same polarity and in parallel with the K^+ battery.

It is possible not only to squeeze out the axoplasm of the giant axon but later to perfuse the axon with fluids of known composition. In the perfused axons K^+ was diluted with a nonelectrolyte such as sucrose. As expected from the Nernst equation, the membrane potential was reduced. With equal concentration of K^+ across the membrane the potential was close to zero.

With a micropipette inserted into a giant fiber and a resting membrane potential of 60 mv, stimulation of the fiber gave rise to action potentials having amplitudes of 90 mv (Fig. 2-12). An action potential larger in amplitude than the resting membrane potential was a new and unexpected finding. On the basis of the Bernstein theory of

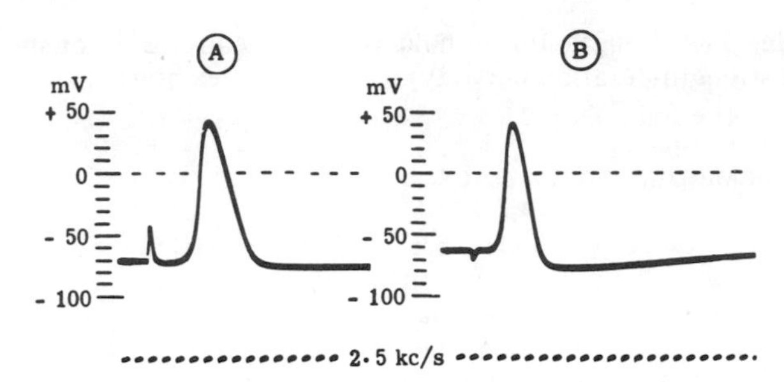

FIGURE 2-12. Action potentials from giant nerve fibers. In A the action potential is recorded from a giant nerve fiber in situ with a microelectrode. The resting membrane potential is 70 mv, inside negative. An action potential shows an overshoot of approximately 35 mv. In B, an internal capillary electrode inserted from one end measures a smaller resting membrane of approximately 60 mv. A hyperpolarization follows the action potential – the positive afterpotential. (From A. L. Hodgkin. *Proc. Roy. Sci. [Biol.]* 148:5, 1958.)

1912, the action potential was believed to be produced by an increased permeability to all ions, and the action potential could not exceed the resting membrane potential because of the ratio of K^+ across the membrane. The excess potential recorded with the inside electrode was called the *overshoot,* and with a resting membrane potential of 60 mv and a 90-mv action potential the overshoot amounted to 30 mv. The explanation of this finding resulted in a new theory of nerve action, the sodium hypothesis, which is now generally accepted as the basis of the action potential.

Sodium ion is high in concentration outside the cell and low inside. The concentration difference tends to make Na^+ enter cells; similarly, negative charge inside the cell attracts the positively charged Na^+ ion. Radioactive tracer studies have shown that Na^+ does in actuality continuously enter the cell. The Na^+ permeability, while low, would eventually allow a sufficient amount of the Na^+ to enter the cell to bring the resting potential to zero if the sodium pump did not eject the Na^+ which enters.

To explain the action potential by the sodium hypothesis: A reduction of the resting potential across the membrane of the nerve to a critical level causes a nonlinear change in the permeability of the membrane to take place. The permeability to Na^+ suddenly increases, bringing in its positive charge within a fraction of a millisecond. The entry of Na^+ accounts for the overshoot of the action potential because of the *equilibrium potential* for Na^+. The equilibrium potential for Na^+ computed from the Nernst equation is that voltage across the membrane which makes the inside 30 to 40 mv positive. It must be emphasized that only several pM of Na^+ move across the membrane during the action potential. The increase of Na^+ permeability does not continue indefinitely because of a process called *inactivation* which begins a few milliseconds later, acting to shut off the entry of Na^+.

Another permeability change then occurs. After a brief lag, K^+ leaves the interior of the cell, returning the membrane voltage to normal, i.e., bringing the action potential to an end.

The net effect at the end of the action potential is that the fiber has gained a tiny bit of Na^+ (3 to 6 pM) and lost an equivalent amount of K^+. The sodium pump will later act to redress the imbalance, as will be discussed in the next section. As far as the net voltage across the membrane is concerned, the total exchange of ions has been balanced. The increase in K^+ permeability can continue for some time, causing a *hyperpolarization* or *positive afterpotential* to appear (Fig. 2-12). The size of the afterpotential is related to the level of the resting membrane potential and the equilibrium potential for K^+. It is larger when the nerve is depolarized below its normal resting membrane potential, as shown in Figure 2-12.

The analysis of the sodium hypothesis was accomplished in the ingenious experiments of Hodgkin, Huxley, and Katz (Hodgkin, 1964) using electrodes inside the axon. By this means the whole of the membrane could be excited and the flow of current through the membrane measured. Upon depolarization, an early brief inflow of current appeared, and this was due to an entry of Na^+. Later the current became reversed, as a result of an outward flow of K^+. When the nerve was immersed in solutions from which Na^+ was removed and replaced with the inert substance sucrose or choline, the early inward portion of the current was eliminated. The computed permeability changes to Na^+ and K^+ are shown in Figure 2-13. They may be conceived of as due to the opening of selective channels in the membrane, first to Na^+ and then to K^+.

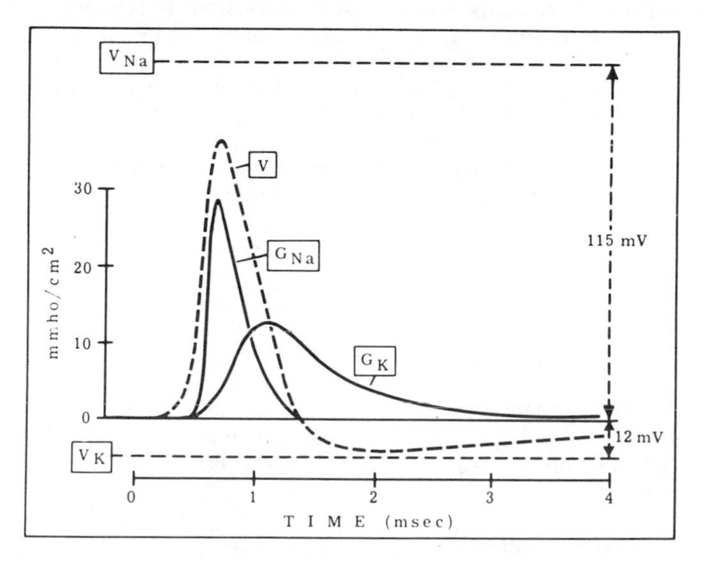

FIGURE 2-13. Computed conductance change in millimho (mmho) per square centimeter to Na (G_{Na}) and to K (G_K) accounting for the action potential (V). Notice that the result of Na^+ entry is to drive the membrane potential to the Na equilibrium (V_{Na}). The membrane at rest is close to the equilibrium potential for K^+ (V_K). (From A. L. Hodgkin. *Proc. Roy. Sci. [Biol.]* 148:5, 1958.)

The electrical resistivity of the membrane is high. As shown in electron micrographs, the membrane is of the order of 75 A thick. Physicochemical studies indicate it is composed in part of lipoproteins, which have a high specific resistivity. Just at the onset of the action potential, the electrical resistance falls drastically, to one-fortieth of its original value. As shown in Figure 2-13, the conductance of Na^+ (G_{Na}) increases, and soon thereafter it is followed by a conductance increase to K^+ (G_K).

Support for the sodium hypothesis has been gained in a number of different ways. Keynes (see Hodgkin, 1964), using radioactive tracers, found that the uptake of Na^+ by nerve or muscle fibers following a repetitive series of action potentials matched the computed inward movements of Na^+ expected per each action potential. A corresponding loss of K^+ was also shown. In the study of the perfused axon, however, an apparent defect in the hypothesis has been seen: Perfusion with low-K^+ sucrose solutions reduced the resting membrane potential, but large resting membrane potentials remained present. The explanation of this phenomenon appears to rest in the use of sucrose solutions having low ionic strength and local fields closely related to the membrane. It is expected that these recent investigations of nerve membrane properties will bring closer an ultimate molecular description of nerve excitation. What is desired is a specification of how active Na^+ permeability increase comes about, and of the inverse process of inactivation, which accounts for refractoriness. Recently, optical studies have shown transient changes directly related to the action potential. Such biophysical techniques appear to indicate molecular rearrangements which could relate to openings of channels first to Na^+ and then to K^+. A number of toxins, particularly tetrodotoxin, act in very low concentrations to interfere specifically with activation of the Na^+ permeability increase and thereby block excitation. Presumably the molecular interactions in the membrane of such substances will lead to knowledge of the molecular events involved in the early phase of excitation.

Calcium is associated in some way with excitability. This is shown with regard to the property of *rhythmicity*. Nerve axons do not usually discharge rhythmically; however, the action potential can be considered as if it were the first discharge of a highly damped rhythmic oscillation. In clinical cases of hypocalcemia, rhythmic discharges of nerve fibers may be observed. Tapping the skin over a nerve in such a patient may cause an abnormal tetanic muscle contraction (Chvostek's sign). When calcium ion is reduced in the medium surrounding isolated nerve fibers, the nerve responds to a stimulus with a long series of oscillations.

The conduction of the action potential is explained on the basis of the electrical properties of the membrane: its high resistance and capacity (Fig. 2-10) distributed along its length. At the region where an action potential has been excited by electrical currents experimentally (or naturally at receptor endings by physical or chemical stimuli [Chap. 3A]) the voltage change across the membrane causes local currents to depolarize the adjoining membrane, bringing it to the critical level for the selective entry of Na^+ and a propagated action potential (cf. Fig. 2-9B).

SALTATORY TRANSMISSION

Through the efforts of Kato and Tasaki and their pupils (Tasaki, 1953) it became possible to isolate single myelinated nerve fibers for study of their physiological properties. Such a fiber is no more than 10 to 18 μ in diameter, and the technique of isolation is one requiring care and patience. An isolated single fiber is more excitable to cathodal current applied at the nodes than to the internodes (Fig. 2-14). This follows from the theory of saltatory conduction, which regards the myelin membranes wrapped around the fiber (Fig. 2-4) as an electrical insulator. A low resistivity is present at the node openings. An applied current, or the electrical currents produced during a spike, can therefore enter and leave only at the nodes. Excitation of a node takes place following depolarization to the critical level when the Na^+ permeability becomes suddenly increased. The entering current, which is due to Na^+, is completed by the loop of return current passing out of the neighboring nodes and back through the external medium to the excited node. The direction of current at the neighboring node is outward across the membrane, i.e., in the direction to cause excitation. When the critical level of depolarization is reached by this outward current, that adjoining node is excited, Na^+ enters there, and the same process is repeated. The excitation at the nodes, instead of occurring continuously along the membrane as is the case in the giant axon fiber and other smaller nonmyelinated

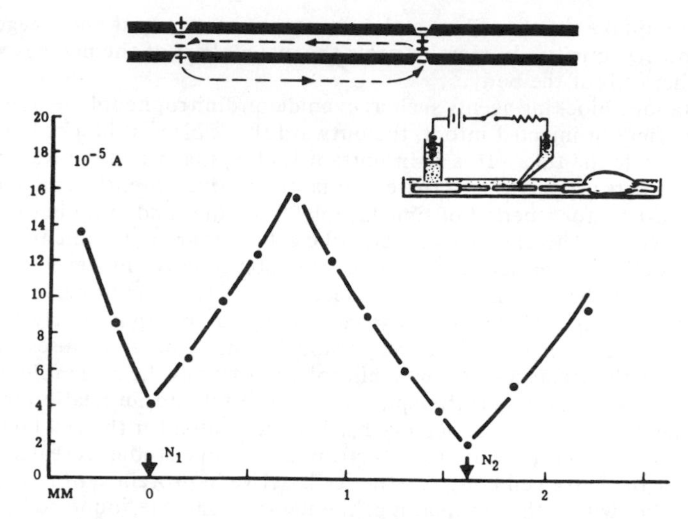

FIGURE 2-14. Transmission in a myelinated nerve occurs via currents passing down along the inside of the fiber and out the adjacent nodes, as shown in the insert above. The internodal portion covered with myelin is less excitable than the nodes. Threshold to electrical currents for a microelectrode placed at various points along a single fiber, as shown in the insert above, is lowest at the two nodes, N_1 and N_2, and is high in the internodal region. (From Tasaki, I., *Nervous Transmission,* 1953. Courtesy of Charles C Thomas, Publisher, Springfield, Illinois.)

axons, is a type of propagation called *saltatory* (L. *saltatio,* "leap"). The local currents passing from an active node to a nearby node do not consume more than a brief period of time. The delay in reaching the critical level for an action potential at the nodes causes a delay which accounts for conduction velocity – a velocity, however, much higher in myelinated nerve than in nonmyelinated fibers. Velocities range up to 100 meters per second and more for the alpha fibers of mammals.

During the propagation of an action potential along a fiber, currents return to the node via an outside path. These external currents do not normally excite the nerve fibers lying alongside, but they are not wholly without effect on nearby fibers, and under special experimental conditions alterations in the excitability of nearlying fibers may be revealed. In some diseases of nerve the myelin is lost, and the result is abnormal nerve excitability and an excitation of nearby fibers – *ephaptic transmission.* In beriberi, a disease caused by a deficiency of the vitamin thiamine, the abnormal sensations may be due to such an alteration.

METABOLISM

After generating an action potential, the nerve has gained a minute amount of Na^+ and lost an equivalent amount of K^+. The process of complete restitution of the nerve after activity requires the pumping out of the Na^+ gained and the recapture of the lost K^+. Otherwise the ionic composition of the fiber would eventually change, with loss of resting potential. The energy-rich phosphate in adenosine triphosphate (ATP) has recently been considered to be involved in such Na^+ pumping action where ATPase in the membrane is likely to be involved. In the mitochondria of nerve, oxygen is utilized and acetyl-CoA is metabolized to build the energy-rich phosphate bonds of ATP. There is a continuous heat production and oxygen uptake in nerve fibers and an increased heat output and oxygen requirement following activity. In most respects the metabolism of the nerve is similar to that of other cells of the body.

If metabolic blocking agents such as cyanide or dinitrophenol are applied to the giant nerve fiber or injected into it, the outward flux of Na^+ is blocked, and Na^+ accumulates inside the fiber. It is an important finding that when the Na^+ pump of giant nerve fibers is poisoned by these agents, the resting potential and action potentials can continue for a period of time lasting 90 minutes and more before block of excitation occurs. This shows that metabolic activities are not immediately related to the production of the action potential. Metabolism must, however, eventually supply energy to the sodium pump and eject the Na^+ entering or nerve potentials will fail. The role of high-energy phosphate in the Na^+ pump is clearly shown when arginine phosphate is injected into the poisoned giant axons and causes Na^+ pumping to resume. In the crustacean arginine phosphate occupies the same role in high-energy phosphate transfer as does creatine phosphate in vertebrates in relation to ATP.

A higher metabolic rate of activity has been indicated for the cell body and dendrites as compared with the axons. Electron microscope studies reveal a structural similarity of the nerve cell body to other cells active in protein synthesis. A number of small bodies within the cytoplasm of the neuron soma are found to be basophilic, staining a deep blue color with aniline dyes. These *Nissl bodies* were shown in electron microscope studies to be made up of a system of fine tubes and channels within the cell, the *endoplasmic reticulum,* with which small particulate structures are associated. These small particles contain *ribonucleic acid* (RNA), which controls protein synthesis within the cell. It has been recognized only in recent years that a high rate of protein synthesis exists in nerve cells. This, coupled with a mechanism of transport of materials outward in the fibers to maintain their function, will be described in a later section.

DEGENERATION AND REGENERATION

When a peripheral nerve is crushed or severed, the part distal to the crush shows the process known as *wallerian degeneration.* The myelin of the fibers begins to bead after a few days and fails to conduct simultaneously all along its length. Over a period of weeks the degenerated myelin is phagocytized.

At present the mechanism of this fast transport system is little understood but it will probably have significance in understanding obscure nerve and muscle diseases. As will be described in Chapter 3, muscle properties are likely to be determined by trophic materials moving from motor nerve fibers into the muscle fibers. Endocrine substances pass from hypothalamic cell bodies in the brain (Chap. 31), no doubt by a similar transport mechanism. The materials are stored in the posterior pituitary.

Soon after peripheral axons are severed, the central ends of fibers above the interruption sprout buds. Renewed fiber growth begins from some of these sprouts, the fibers regenerating down into the distal part of the nerve below the lesion. The process of regeneration of new fibers is similar to the embryonic nerve growth in that unknown directive forces, *neurotropic* stimuli, which are possibly chemical in nature, act on the growing tips of the nerve so that they find their right "addresses." The sheaths left by the distal parts of the transected nerve appear to have a directive influence on regeneration. If the nerve has been interrupted by a crush, the fibers regrow with less time lag than there would be if the nerve had been cut through and a gap left between the cut ends. Some "searching" time is required in the latter case. The surgeon in his repair of nerve lesions assists this seeking process by bridging the gap and suturing the cut ends of the nerve in close apposition. Fibrous growth should also be prevented, for this becomes a source of misdirection of fiber growth. Often tubes of blood vessels or plastic are used as guides for nerve growth. If the outgrowing fibers do not find their way or if the gap is too wide, the fibers turn around, forming a whorl known as a *neuroma*. This may be painful, or it may give rise to a disagreeable sensation such as that of a "phantom limb," when adventitious excitation at the site of the amputation causes the patient to feel that the lost member is still present. In amputations of limbs, procedures to inhibit growth from the end of a cut nerve are necessary to prevent this outcome.

While regeneration of the new fiber process is proceeding, the cell body undergoes the characteristic change known as *chromatolysis*. The staining of Nissl bodies is decreased at this time, and normal stainability returns only slowly and over a period of weeks or months. The chromatolytic reaction is due to dispersion of the Nissl bodies within the soma and to an increased swelling. Later in regeneration RNA and protein content increase greatly; presumably such increases are related to the later increase in diameter of the new axons growing back toward their normal size.

TRANSPORT OF MATERIALS INSIDE NERVE FIBERS

The dependence of a nerve fiber on its cell body, shown by wallerian degeneration and the evidence of a high level of protein synthesis, directs attention to the transport of materials inside nerve fibers. At a constriction of a nerve, materials are found to be accumulated on the central side, much as if they are damned up at the constricted region. Such an effect has been found for particulates including mitochondria and in autonomic nerves containing adrenergic materials (Chap. 7), the particulates containing these substances.

Recently studies with isotopes used to label amino acids have shown an uptake by cell bodies and a rapid incorporation into various components including protein. These materials move down by at least two different mechanisms, as shown by two rates of downflow. One rate is slow – only several millimeters per day. Another more rapid downflow has been shown to move at a rate close to 400 mm per day. This downflow is evidenced by a crest of labeled materials which appear farther down the nerve if more time has elapsed between uptake of the labeled precursor and removal of the nerve for analysis. Such transport requires a supply of energy all along the fiber, and this is supplied by oxidative metabolism.

Such studies appear likely to have much relevance for the normal function of the nerve – the transported protein including enzymes required for the nerve's oxidative metabolism. In addition, trophic materials transported in the nerve fibers carry materials down to and into the muscles to keep their function normal (Chap. 3), as well as to secondary cell structures around sensory nerve terminals which participate in sensory reception (Chap. 3).

REFERENCES

Brink, F., Jr. Nerve Metabolism. In D. Richter (Ed.), *Metabolism of the Nervous System*. New York: Pergamon, 1957. Pp. 187–207.

Conference. Newer properties of perfused squid axons. *J. Gen. Physiol.* 48:1–9, 1965.

Erlanger, J., and H. S. Gasser. *Electrical Signs of Nervous Activity*. Philadelphia: University of Pennsylvania Press, 1937.

Hodgkin, A. L. *Conduction of the Nervous Impulse*. Springfield, Ill.: Thomas, 1964.

Katz, B. *Nerve, Muscle, and Synapse.* New York: McGraw-Hill, 1966.
Nakamura, Y., S. Nakajima, and H. Grundfest. The action of tetrodotoxin on electrogenic components of squid giant axons. *J. Gen. Physiol.* 48:985–996, 1965.
Ochs, S. *Elements of Neurophysiology.* New York: Wiley, 1965.
Tasaki, I. *Nervous Transmission.* Springfield, Ill.: Thomas, 1953.
Young, J. Z. Factors influencing the regeneration of nerves. *Advances Surg.* 1:165–220, 1949.

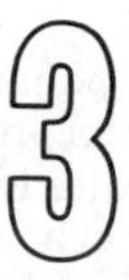

Receptors and Effectors

A. Sensation and Neuromuscular Transmission

Sidney Ochs

SENSATION

Sensations are defined as feelings or impressions produced by stimulation of afferent nerves. This definition is expanded to include those "sensations" known to be present from analysis or experimentation, but not consciously experienced. For example, there is no consciousness of the action of sensory nerves which control the rate of the heart beat, or of sensory impulses from the viscera. Sensations may be categorized as *exteroceptive* (those relating to stimuli from the external world) and *interoceptive* or *proprioceptive* (those from viscera and from somatic structures within the organism). The term *proprioceptive* has more recently been taken to include all sensations arising from within the organism.

Sensations are also categorized by the quality of sensation itself – hence as *epicritic* and the *protopathic* sensations. Epicritic sensations are sharply distinguished, e.g., a pinprick or a point of light. The person receiving such sensory stimuli can localize with a high degree of precision the exact point excited on the surface of the body or excited from the environment. One clinical test for epicritic sensibility is *two-point localization.* The points of a pair of dividers set at various distances are placed simultaneously upon the skin, and the subject is asked whether one or two points can be felt. It is found in general that the localization of two closely spaced points is very highly developed over protruding portions of the body (nose, lips, ears, and fingertips). Protopathic sensations are poorly localized. They include burning types of pain and deep muscle aches and pains which seem to exist over a wide area. A patient experiencing them has difficulty in determining exactly where such pains are localized.

According to Müller's law of *specific nerve energies,* when particular nerve fibers are excited in any way, the sensations aroused are those that would be experienced with normal excitation of the receptors of those fibers. For example, if optic nerve fibers are electrically stimulated, the sensation of light is experienced. Mechanical stimulation produced by pressure applied to the side of the eyeball brings about a visual sensation. Similarly, fibers subserving heat perception will, when adequately excited by any means, give rise to the sensation of heat.

One of the classic experiments which appeared to show a direct relation of a given sensation to a specific receptor found that a particular sensation could be elicited at discrete points over the skin. The points at which cold was experienced appeared to be associated anatomically with cold receptor organs, heat with heat receptor organs, pain with pain receptors, etc. More recently, critical studies of the relation of the presumed receptor organs to the various sensory points within the skin have been made. Studies by Weddell (1961) have shown that, besides the basket-like endings sensitive to touch surrounding the hair follicles in hairy areas of the skin, free nerve terminals are the main type of receptor organ. It appears that the pattern of the termination of the free nerve endings affects the quality of sensation. Thus there may not be a simple *one-to-one relationship* between an end-organ and the reception of a particular sensation; the quality of sensation may depend in part on the overlapping pattern of nerve fibers in the skin. It was once thought that stimulation of the cornea could elicit only pain. There are no specific receptor organs within the cornea and the free nerve endings were supposed to be the receptors for pain. However, with careful examination of the cornea, it has become evident that sensations of touch, cold, and heat can be elicited from the cornea in addition to pain. The C fibers, which constitute many of the nonmyelinated free nerve afferent endings in the skin and other structures, can no longer be considered as purely pain fibers. In experiments involving the skin of the cat, heat evoked action potential responses from C fibers, as did cold and pressure stimulations. The secondary structures associated with a sensory organ, such as the eye or ear, are modifying structures which act to channel the appropriate physical or chemical stimulating agent to the receptor nerve terminals.

THE SENSORY CODE

Receptors do not relay information to the afferent nerve fibers by means of a single action potential. When the appropriate sensory terminations become adequately excited, a repetitive discharge is elicited which passes into the central nervous system (CNS). The repetitive discharge is the "code" representing the applied stimulation. The frequency of the repetitive discharge is related to the intensity.

Rhythmicity of discharge of sensory nerves upon stimulation was first shown by Adrian and his co-workers (Adrian, 1928). One of their early studies was made on sensory fibers known to take origin from muscle. (The sensory organs in muscle will be described in Chap. 5.) If small groups of nerves in muscles are isolated, so that recordings are made from only one or a few afferent nerve fibers, a regular rhythmic discharge is excited in individual sensory fibers upon stretch of the muscle. Other receptors also show a regular repetitive discharge in their fibers upon adequate stimulation (Fig. 3A-1). The rate of sensory discharge increases approximately with the logarithm of the strength of the stimulus.

A logarithmic relationship was found by Fechner and Weber (see Granit, 1956) in their studies of blindfolded subjects who were asked to compare weights placed in the palm of the hand. The ability to estimate small weight increments depended on the weight already present in the palm. The logarithmic relationship was shown in the following formula:

$$\Delta S = \frac{\Delta W}{W}$$

and by integrating:

$$S = K_1 \log W + K_2$$

where ΔS is a just noticeable change in sensation (to weight) to ΔW, an increment of weight. S is sensation, W weight, K_1 and K_2 constants. This logarithmic relationship was generally found to hold for increments of light intensity and of sound intensity.

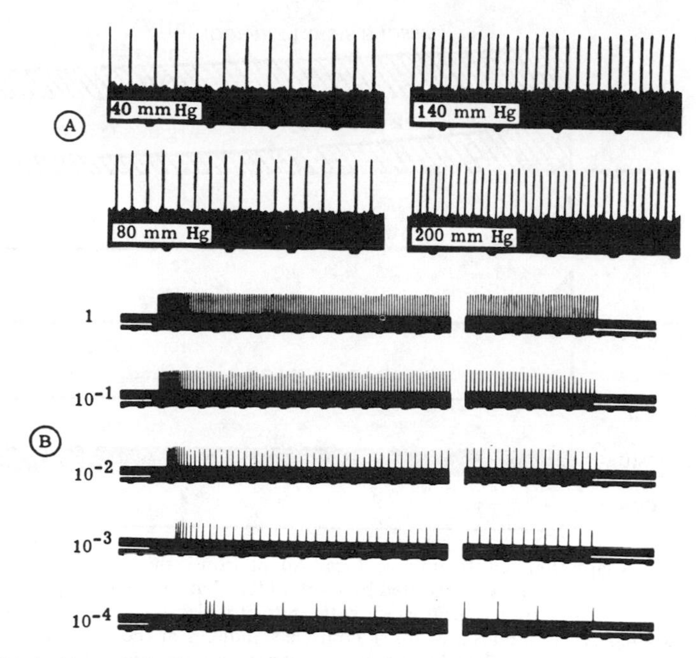

FIGURE 3A-1. Rate of discharge at different stimulus strengths. In A, a single fiber from a carotid sinus pressure receptor is shown with discharge rates at different levels of intracarotid pressure. As the intrasinus pressure (mm Hg) is increased, the discharge rate increases. In B, the discharge from a single optic receptor of the limulus eye is shown at different intensities of illumination. The filled bar underneath signifies the time of stimulation. As the illumination strength is decreased by factors of 10, the discharge rate is seen to decrease. (1A from D. W. Bronk and G. Stella. *Amer. J. Physiol.* 110:711, 1935. 1B from E. F. MacNichol, Jr. In R. G. Grenell and L. J. Mullins [Eds.]. *Molecular Structure and Functional Activity of Nerve Cells.* Washington, D.C.: American Institute of Biological Sciences, 1956. P. 36.)

It is of interest that the rate of discharge found in a sensory nerve bears a similar logarithmic relationship to the degree of stimulation as does the psychophysical relationship of Fechner and Weber. However, there are deviations from this simple logarithmic relationship at the extreme ranges of the stimulus-sensation relationship, as discussed by Stevens (Rosenblith, 1961). A more general and better representation of the relationship, which holds over a wider range of stimulation and for different classes of stimuli, is the power function representation described by Stevens.

Adaptation is an important phenomenon relating sensation to discrimination among several sensory inputs. Although some kinds of stimuli are continually present, attention may not be paid to them. For example, the weight of clothes upon the body is generally neglected unless one is made aware of it. Adaptation can be traced, to some extent, to processes within the receptor. Recording the action potentials from the afferent nerves of various receptors shows that with continued application of a stimulus the frequency of discharge declines. The rate of decline varies according to the type of receptor involved (Fig. 3A-2). Adaptation is only moderate for pressure receptors in muscle, whereas for touch it is quite rapid.

The central interpretation of sensation is controlled only in part by adaptive processes in the peripheral receptors. Adaptation may have some function with regard to total sensory influx insofar as by this means the field of sensory attention

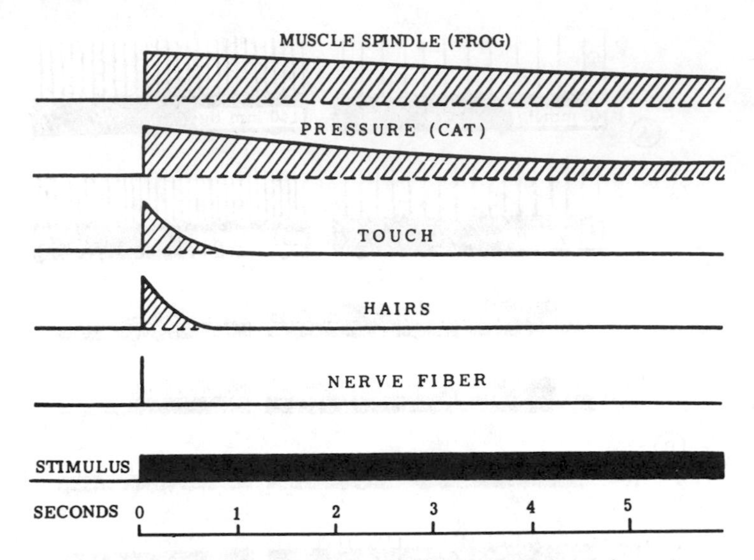

FIGURE 3A-2. Adaptation in different receptors. Adaptation is shown for various sensory receptors to a constant stimulus represented by a solid black bar in the lower part of the figure. The hatched parts of the various receptors show the rate at which the single units of the receptor are discharging. Adaptation is least for the muscle spindle. For the pressure receptor, adaptation is moderate, and for touch and hair receptors adaptation is rapid. The nerve fiber itself responds with a very brief discharge. (From Adrian, 1928.)

becomes "cleared" and open for new sensory information. Higher centers have additional mechanisms to complement these peripheral adaptive effects so that attention can be directed to one rather than another sensory input (see Chap. 9).

TRANSDUCTION

Microelectrode techniques, such as have been used to study nerve properties (see Chap. 2), have been applied to the study of the mechanism underlying excitation of the afferent fiber receptors. A microelectrode may be put within the cell body of a single stretch receptor cell of the crustacean abdominal segments (Fig. 3A-3). This stretch receptor has dendritic portions wrapped around the central region of the muscular portion of the receptor. When the muscular portion is elongated, a deformation of the dendritic branches of the sensory stretch receptor neuron occurs. The deformation of the dendritic membranes causes repetitive activity (Fig. 3A-4). The discharge rate increases when the receptor is stretched to a greater degree. The microelectrode study has revealed the important fact that preceding and concomitant with the repetitive activity there is a depolarization known as a *generator potential.* This long-lasting depolarization of the receptor cell body is produced within the dendrites and spreads to the rest of the cell. The generator potential causes the initial segment of the axon adjacent to the receptor cell body to respond repetitively. The level of stretch controls the size of the generator potential, and in turn the rate of discharge. It has been mentioned that repetitive activity is not usually found in axons unless they are subjected to special conditions such as cathodal depolarization or a lower Ca^{++} level (see Chap. 2). The receptor differs from axons insofar as a repetitive activity is its normal mode of behavior.

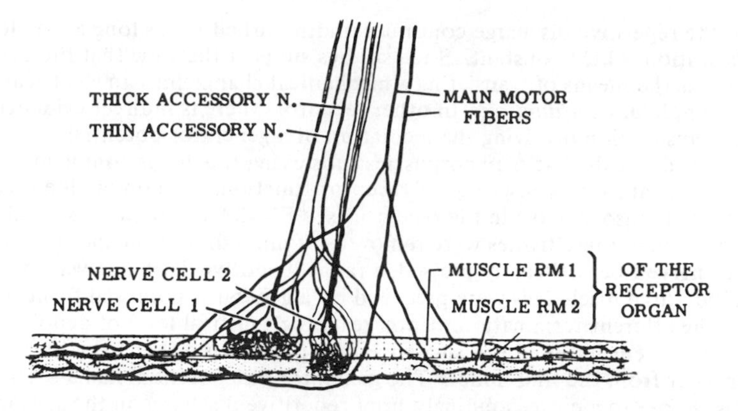

FIGURE 3A-3. Crustacean stretch receptors. Two stretch receptors of the crayfish are shown with the dendritic portions of the cells wrapped around the muscular parts of two different receptors. Nerve cell 1 is from the slow-adapting, and nerve cell 2 is from the fast-adapting, receptor. Movements of the muscular portions of the receptor distort the membrane of the dendritic portion of the receptors and cause repetitive discharge in those sensory afferents. Also shown are motor fibers which act in a feedback system to control the level of sensitivity of the stretch receptors. (From J. S. Alexandrowicz. *Quart. J. Micr. Sci.* 92:190, 1951.)

Two types of stretch receptors are found in the crustacean abdominal segments – the slow- and the fast-adapting. The fast-adapting shows a more rapidly decreasing rate of discharge with maintained stretch. The slow-adapting cells maintain their firing rates for a much longer time (Fig. 3A-4). The generator potentials show a corresponding difference. A long-maintained plateau is seen in the slow-adapting receptor, a faster fall in potential in the fast-adapting receptor. If depolarization is held at a controlled level by a current introduced into the receptor through an electrode, the

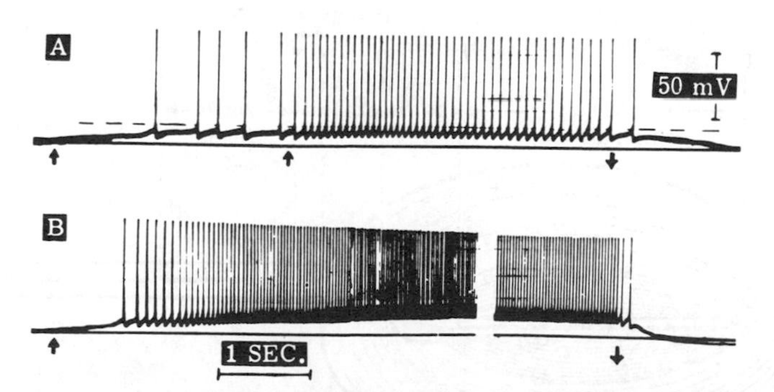

FIGURE 3A-4. Discharge from a single crustacean stretch receptor fiber. In A is shown the discharge of a single slow-adapting stretch receptor recorded from inside the soma with a microelectrode. A stretch is gradually applied, resulting in depolarization, indicated by the shift in the baseline. The rate of discharge from the receptor increases with the level of depolarization. With a maintained stretch a fairly constant level of discharge is maintained until the stretch is released; then depolarization decreases and the rate of discharge of the receptor falls off. In B, a greater stretch gives rise to a higher level of depolarization and a more rapid rate of discharge. (From C. Eyzaguirre and S. W. Kuffler. *J. Gen. Physiol.* 39:87-119, 1955.)

rate of the repetitive discharge continues undiminished for as long as the level of depolarization is held constant. Such studies support the view that the generator potential is the means of translating a mechanical change into an electrical correlate – an example of *transduction.* In other receptors there is indirect evidence for a similar transduction involving the mediation of a generator potential.

The study of the pacinian corpuscles subserving touch has complemented and added significantly to knowledge of receptor function. The onion-like layers around the terminal sensory nerve in the receptor (see Fig. 3A-5) are not essential to its function. When these tissues were removed by microdissection and the naked central afferent nerve fiber delicately pressed, a rapid repetitive discharge was excited. The action potential discharges were preceded by a generator potential from the deformed part of the afferent termination. Upon reaching a critical level of depolarization, this generator current initiated propagated action potentials either from the nonmyelinated terminals or from the first nodes. The generator-like potential had a brief duration and it gave rise to a correspondingly brief repetitive discharge in the sensory fiber.

The deformation of the membrane of the afferent fiber termination increases the ionic exchange across the membrane to produce the generator potential. The supporting structures around this excitable membrane appear to limit the range of possible responses and to protect the ending from excessive stimulation or to allow a response to a deformation in one direction rather than another. This phenomenon is seen in other specialized sensory receptors where the secondary structures enable the sensory fibers to respond to selected aspects of the environment, e.g., the auditory hair cells and associated structures (see Chap. 4B). It is seen in the retina of the eye, where specialized photochemical substances (see Chap. 4A) have evolved to enable the optic nerve cells of the retina to respond selectively to colors as well as to degrees of illumination. The static and dynamic receptors of the semicircular canals, utricle, and saccule enable these sensory organs to signal position and movement of the head in certain directions in space (see Chap. 4C).

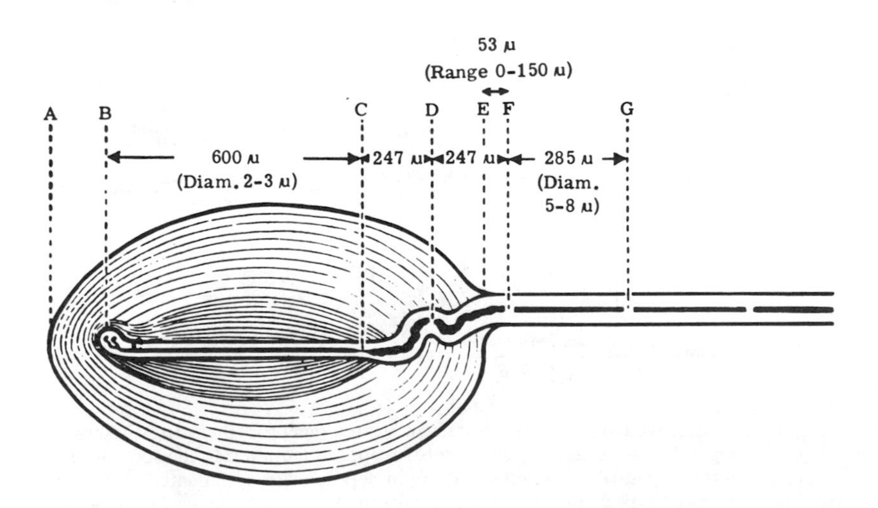

FIGURE 3A-5. The pacinian corpuscle. The letters C and D indicate the nodes of the sensory fiber within the receptor. The sensory fiber is surrounded by lamellated structure. (From T. A. Quilliam and M. Sato. *J. Physiol.* [London] 129:167, 1955.)

In some of the specialized sensory organs, secondary cells have evolved, which give rise to electrical potentials in response to physical or chemical stimulation. These *receptor potentials* are similar to the generator potential. They initiate afferent activity in the sensory afferent nerve terminals of the receptor organ.

PAIN

Pain has sometimes been considered as being due to an excess of sensory stimulation, but present opinion holds that there are specific fibers for pain. Pain may range from mild disagreeableness to an overbearing sensation from which one must have relief, with all else driven from the mind. Some people believe that itch and tickle are due to milder stimulation of the same afferent fibers subserving pain. An argument against this view is that *pruritus* (severe itch) may appear as a separate symptom without the sensation of pain.

Pain is often described by its quality, e.g., as dull or burning. The terms used to describe various pains make a rather lengthy list. The difference between the epicritic or sharp type of pain and the duller, less easily localized protopathic pain can be demonstrated by a simple experiment. Touching the skin of the toe with a hot match ember elicits a sharp first pain followed after a few moments by a duller second pain. This double-pain sensation has been interpreted to mean that the epicritic pain is carried in faster-conducting fibers than the protopathic pain. If the skin of the lower leg is similarly excited, the time between the two pain sensations is decreased, and it is decreased even more if the thigh is stimulated. Such a result is in accord with other data that slower-conducting C fibers carry the protopathic pain sensations and the delta group of A fibers carries the faster pain sensations.

Experimental attempts to determine pain thresholds, e.g., by graded radiant heat, have been unsuccessful because subjective factors are probably the most important features in such experiments with pain. Little understood are the central changes responsible for the alleviation of pain by analgesics.

One characteristic of pain related to central nervous processes is its tendency to *irradiate* and to give rise to *referred pain.* For example, pathological heart changes may be experienced by the patient as a pain of the left upper arm or a pain passing down the left arm and hand. Kidney disease may be felt as a painful back injury, and ureteral trauma produced in the passing of a kidney stone may be interpreted, because of irradiation, as a severe pain of the flank. Pain from a stomach ulcer may be referred to the right shoulder. These misdirections of pain sensation appear to be due to excitation of a common pool of neurons within the central nervous system from two different afferent sources. According to this theory, pain is a functional "spillover" of activity from one afferent input to affect those neurons usually excited by the other set of afferent inputs. Some of the general features of pain plus electrophysiological investigation of transmission in the spinal cord led Melzack and Wall (1965) to develop the "gate" theory. It is thought that intermediary cells in the substantia gelatinosa of the cord can control which of several incoming sensory impulses will ascend the cord to the brain and eventually reach consciousness. The afferent pattern is in this theory an important determinant.

The referral of one sensation to the other set of afferent terminations may also lead to an enduring state of sensitivity. For example, a painful procedure such as the drilling of teeth may give rise to pain long after the actual cause has ceased. The functional change, producing a continued painful state, may exist at various levels within the central nervous system. Transecting the nerve from a site originally giving rise to the pain, e.g., a neuroma resulting from amputation of a limb (see Chap. 2), is usually not permanently effective. Spinal cord section of the afferent tracts may be required, or frontal lobotomy, which is sometimes necessary for intractable pain. The factors of attention and attitude at the time of injury are most important in determining the degree of pain experienced. In World War II, as had been noted in previous wars, a sizable proportion of seriously wounded patients did not experience any pain at the time of injury, probably because of the stress of battle. According to the gate theory, descending influences from the brain control the gate cells to determine which types of ascending impulses will reach the brain.

NEUROMUSCULAR TRANSMISSION

The concept that transmission between a motor nerve and the muscle it innervates is affected by the release of a chemical agent from the nerve was advanced through the research of Dale and his co-workers, who indicated that the transmitter agent is *acetylcholine* (ACh). This substance, released from the nerve endings when motor nerves are fired, moves across the gap between the nerve and muscle membrane at the end-plate to excite the muscle membrane and, in turn, a propagated action potential along the muscle fiber. The classic proof of a transmitter agent was given by Loewi, who found that a substance was released into the fluid medium following vagal nerve stimulation of the frog heart. This "vagus stuff" caused a slowing of the beat of another heart. The vagus stuff was later identified as acetylcholine (see Chap. 7). Acetylcholine is labile, and if blood is present, the cholinesterase present breaks down the acetylcholine to an inactive form. If anticholinesterase agents, such as *eserine* or *neostigmine,* are present, ACh can then be detected.

Electron microscopy of the end-plate region shows that there is no continuity of the nerve membrane with the muscle membrane. The nerve fiber occupies a channel within the muscle membrane, with a cleft between them of approximately 300 to 500 A (Fig. 3A-6). Vesicular bodies approximately 500 A in diameter are found within the terminal portion of the nerve. It is believed that they contain the transmitter substance ACh. Mitochondrial bodies are also seen within the nerve endings. Since mitochondria are present wherever metabolic activity is high, energy sources might well be associated with discharge and a possible resynthesis of ACh and its reentry into the vesicles.

THE END-PLATE POTENTIAL

When a microelectrode is placed inside the muscle cell close to an end-plate and the motor nerve is excited, there is very little evidence of action currents flowing from the nerve terminal to the muscle. This lack of effective flow of current indicates that the action current of the nerve endings is not the effective agent to excite the postsynaptic membrane. There is instead a delay, and then a special electrical process is seen before the propagated action potential in the muscle. This is a depolarization which ordinarily reaches an amplitude of 30 to 50 mv or more and is localized to the end-plate region; it is called the *end-plate potential* (EPP) (Fig. 3A-7). There is a fundamental difference in the ionic mechanisms underlying the EPP and those of the propagated action potential. The action potential is brought about by a brief increased permeability first to Na^+ and then after an interval to K^+, as described for the giant nerve fiber in Chapter 2. The EPP, on the other hand, is a generalized depolarization resulting from simultaneous increased permeability to Na^+ and K^+ caused by ACh.

The EPP also differs from the usual spike potential in that it is not an all-or-none response. It diminishes in amplitude with distance from the region of the end-plate. Curare (d-tubocurarine when purified) attaches in some as yet unknown manner to the receptor substance present in the membrane on the muscle side of the end-plate

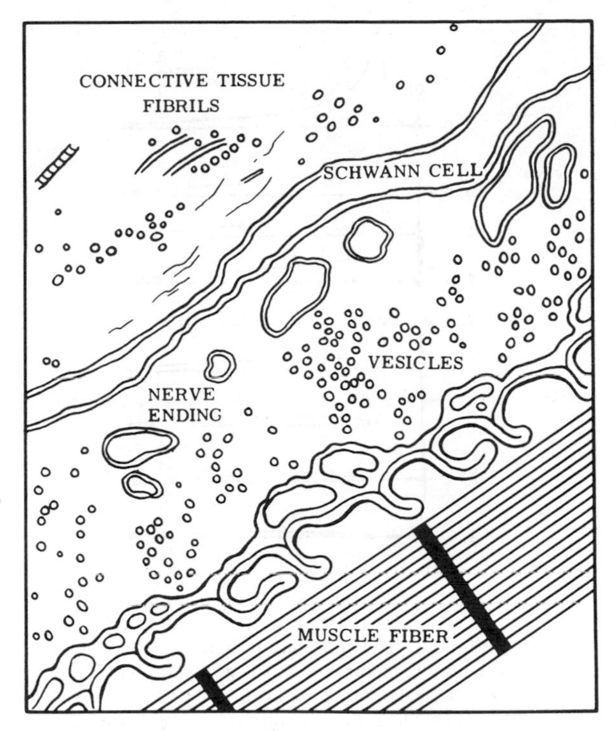

FIGURE 3A-6. Neuromuscular junction. A longitudinal section of a nerve fiber within a tunnel in the muscle surface is diagrammed from an electron micrograph. The nerve fiber is approximately 1.5 μ in diameter. Within the nerve terminal two types of particulate structure are shown, mitochondria and small vesicles believed to contain the transmitter substance acetylcholine. (From R. Birks, H. E. Huxley, and B. Katz. *J. Physiol.* [London] 150:130, 1960.)

region. The ACh released from the nerve ending cannot affect those receptor sites which are occupied by d-tubocurarine or its derivatives. The EPP is therefore reduced in size, the extent of reduction depending upon the amount of d-tubocurarine present. When the EPP is reduced below a critical level of approximately 30 to 50 mv, the EPP depolarization is not large enough to excite a propagated action potential in the adjoining membrane. d-Tubocurarine therefore does not bring about its blocking action by effecting block of the propagated action potential. It blocks by its effect on the amplitude of the EPP.

Most of the relatively long time course of the EPP, approximately 15 msec, is due to a passive distribution of current along the adjoining muscle membrane. Such electrotonic distributions of potential are similar to the local currents discussed in conjunction with nerve membrane (see Chap. 2). The amplitude of the EPP becomes smaller as the distance from the end-plate increases (Fig. 3A-7) because the charge at any point is determined by the electrical characteristics of the membrane of the muscle, i.e., its resistivity and capacitance and the resistivity of the muscle fiber.

The EPP is initiated by the transient increase in permeability to cations produced by the acetylcholine released from the nerve ending. The extracellular fluid is mainly composed of Na^+ and Cl^-, with small amounts of K^+ and lesser amounts of other ions. As the permeability is increased, the cations move according to their concen-

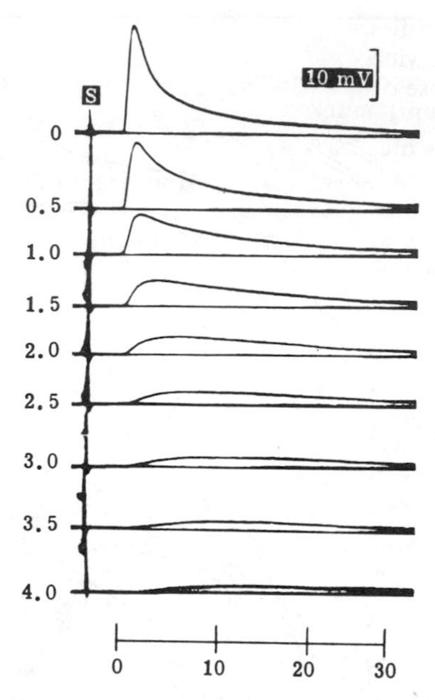

FIGURE 3A-7. End-plate potential. The end-plate potential (EPP) is recorded with a microelectrode from just under the muscle membrane in the end-plate region. It is a depolarization which occurs after a latency and is relatively long lasting. The height of the EPP decreases with distance, as shown by the numbers indicating in millimeters the distance of the recording microelectrode from the end-plate. (From P. Fatt and B. Katz. *J. Physiol.* [London] 115:320, 1951.)

tration and voltage gradients toward a zero membrane potential. The introduction of ACh through a micropipette in the close vicinity of the end-plate region causes a similar increased permeability and also results in a depolarization which, if it reaches a sufficient level, sets off a propagated action potential. When ACh is released from a microelectrode inserted just under the end-plate membrane, it is ineffective. Apparently, only the outer surface of the membrane of the end-plate contains the receptor substance responsive to ACh. Acetylcholinesterase is also present at the end-plate, acting to terminate the depolarization produced by ACh. Depolarizing blocking agents, e.g., choline, succinylcholine, and decamethonium, which are similar in structure to acetylcholine and not acted on by the cholinesterase, produce a persistent depolarization of the end-plate, thereby preventing transmission. The mode of action of these blocking agents therefore differs from that of curare, which does not produce depolarization at any stage during its blocking duration.

Besides the curariform and the depolarizing groups of neuromuscular blocking agents, a special type of neuromuscular block is produced by botulinus toxin. It acts upon the membrane of the motor endings, interfering with the mechanism of release of the transmitter substance. The toxin is extremely potent, and when injected into the circulation it is selectively taken up by motor nerve endings. The toxin also has an affinity for cholinergic autonomic nerve endings, causing a block of vagal inhibition of the heart. The very potent substance tetrodotoxin causes a block of action potential propagation in nerve terminals and thus effects motor responses.

A continual irregular discharge of small potentials, several millivolts in amplitude, has been recorded in the vicinity of the end-plate. These have a duration and other properties similar to those of the EPP and have been termed *miniature end-plate potentials* (MEPP). MEPP's, like the EPP itself, are blocked by curare. The rate of discharge of the MEPP is increased greatly by depolarization of the motor nerve terminals, and it is believed that when an action potential invades the nerve terminal the discharge of MEPP becomes synchronized to give rise to the larger-sized EPP.

Careful searching along the length of the nerve terminals of the neuromuscular junction with an electrode has disclosed that MEPP's arise from active synaptic spots. The propagation of an action potential along the terminals was shown to synchronously release a large number of MEPP's. The variable latency and number of MEPP's released spontaneously can be described by the probability that a unit or packet consisting of several thousand molecules of ACh will be discharged at the active spot. This was determined by bringing to the active spot a micropipette containing ACh and releasing known amounts of the agent to produce depolarizations. By this means it was also shown that there is very little latency for the applied ACh to produce depolarization, whereas there is a range of approximately 0.5 to 2.5 msec for the release of a packet of ACh following nerve activity. The irreducible latency of 0.5 msec represents the time taken for the release of transmitter from the terminal to become effective at the receptor.

An attractive theory is that each vesicle in the motor nerve terminal contains a packet of several thousand molecules of ACh and gives rise, on discharge, to a single MEPP. The simultaneous release of a number of vesicles from the nerve terminal is brought about by the action potential. Calcium was shown to be required for the release of ACh from the terminals. The most likely hypothesis is that the nerve action potential changes the nerve membrane properties, and in conjunction with Ca^{++} the probability of release of vesicles is greatly increased for a brief time.

Repetitive stimulation changes the probability of release of vesicles from the active sites, as is seen following a period of high-frequency stimulation of the motor nerve. *Post-tetanic potentiation* (PTP) is observed as an increase in the size of the EPP for a short time after a period of stimulation. In a partially curarized muscle, in which the EPP has been reduced in amplitude, an augmentation of the second EPP soon after a first EPP may produce a propagated spike.

Neuromuscular transmission is poor in the muscle disease *myasthenia gravis* possibly because of a failure of mechanisms controlling the amount of acetylcholine release. EPP's are of lower than normal amplitude, and transmission may fail because critical levels are not always reached. Anticholinesterase agents such as eserine or neostigmine can improve neuromuscular transmission by prolonging the duration of the EPP and may thereby permit critical levels to be reached.

TROPHIC INFLUENCE OF NERVE ON MUSCLE

When the motor nerve to a skeletal muscle is transected, the muscle after a while undergoes a series of profound biochemical and functional changes. A period of increased excitability and sensitivity to injected acetylcholine follows, the muscle showing *denervation hypersensitivity*. *Fibrillation,* large irregular discharges, occur at this time. Metabolism is abnormal. Then the muscle slowly shrinks over a period of weeks and months, an event called *denervation atrophy*.

Recent studies have shown that, whereas normally only the end-plate region is highly sensitive to acetylcholine, the area of sensitivity to acetylcholine spreads slowly outward from the end-plate region of a denervated muscle, so that in a matter of days or weeks, depending on the species, most or all of the membrane becomes sensitive to applied acetylcholine. If the muscle becomes reinnervated, this increased area of hypersensitivity slowly becomes reduced in size again until only the region of the end-plate shows the usual high sensitivity to acetylcholine. An interpretation of this phenomenon is that a trophic substance present in nerve passes into the muscle in the end-plate region and then moves out into the muscle beyond the end-plate region. Interruption of the nerve stops the continual supply of this trophic substance, which is being continually carried down the nerve fiber by axoplasmic transport (Chap. 2).

Following nerve interruption or in animals poisoned with botulinum, the continual low level of discharge of ACh represented by the MEPP is blocked. The muscle later develops the hypersensitization of denervation. This suggests that ACh is the trophic material. However, the failure of ACh applied to denervated muscle to prevent hypersensitization in vitro and other evidence indicate that some as yet unknown substance is the trophic agent.

There may be a number of different agents carried from the nerve into the muscle. In the mammal, white muscles have a faster twitch duration than red muscles. The red muscles are of the tonic group (see Chap. 5), and their longer twitch duration is related to their more prolonged activity in maintaining posture. When the motor nerve to the fast muscles of the leg of the cat is cut and crossed with the cut nerve of a slow muscle of the leg, the muscles innervated by their new nerve supply take on new properties. The fast muscles became slow contractors and the slow muscles tend to become fast. This experimental result is explained by the concept that substances within the nerve fiber flow down into the muscle to determine the functional properties of muscle, e.g., twitch duration.

Further study of the control of muscle properties by nerve may be expected to reveal the causes of diseases of muscle which have their origin in changes of their nerve supply.

REFERENCES

Adrian, E. D. *The Basis of Sensation.* New York: Norton, 1928.

Davis, H. Some principles of sensory receptor action. *Physiol. Rev.* 41:391–416, 1961.

Eccles, J. C. *The Physiology of Synapses.* New York: Academic, 1964.

Eyzaguirre, C. *Physiology of the Nervous System.* Chicago: Year Book, 1969.

Granit, R. *Receptors and Sensory Perceptors.* New Haven, Conn.: Yale University Press, 1956.

Gray, J. A. B. Mechanical into electrical energy in certain mechanoreceptors. *Progr. Biophys.* 9:286–324, 1959.

Gutmann, E. (Ed.). *The Denervated Muscle.* Prague: Publishing House, Czechoslovak. Acad. Sci., 1962.

Gutmann, E., and P. Hnik (Eds.). *The Effect of Use and Disuse on Neuromuscular Functions.* Prague: Publishing House, Czechoslovak. Acad. Sci., 1963.

Katz, B. The transmission of impulses from nerve to muscle and the subcellular unit of synaptic action. *Proc. Roy. Soc. [Biol.]* 155:455–477, 1962.

Katz, B., and R. Miledi. Propagation of electric activity in motor nerve terminals. *Proc. Roy. Soc. [Biol.]* 161:453–482, 483–495, 496–503, 1965.

Kuffler, S. W. Synaptic inhibitory mechanisms. Properties of dendrites and problems of excitation in isolated nerve cells. *Exp. Cell Res.* 5 (Suppl.):493–519, 1958.

Livingston, W. K. *Pain Mechanisms: A Physiologic Interpretation of Causalgia and Its Related States.* New York: Macmillan, 1947.

Melzack, R., and P. D. Wall. Pain mechanisms: A new theory. *Science* 150:971–979, 1965.

Rosenblith, W. A. (Ed.). *Sensory Communication.* New York: Wiley, 1961.
Singer, M., and J. P. Schade. Mechanisms of Neural Regeneration. In M. Singer and J. P. Schade (Eds.), *Progress in Brain Research,* Vol. 13. New York: American Elsevier, 1964.
Sweet, W. H. Pain. In J. Field (Ed.), *Handbook of Physiology.* Section 1: Neurophysiology, Vol. I. Washington, D.C.: American Physiological Society, 1959. Pp. 459–506.
Thesleff, S. Effects of motor innervation on the chemical sensitivity of the skeletal muscle. *Physiol. Rev.* 40:734–752, 1960.
Weddell, G. Receptors for Somatic Sensation. In M. A. B. Brazier (Ed.), *First Conference on Brain and Behavior.* Washington, D.C.: American Institute of Biological Sciences, 1961. Pp. 13–48.

Receptors and Effectors

B. Effectors: Striated, Smooth, and Cardiac Muscle

Kalman Greenspan

STRIATED MUSCLE

One of the most important effector systems in the mammalian organism is the striated musculature. The primary function of this musculature is reaction to changes in the external environment that will effect the rapid contracting of striped skeletal muscles indirectly via the central nervous system, producing a movement of the whole or parts of the body. Musculature contraction initiates a reversible chain of events involving structure and associated electrical, chemical, and thermal changes.

STRUCTURE

A skeletal muscle consists of long cylindrical muscle fibers which vary in length from 1 to 40 mm and in diameter from 10 to 100 μ (Fig. 3B-1). Each muscle fiber constitutes a single skeletal muscle cell and is in turn composed of many smaller units called *myofibrils,* which lie in the sarcoplasm. Myofibrils range in diameter from 1 to 2 μ. Each muscle fiber or cell is enclosed by a rather structureless membrane called the *sarcolemma.* The muscle fibers themselves are attached by their fibrous ends to other muscle cells and eventually through perimysium and epimysium to tendons which insert into the bone of the skeleton.

Under the light microscope, individual muscle fibers are seen to have a striated or banded structure. Under the electron microscope, the banding of the myofibrils is more clearly seen as alternate dark (A band) and light (I band) areas. These bands were long considered to have greater (anisotropes) and lesser (isotropes) molecular complexity when viewed with polarized light. The molecular basis for the banded appearance is readily noted after preparation under the electron microscope (Fig. 3B-2). In the center of each A band is a less dense area called the H zone in which there may, on occasion, be another set of filaments, called the S fibrils.* Each I band is bisected by a narrow, somewhat darker line called the Z line (or Z membrane). Every myofibril is therefore made up of units that encompass all the material between two Z lines. These units, called *sarcomeres,* are approximately 2.3 μ in length and repeat themselves in a specific pattern in each myofibril. In its fully relaxed

*Some authors refer to the H zone as being composed of a dark M line adjacent to which, on each side, are lighter areas called L lines. This M-L combination is then referred to as the pseudo-H zone.

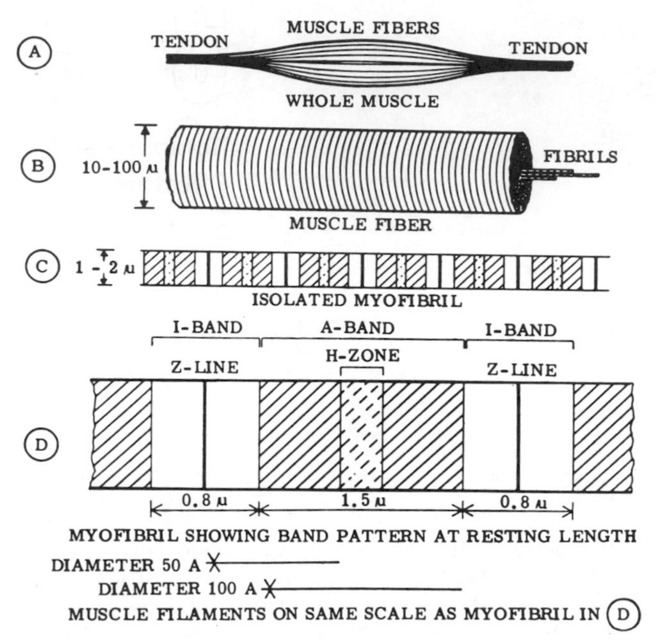

FIGURE 3B-1. Structure of muscle. In A the whole muscle is shown, composed of muscle fibers. These fibers range from 10 to 100 μ (B). Within the muscle fibers, fibrils are found (C). The myofibril shows the striated appearance typical of striated muscles. In the electron micrograph the structure of the myofibril is shown to be composed of an I band and an A band (D). Within the I band a Z line is found, and the sarcomere is taken to extend from Z line to Z line. (From H. E. Huxley. In J. Brachet and A. E. Mirsky [Eds.]. *The Cell.* New York: Academic, 1960. Vol. 4, p. 369.)

state each A band is about 1.5 μ long, while the I band is about half this length (Fig. 3B-1D). In addition, the myofibrils are closely connected with an intracellular system called the *sarcoplasmic reticulum* (SR). The system consists of two components: One is a longitudinal series of tubes within the sarcomeres but not connected with the outside of the cell; hence the longitudinal sacs are embedded in the A and I bands. These tubes in turn embrace the second component of the SR, a transverse tubular system *(T system)* that passes into a myofibril alongside the Z line. The T system connects with the sarcolemma and in fact appears to consist of invaginations of the cell membrane. Thus, a given sarcomere is covered by a longitudinal and a transverse system of tubules forming *triads,* which may play a role in the coupling of the excitation-contractions events (see below).

The fibers may be arranged in a parallel or in a pennate fashion. These are essentially the basic arrangement of patterns, although various combinations of the two exist. The fiber arrangement is significant in that it is related to skeletal muscle function. The force a muscle can exert is a function of its cross-sectional area. Muscle arranged in a pennate fashion holds more fibers per unit volume than does muscle arranged in a parallel fashion, and thus will exert a greater force per gram of tissue. However, the distance which pennate-arranged fibers shorten is much less than that for parallel-arranged fibers, but the former induce short, powerful movements, while the parallel fibers move over a comparable distance more rapidly.

FIGURE 3B-2. Electron micrograph of striated muscle. The Z lines are shown as the very dense thin lines within the light-appearing I bands. Within the sarcomere extending between the lines, the denser A band can be readily discerned. Myosin composes the myofilaments of the A band. Within the A band the lighter H zone can be made out. (From H. E. Huxley. In J. Brachet and A. E. Mirsky [Eds.]. *The Cell.* New York: Academic, 1960. Vol. 4, p. 371.)

ELECTRICAL ATTRIBUTES

The electrical properties of skeletal muscle parallel those properties in nerve fibers. The ionic distribution is such that the concentration of potassium (K^+) is greater inside the cell, while the predominant cation outside the cell membrane is sodium (Na^+). Under conditions of rest the cellular membrane is highly permeable to K^+ and only slightly permeable to Na^+. The magnitude of the resting potential of a mammalian skeletal muscle is about 90 mv, the inside being negative with respect to the outside. During activity, this transmembrane potential is abolished and reverses to about +30 mv (overshoot), indicating a change in membrane permeability during cell activity (the membrane is now highly permeable to Na^+); indeed, as in nerve, it is the entry of Na^+ which is responsible for membrane depolarization. The overshoot appears to be dependent upon Na^+ but may be due as well to other ions such as Ca^{++}. The unit of recording, as in nerve, is the spike potential, the duration being about 2 msec. Absolute and relative refractory periods are also observed, their duration being roughly proportional to spike duration. The contour of the muscle action potential, therefore, is characterized by a fast rise (depolarization), then an abrupt fall to a point, after which there is a slower decline with the characteristic positive and negative afterpotentials. The afterpotentials seem to be chemically dependent, since they are markedly affected by temperature and metabolic inhibitors. The conduction velocity of a skeletal muscle impulse is much slower than that for myelinated nerve fibers, being about 3 to 6 meters per second, depending upon the type of muscle.

The significant aspect of the muscle action potential is that the electrical response always occurs before, and is completed by, the time the mechanical response begins. Thus, under normal conditions a muscle which contracts has an associated action potential. However, under certain abnormal conditions a muscle can and will shorten without a preceding action potential. A muscle so shortened is in a condition of *contracture,* which is usually a reversible type of activity, but during such an occurrence the relaxing phase of the mechanical response is extremely slow. Furthermore, a contracture is not conducted from one end of the muscle to the other, and its production involves a direct excitation of the contractile mechanism. Contracture has been associated with a depletion or complete loss of adenosine triphosphate (ATP) content. A type of contracture which is irreversible is *rigor mortis.*

MECHANICAL PROPERTIES

Following the electrical event there is a period of quiescence called the *latent period,* after which the muscle contracts and develops tension. Thus, just prior to the development of tension, the muscle seems to relax so that during this latent period it shows a decrease in tension, called *latency relaxation,* representing the first or initial mechanical event in muscle contraction. It is believed that latency relaxation is due to the series elastic component, which undergoes a sort of stress relaxation at the onset of contraction. These elements may be the S filaments, which connect the ends of the opposite thin filaments to each other through the zone of H, as discussed by Huxley. Following latency relaxation there occurs a typical mechanical response characterized by tension development which reaches a peak and is then followed by a decrease in the developed tension or force (muscle relaxation). Such a single mechanical response is called a *twitch*, and its duration is of the order of 100 to 200 msec, depending on the animal species and type of muscle. The interval between the beginning of the electrical response and the peak of the tension curve is called *contraction time.* It is a measure of the speed of action of the muscle and is directly related to the function subserved by the muscle. Thus, the external muscles of the eye have fast contracting times and are extremely rapidly acting muscles. On the other hand, the muscles used in the maintenance of posture have long contracting times and accordingly are slowly acting effectors. In addition, muscle activity is dependent upon the number of fibers responding to a nerve impulse. Since there are about one-quarter of a billion individual muscles, but only 400-odd thousand myelinated nerve fibers in all of the ventral roots of the human spinal cord, it is apparent that each muscle fiber is not supplied by an individual motor nerve fiber. Actually, the axon of each motoneuron branches many times and thereby can innervate a number of muscle fibers. The total number of muscle fibers innervated by a single axon of a motoneuron and its branches constitutes a *motor unit.* This complex represents the unit of muscular activity. Furthermore, the number of muscle fibers in one motor unit is represented by the *innervation ratio,* which indicates the size of the motor unit. The physiological significance is that the smaller the ratio, the greater the delicacy of muscle action. Coarse activity results from a unit with a large innervation ratio.

TYPES OF MUSCLE CONTRACTION

All striated muscle when stimulated will contract and develop tension. However, some will also shorten and produce movement. Since weight is carried through a distance – as in running, walking, or lifting an object – work is performed. A muscle which contracts and shortens under a constant load is said to contract *isotonically.* Since the external force which opposes this muscle is less than the force being exerted by the muscle, the function of an isotonically contracting muscle is one of acceleration. On the other hand, when a muscle exerts tension but does not shorten, the contraction is said to be *isometric.* Such a contracting muscle operates during the maintenance of posture and equilibrium, and in holding an object. Since it opposes the force of gravity, its function is one of fixation. A muscle contracting isometrically does not perform work. A third type of contraction can occur in which the tension developed is less than the opposing force, so the muscle is stretched or *lengthened* during contraction. In this case physical work is being done on the muscle in stretching it, and energy is supplied to the muscle. The muscle itself does not do any external work. The function of such contracting muscles is for deceleration, as in checking the forward velocity of a leg as the limb reaches its limit during walking or running. In all three types of contraction the muscle develops tension or exerts a force, but, depending upon the type of contraction, the muscle may do positive work, no work, or negative work.

In both isometric and isotonic contractions the mechanical event far outlasts the electrical response; a second stimulus applied during a twitch produces a second contraction superimposed upon the first. The tension thereby developed is greater than in the single twitch and is known as *summation of contraction* (Fig. 3B-3). This response is a property of the contractile elements, since summation can be

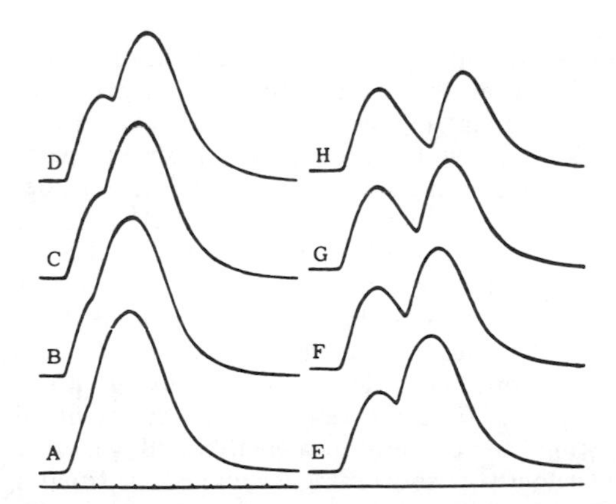

FIGURE 3B-3. Summation of muscle twitches. Two muscle twitches are excited with a double stimulation at different times between shocks. From A at the bottom through H, as the time between the two shocks is increased, the twitch height can be resolved into two separate individual twitches. The addition of the twitch responses is due to the fact that the contractile mechanism of the muscle shows summation, unlike the refractoriness of an action potential in a nerve or muscle membrane. (From S. Cooper and J. C. Eccles. *J. Physiol.* [London] 69:379, 1930.)

produced in a single fiber. Twitch summation can occur because the mechanical response does not have a refractory period like that present in the electrical response. If a series of stimuli is delivered, the muscle twitches combine into a *tetanic* response, which can be as much as four times as great as the amplitude of a single twitch. At high frequencies of stimulation the tetanic contraction is smooth *(fused or complete tetanus)* and does not show the individual twitches seen in *incomplete tetanus,* which results from lower frequencies of stimulation. Thus, the frequency of stimulation will determine the degree of tetanus, and muscles differ as to their *tetanizing frequency.* It should be pointed out that when a muscle is stimulated through its nerve, and when the nerve impulses are so rapid that successive action potentials begin to fall into the refractory period of preceding muscle action potentials, tetanic summation does not occur. The muscle might respond to the first nerve impulse and then not to the succeeding impulses (Wedensky inhibition).

The rate of discharge of motoneurons is normally low, and in the muscles of a motor unit incomplete tetanus occurs or minute "twitches" are produced by each motor unit. Therefore, the smoothness of muscular movements comes about by the operation of a number of motor units which are discharging in an asynchronous manner so that some motor units are contracting while other units are simultaneously relaxing.

CONTRACTILITY (INOTROPY)

The above section has described the types of contraction any muscle can undergo. It has not, however, defined the concept of and the way(s) in which a muscle can change its contractility or inotropy. Muscular performance appears to be the product of at least three principal variables inherent in the constitution of the muscle fibers. A given muscle may show an increase or a decrease in contractility (positive or negative inotropy), by affecting its (1) force-velocity relation, (2) length-tension relation, (3) active state, or (4) via a combination of these relationships.

Force-Velocity Relationship

A muscle which contracts isotonically with tetanic stimulation not only is shortening but is exerting the maximum tension it can produce. The work this muscle performs, as well as the speed of shortening, depends upon the load it is working against (Fig. 3B-4). Hence, with zero load, it will shorten most rapidly but perform no work. With increasing loads, the speed with which the muscle shortens decreases exponentially. Thus, the force generated in contraction is inversely proportional to the velocity of shortening of the contractile elements. Muscle appears to possess the capacity to alter the force-velocity relation independently of resting fiber length. Figure 3B-4 illustrates the basic force-velocity relation. It is to be noted that the speed of fiber shortening is greatest at the onset of contraction, at which point no load is encountered by the contractile elements and therefore no force is generated. At this juncture the velocity of fiber shortening is maximal (labeled V_{max}). Subsequently, as the contracting fiber encounters a greater load, generating an appropriately greater force, the velocity of shortening is reduced. Ultimately, with the expression of maximal force (P_0), contractile element shortening ceases.

Several studies have demonstrated that the force-velocity relation can be altered by a number of inotropic interventions – increased frequency of contraction, for example. Such enhancement of the contractile state may be characterized *primarily* by an increase in the maximal rate of fiber shortening (V_{max}), with or without a change in the maximal force generated. In either case, however, the time required to achieve maximal shortening of the fiber, maximal force generation, and maximal

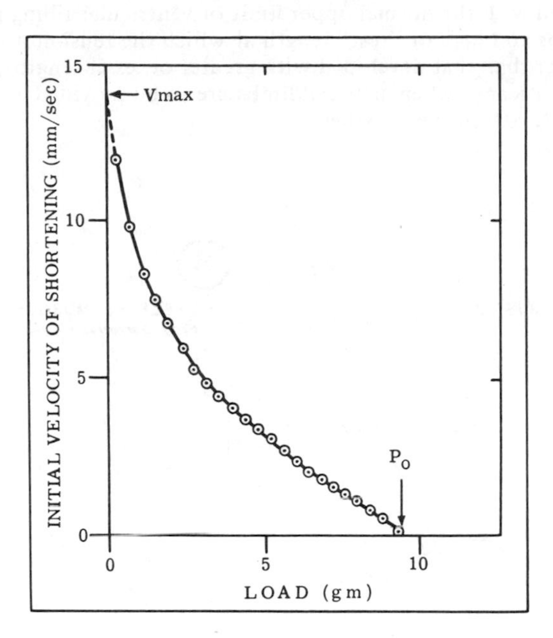

FIGURE 3B-4. The velocity of contractile shortening, which is inversely related to the force the muscle develops or the load it moves. Thus at zero load (no force developed) the velocity of shortening is maximal; this is extrapolated and called V_{max}. The type of shortening is isotonic, and the muscle will shorten as it starts to exert a force, but the velocity of shortening will diminish at maximal force generation (P_0). At that point (P_0) the velocity of contractile shortening is zero and the contraction is isometric. (After A. V. Hill. *Proc. Roy. Soc. [Biol.]* 126:136, 1938.)

rate of force development is consistently abbreviated. Moreover, the force-velocity relation of all muscle, as characterized by V_{max}, may vary quite independently of the resting length of the muscle fibers. Conversely, alterations in the resting fiber length may alter the maximum force generated in contraction without altering V_{max} (see below). An interesting observation is that the more rapid the contraction, the greater the energy requirement. This indicates that when a muscle is contracting more slowly, the energy that is mobilized has a longer time to be transformed into energy for contraction, and less energy will be dissipated as heat.

Length-Tension Relationship

Under any given situation a muscle contracting tetanically exerts its maximum tension. However, the amount of tension that is developed depends upon a number of factors, one of which is the length of the muscle at the time of stimulation. Thus, if a muscle is set at *rest length* (identified as L_{max} and defined as the length it normally assumes in the body under conditions of rest or when maximally extended in the body), the tension it develops during tetanus is greater than that developed with longer or lesser length than L_{max}. This indicates that a muscle is able to exert a greater tension as its length is increased, if the resting length of the muscle is not exceeded (Fig. 3B-5). (In cardiac muscle, L_{max} seems to be the "resting" muscle length associated with the normal upper limit of ventricular filling pressure.) Since every muscle has an L_{max} or "rest" length at which the tension it develops during tetanus is greater than that developed with greater or lesser lengths, this has important clinical significance when artificial limbs are to be provided and a muscle length is to be chosen to attach the prosthesis.

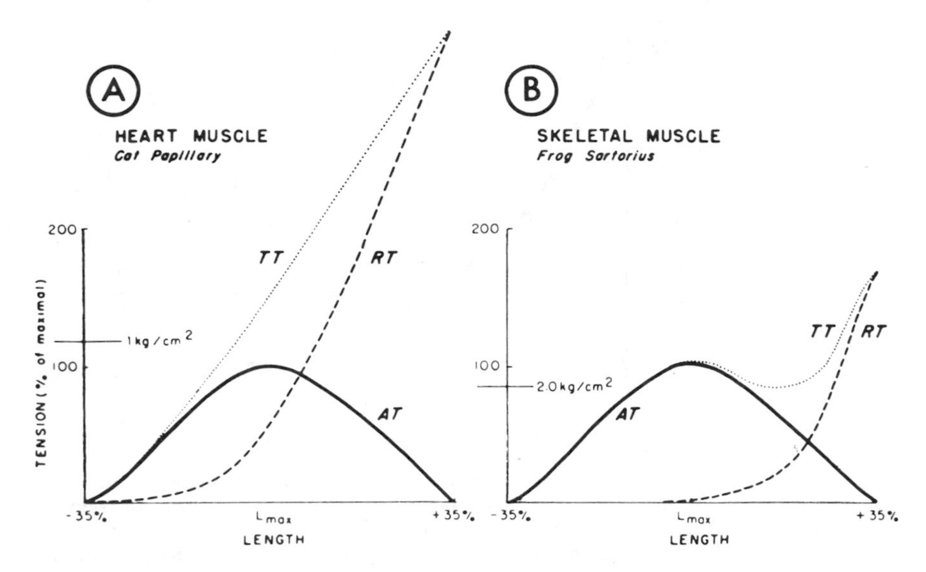

FIGURE 3B-5. Relation between length and tension for heart (A) and skeletal muscle (B). Tension has been related to percentage change in initial length. Although the absolute maximal developed tensions found at L_{max} are different in heart and skeletal muscle, the shapes of the curves are quite similar. *RT* = resting tension; *AT* = actively developed tension; *TT* = total tension or AT + RT. (From D. Spiro and E. H. Sonnenblick. *Circ. Res.* 15 [Supp. 2]:14–37, 1964.)

Active State: Relationship to Twitch and Tetanus Tension

The question arises as to why the tension or force developed by a muscle during complete tetanus is far greater than that seen for a single twitch. From the preceding discussion it is obvious that the tension produced during a twitch will be greater with longer length (up to L_{max}). However, if a muscle at a given length is stimulated to contract and then suddenly stretched at various times during, and up to the peak of, the twitch, the tension developed is greater at the new stretched length than it would have been had the muscle at the onset been stimulated at that new length. This indicates that, at a time of tension rise, sudden stretch somehow draws out earlier the available force of which the contractile element is capable. In addition to contractile elements, a muscle has elastic elements which are in parallel and in series with the contractile elements. The elastic elements are probably not taut. Hence, the contractile machinery must, before it reflects its activity, take up the slack of the elastic elements, which requires some time. Reducing the slack time should result in the more rapid development of tension, and this is what quick stretch seems to do. The greater tension seen following quick stretch indicates that the energy of the twitch is made available much sooner than is ever indicated by a recording of a tension curve. It also means that the maximal tension which a muscle is capable of producing, for a given length, is never attained in a twitch.

The tension which the contractile elements are capable of producing, called *active state,* is quickly developed following a stimulus, is maintained for a time, and then declines exponentially. Therefore, with a single stimulus the contractile elements develop their active state at once but maintain this intensity for only a very short time. In order for the muscle to develop tension, the contractile elements must first stretch elastic elements. The stretching takes time, and furthermore, the heavier the load, the longer it takes. By the time the slack is taken up, then, the active state has started to decline and the tension which does develop at the peak of the twitch is less than the theoretical maximum. When the active state duration is prolonged (as with epinephrine or anions), the twitch tension is greater. Another way to prevent the active state from decaying is a second, third, or more stimuli immediately following the first. This should produce a twitch of greater magnitude (Fig. 3B-3). Physiologically it is called summation of two or more contractions. Therefore, if two or more stimuli maintain the active state, a series of successive stimuli applied while the contractile elements are in their maximal state of activity should maintain the condition until fatigue sets in. This is exactly what occurs with a tetanizing current of stimulation, and the tetanus tension developed will be the maximal activity of which the muscle is capable *at that moment,* depending upon its metabolic state, length, etc.

Since onset of the active state is a result of cell depolarization, it has been suggested that the active state decay may be initiated by repolarization of the action potential. Any factor which delays repolarization also prolongs the duration of the active state. In turn, it is interesting to note that substances (such as choline and tetraethylammonium) which prolong the active state are also effective in delaying repolarization of the action potential.

Thus, there is a third variable which has a bearing upon the expression of either of the two above-described muscle functions: the duration of the active state of the contractile elements. The active state may be described as the *existence* of force generation or active shortening of the contractile elements. It is apparent that this so-called active state possesses both intensity and duration. The intensity is characterized by the magnitude of force generated (P_0) and the velocity of contractile shortening relative to the force generated. These characteristics of the intensity of the active state have been described above with regard to the length-tension and force-velocity relations. The duration of the active state merits further consideration. In vitro studies have demonstrated that it is directly proportional to the time required to reach peak isometric tension; thus an objective criterion is available for the assessment of active state duration. It has also been noted that the duration of the active state tends to vary inversely with the maximal velocity (V_{max}) of contractile shortening. There is a theory that the duration of the active state may limit the maximum force generated by stimulation at any resting fiber length or inotropic state. Presumably, the resting fiber length determines the maximum force generated by contraction, but this may not be achieved because of the limited active state duration. Prolongation of the active state (for example, as a consequence of hypothermia) may permit full expression of the force-generating capacity inherent at that resting fiber length (and at that force-velocity state).

Interaction of Factors

It has become increasingly apparent that there are three main determinants upon which the functional characteristics of muscle fibers are predicated. The first is the length-tension principle, which dictates that the maximum force that may be generated is directly related to the resting length of the muscle fibers. The second is the force-velocity curve, which expresses an inverse relation between the velocity of contractile shortening and the force generated or load borne by the contracting fibers. The inotropic or contractile state of the muscle is characterized by the *maximal* velocity of fiber shortening (V_{max}), which occurs at a time when the contractile elements are unencumbered by a load or resistance to shortening. It has also been demonstrated that the aforementioned basic factors may vary independently of one another, rendering any assessment of the state of myocardial function more complex. In addition, in reviewing the determinants of muscle performance one must consider a third variable which has a bearing upon the expression of either of the two other described muscle functions: duration of the *active state* of the contractile elements.

An assessment of muscle performance must therefore conclude that the end result of contraction stems primarily from the interaction of these three factors. It is also wise to remain cognizant of the fact that resting fiber length, the force-velocity state, and active state duration may vary independently, influenced by a variety of circumstances (although, in general, V_{max} and active state duration tend to bear an inverse relation to each other). An enhancement of the inotropic state, without change in resting fiber length or active state duration, may permit expression of the maximum force generation capacity (P_0) inherent at that fiber length and previously inexpressible because of the limitation of active state duration. Moreover, an increase in resting fiber length alone may permit greater force generation without change in the contractile state of duration of the active state. Finally, an alteration in active state duration may limit or permit full expression of the force-generating capacity inherent at an unchanging V_{max} and resting fiber length.

THERMAL EVENTS OF CONTRACTION

During the events resulting from excitation and contraction a number of chemical reactions occur within the muscle fibers causing liberation of energy. Associated with this energy liberation is the production of heat, the study of which has revealed to a great extent what happens during muscular activity. In the inactive muscle chemical reactions take place which maintain the integrity of the cells. From these processes the muscle liberates a *resting heat.* When activated, the muscle liberates heat in excess of the resting base level. The rapid outburst of heat in the case of tetanic contraction is called *maintenance heat* but in the case of a single twitch is described as *activation heat.* Activation heat is seen whether the muscle contracts isotonically or isometrically and represents energy expenditure for maintenance of the active state. However, if the muscle is permitted to shorten under a load, then more heat is liberated, the magnitude increasing with the degree of shortening. The amount of heat liberated during shortening is not dependent upon the load of the muscle; it is independent not only of the work done by the muscle but also of the speed of muscle shortening. It is solely dependent upon the amount of shortening. The fact that *shortening heat* is not related to the amount of work accomplished means that more energy can be usefully utilized when the muscle does work. Thus, a muscle which does not shorten but develops an equal amount of tension will obviously liberate less energy. The muscle when it does work can therefore regulate and mobilize its energies depending upon the amount of work to be done. This concept of more work-energy mobilization (Fenn effect) dispelled the prior notion that every muscle had a fixed quantum of energy regardless of the degree of work. At a fixed level of shortening and with greater amounts of work, there is not only a greater liberation of energy but a reduction in wasteful (except for temperature maintenance) heat loss. This means that the contractile process is utilizing the energy more *efficiently,* or perhaps that there is less loss through friction during muscular work.

Subsequently, when the muscle contraction is over, relaxation begins, and as the load is lowered there is a further liberation of heat. This *relaxation heat* is produced as a result of the muscle's being stretched by the load. It should be recalled that at the onset of contraction, energy was required to stretch the elastic elements. This energy is stored during elastic stretch and released as relaxation heat during muscle relaxation. Relaxation does not appear to involve any energy utilization, and all the heat lost during this phase comes from the energy stored in the contraction phase. Hence, muscle relaxation appears to be a purely passive phenomenon. However, the magnitude of the relaxation heat depicts exactly the work done on the muscle by the load as it falls and stretches the muscle. If the weight is removed at the peak of shortening and the muscle permitted to relax under no tension, no relaxation heat is liberated. This is of course paradoxical since the question may be raised as to where the energy stored in the elements went upon removal of the load. No data are currently available to explain the observation.

The heat liberated during the active phase of contraction (activation, shortening, and relaxation heat) is collectively known as the *initial heat.* It is not dependent upon oxygen supply and is independent of glycolysis. Initial heat may be related to degradation of substances rich in high-energy phosphate.

Finally, after the muscle has returned to its original physical condition, an additional amount of heat is liberated. This *recovery* or *delayed heat* is concerned with refueling of the energy that has been used during the active contraction period. The recovery is dependent upon oxygen and hence is essentially an oxidative manifestation.

MUSCLE CONTRACTION

The exact mechanism involved in activity of the contractile elements when the cell membrane becomes excited has not been completely elucidated. Many theories have been postulated, only to fall by the wayside as new information is gathered. The currently accepted theory is the *interdigitation* or *sliding of myofilaments.* The muscle contains two main proteins, *myosin* and *actin.* The addition of certain chemicals can selectively remove these proteins from the muscle, leaving behind a fibril containing intact Z lines which are joined together by material that maintains structural integrity. When the myosin filaments are removed, the A band disappears and there is left behind a material that stretches from the Z line to the rim of what was the H zone. At this stage, removal of the actin filaments results in a disappearance of the I band and leaves behind a ghost fiber with only Z lines. Therefore it seems evident that the myosin filaments reside in the A band, whereas the actin filaments reside in the I band. The actin starts at a Z line and goes through half of an I band, through an A band to terminate at the H zone. On the basis of electron microscopic studies, these proteins lie in a longitudinal arrangement and are spaced a few hundred angstroms apart. Furthermore, the myosin filaments are about 100 A thick and 450 A apart, and extend through the A band. The actin filaments are about half as thick and extend, as mentioned, from a Z line to the H line. The actin or thin filaments are arranged in a hexagonal manner and in turn each actin filament as it courses through the A band is surrounded by three thick or myosin filaments (Fig. 3B-6). The myosin filaments are thus arranged in a trigonal manner. The above information suggests that a muscle is made up of two sets of filaments, which overlap in the A band. Reference to Figure 3B-6 indicates in diagrammatic fashion the changes which occur in the filament arrangement under conditions of stretch and contraction as compared to a rest condition. Under any condition the A band

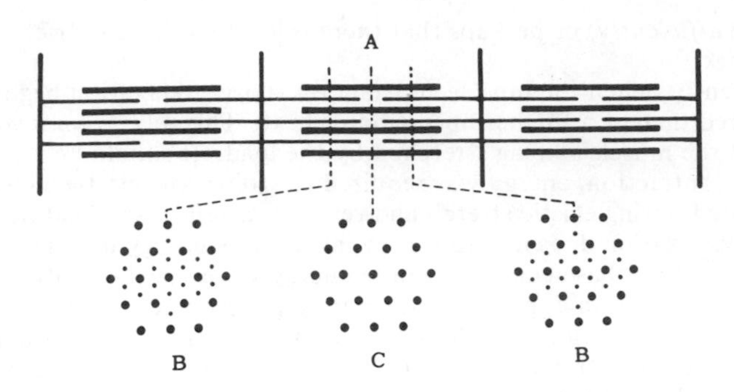

FIGURE 3B-6. Myofilaments of a muscle. In A the dashed lines indicate where a cross section was taken across the various portions of the A band of the sarcomere. The section across the middle of the sarcomere, the H zone, shows the position of the thick filaments (C). The myosin-containing myofilaments are the thicker. In sections on either side (B), where both the thick and the thin filaments are cut through, the position of the thinner myofilaments with relation to the thick ones is shown. According to the sliding filament theory, it is the relative movement of the two myofilaments which constitutes contraction. (From H. E. Huxley. In J. Brachet and A. E. Mirsky [Eds.]. *The Cell.* New York: Academic, 1960. Vol. 4, p. 375.)

does not change its length. Furthermore, the H zone length increases or decreases with the I band length; however, the distance from one H zone to another H zone does not change. Since this distance is exactly the length of the thin filaments, it is indicative that these filaments too do not alter their length. The conclusion reached from these observations is that when a muscle changes its length the two sets of filaments slide past each other. With extreme shortening, the proteins do meet and may overlap (Fig. 3B-7). Upon relaxation of the muscle, the two sliding elements again withdraw to the spacing of the relaxed muscle.

Just how, on a molecular basis, the movement takes place is at present a matter of research. In electron micrographs, side projections can be seen along the myofilaments. Huxley (1958) believes a "catch" mechanism occurs in which side extensions move the myofilaments past one another by a ratchet-like process (Fig. 3B-8). The energy of contraction comes from adenosine triphosphate, which contains energy-rich phosphate bonds (see Chap. 27). Somehow, ATP attaches to the myosin, which has been shown to act as an ATPase to release thereby the high-energy phosphate of the ATP. The chemical energy in the high-energy phosphate bond is in some manner translated to contractile proteins and thereby provides the mechanical action of the muscle. The details of this coupling of chemical bond energy to mechanical action are not clear at present.

As mentioned above, the link uniting the events between the initial excitation and the terminal event of contraction is still obscure. Recent morphological and physiological studies have demonstrated that the Z lines are made up of two parts – one being composed of the Z line of contractile elements, and the other being the central part of the triads, the transverse tubes or T system, which appear to be continuous with the plasma membrane of the cell. It has been suggested that the T system represents the pathway for current flow and hence plays a role in excitation-contraction coupling. In addition, calcium ion plays an important part in contraction. In fact, calcium in bound form is stored in saclike structures located near the transverse tubules, the membrane of which is permeable to the unbound form of the cation. The excitation-contraction coupling can be very briefly outlined as follows. Upon cell depolarization, the current flowing through the T system somehow

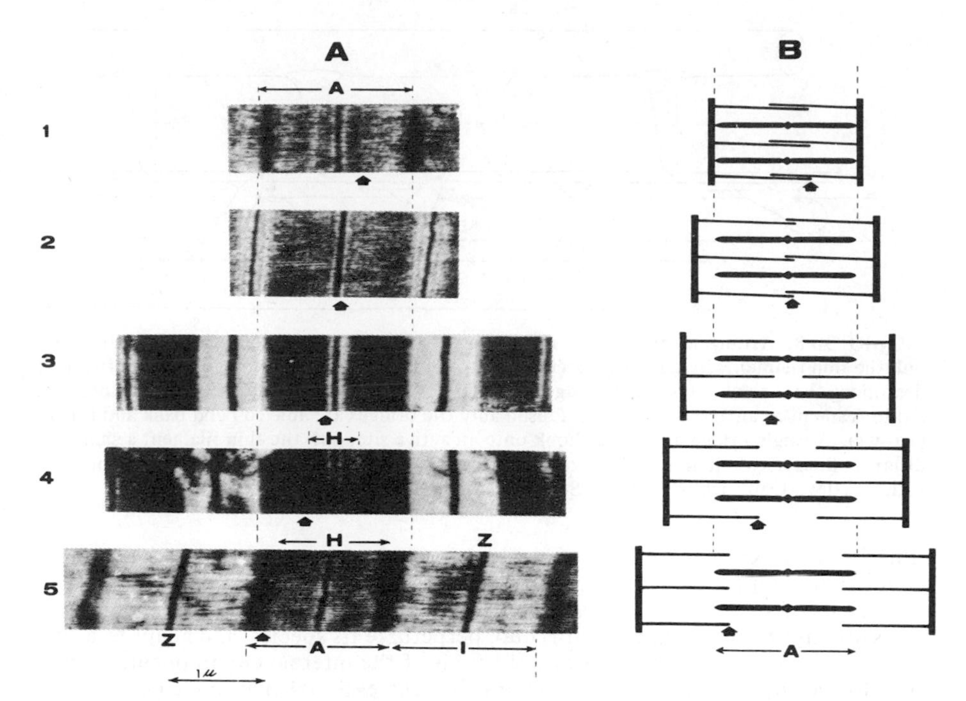

FIGURE 3B-7. Relation between sarcomere length and band patterns in skeletal muscle (frog sartorius). Panel A (left) shows the band patterns as seen electromicroscopically, and panel B (right) the disposition of the thick and thin filaments that create these patterns. The vertical arrows in both panels denote the ends of the thin filaments that insert at the Z line at the left. Panel A (3) represents the sarcomere at the apex of the length-tension curve – that is, at L_{max}. In (1) and (2), sarcomere length has been progressively decreased, whereas in (4) and (5) it has been progressively elongated. Throughout, the A band remains constant in width. The placement of filaments to provide for maximum overlap is shown in B (3). (1) shows the sarcomere pattern near L_0 in the shortened muscle (L_0 is defined as that muscle length at which active tension approaches zero as the muscle is allowed to shorten); the I band has disappeared, and a secondary dark band has been formed at the center of the sarcomere termed the *C contraction band,* which is due to the passage of thin filaments through this area as in B (1). In A (4) and (5), an expanding H zone has appeared owing to the withdrawal of the thin filaments from the A band, as shown diagrammatically in B (4) and (5). (From Braunwald et al., 1968.)

causes the sacs to release free calcium. The transport of calcium into the T system may be linked with the breakdown of a high-energy phosphate bond. The ionized calcium diffuses to the active sites on the actin and myosin filaments, initiating the sliding of these filaments. With relaxation some factor removes the calcium from the active sites. The calcium then actively diffuses out, to reaccumulate in the lateral sacs of the sarcoplasmic reticulum.

SMOOTH MUSCLE

The primary function of muscle is mechanical – to generate a force. This force may be utilized to perform many activities; and in smooth muscle, contraction is respon-

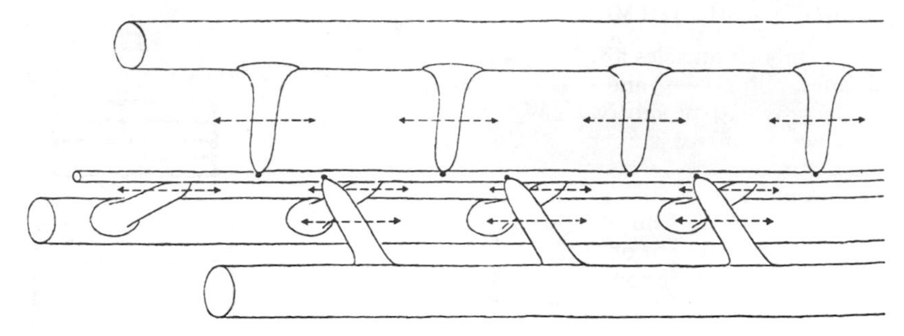

FIGURE 3B-8. Arrangement of cross-bridges suggests that they enable the thick filaments to pull the thin filaments by a kind of ratchet action. In this schematic drawing one thin filament lies among three thick ones. Each bridge is a part of a thick filament, but it is able to hook onto a thin filament at an active site (dot). Presumably the bridges are able to bend back and forth (arrows). A single bridge might thus hook onto an active site, pull the thin filament a short distance, then release it and hook onto the next active site. (From The contraction of muscle, by H. E. Huxley. Copyright © 1958 by Scientific American, Inc. All rights reserved.)

sible for the mixing and moving of ingested food, the maintenance and regulation of circulation, expulsion of the fetus, regulation of light admitted to the retina, etc. The end results of smooth muscle contraction are extremely important to the organism in its attempt to maintain the status quo, perpetuate its species, and adapt to a constantly changing environment. Control here is of the internal environment, in contrast to skeletal muscle control, which involves the external environment. Both effector systems, however, act by contracting to preserve the integrity of the organism.

The classification of muscle into skeletal, smooth, and cardiac muscle types is dependent upon the embryological, histological, and physiological functions of these effectors. Embryologically, smooth muscle is derived from the mesenchyme, while skeletal muscle originates from mesodermal somites. Histologically, the smooth muscle fibers are smaller than skeletal fibers and contain fewer myofibrils per unit. A distinct sarcolemma is absent in most smooth muscle, although one can be observed in uterine fibers which are about 150 A thick. In addition, there seems to be no differentiation between the A and I bands in smooth muscle.

Metabolically, smooth muscle cells under rest conditions utilize less oxygen than do skeletal muscle cells. Oxygen is required for the continual maintenance of tonus, but when activity increases there is very little increase in oxygen consumption. Under anaerobic conditions, smooth muscle shows an increase in lactate formation and a decrease in glycogen content. Phosphocreatine is present in smooth muscle, decreasing during activity and increasing during an aerobic recovery period. Such recovery does not occur under anaerobic conditions.

Generally, the metabolism of smooth muscle is similar to that of skeletal and cardiac muscle. In all three muscle types, the metabolism is primarily aerobic. They contain the same enzyme systems and utilize the same substrates, and their major energy source is from the Krebs cycle. Functionally, muscle can be classified according to (1) nerve supply to the fibers, (2) nature of the chemical transmitter, and (3) electrical and mechanical attributes of the tissue. Such criteria can be utilized to divide smooth muscle into two classes: (1) the multiunit smooth muscle and (2) visceral smooth muscle. In this section, only the electrical and mechanical properties will be discussed, since the innervation and the chemical transmitters affecting smooth muscle are described in Chapter 7.

THE MULTIUNIT SMOOTH MUSCLE

Multiunit smooth muscles are exemplified by the ciliary muscle, iris and nictitating membranes, pilomotors, and many blood vessels. Their activity is largely controlled by discharge from true motor nerves. They are very similar to striated muscle; however, the nerve control is not as precise as that of skeletal muscle. The resting potential of a multiunit muscle cell is on the order of −50 mv. A single threshold stimulus results in depolarization, followed by a long delay and contraction similar to a twitch. However, the time course is greatly prolonged. Repetitive stimulation or increasing the strength of stimulation results in summation and tetanic contractions. It appears that following nerve stimulation there is a release of norepinephrine, which is responsible for muscle depolarization. The nerve distribution of the multiunit smooth muscle is such that two or more nerve fibers innervate each muscle fiber so that an overlap exists. When stimulus strength is increased, more fibers are activated and a greater amount of the mediator is released, resulting in greater contractions. When the frequency of stimulation is increased, a faster release of mediator occurs which affects a greater area of muscle.

VISCERAL SMOOTH MUSCLE

The visceral smooth muscle behaves as though it were a single unit or multiple of large units. The individual fibers are believed to be in a functional continuity and hence can be considered to form a syncytium. They are innervated by the fibers of the autonomic nervous system. However, these muscle fibers are not dependent upon the nerves of their rhythmic activity. The role of the nerves is one of regulation, not initiation, of visceral smooth muscle activity.

Electrical Properties

Essentially, the distribution and permeability of K^+ across the visceral smooth muscle membrane is similar to that in most viable cells. The smooth muscle cell has a much higher Na^+ content, and this ion can readily cross the cell membrane. Intracellular recordings have been made on these extremely small cells and a wide scatter of resting potentials (maximum polarization) obtained. A resting value of about 60 mv is generally accepted, the inside being negative with respect to the outside of the cell. The exact value differs depending upon the type of smooth muscle tissue and whether the tissue is under the influence of hormones, stretch, activity, etc. In general, the resting potential of visceral smooth muscle is lower and less stable than that of skeletal muscle. A characteristic finding is the small oscillation or fluctuation of the resting potential, usually subthreshold in nature. When these *slow waves* or "fluctuations" reach a threshold potential, they give rise to a *spike potential.* Such cells showing these local depolarizations are believed to be "pacemaker" cells possessing the property of automaticity (see Chap. 14). In addition, spike potentials as well as plateau-type action potentials (see Chap. 14) have been recorded from cells not showing the slow wave, and these action potentials may be elicited from conducted impulses. In either case, the rate of depolarization is extremely slow. Furthermore, the overshoot is usually lower than that observed in skeletal muscle action potentials. The duration of the action potential varies from tissue to tissue depending upon the type. The spike potential duration varies from 5 to 40 msec, while the plateau type may last as long as 400 msec. The usual refractory periods and afterpotentials are present in both types of action potential.

To a large extent, the basis of the resting and action potentials of visceral smooth muscle can be explained on the ionic hypothesis (see Chaps. 1 and 2). The usual logarithmic relationship between the external K^+ concentration and the resting potential holds true for higher concentrations. Raising extracellular K^+ results in a slight depolarization and an increase in smooth muscle activity, whereas removal of K^+ results in a slight hyperpolarization. However, other ions may be involved in the maintenance of the electrical gradient across the cell membrane. Thus, substitution of Cl^- with other anions causes depolarization; this anion may play a part in maintaining the electrical gradients across visceral smooth muscle cells.

The ionic hypothesis postulates that the action potential is a result of increased membrane permeability to Na^+. This does appear to be the case for skeletal and cardiac muscle. However, the relationship between Na^+ and the electrical events of smooth muscle is not clear. The fact that an overshoot is observed indicates that Na^+ conductance is augmented during the rising phase. Yet for many types of smooth muscle, replacement of 90 percent of the Na^+ appears to have very little effect on the spike potential. Nevertheless, the Na^+ ion cannot be totally exchanged for a substitute, since in the complete absence of Na^+ all activity eventually ceases. Pacemaker activity of smooth muscle appears to be influenced by variation in Na^+ concentration; it is abolished when the muscle is placed in a sodium-free solution. What may be important for the spike potential is the concentration of calcium. This ion appears to act as a moderator of the membrane permeability to Na^+. In a solution containing excess Na^+, the removal of Ca^{++} results in oscillatory activity of the muscle. Excess Ca^{++} produces a regular pattern of activity. Although spike activity is still maintained in a low Na^+ solution, the removal of Ca^{++} results in a cessation of this activity. It is therefore highly suggestive that pacemaker activity is dependent upon Na^+, whereas the key substance for spike activity may be Ca^{++}. However, the possibility must be entertained that the Ca^{++} spikes are operative only under adverse conditions (such as during Na^+ removal), and that under normal conditions the spike is still dependent upon Na^+. On the other hand, the recording of spikes under low Na^+ conditions need not be interpreted as a nondependency of the spike upon this ion. Conceivably, the tissue is not completely sodium-depleted even in the sodium-free solution.

Factors Affecting Membrane Excitability

Smooth muscles comprise the visceral hollow organs of the body. These muscles are constantly under the influence of a variety of factors, singly and in combination, which directly affect their excitability and activity. The organs of the visceral system, by the very nature of their makeup and function, are continuously being regulated by hormones, autonomic transmitter agents, stretch, and tension. In turn, the response of the tissue will depend on predisposing factors such as the metabolic condition of the tissue and the state of tonus (see below).

It should be recalled that stretch, hormones, and the autonomic transmitter agents are normal stimuli for smooth muscle. Thus, acetylcholine (ACh), histamine, stretch, and estrogen (for some muscles) increase excitability, and the frequency of spike activity is preceded by a marked depolarization. The effect of ACh, however, can be blocked by atropine, resulting in hyperpolarization and relaxation. Further, high concentration of K^+ will antagonize the effect of ACh by depolarizing the membrane to below threshold, thus rendering it inexcitable. The mechanism of ACh depolarization seems to be similar to that for skeletal neuromuscular function. That is, ACh causes a nonspecific increase in membrane permeability to all cations. This conclusion is based upon the observation that (1) ACh will depolarize when the solution is free of Na^+ but it does not increase spike activity; (2) excess Na^+ enhances the ACh effect; (3) ACh fails to depolarize in calcium-free solution but is potentiated by excess Ca^{++}. These findings are indicative of an increased permeability to Na^+ and Ca^{++} and also support the concept that both Na^+ and Ca^{++} are essential for smooth muscle excitability and development of spike activity. In addition, the smooth muscle membrane becomes unstable and depolarized by any factor which decreases the metabolic rate of the tissue. Thus, sudden cooling, removal of energy-rich compounds, and inhibitors of the sodium pump (such as the digitalis compounds) result in increased activity preceded by depolarization.

On the other hand, any factor or agent which increases the metabolic activity of the tissue will cause hyperpolarization, reduction or loss of excitability, and cessation of activity. Thus, epinephrine, progesterone, addition of energy-yielding compounds, or raising the temperature will stop activity and increase the transmembrane electrical gradient. Interestingly, the increase in membrane potential due to epinephrine does not lead to inhibition of activity, since the spontaneous discharge stops before hyperpolarization. The degree of hyperpolarization appears to be related to the potential level at the time of drug action. The closer the potential is to the K^+ potential, the less the degree of hyperpolarization. Furthermore, epinephrine is able to hyperpolarize in potassium-free solutions even at a time when the potential level is high. This indicates that hyperpolarization may be initiated by the ability of the catecholamines to accelerate the extrusion of Na^+. It is worthwhile to refer to a review by Daniel (1964), who has attempted to unify all the various factors which affect smooth muscle activity into a mechanism that is dependent upon the ratio of bound-to-free Ca^{++} within the membrane. Daniel thinks that Ca^{++} is bound to membrane anionic sites, and the factors which alter muscle activity do so by inducing a change in the bound-to-free Ca^{++} ratio. It should be pointed out that the factors (ACh, stretch, etc.) which result in depolarization and increased activity of smooth muscle also produce an increase in muscle tension. Associated with this tension is an increase in K^+ efflux from the cell. During normal muscle relaxation or relaxation induced by epinephrine there is an enhanced uptake of K^+ by the cell. Thus, during muscle activity there is a K^+ efflux, and during relaxation there is a K^+ influx. These processes normally take place during activity and recovery from activity. Therefore, the depolarization and hyperpolarization resulting from the agents discussed above can be explained on the basis of the outward and inward movement of K^+. Essentially, these substances (ACh, stretch, epinephrine, etc.) change the ratio of K^+ concentration inside and outside the cell by either inhibiting or stimulating the active transport system. Invoking this mechanism, however, would require a rather large net loss or gain of K^+. An alternative explanation has been postulated. In smooth muscle, an electrically neutral pump (Na^+-K^+ coupled) may not be functioning, but rather a sodium pump alone (see Chap. 1). In such a system, any factor increasing the metabolic rate would increase the electrogenic extrusion of Na^+ and raise the membrane potential, resulting in hyperpolarization (which would accelerate K^+ influx). On the other hand, factors which decrease the metabolic rate, as well as ACh, stretch, histamine, etc., cause a decrease in Na^+ extrusion by inhibition or slowing down of the Na^+ pump, resulting in a decrease in membrane potential *(which would accelerate a K^+ efflux)* and depolarization.

The possible existence of an electrogenic pump in smooth muscle is a question still unresolved.

Mechanical Attributes

The characteristic mechanical response of skeletal muscle is a twitch which occurs fairly rapidly following a nerve stimulus. The mechanical response of visceral smooth muscle is a slow, sustained contraction followed by a slow relaxation. Moreover, the contractions occur spontaneously and in a rhythmic fashion. This sustained and powerful contraction can be maintained over a long period of time and is preceded by a spike potential (or plateau-type action potential), which in turn arises from one of the slow fluctuations that has attained a threshold potential.

In addition to rhythmic contractions, smooth muscle possesses the mechanical property of *tonus.* This is a condition of persistent, sustained activity which is independent of extrinsic and intrinsic nerve activity. Tonus is therefore myogenic in origin. (It differs from skeletal muscle *tone,* which is a basic level of activity initiated and maintained by motor nerve activity.) Tonus is a biological property of smooth muscle itself, and although it represents a basic level of activity, it sets the stage for the type of mechanical response the muscle produces when stimulated. An example of how tonus affects smooth muscle response can be seen when some stimulating agent is utilized to cause contraction. If tonus is high, the agent will have very little effect in further inducing a positive muscle response. On the other hand, the very same agent will have a profound effect if it is acting on a smooth muscle in which tonus is low. Thus, if tonus is low, vagal stimulation will initiate intestinal contraction if the muscle was previously quiescent, or the vagal effect would further augment an existing activity. But if gastrointestinal tonus is high, vagal stimulation will result in a diminution of activity. Similarly, a vasodilating agent will have much less influence (if any) on the blood vessels if the tonus of its muscle is low (already dilated) than it would have if the vessel were constricted (high tonus). A detailed account of neurogenic and humoral control of the various types of smooth muscle appears in Chapter 7.

Variations in tonus from its inherent level can occur spontaneously, for tonus is subject to many outside influences such as autonomic nerve activity, change in the physical environment, and chemical agents. Thus, alkalinity and low temperature increase tonus, while heat and acidity decrease it. True tonus change implies a change in fiber length without a change in the magnitude of the contractions. The ability of smooth muscle to change its tonic condition means that these muscles can exist at different lengths under equal degrees of tension.

The exact mechanism for the development of tonus is not understood, but it might be thought of as an asynchronous, basic level of contraction and relaxation of different muscle bundle components in such a manner that contractility remains unchanged but the length does change, as a result of the out-of-phase change between muscle units. Hence, as one component is contracting the other unit or units are relaxing and the base tension remains constant. Although tonus is myogenic in origin, it has been suggested that inherent tonus is due to gradual release, within the smooth muscle, of such substances as catecholamines, ACh, histamine, and 5-hydroxytryptamine. However, blocking agents of these substances, such as Dibenamine, atropine, and antihistamine, are without effect on smooth muscle tonus.

In addition to tonus, smooth muscle exhibits another property called *plasticity,* best observed when smooth muscle is stretched. Smooth muscle can be stretched slowly to new lengths without a permanent increase in tension. The first response to stretch is an increase in tension, but there is a decline to initial tension by the time the new length has been reached. This *release of tension* is due to the plasticity of the muscle. It is as if the muscle fibers rearrange themselves, with a resulting tension decrease (from the initial increase back to the prestretch tension) with a change in length. This is an extremely important property since it enables the hollow organs, which are encircled by smooth muscle, to serve as very effective reservoirs. The volume of the organ can thus be appreciably increased (and hence the length increased) with a minimum change in pressure. Eventually, however, a length is attained in which the tension does increase and is maintained, and for many organs this is the stimulus for the initiation of contractions which result in the evacuation of the reservoir. At greater lengths, the tension of contracting smooth muscle declines and is very similar to the tension-length relationship observed for skeletal muscle. It is interesting to note that the dependence of force of contraction upon muscle length may be a fundamental property of all contractile tissue.

Similarly, as in skeletal muscle, the force-velocity of shortening relationship holds true for smooth muscle. Thus the rate of shortening is fast when the muscle contracts against a small volume and becomes slower with increasing volume. Yet the work-tension relationship for smooth muscle is unlike that of skeletal muscle in that it is highly susceptible to environmental changes. The work performed by smooth muscle increases as the load is increased, until a work plateau is attained. However, the administration of *estrogen to uterine muscle* shifts the curve up (more work for the same load as compared to the same tissue without hormone administration). On the other hand, progesterone shifts the curve below normal. Generally, then, the determinants of smooth muscle contractility appear to be similar to those described for the striated musculature.

CARDIAC MUSCLE

This section deals with the structure of the myocardium and its properties of excitability and contractility. The electrophysiological properties of automaticity, rhythmicity, and conductivity will be described in detail in Chapter 14.

STRUCTURE

Cardiac muscle structure is such that it may be regarded as part skeletal and part smooth muscle. It is striated in structure and syncytial in function. Electromicroscopic studies have shown that cardiac cells are separated by *intercalated discs* which in turn join the sarcolemma. The sarcolemma is composed of two layers, 200 to 300 A thick: an outer one (basement membrane) which runs uninterrupted and an inner one (plasma membrane) which joins the intercalated discs, as well as other tubular structures, the *sarcoplasmic reticulum.* The latter appears to connect with the Z lines of the sarcomeres. Hence, cardiac cells are structurally separated as units, although from a functional point of view the separations do not seem to represent a high-resistance barrier for conduction of the electrical impulses (Fig. 3B-9).

The heart is essentially composed of tissue that contains (1) true contractile cells, (2) specialized conducting cells, and (3) nodal or automatic (pacemaker) cells. The true contractile cells are present in the atria and ventricles. The atria are thin-walled structures separated from the thicker ventricles by the fibrous valves.

The nodal tissue is composed of small fibers which contain only a few myofibrils and which are located in two areas of the heart. One such area occurs at the union of the superior vena cava with the right atrium and extends as a curved band down toward the mouth of the inferior vena cava. This *sinoatrial* (SA) *node* is composed of muscular tissue, but the fibers are more elongated, less heavily striated, and contain less muscle glycogen than do common contractile fibers. The fibers radiate out from the node in all directions and merge with the common atrial fibers. This is the site from which depolarizing impulses originate, and the area of the node is rich in nerve fibers from the vagus and sympathetic nerves. The human SA node measures 25 to 30 mm in length and is 2 to 5 mm thick.

The second nodal area is located near the interatrial septum at the junction of atria with ventricles. This *atrioventricular* (AV) *node,* or the node of Tawara, is a small nodule of modified muscular tissue. It measures 5 mm in length and 2 to 3 mm in width in the human heart. In some ways it resembles the tissue at the sinoatrial node and contains elongated, less heavily striated fibers with a lower glycogen content. The node continues into the *bundle of His* or *common bundle,* which passes through the fibrous skeleton at the ventricular base.

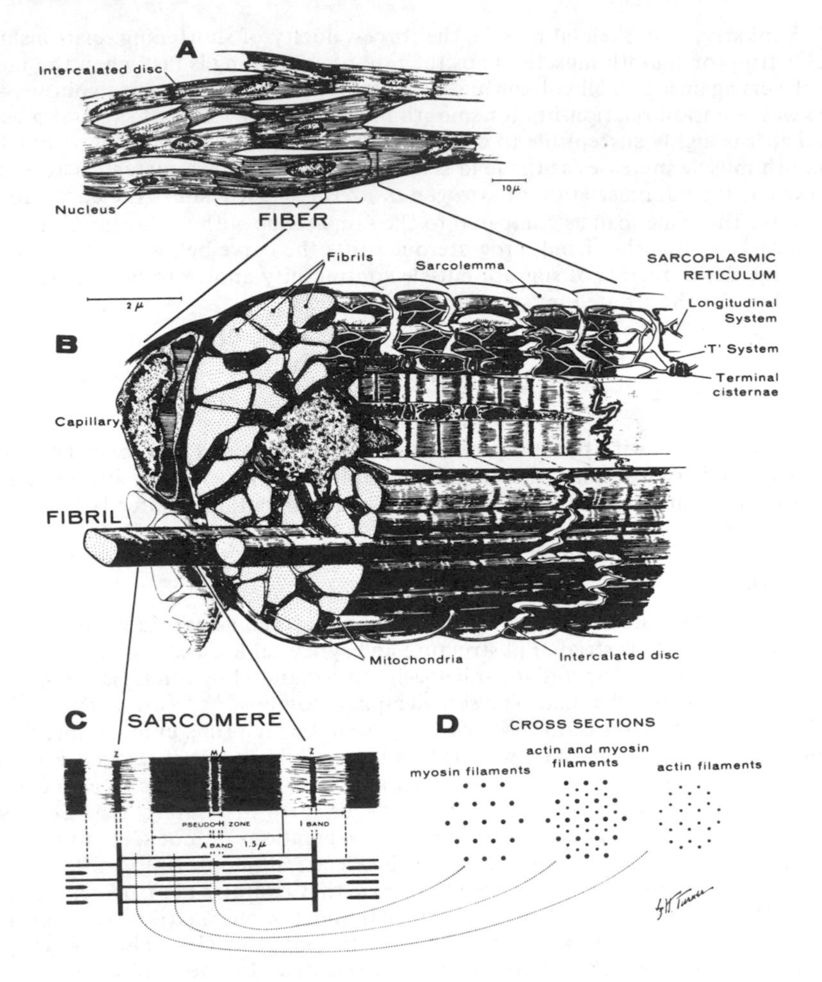

FIGURE 3B-9. The microscopic structure of heart muscle. A. Myocardium as seen under the light microscope. Branching of fibers is evident, each containing a centrally located nucleus. B. A myocardial cell or fiber reconstructed from electron micrographs, showing the arrangement of the multiple parallel fibrils that compose the cell and of the serially connected sarcomeres that compose the fibrils. *N* = nucleus. C. An individual sarcomere from a myofibril. A diagrammatic representation of the arrangement of myofilaments which make up the sarcomere is shown below. Thick filaments, 1.5 μ in length, composed of myosin, form the A band, while thin filaments, 1.0 μ in length, composed primarily of actin, extend from the Z line through the I band into the A band, ending at the edges of the H zone. An H zone exists in the central area of the A band where thin filaments are absent. An overlapping of thick and thin filaments is seen only in the A band. D. Cross sections of the sarcomere, showing the specific lattice arrangements of the myofilaments. In the center of the sarcomere (left) only the thick (myosin) filaments arranged in a hexagonal array are seen. In the distal portions of the A band (center) both thick and thin (actin) filaments are found with each thick filament surrounded by 6 thin filaments. In the I band (right) only thin filaments are present. (From Braunwald et al., 1968.)

The third type of cardiac tissue comprises the *specialized conducting system* of the heart and is composed of fibers which are much larger in diameter (50 to 70 μ) than the contractile fibers (15 μ). This system is made up of a bundle of His which is continuous with the AV node and which breaks up into two branches that run down the right and left side of the interventricular septum beneath the endocardium. The bundles, along their length, give off many branches which penetrate both ventricles and comprise the peripheral *Purkinje network.*

Excitability and Contractility

One fundamental property of the myocardium is irritability or excitability. Heart muscle has the ability to respond to a stimulus of adequate strength and duration. This response consists of the generation of a propagated action potential and a mechanical contraction. Unlike nerve and muscle action potentials, which are measured in milliseconds, the electrical spike of certain cardiac tissue actually plateaus in the depolarized state and is greatly prolonged. The duration of the action potential in the human ventricle is 0.3 second, and that in the atrium is 0.15 second.

Because a depolarized tissue may not respond to another stimulus, it is obvious that cardiac tissue has a prolonged refractory period. The length of the refractory period varies among the fiber types of the heart. For all the fiber types, however, reexcitation cannot occur during the fully depolarized state, no matter how strong the second stimulus. This time period constitutes the *absolute refractory period.* Briefly thereafter the fiber can be reexcited, but it requires stimulus strength of much greater intensity than is normally needed. This period of time constitutes the *relative refractory period* of the action potential. Reexcitation during the relative refractory period results in an action potential which is much lower in magnitude and which is conducted at a slower rate. The mechanical response elicited by the reduced action potential is also altered, despite the fact that membrane excitability has fully returned, and indicates that recovery of contractility has not occurred.

As the second stimulus is applied later, the premature contraction will be greater. In any event, when a premature contraction is produced, regardless of its strength, there follows a delay or *compensatory pause* before the next normal contraction occurs. The pause is a result of the heart's being made inexcitable by the premature beat and the fact that the next normal sinus rhythm is blocked. However that may be, during normal cardiac activity each contraction is preceded by electrical depolarization of the heart. It should be emphasized that the *total refractory* period (absolute plus relative) does not outlast the muscular contraction. The relative refractory period is over by about 75 msec before tension in the muscle has decreased to its original level. Consequently, reexcitation can and does result in a fusion of contraction, but the total force of contraction is less than normal. True summation of contraction or tetanus, with repetitive stimulation, does not occur in cardiac muscle because of the prolonged time required for full recovery of contractility.

The heart is a muscular pump and functions by contraction of its fibers. The contraction, or *systole,* as it is commonly called, is all-or-none and not tetanic. It is not graded by changing the number of active fibers, but it can be graded through direct adjustments upon the contractile mechanisms by mechanical, humoral, and neural factors. As mentioned before and will be emphasized in other chapters, the myocardial contraction may be graded or altered in regard to velocity of contraction, strength of contraction, and extent of contraction. The velocity of cardiac contraction is less than that of skeletal muscle but greater than that of most smooth muscles. As far as the heart action is concerned, relaxation, or *diastole,* is as important as contraction. Prolonged tetanic contraction would be detrimental to the organism, for effective pumping can be brought about only by adequate relaxation and refilling after each contraction.

The force of cardiac contraction is related not only to recovery of electrical excitability but to recovery of contractility as well. In addition, the force is related to the regularity and frequency of stimulation, muscle length (or diastolic volume against which the heart contracts), and a variety of neural and humoral factors. Functionally, the normal heart behaves as a syncytium or, in a sense, as one cell. Like a single nerve or muscle fiber, this single cell responds as a whole or not at all, in an all-or-none fashion, a fundamental property that is of utmost importance in any attempt to describe myocardial behavior. Because of the heart's syncytial nature, fiber recruitment or spatial summation is eliminated as a mechanism of gradation of contraction. However, to say that the heart behaves in all-or-none fashion should not imply that the contractile response is always the same. The magnitude of the response may be greatly altered by varying the prevailing conditions. An example is *treppe* or "staircase." When an isolated, quiescent frog heart is induced to beat by constant, maximal stimuli, each successive beat increases in magnitude. It has been suggested that the phenomenon of treppe may be due to a shift of the potassium ion out of the cell, producing a more favorable environment for the actin-myosin activity. Another phenomenon observed in isolated mammalian hearts is that of *post-tetanic potentiation* (i.e., when the frequency of stimulation is suddenly decreased, the contractions that follow are stronger, subsequently declining to a level that is "normal" for that slower frequency of stimulation). The initially greater contraction may be due to the longer recovery time of contractility that results from the reduced frequency of stimulation.

Both treppe and post-tetanic potentiation are effects produced from isolated hearts and are not observed in the intact, in vivo myocardium. However, an analogous situation does exist for the in situ heart, known as *postextrasystolic potentiation.* It was mentioned above that a compensatory pause follows a premature beat or PVC (premature ventricular contraction). The initial contractions which follow the pause induced by the PVC are greater than the normal contractions. This potentiation might be explained on the basis of the greater volume of blood which entered the ventricles during the compensatory pause and which stretched the fibers to result in greater contraction (Frank-Starling's law of the heart). A more detailed discussion of this length-tension relationship is presented in Chapter 15. Suffice it here to say that this relationship indicates that within certain limits the energy of contraction is proportional to the initial or resting length of the fiber. The relationship appears to be a fundamental property of most contractile tissue. Perhaps the law should be amended with the following limitation: When nervous, endocrine, or metabolic factors play a role, Starling's law may not be obeyed. At least, this holds insofar as the concept was conceived, since under the influence of the above factors the heart may be operative at a different Frank-Starling curve. However, the important thing is that when these controlling factors are completely or even partially absent, the heart has the remarkable ability to adjust itself by pumping more blood out if more blood has returned and filled it to greater dimensions.

The applicability of this law to the postextrasystolic potentiation is open to question since:

1. Potentiation can be produced in isolated papillary muscle in which the fiber length is unchanged despite the postextrasystolic pause.
2. There is no linear relationship between the degree of potentiation and the filling time of the ventricle with blood.
3. Studies in which the filling volumes of blood were varied showed greater potentiation with no filling than during normal cardiac filling.

Hence, the mechanism for postextrasystolic potentiation remains obscure. Nevertheless, it may be stated that this phenomenon appears to be a function of enhanced contractility. Much as neural or humoral factors may alter the contractile properties of the heart (a shift to a "new" Starling curve [see Chap. 15]), so does this postextrasystolic pause appear to enhance contractility.

Augmentation of cardiac contraction, associated with abrupt rate change, has been noted in a variety of mammals as well as the human. It occurs under a number of circumstances and is manifested by postextrasystolic, poststimulation, and rest potentiation. It has, at one time or another, been variously attributed to an effect of the premature beat per se, to increased diastolic filling associated with the compensatory pause, and to an imbalance of positive and negative inotropic substances engendered by each excitation. These types of contractile augmentation need not be associated with diastolic filling alone, because, as mentioned above, potentiation has been demonstrated in vitro, and even without change in resting fiber length.

However, a fundamental relationship has been observed between cycle interval and strength of contraction, which appears to be common to all varieties of potentiation produced by abrupt rate change. This denominator, common to all three types of contractile augmentation, is constituted by the relative prolongation of the cycle length – that is, interval prolongation relative to the previous cycle length and preceding the potentiated contraction. Although the temporal course of the interval-dependent contractile alterations has been ascertained, the physiological mechanism of such enhancement has not been further clarified.

Most studies of these inotropic phenomena have suggested that the interval-dependent potentiation is mediated through changes in excitation-contraction coupling and is not associated with alteration in the excitatory process itself. Cardiac action potential changes concomitant with increased contraction either have not been demonstrated or have been assumed to be of no significance; neither the action potential duration nor its amplitude has been found to increase with augmentation of contractility. Nevertheless, studies in the author's laboratory have demonstrated a characteristic action potential change following a sudden prolongation of the cycle length. Following a relative prolongation of the cycle length, both shortening of phase 2 of the cardiac action potential and potentiation of contractility are observed. In contrast, a relative abbreviation of cycle length is associated with both depression of contractility and widening of phase 2 of the action potential. These interval-dependent alterations in contractility are reflected by postextrasystolic, poststimulation, and rest potentiation, as well as by depression of contractility of the premature beat itself associated with a premature stimulus. In each circumstance, the extent of action potential alteration correlates with the magnitude of contractile change, and both phenomena appear to be a function of the relative degree of cycle length alteration, rather than of the absolute length of any single interval. Since these results were obtained from isolated segments of ventricular muscle undergoing isometric contraction, any reflex activity is therefore precluded and the contribution of muscle length variation (i.e., the Frank-Starling relation) is excluded. The variations in contractile strength could, then, be attributed to alteration in the inherent speed of contractile shortening. The electrophysiological correlate of the inotropic state of the myocardium appears to have been identified since the electrophysiological finding is consistent and uniform with all types of inotropic changes (i.e., due to digitalis, pentobarbital, nicotine, catecholamines, etc.).

In addition, evidence has been obtained in the author's laboratory which demonstrates a direct relation between postextrasystolic T wave alterations and the magnitude of the postextrasystolic contractile alternans in the "normal" anesthetized animal as well as under the influence of positive and negative inotropic agents. This relation is observed consistently in the electrogram recorded directly from the ventricular epicardium, while only the most prominent contractile alterations are associated with discernible T wave alterations in the accompanying lead II of the conventional electrocardiogram (ECG). It is suggested, as a consequence, that the postextrasystolic T wave change, as well as rate-induced electrical alternans of the T wave, is a correlate of the inotropic state of the myocardium altered by abrupt rate change. The relative infrequency with which such T wave changes are observed in the conventional ECG appears in part due to the insensitivity of the surface ECG, resulting in the recording of complexes of a lower voltage. Conversely, the frequency with which postextrasystolic T wave changes occur in diseased hearts may reflect, in part, a relatively greater interval-dependent contractile change, associated with which there occurs a relatively more prominent T wave alteration.

REFERENCES

Braunwald, E., J. Ross, and E. H. Sonnenblick. *Mechanism of Contraction of the Normal and Failing Heart.* Boston: Little, Brown, 1968.

Burnstock, G., M. E. Holman, and C. L. Prosser. Electrophysiology of smooth muscle. *Physiol. Rev.* 43:482–527, 1963.

Daniel, E. E. Effect of drugs on contractions of vertebrate smooth muscle. *Ann. Rev. Pharmacol.* 4:189–222, 1964.

Eichna, L. W. (Ed.). Proceedings of a symposium on vascular smooth muscle. *Physiol. Rev.* 42 (Pt. II):1–365, 1962.

Greenspan, K., R. E. Edmands, and C. Fisch. The relation of contractile enhancement of action potential change in canine myocardium. *Circ. Res.* 20:311–320, 1967.

Huxley, H. E. The contraction of muscle. *Sci. Amer.* 199:67–82, (Nov.) 1958.

Sperlakis, N., and C. L. Prosser. Mechanical and electrical activity in intestinal smooth muscle. *Amer. J. Physiol.* 196:850–856, 1959.

Special Senses

A. The Eye

Paul C. Johnson

THE EYE performs two separate but closely related functions. First, the eye is an optical instrument which collects light waves from the environment and projects them as images onto the retina. Second, the eye is a sensory receptor which responds to the images formed on the retina and which sends sensory information to the visual areas of the brain. Before considering in detail the functions of the eye, it is necessary to consider some of the properties of light and optical systems.

PHYSICAL PROPERTIES OF LIGHT

The visible light spectrum ranges from 400 to 800 mμ, with blue light having the shortest, and red the longest, wavelength. Immediately adjacent to the visible spectrum are the areas of ultraviolet and infrared radiation. Although not perceived by the eye as such, radiation in these areas is biologically important. For example, ultraviolet radiation is the cause of the tanning and burning effects of the sun on the skin. Infrared radiation has heating effects on the body.

From the time of Newton, scientists have debated whether light is a stream of corpuscles having the properties of matter, or a form of wave reflected or emitted from material objects. Careful experimentation has shown that it actually possesses properties of both matter and waves, and consists of electrical and magnetic energy which travels in packets, or *photons,* at a speed of 186,000 miles per second.

The wavelike properties of light are of particular significance in considering the optical function of the eye. These properties enable light to be focused, reflected, and refracted like other waves. One important property of wave phenomena is *Huygens' principle,* which states that *any point on a wave front may be regarded as a new source of light.* For example, if light passes through a narrow slit, the rays fan out when they leave the slit as if that slit were a new source of light. If a screen is placed beyond the slit to intercept the rays, they form a central image on the screen with less intense secondary images of the slit on either side of the central image (Fig. 4A-1A). This formation of a *diffraction pattern* is a consequence of the wave nature of light. It is of practical importance because diffraction or breaking up of the light limits the sharpness of the image. As a result, when two objects are very close together, the diffraction patterns overlap, and it may be difficult to determine whether the objects are actually separated.

REFRACTION AND IMAGE FORMATION

When light rays strike an object they are either reflected, absorbed, or transmitted through it. Even rays that are transmitted do not escape some alteration. The entering light rays bend to an extent depending upon the angle at which they strike the surface, and upon the *refractive indices** of the surrounding medium and of the object. For reference, the refractive index of a vacuum is taken as unity, and that

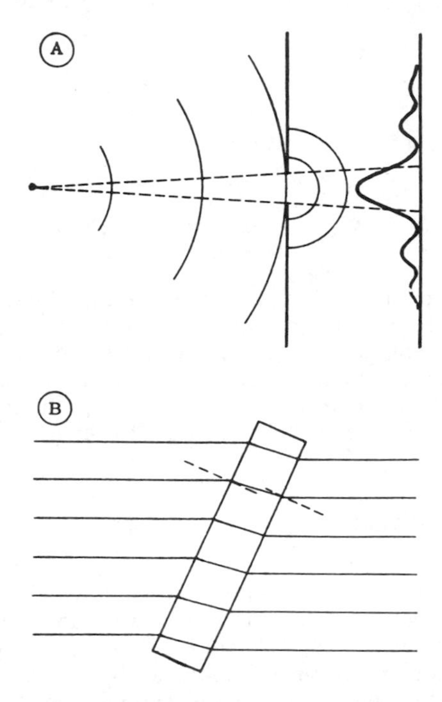

FIGURE 4A-1. A. Diffraction pattern formed by light rays passing through a narrow slit. Dashed lines indicate the path which light rays would follow if diffraction did not occur. B. Bending of light rays at the points of entrance and emergence from a medium with a higher refractive index.

of air is practically the same. As light rays pass from one medium into a second having a higher refractive index than the first, the rays are bent toward a perpendicular line drawn through the surface at the point entered (Fig. 4A-1B). The greater the difference in the refractive index, the greater the degree of bending. Conversely, if the rays leave a medium having a high refractive index for a low one, they are bent away from the perpendicular. Finally, if light rays strike perpendicularly to the surface, they are not bent, regardless of the refractive indices of the media involved.

The property of refraction is basic to the operation of lenses, as may be seen in Figure 4A-2A. The central rays from the point source of light strike almost

---

*The refractive index of a medium is defined as the ratio between the sine of the angle of incidence and the sine of the angle of refraction. It is due to a difference between the velocity of light in the medium and in a vacuum.

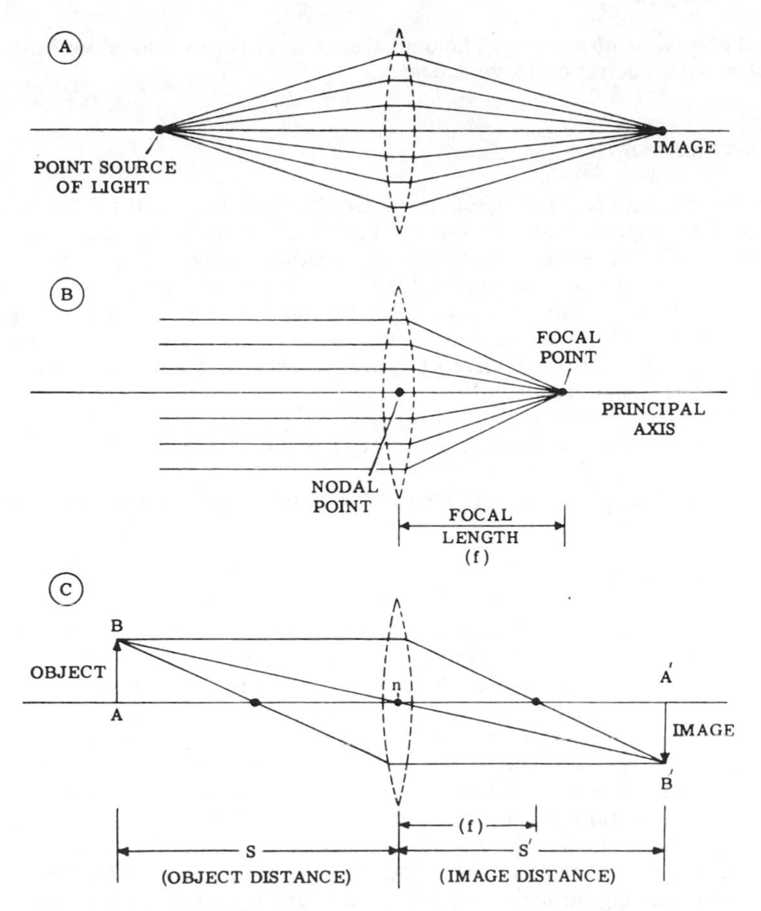

FIGURE 4A-2. A. Formation of an image by a spherical lens, showing the greater deviation of peripheral rays. B. Illustration of focal length as the distance from the lens at which parallel rays are brought to focus. C. Construction of an image when lens strength, object size, and object distance are known.

perpendicularly to the lens surface and are bent only to a small degree, whereas the peripheral rays strike the surface at an acute angle and bend considerably more. If the surface of the lens is spherical, the transmitted rays intersect at a point on the other side of the lens to form an image of the point source. The lens illustrated is a converging, convex spherical lens. The surface of such a lens is a portion of a sphere; i.e., it has the same curvature in all planes. One or both surfaces may be spherical. A lens which causes light to diverge is constructed by making the spherical surfaces concave rather than convex.

The light rays passing through the peripheral portion of a spherical convex lens do not focus at exactly the same point as those passing through the center. This phenomenon is called *spherical aberration.* An additional limitation is the fact that the refractive index varies according to the wavelength of light used, blue light being bent more than red. Thus, if white light is used, the rays of different wavelength form separate images, and an expanded image is produced. This characteristic is termed *chromatic aberration.* The optical system of the eye possesses both spherical and chromatic aberration to some degree.

GEOMETRICAL OPTICS

Geometrical optics is the study of the properties of lenses and the formation of images. Two important applications of this branch of optics are the determinations of the strength of a lens and of the size and position of the image it forms. When parallel rays of light fall upon a lens, they come to focus at a point called the *focal point,* as in Figure 4A-2B. The *focal length* is the distance from the focal point to the lens. The focal length is a convenient measure of the strength of a lens. In physiological optics, the lens strength is usually expressed in *diopters,* the reciprocal of the focal length expressed in meters.

$$\text{Diopters (D)} = 1/\text{focal length (meters)}$$

For example, if a lens has a focal length of 0.1 meter, its power is 10 diopters (10 D).

IMAGE FORMATION

The position at which an image is formed by a lens can be determined geometrically by tracing the path of several rays through the lens, as shown in Figure 4A-2C. The determination requires only that object size, object distance, and the focal length of the lens be known. There are three rays whose paths can be readily traced. First, a ray which leaves a point source (the end of the arrow) and passes through the center of the lens is bent as it enters the lens and again as it leaves. Because the two surfaces of the lens are parallel at the center, there is little deviation and the ray passes on, its direction essentially unchanged. Second, a ray which proceeds parallel to the principal axis is bent sufficiently to pass through the focal point on the principal axis on the image side of the lens. As this ray proceeds further, it intersects the ray passing through the center of the lens and an image is formed at the intersection. Third, a ray which passes through the focal point on the object side of the lens is bent just enough to meet the other two rays at their point of intersection.

If the lens is made stronger or the object is moved farther from the lens, the image is formed closer to the lens. The object distance, image distance, and lens strength are therefore interrelated. This relation is expressed by the *lens formula:*

$$\frac{1}{S} + \frac{1}{S'} = \frac{1}{f}$$

where S is the object distance, S' is the image distance, and f is the focal length.

Example 1. If an object 20 cm from the lens forms an image 20 mm from the lens, what is the lens strength? First, all values should be expressed in meters. The lens strength can then be obtained in diopters, as follows:

$$\frac{1}{0.2} + \frac{1}{0.02} = \frac{1}{f}$$

$$\frac{1}{f} = 55 \text{ D}$$

Example 2. If an object is placed 40 cm from a lens of 50 D strength, what is the image distance?

$$\frac{1}{0.4} + \frac{1}{S'} = 50$$

Multiplying by 0.4S':

$$\frac{0.4S'}{0.4} + \frac{0.4S'}{S'} = 50 \times 0.4S'$$

$$S' = 0.021 \text{ meter}$$

If a point source of light is placed at the focal point of a lens, the rays are parallel after passing through the lens and do not come to a focus (or focus only at an infinite distance from the lens). If the source is moved away from the focal point, an image is first formed at a great distance from the lens, gradually moving closer as the source retreats from the lens. When the source is an infinite distance from the lens, the light rays falling on the lens are parallel and the image forms at the focal point. This can be shown by the lens equation. Since S is infinite, the equation becomes

$$\frac{1}{S'} = \frac{1}{f}$$

For any optical system there is a point beyond which the object can be considered for all practical purposes to be at infinity. For the eye this distance is 6 meters (20 feet). Any further movement beyond that point has little effect on the image distance.

IMAGE SIZE

The size of the image is directly proportional to the size of the object and the ratio of the image and the object distances. For example:

if AB = object length

and $A'B'$ = image length

then $$\frac{A'B'}{AB} = \frac{S'}{S}$$

These relations can be seen from Figure 4A-2C, and follow from the fact that ABn and A'B'n are similar triangles.

OPTICAL FUNCTION OF THE EYE

The basic optical principles outlined above can be applied to the function of the eye. The structure of the eye is shown in Figure 4A-3. Light rays striking the eye first enter the *cornea,* which has a high degree of curvature and a refractive index of 1.33 compared with 1.00 for air. These two factors combine to cause a considerable bending of the light rays as they enter the cornea. The resting eye has a power of 67 D. Most of this optical power (45 D) is in the cornea. When one swims underwater the refractive power of the cornea is lost, since the water also

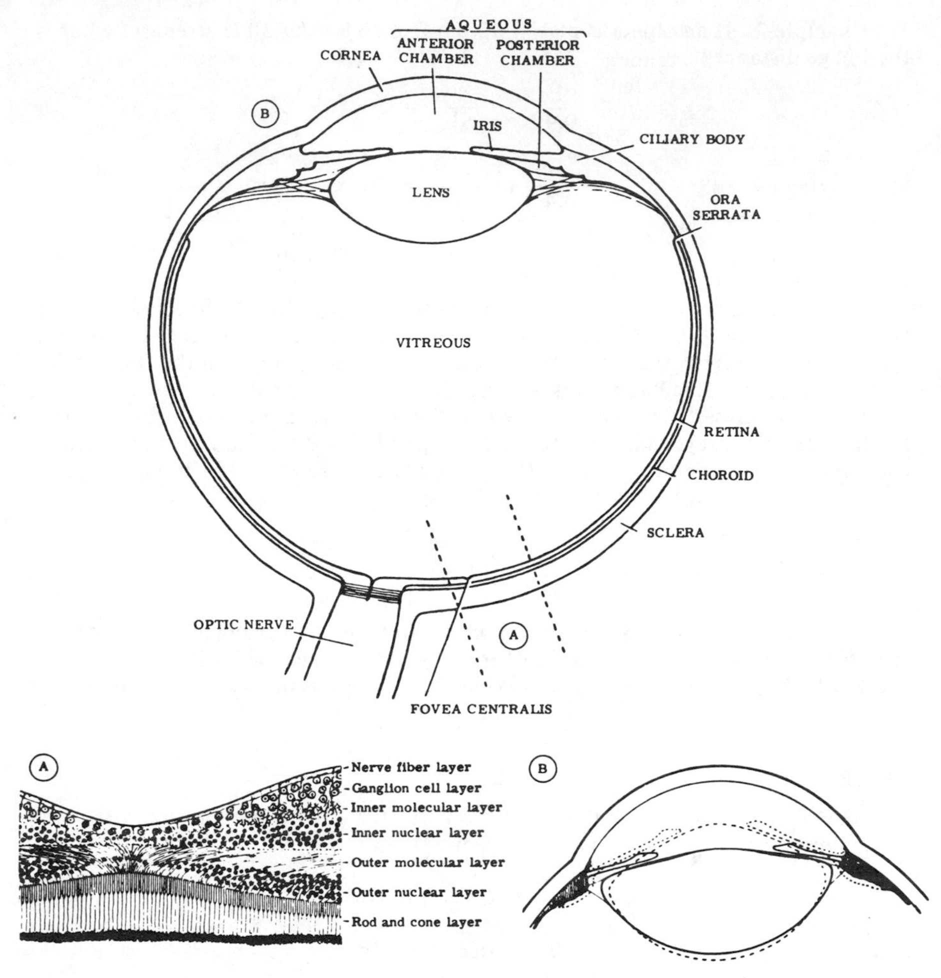

FIGURE 4A-3. Structure of the eye. Details of regions A and B in the upper figure are shown. A. Fine structure of the retina. B. Changes in lens and associated structures with accommodation for near vision (accommodation indicated by dashed lines). Note the forward movement of the lens anterior surface in B. (From F. R. Winton and L. E. Bayliss. *Human Physiology* [5th ed.]. Boston: Little, Brown, 1962. P. 520.)

has a refractive index of 1.33. The rays pass through the *anterior chamber* (refractive index 1.33) without further alteration but are bent when they enter the *crystalline lens* of the eye. The lens has a graded refractive index which varies from 1.41 at the center to 1.39 at the periphery. This graded index decreases the power of the lens, making it equivalent to one having a uniform index of 1.39. The power of the lens is approximately 20 D at rest but may be increased by as much as 12 D by accommodation.

Accommodation is the process by which the refractive power of the eye is modified for viewing nearby objects. It is produced by a thickening of the center of the lens (mostly at the anterior surface), as shown in Figure 4A-3 (lower right corner).

The lens has an elastic capsule which tends to make this structure assume a spherical shape. However, this tendency is opposed by the tension within the *sclera* or outer coat of the eyeball. The tension is due simply to the intraocular pressure, normally 20 mm Hg, and is transmitted to the lens by the ring-shaped *ciliary muscle,* which encircles the lens, and by the *suspensory ligaments,* which are attached to the periphery of the lens. The tension applied to the lens by the ligaments flattens it by a stretching process. However, when the ciliary muscle contracts, its diameter decreases, thereby decreasing the tension on the suspensory ligaments and allowing the lens to assume a more spherical shape. Thus accommodation depends upon the elastic properties of the lens capsule

Associated with accommodation of the lens is a constriction of the sphincter-like *pupillary muscle,* which eliminates rays passing through the more peripheral portions of the lens (Fig. 4A-3B). This decreases spherical aberration, which might otherwise be a limiting factor in near vision. When one observes a nearby object, where accommodation is required, the axis of the eye is shifted by the *extrinsic muscles* to train both eyes on the object.

For most purposes the optical properties of the compound lens system of the eye may be considered equivalent to those of a single refractive surface, the optical center or nodal point of which is situated 5 mm behind the anterior corneal surface and 15 mm in front of the *retina* (called the reduced eye of Listing). It is possible with this simplifying assumption to apply the lens equation to the eye. The normal power of the eye and the power of accommodation may be readily determined. For example, in the normal (emmetropic) eye at rest, distant objects are brought to focus on the retina. If the object distance in this case is considered to be infinite, the equation becomes

$$\frac{1}{\infty} + \frac{1}{S'} = \frac{1}{f} \quad \text{or} \quad \frac{1}{S'} = \frac{1}{f}$$

Since the image is focused on the retina, $S' = 15$ mm; therefore

$$\frac{1}{.015} = \frac{1}{f}$$

giving 67 D as the optical strength of the resting eye.

The maximal optical strength of the eye can be determined by measuring the shortest distance at which an object may be seen distinctly. This is called the *near point* of vision. In the young adult this distance is about 10 cm. The maximal strength is then

$$\frac{1}{0.10} + \frac{1}{.015} = \frac{1}{f}$$

$$\frac{1}{f} = 77 \text{ D}$$

The difference in strength between the resting and the maximally accommodated eye is the *power of accommodation.* This is about 12 D in children and 10 D in young adults. With increasing age, the elasticity of the lens decreases, thereby reducing the power of accommodation and causing the near point to recede (Table 4A-1). This decrease in accommodation with age is known as *presbyopia.* It is an inevitable result of the aging process. Between ages 40 and 50 the near point recedes beyond arm's length, making reading glasses necessary.

TABLE 4A-1. Effect of Age on Amplitude of Accommodation and Near Point

Age	Amplitude (D)	Near Point (cm)
10	11.3	8.8
20	9.6	10.4
30	7.8	12.8
40	5.4	18.5
50	1.9	52.6
60	1.2	83.3
70	1.0	100.0

OPTICAL ABNORMALITIES

Two of the most common optical defects result from an abnormal size of the eyeball. These conditions are hyperopia, or farsightedness, and myopia, or nearsightedness. *Hyperopia* is characterized by an abnormally short eyeball with a decreased distance from lens to retina. Much less frequently, it is due to insufficient refractive power of the optical system. This defect causes the image of a distant object to be formed behind the retina in the resting eye (Fig. 4A-4). The hyperopic individual can focus the image on the retina by partially accommodating the lens. This, of

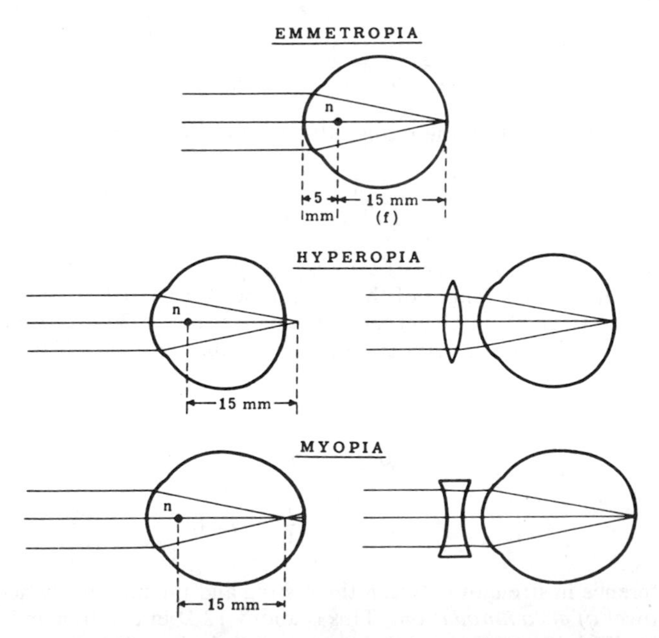

FIGURE 4A-4. Optical dimensions of the normal and abnormal eye. In the emmetropic eye the nodal point of the lens system is 15 mm before the retina. In the hyperopic eye the nodal point is less than 15 mm before the retina, causing the focal point to fall behind the retina. This is corrected by a spherical converging lens placed before the eye. In the myopic eye, rays focus before the retina. This is corrected by a spherical diverging lens.

course, is not a normal situation and leads to eye fatigue. Because some accommodation is required even for distant viewing, the near point is more distant than normal, and near vision is deficient – hence the descriptive term *farsightedness.* The hyperopic condition can be remedied by placing the appropriate convex spherical lens before the eye (Fig. 4A-4).

Myopia is characterized by an abnormally long eyeball with an increased lens-to-retina distance. Occasionally this condition is produced by an abnormally great curvature of the cornea or lens. This defect causes the image of a distant object to be formed before the retina in the resting eye (Fig. 4A-4). There is no simple way by which the defect can be compensated, since the normal process of accommodation only aggravates the deficiency. Vision of near objects is not impaired, however, and the near point is closer than normal. Myopia can be remedied by placing the appropriate concave spherical lens before the eye (Fig. 4A-4).

Astigmatism is a common optical defect which is most often due to an abnormal curvature of the cornea. Normally, the corneal surface is spherical. In astigmatism the surface is ellipsoid or egg-shaped (Fig. 4A-5). This causes the rays traveling in one plane to be bent more drastically than those in another. As a result, the rays in one plane may focus on the retina while those in another do not. Astigmatism can be corrected by a *cylindrical* lens. Such a lens may be thought of as a portion of a cylinder which is cut longitudinally. If such a lens is viewed from above, the curvature is seen in cross section (Fig. 4A-5). Rays traveling in the horizontal plane also "see" the lens as a curved surface and are bent accordingly. Rays traveling in the vertical plane "see" the lens as a rectangular object and are not bent as they pass through it. By use of a cylindrical lens, the rays not focused on the retina may be brought to focus at that point. Thus it is necessary that the curvature of the correcting lens be placed in the same plane as the rays having an abnormal focal point. The longitudinal axis of the correcting lens then is perpendicular to the plane of the astigmatism.

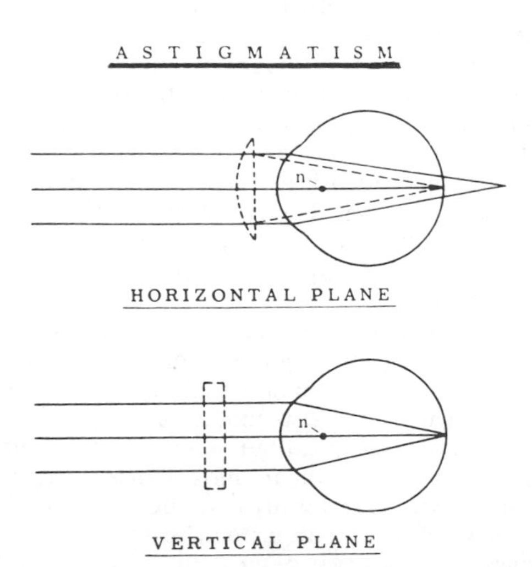

FIGURE 4A-5. Optical features of astigmatism. In the uncorrected eye, light rays (shown by solid lines) focus behind the retina in one plane and on the retina in another. The defect is corrected by a cylindrical lens so placed that the curvature is only in the plane exhibiting the defect.

THE OPHTHALMOSCOPE

An ophthalmoscope is a device used for examination of the optical and physical properties of the eye. It consists essentially of a light source and a mirror or prism which reflects light into the eye and onto the retina. Some of the light is reflected from the retina, which becomes, in effect, a new light source. The light passes out of the eye through the lens and cornea. If the eye is normal and relaxed, the focal point of the lens and cornea coincides with the position of the retina. Consequently light rays reflected from the retina are bent just enough to be rendered parallel as they pass through the lens and leave the eye, whereas if the eye is myopic the rays are convergent.

The degree of abnormality of the eye may be readily determined by a variety of means with the ophthalmoscope. For example, if the parallel rays emerging from the emmetropic eye pass through a +1 diopter lens, an image of the retina is formed 1 meter from the lens. An additional converging lens for the hyperopic eye or diverging lens for the myopic eye must be used to bring the image to this same point. The strength of the additional lenses required is a measure of the abnormality which exists. This method illustrates the principle underlying the use of the ophthalmoscope. At present, more indirect methods which permit rapid precise measurement of the degree of abnormality are utilized.

VISUAL ACUITY

The ability of the eye to detect a separation between two adjacent objects depends upon the ability of the retina to perceive a separation between the images which fall upon it. The separation of two adjacent images on the retina depends upon the adequacy of illumination, the fidelity with which the light rays are transmitted by the optical system, and the diffraction pattern of the image on the retina. In addition, the density of the packing of the retinal receptors is a determining factor in the resolution of images. If the separation between images on the retina is to be perceived, a receptor must be present in the intervening space. Since the diffraction pattern tends to diffuse the edges of the image, the separation may not be clear-cut. There may be, in fact, a "gray" zone between two images, which the retina must recognize as a separation.

In viewing nearby objects, the iris of the eye constricts to block peripheral rays and minimize spherical aberration. The visual area of the cortex is apparently able to compensate in some fashion for chromatic aberration. Thus the resolving power of the eye is determined primarily by retinal "grain" or receptor density and the unavoidable diffraction of rays.

The minimum separation between images that can be detected may be conveniently expressed in terms of the visual angle. This is the angle which the separation subtends in the visual field. The minimal visual angle for the normal eye is approximately 1 minute. The images of objects undergoing close scrutiny are projected onto the *macula lutea* or yellow spot, the portion of the retina where the *cones,* the receptors which subserve detail vision, are most densely packed. In other portions of the retina the visual angle is greater because the cone population is less dense, and for other reasons which will be discussed shortly.

In clinical studies, visual acuity is determined by use of test charts or letters. The subject is situated 20 feet from the chart so that the emmetropic eye will not be accommodated. The Snellen test chart, which is most often used, is composed of block letters so constructed that the width of the trace and the separation between limbs of the letter are standardized. The capital letter E, for example, is so constructed that the width of the horizontal bars is equal to the separation between them. The chart contains lines of various sizes of type. The test is carried out by having the subject read the smallest type possible from the 20-foot distance, each eye being tested separately. For example, if the subject can read at 20 feet what the normal person can read at that distance, his acuity is expressed as 20/20 and corresponds to a visual angle of 1 minute. If the smallest type the subject can discern at 20 feet corresponds to what the normal individual can read at 40 feet, his acuity is expressed as 20/40 and his visual angle is 2 minutes.

STRUCTURE OF THE RETINA

Details of the central portion of the retinal structure are shown in Figure 4A-3A (lower left corner). The retina extends over a large portion of the posterior pole of the eye (roughly 180 degrees). Before the light rays enter the receptor layer, they must pass through layers of blood vessels, nerve fibers, ganglion cells, and bipolar cells. Only in the macula lutea do the light rays have free access to the receptors. Within the macula lutea is an area called the *fovea centralis,* where the receptors are most densely packed. This region contains only *cones,* which are the receptors concerned with fine detail and color vision. Moreover, each cone in this area has a "private line" to the visual cortex. Visual acuity decreases toward the periphery of the retina. There the cones become less dense and there are many cones on a single "party line" to the cortex. Furthermore, the layers overlying the cones tend to scatter and absorb light. As a result, visual acuity at the fovea centralis is twice that just outside the macula lutea and about 40 times that at the retinal border.

Visual acuity and the contrast between an object and its background may be intensified by mechanisms within the retina. When a point source of light is focused on a portion of the retina, some fibers from this area which were previously silent begin to fire. These are called *on* fibers. Other fibers, previously discharging, become silent. These are termed *off* fibers. Still other fibers rapidly adapt to existing light, discharging for a brief period at the beginning and after the end of a light stimulus. These are the *on-off* fibers. Movement of an object across the retinal field leaves a trail of on-off fiber activity in its wake. The fact that the on-off fibers are more numerous than the other two types may account for the great ability of the eye to detect small movements. Normally, the eye does not fix upon an object persistently but scans an area by movements which are rapid and small in amplitude. Apparently this scanning action helps maintain visual acuity, since the image begins to fade if the movements are blocked. The on-off response may be important in this respect, for the scanning tends to create a halo of on-off impulses at the border of the image. There are also lateral neuronal connections in the retina which may be important in visual acuity. When a small area is illuminated, inhibitory impulses are sent to surrounding nonilluminated areas. This tends to intensify the image.

Although visual acuity is lower in the peripheral regions of the retina than in the macula lutea, night vision is better because of the increased population of rods, which are specifically adapted for night vision. The rods are much more sensitive to low levels of illumination than the cones but have a lower level of visual acuity. (Rod vision is about the same throughout the retina and is one-twentieth as acute as the cone vision at the fovea centralis.) The lower visual acuity is due, at least in part, to the convergence of a number of rods upon a single line to the cortex. There are some 6 million cones, 125 million rods, and 1 million optic nerve fibers in the human retina. The rods and cones in most regions of the retina are therefore connected in large numbers to a single ganglion cell and optic fiber.

ADAPTATION TO ILLUMINATION

If a bright light is shone into the eye, the pupil immediately constricts. This is the *light reflex,* which is initiated at the retina and passes by way of the pretectal region and the ciliary ganglion to cause contraction of the sphincter pupillae muscles. If only one eye is stimulated, the reaction occurs in the pupil illuminated, the *direct light reflex,* and in the opposite pupil as well, the *consensual light reflex.* In tabes dorsalis the light reflex may be lost, while the pupillary reaction during accommodation remains unimpaired. This condition is called the *Argyll Robertson pupil.* The reaction of the pupil to light is usually a temporary phenomenon, the purpose of which is to protect the retina from too intense illumination. As time passes, the retina adapts to the new level of illumination and the pupil returns to its original size. The maximal diameter of the pupil is 8 mm and the minimal is about 1.5 mm. Thus the area, and therefore the amount of light entering, can be changed 30-fold. However, the eye is able to adapt to changes in light intensity of about 10-billion-fold because of changes in the retina.

The reactions of the retina to changes in light intensity may be more readily understood by considering the adaptation to darkness. If an individual enters a dark room, there is a period during which he perceives nothing followed by one in which the objects around him gradually become more discernible. If periodic measurements of the light threshold or the just perceptible light stimulus are made, a gradual decrease in this threshold is found, as shown in Figure 4A-6. Both the rods and the cones begin to adapt immediately. The cones initially adapt more rapidly, so that the threshold measured in the early stages of adaptation is essentially that of the cones alone. The cones increase their sensitivity 20-fold to 50-fold in the first five minutes, after which little further adaptation of these receptors occurs. The rods, on the other hand, adapt more slowly but to a much greater degree. After the cones have reached their maximal sensitivity, the threshold of the eye as a whole continues to decrease, because of the adaptation of the rods. Since these adaptations proceed at different rates, the dark adaptation curve for the eye breaks sharply at the point where the rods become more sensitive than the cones. The adaptation of the rods is essentially complete in 25 to 30 minutes, although it may continue slowly

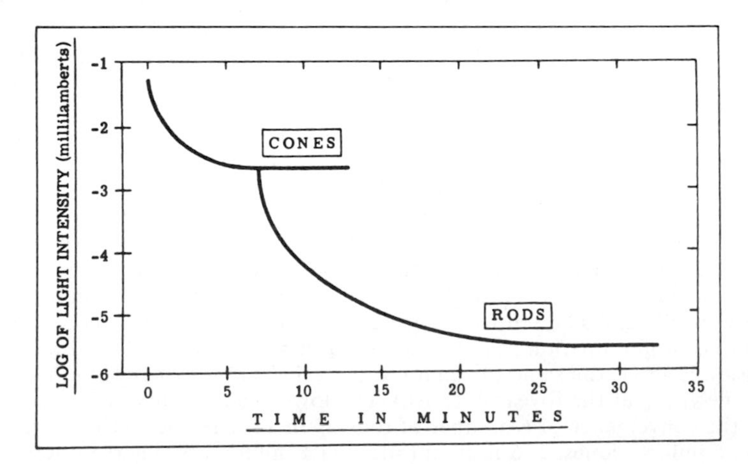

FIGURE 4A-6. Adaptation of rods and cones to darkness. Change in light threshold is shown on a *logarithmic* scale.

for several hours thereafter. In passing from a dark area to a light one, the process is reversed. The two-step change in sensitivity is not seen; the photosensitive material in the rods is immediately bleached, and the cones then determine the light threshold.

The process of adaptation is a manifestation of the properties of the photosensitive chemical substance in the light receptor. The nature of this substance has been studied in much more detail in the rods than in the cones.

SCOTOPIC VISION

The rods contain a reddish purple pigment called *rhodopsin* or *visual purple,* which is bleached upon exposure to light. The bleaching process causes excitation of the receptor by means not yet understood. The bleaching and regenerative processes involve changes in the rhodopsin molecule which can be studied in vivo or in vitro. The first stage in the bleaching process is the formation of orange pigments *(lumi-rhodopsin* followed quickly by *meta-rhodopsin),* which then split into *retinene* and *opsin* (visual yellow) (Fig. 4A-7). At first, the retinene is in the *trans* isomer form. Some of this retinene is converted to the *cis* form by a photochemical process. Cis

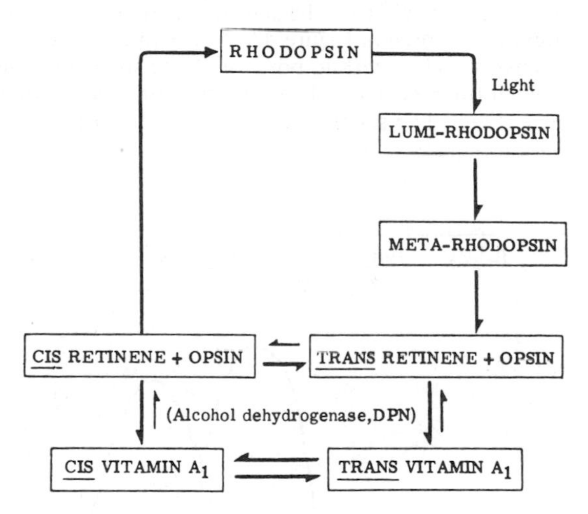

FIGURE 4A-7. Degradation of rhodopsin when exposed to light. Resynthesis from cis retinene and opsin to rhodopsin occurs in dark.

retinene in a solution containing opsin is converted in the dark to rhodopsin. Only one of the several cis isomers (11-cis) is the percursor of rhodopsin. The trans retinene which is not isomerized to the cis form is reduced to vitamin A_1. Before the regeneration of rhodopsin can occur, vitamin A_1 must be isomerized from the trans form to the cis form. This occurs elsewhere in the body by a process not yet understood. Meanwhile, the active form of vitamin A_1 is withdrawn from the blood, to be used in regeneration of visual purple.

Rhodopsin is most readily bleached by light having a wavelength of 500 mμ. Its sensitivity falls on either side of 500 mμ; it is about 40 percent as sensitive to light at the blue end of the spectrum. The sensitivity to light in the red region is very low.

For this reason, wearing red-tinted glasses allows full adaptation of the rods to night vision even while general illumination is normal. Since some cones are stimulated by red radiation, they do not adapt under these circumstances.

PHOTOPIC VISION

Photopic vision is the term for the visual function performed by the cones. The cones, as a group, are most sensitive to light in the region of 550 mμ. The sensitivity of the cones extends throughout the entire visible spectrum, but is much less at the two ends of the spectrum than at the center. This is why yellow seems much brighter than blue or red. This spectral variation in sensitivity appears to be a summated effect of several different types of cones, each of which responds to a limited portion of the spectrum.

COLOR VISION

The most widely accepted theory of color vision is the Young-Helmholtz theory. It is based upon the assumption that there are three receptors, red, green, and blue (violet), which subserve color vision. The sensitivity of these receptors, as deduced from indirect evidence, is shown in Figure 4A-8. It can be seen that, although each has its maximum sensitivity in a certain portion of the spectrum, there is considerable overlap between the receptors. The sensation of blue is due to stimulation of the blue receptor alone. The sensation of blue-green is due to stimulation of both the blue and the green receptors. On the other end of the spectrum, the sensation

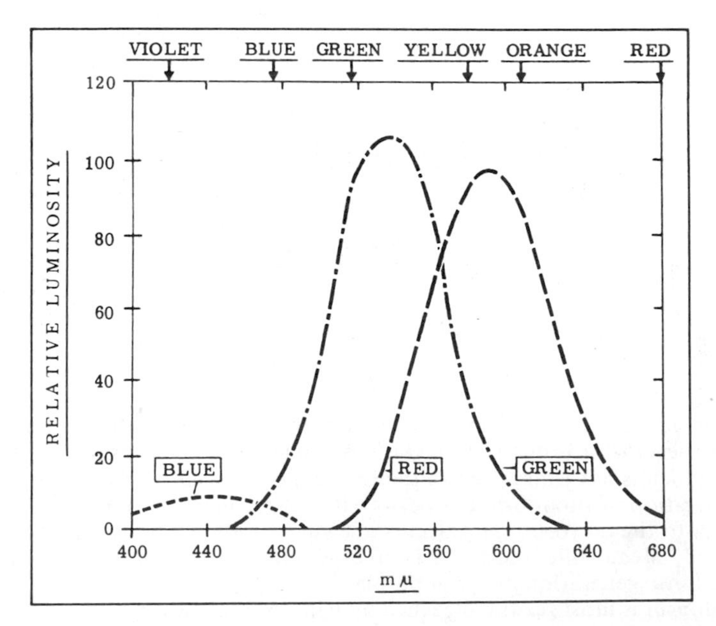

FIGURE 4A-8. Wavelength sensitivity of the retinal color receptors according to the Young-Helmholtz theory.

of red is due only to stimulation of the red receptor. However, shifting toward the green portion of the spectrum results in the perception of the sensations of orange and then yellow because of stimulation of both the red and green receptors. Thus the various hues of the spectrum, plus the extraspectral color purple, may be differentiated on the basis of the three color receptors. If all the receptors are stimulated simultaneously in the correct proportions, the sensation of white is perceived.

For 150 years the Young-Helmholtz theory, first advanced in 1803 by Stephen Young, was without direct substantiation. Indirect evidence from studies of color perception had made it possible to formulate the relationship between the three hypothetical receptors shown in Figure 4A-8. However, in 1964 experiments were performed which showed conclusively that such color receptors do exist. Brown and Wald succeeded in measuring the absorption spectrum of single rods and cones in an excised portion of human retina. They found that the rods display an absorption peak at about 505 mμ (Fig. 4A-9). They also recorded the absorption of three types of cones: a blue-sensitive cone with an absorption maximum at 450 mμ, two green-sensitive cones having absorption maxima at 525 mμ, and a red-sensitive cone with its absorption maximum at 555 mμ. The results of these remarkable experiments are shown in Figure 4A-10. It should be noted that these absorption spectra agree well with the results obtained in color perception studies.

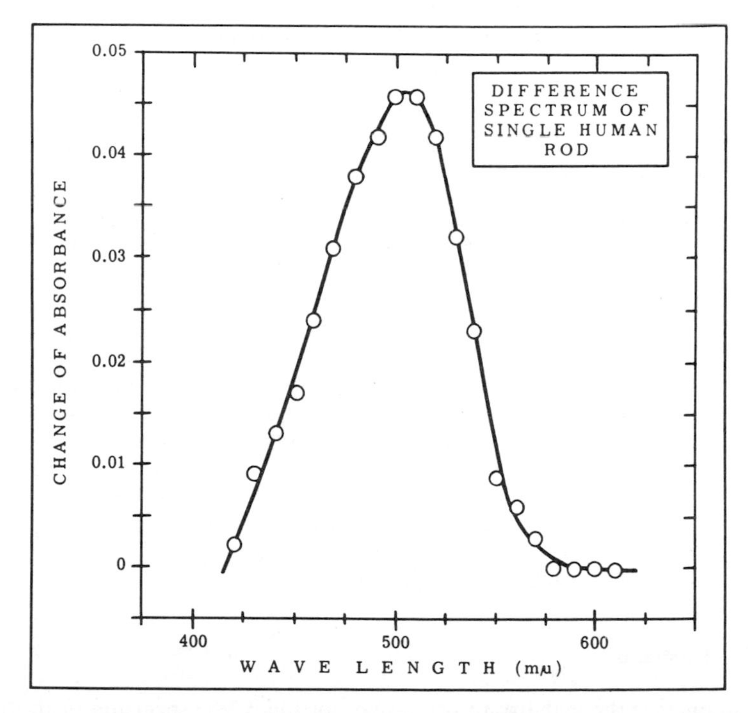

FIGURE 4A-9. The difference spectrum of a single rod in the human retina (parafoveal region). The spectrum was recorded in darkened room using very low light intensities to determine the absorption. The experiment was repeated after bleaching with a flash of yellow light. Curve represents difference between the two absorption spectra. (From P. K. Brown and G. Wald. *Science* 144:45–52, Issue 3614, 1964. Copyright 1964 by the American Association for the Advancement of Science.)

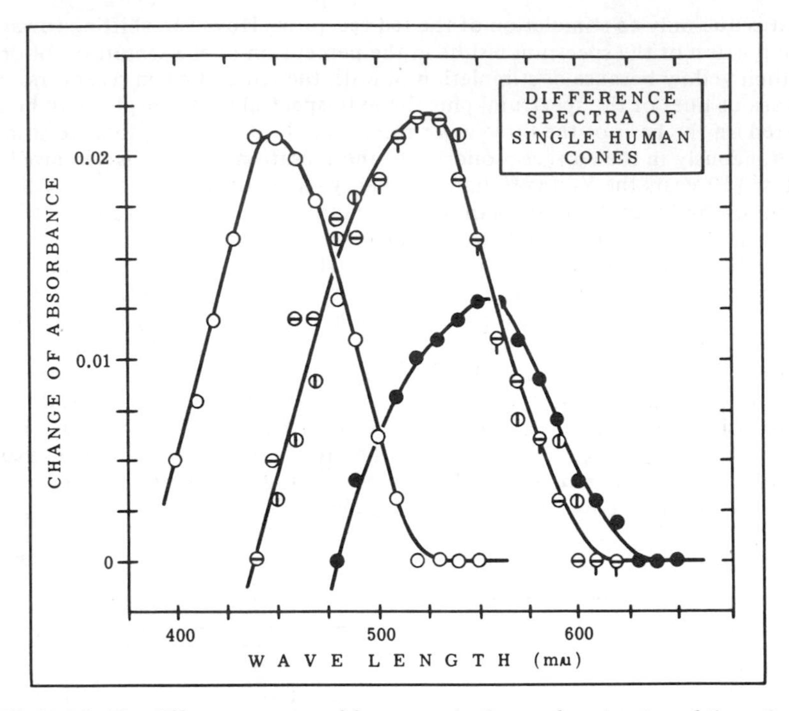

FIGURE 4A-10. The difference spectra of four cones in the parafoveal region of the retina. Difference spectra represent reduction in light absorption after bleaching with flash of yellow light. The three types of receptors have maximal absorbance at 450 mμ (blue sensitive), 525 mμ (green sensitive), and 555 mμ (red sensitive). (From P. K. Brown and G. Wald. *Science* 144:45–52, Issue 3614, 1964. Copyright 1964 by the American Association for the Advancement of Science.)

Equally convincing are the results obtained by MacNichol (1964) on the goldfish retina. Using photometric techniques also, he was able to measure the absorption spectra of more than 100 individual cones and demonstrated that they fell into just three groups, having maxima at 455, 530, and 625 mμ. Since this species is known to be color perceptive, the evidence strongly favors the trichromatic theory of color vision. These findings also indicate that the mode of excitation of cones strongly resembles that of rods, with a pigment being bleached in the receptor when it is presented with an adequate stimulus.

COLOR BLINDNESS

Color blindness is the inability to perceive a portion of the spectrum or to distinguish between colors which the normal person recognizes as being different. This confusion arises between colors which are adjacent in the spectrum. According to the Young-Helmholtz theory, color blindness is due to absence or reduced sensitivity of one or more of the three color receptors. It is present to some degree in 9 percent of males and 2 percent of females.

For example, an individual may be red-blind. This condition is termed *protanopia* and is due to the absence or deficiency of the red receptor. Since the individual has no cones sensitive to red, the red portion of the spectrum is not detected by any color receptor. Apparently the spectral range of the eye is also reduced, since the protanope does not perceive radiation in the red range, particularly beyond 640 mμ. The classic example of lack of red perception is the protanope who appeared at a funeral wearing a bright red tie. The protanope has difficulty distinguishing between adjacent colors in the yellow, green, and orange spectral regions. His perception of blue is normal.

Instead of outright lack of a color receptor, some persons may show a color weakness. For example, if this occurs in the red region, the person requires an abnormal amount of red and green to match an intermediate color such as yellow. This condition is known as *protanomaly.* Protanopia and protanomaly are the most common forms of color blindness.

A less common form of color blindness is deuteranopia, which is due to lack of the green receptor. The entire spectrum appears to the deuteranope to be composed of yellows and blues. In this instance, the green receptor shifts to the red spectral range and gives the sensation of yellow in the yellow, green, orange, and red ranges. Thus the deuteranope is unable to distinguish between adjacent colors in this region. A weakness of the green receptor is termed *deuteranomaly.* The deuteranope makes many of the same mistakes as the protanope in matching colors. However, the two conditions can be distinguished by the protanope's lack of perception of red.

Tritanopia is a rare form of color blindness which is due to lack of the blue receptor. The spectrum is shortened at the blue end for individuals with this defect.

The types of color blindness just described are due to lack of a single type of receptor. Individuals having them are termed *dichromats,* since their entire spectrum is subjectively composed of two colors or combinations thereof. Normal individuals are *trichromats.* In rare instances, two of the receptors are absent; persons with this defect are *monochromats.* They have no perception of color as such. Apparently they detect only one color and gradations thereof. Total color blindness has also been observed. In this case the cones seemingly do not function at all. The spectral sensitivity curve for persons so afflicted is the same as for persons with scotopic vision (maximum at 500 mμ).

VISUAL FIELD

The visual field is the portion of the external environment which is represented on the retina. The extent of this field is limited at most points by anatomical factors such as the supraorbital ridges above and the nose medially. The temporal field is limited only by the orientation of the eye and the sensitivity of the peripheral portions of the retina.

The visual field is also limited to a small extent by the structure of the retina. Since there are no receptors at the exit of the optic nerve from the retina, images falling on this area are not perceived. This is the *blind spot* of the retina. It is situated 10 degrees from the fovea centralis on the nasal side of the retina and encompasses about 3 degrees of visual field.

The extent of the visual field projected onto the retina may be determined by use of the *perimeter.* The perimeter is a metal band shaped as a half-circle, and the subject is situated so that one eye is at its center. If the eye being tested is directed straight ahead at the middle of the band, the two ends of the perimeter are 90 degrees removed from the visual axis. If a small target is moved in from the periphery along the perimeter, the image moves from the peripheral to the central regions of the retina. The subject then indicates when the target is in view. By repeating this test along the horizontal, vertical, and several intervening meridians, one can determine the extent of the visual field. Normally, the visual field for the eye extends 50 degrees upward, 80 degrees downward, 60 degrees nasally, and more than 90 degrees temporally. The field is most extensive for a white target and becomes successively smaller when blue, red, and green targets are used. The extent of the color fields is said to be a relative matter, depending on the brightness of the target.

A reduction of the extent of the visual field occurs in lesions of either the optic nerve, the optic tract, or the visual pathways in the central nervous system. Figure 4A-11 is a diagram of the pathways to the cortex and the effect of various lesions.

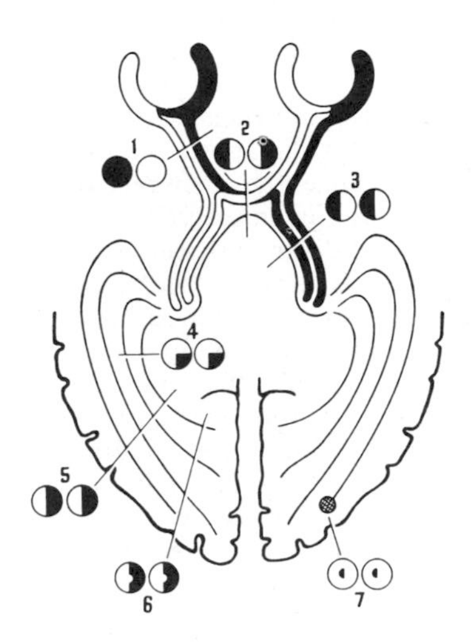

FIGURE 4A-11. The visual pathways, showing sites of interruption of pathways and the resultant abnormalities in visual fields. (1) Optic nerve – blindness on the side of lesion, with normal contralateral field. (2) Optic chiasm – bitemporal hemianopia. (3) Optic tract – contralateral homonymous hemianopia. (4) Medial fibers of the optic radiation – contralateral inferior homonymous quadrantanopia. (5) Optic radiation in the parietal lobe – contralateral homonymous hemianopia. (6) Optic radiation in the posterior parietal lobe and occipital lobe – contralateral homonymous hemianopia with macular sparing. (7) Tip of the occipital lobe – contralateral homonymous hemianopic scotoma. (From D. O. Harrington. *The Visual Fields*, 2d ed. St. Louis: Mosby, 1964.)

Images formed on the retina are inverted and reversed left from right. Thus, an object in the lower left quadrant of the visual field projects an image onto the upper right quadrant of the retina. Also, the portions of the visual field of each eye which overlap are projected onto the same area of the cortex. The detailed structure of the anatomical pathways which carry this information may be better understood from a consideration of the effects of lesions in these paths.

A lesion of the optic nerve, for example, produces blindness in that eye. A lesion at the chiasm involves the fibers from the nasal portion of the retina which cross to join contralateral temporal fibers at this point. This produces a *hemianopia* or blindness in one-half of the visual field of each eye. The blindness is referred to the *visual field* rather than the *retinal field.* Thus, this defect would be termed *bitemporal hemianopia.*

If the lesion occurs in the optic tracts, it interrupts pathways serving the same visual fields in each retina and produces *contralateral homonymous hemianopia.*

Since the lesion is on the side opposite the visual field which it affects, it is termed contralateral. In this instance the lesion is on the right and the impaired portion of the field is on the left. *Homonymous* refers to the fact that the same side of the visual field is affected in each eye.

The fibers in the optic tract synapse at the external geniculate ganglion. Optic radiations (geniculocalcarine tract) from the ganglion pass to the occipital lobe. A lesion in the geniculocalcarine tract, as in the optic tract, may cause contralateral homonymous hemianopia. If the lesion is less extensive, only one quadrant is obliterated and *contralateral homonymous quadrantanopia* results. If the lesion occurs in the posterior parietal lobe and occipital lobe, a *contralateral homonymous hemianopia with macular sparing* results. The sparing of macular vision occurs because macular fibers terminate before the calcarine cortex. A lesion of the macular field on one side causes *hemianopic scotoma. Scotoma* is a generalized term referring to a blindness or weakness in a portion of the visual field.

BINOCULAR VISION

The extent of the visual field for a single eye has been noted previously. Most of the visual field is the same for both the right and the left eye. If both eyes fix straight ahead upon an object, the visual axes of the eyes are directed toward the object and the centers of the two visual fields coincide. All points within approximately 60 degrees of this center are seen by both eyes.

Each point on the visual cortex within the binocular field receives impulses from a point on each retina. These retinal points are called *corresponding points.* Each gives rise to the same sensation in the cortex. The retina-to-cortex connections are so arranged that when the foveas are trained on the same point, so also are the corresponding points on the two retinas. This causes the same information to be sent to the cortex* from all corresponding points on both retinas.

If an object is moved toward an observer, the orientation of the eyes must be shifted so that the foveas remain trained on the object and the corresponding points of the retinas remain matched. This is the act of *convergence,* a reflex rather than a voluntary act. It is accomplished by contraction of an extrinsic striated muscle, the *medial rectus,* which draws the visual axis medially as the object approaches.

Each eye has six extrinsic muscles; they rotate the eyeball on the horizontal, vertical, transverse, and oblique axes. When the eyes are following a moving object, these muscles act in concert to keep the object trained on corresponding points of the retinas. They are capable of moving the eye approximately 50 degrees in any direction from the normal position. If, for some reason, the coordination between these muscles is lost, the images formed in the two eyes no longer fall on corresponding points. A condition known as *diplopia* then occurs, in which a double image is projected onto the cortex. If this condition is chronic, one of the images is suppressed and the corresponding eye suffers deterioration of its performance, a condition known as *amblyopia.*

REFERENCES

Adler, F. H. *Physiology of the Eye,* 3d ed. St. Louis: Mosby, 1956.

Brown, P. K., and G. Wald. Visual pigments in single rods and cones of the human retina. *Science* 144:45, 1964.

---

*Because of the separation of the eyes there are small differences in the shape and pattern of objects in the two visual fields. The closer an object is to the observer, the greater these differences. The observer associates these differences with distance between himself and the object; hence, depth perception depends importantly on these small differences.

Granit, R. Neural Activity in the Retina. In J. Field (Ed.), *Handbook of Physiology.* Section 1: Neurophysiology, Vol. I. Washington, D.C.: American Physiological Society, 1959. Pp. 693–712.

Harrington, D. O. *The Visual Fields*, 2d ed. St. Louis: Mosby, 1964.

Meyer, C. F. *The Diffraction of Light, X-rays, and Material Particles,* 2d ed. Ann Arbor, Mich.: Edwards, 1949.

Graham, C. H. Visual Perception. In S. S. Stevens (Ed.), *Handbook of Experimental Psychology.* New York: Wiley, 1951.

MacNichol, E. F., Jr. Retinal mechanisms of color vision. *Vision Res.* 4:119, 1964.

Wald, G. The Photoreceptor Process in Vision. In J. Field (Ed.), *Handbook of Physiology,* Section 1: Neurophysiology, Vol. I. Washington, D.C.: American Physiological Society, 1959. Pp. 671–692.

Special Senses

B. The Ear

Carl F. Rothe

THE EAR is an organ of exquisite sensitivity and superb design. It is stimulated by energies infinitesimally small, since at threshold it will respond to an energy flux of a millionth of one billionth of a watt per square centimeter. As the function of the ear is described, note the relative ease by which pathological or traumatic changes may occur, and imagine the deleterious effects of calcification of various parts, reduction in compliance, or changes in motion due to inflammation.

Sound is produced and transmitted by the vibratory motion of bodies or air molecules which are displaced from a position of equilibrium in a direction parallel to the direction of propagation, are rapidly restored to this position, overshoot, and again are restored to continue the cycle. Thus sound waves produced by a tuning fork are waves of alternating compression (increased pressure) and expansion (reduced pressure) of air as the tines of the fork swing back and forth. Physiologically, hearing is the subjective interpretation of the sensations produced by vibrations of a frequency and energy adequate to stimulate the auditory apparatus.

The *frequency* of sound is given in cycles per second (cps) or hertz (Hz). The psychophysiological appreciation of frequency is pitch. The natural resonant frequency of a system forced to vibrate depends inversely upon the mass in motion (inertia) and directly upon the restoring force (elasticity) acting to bring the body back to equilibrium. The duration of the vibrations depends on the degree of damping (viscous resistance).

The *quality* of a sound refers to the sensation perceived when one hears a mixture of related frequencies. Although musical instruments such as a flute produce a relatively pure sound of only one frequency, most musical sounds are composed of a fundamental frequency plus various amounts of harmonics or overtones which are integral multiples of the fundamental frequency. Nonintegral multiples are also present in some sounds, as from bells. Such mixture of frequencies, regularly repeated, determines the quality of a musical sound and its distinctiveness. *Noise,* on the other hand, is composed of a random mixture of unrelated frequencies. The *duration* of a sound also helps to make it distinctive. Whereas a plucked violin string continues to produce sound for several seconds, the soft tissue of the body effectively dampens the vibrations of such sound-producing organs as the heart.

The *intensity* of a sound is expressed in physical terms as the *amount of energy transmitted per second* through a unit *area* perpendicular to the direction of travel of the wave, i.e., power flux. The usual units are watts per square centimeter. This intensity is dependent on the fluctuation in air pressure. The psychophysiological appreciation of intensity is proportional to the 0.6 power (i.e., approximately the square root) of the sound intensity. It is also related to the frequency. Our ears are most sensitive to sounds with frequencies between 2000 and 3000 Hz. If the frequency is increased above or decreased below this level, our ears are less sensitive. This difference in sensitivity is marked, in that it requires about 10,000 times more sound power to hear a 100- or a 15,000-Hz sound than it does to hear a 2000-Hz sound (Fig. 4B-1A). The *pressure* fluctuations must be 100 times as great at 15,000 Hz to be heard as well as a 2000-Hz sound, that is, to sound equally *loud.*

To represent the extreme range of the intensity of sound, the *decibel* (db) is used. This ratio is somewhat related to perception of intensity. The just-noticeable

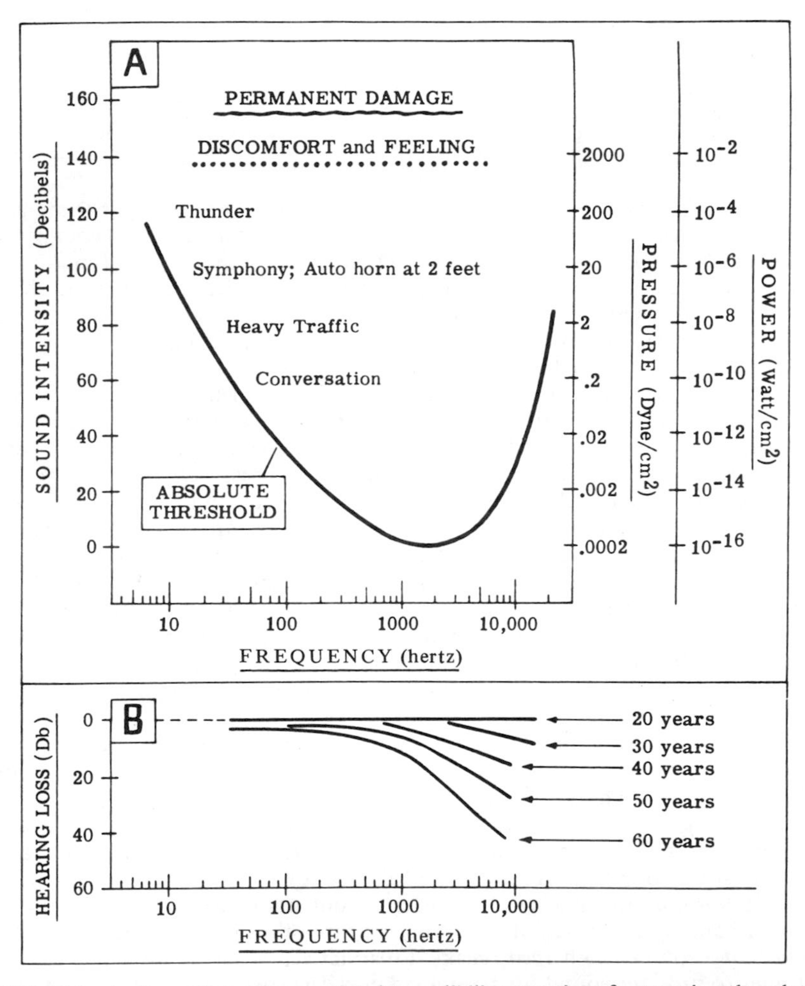

FIGURE 4B-1. A. Sound intensity as related to audibility at various frequencies, the relative pressure, and energy fluxes. Sound intensity that is uncomfortable or damaging is relatively independent of frequency. B. Presbycusis; progressive loss of hearing ability of men for high frequencies at various ages compared with hearing at 20 years. (A. After Licklider, 1951. P. 995. B. Data obtained from C. C. Bunch, *Arch. Otolaryng.* 9:625, 1929, and from J. C. Webster et al., *J. Acoust. Soc. Amer.* 22:473–483, 1950.)

difference of loudness is about 1 db but ranges between 0.3 and 5.0 db depending on the frequency and absolute sound level. A decibel is defined as 10 times the logarithm of the ratio of the intensity of the sound in question to the intensity of a reference sound. It is not an absolute unit but always is related to some reference level. In equation form:

$$N(db) = 10 \log_{10} \frac{I_1}{I_2} = 20 \log_{10} \frac{P_1}{P_2}$$

in which I is the intensity in watts per square centimeter and P is pressure fluctuation in dynes per square centimeter. If intensity is in terms of amplitude or pressure change, instead of power flux, the factor is $2 \times 10 = 20$ instead of 10 because for a given system: $W = E^2/R$. Here W is watts (power), E is the intensity factor (analogous to voltage), and R is resistance. Thus a unit change in power is related to the square (2 times the logarithm) of the potential or pressure fluctuation, assuming a constant resistance. The reference level generally used for intensity is a power flux of 10^{-16} watts/cm^2 and for sound pressure a fluctuation of 0.0002 dynes/cm^2. (Rather than peak-to-peak or simple arithmetic average, the fluctuation is measured as the root-mean-square [r.m.s.] average, i.e., the square root of the arithmetic average of the square of the values at each instant in time.) The reference levels, at about 2000 Hz, are about at the threshold of hearing under ideal conditions. Since a dyne of force is roughly equivalent to 1 mg of weight, the usable range of sensitivity to the pressure fluctuations varies between about 0.2 μg/cm^2 at threshold and 2 gm/cm^2 (1.5 mm Hg) as a result, for example, of a sonic boom. This is a ratio of $1:10^7$ or 140 decibels (20×7), since pressure, not power, is the unit considered. The human voice, during normal conversation, develops about 50 microwatts of sound. This gives a sound intensity at the ear of the listener of about 10^{-10} watts/cm^2 (60 db).

Audiometers are instruments for measuring hearing ability. There are two basic types – the pure-tone audiometer, producing tones of various frequencies and intensity levels, and the speech audiometer, providing for the presentation of speech-testing material by live or recorded voice at different intensities. These instruments are used clinically to measure two basic dimensions of hearing: (1) the threshold of hearing various tones and test words, and (2) how clearly speech can be understood when it is presented at a comfortably loud level. Hearing may be impaired in varying degrees for either one or both of these dimensions.

In the pure-tone audiometer the test frequencies usually are at octave (doubling the frequency) or half-octave intervals, ranging from 125 to 8000 Hz. This frequency range is somewhat less than the capability of the young adult ear, which may perceive frequencies as low as about 20 Hz and as high as about 20,000 Hz. Dogs and some very young children may hear frequencies as high as 40,000 Hz. The thresholds of hearing at the extremes of the frequency range require a high energy flux (Fig. 4B-1A). The intensities on an audiometer are scaled in decibels of hearing threshold level (decibels of *hearing loss* on older audiometers). The zero reference is related to the performance of normal ears rather than to the physicist's standard zero sound pressure level of 0.0002 dyne/cm^2. This threshold reference is more convenient since the ear is not equally responsive at all frequencies. Moreover, the typical normal ear even at the most easily heard frequencies is not quite able to hear sounds as faint as the physicist's zero.

The audiometric zero is based on extensive hearing surveys of young otologically normal ears. It is being changed in the United States because the American standard for audiometric zero (ASA 1951) differs significantly from standards of other countries. The International Standards Organization has proposed worldwide adoption of a single audiometric threshold reference, designated briefly as ISO 1964. The new threshold reference requires the subject to hear considerably fainter sounds in order to obtain an "0 db hearing threshold level." The difference between the reference thresholds varies with different frequencies from 15 db to 6 db. A hearing threshold 25 db above the ISO 1964 reference level (15 db for ASA 1951) in the range of 500 to 2000 Hz is considered to be a slight hearing impairment.

The ability to hear high-frequency sounds suffers with age, as does the ability to focus for close vision. This poor hearing of the elderly, called *presbycusis,* is particularly obvious in men (Fig. 4B-1B). A man 50 years of age normally has a hearing loss of about 20 db (ASA 1951) at 4000 Hz compared with a young adult. This is a ratio of $10^{-20/10}$, which is $10^{-2.0}$ or a loss of 1:100. Presbycusis is primarily a result of hair cell degeneration and loss of neuron population, mainly in the spiral ganglion (first neuron), and so resembles – and indeed in part is possibly due to – hearing impairment from intense noise.

OUTLINE OF THE HEARING PROCESS

In outline, the process of hearing is diagramed in Figure 4B-2A. An anatomy textbook should be consulted for details of the structure of the ear. Sound enters the external auditory meatus and impinges upon the tympanic membrane or ear drum, putting it into motion. The tympanic membrane in turn is coupled to the auditory ossicles, or bones of the middle ear, which transmit the sound to the inner ear, the cochlea. The inner ear of the living animal is curled in the form of a snail; thus the name. The movement of the auditory ossicles sets into motion the oval window, which separates the aqueous perilymph of the inner ear from the air in the middle ear. The motion of the fluid in the scala vestibuli of the inner ear (the vestibule) causes the basilar membrane to move in a pattern determined by the frequency and intensity of the sound. Movement of fluid in the scala tympani in turn moves the round window. The motion of the basilar membrane stimulates sensory elements in the inner ear (the organ of Corti) so that nerve action potentials are transmitted by the auditory nerve to the auditory cortex. We then perceive these impulses as sound. The vestibular apparatus or labyrinth is part of the inner ear and functions to provide sensory inputs for the maintenance of postural equilibrium (see Chap. 4C).

THE EXTERNAL EAR

The external ear consists of the pinna or auricle, the external auditory meatus, and the tympanic membrane. The *meatus* provides a passage for sound to enter the middle and inner ear. These delicate structures are surrounded and protected by the bone of the skull. The meatus prevents the entrance of large insects and objects which might damage the paper-thin (0.1 mm) ear drum. Furthermore, it acts to keep the air moist and near body temperature (±0.2°C), essential conditions if the ear drum is to function adequately. The *tympanic membrane* or ear drum is roughly conical in shape. This provides a degree of rigidity for coupling to the auditory bones.

THE MIDDLE EAR

The middle ear is air-filled and contains the *auditory ossicles.* These bones weigh in all about 55 mg. They are shaped roughly as shown in Figure 4B-2A. The *malleus,* or hammer, is fastened to the tympanic membrane; the *incus,* or anvil, is firmly attached to the malleus and then acts on the *stapes,* or stirrup, which is attached to the oval window of the inner ear.

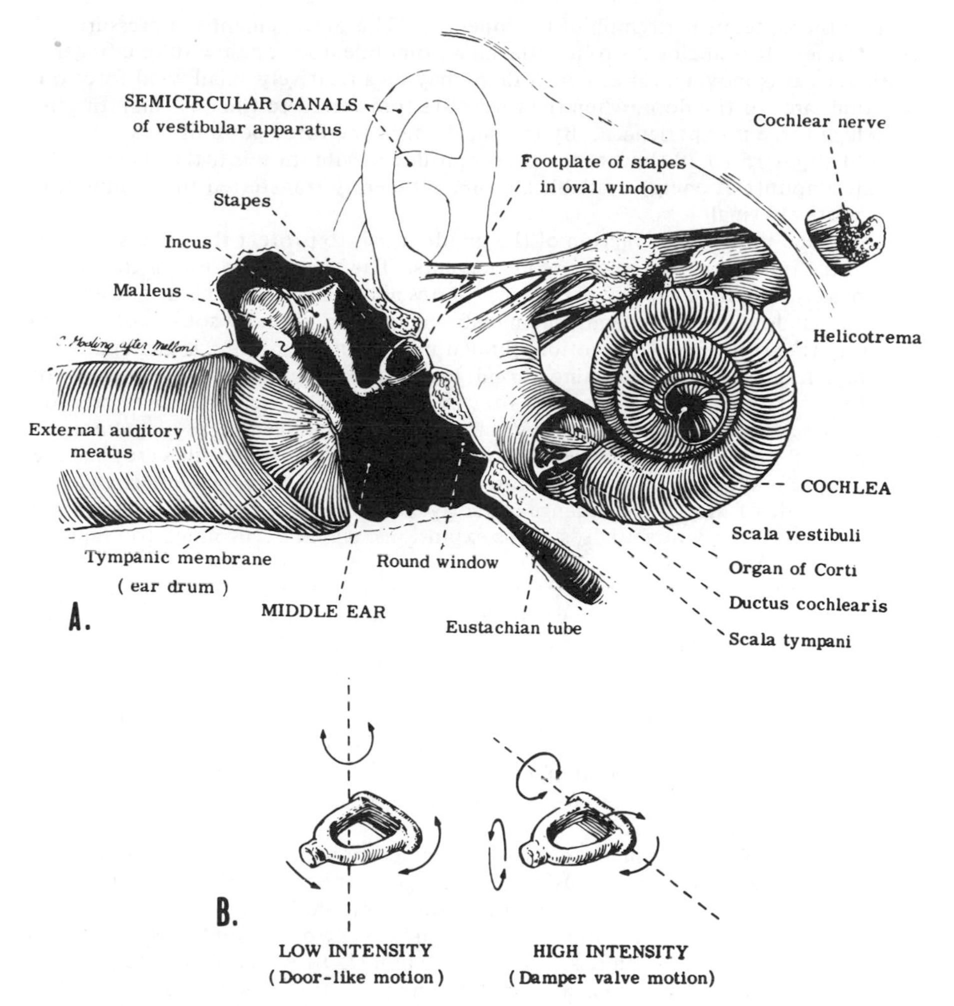

FIGURE 4B-2. A. Diagram of the human hearing apparatus. B. Modes of vibration of the stapes. At high intensities, near the threshold of feeling, the primary mode of rotation is shifted to a horizontal axis, so that as one edge of the stapes goes in, the other comes out, protecting the inner ear. (A. Redrawn from *Dorland's Illustrated Medical Dictionary,* 24th ed. B. Redrawn from *Experiments in Hearing* by Georg von Békésy. Copyright 1960, McGraw-Hill Book Company. Used with permission of McGraw-Hill Book Company.)

The primary function of the middle ear is to efficiently couple the movements of the low-density air to the high-density aqueous medium of the inner ear. If the energy flux is to be transferred efficiently, there must be impedance matching. That is, the relatively high amplitude but low force of movement of the air must be efficiently coupled to the high resistance to movement (inertia) of the fluid of the inner ear, so that the maximum amount of power (force times velocity) is transmitted. The tympanic membrane has an area of about 0.7 cm^2 in man. It is coupled through the bones to a much smaller (0.03 cm^2) oval window. The coupling acts to convert the movements of easily compressible air to the higher forces necessary to put into motion the aqueous perilymph of the inner ear. This arrangement is a pressure transformer. It is analogous to holding an automobile door open with one finger while the car is moving, for although there may be a relatively small wind force on each unit area of the door, when the forces are transferred to the one small fingertip a high force is experienced. By this mechanism the force per unit area is increased about 15 to 20 times while the amplitude of vibration is little changed. The minute amounts of energy available are thus efficiently transferred to the inner ear with relatively small loss.

A second important function of the middle ear is to protect the exquisite structure of the inner ear from excessive movements. The axis of rotation of the malleus and incus is roughly such that the stapes rotates about a vertical axis at one edge of the oval window when the tympanic membrane is acted upon by sounds of moderate intensity (Fig. 4B-2B). This motion, though minute in magnitude, is analogous to that of a door opening and closing. Loud, low-pitched sounds cause the axis of rotation to shift so that the major axis of rotation is horizontal, across the oval window. Under these conditions one edge goes in and the other comes out, as diagrammed in Figure 4B-2B. Excessive movements of the fluid within the cochlea are thus prevented by this short-circuiting procedure. This is a crucial protective mechanism in that the mode of motion can change instantly. It provides a means of protecting the inner ear from transient sounds, such as explosions, which occur much too rapidly for any reflex mechanism.

Another protective device of the middle ear is the reflex action of the muscles, which functions in a manner similar to, and fully as fast as, a blink of the eye. The *tensor tympani* acts to pull the malleus and tympanic membrane into a middle ear, as the name implies. The *stapedius* tends to pull the stapes out of the oval window. The two muscles, acting together, snub low-frequency vibrations and so help protect the inner ear. However, above about 2000 Hz, this mechanism provides but little protection. Furthermore, the response time of this reflex is at least 10 to 20 msec, and about one-sixth of a second of a loud, reflex stimulating tone is required for maximal protection. It is therefore of little value in shielding the inner ear from explosions or blows on the ear.

These two protective mechanisms, in conjunction with the characteristics of the inner ear, provide the tremendous dynamic range of the ear, so that we can perceive, without damage, pressures 10 million times the threshold level.

The *eustachian tubes* act to equalize pressures. The body fluids absorb gases, so that the total gas pressure in tissue is about 60 mm Hg below atmospheric pressure. Consequently, air must periodically enter the middle ear if a partial vacuum and impairment of hearing are not to develop. Air enters or leaves by way of the pharyngeal slits of the eustachian tubes when an individual yawns or swallows.

Chronic *otitis media* from infection of the middle ear may so damage the ossicles and their supporting structures that surgical replacement of the ossicles by a single columella between the tympanic membrane and the oval window is required. Since air conduction is impaired, this disorder is an example of *conduction deafness.* A common cause of hearing disability is *otosclerosis.* This disease causes a significant hearing loss in about 1 percent of Caucasians (twice as frequent in females as in males) but is rare among Negroes. There is destruction and regrowth of bone about the otic capsule. In most instances otosclerosis produces a primary middle ear lesion. The otosclerotic foci invade the stapes footplate, gradually reducing its mobility. In the early stages of the disease only the low frequencies are affected. However, as the invasion of the footplate continues, the high frequencies are also involved. With complete stapes ankylosis all frequencies are about equally affected, and a hearing loss of about 60 to 65 db will be present – the maximum loss which can be imposed by a conductive lesion. Surgical intervention may relieve that portion of the impairment assignable to the reduced mobility of the stapes, but it is not effective at all in reducing impairment attributable to sensory-neural involvement. The surgical procedure of choice used to be fenestration, in which a new oval window was created in the horizontal semicircular canal. Without an impedance-matching transformer system even the best surgical results left a mild hearing deficit of about 25 db. Later techniques were developed which retained the transformer action of the ossicular chain. At the present time the procedure of choice is stapedectomy with the substitution of a prosthesis.

THE INNER EAR

Movement of the stapes causes movement of the perilymph fluid, basilar membrane, and organ of Corti, which in turn triggers neural impulses. Many theories have been proposed to explain frequency analysis by the ear and the triggering of the auditory nerves. The theories presented here are generally accepted and are in accord with most of the data now available.

Place and Volley Theories of Frequency Discrimination

Very-low-frequency pressure waves cause the perilymph to move back and forth through the helicotrema, a minute opening connecting the scala vestibuli to the scala tympani (Fig. 4B-2A). They have little effect on the basilar membrane. At somewhat higher frequencies, for example 30 Hz, the pressure waves tend to short-circuit through the basilar membrane because of the inertia of the fluids; thus they cause it to move back and forth. The movement of the basilar membrane causes distortion of the hair cells, which in turn initiates volleys of neural impulses. Under these conditions frequency discrimination is performed by the cerebral cortex. At these low frequencies, the volleys of neural impulses correspond to the fluctuations in pressure of the incoming sound.

At high frequencies perception of sound frequency is based upon the *place* where the maximal movement of the basilar membrane occurs. Although a pattern of the sound pressure changes may still be seen in the action potential pattern to about 3000 Hz, above about 120 Hz the place discrimination process becomes important. The basilar membrane is relatively massive at the distal or apical end (about 0.5 mm wide); here it has a relatively low stiffness, and because of its distance from the stapes a relatively large mass of perilymph is involved in moving the distal end. Thus low frequencies tend to act here. On the other hand, just inside the oval window and stapes the supporting structure is lighter, narrower (0.04 mm), and more rigid. Most important, there is less fluid to move. High frequencies tend to produce the maximal effect here; that is, the amplitude of the oscillation is greater at this place than farther on in the cochlea (Fig. 4B-3).

Von Békésy (1960) has provided much information concerning the dynamic activity of the inner ear. Using a microscope and microtechniques with extreme care, he bored holes in the walls of the cochlear channels, applied minute specks of silver, and with a stroboscope actually measured the displacement at various places along the membrane when sounds of various frequencies were applied to the ears of cadavers (Fig. 4B-3, right). From this evidence there is little question that the basilar membrane undulates maximally at a certain place along the membrane when stimulated by a specific frequency. The nature of this undulation is shown in Figure 4B-3. This kind of wave form is called a *traveling wave*. Because of the relatively large mass of fluid and low tension on the basilar membrane, the resonance theory of Helmholtz, which pictured the inner ear as being tuned like a harp, is not adequate.

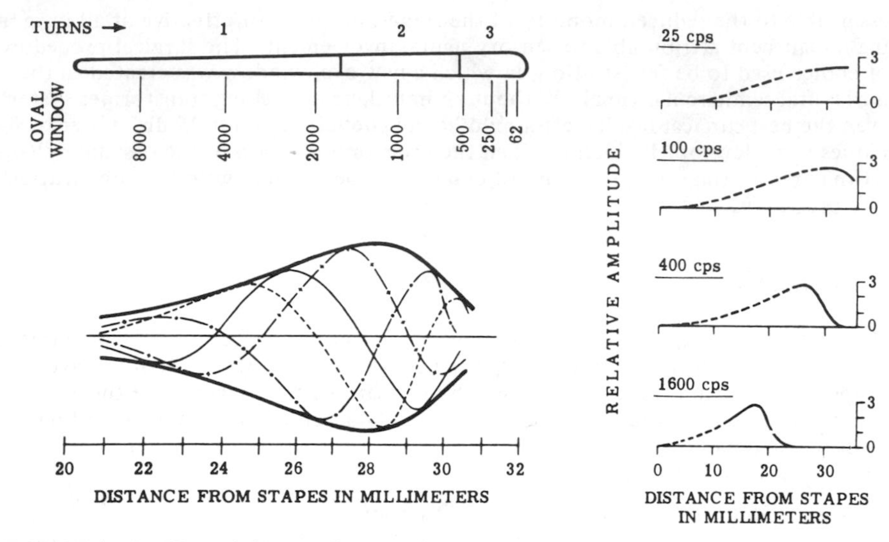

FIGURE 4B-3. (Upper left) Localization of pitch discrimination along the basilar membrane of man is deduced from experiments with guinea pigs. Lesions were produced in the basilar membrane and electrical audiograms were then made. In man there are three turns in the cochlea, with high frequencies acting near the stapes and oval window. (Lower left) Pattern of motion of a traveling wave with maximal amplitude of the envelope of motion at about 28 mm from the stapes. (Right) Relative amplitude of motion of cochlear partition of a cadaver specimen. (*cps* = hertz.) (Upper left: After S. S. Stevens et al. *J. Gen. Psychol.* 13:297, 1935. Lower left, right: From *Experiments in Hearing* by Georg von Békésy. Copyright 1960, McGraw-Hill Book Company. Used with permission of McGraw-Hill Book Company.)

Studies of inner ears damaged by high-intensity sounds of a specific frequency support the theory that sound frequencies act maximally on specific sites along the basilar membrane. Frequencies above about 10,000 Hz will be discriminated rather poorly because of the closeness of the frequency scale on the basilar membrane (Fig. 4B-3, top left). Thus there is mechanical analysis of frequency by localization of the place of maximal displacement of the basilar membrane. Low frequencies are localized at the far end, and high frequencies act next to the inner ear windows.

The *structure* of the inner ear is complex. A cross-sectional drawing of the cochlea is presented in Figure 4B-4. The total diameter is about 3 mm and its volume is about 100 μl (the volume of 2 drops of water!). Pressure waves come in along the top, the scala vestibuli, and cause the basilar membrane to move up and down. The lower drawing is still more highly magnified and shows the hair cells and auditory nerves. The *tectorial membrane* is a rather rigid and massive structure which is in contact with the hair cells. Because of the geometry of the attachments of the tectorial membrane and the organ of Corti, sound-induced vibrations produce a shearing action which causes the hairs to distort the cuticular plates of the hair cells. The movement of these hairs somehow excites the auditory nerves. The amount of movement is exceedingly small. A 3000-Hz sound of just-threshold intensity moves the tympanic membrane back and forth about 10^{-10} cm. The amplitude of movement is much less than the diameter of a hydrogen molecule (2×10^{-8} cm). The basilar membrane moves only a small fraction as much as this. The auditory apparatus is indeed extremely sensitive. The reason we do not hear blood flowing through the vessels of the tympanic membrane is probably because the otherwise

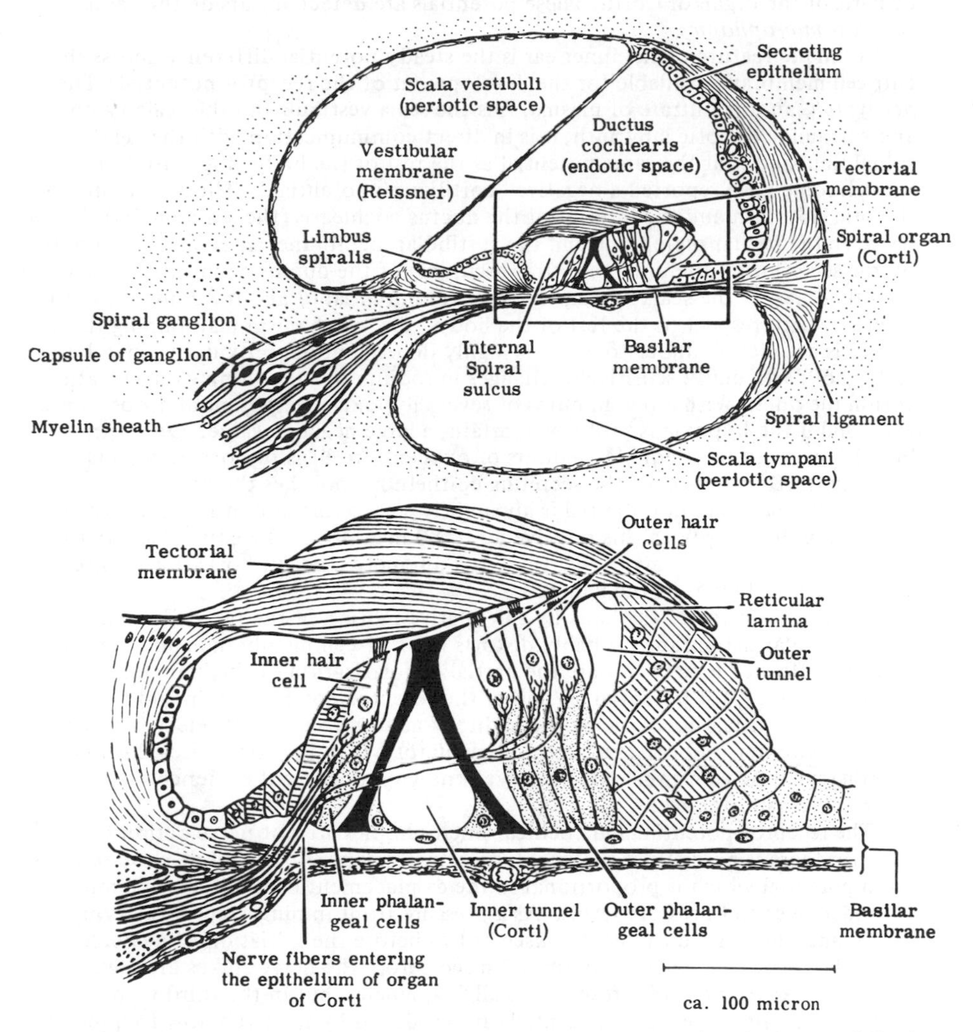

FIGURE 4B-4. (Upper) Cross section of the human cochlea showing the spinal organ of Corti and the three ducts of the inner ear. (Lower) Basilar membrane with organ of Corti at higher magnification. The periodic spaces are filled with perilymph and the endotic space is filled with endolymph. (Modified from A. T. Rasmussen. *Outlines of Neuro-anatomy,* 3d ed. Dubuque: Brown, 1943. Pp. 45, 47.)

audible higher frequencies are damped so that the flow is steady. It is of significance that there are no blood vessels in the organ of Corti. Its nutrient supply is dependent upon the secretory epithelium of the ductus cochlearis.

Initiation of Neural Impulse

The most recent evidence indicates that neural impulses are triggered in some manner by the development of *receptor potentials* resulting from relative movements of parts of the organ of Corti. These potentials are detected outside the ear as *cochlear microphonics.*

A unique feature of the inner ear is the steady potential difference across the hair cell membrane available for the development of the receptor potential. The *perilymph,* an ultrafiltrate of plasma, fills the scala vestibuli and the scala tympani and surrounds the otic labyrinth; it is in direct communication with the cerebrospinal fluid and is at the same potential as the rest of the body (Figs. 4B-2, 4), although some have reported a negative "cortilymph potential." The *endolymph* fills the otic labyrinth and the interior of the ductus cochlearis (the scala media) located between the basilar membrane and the vestibular membrane. It is probably produced by the secreting epithelium (the stria vascularis) of the ductus cochlearis (Fig. 4B-4). This fluid has, in the scala media, an electrical potential of about 80 mv positive (not negative) with respect to the rest of the body. The mechanism maintaining this potential is not clearly understood. It is highly dependent upon oxidative metabolism and is not particularly sensitive to changes in sodium or other ionic concentrations within the duct. With a few minutes of severe hypoxia, the potential drops to near zero, and if the hypoxia is prolonged, attains a large negative value. On return of blood flow carrying adequate amounts of oxygen it returns to normal in a matter of seconds. Destruction of the secreting epithelium abolishes the potential. Thus the positive endolymph potential is apparently a secretory potential rather than a diffusion potential. With this positive 50 to 80 mv outside the hair cells, and the usual negative 20 to 80 mv inside, there is a uniquely high (up to 160 mv) potential across the membranes of the hair cells.

Slight movement of the hairs by the relative motion of the tectorial membrane and the basilar membrane probably distorts the hair cell membrane and opens pores to allow a flow of ions to initiate the partial depolarization of the hair cell membrane – the receptor potential. Whether the receptor potential is, in addition, a generator potential, depolarizing the auditory neuron to excite it (electric transmission), or whether it liberates a chemical mediator which in turn excites the auditory neuron (synaptic transmission), is not yet known, but present evidence suggests a chemical transmitter agent.

The receptor potential and cochlear microphonics are not nerve action potentials, for there is no distinct threshold, no refractory period, no all-or-none response, but a potential which is proportional to the displacement of the basilar membrane at moderate sound pressure levels. As a pressure front impinges upon the tympanic membrane, there is a delay of 0.1 msec or less before the initiation of a cochlear microphonic, then a delay of about 0.7 msec before the nerve spikes are seen. The hair cells next to the stapes respond to all frequencies, but in the third turn of the cochlea (Fig. 4B-3), receptor potentials are produced by relatively low frequencies only, supporting the place theory of frequency analysis.

Although sound of a particular frequency sets into motion a relatively large part of the basilar membrane, some mechanism, probably mediated through the auditory cortex (see review by Grinnell, 1969), sharpens the sensation so that frequencies differing by 2 to 3 Hz can be distinguished in the range of 60 to 1000 Hz when presented separately. Above 1000 Hz, the ability to discriminate between different frequencies is about 2 per 1000. Thus 8000- and 8016-Hz tones, if heard alternately, will be perceived as just different in frequency. If two tones of similar loudness and nearly the same frequency are heard simultaneously, a phenomenon called *beats* will be heard. There will be a variation in loudness which occurs at a frequency equal to the *difference* of the frequencies of the two tones. Beats between two tones can be detected up to a beat frequency of about six per second. Listening for these beats is the technique used by musicians to tune their instruments. If the beat frequency is greater than about six per second, a sensation of dissonance occurs.

Loudness discrimination is possible since sounds of higher intensity cause a greater movement over a wider area of the basilar membrane than do those of low intensity. By moving more hairs to a greater degree, more hair cells are stimulated to excite more auditory nerves and also to increase the nerve impulse frequency to give the sensation of greater loudness. In addition, the inner hair cells (Fig. 4B-4) have a higher threshold and therefore, when stimulated, may add to the sensation of loudness.

Sensory-neural deafness, involving damage to the inner ear or the auditory nerve, may occur at any age as a result of various infections or trauma. Some antibiotics of the streptomycin group are ototoxic and thus cause degeneration of the organ of Corti.

Prolonged exposure over a period of years to high-intensity noise can result in *noise-induced hearing impairment.* Thus hearing conservation measures are recommended if the noise level exceeds 85 db over the range of 300 to 2400 Hz. At this level of noise, conversation is not possible unless the voice is raised and the people are within a few inches of each other. The availability of high-power audio amplifiers for popular music has made permanent hearing impairment by such "noise pollution" a significant problem for many young people. *Acoustic trauma* to the organ of Corti may be caused by a single high-intensity noise of greater than about 140 db. When the sound is a pure tone, histologically detectable damage to a localized spot along the basilar membrane occurs. The maximum loss in hearing ability occurs at a frequency about a half-octave above the frequency of the damaging tone, a site nearer the oval window. Hearing impairment by noise is dependent upon not only the intensity of the noise but also its frequency and duration.

Sound intensity of 180 db, a level attained, for example, close to a jet engine, has a power flux of 10 watts per square centimeter. Since the soft tissues of the body absorb sound, particularly that of high frequency, and convert it to heat, there is a significant and dangerous increase in body temperature under these conditions. Furthermore, the pressure fluctuations are of the order of 200 gm/cm^2 (about 150 mm Hg), and delicate tissues such as those of the ear and brain can be literally torn apart by the vibrations. An energy flux of 1 watt/cm^2 at 20,000 Hz will kill a mouse in one minute.

High-intensity sound can be used to destroy pathological structures in the brain. Ultrasonic frequencies of about 1 million per second, well beyond the audible range, may be focused to impinge on a small area within the body. When such beams of high-intensity sound are focused on a desired spot in the brain, tissue can be selectively destroyed by the heat generated without serious damage to intervening tissue. The echoes from pulses of ultrasound of low intensity are being used to visualize internal structures of the body, such as heart valves, the ventricles of the brain, a fetus.

LOCALIZATION OF SOURCE OF SOUND

The process by which we localize the source of a sound is complex in that several mechanisms are involved. The *time of arrival* of the pressure front, e.g., from a click, provides one mechanism. If the source is to the right of the individual, the sound first stimulates the right ear and then the left ear. By orienting sound sources and having a blindfolded subject indicate the direction of the sound source, or by using two earphones and an electronically produced time-delay for one of the ears, it has been found that the normal individual can distinguish differences in the apparent location of the source when the arrival times of two clicks are spaced as closely as about 0.01 msec. If the clicks are more than about 2 msec apart, the sensation is not that of localization but of hearing two separate sounds. The 2-msec minimum separation is important. In the interpretation of heart sounds, for example, if the aortic and pulmonary valves close within an interval less than 2 msec, the sound will be perceived as a single sound and not as a split sound.

With low frequencies, there is localization based on the *phase* relationship. This is closely similar to the time-of-arrival mechanism but involves a smoothly changing type of sound (sine wave) rather than sharp waves such as are associated with clicks.

Intensity of sound at each ear provides another important mechanism, particularly at high frequencies. The head tends to shadow the sound, and since we can distinguish a difference in intensity of about 1 db, this ability furnishes a mechanism for localization of sound. At high frequencies the difference in intensity, because of the shadowing effect, is as high as 30 db; thus, rather fine localization is obtainable. This difference in intensity of the high frequencies is of prime importance in the appreciation of streophonic sound. Interestingly enough, at about 3000 Hz, the frequency of maximal sensitivity of the human ear, neither the time of arrival nor the intensity mechanism is particularly efficient; hence, these frequencies are difficult to localize. The human voice, having transient, clicklike sounds and lower frequencies, is easily localized.

REFERENCES

Beranek, L. L. Noise. *Sci. Amer.* 215:66–76, December, 1966.

Davis, H. Biophysics and physiology of the inner ear. *Physiol. Rev.* 37:1–49, 1957.

Davis. H. Excitation of Auditory Receptors. In J. Field (Ed.), *Handbook of Physiology.* Section 1: Neurophysiology, Vol. I. Washington, D.C.: American Physiological Society, 1959. Pp. 565–584.

Davis, H., and S. R. Silverman (Eds.). *Hearing and Deafness,* 3d ed. New York: Holt, 1970.

Glorig, A., and H. Davis. Age, noise and hearing loss. *Ann. Otol.* 70:556–571, 1961.

Goldstein, M. H., Jr. The Auditory Periphery. In V. B. Mountcastle (Ed.), *Medical Physiology,* 12th ed. St. Louis: Mosby, 1968. Chap. 64.

Grinnell, A. D. Comparative physiology of hearing. *Ann. Rev. Physiol.* 31:545–580, 1969.

Jerger, J. F. (Ed.). *Modern Developments in Audiology.* New York: Academic, 1963.

Katsuki, Y. Comparative neurophysiology of hearing. *Physiol. Rev.* 45:380–423, 1965.

Licklider, J. C. R. Basic Correlates of the Auditory Stimulus. In S. S. Stevens (Ed.), *Handbook of Experimental Psychology.* New York: Wiley, 1951. Chap. 25.

Mountcastle, V. B. Central Neural Mechanisms in Hearing. In V. B. Mountcastle (Ed.), *Medical Physiology,* 12th ed. St. Louis: Mosby, 1968. Chap. 65.

Rosenblith, W. A. (Ed.). *Sensory Communication.* Cambridge, Mass.: M.I.T. Press, 1961.

von Békésy, G. *Experiments in Hearing.* New York: McGraw-Hill, 1960.

Wever, E. G., and M. Lawrence. *Physiological Acoustics.* Princeton, N.J.: Princeton University Press, 1954.

Special Senses

C. The Ear: Vestibular Apparatus

Julius J. Friedman

ORIENTATION of the body during movement is maintained in part by reflex activity (acceleration reflexes) which originates from the *labyrinths* (vestibular apparatus), the nonacoustical part of the inner ear. These reflexes adjust the position of the head with respect to the horizontal plane and gravity and reorient the body to effect a normal postural alignment. In addition, man uses other sensory information to maintain orientation with respect to the environment. This information includes visual perception, proprioceptive impulses from the body musculature, and external sensations of relative motion and position. The vestibular apparatus consists in part of the *semicircular canals,* which are concerned with *angular* movement of the head. The horizontal (lateral) are the *nongravity* set of canals, concerned with turning movements around a vertical axis, such as in waltzing. The vertical are the *gravity* set of canals, concerned with movements around a horizontal axis, such as in leaping and landing and in protection against falling. The other parts of the vestibular apparatus are the otolithic organs, the *utricle* and *saccule,* which are stimulated by slow tilting (gravity), centrifugal force, and linear acceleration.

STRUCTURE OF THE LABYRINTH

The horizontal (lateral), anterior (superior), and posterior (inferior) membranous semicircular canals are enclosed in the *bony labyrinth* (Fig. 4C-1). These canals are continuous with the utricle, and both are filled with a fluid, *endolymph.* Near the point of junction with the utricle, each canal is equipped with an enlarged region called the *ampulla.* The ampulla contains the receptor organ, *crista ampullaris* (Fig. 4C-2), a ridge of columnar epithelial hair cells possessing hair filaments which are embedded in a gelatinous mass called the *cupula.* The cupula projects from the crista to the opposite wall of the ampulla and represents a movable partition which can be distorted by movements of endolymph within the canals. The hair filaments protruding from each sensory hair cell are composed of 40 to 70 *stereocilia* and one *kinocilium.* The kinocilia of the cells of the horizontal crista are always located at the peripheral margin of the cells on the side toward the utricle, whereas the kinocilia of the cells of the posterior and anterior cristae are on the canal side (Fig. 4C-2). These receptors respond bidirectionally. Displacement of the stereocilia toward the kinocilium depolarizes the cell, increasing its discharge frequency; displacement of the stereocilia away from the kinocilium hyperpolarizes the hair cell, decreasing its discharge rate.

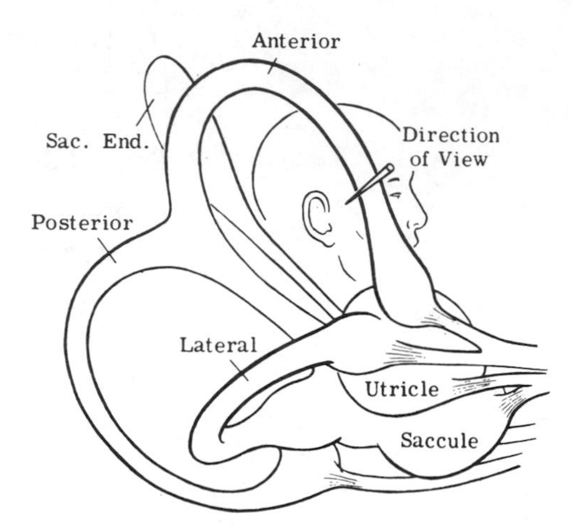

FIGURE 4C-1. Structural relations of the human vestibular apparatus. (From M. Hardy. *Anat. Rec.* 59:403, 1934.)

The utricle and saccule are situated anterior to the semicircular canals and communicate with each other through the independent utricular and saccular, and the common endolymphatic, ducts. The receptor organs of the utricle and saccule are called *maculae* (Fig. 4C-2) and are structurally similar to the cristae. However, the stereocilia and the kinocilia of the utricular maculae are oriented in a more complex pattern. Kinocilia of these sensory organs are pointed in both directions, providing both hyperpolarization and depolarization with a displacement in either direction. The depolarization effect must exceed that of hyperpolarization since displacement increases the discharge frequency. Embedded in the hair cell filament membrane of this receptor are numerous small particles of calcium carbonate called *otoliths*. The macula of the utricle is oriented horizontally with the head in the erect position, whereas that of the saccule is oriented vertically.

The nerve endings in the cristae and maculae communicate with the central nervous system via the vestibular ganglion and vestibular branch of the eighth nerve.

FUNCTION OF THE SEMICIRCULAR CANALS

MODE OF STIMULATION OF THE SEMICIRCULAR CANALS

Under resting conditions there is a balanced basal discharge of impulses from the ampullae on both sides of the head. Deviation of either cupula from the resting position will create an imbalance and produce the sensation of rotation. For example, an animal lacking one vestibular apparatus may experience the sensation of rotation whether or not the ampulla is compressed or decompressed, because of the gradient established by a deviation from resting balanced condition. With two vestibular systems present, the intensity of sensation is increased, owing to the greater differential between the ampullae on either side.

The crista of a particular canal is stimulated when the head is rotated in the plane of the canal. Because of the inertia of the endolymph, movement of this fluid lags behind movement of the canal throughout the period of rotational

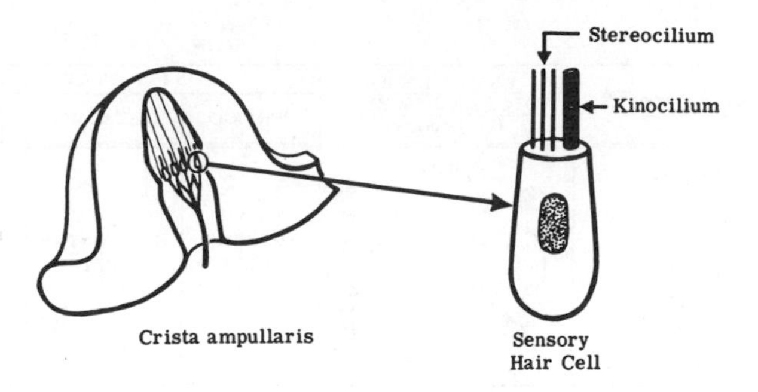

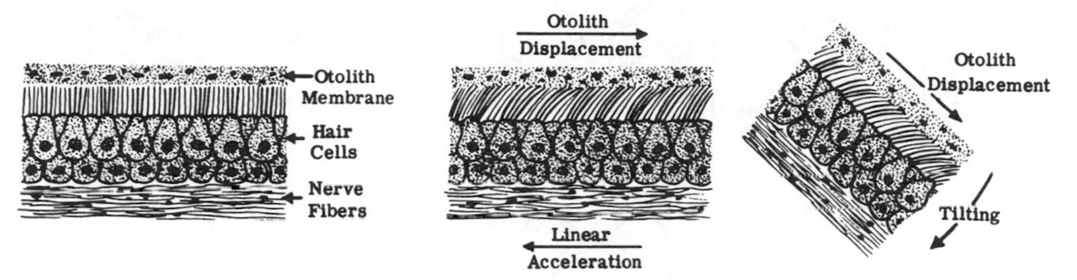

FIGURE 4C-2. Diagrams of the vestibular sense organs. (Above) Crista ampularis. (From A. Flock and J. Wersall. *J. Cell Biol.* 15:21, 1962.) (Below) Macula. (From Fischer, 1956. P. 17.)

acceleration. Since this situation is equivalent to a reverse flow of endolymph, the endolymph is displaced during acceleration in a direction opposite to rotation. Thus, with horizontal rotation to the right, endolymph is displaced to the left, thereby compressing the right and decompressing the left ampullae (Fig. 4C-3). This movement approximates the stereocilia and kinocilia of the right and separates those of the left cristae. Consequently, the frequency of afferent impulses from the right ampulla increases, whereas that from the left ampulla decreases, with the result that the sensation of rotation to the right is perceived. It should be noted that the *sensation of rotation is in the direction opposite to the direction of the endolymph displacement.* This is true of rotation in all planes. As the speed of rotation is held constant and as acceleration becomes zero, the endolymph lag disappears. Owing to its elasticity, the cupula returns to the resting position and the discharge of impulses from the ampullae is restored to the basal frequency. In this manner, the sensation of rotation declines as angular acceleration approaches zero. With deceleration, the inertia of the endolymph causes it to be displaced in the direction of rotation, thereby compressing the left and decompressing the right ampullae and producing a postrotatory sensation of rotation in the opposite direction (Fig. 4C-3).

In an ordinary head movement more than one canal is stimulated on each side; the three pairs of canals will always give rise to a combination of endolymphatic streams the resultant of which will flow in the "plane of rotation." With transverse rotation down and to the right, endolymph is displaced upward and to the left, thereby compressing the left anterior and posterior ampullae and decompressing the right anterior and posterior ampullae. Thus the stereocilia and kinocilia

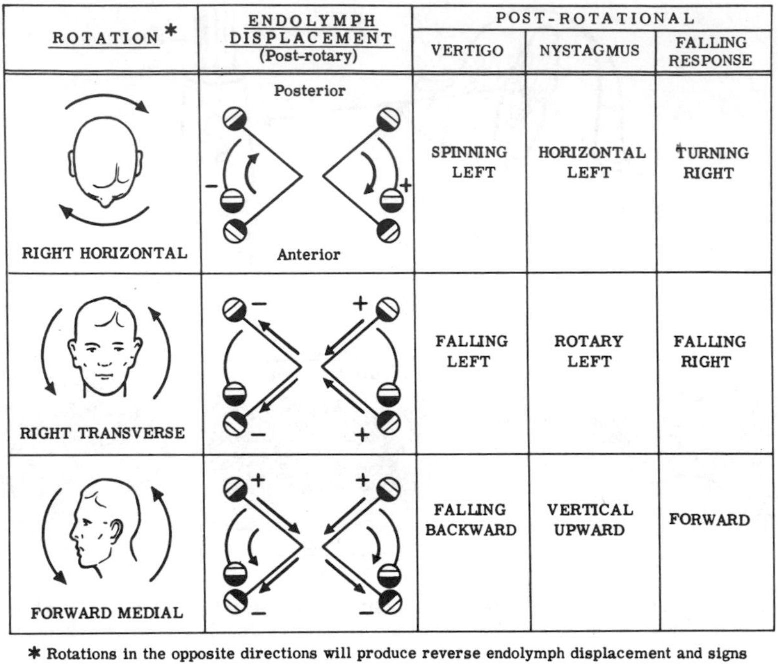

ROTATION*	ENDOLYMPH DISPLACEMENT (Post-rotary)	POST-ROTATIONAL		
		VERTIGO	NYSTAGMUS	FALLING RESPONSE
RIGHT HORIZONTAL	Posterior / Anterior (− +)	SPINNING LEFT	HORIZONTAL LEFT	TURNING RIGHT
RIGHT TRANSVERSE	(− + − +)	FALLING LEFT	ROTARY LEFT	FALLING RIGHT
FORWARD MEDIAL	(+ + − −)	FALLING BACKWARD	VERTICAL UPWARD	FORWARD

* Rotations in the opposite directions will produce reverse endolymph displacement and signs

FIGURE 4C-3. Effects of rotation in the three main axes on the postrotatory endolymph displacement. The dark segment and thin line of each crista respectively denote the relative position of the kinocilia and stereocilia. Rotations in the opposite directions will produce reverse endolymph displacement and signs.

of the right ampullae are brought closer together, and the discharge frequency increases while the opposite events take place in the left cristae. This elicits the sensation of rotation down and to the right. With deceleration, the endolymph is displaced in the direction of rotation, compressing the right and decompressing the left anterior and posterior ampullae and causing the postrotatory sensation of rotation up and to the left. Medial rotation forward and down produces endolymph displacement upward and backward, compression of the posterior and decompression of the anterior ampullae, and a sensation of rotation forward and downward. With cessation of rotation, endolymph is displaced in the direction of rotation (forward and down), thereby compressing the anterior and decompressing the posterior ampullae.

EFFECTS OF LABYRINTH STIMULATION

Stimulation of the semicircular canals produces the subjective manifestations of vertigo, nausea, and other autonomic nervous responses, in addition to the objective manifestations associated with changes in tonus of eye muscles (nystagmus) and of the antigravity muscles (falling reaction).

Vertigo

Vertigo is the sensation of spinning with respect to the environment and may cause a person to lose balance and fall. The direction of the sensation of spinning depends on which semicircular canal is stimulated. *In every case, the vertigo is oriented in the direction opposite to the endolymph displacement.* Therefore, the vertigo is in the direction of rotation during the period of rotation, whereas that experienced during the postrotatory state is in the direction opposite to the original rotation.

Autonomic Responses

Autonomic responses resulting from the vertigo include nausea and vomiting, pallor, perspiration, and – with intense stimulation – cardiac, vasomotor, and respiratory influences which can lead to hypotension and hyperpnea.

Alterations in Muscle Tone

Stimulation of the semicircular canals produces an increased muscle tone on the ipsilateral side and a decreased tone on the contralateral side. As a consequence of these changes in muscle tone, one experiences nystagmus, past pointing, and falling reactions. *Nystagmus* is an oscillatory eye movement consisting of a slow and a fast component. The slow component is always in the direction of the flow of the endolymph displacement and is a result of a reflex from the cristae via the vestibular nuclei in the brain stem to the eye muscles. The fast component, the recovery phase of eye rotation, which unfortunately has been designated to define the direction of a nystagmus, is due to an efferent central nervous component of the vestibular reflex from the reticular formation in the brain stem, not from the vestibular nuclei. Rotation in the horizontal plane produces a horizontal nystagmus, medial rotation produces a vertical nystagmus, and transverse rotation produces a rotatory nystagmus. *Past pointing* and *falling reactions* are due to a change in tonus of the antigravity muscles wherein the tone of the muscles increases in the side toward which the endolymph is displaced and decreases on the opposite side. Thus, if the ground gives way under a person's right foot, he and his head lurch to the right, shifting his endolymph to the left. This reflexly causes an immediate extension of his right arm and leg and flexion of his left arm and leg, accompanied by deviation of his eyes to the left. These are protective righting reflexes.

CALORIC STIMULATION

Douching the ear with warm or cool water produces convection currents in the endolymph which cause crista stimulation. This approach has an advantage over stimulation by rotation in that the vestibular system of each ear can be tested separately. The technique is used widely in diagnosis of vestibular disorders. Douching each individual ear at suitable intervals with water at 30°C and then at 44°C is a method used to ascertain canal paralysis, paresis, and a phenomenon called *directional preponderance.* The latter means there is a tendency for the nystagmus elicited by both ears to beat more (longer) in one direction (right or left) than in another. This is sometimes indicative of central nervous system disease.

When the head is held backward at 60 degrees, the horizontal canals are brought into vertical position. The douche causes a greater change in the temperature of the endolymph in the part of the canal lying near the external auditory meatus than in the more deeply situated parts. With a cold douche, the currents created move away from the ampulla; with a warm douche, they move toward it. The convection currents thus set up stimulate the cirsta, and *horizontal nystagmus* and vertigo result. With the head upright, both sets of vertical canals are stimulated, and *rotatory nystagmus* results, away from the cold-syringed ear and toward the warm-syringed ear.

FUNCTIONS OF THE UTRICLE AND SACCULE

The utricle is the most important gravity organ in the labyrinth. Its maculae respond to slow tilting, linear acceleration, and centrifugal force. With the head in the erect position, the utricular maculae are in the horizontal plane, and the otoliths exert uniform pressure down upon the hair cells. In this case there is no distortion of the hair filaments, and the discharges from both right and left utricles are in balance, the hair cells firing at the rate of 10 to 20 per second. Linear acceleration or tilting the head causes the otoliths to be displaced because of inertia and the force of gravity. This displacement produces distortion of the hair filaments and elicits a change in discharge frequency of the hair cells. More specifically, if the left utricle is tilted to the left, the discharge rate will increase, reaching a maximum in the left side down position, whereas a tilt to the right decreases the discharge of the left utricle to a minimum in the left side up position. Complementary changes occur on the opposite side; these amplify the imbalance between the right and the left utricular discharge. The function of the saccule is obscure and still relatively unknown.

REFERENCES

Fischer, J. *The Labyrinth: Physiology and Functional Tests.* New York: Grune & Stratton, 1956.

Gernandt, B. E. Vestibular Mechanism. In J. Field (Ed.), *Handbook of Physiology.* Section 1: Neurophysiology, Vol. I. Washington, D.C.: American Physiological Society, 1959. Pp. 549–564.

5

Reflexes and Reflex Mechanisms

Sidney Ochs

DESCARTES first clearly defined the basic behavior pattern of the reflex, giving as an example a foot placed near a fire, which when painfully stimulated is quickly withdrawn. (In modern terms this is called a *flexor withdrawal reflex.*) Involved in reflex activity are: excitation of sensory receptors, conduction over afferent nerve fibers, and finally, after specific central nervous system (CNS) activity, excitation of the motor nerves leading to the musculature to give an appropriate contraction.

Reflex responses can be described as *machine-like* in character, which implies reproducibility of response. They have also been called *purposeful,* since reflexes are generally protective to the organism. For example, when a beam of light is shined into the eye, a constriction of the pupil occurs. This reflex response helps prevent the retina from being subjected to intense illumination. Similarly, muscular control of the ear drum acts to decrease movement of the middle ear and thus to prevent excessive sounds from being transmitted to the inner ear. Another example is the *pinna reflex,* found in the cat and the dog. When a foreign object enters the outer ear, a series of reflex ear twitches act to dislodge it. The purposive nature of reflexes is further demonstrated in the spinal frog. Pithing the brain of this animal results in a short period of spinal shock (see following section). After recovery, when an irritant acid solution is applied to one flank, the leg on that side is brought up and wiping movements are performed. If the leg is held, the other leg performs the wiping movements. The direction of the limb toward the spot of excitation on the flank is an example of *local sign.* This implies that stimuli which enter the nervous system have sufficient localizing information so that the resulting reflex actions are directed to the appropriate site.

In the last century, the purposiveness of reflexes was taken to indicate that there was some kind of primitive consciousness in the spinal cord which directed reflex activity. Sherrington (1906), however, pointed out that the apparently purposive nature of reflex responses represents a selection, during evolution, of responses which have survival value. Certain reflexes damaging to the organism, rather than protective, may be elicited, indicating the lack of conscious direction.

Some actions are partially voluntary and partially reflex. For example, a mixture of volition and reflex activity is involved in the act of swallowing. Food in the mouth may be voluntarily rejected, but once it has passed beyond the fauces, reflexes coordinated by a swallowing center in the medulla are triggered.

A practical use of reflexes is illustrated in anesthesiology. Touching the eyelids causes a reflex closure, the *eyelid reflex,* which is lost with moderately deep anesthesia. Touching the cornea of the eye causes a *corneal reflex,* a blinking and covering of the cornea. This reflex is diminished or lost when the brain stem has been depressed to a perilous degree. The *pupillary light reflex* is one of the last reflexes to disappear in deep anesthesia, along with respiratory and cardiovascular reflex control mechanisms having their centers in the medulla. Another reflex of much value for judging whether the depth of anesthesia is satisfactory for operative manipulation is the response produced by pinching the skin (one example of *nociceptive,* or injury-provoking, stimulation). This evokes a flexor withdrawal reflex in which the limb flexes away from the site where the stimulus is applied. When this reflex is absent, anesthesia is generally considered sufficiently deep to permit operative procedures.

As will be discussed in more detail in a later section, the *stretch reflexes* of skeletal muscle are of great importance for normal posture and locomotion, and these have been used to form our basic concepts of the reflex mechanism. A stretch reflex occurs as a result of a brief extension of the muscle, to give rise to a quick reflex contraction of the muscle. Or, as a result of a maintained stretch placed on the muscle, a maintained tonic reflex tension is produced in the muscle. Muscles are termed *flexors* or *extensors* depending on whether they flex or extend a limb at a joint. Again, as will be discussed more fully later on, there is a special relationship of flexor and extensor muscles in stretch reflex responses in that when one muscle is excited reflexly, the other is inhibited, and vice versa.

The various reflexes are so interrelated that in the intact animal the result is one continuous smooth and well-directed behavior pattern, as each reflex succeeds and merges with the next in rapid sequence. This interrelation of one reflex with another is demonstrated in locomotion. An animal made spinal some weeks previously and allowed to recover shows a high level of reflex excitability (see below). When it is laid down so that the legs are moderately flexed and a hand is gently pressed upward against the toes, the slight spreading of the toe pads suddenly results in a powerful downward thrust of the leg, the *extensor-thrust reflex.* During locomotion, the limb is flexed and brought up from the ground. After the body of the animal has carried it forward, the limb is extended and comes into contact with the surface. When this happens, the extensor thrust is excited and the limb is converted into an extended rigid column to give a pole-vaulting effect to the body, carrying it forward. The reflex is soon after inhibited, permitting the leg to flex, and the cycle of flexion and extension is repeated.

The reflex relationship of the two hind limbs is shown in the *crossed reflexes.* When the spinal animal suspended in a body harness with legs pendant receives a noxious stimulation to one limb, a flexor withdrawal is produced. The contralateral hind limb then may show a *crossed extension.* Forceful flexion of one leg also causes reflex extension of the contralateral limb. Some animals do not walk or run but hop or gallop with both hind legs flexing and extending together. Rabbits are hopping animals, and in this species reflex flexing of one hind limb usually causes a crossed flexion of the contralateral hind limb rather than the crossed extension usually seen in the cat and the dog. Another type of response associated with locomotion, the *walking reflex,* may be observed when reflex excitability is high. Noxious stimulation of one leg of a spinal dog to cause flexion may not only also produce a crossed extension of the opposite limb but be followed later by an extension of the originally reflexly flexed leg. This proceeds in an alternating fashion with a fairly constant rhythm, so that there is a cyclic pattern of flexions and extensions of the two hind limbs which may persist for several minutes.

Walking and running are activities of the quadruped which require all four legs. When one leg is stimulated, not only does the leg of the contralateral side show crossed extension, but the foreleg may extend on the side on which the hind limb is stimulated to flex. Furthermore, the opposite foreleg may become extended. Such patterns are referred to as *reflex figures.* The reflexes elicited by these stimulations indicate that one is exciting parts of reflex patterns within the spinal cord controlling normal locomotory behavior of the four-legged animal. These local or lower spinal reflexes are modulated by control centers in the brain, as will be described in the next section and in Chapter 9.

LOWER AND UPPER MOTOR CONTROL

The basic machinery for the most fundamental types of reflex mechanisms is found within the spinal cord and the brain stem. In local reflexes a complex interplay of control is present. Sherrington (1906) used such terms as *prepotency* and *dominance* to describe the interaction of stimuli and reflexes. Noxious stimulation is usually prepotent. Its prepotent nature is demonstrated in competition with the *scratch reflex.* To excite the latter, a stimulus is used to imitate an insect crawling on the skin. This gives rise to a rhythmic scratching movement of the limb, directed toward the stimulus. If the scratch reflex is induced and then the skin is excited with a nociceptive stimulation, the scratch reflex ceases and the more prepotent flexor withdrawal response to the noxious stimulation takes command of the reflex channels. Only a relatively limited number of lower motoneuron cells are available to respond to reflex action, and some mechanisms dominate to switch command from one reflex to the other. Sherrington (1906) applied the term *final common path* to the motoneuron cells whose axons innervate the muscles of the body. All the varieties of reflex activities and the resulting complex behavior which animals display must somehow be channeled to and funneled through the relatively few cells of the final common path.

If, as a result of compression asphyxiation or neurotropic diseases such as poliomyelitis, the final common path motoneurons are destroyed, there is a total loss of reflex excitability and later a wasting of the muscles innervated by these cells (Chap. 3A). This is part of the condition termed a *lower motor lesion.* If the spinal cord is transected above the level of the motoneurons so that the latter are not damaged, then, with the descending connections from higher motor centers broken because of this *higher motor lesion,* two distinct changes in motor behavior develop. Immediately upon the making of such a section, a diminution or elimination of motor reflexes occurs, and a lessened muscle tone, the state of *spinal shock.* A gradual recovery of tone and reflexes takes place over a period of minutes or months, depending on the phylogenetic position of the animal. The shock is not due to the immediate effects of the lesion, i.e., a possible stimulation produced in cutting across nerve fibers. This was shown by Sherrington when, after the recovery from spinal shock, a second cut was made below the level of the first and no further period of shock ensued. The degree of spinal shock is more profound and the recovery time more prolonged, the higher the animal is in the phylogenetic series. In the frog, much of the reflex function returns within five minutes. In the dog and cat, reflexes begin to return within several hours, full recovery taking several weeks. In the monkey, months are required for recovery, and in man it may take many months before even a limited recovery of reflexes. These differences in the time course of recovery from spinal shock in the different species are due to *encephalization,* i.e., the greater dependence of the lower motor centers upon higher motor control mechanisms present in the brain of the higher species.

Not only are the somatic reflexes depressed during spinal shock, but one finds a similar depression of the visceral reflexes. Some of these higher visceral control mechanisms will be outlined later in the chapter, but because it is an important example of reflex loss, mention is made here of the effect of spinal shock on urinary bladder function. A reflex emptying of the bladder occurs when it is filled by urine to a certain level and stretch receptors in the musculature of the wall are activated. The reflex control mechanism is localized within the lower spinal cord but controlled in part by higher centers. In the normal animal, voluntary inhibition of this reflex level by upper brain mechanisms is possible, so that other ongoing behavior is not interfered with by the requirements of this function. Upon spinal cord transection, the upper influences are interrupted and the threshold for the local spinal reflex becomes greatly elevated. In such an event, the bladder may attain a considerable size before the elevated threshold of reflex micturition is reached and bladder emptying is finally activated. For the preservation of spinal animals and in the clinical management of human spinal cases, therefore, either the bladder must be drained or the urine must be manually expressed. Such procedure is continued for some days for animals, and for weeks for man, until finally excitability increases and reflex emptying occurs at moderate filling volumes. This is the *automatic bladder* condition. Other visceral reflex mechanisms, those of vasomotion and sweating, are also depressed and gradually recover after spinal cord section.

Spinal shock appears to be due to the removal of upper influences which act on lower motor mechanisms in a mainly excitatory fashion. Removal of such excitation results therefore in a reduced level of a local excitatory state in the lower motor centers and a depression of spinal reflexes and tone. Possibly, spinal shock may also be due to the removal of a local inhibition of reflex control. In this view, loss of upper control results in an excessive inhibition within the lower motor mechanisms.

Later in the course of recovery from spinal shock, somatic and visceral reflexes may increase in their excitability far beyond normal levels, so that a mild stimulation can excite a very great response, a *mass discharge.* Profuse sweating, flushing, urination, defecation, and flexor or extensor motor spasms of the limbs are seen during mass discharges in spinal man. It is possible that sprouting of local entering afferent fibers and interneurons acting on motoneurons is responsible for these later manifestations of exaggerated neural activities. Or the properties of the remaining cells may have changed so that their thresholds to activations are abnormally low.

THE FLEXOR REFLEX AND SOME GENERAL PROPERTIES OF REFLEX POOLS

The flexor reflex acts to protect against a potential trauma caused by a painful stimulation, i.e., a nociceptive stimulation. Stimulation of the skin of the limb causes a contraction of the limb muscles with a movement of the limb, or even the entire body, away from the stimulation. Flexor reflexes are widespread, engaging many muscles of a limb, and if they are strong enough, a spread to other limbs occurs. For example, a moderate nociceptive stimulation of the lower foot may cause only the foot to be withdrawn. If a stronger stimulation is used, the lower leg is also flexed. With still stronger stimulation, the upper leg is flexed as well. Furthermore, other limbs may also enter into the reflex when very strong excitation is used. The *Babinski reflex,* routinely looked for during clinical examination, is usually in adults a pathological flexor withdrawal to nociceptive stimulation. The bottom of the foot is stroked along the outer edge, and normally the response is a downward movement of the big toe (plantar flexion). In the Babinski reflex, an upward movement of the big toe takes place, often accompanied by fanning of the other toes. If reflex excitability is high, the foot may be everted from the site of stimulation and the whole limb withdrawn. The presence of the Babinski reflex has been associated with an upper motor lesion.

When a brief electrical stimulus is applied to the central end of a cut muscle nerve, a brief reflex contraction is induced in other flexor muscles of that limb. The extensive distribution of response of flexor muscles in response to such stimulation of the nerve of one muscle is brought about by widespread synaptic connection found within the spinal cord. Two classes of afferent fiber synaptic termination and distribution in the cord were shown by Ramón y Cajal (1952). In one type (Fig. 5-1A), afferent fibers synapse directly on the motoneurons and have a restricted distribution to the peripheral musculature. Another class of afferent fibers (Fig. 5-1B) connect with interneurons, which in turn have widespread branching synaptic contact with other interneurons, and eventually motoneurons. The stretch reflexes are subserved by the *circumscribed* type of connection and the flexor withdrawal reflex by the widespread or *diffuse* type.

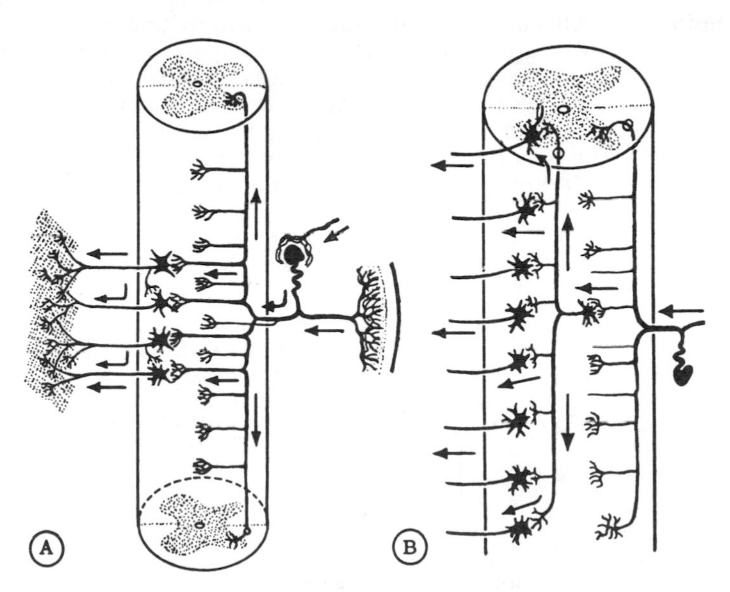

FIGURE 5-1. Diffuse and circumscribed distribution of afferent fibers terminating in the spinal cord. In A, a circumscribed monosynaptic type of termination is shown. The fibers synapse directly upon the motor horn cells, usually within the spinal cord segment of entry of those fibers. In B, a diffuse transmission within the central nervous system is shown. Sensory fibers terminate upon interneurons within the spinal cord which, by multiple branches, engage a large number of motoneurons that in turn are distributed to many muscles of the peripheral musculature. (From S. Ramón y Cajal. *Histologie du système nerveux.* Madrid: Consejo Superior de Investigaciones Cientificas, 1952. Vol. 1, pp. 531, 532.)

The fibers make contact with other neurons by enlarged endings, the synaptic endings (boutons) present on nearly all neurons – in particular the motoneurons. They are numerous over the membrane surface including both the somas and the dendrites. A number of different neurons have boutons terminating on the motoneurons to excite or inhibit their discharge, as is shown by interaction studies made by stimulating two afferent nerves entering the spinal cord. These are commonly selected from the various nerves innervating the limb muscles or the nerves which are purely sensory (those innervating the skin or limb joints). The nerves are cut distally, mounted on electrodes, and kept protected from drying or being electrically shorted by body fluids. Each afferent nerve has an input which eventually can activate a number of the motoneurons present in the available pool of motoneurons innervating the peripheral limb musculature to give rise to a flexor reflex. The type of effect observed depends on whether a weak or strong stimulus is used to excite the afferent nerves. With weak stimuli, not all the cells of the motoneuron pool are excited sufficiently to give action potentials. The cells which do not fire at low stimulation strength may be *subliminally excited.* When two afferent nerves are stimulated simultaneously with shocks of low intensity, so that each produces subliminal excitation in a group common to each, as shown in Figure 5-2A, there is a *summation* of subliminal excitability and a response in that group of neurons which would not ordinarily be excited. In other words, the result is a discharge of more than the sum of each given separately. This is called *facilitation.*

If the shocks to the afferent nerve inputs are each made very strong, the group of cells previously subliminally excited are then excited to discharge (Fig. 5-2B). In this case the response to stimulation of two afferent inputs together is less than the sum of each of the responses given separately. This demonstrates the phenomenon of *occlusion.* The presence of multiple synapses over the motoneuron gives rise to different varieties of interaction. When two inputs from different sources are arranged to occur at different times, there are *spatiotemporal* interactions. Two stimuli presented to the same afferent nerve at different times demonstrate *temporal*

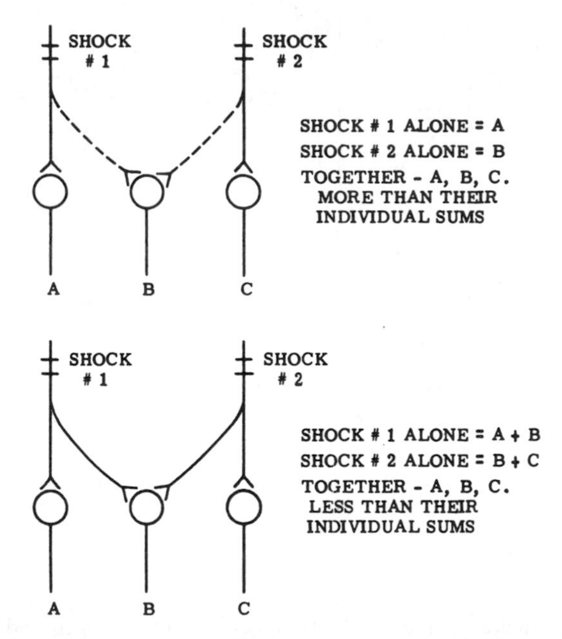

FIGURE 5-2. Facilitation and occlusion. Excitation of one sensory nerve, #1, excites a group of neurons, A; and a collateral branch, shown by the dashed line, subliminally excites a group represented by B. Similarly, excitation of sensory nerve #2 excites neuron group C and subliminally group B. The sum of the two shocks presented separately is A + C. If #1 and #2 are excited together, the two subliminal excitations of group B summate, and the resulting discharge of A, B, and C together is more than the sum of the individual shocks. In the lower half of the illustration, occlusion is shown. A shock to #1 excites neuron groups A and B. A shock to #2 excites groups B and C. A shock to #1 and #2 together, because of their common termination on B, is less than the sum of their individual responses – occlusion.

interaction. If the strength of the first shock is low enough in magnitude so that only relatively few cells are fired, with many subliminally excited, later test shocks may be facilitated *(temporal facilitation).* An enduring excitatory change in the motoneuron membrane explains the central excitation state. If stronger conditioning shocks are used, an *inhibition* may persist for up to several hundred milliseconds. The long-lasting inhibitions are brought about in part by an effect on the postsynaptic neuronal membrane and in part by long-enduring changes at the presynaptic terminations, as will be noted later on.

STRETCH MYOTATIC REFLEXES – THE MONOSYNAPTIC REFLEX SYSTEM

A sudden stretch of a limb muscle gives rise to a reflex contraction of that same muscle. The classic example of such a stretch, or *myotatic,* reflex is the *patellar reflex,* or *knee jerk.* When the patella is given a light tap, the quadriceps muscles are abruptly stretched, thereby exciting stretch receptors and a reflex extension of the lower leg. The reflex nature of the response is shown by cutting the appropriate spinal cord dorsal roots through which the afferent fibers from that muscle enter the spinal cord, in which circumstance the sudden stretch of the muscle no longer elicits a reflex contraction. The stretch reflexes are also elicited by means of a maintained stretch on the muscle and are shown by the resulting tension increase. The reflex nature of the stretch reflex is shown by the sudden drop of tension which occurs when the dorsal roots are cut.

Muscles usually operate in pairs for the movement of a limb, so that when one set of muscles, the *agonistic,* is contracting, the opposing *(antagonistic)* muscles are relaxing. This coordinated opposed grouping of agonist and antagonist muscles is called *reciprocal innervation.* In order to demonstrate the reflex properties of a reciprocally innervated set of two such muscle groups, e.g., of biceps and triceps muscles, strings are tied to the cut ends of the tendons of those muscles so that stretch can be placed on them and tension recorded. A stretch of one muscle elicits a reflex contraction in that same muscle, the other muscle showing *inhibition* by its relaxation. Reflex inhibition is mediated by inhibitory discharges on the motoneurons innervating the antagonistic muscle of the reciprocal pair. If the opposite muscle were similarly stretched, the other muscle would show inhibition again by inhibitory influences on its motoneurons in the CNS.

Both the fast stretch reflexes excited by quick stretch of a muscle and the slow myotatic reflexes produced by a maintained stretch are excited by *spindle receptors* present among the skeletal muscle fibers. A more detailed description of these muscle receptors will be given in a later section of this chapter, but it is noted here that just a small elongation of the spindle receptors is sufficient to excite them. The afferent fibers from the spindles enter the cord with their synaptic terminal endings directly on the motoneurons. For stretch receptors the distribution of the entering fibers is of the circumscribed type shown in Figure 5-1A. Only one synapse between afferents and motoneurons is involved. To study the cord connectivity, the dorsal root fibers are electrically stimulated with a brief impulse in order to elicit a reflex discharge recorded in the ventral roots, as shown in Figure 5-3.

The central end of the cut dorsal root is placed on stimulating electrodes, and the central end of the cut ventral root of that segment on recording electrodes. After the artifact signifying the time of stimulation and a latency of approximately 1.5 msec, a spikelike response known as the *monosynaptic response* is recorded in the ventral root (Fig. 5-3). This is followed by an irregular discharge lasting approximately 15 msec, the *multisynaptic* or *polysynaptic discharge.* As indicated in Figure 5-3, the monosynaptic response is excited by the fibers synapsing directly

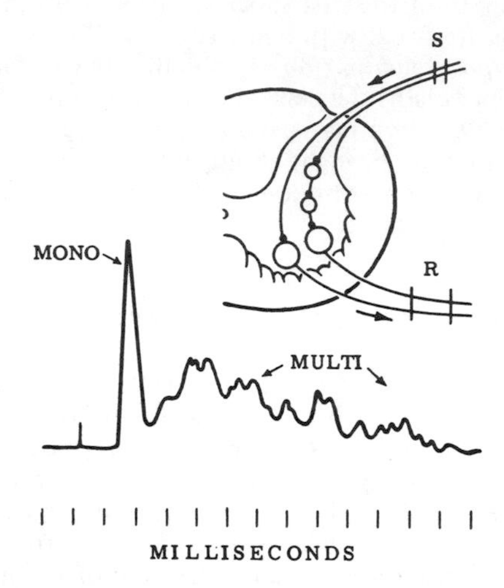

FIGURE 5-3. Electrical reflexes of the spinal cord. The diagram in the upper portion of this figure shows a stimulus (S) of afferent fibers in the dorsal root which terminate directly on the motoneurons to give the monosynaptic (mono) responses recorded in the ventral root; terminations on interneurons give rise to multisynaptic (multi) discharges. Electrical reflex response patterns are shown in the lower part of the figure.

on the motoneurons, the multisynaptic response after a number of intervening neurons are activated before terminating on motoneurons.

Evidence for the concept that the monosynaptic response involves only one synapse has been obtained from the studies initiated by Lorente de Nó on the oculomotor nucleus and by Lloyd, Renshaw, Eccles, and others on the spinal cord. Such studies have shown that part of the monosynaptic latency seen after dorsal root stimulation is due to the time for conduction of the impulses in the afferent fiber branches, referred to as the *reflexomotor collaterals* by Ramón y Cajal, which terminate directly on the motoneurons; part is due to conduction time in efferent fibers; and part is due to processes underlying synaptic transmission at the motoneurons.

As previously noted, synaptic transmission is affected by a large number of boutons ending over the surface of the motoneurons. Because of deficiencies in the usual staining techniques, bouton endings cannot always be found along the length of dendrite branches of the motoneurons. However, electron microscope studies have shown that they extend all along the dendrites to their furthermost branches. Synapses on the motoneurons are made not only by the locally supplied afferent inputs, as well as from numerous close-by interneurons, but in addition by fibers descending in tracts from the upper brain regions and brain stem. In primates, including man, some of the upper tract fibers synapse directly on the motoneurons besides synapsing on interneurons which in turn synapse on motoneurons. If a brief electrical impulse is delivered through the tips of the needle electrodes inserted into a population of motoneurons, a double response is recorded in the ventral root (Fig. 5-4). The first spike (m) can be accounted for as a direct excitation of the motoneurons or of their axons. The m response is followed after an additional latency of 0.5 to 1.0 msec by an s response due to excitation of afferent fibers which synapse on motoneurons. The time between the m and the s responses cannot be reduced

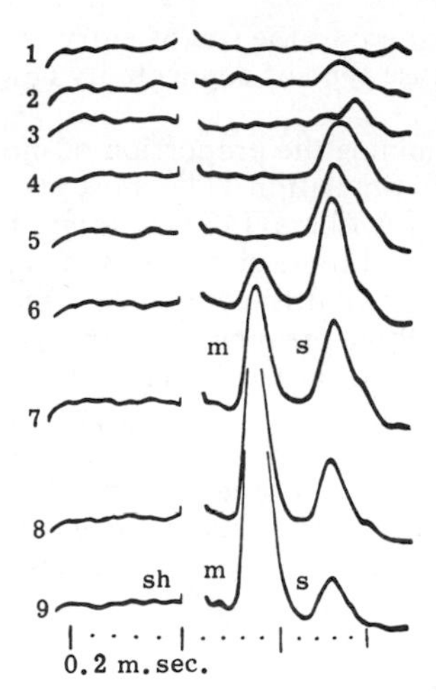

FIGURE 5-4. Focal stimulation within the central nervous system. With a stimulating needle electrode in the oculomotor nucleus the motoneurons are excited, as are afferent branches terminating on another group of motoneurons. As the strength of the shock is successively increased (shown in the traces from above downward), a small longer latency response is seen, corresponding to a monosynaptic response with synaptic (s) delay. As the strength is further increased, an earlier m response is seen which is due to a direct excitation of the motoneurons. Notice that as the size of the direct response increases, it occludes and reduces the synaptic response. Even at the strongest shock levels synaptic delay is no shorter than 0.5 msec. Sh is the shock artifact. (After R. Lorente de Nó. *J. Neurophysiol.* 2:409, 1939.)

below 0.5 msec no matter how strong the stimulation. The synaptic delay time is therefore shown to range from 0.5 to 1.0 msec.

As the strength of a dorsal root stimulus is gradually increased, more dorsal root fibers and more synapses on the motoneuron cells are excited. The size of the resulting monosynaptic response in the ventral root increases in sigmoidal fashion with increasing stimulus strength. When the stimulus is weak, there is a subliminal activation of a large number of *fringe* cells. When the stimulus strength is increased, a larger number of neurons are fired and the fringe group is smaller. However, even with a maximal response a significant fringe group exists, as is shown by the phenomenon of *post-tetanic potentiation* (PTP). The root is tetanically stimulated for several minutes, and monosynaptic responses are thereafter increased above control levels and then gradually return to control levels within a few minutes. The mechanism of synaptic transmission will be discussed later in this chapter. However, to anticipate that discussion, the chemical mediator involved in synaptic excitation appears to be mobilized by the tetanic activation with a temporarily increased probability of release of transmitter.

The more prolonged electrical discharge which appears after the monosynaptic response, the multisynaptic response, is believed to be due to repetitive activity in chains of interneurons. These interneurons terminate onto motoneurons. The motoneurons which give rise to multisynaptic discharges are distributed mainly to flexor muscles, while the monosynaptic response is restricted to the same muscle in which the spindle afferents originate. The response is seen in the ventral root of the spinal cord segment in which the stimulated afferent fibers enter, but the multisynaptic discharges are recorded from the ventral roots of a number of segments up and down the spinal cord lateral to the site of entry of the stimulated dorsal root. This further shows the diffuse type of connectivity (Fig. 5-1B) involved in the multisynaptic type of discharge.

A variable factor determining the proportion of monosynaptic to multisynaptic responses is the strength of stimulation. The most excitable afferent fibers, the largest-diameter subgroup of A fibers (IA) take origin from the spindle stretch receptors within the muscles, and these give rise to the monosynaptic response. Fibers of smaller diameter give rise to multisynaptic discharges. These fibers originate from receptors of various types. By depressing the activity of the spinal cord, either with barbiturate or by using weak shocks to afferent nerve fibers, one can obtain monosynaptic responses in relative isolation. If the nerve selected for afferent stimulation is purely a skin sensory nerve (e.g., the sural nerve), the reflex response in the ventral root is typically a multisynaptic discharge with no evidence of monosynaptic discharge. When studying synaptic processes in motoneurons it is advantageous to arrange the experiment so that monosynaptic responses alone are obtained (see below).

FACILITATION AND INHIBITION

Muscles acting at a joint in concert, for example, the lateral and medial gastrocnemius muscles, are *synergists.* Facilitation may be clearly demonstrated by using the monosynaptic response obtained by stimulation of synergistic nerves. The central end of the medial gastrocnemius nerve is excited by a conditioning stimulus, and either concurrently or at various intervals thereafter a test stimulus is delivered to the central end of the lateral gastrocnemius nerve. The conditioning stimulus may be adjusted so that only a few motoneurons fire, and the effect is a just liminal excitation. When stimuli are delivered concurrently to both the medial and lateral gastrocnemius nerves, the test response may be doubled. As the time between the two volleys is gradually increased over a period of approximately 14 msec, the facilitating effect gradually falls to zero (Fig. 5-5A).

Because facilitation is seen when the two nerves are stimulated concurrently, and there is no time for an intervening synapse, this is termed a *direct facilitation.* Afferent fibers pass to the synergistic motoneurons, to cause the latter to be subliminally excited. With concurrent delivery of stimuli from two synergistic nerves, the addition of subliminal excitatory activity in the fringe group common to those synergists results in firing (cf. Fig. 5-2A). Reversing the order of test and conditioning stimulations presented to the synergistic pair of nerves gives rise to the same pattern of facilitation.

Because of the simplicity of the monosynaptic response, i.e., one presynaptic fiber and one postsynaptic element, a similar direct system for inhibition on reciprocally paired inputs might be expected. To a considerable extent this is the case, with one reservation to be more fully discussed in the next section: an *inhibitory interneuron* is interposed. To show inhibition, a reciprocal pair of muscle nerves is selected. The central end of one nerve (e.g., that to the quadriceps) is stimulated to give rise to a monosynaptic response in the ventral root. This test monosynaptic response is preceded by a conditioning shock to the central end of the nerve of its antagonistic muscle (in this case the semitendinosus). When conditioning and test stimuli are delivered concurrently, very little inhibitory effect is seen, but with a small increase in the interval, the test response which is delivered later becomes diminished (Fig. 5-5B). The resulting curve of inhibition found with different times between the two stimuli shows a peak effect at an interval of 0.5 msec between the conditioning and test volley. As the time between the two stimuli is further increased, the test response gradually regains its normal size within

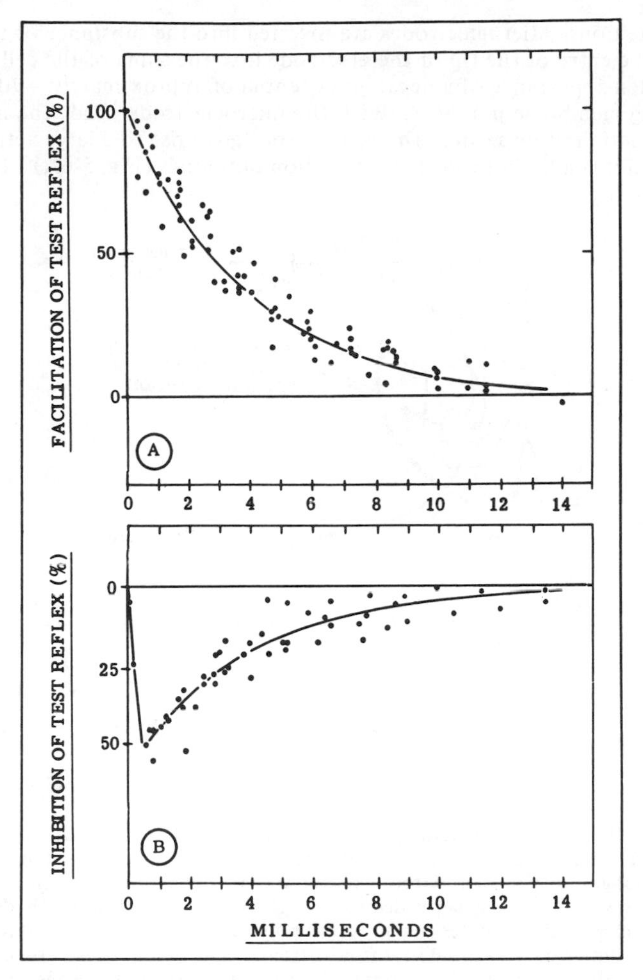

FIGURE 5-5. Direct facilitation and "direct" inhibition. The monosynaptic response is used as a test response and is preceded at the time intervals shown on the abscissa by a conditioning shock. In A the conditioning shock is a volley in a synergistic nerve. In B the conditioning volley is to an antagonistic nerve. Both facilitatory and inhibitory effects decline gradually to control levels in approximately 15 msec. (From Lloyd, 1946.)

approximately 15 msec. An explanation of facilitation and inhibition requires an understanding of the synaptic mechanisms involved.

MECHANISMS UNDERLYING SYNAPTIC TRANSMISSION

The preceding discussion concerned a population of neurons albeit of relatively simple form as shown by the monosynaptic responses. A deeper analysis requires events in individual motoneurons to be recorded.

Microelectrodes were first successfully used for this purpose by Eccles (1964, 1965) and his colleagues to study the mechanism of synaptic transmission in spinal cord motoneurons. Microelectrodes are inserted into the substance of the spinal cord until the entry of the tip of the electrode into the soma of the cell is indicated by the sudden appearance of a negative potential of approximately −70 mv. This is the resting membrane potential. With the microelectrode inside the motoneuron soma, stimulation of its axon in the ventral root gives rise to a large action potential with overshoot similar in form to other action potentials (Fig. 5-6A). The latency

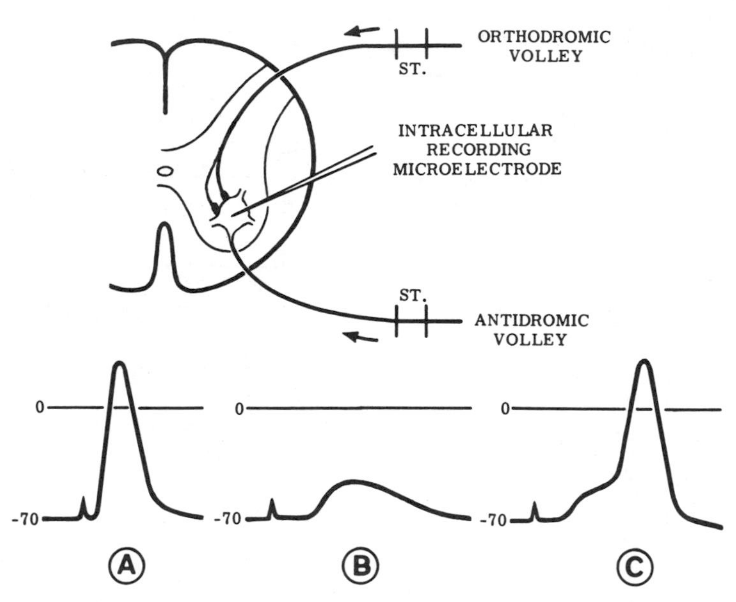

FIGURE 5-6. Microelectrode recording of responses from motoneurons. A recording microelectrode is shown inside a motoneuron in the top diagram. Stimulation of the cell's fiber in the ventral root gives rise to an antidromic response (A). Stimulation orthodromically with a weak shock gives rise, after a synaptic delay, to an EPSP (B). Stimulation orthodromically with a stronger shock gives rise to EPSP and an action potential (C). The response recorded in the soma upon excitation of the ventral root results from conduction in the reverse direction and it is therefore called an *antidromic response* (Gr. *dromos,* "running"). This is distinguished from propagation in the usual direction, i.e., *orthodromic* excitation via afferent nerve excitation.

of the antidromic response is very short, as expected from the short conduction distance between ventral root and soma. If the dorsal root is stimulated with a weak excitation, the response to this orthodromic activation (recorded with a microelectrode) is a depolarizing potential of the motoneuron membrane appearing with a latency of 0.3 to 0.5 msec (Fig. 5-6B). This *excitatory postsynaptic potential* (EPSP) lasts approximately 15 msec and has a form similar to the end-plate potential (EPP) recorded in the muscle (Chap. 3A). It increases in size as the volley to the afferent fibers is increased in strength. When the orthodromic volley and the EPSP become large enough, a critical level of depolarization is reached and an action potential is then excited in the motoneuron (Fig. 5-6C). This propagates down along its axon to excite the muscles innervated by that motoneuron.

The EPSP size is related to the number of presynaptic fibers activating bouton endings on the membrane of the motoneuron. The depolarization produced by activity of each bouton adds up to give rise to the EPSP. The EPSP accounts for the subliminal excitatory changes found by interaction studies previously discussed. The EPSP produced by a conditioning volley can summate with a second EPSP produced by a volley in synergistic nerves. When both afferent nerves are stimulated together, the EPSP becomes larger until finally a critical size is attained and an action potential is fired. The facilitating effect shown by the summation of EPSP's over a period of approximately 15 msec is also the time course of the duration of the EPSP itself (Fig. 5-7).

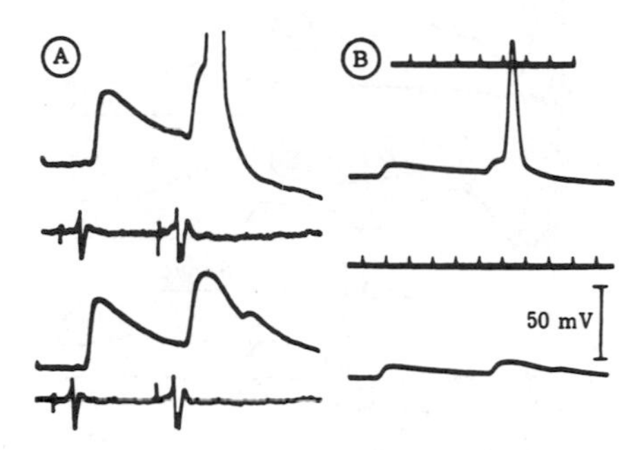

FIGURE 5-7. Synaptic potentials of motoneurons. Two excitatory postsynaptic potentials (EPSP) recorded from inside a motoneuron in response to two successive weak shocks are shown in A at high and in B at lower amplification. The EPSP's are smaller and longer-lasting depolarizing potentials. The first EPSP of the upper traces is ineffective, but the second, occurring a short time afterward, excites an action potential. In the lower trace, EPSP's only are seen. A small shift in excitability caused the second EPSP to be ineffective. (From J. C. Eccles. *Neurophysiological Basis of Mind.* London: Oxford University Press, 1953. P. 133.)

If an antagonistic nerve to the motoneuron is stimulated, one sees a smaller hyperpolarizing type of response, an *inhibitory postsynaptic potential* (IPSP) in the cell (Fig. 5-8). The effect of the inhibitory excitation is to reduce membrane potential below the critical level for firing and so block excitation of the cell. During inhibitory action, an EPSP may remain present while the propagated spike is blocked, the total level of depolarization required for firing being reduced below the critical level (Fig. 5-8). The IPSP also increases the ionic conductance of the membrane, which likewise acts to reduce the effectiveness of an EPSP to excite an action potential.

The delay found for the IPSP is greater than that for an EPSP. As already indicated, there is an additional intercalated interneuron on the inhibitory line. Branches from the excitatory fibers send collaterals onto small inhibitory cells in the intermediate part of the spinal cord. These cells, in turn, have short axons which synapse on motoneurons to give rise to an IPSP in the motoneurons. The synaptic delay caused by transmission through these intervening cells accounts for the maximum inhibition seen at 0.5 msec after the conditioning stimuli in the interaction studies shown in Figure 5-5B.

Excitation of an action potential in the motoneuron does not normally occur on the cell body membrane. Rather, evidence from a variety of cells indicates that the propagated action potential originates from the axon just distal to the cell body, at the nonmyelinated *initial segment* (IS). The soma membrane is actually less excitable than the initial segment portion of the cell. The lower excitability of the

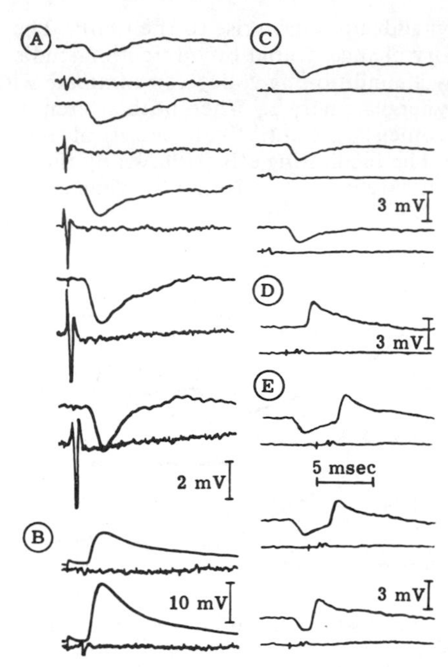

FIGURE 5-8. Inhibitory postsynaptic potentials. The hyperpolarizing inhibitory postsynaptic potentials (IPSP) recorded with a microelectrode from within motoneurons are shown at increasing strengths of stimulation in A. The action potential of the afferents in the dorsal root volley is shown on the lower traces. As the strength is increased, the size of the IPSP increases. The IPSP's are to be compared with EPSP's excited in these same motoneurons (B). In C, inhibitory volleys are set up, and then (D) an excitatory volley; finally, in E, an excitatory volley follows soon after an inhibitory volley. The IPSP decreases the resultant depolarization brought about by the EPSP to below critical level for firing. (From J. C. Eccles. *Neurophysiological Basis of Mind.* London: Oxford University Press, 1953. P. 155.)

soma compared with the initial segment is shown by the failure sometimes seen with antidromic invasion of the soma where one is left with a small *initial segment potential.* Other evidence was obtained from the directly visualized stretch receptors of crustaceans (Chap. 3A). With recording electrodes placed successively at the soma and at the initial segment, propagated responses were seen to be initiated first at the initial segment. The sequence of events during normal orthodromic excitation is therefore as follows: An EPSP excited in the soma by synaptic activity results in a flow of current outward through the initial segment of the cell; excitation of an action potential follows, which then propagates down the axon to the muscle, and at the same time back into the cell body to give rise to a *soma-dendrite* action potential.

Excitation at the initial segment is reasonable if it is considered that this is a common site on the cell where the depolarizations of the EPSP are tallied up. If a large number of excitatory synapses are fired, a larger flow of outward current passes through the membrane of the initial segment. If at the same time, or just previously, IPSP activity is excited, the resulting hyperpolarization of the initial segment keeps the initial segment below the critical level for the firing of a propagated action potential.

SYNAPTIC TRANSMITTERS

There is an irreducible latency between activation of the afferent nerve terminals and the beginning of an EPSP. This is taken to be evidence that a transmitter substance is released from the terminals to affect the postsynaptic membrane rather than electrical currents from the presynaptic terminals effecting synaptic transmission. However, the transmitter substances giving rise to the EPSP and IPSP are as yet unknown. The transmitters for EPSP and IPSP are different because of the different electrical actions they produce on the postsynaptic membrane of the motoneurons, depolarization and hyperpolarization respectively.

The ionic permeability increase during an EPSP is generalized, with mainly Na^+ and K^+ carrying the membrane toward a zero equilibrium potential. The inhibitory transmitter substance, on the other hand, produces a more selective permeability increase to K^+ and to Cl^- with a resultant equilibrium potential at a higher membrane potential, i.e., from −70 to −80 mv. These different ionic permeabilities during EPSP and IPSP activation were shown by the use of double-barreled microelectrodes inserted into motoneurons. One barrel recorded the EPSP or IPSP; the other was used to pass different ions into the cell. By this means it was shown that the inhibitory transmitter permitted only ions of smaller overall size to pass from the soma, apparently because a smaller pore opening is produced in the membrane as compared to the wider pores opened by the excitatory transmitter. Another interesting difference between the action of the transmitter agents is that the convulsant agent strychnine will block the IPSP without affecting the EPSP. Injection of substances into the blood which could augment or block a cholinergic synapse, e.g., ACh, was ineffective on the EPSP and IPSP. That this was not due to a barrier preventing their entry into the CNS (a blood-brain barrier) was indicated by the effect of cholinergic and cholinergic-blocking agents on a special class of neurons in the spinal cord, the Renshaw cells, to be described below. Some recent candidates for transmitter substances are amino acids normally found in neurons. Glutamic acid has been considered an excitatory transmitter, and gamma aminobutyric acid (GABA) and glycine are thought to be inhibitory transmitters. Recent evidence favors glycine (Aprison and Werman, 1968).

RENSHAW CELL INHIBITION

A type of inhibitory interneuron has been found related to the motoneurons within the ventral horn portion of the spinal cord, the *Renshaw cells* (Fig. 5-9). Collateral fibers from the motoneurons synapse on the Renshaw cells, which in turn synapse onto a number of motoneurons in their immediate vicinity to give rise to a series of repetitive IPSP's in the motoneurons. The repetitive discharge of the Renshaw cell therefore results in a long-maintained inhibitory effect on the motoneurons.

The functional significance of the inhibition brought about by Renshaw cell activity is as yet not completely determined. It may function to sharpen the effect of those motoneurons which have fired by inhibiting other motoneurons which have not been activated. Apparently its effects are greatest on the tonically active motoneurons which supply muscles important in maintaining posture for long periods of time.

The Renshaw cells have further theoretical significance. The motoneuron collateral fiber which ends upon the Renshaw cell is split from the main axon, which releases acetylcholine at the neuromuscular junction. This suggested that ACh is also the transmitter agent acting on the Renshaw cells. Injection of substances into the blood stream which are known to excite or block ACh transmission at the

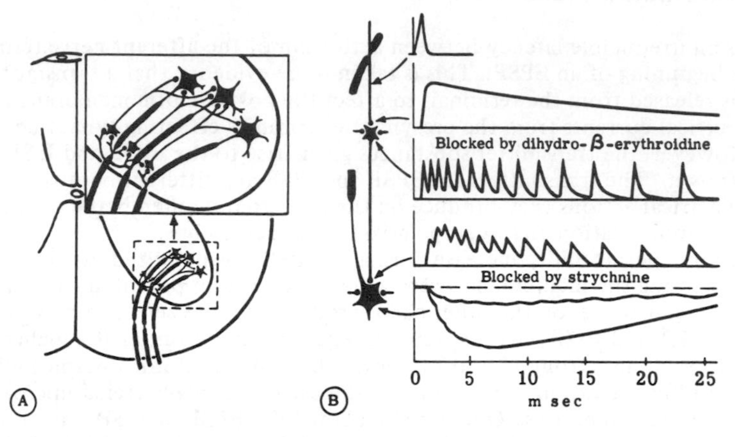

FIGURE 5-9. Renshaw cell activity of the spinal cord. In A, the ventral horn of the spinal cord is shown, and in the inset an enlarged view. From the axons of three motoneurons, collaterals can be seen taking origin from the first node. These synapse on smaller cells (Renshaw cells), which in turn send axons to the surrounding motoneurons and branches back to the same motoneuron. In B the collateral from the motor horn cell axon is shown at the top synapsing on a Renshaw cell. The Renshaw cell, when it discharges, does so at a very rapid rate and gives rise to a series of discharges which can be blocked by dihydro-β-erythroidine. The Renshaw cell in turn synapses on motoneurons, and at this point inhibitory mediator substance is released in successive summated IPSP's. This IPSP can be clocked by strychnine. (From J. C. Eccles et al. *J. Physiol.* 126:542 and 557, 1954.)

neuromuscular junction also excites or blocks synaptic activation of the Renshaw cells. The substance dihydro-β-erythroidine, which can pass the blood-brain barrier and which has a curariform activity similar to that of d-tubocurarine, was shown to block Renshaw cell activity (Fig. 5-9). Conversely, ACh and cholinomimetic substances are excitatory, and they prolong the discharge. The findings are taken as evidence that the collateral from motoneurons synapsing on the Renshaw cell releases ACh as the synaptic transmitter.

THE GAMMA SYSTEM

As has already been noted in Chapter 3A, the muscle spindle receptor is excited by stretch to give rise to volleys passing into the spinal cord via the dorsal roots and reflexomotor collaterals of Ramón y Cajal. These end directly on the motoneurons to excite them. The large-diameter afferent fiber looped around the central portion of the spindle receptor is the *primary receptor,* or *annulospiral ending.* Elongation of the spindle deforms the sensory terminations, thereby exciting afferent discharges in its fibers with a rate proportioned to the degree of stretch (Chap. 3A).

The spindle receptor has a complex structure. Small motor fibers synapse on the muscular parts of the spindle receptor present on either side of the central receptor portion. The small motor fibers fall into the gamma range of A fibers and are therefore referred to as *gamma motor* or *fusimotor* fibers. When the gamma motor fibers are fired, the muscular portions of the spindle receptor contract and stretch out the central portion of the receptor. This has the same effect as a pull on the whole muscle, and the spindle receptor is thereby excited. With a gamma discharge present, a smaller stretch of the spindle is required to excite an efferent discharge. When the gamma system is less active, or if it is stopped altogether, a greater stretch must be placed on the spindle to cause it to discharge. Because the spindles are anatomically in parallel with the contracting muscles, as shown in Figure 5-10A, a twitch which contracts the extrafusal muscle (muscle fibers other than the spindles are referred to as extrafusal) will cause a relaxation of the stretch receptor by *unloading* the tension on it. This relaxation of the stretch receptor during a twitch contraction results in a diminution of the discharge in the spindle afferent fibers known as the *pause.* A fiber isolated from among the fibers of the dorsal roots can be identified as coming from a spindle stretch receptor by the appearance of the pause during a muscle contraction. Stimulation of gamma activity, if great enough, may, however, overcome the pause and cause the stretch receptor to discharge at a fairly high rate during the twitch response.

The gamma system is an example of a feedback mechanism. Activity in the CNS by exciting or inhibiting synapses on the cell bodies of the gamma fibers can modify their discharge, and in turn the sensitivity of the spindle receptor. The input from the spindle receptor has an action to modify not only motoneurons but the gamma cells as well, thereby completing the feedback loop. An excessive involvement of this feedback system appears to underlie the pathological state of spasticity, possibly through an action of higher brain centers on gamma cell bodies to increase

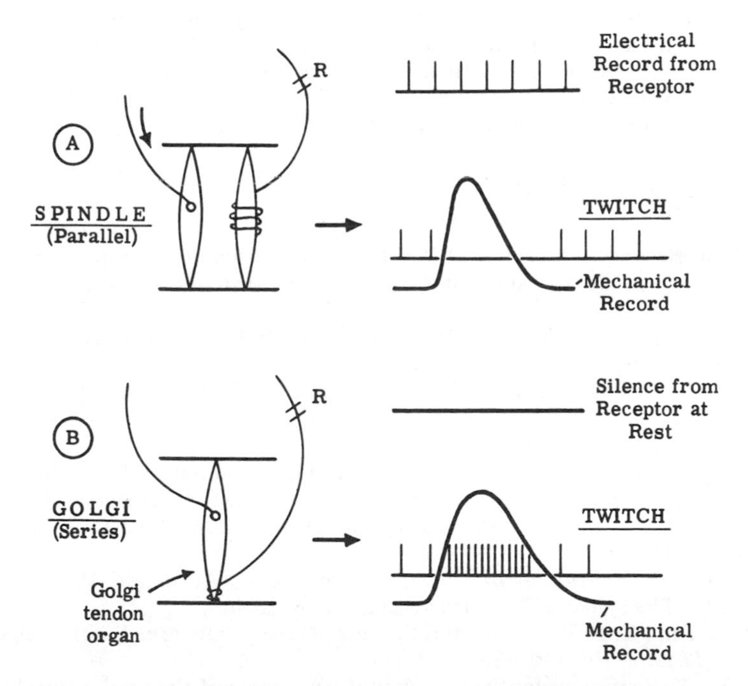

FIGURE 5-10. Muscle receptor organs. Two main types of muscle receptor organs are pictured, the spindle stretch receptor (A) and the Golgi tendon organ (B). The spindle is placed in parallel with the muscle fibers. When the muscle is caused to contract, the electrical discharge recorded from a spindle receptor shows a decrease, the "pause." The Golgi tendon organ is in series with skeletal muscles, and contraction of the muscle during a twitch causes a speeding up of the discharge rate during the time of the muscle contraction.

their excitability. A condition of exaggerated hypertonus and increased reflex excitability known as *decerebrate rigidity* is produced by transecting the brain stem of a dog or cat at the level of the midbrain between the colliculi. The position of the animal in decerebrate rigidity is one of extreme hyperextension of the limbs. Usually the head is also arched back and the tail is elevated; this condition is known as *opisthotonus.* The whole posture in this state has been called a caricature of standing. If in such a decerebrate preparation the dorsal roots are cut, the limb immediately becomes flaccid, proof that the decerebrate state is produced by reflex hyperactivity of stretch reflexes. The inference that this state is due to an augmented gamma motor activity was shown by the early appearance of increased discharges in recordings taken from thin fiber slips from the ventral roots and identified as gamma motor fiber discharges.

If the leg of an animal showing a decerebrate rigidity is forcibly flexed, it resists the applied flexion until a certain point is reached when the resistance seems suddenly to melt away. This phenomenon is known as the *knife-clasp reflex* or the *lengthening reaction.* It is believed that, at this level of stretch, sensory afferents from the tendon regions of muscles, the *Golgi tendon organs,* reach threshold, and these have an inhibitory effect on the motoneurons. The Golgi tendon organs, unlike the spindles, are placed in a series with respect to the skeletal muscle (Fig. 5-10B). A single sensory nerve fiber isolated in the dorsal root can be identified as coming from a Golgi tendon organ if it shows an increased repetitive activity during a twitch of the muscle (Fig. 5-10). Another identifying characteristic is the lack of effect of gamma motor activity on the discharge coming from the Golgi tendon organ.

Recently it has been discovered that boutons in the CNS are found terminating on other afferent terminals. By their action these *presynaptic endings* can inhibit the discharge of the afferent terminal on which they synapse. The relation of *presynaptic inhibition* to information transfer is at this time not well understood.

REFERENCES

Aprison, M. H., and R. Werman. A combined neurochemical and neurophysiological approach to identification of central nervous system transmitters. *Neurosci. Res.* 1:143–174, 1968.

Barker, D. (Ed.). *Symposium on Muscle Receptors.* Hong Kong: Hong Kong University Press, 1962.

Boyd, J. A., C. Eyzaguirre, P. B. C. Mathews, and G. Rushworth. *The Role of the Gamma System in Movement and Posture.* New York: Association for the Aid of Crippled Children, 1968.

Creed, R. S., D. Denny-Brown, J. C. Eccles, E. G. T. Liddell, and C. S. Sherrington. *Reflex Activity of the Spinal Cord.* London: Oxford University Press, 1932.

DeMyer, W. D. *Technique of the Neurologic Examination.* New York: McGraw-Hill, 1969.

Eccles, J. C. *The Physiology of Synapses.* New York: Academic, 1964.

Eccles, J. C. The synapse. *Sci. Amer.* 212:56–66, 1965.

Hunt, C. C., and E. R. Perl. Spinal reflex mechanisms concerned with skeletal muscle. *Physiol. Rev.* 40:538–579, 1960.

Leksell, L. The action potential and excitatory effects of the small ventral root fibers to skeletal muscle. *Acta Physiol. Scand.* 10 (Suppl. 31):1–83, 1945.

Liddell, E. G. T. *The Discovery of Reflexes.* New York: Oxford University Press, 1960.

Lloyd, D. P. C. Facilitation and inhibition of spinal motoneurons. *J. Neurophysiol.* 9:421–438, 1946.

McLennan, H. *Synaptic Transmission.* Philadelphia: Saunders, 1963.

Mountcastle, V. B. (Ed.). *Medical Physiology,* Vol II. St. Louis: Mosby, 1968.

Ochs, S. *Elements of Neurophysiology.* New York: Wiley, 1965.

Ruch, T. C., and H. D. Patton (Eds.). *Physiology and Biophysics,* 19th ed. Philadelphia: Saunders, 1965.

Sherrington, C. *The Integrative Action of the Nervous System.* Cambridge, Eng.: Cambridge University Press, 1906. Rev. reprint, New Haven, Conn.: Yale University Press, 1947.

Wiersma, C. A. G. Neural transmission in invertebrates. *Physiol. Rev.* 33:326–355, 1953.

Properties of the Cerebrum and Higher Sensory Function

Sidney Ochs

THE TERM *encephalization* refers to the increased power of command which the more headward parts of the central nervous system exert on the lower reflex mechanisms. An increased mass and complexity of brain structure are found in the higher phyla, along with an increased range of function. This does not mean, however, that the lower centers have declined in complexity. The cerebral cortex made its appearance later in phylogenetic development, and on this basis it has been taken to be the seat of intelligence. But it will be seen that subcortical centers are important as well in the exercise of higher functions.

Clinical observations indicating that certain functions could be localized within the cerebral cortex were confirmed when Fritsch and Hitzig in 1870 showed that electrical stimulation of certain regions of the cortex gave rise to motor responses (Chap. 8). Later, various other areas were found to receive sensory inputs. Other areas in the cortex, designated as associational, were believed to control higher functions (Chap. 9).

The concept of strictly separable motor and sensory centers in the cortex has been modified with the realization of the existing complex interrelations between sensory and motor centers. The term *sensorimotor* cortex is often used to stress this interrelation. We shall for didactic reasons treat them separately. The electrical properties of the cerebrum and their relation to its function – the changes during sleep and wakefulness and sensory reception in the cortex – will be discussed in this chapter and motor control in Chapter 8.

THE ELECTROENCEPHALOGRAM

In 1875 Caton first discovered a continuous electrical activity in the exposed cerebrum of animals. In 1929 Berger, using the string galvanometer, showed that similar electrical activity could be recorded through the intact human skull and that it changed with sleep or altered consciousness. The availability of improved electronic instrumentation since then has allowed the *electroencephalogram* (EEG) to become a valuable clinical tool in the study of epilepsy, brain tumor localization, and diagnosis of other neurological conditions. The study of EEG activity has also advanced general knowledge of brain function and has been used as a tool for showing brain changes related to higher function (see Chap. 9).

To record the EEG, electrodes are placed over the surface of the skull in a regular pattern using *monopolar* or *bipolar* recording. Monopolar recording is accomplished with an active probe or electrode over an active site with an *indifferent electrode* placed on inactive tissue at a distance. The ear lobes are usually used as the site for the indifferent electrode, because electrical activity from the heart (the electrocardiogram) would interfere if it were placed on the body. The indifferent electrode does not "see" electrogenic activity if it is placed at a sufficiently remote distance, although in practice a small amount of electrical activity is usually recorded. Bipolar recording is accomplished with two electrodes over active electrogenic sites, and the potential at any instant is the algebraic resultant of voltage which appears under each electrode. Figure 6-1 shows an EEG taken from a human. A fairly regular 8 to 12 per second series of waves may be readily distinguished in some portions

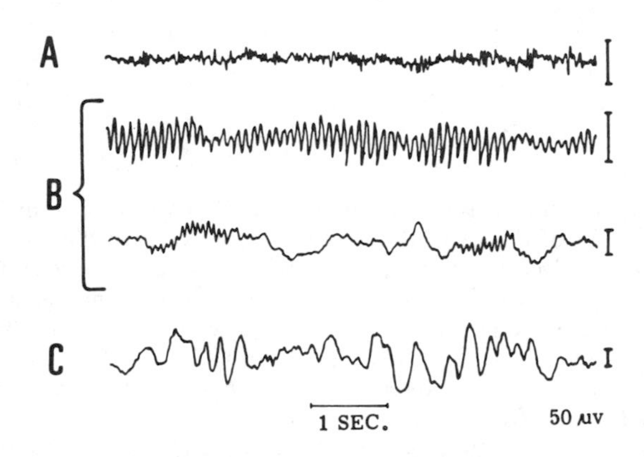

FIGURE 6-1. EEG during different states of sleep and wakefulness. The EEG of the alert state (A) shows a low-amplitude, fast *beta* wave type of activity. In the drowsy or light sleep state (B), *alpha* activity is seen. There may be periods in the drowsy state when a mixture of alpha and slower wave activity is seen. In deep sleep (C), large-amplitude, slow *delta* waves are predominant. (From W. Penfield and T. C. Erickson. *Epilepsy and Cerebral Localization.* Springfield, Ill.: Thomas, 1941. P. 401.)

of the records. These are known as the *alpha waves,* although the term *Berger rhythm* is sometimes used in honor of their discoverer. The alpha rhythm is usually present in an individual who is relaxed, with eyes closed, and in a quiet room. If the subject is asked to open his eyes, the alpha rhythm is abruptly *blocked* and replaced with the smaller-amplitude and faster-frequency waves. The same blocking of the alpha rhythm can occur if the subject is asked to visualize a scene with his eyes closed, if a sudden sound is made, or if the subject is otherwise *alerted* to some suddenly produced sensory stimulus in the environment.

The various rhythms present in the EEG may be grouped according to the dominant frequencies found. Two additional frequency groups are designated as the *beta* wave group, 18 to 30 cps, and the *theta* group, 4 to 7 cps. These have some relation to the state of brain function. Rhythms of a still slower frequency range, below 4 per second, are called *delta waves.* They are usually larger in amplitude than the alpha waves and are seen during sleep.

The wave form of the EEG may be difficult to analyze into dominant frequencies by simple inspection. Automatic frequency analyzers have been used to measure the EEG spectrum, with a type of Fourier wave analysis. The frequency analyzers assess the relative amount of activity in selected divisions of the frequency spectrum by means of a series of filters. The relative amount of activity present in the various frequency bands within a short sampling time is determined.

If the blood supply to the brain is suddenly interrupted, the EEG activity disappears within 15 to 20 seconds, with loss of consciousness. If the blood supply to the brain is returned within approximately 5 minutes, recovery is generally complete. (See Circulation and Metabolism, below.)

The EEG waves are believed to reflect activity within the dendrites of pyramidal cell neurons of the cerebral cortex (Fig. 6-2). A large number of the apical dendrites of the pyramidal cells are found stacked vertically in the cortex, and their finer branches are numerous in the upper (molecular) layer of the cortex. Many neuronal elements synchronous in their discharge are required to cause sufficient current flow and thereby produce the relatively large EEG voltages found. Voltages present at the surface of the cerebral cortex as a result of current flow in the apical dendrites are explainable in part on the basis of general electrical principles. If a polarization

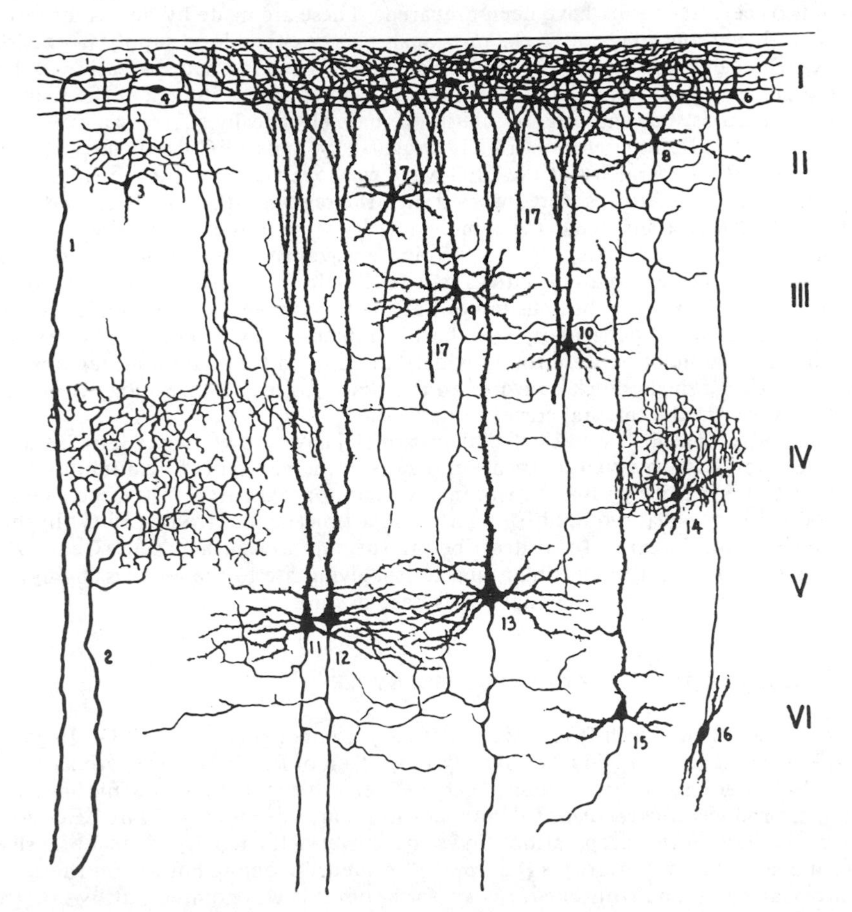

FIGURE 6-2. Section through the cerebral cortex with characteristic cells through all the layers I-VI. Pyramidal cells 7, 10, 11, 12, 13 have apical dendrites extending up to the molecular (1st) layer, where they arborize profusely. Other cells include short axon cell, 3, and Golgi type II cell, 14. (From H.-T. Chang. *J. Neurophysiol.* 14:1, 1951.)

difference exists between the upper and lower shafts of the apical dendrites, current will flow in the extraneuronal spaces from polarized to depolarized parts of the cell. This flow of current gives rise to the potentials recorded on the surface. Current flow in a three-dimensional conducting volume giving rise to potentials on its surface, as is the case with the brain, can be complex. (The student is referred to books dealing with volume conductor theory.) For the purposes of this discussion it will suffice to consider only the simplest principles involved. Areas of depolarization on the neuron membrane are called *sinks* because current flows into them from polarized parts of the membrane called *sources.* A large number of cells must be synchronized in their depolarization to give rise to the relatively large voltages recorded from the brain. In the production of the series of repeated individual waves characteristic of the EEG, a further requirement is that the synchronization of discharge be repeated in a regular sequence.

A propensity for rhythmic response appears to be a fundamental property of cells in the cerebral cortex. In order to study the origin of rhythmicity, isolated islands of cerebral cortex have been prepared. These are made by means of cortical cuts produced some days or weeks previously, leaving the blood supply to the island intact while all neuronal connections are severed. When the neurons inside such an island are electrically stimulated, rhythmic responses are excited. Yet the neurons in the cortical islands without stimulation remain electrically silent. An external source of activation therefore normally triggers rhythmic EEG-like behavior. A pacemaker for rhythmicity of the EEG wave was located in the thalamus. Electrodes insulated but for their tips were inserted into the regions, and stimuli at a rate close to that of the naturally occurring spindle activity were found to give rise to typical *spindle waves* in the cortex (Fig. 6-3). Spindle waves show *recruitment* in that successive waves increase in amplitude as more cells add their discharge to the group giving rise to the wave. The cells fall out at the tapering end of the spindle. Spindles are seen in sleep and particularly with barbiturate anesthesia. The spindles excited by thalamus stimulation were found widely distributed throughout all regions of the cortex. Also, a single shock delivered to the thalamus may excite a long train of spindle waves in the cerebral cortex (Fig. 6-3).

Recently single-cell studies of thalamic nuclei have shown that, following a discharge, a long-lasting hyperpolarization occurs in the neuron before a rebound of negativity takes place with a spike at intervals of approximately 100 msec. This period is close to that required for a pacemaker function controlling EEG rhythms. Inhibitory cells similar to those described in the cord as Renshaw cells (Chap. 5) have been proposed as the cellular mechanism giving rise to the long-lasting hyperpolarizations in thalamic cells that act as pacemakers.

ALERTING AND THE SLEEP-WAKEFULNESS CYCLE

During sleep there is a shift to lower-frequency components of the EEG. In general the EEG from alert individuals is largely composed of fast beta wave activity.

By transecting the brain stem of cats between the colliculi of the midbrain Bremer produced the *cerveau isolé* preparation (cut B in Fig. 6-4). The EEG taken from the brain of this preparation shows the slower delta activity typical of a sleeping animal. This cut interrupts the upward influences coming both from the specific sensory afferents and from the *reticular formation.* It was pointed out by comparative anatomists many years ago that the reticular formation, composed of cells with interspersed fibers, could have an integrative function. Collaterals branching off from the various sensory tracts passing up to the cerebral cortex relay impulses into the reticular formation on their passage by this brain stem region. These collateral

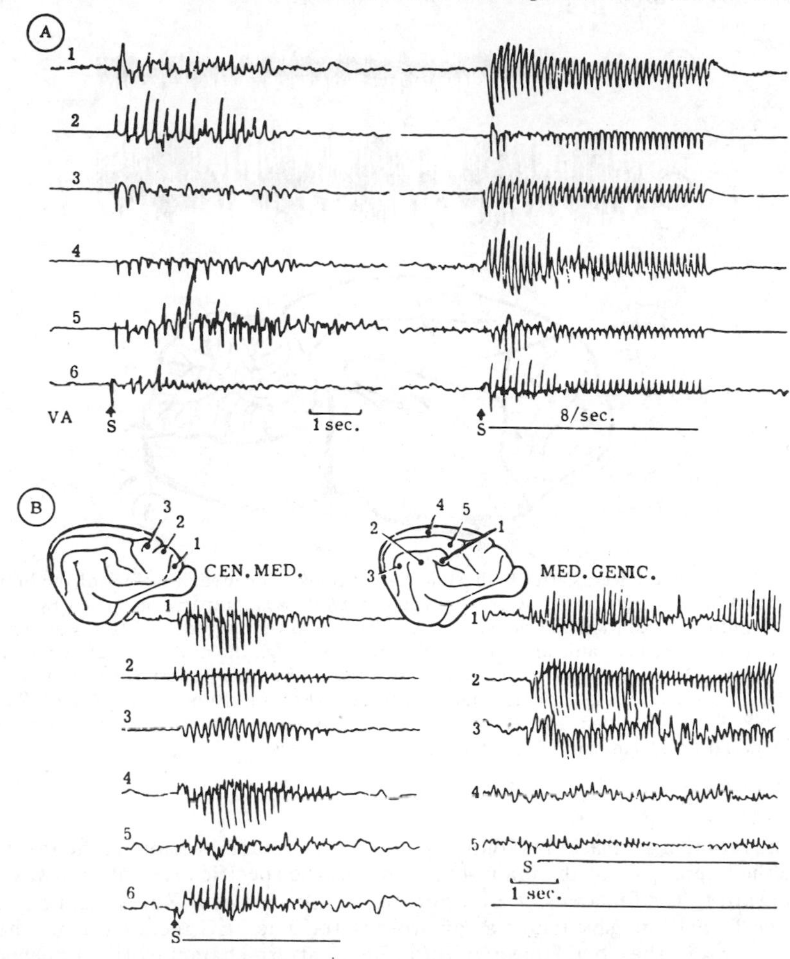

FIGURE 6-3. Recruiting waves excited by stimulation of various nonspecific thalamic nuclei. In A, to the right, a train of 8-per-second stimuli was initiated. The evoked responses are large and follow the rate of stimulation. To the left, a single stimulus was delivered and a train of repetitive responses was excited. In B, nuclei of the nonspecific group are excited with single shocks, as in A, with trains of stimulation. Recruiting waves are apparent in the various regions of the cat's cortex. (From J. Hanberry and H. H. Jasper. *J. Neurophysiol.* 16:256, 1953.)

inputs were demonstrated by recording *evoked responses,* i.e., a grouped discharge of neurons in the reticular formation by means of a peripheral sensory stimulation. Evoked responses were elicited in response to all the various sensory excitations, e.g., sound, skin touch, or a light flash to the eye. The evoked responses recorded in the reticular formation to different sensory sources also showed occlusion. A second evoked response was reduced in amplitude if it followed too soon after a first such response whether elicited by excitation of the same or a different sensory modality. The occlusion suggests that the different types of excitatory inputs on entering the reticular formation become "added up" on cells which receive inputs in common from all these sensory inputs. In this way the reticular formation reflects the general level of sensory excitation.

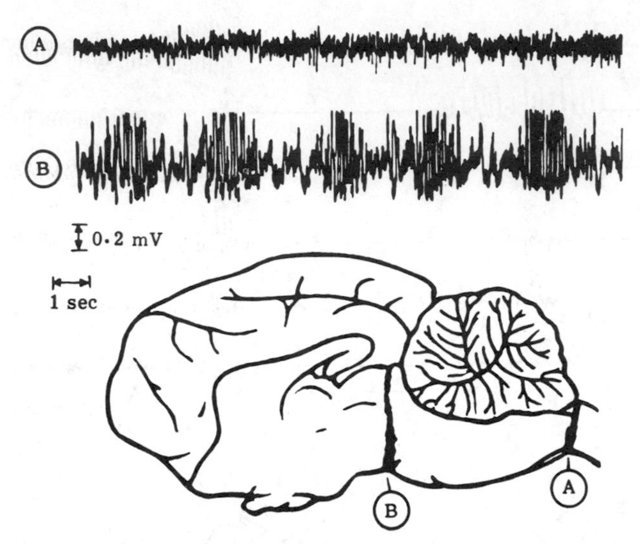

FIGURE 6-4. EEG patterns after brain stem transections. The medial surface of the brain is represented in the diagram. Cut A produces an *encéphale isolé* preparation. The EEG record taken from this preparation and shown in line A has the low amplitude and fast wave activity characteristic of the alert animal. A transection made through the midcollicular region of the brain stem, cut B, produces a *cerveau isolé* preparation. The EEG pattern (line B) in this case is of the sleep type, with larger-amplitude slow waves. (From F. Bremer. *Bull. Acad. Roy. Med. Belg.* 2:68, 1937. Modified from M. A. B. Brazier. *The Electrical Activity of the Nervous System.* New York: Macmillan, 1960. P. 212.)

The reticular formation is found mainly in the medial portion of the medulla, while the lateral part of the brain stem contains the specific afferent pathways passing upward to the thalamus. Lesions of the middle part of the medulla, i.e., the reticular formation, gave rise to a comatose state. The EEG recorded from the animals showed the slow, large-amplitude EEG pattern characteristic of deep sleep. Lesions placed laterally in the brain stem, in the tract containing the ascending specific sensory pathways, produced animals which remained behaviorally in a waking state and had an EEG pattern typical of an alert animal.

Alerting or *arousal* refers to the change seen in the EEG of a drowsing animal when it is suddenly presented with a brief sensory stimulation – a click of sound, a pinch of the skin, etc. In this event the slow wave activity is abruptly replaced by a low-amplitude pattern of fast EEG activity which may persist for some time after the stimulus (Fig. 6-5A). The pattern of arousal or *desynchronization* is widespread in the cortex. A similar period of alerting of the EEG could be obtained following electrical stimulation of the reticular formation through electrodes inserted into its substance (Fig. 6-5B).

When a transection was made below the brain stem, the *encéphale isolé* preparation (cut A in Fig. 6-4), the EEG pattern (EEG record A in Fig. 6-4) was one of wakefulness. The difference in this case is due to sensory inputs coming into the brain stem from the trigeminal nerve innervating the head. These inputs into the reticular formation activate it and thence the cortex to produce the wakeful state. Destruction of the trigeminal nerve afferents in an encéphale isolé preparation promptly brings about the slow wave EEG pattern typical of sleep.

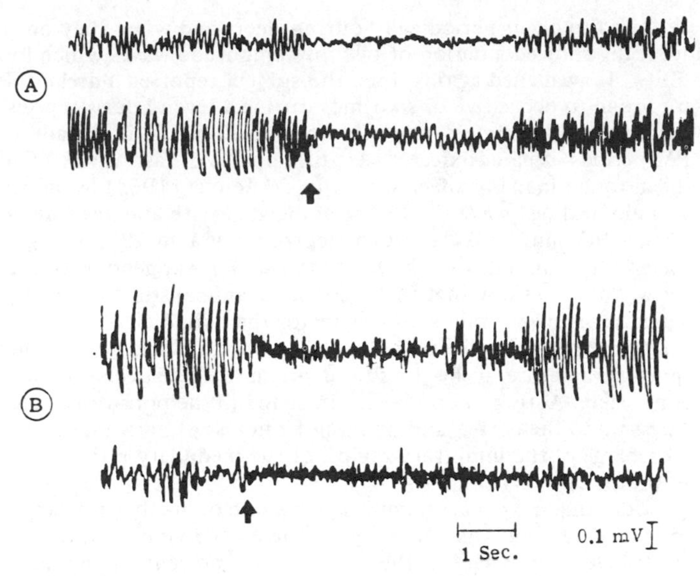

FIGURE 6-5. Alerting responses in the brain. In A the EEG obtained from the motor cortex and visual cortex is shown. After a sudden sensory stimulation (arrow), which may be a noise, a touch, or a flash of light, the EEG pattern changes to the alert type. Alerting or desynchronization of EEG lasts for a brief time before return of the regular alpha or spindle type of record characteristic of the drowsy animal. In B, a brief electrical stimulation is delivered to the reticular formation, and it has a similar alerting effect on the EEG of the cortex. (From F. Bremer. In G. E. W. Wolstenholme and M. O'Connor [Eds.], *Ciba Foundation Symposium on the Nature of Sleep.* Boston: Little, Brown, 1960. Pp. 32, 38.)

From the reticular formation, impulses are carried to the cerebral cortex by fibers via a path yet to be determined. The overall effect of reticular system activation on the cortex is based on a study of evoked responses. As will be seen farther on, the surface-negative type of evoked response is nonsummating, and the effect of reticular formation stimulation to decrease its amplitude can be considered a form of occlusion. Therefore, it is believed that the input of the reticular formation is excitatory to the cortex.

The state of sleep is generally thought to be due to diminished corticopetal sensory influences, and waking to their increase. Sleep, however, may also be actively excited, as Hess some time ago showed by means of an electrical stimulation of thalamic nuclei of the cat via chronically implanted electrodes. On stimulation, the animal shows diminished motor activity, seeks out a resting place, then curls up and behaviorally appears to fall into a natural state of sleep. Other brain stem regions may be involved in active sleep. Recently, injection of ACh into the brain stem by chronically implanted cannulas was shown to produce a behavioral change similar to normal sleep. Regions for sleep induction and pathways from them were traced by Hernández-Peón and his colleagues. These sleep-inducing or *hypogenic* regions are considered to interact with arousal from the reticular formation to bring about the states of sleep and wakefulness.

Until lately it was assumed that sleep was a rather homogeneous state. A recognition of periodic alterations occurring during sleep has come from the work of Kleitman, Aserinsky, and Dement (see Jouvet, 1965). Rapid eyeball movements

(REM) were found to occur periodically during sleep, and when they occurred, the EEG paradoxically showed a period of fast low-amplitude waves which looked like an alerted EEG. If awakened at this time, the subject reported a dream. REM periods were found to occur five or six times during a seven-hour sleep session, and to occupy approximately 20 percent of the total sleep period in the adult.

The REM periods of paradoxically alert-like phases of fast wave EEG during sleep have been found in many species. In the cat Jouvet (1965) found that the periods of paradoxical fast wave EEG were associated with a marked decrease in the tonus of the neck muscles as recorded electromyographically. Using the decreased muscle tone as an index of the *paradoxical phase* appearing during REM sleep, Jouvet was able to show that the decorticate animal still had these periodic episodes, indicating a lower brain stem origin for the periodicity. When lesions were made in the upper pons, the periodic changes were blocked. The rhombencephalon appears, therefore, to be the site of origin of these cycles of alert-like EEG activity during sleep. Activity periodically initiated in the pons causes alerting influences to pass up to the cortex and, judging from loss of neck muscle tonus, a downward discharge of the inhibitory region of the medullary reticular formation to the lower motor centers (Chap. 8).

These periodic changes occurring during sleep are of great significance. As already mentioned, dreaming is associated with them. Individuals who are wakened during the fast wave periods seen in the EEG so as to prevent paradoxical sleep from occurring have hallucinations and other behavioral changes. In animals such deprivation also leads to pathological behavior. The implications for psychiatry are apparent.

DIRECTLY EVOKED RESPONSES

The EEG represents a complex temporal and spatial interplay of neuronal activity. To further its analysis, it is possible to synchronize the discharge of a population of cortical neurons. Such evoked responses may be achieved by a volley in a sensory tract leading to a cortical sensory area, as will be described later on in this chapter. Another approach is to stimulate directly the exposed surface of the cortex. When the cortical surface is stimulated with a single brief pulse of current, a characteristic surface-negative response is recorded a few millimeters away. This *superficial response* or *direct cortical response* (DCR) increases in amplitude as the strength of stimulation is successively increased (Fig. 6-6). The negative wave DCR (N wave DCR) shown in Figure 6-7 is also seen to become shorter in duration and to be followed by a positive wave. Axons pass tangentially in the cortex on apical dendrites of pyramidal cells, and the response is generated in the dendrites. The DCR's show an occlusive interaction when two different sites on the cortex are stimulated at different times. The occlusion-like interaction suggests that the electrogenesis of the response in the apical dendrites is not that of an EPSP. The view derived from such studies is that the N wave DCR is a *graded potential,* one which does not show the summation expected of an EPSP but which is not fully propagated down the axon of the cell.

SENSORY AREAS AND EVOKED RESPONSES

Sensory inputs of various modalities are localized to specified regions in the cerebral cortex. For example, visual input is localized to a part of the occipital cortex in the primate.

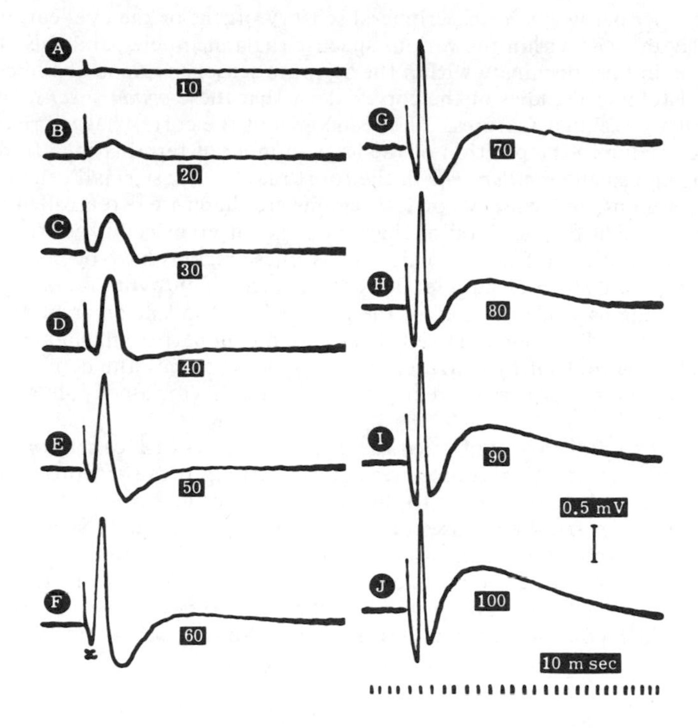

FIGURE 6-6. Direct cortical responses (dendritic potentials). The responses elicited by direct stimulation of the surface in recording a few millimeters away are shown at successively increasing strengths of stimulation. The response is a slow wave response, broader at the weaker stimulus strength, and at the higher strengths of stimulation is followed by a longer-lasting second negative wave. (From H.-T. Chang. *J. Neurophysiol.* 14:1, 1951.)

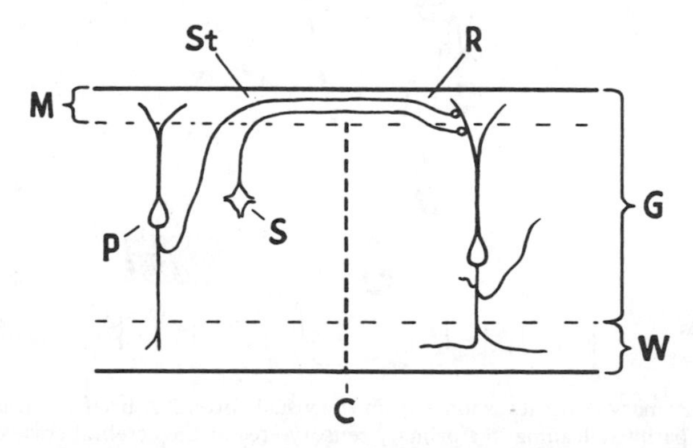

FIGURE 6-7. Hypothesis of neuronal connections underlying the N wave DCR. Stimulation (St) of axons in the molecular layer (M) inferred by passage over a cut (C) of all other layers. They may arise from pyramidal cell (P) collaterals or stellate cells (S). In the recording site (R) synapse of these axons on apical dendrites of pyramidal cells is shown, the latter generating the response. (From S. Ochs and H. Suzuki. *Electroenceph. Clin. Neurophysiol.* 19:230, 1965.)

Nerve fiber pathways from peripheral sensory organs of the eye, ear, and skin have a relay on cells within the various specific thalamic nuclei, and the fibers of these nuclei in turn terminate within the separate sensory regions of the cerebral cortex. Histological studies of the cortex show that these *primary sensory areas* have distinctive cellular features. The visual area of the cortex has a striate appearance which is due to the position of the large number of terminating afferent sensory fibers and small *granular cells* in the fourth layer. These small cells predominate in sensory regions, and sensory cortical regions are therefore often called granular cortex. Because of the variation of fibers and the different cell types found in the different layers of the different cortical areas, these regions have been characterized and given separate letter or number designations. Such *cytoarchitectonic* maps, however, have in the past been carried to the point where too fine an areal division of the cortex was made. Differences in cell shapes are due in part to mechanical distortions produced by the folds of the surface. For these reasons only the distinctly different cytoarchitectural regions are currently considered as having a possible relation to function.

Electrical studies have confirmed the general features of localization of the visual, auditory, and somesthetic receptor areas in the cortex. If the receptor organs are briefly stimulated by a light flash to the eye, a click to the ear, or a touch to the skin, a characteristic electrical response can be detected in the primary sensory area of the cortex to which the fibers of each sense modality project. A more consistent response is seen when a brief single shock is delivered to a sensory projection tract leading to the primary receptor area. This *primary cortical response* consists of a series of small fast waves which continue on to a slower 10- to 20-msec positive wave, followed in turn by a slow negative wave lasting 10 to 20 msec (Fig. 6-8).

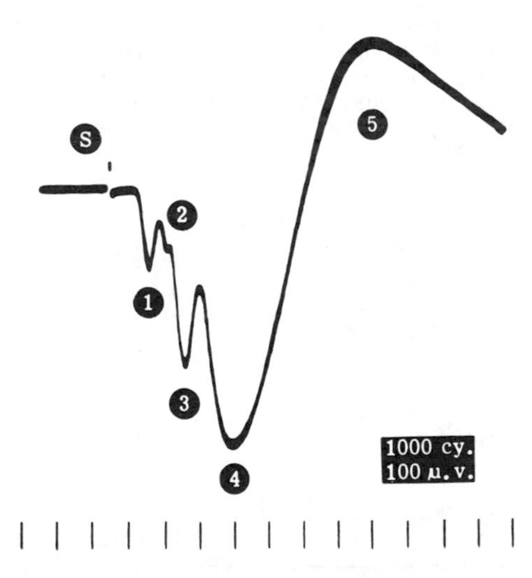

FIGURE 6-8. Primary cortical response from the visual cortex. A brief electrical stimulation of the afferent pathway leading to a primary sensory area of the cerebral cortex (visual cortex in this case) gives rise to a series of fast spikelike waves (1, 2, 3) inscribed on a slow positive wave (4) which is succeeded by a slow negative wave (5). Spike wave 1 – and possibly 2 – is due to the activity of the specific afferent endings within the visual cortex. Waves from 3 on are due to intracortical activity. (From L. I. Malis and L. Kruger. *J. Neurophysiol.* 19:175, 1956.)

The same complex response is seen in each of the corresponding primary sensory areas when its specific sensory relay nuclear group (somesthetic, visual, or auditory) is stimulated. The first of the brief spikelike waves is the sign of activity in the entering specific afferent fibers.

The specific afferent fibers synapse on granular cells in the fourth layer of the cortex. In turn, the granular cells synapse on other neurons and eventually on pyramidal cells to give rise to the slower wave components of the response. The positive portion of the response is due to excitation of elements activated in deeper layers of the cortex, probably on the soma and lower dendrites of pyramidal cells. The negative portion of the sensorily evoked response is due to a later activation and depolarization of the apical dendrites near the surface. It is likely that the same electrogenesis is involved as that described for the N wave DCR. This is indicated by the fact that the negative phase of the sensorily evoked response shows occlusive interaction with the N wave DCR, and, as previously mentioned, occlusive interaction indicates that a common element in (in this case) the apical dendrites is causing both responses.

Electrical stimulation of the primary sensory cortex of man under local anesthesia gives rise to sensations experienced as either visual, auditory, or somesthetic, depending on the sensory area stimulated: for the visual area, optical; for the auditory area, sound; and for the somesthetic cortex, sensations referred to the skin of the body. Sensory areas relating to olfactory, gustatory, and alimentary sensations are also known. Tumors of the uncus, for example, may be associated with disturbing sensations of smell.

Within each of the specific sensory regions of the cortex there is an internal spatial organization, i.e., a topographical localization of parts of the receptor field to parts of the sensory area, as shown for several species in Figure 6-9. The correspondence of a point within the sensory field to a point of the cortex is referred to as *point-to-point representation.* This correspondence was shown experimentally in 1942 by Marshall and Talbot, who used a point source of flashed light brought to a given locus in the visual field of an animal while exploring the cortex with a monopolar electrode. A point on the visual cortex was found where the evoked response had the greatest response amplitude. Reducing the strength of the light and using a small point source allowed the area of the cortex representing the point stimulated in the visual field to be localized to within less than a millimeter.

The topographical relationship within a sensory area may be shown by a figurine, as in Figure 6-10. Such representation more readily shows the extent of the area given over in the cortex to a sensory modality which has a relation to a species' behavior. For example, in the primate, the thumb area in the somesthetic region is large, as is the face region, especially the mouth and tongue. This is also the case in man, and the area of these regions conforms with their importance. Another example is the large extent of the cortical somesthetic representation for the snout in the pig.

Point-to-point localization in a sensory region should not be considered to mean that fibers are activated in a direct line from points within the sensory organ to the cortex, much as in a television receiver the beam represents a scene point for point by the intensity at each point on the surface of the screen. Studies of the retina have shown that when a pinpoint source of light falls on the retina, a relatively large number of cells may be discharged, and cells in the surrounding area are inhibited as well. Retinal cells, therefore, show complex and widespread excitatory and inhibitory effects, and the same sort of surrounding inhibition (and excitation) occurs at all points within the central nervous system. The complexity increases along the sensory pathways upward to and in the cortex. Within the cortex each single specific afferent fiber synapses on a large number of neurons within the field of its

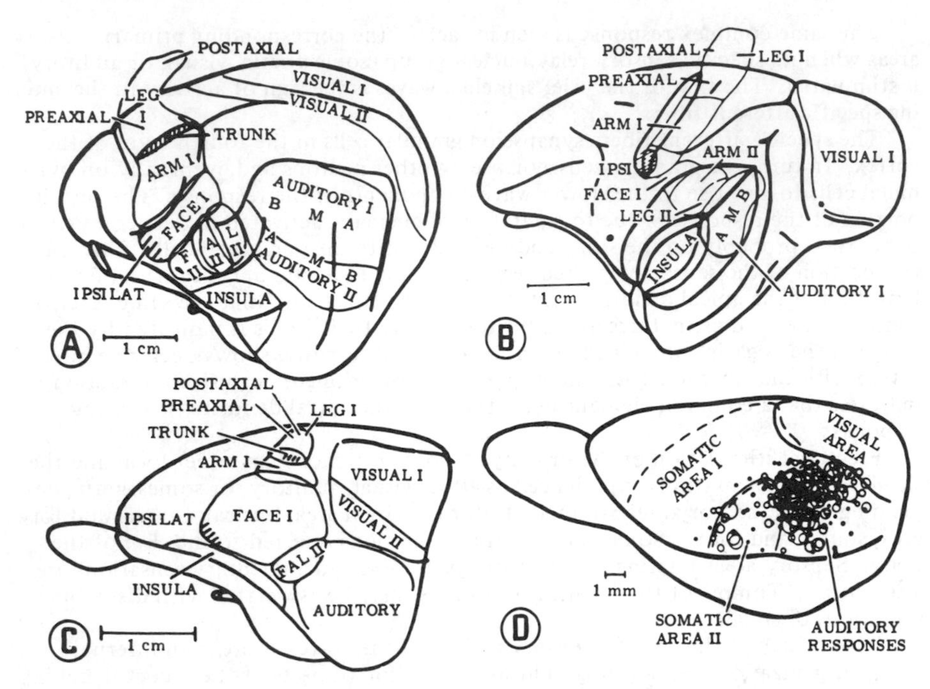

FIGURE 6-9. Sensory maps of several animal species. A is a sensory map for a cat. Two visual areas are shown, and the somesthetic area I is represented by leg, arm, and face areas. A second somesthetic area II is also shown, and two auditory areas as well are found. B is a corresponding map for the monkey, C for the rabbit, and D for the rat. Notice that in general the relative positions of the main sensory areas have a similar pattern over the surface of the brain. (From C. N. Woolsey and D. H. Le Messurier. *Fed. Proc.* 7:137, 1948.)

multiple branching terminations. These afferents extend laterally, and through their numerous branching terminations and synaptic endings a relatively large volume of cortex is involved by each termination. The "grain" of the cortical region excited by a single afferent fiber would therefore be much too coarse to provide an appreciation of fine detail if each afferent fiber represented one point of a point-to-point representation and if no other mechanism of discrimination between points in the visual field were operating. Some additional process must come into play to "sharpen" the representation of the points of its surface. This sharpening could come about in part by the surrounding inhibitory effect and in part by the overlapping of terminations of several afferents ending within a cortical region which could be required for adequate excitation of a smaller number of cortical cells.

Topographical localization within sensory regions is not only spatial, for it is also seen in the auditory cortex, where different sound frequencies are localized to specific places. The spatial localization of frequencies in the cortex is related to appreciation of temporal patterns of sound.

Sensory afferent projections to other regions than the primary sensory areas have been found in the cortex. These *secondary sensory areas* also have a topographical organization with point-to-point representation of the peripheral sensory field (Fig. 6-9). The function of the secondary receptor areas and their relation to the first sensory areas are unknown.

FIGURE 6-10. Figurine representation of the somesthetic area of the dog. A body representation of tactile areas of the dog, somatic area I, is shown determined by evoked potential techniques. A second somesthetic area is also found which is a smaller duplication of the first, somatic area II. Note the greater disproportionate size of the snout region in this species. (From C. N. Woolsey and D. H. Le Messurier. *Fed. Proc.* 7:137, 1948.)

SINGLE CELL RESPONSES

In recent years the use of microelectrodes to explore the activity of single units within the central nervous system has aroused increasing interest. The activity of single cells of the sensory regions of the cortex has been sampled during sensory stimulation. Microelectrodes used for such unit cell studies are nearly as small as microelectrodes used to study transmembrane potentials from a single cell (see Chaps. 2 and 3). The tip is passed close to a spontaneously active cell or a cell which can be activated by sensory stimulation. Because of the small size of the recording tip there is a rapid decrement of potential with distance from the discharging surface, and this allows the activity of a nearby cell to be recorded in relative isolation. Other more distant cells usually contribute too small a voltage to be recorded. The technique has the disadvantage that a statistical survey must be made to determine what a given population of neurons is doing in any small area. As an example, during an evoked response to a light flash some neuronal units increase in their rate of firing, others decrease, and still others are not affected. These various types of unit response indicate that much of the function of the primary receptor areas underlying vision is complex.

Single neurons in the cortex of the somesthetic area have been found which respond either to movement of hairs of the body surface or to rotation of limb and digit joints. On recording successively downward from the surface, neurons were found to respond either to the one or to the other type of stimulation. The neurons responding to a given type of sensory input are arranged vertically in columns. The regions of the skin surface from which the column of similar responses was elicited were rather discrete and generally contralateral. In the secondary sensory areas of the cortex, the peripheral body surface area from which excitation could be obtained was more widely spread and in addition could be obtained from bilateral areas on the surface.

In the thalamus, where a relay is made between the tracts carrying sensory information from the periphery to the cortex, the ventrobasal region was found to contain neurons corresponding to the discrete type of peripheral sensory organization, and in the posterior nuclear group neurons were found responding to the widespread and bilateral skin stimulation.

In the visual cortex, Hubel and Wiesel (1962) found a similar system of neurons having a columnar organization. Units in the visual cortex were found to respond to slits of light which were oriented at a certain specific angle. For those neurons, an orientation in a different direction brought no response. Upon recording from cells successively down in the cortex, the cells were found to respond to a given orientation in a slitlike fashion all along that vertical column of cortex. Cells in nearby columns were found responding to slits of other orientations or to other sensory stimuli – for example, to the movement of a slit of light in one certain direction, or to different-shaped contours in the complex type of cell (Fig. 6-11).

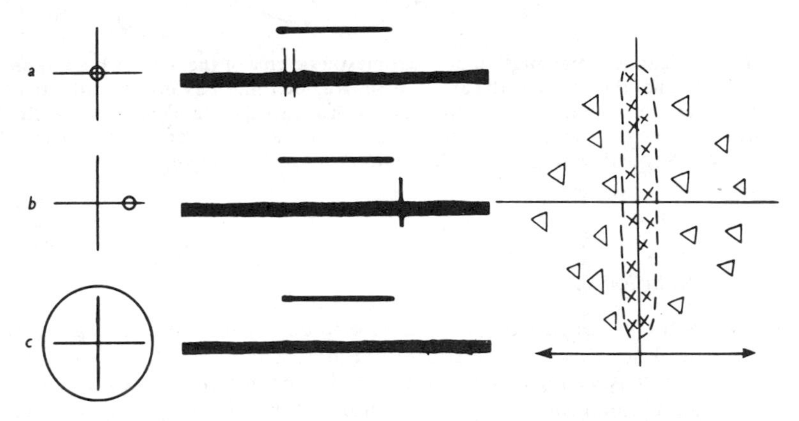

FIGURE 6-11. Responses of a unit to a stimulation with circular spots of light in the visual field. The receptive area is activated by: a, 1° spot in the center; b, same spot displaced 3° to the right; c, 8° spot covering a large part of the receptive field. (From Hubel and Wiesel, 1959.)

In the excitatory discharges described, the visual cells gave rise to a discharge at the onset, *on* response. In addition, other cells discharged at the *off* response, as well as *on-off* cells to both. Excitation and inhibition were discovered to be functionally related, so that in the region of the cortex surrounding a group of excitatory units, e.g., of on units, one finds an inhibitory discharge of off units. The two primary visual areas in each hemisphere are also connected for binocular vision, and unit recording which is shown when a neuron responding to a given type of slit orientation will, in a fair proportion of cases, respond also to stimulation of the eye on the other side with slits of the same orientation.

A simplification of neuronal networks is obtained by study of the sensory systems of lower organisms. The *ommatidium* of the *Limulus* is a small cluster of light receptor cells, a number of these forming the compound eye. An *eccentric* cell is found in close apposition to cells in the ommatidium, and discharges in its axon have been recorded with a rate determined by the degree of light shone into the ommatidium. There are collateral interactions between the axons of the eccentric cells such that discharges in one will inhibit those of the other around it. This form of *lateral inhibition* is also found in the sensory systems of higher forms.

STEADY POTENTIALS AND SPREADING DEPRESSION

The relation of apical dendrites of pyramidal cells to the production of EEG waves has been discussed. Another aspect of pyramidal cell activity and the functional state of the cortex is the *steady potential,* which may be recorded from the surface of the cerebrum. A DC coupled amplifier is required to record the steady potential levels from the surface. For this reason these potentials have also been called *DC potentials.* Changes in the steady potential have been related to differences in polarization of the upper and lower portions of the apical dendrites of pyramidal cells, which could reflect the summed synaptic excitation present on upper and lower portions of the pyramidal cells. The relation of the steady potential to activity was shown by a negative shift in the steady potential recorded when nonspecific thalamic nuclei with axons presumably terminating on upper apical dendrite sites were stimulated.

The steady potential is also related to metabolism. If the blood supply to the brain is suddenly interrupted, a latent period of approximately two to five minutes follows, and then a rapid shift of the steady potential to a surface negativity takes place (Fig. 6-12). At the time of this negative shift, the electrical resistance of the cortex increases markedly. The negative change in steady potential and the increased electrical impedance of the cortex were related to the loss of normal ion permeability

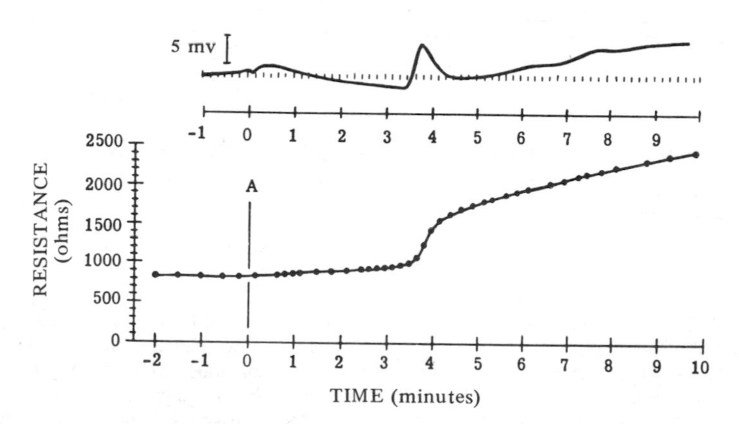

FIGURE 6-12. Asphyxial changes in the cerebral cortex. In the upper curve, the steady potential is recorded; in the lower curve, the electrical resistivity of the cortex. At A, the brain is asphyxiated by sudden deprivation of its blood supply. A latent period of several minutes ensues before a rapid increase in resistivity is seen. At the same time a negative depolarization of the brain takes place. (From A. van Harreveld and S. Ochs. *Amer. J. Physiol.* 187:184, 1956.)

of the apical dendrites, and to an entry of Na^+, Cl^-, and water into the cells from the extracellular space. Evidence for this is a measured swelling of the apical dendrites. The process heralds the irreversible destruction of the cortical neurons. Consequently it is imperative that oxygenated blood be supplied to the brain before this change takes place; otherwise, cortical functions are lost, and in humans a permanent loss of consciousness results.

Some aspects of the changes seen after *asphyxiation,* described above, are similar to those in the phenomenon discovered by Leão in 1944 and termed *spreading depression* (SD). Spreading depression is elicited by mechanical, chemical, or electrical stimulation of the surface of the brain and characterized as a slowly spreading

(2-5 mm per minute) depression of the EEG and of evoked responses. It moves through the cortex without regard to the function of the area involved: sensory, motor, or associational. In the area involved, an increased permeability occurs in the apical dendrites of the pyramidal neurons, causing a negative shift of the steady potential and an increased electrical resistivity (Fig. 6-13). Apparently, glutamate is released from the involved cells which diffuses to nearby cells and excites SD in them. The apical dendrites of the involved neurons show evidence of swelling and entry of Na^+, Cl^-, and water into them similar to that of the asphyxial change. However, unlike the asphyxial change, recovery from the effects of SD takes place

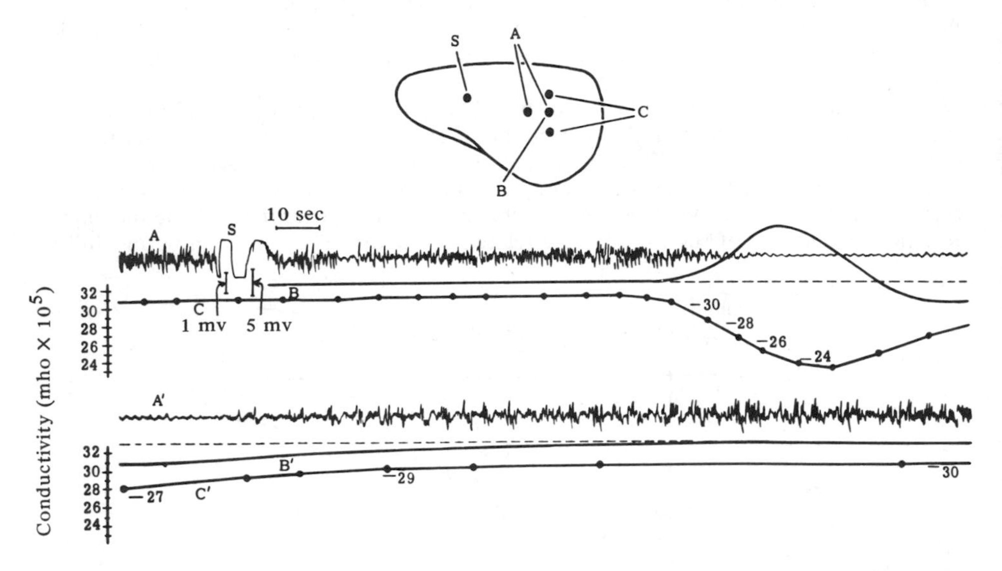

FIGURE 6-13. Spreading depression of the cerebral cortex. The EEG is shown in line A, the steady potential of the surface of the brain in B, and on line C the electrical conductivity (inverse of resistivity) of the cerebral cortex. These records are continued below as A', B', and C'. The electrode positions on the surface of the brain are shown in the diagram. A brief intense stimulation of the brain is employed at S to start spreading depression. As depression spreads out and occupies the region of the recording electrodes, the EEG shows a diminution, and at the same time a negative variation is seen in the steady potential level of the brain. A decrease in conductivity also occurs at this time. Recovery occurs some minutes later. (From A. van Harreveld and S. Ochs. *Amer. J. Physiol.* 189:159, 1957.

in a few minutes. An increased metabolic activity is found after SD, presumably the activity of the Na^+ pump in the cells required to restore normal conditions. Spreading depression occurs readily in the rat, rabbit, and guinea pig, but in higher species it is more readily seen after cortical exposure, or after application of Ringer's solution with augmented K^+. It may, however, occur pathologically in man following concussion or in certain types of *convulsive states.* Convulsions are due to abnormal synchronous activity of neurons leading usually to large spikelike activity recorded from the cortex.

The term *spreading depression* may be rather misleading because a small amount of convulsive spike activity is commonly seen; under certain conditions, a prolonged convulsive discharge occurs rather than depression. Convulsive discharges probably take place in neurons in the lower part of the cortex which have been *released* by the presence of SD in the upper layers. Perhaps these convulsive charges are responsible for the clinical condition in man known as *jacksonian march,* a convulsive spread in the motor cortex at the slow rate usual for SD, with a successive involvement of the regions of the motor cortex and their corresponding musculatures (see Chap. 8). The convulsive state is properly a clinical subject, but from a physiological point of view there is much of interest in the abnormal synchronization of neuron activity involved in the various convulsive disorders. In *grand mal* epileptic attacks there is a train of large convulsive spikes, and in the body a tonic extension of the lower limbs and flexion of the upper limbs followed by a period of clonic jerking movements. In the epileptic condition known as *petit mal,* consciousness is lost without much motor involvement. Petit mal is triggered from the nonspecific thalamic nuclei, mentioned above as the EEG pacemaker area, giving rise in the cortex to a spike-and-dome type of abnormal wave discharge which is synchronized in both hemispheres.

NEUROTRANSMITTERS

Electron micrographs of the central nervous system (CNS) reveal structures resembling the presynaptic terminals present at the neuromuscular junction (Chap. 3A). They contain vesicular structures approximately 500 A in diameter and mitochondria. In favorable situations in the cerebral cortex, it has been shown that these terminals are synaptically apposed to spinelike projections from the shafts of the apical dendrites of pyramidal cells, and typical synaptic junctions are found on the somas similar to those seen on spinal cord motoneurons (Chap. 5).

A further advance on the functions of the transmission affected by the synaptic structure has become possible through the technique of subcellular fractionation. The brain is homogenized in an isotonic sucrose solution and subjected to centrifugation at different speeds. One fraction can be selected containing structures identical to the presynaptic terminals. Vesicles, mitochondria, and in some cases the part of the postsynaptic membrane torn off in separation are seen. Apparently the stems of the bouton endings become pinched off to form these *synaptosomes.* In further treatment of the synaptosome, the membranes around them are ruptured with the release of the vesicles. Pharmacological and biological studies of the vesicles have shown that ACh, serotonin (5-HT), and possibly other transmitter substances are present in them.

In CNS tissue there is a very high concentration of *glutamic acid* and *gamma aminobutyric acid* (GABA). The enzyme *pyridoxine decarboxylase* will transform glutamic acid to GABA. It is of great interest that, by means of the microiontophoretic technique, very small amounts of glutamic acid or GABA released in the immediate vicinity of neurons in the cortex can alter their discharge rate. It has been noted in this way that glutamic acid excites a rapid increase in the rate of discharge of most neurons, whereas GABA will markedly inhibit their discharge. These effects are found with extremely low concentrations, suggesting that these substances are excitatory and inhibitory transmitter substances. Alternatively they are *modulators,* acting to control the excitatory state of cortical neurons in a more general fashion.

CIRCULATION AND METABOLISM

The utilization of oxygen by the brain is surprisingly high relative to that of the rest of the body. It amounts to approximately 3.3 ml per 100 gm per minute and roughly 20 to 25 percent of the total oxygen uptake of the body. Glucose is the metabolite utilized, as indicated by a respiratory quotient (RQ) close to 1.0.

A close dependence of brain function on oxidative metabolism is shown by the rapid alteration of function on acute cessation of blood supply. Consciousness is lost in approximately 10 to 15 seconds with a loss of the EEG, an *isoelectric* pattern. If blood circulation is not returned within approximately five minutes, there is a widespread depolarization of the cortex much like that seen in spreading depression but of larger amplitude. At this time there is a significant shift of Na^+ and Cl^- ions and water from the extracellular space into the cells. Irreparable neuronal damage begins with periods of asphyxiation of this duration and entails the possibility of a permanent lack of memory or consciousness. It is possible, however, to stop the circulation for times longer than five minutes by reducing the temperature of the brain through cooling the body. In this way brain surgery may be extended for a period of 20 to 30 minutes or longer.

The maintainance of adequate circulation to the brain is accomplished by a number of important reflex mechanisms outside the brain (Chap. 17). The fine control of circulation within the brain apparently depends not on local neural control but on metabolic mechanisms, primarily by variation in P_{CO_2} level, which in turn is determined by local metabolism. Increase in P_{CO_2} increases cerebral blood flow, and vice versa (Fig. 6-14). The use of a Fick technique (Chap. 11) for estimation of cerebral blood flow gives values of approximately 60 ml per 100 gm per minute, and this flow does not show much variation with changing conditions such as sleep or waking. Adequate blood flow is maintained in fairly constant fashion as systemic blood pressure varies from approximately 70 to 150 mm Hg and more; i.e., it shows autoregulation. Low blood pressures are dangerous, however, for with

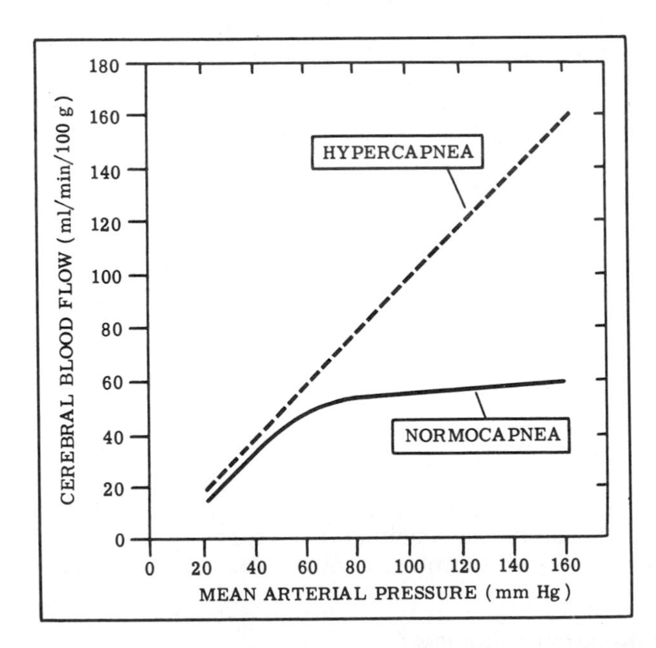

FIGURE 6-14. Pressure-flow relationship in the cerebral circulation. Autoregulation of blood flow is present at normal arterial blood levels of CO_2 (normocapnea) but is abolished by high arterial blood levels of CO_2 (hypercapnea). Then CBF varies as a linear function of mean arterial blood pressure. (From R. M. Berne and M. N. Levy. *Cardiovascular Physiology.* St. Louis: Mosby, 1967. P. 221.)

decreases to or below a level of approximately 40 to 50 mm Hg, as may occur in shock (Chap. 18), cerebral blood flow becomes inadequate for proper oxygen uptake and metabolism. Coma and permanent loss of cerebral function may occur if this is not soon reversed.

A more subtle alteration of metabolism is possible. As noted earlier, neurons have a high rate of protein synthesis which supplies materials carried outward in the fibers by transport mechanisms necessary to maintain normal nerve function and synaptic transmission. At present not enough is known about the effects of various diseases on this transport function. There is, however, a growing awareness of its role. For example, brain damage can occur in cases of protein deficiencies in the diets of infants, with a resultant subnormal intelligence later in life.

A large portion of the cells in the CNS and in the cortex are neuroglial (glial) cells. The exact function of the glial cells is not understood. According to one theory, the glial cells control the transport of materials from the blood capillaries to the neurons. This role has classically been that of the *blood-brain barrier,* usually ascribed to capillaries, which controls entry of substances to the extracellular space and thence to cells. Part of the evidence for such an intermediating role for glia comes from electron microscopic studies which show that there is very little extracellular space present in central nervous tissue, possibly too little for the diffusion of the materials from the blood capillaries to the neurons. However, the size of the extracellular space is likely to be similar to that found in other tissues – about 20 percent. This has been determined from electrical impedance studies and the increase in impedance upon asphyxiation when Na, Cl, and water enter the cells with a corresponding reduction of the extracellular space.

The seeming lack of an extracellular space noted in the usual electron micrographs (and in turn an anatomical evaluation of the size of the extracellular compartment) is not a settled question. It is possible that the apparent lack of extracellular space is due to the previously discussed translocation of Na, Cl, and water into cells during the time the tissue is removed and prepared for electron microscopic examination. An extracellular space with a value closer to that obtained from measurement of cortical impedance has recently been shown in electron micrographs by van Harreveld (1966) using freeze-substitution, and it is expected that this technique may help resolve not only the problem of the size of the extracellular compartment but the function of glial cells.

REFERENCES

Brazier, M. (Ed.). *Brain Function.* UCLA Forum in Medical Sciences No. 1. Berkeley: University of California Press, 1963.

de Robertis, E. *Histophysiology of Synapses and Neurosecretion.* New York: Macmillan, 1964.

Hernández-Peón, R. Central neurohumoral transmission in sleep and wakefulness. *Progr. Brain Res.* 18:96–117, 1965.

Hill, D., and G. Parr (Eds.). *Electroencephalography,* 2d ed. London: MacDonald, 1963.

Hubel, D. H., and T. N. Wiesel. Receptive fields of single neurons in the cat's striate cortex. *J. Physiol.* 148:574–591, 1959.

Hubel, D. H., and T. N. Wiesel. Receptive fields of single neurons in the cat's striate cortex. *J. Physiol.* 160:106–154, 1962.

Jouvet, M. Paradoxical sleep – a study of its nature and mechanisms. *Progr. Brain Res.* 18:20–62, 1965.

Kety, S. The Cerebral Circulation. In J. Field, H. W. Magoun, and V. E. Hall (Eds.), *Handbook of Physiology.* Section 1: Neurophysiology, Vol. III. Washington, D.C.: American Physiological Society, 1960. Chap. 61.

Krnjević, K. Iontophoretic studies on cortical neurons. *Int. Rev. Neurobiol.* 7:41–98, 1964.

Lorente de Nó, R. Cerebral cortex: Architecture, intracortical connections, motor projections. In J. F. Fulton (Ed.), *Physiology of the Nervous System,* 3d ed. New York: Oxford University Press, 1949.

Magoun, H. W. *The Waking Brain,* 2d ed. Springfield, Ill.: Thomas, 1963.

McIlwain, H. *Biochemistry and the Central Nervous System,* 3d ed. London: Churchill, 1966.

Ochs, S. The nature of spreading depression in neural networks. *Int. Rev. Neurobiol.* 4:1–69, 1962.

Ochs, S. *Elements of Neurophysiology.* New York: Wiley, 1965.

Ochs, S., and F. J. Clark. Tetrodotoxin analysis of direct cortical responses. *Electroenceph. Clin. Neurophysiol.* 24:101–107, 1968.

Penfield, W. G., and H. H. Jasper. *Epilepsy and the Functional Anatomy of the Human Brain.* Boston: Little, Brown, 1954.

Phillis, J. *The Pharmacology of Synapses.* New York: Pergamon, 1970.

Ross Conference. *Brain Damage in the Fetus and Newborn from Hypoxia or Asphyxia.* Columbus, Ohio: Ross Laboratories, 1967.

Rossi, G. F., and A. Zanchetti. The brain stem reticular formation. *Arch. Ital. Biol.* 95:195–435, 1957.

Sholl, D. A. *The Organization of the Cerebral Cortex.* London: Methuen, 1956.

van Harreveld, A. *Brain Tissue Electrolytes.* Washington: D.C.: Butterworth, 1966.

von Bonin, G. *Some Papers on the Cerebral Cortex.* Springfield, Ill.: Thomas, 1960.

7

Regulation of Visceral Function

A. The Autonomic Nervous System

Carl F. Rothe

HOMEOSTASIS, the constant and optimal internal environment of the body, is maintained in large part by the actions of the autonomic nervous system, for it is this part of the nervous system which provides a fine control of the visceral or internal functions of the body. It is generally involuntary and acts on the internal effectors such as (1) nonstriated (smooth) muscle, as in the intestine; (2) cardiac muscle; (3) exocrine glands, e.g., the sweat and salivary glands; and (4) some endocrine glands. The endocrine system also participates in the control of the visceral function of the body, but it is generally slower and acts through the release of internal secretions (hormones) which are transported by the cardiovascular system. The autonomic nervous system, along with the endocrine system, thus aids in keeping the internal environment of the body at such a composition and temperature that cellular life may proceed optimally.

CHARACTERISTICS OF THE AUTONOMIC NERVOUS SYSTEM

Anatomically, the autonomic nervous system is the efferent pathway linking the control centers in the brain and the effector organs such as smooth muscle and secretory cells. However, physiologically, control of visceral function must include sensors, afferent pathways, and central control centers as well. The sensory afferents from the viscera in the vagus and splanchnic nerves, for instance, serve both the somatic and the autonomic nervous system. Other sensors, such as of blood plasma osmolarity and carbon dioxide partial pressure, are located within the cells of the central nervous system itself. The autonomic nervous system differs from the somatic in that the motor neurons which come into immediate functional relationship with the effector cells lie completely outside the central nervous system. These motor neurons are called postganglionic, since they synapse with the preganglionic fibers coming from the spinal cord in ganglia (clumps of neuron cell bodies) located in chains beside the spinal cord or in the organ innervated. The adrenal medullae are an exception, for postganglionic neurons are not present. The adrenal medullary tissue is of the same embryological origin as postganglionic tissue, however, and the gland functions in a manner analogous to that of postganglionic fibers by releasing catecholamines (see below).

The autonomic nervous system may be divided, both functionally and structurally, into the *sympathetic division,* with neurons leaving the spinal cord from

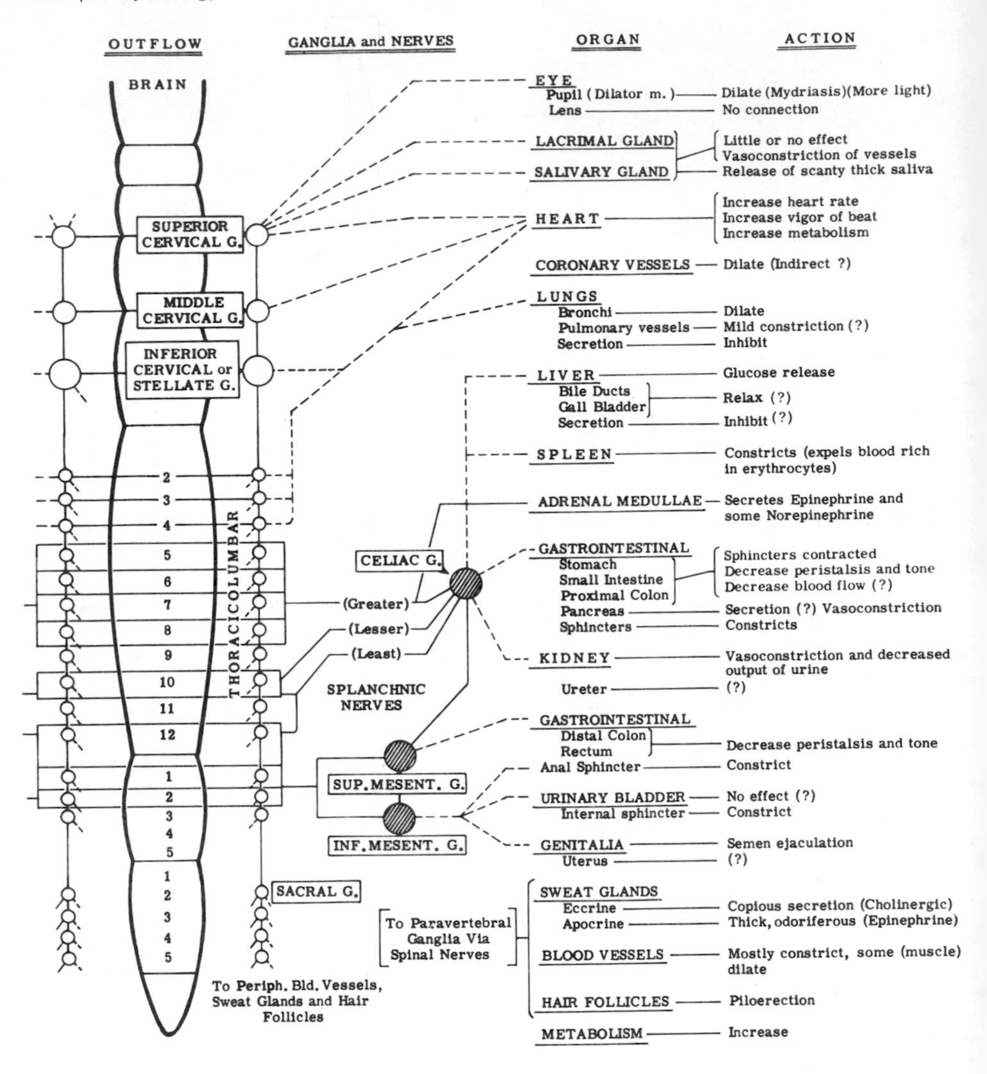

FIGURE 7A-1. Sympathetic division of the autonomic nervous system. Cholinergic fibers, ——; adrenergic (postganglionic) fibers, - - - -. The celiac, superior, and inferior mesenteric ganglia are called prevertebral (or paravertebral) ganglia, since the synapses (ganglia) are not in the sympathetic trunk along the spinal column.

the thoracic and lumbar segments (Fig. 7A-1), and the *parasympathetic division,* with neurons leaving from the cranial and sacral segments (Fig. 7A-2).

THE SYMPATHETIC DIVISION

The sympathetic division tends to act in a widespread manner to prepare the body for emergencies and vigorous muscular activity. Life is generally possible without

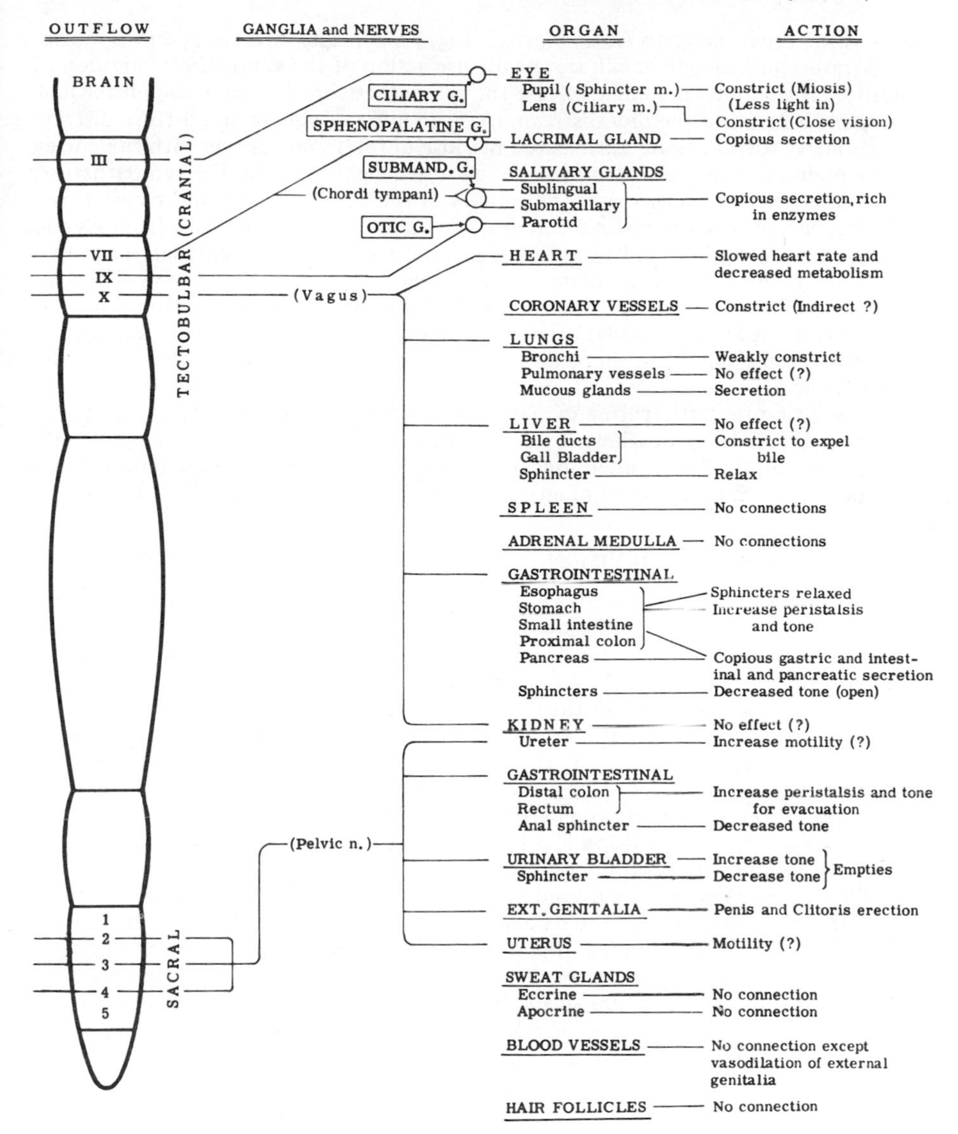

FIGURE 7A-2. Parasympathetic division of the autonomic nervous system (all nerve fibers are cholinergic).

the sympathetic division; but the animal must be sheltered, for the sympathectomized animal is much less resistant to extremes of environmental temperature, hypoxia, and other forms of stress. The cardiovascular system is no longer finely adjusted to provide for prolonged or severe exertion. The animal tends to be weak and apathetic.

The synapses are characteristically located either in the sympathetic paravertebral chain on each side of the spinal cord throughout its length, or in special ganglia such as the celiac ganglion *(solar plexus)* (Fig. 7A-1). The interconnections between the paravertebral ganglia facilitate the diffuse action of the sympathetic division. In addition, the adrenal medullae release the sympathetic mediators, epinephrine and norepinephrine, into the blood stream for even more complete and diffuse distribution.

Homeostatic needs are anticipated because of interconnections with the cortex of the brain. For instance, the mere thought of a fight activates the sympathetic division, which in turn increases the activity of the cardiovascular and respiratory systems, curtails gastrointestinal activity, causes the release of glucose from glycogen stores, channels more blood to muscles, and starts sweating for removal of excess heat. The pupils of the eyes dilate, the eyes tend to protrude, and the eyelids widen. In many animals, the hairs of the back and tail bristle. All these reactions are in anticipation of vigorous muscular activity. The sympathetic division thus acts in emergencies to adjust the organism to an adverse environment or a rapid change in internal requirements.

In addition to participating in massive discharge for emergency situations, the various parts of the sympathetic division function in everyday life in independent and discrete ways. The connections of the sympathetic division with various organs and its actions on them are summarized in Figure 7A-1 and bear careful study.

THE PARASYMPATHETIC DIVISION

The parasympathetic division acts more discretely on individual organs or regions. The abdominal viscera are innervated by preganglionic neurons which leave the cranial part of the cord and travel via the vagus nerves. The synapses are generally located *within* the innervated organ. The vagi contain most of the parasympathetic efferent fibers, although they consist mainly of sensory afferents which participate in visceral reflexes, such as those from the abdomen and the pressure and chemical receptors located in the aortic arch and the heart. There are also a few somatic efferent fibers. Dogs, in addition, have in the cervical vagus trunks some cervical postganglionic sympathetic nerve fibers to the head. The sacral part of the parasympathetic division is primarily concerned with the emptying of the pelvic organs and with reproductive functions.

Whereas the sympathetic division is mainly involved with homeostasis during voluntary muscular activity, the parasympathetic division is involved with restorative vegetative functions such as digestion and rest. Imagine an old man sleeping after eating. His heart rate is slow, his breathing is noisy because of bronchial constriction, and his pupils are small. Saliva may run from the corners of his mouth, and rumbles from his abdomen reveal much intestinal activity.

The mediator that is released at the parasympathetic nerve endings is the same as in the somatic nervous system and ganglia – acetylcholine. The actions on various organs are presented in Figure 7A-2 and should be studied.

THE HYPOTHALAMUS

The hypothalamus (Chap. 8) is the center of interrelating the visceral and somatic functions of the body. Not only are motor actions accompanied by widespread and complex visceral responses, but visceral activity modifies somatic reactions. For instance, during digestion the increased blood flow through the intestine tends to reduce the muscular capacity for work. Such interdependence of visceral and somatic functions is implied in the concept that central control of both functions is exerted through neural mechanisms that are located at common levels in the spinal cord, brain stem, diencephalon, and cerebral cortex; and that both have a common sensory inflow.

If the spinal cord is sectioned at about the first thoracic vertebra, there is an immediate depression not only of somatic but also of autonomic reflexes. Blood pressure drops as a result of loss of peripheral vasoconstrictor tone, sweating is absent, and the body temperature is poorly regulated and tends to approach the environmental temperature. The evacuation reflexes for the bladder and bowel are absent, as are the sexual reflexes. With the passage of time, many of these reflexes return, in part, with the further development of the spinal segmental reflexes; but regulation of blood pressure and of body temperature is severely limited, and micturition, defecation, and sexual reflexes are incomplete.

The decerebrate preparation responds in a similar fashion except that many cardiovascular centers located in the medulla are intact, and cardiovascular regulation is therefore more nearly normal.

If the hypothalamus is left intact by sectioning the brain above the diencephalon, removing only the cerebral cortex, the animal has normal vegetative regulation of homeostasis, but the adjustments appropriate for the anticipation of muscular activity are missing.

As discussed in Chapters 8 and 9 and other appropriate chapters, the hypothalamus is concerned with the regulation of (1) the cardiovascular system, (2) respiration, (3) body temperature, (4) body water and electrolyte concentrations, (5) food intake, and gastric and pancreatic secretion, (6) hypophysial secretion and sexual and maternal behavior, and (7) emotional states.

RECIPROCAL ACTION

Most of the visceral organs have a dual, antagonistic innervation, in that the actions of the two divisions are diametrically opposed. For instance, the sympathetic division increases cardiac activity, and the parasympathetic division decreases it. The parasympathetic division acts to enhance and accelerate visceral functions, such as increased gastrointestinal motility, while the sympathetic division generally acts to constrict visceral blood vessels and to inhibit visceral function such as digestion and elimination.

An exception to the generalization of reciprocal action is the control of secretory function of the salivary glands and pancreas; both divisions seem to stimulate secretion. In addition, not all organs have this dual innervation. The sympathetic division has exclusive innervation of the adrenal medullae, spleen, pilomotor muscles, sweat glands (although the postganglionic fibers are cholinergic), and probably the vasomotor muscles of the blood vessels of the viscera, skin, and skeletal muscle, if not all blood vessels. The parasympathetic division innervates the ciliary and the sphincter muscles of the eye. Functionally, the pupil of the eye has an antagonistic dual innervation; structurally, however, the dilator muscles have sympathetic innervation whereas the constrictor muscles have parasympathetic innervation.

TONE

The autonomic system generally maintains a "tone," a basal level of activity, which then may be either increased or decreased by central control. For example, the flow of blood through an innervated muscle is about one-third to one-half that observed following interruption of sympathetic nerve supply to the muscle, indicating that a sympathetic tone is acting under normal conditions to constrict the vessels. Because of this vasomotor tone, the blood vessels can be either dilated by a decrease in sympathetic division activity or further constricted by an increase in activity.

The parasympathetic division, by way of the vagus, provides a "vagal tone" to the heart. If the vagi of an animal under basal conditions are sectioned, the heart rate increases as a result of removal of this inhibiting effect. Parasympathetic tone is also present in the intestinal tract; it may be interrupted by sectioning the lower vagi. This causes a serious and prolonged, but not permanent, gastric and intestinal dysfunction.

The presence of dual innervation and the possibility of either increasing or decreasing the tone permit a wide range of control.

MEDIATORS

The chemical mediators released at the nerve endings of the autonomic nerve fibers act on *receptors* to produce, in turn, an effector response. A receptor may be conceived as that molecular structure with which a single molecule of mediator or drug interacts. The site may be either an enzyme or the structural configuration of the cell membrane.

The mediator released by the preganglionic fibers of both divisions is acetylcholine. On the other hand, it is difficult to generalize concerning the postganglionic mediator (Fig. 7A-3). The parasympathetic postganglionic fibers are all cholinergic; that is, they release acetylcholine as the transmitter agent. The mediator substance released at most of the postganglionic sympathetic nerve endings is *norepinephrine.* Such fibers are *adrenergic. Epinephrine* and some norepinephrine are released into the blood by the chromaffin cells of the adrenal medullae. Norepinephrine is also called levarterenol, arterenol, or noradrenalin. Epinephrine is norepinephrine plus an N methyl group, and is also called methylarterenol or adrenalin. *Isoproterenol,* a synthetic catecholamine, is norepinephrine with an N isopropyl group. These compounds as a class are called catecholamines. In

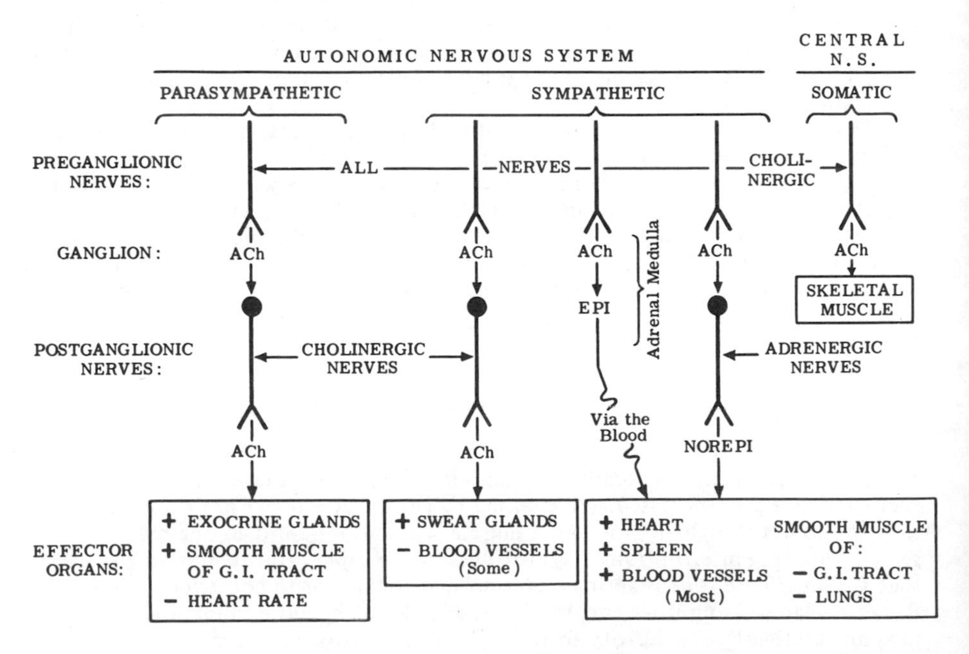

FIGURE 7A-3. Classification of nerves in terms of transmitter or mediator released and location of cell body and synapse; + means stimulatory, − means inhibitory.

addition, some postganglionic sympathetic fibers, such as those going to the sweat glands, are cholinergic, releasing acetylcholine. The best evidence for this conclusion is that their activity is blocked by atropine, a drug which blocks the action of acetylcholine on these cells.

The *mode of action* of acetylcholine and the catecholamines on excitable membranes, such as those of the intestinal smooth muscle and the heart, is apparently that of modifying the rate of polarization and depolarization, thus changing the frequency of spontaneous spike discharge which is the basis of the autorhythmicity characteristic of these tissues. Acetylcholine increases and epinephrine decreases the frequency of spontaneous spike discharge of intestinal smooth muscle. The fact that each spike is followed by a slight contraction explains the effect of these mediators on this tissue. On the other hand, acetylcholine slows or stops the heart beat by hyperpolarizing the pacemaker cell membrane and slowing depolarization, both probably due to an increased permeability of the resting membrane to potassium. The net effect is opposite to that of acetylcholine on intestinal smooth muscle or on the neuromuscular junction, for here depolarization is enhanced. Epinephrine accelerates the heart by increasing the rate of depolarization of the pacemaker during diastole. Thus the response of a particular end-organ to autonomic nervous system activity depends upon the individual makeup of that organ.

Epinephrine and norepinephrine generally act in the same manner, in that both tend to constrict the blood vessels, stimulate the heart, and inhibit the viscera. However, the blood vessels of skeletal muscle are constricted by norepinephrine and, under some conditions, dilated by epinephrine. This is because epinephrine acts in three ways: (1) *Direct, excitatory.* (Other terminology includes *alpha response* and *alpha adrenergic receptor stimulation.*) Epinephrine tends to increase the permeability of the membrane to sodium ions or other ions. The resulting depolarization initiates contraction, viz., vasoconstriction. (2) *Metabolic.* (Also called *beta response* or *beta adrenergic receptor stimulation.*) Epinephrine also acts to increase cellular metabolism. Conversion of glycogen to glucose is speeded; there is increased release of free fatty acids, and oxygen consumption increases. (3) *Direct, inhibitory.* Although vasodilation also occurs in response to increased metabolism by way of local control of the tissue blood flow (Chap. 17), epinephrine appears to have a direct vasodilator effect mediated by beta adrenergic receptors. The sodium pump activity is increased. This stabilizes the membrane by increasing the polarization, and so the muscle relaxes, viz., vasodilation or intestinal smooth muscle relaxation. The observed action of epinephrine is, therefore, the net result of these actions (Fig. 7A-4). In the intestines, since the direct excitatory effect is weak, relaxation is usually seen.

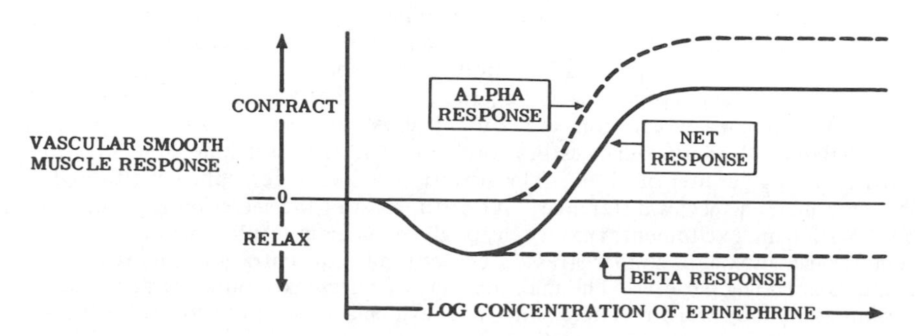

FIGURE 7A-4. Net effect of various concentrations of epinephrine on the vasculature. (After suggestion of D. O. Allen.)

Norepinephrine causes much less beta response than does epinephrine and so is primarily excitatory (alpha response). On the other hand, isoproterenol induces a beta response (increased metabolism, vasodilation, and increased vigor of cardiac contraction). A complication to this classification must be mentioned: Some smooth muscle may be excited even though it is depolarized by high potassium concentrations.

The mediators must be removed from the site of action if they are to be effective under dynamic conditions. *Acetylcholinesterases* destroy acetylcholine by hydrolysis – normally within less than a minute if acetylcholine is introduced into the blood stream. The enzymes are also found in high concentration around nerve endings. The catecholamines are relatively stable in blood but are rapidly inactivated in body tissues. The major mechanism for catecholamine inactivation is the active uptake of the catecholamine by the postganglionic sympathetic nerve endings.

THE ADRENAL MEDULLAE

The release of epinephrine into the blood provides a mechanism for spreading the effects of generalized sympathetic division activity to all cells of the body. These effects are primarily on metabolic processes. Adrenal medullary secretions are important factors in the homeostatic control of blood glucose concentration and release of free fatty acids from adipose tissue during conditions of stress such as violent muscular activity, hypotension, asphyxia, and hypoxia. Epinephrine secreted by the adrenals facilitates synaptic transmission, increases the heart rate and strength of contraction, and increases metabolism. Although the secretions stimulate the cardiovascular system, the direct adrenergic innervation can produce still higher maximal levels of activity.

The adrenal medullae, like the rest of the sympathetic division, are not essential for life if emergency demands are minimized. In fact, it is difficult to show deficiences under nonstressful conditions following removal of these organs. However, a reduction in glycogenolysis and a reduced stress tolerance are seen. Functionally, the adrenal medullary cells are equivalent to sympathetic postganglionic nerve cells. Their cholinergic preganglionic innervation and their embryological origins are similar. The cells are stimulated to secrete under the same conditions of stress as those affecting the adrenergic postganglionic neurons. In both, the catecholamines are held in minute granules and so are protected from enzymatic degradation. The neuroadrenomedullary and ganglionic junctions respond similarly to ganglionic blocking or facilitatory agents. The one clear difference is the secretion of epinephrine by the medullae and, on the other hand, norepinephrine by the adrenergic fibers. The endocrine secretion of the adrenal medullae supplements the neural action of the sympathetic division activity. In addition to the adrenal medullae, similar chromaffin cells secreting small amounts of epinephrine are located adjacent to the paravertebral chain of sympathetic ganglia, near the carotid sinus, heart, and liver.

Adrenal medullary secretions are controlled by centers located in the posterior hypothalamus. Under normal resting conditions the catecholamine concentration is 0.1 to 1.0 μg per liter of blood. The secretion is mainly epinephrine, although there are marked species differences. At birth, most of the secretion is norepinephrine. With pain, excitement, anxiety, hypoglycemia from insulin injection, cold, hemorrhage, anoxia, or similar stresses, the systemic concentrations increase to as much as 100 μg per liter. The maximal rate of release is about 2 μg per minute per kilogram body weight. There is no clear evidence of separate control of norepinephrine and epinephrine secretion.

Because of the metabolic stimulation and glycogenolytic actions, epinephrine from the adrenal medullae has a particular value in homeostasis in response to decreases in body temperature. Cardiac stimulation and facilitation of synaptic transmission by epinephrine are remarkable during hypothermia.

THE PINEAL GLAND

The pineal gland is a neurosecretory gland analogous to the adrenal medullae and neurohypophysis. It is innervated by sympathetic preganglionic nerves and not with fibers originating in the brain. *Melatonin,* a hormone secreted by the pineal gland, acts on the ovaries to inhibit the estrus cycle. Stimulation of the retina by ambient light provides information which is transmitted by way of sympathetic nerves to inhibit the synthesis of melatonin. Light thus inhibits the inhibition of gonadal activity. The role of the pineal gland and melatonin in the timing of the menstrual cycle, or in other visceral functions which may be related to the ambient lighting conditions, is not as yet clear.

EFFECTORS – SMOOTH MUSCLE

Smooth muscle is the effector organ of much of the autonomic nervous system. *Multiunit* smooth muscle has innervation similar to skeletal muscle. Little or no spontaneous or rhythmic activity is present. The nerve supply is excitatory. Examples include the muscle of the iris, the piloerector and nictitating membrane muscles of animals such as the cat, and probably vascular smooth muscle. *Visceral* smooth muscle, on the other hand, contracts spontaneously and rhythmically in response to the local chemical environment or to stretch. This activity is modified by the innervation from the autonomic nervous system, which either inhibits or excites further activity.

In most, if not all, types of smooth muscle under normal conditions, electrical activity of the muscle membrane precedes contraction (Chap. 3B). A prolonged contraction comes about as a result of repetitive action potentials. The evidence is not certain for all organs because of the minute size of the smooth muscle cells, most of which have diameters of the order of 1 μ or less. In the case of visceral smooth muscle and the heart, the spontaneous contraction results from an unstable resting membrane potential of the pacemaker cell, which upon reaching threshold initiates an all-or-none action potential that is transmitted from cell to cell. Stretching visceral smooth muscle in many cases depolarizes the membrane to initiate an action potential. Sodium and potassium are important in determining the resting membrane potential, while sodium and calcium are involved in the action potential (Chap. 3B). The smooth muscle cell membrane is apparently much more permeable to sodium at rest than that of neurons or skeletal muscle, because cellular metabolism and the sodium pump activity are important factors in determining visceral smooth muscle tension.

EFFECTS OF THE AUTONOMIC NERVOUS SYSTEM ON SPECIFIC ORGANS

THE EYES

The adjustment of the eyes to varying light conditions and to focusing at various distances is controlled by the autonomic nervous system. Sympathetic activity to the radially oriented muscle of the iris causes it to dilate and to allow more light to enter the eye. Neural impulses of the parasympathetic division cause the sphincter muscles of the iris to contract, and thereby cause constriction of the pupil. During excitement the pupils of the eye are characteristically dilated *(mydriasis)*, whereas poisoning with anticholinesterase compounds (see below) causes a characteristic pinpoint pupil *(miosis)*.

Focusing of the lens of the eye is controlled almost exclusively by the parasympathetic division. The lens at rest is focused for distant vision and is relatively flat, but excitation by the parasympathetic division causes the ciliary muscles to contract and make the lenses more spherical for accommodation for near vision.

THE HEART

As discussed in the chapters on circulation, vigorous muscular activity is accompanied by an increased sympathetic division activity, causing the heart to beat faster and to contract more vigorously so that more blood is pumped. At some later time, the parasympathetic division, by depressing cardiac activity, serves a restorative function.

SMOOTH MUSCLE OF THE SYSTEMIC BLOOD VESSELS

The blood vessels of the abdominal viscera and skin are generally constricted by generalized sympathetic activity. On the other hand, the vessels of the heart and active muscles are dilated either directly or indirectly. Blood vessels other than those to the external genitalia have no parasympathetic innervation. Sympathetic cholinergic nerves for active vasodilation of skeletal muscle act to increase the flow of blood even before contraction starts. The cutaneous vasodilation enhancing heat loss probably results from a decrease in the sympathetic vasoconstrictor tone. Increased blood flow of salivary glands following stimulation via the parasympathetic division is due to the increase in metabolism of the organs and the release of *bradykinin,* a polypeptide which is a potent vasodilator on the one hand and an intestinal and uterine smooth muscle stimulant on the other. The evidence for a *direct* vasodilating action by the parasympathetic or sympathetic division has been uncertain: Much, but not all, of the data can be explained either by a reduction in sympathetic vasoconstrictor tone or by the release of vasodilator materials resulting from the enhanced metabolism brought about by increased activity of the tissues, e.g., heart, muscle, or glands (see Chap. 17).

If the flow of blood to the skin, intestinal tract, and kidneys is reduced (for example, during exercise and hemorrhage), blood flow is available for diversion to active muscle, the heart, and the brain. On the other hand, muscle blood flow is decreased during asphyxia.

SPLEEN

The contraction of the spleen and subsequent discharge of blood rich in erythrocytes is another effective mechanism for maintenance of homeostasis during physiological stress, such as exercise, hemorrhage, and anoxia. The erythrocyte storage function of the spleen is of minor importance in man.

RESPIRATORY TRACT

The autonomic system apparently has a minor effect on the functioning of the lungs. Consistent with the concept of the importance of the sympathetic division for voluntary muscular activity, sympathetic division stimulation causes the bronchi to dilate, thus allowing air to enter more easily. The parasympathetic division acts in an opposite manner, and also stimulates the secretion of mucus.

EXOCRINE FUNCTION

Most of the exocrine glands of the body are stimulated to secrete as a result of parasympathetic division activity. An exception is the mammary gland; milk secretion is under hormonal control. The *nasal* and *lacrimal glands* secrete mucus and tears, thereby providing a protective function. The *salivary glands,* when stimulated by the parasympathetic division, supply a copious secretion for mastication and digestion. The *glands of the stomach* and *pancreas* are stimulated by parasympathetic division impulses and provide an increased secretion of pancreatic and digestive juices for digestion. Not all control of these secretions is neural, however.

The sympathetic division has little or no effect upon exocrine secretion except that its activity generally causes vasoconstriction, which tends to reduce secretion. Exceptions are the effects on the salivary glands and pancreas. Sympathetic stimulation causes a small but definite increase in flow of saliva, which is thicker and more viscous than that following parasympathetic division activity. Splanchnic nerve stimulation (sympathetic) causes some pancreatic secretion and changes in the zymogen granules of the acinar cells of the pancreas.

The *eccrine* type of *sweat gland,* producing copious watery sweat *(hidrosis),* is innervated by sympathetic nerves with no connections from the parasympathetic division. Their innervation is by cholinergic fibers, an exception to the general rule that sympathetic postganglionic fibers are adrenergic.

The *apocrine glands,* located primarily in the axillary and pubic regions, secrete a thick material which is decomposed by skin flora to produce characteristic body odors. Although closely related to the eccrine glands, they respond to adrenergic stimulation. In fact, there is evidence that they are not innervated, but respond to circulatory epinephrine only. Whereas the eccrine glands are associated with heat loss, the apocrine glands are associated with fear, anxiety, and vigorous muscular or sexual activity.

SMOOTH MUSCLE OF THE GASTROINTESTINAL TRACT

Although the gastrointestinal system has its own intrinsic neural control, both divisions of the autonomic nervous system affect the gastrointestinal activity. Parasympathetic division activity generally increases gastrointestinal motility. The role of the sympathetic division is primarily an inhibiting one seen only in some diseases or during stress.

The esophagus is supplied by parasympathetic nerves. The involuntary parts of the process of swallowing are mediated by the parasympathetic division, which initiates a wave of constriction down the musculature of the esophagus.

SMOOTH MUSCLE OF THE PELVIC VISCERA

The parasympathetic system is involved in contraction of the bladder and lower colon for emptying of these organs and is essential for erection of the penis and clitoris. The effect on the uterus is variable and depends on its physiological state. The sympathetic division has an inhibiting effect on urination and defecation. An intact sympathetic reflex arc is not essential for emission but is essential for the ejaculation of semen. The sympathetic division is apparently not essential for gestation.

RESPONSE TO DENERVATION

Immediately following sectioning of sympathetic or parasympathetic nerves, the denervated organ loses most of its tone and activity, but not permanently. Compensation develops which is primarily an increased sensitivity to circulating chemical agents. Denervation supersensitivity is most pronounced if the postganglionic nerves are severed. This was first studied in connection with *Horner's syndrome,* which results from interruption of the sympathetic division supply to the face. As a result of lack of sympathetic tone the pupils are constricted (miosis), the eyelids droop *(ptosis),* and the face is flushed. With the passage of time, however, these symptoms tend to disappear. If one side of the face has the preganglionic sympathetic supply sectioned and the other side the postganglionic fibers, at first the effects are symmetrical, but with time the side with the postganglionic innervation cut shows a larger pupil (return of sympathetic tone) than does the other side. The discrepancy is intensified when the animal is frightened. These phenomena are explained as follows: With destruction of the postganglionic fibers, the major inactivation mechanism for catecholamine (i.e., uptake by nerve endings) is removed, and so the receptors are exposed to stimulating concentrations of blood-borne catecholamines. Following preganglionic section, the neural input is blocked and circulating catecholamines in the vicinity of the receptors are readily inactivated by the postganglionic endings nearby. Some supersensitivity develops but of lesser magnitude. Denervation supersensitivity also follows parasympathetic nerve sectioning. Even skeletal muscle shows an increased sensitivity to the mediator after denervation, for sectioning of the motor nerve causes the motor end-plates to become hypersensitive to acetylcholine.

Since preganglionic denervation causes less sensitization than does postganglionic denervation, it provides a surgical tool. For example, *Raynaud's disease* involves painful paroxysmal cutaneous vasospasms (usually in the fingers and toes) following exposure to cold or emotional stress. The vasospasm is so intense that gangrene sometimes results from the lack of circulation. If the postganglionic fibers are sectioned, it is only a matter of time before the blood vessels respond excessively to circulating epinephrine. Preganglionic denervation, however, eliminates the neural constrictor influence without the development of the marked hypersensitivity to the circulating epinephrine.

TRAINING THE AUTONOMIC NERVOUS SYSTEM

Recent studies (e.g., Miller et al., 1970) have supported the concept that the autonomic nervous system can be taught and that people who are in the early stages of hypertensive disease can possibly be trained to reduce their blood pressure. Such visceral learning would be of great therapeutic significance and is the subject of continuing research.

DRUGS ENHANCING OR DEPRESSING AUTONOMIC FUNCTIONS

Drugs which mimic or block sympathetic or parasympathetic division activity provide a tool for a better understanding of normal physiological processes and may be used to reverse or counteract some diseases. Pharmacology texts should be consulted for details.

PREGANGLIONIC SIMULATING DRUGS

Preganglionic simulating agents, e.g., acetylcholine in very high concentrations and nicotine, mimic the activity of the preganglionic neurons and thereby stimulate postganglionic neurons of both divisions. Since nicotine and "nicotinic" drugs such as carbachol stimulate not only sympathetic and parasympathetic postganglionic fibers but also skeletal muscle, it is thought that the receptor system at the postganglionic neurons is similar to the system at the neuromuscular junction.

PARASYMPATHOMIMETIC OR CHOLINERGIC DRUGS

Because acetylcholine is so rapidly destroyed in blood after injection, it does not cause the same effects throughout the body as parasympathetic division activity. However, drugs such as methacholine, muscarine, and pilocarpine are less rapidly inactivated and have effects in the body similar to those of the acetylcholine which is released from parasympathetic endings. These drugs also simulate the activity of sympathetic *cholinergic* fibers; e.g., profuse sweating is seen following administration.

The parasympathetic effects may be potentiated if acetylcholine destruction is inhibited. An agent such as eserine or prostigmine inhibits acetylcholine hydrolysis and thus potentiates parasympathetic activity. Because of this anticholinesterase activity, transmission at sympathetic ganglia and neuromuscular junctions is also enhanced. The poisoning action of most organic phosphorus pesticides is by way of anticholinesterase activity.

The action of the alkaloid muscarine on viscera such as cardiac muscle, exocrine glands, and smooth muscle is similar to that of acetylcholine. The effects are termed the *muscarinic actions* of acetylcholine. Muscarine has little effect on skeletal muscle or ganglionic transmission. *Nicotine,* on the other hand, affects autonomic ganglia – both parasympathetic and sympathetic – and skeletal muscle much as acetylcholine does. Thus, the nicotinic actions of acetylcholine may be differentiated from the muscarinic actions.

SYMPATHOMIMETIC OR ADRENERGIC DRUGS

Sympathomimetic or adrenergic drugs mimic the effects of sympathetic division discharge. Drugs such as ephedrine, amphetamine, and isoproterenol have effects similar to the action of epinephrine and norepinephrine, which are natural hormones. They differ, however, in potency at various effector sites, in mode of action, and in the duration of their activity.

BLOCKING AGENTS

There are two general types of blocking agent used for research and therapy. *Depolarizing* agents, such as nicotine in high concentration, block by maintaining depolarization of the excitable membrane. At first the effector organ is stimulated, but then blockage of transmission occurs. *Antidepolarizers* generally block by competing with acetylcholine for the receptor sites on the postganglionic or postjunction membrane. Hexamethonium and tetraethylammonium act in this manner at the ganglia, d-tubocurarine has a similar blocking action at the skeletal muscle neuromuscular junction, and atropine blocks at the visceral sites.

At the ganglia the nicotinic effects of acetylcholine are blocked by agents such as tetraethylammonium and trimethaphan camphorsulfonate, but atropine or curare is relatively ineffective. Ganglionic blocking agents are used primarily to block sympathetic activity, as in reducing hypertension, but also act on the parasympathetic division ganglia.

At the neuromuscular junction of skeletal muscle the receptor agent for acetylcholine is different from that in smooth muscle, for here atropine is not effective, while d-tubocurarine is. Succinylcholine is an example of a depolarizing blocking agent, whereas curare reversibly binds the receptor.

At the visceral effector organs, the muscarinic effects of acetylcholine are blocked by atropine and scopolamine. Atropine blocks not only the excitatory effects of acetylcholine, such as those on the intestine or exocrine glands, but also the inhibiting effects such as in the heart.

The direct excitatory (alpha) responses of sympathetic effector cells to agents that cause vasoconstriction are blocked by drugs such as Dibenamine and Dibenzyline (phenoxybenzamine). After administration of these drugs, epinephrine dilates certain blood vessels – the constricting effects are blocked. The drugs do not block the beta receptors, for catecholamines continue to stimulate metabolism and inhibit the gastrointestinal tract.

The beta response to epinephrine is blocked by propranolol. Use of such agents aids in studying the action of catecholamines.

Reserpine appears to block the sympathetic system, but in fact it acts to reduce the amount of catecholamines in the brain and adrenal medullae available for subsequent release after neural stimulation. Indeed, the catecholamines in the heart and blood vessels approach zero in animals treated with high doses of reserpine.

REFERENCES

Ahlquist, R. P. Effects of the Autonomic Drugs on the Circulatory System. In J. Field (Ed.), *Handbook of Physiology.* Section 2: Circulation, Vol. III. Washington, D.C.: American Physiological Society, 1965. Pp. 2457–2677.

Burn, J. H. *Autonomic Nervous System.* Oxford, Eng.: Blackwell, 1963.

Goodman, L. S., and A. Gilman (Eds.). *The Pharmacological Basis of Therapeutics,* 3d ed. New York: Macmillan, 1965.

Hillarp, N.-Å. Peripheral Autonomic Mechanisms. In J. Field (Ed.), *Handbook of Physiology.* Section 1: Neurophysiology, Vol. II. Washington, D.C.: American Physiological Society, 1959. Pp. 979–1038.

Ingram, W. R. Central Autonomic Mechanisms. In J. Field (Ed.), *Handbook of Physiology.* Section 1: Neurophysiology, Vol. II. Washington, D.C.: American Physiological Society, 1959. Pp. 951–978.

Kuntz, A. *The Autonomic Nervous System.* Philadelphia: Lea & Febiger, 1953.

Malmejac, J. Activity of adrenal medulla and its regulation. *Physiol. Rev.* 44:186–218, 1964.

Miller, N. E., L. V. DiCara, H. Solomon, J. M. Weiss, and B. Dworkin. Learned modifications of autonomic functions: A review and some new data. *Circ. Res.* 27 (Suppl. I, Hypertension):I-3 to I-11, 1970.

Root, W. S., and F. G. Hofmann (Eds.). *Physiological Pharmacology,* Vols. III, IV. New York: Academic, 1967.

Spector, W. S. (Ed.). *Handbook of Biological Data.* Philadelphia: Saunders, 1956. Pp. 305–311.

Von Euler, U. S. Autonomic Neuroeffector Transmission. In J. Field (Ed.), *Handbook of Physiology.* Section 1: Neurophysiology, Vol. I. Washington, D.C.: American Physiological Society, 1959. Pp. 215–237.

White, J. C., R. H. Smithwick, and F. A. Simeone. *The Autonomic Nervous System,* 3d ed. New York: Macmillan, 1952.

Wurtman, R. J., and J. Axelrod. The pineal gland. *Sci. Amer.* 213:50–60, 1965.

7

Regulation of Visceral Function

B. Homeostasis and Negative Feedback Control

Carl F. Rothe

EACH CELL of the body requires an environment which supplies nutrients and removes metabolic wastes. The concept of *a constant and optimal internal environment* as a necessity for normal function was first formulated by Claude Bernard a century ago. Cannon in 1929 further developed the concept of this condition, which he called *homeostasis,* and emphasized the role of the autonomic nervous system. One of the cardinal principles of physiology is that homeostatic mechanisms operate to counteract changes in the internal environment which are induced either by changes in the external environment or by activity of the individual. Thus, disturbances of the internal environment as the result of exercise, nutritional imbalance, trauma, and disease are minimized.

An example of homeostasis is the control of body temperature. If the internal temperature drops, homeostatic mechanisms act to reduce heat loss from the body and to increase heat production. Consequently, these mechanisms limit a decrease in body temperature and so maintain this variable relatively constant. On the other hand, a cold-blooded animal does not possess homeostatic temperature control systems; its body temperature thus tends to be similar to that of its environment.

Homeostatic mechanisms act to minimize the difference between the actual and optimal responses of a system, and are therefore biological examples of *negative feedback control.* In systems of this type the level of the controlled variable is sensed, and action is taken which opposes any change from the desired level. If the response increases, a signal is fed back to an effector mechanism in a negative or inhibitory manner so that the subsequent response is reduced. On the other hand, a decrease in response elicits a subsequent increase. A familiar example of a negative feedback control system is the thermostatic control of room temperature. This is illustrated in Figure 7B-1A. Room temperature is measured by a temperature-sensing element in a thermostat and compared with a reference (desired or optimal temperature) in such a manner that an *error signal* develops when discrepancies exist. If the room is too hot, the furnace is turned off by the action of the thermostat, and if it is too cold, the furnace is turned on. The error signal from the thermostat is used to adjust the system automatically so as to minimize any deviation between the measured temperature and the desired temperature. If the system is properly designed, room temperature can be held nearly constant despite wide fluctuations in outside temperature.

Homeostatic feedback mechanisms of mammals are exceedingly complex and interrelated but are generally amenable to analysis using the approach of engineers.

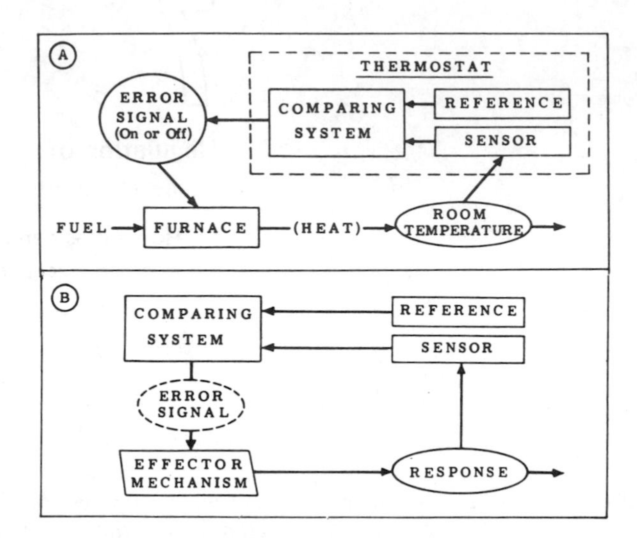

FIGURE 7B-1. Negative feedback control system. A. Schematic diagram of components of a thermostatically controlled furnace. B. Components of a negative feedback control system.

These workers, utilizing the principle of negative feedback control, have made great advances in the design of control systems used in the operation of devices such as automatic airplane pilots, missile guidance systems, computers, and servomechanisms for industrial automation. The primary purpose of using negative feedback in these mechanisms is to provide high accuracy and stability of function in spite of changes in the external environment or changes in the systems themselves. This same general principle acts in mammals to keep constant and optimal such diverse variables as body temperature, blood pressure, blood sugar concentration, electrolyte concentrations, muscle tone, and blood carbon dioxide levels, to mention only a few. The human body has a large number of negative feedback systems, all perfected through evolutionary development. The autonomic nervous system is an important part of most homeostatic mechanisms. To attain a better understanding of the physiology of the normal human, it is necessary to understand the fundamental characteristics of these negative feedback control systems.

SYSTEMS

The concept of a system – a set of components which act together, and which can be treated as a whole – has been found to be highly useful for the study of dynamic, complex biomedical situations. With a system, *inputs* are those stimuli, forcings, or disturbances which act on it, while the *output* or response is the consequence of such inputs. Furthermore, the relationship between output and input is the "law" which describes the system. It is usually written as a mathematical equation. In some cases, such as that describing the relationship between the input forcing of a transducer and the resulting pen deflection of a recorder, the equation is simple; in others, such as those predicting changes in cardiac function in response to changes in pressure at the pressoreceptors, the relationship is so involved that complex differential equations are required to provide even an approximation. If the purpose of the analysis is to understand and predict the behavior of complex systems, the output-to-input relationship written in mathematical form with carefully defined terms is an effective form for the theory describing the system.

SYSTEMS ANALYSIS SIMULATION AND MODELS

As our understanding of biological systems becomes more extensive, it becomes more difficult to organize the information, to check it for consistency, and to predict the consequences of various stimuli or changes of parameters of the system. *Mathematical biology, systems analysis,* and *simulation* are different words for the general approach of (1) describing our understanding through models written in the form of mathematical equations and (2) studying the models with computers to check for internal consistency and for agreement with available experimental data. As an example, both the plasma osmolality and the transmural pressure of the left atrium are important variables in the control of the secretion of antidiuretic hormone, which in turn acts on the kidney in the control of blood volume. What is the relative importance of each, under various conditions? How fast does the system respond? Is there a synergistic response? A threshold? Saturation of response? A mathematical model may provide a concise and accurate summary of thousands of experiments and may permit accurate predictions under a wide variety of conditions.

A *model* is a representation of reality, be it a stereotype, archetype, caricature, syndrome, diagram, prejudice, equation, or verbal description. In comparison to verbal models, mathematical models are more precise, less ambiguous, and more readily permit quantitative verification. The use of computers facilitates the study of multifactor interaction and dynamic (i.e., transient or moment-by-moment) responses. In the process of developing and testing the model – called *simulation* – new experiments are suggested, for gaps in knowledge often become obvious. Not only do models provide a concise summary of pertinent observations, but for many they provide a more satisfying description of the system than catalogues of cases, tables of data, or graphs.

CHARACTERISTICS OF HOMEOSTATIC MECHANISMS

The response of a system may be controlled by two general approaches: (1) *Open loop,* or control based on *compensation* for the effect of each source of disturbance. The quantitative effects of all the important disturbances on the system must be determined. The magnitudes of the disturbances are measured, and the system is then adjusted accordingly. Aiming a gun, after taking into account wind and distance, is an example of this approach. Its effectiveness depends upon precise knowledge of all disturbing variables, both as to their influence and their magnitude and as to their interactions. In addition, the mechanism itself must be precise and stable, features not characteristic of biological systems. (2) *Closed loop.* In this approach, control is attained by measuring or sensing the response and feeding it back to the effector mechanism. Knowledge of the source and magnitude of the disturbance is immaterial – providing they are not beyond the range of control. The effector mechanism may be unstable and imprecise in constructional details – a hallmark of living organisms. With a closed loop, close and reliable control is attainable. A control system in which the output or response is held constant is often called a *regulator,* while one in which the response is made to follow the pattern of a varying input signal is called a *servomechanism.* Control – in the sense of negative feedback control – implies that the response of the total system will be held near certain desired values even though disturbing inputs to the system fluctuate.

The operation of a negative feedback control system may be broken into several separate functions (Fig. 7B-1B). First, the magnitude of the controlled variable must be *sensed* by some means (the sensor). This magnitude must then be *compared* with a reference value which is considered optimal and is often a part of the genetic make-up of the biological organism. The difference between the reference value and the actual value is the *error signal.* This error signal is *fed back* to the *effector mechanism* in such a manner (negative) as to *minimize the error.*

Regulation of the amount of food eaten is an example of a complex biological control system. Hunger is the error signal. The level of intake needed to satisfy the hunger is widely variable, depending in part on the physical activity, psychological status, and general health of the individual. If the food intake is inadequate, the hunger increases, the individual eats more and thereby reduces the hunger – a negative feedback system. Negative feedback control systems also operate at the molecular level, controlling enzymatic reactions. Here all facets may be combined into a system in which an increase in the concentration of the product of a chemical reaction inhibits the reaction itself so that the product concentration is held relatively constant.

Comparing the response of the system to the input or reference by subtraction is one way to obtain the error signal. A different approach is to divide the reference signal by a signal representing the response. As applied to an amplifier, this type of control provides an automatic gain control. Although potentially of wide importance in studying biological systems, especially those involving enzymatic actions, this form of description has not been widely exploited as yet.

THE REFERENCE LEVEL

The reference level (or set point) is the desired condition. In the control of room temperature, the reference is the temperature setting of the thermostat. In the healthy individual, it is usually the optimal level of the variable, and aids in defining what is meant by normal. With a control system, the regulated variable may be changed by changing the reference level. Many pathological conditions may be traced to such a change, although the details of how and why are far from being solved. For example, fever is primarily a change in the reference level for the control of body temperature. Hypertension is possibly a change in the reference level for the control of blood pressure. When the reference is changed, the controlled function then fluctuates about this new level.

In most biological systems there is no clearly defined arithmetic comparator for the generation of an error signal by subtracting a value representing the response from a reference value. Thus most biological systems function around an *operating point* determined by the characteristics of the chemical and physical structures of the system.

FEEDBACK FACTOR

The feedback factor has been variously defined, but it generally means the ratio of the response to a disturbing influence of an uncontrolled system to that of the controlled system.

$$\text{Feedback factor} = \frac{\text{Uncontrolled response}}{\text{Controlled response}}$$

It is a measure of the effectiveness of the control system. As an example, the control of body temperature in mammals is excellent; indeed, such animals are called homeotherms. The body temperature of a group of subjects in a room with the temperature at 40°C was 37.8°C, while that of the same persons under similar conditions but at 2°C was 37.2°C. The feedback factor for these people was

$$\frac{40°\text{C} - 2°\text{C (uncontrolled response)}}{37.8°\text{C} - 37.2°\text{C (controlled response)}} = \frac{38}{0.6} = 63$$

The blood pressure is said to be regulated with a feedback factor of 3. This means that an influence which would cause only a 20 mm Hg change in the normal animal would cause a 60 mm Hg change in the animal without control. Present research suggests that it may be as high as 10. Hypertension might be explained by some factor causing an increase in blood pressure in an animal which has a feedback factor too low to stimulate adequate compensation.

It is generally quite difficult to obtain a completely uncontrolled response of a biological system since it is hard, and in most cases probably impossible, to remove all the regulatory features without damaging the normal, basic biological mechanisms. The concept of feedback factors is of value, however, in assessing the importance of parts of a biological control system or in testing the adequacy of a homeostatic mechanism. For example, if before anesthetization the change in blood pressure of a dog in response to a temporary loss of blood volume is −2 mm Hg and after anesthetization the response to a similar loss is −20 mm Hg, it can be said that the efficiency of the control of blood pressure is reduced by a factor of 10 (20/2) by the anesthetic agent. For use as a diagnostic approach to disease, standardized stresses or test situations must be developed.

MODES OF CONTROL

There are several modes of control used in engineering practice which have been found in biological systems: *Proportional.* In this form the error signal driving the effector organ or actuator is directly proportional to the magnitude of the error. Here the error will be relatively large if the external disturbances are large. *Integral.* To eliminate the steady-state error, the error may be added up as time goes by (integrated) to develop an additional signal to reduce the overall error. *Derivative.* To anticipate the degree of corrective action needed, the rate of change of error (the first derivative) can be used to develop a signal to drive the effector system. This assumes that a large disturbance will produce initially a faster-changing response than a small disturbance. Control of blood pressure and body temperature are rate dependent. Because all physical systems have energy-storing features, e.g., inertia and compliance, the response never exactly follows the stimulus, and so perfect control during transient changes, even with these more complex modes, is impossible. *On-Off.* This is the familiar type of control used with automatic appliances such as home furnaces, refrigerators, and pumps. When the controlled variable exceeds set limits, the unit is turned on or off. Though it is discontinuous, this type of control can be highly effective, especially if used as a supplement to open-loop control.

LIMITS OF CONTROL

A control system will perform accurately only within certain limits of change in components or energy supply. If the energy supply is markedly reduced or some component is seriously damaged, the effector mechanism may operate at maximal, yet inadequate, levels. With a serious continuing loss of blood, the blood pressure may be maintained for a while by compensatory mechanisms, but a point will be reached at which the limits of compensation have been exceeded. Then the pressure will fall, often in a fulminating manner. Another example of an overdriven control system is seen when insufficient iodine in the diet limits the production of thyroid hormone. A feedback mechanism involving the pituitary gland causes hypertrophy of the thyroid (simple goiter), which may lead to irreversible damage of this gland if the iodine deficiency is not relieved.

When a biological feedback system is driven to its limits because of malfunction or external stress, the ideal form of treatment is to find and correct the cause of the change rather than attempting compensatory changes which might lead to further overloading of the system and possible fulminating failure. Defective function of any link in the homeostatic system is ordinarily reflected in some degree of hyperfunction or hypofunction of other elements in that system. This is why symptoms are commonly clustered into characteristic *syndromes.* Because the various parts of a homeostatic system are connected functionally into a closed loop, the chain of cause and effect among symptoms is often difficult to discover unless studied in the light of overall system operation.

POSITIVE FEEDBACK, RESPONSE TIME, AND STABILITY

If the effect of the feedback is to increase, rather than decrease, the deviation from the desired level, the feedback is said to be positive. Positive feedback may be defined as a situation in which an *increase* in response, when fed back to the control mechanism, causes a still *further increase* in response. Positive feedback occurs if the feedback factor is less than 1, i.e., if the "controlled" response to a disturbing influence is greater than the "uncontrolled" response with no feedback. Positive feedback, if not limited, causes a progressive change in response so that eventually the system operates at a maximum or minimum level – a vicious cycle.

If the thermostat for the furnace were connected in reverse, so that an increase in temperature turned the furnace *on,* the result would be an example of unlimited positive feedback. Biological examples of unlimited positive feedback are relatively rare, since the condition results in extremes of response. The response of a nerve to an adequate stimulus is a good example, however. If the stimulus to nerve action is inadequate, there is no response; but if the stimulus is adequate, once the response of the nerve starts, the process reinforces itself (positive feedback) so that the nerve fiber responds maximally. Only later is the process reversed.

Positive feedback can lead to death. For example, if the heart starts to fail significantly, less blood is pumped, the blood pressure falls, and there is consequently a lesser supply of blood to the heart itself. This weakens it *further,* leading to a further decrease in blood pumped and so, eventually, to death.

The *response time* of a system is a measure of the time required for it to respond to a change in conditions, and is defined, for the simplest system, as 63.2 percent $(1-1/e)$ of the time taken to achieve the final total change in response. Following a sudden change in conditions, the error signal even in an efficient control system is relatively large at first, but then it declines toward zero as the desired level is again approached.

Because of the interaction between energy-storing devices, oscillations may occur in response to a sudden change in input or to disturbances. The transport time of materials through the vascular system, and the diffusion of products from sites of synthesis are conducive to oscillations. In most control systems, oscillations are a sign of malfunctioning or poor control. The blood pressure sometimes oscillates in an animal with a failing cardiovascular system. Predicting the conditions for system stability, i.e., lack of sustained oscillations, is a major part of the rather complex science of control system design. In biological control systems, some of the components are not even known, so no complete description of the characteristics of the systems can be given. This, then, is an area for further biophysical research.

For accurate control, a high feedback factor is essential, so that the effects of changes in the metabolic energy supply, parts of the system, or environment will be reduced toward zero. Such systems, however, tend to be unstable and to oscillate, especially if the feedback system has a relatively long response time, i.e., is sluggish. The response of the somatic nervous system is speedy; therefore, accurate control of posture by rapidly responding muscles is possible. In some disorders, however, the feedback system is sluggish or the feedback factor (gain) is too high, and so there is tremor on attempting to make fine movements.

BIOLOGICAL CONTROL SYSTEMS

Many biological variables are controlled directly. For example, the amount of carbon dioxide in the arterial blood is controlled by homeostatic mechanisms which regulate the degree of lung ventilation so that the proper amount of carbon dioxide is exhaled. Other variables, such as the amount of oxygen in the arterial blood, tend to be regulated in an indirect manner, for the arterial oxygen tension is dependent largely on the amount of oxygen in the air and the degree of lung ventilation, which in turn is controlled directly by the amount of carbon dioxide in the blood. Finding the primary variables and the interrelationships of these homeostatic mechanisms is one of the challenges of modern physiology.

It must be emphasized that a negative feedback control system is not highly dependent upon the accuracy and stability of its components. An individual's heart may be severely damaged by disease, yet the blood pressure may be normal if the person does not strenuously exert himself. There are marked differences in the dimensions of the organs of people and in their functional reserves between birth and old age. Yet these differences do not preclude normal function.

Finally, biological homeostatic mechanisms often possess several feedback systems which act on the same variable, providing constancy and reliability to the organism (yet complexity to the physiologist and student). As an example, there are several types of sensors and effector mechanisms for the control of blood pressure.

The autonomic nervous system is an essential part of most mammalian control systems. As better understanding of the various homeostatic mechanisms and sites of failure is obtained, more effective therapy for various pathological conditions may be devised.

In conclusion: "The essence of physiology is regulation. It is this concern with 'purposeful' system responses which distinguishes physiology from biophysics and biochemistry. Thus, physiologists study the regulation of breathing, of cardiac output, of blood pressure, of water balance, of body temperature and of a host of other biological phenomena" (Grodins et al., 1954).

REFERENCES

Adolph, E. F. Early concepts of physiological regulations. *Physiol. Rev.* 41:737–770, 1961.

Apter, J. T. Biosystems Modeling. In M. Clynes and J. Milsum (Eds.), *Biomedical Engineering Systems.* New York: McGraw-Hill, 1970. Chap. 5.

Bailey, N. T. J. *The Mathematical Approach to Biology and Medicine.* New York: Wiley, 1967.

Blesser, W. B. *A Systems Approach to Biomedicine.* New York: McGraw-Hill, 1969.

Cannon, W. B. Organization for physiological homeostasis. *Physiol. Rev.* 9:399–431, 1929.

Defares, J. G. Principles of Feedback Control and Its Application to the Respiratory System. In J. Field (Ed.), *Handbook of Physiology.* Section 3: Respiration, Vol. I. Washington, D.C.: American Physiological Society, 1964. Chap. 26, pp. 649–680.

Grodins, F. S. *Control Theory in Biological Systems.* New York: Columbia University Press, 1963.

Grodins, F. S., J. S. Gray, K. R. Schroeder, A. L. Norins, and R. W. Jones. Respiratory responses to CO_2 inhalation. A theoretical study of a nonlinear biological regulator. *J. Appl. Physiol.* 7:283–308, 1954.

Machin, K. E. Feedback theory and its application to biological systems. *Sympos. Soc. Exp. Biol.* 18:421–445, 1964.

Milsum, J. H. *Biological Control Systems Analysis.* New York: McGraw-Hill, 1966.

Riggs, D. S. *Control Theory and Physiological Feedback Mechanisms.* Baltimore: Williams & Wilkins, 1970.

Warner, H. R. Simulation as a tool for biological research. *Simulation* 3:57–63, 1964.

Wiener, N. *Cybernetics, or Control and Communication in the Animal and the Machine,* 2d ed. New York: Wiley, 1961.

Yamamoto, W. S., and J. R. Brobeck. *Physiological Controls and Regulations.* Philadelphia: Saunders, 1965.

Higher Somatic and Visceral Control

Sidney Ochs

MOTOR AREAS OF THE CEREBRAL CORTEX

Fritsch and Hitzig in 1870 first showed that electrical stimulation of certain parts of the cortex produced movements of specific portions of the peripheral musculature. Since then others have shown, in an increasingly refined way, the correspondence of points within such *motor areas* to the peripheral musculature. In Figure 8-1, the motor area of the human brain is labeled with the body musculature excited from the various cortical sites. A general similarity or homology of motor areas and their muscle control patterns within the cerebral cortex has been found for the different species. A relationship of motor function to the underlying cellular structure was indicated for the area of the brain numbered 4 by Brodmann (1909) (Fig. 8-1), which contains large *pyramidal tract* (PT) cells, known also as *Betz cells.* Some of the evidence obtained supports the possibility that it is the large PT cells which are electrically stimulated by applied electrical currents to give responses in the different limb muscles. However, there is not an exact correspondence of motor control of a given musculature to a specific PT cell population; the motor area is known to extend beyond area 4. Also, as shown anatomically, axons from a given part of the motor cortex are widely distributed to many motor nuclei of the lower motor centers – although with a predominant convergence onto one group of motoneurons innervating a particular muscle.

With the use of a constant level of anesthesia, and other control of the stability of the experimental preparation, a fairly consistent and detailed representation of the peripheral musculature in the motor region may be found. The motor area determined in this way by Woolsey and his co-workers (1952) was called the *precentral motor area.* Figure 8-2 shows the result obtained from a *Macaca mulatta* brain, where a map is constructed from those points of the cortex having the lowest threshold for responses in the various muscles represented. As previously indicated, the correspondence of motor areas with the various cytoarchitectonic regions of the cortex is only approximate, with portions of the motor cortex falling outside area 4. In area 6 of Brodmann, larger muscle groups of the back and the upper limb muscles are represented. From area 4, a control of finer muscles is found, those of the lower arms, the digits, tongue, and face. These areas of finer muscle control are disproportionately large with respect to the control areas of other parts of the musculature. Therefore, lesions of the motor cortex are more destructive of the finer and complex manipulative movements of the digits, while the grosser movements of the axial

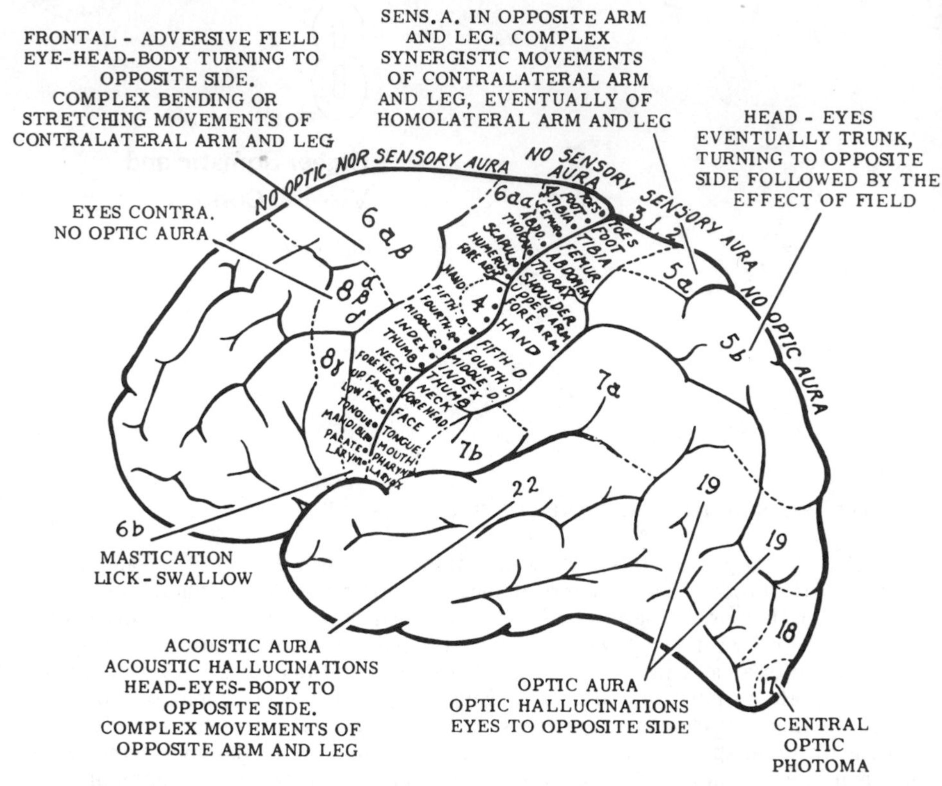

FIGURE 8-1. Motor areas of the human brain. The motor areas of the brain are in front of the central sulcus. The various parts of the peripheral musculature excited by stimulation in that region are labeled. Various other regions in different areas of the brain also can give motor responses as indicated. Notice the correspondence of the motor area to the sensory regions posterior to the sulcus. (From W. Penfield and T. C. Erickson. *Epilepsy and Cerebral Localization.* Springfield, Ill.: Thomas, 1941. P. 401.)

musculature or of muscles of the upper limbs remain relatively less affected. Recent evidence has shown that in man and primates in general, pyramidal tract fibers controlling fine movements end directly on spinal motoneurons. However, a large share of the control is apparently also subserved in tracts other than the pyramidal, in *extrapyramidal* tracts.

A greater susceptibility to disruption of fine control is characteristically seen in patients who have had brain strokes with lesions in the internal capsule interrupting a large part of the fiber downflow from the motor cortex. The patient so affected has a disturbance of posture and movement. The position of the limbs is said to show a reversion to primitive patterns of motor control – flexion of the upper arms and hyperextension of the lower legs with increased reflexes, i.e., *spasticity*. The lower motor centers and brain stem have been *released* by the capsular lesion, and a positive Babinski sign is present (Chap. 5), but it is in connection with fine complex movements that the disturbance is most apparent. The attempt to move the fingers – opening and closing them or approximating the fingers – is difficult or not possible. The upper limb, however, may be moved. In the lower limb this pattern of defect gives rise to the spastic gait typical of a stroke patient. The leg is held stiffly

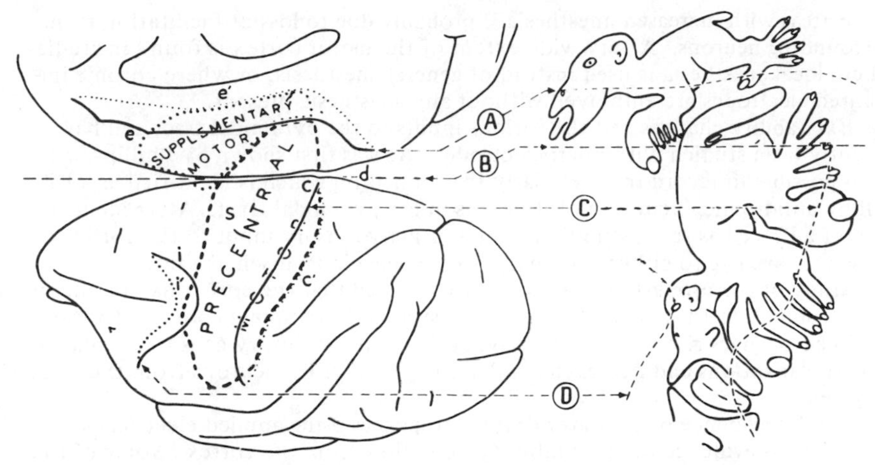

FIGURE 8-2. Motor areas of the monkey brain. To the left, the precentral motor area is shown with dashed lines, and to the right a map of the body musculature represented within the area. In addition to this primary motor area, a smaller supplementary motor area can be found. Much of it is hidden from view on the medial aspect of the brain. For both the primary and secondary motor areas the thumb area is relatively very large, as are the areas for the digits of hand and foot. The face is quite large, with a large tongue area. (From Woolsey et al., 1952.)

extended, and in walking it is swung outward and forward. The rhythmic flexion and extension of the limbs which is part of the normal walking pattern is lost. In part, the earlier attempts to assign a coarse control function to area 6 were due to unrecognized involvement with a distinct second motor area, the *supplementary motor area,* which can also be represented by a second map (Fig. 8-2). Part of this supplementary motor area is turned around and hidden from view on the medial surface of the hemisphere. The responses to stimulation of the supplementary motor area give rise to more generalized movements affecting large muscular groups. Lesions in this region are therefore more likely to produce spastic phenomena. The efferent pathway from the supplementary area projects separately into the pyramidal tract and does not relay through the precentral motor area of the cortex. Cutting the pyramidal tract by itself does not cause spasticity (Wiesendanger, 1969).

Interconnections between associational areas and motor areas permit other parts of the brain to affect motor control. The sensory receptor areas within the cortex have close connections to the motor areas, and electrical stimulation of parts of the cortex outside the "true" motor area may be compounded of sensory and associational cortex stimulation, with a secondary excitation of the motor neurons. Note the correspondence of peripheral representation in somesthetic areas 1, 2, and 3 with motor control regions in Figure 8-1. In some species there is more overlap in the musculature, and one considers a *sensorimotor* cortex rather than separate motor and sensory regions.

A region from which motor responses are obtained probably represents a complex interplay of neurons having facilitatory and excitatory effects on the motor outflow. Anesthesia, which might be expected to have an effect on the excitation, has been shown by Liddell and Phillips (1951) to decrease the size of the area from which a motor response in a given muscle can be obtained on electrical stimulation. Studies indicate that the organization of the motor cortex consists of widely overlapping areas of muscle representations instead of pointlike areas. The shrinkage of these areas with increased anesthesia is probably due to loss of facilitation from surrounding neurons. A very wide extent of the motor cortex is found in studies where local anesthesia is used instead of general anesthesia, or where chronic implanted electrodes are employed without any anesthesia present.

Excitability changes and the various inputs to the pyramidal tract cell have recently been studied with microelectrodes. As was first shown by Phillips, it has been possible to record intracellularly from a neuron which is identified as a PT cell by antidromic excitation of its axons in the pyramidal tract. After identification of a PT cell as such, stimulation of a primary sensory input to the cortex was shown to give rise to either a discharge or a change in its discharge rate.

Study of the properties of PT cells has revealed that the cortex has an inhibitory system similar to that of the Renshaw cell system in the spinal cord. Activation of a PT neuron gives rise by collaterals synapsing onto inhibitory cells to long-lasting hyperpolarizations and decreased excitability in other PT neurons from the inhibitory cells.

Besides evidence of excitatory motor responses using applied electrical stimulation, there is evidence for an inhibitory downflow from the cortex. Some of the cortical inhibition is reciprocal. Along with an excitation of one set of muscles there is a reciprocal inhibition of the antagonistic muscles of that limb. Inhibition from the cortex is also found in some cases to be more general with inhibition of all muscle groups having motoneurons at the lower brain stem or spinal level.

At one time it was believed that there were special inhibitory regions in the cortex, called suppressor strips, which gave rise to a general inhibition of the peripheral musculature. The suppressor strip areas were believed to exist near the motor area, and also to send fibers to, and have action on, the reticular formation. (The motor control function of the reticular formation is described below.) More recently, it has come to be recognized that suppression is in all likelihood part of the phenomenon of spreading depression (Chap. 6). The onset and termination of spreading depression are much too slow for the rapid control expected of an inhibitory mechanism. Spreading depressions last typically for from 10 minutes up to one-half hour or more, while inhibition from the cerebral cortex occurs within seconds and stops quickly after stimulation of an inhibitory site.

RETICULAR FORMATION CONTROL

The upward spread of sensory activation exerted by the reticular formation on the cerebral cortex was referred to in Chapter 6. Historically, the motor control exerted by the reticular formation on lower motor centers was discovered first. Magoun and Rhines (1947) showed that stimulation within certain areas of the reticular formation could result in a facilitatory effect on lower motor centers, whereas stimulation in other reticular formation areas gave rise to inhibitory effects. This is shown in Figure 8-3. While a reflex, the knee jerk, was being elicited continuously at regular intervals, stimulation through electrodes stereotaxically placed in the reticular formation of the lower medial part of the medulla gave rise to a rapid inhibition. On cessation of stimulation, the reflexes recovered within a few seconds. A similar effect of reticular formation stimulation on cortically evoked limb movements was also found from this region. If stimulation was administered to the upper lateral portion of the reticular formation, a facilitation of knee jerk reflexes was found (Fig. 8-4), as well as of cortically evoked lower limb movements. By this means, the presence of facilitatory and inhibitory regions within the reticular formation could be plotted (Figs. 8-3 and 8-4). Loss of the normal inhibitory control descending from the cortex and other subcortical centers acting on the inhibitory

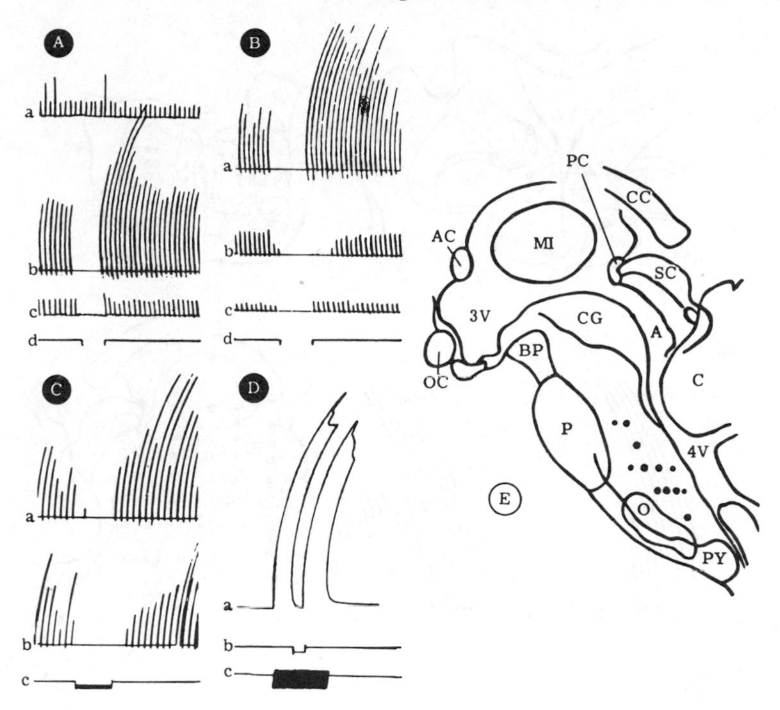

FIGURE 8-3. Inhibitory regions in the reticular formation. To the right, in E, is a sagittal view of the brain stem, with dots in the lower portion indicating regions effective for causing inhibitory effects. These effects are shown in A and B as an inhibition of the knee jerks at the time of reticular formation stimulation (see lower segment lines). In C and D, stimulation to the inhibitory regions is seen to be effective for inhibiting cortically evoked movements. (From H. W. Magoun and R. Rhines. *J. Neurophysiol.* 9:166, 169, 1946.)

regions of the reticular formation centers could result in an unopposed excitatory influence descending onto lower motor centers from the facilitatory reticular formation center. Or, a loss of discharge from the inhibitory reticular formation could also give rise to a release of lower motor centers and the exaggerated reflex activity associated with spasticity. A high control of the spinal cord gamma motoneurons (Chap. 6) has been located in the reticular formation, and exaggerated gamma discharge has been associated with the spastic-like condition of decerebrate rigidity (Chap. 5). An increased outflow of gamma excitation may very well occur in the stroke victim as a result of loss of higher inhibitory influences acting on the reticular formation. However, not all the excitatory downflow from the reticular formation takes place on the gamma motor neurons of the spinal cord. When a monosynaptic reflex is elicited by stimulation of the central end of cut dorsal roots, the gamma neurons are excluded from the response. Using this system, it has been shown that stimulation of the reticular formation causes facilitation or inhibition of the monosynaptic response, depending on the site stimulated within the reticular formation. This would indicate, as Lloyd demonstrated in 1942, that the fibers from the reticular formation terminate on the interneurons, which in turn end on the motoneurons to modify their excitability.

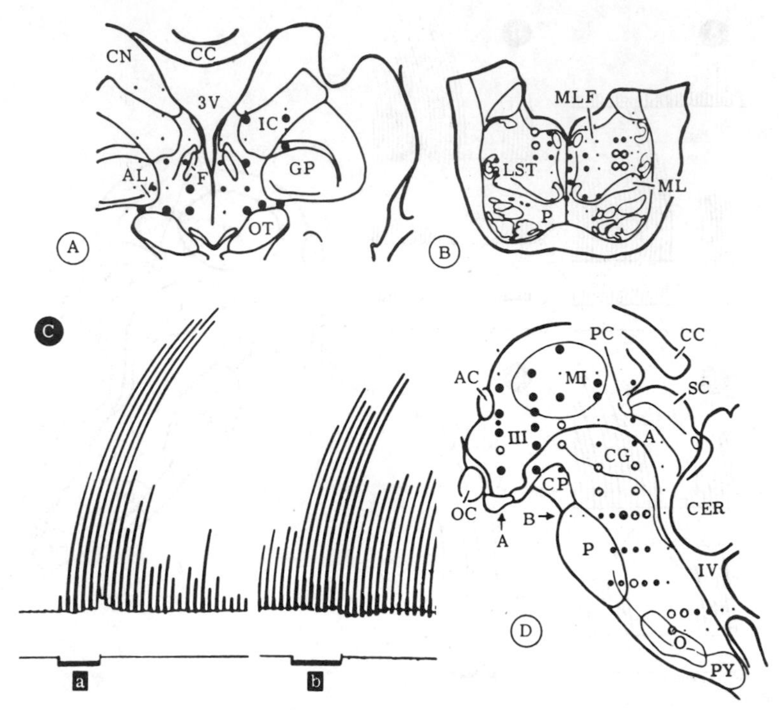

FIGURE 8-4. Facilitatory regions of the reticular formation. D is a sagittal view of the brain stem. A and B are cross sections of the reticular formation, with dots representing the areas from which facilitatory effects are obtained. In general, these facilitatory regions are higher in the brain stem and more lateral than is the inhibitory region. In C, cortically evoked responses are facilitated during the period of stimulation of the facilitatory region of the reticular formation (signal a on lower line). At the right, a knee jerk reflex is similarly facilitated during the period of stimulation of the reticular formation (signal b on lower line). (From R. Rhines and H. W. Magoun. *J. Neurophysiol.* 9:222, 220, 221, 1946.)

Studies have been cited showing an inhibitory action from one region and facilitation from other regions of the reticular formation. However, functional control from each of these reticular formation regions is not "global" under all conditions of study. Mixed facilitatory and inhibitory effects have been found under certain experimental conditions. Anesthesia probably causes a simplification of what in reality is a more complex control of reflex responses from the reticular formation. Electrodes have been chronically implanted into the brain stem reticular formation so that stimulation can be delivered through a cable attached to the otherwise free-moving and unanesthetized animal. In this case, one finds coordinated movements of the limbs elicited in response to excitation in the reticular formation rather than simply a generalized inhibition or facilitation from the separate areas identified in acute experiments. Present knowledge of motor control from the reticular formation region appears to be that some types of motor behavior, particularly those controlling the tonic or postural states, involve an underlying generalized inhibition or facilitation mechanism. In addition, a more complex organization is actuated from the same region, an organization probably related to phasic activity initiated from the cerebral motor cortex or subcortical motor areas.

CEREBELLAR CONTROL

The cerebellum, or "lesser brain," is situated behind the cerebrum in the higher species. Its removal results in gross impairment of motor coordination and body equilibrium. An animal with its cerebellum ablated shows an inability to adjust the position of its limbs with respect to a goal. It overshoots or undershoots its goal when, for example, reaching for food: a condition called *dysmetria.* Movements, instead of occurring smoothly, are disjointed or jerky, indicating *asynergia.* A cerebellar tremor is also seen during a voluntary movement. The tremor is coarse and becomes greater as the goal is being approached: an *intention* tremor. Decerebellated animals appear weaker *(asthenic);* the weakness of the muscle is described as a diminution of tone, or *hypotonia.*

Removal of different parts of the cerebellum has different effects upon motor behavior. The newer posterior and lateral portions, the *neocerebellum,* are joined with the cortex by extensive afferent and efferent fiber connections. This part of the cerebellum is associated with cortically controlled phasic movements. The phylogenetically older portion of the cerebellum, the *paleocerebellum,* is located anteriorly and medially in the cerebellum and is concerned with tonic postural mechanisms as well as with locomotion which is more automatic in nature. The flocculonodular portions of the cerebellum, part of the *archicerebellum,* are found laterally and are concerned with body equilibrium. Nerve impulses from the vestibular sensory receptor organs (semicircular canals) give information as to the position of the head with respect to spatial orientation (Chap. 4). The afferent fibers of these receptors terminate in the flocculonodular portion of the cerebellum. Ablation of either the vestibular organs or the flocculonodular lobes causes impairment of orientation and loss of equilibrium. A human so damaged falls if asked to close his eyes, because he thereby loses the remaining visual cues for equilibrium. As has been shown by Snider and his colleagues, sense modalities such as vision and sound are represented in topographical fashion, as are sensory inputs from the skin and from muscle receptors (Fig. 8-5).

Analysis of the function of the cerebellum shows that it acts as a feedback control mechanism to smooth and integrate motor activity initiated elsewhere in the brain. How the interplays of the various sensory inputs to the cerebellum come to be compared with one another – the *comparator* function – and how actions are integrated and adjusted with respect to incoming sensory information will eventually be known from studies recently initiated by Eccles, Ito, Llinas, and their colleagues. Their microelectrode studies have shown a relatively simple connectivity for the neurons of the cerebellar cortex (Fig. 8-6).

The cells of the cerebellar cortex are arranged in three layers. At the bottom of the outer molecular layer are the large cell bodies of the Purkinje cells. These have large ramified dendrites which spread fanwise at right angles to the folia of the cerebellum in the molecular layer. Also in this layer are the parallel fibers which course through the dendrites in the direction of the folia and make synaptic contact with them. The parallel fibers originate from granular cells in the middle layer. These have axons ascending to the branch and become the parallel fibers. The deepest layer is the white matter, representing fibers entering and leaving the cerebellar cortex.

There are two sensory inputs to the cerebellar cortex. One, originating from the inferior olivary nucleus of the brain stem, ascends into the molecular layer to intertwine on and synapse on the dendrites of the Purkinje cells. These are the *climbing fiber* afferents. The other afferent input is made up of the *mossy fibers,* representing sensory inputs from the spinal cord, brain stem, and upper motor areas of the brain. These end in bulbous *glomeruli* – regions of synapse where

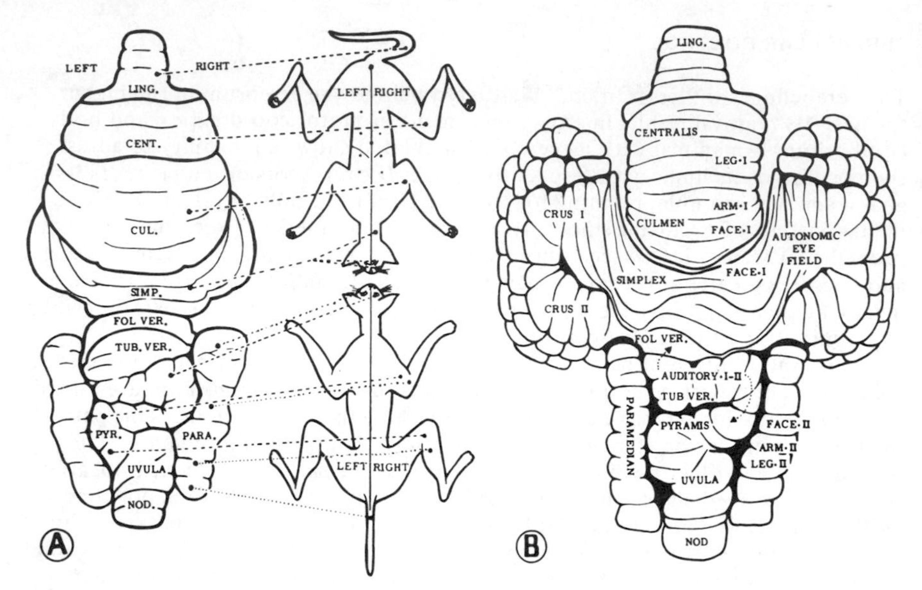

FIGURE 8-5. Somatotopic localization in the cerebellum. In A the localizations of motor functions found by stimulating areas over the cerebellar surface are shown. In B, the localization within the cerebellum of various sensory inputs using evoked response techniques is shown. (From J. L. Hampson et al. *Association for Research in Nervous and Mental Disease, Proceedings* 30:304, 1952, and from J. L. Hampson, *J. Neurophysiol.* 12:47, 1949.)

granule cells are excited. These inputs and the intrinsic inhibitory interneurons of the cerebellar cortex, the *stellate* and *basket cells,* eventually contribute to a complex control of excitation and inhibition acting on the Purkinje cell. This type of neuron represents the final common path from the cerebellum. Its axons are inhibitory to deeper relay nuclei of the cerebellum (fastigial, interpositus, and dentate), which in turn synapse on brain stem nuclear groups, vestibular, reticular formation, and red nucleus groups with their tracts descending to lower spinal reflex centers. The outflow from the dentate goes to the ventrolateral nucleus of the thalamus and thence back to the sensorimotor cortex of the cerebrum to regulate its motor control functions.

BASAL GANGLION CONTROL

An important but little understood group of nuclei in the brain stem is referred to as the *basal ganglia:* caudate nucleus, putamen, and globus pallidum (the striate body); subthalamic nucleus or Luys' body; substantia nigra; and red nucleus. These have a motor role but a clear understanding of their function is not yet in hand.

Stimulation of the caudate and other basal ganglion structures produces complex stereotyped movements – a turning of the body or a sudden arrest of ongoing movement. Inhibitory effects on lower reflexes have been obtained much like that from the inhibitory regions of the reticular formation. *Parkinson's disease* is characterized by a rapid alternating tremor of extensor and flexor muscle groups. The destruction of the efferent outflow of the globus pallidus by electrocoagulation or by focused high-frequency sound waves has resulted in dramatic relief.

A deficiency of the transmitter dopamine in the nigrostriatal fiber system has

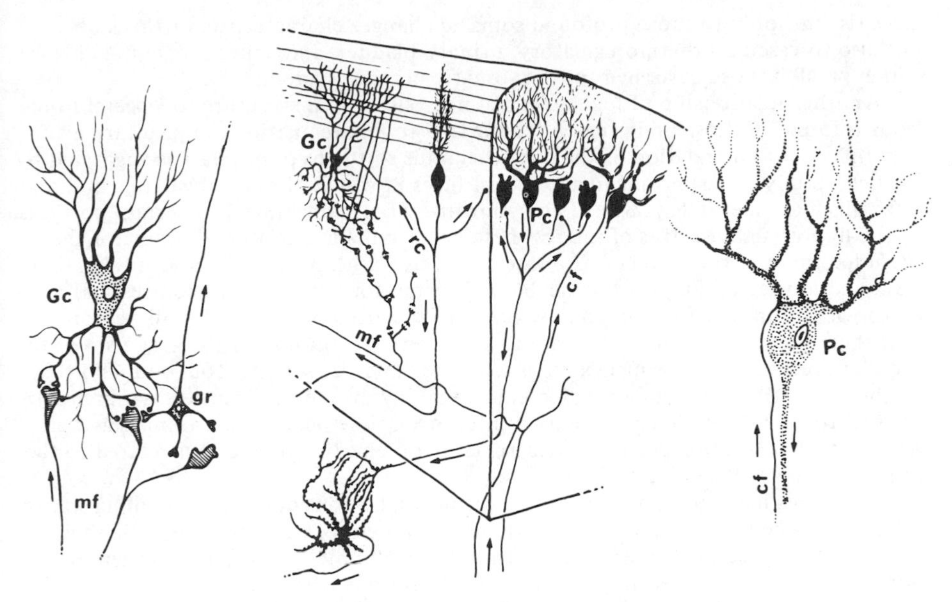

FIGURE 8-6. Details of cerebellar neurons. Left – afferent mossy fibers (mf) terminate on granule cells (gr) of the granular layer and on Golgi cells (Gc). Right – afferent climbing fibers (cf) terminate on dendrites of Purkinje cells (Pc). Middle – Purkinje cells have an extensive dendritic arborization across the folium in the molecular layer, where parallel fibers thread through and synapse on them. Efferent outflow is from the Purkinje axon. (After C. A. Fox. In E. C. Crosby, T. H. Humphrey, and E. W. Lauer [Eds.], *Correlative Anatomy of the Nervous System.* New York: Macmillan, 1962; and from J. C. Eccles, *Perspect. Biol. Med.* 8:289, 1965. Copyright © 1965, The University of Chicago Press.)

been related to parkinsonism and in many cases helped by the administration of the dopamine precursor L-dopa.

There is evidence that the caudate may also have some role in sensory transfer, acting to suppress evoked responses to sensory stimulation of the associative cortex.

HIGHER CENTERS OF VISCERAL CONTROL

In later chapters the physiology of various visceral systems – cardiovascular, respiratory, genitourinary, digestive, and others – is thoroughly dealt with. It will be of value, however, in this section to note briefly the mechanisms of upper visceral control located in the central nervous system and their relation to visceral systems and to the endocrine system of general homeostatic regulation.

The peripheral part of the autonomic nervous system, with its cell bodies located in the intermediary portions of the gray matter of the spinal cord, is discussed in Chapter 7A. This system, like the somatic nervous system, has afferent fibers entering the spinal cord which synapse eventually upon cell bodies within the intermediary parts of the gray matter of the spinal cord, which in turn give rise to efferent fibers emerging from the ventral roots. The autonomic motor nerves are distributed in a more or less diffuse fashion to the heart, various smooth muscles, and glandular structures throughout the body. Like the somatic nervous system, the local cord reflex centers controlling the autonomic visceral functions are in turn controlled by higher brain centers. In spinal shock, as mentioned, transection of the spinal cord gives rise not only to those profound somatic changes characteristic of this state but also to vascular, thermoregulatory, urinary bladder, and other urogenital changes which parallel those described for the somatic nervous system.

Another relationship of higher central nervous system structure to visceral functions is found in those brain regions connected to and regulating the pituitary gland (hypophysis). The relation of the hypothalamus to the hypophysis was suggested by Frölich's study of a synrome in adolescent boys in which obesity, dwarfing, and sexual infantilism are met (Chap. 31). This syndrome was finally traced to certain portions of the hypothalamus. It is of interest here to describe the control this area of the brain has upon visceral functions, those connected with food and water intake, body fluid and ion regulation, cardiovascular and respiratory control, gastrointestinal regulation, genitourinary function, and sexual and maternal behavior. It will be noted that these visceral functions all relate to the internal economy of the organism or to broader events in the organism's life, such as sex and the struggle for food, and that all these functions have an attending emotional involvement. Emotion is a predominant factor in human behavior, and by inference in the behavior of animals as well. Although closely connected with visceral control, emotional reactions related to the hypothalamus will be discussed separately (Chap. 9).

The hypothalamus may, for present purposes, be considered a controlling station for most of the visceral functions that have been mentioned, serving as a "high level" reflex control system. Other regions of the brain (limbic system) have, as will be noted in the next chapter, important interconnections with the hypothalamus and with the cortex. The limbic system and the neocortex can also therefore exert, at a still higher level, a regulation of visceral functions.

ROLE OF THE HYPOTHALAMUS IN VISCERAL CONTROL

In the preceding chapter, attention was directed to the importance of the hypothalamus as a higher regulating center for the autonomic nervous system. The purpose in this section is to develop this aspect of neural control. The areas to be discussed are (1) cardiovascular control, (2) respiration, (3) temperature regulation, (4) water and electrolyte balance, (5) control of food intake, and (6) sexual and maternal function.

CARDIOVASCULAR CONTROL

An animal in a state of spinal shock shows various degrees of loss of vasomotor activity in those parts of the nervous system below the transected level. The peripheral vessels are dilated, blood pressure is reduced, and response to reflex vasomotor activity is lessened. Vasodilation follows from the loss of normal reflex vasomotor tone acting upon the vessels. If the transection of the brain is made forward of the hypothalamus, the control of the nervous system is relatively normal insofar as vasomotor activity is concerned, and blood pressure is not affected.

Anesthesia, if very deep, causes a similar state because of its central action with a loss of reflex adjustments. The anesthetic agent chloroform sensitizes the heart to adrenergic action, leading to irregularities of rhythm, extrasystoles, and fibrillation. Cardiovascular control originates from the hypothalamus with a pathway traced from the hypothalamus and descending in the brain stem via sympathetic nerves which exit from the spinal cord at the thoracic levels to innervate the heart. The release of adrenergic transmitter leads to fibrillation if agents such as chloroform are present.

Parts of the brain higher than the hypothalamus also have been shown to have vascular effects. Stimulation of regions of the cingulate and neocortex can give rise to localized vasodilations and constrictions of skin, muscle, and body organs.

RESPIRATION

As indicated in Chapter 21, there is much evidence that a fundamental pacemaker activity underlies the rhythm of respiration. This rhythm is also controlled by higher nervous centers, so that voluntary control is possible. Higher centers in the brain, afferent discharges from the lungs and body, and also the oxygen, carbon dioxide content, and pH of the blood all affect the rhythm of respiration. But the nature of the rhythmic changes in cells leading to periodic discharge is at present little understood. One would like to know what series of molecular events causes the rhythmic changes in those cells. A similar problem relates to the pacemaker cells of the heart (Chap. 14) and EEG activity. The generator potential of sensory nerve endings related to rhythmic discharge (Chap. 3) is an analogous system whose study may eventually lead to an understanding of pacemaker activity.

Evidence exists (Chap. 21) that the hypothalamus has an excitatory effect on the inspiratory neurons in the medulla. Higher control by the cerebral cortex is indicated by the modifications of respiration that occur with speech, chewing, swallowing, and smelling. Stimulation of the cingulate cortex and other parts of the limbic system also has been found to affect respiration.

TEMPERATURE REGULATION

The relation of the hypothalamus to temperature regulation is shown when, upon local heating of small regions within the hypothalamus, peripheral changes are found in the body acting in the direction of removal or loss of body heat from the animal (Chap. 28). Conversely, cooling of the blood passing to the hypothalamus will cause changes in the direction of conserving body heat or increasing heat production by shivering. Shivering results in an increase of body warmth, and electrical stimulation of the posterior hypothalamus is effective in producing shivering. Electrodes placed in the hypothalamus show potential changes indicative of neural activity during heating or cooling of the blood or of the cells in this region. If the outflow from the hypothalamus is cut or the hypothalamus destroyed, heat-regulatory mechanisms are interfered with. In such cases an animal may be changed from a *homeothermic* to a *poikilothermic animal.* In a cold environment the animal will show a reduction in body temperature, with a fall toward that of the external temperature. Conversely, if the animal is placed in a warmer than usual environment, body temperature will rise. In other words, the homeostatic reflex adjustments which are mediated by the hypothalamus in the homeotherm are interfered with or lost.

Sweating is part of the hypothalamic regulation of body temperature in some species, including man, and operates in conjunction with the aforementioned control of the peripheral vascular bed, which in respect to temperature control provides appropriate vasoconstriction to aid the body's heat retention or vasodilation to assist its heat loss.

WATER AND ELECTROLYTE BALANCE

Receptors sensitive to changes in osmotic pressure of the blood, *osmoreceptors,* have been localized within the anterobasal part of the hypothalamus. Injection of hypertonic fluid into the arterial supply of the hypothalamus causes the release of antidiuretic hormone (ADH) from the posterior pituitary (neurohypophysis) after reflex excitation of the osmoreceptors of the hypothalamus (Chap. 31). An increased discharge of neural units within the supraoptic nuclei of the hypothalamus is also found following injection of small amounts of hypertonic fluids into the carotid arteries. Electrical stimulation within the hypothalamus of goats has been shown to lead to excessive thirst and excessive intake of water to the point where death can ensue.

From cells in the anterobasal region of the hypothalamus (supraoptic and paraventricular nuclei), tracts pass down into the neurohypophysis. It is believed that ADH is formed in the cells of these hypothalamic nuclei. The ADH moves down inside the fibers of the tracts by axoplasmic transport (Chap. 2), to be stored and later released as required from the pituicytes of the neurohypophysis. This hormone is required for the normal return of water from the kidney tubules to the blood (Chap. 22). In the absence of ADH, polyuria results.

CONTROL OF FOOD INTAKE

Within the hypothalamus there are areas which on stimulation cause an animal to seek food, and other regions which stop the satiated animal from continuing to eat. Lesions made in the medial part of the hypothalamus will make an animal eat excessively; i.e., it shows *hyperphagia.* Such an animal may more than double its normal weight. On the other hand, lesions in the lateral part of the hypothalamus will cause a reduction of food intake, *hypophagia.* If the lesions are complete, *aphagia* may result. Electrical stimulation of the medial nuclei makes an animal eat less, and stimulation of the lateral hypothalamic nuclei causes increased or continued eating. The medial nuclei are therefore regarded as a *satiety center,* the lateral nuclei as a *hunger center.* These two centers within the hypothalamus receive appropriate sensory inputs when hunger is present or after sufficient food is taken, and together they regulate the intake of food with relation to body needs. Receptors responding to glucose levels seem to be one important factor in the stopping of eating by exciting the satiety center. An increase in the level of food intake over a long enough time causes *obesity.* The deposition of fat in excess amounts in obesity is related to later degenerative changes in the cardiovascular and other systems and is associated with a shorter life-span. Obesity may also be brought about through influences from the higher centers (limbic, brain, and cortex) acting upon the food intake mechanisms of the hypothalamus. Some pharmacological agents appear to act on the hypothalamus to decrease the sensitivity of the food intake center, i.g., Dexedrine. This adrenergic drug apparently depresses the sensitivity of the food intake center and produces a decrease in the desire to eat. Higher centers may also be involved. In the cephalic phase of digestion (Chap. 26), the sight and smell of food stimulate the intake of food, and conditioned reflexes associated with food and eating are of great importance with regard to the level of food intake or the desire to eat.

REGULATION OF SEXUAL AND MATERNAL BEHAVIOR

The implantation of minute amounts of estrogen within the hypothalamus has shown that this region controls behavior patterns concerned with mating. Other limbic areas controlling overt behavior of this function will be discussed in Chapter 9.

The secretion of gonadotropic hormones by the adenohypophysis in both the male and the female is under the control of the hypothalamus. Lesions placed in the anterior hypothalamus prevent the release of luteinizing hormone, and a condition of constant estrus results. Lesions placed in the posterior tuberal area cause gonadal atrophy. When the adenohypophysis is removed from its normal site in the sella turcica and transplanted to the kidney capsule or anterior chamber of the eye, it is unable to secrete gonadotropins in amounts adequate to maintain gonadal activity. The control of hormonal release which the hypothalamus exerts over the adenohypophysis occurs through passage of hormones *(releasing factors)* in the rich vasculature found around the adenohypophysis, the pituitary portal circulation (Chap. 31).

REFERENCES

Anand, B. K. Nervous regulation of food intake. *Physiol. Rev.* 41:677–708, 1961.

Brooks, C. McC., J. L. Gilbert, H. A. Levey, and D. R. Curtis. *Humors, Hormones, and Neurosecretion.* New York: State University of New York, 1962.

Eccles, J. C. Functional meaning of the patterns of synaptic connections in the cerebellum. *Perspect. Biol. Med.* 8:289–310, 1965.

Eccles, J. C., M. Ito, and J. Zentagothai. *The Cerebellum as a Neuronal Machine.* New York: Springer, 1967.

Fulton, J. F. *Physiology of the Nervous System,* 3d ed. New York: Oxford University Press, 1949.

Hardy, J. D. Physiology of temperature regulation. *Physiol. Rev.* 41:521–606, 1961.

Harris, G. W. *Neural Control of the Pituitary Gland.* London: Arnold, 1955.

Hoff, E. C., J. F. Kell, Jr., and M. N. Carroll, Jr. Effects of cortical stimulation and lesions on cardiovascular function. *Physiol. Rev.* 43:68–114, 1963.

Jung, R., and R. Hassler. The extrapyramidal motor system. In J. Field (Ed.), *Handbook of Physiology.* Section 1: Neurophysiology, Vol. II. Washington, D.C.: American Physiological Society, 1960. Pp. 863–927.

Lassek, A. M. *The Pyramidal Tract.* Springfield, Ill.: Thomas, 1954.

Liddell, E. G. T., and E. G. Phillips. Overlapping areas in the motor cortex of the baboon. *J. Physiol.* 122:392–399, 1951.

Llinas, R. R. (Ed.). *Neurobiology of Cerebellar Evolution and Development.* Chicago: American Medical Association, 1969.

Magoun, H. W., and R. Rhines. *Spasticity: The Stretch-Reflex and Extrapyramidal Systems.* Springfield, Ill.: Thomas, 1947.

Penfield, W. *The Excitable Cortex in Conscious Man.* Springfield, Ill.: Thomas, 1958.

Preston, J. B., and D. G. Whitlock. Intracellular potentials recorded from motoneurons following precentral gyrus stimulation in primate. *J. Neurophysiol.* 24:91–100, 1961.

Snider, R. S., and A. Stowell. Receiving areas of the tactile, auditory and visual systems in the cerebellum. *J. Neurophysiol.* 7:331–357, 1944.

Spiegel, E. A., and H. T. Wycis. *Stereoencephalotomy.* Vol. II, Clinical and Physiological Applications. New York: Grune & Stratton, 1962.

Wiesendanger, M. The pyramidal tract. *Ergebn. Physiol.* 61:72–136, 1969.

Woolsey, C. N., P. G. Settlage, D. R. Meyer, W. Sencer, T. P. Hamuy, and A. M. Travis. Patterns of localization in precentral and "supplementary" motor areas and their relation to the concept of a premotor area. *Association for Research in Nervous and Mental Disease, Proceedings* (1950) 30:238–264, 1952.

Higher Nervous Functions

Sidney Ochs

HIGHER NERVOUS FUNCTIONS may be understood as those neural mechanisms relating to complex behavior, a resultant of interactions with the environment whereby an animal learns to cope with the many kinds of stimuli or signals indicating possible success or failure in attaining goals. It can be appreciated that this field of study is a difficult one. Often, inferences from subjective experience must be used in an attempt to augment limited objective observations made in the laboratory. The subject will be divided into *emotional* and *intellectual* behavior, with full recognition of the difficulty of such separation into categories.

EMOTION AND THE LIMBIC SYSTEM

Emotion is something individuals are aware of throughout the course of their lives. Emotion has two aspects: internal awareness and external display. The release of emotion appears at times to be almost a reflex-like expression of that state. For example, we may have a feeling of sorrow or depression, but a stronger sorrow may lead to uncontrollable weeping and sobbing. We may be amused or give way to sudden laughter. We may feel pleasure or enter a state of ecstasy. Displeasure and anger in the extreme can turn to rage. These human experiences have some parallels in experimental investigation of the emotions in animals. We ascribe emotional states to those displays in the animal corresponding to the behavior seen in man, usually the sudden reflex-like changes.

The similarity of brain structures in the human and the animal suggests generalizations for the localization of higher nervous function in brain regions having a similar form. For the most part the hypothalamus and the *limbic system* are involved. The limbic system, as described by Broca in 1878, is a group of nuclear regions and their interconnection found around the brain stem on the medial and ventral aspect of the brain (see below). Although the hypothalamus is sometimes included in the limbic system because of important connections made by the fornix carrying impulses from the limbic structures to the hypothalamus, this discussion will keep them separate.

Starting then with the hypothalamus: It was recognized that an extreme state of angry behavior known as *sham rage* can be produced in animals by amputation of the brain in front of the hypothalamus. Such an animal shows periodic displays of rage, with claws extended, jaws open, and muscles tense. Upon provocation, or even gentle restraint, the animal growls, thrashes about, lunges, and snaps. The rage display is not directed specifically to the source of provocation because the higher centers are lost. Sympathetic discharge from the hypothalamus is shown by the increased heart rate and blood pressure, pupillary dilation, hair bristling, and salivation (Chap. 7). An augmented sympathetic discharge is not seen if the hypothalamus has been destroyed. If, in the monkey, a cut is made in the base of the brain just anterior to the hypothalamus, a rage state is produced, but in this case the animal can focus its attacks. The extreme ferocity of this preparation can be appreciated from Fulton's graphic description of his experimental animals (1951). Some components of a rage reaction can be elicited from yet lower brain stem structures. A chronic decerebrate preparation shows some aspects of ragelike behavior referred to as *pseudoaffective*. However, while some of the components of rage display are evoked from lower brain structures, the more organized response requires the hypothalamus.

Localized electrical stimulation of the hypothalamus using chronically implanted electrodes can clearly bring about a display of ragelike behavior, with signs of sympathetic discharge also present. Some experiments show that sham rage elicited by such electrical stimulation of the hypothalamus does not have a subjective component; an animal may purr and even respond to petting between displays and during such stimulation. Stimulation directly in the hypothalamus and conditioning (see following section) would suggest a subjective aspect to the electrically elicited rage. While all of the display may not be conditioned, animals could be conditioned to avoid hypothalamus stimulation. It would appear that the hypothalamus acts as a control center for the outward expression of rage responses without necessarily being the place where emotions as such are experienced.

Papez (1937) related limbic brain structures (Fig. 9-1) to emotions by theorizing that afferent activity passes from the thalamus to the hypothalamus, then via mammillary bodies to the anterior thalamic nucleus, and from there to the cingulate cortex. The cingulate cortex was supposed to be the site of the subjective appreciation of the emotional states not only of rage but of sexual and feeding behaviors. This theory has been significant in providing impetus to investigations of the limbic brain structures with regard to their suggested connections with emotions.

The ablation studies of temporal lobes of monkey brain by Klüver and Bucy (1939) also directed attention to the relation of limbic brain structures to behavior. In their work, a number of limbic structures were removed, including the amygdaloid nuclei. Monkeys with both temporal lobes ablated showed bizarre behavior patterns. They lost their normal fear of objects such as snakes and other similar genetically determined "fear" objects. They seemed to "explore" their environment orally, putting edible and inedible objects alike into their mouths, no matter how abnormal or repulsive these objects might be with respect to the usual behavior for the animal. They would not necessarily swallow them, spitting out inedible objects and eating food. They appeared to show lack of recent memory by returning soon after to an object previously spit out and again putting it into their mouths. The picking up of any and all objects for oral exploration might suggest a lack of visual discrimination. However, temporal-lobe-ablated animals can learn to discriminate visually between symbols, e.g., a square versus a circle, when these are presented simultaneously in a standard visual discrimination experiment. Apparently the character of the objects is not recognized, or they cannot inhibit oral exploration of them. Animals with lesions in the amygdaloid nuclei of the temporal lobes also show a steady gain in weight. Their level of food intake seemed to have been raised by the lesion, or they became less active and therefore accumulated fat (Chap. 8).

Temporal lobe lesions give rise to a peculiar altered sexual behavior. The animals take other species as sex partners or even inanimate objects and in general show an increase in sexual activity. Other limbic system structures are related to sexual activity. The retrosplenial cortex in the rat is homologous to the cingulate cortex,

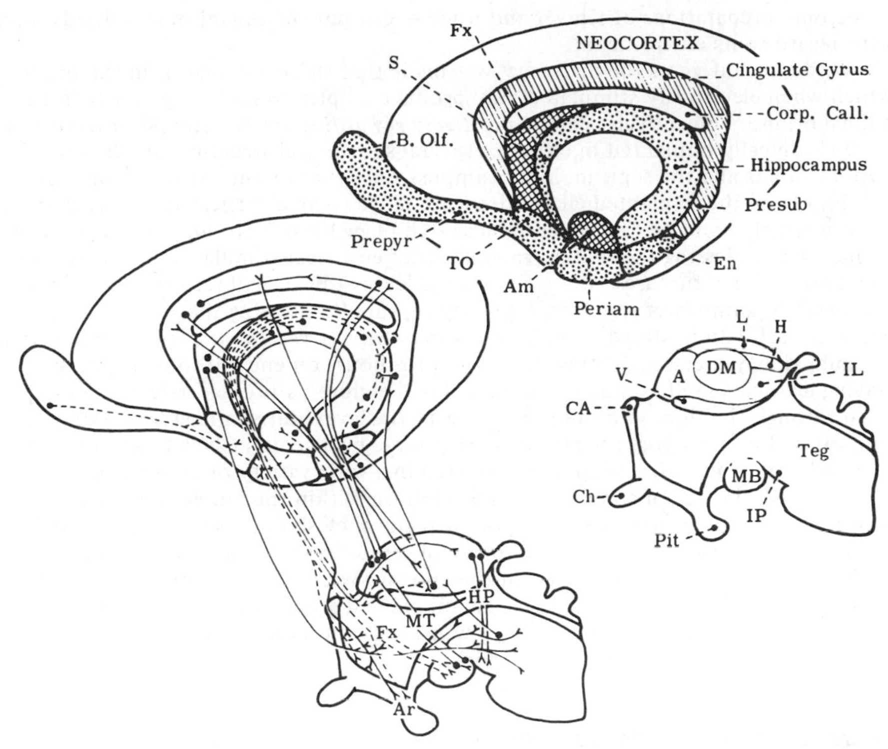

FIGURE 9-1. Limbic system shown at the right, with upper diagram representing the medial aspect of the brain, lower diagram the medial aspect of the brain stem. The corresponding pair of diagrams at the left show their regional interconnections. Three types of region are distinguished: (1) the *paleocortex,* represented by stippling, including as its chief structure the hippocampus and the olfactory structures, olfactory bulb (B. Olf.), and prepyriformis (prepyr); (2) the *juxtallocortex,* shown by slanted lines, which includes the cingulate cortex and presubiculum (presub); and (3) the *brain stem* structures, cross-hatched in the upper diagram – the septum (S), and amygdaloid complex (Am). These regions are related to one another and to brain stem regions as shown at the left. Important efferent connections are made from the hippocampus to the mammillary bodies in the hypothalamus via the fornix (Fx) and from the mammillary bodies to the anterior thalamic nucleus via the mammillothalamic tract (MT). Connections from the septum to lower brain stem centers are carried by the medial forebrain bundle, which is not labeled. (From J. V. Brady, "The Paleocortex and Behavioral Motivation," in H. F. Harlow and C. N. Woolsey, editors, *Biological and Biochemical Bases of Behavior* [Madison: The University of Wisconsin Press; © 1958 by the Regents of the University of Wisconsin], plate 23.)

a limbic brain structure. If it is destroyed, mother rats do not retrieve their pups from danger. Female rats so damaged show an abnormally increased responsiveness to the male. In the male rat, electrical stimulation of cingulate cortical regions can give rise to erection.

The two constellations of behavior (feeding and sex) have been referred to as "preservation of the self and preservation of the species." Obviously, these two biological goals are required for carrying on animal life. Aggressive or placid emotive states connected with hunger and satiation are of biological significance with regard to the organisms's drive toward or inattention to those goals. Certainly the altered behavior produced by a defect of this system – e.g., the bilateral temporal lobectomy preparations of Klüver and Bucy – can put the organism at a disadvantage with regard to its environment.

Within recent years the discovery was made that there are regions in the brain which when electrically stimulated may produce a "pleasurable" experience in the animal or, more objectively, can be *positively rewarding* for the animal. Electrodes were chronically implanted in either of several subcortical structures of the rat, the hypothalamus and the septum, hippocampus, and other limbic brain regions. Connections from the implanted electrodes were made to an electrical stimulating apparatus which could be controlled by a switch placed inside the animal's box. The animal, when pressing the switch, causes a brief electrical stimulation to be delivered to its own brain structures. Once the animal learns this, it will return to the switch and begin to administer *self-stimulation* by repeatedly pressing the switch. Apparently, the self-administered stimulation is so *positively reinforcing* (pleasurable) that the animal may incessantly press the switch for hours on end without stopping. Having learned to self-stimulate, an animal will endure painful electrical shocks applied through the floor grid in order to get to the lever and administer self-stimulation. It will prefer self-stimulation to other biological goals such as food or mating. Such self-stimulation has been demonstrated in a wide variety of animal species.

There is evidence of some special distribution within the limbic structures of positively rewarding sites. However, other regions, close to these same structures, presumably give rise to painful or unpleasant sensations when electrically excited, as shown by the *aversive* responses of the animal. When the electrodes are inserted into such *negatively reinforcing* sites, an animal will stimulate itself once and not press the switch again. Furthermore, it will exert considerable effort to prevent excitation of negatively rewarding regions.

LOCALIZATION OF LEARNING IN THE BRAIN

In the previous section the relation of parts of the brain to fundamental biological goals which are emotionally charged was emphasized. The alteration of innate behavior with respect to experiences in the environment, i.e., learning, will now be discussed. In their well-known studies, Pavlov and his colleagues investigated what is now known as *classical conditioning.* An animal is conditioned according to the Pavlovian technique by first receiving a sensory stimulus to which it will respond in an unlearned manner – the *unconditioned stimulus.* Food juices placed in the mouth of a hungry dog by means of an implanted cannula represent such a stimulus. This unconditioned food stimulus gives rise to an *unconditioned response,* in this case salivation. The measure of the response is the amount of saliva produced. If, at the time that the unconditioned food stimulus is being presented (or a short time beforehand) a bell is rung, the animal after a number of such pairings will salivate in response to only the ringing of the bell.

The sound of the bell is termed the *conditioned stimulus,* and salivation in response to it is a *conditioned response.* According to Pavlov, some new neural connections are made in the cerebral cortex after conditioning has occurred, so that the conditioned stimulus can take the place of the unconditioned stimulus with which it has been paired. The basis of Pavlov's belief was a report of an inability to condition decorticated dogs. Subsequent investigations have shown that simple types of classical conditioning can occur in the decorticate animal, and this finding directed attention to subcortical regions which may be involved in learning. Stimulation of the centrencephalic system localized in the nonspecific thalamic nuclei causes a petit mal type of widespread convulsive discharge synchronized in both hemispheres with a loss of consciousness (see Chap. 6). On this basis the centrencephalic system was regarded by Penfield (1958) as Hughlings Jackson's "highest level of consciousness."

It is possible to use direct stimulation of subcortical structures as an unconditioned stimulus, and peripheral stimulation or a sensory stimulus, such as a sound, as the conditioned stimulus to cause conditioning. A deviational response (a turning of the head) could be obtained from limbic structures and was conditioned by sound. From the septum sexual responses as the unconditioned responses could be conditioned to a tone. The display of rage obtained from some subcortical structures could not, however, be conditioned.

In contrast to the simple type of classical conditioning, a greater involvement of the cortex in learning is shown by use of *instrumental* or *operant conditioning.* In instrumental conditioning, an animal makes some response which at first may be accidental or part of the animal's normal repertoire of activity. The experimenter arranges conditions so that when the desired response is made, the animal is immediately rewarded with food or some other reward. The animal soon learns to discriminate between rewarded responses and other kinds of behavior and will, if sufficiently motivated, consistently perform those responses which have been so *reinforced.* In Skinner's type of experimental arrangement, an animal is placed in a small box with a small bar projecting from one wall. In moving about inside the box, the animal will occasionally strike the bar. If, when a hungry rat strikes the bar, a food pellet is delivered into a food hopper within the box, the animal learns to repetitively strike the bar for food. If reinforcement is removed, the animal gradually decreases its rate of responding until no more responses are made, a phenomenon known as *extinction.*

Various schedules of reinforcement have been described by Skinner and Ferster (1957). The animal may be required to press the bar a fixed number of times to obtain reinforcement. Or, the animal may learn that reinforcement will take place only when the bar is pressed after a fixed interval of time. These and other schedules are of interest insofar as they have been shown to be differentially affected by drugs; and cortical ablation or spreading depression present in both cortices will prevent the learning or execution of an instrumental response.

While recent studies have shown the possibility of subcortical conditioning, the importance of the cortex for higher functions should not be lost sight of. It is possible, as was shown by Doty and his colleagues, to stimulate the cortex directly for use as a conditioned stimulus in order to elicit a conditioned response. The cortex appears to be critically involved in more complex types of learning tasks. Attention has long been directed to the frontal lobes, an associational region thought to be connected with learning. Animals with frontal lobes ablated show a defect of *delayed response* performance. In a delayed response, an animal is allowed to see a food bait placed under one of two specially marked covers. Then, after a delay, the animal is permitted to choose one or the other of the covers to obtain the bait. Successful performance is measured by the proper choice of the cover over the food bait. Such delayed response behavior was greatly interfered with by frontal lobe lesions. Intense electrical stimulation or convulsive activity induced in the frontal cortex was also effective in interrupting delayed responding. Similar disruptive activity induced in other cortical areas was ineffective. Animals with their frontal lobes ablated do not show a loss of ability to discriminate one object from another if both objects are presented simultaneously. An interpretation of this result is that the frontal cortex is related to recent memory – or more likely that frontal-ablated animals are more distractible while they must wait before choosing. They may not have a sufficiently prolonged interest in the test object during the time it is out of sight. Once learning has been achieved, disruption of the activity of the frontal lobes is ineffective in interrupting delayed responding. This demonstrates a fundamental difference in the mechanisms responsible for the *acquisition* of learned behaviors and for their *retention.* Retention, i.e., memory mechanisms, will be discussed in a later section.

EEG CORRELATES OF BEHAVIOR

Changes in the electroencephalogram (EEG) have been correlated with a classic conditioning procedure using the alerting reaction (Chap. 6). A light shone in the eye gives rise to the usual period of low-amplitude fast waves (Fig. 9-2B). Also required to show EEG conditioning is a tone signal which does not by itself produce alerting.

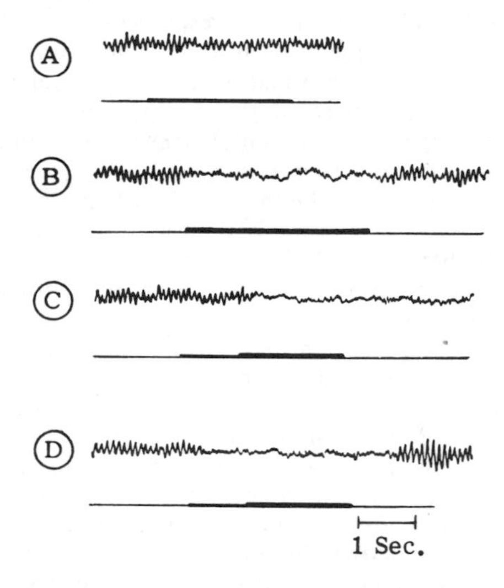

FIGURE 9-2. Conditioning of the EEG. This EEG of a human shows in A a record of a resting EEG with no alerting response to a tone signal (time of presentation of tone shown by thickening of the line underneath). In B, alerting to a light stimulus is seen (light on at time shown by heavier line underneath). In C, the tone signal precedes the light and no conditioning has yet occurred. After a number of such pairings, conditioning is seen in D when alerting occurs at the onset of the *tone* signal. (From F. Morrell. *Neurology* 6:329, 1956.)

This is arranged by *habituation* of the alerting response to the sound signal. A first presentation of a sound stimulus produces alerting, and with repeated presentations there is a diminished alerting response in the EEG. Eventually the stimulus no longer produces alerting, and the animal or human subject has become habituated to that specific stimulus. If the sound is altered to another tone close in frequency to the first, it will produce alerting. In the example of Figure 9-2A, the response to a sound stimulus had been habituated and no longer caused alerting. Alerting was caused in B by presentation of the light signal, and also in C when, early in the course of conditioning, the light signal was preceded by a tone as a conditioning stimulus. After a number of such pairings, conditioning occurred to the sound stimulus, as shown in D. Notice that alerting occurred in response to the conditioning signal and preceded the onset of the light stimulus, showing that alerting to the sound signal had taken place.

An interesting type of EEG conditioning is shown by the use of a light flashed at the rate of 7.5 times per second as the unconditioned stimulus, the flashing light exciting a series of evoked waves in the cortex. A tone signal to which the animal was habituated was the conditioned stimulus and was presented a short time before the light stimulus. After a number of pairings the tone signal elicited conditioned responses at the same frequency as that of the light flashes used as the unconditioned stimulus. This conditioned response was seen in a number of cortical regions. *Discrimination* was shown when in a later stage of conditioning the conditioned response became localized to the visual cortex (the unconditioned primary receptor areas). Habituation in the other cortical areas may have occurred during the restriction to the visual field.

The limbic cortex, and in particular the hippocampus, has a connection with the learning processes. When the cortex shows the faster, smaller-amplitude waves of alerting, the hippocampus gives rise to large waves at the rate of 5 to 7 per second. During conditioning, EEG changes were registered from the hippocampus via chronically implanted electrodes, and a characteristic shift to a lower frequency was found to take place during learning. The electrical correlate of learning suggests that the hippocampus is involved in the early process of learning.

MEMORY MECHANISMS

Intellectual ability is based on memories, to which the individual is continually adding. It must be admitted that little is known of the neuronal basis underlying memory, although some general statements may be made. Memories can exist for years, perhaps a lifetime, and thus would appear to depend on some permanent cellular changes. Memories persist after normal cerebral activity has been temporarily stopped, either by a degree of anesthetization sufficient to decrease all EEG activity of the cerebrum or by cooling to a low enough body temperature so that EEG activity is absent. All this supports the conclusion that memory stores are due to some permanent structural change of the neurons rather than being the result of a continuing neuronal activity.

Another general observation concerning memory is that recent memories are less stable than older ones. Clinical experience indicates that a blow upon the head may produce a *retrograde amnesia,* i.e., a loss of memory extending back to include a period of several hours or more before the blow. The susceptibility of recent memories to interruption of neural activity was experimentally shown by application of electrical shock to the heads of rats at different intervals after a conditioning stimulus. Animals electroshocked at too short a time after a learning session do not learn conditioned responses even when numerous conditioning and shock sessions are given. If the interval between the conditioned experience and the administration of the electroshock is lengthened to an hour or more, then even though a great amount of electroshock is given, these animals are conditioned as well as unshocked animals. The destructive influence of electroshocks on recent memory shows a curve of decreasing effect: It is greatest when shocks are given just after the training session, and the effect diminishes in regular fashion as more time between training sessions and shock is allowed. A similar destructive effect on conditioned learning was found using quickly induced anesthesia or a rapid cooling of the brain after conditioning sessions. Therefore, two types of memory are distinguished: that level of neuronal activity responsible for recent memories, and molecular changes in neurons responsible for long-term memories.

In an attempt to find the place in the brain where long-term memories are stored, Lashley (1950) studied the effects of various cortical lesions on the retention of a learned response. From this work on the rat it appeared to Lashley that memories or engrams could not be specifically located and that the amount of cortex damaged was related to the degree of memory loss. On the other hand, Penfield has indicated the importance of a certain part of the brain, the temporal lobe, with respect to storage and retrieval of memories in man. Electrical stimulation of the temporal cortex of conscious patients under local anesthesia gave rise to vivid memories. Stimulation of cortical areas outside the temporal lobe did not elicit memories. Penfield considered such responses from the temporal lobe to be support for its *interpretative function.*

The memory responses obtained by electrical stimulation of the temporal region might be a misinterpretation or altered interpretation of present experience with respect to past experiences. Thus, a mother under the influence of electrical stimulation in this region appeared to be seeing her child present in the operating room with all the adjoining sensations and sounds of an actual experience which had occurred previously. It should be noted that such hallucinatory effects may also have taken place in these patients as a result of the excitation produced by a tumor in the temporal region. These are possibly related to *auras* – the sensory alterations or hallucinations preceding a generalized epileptic attack.

The fact that electrical stimulation of the temporal cortex can excite a past memory suggests that neuronal activity in the normal temporal lobe cortex is somehow connected with a memory. The memory does not exist in this part of the cortex per se, but, as suggested by Penfield, it can trigger other regions, e.g., the centrencephalic region in the thalamus. A subcortical focus was indicated by the fact that after an ablation of the temporal cortex, stimulation of the underlying fibers was also effective in eliciting memories.

Myers and Sperry advanced the study of learning by use of their *split-brain* preparation (Fig. 9-3). To produce a split-brain cat or monkey, the optic chiasm–corpus callosum tract connecting the two hemispheres as well as the crossing fibers of the anterior commissure are cut in the anteroposterior plane. The visual information input from each eye can therefore pass only to the hemisphere of the corresponding side. With one eye covered and the other eye open, the animal is taught a visual discrimination in the hemisphere on the side on which the eye is open. After training is completed on that side, the eye of the opposite side is uncovered and the eye on the trained side covered. The animal is then unable to perform the visual discrimination to which it had previously been trained. The habit remains localized to the brain on the side on which the eye was open during training. Use of this technique makes it possible for two habits to be laid down, one on each side. Even opposing habits may be trained into the two sides. On one side the animal may learn to respond to a square and not to a circle; on the other side, to a circle and not to a square. With both eyes open the animal responds positively to one or the other symbol. At times one habit dominates, at times the other; the animal does not respond partially to the habits laid down on each side.

If the optic chiasm is cut in the midline as usual but the corpus callosum remains uncut, sensory information is channeled to the other side. This was shown when, after such animals were trained with one eye open, the other covered, and then later the previously open eye was closed and the other opened, the animal readily recognized the visual stimulus and responded to it.

By the use of the phenomenon of spreading depression (Chap. 6), one or both cerebral cortices may be functionally incapacitated for several hours. During such functional and temporary decortication Bureš and Burešová (Bureš, 1959) have shown that animals may learn a simple type of classical conditioned response. However, spreading depression will prevent an acquisition of complex conditioning. This was also shown using instrumental conditioned responses. With spreading

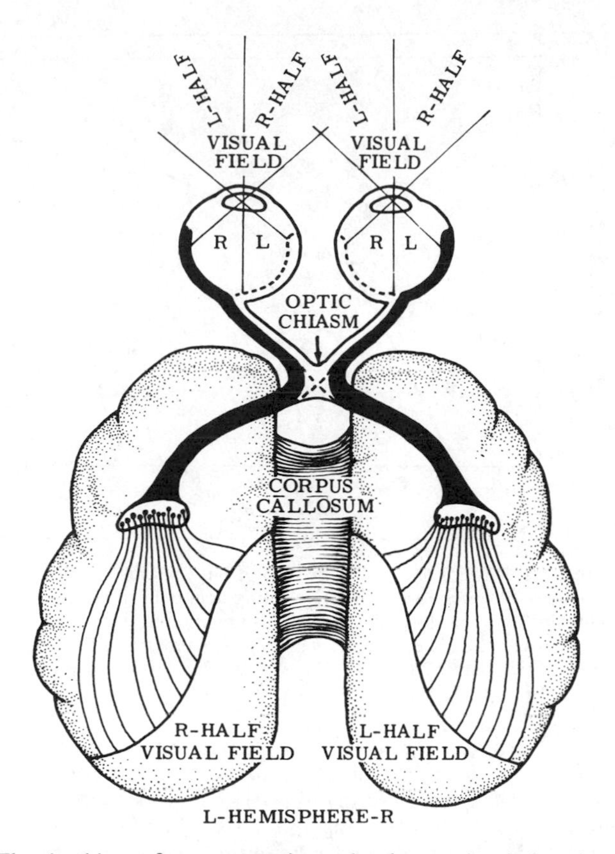

FIGURE 9-3. The visual input from one eye is restricted to one hemisphere by cutting the optic chiasm in the midline. The corpus callosum is also sectioned to produce a split-brain animal. (From R. W. Sperry. *Science* 133:1749–1757, [June] 1961. Copyright 1961 by the American Association for the Advancement of Science.)

depression present in one hemisphere, an animal will learn an instrumental response on the side not occupied with spreading depression. The localization of the learned response to the cortex which was not depressed during the training sessions was shown when later the trained cortex was depressed. The animal then acts as an untrained animal would. Such *lateralization* of the learned response to one hemisphere may remain for a long time (Fig. 9-4), even though the interhemispheric connections present in the corpus callosum might be expected to transfer the habit from the trained side to the other side without spreading depression in the interval between trials.

Transfer can occur after lateralization of a learned response to one side has been achieved. Transfer will occur if with no spreading depression present an animal is permitted to press the bar and receive reinforcement. A high rate of responses is then seen after depression of the trained side, whereas previously the animal did not respond at all. It is necessary, however, that the animal not only press the bar but also receive a reinforcement before transfer is effected. The neuronal mechanisms underlying transfer of learned responses remain as a fascinating problem for future studies.

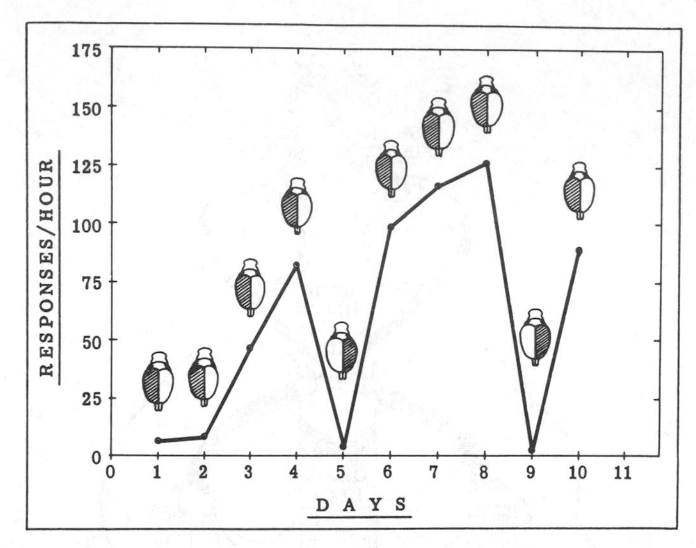

FIGURE 9-4. Unilateral learning shown by the rate of bar pressing. With spreading depression (SD) on the hemisphere of one side (hatched), learning occurs in the open side on days 3 and 4. With SD on the previously open side, days 5 and 9, no responses; with SD on the side originally depressed, responses on days 6, 7, 8, and 10. (After I. S. Russell and S. Ochs. *Science* 133:1077–1078, [April] 1961. Copyright 1961 by the American Association for the Advancement of Science.

Attempts have been made to show the sites where memories are stored by means of lesions placed in various limbic and other subcortical structures. Lesions in the septum and in the pathways connecting hypothalamus and hippocampus to other regions will produce a defect of learning and a loss of conditioned responding. A number of other subcortical regions have been implicated. Further analysis is required to determine whether memory or motivation is involved by such lesions.

Much is still to be learned about memory. If all memories are present subcortically, does the cortex play a role in retrieving them? If part of the memory is laid down in one site and part in another, how are they gathered together to give rise to a whole memory sequence? Finally, much evidence suggests a molecular basis for permanent memories. Changes in the ribonucleic acid (RNA) composition of neuron cell bodies, and in turn their synthesis of specific proteins, may be responsible for memories. But how is neuronal activity transformed into RNA changes, and in turn specific proteins? How do such proteins later effect a neural discharge of a given pattern? The importance of these problems for clinical medicine is apparent, but as yet too little is known. The reader is referred to the cited works for further information.

REFERENCES

Brady, J. V. The Paleocortex and Behavioral Motivation. In H. F. Harlow and C. N. Woolsey (Eds.), *Biological and Biochemical Bases of Behavior.* Madison: University of Wisconsin Press, 1958.

Bureš, J. Reversible Decortication and Behavior. In M. A. B. Brazier (Ed.), *The Central Nervous System and Behavior.* New York: Josiah Macy, Jr. Foundation, 1959. Pp. 207–248.

Doty, R. W., and C. Giurgea. Conditioned Reflexes Established by Coupling Electrical Excitation of Two Cortical Areas. In A. Fessard, R. W. Gerard, and J. Konorski (Eds.), *Brain Mechanisms and Learning: A Symposium.* Springfield, Ill.: Thomas, 1961. Pp. 133–151.

Fulton, J. F. *Frontal Lobotomy and Affective Behavior: A Neurophysiological Analysis.* New York: Norton, 1951.

Gaito, J. (Ed.). *Macromolecules and Behavior,* 2d ed. New York: Appleton-Century-Crofts, 1971.

Herrick, C. J. (Memorial Symposium): Neurophysiology of learning and behavior. *Fed. Proc.* 20:601–631, 1961.

Klüver, H., and P. C. Bucy. Preliminary analysis of functions of the temporal lobes in monkeys. *Arch. Neurol. Psychiat.* 42:979–1000, 1939.

Landauer, T. K. (Ed.). *Readings in Physiological Psychology.* New York: McGraw-Hill, 1967.

Lashley, K. S. In search of the engram. *Sympos. Soc. Exp. Biol.* 4:454–482, 1950.

MacLean, P. D. Psychosomatic disease and visceral brain. *Psychosom. Med.* 17:355–366, 1955.

Miller, N. E. Learning of visceral and glandular responses. *Science* 163:434–445, 1969.

Morrell, J. Electrophysiological contributions to the neural basis of learning. *Physiol. Rev.* 41:443–494, 1961.

Mountcastle, V. B. (Ed.). *Interhemispheric Relations and Cerebral Dominance.* Baltimore: Johns Hopkins University Press, 1962.

Olds, J. Self-stimulation of the brain. *Science* 127:315–324, 1958.

Papez, J. W. A proposed mechanism of emotion. *Arch. Neurol. Psychiat.* 38:725–745, 1937.

Pavlov, I. P. *Conditioned Reflexes.* New York: Dover, 1960.

Penfield, W. Functional localization in temporal and deep sylvian areas. *Res. Publ. Ass. Res. Nerv. Ment. Dis.* 36:210–226, 1958.

Russell, I. S., and S. Ochs. Localization of a memory trace in one cortical hemisphere and transfer to the other hemisphere. *Brain* 86:37–54, 1963.

Russell, R. W. (Ed.). *Frontiers and Physiological Psychology.* New York: Academic, 1966.

Skinner, B. F., and C. Ferster. *Schedules of Reinforcement.* New York: Appleton-Century-Crofts, 1957.

Sperry, R. W. Cerebral organization and behavior. *Science* 133:1749–1757, 1961.

10

Functional Properties of Blood

Julius J. Friedman

UNICELLULAR ORGANISMS are in immediate contact with their external environment. Nutrition is derived directly from and excretion occurs directly into the external medium. The organism is able to survive under these conditions because the volume of the extracellular environment greatly exceeds that of the cell. Consequently, acquisitions from and contributions to the surrounding medium produce no significant changes in concentrations of the various constituents of the medium, and the cell can be considered as being exposed to a constant environment.

In complex multicellular organisms, cells neither gain nutrition directly from the external environment nor excrete waste materials directly into it. Instead, the organisms are equipped with specialized tissues which are concerned with the processes of nutrition and excretion. To provide these facilities to the remainder of the cells throughout the organism, a vehicle of communication between them is necessary.

It is the primary function of blood to provide a link between the various organs and cells of the body. Blood maintains a constant cellular environment by circulating through every tissue and continuously delivering nutrients to the tissues and removing waste products and various tissue secretions from them. Blood consists of a fluid plasma in which are suspended a number of formed elements.

PLASMA

Plasma is a pale amber fluid which contains numerous dissolved materials, including proteins, carbohydrates, lipids, electrolytes, pigments, hormones, and others. The proteins are responsible for many of the characteristics of this fluid. The plasma proteins include *albumin, globulin,* and *fibrinogen,* and each can be fractionated by salt precipitation or by electrophoresis. The proteins are found to be present in the concentrations listed in Table 10-1. A normal electrophoretic pattern of plasma (Fig. 10-1A) reveals that a number of globulin fractions are present. These include α_1, α_2, β_1, and γ varieties. Each fraction apparently contributes to a different characteristic of the plasma.

The proteins contribute significantly to the specific gravity of plasma, which is about 1.028. They are of great functional importance with respect to osmotic pressure, viscosity, and suspension characteristics of the plasma. The proteins also participate in nutrition, coagulation, acid-base regulation, and immunological responses.

TABLE 10-1. Plasma Proteins

Protein	Molecular Weight	Fractionation by Electrophoresis (% wt.)	Fractionation by Salt Precipitation (gm per 100 ml.)
Albumin	69,000	55	4–6
Globulin	80,000–200,000	38	1.5–3
Fibrinogen	350,000–400,000	7	0.2–0.4

A number of disorders manifest profound changes in the concentrations of the plasma proteins which can be detected by electrophoresis (Fig. 10-1B, C, D).

FORMED ELEMENTS

The formed elements of the blood include erythrocytes, leukocytes, and thrombocytes (platelets).

ERYTHROCYTES

Erythrocytes, or red blood cells, are annucleated biconcave discs which are approximately 2 μ thick and 8 μ in diameter and which possess a volume of approximately

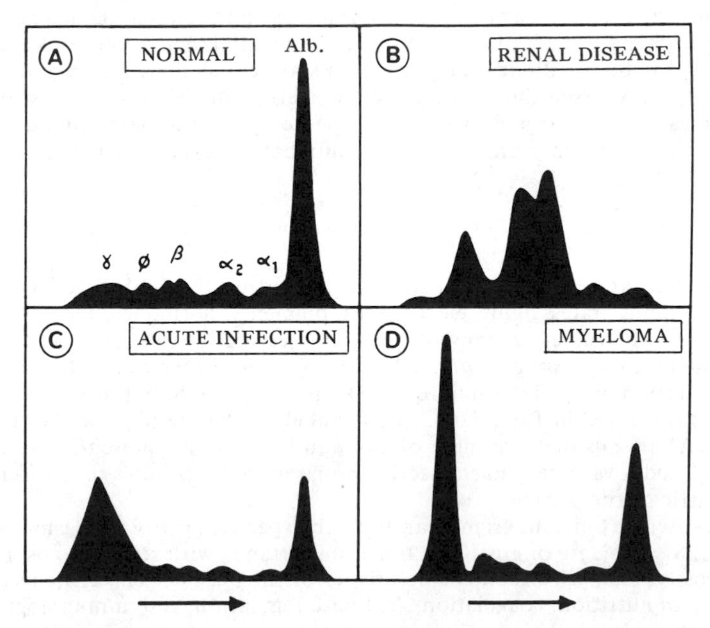

FIGURE 10-1. Densitometric patterns of paper electrophoresis of human plasma.

85 μ^3. The major function of the red cell is to transport oxygen and carbon dioxide. The red cell must be equipped to release its oxygen to the tissues rapidly and take up carbon dioxide. The biconcave shape of the cell is ideally suited to this function. It provides the maximum surface area for diffusion for the volume of the cell.

There are about 4.5 to 6.0 million red blood cells per cubic millimeter of blood. Since the total blood volume is 5 liters, the total number of red blood cells present in the circulation is about 25×10^{12}. The number can be estimated by counting those in a small representative sample of blood with the aid of the microscope and the graduated slide called a hemacytometer.

A unit used to describe the relative red cell content of blood is the *hematocrit ratio.* This ratio is derived by first rendering a blood sample incoagulable and then centrifuging it in a Wintrobe tube at 1500 g (3000 rpm with an arm 15 cm long) for 30 minutes. A Wintrobe tube is merely a heavy-walled, small-volume, uniform-diameter tube graduated from 0 to 10 (Fig. 10-2). Since the specific gravity of red cells is about 1.088, they become packed at the bottom of the tube, and the less dense plasma is situated above the red cells. The white blood cells and platelets possess a specific gravity intermediate between that of plasma and red cells and form the white buff-colored layer immediately above the red cell column. The hematocrit ratio is determined simply by measuring the height of the red cell column and dividing this by the height of the total blood column (Fig. 10-2). The

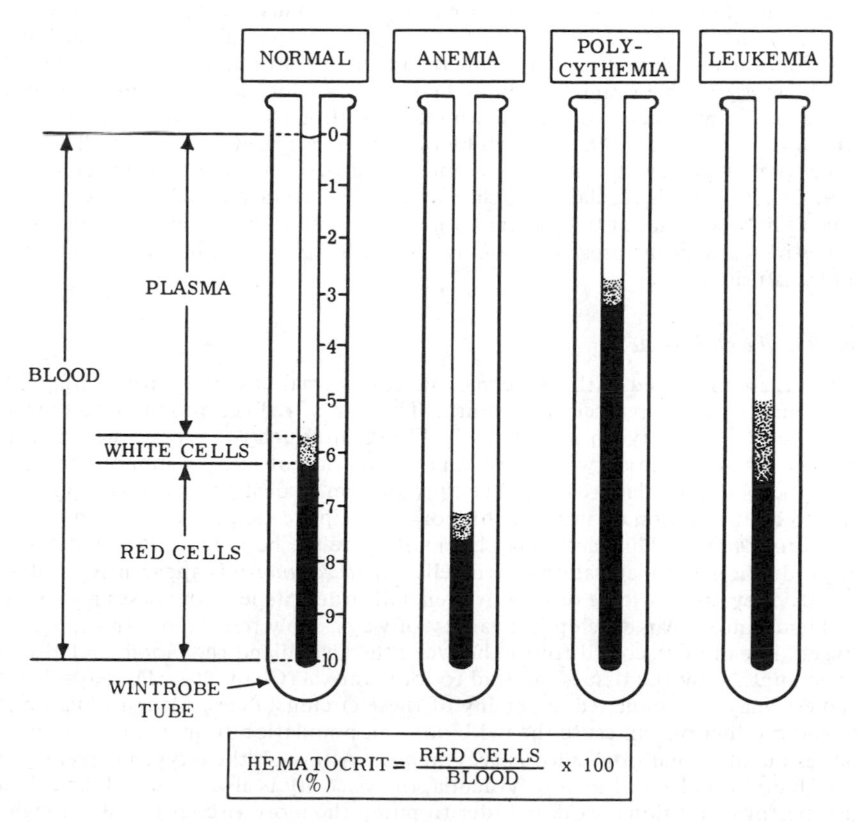

FIGURE 10-2. Determination of hematocrit ratio.

white blood cells are not included in the hematocrit ratio. This ratio is usually multiplied by 100 to provide a hematocrit reading in percent. The venous hematocrit of a normal adult determined in this manner is about 45 percent. However, because the cells are disc-shaped, the small space between adjacent cells is occupied by plasma which must be subtracted from the initial measurement. It has been estimated, by using radioactive plasma, that approximately 4 to 5 percent of the volume of packed cells in the Wintrobe tube is represented by plasma. Thus, the corrected venous hematocrit reading is approximately 43 percent. Care must be exercised in evaluating changes, since the hematocrit varies with changes in sampling site as well as in many clinical conditions (Fig. 10-2).

The hematocrit is not uniform throughout the body. That of venous blood is slightly greater than that of arterial blood, because of fluid and electrolyte shifts. In addition, the hematocrit varies from tissue to tissue. The splenic hematocrit is approximately 70 percent, whereas that of the kidney is only about 20 percent. The values for other tissues are distributed between these two extremes. This variation in tissue hematocrit has been attributed in part to a phenomenon called *axial streaming.* As blood flows through the peripheral circulation, mostly plasma is "skimmed" off into the capillary-venular circulation from the periphery of the flow stream. Since the capillaries and venules contribute significantly to tissue blood volume, the low hematocrit of this segment of the tissue vasculature imparts a low reading to the remainder of the tissue circulation, and tissue hematocrit is therefore lower than either arterial or venous hematocrit. The most notable exception, the spleen, by virtue of its sinusoidal vasculature, possesses the ability to "filter" red cells from the circulation, and this imparts a high hematocrit to the splenic blood.

The most reliable estimate of body hematocrit is derived from independent total plasma and total red cell volume determinations. Body hematocrit calculated as the ratio of total red cell volume to total blood volume is about 35 percent. The ratio of the total body hematocrit to the venous hematocrit, which has been referred to as the *F cells ratio,* is regularly less than unity. The average F cells ratio seems to remain fairly constant at 0.91 under a variety of conditions of cardiovascular stress. It does increase during pregnancy and is susceptible to change in severe hemorrhage and transfusion.

Red Blood Cell Production

Red cells are produced in the bone marrow at a normal rate of approximately 400 to 500 ml of packed red cells per month. The rate of red cell production is probably determined by the oxygen tension of the blood, although the precise mechanism is not clear. Any condition which results in a reduction of oxygen supply to the tissues appears to stimulate red blood cell production. Local tissue anoxia apparently leads to the formation of a specific humoral principle called the *erythropoietic stimulating factor.* This factor may then stimulate the bone marrow to increase the production and maturation of red cells. *Erythropoietin* is apparently produced in the kidney, and in some obstructive renal disorders its level increases markedly.

Tissue anoxia can develop in a variety of ways. Low tension of atmospheric oxygen, impaired gaseous diffusion between the alveoli and the blood, and altered oxygen uptake by the tissues can lead to local anoxia (Chap. 20). Therefore, erythropoiesis may be stimulated under any of these circumstances. Overstimulation of red cell production can cause the red blood cell population to increase excessively and results in a condition called *polycythemia.* Although the oxygen-carrying capacity of blood is enhanced in polycythemia, the viscosity is also increased, and the heart must perform additional work in order to pump the more viscous blood through the circulatory system. To accommodate the added work load, the heart hypertrophies.

Polycythemia and cardiac hypertrophy are most prevalent among people who reside at higher elevations and who are thus chronically exposed to low oxygen tensions.

Red Cell Destruction

The red cell travels about 700 miles during its lifetime and is subjected to considerable mechanical and chemical stress, which eventually leads to its destruction. The life-span of the red blood cell, as determined by radioactive methods, is about 120 days. Approximately 1 percent of the red cells are replaced daily, therefore, and the total red cell volume is replaced every four months. Red cells are primarily destroyed in the spleen. Continuous friction between red cells and the vessel walls and the action of chemical substances in circulation apparently alter the integrity of the red cell membrane, thereby rendering it more susceptible to osmotic stresses. Figure 10-3 presents a *fragility curve,* which is a plot of the percentage of hemolysis against extracellular sodium chloride concentration. The sigmoid character of the plot is thought to represent the hemolytic behavior of cells of different ages. Thus, older cells hemolyze with lower osmotic stress, whereas the younger cells are most resistant to osmotic influence. Fragility tests are often used to distinguish between hereditary and acquired red blood cell disorders.

The number of red cells in the circulation at any one time is determined by the balance between production and destruction. When production exceeds destruction, the red cell volume and hematocrit rise, and polycythemia results. Conversely, when production is less than destruction, red cell volume and hematocrit fall, and *anemia* results.

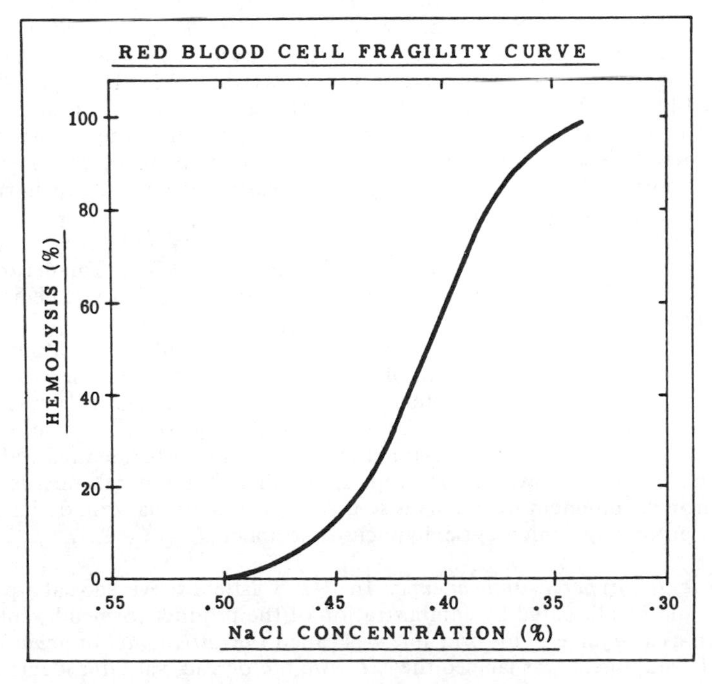

FIGURE 10-3. Red blood cell fragility curve. Effect of varying osmotic environment on red cell survival. (From F. T. Hunter. *J. Clin. Invest.* 19:691–694, 1940.)

Anemia

Anemia refers to a condition in which the red cell population is below normal levels. It may result from a number of causes, including hemorrhage (acute and chronic), maturation deficiency, aplastic bone marrow, and enhanced fragility. Each situation influences both the size and hemoglobin content of the red cell in a characteristic manner; these two parameters are used to classify anemia. The cell of normal size is *normocytic.* A cell larger than normal is *macrocytic,* while one smaller than normal is *microcytic.* The extent of microcytosis or macrocytosis is reflected in the mean corpuscular volume (MCV)

$$\text{MCV} = \frac{\text{Hematocrit ratio} \times 1000}{\text{Red cell count in millions/mm}^3}$$

Similarly, a cell which possesses a normal concentration of hemoglobin is said to be *normochromic.* Red cells with subnormal hemoglobin concentration are *hypochromic,* and those with an above normal concentration are *hyperchromic.*

The degree of hemoglobin alteration is reflected in the mean corpuscular hemoglobin (MCH) determination.

$$\text{MCH} = \frac{\text{Hemoglobin (gm/1000 ml)}}{\text{Red cell count in millions/mm}^3}$$

Normocytic, normochromic anemia. Although red blood cells can be of normal size and contain the normal concentration of hemoglobin, anemia may exist if the number of red cells in the circulation is below normal. This condition arises with acute hemorrhage. As the result of hemorrhage, both plasma and red cells are lost from the circulation and the red cell volume is reduced. The plasma volume is restored in a short time by compensatory responses (Chap. 12). However, the process of red blood cell production is slow, and the circulation remains deficient in red cells for some time. Since the bulk of the cells present in the circulation during this period are the same ones that were present prior to hemorrhage, the cells are essentially normal. Consequently, the anemia resulting from acute hemorrhage is normocytic and normochromic.

Microcytic, hypochromic anemia (iron-deficiency anemia). Iron-deficiency anemia often occurs in women of childbearing age or during chronic hemorrhage, and in infants.

During chronic hemorrhage, red cell loss from the circulation continues over a period of time, and the replacement of these cells must be maintained throughout this period. The persistent and extended stimulus to erythropoiesis imposes a continuous demand on the iron stores of the body and leads to a condition of iron deficiency. Consequently, the hemoglobin content of newly formed red cells is lower than normal (hypochromic) and the cells are smaller than normal (microcytic).

The iron requirement of infants is so great that nutritional iron deficiency often develops, producing a microcytic, hypochromic anemia.

Macrocytic, hyperchromic anemia. In 1929 Castle discovered that a particular anemia could be alleviated by administration of the product formed by incubating meat with gastric juice. This principle was called the *antianemic* or *hematinic factor.* The meat component was labeled the *extrinsic factor* and was later identified as vitamin B_{12}. The gastric juice component was called the *intrinsic factor.* Vitamin B_{12} contained in ingested meat combines with the intrinsic factor to form the hematinic factor, which is absorbed through the intestinal wall and stored in the liver. In the absence of either of these factors, the maturation of red cells is altered and excessively large red cells (macrocytes) are formed in reduced number. These cells contain a greater than normal amount of hemoglobin, owing to their increased size, and therefore are hyperchromic. The clinical condition which characterizes these changes is called *pernicious* or *addisonian anemia.*

Pernicious anemia may occur under any circumstances which interfere with any phase of the incorporation or utilization of the antianemic factor. Consequently, a reduction in extrinsic factor due to dietary deficiency, reduced gastric juice secretion, reduced intestinal absorption, or decreased liver function may lead to pernicious anemia. Because the liver is the storage site of the hematinic factor, either the addition of liver to the diet or administration of liver extract is generally sufficient to alleviate the condition.

Other megaloblastic anemias, which resemble pernicious anemia but have an obscure pathogenesis, occur in tropical sprue and pregnancy, as well as in conditions of folic acid and vitamin C deficiencies, and are referred to as *nutritional anemias.*

Aplastic anemia. Occasionally a subject with malfunctioning bone marrow is encountered; his capacity to produce normal cells is absent. In acute conditions, such as those due to excessive exposure to radiation, the cells in circulation are essentially those which were present prior to exposure and are therefore normocytic and normochromic. However, since destruction proceeds independently of production, the red cell count progressively falls. Aplastic bone marrow can also occur spontaneously.

Hemolytic anemia. Anemia may result when enhanced red cell hemolysis takes place because of chemical agents, hereditary defects of cell formation, enhanced reticuloendothelial destruction, and hypersplenism. Venoms, bacterial toxins, and some drugs produce hemolysis by means of lysis of the cell membrane. *Familial hemolytic anemia* is a hereditary condition characterized by the production of small spherical red cells called microcytes or spherocytes. These cells are normochromic but hemolyze readily because their shape imparts great tension to the membrane during the deformations encountered by the cell as it passes through minute vessels. *Sickle cell anemia,* particularly prevalent among Negroes, represents another hereditary abnormality in which the red cells are shaped like sickles or crescents and therefore are highly susceptible to osmotic stress. In addition, changes in the nature of the hemoglobin in these cells modify their binding of oxygen and impair gaseous exchange.

LEUKOCYTES

The major function of leukocytes is to protect against the invasion of bacteria and infection. These cells are capable of passing through the vascular endothelium and entering the tissue spaces (diapedesis) in accordance with the local needs. They are apparently attracted by chemical substances (chemotaxis) released by the bacteria and can engulf and digest the foreign substances by means of ameboid movement (phagocytosis). Normally there are approximately 5000 to 7000 white blood cells per cubic milliliter of blood. This total is comprised primarily of polymorphonuclear neutrophils (60 percent) and lymphocytes (35 percent), the remaining 5 percent consisting of polymorphonuclear eosinophils, basophils, and monocytes. The leukocytes form part of a broader system which provides protection against infection. This is the *reticuloendothelial system;* it includes all phagocytic cells, fixed and mobile, which are primarily situated in the liver, spleen, lymph nodes, lung, and gastrointestinal tract.

PLATELETS

Platelets are small, colorless bodies, ranging in size from 2 to 4 μ in diameter, which are intimately associated with coagulation. Contained within or adsorbed on the platelets are a number of factors of coagulation. When tissue injury occurs, the platelets may disrupt at the site of injury, releasing their contents.

There are about 150,000 to 350,000 platelets per cubic milliliter of blood. They have a life-span of about one week.

The reader is referred to standard histology textbooks for the details of the morphology and histology of the formed elements and associated organs.

BLOOD VOLUME

The total blood volume in a subject is best determined by estimating both the plasma and the red cell volumes independently and adding them together. These volumes are estimated by means of a dilution principle wherein a known quantity of indicator (Q) is introduced into an unknown volume. Sufficient time is allowed for complete mixing, and the concentration of indicator (C) in a representative sample is determined. The volume of dilution (V) is then calculated as:

$$V = \frac{Q}{C}$$

PLASMA VOLUME

Plasma volume (PV) is estimated by injecting into the circulatory system about 5 microcuries of radioactively iodinated (^{125}I) serum albumin (RISA). If albumin distribution were restricted to the vascular system, a sample after 5 to 10 minutes of vascular mixing would be adequate for the determination of indicator concentration. Unfortunately, albumin leaves the circulation slowly (about 10 percent per hour) so that it is necessary to obtain at least two samples – 10 and 20 minutes after injection – and extrapolate from these two points to zero time. This procedure provides an estimate of the mixed indicator concentration in the plasma before any extravascular loss occurred.

An average plasma concentration of indicator in a 70-kg male would be about 0.0015 microcurie per milliliter of plasma. Dividing this concentration into the dose administered provides an estimate of about 3200 ml of plasma.

RED BLOOD CELL VOLUME

Red blood cell volume (RCV) is generally estimated with red cells labeled with radioactive chromium (^{51}Cr). The procedure employed is as follows: Approximately 20 to 30 ml of blood is withdrawn from the subject and mixed with 5 ml of an anticoagulant acid citrate dextrose (ACD) solution. A dose of 15 to 20 microcuries of sodium chromate (^{51}Cr) is added to the blood and gently agitated at room temperature for 30 to 45 minutes. This provides 90 percent binding of ^{51}Cr. Tagging is terminated by adding 50 to 100 mg of ascorbic acid, which reduces the chromate.

The labeled blood is injected intravenously and allowed to mix in the circulation for about 10 minutes. A sample is obtained and corrected for unlabeled ^{51}Cr, and the concentration is determined. Average concentrations are about 0.011 microcurie per milliliter of red cells. Dividing the corrected concentration into the corrected dose provides an estimate of 1800 ml of red cells for a 70-kg adult male.

The times allowed for vascular mixing apply to a subject under reasonably normal cardiovascular conditions. In cases of cardiovascular stress where tissue blood flow is severely altered, longer mixing times will be necessary.

TOTAL BLOOD VOLUME

Total blood volume (BV) is calculated by the expression

$$BV = PV + RCV$$

Thus the total blood volume of a normal 70-kg adult male is about 5000 ml. The blood volume of a normal adult female is about 4500 ml.

It is possible to obtain an estimate of total blood volume from a single determination of plasma or of red cell volume and corrected venous hematocrit according to the expressions

$$BV = \frac{PV}{(100 - \text{Hematocrit}) \times \text{F cells}}$$

$$BV = \frac{RCV}{\text{Hematocrit} \times \text{F cells}}$$

Total blood volume varies under circumstances involving the integrity of the vascular system, the state of water balance of the individual, and the state of physiological activity. These will be discussed in detail in Chapters 18 and 23.

Changing posture from a lying to a standing position will result in a fall in blood volume due to a reduction in plasma volume. This reflects the change in transcapillary dynamics upon standing, which produces enhanced filtration (see Chap. 12). Exercise produces similar variations in blood volume. Adaptation to low tissue oxygen tensions such as occur at altitude and with physical training increases red cell and blood volume. In pregnancy both plasma and red cell volume increase.

The blood volume is not uniformly distributed throughout the body. Approximately 80 percent of the total blood volume is contained within the low pressure system (veins, right heart, and lungs). Because of the high distensibility of the pulmonary vasculature, and the large mass of muscle, these two tissues are prominent blood reservoirs which participate in large volume shifts during conditions of circulatory stress. About 50 percent of a 400 to 1000-ml hemorrhage is provided by the intrathoracic compartment. During orthostasis, approximately 700 ml is translocated from the thorax to the legs.

Changes in blood volume distribution are manifested by either passive or active adjustment of the capacitance vasculature. Passive behavior depends on the elastic recoil of the vessel system, whereas active behavior is primarily related to venomotor tone.

BLOOD GROUPS AND TRANSFUSION

In conditions of anemia or hemorrhage, it is often desirable to transfuse additional volumes of blood into the circulation in order to improve the hemodynamics of the system. When this course of treatment is indicated, care must be exercised in selecting the blood to be transfused, since not all blood is compatible. Numerous reactions may arise from incompatibilities in blood. These incompatibilities reflect variations in principles present in the cells and in the plasma of different individuals.

Blood has been classified, according to its antigenic activity, as types O, A, B, and AB. The specific representation in the blood of an individual is determined genetically, so there must be a double representation, one derived from each parent. Consequently, the genotypes of the blood groups are OO, AA, AO, BB, BO, and AB. Present on red cells are specific chemical principles called *agglutinogens,* which determine the antigenic behavior of the cells. When serum possesses a suitable antibody, *agglutinin,* agglutination of the red cells takes place.

Fortunately, when an agglutinogen is present in the blood, the antagonistic agglutinin is absent. Conversely, where the agglutinin is present, the agglutinogen is absent. Type A blood contains agglutinogen A and anti-B or beta agglutinin. Similarly, type B blood contains agglutinogen B and anti-A or alpha agglutinin. Type O blood does not usually contain any agglutinogen, and both alpha and beta agglutinins are present, whereas type AB blood possesses both agglutinogen A and agglutinogen B; consequently no agglutinins are present. The blood groups, their constituents, and their percentage distribution are listed in Table 10-2.

When type A blood is transfused into a type A recipient, no blood reaction occurs because no antagonistic blood principles are present. However, if type A blood is transfused into a type B or a type O recipient, two antigen-antibody reactions are

TABLE 10-2. Blood Group Constituents

Group	Agglutinogen	Agglutinin	Percent of Population
O	–	α, β	46
A	A	β	42
B	B	α	9
AB	A + B	–	3

possible. The agglutinins of the donor may react with the agglutinogen of the recipient to produce agglutination of recipient cells, or agglutinogens of the donor may react with agglutinins of the recipient to produce agglutination of donor cells. The former reaction is unlikely, because the addition of small quantities of donor agglutinins into a relatively large volume of the recipient's blood provides sufficient dilution to render the donor agglutinins ineffective. Transfusion reaction therefore involves the agglutination of donor cells by recipient agglutinins.

Since type O blood possesses no agglutinogens, type O cells cannot be agglutinated by any of the agglutinins. Consequently type O blood can be transfused into any other type without reaction, and type O persons are referred to as "universal donors." Conversely, type AB blood is without agglutinins and will not agglutinate any type of donor blood. Persons with type AB blood are referred to as "universal recipients."

There are a number of other factors in the blood which can lead to low-level incompatibilities or sensitization. The most prominent of these, the *Rh factor,* is present in approximately 85 percent of the population. When it is present, the blood is Rh-positive; when it is absent, Rh-negative. Antibodies to the Rh factor do not normally exist in the blood but arise following the exposure of individuals with Rh-negative blood to the Rh antigen. An Rh-negative mother bearing an Rh-positive child may become sensitized and develop Rh antibodies. These antibodies can then diffuse into the fetal circulation, to produce an Rh reaction and death (erythroblastosis fetalis). Usually a number of exposures are required to establish an antibody titer sufficiently high to produce fetal death.

Other factors, such as M and N, are occasionally considered in blood typing, particularly in circumstances of disputed parenthood.

HEMOSTASIS

The rupturing of a blood vessel results in blood loss which, if unchecked, would eventually lead to hemorrhagic shock and death. During normal activity, minor accidents frequently occur in which minute blood vessels are ruptured, yet blood loss is minimal. When a blood vessel is ruptured, hemorrhage is limited by a reduction of blood flow to the site of injury by means of local vasoconstriction, by platelet aggregation, by coagulation, and by clot retraction.

VASOCONSTRICTION

Numerous influences can produce vasoconstriction (Chap. 17). With vascular injury, platelets release serotonin, which may provide the stimulus for constriction of small arteries and arterioles. By reducing blood flow from the severed vessel, vasoconstriction enables platelet aggregation and coagulation to proceed with effectiveness.

PLATELET AGGREGATION

Adenosine diphosphate (ADP) released from damaged cells causes platelet changes which result in increased adhesiveness and aggregation. It appears that the platelet plug organizes in association with the positive electrical charge of collagen at the site of injury.

COAGULATION

Coagulation represents the interaction of platelet, plasma, and tissue factors which culminates in the formation of a fibrin meshwork at the site of vessel rupture. While numerous theories have been introduced to account for this phenomenon, there is still considerable uncertainty about it owing to the limited availability of purified factors.

A recent scheme of coagulation proposed by Seegers places prothrombin in a pivotal position. *Prothrombin* is a glycoprotein of molecular weight 68,000 which is formed in the liver. It is thought by Seegers to possess the characteristics attributed to other factors of coagulation, namely, factors VII, IX, and X, because all depend on vitamin K for synthesis, are heat stable, and are absorbable from oxalated plasma. Further, prothrombin is represented as a loose complex of precursors – in particular, *autoprothrombin III* and *prethrombin.*

Coagulation occurs in a sequence of three phases: (1) conversion of autoprothrombin III to autoprothrombin C, (2) conversion of prethrombin to thrombin, and (3) conversion of fibrinogen to fibrin.

Phase I: Conversion of Autoprothrombin III to Autoprothrombin C

Coagulation may be initiated by means of tissue damage, wherein tissue fluids gain access to the circulation, or it may take place without accompanying tissue damage.

The sequence of events following tissue damage is referred to by some as *extrinsic* coagulation, whereas coagulation without the participation of tissue factors is regarded as *intrinsic.*

The active principle in tissue is *thromboplastin,* which is released or activated upon damage of tissue.

Tissue thromboplastin apparently acts on or with Ca^{++} to remove peptides from autoprothrombin III to form autoprothrombin C.

The intrinsic sequence of coagulation is initiated by means of platelet alterations produced by either ADP, collagen, Hageman factor, electrical charge, or thrombin which cause the release of platelet factor 3, a lipoprotein. This lipid acting along with a platelet cofactor (factor VIII – the hemophilic factor) and with Ca^{++} converts autoprothrombin III to autoprothrombin C. See Figure 10-4A.

Phase II: Conversion of Prethrombin to Thrombin

Since blood clots rapidly in the presence of *thrombin,* significant amounts cannot be present, as such, in the circulation. Instead it is represented by a precursor, prethrombin. Prethrombin can be converted to thrombin by autoprothrombin C alone; however, the rate of conversion is substantially accelerated by the presence of a lipid, Ca^{++} and Ac-globulin (factor V). See Figure 10-4B.

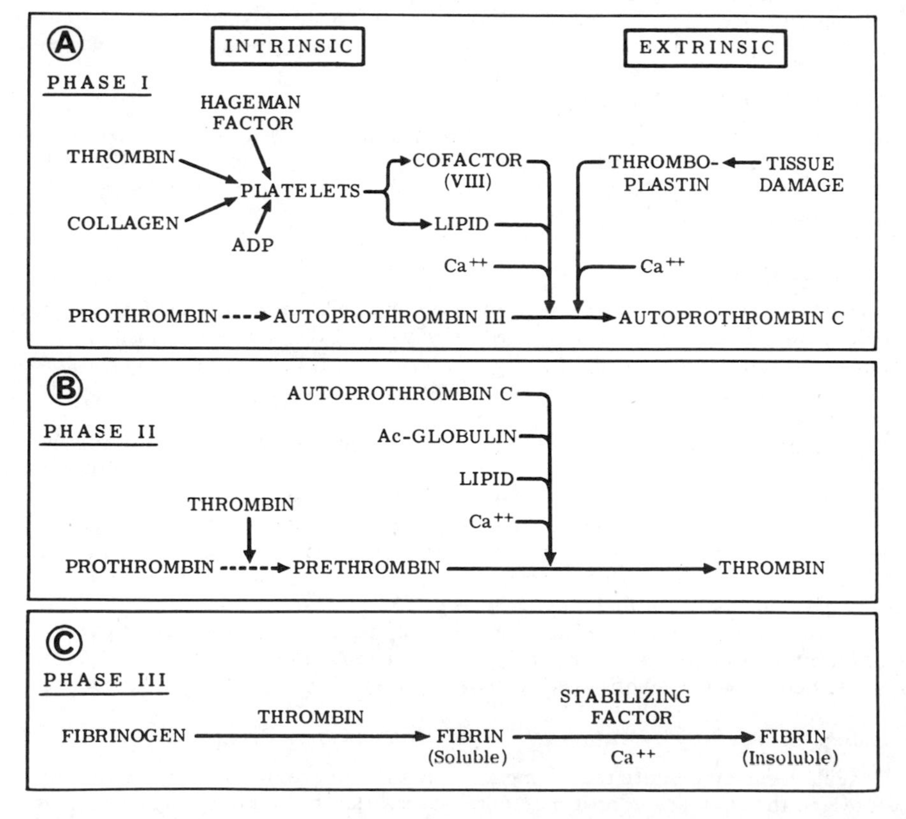

FIGURE 10-4. Schematic representation of the mechanism of coagulation. (From Seegers, 1969.)

Phase III: Conversion of Fibrinogen to Fibrin

Fibrinogen is converted to fibrin by thrombin. Thrombin partially proteolyzes fibrinogen by splitting off two peptides. The fibrin monomer thus formed is unstable and dissolves easily in urea. However, in the presence of calcium ions and *fibrin-stabilizing factor* (factor XIII) derived from plasma transglutaminase, fibrin becomes cross-linked to form an insoluble clot. See Figure 10-4C.

There are numerous synonyms for the coagulation factors cited above, many of which are included in Table 10-3.

TABLE 10-3. Common Synonyms of Blood-Clotting Factors

Factor	Synonym
I	Fibrinogen
II	Prothrombin
III	Thromboplastin (tissue)
IV	Calcium ion
V	Proaccelerin, labile factor, plasma accelerator globulin (AcG)
VI	Discontinued
VII	Proconvertin, stable factor, precursor of serum prothrombin conversion accelerator (pro-SPCA)
VIII	Antihemophilic factor (AHF), antihemophilic globulin (AHG), thromboplastinogen, plasma thromboplastic factor (PTF), platelet cofactor I
IX	Plasma thromboplastic component (PTC), Christmas factor, platelet cofactor II
X	Stuart-Prower factor
XI	Plasma thromboplastin antecedent
XII	Hageman factor
XIII	Fibrin-stabilizing factor

Clot Retraction

After the clot forms, it retracts and extrudes fluid identical with plasma except for the absence of fibrinogen and factors VII and IX. This fluid is *serum*. Platelets are necessary for clot retraction since the process does not occur in platelet deficiency. The platelets apparently contain a contractile protein, similar to actomyosin, which contracts when treated with adenosine triphosphate (ATP). The function of clot retraction has been considered to be that of drawing the wound surfaces together, recanalizing thrombosed vessels, or promoting more effective thrombosis.

Fibrinolysis

Blood clots are not permanent. After some time the clot is dissolved or lysed. The process of *lysis* involves the conversion of an inactive precursor, *profibrinolysin* (plasminogen), to the active form *fibrinolysin* (plasmin). Stress, anoxia, or pyrogens result in the release of *urokinase,* which elicits the transformation of profibrinolysin to fibrinolysin, which degrades the fibrin meshwork, disrupting the clot.

There is evidence suggesting that a balance may exist between fibrin clot formation and lysis. Interruption of coagulation by heparin administration results in an elevation of profibrinolysin concentration. This mechanism has also been considered responsible for the maintenance of normal capillary flow by preventing microthrombus formation.

NATURAL INHIBITORS OF COAGULATION

Since tissue breakdown and platelet destruction probably occur under normal circumstances, and since some platelets are probably damaged and thromboplastin probably enters the circulation to form thrombin, it would seem possible that coagulation in vivo might occur. Fortunately, there are inherent inhibitors which control clotting at every stage. These include *antithromboplastin, heparin,* and *antithrombin.* Consequently, when small quantities of thromboplastin are released from tissue or formed in blood, inactivation by antithromboplastin occurs. However, even in situations of low-level release, the concentration of thromboplastin in the plasma may become sufficiently great to form autoprothrombin C. Once again coagulation is prevented, this time by virtue of the presence of heparin.

Heparin is a polysaccharide produced primarily in the basophilic mast cells distributed throughout the pericapillary tissue. It is available commercially and serves as one of the major clinical anticoagulants. Heparin prevents coagulation in a variety of ways. Apparently it can inhibit prothrombin conversion, and possibly thrombin. Coagulation is also inhibited by the presence of antithrombin, which prevents the conversion of fibrinogen to fibrin. Antithrombin appears to be a complex of substances which are potentiated by heparin. In addition to this function, antithrombin is believed to clear thrombin from the blood once clotting is completed. Otherwise, it is conceivable that thrombin would progressively accumulate and produce widespread coagulation. With the presence of these anticoagulant factors, obviously coagulation does not normally occur spontaneously but requires adequate tissue damage to take place.

COAGULATION DEFICIENCIES

A deficiency of coagulation and a tendency toward bleeding may result from a deficiency of any of the principles cited in Figure 10-4 except the Hageman factor. Deficiencies may arise in several ways. A dietary or absorption deficiency of vitamin K will result in a low blood prothrombin level (prothrombinemia). A reduced platelet count (thrombocytopenia) will lead to deficiency of cofactors. Since fibrinogen, prothrombin, and probably other factors in coagulation are formed in the liver, liver disease which alters prothrombin or fibrinogen synthesis will result in coagulation deficiencies. The classical hereditary condition of *hemophilia* results from the absence of antihemophilic factor (AHF), which leads to coagulation deficiency and a bleeding tendency. This effect is an inherited, sex-linked abnormality of males. In each case the replacement of the missing factor is usually sufficient to restore coagulation. Coagulation may also be impeded by excessive concentrations of fibrinolysin such as are seen in fibrinolytic purpura.

COAGULATION EXCESSES

When the concentration of any of the inherent anticoagulants is low, when endothelial damage is great, or when blood is stagnant, coagulation in vivo may occur. A clot which remains fixed in a vessel is called a *thrombus,* and the condition, *thrombosis.* As blood flows by a thrombus, fragments *(emboli)* may break away and circulate throughout the vascular system. An embolus continues to circulate until it is confronted with a vessel possessing a cross-sectional area smaller than its own; it then becomes lodged in the vessel. This effectively occludes the vessel and deprives the tissue supplied by the vessel of adequate nutrition. When the area supplied is extensive or is situated in a vital organ, death can result. Usually the occluded region, if restricted in size, becomes invaded by a collateral circulation, and this may restore an adequate blood supply. When collateral circulation is inadequate, the occluded area degenerates to form an *infarct.*

ANTICOAGULANT AGENTS

The most prominent anticoagulant agents administered clinically in circumstances of excessive coagulation in vivo are heparin and *Dicumarol,* a drug which acts in the liver to prevent prothrombin formation by antagonizing the action of vitamin K. Because calcium ion is intimately associated with both the formation of the converting enzyme and the conversion of prothrombin, effective removal of calcium ion from the plasma prevents coagulation. However, the use of calcium precipitants or complexing agents is restricted to in vitro anticoagulation because a decreasing of the calcium ion concentration leads to cardiac arrest. *Sodium citrate* is a compound which removes calcium ion from solution by forming a poorly dissociating calcium salt. This substance is used routinely in blood banks as an anticoagulant. *Oxalates* are also occasionally used as in vitro anticoagulants.

COAGULATION TESTS

Numerous tests are in use to assess the effectiveness of the coagulation system. These include clotting time, prothrombin consumption time, thromboplastin generation, thrombin generation time, and others. The clotting time is the most comprehensive assay, since it is sensitive to deficiencies of most of the coagulating principles. The other tests are used to identify the coagulation deficiency more specifically. Details of the tests are to be found in any standard hematology text.

REFERENCES

Biggs, R., and R. G. MacFarlane. *Human Blood Coagulation and Its Disorders,* 3d ed. Philadelphia: Davis, 1962.

MacFarlane, R. G. (Ed.). *The Functions of Blood.* New York: Academic, 1960.

Putnam, F. W. (Ed.). *The Plasma Proteins.* New York: Academic, 1960.

Ratnoff, O. G. *Bleeding Syndromes.* Springfield, Ill.: Thomas, 1960.

Seegers, W. H. Blood clotting mechanisms: Three basic reactions. *Ann. Rev. Physiol.* 31:269–294, 1969.

Wintrobe, M. M. *Clinical Hematology.* Philadelphia: Lea & Febiger, 1961.

Fluid Dynamics

Carl F. Rothe

THE RATE of blood flow through tissues is a vital determinant of the adequacy of cellular nutrition and waste removal. The physical factors which govern this flow are the same as those modifying fluid flow through any living or nonliving system. A thorough understanding of cardiovascular function therefore requires the application of these principles of fluid dynamics. Once these concepts are understood, the effects of disease on the function of the cardiovascular system can be appreciated more fully, for one may reasonably predict the consequence of various changes. Most of the principles are summarized by the equations given in this chapter. The same equations, with differences in parameters or added terms, are applicable to flow in any system, e.g., air flow in the lungs.

FACTORS INFLUENCING FLOW

The rate of flow of a fluid is proportional to the driving pressure gradient. Expressed mathematically, this is:

$$\dot{Q} = G\,\Delta P = \frac{1}{R}\,\Delta P \tag{1}$$

The flow, $\dot{Q}$, is expressed in units of volume per unit of time, e.g., milliliters per minute. The *perfusion pressure*, or *effective pressure gradient* along the vessel in question, $P_1 - P_2$ or ΔP (Fig. 11-1A), is expressed in millimeters of mercury (mm Hg or torr), or newtons per square meter. (The unit of pressure in the System International, n/m^2, is derived from the basic meter, kilogram, and second. It is a rather small unit, since 1 mm Hg = 133.3 n/m^2.) The *conductance,* G, is the proportionality constant of equation 1 and is expressed in flow units divided by pressure units, e.g., milliliters per minute per millimeter of mercury. The reciprocal of conductance is *resistance,* R. The resistance opposing the flow of fluid cannot be measured directly but is calculated from simultaneous measurements of the flow and pressure gradient using equation 1. It is expressed as the units of pressure divided by the units of flow. The equation $\dot{Q} = P/R$ is fundamentally the same as Ohm's law of electrical circuits: $I = E/R$. There is no commonly accepted resistance unit analogous to the ohm. Note that if the pressure gradient increases, the flow increases, whereas if resistance increases, the flow decreases.

Flow is proportional to the pressure gradient, which in the case of a tissue vascular bed is the arterial minus the venous pressure. Because the venous pressure is much less than the arterial pressure in the systemic circuit, the venous pressure can often be neglected in calculating overall tissue vascular resistance. However, measurement of the resistance to flow through a low-pressure circuit, such as the pulmonary circulation, requires a knowledge of the pressure in the pulmonary veins as well as in the pulmonary artery.

POISEUILLE'S EQUATION

Poiseuille's equation describes in more detail the factors determining the resistance to flow (Fig. 11-1A). Poiseuille's equation is:

$$\dot{Q} = \Delta P \times \frac{\pi r^4}{8L} \times \frac{1}{\eta} \qquad (2)$$

Here r is the radius of the vessel or pipe, L is its length, and η is the coefficient of viscosity. Note that flow will increase if the pressure gradient or radius is increased or if the length or viscosity is decreased.

There are several important consequences from the mathematical derivation of Poiseuille's equation. The primary assumption, attributed to Sir Isaac Newton, is that fluid flows as infinitely thin layers or laminae sliding past each other giving *laminar* flow. The force per unit area *(shear stress)* tending to impede this sliding is directly proportional to the relative velocity between two adjacent layers *(shear rate).* The shear rate is a velocity gradient and is the longitudinal displacement per second divided by the distance between layers. Hence, the dimensions are: cm/sec per cm or sec^{-1} (the units of length cancel). The ratio of shear stress to shear rate is called the viscosity of the fluid. Like friction, there is dissipation of energy to heat because of the relative motion of the molecules past each other. If the viscosity is independent of shear rate, the fluid is said to be *newtonian.* Blood and many colloidal suspensions may not be newtonian, for at low shear rates (less than 100 sec^{-1}, i.e., 10 cm/sec per millimeter of thickness) blood viscosity becomes sensitive to shear rate, as will be described below.

From the theoretical derivation and from actual measurement, the fluid at the wall of a blood vessel is stationary – the velocity is zero (Fig. 11-1B). However, the velocity gradient at this point is maximum. At the center of the tube the velocity is maximum whereas the velocity gradient is zero. Flow is equal to the cross-sectional area times the mean velocity. Since the cross-sectional area is proportional to the square of the radius and since, from the theoretical derivation, the velocity is proportional to the square of the distance from the axis, that is, the square of the radius, flow is proportional to the fourth power of the radius of the tube.

There are two types of use for an equation such as this: (1) The equation *describes* the characteristic of a physical *system* (Chap. 7B). It is the law relating input to output of the system; i.e., it describes how the variable Q is dependent on the various independent variables ΔP, r, η, and L. The equation is considered to be valid and the assumptions used in arriving at it are not disproved if, after each of the independent variables and the flow have been measured, the predicted flow equals the measured flow. Statistical analyses are used to estimate whether the discrepancies between measurement and prediction are likely to be real or a result of random variation. (2) The equation may also be used to *estimate* one of the independent variables. Assuming that the equation is valid under the conditions used, if flow and all but one other variable are measured, the value of this other variable may be calculated. The commonest way to determine viscosity is to measure the

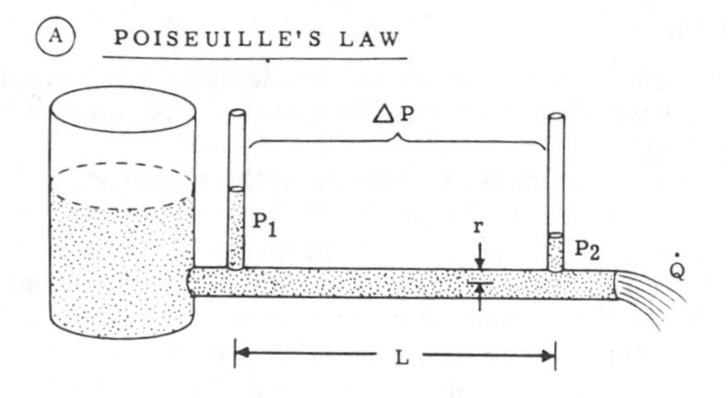

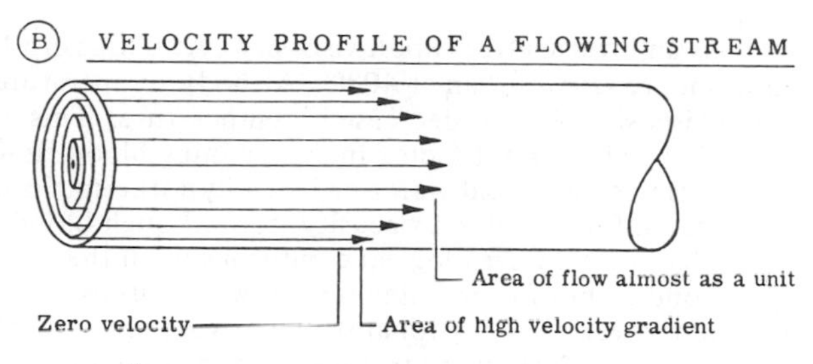

FIGURE 11-1. A. Flow ($\dot{Q}$) is proportional to the pressure gradient (ΔP), directly proportional to the vessel radius to the fourth power (r^4), and inversely proportional to the length of the resistance element (L). B. The velocity profile of the flowing stream is parabolic, with the highest velocity in the center.

rate of flow of the fluid through a tube of standardized dimensions using a standardized pressure.

Note that the flow is inversely proportional to the viscosity (η) of the fluid. Syrup, for example, is a highly viscous fluid, whereas water has a relatively low viscosity. The viscosity of air is much less than that of water, but it is not negligible; it is about one-thirtieth of that of water at body temperature.

Since flow through a resistance vessel is directly proportional to the fourth power of the radius of that vessel, the resistance is markedly influenced by the radius of minute blood vessels such as the arterioles. For example, at a constant pressure gradient, a 19 percent increase in arteriolar radius permits a 100 percent increase in flow. Physiologically, resistance to the flow of blood is determined primarily by the caliber of the small arteries, arterioles, and precapillary sphincters, which in turn is dependent on changes in the vasomotor tone. This vasomotor tone is the state of tonic contraction of the smooth muscles of these resistance vessels (see Fig. 11-3A and Chaps. 12 and 17).

The *total peripheral resistance* (TPR) is the resistance to flow of blood of the systemic vasculature as a whole. It is the aortic pressure minus the central venous pressure divided by the total flow through the system (the cardiac output). It decreases in exercise and is increased in essential hypertension. Since the circulation consists of many parallel branches, the calculated value of the total peripheral resistance is less than that of any resistance through a particular organ or region. When using total peripheral resistance as a diagnostic tool, one should remember that pathological conditions causing an increase in resistance to flow in one organ may be compensated by opposite changes in other organs, so the total peripheral resistance may remain constant.

VISCOSITY OF BLOOD

Plasma is about 1.8 times as viscous as water. Whole blood has a variable viscosity of between 2 and 15 times that of water. The causes of this range of blood viscosity are physiologically important.

The *hematocrit ratio* has a marked effect upon the viscosity and flow of whole blood, as may be seen in Figure 11-2A and B. If the hematocrit were zero, the viscosity would be that of plasma, but as the proportion of blood which is cells increases, the viscosity of the blood increases. Beyond the normal hematocrit ratio of about 0.40, the blood becomes much more viscous. Thus, if the concentration of erythrocytes is abnormally large, as in polycythemia, the flow tends to be diminished because of the increased viscosity of the blood. In contrast, in anemia the proportion of cells is so reduced that the flow tends to increase because of the decrease in viscosity.

Temperature has an inverse effect upon viscosity. At 0°C, both blood and water are about 2.5 times more viscous than at 40°C. At body temperature, there is a 2 percent increase in viscosity per 1°C decrease in temperature. This increase in viscosity of cold blood is an important factor in determining blood flow when extremities are cooled by immersion in cold water or when hypothermia is used in surgery.

Flow velocity may influence blood viscosity, for although blood behaves as a newtonian fluid under normal physiological conditions in all the vasculature, under some pathological conditions of nearly stagnant flow the viscosity of blood increases. Since flow is then not linearly proportional to the driving pressure, the blood is *non-newtonian* and the ratios of shear stress to shear rate are called the *apparent viscosities.* Because the geometry of the minute blood vessel network is so complex that calculation of viscosity from pressure-flow data and the Poiseuille equation is impossible, the viscosity of blood is compared to that of water – a fluid of known characteristics. The *relative viscosity* of a fluid is thus the ratio of the flow of water to that of the fluid under the same conditions of tubing size and geometry, temperature and pressure gradient. Such a comparison also provides an estimate of the apparent viscosity.

A major cause of the increased viscosity of blood at abnormally low shear rates is the tendency for erythrocytes to aggregate into rouleaux (i.e., they resemble a stack of coins). Since force is required to break up the rouleaux so that flow can continue through the capillaries, less driving force is available to overcome the viscous friction. As the shear rate is increased, the apparent viscosity decreases toward an asymptote. Fibrinogen is necessary for this anomalous rheological behavior of whole blood. Although normal blood does not undergo rapid aggregation, the tendency toward cell clumping may be important under conditions such as traumatic or burn shock, when the blood tends to *sludge.*

FLOW IN MINUTE VESSELS

If blood is pumped through very small tubes such as arterioles, the apparent viscosity is decreased, as shown in Figure 11-2A and C. Blood thus behaves more like water when flowing through small tubes than when pumped through larger tubes. (With either fluid, however, the flow in small tubes is small!) This phenomenon is called the *Fahraeus-Lindqvist,* or *sigma,* effect. It is caused by the erythrocytes flowing through vessels only a few times larger in diameter than the diameter of these particles.

One explanation may be derived from the calculation of Poiseuille's equation by summating the layers of fluid which are as thick as erythrocytes, instead of assuming that the layers are infinitely thin as in the classic derivation. The complex equation thus derived more accurately predicts the pressure-flow relationship. In large tubes (> 0.5 mm) the viscosity of blood is not dependent upon the size of the

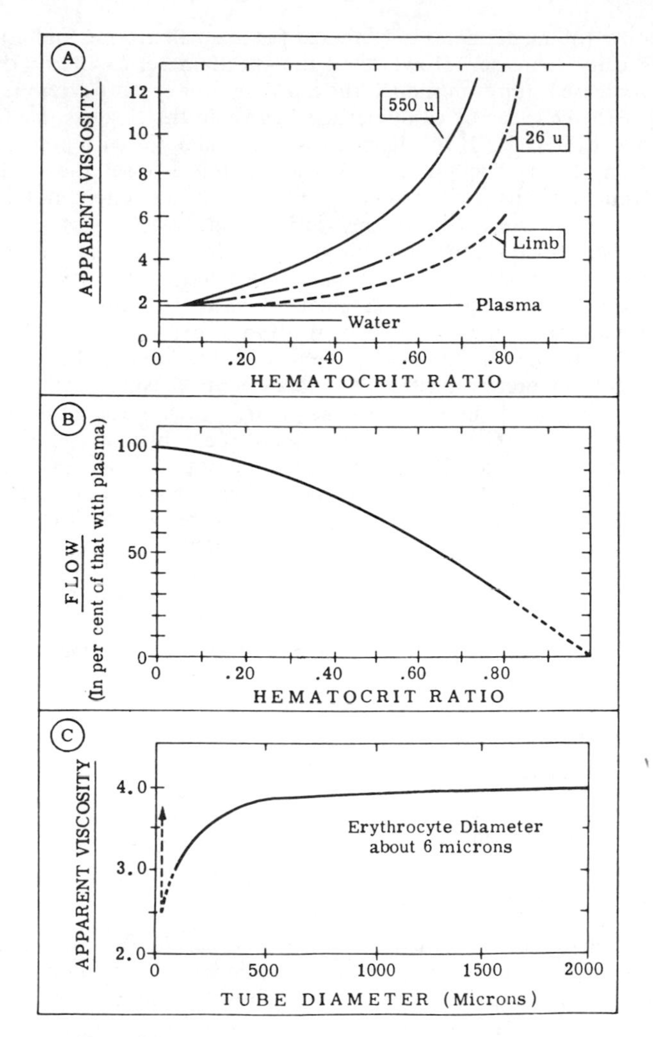

FIGURE 11-2. A. Effect of the concentration of erythrocytes (hematocrit ratio) on the apparent viscosity of blood perfused through tubing of various diameters and an isolated dog hind limb. There is less effect of hematocrit ratio on viscosity if the blood flows through small, 26 μ diameter tubes than if the flow is through larger, 550 μ, glass tubes. (Data from Bayliss, 1952.) There is even less effect when blood flows through a hind limb. (Data from Whittaker and Winton. *J. Physiol.* 78:357, 1933.) The reduction in viscosity is largely a consequence of the sigma effect and axial accumulation of the erythrocytes. B. As the hematocrit ratio of blood is increased, the flow of blood through isolated tissue decreases, other factors being constant. Note the relatively small effect at low hematocrit ratios. Active changes in vasomotor tone as a result of a deficit or excess of nutrient inflow or metabolite removal complicate quantitation of these effects. (Derived from data of Whittaker and Winton. *J. Physiol.* 78:357, 1933.) C. The effect of tube diameter on the apparent viscosity of blood. Experimental data fit this theoretical curve based on the sigma effect on flow of fluid layers of finite thickness equal to the diameter of the erythrocyte. (Data from Kumin and Haynes in Burton, 1965.)

tube, but as the tubing diameter is decreased (as in small arteries and arterioles) the apparent viscosity decreases. However, in the 4 μ diameter capillaries the apparent viscosity is increased, for even though the 8 μ diameter erythrocytes are readily distorted so that the cells can be easily perfused through the tissue, some force is required to distort the cells. If the diameter is yet smaller, as with precapillary sphincter constriction, the erythrocytes occlude the resistance vessel, the apparent viscosity becomes infinite, and the flow ceases (Fig. 11-2C). This phenomenon is superimposed on the general effects of vasoconstriction (Fig. 11-3A).

Another aspect of the anomalous viscosity of blood and a partial explanation of the Fahraeus-Lindqvist effect is the *axial accumulation* and *streaming* of the red blood cells, which leaves a cell-free marginal zone a few microns thick along the wall of the vessel and an increasing concentration of cells toward the center of the flow stream. Erythrocytes tend to migrate toward the middle of the flowing stream in small tubes at moderate velocities. The less viscous plasma is left at the wall, where the greatest velocity gradient exists. Consequently, such blood approaches the viscosity of plasma alone. The explanation of axial accumulation follows in part from Bernoulli's equation (Eq. 5). Because of the velocity profile (Fig. 11-1B) across the minute vessel, blood moves past the center-oriented surface of an erythrocyte faster than along the outer surface, since the erythrocytes are held back by the slower-moving blood along the wall. The higher velocity toward the center tends to reduce the lateral pressure, following Bernoulli's equation. As a result, the particle tends to move toward this area of lower pressure, i.e., toward the axis of the tube. This reduction in apparent viscosity is particularly noticeable with abnormally high hematocrit ratios (Fig. 11-2A). Axial accumulation, by permitting plasma skimming into side branches, is a factor in determining the distribution of erythrocytes and plasma in the capillary bed (Chap. 12).

CONTINUITY EQUATION

The continuity equation states that the velocity of flow times the cross-sectional area of a vessel is at all points constant:

$$v_1 \times A_1 = v_2 \times A_2 \tag{3}$$

Here v is the velocity and A is the cross-sectional area. As the cross-sectional area of the vascular system increases, the velocity must decrease since the mass of blood moving in the body is constant (Fig. 11-3B). Thus, in a human aorta of about 2 cm^2 in area, the mean velocity is about 50 cm per second, but in the 4 μ diameter capillaries it is only about 0.1 cm per second because the *total* cross-sectional area in man is over 1000 cm^2 (1 square foot).

EQUATION OF MOTION

When a pressure or force is applied to a system having mass, it does not start moving instantly. Inertia must be considered. Newton's laws of motion may be applied to a flowing stream to obtain an estimate of the distribution of the force, since the algebraic sum of the forces is zero (d'Alembert's principle). The basic equation applied to a mechanical system, such as a weight suspended on a spring and moving in a viscous liquid is

$$Ma + Rv + Sx = F \tag{4a}$$

Here F is the applied force; x is the displacement of the spring from the neutral or zero force position; S is the parameter defining the stiffness of the spring (force/unit

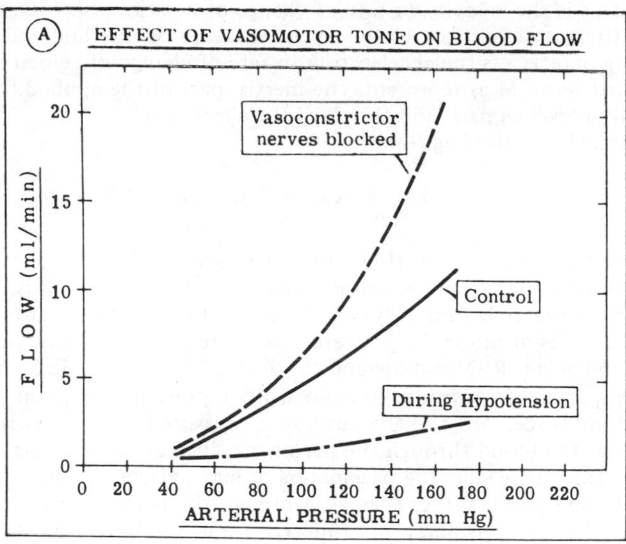

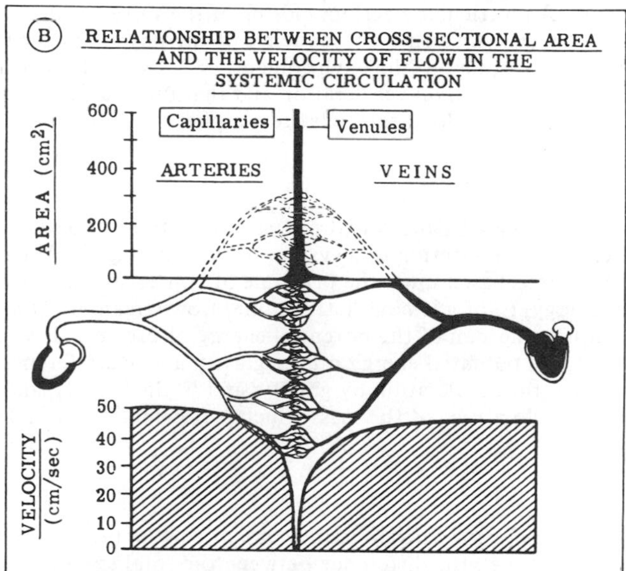

FIGURE 11-3. A. At a given pressure, the flow of blood through tissue is controlled primarily by the vasomotor tone. If impulse conduction along vasoconstrictor nerves is blocked, the flow is increased. Conversely, during severe hypotension, as a result of hemorrhage, there is an intense, neurally induced vasoconstriction of the blood vessels in vascularly isolated muscles of dogs, so that the flow is reduced even though the perfusion pressure may be normal. In addition to this neural control of vasomotor tone, there is an important intrinsic control of blood flow to maintain adequate nutrition and waste removal. B. The relationship between the cross-sectional area of the vascular system at various sites and the velocity of flow in the systemic circulation of a dog, demonstrating the continuity equation. (In the capillaries the flow is less than 0.1 cm per second. The area of the dog aorta is about 1 cm^2.) (B after R. F. Rushmer. *Cardiovascular Dynamics,* 3d ed. Philadelphia: Saunders, 1970. P. 8.)

displacement); v is the velocity or rate of change of position, (dx/dt); R is the resistance or coefficient of friction related to the viscosity of the fluid and the geometry of the moving object; a is the acceleration or rate of change of velocity; and M is the mass. The first term (M a) represents the inertial part of the applied force, the second the frictional or viscous part, and the third the elastic part.

When applied to a flowing fluid,

$$M \ddot{Q} + R \dot{Q} + S Q = P \tag{4b}$$

where $\ddot{Q}$ is the rate of change of flow or acceleration of the bolus of fluid, $\dot{Q}$ is the flow, Q is the distending volume, and P is the applied pressure. Thus, the pressure developed in the ventricle during systole is distributed into three parts acting to: accelerate the bolus of blood, $M \ddot{Q}$; overcome the resistance to flow through the aortic valve and aorta, $R \dot{Q}$; and distend the aorta, S Q. At the peak of ventricular pressure, the inertial and viscous terms normally represent only about 5 percent each of the total force. Most of the developed pressure is acting to distend the aorta and move the blood through the periphery. The above analysis is only an approximation, for it considers the parameters of mass, viscosity, and stiffness as lumped at discrete points rather than as continuously distributed throughout the system. Nonetheless, consideration of inertia, for example, is crucial in understanding the dynamics of ventricular ejection, closing of the valves, and the resonant behavior of fluid systems. A further consequence of the physics of the system is that energy is dissipated, i.e., lost to the environment, *only* by the frictional term. The energy stored as momentum on acceleration and that used to distend the aorta are available later to drive the blood through the periphery.

BERNOULLI'S EQUATION

Rather than considering a balance of forces, as in equations 4a and 4b, one may obtain another useful description of a system by considering a balance of energy. Bernoulli's equation is based upon the principle of conservation of energy and states that the total energy, E, of a laminar (streamline) flow stream *without resistance* is constant and equals the sum of the potential energy of pressure, P V (here, V is volume of fluid); the potential energy due to gravity above a reference plane, m g h, (here, m is mass; g, the acceleration by gravity; and h, the height); and the kinetic energy due to the velocity, v, of the mass which is flowing, $\frac{1}{2}$ m v^2. Thus:

$$E = P V + m g h + \tfrac{1}{2} m v^2 \tag{5}$$

The driving force or *effective pressure gradient* for the propulsion of blood is the total energy gradient, i.e., the difference between the total energy, as expressed in equation 5, at the upstream point of measurement and that at the downstream end. Thus, the effective pressure gradient includes not only the usual *pressure* type of potential energy, but also hydrostatic or *gravitational* potential energy and the *inertial* or kinetic energy of the moving mass. The hydrostatic factor is important in understanding the factors determining the flow to and from the legs (Chap. 16). The inertial factor accounts for over half of the driving force in the pulmonary artery or great veins during vigorous exercise. During one part of the ejection phase of the cardiac cycle, the inertial term is great enough so that flow continues even though the pressure gradient is reversed (Chap. 13). When the rate of flow changes, as in the arteries, the resistance to flow, now called *impedance,* includes this inertial term. The pattern of flow then lags behind that of the pressure. The mathematics describing the relationship between the driving arterial pulse and the resulting flow becomes complex (e.g., Attinger, 1964, 1968; Noordergraaf, 1969).

A consequence of the Bernoulli principle is that if the velocity of flow increases because of a constriction in a tube or because blood is being rapidly propelled from the heart, more of the total energy is in the form of kinetic energy (velocity) and less is in the form of potential energy (pressure). Thus the lateral pressure tending to distend a blood vessel is decreased at a constriction, since the constriction causes a higher velocity. The amount of energy in the form of kinetic energy is then higher; hence, the potential energy in the form of distending pressure must be lower. Under most physiological conditions this factor is minor. However, it does help explain the closure of the valves of the heart, for blood rapidly flowing past partially closed valve leaflets tends to suck them together.

The orientation of a catheter to measure the lateral pressure of a flowing stream is important, for if it is directed upstream against the direction of flow it will give a somewhat higher value than one directed at right angles to the flow stream. In the latter case, as the flow impinges on the cannula, the kinetic energy due to the inertia of the blood is converted to potential energy and is added to the lateral pressure. In the larger systematic arteries this "end-pressure" or kinetic factor is minor, but in the pulmonary vessels with high flow velocities and low lateral pressures, this factor becomes significant.

TRANSMURAL PRESSURE

The *transmural pressure,* P_T, is the difference between the pressure inside, P_i, a vessel and the pressure outside, P_o, the vessel:

$$P_T = P_i - P_o \tag{6}$$

Often the outside pressure is atmospheric and so may be considered zero.

The transmural pressure across a resistance vessel is an important factor determining the resistance to flow, for as the pressure increases, the vessels are passively distended and allow an even higher flow than would be found with constant vessel size (Fig. 11-4B). Contraction of vascular smooth muscle in conjunction with the elastic elements counteracts the distending force and so regulates the flow of blood to the tissue. This phenomenon explains in part the nonlinear pressure-flow characteristics of blood through a living tissue (Fig. 11-3A).

COMPLIANCE

Hollow organs and tubes show an increase in volume with an increase in transmural pressure. This property is called *distensibility.* Compliance, C, is a general term describing the change in dimension (strain) resulting from a change in transmural pressure (stress). It is usually defined as

$$C = \frac{\Delta V}{\Delta P} \tag{7a}$$

where ΔV is the change in volume in response to a change in pressure, ΔP. The coefficient of stiffness, S, of the third term in equation 4b represents the reciprocal of compliance. The histological structure of most biological blood vessels – and indeed most tissues – is such that with distention, the compliance decreases, i.e., they become stiffer. Furthermore, blood vessels contain a volume at zero transmural pressure. Therefore, it is important to specify the volume and transmural pressure at which a compliance measurement is made. *Capacitance,* a term borrowed from electronics, implies a storage function, suggesting the ratio of the total volume contained to the transmural pressure. Some workers equate it with compliance. Because a long vessel would have a larger change in volume for a given pressure increase than a shorter vessel, a *coefficient of volume distensibility,* κ, is defined as the fractional change in volume per unit change in pressure:

$$\kappa = \frac{\Delta V}{V \cdot \Delta P} \qquad (7b)$$

LAPLACE EQUATION

Because blood flow is proportional to the fourth power of the radius of the arterioles and other "resistance" vessels, small changes in vessel dimensions may exert profound effects on blood flow. Therefore, the factors influencing vascular dimensions, such as transmural pressure and wall tension, are important contributors to the state of vascular resistance. The Laplace equation states that the tension, T, in the wall of a vessel required to maintain a given radius is proportional to the product of the *transmural pressure,* P, and the *radius,* r.

$$T = Pr \qquad (8)$$

A derivation of the Laplace equation provides a valuable insight into simple mathematical reasoning. If the wall of the vessel is to be stationary, the *distending force* pushing the vessel wall out must be balanced by a *restraining force* produced

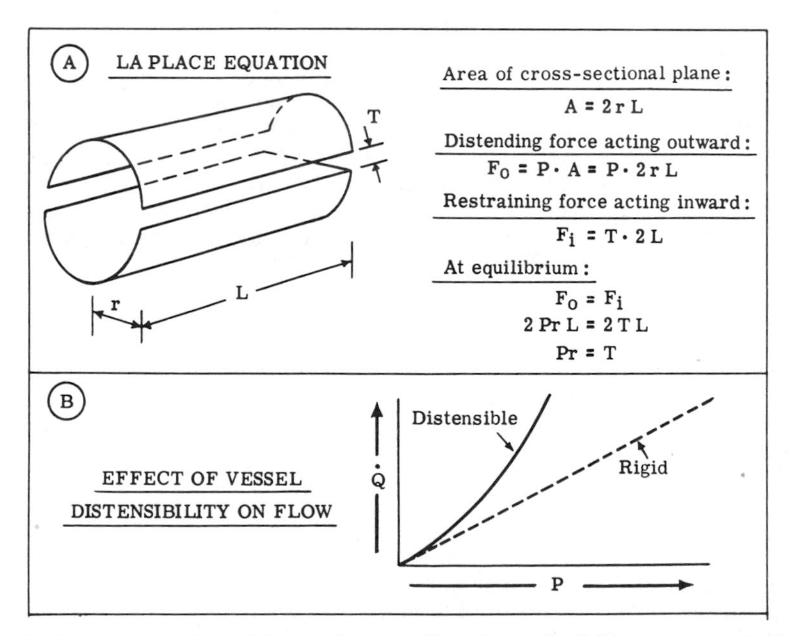

FIGURE 11-4. A. Derivation of the Laplace equation. B. As the driving pressure in a distensible vessel is increased, the flow increases more rapidly than that in a rigid vessel of the same resting diameter, because both the pressure gradient and the radius are increased by increases in the driving pressure.

by the tensions of the fibers in the wall. The distending force is equal to the area of a longitudinal sectional plane times the transmural pressure. The distending force, in the case of a cylinder, equals $P \times 2rL$, where P is the pressure in grams per square centimeter, r is the radius, and L is the length of the cylindrical vessel (Fig. 11-4A). The restraining force is equal to $T \times 2L$, where T is the tension per unit length and L is the length. The factor 2 is used because both edges of the vessel are acting to hold it together. In this simple derivation the walls of the vessel are assumed to be infinitely thin, an assumption which is not valid for calculating wall tensions of small arteries, for example. Equating these two expressions gives the relationship $T = Pr$. (For a sphere, the equation becomes $T = P\,r/2$.) In a larger vessel the forces in the wall required to keep it from distending at a given pressure must be higher than in a smaller vessel. For the heart to eject blood at a given pressure, the myocardial tension must be greater in a large heart than in a small heart (Chap. 15). In aortic aneurysm the walls of the aorta are damaged, so the diameter becomes much larger than normal. This is exceedingly serious because the enlarged vessel is much more likely to rupture, owing to the high tensions in its walls, even at normal arterial pressures. The capillaries, because of their minute size, have wall tensions of less than 10 mg per centimeter of vessel length and so remain intact in spite of wall thicknesses of about 1 μ.

TURBULENCE

Blood flow through vascular channels is normally streamline in nature. Under special conditions turbulent fluid flow, with swirls or eddies, may develop. With the development of turbulence, flow is approximately proportional to the square root of the pressure drop, so a given increase in pressure results in less than a proportionate increase in flow:

$$\dot{Q} \simeq k\sqrt{\Delta P} \tag{9}$$

This occurs because the potential (driving) energy is dissipated in the swirls and eddies as heat. If turbulence is of sufficient magnitude, audible sound is produced like the hiss of air from a compressed-air outlet.

If the ratio of inertial to viscous forces of a flowing stream exceeds a given value, a slight disturbance in a streamline, laminar flow pattern will lead to the development of a randomly oriented turbulent pattern. Turbulence is likely when the ratio, called Reynolds' number (Re), exceeds a critical value of about 2000. Reynolds' number may be computed for a cylindrical conduit with the following equation:

$$Re = \frac{\rho\, v\, D}{\eta} \tag{10a}$$

where ρ is the density of the fluid in $gm/cm^3 \cdot g$ (here g is the gravitational constant, 980 cm/sec^2), v is the average velocity in cm/sec, D is the diameter of the vessel in cm, and η is the viscosity in $gm \cdot sec/cm^2$. Other factors being constant, an increase in velocity is likely to result in turbulence. Writing the equation in terms of flow, $\dot{Q}$, rather than velocity gives

$$Re = \frac{4\rho\dot{Q}}{\pi D \eta} \tag{10b}$$

Since blood density is reasonably constant, the factors tending to cause turbulence are an increase in the rate of blood flow, such as occurs in exercise; a decrease in blood viscosity, as seen in anemia; or a decrease in the radius of the blood vessel, as occurs with stenotic valves or coarctation of a blood vessel such as the aorta.

The streamline flow of fluid may also break into organized swirls of turbulence and produce *aeolian tones* at flow velocities lower than those causing fully developed random turbulence. The term comes from the singing aeolian harps placed in the wind along the coast of the Mediterranean Sea many centuries ago. *Aeolian* means "born of the wind." As a fluid flows past an obstacle in the flow stream, a slight disturbance causes the flow wake to fluctuate at first toward one side. Inertial forces continue the stream into a higher-velocity area, which then forces it back in the opposite direction, where it overshoots and then is forced to swing back to repeat the cycle. The fluttering of a flag in a stiff breeze is produced by the same mechanism. Described another way, the fluid tends to follow the curvature of the obstacle, and so curves in a swirl around it to form a vortex. This is an area of low pressure; hence, the stream tends to flow into that area from the opposite direction. The cycle is repeated indefinitely with swirls of opposite direction being set up, vortices shed and carried downstream. Each time the wake of the flow stream swings back and forth, it changes the lateral pressure on the wall of the vessels.

The studies of Bruns (1959) have shown that the frequency and location of cardiovascular sounds under physiological conditions are explainable by this process. Note that the fluctuating wake has a maximal component of fluctuations at right angles to the flow stream. If lateral fluctuations of pressure from the oscillating flow stream are of sufficient magnitude and frequency, audible sound is produced. In this kind of turbulence, relatively more sound is produced with a given driving pressure than in the randomly oriented flow pattern of classic Reynolds' turbulence. It is probably the source of most of the murmurs heard in the cardiovascular system and can occur under conditions in which Reynolds' number is much less than 2000.

SUMMARY

1. Factors regulating flow are the pressure gradient along the vessel, the viscosity of the fluid, and the physical dimensions of the resistance vessel.
2. An increase in transmural pressure, by distending the resistance blood vessels, tends to increase flow irrespective of its concomitant effect on the driving pressure.
3. With constant pressure gradient, the smooth muscle tension (vasomotor tone) acts to reduce the caliber of the resistance vessels and so reduces flow.
4. The apparent viscosity of blood is increased (hence the flow decreases if other factors remain constant) as a result of increased hematocrit ratio, decreased temperature, turbulence, an increase in minute vessel size up to about 0.5 mm, or erythrocyte aggregation if flow rate is reduced toward zero.

MEASUREMENT OF BLOOD FLOW

Understanding of the performance of the circulatory systems requires a measurement of the flow of blood. In measuring flow, two sources of error must always be considered: errors in the measurement itself, and deleterious effects of the measurement process upon the functioning of the organism. In cardiovascular research, if highly accurate results are to be obtained, measurements must be made without disrupting the organism by concomitant surgery, anesthesia, pain, impeding the normal blood flow, inducing blood clotting, or causing uncompensated loss of blood. Much has been learned about the cardiovascular system, however, without such stringent requirements being placed on the measuring process.

The simplest and fundamentally the most accurate method of measuring blood flow is to collect a known volume of vascular outflow over a measured time interval by means of a graduate and stopwatch. This method, although simple, is obviously of no value for measuring cardiac output because the removal of the blood would seriously affect the animal. However, it is of value in measuring the flow through small amounts of tissue.

The flow of blood from small segments of tissue may be measured by cannulating a vein and counting the number of drops of blood per minute leaving the cannula. If the volume of these drops is known, the flow may be determined. The maximum flow measurable is limited to about 10 ml per minute. Furthermore, the size of the drops is dependent on a multitude of variables; thus, reliable measurement of complex fluids such as blood is difficult.

To measure the pattern of pulsatile blood flow, electromagnetic and ultrasonic flowmeters are available. The probes of these instruments can, if desired, be implanted in an experimental animal so that painless measurements can be made at a later date without anesthetics. The artery does not have to be cannulated since the devices detect the flow through the walls of intact blood vessels.

The *electromagnetic flowmeter* is based upon the principle of an electric generator. If a conductor is moved through a magnetic field, an electrical potential is developed in the conductor which is proportional to the length and velocity of the conductor. Blood going through a magnetic field forms a single conductor; consequently, electrodes can be placed on both sides of the blood vessel to pick up the potential developed.

The *ultrasonic flowmeter* is based upon the principle that the velocity of sound along a flowing stream in relation to a stationary point is higher going downstream than going upstream. One-half the difference between the two velocities is the stream velocity. By the use of techniques similar to those used for radar and sonar, a pulse of sound is produced and the time required for it to travel downstream is compared with the time for a similar pulse of sound to be transmitted upstream.

Indicator dilution is another class of noncannulating flow measurements, which has the additional advantage of requiring only minimal surgery. It is used particularly for determinations of cardiac output. These chemical methods of measuring blood flow involve injection of an indicator into the blood stream and are based on the principles of material conservation and indicator dilution. If proper conditions are met, they are fully as accurate as the direct methods.

The *material conservation principle* states that if a known amount of a substance is carried by a flow stream to a particular region, it must either (1) leave the region, (2) accumulate in the region, (3) be converted to some other material, or (4) be excreted by some route other than the flow stream.

The *indicator dilution principle* is based on the fact that an amount of a substance, A, mixed uniformly in a volume, V, has a concentration, C, such that by definition of concentration

$$C = \frac{A}{V} \qquad (11)$$

This *dilution method* (see also Chaps. 10 and 23) is used to estimate the volumes of various fluid compartments in the body by rearranging the equation to $V = A/C$. A material is used which equilibrates within the entire fluid compartment in question. Any losses of indicator from the fluid compartment are accounted for by the material conservation equation.

Indicator dilution can also be used to measure flow. Mean flow, $\dot{Q}$, is volume, V, moved per unit time, t:

$$\dot{Q} = \frac{V}{t} \qquad (12)$$

This volume of fluid passing a given point in a unit of time may be estimated by the dilution principle, by substituting V = A/C into equation 12. If the amount of indicator added or removed by an organ is known, and is divided by the change in concentration of a substance as it passes through the organ, the flow can be calculated, as indicated below. The material conservation principle is used to account for the indicator used.

Fick Method (Constant Infusion or Removal)

The blood flow through an organ can be determined by measuring the amount of a substance removed by the organ per minute and dividing by the change in concentration of the substance in the blood as it goes through the organ. For *cardiac output,* for example, measurement of the flow of blood through the heart and thus through the lungs is desired. The transport of oxygen by the lungs provides a convenient indicator agent. The volume of oxygen consumed per minute, A/t, may be measured by means of a spirometer. Some technique is then used to determine the change in oxygen concentration. Oxygen consumption is measured over an interval of 5 to 10 minutes. The concentration of oxygen in the venous blood coming to the right heart and lungs, C_V, and the concentration of oxygen leaving the lungs and left heart in the arterial blood, C_A, are measured to give the change

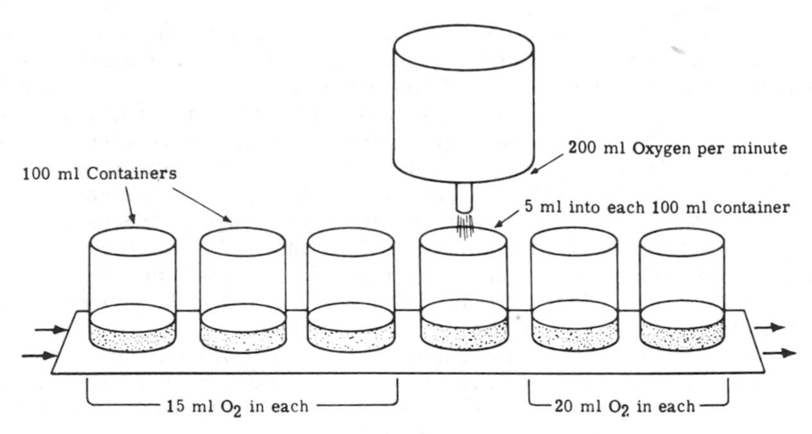

Number of containers receiving 5 ml each (20 ml - 15 ml = 5 ml) which must pass in one minute to remove the 200 ml of O_2

$$\dot{N} = \frac{200}{5} = 40, \text{ 100 ml containers per minute}$$

If each of these were 100 ml of blood: 40 x 100 = 4000 ml

$$\text{Flow, } \dot{Q} = \frac{\text{Amount } O_2\text{/minute}}{\text{Concentration change}} = \frac{A/t}{C_A - C_V} = \frac{200 \text{ ml/minute}}{\dfrac{20 \text{ ml } O_2}{100 \text{ ml Blood}} - \dfrac{15 \text{ ml } O_2}{100 \text{ ml Blood}}}$$

$$\dot{Q} = \frac{200 \text{ ml } O_2\text{/minute}}{5 \text{ ml } O_2\text{/100 ml Blood}} = 40 \times 100 \text{ ml Blood/minute} = 4 \text{ L/minute}$$

FIGURE 11-5. Derivation of the Fick equation for the measurement of cardiac output or blood flow.

in concentration, ΔC. This change in concentration is the dilution. The *Fick equation* is:

$$\dot{Q} = \frac{A/t}{\Delta C} = \frac{A/t}{C_A - C_V} \tag{13}$$

The technique and calculations are shown schematically in Figure 11-5. The problem to be solved is: If each 100 ml of blood going through the lungs takes up 5 ml oxygen, how much blood must flow per minute to take up 200 ml of oxygen?

Arterial blood samples are obtained by arterial puncture. The major problem with this method is obtaining a valid sample of mixed venous blood. This blood must be representative of all the venous blood. Since the kidneys usually extract only about 2 to 3 volumes of oxygen per hundred volumes of blood, the brain about 6, and the heart and active muscle 10 to 15, the sample must be obtained after mixing. Mixing is adequate only at sites beyond the right ventricle. Cardiac catheterization is thus routinely used to obtain these samples. The catheter is inserted into an arm vein and advanced through the veins into the right heart and on through the pulmonary valve into the pulmonary artery. The procedure is safe, but to the patient it is a rather heroic one, so the apprehension tends to give abnormally high values.

Because of the problems in obtaining the concentration of indicator in mixed venous blood, various *indirect methods* have been developed in the past to bypass the necessity of obtaining a mixed venous blood sample. For example, it has been assumed that a sample of air taken at the end of exhalation is in equilibrium with the alveolar gases in the lungs, which in turn are assumed to be in equilibrium with the arterial blood. After rebreathing several times into a small bag, the gas at the end of an exhalation is assumed to be in equilibrium with the *venous* blood. The indirect Fick methods have not been particularly reliable, however, because of errors in these assumptions.

The direct Fick is the standard method of determining the cardiac output in man but is also subject to errors, particularly during severe exertion. It is a *determination* of mean flow, since the oxygen consumption is averaged over a 5- to 10-minute period. The venous sample should be taken continuously during this period if accurate results are desired. It is not adequate for measuring rapid changes in cardiac output, since it is impossible to measure the true oxygen consumption over short intervals of time because of the variable amount of air held in the lungs from breath to breath during such changes. Repeated determinations using the Fick and dye dilution methods (discussed below) agree within a range of $\pm$ 15 percent. Much of this variation is probably due to true minute-by-minute variation in the cardiac output.

The indicator dilution method, using the Fick equation, provides a convenient tool for the measurement of mean *organ blood flow* if the indicators used are removed from the blood by these organs. The liver removes bromsulphalein (BSP), and the kidneys remove p-aminohippurate (PAH). The brain and heart absorb nitrous oxide to provide arteriovenous (A-V) differences that enable flow to be measured, assuming that the amount of material accumulated is related to the solubility of the gas in the tissue. The pattern of "washout" of radioactive xenon or krypton may also be used to calculate flow.

Stewart-Hamilton Method (Single Injection)

The Stewart-Hamilton method is based upon the rapid injection of a definite, known amount of indicator. If the amount injected and the mean concentration over a given time (the minute-concentration) are measured, the flow can be calculated. It is in many ways more convenient than, and is fully as accurate as, the direct Fick method. In practice, for cardiac outputs, a known amount of an indicator such as Evans blue, indocyanine green, or albumin labeled with ^{131}I is injected into a vein. It is diluted and mixed with the blood flowing through the heart and lungs. The resulting concentration of dye in the arterial blood is then measured and the cardiac output calculated.

To understand this procedure, first imagine a flowing stream (Fig. 11-6A). If the total flow containing an added amount of dye, A, is collected over a definite period of time – for example, one minute – and the dye is then mixed throughout the total collected volume to give a concentration, C, the volume can be calculated from the basic dilution equation, V = A/C. This is the volume per minute, i.e., the rate of flow.

Next, imagine that the dye is uniformly mixed across the flow stream and that a definite proportion, assume one-tenth, of the flow is collected as in Figure 11-6B. Starting before the dye appears and ending after the dye has cleared the sampling site, a volume containing dye is collected for one minute and is thoroughly mixed so that the mean concentration may again be determined. If, in the sample, the proportion of the total amount of dye is exactly equal to the proportion of the total flow collected over one minute, i.e., one-tenth of each, the total flow can again be calculated. Since the proportionality constants for the dye and the volume can be canceled (Fig. 11-6B), it is not even necessary to know what proportion of the flow is collected. This proportion must, however, be constant throughout the collection, and the dye must be uniformly mixed across the flow stream.

Finally, instead of collecting a sample over a given time interval and mixing, individual samples may be analyzed or a continuous measure of the indicator concentration may be made with a densitometer (Fig. 11-6C). The mean concentration is mathematically calculated (Fig. 11-6D) over the time interval, usually one minute. The *Stewart-Hamilton indicator dilution equation* is

$$\dot{Q} = \frac{A}{\int_0^\infty C dt} = \frac{A}{\Sigma C \Delta t} \tag{14}$$

where Δt is the fraction of the unit time interval.

The concentrations at various fractions of a minute are added together to give $\Sigma C \Delta t$, the *minute-concentration,* a term somewhat analogous to the *respiratory minute volume.* Usually, after less than a minute the concentration has declined to so near zero that in actuality no further additions are made to the sum, called the integral. Because fractional units of time are used, this is a mean concentration averaged over one unit of time – a minute usually.

In practice, a known amount of dye is injected into the blood stream. It must mix across the total flow stream. For cardiac output determinations, this occurs in the right or the left ventricle. A representative fraction of the cardiac output containing changing concentrations of dye is obtained by taking a continuous sample of arterial blood, as shown schematically in Figure 11-6C. The concentration of dye is analyzed and plotted (solid curve, Fig. 11-6E) over an interval long enough for a complete passage of dye through the circulatory system. In calculating the mean concentration of dye over a certain period (the minute-concentration), one must calculate the area under the curve either by measuring with an instrument called a planimeter or by summation of instantaneous values. The latter consists of finding the concentration during the first second and adding it to that at the second second, adding to the third, and so summating throughout the whole minute. Dividing this sum by the number of samples measured during the minute, 60, gives the minute-concentration. During the first few samples after the indicator

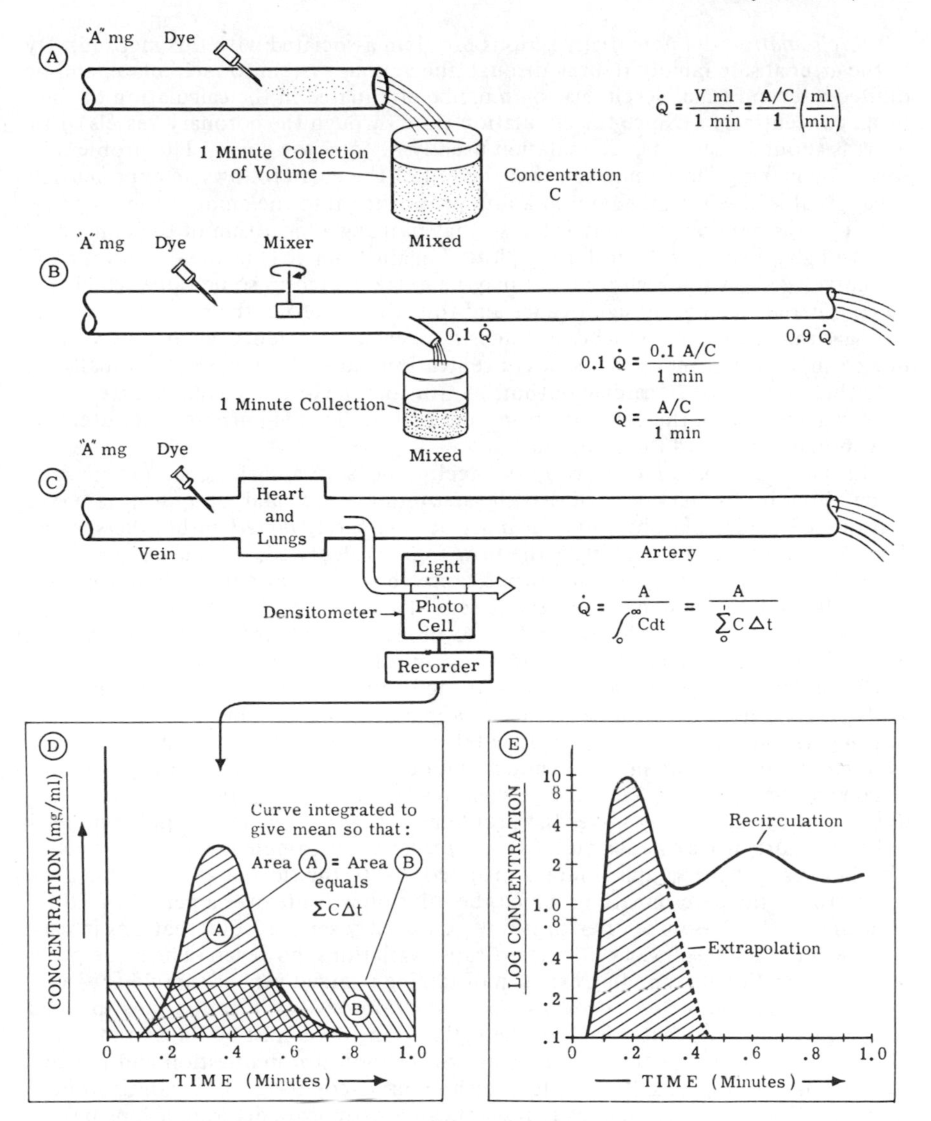

FIGURE 11-6. Derivation of the Stewart-Hamilton indicator dilution equation for the measurement of cardiac output by the dye dilution method. A. Total flow stream collection. B. Collection of a fraction of the flow stream with no recirculation. C. Measuring dynamic concentration changes with a densitometer. D. Record of dye concentration and the derivation of the concentration-time factor, which is usually the mean concentration during 1 minute, i.e., the minute-concentration. E. Effect of recirculation and the method of extrapolation using a semilogarithmic plot of dye concentration.

injection, the concentration will be zero. If there were no recirculation of dye, there would be no dye in the last samples and so no additions to be made to the sum. This situation is analogous to the calculation of the daily rainfall over one year. Some days are without rain, and for these zero amount of rain is added in the calculation.

Recirculation is a potentially serious problem associated with this method. Dye in the arterial side rapidly returns through the venous system to recirculate, and be diluted a second time. Recirculation must be eliminated in the calculation of the mean concentration. Since the circulation time through the coronary vessels of the heart is about 15 seconds, recirculation usually distorts the curve. The problem is solved by making the assumption that the tail of the curve follows an exponential decay; that is, dye is washed out at a rate proportional to the amount that is present. This assumption is usually valid, so that plotting a logarithm of the concentration will give a straight downslope if plotted against time. Thus the downslope of the curve, as shown in Figure 11-6E, may be extrapolated to some value which can be considered to be of no significance and thus called zero in the summing process. (A logarithmic curve approaches but does not reach zero; hence, an arbitrary baseline of about 1 percent of the peak concentration must be established.) Finally, note that an increase in cardiac output, by diluting the dye to a greater extent, results in a decrease in the concentration-time factor. Special-purpose computers are now available to speed the computation.

In the determination of flow by this technique, several major assumptions are made. Conditions must be such that the assumptions are valid, or appropriate corrections are made. (1) The indicator must be adequately *mixed* within the system. The flow at the point of mixing is the flow that is calculated. This usually presents no problem in measuring cardiac output, since the ventricles mix the blood vigorously. It does preclude, however, the possibility of injecting dye in the arch of the aorta and sampling at the femoral artery to obtain accurate information about aortic blood flow. (2) There must be no *loss* (or gain) of indicator. The dye must not be lost in tubing, absorbed on artery walls, or be lost by passing through the capillary walls. If indicator is lost, there will be an apparently higher diluting volume, and thus higher indicated flow. (3) In flow determinations, corrections must be made for *recirculation.* There must be enough data on the downward side of the curve to obtain an accurate estimate of the downslope for the extrapolation. Injection of dye near the heart usually solves the problem, but in some clinical conditions, such as heart failure, it may be serious. (4) A *representative* sample must be obtained. This implies that the sampling rate is proportional to the flow at all moments during the curve for the determination. Since the dilution is made over a period of 10 to 20 heart beats, this source of error is not particularly serious. Fluctuations in cardiac output from respiratory efforts do cause variations, however. The same problem of proportionate sampling exists in obtaining a mixed venous sample for the direct Fick method. (5) An *accurate determination* of concentration must be made. If a single calibration factor is used, the calibration curve must be linear and pass through zero. The detector must respond to the indicator in question and not to extraneous indicator – a difficult feat with external counting when radioisotopes are used. Good practice dictates that duplicate or triplicate determinations be made.

Both the dye dilution and Fick methods measure the mean flow over a period of many seconds to several minutes. Thus, there are variations from one determination to the next because of variations in cardiac output from moment to moment. The respiratory cycle is a particularly important source of this type of error since the cardiac output often fluctuates in response to changes in intrathoracic pressure and respiratory center activity.

A discussion of the factors influencing cardiac output in man is presented in Chapter 15.

REFERENCES

Attinger, E. O. (Ed.) *Pulsatile Blood Flow.* New York: Blakiston Div., McGraw-Hill, 1964.

Attinger, E. O. Analysis of pulsatile blood flow. *Advances Biomed. Engin. Med. Phys.* 1:1–59, 1968.

Bayliss, L. E. Rheology of Blood and Lymph. In A. Frey-Wyssling (Ed.), *Deformation and Flow in Biological Systems.* New York: Interscience, 1952.

Bruns, D. L. A general theory of the causes of murmurs in the cardiovascular system. *Amer. J. Med.* 27:360–374, 1959.

Burton, A. C. Hemodynamics and the Physics of the Circulation. In T. C. Ruch and H. D. Patton (Eds.), *Physiology and Biophysics.* Philadelphia: Saunders, 1965. Chap. 27.

Burton, A. C. *Physiology and Biophysics of the Circulation.* Chicago: Year Book, 1965.

Eskinazi, S. *Principles of Fluid Mechanics.* Boston: Allyn & Bacon, 1962.

Glasser, O. (Ed.). Circulatory System. In *Medical Physics.* Chicago: Year Book, 1960. Vol. 3, pp. 119–193.

Green, H. D. Circulation: Physical Principle. In O. Glasser (Ed.), *Medical Physics.* Chicago: Year Book, 1944. Vol. 1, pp. 208–232.

Hamilton, W. F. (Ed.). *Handbook of Physiology.* Section 2: Circulation, Vols. I and II. Washington, D.C.: American Physiological Society, 1962, 1963. See: A. C. Burton, Physical Principles of Circulatory Phenomena: The Physical Equilibria of the Heart and Blood Vessels, Chap. 6, pp 85–106. L. E. Bayliss, The Rheology of Blood, Chap. 8, pp. 137–150. W. F. Hamilton, Measurement of the Cardiac Output, Chap. 17, pp. 551–584. K. L. Zierler, Circulation Times and the Theory of Indicator-Dilution Methods for Determining Blood Flow and Volume, Chap. 18, pp. 585–615. H. D. Green et al., Resistance (Conductance) and Capacitance Phenomena in Terminal Vascular Beds, Chap. 28, pp. 935–960. K. Kramer et al., Methods of Measuring Blood Flow, Chap. 38, pp. 1277–1324.

Noordergraaf, A. Hemodynamics. In H. P. Schwann (Ed.), *Biological Engineering.* New York: McGraw-Hill, 1969. Pp. 391–545.

Replogle, R. L., H. J. Meiselman, and E. W. Merrill. Clinical implication of blood rheology studies. *Circulation* 36:148–159, 1967.

Stacy, R. W., D. T. Williams, R. E. Worden, and R. O. McMorris. *Essentials of Biological and Medical Physics.* New York: McGraw-Hill, 1955. Chaps. 30–31.

Whitmore, R. L. *Rheology of the Circulation.* New York: Pergamon, 1968.

12

Microcirculation

Julius J. Friedman

THE CARDIOVASCULAR SYSTEM is organized toward the maintenance of a homeostatic environment for the tissues of an organism. Cardiac and pipherovascular functions are coordinated to transport blood to and from the capillary-venular networks where exchange of nutrients and cell products between blood and tissue fluids takes place. Examination of the anatomical and functional organization of the microcirculation unit should provide an insight into the capability of this system to meet the needs of the tissues.

ANATOMICAL ORGANIZATION

A microcirculatory unit consists of serially and parallel arranged blood vessels including *arterioles, metarterioles, precapillary sphincters, capillaries, arteriovenous anastomoses, venules*, and *collecting venules*. The dimensions and structural constituents of these blood vessels are presented in Figure 12-1.

The precise arrangement of these microvascular elements differs in different organs. In the mesoappendix of the rat they are arranged in a manner diagramed in Figure 12-2. Blood flow enters the microcirculatory unit by means of the *arteriole,* which finally terminates as terminal arterioles or *metarterioles.* The metarteriole differs from the arteriole in that the smooth muscle cells in the vessel wall are discontinuous.

Numerous capillaries originate from single metarterioles and terminal arterioles, and they branch and anastomose repeatedly to produce a great increase in surface area and reduction in blood flow velocity. At the entrance of most capillaries there exist segments of vessels well provided with smooth muscle. These are the *precapillary sphincters,* and they regulate the effective surface area of the *capillary-venular* network. Capillaries come together to form venules, which merge further to form *collecting venules* to drain the microcirculatory unit. While it is functionally correct to consider the microcirculation as consisting of units, extensive arteriolar and venular anastomoses provide elaborate collateral circulation. In some tissues, the arteriolar system is organized as an arcade network which may act to establish an isopressure perfusion source for all the microcirculatory units.

FUNCTIONAL ORGANIZATION

Functionally, the series and parallel coupled components of the microcirculation may be classified as *resistance vessels, exchange vessels, shunt vessels,* and *capacitance vessels.*

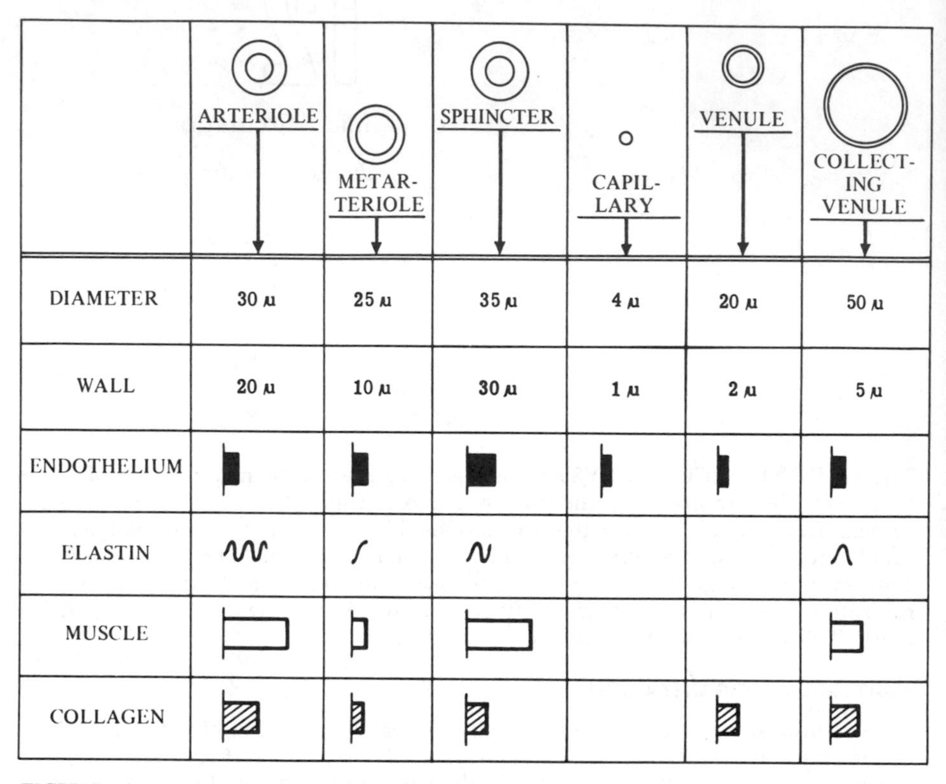

	ARTERIOLE	METARTERIOLE	SPHINCTER	CAPILLARY	VENULE	COLLECTING VENULE
DIAMETER	30 μ	25 μ	35 μ	4 μ	20 μ	50 μ
WALL	20 μ	10 μ	30 μ	1 μ	2 μ	5 μ
ENDOTHELIUM						
ELASTIN						
MUSCLE						
COLLAGEN						

FIGURE 12-1. Dimensions and structural constituents of the various vessels of the periphero-vascular unit. (Modified from A. C. Burton. *Physiol. Rev.* 34:619–642, 1954.)

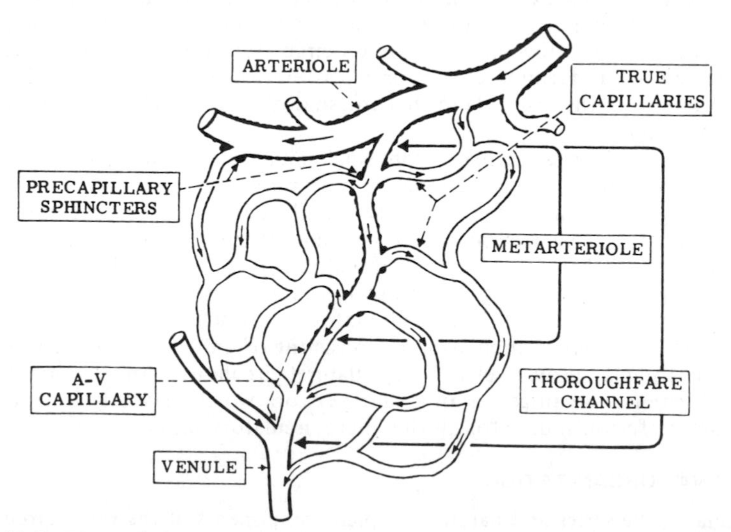

FIGURE 12-2. Diagram of a microcirculatory unit. (After R. Chambers and B. W. Zweifach. *Physiol. Rev.* 27:436–463, 1947.)

RESISTANCE VESSELS

Resistance to flow is manifested by all blood vessels which may essentially be classified as precapillary and as postcapillary resistance elements. The *precapillary resistance elements* include small arteries, arterioles, and precapillary sphincters. The small arteries and arterioles are the major variable resistance elements of the microvascular unit which determine the extent of total tissue blood flow. The precapillary sphincters, on the other hand, by adjusting the number of capillaries open, determine the distribution of capillary blood flow, the extent of capillary flow velocity, capillary surface area, and the mean extravascular diffusion distances. In short, precapillary sphincter activity exerts a great influence on the nature of transcapillary exchange.

Postcapillary resistance vessels include the muscular venules and small veins. Although they manifest only small changes in resistance, their strategic position, immediately postcapillary, enables them to markedly influence capillary pressure. Substantially, it is the *ratio* of precapillary to postcapillary resistance that determines capillary hydrostatic pressure.

EXCHANGE VESSELS

Exchange between the vascular and extravascular compartments occurs primarily across capillaries and venules. These vessels are uniquely suited for exchange by virtue of their high surface-area-to-volume ratio and their thin wall. Moreover, exchange vessels are usually within 20 to 50 μ of tissue cells so that diffusion distances are minimal. Transvascular exchange is different in different segments of the exchange network in the same tissue as well as in different tissues because permeability of the capillary-venular circuit is not uniform. Within a single tissue the venous end of the exchange vasculature is more permeable to water and to solute than is the arterial end. In different tissues the structure of the capillary wall possesses different characteristics which influence their relative permeabilities.

SHUNT VESSELS

All elements which bypass the effective exchange circulation of a tissue serve as shunt vessels. They include arteriovenous anastomoses, preferential (thoroughfare) channels, and even capillaries when they receive blood flow greatly in excess of their exchange capacity. Shunts of varying dimensions have been described in most tissues. Except for those in the skin that subserve temperature regulation, their precise functional significance is unclear. When effective shunts exist, some fraction of the total flow to an organ or region will not participate in exchange. With shunting, redistribution of blood flow may occur within the tissue without any apparent change in total tissue blood flow, and variations in total blood flow need not be associated with comparable variations in tissue nutrition.

CAPACITANCE VESSELS

Because of their great distensibility, the veins are the major capacitance elements of the vascular system. The distribution of blood volume within the venous circulation is somewhat uncertain. Current evidence indicates that the venular and small venous division constitutes the major functional capacity of the vascular system. The total system is estimated to contain about 70 percent of the total blood volume.

FUNCTIONAL ACTIVITY OF THE MICROCIRCULATION

NORMAL BLOOD FLOW

In contrast to the laminar flow seen in small arterial and venous vessels, laminar flow characteristics in the microcirculation are less evident. Red cells flow through capillaries in single file at varying rates and are usually distorted because of a restricted vessel lumen. Arteriolar flow velocity has been estimated at 4.6 mm per second, and venular flow velocity at 2.6 mm per second.

Red blood cell velocity in the capillary is estimated to be about 1 to 2 mm per second. Velocity of flow through a segment of capillary does not necessarily indicate the transit time of red blood cells through the capillary circuit. Capillary flow is intermittent and often reverses direction, so that the effective red cell transit time is probably in excess of the estimated 1.0 second.

INTERMITTENCY OF BLOOD FLOW

Direct examination of the microcirculation reveals that flow is very intermittent. Capillary flow may vary in rate and direction from one moment to the next. The intermittency probably reflects changes in the pressure gradient through a particular vessel due to sphincter action or to transient plugging upstream or down. It does not represent active capillary activity, since capillaries do not contract.

Flow in any single capillary depends primarily upon the state of the precapillary sphincters and metarterioles. These elements exhibit alternate phases of contraction and relaxation with a period of about 30 seconds to several minutes. This phasic contractile activity has been termed *vasomotion.* Vasomotion determines the distribution of blood flow between various regions of the capillary bed. Metarteriolar vasomotion is primarily concerned with main channel flow. The contraction phase of metarteriolar vasomotion is never intense enough to arrest main channel flow. However, the contraction phase of precapillary sphincter vasomotion is complete, and capillary flow is arrested. With augmented vasomotion, the frequency of contractions and relaxations increases, the contraction phases becoming more prominent. Under these circumstances, capillary blood flow decreases. On the other hand, a reduction in vasomotion involves a reduction in the frequency of the cycle with the dilator phase predominating, and capillary blood flow increases.

Vasomotion is the response of vascular smooth muscle to local chemical, myogenic, and neurogenic influences. However, since vasomotion is seen in denervated regions, the primary influence is probably exerted by chemical and physical influences acting on the vascular musculature. Neurogenic influences may serve to modify the reaction of vascular smooth muscle to the other agents. Consequently, vasomotion is dependent to a large extent upon tissue activity. There is a normal level of vasomotor tone which partially restricts blood flow into the capillary network. As tissue metabolism increases or as a degree of anaerobic metabolism ensues, the enhanced accumulation of metabolic products depresses the peripheral vasculature, thereby reducing vasomotion, and capillary blood flow increases. Conversely, minimal metabolic activity, such as exists in the basal state, provides a low concentration of depressed metabolites, and vasomotion is maintained at a relatively high frequency.

Phasic contractions similar to vasomotion are also seen in arterioles. However, arteriolar vasomotion is probably influenced more by central nervous impulses traveling through the sympathetic outflow to the arterioles than by local humoral influences.

TRANSCAPILLARY MOLECULAR AND FLUID EXCHANGE

Normal tissue function requires that water and soluble material pass between the blood and tissue fluids. This transfer of material across the capillary-venular wall occurs by means of filtration, diffusion, and cytopempsis. The rate and extent of transcapillary movement depend on the nature of the molecule and of the capillary structure.

CAPILLARY STRUCTURE

Electron microscopic examination of capillaries shows them to consist of a single layer of *endothelial* cells. At the outer surface of the endothelial cells is an amorphous mucopolysaccharide matrix called the *basement membrane.* It forms a barrier about 500 A thick which contributes to the permeability characteristics of the capillary. Capillaries have been classified on the basis of the nature of *endothelial perforations* (fenestrations) and the extent of *basement membrane* and *perivascular investment.* Thus, capillaries of muscle, skin, heart, and lung (Fig. 12-3A) have an

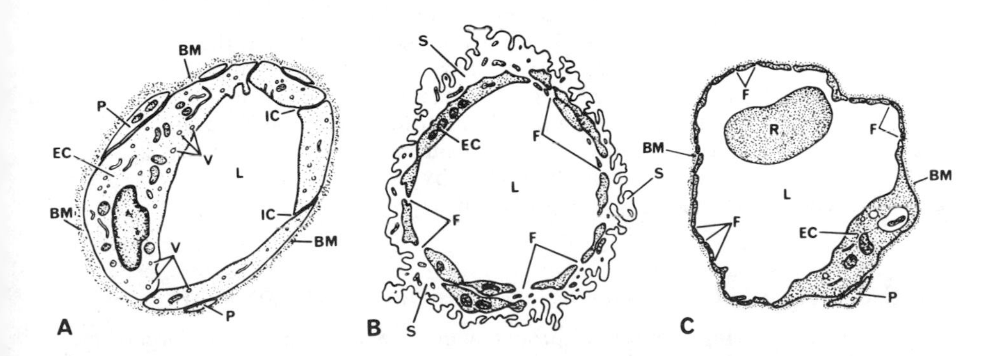

FIGURE 12-3. Diagrammatic representation of the electron microscopic characteristics of capillaries from different tissues. A. Muscle. B. Liver. C. Intestine. Symbols: EC = endothelial cell; IC = intercellular gap; L = vessel lumen; V = vesicle; F = fenestration; BM = basement membrane; P = pericyte; R = red cell; S = extravascular space. (After Bennett, Luft, and Hampton, 1959.)

unperforated endothelium, a prominent continuous basement membrane, and a prominent pericapillary investment. Liver capillaries (Fig. 12-3B) possess large endothelial perforations, a faint discontinuous basement membrane, and no significant pericapillary investment. Intestinal capillaries (Fig. 12-3C) possess thin, fenestrated endothelial walls with a continuous but indistinct basement membrane and only a slight perivascular investment.

CAPILLARY PORES

It was formerly believed that the major transcapillary exchange occurred through pores present in the simple edge-to-edge intercellular gaps of endothelial cells.

However, electron microscopy studies indicate that the junction between the cells is edge to edge only occasionally. More commonly cells overlap irregularly for 0.5 to 1.0 μ. The gap between cells is 90 to 150 A and is usually filled with mucopolysaccharide substance similar to that of the basement membrane. Pores are rarely apparent at this site, although the gap is about twice as large as the estimated pore radius of 40 A. However, particulate matter such as insoluble microspheres, lipid droplets, and leukocytes can pass through the intercellular gap, and physiological evidence supports the concept of functional pores. Moreover, the intercellular gaps constitute about 0.1 percent of the capillary surface, and it has been estimated that about 0.1 percent of the surface area of the capillary is available for water exchange.

The fenestrations, which may be considered as "pores," must certainly be prominent elements in capillary permeability. Liver capillaries, with numerous large fenestrations, possess a relatively high permeability. Intestine and kidney capillaries, which have intermediate-size fenestrations, also have intermediate permeabilities. Muscle, skin, heart, and lung capillaries, which have no fenestrations, have low permeabilities.

CYTOPEMPSIS (VESICULAR TRANSPORT)

Electron microscopy reveals vesicles in endothelial cytoplasm, and inpocketings are present at either margin of the endothelial cell membrane. These have been implicated in the transport of molecules by means of engulfing them at the vascular border, transporting them through the cell, and discharging them at the interstitial surface. The process has been called *cytopempsis.* However, the same type of vesicles is found in numerous cells unrelated to exchange. In addition, experimental evidence indicates that this phenomenon is probably not significant in normal transcapillary exchange. Cytopempsis may represent the method whereby extremely large molecules such as globulins and fibrinogens pass across the capillary wall.

TRANSCAPILLARY MOLECULAR EXCHANGE (CAPILLARY PERMEABILITY)

Transcapillary exchange is a passive process which depends on the diffusion of molecules and on the degree of membrane restriction to particle penetration. The free diffusion of molecules in solution is described by the Fick relationship (Chap. 1). According to this formulation, the rate of diffusion of a quantity of materials (Q) per unit time (t) with a diffusion coefficient (D_S) through a cross-sectional area (A_S) over distance (x) is directly proportional to the concentration gradient (dc/dx).

$$\frac{dQ}{dt} = DA\frac{dc}{dx}$$

Capillary permeability (P_S) may be defined as the degree to which the capillary permits the passage of molecules. The capillary wall is a passive barrier, and the major determinants of molecular penetration are molecular lipid solubility and molecular size or shape and charge. Lipid-soluble molecules utilize the entire endothelial surface for transport in accordance with their oil:water partition coefficients. However, most ionic and molecular constituents of plasma are not lipid soluble, and presumably they must pass through aqueous channels in or between cells in order to enter the extravascular space.

The extent of the gaps and fenestrations and of electrostatic interaction is reflected in the apparent area available for diffusion (A_S) compared to the total area of the membrane (A_m). This may be formulated as follows:

$$P_s = D_s \times \frac{A_s}{A_m}$$

Capillary permeability to a particular solute differs from the free diffusion of that solute to the degree that the effective or apparent area for diffusion differs from the total area of the membrane modified by any electrostatic interaction that may occur between the charged particles and fixed charges in the capillary membrane. The apparent area for diffusion for all lipid-insoluble molecules is considerably less than the total area of the membrane. Consequently, the rate of capillary penetration is substantially less than that of free diffusion. Since there are relatively few large pores, the area of diffusion for large lipid-insoluble molecules is less than that for smaller molecules, and their capillary permeability is therefore considerably lower than for small molecules. The relationships between molecular size and capillary permeability have been established for muscle and are presented in Table 12-1.

Lipid-soluble molecules such as oxygen and carbon dioxide have an area for diffusion that should be approximately equal to the total endothelial surface. Permeability for these molecules should thus be expected to be approximately equal to the

TABLE 12-1. Permeability of Mammalian Muscle Capillaries to Lipid-Insoluble Molecules

Substance	Molecular Weight (gm)	Diffusion (D) (cm^2/sec^{-1})	Molecular Radius (Approx.) (cm)	Permeability (P) (cm/sec^{-1})	P/D* (cm^{-1})
H_2O	18	3.4×10^{-5}	1.5×10^{-8}	0.54×10^{-6}	17.2×10^{-3}
NaCl	58	2.0	2.3	0.33	14.3
Urea	60	1.95	2.6	0.26	13.5
Glucose	180	0.90	3.7	0.090	10.0
Sucrose	342	0.70	4.8	0.050	6.9
Raffinose	504	0.64	5.7	0.039	6.0
Inulin	5,500	0.24	13.0	0.005	2.6
Myoglobin	17,000	0.17	19.0	0.0004	0.3
Serum albumin	67,000	0.085	36.0	0.0000†	0.0

*The ratio P/D is the ratio of the permeability coefficient to the free diffusion coefficient and is equivalent to the effective pore area per unit length of the diffusion path through the capillary wall per square centimeter of capillary surface.

†Serum albumin possesses permeability less than 0.00001.

Source: Data from Landis and Pappenheimer (1963) and from Renkin (1959).

free diffusion coefficient. However, capillary permeability to lipid-soluble molecules is also influenced by the oil:water partition coefficient of each molecule. Therefore, D_{O_2} in serum is less than P_{O_2}. In the case of CO_2, which is more soluble in water and lipid than is O_2, D_{CO_2} is of the same order of magnitude as P_{CO_2}.

Because of variations in the extent of fenestration and the development of the basement membrane of different tissues, it is obvious that capillary permeability will not be uniform throughout the body. Capillaries of liver and intestine are highly permeable and restrict protein passage only slightly; those of muscle, skin, heart, and lung, which are poorly permeable, do restrict protein passage significantly.

Further evidence of nonuniform distribution of capillary permeability between tissues has been obtained by examining the escape of different-size molecules of dextran upon their circulation through various tissues. It was found that liver, with a high permeability, passed a greater ratio of large- to small-size molecules than did muscle, with a low permeability. Intestine, with an intermediate permeability, passed some intermediate ratio of small to large dextran molecules.

TRANSCAPILLARY FLUID MOVEMENT

Net transcapillary fluid movement is determined by the interaction of hydrostatic and osmotic forces acting across the capillary wall, as first described by Starling, and may be formulated as:

$$FM = CFC\ (P_c + \pi_{i.f.} - \pi_p - P_t)$$

FM represents *net fluid movement* and CFC the *capillary filtration coefficient. Capillary hydrostatic* (P_c) and *interstitial fluid colloidal osmotic* ($\pi_{i.f.}$) *pressures* represent the *filtration force* whereas *plasma colloidal osmotic* (π_p) and *tissue hydrostatic* (P_t) *pressures* represent the *absorption force.* FM is positive when the filtration force exceeds the absorption force and fluid is filtered from the circulatory system into the interstitial spaces. Conversely, FM is negative when the filtration force is lower than the absorption force and fluid is absorbed from the interstitial spaces into the circulatory system.

CAPILLARY FILTRATION COEFFICIENT

The capillary filtration coefficient (CFC) may be defined as the volume of fluid filtered in one minute by 100 gm of tissue for a 1 mm Hg change in capillary pressure. CFC reflects the status of water permeability and of surface area of the capillary wall. Capillary membrane water permeability is considered to be reasonably constant so that CFC is determined primarily by capillary surface area.

The filtration of water through the capillary wall of the vessels of the human forearm, measured from the slow change in forearm volume which follows an elevation in venous pressure, is approximately 0.0057 ml/min $\times$ 100 gm of tissue per centimeter of water capillary pressure. Fluid filters from kidney, intestine, and liver at rates considerably higher than those of muscle or skin. However, because of the great mass of muscle and skin, the total body filtration rate resulting from elevations of central venous pressure is approximately 0.0061 ml/min $\times$ 100 gm of tissue per centimeter of water venous pressure. In an adult, an increase in central venous pressure of 10 cm HOH will create a loss of 250 ml of fluid from plasma in 10 minutes.

Most of the water is believed to pass across the capillary wall between cell margins and through fenestrations. In muscles it would appear that these comprise approximately 0.1 percent of the total membrane surface area. In kidney, the glomerular filtration rate is considerably greater, so that the effective surface area available for filtration must be greater.

CAPILLARY HYDROSTATIC PRESSURE

Under normal circumstances, capillary hydrostatic pressure is probably the major determinant of transcapillary fluid movement, since it is probably the most variable component of the transcapillary forces.

Capillary pressure (P_c) is determined by arterial pressure (P_a), venous pressure (P_v), and the ratio of postcapillary to precapillary resistance (r_v/r_a) according to the formulation

$$P_c = \frac{\frac{r_v}{r_a} P_a + P_v}{1 + \frac{r_v}{r_a}}$$

Thus elevations of P_a, P_v, or r_v/r_a will produce elevations of P_c. Since r_a is normally about four times greater than r_v, 80 percent of changes in P_v is reflected back to the capillaries compared to 20 percent of changes in P_a. This apportionment will be modified by changes in either precapillary or postcapillary resistance.

PLASMA COLLOIDAL OSMOTIC (ONCOTIC) PRESSURE

The major contributors to plasma oncotic pressure are the plasma proteins, albumin, globulins, and fibrinogen (Table 12-2). Albumin, the smallest and most concentrated

TABLE 12-2. Osmotic Pressure of Plasma Proteins

Protein	Molecular Weight	Concentration (gm/100 ml)	Osmotic Pressure (mm Hg)
Albumin	68,000	5.0	20
Globulins	200,000	1.5	5
Fibrinogen	500,000	0.5	1

of the plasma proteins, provides the major contribution. In addition, approximately 30 percent of the total effective osmotic pressure results from the unequal distribution of electrolytes in an ionized colloidal system caused by the Gibbs-Donnan phenomenon.

TISSUE HYDROSTATIC PRESSURE

Tissue hydrostatic pressure represents that pressure which develops in the interstitial compartment outside the capillary wall. It is determined by the volume of fluid and the distensibility of the interstitial space. Attempts to measure tissue pressure have been made with needle micropuncture and by subcutaneous capsule implantation. Micropuncture provides estimates of a positive pressure of 0 to 5 mm Hg, whereas the capsular pressures are found to be about 7 mm Hg negative. It has been suggested that the negative capsular pressure is due to a dynamic osmotic balance which develops across the capsule wall as a result of the formation of a semipermeable membrane about the capsule. In either case, tissue pressure normally changes only slightly along the length of the capillary; thus its contribution to normal transcapillary exchange is slight. However, in conditions of lymphatic blockage or increased capillary permeability, tissue pressure can achieve high levels and influence transcapillary dynamics significantly.

INTERSTITIAL FLUID COLLOIDAL OSMOTIC (ONCOTIC) PRESSURE

Tissue colloidal osmotic pressure is provided by plasma proteins which have passed through the capillary wall into the interstitial fluid. Consequently, this pressure varies in accordance with the permeability of the capillary wall to plasma proteins. Muscle capillaries are poorly permeable to protein, as reflected in the lymph protein concentration of about 0.5–1.0 gm per 100 ml. In intestine and liver the lymph protein concentration is about 3–4 and 5–6 gm per 100 ml, respectively.

It is estimated that 1 percent of protein provides about 3 mm Hg osmotic pressure. However, it is difficult to estimate the tissue oncotic pressure from the total protein concentration alone because the interstitial protein is not uniformly distributed. A substantial portion of the interstitial space is occupied by ground substance which consists mainly of hyaluronic acid and chondroitin sulfate. One gram of hyaluronic acid can hold up to 100 ml of water while excluding albumin. Thus, the interstitial protein is distributed in a volume of fluid which is less than the total interstitial fluid volume, and effective interstitial protein concentration and tissue oncotic pressure are probably higher than would be anticipated from data of protein content and volume of the interstitial space.

NORMAL TRANSCAPILLARY FLUID EXCHANGE

Transcapillary exchange of fluid under normal conditions is determined primarily by the relationship of capillary pressure to plasma oncotic pressure. When capillary hydrostatic pressure exceeds plasma oncotic pressure, filtration results. When capillary hydrostatic pressure is less than plasma oncotic pressure, absorption occurs. It was suggested that hydrostatic exceeded oncotic pressure at the arterial end of the capillary, where filtration took place. As blood moved along the exchange vessels, hydrostatic pressure declined progressively until filtration ceased and absorption from the interstitial space to the vascular lumen occurred.

An alternative concept considers the effect of vasomotion on exchange. It was suggested that during the dilator phase of vasomotion, capillary hydrostatic pressure was high and filtration occurred, whereas the constrictor phase of vasomotion produced a reduction in pressure and permitted absorption to take place in the same vessel.

VARIATIONS IN TRANSCAPILLARY EXCHANGE

The net exchange of fluid across the capillary membrane may be influenced by variations in either capillary or tissue hydrostatic pressure, plasma or tissue protein concentration, lymphatic drainage, capillary permeability, or capillary surface area.

ALTERATIONS IN CAPILLARY HYDROSTATIC PRESSURE

Capillary hydrostatic pressure is mainly determined by arterial blood pressure, venous pressure, and precapillary and postcapillary resistance. The effects of changes in these variables on transcapillary fluid exchange are illustrated in Figure 12-4.

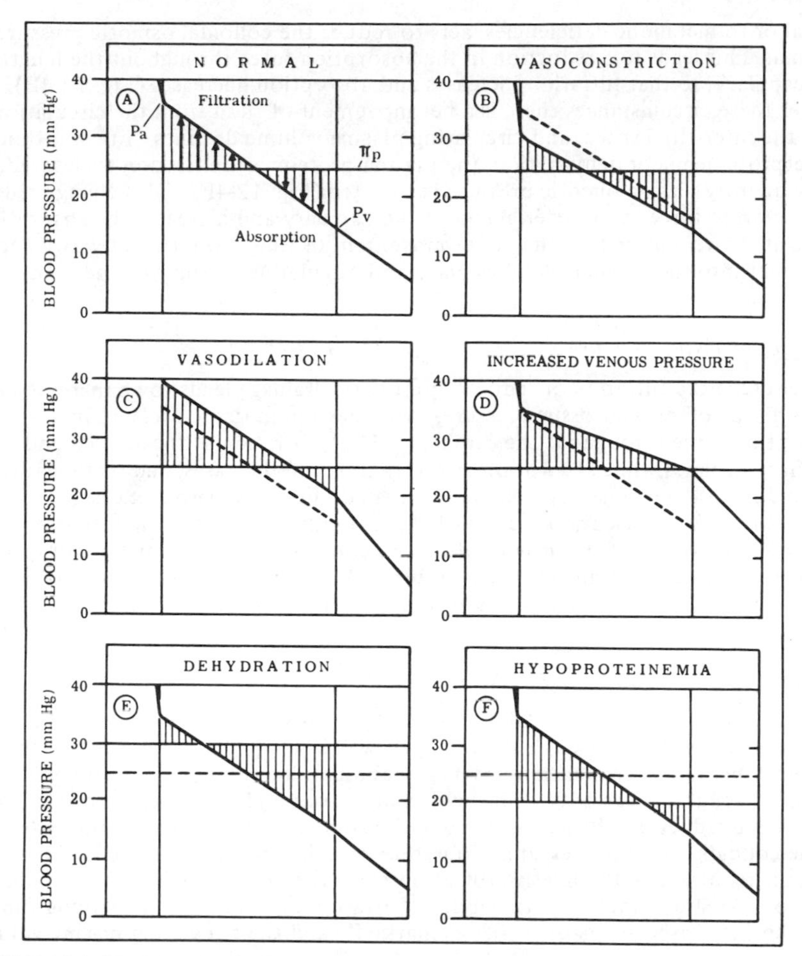

FIGURE 12-4. Variations in filtration and absorption produced by changes in arterial and venous pressures and resistances and plasma colloidal osmotic pressure. The broken lines represent normal levels.

The normal balance between filtration and absorption is presented in Figure 12-4A. A reduction in capillary pressure by means of reduced arterial pressure or increased precapillary resistance (Fig. 12-4B) results in reduced filtration and increased absorption. Increasing capillary pressure by increasing arterial pressure or reducing precapillary resistance (Fig. 12-4C) increases filtration and reduces absorption. Marked increases in filtration and reductions in absorption follow elevations of capillary pressure by increasing venous pressure or postcapillary resistance (Fig. 12-4D).

Changes in Plasma Colloidal Osmotic (Oncotic) Pressure

Plasma colloidal osmotic pressure is primarily determined by the plasma albumin concentration. A reduction of plasma albumin concentration, due either to nutritional or to metabolic deficiencies, acts to reduce the colloidal osmotic pressure of plasma. This leads to a reduction in the absorption force throughout the length of the capillary, so that filtration increases and absorption decreases (Fig. 12-4E). Under these circumstances there is a net movement of fluid from the circulation into the interstitial space, and circulating plasma volume declines. In conditions of hyperproteinemia or dehydration, the plasma protein concentration is elevated, and consequently plasma oncotic pressure is elevated (Fig. 12-4F). This change reduces the filtration force at the arterial end of the capillary and increases the absorption force at the venous end, so that a net movement of fluid from the interstitial compartment into the circulation takes place and circulating plasma volume tends to rise.

Changes in Tissue Pressure

Either enhanced filtration or reduced lymphatic drainage leads to an increase in interstitial fluid volume and tissue pressure. An elevation in tissue pressure increases the absorption force throughout the capillary. Thus, filtration at the arterial end of the capillary declines while absorption at the venous end increases, and further filtration of fluid from the circulatory system is impeded. For this reason, excessive filtration of fluid from the circulation is somewhat self-limiting. However, before tissue pressure achieves a level of this magnitude, filtration must exceed lymphatic drainage for a sufficient time or by sufficient magnitude. When this circumstance exists, interstitial fluid accumulation becomes excessive and edema develops.

CHANGES IN CAPILLARY PERMEABILITY

The osmotic pressure exerted by the plasma proteins is effective only as long as diffusion of the proteins through the capillary wall is adequately restricted. Under the circumstances of increased capillary permeability, albumin molecules are able to diffuse through the capillary endothelium at a more rapid rate than normal. This results in a reduction of the plasma colloidal osmotic pressure and an increase in tissue colloidal osmotic pressure. Consequently, the absorption force in the circulation is reduced and the relative filtration force is enhanced. Therefore, in conditions of increased capillary permeability, filtration of fluid from the circulation into the extravascular spaces increases markedly and the circulating plasma volume declines.

LYMPH AND LYMPHATICS

Large molecules which reach the interstitial space by filtration, cytopempsis, or cellular metabolism and secretion are not removed by the exchange vessels. If they were to accumulate, they would exert sufficient effective osmotic pressure to upset transcapillary fluid exchange, and excessive fluid would accumulate in the interstitial space. Under normal conditions this does not occur.

Filtered fluid and other plasma constituents which accumulate in the extravascular spaces are drained and conducted back to the circulatory system by way of the lymphatic network. In addition to this drainage function, the lymphatic system possesses concentrated areas of reticular endothelial cells at various sites which remove bacteria and foreign material from the lymph circulation. This action is one of the major protective mechanisms of the body against the invasion of harmful agents. The lymphatics also serve as a transport system for a number of materials, such as vitamin K and lipids absorbed from the intestine.

The lymphatic network originates in the tissue spaces as very thin, closed endothelial tubes (lymphatic capillaries), which have the same relation to the tissue spaces as have blood capillaries. However, lymphatic capillaries are far more permeable to large particles than are blood capillaries. Although the lymphatic capillaries consist of endothelial cells, their porosity is apparently so great as to provide little or no resistance to the passage of particles into the lymphatic system. These capillaries are also devoid of any basement membrane, which may account for their high permeability. The lymphatic capillaries converge on one another to form larger lymphatic vessels which possess valves to ensure a unidirectional flow of the lymph.

LYMPH FORMATION

Lymph is defined as the fluid which returns to the circulation from the tissue space by way of the lymphatics. Since lymph originates from plasma, as the difference between filtration and absorption of fluid across the capillary wall, the composition of lymph is very similar to that of plasma, and all influences which modify transcapillary exchange also modify the formation and composition of lymph. Except for variation in protein concentrations and electrolytes associated with the Gibbs-Donnan equilibrium, lymph and plasma are almost identical. The average protein concentration of lymph is approximately 1.5 percent, compared with 6 percent in plasma. In addition, the concentrations of the different protein fractions in lymph differ from those in plasma. Since albumin is the smallest plasma protein, it permeates the capillary endothelium more easily than does globulin. Consequently more albumin, relative to globulin, accumulates in lymph. Evidence of this differential rate of diffusion is found in the comparison of the ratio of albumin to globulin (A/G) in lymph and in plasma. The plasma A/G ratio is about 1.8, whereas the A/G ratio of lymph is 2.5.

In general, the protein concentration of lymph varies inversely with the rate of formation. With increased filtration from the capillaries, more fluid in relation to protein is cleared from the circulation; and although the quantity of protein filtered increases, the protein concentration in the interstitial compartment is reduced. Since capillary permeability and the rate of filtration vary between different regions of the organism, the protein concentration of lymph from these different regions also varies. The protein concentration of lymph from the liver and intestine may be as high as 5 percent, whereas that in lymph from the extremities is approximately 0.5 percent.

LYMPH FLOW

The rate of lymph flow in mammals is very low under normal circumstances. Furthermore, it varies between tissues in accordance with local transcapillary dynamics. The flow of lymph in the thoracic duct is approximately 1.38 ml per kilogram per hour. In a 24-hour period this represents a volume of fluid equal to the plasma volume. In addition, 50 to 100 percent of the circulating plasma protein is returned to the circulatory system during this period by the lymphatics. Therefore, this system is essential in maintaining the circulating blood volume.

Lymph flow is maintained by the influence of active contraction of muscular elements in their walls, and by external forces. Lymph flow is dependent, in part, on the tissue pressure which is generated by the continuous filtration of fluid from the circulation. Lymph flow is aided by compression of the lymphatic channels as a consequence of contraction of neighboring musculature and is enhanced by the negative intrathoracic pressure. Although increased respiratory or muscular activity enhances lymph flow, the major limitation and therefore the major determinant of the flow is lymph formation. Consequently, any circumstance which increases the rate of filtration of fluid from the capillaries also increases the rate of lymph flow. Raising capillary pressure by either arterial vasodilation or venous constriction enhances the rate of lymph flow.

Intravascular infusions of various solutions enhance the formation of lymph and lymph flow in two ways. First, these solutions increase the volume of fluid in the circulatory system and thereby increase arterial and venous blood pressure. This increases capillary pressure and augments filtration. Moreover, the increased volume of fluid in circulation dilutes the plasma proteins, effectively reducing the plasma colloidal osmotic pressure, which enhances filtration and reduces absorption.

Cardiac muscular contraction, the peristaltic action of the intestinal smooth muscle, or voluntary muscular contraction massages the lymphatic channels, and the alternate compression and relaxation exerted by the contracting muscles propel lymph centrally. The valves in the lymphatic channels prevent retrograde flow. Lymph flow is also augmented by increased tissue activity independent of external muscular contraction. The heightened tissue activity results in the accumulation of metabolites, which presumably cause the arterioles and precapillary sphincters to dilate and the capillary endothelium to become somewhat more permeable, so that blood flow into the capillary network increases at the same time that permeability is augmented. Eating results in lymph flows as high as 5.8 ml per kilogram per hour.

EDEMA

Edema is the condition of excess accumulation of fluids in the tissue spaces. It is of functional significance in that the presence of excess fluid in the interstitial space retards the exchange of nutrients and metabolites between cells and plasma. Edema results when filtration of fluid from the circulation exceeds the drainage. It may result from any condition which produces sufficiently increased capillary pressure, decreased plasma protein concentration, and increased capillary permeability. When these changes become excessive, fluid accumulates in the extravascular compartments more rapidly than it is removed by lymphatic drainage, and the extravascular compartment swells and becomes congested.

Although elevation in capillary pressure or reduction in plasma oncotic pressure facilitates filtration and may eventually lead to edema, increased capillary permeability and lymphatic blockage cause the most rapid development of severe edema. With increased capillary permeability, plasma proteins permeate the capillary wall more easily and thereby reduce the plasma oncotic pressure, while at the same time they increase the tissue oncotic pressure. Therefore, the accumulation of fluid in the extravascular spaces is doubly augmented.

Capillary permeability increases under a variety of circumstances. Marked increases in capillary volume and pressure created by large intravascular infusions, large doses of vasodepressant drugs, advanced bacterial inflammation, and the production of toxic metabolic products stretch the capillary wall and increase the dimensions of the intercellular and intracellular openings. In addition, the toxic metabolites, agents such as histamine, and bacterial toxins appear to act directly upon the capillary wall to increase its permeability. In each of the above conditions lymph flow is greatly enhanced. However, the capacity of lymphatic drainage is limited, and when capillary filtration exceeds lymphatic drainage, edema results.

The most exaggerated form of edema, *elephantiasis* of the limbs and scrotum, is caused by blockage of the lymphatics by the filarial organisms. Under these circumstances, lymphatic drainage from the tissue spaces is retarded or interrupted. Nevertheless, filtration of fluid proceeds, and interstitial fluid accumulation becomes progressive. Ultimately, the accumulation of extravascular fluid becomes so great that tissue pressure rises to a level at which it effectively counteracts the hydrostatic pressure in the capillary. To this extent the condition of edema may be said to be somewhat self-limiting. Unfortunately, because of the great distensibility of the extravascular space and because of the large capacity of body cavities in which the various viscera are contained, this point is not achieved until excessive quantities of fluid have been removed from circulation and circulating blood volume declines significantly. Moreover, because of the marked diffusion gradient for protein between plasma and interstitial fluid, protein molecules continue to diffuse from the circulation into the interstitial spaces despite a reduction in gross filtration. This increases tissue oncotic pressure and reduces plasma oncotic pressure, with the result that edema formation may continue even after hydrostatic pressures have equilibrated. Ultimately, hydrostatic pressure and oncotic pressure equalize across the capillary wall and no further net movement of fluid takes place. This point of equilibrium is seldom, if ever, achieved.

Edema also occurs in renal disease and is an outstanding feature of the nephrotic syndrome. In conditions of nephrosis, excessive protein is lost from the circulation into the urine, thereby reducing the plasma protein concentration. The kidney can also contribute to the formation of edema by means of salt and fluid retention. With this retention, the volume of the circulatory and extravascular systems is increased and venous pressure is elevated, a condition which facilitates the accumulation of fluid in the extravascular space. A more extensive discussion of fluid balance appears in Chapter 23.

REFERENCES

Bennett, H. S., J. H. Luft, and J. C. Hampton. Morphological characteristics of vertebrate blood capillaries. *Amer. J. Physiol.* 196:381–390, 1959.

Fulton, G. P., and B. W. Zweifach (Eds.). *Factors Regulating Blood Flow.* Washington, D.C.: American Physiological Society, 1958.

Landis, E. M., and J. R. Pappenheimer. Exchange of Substances Through the Capillary Walls. In W. F. Hamilton (Ed.), *Handbook of Physiology.* Section 2: Circulation, Vol. II. Washington, D.C.: American Physiological Society, 1963. Pp. 961–1034.

Mayerson, H. S. The Physiologic Importance of Lymph. In W. F. Hamilton (Ed.), *Handbook of Physiology.* Section 2: Circulation, Vol. II. Washington, D.C.: American Physiological Society, 1963. Pp. 1035–1073.

Pappenheimer, J. R. Passage of molecules through capillary walls. *Physiol. Rev.* 33:387–423, 1953.

Renkin, E. M. Capillary Permeability and Transcapillary Exchange in Relation to Molecular Size. In Reynolds and Zweifach, 1959. Pp. 28–36.

Reynolds, S. R. M., and B. W. Zweifach (Eds.). *The Microcirculation: Factors Influencing Exchange of Substances Across Capillary Wall.* Urbana: University of Illinois Press, 1959.

Ruszynák, I., M. Földi, and G. Szabó. *Lymphatics and Lymph Circulation.* New York: Pergamon, 1960.

Wiederhielm, C. A. Dynamics of transcapillary fluid exchange. *J. Gen. Physiol.* 52:29–63, 1968.

Yoffey, J. M., and F. C. Courtice. *Lymphatics, Lymph and Lymphoid Tissue.* Cambridge, Mass.: Harvard University Press, 1956.

13

The Heart as a Pump: Mechanical Correlates of Cardiac Activity

Ewald E. Selkurt
and Robert W. Bullard

THE FUNCTION of the circulatory system is to maintain an optimum environment for cellular function. This optimum environment may involve concentrations of nutritive, hormonal, and waste materials; tensions of respiratory gases; and temperature. Because of continuous cellular activity an optimum environment may be maintained only by an uninterrupted flow of blood to the tissues. To maintain this continuous flow, then, is the role of the circulatory system; and it is the heart that serves as the pump to maintain the flow. The properties of heart muscle have been discussed in Chapter 3. This chapter will deal with the specific anatomical and physiological properties of the heart that fit it for the role of a pump for the circulation of blood.

FUNCTIONAL ANATOMY OF THE HEART

A logical starting point in following the functional anatomy is the so-called *fibrous skeleton* of the heart. This is the dense connective tissue septum between the atria above and the ventricles below, and it is mainly composed of the four valve rings. Through this skeleton passes the bundle of His, made up of specialized conductile tissue derived from muscle. The bundle of His is the only structure maintaining the syncytium between atria and ventricles, or the only site of atrioventricular protoplasmic continuity.

ATRIAL ANATOMY

Above the fibrous skeleton sit the thin-walled, cup-shaped right and left atria. It is believed that the musculature of the atrial walls consists of two thin sheets: (1) a superficial sheet common to both atria, and (2) two deep sheets, one for each atrium, with the fibers more or less perpendicular to the superficial sheet. The thin-walled atria do very little work. They simply pump through the valve rings into the ventricles a small portion of the total blood returning from the veins.

VENTRICULAR ANATOMY

The heavy-walled ventricles must pump a volume of blood from the low-pressure venous system into the high-pressure arterial distribution system. Earlier anatom-

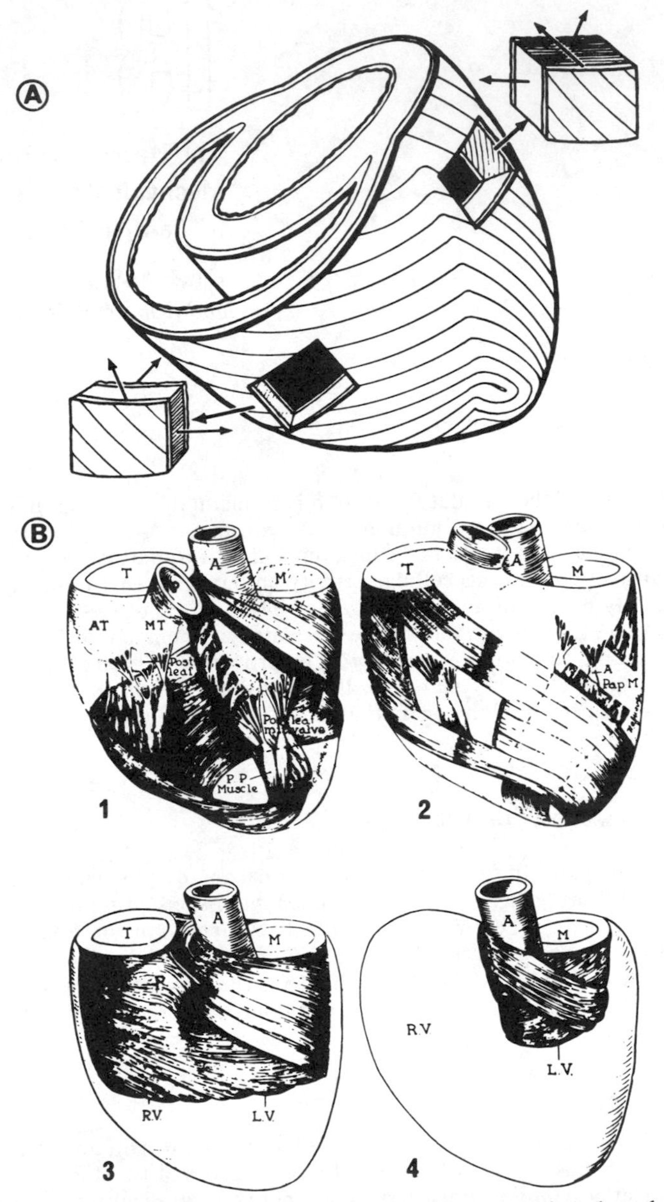

FIGURE 13-1. A. The ventricular anatomy. Note that the crescent-shaped, pocket-like right ventricle is more or less supported by the heavier-walled left ventricle. The blocks of tissue show three bands of muscle. The heavy, closed, spiraled constrictor band is characteristic of the left ventricle. B. Suggested arrangement of "scrolls" of ventricular muscle bundles. (1) The superficial bulbospiral muscle as seen from the front of the human heart. A = aorta; M = mitral orifice; P = pulmonary artery; T = tricuspid orifice; AT = anterior leaflet of tricuspid valve; MT = medial leaflet of tricuspid valve. A V-shaped section is cut from those fibers encircling the left ventricle subendocardially, so that the mitral valve may be seen. A similar band on the right is not sketched in. (2) The superior sinospiral muscle as seen from the anterior

ical studies have apparently demonstrated separation of the ventricular musculature into four bands: (1) the superficial bulbospiral muscles, (2) the superficial sinospiral muscles, (3) deep sinospiral muscles, which encircle both ventricles, and (4) deep bulbospiral muscle fibers, which encircle the left ventricle (Fig. 13-1B). (*Bulbo* indicates an origin at the mitral valve ring, and *sino* indicates an origin at the tricuspid valve ring.)

From a functional point of view, Rushmer has interpreted (Fig. 13-1A) the ventricles as being formed of two sets of myocardial bundles: the internal and external layers of spiral muscle, and the thicker ventricular constrictor muscles. The internal and external investments of the ventricular chambers are presumably composed of the same muscle bundles spiraling in opposite directions, which tend to oppose each other on contraction (see direction of arrows). As a result, the heart does not rotate appreciably with constriction.

Recent anatomical and functional studies (Armour and Randall, 1970) in nine different mammalian species, including dog and man, have challenged the older concepts of cardiac architecture as composed of "scrolls of muscle bands." Their structural analysis, which allows easier understanding of myocardial mechanics, stresses three basic characteristics: (1) The ventricle is made up of a continuum of interdigitating fibers which do not sharply delineate into separate muscular bands. (2) Deep fibers are oriented at right angles to epicardial fibers with a transition of angular change between them, but with a "principal fiber direction" consisting of the bulk of the ventricular fibers which is generally oblique to the vertical base-to-apex axis. (3) Epicardial fibers originate from the aortic ring area, descend, and at the apex enter deeply to form a major part of the papillary muscles, reinserting into the mitral ring via the chordae tendineae. However, a significant number of muscle strands do not pass from base to apex but swing deeply from the epicardium toward the endocardium at all vertical levels.

The papillary muscles have been discerned by Armour and Randall to have a great influence on cavity geometry as they change in size between diastole and systole. In diastole, they are relatively small and delineate a large submitral inflow region (Fig. 13-2), the so-called inframitral volume tract. During systole, the bulging of the papillary muscles closes off this region and accentuates the cylindrical outlet channel up to the aortic root – the outflow tract. The right ventricular papillary muscle contraction shows comparable changes in the inflow region and outflow tracts.

VENTRICULAR ACTION

By the use of a technique known as biplane high-speed cinefluorographic angiocardiography, the pumping action of the right and left ventricles can be studied. In this technique a radiopaque (x-ray absorber) substance is injected into the blood stream, and x-ray movies are taken at the rate of 60 frames per second with the opaque blood outlining the ventricular chambers. The studies have been extended

---

surface of the heart. Again the subendocardial layer has been cut through in order to show deeper structures. The window in the right ventricular wall shows the fibers from the trabeculated area running up to the anterior and medial leaflets of the tricuspid valve. In both of these superficial muscles, blood vessels follow the muscle strands as they encircle the apex. (3) The deep sinospiral muscle as seen from the front. Note the division of the muscle at the posterior interventricular sulcus, with fibers passing anteriorly to form most of the basal two-thirds of the septum; these septal fibers lie just distal to the band of the left head of origin at the base of the aorta. (4) The deep bulbospiral muscle, a powerful sphincter encircling the left ventricular base and enclosing both the aorta and the mitral orifice within its sweep. (A from Rushmer, 1970. B from J. S. Robb and R. C. Robb. *Amer. Heart J.* 23:455–467, 1942.)

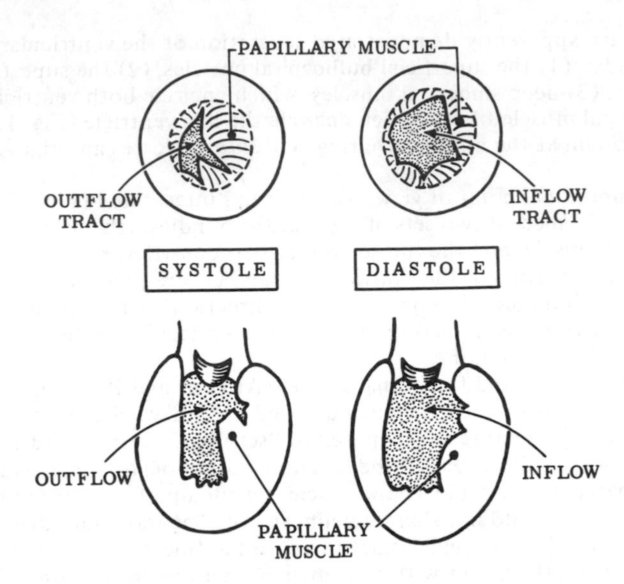

FIGURE 13-2. Cross-sectional dimensions of left ventricle in diastole and systole, illustrating variations of internal configuration caused by papillary muscles. In diastole, papillary muscles protrude minimally, the cavity being relatively globular. In systole, papillary muscles, especially anterior papillary muscle, bulge into cavity leaving a smooth-walled infra-aortic outflow tract. The dashed lines (upper figures) indicate a region of rapid fiber strand transition from parallel to near vertical in the endocardial papillary muscle portion. (From Armour and Randall, 1970.)

by the sewing of metal markers (tantalum screws and lead beads) on the ventricular walls so that their movement can be followed with x-ray movies. Other techniques for measuring changes in the heart size and volume of intact, unanesthetized dogs have employed differential transformers sewed within the chambers of the heart to measure changes in diameter (Rushmer, 1953, 1956, 1970) and inductance gauges using mercury in rubber as a single-turn coil to detect changes in cross-sectional area (Hawthorne, 1961). By means of high-frequency sound (ultrasonics) and the techniques of radar, the distances between a transmitter and a receiver crystal have been used to measure various cardiac diameters (Rushmer, 1970). These techniques have given valuable information about changes in the shape of the ventricles during the heart beat.

CHANGES IN LEFT VENTRICULAR LENGTH

The first significant dimensional changes occur during the *isovolumetric* (isovolemic) *contraction* (IVC) *phase* or the phase prior to the ejection of blood. No change in volume occurs, but there is a shortening in the base-to-apex length *(inflow tract)* (Fig. 13-3, curve 3). The circumflex-to-apex length (curve 2, "L_1") also shortens. The aorta-to-apex dimension (curve 1), the *outflow tract,* shows a lengthening of a couple of millimeters, however.

During *ejection,* the length shortens in all dimensions (1, 2, 3). Late in ejection, the base-to-apex length (3) increases slightly. *Isovolumetric* (isovolemic) *relaxation* (IVR) and the *rapid filling* (RF) *phase* are associated with a sudden sharp increase in length of all dimensions into the phase of *reduced filling* (see Fig. 13-7).

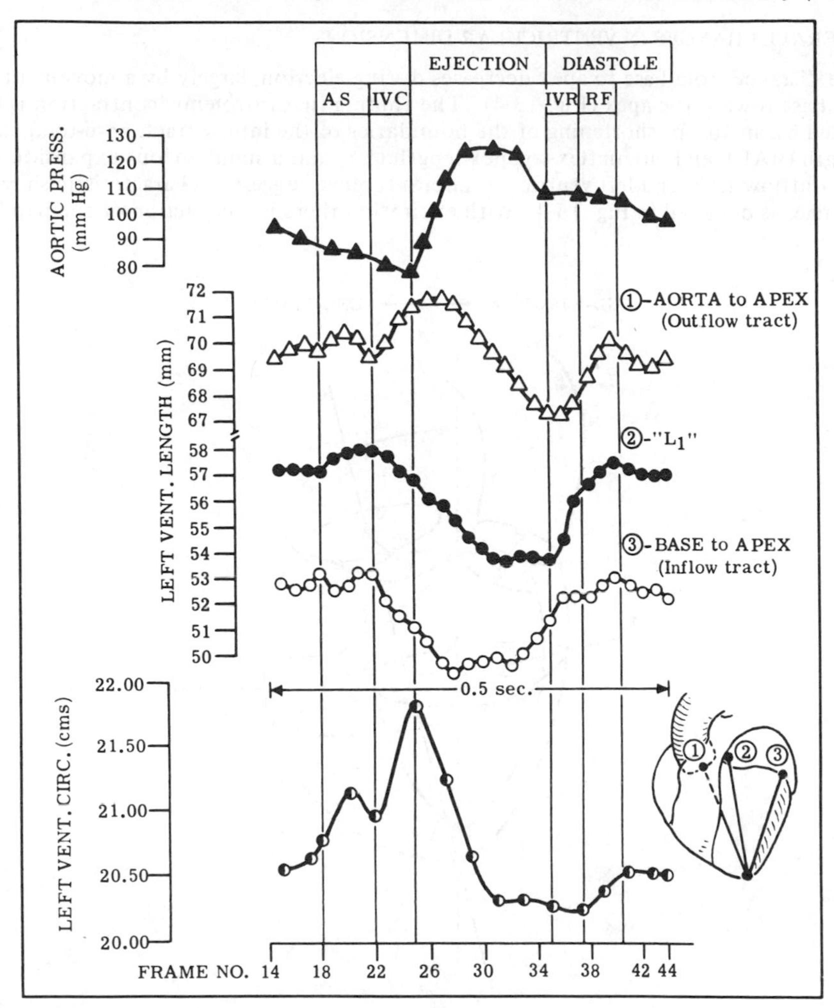

FIGURE 13-3. Plot of the simultaneous changes in canine left ventricular length: aorta-to-apex length, AAL (curve 1), the outflow tract; circumflex-to-apex length, L_1 (curve 2); and base-to-apex length (curve 3), inflow tract. Bottom curve shows change in left ventricular circumference (LVC) measured by a mercury silastic gauge. Insert of heart (lower right-hand corner) shows the dimensions measured for length curves 1, 2, and 3 (measurements made by electromagnetic induction changes and radiopaque markers using biplane cinefluorography). AS = atrial systole; IVC = isovolumetric or isovolemic contraction; IVR = isovolemic relaxation; RF = rapid filling phase. (From Hinds, Hawthorne, Mullins, and Mitchell, 1969.)

LEFT VENTRICULAR CIRCUMFERENCE

Left ventricular circumference (LVC) definitely increases during IVC, with the result that the heart increases in the transverse diameter, to result in a bulging effect. Thus, the heart actually becomes more spherical during this phase.

With ejection, the transverse diameter reduces sharply, remaining constant during IVR. As with length, LVC increases during RF.

OVERALL CHANGES IN VENTRICULAR DIMENSIONS

The distance from base to apex decreases during ejection, largely by a movement of the base toward the apex (Fig. 13-4). The finding that isovolemic contraction is initiated by an abrupt shortening of the boundaries of the inflow tract, base-to-apex length (BAL), and circumflex-to-apex length (L_1) and a simultaneous expansion of the outflow tract and left ventricular circumference suggests a characteristic shape change, as depicted in Fig. 13-4. With excitation there is contraction of the papillary

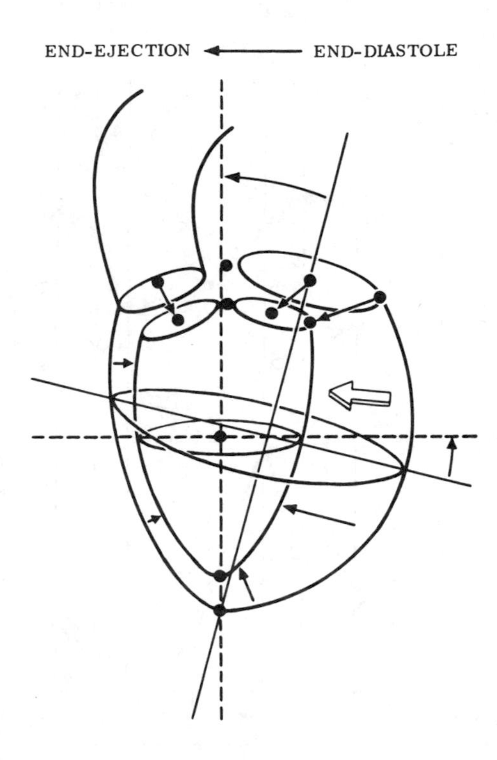

FIGURE 13-4. Diagrammatic summary of external dimensional changes on the movement of the left ventricle. (From Hinds, Hawthorne, Mullins, and Mitchell, 1969.)

muscles and the trabeculae carneae, with associated thickening of these muscles. Comment has been made earlier concerning these changes (Fig. 13-2). The resulting opposition of the muscles at their bases causes an internal translocation of intraventricular volume from the more apical region to the basal part of the left ventricular cavity, poised, as it were, ready for ejection into the outflow tract.

During ejection, there is a constant shifting of the center of the mass from the inflow to the outflow tract (Fig. 13-4). The major movement is an overall base-to-apex shortening.

The finding that all length dimensions show a sharp increase associated with very little change in left ventricular circumference during IVR suggests that this phase is attended by decreases in wall thickness at the time when intraventricular pressure is falling sharply (see Fig. 13-7). This phenomenon leads to the interesting concept that during IVR the ventricle is "prepared" for its transition from systole to diastole (Hinds et al., 1969). It appears that with a constant intraventricular volume there is a resultant sudden increase in surface area by the change from a small to a larger ellipsoid, so that the intracavitary pressure and wall tension fall, to allow the phase of rapid filling to proceed at a low atrioventricular pressure gradient.

During RF, the dominant change is one of left ventricular circumference expansion, with a shift to center of mass from the outflow tract back to the inflow tract, so that the ventricle is oriented for filling.

Reduced filling *(diastasis)* is characterized by very small gradual changes in dimensions and in intracavitary pressure.

At *atrial systole* (AS), aorta-to-apex length (AAL) increases slightly with no change in L_1 (2), and BAL (3) decreases slightly. LVC increases measurably. This change reflects the effect of the forceful atrial contribution to a passively filled ventricle, causing some change in its shape during AS.

PUMPING ACTION OF THE HEART

There appear to be three distinct actions by which blood is ejected from the right ventricle (Rushmer et al., 1953). Part of the pumping action is brought about by the movement of the base toward the apex, which decreases the contained volume of blood. A second action has been described as the "bellows effect." A bellows consists of two planes with a volume of fluid between them. A slight movement of the planes toward each other displaces a large volume of fluid but with a fairly low mechanical advantage for building up pressure. The right ventricle has been looked upon as a pocket hanging from the heavier-walled left ventricle. It appears that the free wall of the right ventricle is moved toward the interventricular septum by muscle contraction, thus simulating the bellows action. A third action may not be important normally but in some patients could mean the difference between life and death. It has been noted that rather extensive muscular damage can occur in the right ventricular wall without causing death, and also that cauterization of the entire right ventricle of the dog is compatible with life. This third action is called *left ventricular aid.* As the constrictor muscles of the left ventricle contract, the curvature of the septum increases and tends to pull the free right ventricular wall toward the septum, thus displacing some blood from the chamber. This effect of increasing the curvature is analogous to the tendency of tight pants to split when the wearer bends over.

Under the influence of sympathetic nerve stimulation, right ventricular systolic pressure can exceed 70 mm Hg (Pace et al., 1969). The sinus of the right ventricle is like the left ventricle in architecture and can be viewed as a pressure generator. The conus (entry into the pulmonary artery) can be considered a pressure regulator, attenuating the effect of marked increases in intraventricular pressure on pulmonary artery pressure under circumstances of sympathetic nerve stimulation.

There are two distinct pumping actions in left ventricular systole. The first, again, is the shortening of the chamber as the base moves toward the apex. The second is a reduction in transverse diameter by the shortening of the constrictor fibers. The ventricle is most accurately described as an ellipsoid with a complex relationship of the stress/pressure ratio to wall thickness (Sandler and Ghista, 1969). However, for simplicity's sake, the left ventricle may be thought of as a cylinder with a cone at the apex end. Because the volume of a cylinder is equal to $\pi r^2 h$, a decrease in the radius displaces a volume proportional to the second power of the initial radius, minus the second power of the final radius, while a decrease in height displaces only a volume directly porportional to the height decrease. The squeezing action of the heavy constrictor band being very effective in ejecting blood against a high resistance, the band is an efficient force pump.

When the right and left ventricles are compared as to function, the chief difference is that the left ventricle must pump a volume of blood through the high-resistance systemic circulation while the right ventricle pumps this same volume through the much shorter low-resistance pulmonary circulation. This difference in function is perfectly reflected in the difference in structure.

The thin-walled structure of the atria is a reflection of the low work demands they are required to meet. Nature exemplifies this relationship of structure to function when a condition arises in the pulmonary circuit which increases resistance to blood flow. In this condition the right ventricle may hypertrophy or increase its muscular mass and assume the architecture of the force pump.

ARRANGEMENT OF THE VALVES

Effective pumping action by the heart demands unidirectional flow from the venous side to the arterial side. The unidirectional flow is the function of the four cardiac valves located in the fibrous skeleton of the heart at the entrances and exits of the two ventricular chambers. The aortic and pulmonary valves, or semilunar valves, located at the exits of the left and right ventricles, are perhaps the simpler pair. They act to prevent the return of blood from the arterial system. Both valves consist of three tough but flexible flaps or cusps attached symmetrically around the valve rings to form a triangular orifice when they open during ventricular systole. The position of the three aortic valve cusps during their open phase is extremely important. A widening of the aorta just above the valve ring is effected by three outpouchings known as the sinuses of Valsalva. Hydraulic studies have shown that, when a fluid is forced through an orifice into a wider chamber, eddy currents may be generated. In the aorta eddy currents appear to circle behind the cusps and exert a force which keeps them located in the stream. If they were forced back to the aortic wall, blockage of the coronary orifice could occur and cause serious myocardial impairment. Eddy currents also hold the valve cusps in a position ready for a rapid closure (Fig. 13-5).

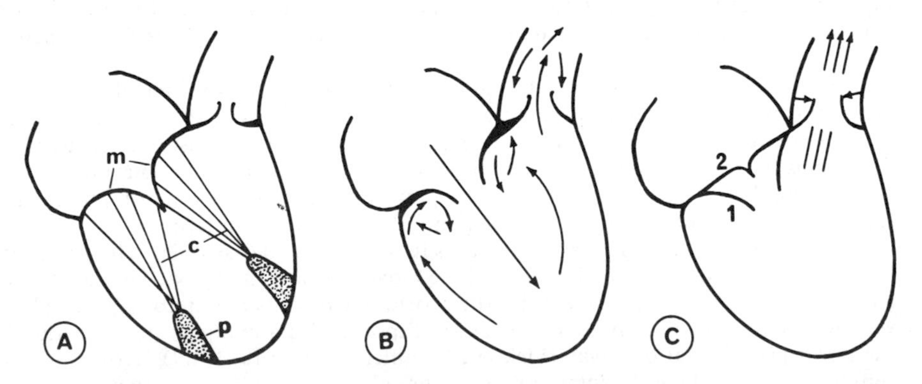

FIGURE 13-5. Mechanisms concerned in movement of the atrioventricular and aortic valves, (A) illustrating the action of chordae tendineae, (B) the importance of turbulent flow and eddy currents in valve closures, and (C) the suction created by sudden cessation of a jet. (After C. J. Wiggers. *Physiology in Health and Disease.* Philadelphia: Lea & Febiger, 1949. P. 583.)

The two atrioventricular valves (the tricuspid on the right side and mitral on the left side) act to prevent the ventricles from ejecting blood back into the atria. Both valves contain two large primary cusps attached completely around the valve ring. The tricuspid has, in addition, a fairly prominent secondary cusp. The total area of the cusps is much greater than that of the orifice they guard, which ensures a snug and leakproof fit upon closure. To the lower side of the cusps are attached the chordae tendineae leading to the papillary muscles on the ventricular walls. The papillary muscles function through the guylike chordae tendineae to prevent eversion of the flexible cusps into the atrium. The attachment of the chordae tendineae to papillary muscles is such that contraction tends to hold the cusps together rather than pull them apart. The prevention of eversion makes the ejection phase more efficient, since a bulging valve would, in a practical sense, be as detrimental as a leak of the same volume as the bulge.

Recent observations utilizing cineangiographic techniques in unanesthetized dogs indicate that the papillary muscles contract isometrically during ventricular ejection (Karas and Elkins, 1970). The critical distance between the free margin of the valve and the apex is kept constant by the chordae tendineae and papillary muscle so that the valve can open and close appropriately. If the papillary muscle shortened during ventricular contraction, it would prevent apposition of the leaflets and hence allow regurgitation.

VALVE ACTION

Because the valves ensure unidirectional flow from the venous side to the arterial side, they passively open whenever the pressure in the chamber nearest the venous side exceeds the pressure in the chamber or vessel nearest the arterial side. There are possibly two mechanisms involved in valve closure. The most obvious would be that brought about by the backflow of the blood. For example, closure of the mitral valve would occur as the left ventricle started to contract. This mechanism of closure requires some regurgitation of blood in a countercurrent direction. Another mechanism is present and may be readily seen under special circumstances. In pathological conditions where there is a retardation of the conduction of excitation between atrium and ventricle, the atrioventricular valves have been seen to close immediately after atrial systole, open again in the delay time, and close again with ventricular contraction. The first closure is apparently due to the eddy currents exerting a force in back of the open cusps. When atrial systole stops, this force closes the valve. The second closing is, of course, due to the backflow resulting from ventricular contraction swinging the valve closed. How important the eddy currents are in normal closure has not been completely decided. It is obvious that they act to hold the valves somewhat out into the stream in a position ready for rapid closing. One idea is that the eddy currents may start the closure at the end of atrial systole and that ventricular systole completes the closure. This is done so rapidly that it appears to be one movement. Because the backflow mechanism would therefore need to move the valve cusps only a short distance, the possibility of inefficient regurgitation is greatly reduced.

A brief description of the functional properties of cardiac tissue and anatomical actions of myocardium and valves has been presented. The following section is a consideration, in some detail, of the precise time relationships of these actions with volume and pressure changes.

PRESSURE AND VOLUME EVENTS OF THE CARDIAC CYCLE

METHODOLOGY

The events that are to be recorded occupy less than 1 second in the course of a normal heart beat. Therefore, the pressure recording system should record changes with correct phasic relationships. Reasonable reproduction of a wave form can be obtained by using a manometer system which has a uniform response to the tenth harmonic of its fundamental frequency. With the heart rate, which may go as high as 240 beats per minute, the pulse frequency is 4 per second and the tenth harmonic would be 40 cycles per second. With such a frequency response, rapid changes in pressure can be accurately recorded. Optical manometers, and electrical pressure transducers whose impulses are amplified and recorded by galvanometers, meet the frequency response requirements.

The voltages from the transducers are usually amplified to drive galvanometers which record with light (photokymograph), ink, or a heated stylus melting white wax placed over black paper. The electrical fluctuations may also be observed with a cathode-ray oscilloscope. An earlier section has discussed several methods for recording volume changes of the heart.

Volume change of the ventricles of the heart, as such, has also been measured by the classic technique employing the Henderson cardiometer. This is a thistle-shaped glass chamber having a capacity somewhat larger than the volume of the ventricles of the heart, with a perforated rubber membrane covering the open end. The perforation is of a size that permits the ventricles of the heart to be slipped into it in such a position that the membrane is snugly applied to the atrioventricular (AV) groove. This air-filled system is connected to a recording tambour and lever, optical manometer, or electronic recording device. Changes in volume of the heart cause variations in pressure within this system which in turn are inscribed on the recording paper by the recording system.

SIMULTANEOUS RECORDING OF PRESSURE AND VOLUME CURVES

Simultaneous recording of as many as three events gives insight into the rapid variations in pressure and volume of the heart and supplies direct evidence about the factors which are concerned with the opening and closure of the valves and, in turn, filling and emptying of the heart. This work has been done in classic experiments by Wiggers in open-chest dogs (1952). Atrial pressures have been measured by direct insertion of cannulas through appendages of the atrium or by entering the atrial chamber via the jugular vein and superior vena cava. Aortic pressure variations have been measured by insertions of the cannula via the carotid artery. Ventricular pressure variations have been measured by direct puncture of the ventricular wall, care being taken not to injure the coronary vessels. Studies on the dog have yielded much valuable information on the events of the cardiac cycle.

The modern era in the study of circulatory dynamics in man began about 1941 when Cournand and Ranges demonstrated the practicability and safety of catheterization of the right side of the heart. Their studies were built upon Forssmann's pioneer demonstration in 1929 of the feasibility of this technique. The ability to record pressure pulses in the chamber of the right heart and pulmonary artery, and the opportunity to sample blood from these areas, furnished the basis for precise and detailed study of the lesser circulation in the normal and in various pathological states. Thus, catheterization of the right heart permits evaluations of the pulmonary and systemic blood flow (cardiac output by the Fick method), pulmonary vascular tension and resistance, and congenital and other lesions of the right heart.

However, the need for information about the function of the less accessible left side of the human heart became apparent. A powerful stimulus to the development of methods for more precise characterization of the circulatory dynamics of the left side of the heart was furnished when surgical procedures for the correction of rheumatic valvular lesions were found to be feasible. The availability of methods for catheterization of the left side of the heart and the aorta has given valuable information, in the normal and pathological states, on simultaneous pressures measured in the left atrium, the left ventricle, and the central aorta. It has become possible to make accurate appraisal of rapid hemodynamic changes across the cardiac valves during exercise, changes in posture, and injection of drugs or contrast media.

SEQUENCE OF EVENTS IN THE CARDIAC CYCLE

Consideration of the heart as a pump focuses attention on the action of the left ventricle and the play of the valves at the entrance to and the exit from this chamber as they are influenced by the changing gradient of pressure during a cardiac cycle. The stages shown in Figure 13-6 serve to define the phases of systole in relation to valve action (isovolumetric contraction and ejection), and diastole (isovolumetric relaxation and filling: rapid, reduced [diastasis], and atrial systole). With this introductory survey, attention is next directed to the more complex interrelationships of Figure 13-7.

The upper part of Figure 13-7 shows the events in the left ventricle, aorta, and

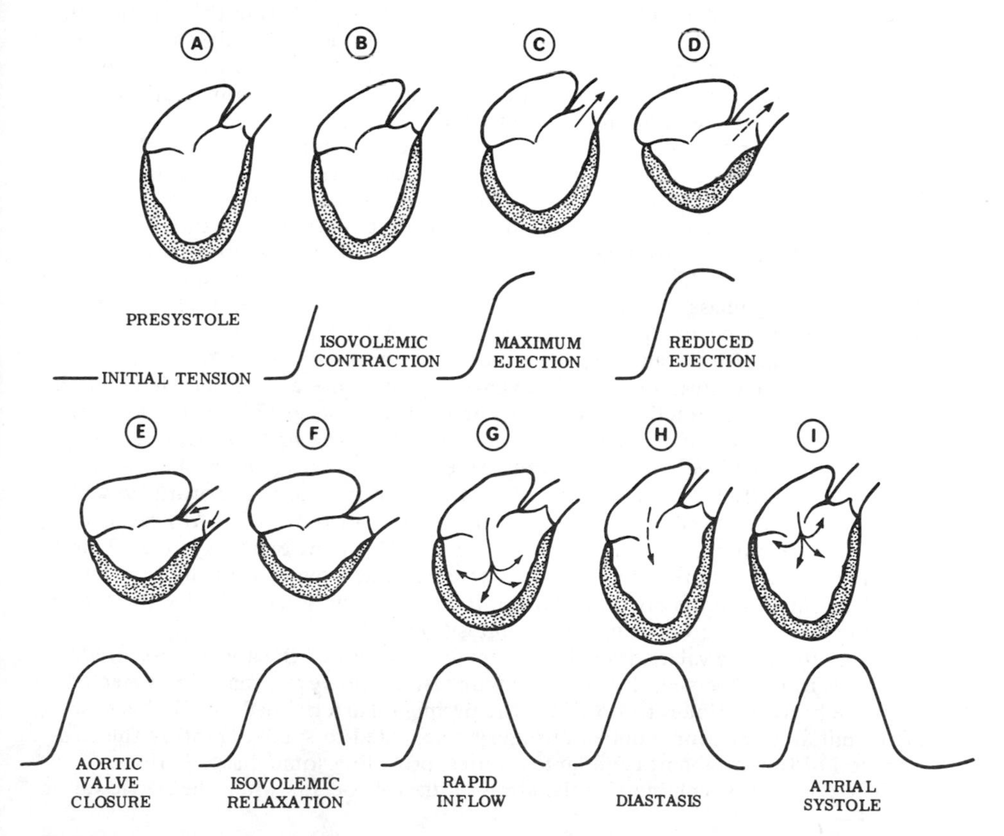

FIGURE 13-6. Diagrams illustrating successive phases of the cardiac cycle and the corresponding development of the left ventricular pressure curve. (After C. J. Wiggers. *Physiology in Health and Disease.* Philadelphia: Lea & Febiger, 1949. P. 654.)

left atrium. The pressure begins to rise rapidly to initiate *ventricular systole* coincidental with the peak of the QRS complex, the excitation wave for ventricular contraction, of the electrocardiogram (ECG) (top of figure). The rapid increase in pressure developed by the ventricle soon exceeds the pressure in the left atrium, and the AV valves close, to begin this phase of the cycle. The aortic valves remain closed until the pressure in the left ventricle exceeds that in the aorta, at which time the aortic valves open and ejection by the heart in systole begins. Although the orientation of the musculature of the heart may change during this phase, reference to the volume curve of the heart shows that its volume does not change; hence the name of this phase. Most of the ventricular ejection occurs during the phase of *rapid ejection,* as indicated in the volume curve of the heart. Left ventricular volume declines more gradually during the phase of *reduced ejection* and reaches a minimum during the *isovolumetric relaxation* phase. The left atrial volume has reached its minimum during *atrial systole* and is refilling during ventricular ejection. The pressure of the ventricle rises very rapidly at first, that in the aorta being just below it. The output into the aorta initially exceeds runoff into the peripheral arteries. However, the pressure begins to flatten off and reaches a summit as output and runoff reach equilibrium. The *reduced ejection* phase then begins. Since runoff from the aorta now exceeds output of the heart, the pressure in the common chambers begins to decline. When the ventricle starts to relax, the decrease of pressure within the ventricle initiates the closure of the aortic valves. The closure of the aortic valves is preceded by a sharp dip, the *incisura,* followed by a small secondary rise in aortic pressure. For ease of analysis, some authorities use the incisura as a convenient indication of the end of systole and the beginning of diastole of the ventricle.

Instantaneous flow in the ascending aorta of dogs can be metered with an electromagnetic flowmeter. Such a flow curve, adapted to values expected in man, occupies the middle of Figure 13-7. Pressure transients across the aortic valves have been measured through needles inserted into the left ventricular chamber and the ascending aorta. The recordings of differential pressure show a positive value for differential pressure (LVP-AP) reaching a maximum differential of Ca + 14 mm Hg during the rising phase of aortic flow (ventricular outflow). Then LVP-AP reverses itself during reduced ejection, to reach a negative value late in systole.

When the pressure gradients at the root of the aorta are compared with the intraventricular pressures, it is seen that ventricular pressure exceeds aortic pressure only during an early fraction (average 33 percent) of systole. This is the time of greatest acceleration of aortic flow and coincides with the phase of *rapid ejection.*

During *reduced ejection,* aortic flow decelerates, and pressure within the ventricle falls actually below that in the root of the aorta, although forward flow continues.* Toward the end of systole, the differential pressure (LVP-AP) reverses itself, to reach a negative value of about −15 mm Hg at the end of systole. This "reversed gradient" initiates a brief interval of retrograde flow toward the ventricular chamber, which is probably important in the mechanism of aortic valve closure, aiding in bringing the valve cusps into apposition.

When the aortic valves have closed, flow in the root of the aorta is essentially zero, as is to be expected. It should be emphasized, however, that blood ejected from the heart continues to flow into the peripheral arterial bed mainly because of the initial momentum applied during systole, aided to some extent by the elastic recoil of the distal aorta and large arteries, providing some diastolic flow.

The phase of isovolumetric relaxation of diastole occurs when the aortic valves

---

*This finding in late systole implies that at this time blood is flowing out of the left ventricle under its own momentum and that the aorta is a pressure source for the ventricle, the so-called momentum hypothesis (I. M. Noble. *Circ. Res.* 23:663–670, 1968).

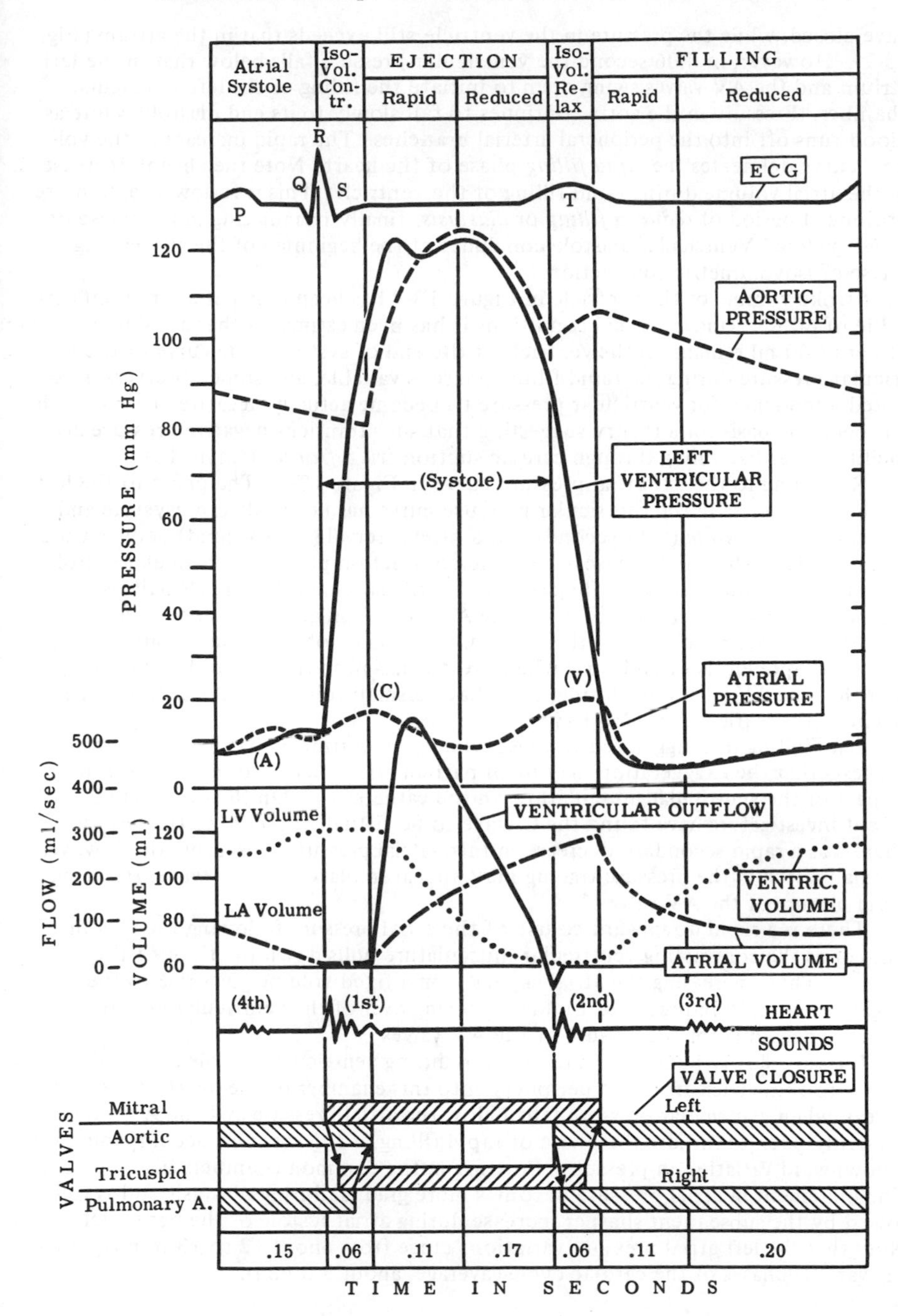

FIGURE 13-7. Pressure and volume events of the left heart during the cardiac cycle.

have closed, while the pressure in the ventricle still exceeds that in the atrium (Fig. 13-7). However, in 0.06 second the ventricular pressure falls below that in the left atrium and the AV valves swing open to initiate the filling of the left ventricular chamber. Pressure in the aorta continues to fall slowly to its end diastolic value as blood runs off into the peripheral arterial branches. The rapid increase in the volume curve delineates the *rapid filling* phase of the heart. Note the abrupt decrease in the atrial volume during rapid filling of the ventricle. This is followed by a more prolonged period of *reduced filling* or *diastasis,* finally terminating in the phase of *atrial systole.* Ventricular diastole concludes at the beginning of the succeeding phase of isovolumetric contraction.

Stroke volume of the ventricle in Figure 13-7 has been taken at a representative value of 60 ml. Under normal conditions it has been estimated that a residual volume of about 60 ml remains in the ventricle at the end of systole. The curve of the ventricular pressure during the rapid filling phase is variable, and some observers have noted a tendency for ventricular pressure to become actually negative at times. This has been the basis for a theory suggesting that such transient negative pressure also facilitates cardiac filling through cardiac suction *(vis a fronte)* (Chap. 16).

Returning to the left atrial pressure curve in Figure 13-7: The phase to the left of the sudden rise in the ventricular pressure curve has been called presystole and includes *atrial systole* (0.11 second) and a brief interval (0.04 second) between the end of atrial systole and the isovolumetric contraction phase of ventricular systole. Atrial systole concludes the filling of the ventricular chambers. The small rise in pressure during atrial systole is called the A wave. A small dip precedes a second rise, the C wave, which falls in the isovolumetric phase of ventricular contraction. The C component is variable and when present has been thought to be due to the fact that the AV valves, in closing, press back a small volume of blood contained between the leaflets (Wiggers, 1952).

The finding of exaggerated C waves when catheterization has been employed suggests that the exaggeration may be in part an artifact created by mechanical impact of the ventricular musculature on the catheter tip. On the other hand, different investigators have found the C wave to be virtually absent, particularly when there was a rapid secondary decrease in intra-atrial pressure, obscuring the C wave. A small notch in the pressure tracing may appear in place of the C wave, resulting from closure of the AV valves.*

There is a rapid downward course of the atrial pressure following the C component as the contracting ventricular musculature pulls down the floor of the atrium. Thus, increasing the atrial capacity on a fixed volume gives rise to the negative variation in pressure. Sudden crossing over of the ventricular and atrial pressure accounts for the closure of the AV valves.

Pressure slowly builds up in the atrium during ventricular systole as blood continues to come back from the periphery into this chamber of the heart. It reaches a peak when ventricular pressure falls below the atrial pressure and the AV valves suddenly open to initiate the phase of rapid filling of the ventricle accompanied by a downward variation in pressure. Pressure in the common chambers begins to rise. The increment in atrial pressure becomes more gradual during diastasis and is followed by the subsequent sharper increase during atrial systole of the next cycle. Note that the left atrial pressure variations range from about +2 to 13 mm Hg during various phases of the cardiac cycle (average, about 8.0 mm).

---

*Luisada has recommended that the peak or notch be designated by the symbol AV (atrioventricular) rather than C, since that letter was originally used to describe the C wave of the venous tracing, in which C was employed because the wave was caused by impact from the carotid artery (Chap. 16).

Below the variations in pressure associated with left ventricular activity in Figure 13-7 appear the times of opening and of closing of the valves of the left and right sides of the heart. As regards the order of valve closure, the mitral valve closes before the tricuspid, and the pulmonary valve opens before the aortic. At the end of systole the reverse occurs, i.e., the aortic valve closes earlier because of the higher pressure in the aorta than in the pulmonary artery but the tricuspid valve opens before the mitral because the pressure in the right ventricle has much less of a range over which to change before the valve may open. As a consequence, the phases of isovolumetric contraction and relaxation are briefer in the right ventricle than in the left, 0.04 second instead of 0.06, and the ejection phase is somewhat longer. The peak pressure developed in the right ventricle is about one-fifth that generated by the left ventricular contraction at its systolic peak. End diastolic pressure averages 3 mm Hg. The average peak systolic pressure is 24 mm Hg.

Pulmonary artery systolic pressure averages 22 mm Hg and diastolic about 9 mm Hg. The mean pulmonary arterial pressure averages 15 mm Hg.

The right atrial pressure variations are similar to those in the left atrium. Atrial systole on the right side begins somewhat earlier than on the left because of the slightly later excitation of the left atrial musculature. As with the left atrium, the C component may show considerable variability. The right atrial pressure has been found to vary between −2 and +5 mm Hg during the cardiac cycle. The lowest values have been obtained during the rapid ejection phase of the ventricle, the highest during atrial systole.

The relationship of the heart sounds to the cardiac cycle is indicated by the phonocardiogram tracing (heart sounds) in Figure 13-7. This shows that the *first sound* is related to the beginning of ventricular systole; it is caused by the closure of the AV valves. The *second sound* is caused by the closure of the semilunar valves in the aorta and pulmonary artery. The *third sound* is associated with the rapid filling phase of the heart. The *fourth sound* is related to atrial systole.

THE HEART SOUNDS

AUSCULTATION OF THE HEART

It is recommended that the student review the physics of sound and sound production before continuing in this section. The mechanisms of hearing of the human ear should also be reviewed (Chap. 4B).

Although an instrument like a tuning fork produces a pure sound, i.e., a sound with but one frequency, most natural sounds are composed of various frequencies or overtones which combine to determine the quality of the sound. Vibrations from the heart are composed of unrelated frequencies with very brief durations. Since the mass of tissue is large in relation to its elasticity, the vibrations occur at low frequencies. The vibratory motion tends to die out as the original energy imparted is dissipated. In this manner, the soft tissues of the body quite effectively damp the vibrations of the internal structures such as the heart, so that they typically consist of relatively few vibrations.

The audible components of the heart sounds and murmurs are in the frequency range between 30 and 250 cps. The normal first and second heart sounds range from 60 to 100 cps, and the third and fourth heart sounds, when heard, are usually below 40 cps. Even the so-called high-pitched murmurs may not greatly exceed 300 cps. The auditory range at normal intensities lies between 20 and 16,000 cps, but the maximal sensitivity of human audition lies within the so-called speech range of about 1000 to 2000 cps. Advantage is gained by making use of phonocardiography.

This may reveal sounds of a frequency inaudible or barely audible to the ear and supplies the temporal relationship between the heart sounds and the mechanical events of the cardiac cycle which is important in interpretation of the significance of sounds and murmurs.

ORIGIN OF HEART SOUNDS

First Heart Sound

The first heart sound is associated with the following dynamic phenomena: initiation of ventricular systole, closing of the AV valves, and opening of the semilunar valves. The first sound is lower in pitch and longer in duration than the second heart sound because the larger mitral and tricuspid valves close at low pressure compared with the aortic and pulmonic valves. Several authors have suggested a muscular component of the first heart sound. Although some vibrations can be recorded from contracting skeletal and cardiac muscle strips, their intensity is probably not sufficient to contribute to the sound heard on direct auscultation. It is more probable that the sudden closure of the AV valves as the result of abrupt pressure change causes vibrations in the cardiac structures which, when transmitted to the tissues of the body, are picked up from the precordium by the stethoscope.

As many as four components have been identified in the first heart sound. The first component is noted at the onset of ventricular contraction caused by the surging of blood toward the AV valves to initiate their closure. This induces a low-frequency sound of low intensity. The second component results as pressure rapidly increases in the ventricular chamber and the AV valves close and are set vibrating. Overstretching of the valves causes a recoil back toward the ventricular wall, accounting for the ensuing recoil of the contracting ventricular myocardium. The third component may involve oscillations of blood between the distended root of the aorta and the ventricular walls. The fourth component may represent vibrations due to turbulence in blood flowing rapidly through the ascending aorta and pulmonary artery. Indeed, all may be damped oscillations resulting from the primary event – valve closure.

Second Heart Sound

The second sound complex is caused mainly by the closing of the semilunar valves and the resulting vibrations of the heart and chest wall. More specifically, blood in the roots of the aorta and pulmonary artery rushes back toward the ventricular chambers, but this movement is abruptly arrested by closure of the semilunar valves. The momentum of the moving blood overstretches the valve cusps, and the recoil causes oscillations to occur in both the arterial and ventricular cavities.

Third Heart Sound

The third sound occurs beyond the point of atrial and ventricular pressure crossing, which concludes isovolumetric relaxation, and it occurs in the time of transition between rapid filling and reduced filling. When intraventricular pressure drops below atrial pressure, the AV valves swing open before a mass movement of blood into relaxed ventricular chambers. Inflow is arrested suddenly in the transition between rapid filling and diastasis. The momentum of the moving mass of blood produces low-frequency vibrations because the chamber walls are relaxed. Such vibrations are more likely to occur when the rapid filling phase terminates abruptly. The sound is not often heard in normal adults but may be heard more often in children; it is of low pitch and intensity.

Fourth Heart Sound

The fourth sound occurs at the time of atrial contraction and is related to the forceful ejection of blood into the ventricles, which are already distended. Almost always inaudible in normal persons, it is of low pitch and intensity when heard.

SPLITTING OF HEART SOUNDS

Some authorities believe that the first heart sound has two components, a mitral and a tricuspid. They base their opinion on the fact that closure of these valves is often slightly asynchronous, the mitral sound being heard first and then the tricuspid, because the left ventricular contraction precedes the onset of the right by a brief interval.

The aortic and pulmonic valves, also, do not always close at the same time, because of the different pressure and flow relationships in the aorta and pulmonary artery. Therefore, splitting of the second sound is possible in both normal and pathological circumstances. The aortic component precedes the pulmonic component and is particularly noticeable during inspiration. During inspiration, the effective pulmonary venous filling pressure remains unchanged, since the pulmonary veins, left atrium, and left ventricle are subject to the same intrathoracic pressure change. But there is increased filling of the right side of the heart, secondary to decreased intrathoracic pressure below that of the extrathoracic great veins, with resulting delay of ejection of blood from the right ventricle and hence a later closure of the pulmonic valve.

MURMURS

Murmurs are audible vibrations produced by the flow of blood. They originate in either the heart or the great vessels as a result of turbulent blood flow which produces a noise recognized as a murmur.

Blood flow through normal vascular channels is described as laminar or streamline flow and is silent. Turbulent flow may develop under special circumstances. The factors which determine whether flow is laminar or turbulent are the radius of the channel and the velocity, density, and viscosity of the fluid medium. Since blood viscosity and density are reasonably constant, the factors usually concerned are the velocity of flow and the radii of the parts of the cardiovascular system. How these factors are related to the production of random turbulence (Reynolds') and audible sounds should be reviewed (Chap. 11).

It is probable that some turbulence exists at specific points even normally, e.g., in the aorta or pulmonary artery just beyond the valve, but the noise produced is usually not intense enough to be heard with the stethoscope. Such minimal turbulence can be increased and result in an audible murmur when the velocity of blood flow is increased as in fever, exercise, pregnancy, or hyperthyroidism.

But when the caliber within parts of the cardiovascular system is reduced by disease, a murmur may result at normal or even decreased flow. Also, turbulence tends to occur at reduced velocity where fluid flows into a channel of much larger diameter – for example, blood flowing through a stenotic aortic valve into a normal aorta. Likewise, a murmur is produced by regurgitation of blood through an incompetent valve. In a tube of uniform caliber, inserting a local obstruction produces turbulence at a much lower velocity of flow (see Chap. 11, Turbulence).

Another mechanism by which murmurs may be produced is vibration of a structure within the heart, such as a ruptured valve cusp or chorda tendinea. This usually causes a murmur which is musical in quality since the process is similar to the vibration of a violin string.

SYSTOLIC MURMURS

Systolic murmurs are those which originate with or after the first heart sound and end with or before the second heart sound. They have been classified into *ejection* and *regurgitant* murmurs. The ejection murmur is one with midsystolic accentuation, such as that heard in aortic and pulmonic stenosis, and is associated with ejection of blood in the normal direction through a stenotic valve. A regurgitant systolic murmur is associated with regurgitation of blood through an insufficient valve. Regurgitant murmurs begin immediately after the first sound and extend throughout systole, with fairly constant intensity.

Mitral and Tricuspid Regurgitation

The murmurs of mitral and tricuspid regurgitation (Fig. 13-8), whether the valve is insufficient and diseased or incompetent because of a dilated annulus, are similar in quality because the mechanism is the same. When the pressure in the ventricle exceeds that in the corresponding atrium, the mitral and tricuspid valves close, producing the first heart sound. If closure is not complete, blood is forced into the atrium under pressure, causing turbulence and therefore a murmur. Usually the murmur begins immediately after the first sound and extends throughout systole, ending with the second sound, although in *mild* mitral insufficiency the murmur may end before the aortic component of the second sound.

Aortic Stenosis; Pulmonic Stenosis (Ejection Murmurs)

The murmurs of aortic stenosis (Fig. 13-8B), both congenital and acquired, and of pulmonic stenosis, which is usually congenital, have a diamond shape when visualized on a phonocardiogram. Since the intensity of the murmur is related to ejection of blood, the murmur is loudest when the ejection is at its maximum, i.e., near midsystole. When stenosis is severe, the ejection of blood is delayed and maximum accentuation occurs later in systole.

Atrial Septal Defects

With atrial septal defects (Fig. 13-8C) and with anomalous pulmonary venous drainage, a systolic murmur is frequently present in the pulmonic area, due to the proportionally large flow of blood past the pulmonic valve into an artery which is often dilated. The murmur so produced usually has an early systolic accentuation, although at times it sounds much like the murmur of pulmonic stenosis.

Ventricular Septal Defects

With ventricular septal defects (Fig. 13-8D), on the other hand, the murmur is related to the passage of blood through a small defect which acts much like a stenotic valve. The resulting murmur, best heard in the third and fourth left interspaces at the sternal border, is loud and harsh and often has a midsystolic accentuation.

Coarctation of the Aorta

The murmur of coarctation of the aorta is also produced by the flow of blood through a constriction, but, since the narrowing is some distance from the heart, the murmur

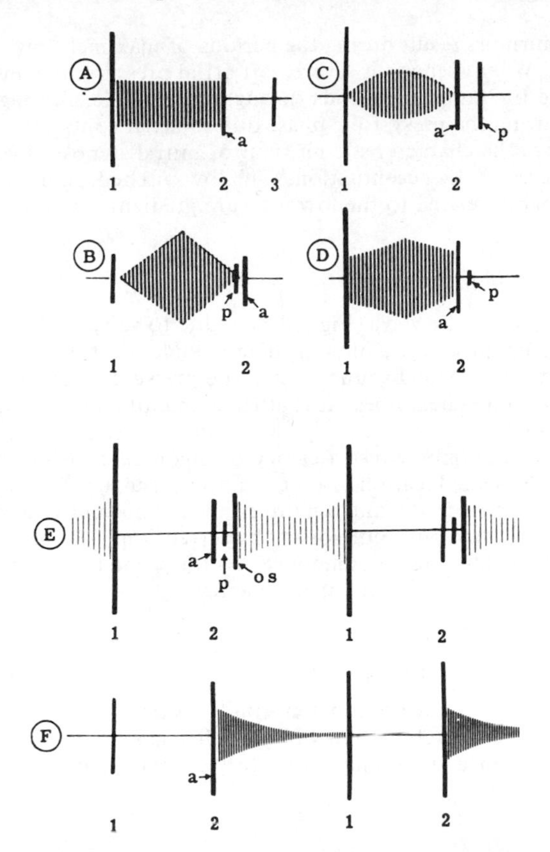

FIGURE 13-8. Phonocardiograms of systolic murmurs (A–D) and diastolic murmurs (E–F). A. Murmur of mitral regurgitation. B. Aortic stenosis. C. Atrial septic defect. D. Ventricular septal defect. E. Mitral stenosis. F. Aortic insufficiency. a = aortic component of second sound; p = pulmonic component; 1, 2, 3 = first, second, third heart sounds; os = opening snap of mitral valve. (From V. Hardman and J. S. Butterworth. *Mod. Conc. Cardiovasc. Dis.* 30:653, 655, 1961.

begins some time after the first sound and often extends past the second sound into early diastole. There is often a diamond-shaped configuration, although the intensity of the murmur is variable. Dilated intercostal vessels, due to collateral circulation, also contribute to the murmur.

DIASTOLIC MURMURS

Mitral and Tricuspid Stenosis

At the end of systole, ventricular pressure falls below atrial pressure, and the mitral and tricuspid valves open. Then comes the period of rapid blood inflow into the relaxing ventricles, followed by reduced inflow. In the presystolic period, with normal sinus rhythm, atrial contraction forces more blood into the ventricles. When the mitral or tricuspid valve becomes stenotic (Fig. 13-8E), the flow of blood is impeded and murmurs result during the periods of maximal flow, i.e., early diastole and presystole. When stenosis is severe, left atrial pressure remains elevated throughout diastole and the murmur extends throughout diastole, although there is still an accentuation during the presystolic phase due to atrial contraction. When there is atrial fibrillation, the characteristic murmur of mitral stenosis does not disappear, but it loses its presystolic accentuation. The low-pitched, rumbling quality of this murmur is probably related to the low pressure gradient across the valve.

Aortic and Pulmonic Insufficiency

Insufficiency of the aortic valve (Fig. 13-8F), due to valvular disease or to dilation of the valvular ring, produces a blowing decrescendo diastolic murmur which begins immediately after the second sound. Since the pressure gradient is greatest in early diastole, the murmur is heard best at that time and diminishes in intensity as the gradient decreases.

The murmur of pulmonic insufficiency or incompetence has the same quality and is heard in the same location as that of aortic insufficiency. It is indistinguishable by auscultation from the much more common murmur of aortic insufficiency. However, it is associated with pulmonary hypertension, and there is usually a marked accentuation of the pulmonary component of the second sound to aid in differentiating it from the murmur of aortic insufficiency.

CONTINUOUS MURMURS AND SOUNDS

Continuous murmurs are heard in patent ductus arteriosus and arteriovenous fistulas. The murmur of patent ductus is most intense during systole, when the pressure gradient between the aorta and pulmonary artery is greatest and flow through the shunt is most rapid.

REFERENCES

Armour, J. A., and W. C. Randall. Structural basis for cardiac function. *Amer. J. Physiol.* 218:1517–1523, 1970.

Brecher, G. A., and P. M. Galletti. Functional Anatomy of Cardiac Pumping. In W. A. Hamilton and P. Dow (Eds.), *Handbook of Physiology.* Section 2: Circulation, Vol. II. Washington, D.C.: American Physiological Society, 1963. Pp. 759–798.

Hawthorne, E. W. Instantaneous dimensional changes of the left ventricle in dogs. *Circ. Res.* 9:110–119, 1961.

Hinds, J. E., E. W. Hawthorne, C. B. Mullins, and J. H. Mitchell. Instantaneous changes in the left ventricular lengths occurring in dogs during the cardiac cycle. *Fed. Proc.* 28:1351–1357, 1969.

Karas, S., and R. C. Elkins. Mechanism of function of the mitral valve leaflets, chordae tendineae and left ventricular papillary muscles in dogs. *Circ. Res.* 26:689–696, 1970.

Luisada, A. A. (Ed.). *Cardiology, An Encyclopedia of the Cardiovascular System.* New York: McGraw-Hill, 1959. Vol. I.

McKusick, V. A. *Cardiovascular Sounds in Health and Disease.* Baltimore: Williams & Wilkins, 1958.

Moscovitz, H. L., E. Donoso, I. J. Gelb, and R. J. Wilder. *An Atlas of Hemodynamics of the Cardiovascular System.* New York: Grune & Stratton, 1963.

Pace, J. B., W. F. Keefe, J. A. Armour, and W. C. Randall. Influence of sympathetic nerve stimulation on right ventricular out-flow tract pressures in anesthetized dogs. *Circ. Res.* 24:397–407, 1969.

Pedersen, A. *The Venous Pressures in the Pulmonary Circulation.* Copenhagen: Nyt Nordisk Forlag, Arnold Busck, 1956.

Rushmer, R. F. Pressure-circumference measurements of the left ventricle. *Amer. J. Physiol.* 186:115–121, 1956.

Rushmer, R. F. *Cardiovascular Dynamics,* 3d ed. Philadelphia: Saunders, 1970.

Rushmer, R. F., D. K. Crystal, and C. Wagner. The functional anatomy of ventricular contraction. *Circ. Res.* 1:162–170, 1953.

Sandler, H., and D. H. Ghista. Mechanical and dynamic implications of dimensional measurements of the left ventricle. *Fed. Proc.* 28:1344–1350, 1969.

Starling, E. H. *The Linacre Lecture on the Law of the Heart.* London: Longmans, 1918.

Wiggers, C. J. *Circulatory Dynamics.* New York: Grune & Stratton, 1952.

Zimmerman, H. A. (Ed.). *Intravascular Catheterization.* Springfield, Ill.: Thomas, 1959.

14

Cardiac Excitation, Conduction, and the Electrocardiogram

Kalman Greenspan

THE EXCITATION or stimulus for the heart arises from within the cardiac muscle itself. How and where this excitation occurs and how it is conducted throughout the entire myocardium are discussed in this chapter. The electrical events of the cardiac cycle as manifested in the electrocardiogram and a simplified theory of electrocardiographic interpretation are presented.

RESTING AND ACTION POTENTIAL

The typical transmembrane potential of an inactive myocardial fiber (Fig. 14-1A, B, C) varies between 80 and 100 mv, the interior of the cell being negative with respect to the cell's exterior. When the cell is excited, however, the potential difference is quickly lost and in fact becomes opposite in sign (+20 to +40 mv), indicating that a reversal of polarity has occurred. This sudden initial upstroke (depolarization) is designated as phase 0 and includes the overshoot. Soon after the reversal of polarity the restorative processes start (repolarization), and the potential difference starts to decline toward its resting value. This is characterized by fast repolarization (phase 1), which then slows down and results in a plateau (phase 2). Following the plateau there is once again a fairly rapid wave of repolarization (phase 3), until the resting transmembrane potential or diastolic period is attained (phase 4). The time course of the cardiac action potential will differ also according to the fiber type; atrial cells (Fig. 14-1A) show a less prominent plateau. Action potentials obtained from the ventricular myocardium (Fig. 14-1B) and from the Purkinje network (Fig. 14-1C) located in and ramifying throughout these chambers show a more prolonged plateau. However, phase 0 is similar for atrial, ventricular, and Purkinje fibers, being rapid in upstroke.

Quite the contrary is seen in the configuration of the action potentials obtained for the fibers in the sinoatrial (SA) node (Fig. 14-1D) and the atrioventricular (AV) nodes (Fig. 14-3A). Phase 4 demonstrates a slow continuous diastolic depolarization so that during diastole the SA membrane potential progressively becomes less negative. The magnitude of the transmembrane "resting" potential is about −70 mv. Furthermore, phase 0 is not rapid; the overshoot (phase 1) is much reduced and in many instances is not present. Phases 1 and 2 blend with phase 3, and the entire process of repolarization is slow. Following the repolarization wave there is a progressive decline in the magnitude of the resting potential (an incline in the phase 4

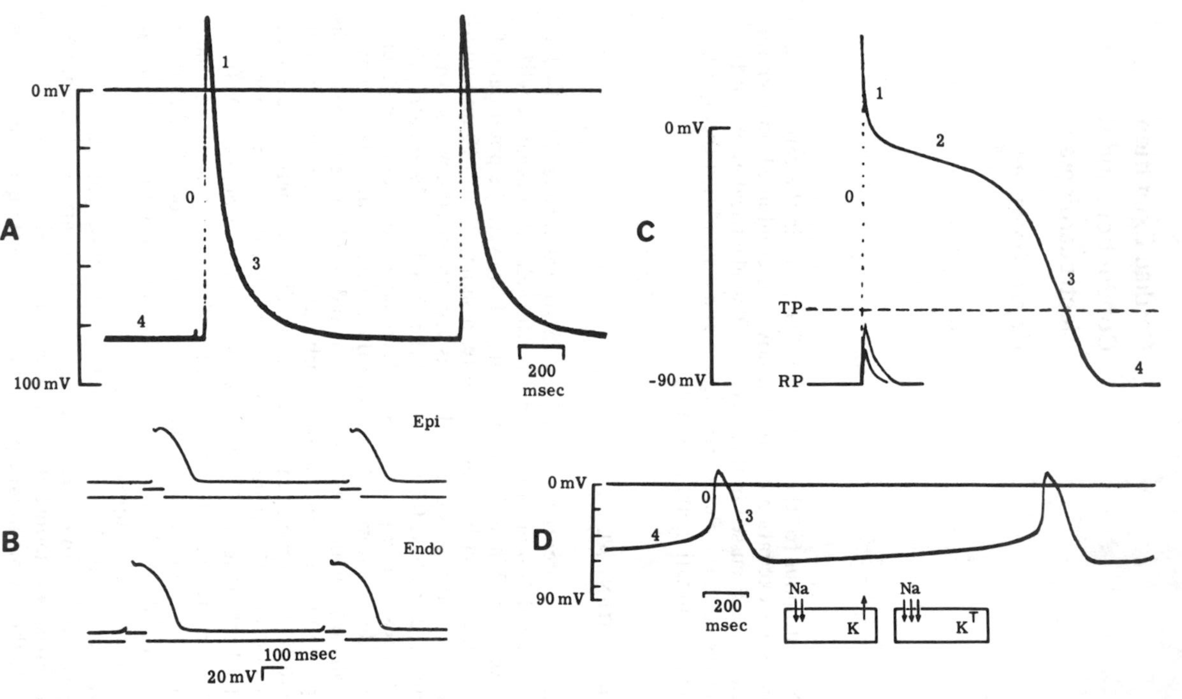

FIGURE 14-1. Records of membrane potentials obtained from: A. Atrial contractile fiber. B. Ventricular epicardial (Epi) and endocardial (Endo) surfaces. C. Purkinje fiber. D. Sinoatrial pacemaker cell. Note the differences in rapid depolarization (phase 0), the plateau (phase 2), and repolarization (phase 3) between these cardiac cells. C. TP = threshold potential; RP = resting potential. Application of progressively increased cathodal stimuli to cardiac Purkinje fiber. With a threshold stimulus (S_3), the all-or-none action potential is produced. D. In automatic fibers progressive spontaneous self-depolarization occurs during diastole. Excitation occurs when the threshold level of depolarization is reached. Below, in diagrammatic form, are the possible ionic fluxes that could account for the continuous slow diastolic depolarization.

slope) so that on reaching its threshold potential the cell becomes excited in a self-sustaining manner, resulting in full depolarization. All cells potentially capable of developing intrinsic automaticity* exhibit this characteristic sign of slow diastolic depolarization. The time course of these action potentials has great physiological significance, in that cardiac rate and rhythmicity are dependent upon the magnitude of the resting potential, the level of the threshold potential, and the slope or steepness of the diastolic depolarization. In turn, the slow diastolic depolarization is probably responsible for the property of automaticity, while the level of membrane potential will determine excitability and conduction. The mechanism responsible for the slow diastolic depolarization has not been clearly defined, although there is some evidence of a concomitant progressive increase in membrane resistance during diastole which may be indicative of a decreased membrane permeability to K^+.

PHYSIOLOGY OF PACEMAKER ACTION

Following depolarization of the primary automatic cell in the sinoatrial node, the current so generated flows longitudinally and lowers the transmembrane potential of adjacent areas. When threshold level is attained, these areas in turn depolarize, and self-sustaining, propagated action potentials spread over the entire myocardium. There is, however, a preferential sequence of activation (Fig. 14-2). Activity is first

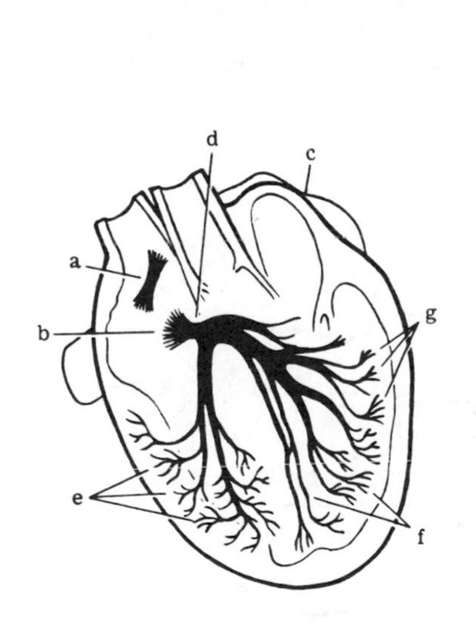

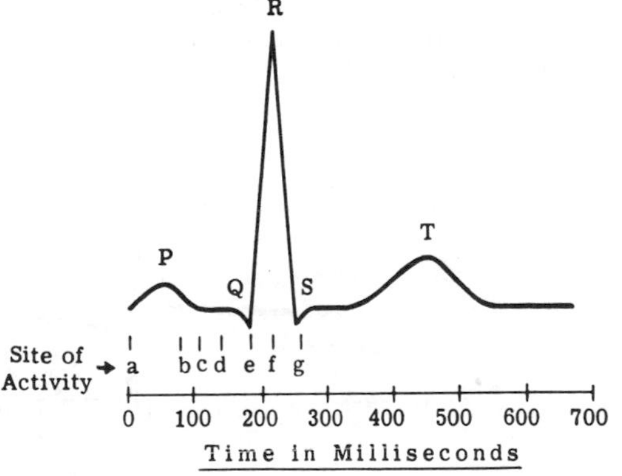

	Site	Time for Excitation (Milliseconds)
a	Sinoatrial node	0
b	Atrioventricular node	66
c	Remote atrial surface	100
d	Bundle of His	130
e	Anterior surface, Rt. ventricle	190
f	Apical surface	220
g	Posterior basal area	260

FIGURE 14-2. Conduction pathways of the heart. The times, in milliseconds, following pacemaker stimulation required for excitation to occur are shown for various regions. An electrocardiogram is shown correlated with the conduction sequence.

*A myogenic property of certain cardiac cells defined as an inherent ability to develop spontaneous depolarization. Such cells are also called pacemaker cells.

initiated in the vicinity of the SA node. The velocity of conduction through pacemaker cells is difficult to measure since it increases with distance from the primary pacemaker cell. In the rabbit heart, the earliest activity was detected in the wall of the superior vena cava, a few millimeters away from the crista terminalis, and the activity propagated radially at an extremely low velocity of 0.05 meter per second. As propagation approached the crista terminalis, conduction velocity became increasingly faster. In fact, activity from the crista terminalis spreads into the atrial roof and down to the coronary sinus at velocities ranging from 0.5 to 1.0 meter per second. From the atria, activity spreads through the atrioventricular node, where after some delay the sequence of activation is through the His bundle, bundle branches, peripheral Purkinje fibers, and then ventricular tissue proper.

Of great significance is the delay in the spread of excitation which occurs in the atrioventricular region. Microelectrode studies of rabbit heart have aided immensely in the elucidation of the mechanism of nodal delay. From a functional point of view, the AV nodal region can be divided into an AN (atrionodal), N (nodal), and NH (nodal-His) zone. Activity approaches the AN boundary region at right angles, and propagation becomes slower. Associated with the reduced conduction velocity is a decrease in the rise time of the action potential when the excitatory wave reaches the narrow, middle N zone; propagation is as slow as that seen in the SA pacemaker (0.05 meter per second). The upstroke of the action potential is slow, and the propagation time consumes about 30 milliseconds of the total AV delay. After the N zone is passed and activity enters the NH zone (Fig. 14-3A), conduction velocity increases (1.0 to 1.5 meters per second) and the action potential starts to assume a configuration almost identical to that seen in the area of the bundle of His (Fig. 14-3B).

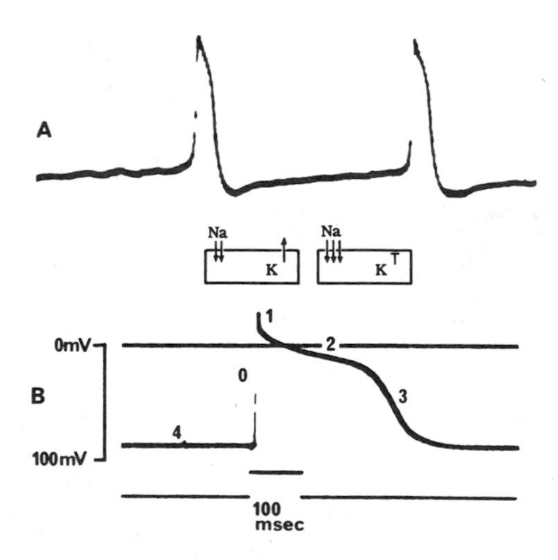

FIGURE 14-3. A. Record of action potentials obtained from rabbit AV node (NH region). Note the appearance of a spike similar to that in a His-type action potential, but the period of diastole (phase 4) shows an indication of continuous spontaneous depolarization. It may, however, represent a period of hyperpolarization that tends to level off. See text for discussion. Below the potentials are the possible ionic fluxes (decrease in K^+ conductance with an increasing Na^+ influx) that could account for the continuous depolarization. B. Action potential recorded from the bundle of His area. Note that there is no spontaneous self-depolarization; this activity requires a stimulus. The action potential is characterized by a rapid upstroke velocity, a sharp phase 1, and a prolonged plateau (phase 2).

It has been postulated that conduction through the AN and NH regions is all-or-none, while the mechanism of propagation within the N layer node is one of decremental conduction. Decremental conduction means that, whatever the cause, there is a gradual decrease in the rising velocity and a gradual diminution in the amplitude of the action potential as the activity propagates over a given path. Associated with these phenomena is a decrease in conduction velocity. Accordingly, the N layer is the region where AV block is most likely to occur, since it is in this zone that conduction velocity is at its minimum. If decremental conduction in the N zone is enhanced, as with acetylcholine (ACh) or vagal stimulation, the action potential amplitude attained may not be sufficient to excite the next zone and the propagation wave will not pass to the bundle of His. The result is AV block.

Thus the potential "dam" for excitation lies in the fact that conduction through the N layer is decremental. However, once excitation passes this zone with a high enough amplitude, the wave will reach the bundle of His. As the activity enters the His area and the bundle branches, there is again an increase in conduction velocity (to 3–4 meters per second) attributable to the greater diameter of these fibers and their infrequent branching. Action potentials from these cells are characterized by a rapid upstroke, high amplitude, and long duration (Fig. 14-3B). Conduction then slows down (to 1 meter per second) as activity enters the ventricular muscle (Fig. 14-1B) presumably because of the smaller fiber diameter and frequent branching.

Generally, impulse transmission in cardiac tissue occurs at different velocities in the various parts of the heart. In any given fiber, however, changes in conduction velocity can be attributed to changes in the resting potential, threshold potential, rate of depolarization, action potential duration, degree of repolarization prior to the arrival of the next impulse, or changes in the core conductor properties of that tissue.

Automaticity in the cardiac cell is demonstrated by a slow continuing depolarization during the diastolic phase (no isoelectric period is observed during phase 4). It can be demonstrated that such automatic cells reside in the SA node (Fig. 14-1D), within the atrium, and in the His-Purkinje system. True atrial and ventricular cells apparently do not meet the electrophysiological and pharmacological criteria for automaticity and hence are not considered as being composed of automatic cells. The cells of the AV node, however, require special consideration. At first glance these cells seem to show spontaneous depolarization (Fig. 14-3A). However, Hoffman and Cranefield (1964) have stated that the AV node does not contain automatic fibers. They base this opinion on the observation that following phase 3 there is a phase of hyperpolarization which rapidly returns to an isoelectric level and remains steady for the duration of phase 4. In addition, a foot appears just prior to the AV nodal action potential upstroke. During rapid heart rates this foot of the nodal action potential merges with the early hyperpolarization, giving the appearance of diastolic depolarization (Fig. 14-3A). Thus, the so-called clinical nodal rhythms probably originate from the transitional fibers located between the NH area and the His area, or from the bundle of His itself. These areas possess automatic cells, and when one of the areas serves as the ventricular pacemaker, electrocardiographic interpretation would be a nodal rhythm.

IONIC BASIS OF ELECTRICAL ACTIVITY

With some reservations, the ionic theory as postulated for the squid axon can be applied to cardiac cells (see Chap. 1). The theory rests essentially upon the concept that any viable membrane possessing the property of excitability can and will alter its permeability to several ionic species. A number of investigators have shown that the intracellular K^+ concentration is 30 to 40 times as much as is present extracellularly. The reverse situation exists for the Na^+ ion. It has also been observed that the membrane at rest is much more permeable to K^+ than to Na^+. Existence of this selective permeability, plus the uneven distribution of these ions across the membrane, allows the cell to act as a battery. The uneven ionic distribution is probably attained through a metabolism-dependent "pump," which actively shifts Na^+ out of the cell interior while transferring the K^+ ion inside.

Two factors appear to be responsible for, and determine the values of, the cardiac transmembrane resting and action potentials. They are the concentration gradients of K^+ and Na, and the relative membrane permeability of these two cations. The equation depicting the contribution of these factors is presented in Chapter 1 (see equation 13). To comprehend the equation fully, one must constantly keep the aspect of permeability in mind. Thus, when the cardiac cell is not excited but at rest, its membrane is relatively more permeable to K^+, and so the transmembrane potential is dominated by the K^+ gradient. With activity, as the heart is depolarized, the permeability to Na^+ increases, permitting some Na^+ to enter the cell and causing further depolarization. This in turn lets the membrane permeability to Na^+ increase further, thus facilitating a greater influx of Na^+. Finally, Na^+ permeability becomes greater than the K^+ permeability, and the membrane potential is hereby determined by the Na^+ gradient. At the peak of the action potential, the membrane starts the restorative process, which involves again a reversal of its permeability to Na^+ and K^+. The first step is a decrease in Na^+ permeability which is known as *inactivation* (P_{Na} begins to fall toward its resting value), and soon after P_K begins to rise. Both these changes allow the K^+ gradient to dominate, and thus determine the transmembrane potential responsible for repolarization. Membrane conductance during the plateau, or phase 2, of the ventricular and His-Purkinje action potentials remains undefined. It has been suggested, however, that during the plateau P_K and P_{Na} are equal, resulting in a temporary maintenance of the potential at or near the zero level.

The ionic theory briefly discussed above postulates that there is an ionic basis for the potential differences existing across the cardiac cell membrane under conditions of rest and activity. In order for a specific ion at any given moment of time to be responsible for a potential difference, it must build up an electromotive force. This is possible, given two conditions:

1. An uneven distribution of that ion (i.e., causing a gradient to exist across the membrane).
2. Greater permeability for that ion than for any other ion at that given moment of time.

For the heart, as for other tissues, evidence indicates that during diastole and during repolarization the specific ion responsible for the potential is K^+. At these times the cardiac membrane permeability is much higher for this cation than for any other positively charged ion. Similarly, during the depolarization the Na^+ ion determines membrane potential. Hence changes in extracellular ionic concentration of K^+ should reflect changes in the amplitude, configuration, and duration of the monophasic action potential. Concomitant changes should occur on the complex of the electrocardiogram. Furthermore, any agent, drug, or hormone which affects membrane permeability will alter its excitability.

EFFECT OF IONS ON THE RESTING POTENTIAL

A prominent effect of increasing the extracellular K^+ concentration is a reduction in the resting potential; the magnitude of the reduction bears an almost linear relation to the amount of K^+ added to the perfusion solution and can be approximately predicted by the Nernst equation (Chap. 1). The depolarization effect of high K^+ concentration has been described for all fiber types of the heart in many species. Structures studied include cat and rabbit atria, and dog SA node, atria, ventricle, and Purkinje fibers, as well as the hearts of cold-blooded animals (frog and turtle). In all these fiber types the initial effect of the reduction of the resting potential by increasing K^+ concentration is an enhancement of excitability. The reason is that the resting potential encroaches upon the threshold potential level so that the stimulus requirement is lower. However, further increase in extracellular K^+ results in a loss of excitability, presumably through inactivation.

An interesting observation was made in relation to the K^+ and Ca^{++} ratio. This ratio appears to be an important component of cardiac function; a similar interaction between K^+ and Ca^{++} exists for the resting potential. Thus, the depolarizing effect of a high concentration of extracellular K^+ can be counteracted by simultaneously raising that of extracellular Ca^{++}, and augmented by lowering the Ca^{++} concentration. With normal K^+ levels the alteration of Ca^{++} to high or low levels does not affect the resting potential. This concept of a K^+ and Ca^{++} antagonism in cardiac tissue is consistent with that observed in other viable cells.

Paradoxically, lowering the extracellular concentration of K^+ results in a resting potential change similar to that of high K^+ concentration. However, this occurs only in mammalian hearts, since in cold-blooded animals the response to severe reduction of extracellular K^+ concentration (0 to 25 percent of normal) is a marked increase in the resting potential, although an increased resting potential in the isolated rabbit heart, perfused with a solution of low K^+ concentration, has been reported. In any event, the depolarization initiated with a low concentration of K^+ in mammalian hearts can be aided by simultaneously raising extracellular Ca^{++} concentration whereas lowering the concentration of extracellular Ca^{++} in the face of low K^+ concentration will restore the resting potential to its normal value. In fact, it has been observed that ventricular fibrillation induced with a low concentration of K^+ can be terminated if the low-concentrate K^+ perfusate is replaced with a solution low in both K^+ and Ca^{++}.

EFFECTS OF IONS ON THE ACTION POTENTIAL – NONAUTOMATIC CELLS

The change in the resting potential due to an alteration in extracellular K^+ concentration would be expected to affect the contour, amplitude, and duration of the action potential. Thus, following exposure to a high concentration of K^+ there is a reduction in maximal upstroke velocity (Fig. 14-4), amplitude, and action potential duration. These effects have been noted for atria, Purkinje fiber, and ventricular tissue of both mammalian and cold-blooded animals. They may result from the reduced resting potential or perhaps they are due to a direct effect of the added K^+.

The reduced upstroke velocity of phase 0 and the decreased amplitude are apparently indirect effects secondary to the reduced resting potential; the observation was made that if the resting potential is restored, as by anodal current, the effect of high-concentrate K^+ on phase 0 and upstroke amplitude is reversed. However, the enhanced repolarization induced by high K^+ concentration is not completely counteracted by anodal current. It appears to be a direct effect of the cation: Infusion of K^+ during the plateau period (phase 2) of the ventricular action potential causes an immediate repolarization. Similar infusion during diastole results only in depolarization without affecting the time course of the action potential. Thus, shortening of the action potential due to high K^+ concentration can occur independently of a reduction in resting potential, and the action of the cation is a direct effect, not

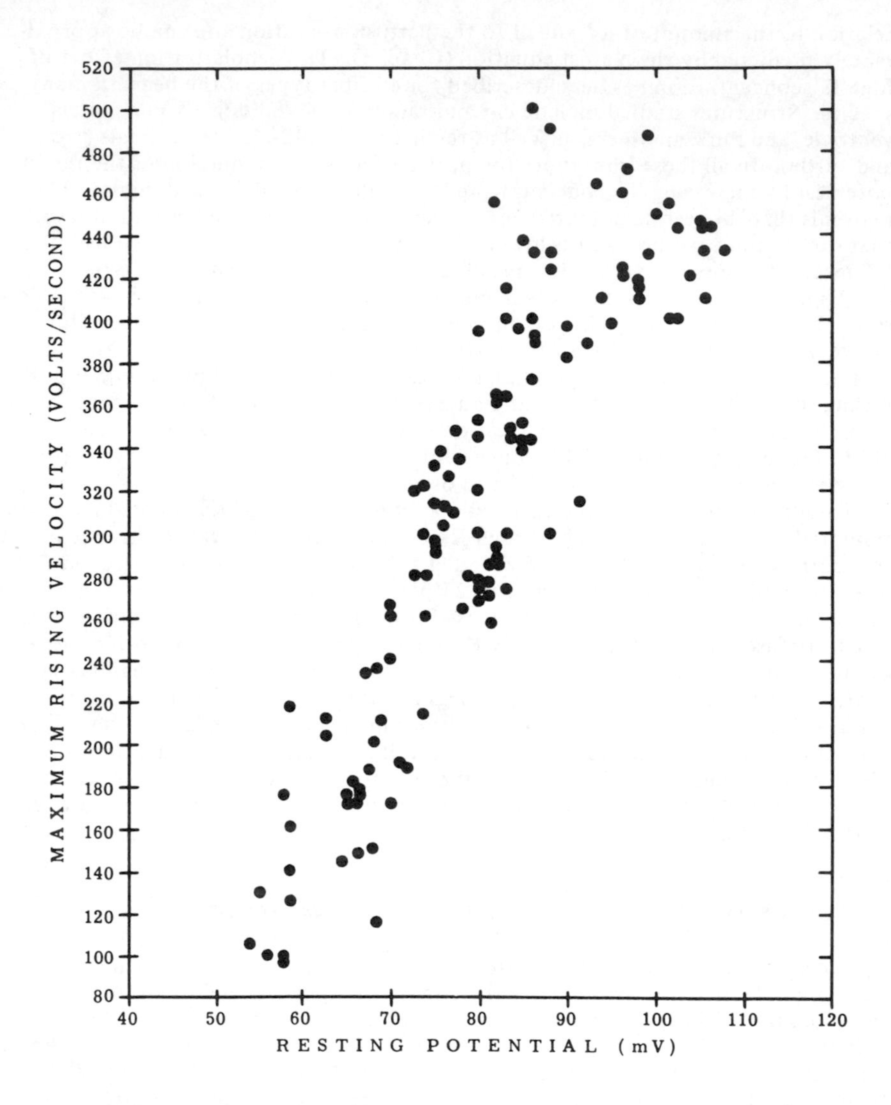

FIGURE 14-4. Relationship between the transmembrane resting potential and the maximum rising velocity in canine Purkinje cells. As the potential is reduced, the maximal rate of rise decreases. The resting potential in the experiment was reduced by exposing the tissue to high concentration of K^+.

a consequence of depolarization. Since during the normal period of repolarization there is an increase in the efflux of K^+, it would seem that the K^+ itself increased K^+ efflux from the cell, presumably by enhancing membrane permeability. Evidence has been advanced that an increase in extracellular K^+ concentration decreases membrane resistance. In fact, this effect is much more pronounced at a membrane potential of -40 mv, the potential level at which the plateau region for sheep Purkinje fiber occurs. Furthermore, both influx and efflux rates of K^+ increase with increasing

K^+ concentrations. The above results are highly suggestive that K^+ itself increases membrane conductance or permeability for K^+.

Perfusion of cardiac cells with solutions containing low K^+ concentration affects the action potential configuration in a manner opposite to that seen with high K^+. Thus, with low K^+ concentration the atrial and ventricular action potentials show an increased amplitude and a prolonged duration. Interestingly, continued perfusion with a low-concentrate K^+ solution results in supraventricular and ventricular ectopic beats, ectopic tachycardias, and finally ventricular fibrillation, which are not terminated by replacement in the normal K^+ solution. However, if the low-concentrate K^+ solution is replaced with a control solution prior to the onset of fibrillation, the result is an immediate cardiac standstill lasting 2 to 30 seconds. Subsequently, the heart starts to beat slowly and within two minutes the rate returns to control values. The interesting electrophysiological observation is that, while in the low-concentrate K^+ solution, the ventricular action potential amplitude and duration are increased. Upon the return to the control solution, however, the action potential amplitude is decreased and the duration is shortened, thereby simulating a "high" K^+ effect. In fact, the amplitude of the action potential after the standstill period levels off at a value intermediate between the low-concentrate K^+ solution and the return to the control solution (i.e., smaller than when in the low-concentrate K^+ solution but higher than the original control amplitude).

Since the phase of depolarization is determined by the movement of Na^+, the changes in action potential configuration can be predicted following changes in concentration of this ion in the extracellular components. Thus, a *reduction* in Na^+ concentration results in a decreased upstroke velocity and a decrease in the amplitude of the atrial action potential. Severe reduction of this ion (to 10 percent of normal) leads to a complete loss of excitability. The ventricular muscle behaves somewhat differently in a low Na^+ medium in that phase 0 is not affected, although the duration is shortened on account of an increase in the steepness of phase 2. *Increasing* the Na^+ concentration has little direct effect on action potential configuration. An indirect effect may be observed as a result of changes in tonicity with the excess Na^+.

Calcium has little effect on the membrane potential and amplitude of the action potential. However, the duration of the action potential is increased in a low-concentrate Ca^{++} medium, whereas it is markedly shortened with Ca^{++} excess. In fact, the ventricular action potential following Ca^{++} excess may resemble an atrial action potential. On the other hand, atrial excitability (unlike ventricular) is notably decreased by Ca^{++} depletion.

EFFECT OF TEMPERATURE ON THE ACTION POTENTIAL OF NONAUTOMATIC CELLS

The electrophysiological properties of atrial and ventricular cells are strongly affected by changes in temperature. Temperature affects the action potential phases directly and indirectly via change in heart rate. With hyperthermia, rate is increased and action potential duration decreased. With cooling, depending upon the severity, the heart rate is decreased, the action potential duration markedly prolonged, and the membrane potential reduced; these changes may ultimately lead to a complete loss of excitability. Temperature change can produce the above alterations directly and independently of rate.

FACTORS AFFECTING RATE AND RHYTHM – AUTOMATIC CELLS

The beat of the heart occurs in an extremely regular, continuous, and rhythmic fashion. It is a result of automaticity, a property present in those cells of the heart which exhibit the electrophysiological pattern of slow diastolic depolarization (Fig. 14-1D). This phenomenon confers upon the cells the ability to depolarize spontaneously, independently of the extrinsic and intrinsic nerve supply. The cells of the heart which exhibit automaticity with the most rapid rate of depolarization, and which are responsible for depolarizing the rest of the heart, are located in the sinoatrial node. Although myogenic in origin, the rate and rhythm may change spontaneously, or they may be influenced by temperature, inorganic ions, and neural and humoral activity. Any factor which changes the rate and rhythm of the heart does so by affecting either (1) the magnitude of the resting potential, (2) the threshold potential voltage which must be attained, and/or (3) the slope of diastolic depolarization. Study of Figure 14-5 reveals that any alteration either of these potentials or of the slope of depolarization alters the time required for reaching threshold and, hence, the rate of stimulus production.

EFFECT OF ACh AND THE VAGUS

Stimulation of the vagus or administration of ACh reduces the heart rate (bradycardia) and, with sufficient intensity or concentration, may lead to sinus arrest. The effects on the automatic cell of such maneuvers are as follows: (1) Repolarization is enhanced. (2) Hyperpolarization takes place: In the lower diagram of Figure 14-5 the potential goes from *a* to *d*. The magnitude of the hyperpolarization depends upon the initial resting potential value; the smaller the resting potential, the greater the

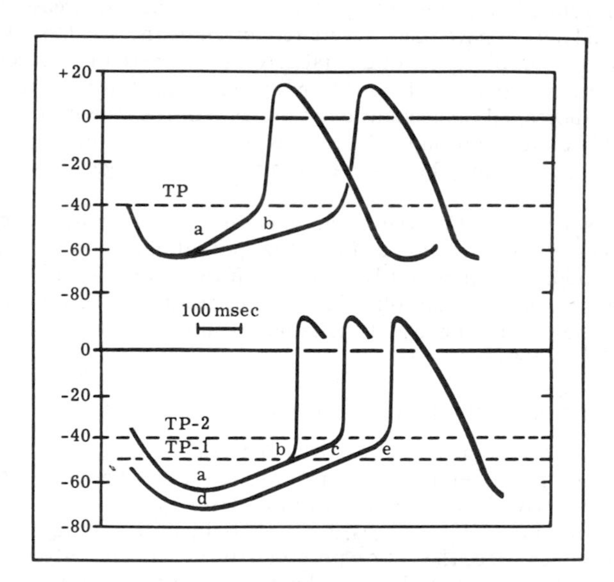

FIGURE 14-5. Diagram representing the important mechanisms responsible for changes in frequency of automatic fiber. The upper diagram shows a decrease in rate caused by a decrease in the slope of phase 4 from a to b and thus an increase in the time required for the transmembrane potential to reach the threshold potential (TP). The lower diagram shows rate changes associated with a change in the level of the threshold potential from TP-1 to TP-2 and an increase in cycle length from a–b to a–c; also shown is a change in rate due to an increase in resting potential (compare a–c and d–e). (From *Electrophysiology of the Heart,* by B. F. Hoffman and P. Cranefield. Copyright 1960, McGraw-Hill Book Company. Used with permission of McGraw-Hill Book Company.)

hyperpolarization produced by ACh. (3) There is a shift of the slope of diastolic depolarization: In the upper diagram of Figure 14-5 the slope goes from *a* to *b*. The net effect of cholinergic activity is an increase in the time required for the fiber to reach its threshold potential. The most prominent effect of cholinergic activity is on phase 4 depression; if severe enough, it can result in cardiac standstill. The mode of action of ACh appears to be via an increase in K^+ permeability, which in turn reduces the loss of diastolic membrane potential and thus leads to a reduction in rate (Fig. 14-6A).

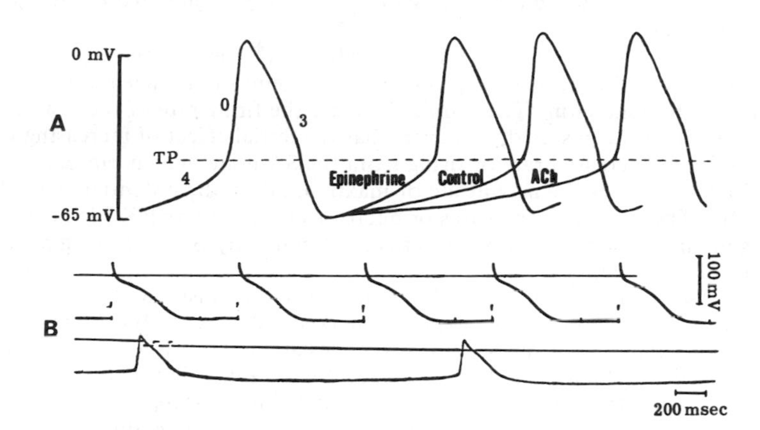

FIGURE 14-6. A. Epinephrine or sympathetic stimulation increases heart rate by enhancing the slope of diastolic depolarization while parasympathomimetic activity or ACh depresses the steepness of phase 4 depolarization to result in slowing of heart rate. B. Purkinje fiber action potentials (upper strip), which are converted to pacemaker type (lower strip) with influence of epinephrine.

CATECHOLAMINES AND SYMPATHETIC NERVE

The administration of epinephrine or norepinephrine, as well as stimulation of the cardioaccelerator nerve, increases the heart rate (tachycardia). The most striking effect is on the slope of phase 4 (Fig. 14-6; Fig. 14-5, in upper diagram *b* to *a*). On latent pacemakers, such as Purkinje fibers, sympathetic activity induces automaticity (Fig. 14-6B) and may lead to multifocal firing and ventricular fibrillation.

TEMPERATURE

Cooling of the SA node reduces spontaneous activity (depression of phase 4), and severe cooling may result in complete standstill. Increasing temperatures increase the slope of diastolic depolarization (Fig. 14-6A, similar to that seen with epinephrine). High temperature may also induce Purkinje fiber to become pacemaker cells (as is also seen with catecholamines – Fig. 14-6B).

EFFECT OF IONS ON AUTOMATIC CELLS

Very few studies have been reported on the effect of K^+ upon the electrophysiological properties of cells possessing the property of automaticity. From the available information it appears that a decrease in extracellular K^+ causes an increase in the slope of diastolic depolarization (probably due to a decrease in P_K), a decrease in the maximum diastolic potential, and a reduction in threshold. (See Fig. 14-5, lower diagram; Low K^+ results in a shift of *d* to *a*; TP-2 to TP-1. In upper diagram slope *b* shifts to slope *a*. See also Fig. 14-1D for ionic fluxes which may be exaggerated with a low K^+ environment.) All of these accelerate activity in the normally automatic cells of the heart. SA node tissue perfused with a K^+-free solution develops multifocal activity. In normally quiescent but latent automatic fibers (Purkinje system), low K^+ concentration tends to initiate pacemaker activity and frequently leads to the production of extrasystoles. Eventually, however, such fibers exposed to an extracellular environment lacking in K^+ will become arrested because of loss of resting potential.

On the other hand, an increase in extracellular K^+ has an opposite effect. The steepness of phase 4 is decreased, probably on account of an increase in membrane permeability to potassium. This would decrease the firing rate of the normal automatic cells. It has been stated, however, that the initial effect of increasing extracellular K^+ is an increased rate of firing from the SA node, which can occur since high K^+ also decreases the magnitude of maximum diastolic potential of the SA nodal cells. The firing rate increases or decreases following high K^+ exposure depending upon whether the cation effect is predominantly on the resting potential or on the diastolic depolarization.

Generally, K^+ affects all the cardiac cells in a similar fashion. However, the sensitivity of the cardiac cells to the cation is markedly different. Within the same heart it appears that the cells of the SA and AV nodes are more resistant to K^+ than are the contractile cells of the myocardium. In turn, the Purkinje fiber network is more sensitive to K^+ than ventricular cells, while the bundle of His is highly resistant to increases of extracellular K^+. The sensitivity to K^+ also differs within the contractile tissue in that conduction fails earlier in the atrial fibers than in ventricular fibers. Of interest is the high resistance of the SA nodal cells to K^+. Increasing the perfusing solution with K^+ to five times greater than normal results in complete suppression of atrial activity, whereas the SA nodal cells show only a slight decrease in the magnitude of the resting potential and in the slope of diastolic depolarization. Only when the K^+ concentration is markedly increased (to 15 times normal) will the spontaneous activity of the pacemaker become arrested.

Other ions, such as Ca^{++}, Na^+, and Cl^-, have little effect on the automatic cells of the SA node unless the ionic changes are severe.

THE ELECTROCARDIOGRAM

In any fundamental consideration of the constitution of the electrocardiogram (ECG) due to differences in the basic electrical properties of different cardiac tissue, the registration of the electrocardiographic complex consists of various wave forms. A representative complex is illustrated in Figure 14-7A. The P wave is constituted by the sum of atrial depolarization. The QRS complex reflects ventricular depolarization. The T wave represents ventricular repolarization.

The relative magnitude of these electrocardiographic components reflects the relative mass of tissue represented. For example, the ventricular mass, which is responsible for the relatively large QRS complex, comprises by far the greatest bulk of the heart. The specialized conducting cells and pacemaker cells (SA node, specialized atrial fibers, AV junctional cells, and His-Purkinje cells) are not graphically registered in the conventional ECG because of relatively small total mass and their distance from the conventional recording electrodes. Atrial repolarization is infrequently observed as it usually occurs simultaneously with ventricular depolarization, the wave form of the latter obscuring the atrial T wave.

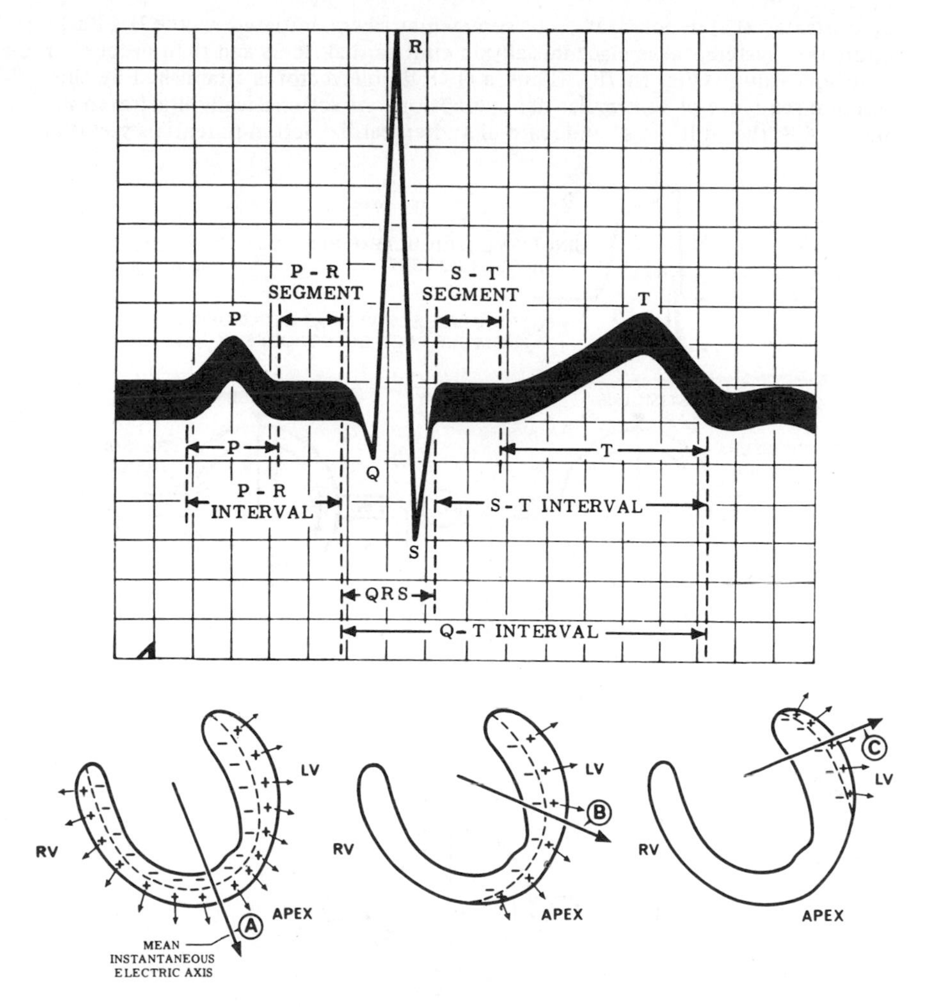

FIGURE 14-7. (Upper) The normal electrocardiogram, showing the time relationships. The distance between the vertical lines is 40 msec. The distance between the horizontal lines is 0.1 mv. (From G. E. Burch and T. Winsor. *A Primer of Electrocardiography,* 5th ed. Philadelphia: Lea & Febiger, 1966.) (Lower) A. The mean instantaneous electrical axis at the beginning of ventricular depolarization. B. The mean instantaneous electrical axis at a later stage of ventricular depolarization. The vector is more to the left and with excitation proceeding toward the epicardium. C. The mean instantaneous axis late in ventricular depolarization. LV = left ventricle; RV = right ventricle. (From Burch and Winsor, 1966.)

Nonetheless, the surface ECG does not closely resemble the single fiber action potential, the summated action potentials of which it is comprised. The source of such a marked disparity becomes evident, however, when one considers that the surface ECG represents an algebraic sum of numerous action potentials, oriented in many directions and happening serially as well as simultaneously. If depolarization and repolarization occurred simultaneously in all fibers, the net electrical difference between two recording electrodes would be zero. But it has been amply demonstrated that depolarization of ventricular fibers, initiated via the His-Purkinje conduction system, takes place initially in endocardial fibers and then proceeds radially to the epicardium (Fig. 14-7B). Thus, a QRS dipole vector is established by this temporal sequence of ventricular depolarization. As schematically illustrated in Figure 14-8, the addition of endocardial and epicardial action potentials yields an

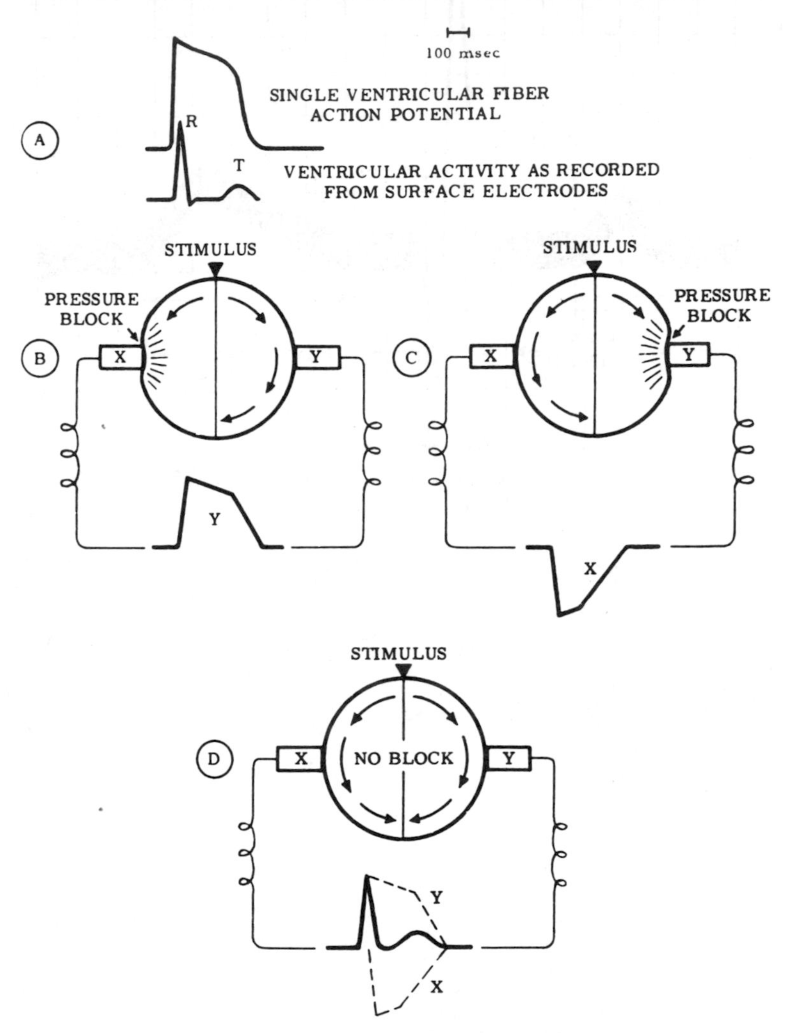

FIGURE 14-8. A. The action potential of a single ventricular fiber and the electrical activity of the entire ventricle are compared. B. Applying pressure to a hypothetical myocardium with an electrode (X) blocks the conduction of the excitation, and this electrode serves as an indifferent electrode. A monophasic action potential is then recorded under the other electrode (Y). C. The situation is reversed and the direction of the deflection is reversed. If electrodes X and Y were symmetrically influenced by the spread of excitation, no deflection would be recorded, as the upward and downward deflections would cancel each other. D. If some asymmetry existed, the recorded deflection would represent the algebraic sum of the two monophasic potentials. The X electrode is recording from the epicardium and the Y electrode from the endocardium. (From K. Jochim, L. N. Katz, and W. Mayne. *Amer. J. Physiol.* 111:177, 1935.)

isoelectric S-T segment. It would seem reasonable, as well, to expect the T wave to be of equal amplitude and opposite direction to that of the QRS if repolarization occurred in the same sequence as depolarization. As shown in this diagram, however, more rapid repolarization of the epicardium than of the endocardium yields an upright T wave in the presence of an upright QRS complex.

The most conventional method of recording this summation of action potentials involves the use of the three standard leads, as conceived by Einthoven in 1908. Viewing the heart as an electrical dipole situated in the center of an equilateral triangle (see Fig. 14-9), one can perceive the derivation of the three standard leads. Lead I is recorded between the right and left arms, producing an upward deflection if the left arm is positive with respect to the right arm (i.e., when the projection of the dipole vector renders the left arm positive with respect to the right arm). Lead II is recorded between the right arm and left leg, yielding an upward deflection if projection of the cardiac vector upon this lead produces positivity in the left leg. Lead III is recorded between the left arm and left leg, yielding an upward deflection

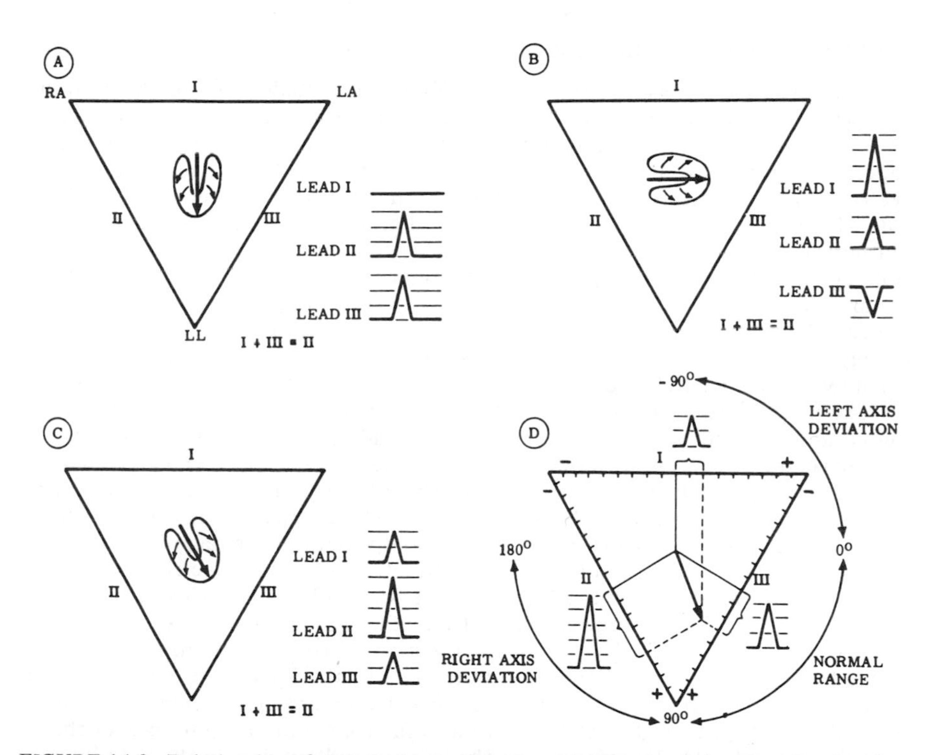

FIGURE 14-9. Relationship of electrical axis of the heart to the recorded potentials. A. The vector of excitation, or axis, is directly vertical downward. The voltage changes seen at each lead are shown. B. The vector of excitation, or axis, is directly horizontal and to the left arm. C. The vector of excitation, or axis, is downward and toward the left. Lead II, which is most parallel to the axis, has the greatest deflection. D. To determine the mean electrical axis from records obtained for any two leads: Count off from the midpoint of each lead in the appropriate direction the number of millimeters of deflection obtained. Drop perpendiculars from each point. The line drawn from the midpoint of the triangle to the point of intersection of the perpendiculars represents the axis. The arrow head represents its direction, the length of the line represents its magnitude, and the degrees represent its orientation.

if the left leg is positive with respect to the left arm. As Einthoven's triangle constitutes an electrical network, physical principles require that the sum of all electrical forces equal zero. Thus, Einthoven's law states that lead I voltage plus lead III voltage equals lead II voltage at any instant of the cardiac cycle. (Einthoven's modification of this basic principle involved a reversal of lead II polarity; thus, lead I + lead III = lead II.)

As implied previously, the summation of cardiac action potentials yields a net or resultant vector, with force and direction. Since the three standard leads reflect the projection of this cardiac vector upon three separate axes, it is apparent that the cardiac vector can be reconstructed from the standard leads (as shown in Fig. 14-9). The orientation of the QRS axis, in the frontal plane, is usually inferior and somewhat to the left. Change in the mean QRS axis can be produced by mechanical alteration of the cardiac position as well as by a wide variety of cardiac and pulmonary diseases which change the position of the heart. The electrical axis of the P and T waves may be calculated in a similar fashion.

OTHER ELECTROCARDIOGRAPHIC LEADS

In the foregoing discussion of the electrocardiogram, consideration has been given only to the three standard leads. Such a limited frame of reference permits an evaluation of the cardiac vector only in relation to the frontal plane (i.e., laterally and superiorly-inferiorly). It should be emphasized, however, that the mean cardiac vector courses anteriorly, as well as to the left and inferiorly, in the average individual. These anteroposterior forces are projected on the horizontal and sagittal planes but not on the frontal plane. Conventional electrocardiography provides information regarding such an anteroposterior component by recording additional leads over the anterior thorax. Furthermore, in addition to recording an anteroposterior component, the precordial leads can disclose the electrical activity of localized areas of the myocardium, to which they are in closer proximity. This latter capacity is provided by the unipolar character of the precordial electrodes.

The unipolar lead should be contrasted with the bipolar leads by which standard limb lead complexes are recorded. The bipolar lead involves the recording of the cardiac vector projection with reference to two electrodes. Thus, at a given moment in the cardiac cycle the bipolar lead records the difference in electrical potential between two leads. In contrast, the unipolar electrode records the electrical potential variation with respect to itself only; the potential variations are recorded with reference to an indifferent electrode which is unaffected by cardiac electrical potential. The indifferent electrode (Wilson's terminal) is formed by connecting each of the three limb electrodes to a central terminal via a 5000-ohm resistor. According to Einthoven's law, leads I + III – II = 0. Consequently, the central terminal is unaffected by cardiac electrical activity, and the precordial electrode functions as a unipolar exploring electrode. The routine clinical electrocardiogram employs six such precordial leads (V_1–V_6), with V_1 located in the fourth intercostal space at the right parasternal area. V_6 is located at the left midaxillary line, in the fifth intercostal space.

Three augmented extremity leads are also routinely recorded. As in the case of the precordial leads, the extremity leads are unipolar in character. The left arm electrode (aVL) records electrical differences between the left arm and the remaining two electrodes. The other augmented extremity leads (aVF and aVR) are constituted in a comparable fashion. The extremity leads complement the standard limb leads by assisting in the determination of the mean electrical axis. They also help delineate the temporal course of the cardiac vector.

THE NORMAL ECG

As described earlier, the components of the surface electrocardiogram (see Figs. 14-7A and 14-10) represent circumscribed areas of depolarization and repolarization. Emanating from the area of the sinoatrial node, or so-called primary pacemaker, the wave of depolarization first traverses the atria, producing the P wave of the surface electrocardiogram, of 80 to 100 msec duration. The P wave vector is usually oriented toward the left arm and inferiorly, producing an upright deflection in the three standard leads (see Fig. 14-10).

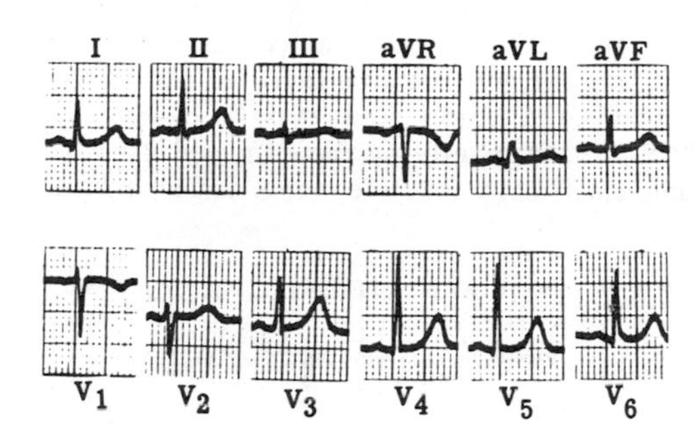

FIGURE 14-10. A normal human electrocardiogram showing a single complex in each of the 12 leads. Note the R wave amplitude progression in the precordial (V_1–V_6) leads.

Following the P wave an isoelectric interval occurs, averaging 80 to 100 msec in duration and reflecting the delay in passage through the atrioventricular junctional area. The QRS complex reflects depolarization of the ventricles in a characteristic sequence. The first vector represents depolarization of the interventricular septal fibers, from left to right, often producing a small Q wave (i.e., an initial downward deflection of the QRS complex) in the three standard leads and V_6, and a small R wave in lead V_1. Subsequently, the principal vector becomes manifest, reflecting depolarization of the bulk of the left ventricle, leftward and inferiorly (along the positive axis of the limb leads). This is inscribed as a tall R wave (i.e., an upward deflection of the QRS complex) in leads I, II, III, and the lateral precordial leads V_4, V_5, and V_6. The precordial lead V_3, anatomically positioned to the right of the bulk of ventricular tissue, records a deep S wave as the wave front is moving away from the electrode. (An S wave refers to a downward deflection following an upward deflection of the QRS complex.)

Terminally, the more basal region of the ventricles becomes depolarized, producing a smaller S wave in leads II and III. The entire course of ventricular depolarization, thus completed, normally consumes 80 to 100 msec.

Following ventricular depolarization, the algebraic sum of ventricular repolarization becomes manifest in the electrocardiogram. The S-T segment corresponds approximately to that portion of the single fiber action potential referred to as the plateau or phase 2 (see Fig. 14-8A and compare to Fig. 14-1). The T wave, in contrast, more closely corresponds to the period of rapid repolarization (phase 3) of the single fiber action potential. As illustrated in Figure 14-1B, the epicardial fibers (the last to depolarize) exhibit more rapid repolarization than do the endocardial fibers. This paradoxical sequence results in the inscription of an upright T wave in the presence of a tall R wave, as previously noted.

CHANGES IN ECG CONFIGURATION

Alterations in repolarization, producing electrocardiographic T wave changes, are the most frequently encountered of ECG abnormalities. While many of these changes are empirically regarded as abnormal, the mechanism by which they occur remains obscure. They are therefore usually referred to as nonspecific T wave changes. Susceptible to a variety of noxious factors, such as ischemia and drugs, or endocrine, metabolic, and electrolyte imbalance, repolarization may be diffusely or locally affected. More rapid repolarization of the endocardium, for example, may result in S-T segment depression and inversion of the T wave (by allowing epicardial repolarization to dominate the summated ECG complex), as can be inferred from Figure 14-8.

Damage to cardiac fibers may impede the course of normal repolarization and so produce a so-called injury current, the area involved remaining electrically negative with respect to the unaffected region. This type of injury results in S-T segment shifts upward in the ECG leads overlying the damaged tissue (Fig. 14-11). As the damaged cells become electrically inactive, the S-T segment subsides. With electrical

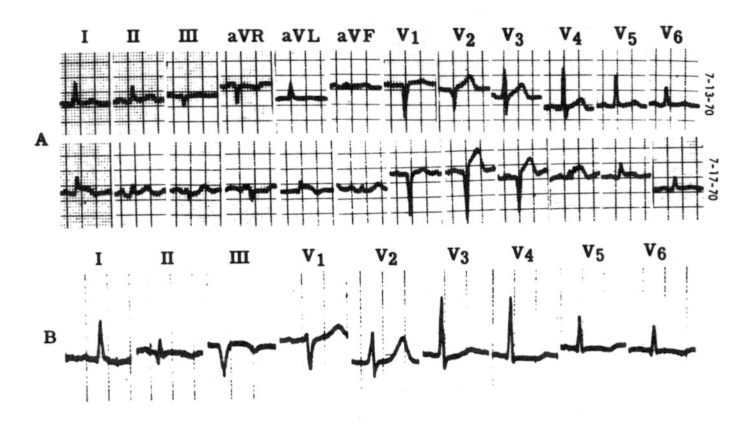

FIGURE 14-11. A. The ECG recorded on 7-13-70 demonstrates a normal ST-T segment configuration. The ECG recorded on 7-17-70 illustrates the progressive ST-T segment elevation (leads I, aVL, V_2, V_3, V_4, and V_5) typical of acute myocardial infarction. B. Healed myocardial infarction is characterized by abnormally wide and deep Q waves (leads II and III).

death of the myocardium, deep wide Q waves develop. Such "electrically silent" areas are manifest by Q waves because the dead myocardium acts as a "window" – the surface electrode records only the residual unopposed forces of depolarization similar to the way an intracavitary electrode would. Thus, the wave of depolarization travels away from the recording electrode. The changes demonstrated in Figure 14-11A are observed clinically in acute myocardial infarction, where the tissue is ischemic but still viable. The development of Q waves correlates with replacement of the dead tissue with fibrous tissue (Fig. 14-11B).

Conduction block within the Purkinje conducting system of the heart produces characteristic alterations of QRS configuration. Such disturbances in conduction may be due to either "functional" (rate dependent, prolonged refractory period, etc.) or pathological lesions of the Purkinje system. This type of injury or lesion causes conduction of the impulse to be delayed as it courses over longer, more circuitous pathways, and via the more slowly conducting ventricular muscle fibers. Lesions of one or the other of the main bundle branches, for example, yield characteristic electrocardiographic patterns, as illustrated in Figure 14-12. Right bundle branch block produces alteration of the QRS complex with widening due to late, unopposed depolarization of the right ventricle, represented by a wide terminal S wave in leads I and V_6 and a secondary R wave (R') in V_1. The left bundle branch block pattern, on the other hand, occurs as the result of a disturbance in the initial course of ventricular depolarization. Normally, septal depolarization takes place from left to right on account of early branching of the left bundle. In the left bundle branch block, however, the septum is depolarized in the opposite direction with loss of the so-called septal q wave frequently found in leads I, II, and V_6. Late depolarization of the left ventricle then produces a widened, notched R wave in these leads.

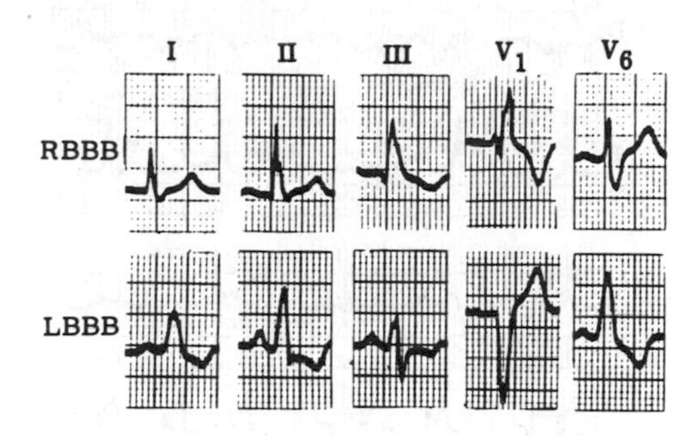

FIGURE 14-12. Right bundle branch block (RBBB) is characterized by right axis deviation, wide S waves in lead I and V_6, and an RSR' in V_1. Left bundle branch block is typified by broad R waves in lead I and V_6 and absence of the small R wave in V_1.

ARRHYTHMIAS

The concept of arrhythmias implies a disruption, transient or prolonged, in the normal sequence of a electrical activation of the heart. The propensity for creating such a disruption is inherent in cardiac tissue, as is evident from a consideration of the electrophysiological properties of various fiber types. As discussed previously, fibers possessing a capacity for automatic behavior are present in virtually every area of the heart. Specialized fibers are found in the atria; lower AV junctional fibers clearly manifest automaticity; and His-Purkinje fibers pervade the ventricular myocardium.

Normally, the more rapid rate of the sinoatrial pacemaker precludes the expression of activity by these more subordinate pacemakers, which have an inherently slower rate of spontaneous diastolic depolarization. Occasionally, however, a secondary pacemaker (ectopic pacemaker*) depolarizes prematurely and at a time when the surrounding tissue is not refractory, and thus may *initiate* depolarization. An impulse originating from specialized atrial tissue is referred to as an atrial premature excitation (APC). Its ventricular counterpart is known as premature ventricular excitation (VPC).† The atrial ectopic beat (Fig. 14-13E) produces a P wave of atrial

*The term *ectopic* refers to an impulse or rhythm initiated by a cardiac pacemaker outside the SA node.

†The terms *APC* and *VPC* have, to date, been used to connote atrial or ventricular premature *contractions*. However, the term *contraction* so used is meaningless when applied to an ECG tracing since the ECG is the recording of the *electrical* events. The latter may or may not lead to the actual mechanical event. Hence *APC* and *VPC* as related to the ECG will here be restricted to defining premature excitations or complexes.

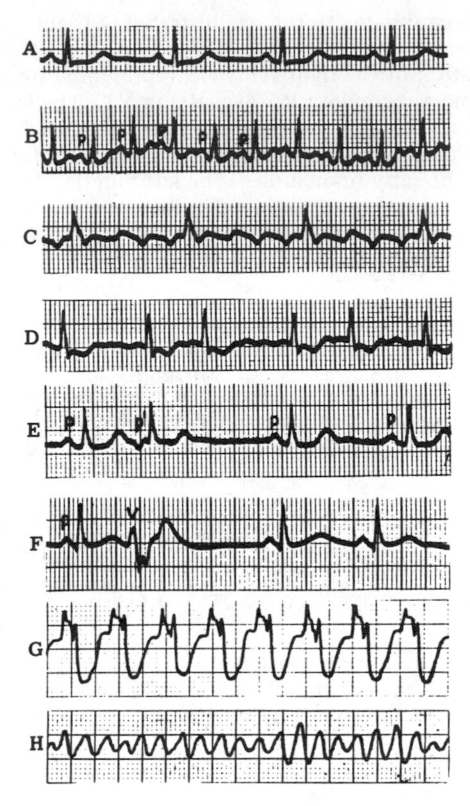

FIGURE 14-13. A. Normal sinus rhythm. B. Rapid atrial tachycardia. C. Atrial flutter with a 4:1 ventricular response. The baseline demonstrates characteristic sawtooth configuration. D. Atrial fibrillation with an irregular ventricular response. The baseline shows no definitive P waves. E. ECG recording of an atrial premature beat (P'). F. Ventricular premature beat (V), followed by a compensatory pause and two normal sinus beats. G. Ventricular tachycardia characterized by wide aberrant QRS complexes and a rapid rate of 150 per minute. H. Ventricular fibrillation characterized by an undulating baseline without definable QRS complexes.

depolarization that may differ in amplitude and configuration from the sinus P; the QRS complex approaches the normal configuration (Fig. 14-13A) since the stimulus traverses the AV node and the normal ventricular conduction system. The premature complex, having depolarized the atrium and SA node region, however, is followed by a slight compensatory pause before reestablishment of sinoatrial excitability. Occasional spontaneous activity may originate from the AV junctional–His bundle area as well, producing similar ventricular excitation, although atrial depolarization here may be obscured by the QRS. Impulses originating from the above area may activate the atrium in retrograde fashion, resulting in a change of the atrial electrical axis, or they may fail to excite the atrium (retrograde).

The premature ventricular complex, originating from one or the other ventricle (Fig. 14-13F), elicits ventricular depolarization over abnormal pathways, producing a widened QRS complex of unusual configuration. This ectopic complex is not ordinarily associated with atrial depolarization; hence the sinoatrial rhythm is not disturbed. The ventricles are refractory to the next sinoatrial impulse, however, and a so-called *fully compensatory pause* results.

Although such occasional ectopic excitations may be observed in normal individuals, under conditions which enhance the automaticity of subordinate atrial pacemaker fibers ectopic activity may increase in rate sufficiently to usurp the function of the SA node. This circumstance may be evident either as frequent atrial premature impulses interrupting the sinus rhythm or as a persistent atrial rhythm, referred to as *atrial tachycardia.* An arrhythmia of this type may vary in rate from 140 to 240 beats per minute (Fig. 14-13B).

Similarly, latent ventricular pacemaker activity may be enhanced to the point of producing ventricular tachycardia (Fig. 14-13G), by usurping SA dominance (i.e., beating more rapidly than the sinus node, and precluding expression of sinus activity). The rate of such a pacemaker may vary from 100 to 150 beats per minute. This electrocardiographic rhythm is characterized by distorted and prolonged QRS (aberrant) waves, not preceded by atrial deflections. Relatively uncommon, ventricular tachycardia is observed primarily in the presence of severe cardiac disease or digitalis intoxication and is associated with deterioration of cardiac output.

Atrial Flutter and Fibrillation

More rapid ectopic pacemaker activity in the atrium may proceed to the point of atrial flutter (Fig. 14-13C). This is characterized by discrete depolarizations occurring at a rate of 200 to 350 per minute. Slow conduction through the AV node and a longer refractory period of this area do not permit a 1:1 ventricular response. Consequently, AV block of varying degree is produced, the ventricles often responding once to each two or four flutter waves. This type of arrhythmia is most commonly observed in rheumatic and coronary heart disease.

Atrial fibrillation (Fig. 14-13D) has an even more rapid (300 to 450 per minute) and, in addition, uncoordinated atrial activation. Atrial contraction is totally ineffective while the ventricular response to such an atrial rhythm is totally irregular (although the ventricular contractions per se are coordinated and effective). The mechanism of arrhythmia is of obscure origin, possibly produced by rapidly firing single or multiple ectopic foci, or by a continuous circus movement* initiated from one focus and perpetuated by a slow conduction velocity (permitting recovery of responsiveness in previously excited areas), a long conduction path, or a short refractory period.

Ventricular Flutter and Fibrillation

Ventricular flutter is an arrhythmia considered by some cardiologists to be a separate entity while others regard it as a phase preceding ventricular fibrillation. Ventricular rhythm during flutter is regular and is characterized by depolarizations that are oscillatory, at a rate of 175 to 250 per minute. The QRS complexes are wide and of large amplitude, and the S-T segment is indistinguishable from the T wave. The ventricular depolarization during flutter is probably from a single ectopic focus. It should be pointed out that ventricular flutter is seldom observed clinically.

Ventricular fibrillation is an arrhythmia (Fig. 14-13H) that is characterized by rapid, uncoordinated ventricular activity and a total irregularity of the QRS complexes. The condition is incompatible with life, because of the ineffectiveness of ventricular contractions, for it is tantamount to ventricular arrest, with total loss

---

*Unidirectional transmission of the impulse traversing in a circuitous manner (reentry) resulting in a continuous reexcitation.

of cardiac output. The mechanism of this type of arrhythmia may be comparable to that underlying atrial fibrillation (although the latter is a *relatively* innocuous rhythm), associated with slowing of conduction velocity and the establishment of a circus movement.

Clinical and experimental experience with exogenous stimulation (i.e., an artificial pacemaker) has provided evidence that ventricular fibrillation may be elicited by a single stimulus at a time when the ventricle is extremely susceptible. This interval is referred to as the vulnerable period and is found toward the end of rapid repolarization (the latter portion of the T wave). These findings have stimulated the development of electrical pacing and defibrillating equipment synchronized to provide an appropriate stimulus well away from the vulnerable period.

CONDUCTION DISTURBANCES

Conduction disturbances may occur in all areas of the heart and may be related to slowing of conduction velocity itself or to an increase in conduction time secondary to a greater tissue mass. The latter results from depolarization traversing a greater mass at the same conduction velocity and may be noted in atrial or ventricular hypertrophy, producing a widening of the P wave or QRS complex respectively.

Conduction disturbances related to a decrease in conduction velocity are perhaps more commonly encountered in clinical practice and most frequently observed in regard to the transmission of depolarizing impulses from atria to ventricles via the AV junctional area. Fibers in the latter region exhibit a resting potential of low magnitude and consequently action potentials of low amplitude. Normally, these low-amplitude action potentials result in slowing of impulse propagation, producing the normal AV conduction delay. This type of conduction, in which action potential amplitude lessens and conduction velocity slows progressively because of successively less adequate action potential generation, is called *decremental conduction.* Thus, when physiological or pathological factors produce a further decrease in action potential amplitude, decremental conduction is enhanced, successive areas of fibers generating lesser and lesser action potentials, to the point of conduction failure. In the latter circumstance the fiber beyond the block remains excitable, but the stimulus is inadequate to effect a response. Illustrations of this phenomenon

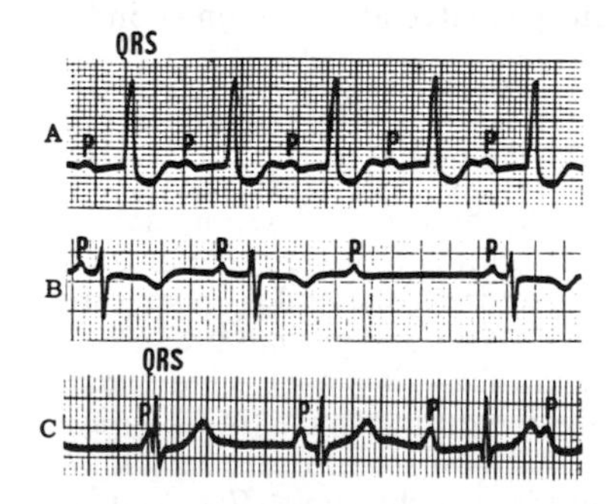

FIGURE 14-14. A. First-degree block characterized by prolonged PR intervals. There are no dropped beats. The maximal normal PR interval is 0.22 second and in this case is 0.31 second. B. Second-degree AV block characterized by a dropped ventricular beat (after the third recorded P). After the dropped QRS complex, PR is normal. C. Third-degree heart block or complete AV block. The P waves and QRS complexes are independent of each other. This is characterized by a simultaneous inscription of the P and QRS complexes of the first beat.

are found in Figure 14-14, in which AV conduction delay is seen to produce an increase in AV conduction time without other disturbance (first-degree AV block), intermittent block of atrial impulses (second-degree AV block), and complete cessation of atrioventricular conduction, with the atria and ventricles beating independently of one another (third-degree AV block).

REFERENCES

Brooks, C. McC., B. F. Hoffman, E. E. Suckling, and O. Orias. *Excitability of the Heart.* New York: Grune & Stratton, 1955.

Burch, G. E., and T. Winsor. *A Primer of Electrocardiography,* 5th ed. Philadelphia: Lea & Febiger, 1966.

Fisch, C., S. B. Knoebel, H. Feigenbaum, and K. Greenspan. Potassium and the monophasic action potential, electrocardiogram, conduction, and arrhythmias. *Progr. Cardiov. Dis.* 8:387–418, 1966.

Fishman, A. P. (Ed.). Symposium on the myocardium. Its biochemistry and biophysics. *Circulation* 24:No. 2, Pt. 2, 1961.

Greenspan, K., and E. Steinmetz. Digitalis and Potassium: Pharmacodynamic Relationship. In L. Dreifus (Ed.), *Mechanisms and Treatment of Cardiac Arrhythmias.* New York: Grune & Stratton, 1966.

Greenspan, K., E. F. Steinmetz, and R. E. Edmands. Clinical and physiological aspects of digitalis and potassium. *Intern. Med. Digest* 1:41–55, 1966.

Hoffman, B. F. Origin of the Heart Beat. In A. A. Luisada (Ed.), *Cardiology.* New York: McGraw-Hill, 1959. Vol. I, Pt. 2, Chap. 4.

Hoffman, B. F., and P. Cranefield. *Electrophysiology of the Heart.* New York: McGraw-Hill, 1960.

Hoffman, B. F., and P. Cranefield. The physiological basis of cardiac arrhythmias. *Amer. J. Med.* 37:670–684, 1964.

Katz, L. N., and A. Pick. *Clinical Electrocardiography.* Philadelphia: Lea & Febiger, 1956.

Lipman, B. S., and E. Massie. *Clinical Scalar Electrocardiography.* Chicago: Year Book, 1959.

Luisada, A. A. *Cardiology Methods.* New York: McGraw-Hill, 1959. Vol. II.

Sodi Pallares, D. S., and R. W. Brancato. The Physiological Basis of the Electrocardiogram. In A. A. Luisada (Ed.), *Cardiology.* New York: McGraw-Hill, 1959. Vol. I, Pt. 2, Chap. 5.

15

Cardiodynamics

Carl F. Rothe

THE FUNCTION of the heart is to propel enough blood into the aorta to maintain the blood pressure at a level sufficient to assure adequate flow of blood to all tissues. To understand these complex relationships, it is necessary to study the components of the system and the relationships between the following variables (Fig. 15-1): *A*. The flow (F) of blood through the capillaries supplying the needs of cells depends upon (1) the difference between the systemic arterial blood pressure and venous pressure (ΔP) and (2) the resistance (R) to the flow. This vascular resistance to flow

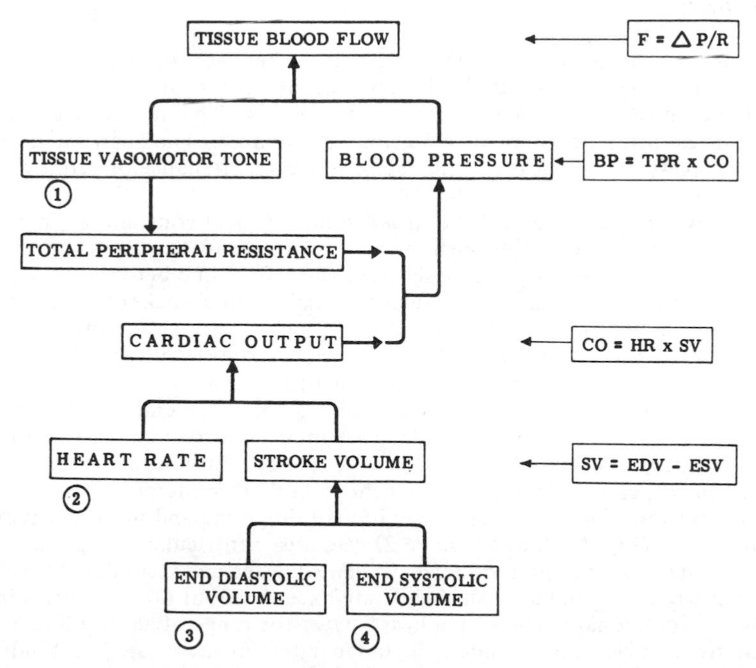

FIGURE 15-1. Cardiovascular system dynamics. The relationships determining tissue blood flow, blood pressure, cardiac output, and stroke volume.

depends on the internal diameter of the arterioles and precapillary sphincters (tissue vasomotor tone), as discussed in Chapters 11, 12, and 17. *B.* The systemic arterial blood pressure is the product of (1) cardiac output (CO) and (2) the summed effects of all these peripheral resistance vessels, i.e., the total peripheral resistance (TPR). (These relationships are analogous to Ohm's law for electrical circuits.) *C.* The cardiac output is the volume of blood pumped by the heart each minute and thus is the product of (1) the volume of each beat, the stroke volume (SV), and (2) the number of heart beats per minute (the heart rate, HR). *D.* The stroke volume is the difference between (1) the volume of blood in the heart at the beginning of systole (the end diastolic volume, EDV) and (2) the amount of blood which remains in the ventricles when the valves close at the end of systole (the end systolic volume, ESV). Four factors, then, are the primary determinants of systemic arterial blood pressure and cardiac output: (1) peripheral resistance to blood flow determined by the tissue vasomotor tone, (2) heart rate, (3) end diastolic volume, and (4) end systolic volume (Fig. 15-1).

These factors are highly related. Some are intrinsic, being dependent on the physical characteristics of cardiac muscle and the vasculature, e.g., the Frank-Starling law of the heart (see below). Other interrelationships are influenced by factors extrinsic to a tissue, e.g., neural and hormonal influences such as those from the pressoreceptor reflex acting by way of the autonomic nervous system on all four variables (Chap. 17). In the normal individual, the fine control of cardiovascular function is neural.

The three factors related to the heart will now be considered in detail. Factors influencing peripheral resistance are considered in Chapters 17 and 18.

HEART RATE

The normal heart rate at rest, in the adult, ranges between 40 and 100 beats per minute and averages about 70. In the newborn, the heart rate is about 135 beats per minute; in the elderly, about 80 beats per minute. The heart of the trained athlete at rest often beats less than 50 times per minute. The heart rate is widely variable, largely because it is a prime factor in the maintenance of cardiovascular homeostasis.

As shown in Figure 15-2, if the stroke volume is held constant, an increase in heart rate causes a directly proportionate increase in cardiac output (curve 1). However, since an increase in heart rate shortens the filling time between beats, the end diastolic volume is decreased. (Consider the analogy of a bucket placed under a running faucet. If the time for filling is reduced by half, only half as much water is collected.) Unless the end systolic volume decreases proportionately to the decrease in end diastolic volume, there will be a consequent decrease in stroke volume. Indeed, at high heart rates, or in a denervated or isolated heart, the end systolic volume does not decrease proportionately to the decrease in end diastolic volume, and the result is a smaller stroke volume (Fig. 15-2, lower curve 2). Without autonomic nervous system control, cardiac performance reaches a peak at moderate heart rates. Any further elevation of heart rate seriously limits filling time and markedly reduces cardiac output (Fig. 15-2, upper curve 2). Because ventricular filling occurs mostly in the first half-second and then tends to plateau after one second, there is little decline in stroke volume until the heart rate exceeds about 60 beats per minute, even in the functionally denervated heart. Over the range of about 60 to 160 beats per minute, cardiac output tends to be independent of heart rate if autonomic nervous system control is blocked, because the stroke volume decreases in proportion to the increase in heart rate above about 60 beats per minute. This limitation of

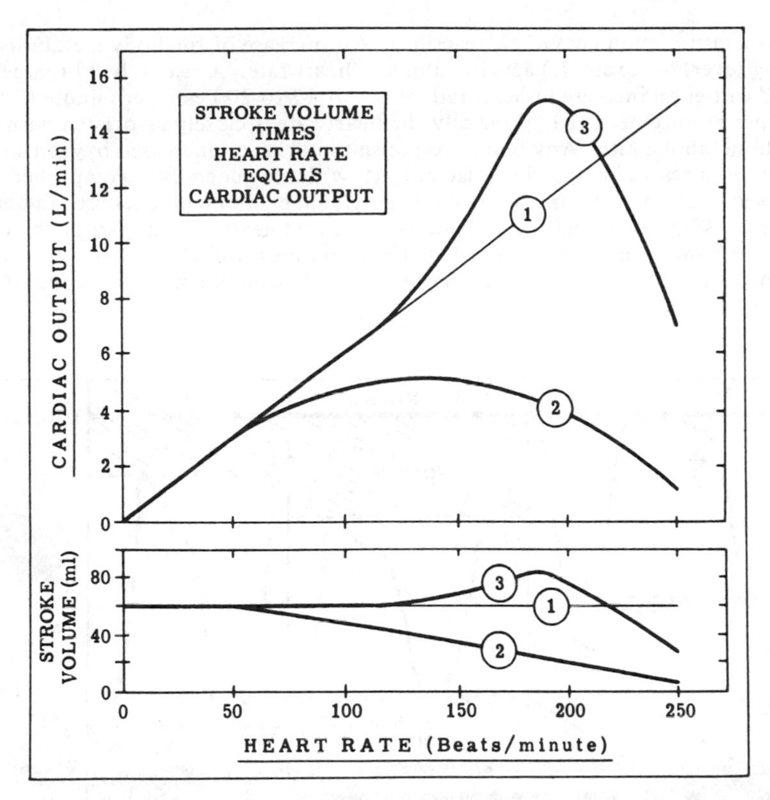

FIGURE 15-2. The effect of heart rate on cardiac output. Since cardiac output is the product of stroke volume and heart rate, if the stroke remained constant, cardiac output would increase linearly with increases in heart rate (curve 1). If the stroke volume declines because of inadequate filling time, there is a peak output and then a decrease (curve 2). As in the normal individual, if the stroke volume is held constant or is increased by compensatory mechanisms such as increased contractility, the output increases to a higher peak before filling time becomes limiting (curve 3).

cardiac output occurs at heart rates only slightly greater than normal if the sympathetic division action on the heart is limited by disease, nerve damage, or drugs. However, under normal circumstances, other mechanisms, mediated primarily by the sympathetic division, act to maintain the stroke volume at a constant level or even increase it (curve 3), e.g., increased vigor of myocardial contraction acting to decrease end systolic volume and thus increase stroke volume. Thus, in the normal individual, the output of the heart per minute tends to increase proportionately as the heart rate increases.

The maximum effective heart rate in man is about 180 beats per minute. Severe *tachycardia,* as seen in atrial fibrillation, is also associated with a decrease in cardiac output, because in this case the time for filling is severely limited by the abnormally high heart rate. In *bradycardia,* following complete heart block, the stroke volume may be large, but the cardiac output is severely limited because of the very low heart rate.

Cooling of the pacemaker cells (during surgical hypothermia, for example) causes a profound reduction in heart rate. This might be expected from the marked effect of temperature upon enzymatic reactions. An increase of the body metabolism (during fever, for example) acts to stimulate heart rate. A rise of body temperature of 1°C causes an increase in heart rate of about 12 to 20 beats per minute (7 to 11 beats per minute per 1°F). Generally the heart rate is closely associated with the overall metabolic rate. Any bodily response requiring an increased oxygen supply usually requires an increased cardiac output, which is generally accomplished by an increased heart rate. Examples include emotional excitement, exercise, and eating. The act of standing upright also elicits an increased heart rate, as part of the homeostatic mechanism maintaining an adequate blood pressure (Chap. 17).

The heart rate is controlled primarily by the autonomic nervous system (Chaps. 7 and 17).

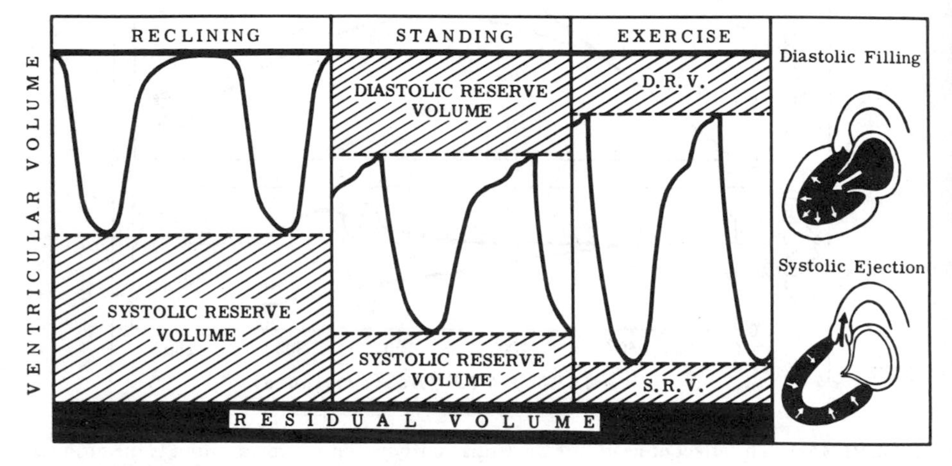

FIGURE 15-3. Ventricular volume under various conditions. (Modified from Rushmer, 1970, p. 512; and from R. F. Rushmer. *Mod. Conc. Cardiovasc. Dis.* 27:473, 1958. By permission of the American Heart Association, Inc.)

The stroke volume, like the tidal volume in respiration, is not as large as the maximum volume of heart. Consequently, there is usually both a diastolic volume reserve or capacity and a systolic volume reserve (capacity) (Fig. 15-3). Discussion of the various factors influencing these volumes follows.

END DIASTOLIC VOLUME

If the heart is to be an effective pump, the ventricles must fill before systole. The adequacy of filling is determined by the filling time, the effective filling pressure, the distensibility of the ventricles, and atrial contraction.

The *filling time* is dependent upon the heart rate. As the heart rate increases, the diastolic period is reduced proportionally more than the systolic period, so that available filling time becomes an important factor at high heart rates. Under normal conditions, however, the filling time is adequate.

The *effective filling pressure* is the pressure gradient between the inside of the ventricles and the pressure outside. It is this pressure gradient which causes the ventricles to distend (Fig. 15-4A). It is closely dependent upon the rate of venous return

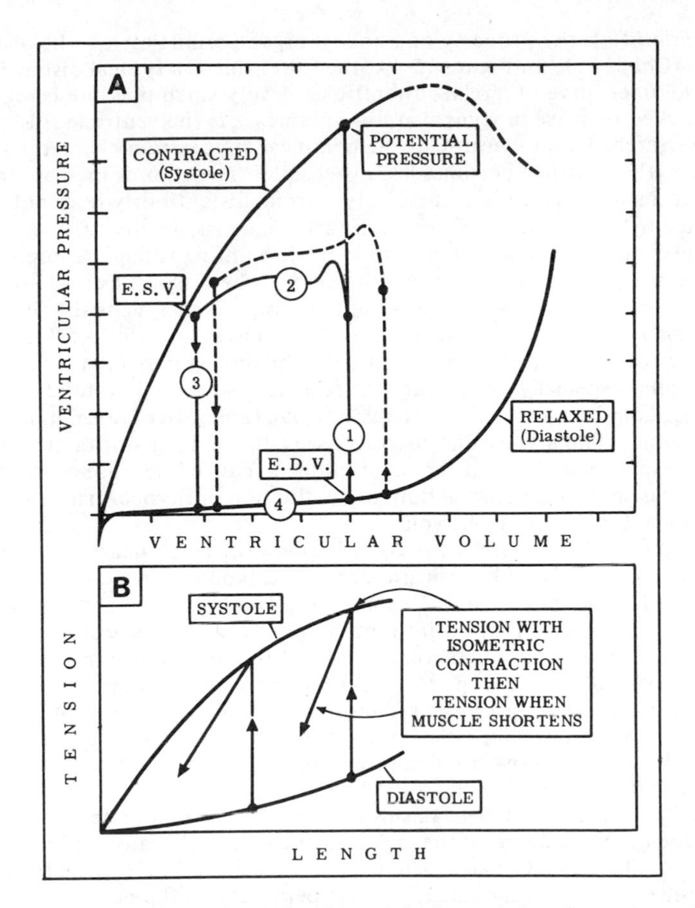

FIGURE 15-4. A. Volume-pressure relationships of the heart. 1, isovolumetric contraction phase; 2, ejection; 3, isovolumetric relaxation; 4, filling. A normal cycle is indicated by a solid line. Systole starts at the end diastolic volume (EDV) and first produces the isovolumetric change in ventricular pressure. It ends at the end systolic volume, after which time isovolumetric relaxation occurs. As predicted by Starling's law or heterometric autoregulation (see below), an increase in end diastolic volume (dotted-line loop) tends to result in an increase in stroke work, since either stroke volume or pressure is increased as a result of the increase in vigor of contraction. B. The effect of shortening on myocardial tension. Muscle at various resting lengths is stimulated to contract isometrically and then allowed to shorten. Note the decrease in tension as a result of viscous forces within the muscle. (A after L. N. Katz. *Circulation* 21:483–498, 1960. By permission of the American Heart Association, Inc. B modified from G. Lundin. *Acta Physiol. Scand.* 7 (Suppl. 20):42, 1944.)

(Chap. 16) and the degree of negative intrathoracic pressure. An increase in the central venous pressure or a more negative thoracic pressure acts to cause an increase in the transmural pressure and thus facilitates filling. Under normal conditions the filling pressure of the heart is adequate and so does not become a limiting factor. However, if the blood volume is reduced – for example, as a result of hemorrhage – or if there is a dilation of the veins and other large volume vessels (venomotor relaxation), the filling pressure drops and there is serious reduction of the cardiac output. Atrial contraction (see below) also contributes.

Distensibility is the property of a hollow organ permitting it to be distended or expanded (Chap. 11). In Figure 15-4A, the lower line is a typical distensibility, or pressure-volume, curve of a relaxed ventricle. A very small pressure is required to cause a marked increase in volume at low volumes. As the ventricle fills, its elastic fibers are stretched, and, because of the nonlinear stress-strain characteristics of these fibers, the ventricle becomes less distensible. Then a unit increase in pressure causes only a small change in volume. Myocardial distensibility does not appear to be modified by sympathetic and parasympathetic division activity.

The mechanisms of distention of the ventricle during filling are important. During the *rapid-filling phase,* the compression of the elastic cardiac tissue by the previous systole tends to cause a recoil of the tissue. This speeds the filling, just as squeezing a rubber bulb and then releasing it causes the bulb to fill rapidly. Most of the filling occurs during this phase in the resting individual. This recoil phenomenon is especially important with relatively small end systolic volumes. Under these conditions actual ventricular suction (a negative transmural pressure) has been found. The myocardial viscosity (see below) is a major factor limiting the rate of rapid relaxation. If the rapid filling is reduced because of low filling pressure, weakened atrial contraction, or a reduction of myocardial recoil, tachycardia will greatly reduce stroke volume.

Atrial contraction becomes especially important at high heart rates, when the diastolic time is limited. When the atria contract, blood is propelled in both directions: toward the veins and into the ventricles. However, the inertia of the blood tends to act as a valve so that a significant amount of blood is propelled into the ventricles to aid in filling. Furthermore, it leads to reduced regurgitation by assisting atrioventricular valve closure (p. 247). During exercise, the amount of blood pumped into the ventricles by the atria can account for as much as 40 percent of the stroke volume. Normally, however, atrial contraction is a minor factor in the filling of the ventricles. Lack of effective atrial contraction, such as occurs with atrial fibrillation, does not jeopardize life, but it does limit exertion.

Normally, homeostatic mechanisms compensate for moderate losses of blood volume which tend to decrease the end diastolic volume because of inadequate filling pressure. These mechanisms include an increased cardiac contractility; less pressure drop due to venous resistance from the periphery to the heart because of the reduced cardiac output and so venous return; and increased venomotor tone (Chap. 16). A human can lose about 1 percent of his body weight of blood without any significant drop in blood pressure. Without the compensatory mechanisms, however, the effects of a decreased filling pressure become highly important. Indeed, a decrease in effective filling pressure (central venous pressure) of only 1 cm of water (about 1 mm Hg) can result in a reduction in cardiac output of nearly 50 percent!

PERICARDIUM

The pericardium provides a limit to the magnitude of the end diastolic volume. Under normal conditions it is not essential in man. The removal of an adherent, dense, scar-filled pericardium following pericarditis provides a significant improvement in the clinical condition of many patients recovering from this disease. The loss of the pericardium does not seriously limit their activity. The function of the pericardium is apparently to check acute overdistention during diastole and to afford a lubricated surface between the beating heart and the lungs. By limiting right ventricular filling, it protects against pulmonary congestion and left ventricular overload. This is important during various cardiac arrhythmias in which prolonged diastolic periods occur. It also aids in keeping the heart and great vessels aligned. Under chronic conditions of myocardial weakness or chronic overloading of the heart, the pericardium slowly distends and cardiac hypertrophy ensues.

If blood or fluids accumulate between the heart and pericardium, filling is impeded. This acute compression of the heart is called *cardiac tamponade.* It is a serious consequence of penetrating wounds of the heart, or of pericarditis, in which there is effusion of fluids from the inflamed membranes.

Cardiac hypertrophy follows a chronically increased load on the heart such as occurs after aortic valvular lesion, mitral insufficiency, or hypertension. The increase in size of the individual muscle fibers (hypertrophy) would be expected to increase the ability of the heart to do the extra work. However, the vascularity of the myocardium usually does not increase proportionately; thus, the ratio between the number of capillaries and the number of muscle fibers decreases as the muscle fibers enlarge and more are formed. In addition, the mean distance which oxygen has to diffuse from the capillaries to the muscle fibers is increased. This limits the effective oxygen supply, and if the hypertrophy is excessive, it is usually followed by cardiac failure.

The cause of hypertrophy is not clearly known, but it often follows a chronic increase in size of the ventricles (dilation). The dilation would be expected to follow increases in filling pressure, regurgitation, or myocardial weakness. Left ventricular heart failure leads to right ventricular hypertrophy because the pulmonary arterial pressure rises as a result of pulmonary congestion and so continuously overloads the right heart (Chap. 18B).

Heterometric Autoregulation – The Frank-Starling Relationship

If cardiac muscle is stretched, it develops greater contractile tension upon excitation. As Starling stated in 1914, "The law of the heart is therefore the same as that of skeletal muscle, namely, that the mechanical energy set free on passage from the resting to the contracted state depends . . . on the length of the muscle fibers." The vigor of contraction is a function of muscle fiber length. Stated another way, stroke volume tends to be directly proportional to diastolic filling; that is, the ventricle tends to eject whatever volume is put into it. Although many investigators, such as Frank in the 1880's, contributed to the study of this mechanism, Starling's formulation was such that it has become known as *Starling's law of the heart.* Sarnoff (1962) has coined the more descriptive term *heterometric autoregulation.* This implies that the performance of the heart is so regulated as to be an effective mechanism for the homeostatic control of circulatory function, by intrinsic (auto) mechanisms which are not neural or hormonal in origin and which are determined by changes (hetero) in myocardial fiber length (metric).

There is no question concerning the validity of this relationship. However, it is but one factor among many controlling cardiac output. The relationship is described graphically in Figure 15-4A. This diagram is similar to diagrams showing the tension developed by contracting skeletal muscle at various initial lengths. At a higher end diastolic volume (dotted-line loop, Fig. 15-4A), the peak isovolumetric pressure (potential pressure with no ejection of blood) which can be developed by the ventricle increases. The increase in end *diastolic* volume, by causing the myocardial fibers to contract with greater vigor, acts to maintain about the same end *systolic* volume. Thus the increase in end diastolic volume is greater than the subsequent increase in end systolic volume, and there is a net increase in stroke volume following increased cardiac filling at a constant pressure load. The increased vigor of contraction results in more external work being done by the heart (see also Fig. 15-5).

As another example, if the total peripheral resistance increases so that the arterial pressure is increased (let us assume from 100 to 150 mm Hg), the systolic ejection is limited by the increased pressure head and the end systolic volume is increased (say from 40 to 50 ml) because the usual amount of blood (80 ml per beat) was not ejected (now only 70 ml). This volume (10 ml) is added to the normal venous inflow (80) to give a larger end diastolic volume (40 + 10 + 80 or 130 ml, instead of the previous 40 + 80 or 120 ml). The larger heart then contracts with more vigor and ejects the same volume per beat (130 – 50 = 80 ml) as before (120 – 40 = 80 ml). Note that the heart is then beating with larger end systolic and end diastolic volumes. Actually, the stroke volume is not perfectly maintained, as in our example, because the venous inflow is limited to a greater degree at the larger size, owing to the decreased distensibility of the heart at the larger volume. Thus the end diastolic volume increases only partially from the 120-ml control end diastolic volume to the 130-ml volume, and so the expected 10-ml increase in end diastolic volume is not achieved.

If the heart is overdistended, the peak pressure tends to decrease (Fig. 15-4A), probably because the sarcomeres are extended so much that the actin and myosin myofilaments do not overlap as effectively (see Braunwald et al., 1968). Then the increase in end diastolic volume is followed by a greater increase in end systolic volume, and a smaller stroke volume is ejected. This may happen in acute congestive heart failure.

From studies with isolated animal hearts, it had long been thought that cardiac output was controlled primarily by changes in stroke volume as determined by the end diastolic volume and the Frank-Starling relationship following increases in venous return. However, the importance of this phenomenon in the intact unanesthetized man has been seriously questioned, since changes in right atrial pressures have been noted without concomitant changes in cardiac output. Recent evidence from human volunteers shows that if the sympathetic division is blocked by drugs, an increase in blood volume causing an increase in right atrial pressure is, indeed, followed by an increase in cardiac output. By using injections of radiopaque dyes in humans and animals, and special transducers to measure the dimensions of the ventricles in dogs, Rushmer and co-workers (1962) have found that the end diastolic volume is normally near maximum at rest in the reclining person or dog. The normal heart is large in the resting, reclining individual; it apparently almost fills the pericardium (Fig. 15-3). Under these conditions a marked increase in heart size in response to increased loads is not possible; hence, most of an increase in cardiac output must take place by means other than increased end diastolic volume and heterometric autoregulation. For example, this can be brought about by increased heart rate and decreased end systolic volume resulting from increased vigor of myocardial contraction (via sympathetic activity).

Standing, excitement, and exertion are often followed by decreases in the diastolic size of the heart as a result of increases in sympathetic division activity (Fig. 15-3). Although the stroke volume increases during maximal exertion, Rushmer (1962, 1970) concludes that the stroke volume remains "remarkably" constant over a wide range of voluntary exertion. Normally, the stroke volume increases only if the total oxygen consumption exceeds about eight times the resting rate. This neural control, which minimizes the importance of heterometric autoregulation (Starling's law), had not been seen to such a great degree in earlier animal experiments for the reason that most anesthetics depress the cardiovascular reflexes acting on the heart and cardiovascular system (Chap. 17). Furthermore, the procedure of thoracotomy, almost always used in these studies, removes the negative intrathoracic pressure, with the result that the heart has a lower effective filling pressure. Thus the heart is not adequately filled during diastole. The end systolic volume of the heart is almost zero under these conditions, and the heart rate is high and relatively constant. It is therefore not surprising that increases in filling pressures distend the heart, giving marked increases in cardiac function explained by Starling's law.

Heterometric autoregulation is without question an important mechanism, in the normal individual, for balancing the outputs of the right and left ventricles. For example, if the right ventricle pumps more blood than the left ventricle, the difference accumulates in the lungs, acts to increase pulmonary venous and left ventricular diastolic pressures, and so increases left ventricular filling. In accordance with Starling's law, the left ventricular output increases and so restores the balance. Heterometric autoregulation is deleterious during changes in body posture in man, for unless other mechanisms act to maintain homeostasis, as the blood pools in the lower extremities there is a profound drop in cardiac output and so in blood pressure (orthostatic hypotension). Likewise, during thoracic surgery, when the effective filling pressure to the heart is reduced, other mechanisms must provide compensation. The Frank-Starling relationship is of value as a defense against heart failure, for the increase in filling pressure (congestion) within certain limits acts to maintain the cardiac output.

END SYSTOLIC VOLUME

The end systolic volume is that volume of blood remaining in the heart at the end of systole at the time when the semilunar valves start to close. At this moment the aortic and ventricular pressure *distending* force is balanced by the *contracting* force produced by the myocardial fibers. The primary factors determining the magnitude of the end systolic volume are the pressure against which the ventricle is pumping and the strength of the myocardial fiber contraction.

PRESSURE LOAD

If the heart is to pump against a higher than previous blood pressure, the myocardial fiber tensions must be greater to expel the same amount of blood – a result of the Laplace relationship. Thus systole will terminate at some higher volume unless the contractility of the myocardium is increased (see below). Conversely, a reduction in systemic blood pressure results in a decrease in end systolic volume.

There is a complex interaction between filling pressure, aortic (outflow) pressure, and cardiac performance (Fig. 15-5). Although the data shown are from dog experiments, the pattern is probably applicable to man. Consider point N as normal (Fig. 15-5A). Increasing ventricular fiber length to point A increases left ventricular (cardiac) output, as predicted by the Frank-Starling relationship. (Ventricular fiber length, ventricular volume, end diastolic pressure, and mean left atrial pressure are generally closely related.) At a constant filling pressure, an increase in mean arterial pressure (pressure load) results in a decrease in cardiac output, point C. Note that at high arterial pressures (200 mm Hg) increasing the filling pressure has relatively little influence on cardiac output, in sharp contrast to the marked influence of filling pressure on cardiac output at low (50 mm Hg) pressure loads. Note, too, that with abnormally low transmural filling pressures, e.g., 4 mm Hg, the cardiac output is low and relatively independent of pressure load.

The rate of doing work is a better index of cardiac function and is more closely related to cardiac reserves in disease states than mere flow output. Figure 15-5B is a three-dimensional presentation of left ventricular work done (per minute) as influenced by filling and output pressure – heart rate, and metabolic, neural, and hormonal influences constant. At a given arterial pressure, an increase in filling pressure leads to increased "work" (the curve B-N-A). If the work per beat is used rather than work per minute (power), the relationship is called a *ventricular function curve* (see below).

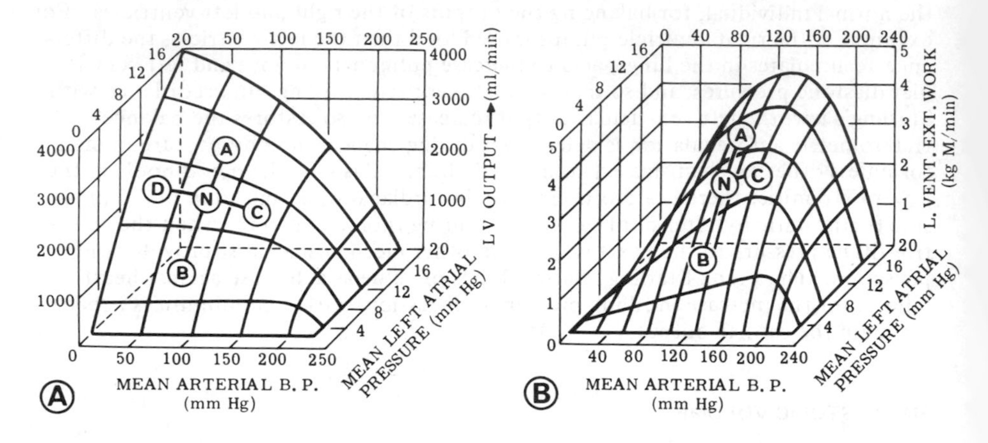

FIGURE 15-5. Effect of filling pressure (mean left atrial pressure) and output pressure (mean arterial pressure) on: A. cardiac output (left ventricular output), and B. heart work per minute (left ventricular external work). Data normalized to a 10-kg dog. (Modified from K. Sagawa. In E. B. Reeve and A. C. Guyton [Eds.]. *Physical Bases of Circulatory Transport.* Philadelphia: Saunders, 1967.)

HOMEOMETRIC AUTOREGULATION

In addition to the increased power output of the heart as a result of increased filling – heterometric autoregulation – the physiology of the heart is such that nearly the same stroke volume is ejected even if there is an increased aortic pressure. This response requires a few beats to develop, but it is not dependent on changes in fiber length (i.e., homeometric) or on neural or hormonal factors (i.e., autoregulation) (Sarnoff and Mitchell, 1962). The relationship is shown in Figure 15-5B, whereby the arterial pressure is increased from 100 mm Hg (point N) to 150 mm Hg (point C). At high mean arterial pressures (greater than 150 mm Hg in these experiments) a reduction in external power output is seen with additional increases in pressure load. When more work is done, the metabolic energy requirements are increased, however. Indeed, the systolic tension may be the principal determinant of not only oxygen consumption (see below) but also homeometric autoregulation.

MYOCARDIAL VISCOSITY

Figure 15-4A, uppermost curve, shows the pressure-volume relationship under isovolumetric, nonshortening conditions. (The curve represents the maximal pressures at various volumes of *isometric* contractions, and each point on it is related to the maximum isometric force [P_0] of a force-velocity curve [Chap. 3B].) The tension developed when strips of myocardium are allowed to shorten is much less than the tension when no shortening is allowed (ejection phase, Fig. 15-4). This loss of force or tension is similar to that which would be seen if the contractile machinery had to drag through a "viscous medium." The force needed to overcome viscous resistance is proportional to the rate of movement; for example, pouring a thick syrup at a very slow velocity requires a low force, but to discharge this viscous material from a container rapidly requires a very high force. The myocardium tends to resist movement in this manner. The myocardial tension develops rapidly during the isovolumetric contraction phase, but as soon as ventricular ejection starts, the myocardial fibers shorten and the effects of the internal viscosity of the myocardium are seen. As a result, the effective intraventricular pressure developed is less than it would be under isovolumetric conditions. If the myocardial fibers did not have to shorten so far or so fast, the proportion of the contractile energy available for the development of useful tension would be higher, and that wasted in viscous friction losses would be lower. Under normal conditions, myocardial contraction is neither isometric nor isotonic, but *auxotonic;* that is, the contraction is against an increasing load, since the aortic pressure increases during the rapid ejection phase.

Less shortening is required by the large heart to eject a certain stroke volume than by the smaller heart. A change of volume (the amount ejected) of a sphere is related to its diameter cubed, and that of a cylinder or cone to the diameter squared, but the circumference (degree of fiber shortening) is directly proportional to the diameter. This relationship holds true under most conditions, especially for the left ventricle, although the geometric relationships of the two ventricles make a quantitative analysis complex. In the larger heart, less muscular tension is lost in overcoming viscous forces and more is available for pumping blood.

ELASTIC FORCES

Another factor which must be considered is the development of elastic forces between the fibers and layers of the heart muscles as shortening occurs. Energy is stored during systole because of tissue compression. During diastole there is a facilitation of filling due to elastic recoil from this compression. This mechanism leads to the diastolic suction which is seen at small end systolic volumes. At normally large ventricular volumes, this factor is minor; consequently, there is more energy available per unit volume for blood ejection.

Thus there are three factors which permit more effective ventricular function at large diastolic volumes: (1) the relationship described by Starling's law of the heart, (2) minimal myocardial viscosity because less shortening is required, and (3) minimal development of elastic forces in connective tissue. However, these factors are opposed by a fourth, which tends to limit severely the work done by the heart at very large volumes.

LAPLACE'S EQUATION

According to the Laplace equation (Chap. 11), the tension of the fibers of a sphere or cylinder is directly proportional to the distending pressure times the radius. Thus, the myocardial tension of the contracting ventricles *required* to expel a unit volume of blood against a given pressure must be greater for a larger heart than for a smaller heart. This factor thus tends to compensate, in the small heart, for the loss of effective tension because of the myocardial viscosity, development of elastic tissue compression forces, and reduced vigor resulting from shorter initial lengths of the myocardial fibers.

MYOCARDIAL CONTRACTILITY

A description of cardiac performance is essential for diagnostic evaluation. *Myocardial contractility* and *vigor of myocardial contraction* are vague terms, connoting the tendency to contract (fiber shortening) as well as the force developed by the muscle fibers to expel blood. As emphasized by Rushmer, such ambiguity of definition tends to "obscure ignorance and impede progress." Much to be preferred is the approach of the physical sciences: the use of descriptive terms which can be accurately measured, described mathematically, and expressed in physical terms. Thus, cardiac performance should be expressed as accurately measured pressure, flow, dimensional changes, power output, and the like. Satisfaction of the purists is precluded at this time, however, by the complexity and relative inaccessibility of this system. Furthermore, the term *contractility* serves a useful purpose by being a conceptual focal point between the many influences on the heart and its performance. Unfortunately, although contractility connotes the ability to contract, many authors (e.g., Sarnoff and Braunwald) restrict the meaning of cardiac contractility to performance or work done *other than* that induced by changes in presystolic fiber length, i.e., factors other than from the Frank-Starling relationship. If the myocardial contractility is increased, either the end systolic volume will be decreased (so increasing the stroke volume) or the same volume of blood will be ejected against a higher pressure. In either case, more work will be performed by the heart with each beat, and more power will be developed.

One approach to evaluation of myocardial contractility is the determination of a *ventricular function curve* (Figs. 15-5B and 15-6). This is the relationship between ventricular stroke work (stroke volume times the change in pressure) and end diastolic fiber length or pressure. The latter is often used because it is easier to measure (cardiac catheterization) and because the curvilinear relationship between the ventricular pressure just before systole and the fiber length is remarkably constant for a given heart. Even the mean atrial pressure has been used to derive satisfactory curves. In the usual method of determining the ventricular function curve, transfusion or hemorrhage is used to provide measured changes in the end diastolic pressure. In addition, cardiac output and heart rate are measured to provide stroke-volume data. The mean aortic pressure and cardiac output furnish sufficient information for evaluating cardiac work, although the integral, during systole, of the instantaneous ventricular pressure times the instantaneous blood outflow provides a more accurate measure. Because such procedures as transfusion and hemorrhage require many seconds, neural and hormonal influences change during the time interval required and thus provide a different type of curve from the one which would be seen with the isolated or denervated heart.

Four features of ventricular function curves should be considered: (1) Under a given set of conditions, increasing the fiber length leads to greater work output (heterometric autoregulation: the basal curve, Fig. 15-6). (2) If there are extracardiac influences, such as sympathetic stimulation, one sees a different curve such that at the same end diastolic pressure more, or less, work is obtained. (3) Small changes in end diastolic pressure of the right ventricle induce large changes in work output. As Guyton (1963) has emphasized, the right ventricular end diastolic pressure is dependent upon the factors influencing the flow of blood to the heart (Chap. 16). Because the left ventricle may pump out only what is pumped to it by the right ventricle, and because of the sensitivity of the right ventricle to filling pressure, potent mechanisms must be available to maintain the optimal filling pressure to the heart. (4) It is important to realize that more work than before can be obtained from a heart which is weakened (i.e., lower function curve) if the filling pressure is increased enough at the same time. Thus, a single point on the curve provides only limited information about the condition of the heart. As the filling pressure is increased, a plateau in response is seen. In the failing heart, still further increases in end diastolic pressure may result in less work than before, even with the same ventricular function curve (Fig. 15-6, bottom curve).

Another approach to evaluating myocardial performance is based on the force-

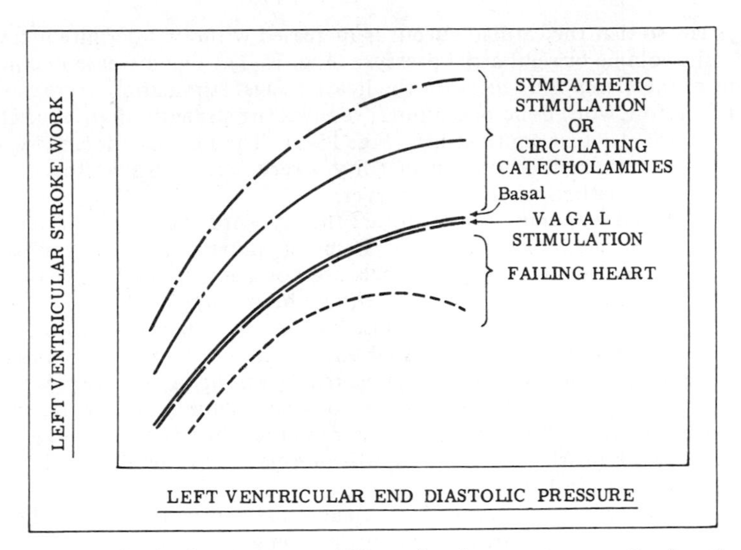

FIGURE 15-6. Ventricular function curve. Effect of various agents on cardiac function curves relating stroke work (stroke volume times mean systolic blood pressure) and left ventricular end diastolic pressure. Under normal conditions there is a small sympathetic division activity giving a higher cardiac function curve than the basal level.

velocity relationship of contracting muscle (Chap. 3B). This concept applied to cardiac muscle has been refined and reviewed by Braunwald et al. (1968). The computation of the velocity of shortening of the contractile elements is based on numerous assumptions concerning geometry, uniformity of the ventricles, stiffness of the series elements, and characteristics of parallel elastic elements in conjunction with the *peak rate of pressure development* in the ventricle. Recent studies by Pollack (1970) raise serious questions about the feasibility of computing the V_{max} of the contractile elements of the intact heart. Although estimation of contractile element shortening is necessary to substantiate our conceptual models of muscle function, measurement theory suggests that the variable actually measured, e.g., peak rate of change of pressure under standardized conditions, had best be used in developing a practical measure of contractility, rather than some derived function based on various tentative assumptions. Rushmer and his colleagues have shown that the peak rate of blood acceleration during ejection (df/dt) is related to ventricular performance. Measuring rates of pressure rise in the ventricle is much easier than developing a ventricular function curve or a force-velocity curve, or even than measuring peak rates of flow under various pressure loads, especially if the subject must be minimally disturbed.

Yet another approach is to suture a strain-gauge arch across a slit in the myocardium to obtain an estimate of the tension developed by the myocardium at this site.

FACTORS INFLUENCING CONTRACTILITY

Myocardial contractility increases markedly in exercise; the heart beats more rapidly and the blood pressure increases. An increase in stroke volume occurs primarily because the end systolic volume is smaller as a result of the sharply increased myocardial contractility. In the normal individual, autonomic nervous effects obscure

Starling's law so that the cardiac output is increased without a definite increase in end diastolic volume or right atrial pressure (Fig. 15-7). The increase in contractility is a result of the *neural influences* on the heart. Vagal stimulation decreases heart rate and therefore work done per minute, reduces the strength of atrial contraction, and decreases ventricular contractility (Fig. 15-6). The maximal depressing effect of the parasympathetic division on contractility is very much less than the augmenting effects of the sympathetic division, however.

The health and *metabolic conditions* of the myocardial cells are an obvious factor determining the strength of contraction. Hypoxia, hypercapnia, and acidosis are depressant. Ischemia from obstruction (occlusion) of a coronary artery leads to rapid deterioration of function because of inadequate blood supply.

The time interval between beats also influences the strength of contraction. A premature depolarization results in a weakened contraction, but the following beat is then more forceful than normal. This *postextrasystolic* potentiation is independent of ventricular filling. It is probably related to the availability and/or transport of calcium between the cell membrane and the contractile machinery. Paired electrical stimuli to the weak heart under some conditions may be of significant therapeutic value (Braunwald et al., 1967).

Inotropes are agents which cause a change in the contractility of the heart. Negative inotropes act on the heart to weaken the strength of contraction. Prime examples are the bacterial toxins released during many diseases, which reduce the ability of the heart to pump blood. Positive inotropes are agents such as digitalis and *circulating catecholamines* which act to strengthen the heart beat. The infusion of norepinephrine in order to supplement the endogenously released catecholamines is of real, but limited, clinical value. Excessively large amounts (more than about 1 μg per kilogram of body weight per minute) may produce cardiac or peripheral damage. In last-resort conditions, the infusions of large amounts of norepinephrine in man may only produce more injury.

In summary, there are five classes of influence on the vigor of myocardial contraction:

1. *The metabolic condition* of the cells, which in turn is dependent on an adequate coronary blood flow, oxygen supply, nutrient supply, and lack of toxins.
2. *Heterometric autoregulation,* the Frank-Starling relationship, or the influence of the length of fibers at the beginning of contraction on cardiac performance.
3. *Central nervous system action,* including the effect of circulating catecholamines and the release of the autonomic mediators at the heart.
4. *Homeometric autoregulation,* the added power output of the heart in response to increases in pressure load without changes in end diastolic fiber length or extracardiac influences.
5. *Interval between beats,* the potentiation following premature ventricular contractions.

RELATIONSHIP BETWEEN CARDIAC OUTPUT AND VENOUS RETURN

Right atrial pressure is an important variable influencing both cardiac output and venous return. Because the cardiac output must equal the venous return over long periods of time and yet may not be equal on a beat-to-beat basis, a study of the interrelationships helps one's understanding of cardiovascular function (see Guyton, 1963, for details).

The curves labeled I-1 to I-3 of Figure 15-7 indicate the cardiac output (right ventricular outflow) which occurs under various conditions at given right atrial pressures. As filling pressure (right atrial pressure) is increased, cardiac output increases. The relationship is also shown in Figure 15-5A with the added variable of arterial pressure. As the level of sympathetic activity is increased (I-3 representing maximum activity versus I-2, a basal level), the cardiac output at a given atrial pressure (e.g., 2 mm Hg) is increased because contractile ability of the myocardium is increased. (Note: The effective filling pressure is about 5 mm Hg greater than the atrial pressure because of the normal negative 5 mm Hg intrathoracic pressure.) This family of ventricular function curves defines the state of the heart.

The curves labeled II-1 to II-3 indicate the venous return (in liters per minute) which occurs at various right atrial pressures. Such curves have been obtained experimentally in dogs, although the technique is difficult and rather traumatic. The point at zero venous return defines one end of the curve. It is measured by fibrillating the heart and rapidly pumping blood from the arterial to venous circuits until the pressures are equal. A steady value must be obtained within three to five seconds before significant fluid shifts or autonomic responses have occurred. With the flow being zero, the pressure, called the *mean circulatory pressure,* is the same throughout the entire vasculature. Furthermore, this pressure is the same as that in the peripheral venules or small veins under normal conditions. Here is located

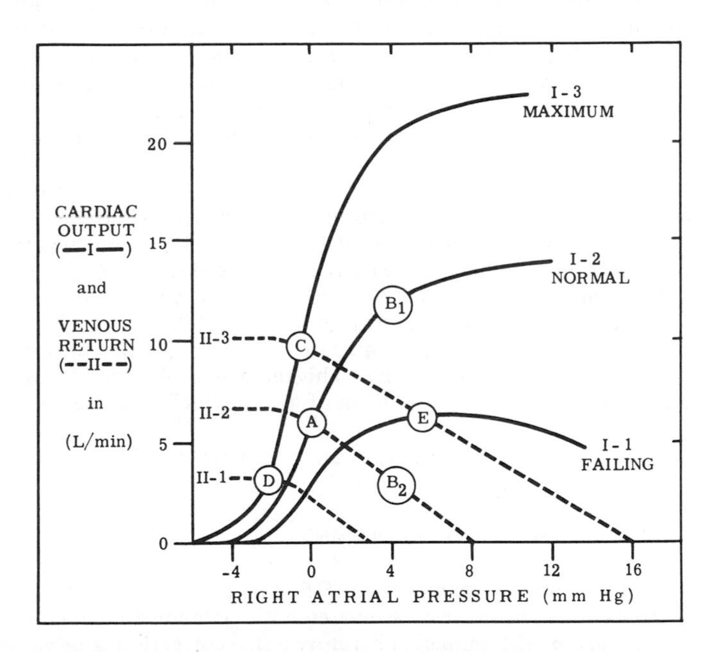

FIGURE 15-7. Relationship between cardiac function and venous return. *Ventricular function curves,* I, relating atrial pressure to cardiac output: 1, failing heart weakened by negative inotropes or ischemia; 2, normal relationship with moderate sympathetic influences; 3, maximal sympathetic influence. *Venous return curves,* II: 1, after hemorrhage and mean circulatory pressure reduced to 3 mm Hg; 2, normal with mean circulatory pressure of 8 mm Hg; 3, with maximum sympathetic influences or after large transfusion of blood. *Operating points:* A, normal; B_1 and B_2, outflow and inflow after sudden increase in atrial pressure; C, at maximum sympathetic activity; D, following hemorrhage; E, with failing heart showing increased central venous (right atrial) pressure.

the greatest portion of the blood. Setting right atrial pressure at a value less than the peripheral venular pressure, a pressure gradient is obtained, and the resulting flow is measured and plotted. At atrial pressures much less than zero, the flow plateaus in value because the vessels entering the chest tend to collapse just outside the chest cavity. Providing a greater negativity does not increase flow, just as sucking harder on a collapsed soda straw does not yield more soda pop. The slope of the venous return curve represents the resistance and is determined by the geometry of the venous vasculature. The zero-flow pressure intercept is determined by (1) the blood volume and (2) the pressure-volume relationship of the vasculature – the capacitance. Capacitance may be reduced to increase the mean circulatory pressure by increasing sympathetic activity which acts on the smooth muscle in the walls of the veins and venules – venomotor tone (Chap. 16).

The steady-state cardiac output and venous return must be equal. This condition will occur at the intersection, point A, of the given cardiac output (I-2) and venous return (II-2) curve. If the filling pressure is temporarily higher than the value at the intersection of the curves, e.g., 4 mm Hg, the cardiac output (point B_1) will be greater than the venous return (point B_2). The volume of blood in the atrium will be rapidly decreased by the increased cardiac output and reduced venous return and so lead to a decrease in atrial pressure until inflow equals outflow. The converse is also true, for if cardiac output is less than venous return, blood accumulates, increasing the atrial pressure and cardiac output and decreasing the rate of venous return. At only one value of atrial pressure – in this case 0 – will inflow equal outflow.

Even though cardiac output is greatly affected by a slight change in atrial pressure (e.g., curve I-3 between 0 and 1 mm Hg atrial pressure), a maximal sympathetic effect on both the heart (I-3) and peripheral venous bed (II-3) results in little or no change in atrial pressure, although the operating point is then at a much higher value, point C.

Note that massive hemorrhage (curve II-1) greatly reduces the mean circulatory pressure and so venous return. Even with massive sympathetic effects on the heart, the cardiac output is greatly reduced, point D. Because the venous return is reduced, the pressure drop from periphery to heart is also reduced ($\Delta P = R\dot{Q}$). The reduction in right atrial pressure with hemorrhage (from 0 to −2 mm Hg) is less than the reduction in mean circulatory pressure (from 8 to 3 mm Hg).

With cardiac failure (curve I-1), an increased mean circulatory pressure is required to maintain the cardiac output (point E). This increase is obtained by sympathetic discharge or, in the long term, by retention of water, which increases the blood volume.

The factors influencing cardiac output are summarized in Figure 15-8.

CARDIAC OUTPUT AND ADAPTATION IN MAN

The cardiac output in normal man is about 5.5 liters per minute. Like many other physical functions of the body, it is associated with metabolism, which in turn is associated with the size of the animal. Therefore a size correction is of value for comparison of various people or animals. Body weight is one factor widely used to determine drug dosage, but the body surface area has been found to be an even better basis for comparison of, for instance, metabolic rate, blood flow, heart rate, size of organs, respiratory rate, and cardiac output. In people or animals of widely different sizes, an empirical equation to convert body weight (in kilograms) to body surface area (in square meters) is:

$$\text{Area} = 0.11 \text{ body weight}^{2/3}$$

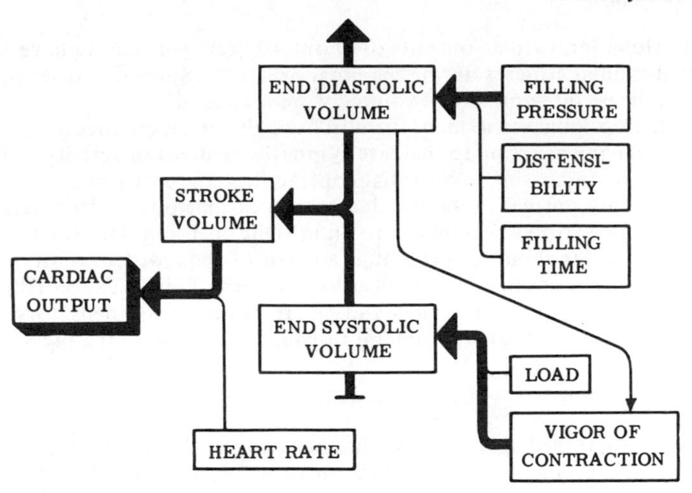

FIGURE 15-8. Factors influencing cardiac output.

In medical terms, an index often means the rate of a physiological function per minute per square meter of body surface area. Thus, the *cardiac index* is defined as the cardiac output in liters per minute per square meter of body surface, and in man it is about 3.5 liters, with values below 2.0 liters or above 5.0 liters definitely abnormal at rest. The determination of the cardiac output in man, as outlined in Chapter 11, provides an important diagnostic tool, especially in cases of valvular lesions. If the cardiac index is below about 2 liters, it is a sign of some form of cardiovascular failure. Tachycardia from atrial fibrillation, or bradycardia from atrioventricular block, often results in seriously reduced cardiac output. The cardiac output is reduced in congestive heart failure and does not increase significantly following muscular effort. In patients in mild heart failure, the cardiac output may be within statistically normal limits, but for a given individual it is lower than it would be if the heart of that individual were unimpaired.

Increases in cardiac output demonstrate compensatory mechanisms in the transport of oxygen to the tissues. In anemia, for instance, blood viscosity is decreased and the oxygen-carrying capacity is reduced. Thus a marked increase in cardiac output might be expected because the blood flows more easily; but more importantly, a greater blood flow is induced as a homeostatic adjustment to maintain the same oxygen transport, although the extraction of oxygen may also be increased. In pregnancy, the blood vessels to the placenta and uterus act as an arteriovenous shunt to lower the total peripheral resistance. An adequate blood pressure is maintained by an increased cardiac output (Chap. 17).

Hyperthyroidism (as in thyrotoxicosis), as well as fever or hyperthermia from high environmental temperature, causes an increase in metabolism which requires an increase in oxygen supply if homeostasis is to be maintained. Vessels dilate in response to hypoxia and so permit an increased flow. The cardiac output must increase if the blood pressure is to be maintained. On the other hand, at hypothermic temperatures of 15° to 20°C, the cardiac output is but 5 percent of normal. If the environmental temperature is high, so that the body temperature tends to rise, more blood is shunted to the skin to allow dissipation of more heat. This shunting of blood leads to an increase in the cardiac output.

The most pronounced increase in cardiac output occurs during severe exertion. The cardiac output of a normal individual walking slowly is about 50 percent greater than basal. However, cardiac outputs of about 40 liters per minute have been measured in well-trained athletes during maximal exertion. Since this is seven times normal, both heart rate and stroke volume were increased.

Hypoxia, emotional excitement, insulin hypoglycemia, cutaneous pain – in fact almost any mechanism acting to increase sympathetic division activity – increase cardiac output by increasing myocardial contractility and heart rate.

About the only normal cause of a decrease in cardiac output from basal flow is the act of changing from a recumbent to an upright position. On standing, the cardiac output decreases about 25 percent as a result of the redistribution of blood and the tendency toward pooling in the lower extremities. This leads to orthostatic hypotension (low blood pressure on standing erect) unless compensatory mechanisms act to increase heart rate, peripheral resistance, and the filling pressure.

CARDIAC WORK

If blood is to be pumped against a pressure head to the tissues, work must be done by the heart to drive a volume of fluid (V) against a pressure head (P). Thus, force (f in newtons) $\times$ distance (d in meters) = work (W in newton-meters) = pressure (P in newtons per square meter) $\times$ volume (V in meters cubed). Consequently, the effective rate of doing work (power developed) is the integral of instantaneous blood pressure times the instantaneous rate of flow during systole. This may be closely approximated as the product of mean aterial blood pressure times cardiac output.

It is informative to calculate the work done per minute by the left ventricle in the normal man. If he pumps 5.5 liters of blood per minute at a pressure of 94 mm Hg, his ventricle will do 5500 cm^3 times 128 gm per square centimeter or 700,000 gram-centimeters of work per minute. (Note that 94 mm Hg = 128 gm per square centimeter, since 1 mm Hg = 1.36 gm per square centimeter.) This 700,000 gram-centimeters of work is 7.0 kilogram-meters of work per minute and is equivalent to lifting three large textbooks 1 yard – a substantial amount of work.

The right heart pumps the same volume as the left heart, but the mean pulmonary arterial pressure is about one-seventh of the mean aortic pressure, and thus the amount of work per minute done by the right heart is one-seventh of that done by the left heart, or about 1 kilogram-meter per minute.

In addition to the potential energy, in the form of pressure-volume work, developed by the heart, metabolic energy is also converted to kinetic energy because of the velocity of movement of the blood. Kinetic energy is one-half the mass moving times its velocity squared. Usually only about 2 percent of the useful work of the heart is in the form of kinetic energy. However, during exercise, when the velocity and rate of flow of blood are higher than normal, the kinetic energy may amount to 25 percent or more of the total useful work. Most of this kinetic energy is reconverted to useful potential energy (pressure times volume) by the distention of the elastic aorta. In effect, the elastic aorta catches the ejected blood and throws it on into the arterial tree.

During exercise, in order to propel more blood to the tissues against the same or higher pressure head, the heart must do more work. The extra work requires an increase in metabolism by the heart. The increase is produced either by a greater utilization of nutrients and oxygen or by an increase in the efficiency of the conversion of these materials to useful work. In man, the biological oxidation of nutrients by 1 ml of oxygen gives about 2 kilogram-meters of energy. Thus the heart requires about 4 ml of oxygen per minute to develop the 8 kilogram-meters per minute of power (1.3 watts) under basal conditions, if it is 100 percent efficient in the conversion of the oxygen and nutrients to the useful work. However, the heart is only about 20 percent efficient. This means that for 20 kilogram-meters per minute of useful work, 100 kilogram-meters per minute of energy must be released. Therefore, the heart of man requires about 4/0.2 or 20 ml of oxygen per minute at rest. Since we use about 250 ml of oxygen every minute, our hearts use about 8 percent of our total oxygen consumption.

CARDIAC EFFICIENCY

Although the oxygen consumption is a good index of the metabolic release of energy under steady-state conditions, the oxygen consumption of the heart is not always closely related to the total useful work done in the intact animal because the efficiency of the heart changes. It has been found by Sarnoff, Braunwald, and coworkers (1958) that the oxygen consumption per minute by the heart is proportional to a tension-time index per minute which is the mean systolic arterial pressure times the duration of systole, times the number of beats per minute. This index is a more exact representation of the relationship noted by Katz: the heart rate times mean blood pressure is highly correlated with myocardial oxygen consumption. Note that the volume of blood moved, i.e., the cardiac output, does not enter this relationship directly.

The relationship between the tension-time index and oxygen consumption may be explained if the following physiological characteristics of muscle are considered: (1) Isometric contraction of muscle requires metabolic energy but results in no useful work since no load is moved. Only tension is developed. The development of a ventricular pressure requires energy even though no blood is ejected and therefore no useful work is done. Useful work is realized only if the heart can eject blood. (2) Heat production by the muscle during contraction is an index of metabolism and is determined primarily by the tension maintained times the duration of maintenance of this tension. This in effect is what the tension-time index measures. (3) When a muscle does shorten under a load, some extra energy is required.

The tension-time index provides a prediction of the relative rate of cardiac oxygen consumption under widely varying conditions. In addition to (1) the myocardial tension and (2) duration of systole, a third factor (3), the level of contractile state, measured as V_{max} (see Chap. 3B), is a determinant of myocardial oxygen consumption. Under special conditions (such as paired electrical stimulation or norepinephrine or calcium infusion), oxygen consumption and the peak velocity of contraction, V_{max}, are increased while the tension-time index is decreased. Thus, the rate of ejection of blood is closely related not only to contractility but also to myocardial oxygen consumption. Indeed, some authors (Braunwald et al., 1968) suggest that V_{max} is the best single predictor of myocardial oxygen consumption.

An increase in *blood pressure* has little effect on cardiac efficiency. The increased work, when the change is in pressure head or load, with a constant cardiac output and heart rate, requires a *proportionate increase in oxygen,* since pressure is a major factor in the determination of oxygen consumption. This is seen clinically in aortic stenosis. In this condition the resistance to flow of blood out of the heart is greatly increased because of the constricted aortic valve, but the cardiac output is held about normal by various compensatory mechanisms. When the pressure load is high, the oxygen consumption by the heart must be high. Overt myocardial hypoxia and heart failure are seen frequently in this condition. Hypertension also presents a serious load to the heart and rapidly leads to cardiac failure if the coronary blood flow is restricted by atherosclerosis.

If the *stroke volume* of the heart increases, with blood pressure and heart rate constant, the efficiency of the heart increases, since the greater volume pumped requires but a *small increase in oxygen consumption.* Systole lasts slightly longer, but the effect is relatively small. In experiments with dogs, the work output can be increased about 700 percent by changes in stroke volume with only about 53 percent increase in oxygen utilization. This is in contrast to increased work by increased pressure, where the oxygen consumption parallels the increased work accomplished. Clinically the heart does much work in mitral insufficiency, where blood leaks back to the atrium with each beat, or in intracardiac and extracardiac shunts where there is a large bypass flow. However, early heart failure is surprisingly rare, considering the enormity of the defect in many cases. Aortic valvular insufficiency is an exception, since there is not only a massive back-and-forth movement of blood through the insufficient valve but also, at the region of the valve, a low blood pressure during most of diastole. Thus the perfusion pressure of the coronary arteries is markedly reduced.

An increase in *heart rate,* blood pressure and cardiac output being constant, acts to decrease the efficiency. An increase in heart rate results in an increase in oxygen consumption without, necessarily, an increase in the amount of work (pressure times volume) done. Since each beat requires a certain amount of energy whether or not useful work is performed, the energy expended increases and the efficiency drops. As a consequence, unless the heart of an athlete adapts to provide a larger stroke volume at relatively low heart rates, the athlete will not excel. Likewise the patient at the limit of cardiac reserve is not aided by excitement, which causes an increased heart rate and thereby a decrease in the efficiency. Increases in heart rate can and do increase the cardiac output, but the price is a decrease in efficiency. Since the normal heart has adequate reserves of coronary blood flow, this is not serious.

In the failing heart, for unknown reasons, the efficiency is decreased. In exercise, with a failing heart, the efficiency appears to be decreased even more. Thus adequacy of coronary flow is particularly important under these conditions. Unfortunately, partial or total occlusion of the coronary arteries is commonly the cause of the cardiac failure in the first place.

NUTRITION OF THE HEART

CORONARY BLOOD FLOW

The sustained ability of the heart to maintain a high blood flow to the peripheral tissues is limited primarily by the cardiac oxygen supply. There are but three mechanisms by which the supply of oxygen to the myocardial cells may be increased: (1) *Increased oxygen extraction.* This is a minor source of additional oxygen for the myocardium, in contrast to peripheral tissue such as skeletal muscle. In skeletal muscle, the arterial-venous difference in the oxygen content of the blood may be increased markedly by decreasing the oxygen content of the venous blood. The heart normally extracts 70 to 90 percent of the oxygen from the arterial blood, leaving only 2 to 6 volumes per 100 volumes of blood in the coronary sinus venous blood – a low reserve. (2) *Increased myocardial efficiency.* The efficiency of the myocardial utilization of oxygen is determined by the type of work done by the heart and is generally independent of the oxygen needs of this organ. (3) *Increased coronary blood flow.* This is the only remaining possibility for an increased supply of oxygen to the heart. Since the heart can incur only a very limited oxygen debt, an increase in the coronary blood flow by dilation of the coronary arterioles is essential.

FACTORS REGULATING CORONARY BLOOD FLOW

Measurement of the coronary circulation is difficult because of the location, short length, and multiplicity of the coronary arteries and veins. Differential pressure flowmeters (orifice and Pitot tube), rotameters, bubble flowmeters, and thermostromuhrs have been used to measure the mean flow in cannulated coronary vessels. Electromagnetic flowmeters are being used to measure pulsatile flow in intact coronary vessels in unanesthetized dogs with the flowmeter chronically implanted, and during cardiac surgery in man. The nitrous oxide method has given reasonably accurate values in unanesthetized humans and dogs. For this method, and the determination of cardiac oxygen consumption, samples of blood are withdrawn from the coronary sinus and aorta using indwelling catheters. (See Chap. 11 for a discussion of flow measurements.)

In man, at rest, the coronary blood flow is about 200 ml per minute (about 70 ml per 100 gm of tissue). From data obtained from dogs, the flow rate can increase by at least nine times during severe stress. The myocardium accounts for about 4 percent of the cardiac output and 8 to 10 percent of the oxygen consumed by an adult at rest.

Metabolism

Myocardial hypoxia has a pronounced effect on coronary blood flow. Hypoxia can increase the coronary blood flow at least five times. The response to anoxia is rapid, since a five-second occlusion of the coronary artery having little or no effect upon the contractility is followed by a definite increase in flow. This sensitivity and rapidity of response to tissue hypoxia is essential, since the increased oxygen required for increased cardiac work must be supplied promptly by increased coronary flow or there will be incipient cardiac failure. Indeed, the coronary blood flow to the heart is closely and directly correlated with the oxygen consumption by this organ, with a correlation coefficient of about 0.90. Although the steps are not completely established, they probably include tissue oxygen tension and adenosine released from the tissue which then acts to dilate the coronary arterioles. An increase in carbon dioxide tension or a decrease in pH has a similar, but lesser, dilating effect upon the coronary blood vessels (Berne, 1964).

Mechanical Regulation of Coronary Flow

Since the coronary arteries are supplied from the aorta and since flow is proportional to pressure, the coronary flow is dependent upon the mean systemic arterial blood pressure. The relationship is not simple, for when the heart contracts, the intramural (intramuscular) pressures can and do exceed the arterial pressures and so occlude arteries and stop the blood flow. The normal contracting myocardium tends to impede flow. Although there is usually an inrush of blood very early in systole, during most of systole the high intramural pressure greatly reduces coronary arterial flow so that only 10 to 40 percent of the total flow occurs during this period, even though the aortic pressure is highest then (Fig. 15-9). The venous discharge is hastened during systole because of the squeezing of the veins.

The anatomical arrangement of the arterial tree of the heart is such that if one of the coronary arteries is suddenly occluded, the pressure beyond the occlusion drops to about 30 mm Hg and contractility of the affected myocardium decreases to useless levels. Collateral flow of about 10 percent of normal is present, but unlike skeletal muscle under similar circumstances, an increase in blood flow to adequate amounts requires several weeks. Thus, sudden coronary occlusion often results in a fatal myocardial infarction, whereas gradual occlusion may be followed by the development of an adequate collateral arterial blood supply.

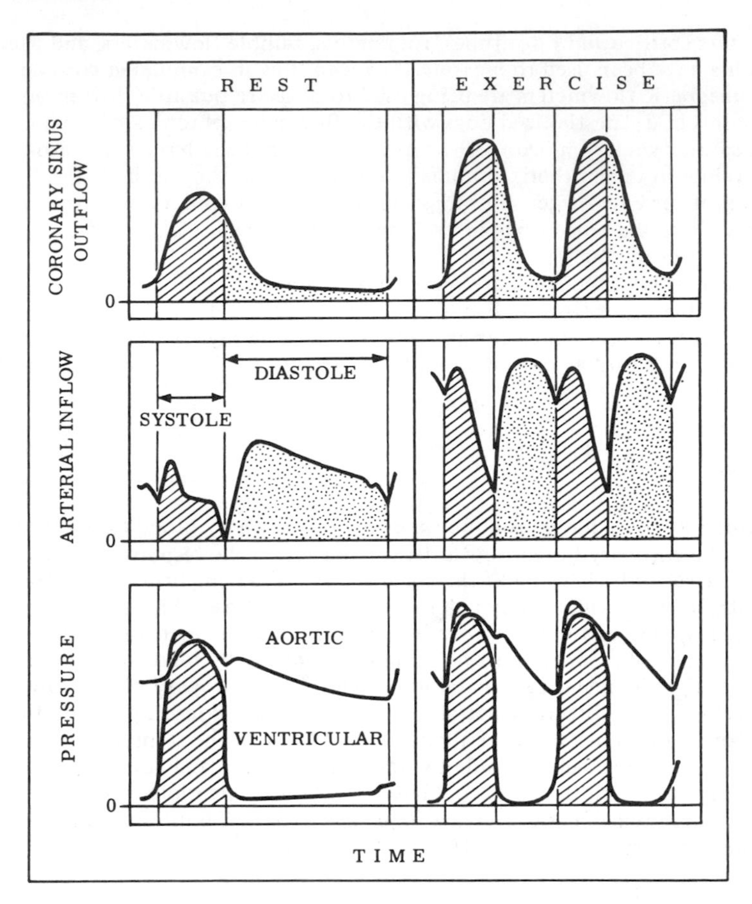

FIGURE 15-9. Effect of myocardial contraction on coronary venous outflow and arterial inflow at rest and during exercise. Note the hindrance to inflow during systole and the arterial regurgitation or backflow. The sharp rise in inflow during systole may be a rebound filling of the compressed arteries. Pressures given for timing reference. (Modified from D. E. Gregg, E. M. Khouri, and C. R. Rayford. *Circ. Res.* 16:105, 1965. By permission of the American Heart Association, Inc.; and from R. F. Rushmer. *Cardiovascular Dynamics,* 2d ed. Philadelphia: Saunders, 1961.)

Neural Control

The innervation of the coronary blood vessels is abundant, but the physiological function of the nerves is not clear, largely because of the difficulty in measuring blood flow under even approximately physiological conditions. Furthermore, it is difficult to differentiate the direct neural effect on the resistance vessels and the concomitant effects of changed metabolism as a result of the neural action. Without question, stimulation of the parasympathetic fibers acts to cause a reduction of the coronary blood flow, but in addition there is a decrease in heart rate and thus increased efficiency, decreased metabolism, and so a decreased oxygen consumption. As a result, one may expect decreased coronary flow, if the oxygen consumption is the primary regulator of this flow. An increase in the activity of the sympathetic nerves acting on the heart causes an increase in flow. Further, the heart rate increases, the contractility increases, more work is done, and thus the metabolism and oxygen consumption rise. The dominant vagal tone in an unanesthetized man acts to keep the heart rate down, the efficiency up, and the coronary blood flow, in consequence, relatively low.

Circulating Epinephrine

Epinephrine causes an increase in coronary blood flow. This is probably a result of the increased metabolism, as occurs in skeletal muscle (Chap. 17). Since epinephrine causes an increased strength of beat, heart rate, and metabolism, the resulting hypoxia probably elicits the arterial dilation. Angina in susceptible patients, following epinephrine injections or sympathetic nervous system action, has been attributed to the increased metabolism with oxygen demands exceeding the supply. Since the increase of blood flow produced by the vasodilating effects of the epinephrine is not adequate to provide for increased oxygen needs, hypoxia and ischemic pain result.

METABOLISM OF THE HEART

Nutrients plus oxygen are required by the heart if it is to do work. As with skeletal muscle, the primary source of energy is the oxidation of glucose. However, the heart is unique in that lactate and pyruvate are readily utilized. In severe exercise, the venous oxygen saturation of active muscle is near zero, anaerobic metabolism is taking place, and lactic acid is released. Under certain conditions the utilization of this lactate by the heart may be greater than the utilization of glucose. It must be remembered, however, that utilization of lactate by the heart requires oxygen. When the heart itself is hypoxic, instead of utilizing lactic acid, it produces it as a metabolite and develops a limited oxygen debt.

Another unusual feature of cardiac metabolism is that fatty acids may be utilized directly. In cases of starvation, the respiratory quotient of the myocardium decreases to about 0.70, indicating that fat, not glucose, is the primary nutrient source of energy. There is no clinical situation in which the cardiac work capacity is limited by the lack of substrate for energy production.

ELECTROLYTES

In addition to the intermediary metabolism, the heart depends upon an optimal electrolyte environment (Chaps. 3B, 13, and 14). The effect of acids on the heart is complex. In dogs, the lethal pH is about 6.0 and is independent of the anion since both hydrochloric and lactic acids have similar effects. At a critically low pH, there is rather sudden, sharp decrease in contractility of the heart. Cardiac arrest in extreme diastole, similar to that caused by potassium, is seen as a terminal event.

REFERENCES

Berne, R. M. Regulation of coronary blood flow. *Physiol. Rev.* 44:1–29, 1964.

Berne, R. M., and L. N. Levy. *Cardiovascular Physiology.* St. Louis: Mosby, 1967.

Braunwald, E., J. Ross, Jr., and E. H. Sonnenblick. *Mechanisms of Contraction of the Normal and Failing Heart.* Boston: Little, Brown, 1968.

Braunwald, E., E. H. Sonnenblick, P. L. Frommer, and J. Ross, Jr. Paired electrical stimulation of the heart: Physiologic observation and clinical implications. *Advances Intern. Med.* 13:61–96, 1967.

Gregg, D. E., E. M. Khouri, and C. R. Rayford. Systemic and coronary energetics in the resting unanesthetized dog. *Circ. Res.* 16:102–113, 1965.

Guyton, A. C. *Circulatory Physiology: Cardiac Output and Its Regulation.* Philadelphia: Saunders, 1963.
Katz, L. N. The performance of the heart. *Circulation* 21:483–498, 1960.
Luisada, A. A. (Ed.). *Cardiovascular Functions.* New York: McGraw-Hill, 1962.
Pollack, G. H. Maximum velocity as an index of contractility in cardiac muscle. *Circ. Res.* 26:111–127, 1970.
Rushmer, R. F. Effects of Nerve Stimulation and Hormones on the Heart; The Role of the Heart in General Circulatory Regulation. In W. F. Hamilton (Ed.), *Handbook of Physiology.* Section 2: Circulation, Vol. I. Washington, D.C.: American Physiological Society, 1962. Chap. 16.
Rushmer, R. F. *Cardiovascular Dynamics,* 3d ed. Philadelphia: Saunders, 1970.
Sarnoff, S. J., E. Braunwald, G. H. Welch, Jr., R. B. Case, W. N. Stainsby, and R. Macruz. Hemodynamic determinants of oxygen consumption of the heart with special reference to the tension-time index. *Amer. J. Physiol.* 192:148–156, 1958.
Sarnoff, S. J., and J. H. Mitchell. The Control of the Function of the Heart. In W. F. Hamilton (Ed.), *Handbook of Physiology.* Section 2: Circulation, Vol. I. Washington, D.C.: American Physiological Society, 1962. Chap. 15.
Sonnenblick, E. H., J. Ross, Jr., J. W. Covell, G. A. Kaiser, and E. Braunwald. Velocity of contraction as a determinant of myocardial oxygen consumption. *Amer. J. Physiol.* 209:919–927, 1965.
Wade, O. L., and J. N. Bishop. *Cardiac Output and Regional Blood Flow.* Oxford, Eng.: Blackwell, 1962.

16

Peripheral Blood Pressures and Pulses: Venous Pressure and Venous Return

Ewald E. Selkurt

ARTERIAL BLOOD PRESSURES are quantitative measurements routinely obtained for diagnostic purposes. In this section the principal physical determinants of arterial blood pressure will be examined. *Systolic pressure* is the highest pressure attained in the aorta and peripheral arteries as a result of the ejection of blood into the aorta by the ventricle. *Diastolic pressure* is the lowest pressure which the gradient of fall reaches during the resting or diastolic phase of the heart. The difference between systolic and diastolic pressure is the *pulse pressure. Mean blood pressure* represents the average pressure attained in the system during the cardiac cycle (Fig. 16-1). Systolic and diastolic pressure must be measured with high frequency–recording optical manometers or electromanometers. Mean blood pressure is not the arithmetic average of the systolic and diastolic pressure, because of the triangular shape of the pulse contour. The mean pressure is computed by measuring the area under the curve from peak to baseline by integration or by the use of a planimeter. If the cycle length is used as a base for construction of a rectangle from the total area, the area divided by this base yields the average height of the rectangle. The height corresponds to the mean arterial blood pressure, when calibrated in millimeters of mercury. (Mean blood pressure can also be measured by a dampened mercury manometer, or by using high-frequency cutoff in the electromanometer.)

The concept of the changes which occur in systolic and diastolic pressure, mean blood pressure, and pulse contour as the blood is ejected into the aorta and passes into the peripheral vascular bed is illustrated in Figure 16-2. It is seen that there is very little change in mean arterial blood pressure as the blood passes down the aorta and the larger arteries. The fall in mean pressure is steeper as the blood enters the small arteries, and the gradient is quite rapid as the blood passes through the terminal branches, the arterioles. It decreases somewhat further in passage through the capillaries. Emerging into the small veins, the blood is at a pressure of about 15 mm Hg, having started out at approximately 90 mm Hg. Pressure then falls further in the return to the heart to enter the thorax and the right atrium in a zone of negative pressure incurred because of the negativity of the thorax.

In passage down the larger arteries, the systolic pressure rises slightly and the diastolic falls. This is the result of reflected waves which amplify somewhat the systolic pressure and depress the trailing portion of the diastolic part of the pulse pressure curve. The mechanism will be discussed below. Finally, the pulse contour changes from that which characterizes the form at the root of the aorta until it reaches the peripheral vessels.

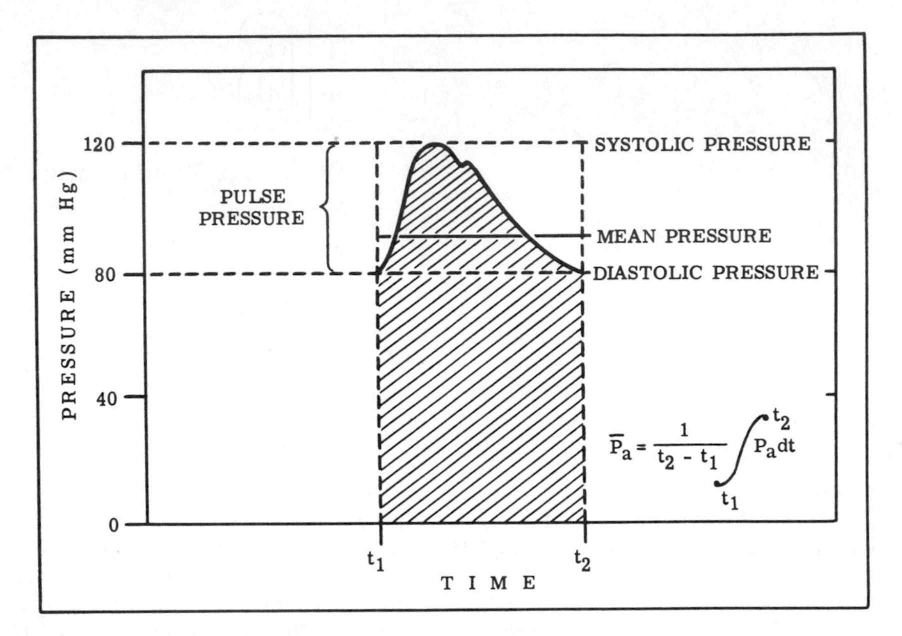

FIGURE 16-1. Arterial systolic, diastolic, pulse, and mean pressures. The mean arterial pressure ($\overline{P}a$) represents the area under the arterial pressure curve (shaded area) divided by the cardiac cycle duration (t_2-t_1).

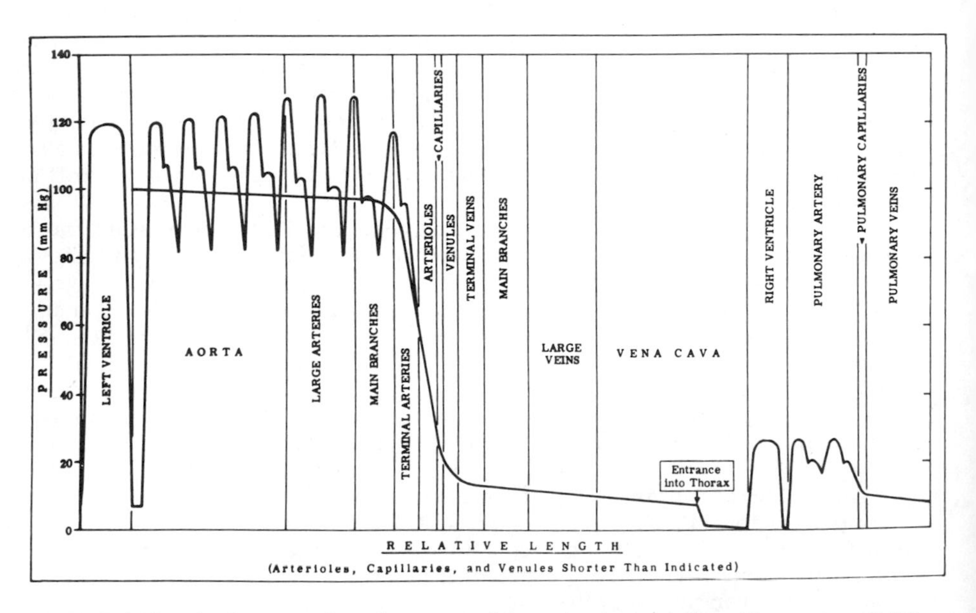

FIGURE 16-2. Blood pressure in various parts of the systemic circulation. The most rapid fall in pressure, associated with loss of the pressure pulse, occurs in the arterioles.

FACTORS WHICH CHANGE SYSTOLIC, DIASTOLIC, AND MEAN BLOOD PRESSURE

The factors which change the mean arterial blood pressure are best considered by an application of Ohm's law to the pressure-flow-resistance relationship in the cardiovascular system. This is the well-known equation

$$F = \frac{P}{R} \qquad (1)$$

which can be rearranged as

$$P = R \times F \qquad (2)$$

It can be readily seen that P, the mean blood pressure, can be altered by variations in peripheral resistance or variations in the flow or, more specifically, cardiac output. Thus an increase in mean arterial blood pressure is the result of an increase in flow (cardiac output) caused by increase in heart rate and/or increase in stroke volume. The latter increases as a result of increased vigor of contraction of the heart or as a result of increased filling. The other factor which increases mean arterial blood pressure is an increase in peripheral resistance, dominated by the setting of the arterioles. A decrease in arterial blood pressure results from opposite effects of the above-mentioned changes.

The factors concerned in the alteration of mean arterial blood pressure (change in heart rate, peripheral resistance, and stroke volume) also modify systolic, diastolic, and pulse pressure. In addition, the elastic properties of the arterial system profoundly influence these pressure parameters.

ROLE OF ARTERIAL ELASTICITY

Before an analysis of the factors which modify pulse pressure is undertaken, the role of the elastic properties of the arterial wall must be understood. One can appreciate this by considering first the static *pressure-volume relationship* of the aorta as shown in Figure 16-3. Successive volumes of fluid were injected under different pressures into the closed-off segments of aortas of humans of different age brackets. After each increment of volume, the static pressure was measured. Curve A (the youngest group) is linear over most of the range (about 75 to 150 mm Hg). However, at each end the curve becomes sigmoidal, and the slope (dV/dP) decreases. At any given point the slope represents the aortic *capacitance* (or *compliance*).

It can also be seen in Figure 16-3 that the curves become displaced downward and slopes diminish with advancing age. At pressure above 80 mm Hg, the capacitance decreases with age, a manifestation of progressive atherosclerosis.

PULSE PRESSURE

If it is assumed that the arterial pressure, P_A, at any moment in time is dependent primarily upon arterial blood volume, V_A, and arterial capacitance, C_A, then it can be shown that the arterial pulse pressure is principally a function of stroke volume and arterial capacitance.

Stroke Volume

The effect of a change in stroke volume on pulse pressure may be analyzed under conditions in which C_A remains virtually constant over the range of pressures under

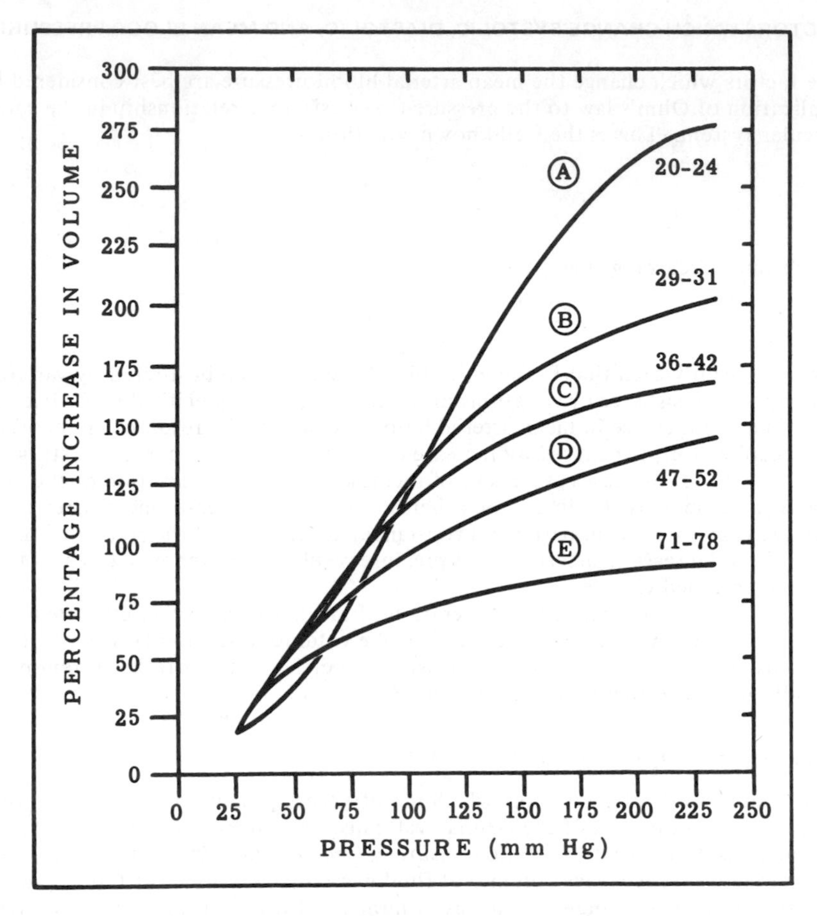

FIGURE 16-3. Pressure-volume relationship for aortas obtained at autopsy from humans in different age groups (see numbers at right end of each curve). (After P. Hallock, and I. C. Benson. *J. Clin. Invest.* 16:595, 1937.)

consideration. In this situation the curve relating P_A to V_A is linear, as in Figure 16-4A. This curve would correspond fairly closely with the curve for the 20–24 year age group in Figure 16-3, especially over the pressure range between 75 and 150 mm Hg.

In an individual with such a $P_A:V_A$ curve, the arterial pressure would oscillate about some mean value ($\overline{P}_A$ in Fig. 16-4A) that depends entirely upon cardiac output and peripheral resistance, as explained above. This mean pressure corresponds to some mean arterial blood volume, $\overline{V}_A$, and the coordinates $\overline{P}_A$, $\overline{V}_A$ define point $\overline{A}$ on the graph. During diastole, peripheral runoff occurs in the absence of ventricular ejection of blood, and P_A and V_A diminish to minimum values, P_1 and V_1, just prior to the next ventricular ejection. P_1 is then, by definition, the diastolic pressure.

During ventricular ejection, there is rapidly introduced into the arterial system a volume of blood which greatly exceeds the peripheral runoff during this same

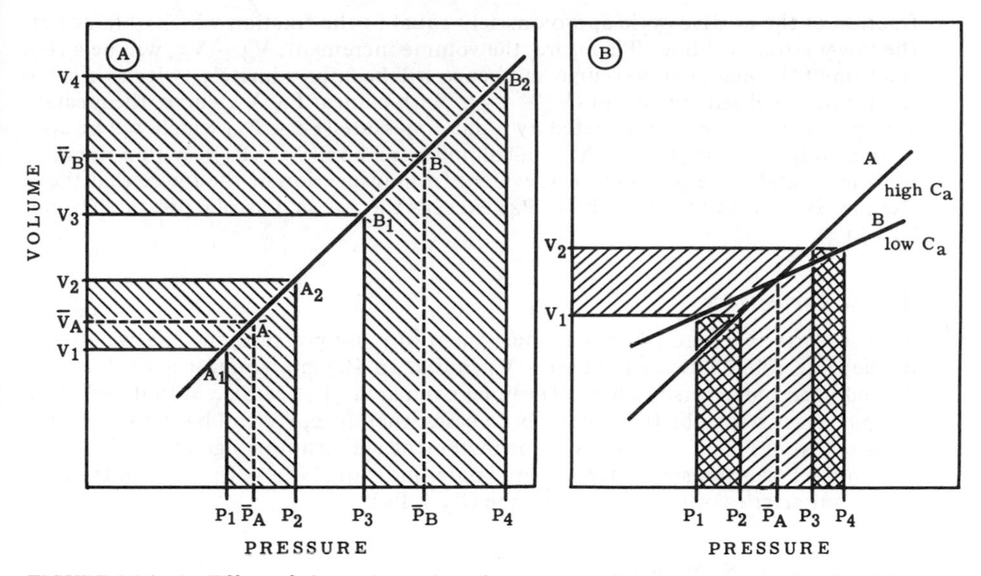

FIGURE 16-4. A. Effect of change in stroke volume upon pulse pressure in a system in which arterial capacitance is constant over the range of pressures and volumes involved. A larger volume (V_4-V_3 as compared to V_2-V_1) results in a greater mean pressure ($\overline{P}_B$ as compared to $\overline{P}_A$) and a greater pulse pressure (P_4-P_3 as compared to P_2-P_1). B. For a given volume increment (V_2-V_1), a reduced arterial capacitance (curve B as compared to curve A) results in an increased pulse pressure (P_4-P_1 as compared to P_3-P_2). (From R. M. Berne and M. N. Levy. *Cardiovascular Physiology*. St. Louis: Mosby, 1967.)

portion of the cardiac cycle. Arterial pressure and volume therefore rise from point A_1 toward point A_2 in Figure 16-4A. As described in Chapter 13, the normal heart discharges most of its stroke volume during the early part of systole, the so-called rapid ejection phase. The maximum arterial volume, V_2, is reached at the end of the rapid ejection phase, and this volume corresponds to a peak pressure, P_2, which is the systolic pressure. The mean arterial pressure is ordinarily somewhat less than the arithmetic average of the systolic and diastolic pressures, as illustrated in Figure 16-1.

The difference between systolic and diastolic pressures ($P_2 - P_1$ in Fig. 16-4A) is called the pulse pressure. The pulse pressure corresponds to some volume increment, $V_2 - V_1$. The increment equals the volume of blood discharged by the left ventricle during the rapid ejection phase minus the volume which has run off to the periphery during this same phase of the cardiac cycle. In the case of a normal heart beating at a normal frequency, this volume increment is a large fraction of the stroke volume (about 80 percent). It will raise arterial volume rapidly from V_1 to V_2 and hence will cause the arterial pressure to rise from the diastolic to the systolic level (P_1 to P_2). During the remainder of the cardiac cycle, peripheral runoff will greatly exceed cardiac injection. The resultant arterial blood volume decrement will cause volumes and pressures to fall from point A_2 back to point A_1.

If stroke volume is now doubled, while heart rate and peripheral resistance remain constant, the mean arterial pressure will be doubled to $\overline{P}_B$ in Figure 16-4A. Thus, the arterial pressure will oscillate about this new value of the mean arterial pressure. A normal, vigorous heart will eject the greater stroke volume during a fraction of the cardiac cycle approximately equal to the fraction which obtained at the lower stroke volume. Therefore, the volume increment, $V_4 - V_3$, will be a large fraction of the new stroke volume and hence will be approximately twice as great as the previous volume increment ($V_2 - V_1$). With a linear P_A:V_A curve, the greater volume increment will be reflected by a pulse pressure ($P_4 - P_3$) which will be approximately twice as great as the original pulse pressure ($P_2 - P_1$). With a rise in both mean and pulse pressures, it is evident from inspection of the figure that the rise in systolic pressure (from P_2 to P_4) exceeds the rise in diastolic pressure (from P_1 to P_3).

Arterial Capacitance

To assess arterial capacitance as a determinant of pulse pressure, the relative effects of the same volume increment ($V_2 - V_1$ in Fig. 16-4B) in a younger individual (curve A) and in an older person (curve B) can be compared. Let cardiac output and total peripheral resistance be the same in both cases; therefore, $\overline{P}_A$ will be the same. It is apparent from B that the same volume increment will result in a greater pulse pressure in the less distensible arteries of the older individual ($P_4 - P_1$) than in the more compliant arteries of the younger person ($P_3 - P_2$).

Total Peripheral Resistance

Let total peripheral resistance (TPR) be increased in an individual with a linear P_A:V_A curve, as shown in Figure 16-5A. If heart rate and stroke volume remain constant, then an increase in TPR will evoke a proportionate increase in $\overline{P}_A$ (from P_2 to P_5). If the volume increments ($V_2 - V_1$ and $V_4 - V_3$) are equal at both levels of TPR, the pulse pressures ($P_3 - P_1$ and $P_6 - P_4$) will also be equal. Hence,

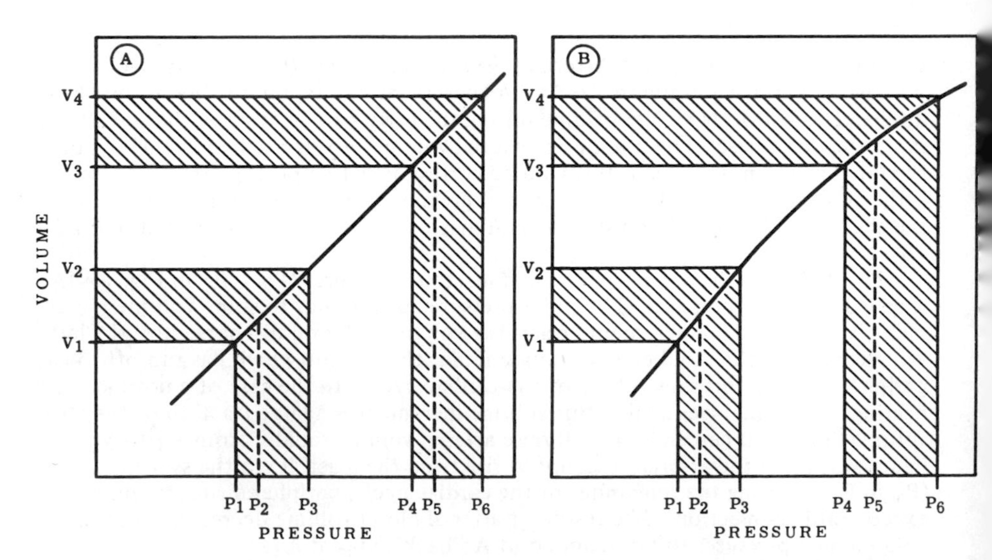

FIGURE 16-5. Effect of a change in total peripheral resistance (volume increment remaining constant) upon pulse pressure when the pressure-volume curve for the arterial system is rectilinear (A) or curvilinear (B). (From R. M. Berne and M. N. Levy. *Cardiovascular Physiology.* St. Louis: Mosby, 1967.)

systolic (P_6) and diastolic (P_4) pressures will have been elevated by exactly the same amounts from their respective control levels (P_3 and P_1).

It is conceivable that with a higher TPR in a steady-state situation a given stroke volume might not be ejected as rapidly as the same stroke volume against a lower TPR. In such case the net volume increment might not be as great for the same stroke volume since a bigger fraction of the peripheral runoff might occur during the longer ejection period. A somewhat smaller net volume increment would result in a proportionately smaller pulse pressure; in Figure 16-5A, $V_4 - V_3$ would be less, and so would $P_6 - P_4$. Under these circumstances an augmentation of TPR would be associated with a somewhat greater rise in diastolic than in systolic pressure.

Chronic hypertension, a condition characterized by a persistent elevation of TPR, occurs more commonly in middle-aged and elderly individuals than in younger persons. The $P_A{:}V_A$ curve would therefore possess the configuration shown in Figure 16-5B (which resembles curves B to E in Figure 16-3), rather than that displayed in Figure 16-5A. The type of curve in Figure 16-5B reveals that C_A is less at higher than at lower pressures. As before, if cardiac output remains constant, an increase in TPR causes a proportionate rise in $\bar{P}_A$ (from P_2 to P_5). For equivalent increases in TPR, the elevation of pressure from P_2 to P_5 will be the same in Figure 16-5A and B. Assuming the volume increment in B at elevated TPR to be equal to the control increment ($V_2 - V_1$), it is evident that the pulse pressure ($P_6 - P_4$) in the hypertensive range will greatly exceed that ($P_3 - P_1$) at normal pressure levels. In other words, a given volume increment will produce a greater pressure increment when the tube is more rigid than when it is more compliant. Hence, the rise in systolic pressure ($P_6 - P_3$) will far exceed the increase in diastolic pressure ($P_4 - P_1$). These changes in arterial pressure closely resemble those seen in patients with hypertension. Diastolic pressure is elevated in such individuals by 10 to 40 mm Hg above the average normal level of 80 mm Hg, and it is not uncommon for systolic pressures to be elevated by 50 to 150 mm Hg above the average normal level of 120 mm Hg.

Heart Rate

Increased heart rate with a *fixed stroke volume* would influence pulse pressure in a manner analogous to the increase in TPR and vice versa. A moment's reflection will lead one to the conclusion that doubling the minute output by a twofold increase in rate against a fixed TPR would produce the same effect as a twofold increase in TPR at a constant output.

Summary

It is apparent, therefore, that the pulse pressure is principally dependent upon volume increment (a function primarily of stroke volume) and arterial capacitance. Stimuli that alter pulse pressure usually do so by changing one or both of these factors. If a change in heart rate, for example, is not accompanied by an alteration in stroke volume, then pulse pressure may or may not be modified, depending upon the value of C_A at the new level of P_A. However, changes in heart rate are usually accompanied by changes in stroke volume, and this in turn will affect pulse pressure in accordance with the principles illustrated in Figure 16-4A.

The foregoing analysis gives some insight into the changes in stroke volume and in systolic and diastolic pressure, mean blood pressure, and pulse pressure to be expected during various stages of cardiac diseases. Although it must be kept in mind that compensatory changes are instituted in the heart action, the following situations will prevail initially. Changes to be anticipated in *aortic* or *mitral stenosis* (narrowing) are similar to those portrayed in Figure 16-4A (going down the pressure-volume line from right to left). The same situation might explain the changes to be anticipated in *mitral insufficiency,* considering that the volume actually ejected from the heart per beat is reduced.

Changes that might occur in *aortic insufficiency* (semilunar incompetence) are a reduction in effective stroke volume but an increase in actual stroke volume, since runoff is by regurgitation both back into the heart and through the peripheral tissue. As a consequence of the increased runoff, the diastolic pressure will decrease greatly. With no homeostatic stimulation to the heart, the systolic pressure will decrease only slightly because of the increase in the actual stroke volume to be pumped. The pulse pressure will thus increase and the mean pressure decrease. With homeostatic stimulation to the heart, a still larger stroke volume is pumped that acts to increase the systolic pressure and mean pressure toward normal. The diastolic pressure remains low, however, with a large pulse pressure – a common characteristic of aortic insufficiency.

The heart may compensate in part on a chronic basis for such alterations and thereby modify the changes that might be predicted. Thus, the ventricle in aortic insufficiency or aortic stenosis might beat more vigorously because of increased stretch of the myocardial fibers, with ultimate hypertrophy, and thus restore mean pressure toward normal. Similarly, with mitral stenosis or mitral insufficiency the contraction of the atrium would be more vigorous. Chronic hypertrophy of the wall of the atrium would tend to restore conditions toward normal.

FORM AND ALTERATION OF THE ARTERIAL PULSE; TRANSMISSION OF PULSE. THE VENOUS PULSE

The pressure pulses produced by the beating of the heart as it ejects blood into the aorta and those that are created in the region of the right atrium during contraction of the heart are transmitted through the elastic vessels toward the periphery. Hence, the pulse is a transient expansion of the vessel as a result of internal pressure changes. There is, therefore, a characteristic arterial pulse which extends to the smallest arteries, where it is finally obliterated. The venous pulse is seen best in the low neck region and is the result of backward transmission of changes in the region of the atrium. Pulse forms can be measured and analyzed by sphygomographs. These can be a button applied to the artery, transmitting the pulsatile variations to a tambour system, or a cup tambour placed over the vessel, recorded by high-frequency recording systems such as an optical manometer or electromanometer.

TRANSMISSION OF ARTERIAL PULSE

The mechanism for transmission of the arterial pulse can be understood with the aid of Figure 16-6. Sudden injection of a quantity of blood into a distensible tube causes a bulging and a rise in pressure at that point (A in the figure). Pressures lower in the tube will not rise until the elevated pressure at point A causes a small amount of fluid to pass along the vessel. This makes the vessel distend at point B while, because of elastic recoil, it begins to shrink at point A. The pressure at B continues to rise because of momentum given to the blood, and that at A continues to fall progressively, building up the pressure at point B. When the pressure is built up at point B, the process is repeated in segment C of the vessel, and so on down the length of the artery. It should be remembered that the movement of the pulse is quite rapid compared with the movement of the blood passing through the vessel. This is clarified by the lower part of the figure, which shows the analogy employing a row of billiard balls. Impact against the first ball moves it against the second one, and so on down

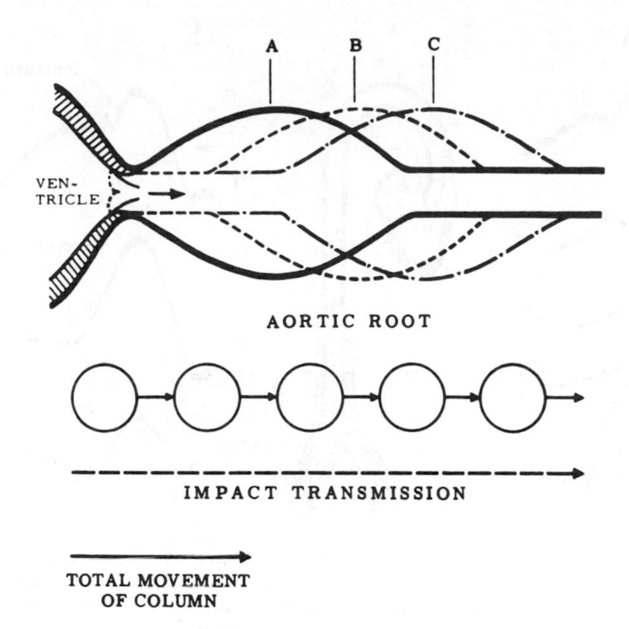

FIGURE 16-6. Mechanism of formation of pulse and pulse transmission.

the line at a rapid rate. The solid arrow (bottom) indicates the total distance moved by the balls. Note that this is much shorter than the distance traversed by the impact along the series of balls (dashed line). In actual fact the pulse wave velocity in the aorta is from 3 to 5 meters per second, while the mean blood velocity in the aorta is only 0.5 meter per second. The velocity of pulse transmission increases in passage through the smaller vessels. It is 7 to 9 meters per second in the subclavian and femoral arteries, and 15 to 40 meters per second in very small arteries of the arms and legs. The speed increases because of the lesser distensibility of the smaller arteries. With arteriosclerotic changes in the vessel, such as occur with old age or with hypertension, the pulse wave velocity speeds up, and it may increase to three times normal.

TRANSFORMATION OF THE PRESSURE PULSE IN TRANSMISSION

In A of Figure 16-7 is shown the pattern of the alteration of the pressure pulse contour from the root of the aorta to the peripheral arteries. Curve I shows modifications which are due to supraposition of free or natural vibrations which result whenever tension is suddenly changed in the arterial system, thereby setting up free vibrations. The best example is found in closure of the aortic semilunar valves. This creates a sharp variation, the *incisura* (labeled c), in the *catacrotic* (falling) limb of the curve. An impact following closure of the aortic valves during isovolumetric relaxation creates a small hump (d) below the incisura on the catacrotic limb. Impacts from the contracting atrium and isovolumetric contraction of the ventricle may modify the anterior (foot) portion of the curve as indicated (a, b), preceding the *anacrotic* (rising) limb of the curve. As this basic pattern moves through the aorta into the larger arteries, changes occur because of the elastic properties of the vessel, the movement of blood, friction encountered, and damping. All variations

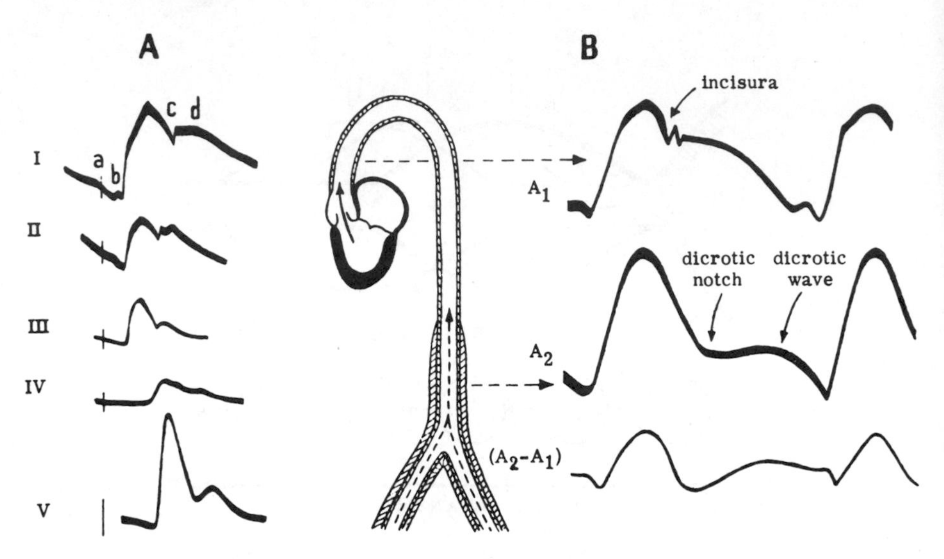

FIGURE 16-7. Transformation of the pulse in transmission. A. Series of human pulse records from: I, subclavian artery pulse; II, lower carotid pulse; III, brachial artery pulse; IV, radial pulse; V, femoral pulse. B. Transformation of pulse from root of aorta, A_1, to abdominal aorta and femoral branches, A_2. $(A_2\text{-}A_1)$ is the calculated *reflected* or *standing wave* in the distal portion of the arterial system. (A from C. J. Wiggers. *Physiology in Health and Disease.* Philadelphia: Lea & Febiger, 1949. B from R. S. Alexander. *Amer. J. Physiol.* 158:289, 1949, in Rushmer, 1955.)

superimposed on the primary curve are obliterated by the damping and frictional phenomena. The bends in the anacrotic and catacrotic limbs tend to disappear, and the pulse skeleton becomes more triangular in form. Finally, major changes in the pulse contour result from reflected waves. When blood is suddenly ejected into the aorta, the pulse, traveling at a speed of 7 to 9 meters per second, reaches the periphery before systole is over. Reflections occur from bifurcation and branches, and even from the peripheral arterioles, which move back up the aorta and meet the oncoming wave during the rest of systole. Standing waves of pressure are created in the descending aorta and peripheral arterial trunks, which tend to augment the peak amplitude of the pulse wave, to obscure the incisura and replace it by a *dicrotic notch,* to amplify the catacrotic limb of the curve to produce a dicrotic wave, and to depress the trailing portions of the pulse wave. An overall increase in systolic pressure and a decrease in diastolic pressure result. Figure 16-7 illustrates how the reflected wave (lowest curve, $A_2 - A_1$) resonates with portions of the oncoming pulse curve (A_1) to amplify certain portions of it and to cancel others. The resulting pulse contour (A_2) resembles that recorded from the femoral, brachial, or radial arteries.

THE CENTRAL VENOUS PULSE

A central venous pulse like that exhibited by fluctuations in the jugular veins represents a retrograde transmission of the pressure variations which originate in the right atrium of the heart. This curve should be reviewed for an understanding of the contour of the central venous pulse as registered by a cup tambour over the vein of the neck. As can be seen in Figure 16-8, the positive and negative variations of the jugular pulse correspond to those which have been observed with direct pressure recording in the atria. The positive waves are labeled a, the result of atrial systole; c, associated with isovolumic ventricular contraction; v, the peak pressure attained just before opening of the atrioventricular (AV) valves. The negative variations are x, the trough between a and c; x', the fall in pressure which is a result of the descent

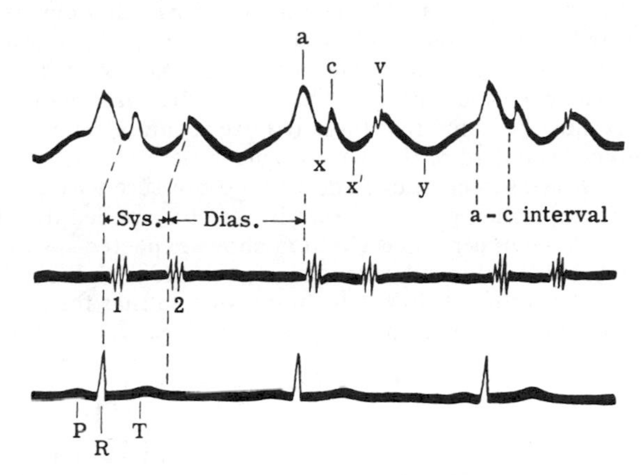

FIGURE 16-8. Simultaneous recording of venous pulse, heart sounds, and ECG (lead II), to demonstrate time relationship. Note necessary correction for delay of propagation of venous pulse. (From C. J. Wiggers. *Physiology in Health and Disease.* Philadelphia: Lea & Febiger, 1949.)

of the base of the atrium during ventricular contraction; and y, the rapid fall coinciding with rapid filling of the heart. One of the major differences from the atrial curve is that the component may be more noticeable, because it has been amplified and exaggerated by impact from the underlying carotid artery, which has been picked up by the superficial recording tambour.

The value of an examination of the central venous pulse trace is that it yields certain information about the cardiac cycle. For example, the interval between the beginning of the a and the c wave gives an approximation of the P-R interval, which should be less than 0.2 second in the normal subject. The time between c and v roughly corresponds to ventricular systole. The time from v to y is an index of the duration of the rapid filling phase of the heart; from y to the beginning of a, the reduced filling phase. Partial or complete heart blocks can be detected by the relationship of the a to the c and v waves. With atrial fibrillation there is no apparent wave. With nodal beats, the a and c waves may summate. In tricuspid insufficiency the systolic wave, the c wave, rises rapidly to a summit upon which vibrations corresponding to the murmur may be recorded. These and other abnormalities of cardiac function may be detected by careful examination of the jugular pulse tracing.

HUMAN BLOOD PRESSURE

The measurement of systolic, diastolic, and mean blood pressure in the human subject by the direct method, i.e., introduction of needles directly into the artery with registration by high-frequency optical manometers of the Hamilton or Gregg type, or alternatively by transducers and electrical manometers, has become commonplace in clinical practice. On the other hand, for routine use and common office practice it is customary to employ the indirect method, using a *sphygmomanometer.* As illustrated in Figure 16-9, the components of the sphygmomanometer are: (1) a compressing bag made of rubber surrounded by an unyielding cuff for application of an extra-arterial pressure; (2) a manometer, usually mercury with a calibrated scale, by which the pressure is read; (3) an inflating bulb or some other device by which pressure can be elevated in the system; and (4) a variable-control exhaust such as a needle valve, by which the system can be deflated either gradually or rapidly. Aneroid manometers for convenience and portability have been preferred more recently to the mercury type. It must be noted that these require accurate calibration at intervals against the reference standard of the mercury manometer.

The principle of measurement of human blood pressure with the sphygmomanometer is illustrated in the figure. Pressure in the cuff applied over the artery in the medial region of the upper arm is elevated above expected systolic pressure in the vessel, thereby obliterating pulse transmission to the periphery. With the use of the exhaust valve, the system is slowly decompressed so that the mercury column drops through the range of expected pressures. The observer needs criteria which

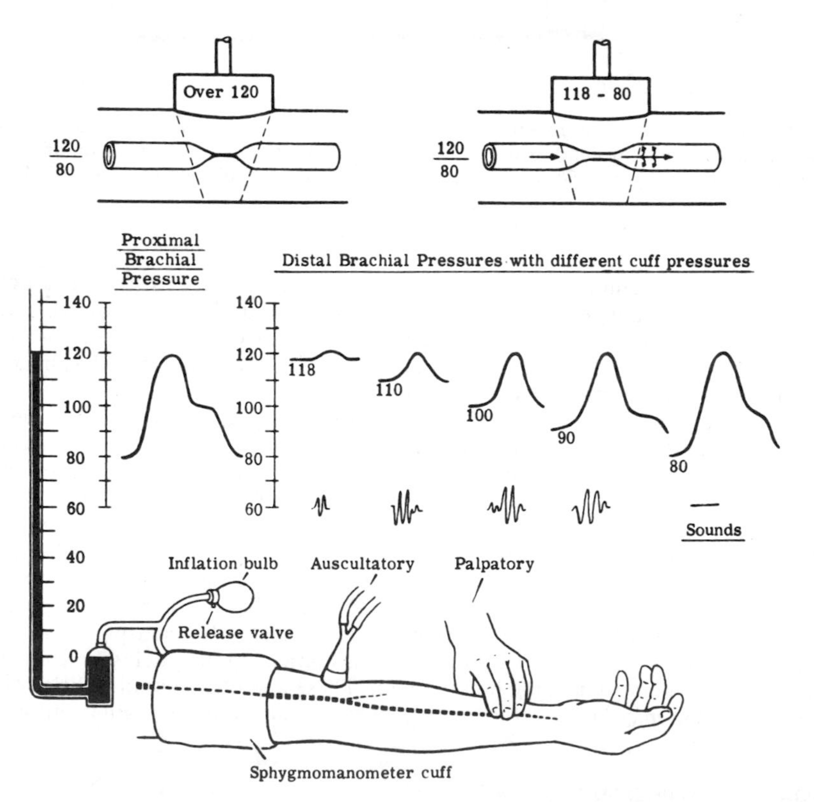

FIGURE 16-9. Principles of sphygmomanometry illustrating the *palpatory* and *auscultatory* techniques for measurement of human blood pressure. (Upper left) Brachial artery completely collapsed at pressure in cuff over 120 mm Hg. (Upper right) Decompression of cuff showing progressive transmission of pulse and accompanying sounds.

tell when the pressure in the manometer system is equal to the systolic pressure and when the level of the pressure in the system matches that of the diastolic pressure to the vessel. The *palpatory method,* originally used by Riva-Rocci, supplies an index for the estimation of systolic pressure. The pressure in the system is inflated to 30 mm Hg above the expected pressure. The cuff system is deflated at a rate of about 2 to 3 mm Hg per heart beat. Continuous palpation of the artery below the cuff gives information of the first appearance of the radial pulse. This approximates systolic pressure. It is obvious that this method will not yield an estimated diastolic pressure. For that, other criteria are needed.

The *auscultatory method* supplies criteria for the estimation of both systolic and diastolic pressure. They are based upon sounds heard by light application of a stethoscope on the artery below the cuff. These sounds were first observed by Korotkow in 1905. In principle, as pressure falls the brachial artery opens momentarily at first, then more and more until it is entirely patent. This progression produces characteristic changes in sounds heard in the stethoscope placed on the brachial artery branches below the cuff in the cubital fossa. The sounds have been divided into several phases as follows: *Phase 1* (10 mm Hg): The artery opens momentarily and the sharp, tapping first sound (systolic pressure) is heard, which becomes progressively louder during the phase. *Phase 2* (10 to 15 mm Hg): The sound becomes softer and may develop a hissing quality similar to a murmur. *Phase 3* (10 to 15 mm Hg): The sound may become louder and have a thudding quality. *Phase 4:* The sound becomes suddenly reduced in intensity and develops a muffled quality during a 5-mm-Hg drop. This muffling has been taken as the criterion for diastolic pressure (Roberts et al., 1953); others have recommended the disappearance of sound as the criterion (Bordley et al., 1951). Although the disappearance of sound is a readily detected end-point for the novice, on theoretical grounds the muffling of sound appears to supply a more accurate criterion of diastolic pressure.

The mechanism for the production of the sounds is as follows. The first sound heard has a sharp, tapping quality, attributed to turbulent flow through the momentarily opened vessel. As the artery remains open for longer intervals, but still partially compressed, additional turbulence develops at the sites of partial constriction to contribute to the development of the Korotkow sounds that are heard in the later phases.

Figure 16-9 presents in diagrammatic form the changes in systolic, diastolic, and mean pressure above and below the point of compression by the cuff of the sphygmomanometer and illustrates graphically the criteria for systolic and diastolic pressure. The left side of the figure shows the situation when the pressure within the cuff is over 120 mm Hg and the artery is entirely collapsed for a portion of its length beneath the cuff. The systolic pressure is approximately 120, the diastolic 80, in the segment of the artery above the point of compression. To the right appears the development of the pulse contour and sounds at varying pressures below the point of occlusion as compared with the proximal brachial pressure. It can be seen that at 118 mm Hg the very top of the proximal pressure pulse curve is transmitted underneath the cuff to yield the first sound and supply the index of systolic pressure. The pressure slowly falls through the indicated range, the character of the sound changing until the artery is open continuously at a level of 80 mm Hg. This is the point of disappearance of sound.

ACCURACY OF THE INDIRECT METHOD OF MEASUREMENT; COMPARISON WITH THE DIRECT METHOD

An important factor to be considered in the influence of the tissue between the cuff and the artery, and with this the width of the cuff. Actually, the length of artery compression is not exactly known. Historically, cuffs 5 cm wide were used until Von Recklinghausen in 1901 showed that readings obtained with cuffs of that width were too high because excessively high pressures were needed to collapse the artery with such a narrow cuff; he recommended, therefore, a 10- to 12-cm width of cuff. On the other hand, if the cuff is too wide, the initial pressure starting out under the cuff has too little force to penetrate the entire length of the occluded vessel until the extra-arterial pressure has fallen to at least 8 mm Hg below the existing systolic pressure in the artery. By making comparisons of direct and indirect methods, recommendations for width of cuff were established, and they vary with the thickness of the appendage used for registration of pressure. Roughly, the cuff width should be 20 percent wider than the diameter of the appendage. It is now recommended that the following cuff widths be used: for infants, 6 cm; for children, 9 cm; for adults, 12 cm; for obese adults, 20 cm. Sometimes the blood pressure is measured in the leg; the recommended width for the thigh is 18 cm.

Even with these improvements in cuff width, comparisons with the direct method (optical manometers, strain gauge transducers) revealed that the auscultatory systolic pressure may be 3 to 4 mm lower than it is when the direct method is used. Comparisons of diastolic pressure by the two methods has yielded somewhat conflicting results. Although Bordley et al. (1951) have recommended the *disappearance* of sound as the criterion for diastolic pressure, the careful studies of Roberts et al. (1953) confirm the *muffling* of the sound (phase 4) as the best criterion. In this connection, one must bear in mind that in certain pathological hemodynamic conditions cessation of sound does not occur even at very low pressure values. Unquestionably, in these circumstances muffling supplies a better criterion of diastolic pressure.

NORMAL VALUES

Systolic and diastolic pressures vary with age. In the newborn they are about 80/45. At 4 years of age they increase to 100/65. The relationship of these pressures to age is shown graphically in Figure 16-10 for a span from 16 to 64 years. The rise in pressure with age is attributed to arteriosclerotic changes in the vessels and loss of elasticity. The increase in pulse pressure is probably the result of loss of elasticity in the central vessels. Below the age of about 35 the pressures in the female are slightly lower than in the male. However, at the age of 40 to 45 both systolic and diastolic pressures in the female rise at a greater rate than in the male. Although no good explanation has been offered for this trend, it has been attributed to the hormonal changes at the menopause.

FACTORS IN PRESSURE VARIATIONS

Racial Factors

It has been observed that blood pressures of Orientals are generally lower than those of Americans and Europeans. These differences are exemplified in Figure 16-11, in which the blood pressure of Londoners is compared with the blood pressures of Fijians and East Indians. Note the significant uptrend in the diastolic and systolic pressure of the European population compared with that of the East Indians and Fijians at age 50 and above. It is not clear whether genetic, environmental, or dietary factors are the basis of the differences. It has been suggested that the high-cholesterol diet of the American and European may be responsible for these changes.

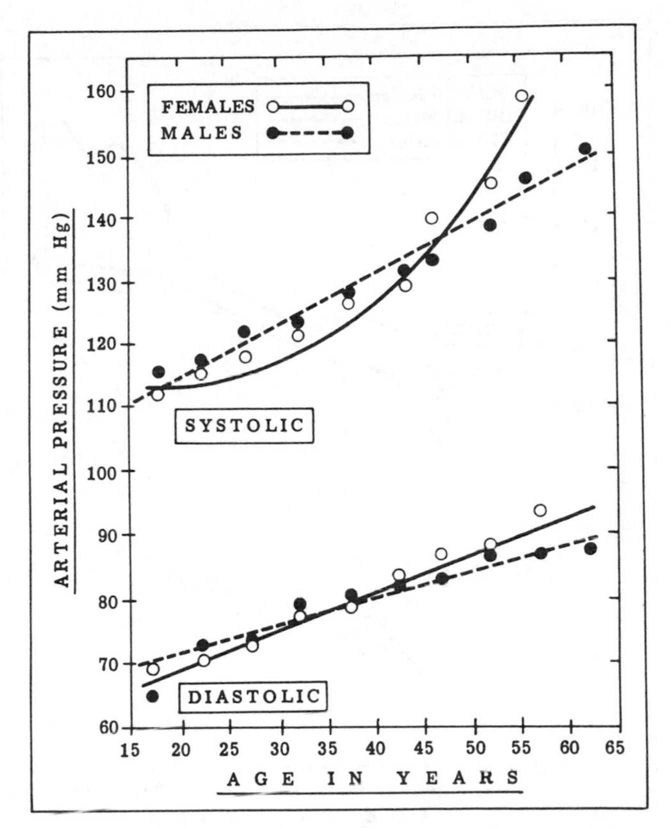

FIGURE 16-10. Relationship of systolic and diastolic pressure to age and sex. (From J. N. Morris. *Mod. Conc. Cardiovasc. Dis.* 30:635, 1961. By permission of the American Heart Association, Inc.)

Other Factors

Emotional factors involving enhanced activity of the sympathetic nervous system have marked influences on blood pressure. These factors must be taken into account when a basal blood pressure reading is desired by the examiner.

Posture has a variable influence on blood pressure. When a person stands, systolic and diastolic pressures both rise approximately 10 mm Hg. There is a slightly greater rise in diastolic than systolic pressure, resulting in a small decrease in pulse pressure. Gravitational effects upon venous return may decrease cardiac output, and compensatory increase in pulse rate may develop.

Exercise has a profound influence on systolic and diastolic pressure – more on systolic than on diastolic. This is related to the degree of exercise. The systolic pressure increases to a greater degree because of the enhanced stroke volume of the heart with severe exercise. Diastolic pressure decreases at first because of dilation of the vessels of the skeletal muscles during exercise. On stopping exercise there may be a sudden fall in blood pressure, particularly systolic, which may be the result of relaxation of the abdominal muscles. A secondary rise follows shortly, which persists for varying periods, depending on the severity and duration of the initial

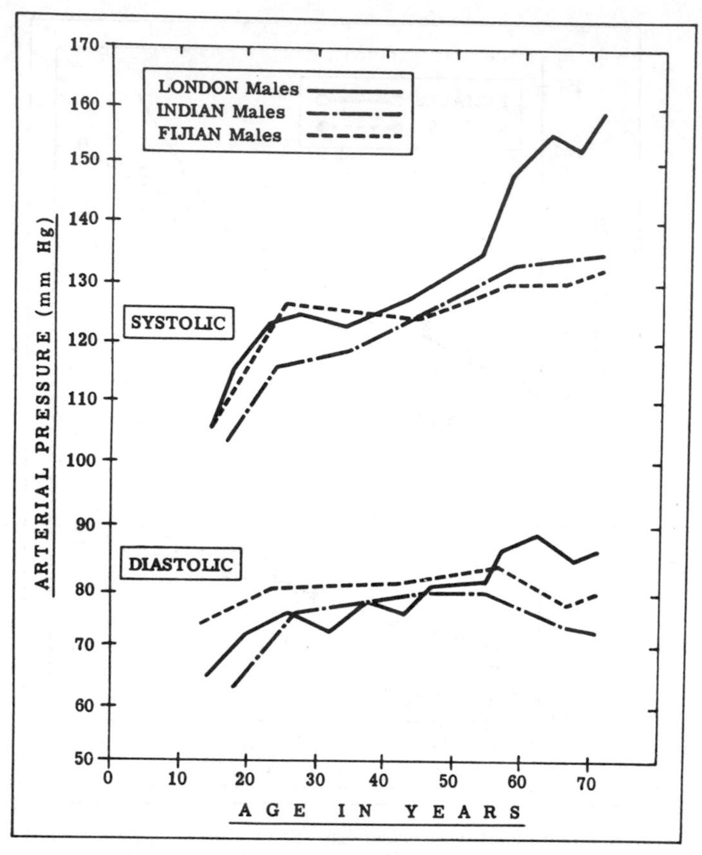

FIGURE 16-11. Racial factors and human blood pressure. (From J. N. Morris. *Mod. Conc. Cardiovasc. Dis.* 30:635, 1961. By permission of the American Heart Association, Inc.)

exercise. *Respiration* has an influence upon systolic and diastolic pressures which may create variations of as much as 10 mm Hg. In simplest form these are decreases with inspiration and increases during expiration.

Summarizing, pressures of 140/90 mm Hg may be regarded as ceilings for blood pressure readings under basal conditions at all ages. Values higher than this must be regarded as hypertension. This topic will be considered in a later section.

TECHNIQUES FOR INDIRECT MEASUREMENT OF BLOOD PRESSURE

1. The patient should be basal or observed during standard conditions. This implies a period of 10 to 12 hours after the last meal and a 30-minute rest before the pressure is taken.

2. The position should be kept standard. Pressure is usually taken with the patient in the sitting position, but many physicians prefer the horizontal, which supplies better relaxation to the patient.

3. The pressure should be measured approximately at the level of the heart. The manometer should be checked to see that the mercury column is at true zero.

The vent at the top of the mercury column should be clear, so as not to impede the movement of the column.

4. The cuff should be evenly applied, with the inflatable part over the artery. It should be snugly wound, but not too tightly or too loosely.

5. The radial artery should be palpated for best placement of the stethoscope.

6. The stethoscope should be lightly applied; heavy pressures may cause abnormal eddy currents and spurious sounds. The stethoscope should not touch the cuff.

7. It is best to use the palpatory method first; this supplies a check for auscultatory gaps, the cause of which is not entirely known.

8. The first few readings should be discarded if the patient appears nervous or apprehensive. Too rapid or too slow decreases in the mercury column give errors in reading. The column should be dropped about 2 mm per second or approximately 2 mm per heart beat. It may be desirable to take the pressure in both arms, particularly if the value does not appear normal.

VENOUS PRESSURE AND VENOUS RETURN

The ultimate force which develops the venous pressure and venous return, the so-called *vis a tergo* (literally, "force from behind"), is derived from the left ventricle. The venous pressure represents what is left after passage of the blood through the arterioles, capillaries, and venules, coming out at about 15 mm Hg (20 cm H_2O). This force is adequate to bring blood back when the body is in the horizontal position (Fig. 16-12) but when it is in the upright position gravity introduces complicating factors. Thus, in the upright position (Fig. 16-13B), under the influence of

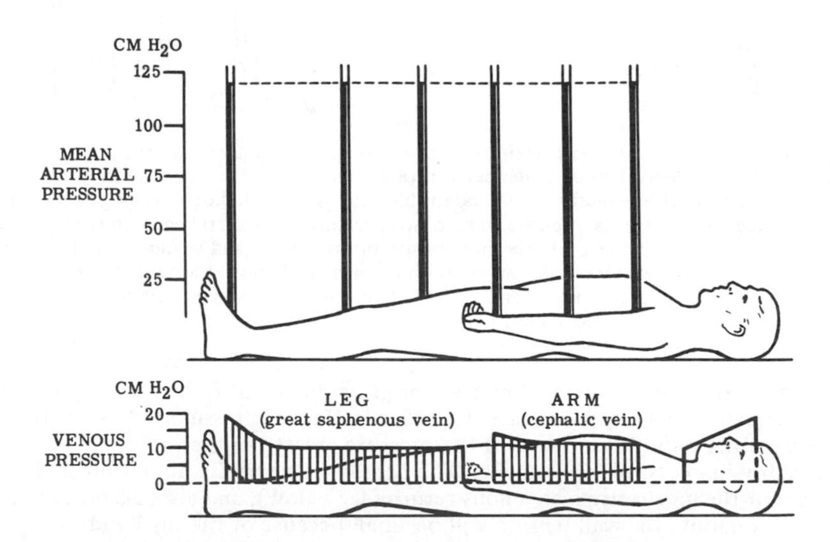

FIGURE 16-12. Mean arterial and venous pressures in reclining subjects. A. The mean arterial pressure diminishes but slightly from the arch of the aorta to the arterial branches, e.g., the radial. This pressure gradient is responsible for the flow of blood through the system. B. The peripheral venous pressure also has a very gradual diminution in pressure from the periphery toward the heart. In the smaller venous branches the pressure gradient is considerably steeper. (After A. Ochsner, Jr., R. Colp, Jr., and G. E. Burch. *Circulation* 3:674–680, 1951.)

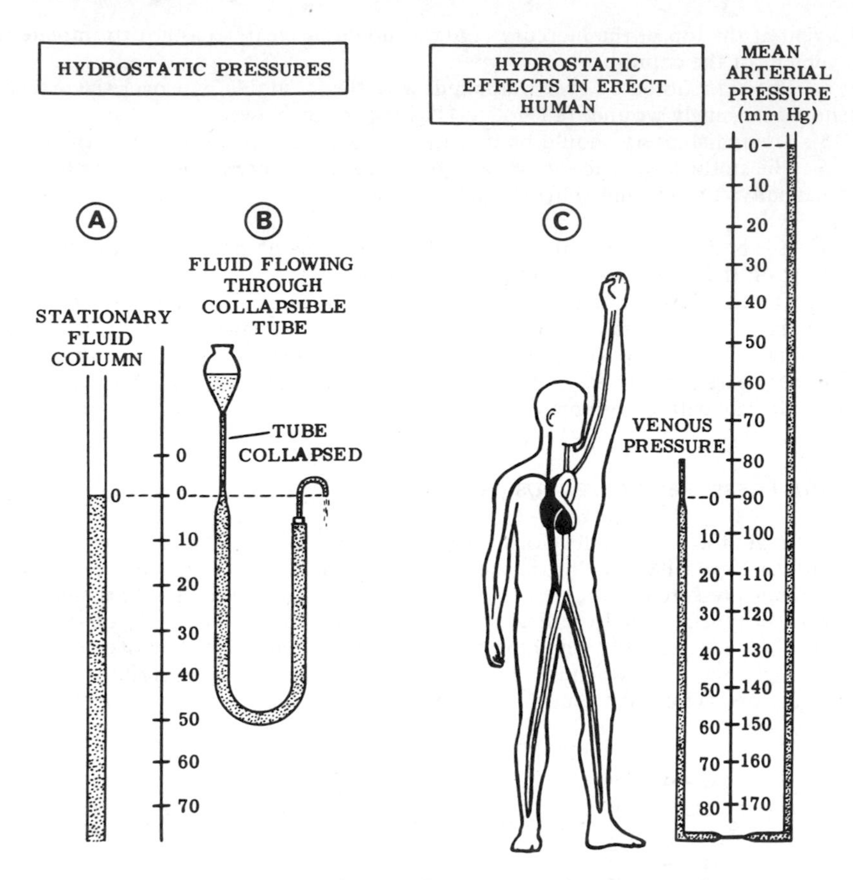

FIGURE 16-13. The nature and significance of hydrostatic pressures. A. The pressure in a column of fluid is dependent upon its specific gravity and the vertical distance from the point of measurement to the meniscus. B. A collapsible tube is distended only so long as the internal pressure exceeds the external pressure. These two pressures are exactly equal in the portion of the tube which is collapsed. C. In the erect position, the arterial and venous pressures are both increased by some 85 mm Hg at the ankle. With the arm held above the head, the arterial pressure at the wrist is about 40 mm Hg and the effective venous pressure is zero down to a level just above the heart. (After Rushmer, 1970.)

the hydrostatic pressure created by the upright columns of fluid, the arterial pressure in the dorsalis pedis artery increases to 175 mm Hg; the pressure is 85 mm Hg in the corresponding vein; this means that the pressure created in the capillaries in the foot is probably in excess of 100 mm Hg. The capillaries normally do not break down because of the mechanisms of venous return (see below), and also, according to the Laplace equation, the wall tension will be small because of the small radius.

In the erect position, there is an immediate shift of blood from the head and upper parts of the body to the abdominal vessels and the vessels of the lower limbs. The veins of the abdomen and lower appendages are normally elliptical in outline and not fully filled in a horizontal position. In the vertical position they increase in fullness, and unless compensatory mechanisms are brought into play, pooling of blood in these areas and excessive capillary filtration may be expected. In man a number of compensating mechanisms are available which minimize excessive development of high venous pressures, filling, and excessive capillary filtration. These include (1) vasomotor reflexes, both arterial and venomotor, (2) the influence of respiration, (3) the massaging action of skeletal muscles, and (4) other mechanisms which aid and favor venous return.

Because of the gravitational effect, the arterial blood pressure may be as low as 50 mm Hg in the arteries of the brain (intravascular pressure, +90 mm Hg −40 mm Hg, hydrostatic pressure). The total venous pressure may be as low as −30 mm Hg (−40 mm Hg hydrostatic pressure +10 mm Hg, normal venous pressure). This net negative hydrostatic pressure results in a negative transmural pressure in veins of the neck and head.

If at any point along the jugular vein the pressure within it equals or is less than the external tissue pressure, the vessel collapses at that level. In an analogy, if a thin-walled tube containing no air is arranged as indicated in Figure 16-13A, the fluid from the reservoir will flow through the tube in response to a pressure gradient. The tube collapses at a level just above that of the outflow tube. Below this level, the internal pressure exceeds the external pressure, and the tube is distended by hydrostatic pressures which increase progressively toward the lower portion of the system. Above the zero level, the pressure within the collapsed tube is equal to the external pressure.

Technically, a free-falling body has no weight because all the potential energy is converted into kinetic energy (movement) or lost as friction (heat). Thus, even though there is fluid flowing through the collapsed portion of the tube, the lateral pressure exactly equals the external pressure. If a normal man assumes a semi-reclining position with his head and trunk oriented about 30 to 45 degrees from the horizontal plane, the lower portion of the jugular vein is distended, but at some point along its course the vein becomes collapsed because the venous pressure equals tissue pressure. This represents the level of zero effective venous pressure.

In the upright position, the pressure in the neck veins falls to 0 mm Hg, and the atmospheric pressure on the outside of the neck causes the veins to collapse all the way up to the skull, particularly with prolonged inspiration. However, the veins of the skull and venous sinuses are in a noncollapsible chamber, and in the standing position a negative venous pressure of about 10 mm Hg exists in the saggital sinus because of the hydrostatic pressure difference between the top and base of the skull.

The cerebrospinal fluid (CSF) pressure and cerebral venous pressure vary together because these fluids are confined within a relatively rigid chamber. Hydrostatic columns in the arteries, capillaries, and veins are balanced by equal changes in extravascular hydrostatic pressure. As a consequence, the cerebrospinal circulation exhibits a stability not exceeded in any other tissue of the body. As was learned in Chapter 6, other mechanisms, both extrinsic (cardiovascular reflexes) and intrinsic (metabolic), contribute to the stability of the cerebral blood flow (CBF).

The heart does not necessarily do more work when the body assumes the upright position from the horizontal. Since the vessels of the body operate on the U-tube principle, with continuously moving fluid a siphon effect is created in the effluent limb of the U system. Work of the heart thus is concerned only with overcoming resistance in the peripheral vascular bed, and this resistance remains essentially the same in all postures.

DETERMINANTS OF VENOUS RETURN

A precise definition of the term *venous return* is difficult because it has been employed in various ways, to mean increased volume flow to the ventricles, increased central venous pressure, increased filling pressure in the ventricles, and so on. The important net result is that cardiac output is ultimately increased by increased venous return, or decreased by reduced venous return.

Guyton (1971) has strongly proposed the *mean circulatory pressure* (also called *mean systemic pressure*) as one of the major factors that determine the rate at which blood flows from the vascular tree into the right atrium of the heart, and, therefore, that control the cardiac output itself. This concept has been discussed thoroughly in Chapter 15, under Relationship Between Cardiac Output and Venous Return, and the reader should review that section to refresh his understanding of the concept.

The mechanism obviously has important connotations with regard to venous return and should be examined as a basis for understanding all venous return mechanisms. Guyton and his associates have depicted the tendency for blood to return to the heart from the circulatory system by a venous return curve which is a plot of blood flow into the right atrium *against* atrial pressure (Fig. 16-14). The experiment was done by replacing the right ventricle with a pump, to vary the rate of removal of flow from the right atrium and to pump it back into the pulmonary artery, completing the circuit via the pulmonary circulation. Flow was measured with a rotameter. The pumping rate and tubing system were adjusted so that a wide range of right atrial pressures was achieved (+8 to −14 mm Hg).

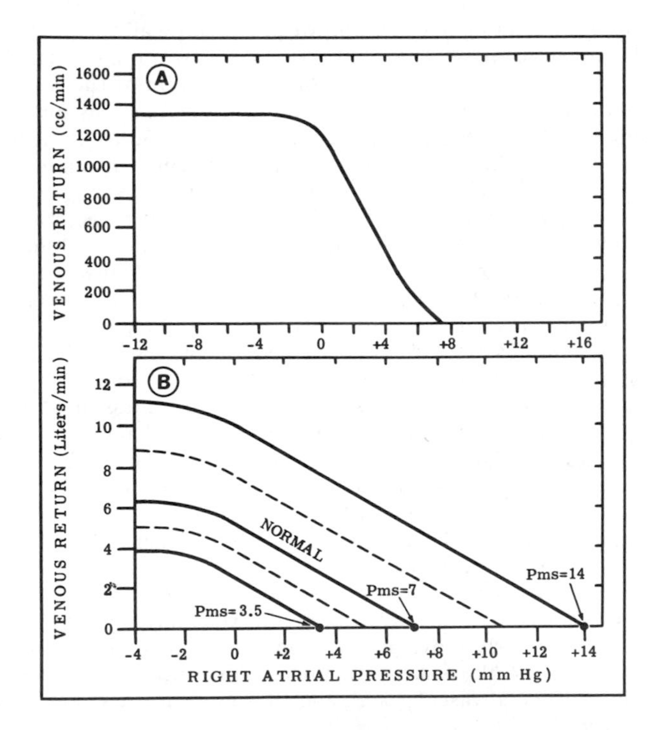

FIGURE 16-14. A. Normal venous return curve for the dog. (From A. C. Guyton, A. W. Lindsey, B. Abernathy, and T. Richardson. *Amer. J. Physiol.* 189:609–615, 1957.) B. Venous return curve adjusted to values observed in man, showing the normal curve when mean systemic pressure (Pms) is 7 mm Hg, and showing the effect of altering the mean systemic pressure. Dashed lines represent effects of intermediate stages of volume depletion or expansion. (From A. C. Guyton. Circulatory Physiology. In Guyton [Ed.], *Cardiac Output and Its Regulation.* Philadelphia: Saunders, 1963. P. 201.)

Figure 16-14A illustrates that venous return reaches a maximum value and remains on a plateau at all right atrial pressures more negative than −2 to −4 mm Hg because of collapse of veins leading into a right atrium at negative pressures. The low right atrial pressure sucks the walls of the veins together and prevents the negative pressure from sucking blood through the veins. This sets pressure in the veins outside the chest at a balance point equal to atmospheric pressure (zero pressure) and prevents any further increase in flow.

As the right atrial pressure rises to positive values, venous return falls, and it reaches zero when right atrial pressure has risen to the mean circulatory pressure, i.e., +7 to 8 mm Hg. This result is due to the distensible nature of the venous system so that any increase in back pressure causes blood to dam up instead of returning to the heart.

When mean circulatory pressure is increased by expanding blood volume, by sympathetic stimulation, and by increased contraction of skeletal muscles (milking out veins, etc.), the curve moves up and to the right. With loss of sympathetic tone, or loss of blood volume, the curve moves down and to the left (Fig. 16-14B). That is, the greater the difference between the mean systemic pressure (mean circulatory pressure) and the right atrial pressure, the greater the venous return, and vice versa (i.e., the "pressure gradient for venous return").

Finally, it should be emphasized that a venous return curve has optimal value only when used in conjunction with a cardiac output curve (Chap. 15).

MECHANISMS FOR INCREASED VENOUS RETURN

One of the major aids to venous return is the influence of respiration. This has been called the *abdominothoracic pump mechanism.* During inspiration there is a decrease in intrathoracic pressure and an increase in intra-abdominal pressure resulting from the downward pressure of the diaphragm and auxiliary muscles of the abdomen. Both events favor forward flow to the heart. In expiration the effect is reversed. The upper portion of Figure 16-15 illustrates the principle involved.

In the lower portion of the figure, the relative importance of the developing negativity of the intrathoracic space and the effect of positive pressure on the large vessels of the abdomen resulting from the descent of the diaphragm and contraction of abdominal muscles are shown from a representative experiment in the dog in which these pressures were measured simultaneously with flow in the superior and inferior venae cavae. To the left is the condition of the closed-thorax animal breathing normally. Note that intra-abdominal pressure rises as intrathoracic pressure falls. Under these circumstances flow in both the superior vena cava and the inferior vena cava is facilitated during the inspiratory phase. To the right is the situation in the open-thorax animal. Intrathoracic pressure rises to atmospheric pressure. With this increase, superior vena cava flow with inspiration is impaired, showing its general dependence upon the intrathoracic negative pressure variations. Enhanced diaphragmatic and abdominal respiratory movements now cause greater variation in intra-abdominal pressure, and it is to be noted that the inferior vena cava flow still rises appreciably during inspiration. Thus, an important mechanism for venous return is the cyclic variation in intrathoracic negativity and the intra-abdominal positive pressure variations during the respiratory cycle.

A second important mechanism favoring return results from the massaging action of skeletal muscles on the veins, aided by the presence of valves in the veins. In effect, the muscles "milk" the blood back to the heart from one venous segment to the next. Continued pulsation of the arteries against accompanying veins, aided by the venous valves, also favors return.

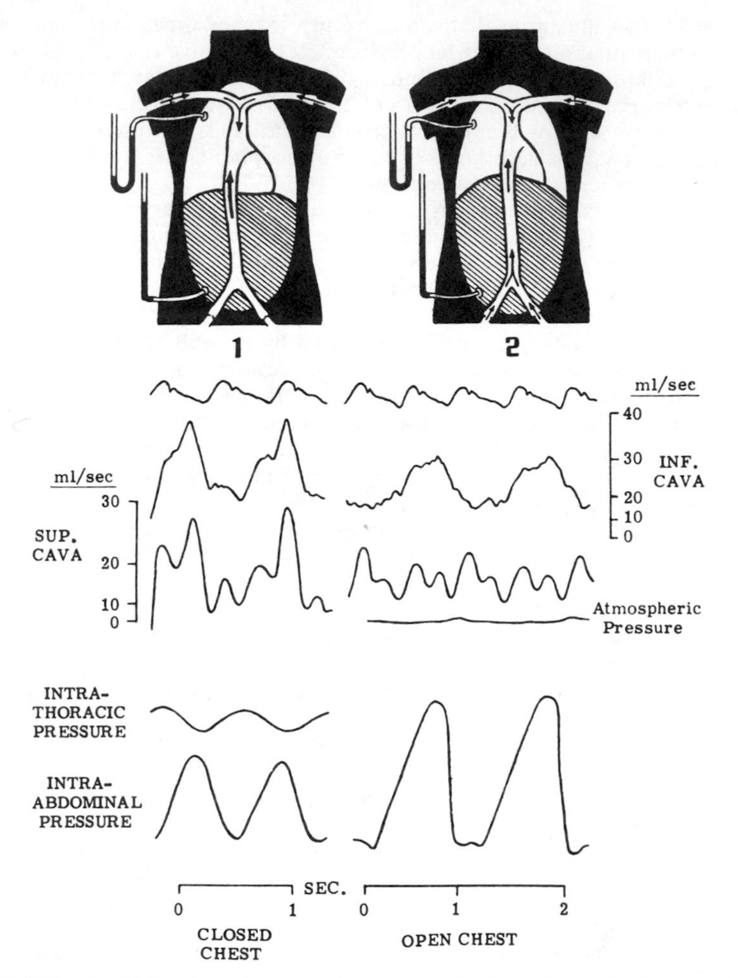

FIGURE 16-15. (Top) The abdominothoracic pump mechanism. 1. Inspiration; accelerated flow into thorax. 2. Expiration; reduced flow into thorax. (Lower half) Inferior vena cava and superior vena cava flow in the dog as influenced by variations in intrathoracic and intra-abdominal pressure. Top trace, aortic pressure; second, inferior vena cava flow (measured with bristle flowmeter); third, superior vena cava flow; fourth, intrathoracic pressure (note, to right, shift upward to atmospheric pressure with open chest); fifth, intra-abdominal pressure (note larger fluctuations to right resulting from greater breathing effort with open chest). (Top from Rushmer, 1970. Lower half from G. A. Brecher. *Venous Return.* New York: Grune & Stratton, 1956. By permission.)

Another mechanism has been called the *cardiac effect.* When the base of the heart drops during contraction of the ventricles, the atria and large vessels are stretched and there is an appreciable dip in atrial pressure (Fig. 13-7). Although sharp drops in inferior vena cava (IVC) and superior vena cava (SVC) flow measured in the dog occur during atrial systole (Brecher, 1956, 1963), a rapid upswing of inflow follows, coinciding with ventricular systole. Brecher believed this was an active drawing or sucking of blood toward the heart during ventricular systole (a manifestation of *vis a fronte,* "force from the front"). Later in the cardiac cycle, when the ventricle relaxed (isovolumetric relaxation), Brecher observed a second acceleration of vena cava flow during ventricular diastole, presumably contributed to initially by the elastic recoil of the ventricular walls. This mechanism also contributed to the vis a fronte. The passive process most likely to be involved is the systolic action on noncontractile elements of the heart which are elastically deformed beyond their equilibrium state, thereby storing potential energy which is released through elastic recoil during diastole. Although often manifested by a phase of negative intraventricular pressure, a negative phase is not essential for the sucking action to occur, in Brecher's view. An analogy of this action is the draining of fluid into an elastic suction bulb previously collapsed.

Atrial systole has two functions, according to Brecher (1963): It (1) ejects some blood into the ventricular cavity and (2) passively enlarges the central venous reservoir ("collapse chamber") by briefly slowing down or stopping atrial inflow (see also Guyton, 1971) and by a back-pressure effect widening the venous reserve chambers, creating a pool from which the next ventricular filling derives its supply. The central veins are particularly suitable for a reservoir function because, through partial collapse of their walls, their contents can change rapidly without change of pressure. Brawley et al. (1966) have substantially duplicated the findings of Brecher but elected a different interpretation. In all their records, acceleration of forward IVC and SVC flow corresponded to decrease in right atrial pressure, and deceleration of caval flow corresponded to increase in atrial pressure. Right ventricular contraction, on the other hand, influenced vena cava flow only as it altered right atrial pressure. Rather than an acceleration of flow with ventricular systole and relaxation, the apparent conclusion was that atrial systole periodically reduced the baseline vena caval flow, which returned to normal during the remainder of the cardiac cycle. Nevertheless, Brecher and Kissen (1958) demonstrated that dog ventricles of approximately normal functional residual capacity could fill by suction at zero ventricular inflow pressure, the basis for the vis a fronte during ventricular diastole. These studies were carried out in open-chest dogs and semi-isolated hearts involving radical surgery. Whether such a mechanism can be extrapolated to the unanesthetized intact mammalian organism remains unsettled.

MEASUREMENT OF VENOUS PRESSURE

Although venous pressure values may be expressed in millimeters of mercury, because of their relatively small magnitude it is customary to give them in millimeters or centimeters of water. The measurement of venous pressure entails careful consideration of the point of zero reference, which has been referred to as the *phlebostatic axis.* This is the pressure in the vein equal to atmospheric pressure at the zero level of the system taken as the base of the tricuspid valve. The anatomical landmark which may be used for the phlebostatic axis is the point of junction between the lateral border of the sternum and the fourth intercostal space. More precisely, the phlebostatic axis is the line of junction between a transverse plane of the body passing through the points of junction of the lateral margins of the sternum and the fourth interspace, and a frontal plane through the sternum to the back. When the body changes its position from vertical to horizontal the reference should be to the *phlebostatic level,* which represents a plane passing through the phlebostatic axis and parallel to the horizon.

Methods for measuring venous pressure are either direct or indirect. Direct techniques involve needle puncture of the vessel and recording of the pressure by electromanometers, or even by means of connections to a water manometer containing anticoagulant solution. An L tube will serve the purpose. Several simple indirect methods provide satisfactory indices of the height of venous pressure. The jugular vein is normally collapsed when the subject is in the erect position, but when venous pressure is elevated, as in congestive heart failure, it may be filled and prominent when directly observed. Another method is as follows: The hand is hung at the side until the veins are well filled; then the hand and arm are slowly raised until the point of collapse of the veins becomes visible. The level of collapse of the vein above the phlebostatic level represents the venous pressure in the veins of the hand. For another convenient index, the examiner is to place one hand of the recumbent subject on the subject's thigh and the other on the table. Under normal venous pressure conditions the veins of the arm and hand on the thigh collapse while those in the hand on the table remain filled. When venous pressure is high, veins of both hands remain filled; when it is low, both collapse.

Consideration of pressure in the veins of the thorax is complicated by the fact that these veins lie in an area of negative pressure and are influenced by the cyclic variations in pressure within the chest during the respiration phases. The effect of this complication is illustrated in Figure 16-16. The effect on the transmural pressure of vessels in the thorax is an increase in the *effective venous pressure* as indicated in the right-hand column of the manometer system. Since the effective venous pressure is important when one considers the venous pressure gradients responsible for the return of blood to the heart, the effect of the intrathoracic pressure must be taken into account. The effective venous pressure is equal to the entering venous pressure minus the intrathoracic pressure. In Figure 16-16B1, the entering venous pressure is 60 mm H_2O and the intrathoracic pressure is −30 mm H_2O; the sum of the minus signs gives a plus value, and the effective pressure is 60 − (−30) or +90 mm H_2O. In phases of respiration when the intrathoracic pressure becomes positive (Fig. 16-16B2), this value is simply subtracted from the entering venous pressure. In this example, 60 mm − 30 mm gives the effective venous pressure of 30 mm H_2O. Effects of changes in intra-abdominal pressure with respiration on the entering venous pressure are ignored in this example.

REPRESENTATIVE PERIPHERAL VENOUS PRESSURES

Average pressure values in mm H_2O in veins of different parts of the body in recumbency are: median basilic (elbow), 97; femoral, 111; abdominal, 115; dorsal metacarpal, 130; great saphenous of ankle, 150 (Burch, 1950). They are normally slightly higher in the male than in the female; for example, 100 mm versus 94 mm H_2O in the median basilic vein. Values are lower in children. Between the ages of 3 and 5 the average value is 46 mm H_2O. Between the ages of 5 and 10 this value increases to 58 mm H_2O.

PATHOLOGICAL CHANGES IN VENOUS PRESSURE

Obstruction of a vein by pressure or by thrombosis elevates venous pressure and ultimately capillary pressure, causing edema fluid to be formed. Such changes may be quite localized, but the same principle applies in pathological states which give rise to rather widespread elevation of venous pressure and to edema because of the resultant increase in capillary pressure. In hepatic cirrhosis there is a marked increase in portal pressure because of the obstruction to flow through the liver. This blocking raises pressure back in the abdominal veins and may be manifested in the esophageal veins by rupture and bleeding. Congestive heart failure is accompanied by an increase in blood volume presumably due to kidney retention of salt and water, and also by increased hemoglobin production by the bone marrow. Some

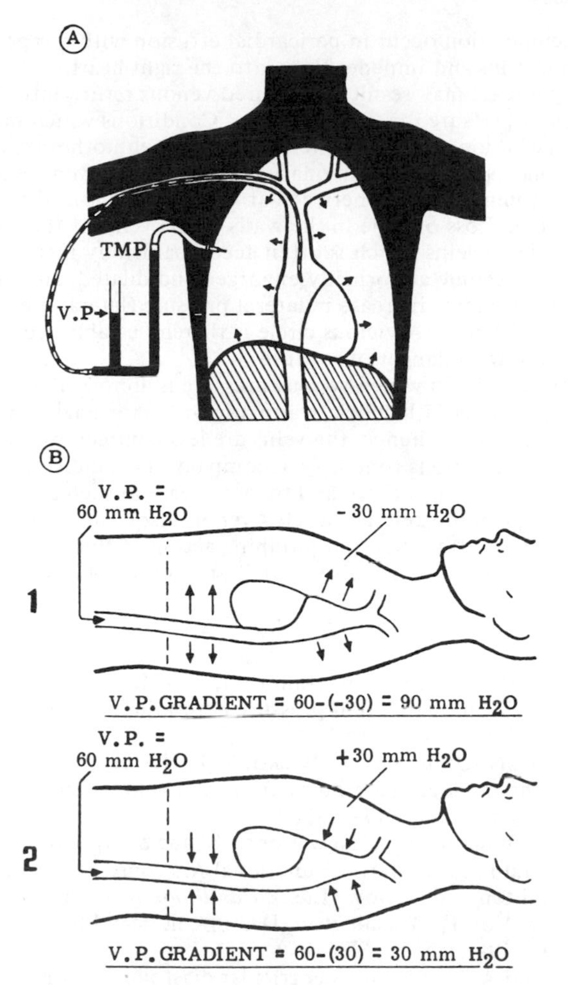

FIGURE 16-16. A. Transmural pressure in the thorax. The central venous pressure, V.P., recorded by catheter approximates atmospheric pressure. However, the *transmural pressure,* TMP, of the veins and atria is greater than the recorded values because of the subatmospheric pressure in the thorax. If the negative intrathoracic pressure is applied to the top of the manometer, the fluid column is elevated by the "suction" of the subatmospheric intrathoracic pressure. The *effective* venous pressure within the chest is indicated by such a manometer. (From Rushmer, 1955.) B. Application of the principle of effective venous pressure.

physiologists believe that increased venomotor tone of small veins accompanies this condition. These factors together produce an elevated systemic venous pressure. This is manifested by an increase in the median basilic vein pressure to 200 to 250 mm H_2O, and it may go as high as 400 mm H_2O, with peripheral edema resulting. The elevation may lead also to congestion of the liver, back pressure in the portal vein system, and accumulation of fluid in the abdomen, known as *ascites.* The same phenomenon operating on the left side of the heart leads to elevated pulmonary vein pressure, which in turn leads to pulmonary edema and impaired respiration and dyspnea. Specialized circumstances producing similar changes in venous pressure and edema accumulation occur in pericardial effusion with tamponade, which compresses the large veins and impedes flow into the right heart.

Pulmonary disease may result in impaired venous return into the thorax, leading to an elevation of pressure in systemic veins. Conditions which might give rise to this are pleural effusion, empyema, hemothorax, pneumothorax, emphysema, extensive pneumonia, and other pulmonary disease. Arteriovenous aneurysms which establish direct communication between arteries and veins lead to localized increases in venous pressure. Loss of tone in the walls of the veins of the leg results in a condition called varicose veins which is often accompanied by incompetence of the valves. The veins become abnormally engorged and dilated, and because of the slowing of flow there is a local increase in lateral pressure (lateral pressure is inversely proportional to velocity). A vicious circle is thereby established which further weakens the veins, with danger of rupture.

Circumstances exist in which venous pressure is abnormally low. This condition is commonly associated with shock, traumatic or hemorrhagic. In shock there is a decrease in blood volume; hence, the veins are less completely filled and pressure falls. A reduction in venous tone may accompany the condition. Shock is accompanied also by less muscle activity and by a decrease in skeletal muscle tone; consequently, the important skeletal muscle "venopressor" mechanism is diminished. A reduction in the contractility and pumping action of the heart – in other words, a reduction in vis a tergo – may also contribute to the lower venous pressure.

REFERENCES

Bordley, J., C. A. R. Connor, W. F. Hamilton, W. J. Kerr, and C. J. Wiggers. Recommendations for human blood pressure determinations by sphygmomanometers. *Circulation* 4:503–509, 1951.

Brawley, R. K., H. N. Oldham, J. S. Vasko, R. P. Henney, and A. G. Morrow. Influence of right atrial pressure pulse on instantaneous vena caval blood flow. *Amer. J. Physiol.* 211:347–353, 1966.

Brecher, G. A. *Venous Return.* New York: Grune & Stratton, 1956.

Brecher, G. A., and P. M. Galletti. Functional Anatomy of Cardiac Pumping. In W. A. Hamilton and P. Dow (Eds.), *Handbook of Physiology.* Section 2: Circulation, Vol. II. Washington, D.C.: American Physiological Society, 1963. Pp. 759–798.

Brecher, G. A., and A. T. Kissen. Ventricular diastolic suction at normal arterial pressures. *Circ. Res.* 6:100–106, 1958.

Burch, G. E. *Primer of Venous Pressure.* Philadelphia: Lea & Febiger, 1950.

Chungchareon, D. Genesis of Korotkoff sounds. *Amer. J. Physiol.* 207:190–194, 1964.

Guyton, A. C. *Textbook of Medical Physiology,* 4th ed. Philadelphia: Saunders, 1971.

Hamilton, W. F. The patterns of the arterial pulse. *Amer. J. Physiol.* 141:235–241, 1944.

Roberts, L. N., J. R. Smiley, and G. W. Manning. A comparison of direct and indirect blood pressure determinations. *Circulation* 8:232–242, 1953.

Rushmer, R. F. *Cardiac Diagnosis.* Philadelphia: Saunders, 1955.

Rushmer, R. F. *Cardiovascular Dynamics,* 3d ed. Philadelphia: Saunders, 1970.

Wiggers, C. J. *Pressure Pulses in the Cardiovascular System.* New York: Longmans, Green, 1928.

Wiggers, C. J. *Circulatory Dynamics.* New York: Grune & Stratton, 1952.

Wiggers, C. J. *Physiology in Health and Disease,* 5th ed. Philadelphia: Lea & Febiger, 1954.

17

Control of the Cardiovascular System

Carl F. Rothe and
Julius J. Friedman

IF TISSUE is to receive the oxygen and nutrients required for metabolism and to have waste products removed, the flow of blood through the tissue must be adequate. This flow through each organ is regulated in part by local and in part by central mechanisms which change the vascular resistance to blood flow in accordance with tissue requirements. For this system to be effective, the systemic arterial pressure should be held reasonably constant by homeostatic mechanisms. With a constant blood pressure, the flow of blood through an organ or tissue is inversely related to its vascular resistance (see Chaps. 11 and 12). To maintain constant systemic blood pressure as the flow through a tissue increases, either the cardiac output must be increased or the flow through some other region of the body must be reduced. Effective treatment of hypertension and many other cardiovascular diseases requires a knowledge of the mechanisms controlling blood pressure.

HOMEOSTATIC CONTROL OF BLOOD PRESSURE

The control of cardiovascular activity to maintain constancy of blood pressure is a classic example of a homeostatic, negative feedback system (Chap. 7B). The *effector mechanism* is dual, in that blood pressure is increased by an increase in either cardiac output or peripheral resistance. Reduction of either of these two factors results in a fall in blood pressure. The blood pressure is sensed by special organs called *pressoreceptors* (baroreceptors) located primarily in the carotid sinuses and the arch of the aorta. Impulses from these receptors are transmitted to the *cardiovascular centers,* located in the brain stem. The procedures by which an operating point, as a reference for comparison, is determined and by which efferent impulses, as error signals, are developed are not as yet understood. The signals, acting to correct deviations in blood pressure, pass by way of the *autonomic nervous system* to the peripheral vasculature and to the heart.

The *arterial pressoreceptor reflex* controlling the systemic arterial blood pressure is described schematically in Figure 17-1. An increase in blood pressure stretches the pressoreceptors in the walls of certain arteries, stimulating them to increase the frequency of discharge along afferent nerves. This acts by way of the cardiovascular centers to *inhibit* sympathetic division action on the heart and peripheral vasculature and to increase parasympathetic division action on the heart by way of the vagus nerves. Cardiac contractility and heart rate decrease, and the degree of vasoconstrictor tone on both the resistance and capacity vessels is reduced. Thus the tetralogy of

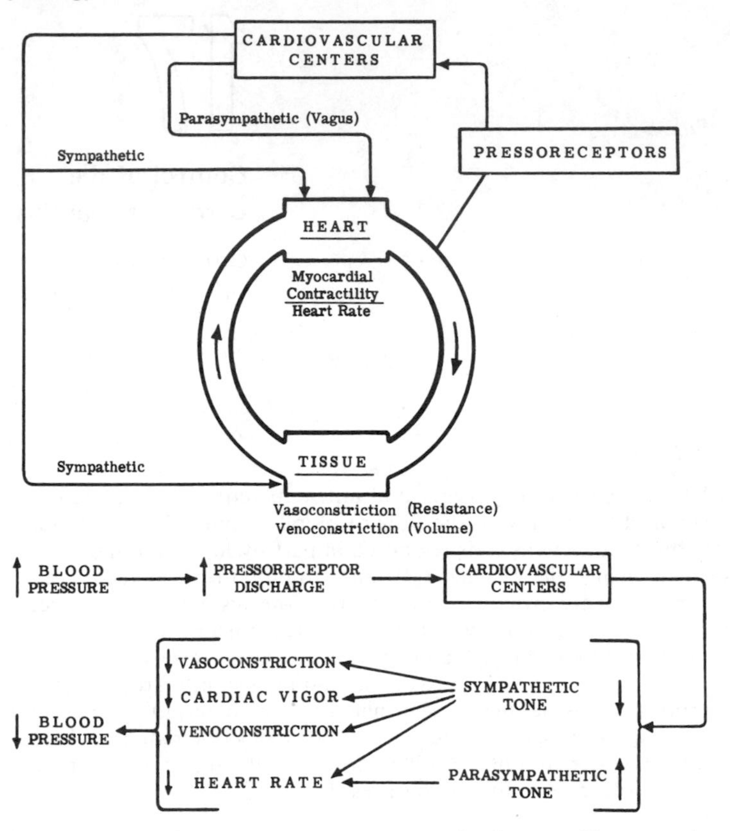

FIGURE 17-1. Operation of the arterial pressoreceptor reflex in controlling systemic arterial blood pressure.

reduced cardiac vigor, bradycardia, vasodilation, and venodilation tends to follow an increase in blood pressure. The combination of these responses results in a decrease in blood pressure. This is a negative feedback response, since the reduction in blood pressure is elicited by an increase above normal.

The arterial pressoreceptor reflex is also important in counteracting a decline in blood pressure. When pressure declines, fewer afferent impulses from the pressoreceptors go to the cardiovascular centers. Vagus tone is decreased, and there is less inhibition of the sympathetic division. As a result, cardiac action and vasoconstriction increase. For example, the blood pressure tends to fall following hemorrhage, and as a result sympathetic division activity is markedly increased. Other names for the arterial pressoreceptor reflex include: carotid sinus reflex, baroreceptor reflex, buffer reflex, and depressor reflex.

The nature of the control system which maintains cardiovascular homeostasis will now be considered in greater detail.

CARDIOVASCULAR CONTROL CENTERS

The neural centers for the control of the cardiovascular system are complex and not completely defined. They are *not* discrete structures but tend to be scattered and intermixed within the neural tissue. The concept of "centers," however, aids in the understanding of the regulatory process. These groups of neurons integrate neural impulses both from peripheral sensors and from the higher brain centers. The most satisfactory information concerning the functional anatomy of these centers has been obtained via electrodes implanted for long periods of time. The effects of electrical stimulation may then be studied in contrast to the lack of activity following ablation of the same areas by electrolytic destruction using much higher currents.

MEDULLARY CONTROL

Neurons in the medulla oblongata are responsible for the integration of afferent impulses and the origination of efferent impulses for the homeostatic control of blood pressure. *Vasoconstrictor centers* in the medulla oblongata are responsible for the neurogenic component of basal vasoconstrictor tone. *Cardiostimulator centers* increase cardiac activity. Normally, however, the cardiostimulator centers are relatively quiescent. Electrical stimulation of these *pressor areas* located in the lateral reticular formation causes an increase in blood pressure by increasing vasoconstriction, heart rate, and probably cardiac vigor (Fig. 17-2). The vasoconstrictor neurons can discharge without afferent stimuli, for vasomotor discharge does not cease and the blood pressure may not decrease even if incoming impulses from peripheral pressoreceptors are stopped. There is a continuous discharge *modified* by impulses from the depressor area, chemoreceptors, and higher centers (Fig. 17-2). The medullary respiratory center neurons are intermingled with those of the pressor area. There are not only similarities in cardiovascular and respiratory response to varying conditions but also interactions between the systems both under normal conditions and following various stresses. Arterial pressoreceptor discharge inhibits not only cardiac and vascular activity but also respiration. The respiratory system may influence the cardiovascular system, as seen by fluctuations in arterial blood pressure *(Traube-Hering waves)* and a complex pattern of heart rate changes *(sinus arrhythmia)* related to respiratory activity.

Cardioinhibitor and *vasodepressor* areas are located in the caudal and medial reticular formation of the medulla in association with the dorsal nucleus of the vagus. Stimulation of these areas results in a reduction in arterial blood pressure. The response is produced by inhibition of the constrictor tone – not activation of vasodilator fibers – and by vagal slowing of the heart (Fig. 17-2).

HYPOTHALAMIC CONTROL

The hypothalamus, being a generalized center of control of the autonomic division, modifies the activity of the bulbar region. There is probably no tonic activity coming from this region, because section at the level of the pons does not necessarily reduce blood pressure. However, redistribution of blood flow and characteristic patterns of cardiovascular response are integrated at this level. The cardiovascular adjustments to emotions such as rage are mediated here. Stimulation of parts of the hypothalamus results in sympathetic division activation, including changes in the blood pressure, cardiac action, and peripheral vascular tone. Centers affecting specific parts of the cardiovascular system are apparently located in the hypothalamus, for highly localized stimulation evokes responses restricted to a single feature of cardiovascular function. Included among these are heart rate, left ventricular size, systolic pressure, and vasodilation of skeletal muscle. Stimulation of certain highly discrete areas results in a cardiovascular response closely similar to that seen during exercise.

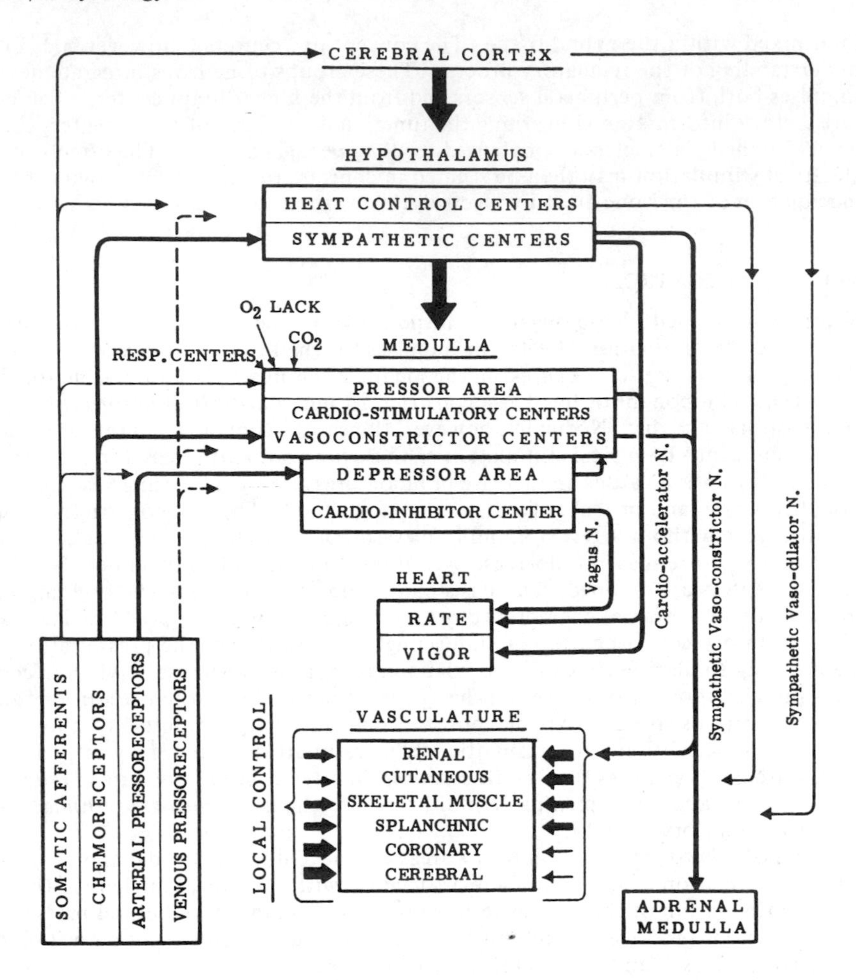

FIGURE 17-2. Control of the cardiovascular system.

The hypothalamus is involved in the distribution of blood flow for the control of body temperature. Lesions in this region impair the ability of an animal to protect itself from extreme environmental temperature, because sweating and cutaneous vasodilation or vasoconstriction no longer occur in response to the appropriate conditions. Temperature control is primarily by way of the sympathetic division, since the parasympathetic division has little direct effect except for that on heart rate.

CEREBRAL CORTICAL CONTROL

Impulses affecting the cardiovascular system originate in the forebrain. Knowledge of these may have an important bearing in the understanding of psychosomatic medicine. The sympathetic vasodilator outflow to skeletal muscle apparently originates in the cerebral cortex and passes through the hypothalamus and medulla, where the efferent discharge pattern may be modified. Electrical stimulation of the cortex may evoke both autonomic and somatic effects, which are usually anatomically and functionally related.

SPINAL CONTROL

Control centers in the spinal cord are apparently of minor importance for the maintenance of an adequate circulation under normal conditions. In cases of circulatory depression the cord is of greater importance. Transection of the thoracic spinal cord causes an immediate and profound fall in blood pressure. However, after a period of time a certain degree of control of the blood pressure may be regained. Once the blood pressure in a spinal animal has returned to near normal levels, total destruction of the cord results in a permanent reduction in pressure. Thus, there are neurons in the cord responding to pressoreceptor impulses or reduced blood flow and hypoxia which initiate impulses along vasoconstrictor fibers. Stimulation of cutaneous receptors by pain or cold induces segmentally arranged vasoconstriction of the intestinal vessels of spinal animals, a phenomenon indicative of the action of spinal centers on the cardiovascular system.

CARDIOVASCULAR SENSORS AND REFLEX CONTROL OF THE CARDIOVASCULAR SYSTEM

Effective cardiovascular control depends on information provided by sense organs which transmit information to the control centers in the brain.

ARTERIAL PRESSURE RECEPTORS

The sensors for the control of the systemic blood pressure by the arterial pressoreceptor reflex are sensitive nerve endings which respond to stretching of the walls of arteries. These pressoreceptors are located not only in the carotid sinuses and arch of the aorta but also along the common carotid arteries. The rate of firing of the pressoreceptors at various pressures has been measured (Fig. 17-3). These receptors are effectively quiescent below pressures of about 60 mm Hg. As the transmural pressure is increased beyond 60 mm Hg, the frequency increases progressively. The *change* in frequency of impulses per millimeter mercury pressure change is maximum at about normal blood pressure. A plateau is reached at about 160 mm Hg. Further increases in pressure do not result in an appreciable increase in the rate of receptor discharge. Consequently, when the blood pressure decreases below about 60 mm Hg or increases above about 160 mm Hg, little further compensatory response is elicited by this reflex system.

The frequency of discharge of the pressoreceptors is also determined by the *rate* of stretch of the receptors as well as by the average magnitude. Thus, there are bursts of activity with each pressure pulse (Fig. 17-3A). In fact, a reduction in pulse pressure with *no change* in mean arterial blood pressure decreases the inhibition on the cardiovascular centers and so results in an increase in cardiovascular activity. In this manner, the *pulse pressure* is an important factor determining the frequency of pressoreceptor impulses going to the cardiovascular centers and consequently the level of sympathetic division influence on the heart and peripheral vasculature. The influence is inversely related to the mean and pulse pressure while the level of vagal activity is directly related to pressoreceptor stretch (Fig. 17-3A). The pressoreceptors apparently do not adapt, but fire continuously at normal blood pressures.

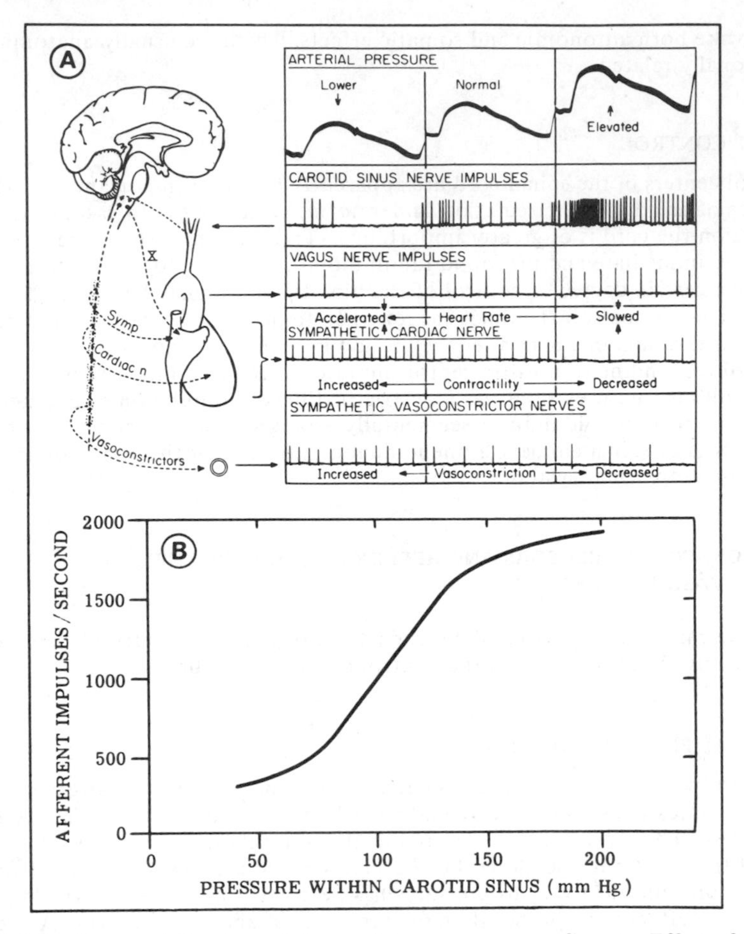

FIGURE 17-3. Neural relationships of the arterial pressoreceptor reflex. A. Effect of arterial pressure pattern on carotid sinus stretch receptor discharge, and impulse rate of efferent nerves. B. Rate of discharge of afferent nerves at various static sinus pressures. (A from Rushmer, 1970. P. 165. B after W. Kalkoff. *Verhandl. Deutsch. Ges. Kreislaufforsch.* 23:399, 1957.)

The role of the pressoreceptors in hypertension is not known. The frequency of impulses is lower than normal in some forms of this disease. It is not known whether the sinus is constricted, less distensible, or less responsive (see Chap. 18A). The surgeon must be cognizant of these receptors, for stimulation of the carotid sinus by probing or pulling can result in profound and sometimes fatal hypotension. Some people have a very sensitive carotid sinus reflex. A tight collar or turning of the neck can stimulate the receptors sufficiently to elicit a response of such magnitude that the blood pressure drops to levels producing cerebral hypoxia and loss of consciousness.

There are other arterial pressoreceptors located along the thoracic and mesenteric arteries. The importance of these receptors in man is not known. Under normal conditions the effects are minimal in the dog. In the cat, however, occlusion of a mesenteric artery results in a marked increase in systemic arterial pressure.

CHEMORECEPTORS

The peripheral chemoreceptors are specialized nerve endings in the carotid and aortic bodies found near the carotid sinus and also in the arch of the aorta. These nerve endings are sensitive to a decrease in oxygen tension, an increase in carbon dioxide tension, or an increase in the hydrogen concentration of the blood. Poisons that inhibit oxidative pathways, such as cyanide or fluoride, stimulate the chemoreceptors. Chemoreceptor stimulation causes an increase in pulmonary ventilation and an increase in blood pressure by peripheral vasoconstriction, but it generally has little effect on heart rate. The cardiovascular responses to hypoxia are complicated by a *lung inflation–systemic vasodilator reflex.* As the depth of respiration is increased because of carotid body stimulation by hypoxia, there is a concomitant tendency for cardiovascular inhibition. These receptors, in conjunction with the neurons of the central nervous system sensitive to hypoxia, hypercapnia, and acidosis, produce complex patterns of responses.

The carotid body with its glomus cells is one of the most vascular tissues of the body. The tissues weigh only about 2 mg in man, but the blood flow through them is about 20 ml per minute per gram of tissue. This is 4 times that of the thyroid, 30 times that of the brain, and about 250 times that of the body as a whole. In addition, they consume oxygen at the rate of about 90 ml per minute per kilogram of tissue, which is about 3 times that of even the brain. The chemoreceptors respond to decreases in oxygen tension and not necessarily to the oxygen content of the blood. For example, in carbon monoxide poisoning if the arterial oxygen tension is maintained, there is little respiratory or cardiovascular stimulation, even though the oxygen content of the blood is markedly reduced. The high blood flow through this tissue is apparently necessary to supply the metabolic needs of the chemoreceptor cells or to remove some stimulatory agent, for a decrease in blood flow through the carotid and aortic bodies, as a result of decreased arterial blood pressure, results in chemoreceptor stimulation. Simultaneous arterial hypoxemia and hypotension are synergistic and thus produce a vigorous response.

The arterial chemoreceptors are normally minimally active and have little effect on the circulation. Only under abnormal conditions, when there is hypoxemia, hypercapnia, or acidosis, are they effective. These receptors are tough enough to respond under adverse environmental conditions so that the cardiovascular and respiratory systems may be stimulated to act to regain an optimal internal environment – an example of the homeostasis.

CAROTID SINUS REFLEX

Occlusion of the carotid arteries on the cardiac side of the pressoreceptors and chemoreceptors produces a carotid sinus reflex. The increase in sympathetic division activity follows from both a reduction of stimulation of the arterial pressoreceptors, because the blood pressure within the sinuses is reduced, and a stimulation of chemoreceptors, because the supply of oxygen to these tissues is decreased. These two receptor systems thus help assure an adequate flow of well-oxygenated blood to the central nervous system. Both sets of receptors are important for the homeostatic compensation following hemorrhage. Incidentally, because of cerebral anastomoses and the vertebral arterial supply, the pressure in the sinuses is not reduced to zero following bilateral carotid occlusions.

CEREBRAL ISCHEMIA

If the flow of blood through the brain is severely restricted, there is a delayed but profound sympathetic division response resulting in intense peripheral vasoconstriction. This central ischemic response acts at blood pressures below those which have no *more* effect by way of the arterial pressoreceptor reflex. The brain becomes hypoxic and hypercapnic below blood pressures of about 50 mm Hg. At these perfusion pressures the chemoreceptors along the carotid arteries are of course also stimulated by the stagnant anoxia. In response, there is a marked elevation of systemic blood pressure to as high as 350 mm Hg. The resulting degree of sympathetic vasoconstrictor discharge gives renal vasoconstriction of such a magnitude that urine flow stops. The mechanism of action probably results from an increase in carbon dioxide and a decrease in oxygen tension in the medullary region of the brain. The removal of carbon dioxide is limited by the reduced flow of blood through the brain. Therefore, carbon dioxide accumulates and acts not only on the cardiovascular centers, to cause vasoconstriction, but also on the respiratory centers, to cause a vigorous increase in respiratory activity.

The *Cushing reflex* is a corollary of the central ischemic response. If the cerebrospinal fluid pressure is greater than the systemic arterial blood pressure, the flow of blood to the brain is stopped, because the fluid pressure acting in the rigid skull occludes the arteries. Under these conditions the central ischemic response produces an increase in blood pressure by peripheral vasoconstriction. This tends to restore cerebral blood flow. A very high blood pressure following head injury or cerebral vascular accident indicates cerebral hemorrhage and the operation of this reflex.

With severely depressed cardiovascular function or serious metabolic acidosis, rhythmic variations in arterial blood pressure may be seen. These fluctuations, called *Mayer waves,* have a period of 15 to 60 seconds and are independent of respiration. As the blood pressure declines, central and/or peripheral chemoreceptors excite reflex vasomotor activity which acts to increase the blood pressure. Chemoreceptor activity then declines, followed by a decrease in blood pressure. The response is oscillatory because of lags in the system. The Mayer waves and Cheyne-Stokes respiration are probably derived from similar mechanisms.

PULMONARY AND CARDIAC RECEPTORS

Right atrial and central vein reflexes following distention of these areas increase heart rate if the vagal tone is high *(Bainbridge reflex).* Nerve impulses along afferent fibers have been recorded in response to changes in pressure in the atria during atrial contraction and also during the distention of the atria during diastole.

Left atrial volume receptors respond to increased transmural pressure. Impulses transmitted to the brain act via the posterior pituitary to reduce antidiuretic hormone secretions. An increase in urine flow, reducing blood volume, follows (see Chap. 23). Increased blood volume, or a more negative intrathoracic pressure, will stimulate these receptors.

The *Bezold-Jarisch* reflex is a bradycardia and hypotension following distention of the ventricles which stimulates proprioceptors. The response is also elicited by injection of veratrine into an artery supplying the left ventricle.

Pulmonary arterial pressoreceptors affect the cardiovascular system in a manner similar to that of the systemic arterial pressoreceptors, but to a much smaller degree; e.g., an increased pulmonary arterial pressure induces bradycardia, hypotension, and hypopnea.

Almost all *sensory nerves* have some connection with the cardiovascular reflex system. Some of these reflexes involve centers in the spinal cord only, since vasoconstrictor and vasodilator reflexes can be elicited in animals with sectioned spinal cords. Generally, a painful stimulus is followed by a rise in blood pressure. The conscious realization of pain, causing anxiety and stimulation of the adrenal medullae, also adds to the response. On the other hand, severe cutaneous pain, painful stimulation of the gastrointestinal or genital tracts, stretching of hollow organs, or the stimulation of deep visceral pain receptors may elicit the opposite response: a fall in blood pressure. However, urinary bladder distention in humans with sectioned spinal cords can cause an increase in systolic arterial pressure to over 300 mm Hg. Emotional stress may be followed by fainting. This is called *vasovagal syncope*, implying a vasodilation of the resistance and possibly the capacitance vessels of the body in addition to a vagal slowing of the heart. Cerebral hypoxia occurs as the blood pressure declines, leading to a loss of consciousness.

NEURAL CONTROL OF CARDIAC OUTPUT

The primary determinants of cardiac output at a given pressure load are the heart rate, filling pressure, distensibility, and contractility (Chap. 15). *Neural impulses* from the cardioregulatory centers in the lower brain act to modify these variables.

The *heart rate* is determined primarily by the balance between the inhibitory effects on the pacemaker of acetylcholine released by the vagus nerves of the parasympathetic division, and the excitatory effects of norepinephrine released by the sympathetic nerve endings (Chaps. 7 and 14). The sympathetic and parasympathetic divisions tend to be antagonistic, since an increase in heart rate, caused by an increase in sympathetic division activity, can be reduced to normal by adequate stimulation of the vagus nerves. In fact, massive vagal stimulation stops the heart for many seconds. A continuous vagal tone reduces the heart rate in the normal resting individual to about 70 beats per minute. Blocking of parasympathetic activity with atropine or sectioning of the vagi results in a marked increase in heart rate. On the other hand, the sympathetic accelerator fibers exert a stimulating influence on the heart, especially during exercise or anesthesia. Sympathectomy by severing the sympathetic outflow in the thorax from T2 to T5 (see Fig. 7A-1) in man is often followed by a reduced cardiac response to exertion.

Myocardial contractility is increased by an increase in the rate of discharge of the sympathetic division nerves going to the heart. An increase in myocardial contractility results in an increase in cardiac vigor: The pressure developed by the ventricles increases; the size of the heart, particularly at the end of systole, tends to decrease; and the rates of change of pressure and size increase. The parasympathetic division has little effect upon ventricular contractility, in contrast to its marked effect on heart rate. The atria are richly innervated by the parasympathetic nerve endings, and there appears to be a decrease in atrial contractility as a result of vagal discharge.

Cardiac distensibility is apparently little affected by the autonomic nervous system, although some experiments indicate that sympathetic division activity may increase ventricular distensibility somewhat and so permit easier filling.

The *filling pressure* of the right heart is determined primarily by the peripheral factors affecting the return of venous blood and the degree of negative intrathoracic pressure (Chap. 16). The degree of constriction of the venous capacitance vessels is important. The filling of the left ventricle is determined primarily by the action of the right heart, the characteristics of the pulmonary bed, and the left atrium.

ADRENAL MEDULLARY INFLUENCE

Catecholamines, released from the adrenal medullae, provide a relatively small, slow-acting, but effective adjunct to the autonomic innervation of the heart. Cardiac output is thereby augmented by means of an increased myocardial vigor and heart rate.

PERIPHERAL CIRCULATION

The peripheral circulation may be considered to be that segment of the circulatory system which is concerned with transport of blood, blood flow distribution, exchange between blood and tissue, and storage of blood.

The various tissue circulations which comprise the peripheral circulation are arranged in parallel, except, for example, the portal components of the hepatic and the pulmonary circulations (Fig. 17-4). Thus the distribution of the cardiac output to these tissues is primarily determined by the relative resistances of their vascular beds. Blood flow and oxygen consumption for the major circulations of the body are presented in Table 17-1. Blood flow distribution is not determined by organ size. The kidneys, which represent 0.5 percent of the body weight, receive 20 percent of the cardiac output, whereas *resting* muscle, which accounts for approximately 40 percent of the body weight, receives only 20 percent of the cardiac output. Tissue blood flow is determined by overall tissue function, which includes normal tissue maintenance as well as specialized tissue function. In this regard, kidney and skin receive blood flow in excess of their oxygen requirement and thus have a low arteriovenous oxygen difference. These flows are related to the respective

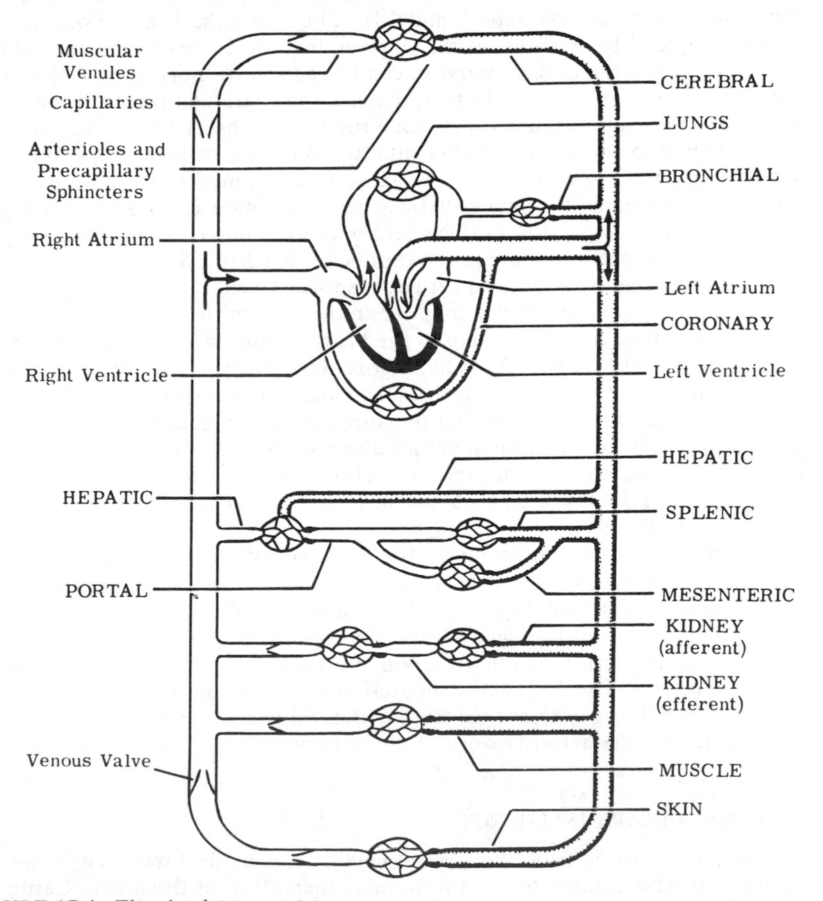

FIGURE 17-4. The circulatory system.

TABLE 17-1. Approximate Distribution of Blood Flow to Organs of a 70-kg Man

Organ	Percent of Total Body Wt.	Blood Flow			AV O_2 Diff.	O_2 Consumption		
		% Cardiac Output	ml/min / 100 gm	ml/min / organ	(vol per 100 ml)	ml/min / organ	ml/min / 100 gm	% Total
Splanchnic	5.7	24	35.0	1400	4.1	58	1.4	25
Skeletal muscle	40.0	20	4.0	1200	6.0	70	0.2	30
Kidneys	0.5	20	350.0	1100	1.7	20	6.0	8
Brain	2.0	15	55.0	750	6.3	46	4.0	19
Skin	10.0	6	0.5	350	1.0	8	0.1	3
Heart	0.5	6	85.0	300	11.4	28	9.0	11
Others*	35.7	8	35.0	900	2.2	20	0.8	4
Total body	100.0	100	85.0	6000	4.5	250	0.4	100

*Other organs include bone, glands, and reproductive organs.

functions of waste clearance and thermoregulation. The coronary and cerebral circulations, on the other hand, receive relatively high blood flows because of their high metabolic rate and oxygen consumptions. This is reflected in the relatively high arteriovenous oxygen difference.

The parallel architectural arrangement provides a means whereby blood flow to a tissue can vary independently of changes in blood pressure. Furthermore, a redistribution of the cardiac output is also possible. This feature of the circulatory system is best exemplified in exercise, when muscle and coronary blood flows are increased while splanchnic, renal, and skin blood flows are reduced.

The flow of blood through tissues is determined by the arteriovenous pressure gradient, the viscosity of blood, and the resistance to the flow of blood (Chap. 11). Since the pressure gradient, blood viscosity, and length of tissue resistance vessels are relatively constant, the caliber of the resistance vessels is the primary determinant of local blood flow.

The flow of blood through tissues is controlled in part by *central* influence manifest by neural and by humoral mechanisms and in part by *local* mechanisms such as oxygen tension, metabolites, intrinsic reflexes, and autoregulation.

Central and local mechanisms continuously interact to modify the distribution of blood flow. Under some conditions, such as voluntary contraction of muscles, the local and central mechanisms tend to act in the same direction to increase flow. When systemic requirements are great, as occurs during inadequate cardiac function, the central mechanisms will predominate over the local mechanisms for tissues such as skeletal muscle. Conversely, when local factors predominate, as in muscular exercise, the local mechanism may override the central influence.

CENTRAL CONTROL OF THE VASCULATURE

Central control of the peripheral circulation is effected through the activity of the sympathetic division of the autonomic nervous system and through the action of adrenomedullary humoral agents.

SYMPATHETIC ADRENERGIC FUNCTION

Vascular Resistance

The primary mechanism of central control of tissue blood flow is provided by the discharge of *sympathetic vasoconstrictor fibers* causing small artery and arteriolar smooth muscle contraction, thereby decreasing lumen diameter and so increasing resistance to flow (Fig. 17-5, curve 1). These vasoconstrictor fibers apparently provide the *only* neural pathway to the vasculature for the *control* of systemic arterial blood pressure. The resting frequency of nerve fiber discharge to skeletal muscle is about 1 to 3 impulses per second. This provides a basal level of vasoconstrictor tone from which adjustment of vascular resistance can occur for the maintenance of circulatory homeostasis. Maximal resistance responses are developed at a frequency of about 10 impulses per second. Inhibition of this basal vasoconstrictor tone results in vasodilation.

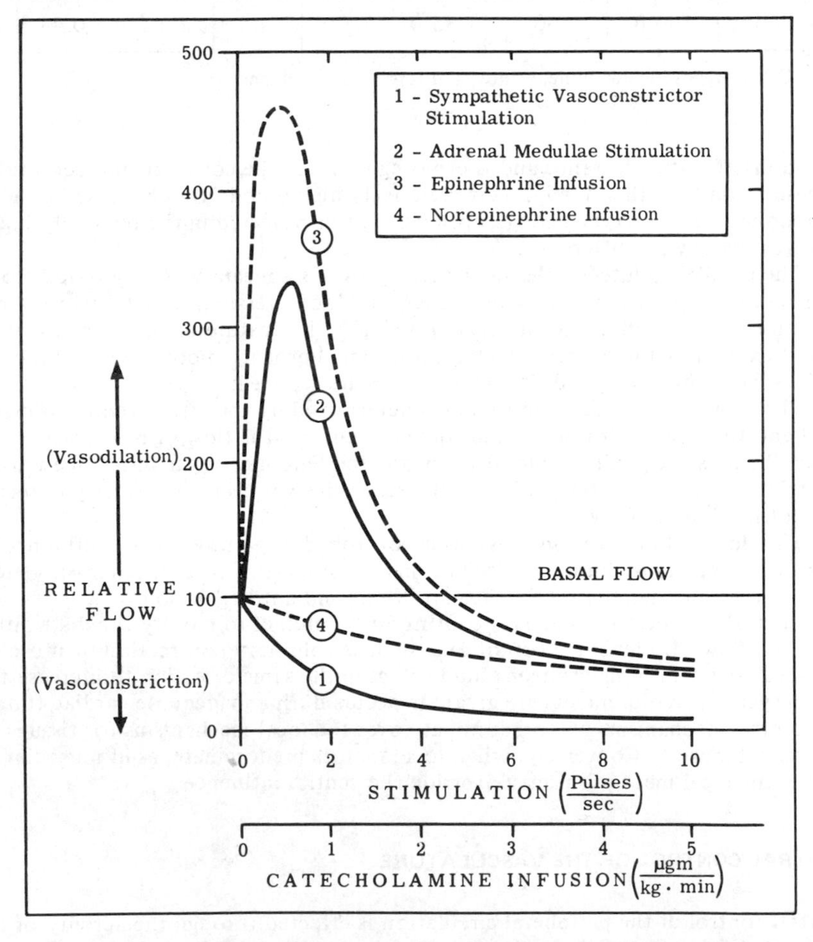

FIGURE 17-5. Effect of adrenergic mechanisms on blood flow through the skeletal muscle vasculature. (Data from Celander, 1954. P. 48.)

Sympathetic vasoconstriction is not the only determinant of the basal vasculature tone, since an intrinsic level of vascular smooth muscle tone is present after denervation. Blockade of the vascular innervation permits a twofold to fivefold increase in flow through skeletal muscle, while skeletal muscular contractions may elicit a further increase to over 20 times the normal flow.

The maximal effects of sympathetic adrenergic stimulation, via vasoconstrictor nerves, on blood flow through various tissues of the body are given in Figure 17-6.

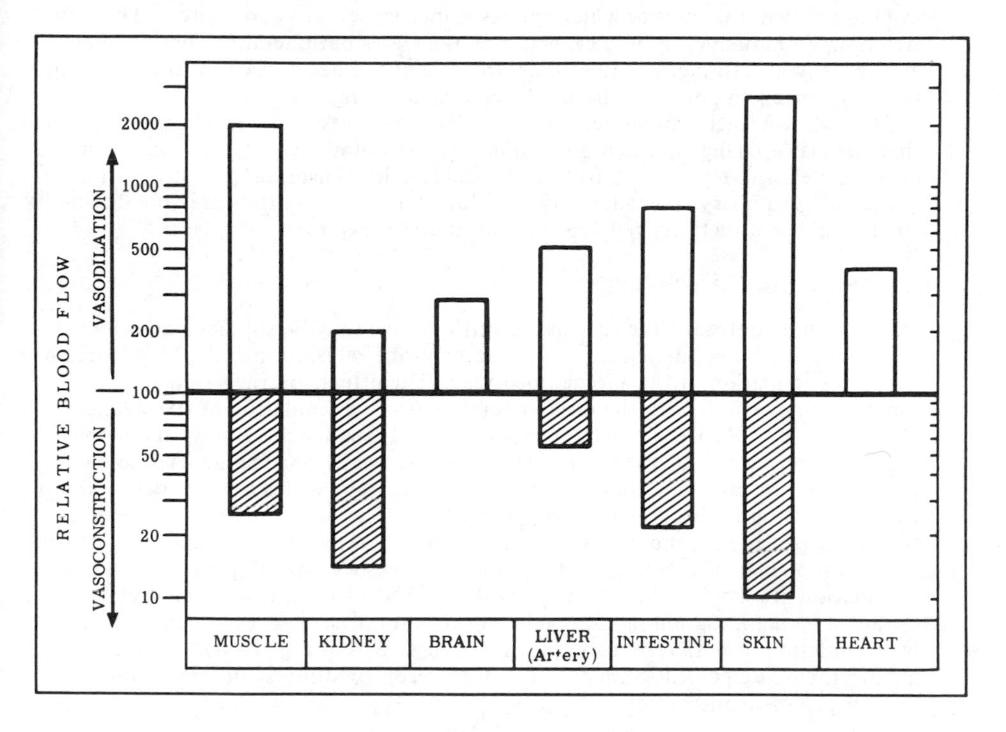

FIGURE 17-6. Maximal sympathetic adrenergic vasoconstriction and maximal metabolic vasodilation on blood flow through various tissues of the body.

Adrenergic stimulation produces vasoconstriction and reduced blood flow in all tissues except the brain and heart. Control of blood flow through these tissues is mediated almost entirely by local factors (see below).

Although the sympathetic vasoconstrictor outflow is somewhat diffuse and has a general effect on the overall peripheral resistance, important changes in the distribution of blood flow may be brought about by these fibers. For example, uncomfortable environmental temperatures, hypoxia, digestion, or muscular exertion results in *discrete* cardiovascular adjustments in the distribution of the cardiac output. Sympathetic vasoconstriction during stress, such as severe hemorrhage and hypotension, can markedly reduce the flow of blood through the skin, skeletal muscle, renal, and splanchnic vascular beds. This blood is redistributed to other more immediately vital tissues. The characteristic cold skin, muscular weakness, and anuria so often seen following hemorrhage are due in part to this redistribution. The cutaneous circulation is almost entirely under neural rather than local control.

Vascular Capacitance

Sympathetic adrenergic activity also increases venular and small vein smooth muscle contraction (venomotor tone) to reduce their capacitance. Capacitance may be defined as the ratio of the volume of blood contained in a vessel to its transmural pressure. In skeletal muscle with constant flow, stimulation is effective in producing less than a 10 percent change in contained volume, whereas in skin over 20 percent change results. Constriction of capacitance vessels occurs at neural discharge rates even lower than the rates at which the resistance vessels are constricted. This constriction, by reducing vascular capacitance, decreases peripheral vascular volume and thereby acts to increase the filling pressure of the heart. Loss of this tone can result in serious hypotension by reducing cardiac filling.

In addition to the direct neurogenic influence on vascular capacitance, it appears that the major change in the blood volume of the venous reservoir occurs by means of passive elastic recoil which follows a reduction in transmural pressure. An increase in precapillary resistance acts to reduce the flow, and thus pressure downstream, and so effectively reduces the transmural pressure.

ADRENOMEDULLARY INFLUENCE

Direct neural control of the vascular smooth muscle may be supplemented by catecholamines such as epinephrine and norepinephrine released into the blood stream through stimulation of the adrenal medullae. The effects on the vasculature are minor in comparison with the direct neural control. Stimulation of the splanchnic innervation to the adrenal medulla over a range of frequencies results in a biphasic change in muscle vascular resistance (Fig. 17-5, curve 2). At frequencies below 3 impulses per second, vasodilation occurs, whereas frequencies from 3 to 10 impulses per second produce progressive vasoconstriction. The basis for this variable effect lies in the fact the adrenomedullary stimulation liberates primarily epinephrine with some norepinephrine. Norepinephrine produces vasoconstriction by alpha receptor stimulation at all doses (Fig. 17-5, curve 4). Epinephrine in low concentrations (below 2 μg per kilogram-minute) acts on skeletal muscle vessels to cause dilation by stimulation of beta receptors (Fig. 17-5, curve 3, and Fig. 7A-4). At doses greater than 2 μg per kilogram-minute, alpha receptor stimulation predominates, resulting in vasoconstriction.

SYMPATHETIC VASODILATOR FIBERS

Sympathetic vasodilator fibers originate in the central nervous system. These apparently innervate only the vasculature of skeletal muscle, external genitalia, and possibly the coronary vessels. The fibers to skeletal muscle, originating in the cerebral cortex, act to increase the flow of blood to muscle during voluntary activity before the local effects of increased metabolism become operative. The enhanced blood flow apparently bypasses the exchange network, as evidenced by reduced O_2 consumption. The functional significance of this system is uncertain. It has no direct effect on the control of blood pressure in response to pressoreceptor activity. Maximal stimulation of these fibers increases skeletal muscle blood flow up to five times the basal flow.

Intra-arterial injections of acetylcholine likewise cause vasodilation. Atropine blocks both the vasodilating effect of acetylcholine injected into an artery and the effects of sympathetic vasodilator fiber stimulation. Sectioning of the sympathetic paravertebral ganglia also blocks the response. It is thus probable that acetylcholine is released at postganglionic cholinergic fibers and acts in some manner to cause vascular dilation.

PARASYMPATHETIC (SECRETOMOTOR) VASODILATOR ACTION

Activation of various exocrine glands (e.g., salivary, sweat) by parasympathetic fiber discharge leads to increased blood flow. However, this does not represent direct vascular influence. Instead, increased glandular activity triggers the release of a proteolytic enzyme which acts on a plasma or tissue globulin to form a potent vasodilator, *bradykinin.* This response, following parasympathetic stimulation, may therefore be referred to as secretomotor vasodilation.

There are no known parasympathetic vasoconstrictor fibers. Although the vagal slowing of the heart is followed by a vasoconstriction of the coronary vessels, this is probably an indirect consequence of the decreased metabolism.

DORSAL ROOT VASODILATION

In addition to the autonomic efferent fibers, stimulation of the skin (or the peripheral end of a sectioned dorsal root fiber) often produces a dilation of the adjacent superficial blood vessels. It is probable that this results from impulses arising from receptor sites (pain) and passing up sensory afferent fibers until a branch in the neuron is reached. Impulses then go back over the branch to a vasomotor ending. For this reason they are called *antidromic vasodilator* impulses. The afferent and efferent parts of this *axon reflex* are formed by the branching of a single nerve fiber. Effective stimulation is probably a result of the release of histamine by tissue trauma. Only in areas rich in pain fibers, such as mucous membrane and skin, does vasodilation occur from this mechanism. The axon reflex may be of functional importance in the development of the *flare* following mechanical trauma to the skin and possibly in the development of inflammation about an infected area. The increase in blood flow would presumably aid the healing process.

LOCAL CONTROL OF TISSUE BLOOD FLOW

PASSIVE RESPONSE TO CHANGES IN TRANSMURAL PRESSURE

In order to appreciate the nature and effectiveness of the local regulation of tissue circulation, the passive behavior of a vascular bed based on the distensible properties of the walls and the transmural pressure must be understood. One can gain some insight into the nature of the vascular bed by examining its *pressure-flow relationship* (Fig. 11-4B). For a rigid pipe system with a newtonian fluid the pressure-flow relationship is linear. That is to say, each elevation in pressure will produce a proportionate elevation in flow. However, the vascular system is a distensible tube network, and elevations in pressure produce greater than proportionate increases in flow. These passive adjustments in the state of the blood vessels are initiated by changes in transmural pressure. Transmural pressure represents the gradient of pressure across the vessel wall. Consequently, transmural pressure may be *increased* by either an increase in intraluminal pressure or a decrease in extraluminal pressure. Then the vessels will dilate and vascular resistance will decrease. Intraluminal pressure can be increased by greater inflow of blood, upstream vasodilation, or downstream vasoconstriction. In addition, placing the tissue below the heart will increase the intraluminal pressure.

Transmural pressure may be *decreased* by either a decrease in intraluminal pressure or an increase in extraluminal pressure. The radius of the vessel thereby decreases, increasing resistance, and consequently reducing blood flow. Intraluminal pressure declines as a result of a decrease in blood flow, upstream vasoconstriction, or downstream vasodilation or of raising the tissue relative to the heart. Extraluminal pressure may increase by means of muscular contraction, abdominal compression, or an increase in tissue pressure resulting from edema. Thus, inflow into skeletal muscle and coronary circulation during contraction is markedly reduced. In the intestine, mechanical compression resulting from distention with luminal contents may also act to retard flow.

An additional explanation of the nonlinear pressure-flow relationship with varying perfusion pressure is that the number of vascular channels perfused may increase with increased pressure. This type of behavior is seen in the skin, pulmonary, and hepatic-portal circulation.

ACTIVE RESPONSE AND AUTOREGULATION OF BLOOD FLOW

Active regulation of blood vessel tone occurs in response to local metabolic factors and to arteriolar transmural pressure variations. It may be modified by extrinsic neurogenic and humoral influences. Active regulation is the control of blood flow related to the function and needs of the tissue and is seen as a deviation from the passive pressure-flow pattern. It is an intrinsic control mechanism, independent of innervation or systemic humoral factors. The term *autoregulation* is commonly applied to this type of control.

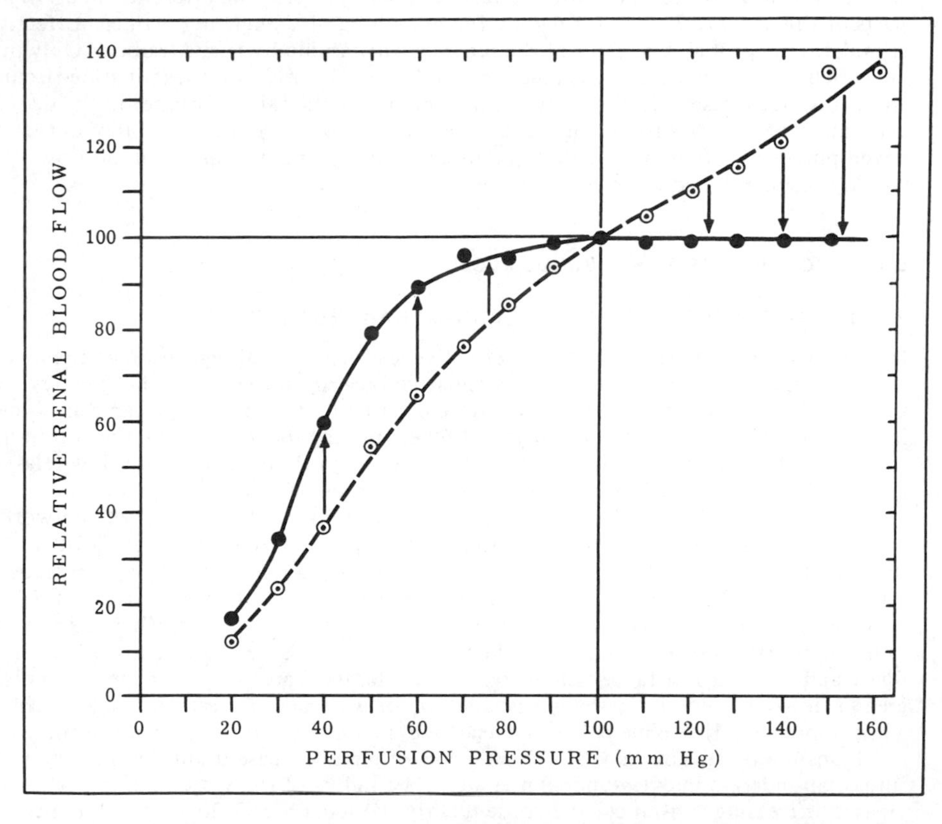

FIGURE 17-7. Autoregulation of renal blood flow. Transient (open circles) and steady state (closed circles) effects on blood flow of elevating and reducing perfusion pressure from 100 mm Hg. (From C. F. Rothe, F. D. Nash, and D. E. Thompson. *Amer. J. Physiol.* 220:1624, 1971.)

Autoregulation of renal blood flow is characterized by the phenomenon illustrated in Figure 17-7. Elevation of perfusion pressure from the control level (100 mm Hg) results in an initial increase of blood flow followed by a reduction of flow to about the previous control level. Since flow is thus maintained relatively constant in the presence of an increased perfusion pressure, vascular resistance necessarily increases. Conversely, reduction of perfusion pressure from the control level produces an initial fall in tissue blood flow which is followed within about 30 seconds by an upward adjustment of flow to the control level. In this case a maintained flow during reduced perfusion pressure reflects a reduction in vascular resistance. Thus autoregulation represents the adjustment of vascular resistance to keep blood flow constant over a range of perfusion pressure. Most tissues exhibit some degree of autoregulation. The mechanism of autoregulation is complex, based on response of the vascular smooth muscle to changes in transmural pressure, intrinsic metabolic factors, or other local determinants of vascular reactivity. A number of theories have been proposed to account for the phenomenon of autoregulation.

Metabolic Theory

There are numerous lines of evidence to implicate metabolic factors in the local control of tissue blood flow. Increased metabolic activity is associated with increased tissue blood flow, in what may be considered *function hyperemia.* Enhanced metabolic activity of muscular exercise causes venous blood oxygen saturation to fall and blood flow to increase markedly (Fig. 17-8). Thus, oxygen consumption, the product of blood flow and the arteriovenous oxygen saturation difference, increases dramatically. The hyperemia during maximal exercise may increase from 1 liter per minute at rest to as much as 20 liters per minute.

In metabolic autoregulation, the local concentration of metabolites (of a vasodilator nature) varies inversely with the blood flow through the tissue. For example, since the arterioles closely parallel their corresponding venules, the accumulation of vasodilator substances with reduced blood flow tends to dilate the resistance vessels and to restore flow. If flow increases, the metabolites are washed out and flow decreases proportionately.

It is presumed that the hyperemias result from alterations of the concentrations of local "metabolic" factors. The capability of metabolites to produce profound vasodilation is illustrated in Figure 17-6. The precise nature of these factors is unclear. Many have been proposed, including oxygen, carbon dioxide, pH, adenosine compounds, histamine, plasma potassium concentration, and plasma osmolarity. Reducing the oxygen partial pressure in the blood (P_{O_2}) or increasing P_{CO_2} certainly has significant influence on cerebral circulation, presumably through a change in local pH. Adenosine compounds and histamine, although potent vasodilators, achieve inadequate precapillary concentrations to account for the hyperemia. Exercise results in elevated venous plasma potassium concentration which could produce vasodilation. Moreover, increased plasma potassium together with reduced P_{O_2} acts synergistically to increase blood flow. Plasma osmolarity is also elevated during exercise. It has been proposed that increased plasma osmolarity causes fluid to leave the cells of small arteries and arterioles to reduce their wall thickness and increase their lumen and thus provide for hyperemia. Inasmuch as all these factors are present to some degree during hyperemias, it is possible that they are collectively responsible for the blood flow patterns seen. The relative significance and level of interactions between them probably vary between tissues as well as between species.

It has been suggested that in the kidney the juxtaglomerular concentration of sodium plays a role as an autoregulatory factor.

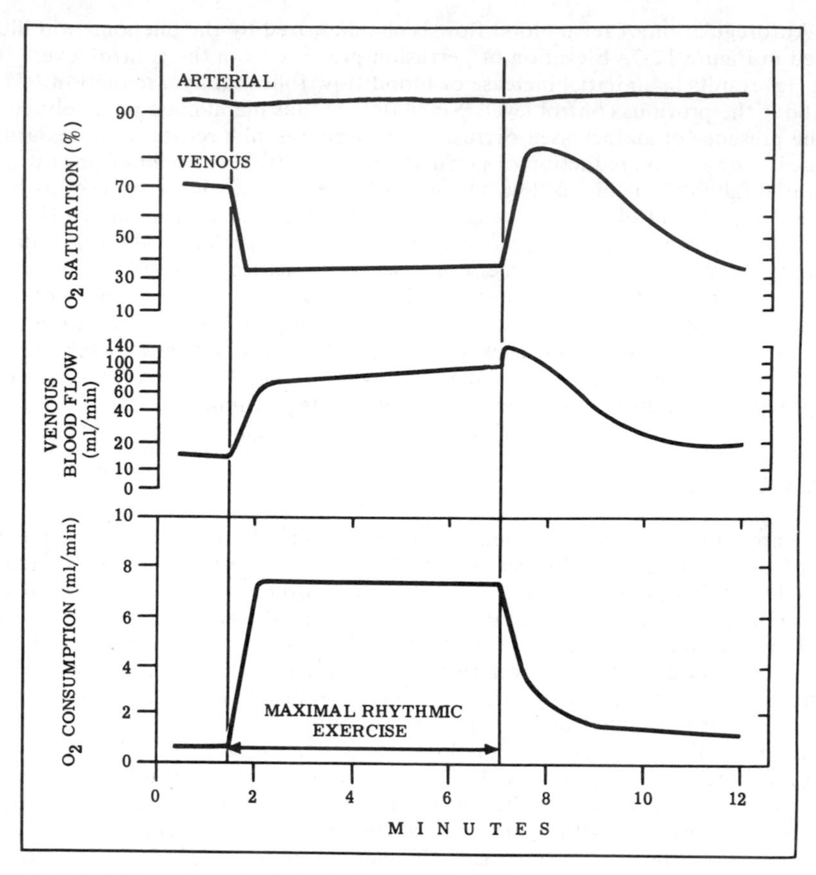

FIGURE 17-8. Effect of maximal rhythmic muscle contraction on venous blood oxygen saturation, blood flow and oxygen consumption. (Data from Kramer et al. *Pflueger. Arch. Ges. Physiol.* 241:717–729, 1913.)

Metabolic autoregulation has been demonstrated as being prominent in tissues with high metabolic activities, such as contracting skeletal muscle, coronary circulation, cerebral circulation, and to some extent the kidney. It is not present in skin.

Additionally, interruption of blood flow to a tissue results, upon restoration of flow, in an increased blood flow, a *reactive hyperemia,* the duration of which is roughly proportional to the duration of cessation of flow. Heart and brain, tissues with minimal oxygen reserves, undergo maximal increases in blood flow. Muscle, intestine, and kidney, with intermediate metabolic rates and oxygen reserves, exhibit intermediate hyperemia. Skin, with a low metabolic rate, and the splanchnic bed show no reactive hyperemia.

Myogenic Theory

Elevating perfusion pressure produces an increase in transmural pressure and therefore wall tension of small arteries and arterioles. Tension on the vascular wall provides the stimulus for visceral smooth muscle contraction, which reduces the diameter, increases the resistance, and impedes flow. The reverse effect occurs when perfusion pressure is reduced. Thus, adjusting vascular resistance in the same direction as the change in perfusion pressure tends to maintain flow at a constant level. Organs in which a myogenic type of autoregulation has been postulated include the kidney, intestine, and hepatic arterial supply.

Other Theories

It has been suggested that, in addition to increasing blood flow, elevating perfusion pressure increases capillary hydrostatic pressure and enhances filtration. This in turn would increase *tissue hydrostatic pressure,* which, by reducing the transmural pressure, should partially collapse the microcirculation to increase vascular resistance and thereby restore blood flow to control. That this sequence is seen in the encapsulated kidney only at excessive tissue pressures, and in the brain with its rigid cranium only when cerebrospinal pressure exceeds arterial pressure, indicates that it is not a physiological phenomenon.

It has been proposed that autoregulation could represent the participation of *neurogenic reflexes.* However, autoregulation has been demonstrated in isolated denervated preparations under influence of local anesthesia. Consequently, this theory appears to be inappropriate.

Summary of Mechanisms of Autoregulation

A single factor is probably not responsible for the autoregulation observed in various tissues. It is likely that the interaction of multiple factors is responsible for autoregulation in any one tissue, and each factor may have a more prominent effect in one tissue than in another. Thus, the metabolic factor is apparently most prominent in highly metabolizing tissues such as exercising skeletal muscle, heart, and brain.

INTERACTION BETWEEN CENTRAL AND LOCAL INFLUENCES

Vascular smooth muscle elements are continuously under central and local influences which may act synergistically or antagonistically to modify vascular tone. The balance between these interactions may be related to the distribution and type of smooth muscle in the various vascular elements. Two types of smooth muscle are present in arterioles. It appears that the smooth muscle at the outer layer of the vessel is of the *multiunit* type and is therefore most responsive to neurogenic influence. The smooth muscle of the inner layer of the vessels is primarily the *visceral* type and is most responsive to transmural pressure and local chemical influence. At the level of the arteriole the smooth muscle distribution favors central neurogenic activity, whereas more distally at the level of the precapillary sphincters, local influences predominate. The smooth muscle of postcapillary resistance vessels seems to be the multiunit type and is therefore under central neurogenic control.

Thus, there exists a framework which can adjust peripherovascular tone to provide for both systemic and local cardiovascular homeostasis. This feature of the control system may be reflected in the effect of sympathetic stimulation on the series vascular elements in muscle. Sympathetic stimulation produces small artery, arteriolar, and venous vasoconstriction to increase peripheral resistance and venous return and thereby contribute to blood pressure regulation. The reduced tissue blood flow which results from the vasoconstriction leads to a reduced transmural pressure as well as an accumulation of metabolites at the precapillary sphincter, both of which produce precapillary sphincter dilation. This provides a more uniform distribution of local blood flow and permits a more complete exchange between the vascular and extravascular compartments, despite the reduced total blood flow.

NEURAL CONTROL OF CARDIOVASCULAR HOMEOSTASIS

Central control of the cardiovascular system is summarized in Figure 17-2. Under normal conditions, the basal neural activity of the cardiovascular centers provides a vagal influence which maintains the heart rate at a relatively low level and some sympathetic, positive inotropic effects on contractility. The heart is relatively large, has a small diastolic reserve, especially if the subject is reclining, and has a large systolic reserve volume (see below and Chap. 15). The sympathetic vasoconstrictor fiber activity acts on the peripheral circulation to reduce flow from the level seen following denervation. In addition, there appears to be a neurogenically induced venomotor tone acting to maintain an adequate filling pressure to the heart.

The four effector mechanisms for the control of arterial blood pressure (Fig. 17-1) are not all used proportionally under all conditions. For example, if the systemic arterial blood pressure is normal, carotid occlusion produces an increase in blood pressure by a peripheral resistance increase with only a slight increase in heart rate and little or no change in cardiac output or capacitance vessel tension. On the other hand, the response to an elevated arterial pressure includes decreased peripheral resistance, heart rate, and cardiac output. Thus the homeostatic mechanisms active in response to elevated pressures appear to be different from those acting in response to reduced pressure.

A differential sensitivity of the medullary neurons controlling the various peripheral vascular beds is also seen. Whereas the skeletal muscle vascular tone is closely related to changes in carotid sinus pressure, there is only a moderate constriction of the intestinal and cutaneous vessels; the renal tone is but little affected, possibly because of its highly effective autoregulation of blood flow. However, under extreme conditions such as severe hemorrhage, renal and splanchnic constriction is marked. Thus, some homeostatic mechanisms may not be activated until relatively extreme changes from the basal level have occurred.

EXTRANEURAL CONTROL OF THE CARDIOVASCULAR SYSTEM

The volume of fluid available to the heart for pumping in relation to the volume of the vascular system provides another factor in the control of cardiovascular function, because an increase in effective blood volume tends to fill the heart more than is normal and so increases cardiac output (see Chap. 15). On the other hand, a decrease in total blood volume tends to decrease cardiac output. However, the passive recoil of the capacitance vessels, associated with precapillary vasoconstriction, provides a powerful extraneural mechanism acting to maintain central venous pressure and cardiac output. The constriction of the resistance vessels results in a decrease in flow, so that the transmural pressure in the capillaries and venules decreases. This leads to fluid reabsorption from the extracellular spaces (Chap. 12) and permits an elastic recoil of the capacitance vessels to redistribute the blood toward the heart. This elastic recoil appears to be a much more important mechanism than neurogenic venoconstriction. Also, a decrease in blood volume, by relieving the stretch of left atrial volume, results in fluid retention by renal activity which acts to restore blood volume (Chaps. 23 and 31).

CARDIOVASCULAR SYSTEM RESERVE

The cardiovascular reflexes provide the widely varying flow of blood needed to meet the metabolic demands of the various tissues of the body, but blood flow to the tissue is limited because of the limitations of the cardiovascular system. A person cannot run at maximum speed (high muscle metabolism) after a full meal (splanchnic bed dilation) on a hot day (skin dilation). Heart disease requires the utilization of cardiovascular reserves to provide even the basal blood flow (Chap. 18B).

Control of the cardiovascular system utilizes four primary cardiovascular reserves in the maintenance of circulatory homeostasis: venous oxygen reserve, heart rate reserve, systolic volume reserve, and diastolic volume reserve.

VENOUS OXYGEN RESERVE

During normal resting conditions, about 4 ml of oxygen is extracted from the arterial blood per 100 ml of blood flow through the tissue (Table 17-1). Thus, a reserve of about 16 volumes of oxygen per 100 volumes of blood (16 volumes percent) is in venous blood. During severe exertion, when the oxygen utilization may be over 10 times the basal level, the extraction of oxygen from the blood is increased to over 12 volumes percent. Active skeletal muscle and cardiac muscle remove at least 75 percent of the oxygen from blood, leaving less than 5 volumes percent of oxygen in the venous blood.

A person with heart failure has a decreased cardiac output, but adequate amounts of oxygen can be transported to tissues because most cells can extract a larger than normal fraction of the oxygen from the blood by increasing the oxygen arteriovenous (A-V) difference. Unfortunately, following cardiac failure, the reduced flow to the peripheral tissue is not uniformly distributed. The blood flow through heart muscle and the brain tends to be maintained because there is little neurally induced vasoconstriction in response to the fall in blood pressure following the decrease in cardiac action. Peripheral vasoconstriction causes a disproportionate reduction flow of blood through the kidneys. Whereas the oxygen extraction of skeletal muscle and intestine can increase to about 12 to 15 volumes of oxygen per 100 volumes of blood, the renal extraction is rather constant. As renal blood flow is reduced, the oxygen consumption is reduced and function impaired. Thus, although an increase in A-V oxygen difference provides a source of oxygen to tissues, its benefits are limited in the patient with cardiac failure because of the encroachment upon renal function. In the normal individual, however, this reserve provides for a twofold to threefold increase in oxygen utilization by active muscle (Fig. 17-8).

HEART RATE RESERVE

During moderate exercise the heart rate is increased by the cardiovascular control system and the stroke volume is held relatively constant. This provides for an increase in cardiac output. However, the heart rate increase limits the filling time; hence, the venous filling pressure or the effective distensibility of the heart must be increased by homeostatic mechanisms, or the end systolic volume must be decreased parallel to a decrease in end diastolic volume, if the stroke volume is to be maintained.

An increase in heart rate does not provide a particularly useful reserve for the cardiac patient, because a rapidly beating heart is less efficient in the use of oxygen than a slowly beating one (Chap. 15). If the heart failure is primarily a result of inadequate coronary blood flow in the first place, this loss of efficiency becomes quite significant. Furthermore, an increase in heart rate tends to limit the coronary blood flow (Chap. 15). Thus, an increase in heart rate is not an efficient means of increasing cardiac output by the failing heart, although it is a highly important mechanism for the normal individual, providing a twofold to threefold increase in cardiac output.

SYSTOLIC VOLUME RESERVE

In the normal individual the stroke volume can be increased from about 70 ml to over 100 ml per stroke. Athletes with cardiac outputs of over 30 liters per minute and heart rates of about 200 beats per minute must have stroke volumes of about 150 ml. This increase in stroke volume during *maximal* exertion by well-trained individuals is provided primarily by a decrease in end systolic volume (see Fig. 15-3). There is a large amount of blood in the heart of a resting individual at the end of systole. This provides a cardiovascular reserve but is available only if the myocardial contractility increases so that the added volume can be ejected. The loss of myocardial contractility in cardiac failure is such that sympathetic activity from the control system is relatively ineffective in utilizing this reserve.

DIASTOLIC VOLUME RESERVE

As the central venous pressure increases in heart failure, the heart becomes distended. The increase in diastolic volume, produced by stretching the myocardial fibers, tends to cause the release of more energy in accordance with Starling's law of the heart, so that more blood is pumped. Furthermore, the heart beats more efficiently because of the decrease in viscous losses and development of internal tensions (Chap. 15). As fluids accumulate, the failing heart becomes larger and more distended. Since even the normal resting reclining individual has a relatively large heart, and therefore a small diastolic volume reserve, there is little reserve for heart failure.

Thus, the person with cardiac failure has severe limits on cardiovascular system reserves. Whereas normally the heart rate is increased and the stroke volume is maintained in response to neural stimuli, the failing heart cannot respond satisfactorily to increased demands, because of myocardial weakness.

CARDIOVASCULAR RESPONSE TO EXERCISE

An increase in oxygen consumption by a highly trained athlete to 21.4 times normal, during sustained maximal exertion, has been recorded. Under these conditions the oxygen extraction increased over 3 times, the heart rate about 3 times, and the stroke volume about 2 times. The cardiovascular reserves utilized during exercise are made available primarily by the changes in the pattern of the autonomic nervous system discharge. Skeletal muscle contraction, initiated in the cerebral cortex, is paralleled by an autonomic nervous system response acting by way of the cardiovascular centers to cause (1) a vasodilation of the blood vessels of the muscle, (2) an increase in adrenal medulla activity, (3) a venous constriction acting to increase the filling pressure to the heart, (4) some splanchnic vasoconstriction shunting blood from the splanchnic area to tissues with increased metabolism, (5) an increase in heart rate, (6) possibly an increase in cardiac distensibility, and (7) an increase in myocardial contractility.

REFERENCES

Abramson, D. I. (Ed.). *Blood Vessels and Lymphatics.* New York: Academic, 1962. Chaps. 8, 10, 11, 16, and 17.

Bard, P. Anatomical organization of the central nervous system in relation to control of the heart and blood vessels. *Physiol. Rev.* 40(Suppl. 4):3–26, 1960.

Celander, O. The range of control exercised by the sympathico-adrenal system. *Acta Physiol. Scand.* 32(Suppl. 116):1–132, 1954.

Chien, S. Role of the sympathetic nervous system in hemorrhage. *Physiol. Rev.* 47:214–288, 1967.

Fishman, A. P., and D. W. Richards (Eds.). *Circulation of the Blood: Men and Ideas.* New York: Oxford University Press, 1964. Chap. 7.

Folkow, B., C. Heymans, and E. Neil. Integrated Aspects of Cardiovascular Regulation. In J. Field (Ed.), *Handbook of Physiology.* Section 2: Circulation, Vol. III. Washington, D.C.: American Physiological Society, 1965. Chap. 49, pp. 1787–1824.

Folkow, B., and S. Mellander. Veins and venous tone. *Amer. Heart J.* 68:397–408, 1964.

Folkow, B., and E. Neil. *Circulation.* London: Oxford University Press, 1971.

Green, H. D., and J. H. Kepchar. Control of peripheral resistance in major systemic vascular beds. *Physiol. Rev.* 39:617–686, 1959.

Green, H. D., C. E. Rapela, and M. C. Conrad. Resistance (Conductance) and Capacitance Phenomena in Terminal Vascular Beds. In W. F. Hamilton (Ed.), *Handbook of Physiology.* Section 2: Circulation, Vol. II. Washington, D.C.: American Physiological Society, 1963. Chap. 28.

Heymans, C., and E. Neil. *Reflexogenic Areas of the Cardiovascular System.* London: Churchill, 1958.

Johnson, P. C. (Ed.). *Autoregulation of Blood Flow.* (*Circ. Res.* 15 [Suppl. 1].) New York: American Heart Association, 1964.

McDowall, R. J. S. *The Control of the Circulation of Blood,* Vols. 1 and 2. London: Dawson, 1938, 1956.

Mellander, S. Comparative studies on the adrenergic neuro-hormonal control of resistance and capacitance blood vessels in the cat. *Acta Physiol. Scand.* 50(Suppl. 176):1–86, 1960.

Mellander, S. Systemic circulation: Local control. *Ann. Rev. Physiol.* 32:313–344, 1970.

Mellander, S., and B. Johanson. Control of resistance, exchange and capacitance functions in the peripheral circulation. *Pharmacol. Rev.* 20:117–196, 1968.

Rushmer, R. F. *Cardiovascular Dynamics,* 3d ed. Philadelphia: Saunders, 1970.

Sagawa, K. Overall circulatory regulation. *Ann. Rev. Physiol.* 31:295–330, 1969.

18

Pathological Physiology of the Cardiovascular System

A. Hypertension

Julius J. Friedman

ARTERIAL HYPERTENSION is a condition of sustained elevated systemic arterial blood pressure. Since arterial pressure is influenced by many factors, including age, sex, body type, race, body temperature, and emotional state, the participation of these factors must be discounted before the diagnosis of hypertension can be made. The minimum level of systemic arterial pressure considered to be hypertensive has been arbitrarily set at 140/90 mm Hg. While hypertension usually involves elevations in mean and pulse pressures, the rise in diastolic pressure is clinically regarded as the critical criterion, suggesting that hypertension is due primarily to an increased peripheral resistance.

Hypertension is a serious cardiovascular disease. It is responsible for approximately 10 percent of the deaths in people over 50 years of age. However, many of the people in this age group may have blood pressures as high as 170/90 without displaying any symptoms of hypertension, and approximately 75 percent of these individuals die of diseases which may have caused their hypertension.

In approximately 30 percent of the cases of hypertension, the elevated blood pressure is a clinical sign of a specific disease, e.g., renal disease, and not a disease entity in itself. Such conditions of elevated blood pressures are referred to as *secondary hypertension.* In the remaining 70 percent the development of hypertension cannot be attributed to any known origin and probably represents a specific disease state. This form of hypertension is called *primary* or *essential hypertension.*

Hypertension is an important clinical disorder because it leads to organic alterations of the heart, brain, kidneys, and arterial vasculature. In hypertension, the heart must pump blood into the arterial system against a higher than normal level of pressure. Consequently, the heart must perform additional work, and therefore it hypertrophies. Hypertrophy may eventually lead to myocardial failure. The cerebral vasculature does not possess the tissue support found in other regions of the body, and cerebral hemorrhages in hypertensives are not uncommon. Renal insufficiency, which results from arterionephrosclerosis, is a frequent complication of hypertension. The continuous added stress placed upon the arterial system in hypertension ultimately leads to sclerosis of the arterial wall. This alteration of the vascular system may modify tissue blood flow and thereby cause disruption of tissue function.

CLASSIFICATION OF HYPERTENSIVE VASCULAR DISEASE

To gain an insight into the nature of essential hypertension, the various diseases which cause an elevation in blood pressure have been studied both clinically and experimentally. These include cardiovascular, neurogenic, endocrine, and renal disorders.

CARDIOVASCULAR HYPERTENSION

Systemic arterial blood pressure is determined by cardiac output and peripheral resistance. Consequently, a variety of changes in the cardiovascular system itself can result in an elevated systemic arterial pressure. Arteriosclerosis of resistance vessels reduces the size of the vascular lumen and thereby increases peripheral resistance. Coarctation of the aorta also increases the resistance to blood flow, and thus blood pressure. Hyperthyroidism, by increasing cardiac output, produces an increase in blood pressure. These conditions lead to hypertension by increasing cardiac output and/or peripheral resistance. Cardiovascular hypertension can usually be alleviated by treating the underlying causes.

NEUROGENIC HYPERTENSION

The centers which regulate and integrate cardiac and vasomotor activity are situated in the brain. Consequently, impairment of cerebral function in such a manner that cardiac and vasomotor activity are enhanced produces an elevation of blood pressure. Cerebral function may be impaired by cerebral ischemia due to either cerebral arterial occlusion or severely increased intracranial pressure, or by tumor development in various regions of the brain. In each case the resulting hypertension may be relieved by appropriate treatment of the organic disturbance.

Another form of neurogenic hypertension may be initiated by carotid sinus baroreceptor denervation. With such denervation, other baroreceptors outside the carotid sinus region must regulate pressure, and these are not as efficient as the carotid receptors. This leads to a sustained elevation of blood pressure by allowing increases in cardiac activity and vasomotor tone to develop with little regulation.

ENDOCRINE HYPERTENSION

The adrenal medulla contains chromaffin cells which secrete the catecholamines, epinephrine and norepinephrine. These cells may develop tumors (pheochromocytoma), which secrete excessive quantities of catecholamines into the circulation, either continuously or periodically; the blood pressure then is elevated by increased cardiac output and peripheral resistance. Surgical removal of the tumorous tissues relieves the condition and usually restores blood pressure to normal levels.

The adrenal cortex has also been implicated; however, the specific role of this tissue and its secretions in the pathogenesis of hypertension remains obscure. Many cases of hypertension can be relieved by restriction of the sodium content in the diet; a high intake of sodium tends to aggravate the condition. Aldosterone, a substance secreted by the adrenal cortex, stimulates the kidney to retain sodium ion and produces hypernatremia. In conditions of adrenal hyperplasia with primary aldosteronism, hyponatremic, hypokalemic alkalosis with volume expansion develops and the blood pressure becomes elevated. This condition is reversible, and removal of the adrenal gland usually restores the blood pressure and electrolyte pattern to normal levels.

In malignant hypertension, aldosterone levels are increased and hypokalemia develops; however, a hyponatremic condition exists, due to volume expansion. In this case the hyperaldosteronism is a secondary manifestation of the hyponatremia. Adrenalectomy in malignant hypertension is of questionable value.

In almost all forms of chronic experimental hypertension, aortic smooth muscle contains an elevated concentration of sodium and potassium ions and an increased volume of water, which may contribute to increased blood pressure in a number of ways. Ion and water accumulation in the walls of resistance vessels could cause swelling of the walls and encroachment upon the lumen, thereby increasing peripheral resistance and blood pressure. In addition, alterations of the ionic composition of vascular smooth muscle could, by disturbing the membrane potential, affect the sensitivity of resistance vessels to circulating vasoconstrictor substances and autonomic nervous influences. Also, ionic changes in the muscular elements of the vascular wall may alter the state of actomyosin to produce an increased contractility. Thus, by producing changes in the physical state of resistance vessels and by increasing excitability and contractility of vascular smooth muscle, alterations in electrolyte balance could contribute significantly to the development of hypertension.

RENAL HYPERTENSION

Permanent hypertension may be induced by a variety of manipulations which effectively reduce either renal blood flow, functional renal tissue, or tubular sodium delivery. Reduction of renal blood flow by arterial occlusion or by renal compression consistently produces hypertension. The relationship of the kidney to hypertension is independent of nervous activity because complete renal denervation does not prevent the development or modify the extent of hypertension. Therefore, renal hypertension is attributed to a humoral mechanism. As early as 1898 it was recognized that an extract of kidney cortex was capable of producing an increase in blood pressure when injected into a normal subject. The principle responsible for this effect is *renin*. However, renin itself has no vasoactive properties. It is a proteolytic enzyme which acts upon a plasma protein, alpha-2 globulin, to produce a polypeptide, angiotensin I, which is also vasoinactive. A converting enzyme, present in plasma, then converts angiotensin I to the active form angiotensin II (formerly known as angiotonin or hypertensin). Angiotensin II produces widespread arteriolar vasoconstriction. There is some evidence that angiotensin also constricts venous elements.

SELF-PERPETUATING NATURE OF HYPERTENSION

Hypertension is characterized by an increase in pulse pressure as well as mean blood pressure. Normally, one would expect an increased peripheral resistance to cause pulse pressure to decline (Chap. 16). The elevated pulse pressure in hypertension is attributed to a reduction of large artery distensibility, which is a secondary rather than a primary effect. It reflects, in part, the progressive sclerotic process in hypertension. The sclerotic process is not restricted to the large arteries. When it involves the resistance vessels, their lumina are permanently reduced and peripheral resistance remains elevated, even after extensive treatment including radical sympathectomy. In this manner, the condition of hypertension enhances and perpetuates its own development, an example of positive feedback.

ESSENTIAL HYPERTENSION

Essential hypertension refers to the sustained elevation in systemic arterial blood pressure for which there is no discernible origin. The hypertension may be "benign" in that it may develop slowly and progressively over many years, or it may be "malignant" and develop rapidly in a brief period of time. Attempts to identify the cause of essential hypertension have been fruitless. After exclusion of hypertension due to renal disease, adrenal dysfunction, or cardiovascular and neurogenic alterations, no reasonable explanation is obvious at this time.

Subjects with essential hypertension consistently exhibit an elevated peripheral resistance which is rather uniformly distributed throughout the body and which is not a result of changes in central vasomotor activity. Investigators have considered the participation of altered pressure regulation effected by a "resetting" of the baroreceptor mechanism. While the regulatory mechanism of the carotid sinus is active in both normotensive and hypertensive patients, the pressoreceptors of hypertensives fire intermittently, and at lower frequencies, at pressures which elicit continuous, higher-frequency firing in normotensives. Thus, the carotid sinus pressor mechanism may, by becoming adapted (set) to a new, elevated pressure level, perpetuate a higher pressure.

An alternative interpretation of altered pressoreceptor activity concerns the organic alteration of the carotid sinus wall. Sclerosis of the carotid sinus wall reduces its distensibility and limits the deformation of the receptors within the sinus wall. Consequently, larger pressure changes are required to activate the pressoreceptor mechanism, and over a wide range the pressure would vary as though the sinus were denervated. Whether this alteration precedes essential hypertension or is caused by hypertension is not clear.

There is evidence from studies of sympathetic blockade that the sympathetic nervous system contributes to the hypertensive state through an increased reactivity of resistance vessels. Patients with essential hypertension exhibit more marked pressor responses to sympathetic influences, both neurogenic and humoral, than do normal subjects. Whether these reactive characteristics are due to metabolic or structural alterations is uncertain. However, the tendency toward hyperactivity is apparently genetically determined, because hypertensive patients almost always possess a family history of the disease. This hyperreactive tendency has been used to identify potential hypertensives. A commonly employed test is the *cold-pressor test,* which involves the measurement of blood pressure before and during the time a subject places a hand in ice water. This procedure produces only a slight rise in blood pressure in normal subjects. However, an elevation in systolic and/or diastolic pressure of 20 mm Hg or more is regarded as indicating a tendency toward hypertension.

An increasing body of evidence suggests that the interrelationship between the kidney and the adrenal cortex is significant in severe hypertension. Renal damage releases renin, which results in the formation of angiotensin, and this agent stimulates the secretion of aldosterone by the adrenal glomerulosa zone (Chap. 35). It may be that angiotensin, acting on vessels made hyperreactive by electrolyte alterations, is a significant feature of essential hypertension. However, at the present time it is not possible to define the exact nature of essential hypertension.

In summary, it is unlikely that any one of the factors cited above is the sole causative agent. In all probability essential hypertension involves the participation of a number of factors to varying degrees in different cases. This multiple-factor concept is particularly attractive in view of the conflicting experimental evidence and unpredictable pharmacological responses of the disease. Furthermore, there are probably many cases of hypertension classified as "essential" which may be due to other causes. They may represent either diseases not yet identified, an imbalance in the relationship of endogenous vasoactive substances, a modification of the participation of regulatory factors in the maintenance of normal blood pressure, or a vasoactive substance which has not yet been characterized. Figure 18A-1 is a diagram relating the various participating systems to each other.

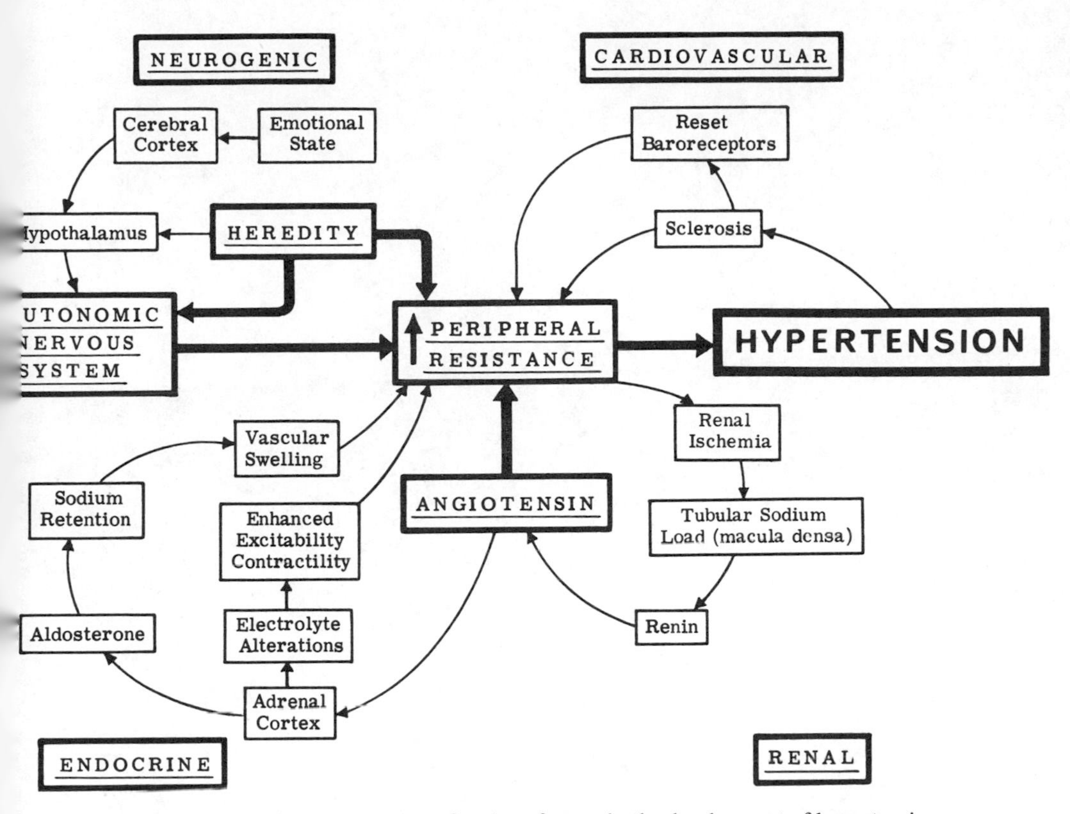

FIGURE 18A-1. Possible participation of various factors in the development of hypertension.

REFERENCES

Bock, K. D., and P. T. Cottier (Eds.). *Essential Hypertension.* An International Symposium sponsored by CIBA, Limited, June, 1960. Berlin: Springer, 1960.

Freis, E. E. Hemodynamics of hypertension. *Physiol. Rev.* 40:27–54, 1960.

Page, I. H. *Arterial Hypertension.* Philadelphia: Davis, 1960.

Page, I. H., and J. W. McCubbin. The Physiology of Arterial Hypertension. In W. F. Hamilton and P. Dow (Eds.), *Handbook of Physiology.* Section 2: Circulation, Vol. III. Washington, D.C.: American Physiological Society, 1965. Chap. 61, pp. 2163–2208.

Rodbard, S., L. A. Sapirstein, and J. E. Wood, III (Eds.). *Hypertension.* Proceedings of the Council for High Blood Pressure Research, American Heart Association, Nov., 1964. New York: A.H.A., 1965.

Skelton, F. R. (Ed.). Hypertension – Chemical and Humoral Factors. Proceedings of the Council for High Blood Pressure Research, American Heart Association, Nov., 1960. *Circ. Res.* 9:715–805, 1961.

Tobian, L. Interrelationship of electrolytes, juxtaglomerular cells and hypertension. *Physiol. Rev.* 40:280–312, 1960.

18

Pathological Physiology of the Cardiovascular System

B. Congestive Heart Failure

Ewald E. Selkurt

HEART FAILURE may be simply defined as a state in which the heart fails to maintain an adequate circulation for the needs and demands of the body despite what appears to be satisfactory filling pressure. When this failure is accompanied by an abnormal increase in blood volume and interstitial fluid, the condition is known as *congestive* heart failure. In Chapter 17, four types of cardiovascular reserves were discussed, which should be reviewed in terms of their role in the failing heart. These were (1) the venous blood oxygen reserve, (2) the heart rate reserve, (3) the systolic volume reserve, and (4) the diastolic volume reserve. As was explained, each has limitations as a reserve for the heart in failure. Perhaps the most valuable of these reserves is the diastolic volume reserve, which acts by way of increased filling pressure. In early stages of heart failure these reserves may be adequate; accordingly, persons at rest or engaged in activity requiring minimal exertion manifest minimal symptomatology. However, increasing venous pressure distends the ventricle beyond the point of critical diastolic stretch, so that ultimately the contractile force declines and the output falls. The diminished cardiac reserve shows up during increased exertion and is evidenced by breathlessness, palpitation (perceptible heart beat), and fatigue with degrees of exercise that were formerly tolerated quite easily. As the condition progresses, these symptoms appear even at rest and are involved in several of the sequelae which result from heart failure and become a part of the overall clinical syndrome.

CAUSES OF HEART FAILURE AND ITS PHYSIOLOGICAL CONSEQUENCES

Circulatory failure occurs for a number of reasons which can be only briefly outlined here. They may be grouped into several categories as follows: First, failure may occur because of *interference with systemic venous return* due to (1) factors usually remote from the heart, as in hemorrhage and shock or the type of peripheral vascular collapse noted in syncope, or (2) factors in the heart region which impair venous return, such as pericardial tamponade, constrictive pericarditis, and tricuspid stenosis. Venous congestion is usually a feature of the latter group. Second, failure may result from *interference with the pumping or filling of the ventricle* due to anatomical abnormalities, either inherited or acquired through disease such as rheumatic fever. This type of failure is seen in semilunar valvular stenosis or insufficiency, mitral stenosis or insufficiency, and tricuspid insufficiency, and is also associated with systemic arterial hypertension and pulmonary arterial hypertension. Third, *primary diseases of the myocardium* lead to failure. This happens with myocardial infarcts created by coronary occlusion. Last, *chronic stimulus to high output* of one or both ventricles may ultimately cause failure. This happens with drastic reduction in peripheral vascular resistance, as in beriberi, systemic arteriovenous (A-V) fistula, Paget's disease (in which increased blood flow through the affected bone acts as an A-V fistula), anemia, and some types of congenital cardiovascular malformations – for example, those that have large left-to-right shunts. Despite the high output in the last group of examples, the blood supply to the various tissues and organs is inadequate, particularly in terms of the metabolic needs. On this basis, hyperthyroidism may lead to high-output failure. In all categories, circumstances occur which lead ultimately to failure of cardiac reserves.

PATHOLOGICAL CHANGES

The pathological changes characteristic of failure of the heart include hydropic degeneration of the myocardial fibrils followed by necrosis, and fibrosis, particularly when associated with degenerative infections and metabolic heart disease. Ultimately, however, the defect must be sought in the ability of the myocardium to utilize energy and to perform work.

METABOLIC CHANGES

From a biochemical point of view, the heart muscle cell failure could result from abnormalities in *energy production,* in *conservation,* or in *utilization* of the energy. Figure 18B-1 summarizes the steps involved in the transfer of foodstuffs to cardiac

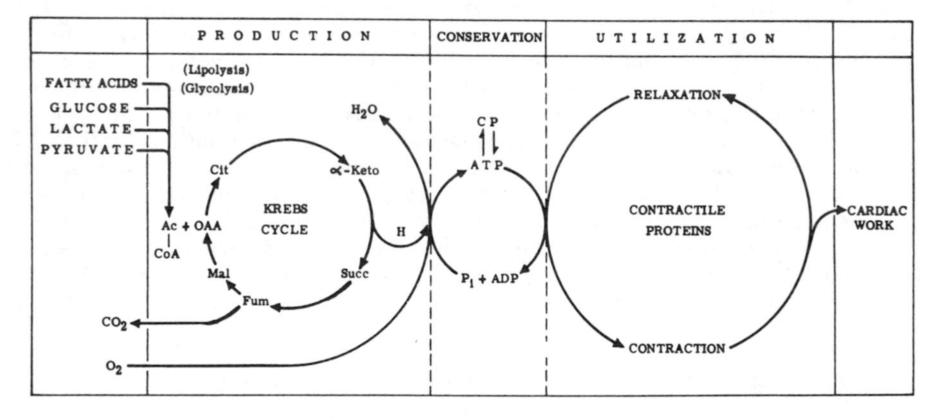

FIGURE 18B-1. *Production.* Metabolism of precursors to Ac-CoA (acetyl coenzyme A), which condenses with OAA (oxalacetate) to form citrate. This goes through the steps of the Krebs cycle to yield 8 hydrogen atoms or electrons ($H = H^+ + e$). These represent the energy content of the acetyl, which contributes 4 hydrogen atoms, plus 4 from the water added to the cycle intermediates in effecting the oxidation of the fragment. This step yields "energy-rich" electrons for the transport of oxygen. *Conservation.* This is associated with the electron flow along the hydrogen transport chain, which is depicted as coupling the phosphorylation of adenosine diphosphate (ADP) to the oxidation of the cytochrome enzymes. *Utilization.* The high-energy phosphate bond of adenosine triphosphate (ATP) is channeled into the contractile mechanism and results in the performance of mechanical work. (From *Annals of The New York Academy of Sciences,* Volume 72, Article 12, Fig. 1, pg. 467, R. E. Olson & D. A. Piatnek. © The New York Academy of Sciences; 1959; Reprinted by permission.)

work. Substrate metabolism has been discussed in Chapter 15. In so-called low output failure, resulting from arteriosclerotic hypertensive heart diseases or from valvular defects, some authorities believe there is no defect in energy production or conservation. In a study of congestive heart failure in both man and dog, no significant alteration of oxidative metabolism was found by Olson and Piatnek in 1959. More specifically, the distribution of phosphorus among the high-energy phosphate bonds have revealed no significant change from normal. Nor does it occur with thyrotoxicosis (often associated with so-called high-output failure), although in this condition a shift toward utilization of fatty acids from glucose, lactate, and pyruvate has been noted. Such a shift in metabolism could not be a reason for myocardial failure, however.

Schwartz and Lee (1962) have challenged the adequacy of levels of energy-rich phosphate or of oxygen and substrate uptake as indices of the rate of energy liberation or storage. They suggest that normal concentration of high-energy phosphates could exist in the face of reduced rate of energy liberation and/or storage if the rate of energy utilization is also decreased. Such heart muscle preparations, however, may show a definite impairment of mitochondrial oxidative phosphorylation and contractile force. Studying mitochondria from failing guinea pig hearts (produced by aortic constriction), they found them clearly depressed with respect to oxidative phosphorylation. Specifically, phosphorylation associated with oxidation of glutamate, succinate, and α-ketoglutarate was uncoupled. The defect probably resided in the phosphorylation step associated with the electron transfer between cytochrome C to oxygen.

Although several other investigators support the conclusions of Schwartz and Lee, recently confirmed by the same laboratory (Lindenmayer et al., 1968), that mitochondrial oxidative phosphorylation is impaired in heart failure, this is not universally accepted. In support of Olson and Piatnek, Stoner et al. (1968) found no impairment in the stressed dog heart (combined tricuspid insufficiency and pulmonary stenosis, pulmonary insufficiency, aortic stenosis, aortic insufficiency, for periods of 332 to 608 days). Chidsey et al. (1966) also found no deficiencies in the energy-transforming capabilities of mitochondria isolated from human myocardial tissues removed at the time of cardiac operations. They concluded that if a biochemical abnormality is responsible for defective ventricular function in such patients, it would appear to involve utilization of energy in the contractile process. Pool et al. (1969), studying right ventricular papillary muscles from cats with heart failure, found the contractile properties severely depressed, to 13 percent of the work of normals, and with this, reduction of use of energy to 7 percent of that used by normal muscle, but with no indication of inefficiency in the direct conversion of chemical energy to mechanical work. Rather, the reduced utilization of $\sim$ P was in relation to reduction in contractile element work.

This important controversy will require solution. Although species differences and differences in techniques may be at the basis of the disagreement, very possibly the degree and duration of stress accounts for the disagreement. Meerson (1962) has defined three significant developmental stages in the failing heart: (1) transient breakdown stage, characterized by contractile insufficiency with deficiency of certain enzymes; (2) protracted stage of relatively stable hyperfunction, characterized by the absence of cardiac insufficiency, by hypertrophy, and by adequate oxidative phosphorylation; and (3) protracted stage of progressing cardiosclerosis and gradual exhaustion of the hypertrophied heart, characterized by disturbances in protein synthesis and sustained hypoxia. It is suggested that stage 3 might be characterized by or caused by an aberration of mitochondrial function(s).

Katz concluded in 1960 that low-output failure does not appear to be related to defects in energy utilization but has suggested rather that muscle tension is not as productive of external work as the normal heart. Olson and Piatnek suggest a defect in the contractile mechanism itself, and this view has been supported by some authorities. They have proposed that an abnormal stable aggregate of myosin is formed in dogs with experimental cardiac failure and that this prevents the formation of actomyosin with normal contractile properties, leading to reduced contractility. But Davis and his associates (1960), also studying possible alteration in cardiac myosin in dogs with chronic congestive heart failure, could give no support to the view that altered contractile protein constitutes the biochemical defect. They found the same value for myosin (molecular weight, 500,000) in the failing ventricle as in the normal heart. In another direction, Chidsey et al. (1964) have investigated catecholamine metabolism in congestive heart failure in man and experimental animals and found a profound reduction in norepinephrine stores. They infer that this reflects impaired function of the cardiac sympathetic nerves, compromising the heart action further.

In certain other pathological conditions the possibility remains that the defect resides in energy production or conservation. For example, in hemorrhagic shock and myocardial infarction, general and local anoxia leads to defective energy production. In anemia, myocardial failure of the high-output type could result because insufficient oxygen is drawn in for substrate metabolism. Likewise, in beriberi heart disease, also associated with high-output failure, a deficiency in thiamine pyrophosphate (cocarboxylase) produces a breakdown of certain decarboxylation reactions that results in interference with normal myocardial energy liberation and conservation.

HEMODYNAMIC ALTERATIONS

It has been customary in the past to consider the hemodynamic alterations in terms of two general aspects of heart failure classified respectively as *forward failure* and *backward failure.* Forward failure implies alterations and sequelae which are the consequence of reduction in cardiac output and peripheral blood flow. The concept of backward failure is concerned with loss of contractile power and distention of the ventricle with ultimate inability for maintaining cardiac output. Venous pressure rises, and with it serious consequences develop in terms of capillary fluid filtration and the formation of edema fluid (Chap. 12).

On a functional basis it seems unrealistic to categorize the consequences of heart failure into these two classifications. Several manifestations of congestive heart failure illustrate this point. For example, the elevated venous pressure which has characteristically typified backward failure involves the accumulation of extra fluid interstitially and an increase in blood volume. The increased fluid volume can best be explained by invoking changes in renal function which result rather directly from forward failure, i.e., reduced cardiac output. Mechanisms will be discussed later. Furthermore, elevated venous pressure is also in part the result of enhanced venomotor activity also brought about by reduced cardiac output. On the other side, the edema which characterizes congestive heart failure cannot be the direct result of reduced cardiac output and impaired systemic blood flow, as was formerly thought to be the case, leading to increased capillary permeability resulting from the presumed hypoxia. This would indeed be a great oversimplification, because neither edema nor venous congestion is produced by the severe anoxia which often accompanies pulmonary disease or cyanotic congenital heart disease, or occurs at high altitude. Evidently other factors are involved. In conclusion, it seems most realistic to emphasize wherever possible the interrelations of the "forward" and "backward" aspects of heart failure to explain the composite picture.

For convenience, our discussion of the hemodynamic changes will begin with a consideration of the consequences of left ventricular failure. Cardiac output is usually less than 4.5 liters per minute. This results in alterations of territorial flow.

Renal blood flow in man decreases from a normal of about 1200 ml to as low as 350 ml per minute. This has ultimate consequences on renal function which will be further discussed. Splanchnic blood flow, including blood supply to the liver, is reduced, usually decreasing in proportion to the reduction in cardiac output. Ultimately, this too impairs hepatic function and is also the basis for gastrointestinal manifestations such as anorexia. The flow to the extremities is reduced and may be the basis of fatigue accompanying heart failure. Cerebral blood flow is usually normal, at least in the early phases of heart failure, as is coronary blood flow unless changes in the coronary vasculature have occurred.

When the left ventricle fails to discharge adequately, the output falls and the residual diastolic volume and diastolic pressure rise, as ultimately does the pulmonary venous pressure. If the right ventricle pumps normally, a redistribution of blood can be expected to occur with more in the pulmonary circulation and less in the systemic circulation. The result is pulmonary venous congestion, which leads to a further decrement in left ventricular output. Actually, this is compensated for in several ways. For example, with increased diastolic stretch, the left ventricle pumps somewhat better. Redistribution of the blood volume may eventually lower the output of the right ventricle, which can pump only what it receives.

PULMONARY CONGESTION

If pulmonary venous pressure rises beyond about 30 mm Hg, edema fluid begins to form and accumulate in the lung tissue and in the alveoli, impairing oxygen exchange. Engorgement of the lungs with blood and the accumulation of edema fluid seriously impair the mechanism of ventilation and increase the work of breathing, leading to undue breathlessness. Fluid accumulation leads to sounds heard by auscultation which are called *rales.* The lung tissue may become tough and lose its resilience because of connective tissue proliferation in the parenchyma. Alveolar membranes become thickened and edematous, further increasing the distance between the alveolar air and blood. *Vital capacity* (Chap. 19) is reduced not only by the extra blood and interstitial fluid in the lungs but also by the usual cardiac hypertrophy which occurs. Finally, there may be accumulation of excessive amounts of mucus and secretions in the small bronchioles and airways, further impeding air exchange. Thus, *dyspnea* is commonly seen. This condition is defined as rapid, shallow breathing of which the patient is conscious, as distinguished from *tachypnea* (rapid breathing) or *hyperpnea* (deep breathing). Dyspnea is not due solely to decreased oxygen and increased carbon dioxide of the blood; it is due in large part to stimulation of vagus nerve stretch receptors in the lung.

Further manifestations of abnormal respiration are noted, perhaps of more serious consequence. These include *orthopnea, paroxysmal nocturnal dyspnea,* and *Cheyne-Stokes respiration.* Orthopnea is difficulty in breathing in the recumbent position as contrasted with the erect. It probably results from the shifting of blood from the dependent parts of the body to the vessels of the lung, exacerbating the subjective sensations and increasing the respiratory effort, probably because of enhanced reflex activity from the lung receptors. Paroxysmal nocturnal dyspnea is similar to this but occurs at night. It is also associated with the redistribution of blood which occurs during recumbency, plus depression of the central nervous system centers during sleep. The subject awakens with a feeling of suffocation, sits up gasping for air, and exhibits excessive pallor of the skin and profuse sweating. The skin may appear cyanotic because of the reduced hemoglobin of the circulation. Coughing and wheezing accompany the attack. The entire episode may last 10 to 20 minutes.

Basically, Cheyne-Stokes breathing presumably results from the depression of the respiratory center which is a consequence of the impaired blood flow following decreased cardiac output. Local vascular disease may contribute to this condition in the elderly. The depressed respiratory center is presumed to be insensitive to normal carbon dioxide tension but may still respond to raised carbon dioxide tension of the blood, and reflexly to anoxia (Chap. 21). With normal P_{CO_2} and P_{O_2}, breathing stops. During apnea the arterial P_{O_2} falls and the P_{CO_2} rises, which directly and indirectly stimulates the respiratory center to renew the breathing. However, the hyperpnea leads to a blow-off of carbon dioxide and improvement of the oxygen tension, with the result that apnea occurs again. This is repeated cyclically. A more detailed discussion of periodic breathing is deferred to Chapter 21.

SYSTEMIC CONGESTION

When failure extends to the right side of the heart, other sequelae become manifest. These may be the indirect consequence of left heart failure and the resulting pulmonary congestion and pulmonary artery hypertension, which lead to ultimate impairment of right ventricular function. They may be more directly the consequence of primary lung disease with pulmonary hypertension or pulmonic valvular stenosis. Pulmonary embolism may lead to the symptomatology, as may certain types of atrial septal defects.

The immediate clinical manifestation consequent to this change is elevation of systemic venous pressure (values of 25 to 40 mm H_2O). Venous engorgement is evidenced by the concomitant distention of the jugular veins. These are distended even in the erect position, which is not the case in the normal individual (Chap. 16). The elevation of venous pressure results not only from the inability of the right heart to pump adequately but also, in chronic heart failure, from the increase in blood volume – an increase both of the erythrocytes (hypoxemic stimulation of bone marrow) and of plasma volume (fluid retention). Unquestionable evidence has recently been gained that this is the case by using tagged albumin (^{131}I) and tagged erythrocytes (^{57}Fe and ^{59}Fe) simultaneously. In addition, several investigators believe that there is enhanced venoconstriction, a reflex response to reduced cardiac output. Ganglionic blocking agents (Dibenamine, Arfonad) which decrease sympathetic nerve activity cause significant reductions in venous pressure in the congestive heart failure patient, while only negligible changes in venous pressure occur in the normal subject given the drugs. This has been taken as presumptive evidence of increased venomotor activity in the heart patient.

A consequence of the elevated venous pressure is that the liver is engorged and becomes tender and painful in response to pressure. Ultimately, as a combined effect of venous congestion and reduced inflow, this leads to *cirrhosis* (fibrotic degeneration) of the liver, evidenced by increased bilirubin in the blood and jaundice. More specifically, decreased indocyanine green and bromsulphalein (BSP) clearance result, indicating reduced hepatic blood flow. Splenomegaly is also seen. Peripheral venous congestion, i.e., in the skin, leads to the appearance of cyanosis; this occurs partly as a result of dilation of venules but also because of the probably impaired oxygenation of the arterial blood and greater oxygen extraction by the tissues. Thus, reduced hemoglobin in cutaneous vessels exceeds the 5 gm per 100 ml of blood necessary for the appearance of visible cyanosis.

MECHANISM OF PERIPHERAL EDEMA

A number of factors are set into play in congestive heart failure which favor the accumulation of fluid in the interstitial spaces of various parts of the body, leading to edema. This is first manifested by an increase in the subject's weight – a gain of 10 to 20 pounds – before any objective signs are noted. Then the ankles and dependent parts begin to swell, and pitting edema is noted. Pitting edema is detected by a firm finger pressure applied to the edematous region. Fluid is expressed, leaving a depression which slowly refills. In the bedridden patient, sacral edema is prominent. Accumulation of the edema fluid in the lung bed *(hydrothorax)* has been discussed above. Fluid accumulation also occurs in the abdomen. This is called *ascites.* Peripheral edema fluid is low in protein, about 0.5 gm per 100 ml, but ascitic fluid may approach the protein content of plasma – 5 to 6 gm per 100 ml.

The production and accumulation of edema fluid in congestive heart failure is a complicated process and is not simply the result of elevation in venous pressure leading to increased capillary filtration pressure. Since intake of fluid is apparently normal, accumulation of fluid must be the result of abnormal retention; hence, the kidney plays a paramount role. An accumulation not only of water but of salt, predominantly sodium chloride, to maintain the isotonicity of the fluid, is involved in this mechanism. Thus, although the first step in the formation of edema fluid is increased capillary filtration, a mechanism for overall accumulation of water and contained salts must be provided.

The mechanism for renal retention of salt and water has not been completely settled. Until the student becomes more familiar with renal physiology (Chap. 22), only brief details can be given. It has been noted that glomerular filtration is usually reduced in congestive heart failure, and some authorities feel that this may be important for the retention of salt and water. Simply stated, when the load of sodium salts (NaCl and $NaHCO_3$) and water offered to the renal tubular cells is reduced, more complete reabsorption of the smaller load occurs. Although such a mechanism can be experimentally demonstrated, it perhaps oversimplifies the situation in congestive heart failure. For one thing, edema fluid accumulates frequently in the absence of measurable decrement in glomerular filtration, and edema fluid can be lost without apparent change in filtration rate. Some authorities believe that the tubular processes of sodium reabsorption are enhanced. Basically, this involves the hormone aldosterone, which greatly potentiates sodium reabsorption by the renal tubular cells. There is evidence that increased excretion of aldosterone occurs in the urine of patients with congestive heart failure, taken as evidence of increased activity of this hormone in the kidneys. The increased appearance in the urine is the result both of increased production and release of the hormone by the adrenal cortex (Chap. 35), and of decreased destruction by the liver, so that nonmetabolized fractions appear in the urine. The decreased metabolic activity of the liver is a consequence of the impaired function of the hepatic cells resulting from impaired circulation, as discussed above. However, continued injection of aldosterone in the normal experimental subject, although it may cause transient sodium retention, is followed by later readjustment of renal mechanisms, so that sodium and attendant water is again lost to restore homeostasis. Thus, hyperaldosteronism itself cannot be the sole mechanism involved but is accepted as an important contributing factor. Possible mechanisms for triggering increased release are discussed in Chapter 35. Davis has suggested that an additional factor, or factors, of a humoral or hemodynamic nature, distinct from those identified at present, is involved in congestive heart failure which potentiates the aldosterone influence. The possibility that a sodium-"losing" hormone is involved in the overall homeostasis has been entertained by some investigators.

It has been suggested that enhanced amounts of antidiuretic-like substances are active in congestive heart failure, again on evidence of increased excretion of such substances in the urine. This should favor water retention by the kidney tubules. It is not known whether this activity represents increased output of antidiuretic hormone (ADH) by the hypothalamo-neurohypophysial (HNS) axis (Chap. 31) or whether such a substance is metabolized to a lesser degree by the liver, as with aldosterone. What the trigger mechanism for enhanced ADH release in congestive heart failure might be is not known, but it could lie in the province of control by volume receptors and osmolar receptors.

In conclusion, primary mechanisms involved in congestive heart failure have been presented, and the sequelae which are consequences of the primary derangements have been discussed. Although the physician is concerned with improving the well-being of his patient by allaying the sequelae – e.g., removing edema fluid by administering diuretics to institute renal loss of salt and water – he continually strives to correct, insofar as possible, the basic derangements. An illustration is his use of drugs such as digitalis to improve the myocardial pumping action.

REFERENCES

Blumgart, H. L. (Ed.). Symposium on congestive heart failure. *Circulation* 21:95–128, 218–255, 424–443, 1960.

Burch, G. E., and C. T. Ray. A consideration of the mechanism of congestive heart failure. *Amer. Heart J.* 41:918–946, 1951.

Chidsey, C. A., G. A. Kaiser, E. H. Sonnenblick, J. F. Spann, Jr., and E. Braunwald. Cardiac norepinephrine stores in experimental heart failure in dogs. *J. Clin. Invest.* 43:2386–2393, 1964.

Chidsey, C. A., E. C. Weinbach, P. E. Pool, and A. G. Morrow. Biochemical studies of energy production in the failing human heart. *J. Clin. Invest.* 45:40–50, 1966.

Davis, J. O. The Physiology of Congestive Heart Failure. In W. F. Hamilton and P. Dow (Eds.), *Handbook of Physiology.* Section 2: Circulation, Vol. III. Washington, D.C.: American Physiological Society, 1965. Chap. 59, pp. 2071–2122.

Davis, J. O., W. R. Carroll, M. Trapasso, and A. Yankopoulos. Chemical characterization of cardiac myosin from normal dogs and from dogs with chronic congestive heart failure. *J. Clin. Invest.* 39:1463–1471, 1960.

Lindenmayer, G. E., L. A. Sordahl, and A. Schwartz. Reevaluation of oxidative phosphorylation in cardiac mitochondria from normal animals and animals in heart failure. *Circ. Res.* 23:439–450, 1968.

Meerson, F. Z. Compensatory hyperfunction of the heart. *Circ. Res.* 11:250–258, 1962.

Meerson, F. Z. The myocardium in hyperfunction, hypertrophy and heart failure. (Monograph 26, American Heart Association.) *Circ. Res.* 25(Suppl. II):1–163, 1969.

Pool, P. E., B. M. Chandler, J. F. Spann, Jr., E. H. Sonnenblick, and E. Braunwald. Mechanochemistry of cardiac muscle: IV. Utilization of high-energy phosphates in experimental heart failure in cats. *Circ. Res.* 24:313–320, 1969.

Rushmer, R. F. *Cardiovascular Dynamics,* 3d ed. Philadelphia: Saunders, 1970.

Schwartz, A., and K. S. Lee. Study of heart mitochondria and glycolytic metabolism in experimentally induced cardiac failure. *Circ. Res.* 10:321–332, 1962.

Sodeman, W. A. *Pathologic Physiology.* Philadelphia: Saunders, 1961.

Stoner, C. D., M. M. Ressallat, and H. D. Sirak. Oxidative phosphorylation in mitochondria isolated from chronically stressed dog hearts. *Circ. Res.* 23:87–97, 1968.

Wiggers, C. J. Dynamics of ventricular contraction under abnormal conditions. *Circulation* 5:321–348, 1952.

Wood, P. *Diseases of the Heart and Circulation,* 2d ed. Philadelphia: Lippincott, 1956.

18

Pathological Physiology of the Cardiovascular System

C. Physiology of Shock

Ewald E. Selkurt

THE TERM *shock* is defined as an abnormal physiological condition resulting from inadequate propulsion of blood into the aorta, and therefore inadequate flow of blood perfusing the capillaries of various tissues and organs. It is typified by deterioration of cellular functions, particularly of vulnerable organs such as the kidney, liver, and heart. It is further characterized by a progressive failure of the circulation eventuating in an *irreversible state.*

CLASSIFICATION OF TYPES OF SHOCK; CAUSES AND MANIFESTATIONS

Circulatory shock describes a syndrome that is characterized by protracted prostration and hypotension, pallor, coldness and moistness of the skin, collapse of superficial veins, reduced sensibilities, and suppression of urine formation. It is important to recognize that all these symptoms may be present without development of the potentially irreversible physiological condition to which the term should be limited. For this reason some authorities have recommended two subdivisions: primary shock and secondary shock. Primary shock is essentially a bout of acute hypotension. It tends to be transient, and there is usually a natural tendency to recovery. Secondary shock is typified by deterioration of cellular function and progressive failure until an irreversible state is reached, culminating in death. In arriving at a diagnosis, it is therefore important to relate the existing clinical signs and symptoms to the antecedent events, gauging the probabilities as to whether they indicate development of true shock or are merely a temporary circulatory derangement. Although better clinical criteria could be desired, progressive reduction of arterial pressure and pulse pressure measured at frequent intervals constitutes the best available evidence as to the existence of a physiological state of shock. Collapse of the veins is an important sign.

SYMPTOMATOLOGY

Clinical symptoms are referable to the skin, the mucous membranes and the countenance, the neuromuscular system, the circulatory and respiratory systems, and the metabolism. The skin is pale and cold because of peripheral vasoconstriction. This is accompanied by sweating because of massive stimulation of the sympathetic division of the autonomic nervous system. The mucous membranes especially tend to be bluish as a result of the enhanced oxygen extraction from the blood (stagnant anoxia). The countenance appears haggard and drawn, in part because of the removal of interstitial fluid from the skin due to hypotension with resultant reduction in capillary hydrostatic pressure which normally opposes the inward-drawing force of the plasma proteins. The reflexes tend to be depressed and responses to noxious stimuli reduced. The blood pressure is decreased, and the pulse is rapid and thready as a result of reduced cardiac output. The respiration is rapid and shallow, on account of both reflex and chemical stimulation. Body temperature may be decreased, partly because of decreased general metabolism resulting from serious reduction in oxygen supply. All these conditions are symptomatic of the loss of circulating blood volume.

PREDISPOSING CAUSES

In most cases there is a loss of blood or plasma fluid with a consequent prolonged period of hypotension. Some of the more important causes are (1) hemorrhage; (2) extensive edema or serous effusion due to mechanical, chemical, or thermal capillary damage; (3) seepage of wounds or burn surfaces; (4) abstraction of fluids through excessive sweating, vomiting, diarrhea, or drainage of secretions after surgical operations; (5) disturbances of electrolyte and water balance, causing withdrawal of water from the blood stream; and (6) unusual engorgement of the capillaries with blood, reducing effective blood volume. The last is frequently the causative factor in the shock that occurs during overwhelming infections, such as those from gram-negative bacteria.

STAGES OF SHOCK

INITIAL OR HYPOTENSIVE STAGE

Primary loss of blood or fluid leads to a disparity between the circulating blood volume and circulatory capacity. This results in reduction of venous return and in an acute decrease in cardiac output, with reduction in arterial and pulse pressures and ultimate impairment of circulation and lowered oxygenation of the tissues of the body. Compensatory mechanisms are instituted rapidly, as follows: Hypotension acts via the carotid sinus and aortic arch baroreceptors to cause reflex vasoconstriction and is supplemented by the action of catecholamines released in increased amounts which augment peripheral arteriolar and venous tone. The same reflexes augment the cardiac vigor and rate. The fall in blood pressure also reflexly increases the depth and rate of respiration, which favors venous return and ultimately cardiac filling – the so-called abdominothoracic pump mechanism discussed in Chapter 16. Increased activity of venopressor mechanisms may contribute to venous return. Further repletion of blood volume results from an increase in plasma volume through the reduction of capillary pressure, which leads to reabsorption of tissue fluid and ultimately plasma replacement. Thus, hemodilution is often noted during this stage.

PROGRESSIVE OR IMPENDING STAGE

The arterial pressure, previously stabilized by compensatory mechanisms, may decrease slowly and progressively. There is no further vasoconstriction; in fact, total peripheral resistance (TPR) may begin to lessen. The low arterial pressure together with vasoconstriction exerts further deleterious effects on the various organs and tissues because of reduction in capillary flow. The generalized tissue hypoxia leads to impairment of enzyme systems and to anaerobic metabolism, with an accumulation of lactates and pyruvates and with failure of resynthesis of organic phosphates. Reduction of alkali reserve occurs, and there is migration of potassium from tissue cells. The vital organs are variously affected. This stage has been called *oligemic shock.*

ORGANIC ALTERATION IN THE PROGRESSIVE STAGE

Kidney

Ischemia and the resultant stagnant anoxia result in failure of glomerular filtration, and the reduced blood flow causes eventual tubular damage, evidenced by impaired extraction of para-aminohippuric acid (PAH). These renal alterations lead to anuria and retention of metabolic products (urea, creatinine, uric acid). Furthermore, they contribute to an upset in the acid-base balance in the direction of metabolic acidosis because of failure of the normal mechanisms for hydrogen ion secretion by the kidney, and the accumulation in the blood of phosphates and lactates. When crushing injuries to the limb accompany shock, there is more profound renal injury because the release of myohemoglobin and products of muscle autolysis contribute to renal failure; renal tubules, particularly the distal and collecting, are blocked by debris and hemoglobin deposits. The term *lower nephron nephrosis* has been applied to this situation. A serious consequence of shock is that delayed renal failure may ultimately lead to death in uremia even though the patient may have been treated with transfusion, with adequate recovery of blood pressure. In any event, even after treatment the kidney may for a prolonged period show residual damage characterized by loss of concentrating power, evidenced in excretion of dilute urine of low specific gravity. This loss of concentrating ability is largely due to the loss of the gradient of osmolality in the kidney (Chap. 22), due to "washing out" of the high papillary concentration needed for urinary concentration by the antidiuretic hormone (ADH) mechanism. The washout results because of marked reduction or cessation of glomerular filtration, with persistence of blood flow through the medulla (Selkurt, 1969).

Liver

The circulatory anoxia impairs the normal mechanisms of metabolic turnover of such substances as pyruvates, lactates, and amino acids released from traumatized tissue, by disorganizing the hepatic cell enzyme systems. The reason for the accumulation of the amino acids in the blood is impairment of the liver's normal ability to form urea. There may be early glycogenolysis and glycogen depletion and hyperglycemia, but later impairment of gluconeogenesis and ultimate hypoglycemia. In connection with the disorganized enzyme systems, it has been observed that the adenosine triphosphate (ATP) and adenosine diphosphate (ADP) concentrations are reduced in the shock liver.

Heart

Ultimately the myocardium may become depressed, because of reduced oxygen supply and possibly accumulation of toxic materials. A reduced coronary flow because of hypotension occurs despite compensatory dilation of the vascular supply of the myocardium.

Brain

The blood supply to the brain appears to be reasonably well maintained because of compensatory vasodilation, and at the sacrifice of flow to the other organs. For example, the blood flow to the skin, kidneys, and splanchnic bed is diverted by more intense vasoconstriction to the brain, pulmonary, and coronary circulations. Brain samples indicate that the metabolic condition appears to be good in terms of ADP, glycogen, and phosphocreatine contents. However, occasionally death by respiratory failure occurs during this stage, suggesting greater susceptibility of the respiratory centers.

IRREVERSIBLE STAGE

There is further decrease in cardiac output. Blood pressure may be very low and pulse pressure may be barely detectable. Respiration becomes depressed. Cardiac deceleration may follow, because of default of reflexes or weakening of the myocardium. In some instances the total peripheral resistance begins to decrease, indicating relaxation of compensatory vasoconstriction. This may be the result of failure of the vasomotor centers due to prolonged hypoxia, but more probably is caused by the production and release of vasoactive substances in the blood which directly or indirectly depress blood pressure or heart action or both. It is of extreme significance that large infusions of blood and blood substitutes given at this time may prove to be of temporary benefit only. With transfusion, blood pressure and other hemodynamic alterations may be reasonably well restored toward normal for a time, but these restorations are likely not to be maintained. The state is called *normovolemic shock* and may progress into irreversible failure and death.

A major manifestation of the irreversible stage is the observation in experimental animals that the capillaries, particularly in the splanchnic bed, dilate by relaxation of the precapillary sphincters. Thus, there is an eminent possibility that stagnant blood will be impounded in these areas. The capacious pulmonary vascular bed does not appear to be an important site of pooling, however (Abel et al., 1967). Furthermore, release of proteases, histamine, and other substances during the hypotensive phases seems to favor increased capillary permeability, which may allow fluid to escape into the extravascular spaces. There is evidence that fluid moves from the interstitial compartment into cells. This may be responsible for the hemoconcentration that is seen in the terminal phases of shock. In the dog, in particular, the intestinal bed shows marked congestion of the capillaries and venules with extreme extravasation of fluid and even blood into the intestinal lumen; hence, significant amounts of transfused blood and fluid may ultimately be lost into the intestinal lumen and be discharged from the body as bloody diarrhea. However, this mechanism is not important in man and the monkey.

Experimental investigation indicates that myocardial depression contributes to and even accelerates the fatal end. This is indicated by a decrease in cardiac output despite terminal elevation of ventricular filling pressure, suggesting weakening of the myocardium; by a lessened response to equivalent states of diastolic distention when infusion is given; and by alterations of the S-T segments of the electrocardiogram.

EXPERIMENTAL PRODUCTION OF SHOCK

The experimental approaches to the study of the basic mechanisms of shock have been many and varied, and space does not permit a detailed description and evaluation of the numerous techniques. Generally, these are designed to simulate the development of shock as it is caused in man. One of the most commonly employed is hemorrhage, illustrated in Figute 18C-1.

The dominant effect of hemorrhage (curve 1) is the reduction in cardiac output that results (curve 3) because of reduced filling pressure (curve 4). Mean blood pressure declines (curve 2), with decrease in pulse pressure, reflecting reduced stroke volume. Reflex stimulation of respiration results (curve 5), supplemented by chemical stimulation from developing acidosis later in the oligemic phase (curve 13). Circulatory compensatory mechanisms include increase in heart rate, particularly if the control rate is slow (curve 6), and enhanced vasoconstriction, so that total peripheral resistance increases markedly at first (curve 11). The increase in TPR is both neurogenic and humoral. Regional and organ changes in vascular resistance vary. The resistance actually *decreases* in the coronary circulation (curve 7) and also in the cerebral circulation (not shown); it shows only minor increase in the splanchnic circulation (curve 8), but increases significantly in the kidney (curve 9) and muscles (curve 10); the dashed line shows the relationship with cold block of the nerve). Later, in oligemic shock, TPR declines somewhat. The apparent reduction of plasma protein concentration (curve 12) suggests that a supplementary compensatory mechanism is the influx of interstitial fluid into the capillaries.

Following transfusion, cardiac output and arterial pressure are temporarily restored, then begin a progressive downward trend. Cardiac output declines because of decrease in filling pressure. Although heart rate is restored almost to control, there is a terminal slowing.

Respiratory rate slows somewhat at first, but increases later in normovolemic shock, reflecting in part the underlying disturbance in acid-base balance. Susceptible animals may die abruptly during this phase (see arrow, curve 5), because of respiratory failure. Respiratory changes are, however, quite variable; a slowing late in normovolemic shock is often observed.

TPR returns toward control following transfusion but then begins a secondary rise. Coronary resistance returns to control but may show a decrease terminally. Renal resistance is temporarily restored toward control level, then shows a progressive increase. Splanchnic resistance shows a phase *below* control which has not been satisfactorily explained, then gradually rises back to the control value. Muscle resistance remains high, although it manifests a downward trend later in normovolemic shock.

Several conclusions can be drawn. Failure of the circulation is not the result of failure of vascular reflex compensatory mechanisms, in view of the well-maintained TPR. Myocardial failure cannot be implicated as an important factor early in normovolemic shock. When extra fluid is given to restore filling pressure, cardiac output is restored. Thus, continued loss of effective circulating blood volume is the reason for the gradual decline in blood pressure. Late in normovolemic shock, despite an attempt to maintain coronary circulation by compensatory dilation during hypotension, flow is impaired because of the reduced cardiac output and blood pressure. Alterations in myocardial metabolism have been demonstrated as a consequence of the impaired flow. Moreover, toxic products of a vasodepressor nature developed in shock probably have an influence on the depression of the myocardium.

Other methods of producing shock include crushing and contusion of muscles, fractures of bones, burns and scalds, intestinal obstruction, and prolonged intestinal exposure and manipulation. Less directly, prolonged ischemia of the limbs, induced by tourniquets and followed by release, or ischemia and release of the intestinal circulation are used for this purpose. Anoxia of the splanchnic bed, experimentally induced by ischemia, has received considerable attention lately as a method of producing shock, for several reasons. In addition to the products of ischemia (see below), the intestinal bed is the site of toxic amines, e.g., tyramine, which might

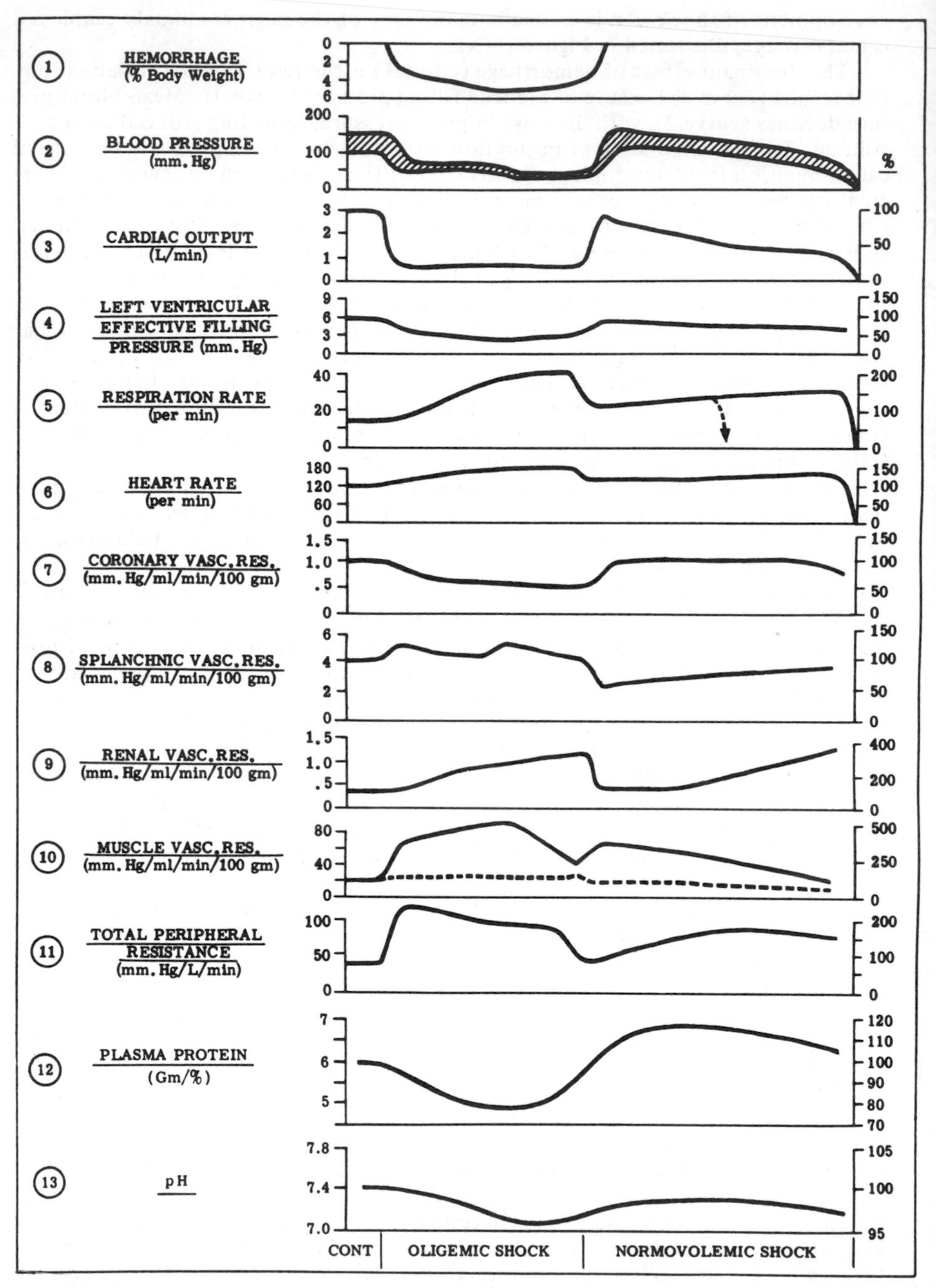

FIGURE 18C-1. Cardiodynamic trends in experimental hemorrhagic shock in dogs. Values to left are in absolute units, to right in percentage change from control *(cont)* (set at 100 percent).

be released into the general circulation with deleterious effects on the cardiovascular system. Ischemia of the intestine created by clamping of the superior mesenteric artery for an hour or two, then release, leads to irreversible shock with much of the symptomatology characteristic of this condition. The organism becomes particularly susceptible because of concomitant impairment of normal liver function, which therefore cannot detoxify the products that may reach the portal circulation. Experimental evidence exists that vasodepressor substances do pass into the systemic circulation as a consequence of intestinal hypoxia, not only with ischemic shock but during hemorrhagic shock.

Some investigators have placed particular emphasis on the role of bacterial endotoxins. In their view, bacteria or their products, released from the intestine during shock, are not screened by the impaired liver and reach the systemic circulation to cause cardiovascular collapse. A lipopolysaccharide which is believed to be the endotoxin involved has been isolated from the plasma of the shock animal. Very small amounts of this substance when injected into the experimental animal lead to shock symptomatology, particularly when the animal has been weakened by a period of moderate hypotension which by itself is not sufficient to cause shock. During the course of hemorrhagic hypotension, the reticuloendothelial system (RES) becomes depressed, with the result that the antibacterial and antitoxic defense mechanisms of the body are impaired (Blattberg and Levy, 1962). Endotoxins from the normal bacterial flora of the intestine constantly enter the intestinal circulation. Ordinarily, they are inactivated by the RES, principally in the liver. When the RES is severely depressed, the endotoxins invade the general circulation. Endotoxins produce a form of shock resembling in many respects that produced by hemorrhage. Therefore, depression of the RES leads to an intensification of the hemodynamic changes caused by blood loss. Sterilization of the intestine by means of antibiotics significantly reduces the mortality from certain standard shock-provoking procedures, including hemorrhage.

The hemodynamic changes have been studied by intravenous injection of the endotoxins into dogs. These endotoxins (derived from *Brucella melitensis* or *Escherichia coli*) cause within one minute of injection a sharp fall in arterial blood pressure and an increase in portal venous pressure. The liver and intestines become engorged with blood, which suggests that the fall in arterial pressure and the elevation of portal pressure may result from pooling of blood in the splanchnic venous bed. These changes lead to an impaired venous return to the heart and reduced cardiac output. One consequence of sequestration of blood in the intestine is impaired circulation in the myocardium; thus, myocardial failure may be secondary to the primary event. Additional factors include capillary injury with resultant excessive loss of fluid into interstitial spaces. These changes together are implicated in the progression of shock to irreversibility.

The relationship of endotoxin shock to hemorrhagic shock is not clear at present. Nor is it yet known how the theory of endotoxin shock will fit ultimately into the picture of clinical shock presented by the injured human subject. Unfortunately, the physiology of shock in the experimental animal, such as the dog, differs in significant ways from that in the primate. Much more work needs to be done in the latter species.

THEORIES OF THE MECHANISM OF SHOCK

Several theories of the causes of irreversibility of shock have been based upon findings in experimental animals, such as the dog, rat, and monkey. A key mechanism is fluid loss from the effective circulation. This can result directly, as by hemorrhage, or indirectly by a loss of fluid resembling plasma, such as occurs in burns. There may be local fluid loss at the site of trauma, or it may be caused by prolonged vomiting and diarrhea. The loss of circulating volume triggers a series of mechanisms (described above) which lead to the secondary sequelae of reduced cardiac output, reflex vasoconstriction, and overall impairment of flow in the various tissues and organs, with consequent secondary effects due to deranged metabolism.

NEUROGENIC FACTORS

The role of the nervous system has been given much attention. The effect of afferent impulses from the sites of trauma, burns, etc., on the vasomotor and respiratory systems has been explored, together with the possibility that failure of the ensuing reflex vasoconstriction accounts for the late irreversibility of shock. Current evidence indicates that total peripheral resistance is well maintained throughout the course of oligemic shock (reduced blood volume) and through most of normovolemic shock (following transfusion). Sympathetic nerve impulse discharge remains high, in the main, and catecholamine output is never decreased (Fig. 18C-2). Thus, progressive fall in blood pressure after transfusion must result in large measure from other factors. Among these, fluid loss has to be given serious consideration.

CHANGES IN BODY FLUIDS

The problem of fluid loss can be considered from the broad point of view of the application of known techniques for the measurement of vascular volume, interstitial volume, and intracellular water during shock, to see whether significant alterations in these volumes indeed occur. Closely related is the consideration of specific organs that might be involved in such disruption by virtue of being particularly susceptible, such as the intestine, or perhaps important because of their mass, such as skeletal muscle.

Many studies have been done in terms of the changes in plasma volume and red cell mass in the post-transfusion phase of hemorrhagic shock. Most of them have been done on the dog, but more and more data are accumulating for man and other primates. Very typically in the dog, the plasma volume and red cell volume return to normal immediately after transfusion, but both plasma volume and red cell volume later decrease progressively.

During oligemic hypotension, blood is diluted by fluid influx (Fig. 18C-2). This is restored on transfusion, but late in the post-transfusion phase protein concentration and hematocrit value appear to increase, presumably because fluid is lost again. Although plasma loss and hemoconcentration are generally found during the late oligemic period and after retransfusion in the dog, it is important to note that hemoconcentration is not found in man, even after severe hemorrhage. If anything, the findings indicate that in man subjected to hemorrhagic shock there is hemodilution rather than extravasation of plasma fluid. Admittedly, this is often complicated by the therapeutic regimen. In a critical experiment done in another species, the sheep, Gillett and Halmagyi (1966), by the use of ^{51}Cr-labeled cells and ^{131}I-labeled plasma for determination of total blood volume, found it virtually unchanged in irreversible shock. Such findings emphasize that fluid loss alone, when it occurs, need not be the sole factor responsible for circulatory failure in shock.

Other relevant questions pertain to the changes in extracellular and intracellular fluid volumes. Extracellular water (ECW), as measured by distribution of inulin or ^{22}Na or ^{35}S sulfate measurements, interestingly shows a decrease in the late hypotensive and late post-transfusion phases in the splenectomized dogs. The typical finding was a small reduction in ECW in the post-transfusion phase. The magnitude

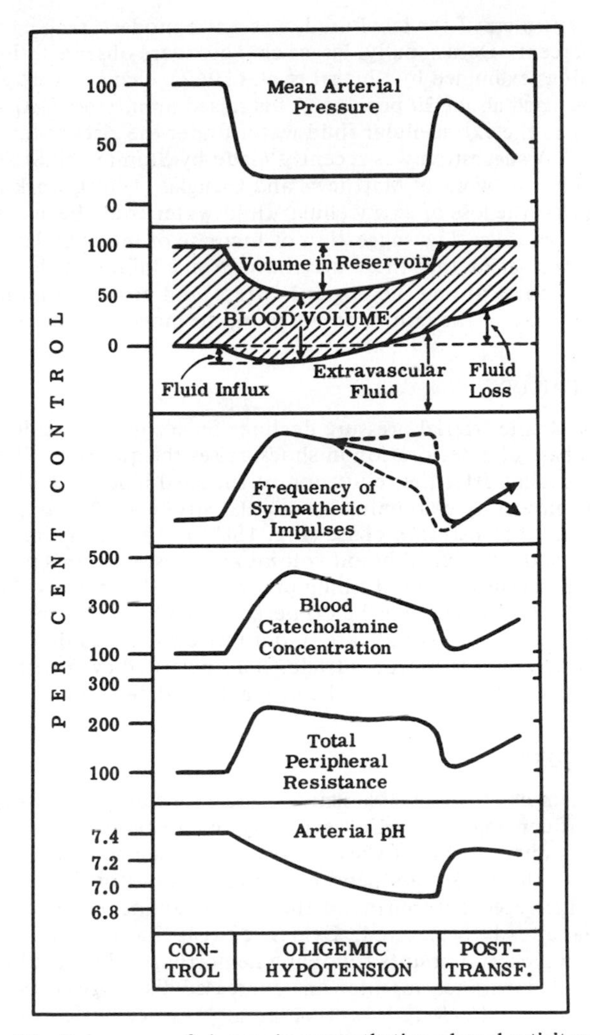

FIGURE 18C-2. The time course of changes in sympathetico-adrenal activity and circulatory functions in constant-pressure hemorrhagic shock (retransfusion after three to four hours at 40 mm Hg) in the dog. These data were taken from different studies. Because of variations in experimental procedures, the time relationships are not quantitative. Ordinate scales are used only to give an approximate estimation of magnitude of changes and are not exact. In the second panel, the remaining blood volume is indicated by the shaded area. Early in oligemic hypotension a portion of the extravascular fluid enters the circulation and becomes part of the blood volume, whereas the reverse occurs later. The initial, transient increase in sympathetic activity during the first few minutes after hemorrhage is not shown. After prolonged oligemic hypotension and after retransfusion, the changes in frequency of sympathetic impulses may take different time courses. There are no experimental data with simultaneous measurements to show whether or not the different time courses are associated with corresponding variations in other parameters, e.g., total peripheral resistance. Total peripheral resistance generally falls toward the control level after an initial rise. This decrease is parallel to the progressive development of acidosis and gradual decline of sympathetico-adrenal activity. (From S. Chien. *Physiol. Rev.* 47:252, 1967.)

of reduction, regardless of the label used, was quite modest, ranging from about 5 percent to 8 percent. Occasionally, larger changes were observed. For example, in several of the dogs examined by Shizgal et al. (1967), changes were greater than 10 percent and averaged about 20 percent. This raised an interesting question as to what happened to the extracellular fluid water under the circumstances of "irreversible shock." A suggestion was recently made by Slonim and Stahl (1968), using rats, supported by the work of Matthews and Douglas (1969), working with splenectomized dogs, that the loss of extracellular fluid water could be accounted for by movement into the cells. The suggestion of increase of intracellular water volume has led to the speculation that there may have been a failure of the sodium pump as a result of the hypoxia during the oligemic stress, and possibly that now water moved into the cells because of the failure of the pump to move sodium out.

MYOCARDIAL FAILURE

The fact that systemic arterial pressure declines on account of inadequate cardiac output in the phase of deterioration in shock raises the question: Why does the cardiac output decline after the reinfusion of the shed blood? Rothe (1970) has searched for an answer by examining ventricular stroke work in dogs during hemorrhage and the post-transfusion phase (Fig. 18C-3). In some animals, massive overtransfusion (twice their original blood volume) was resorted to, supplying an adequate volume and ensuring a good filling pressure. In the animals with more severe stress (hypotension prolonged until a 20 percent to 40 percent uptake of shed blood from the arterial reservoir was indicated), definite evidence of depression of cardiac function was observed in the left ventricle. This persisted even with infusion of 160 ml per kilogram of blood, twice the original bleeding volume.

HUMORAL FACTORS

The humoral theories of the mechanism of shock are concerned with the formation of substances influencing the cardiovascular system or release of such substances from the site of injury. Some of these vasoactive substances may be looked upon as participating in the earlier compensatory mechanisms, for they are vasopressor in action. The increased outpouring of the catecholamines (epinephrine and norepinephrine) has already been cited. Evidence exists that renin or a renin-like substance appears in greater amounts following hemorrhage. This could lead to formation of the pressor substance angiotensin. Increased discharge of vasopressin has been reported. Serotonin (5-hydroxytryptamine), vasoconstrictor in action, may be involved.

Other substances may be toxic to the cardiovascular system and contribute to later circulatory collapse. Some examples are potassium ion, which has a depressing effect on the myocardium, and histamine, which is deleterious to the peripheral circulation. Other products of tissue damage which are vasodepressor in action have been studied. These include the adenosine compounds and vasoactive polypeptides such as bradykinin. Lefer (1970) has proposed a myocardial depressant factor (MDF), probably a polypeptide or glycopeptide of low molecular weight (ca 1000). The sequence of events by which he believes this substance operates to contribute to circulatory failure in shock is illustrated in Figure 18C-4.

Cellular aggression (e.g., anoxia) releases cellular proteases (lysosome breakdown) and activator substances which produce in the plasma new active substances: plasma proteases, serotoxin, and plasma kinins. The kinins are formed by the proteolytic action of the proteases on certain plasma protein precursors, splitting off the vasoactive peptides (e.g., bradykinin and MDF). The toxic, shock-producing principle

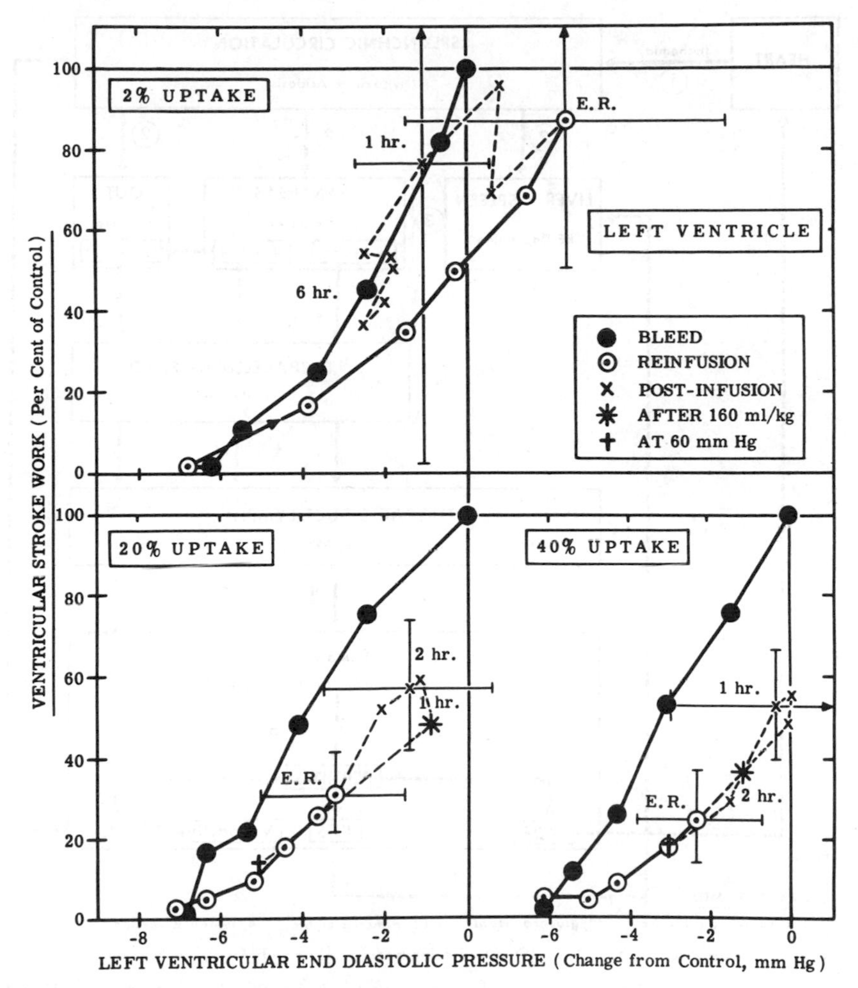

FIGURE 18C-3. Left ventricular function curves of dogs hemorrhaged to 35 mm Hg until 2, 20, or 40 percent uptake of maximum shed volume of blood. Ninety-five percent confidence intervals given at end of reinfusion (E.R.) and one hour after reinfusion. Animals supported with massive transfusions to maintain central venous pressure within 1 mm Hg of control value. (From C. F. Rothe. *Amer. J. Physiol.* 210:1347, 1966.)

serotoxin is also probably produced by a proteolytic step. It differs from other kinins in that it needs the intermediation of histamine release.

The combined action of proteases and vasodilator kinins may explain the increase in capillary permeability in shock, for the proteases weaken the endothelium by their protein-digesting capabilities, while the kinin dilator action further promotes capillary filtration.

Not only do such products of cellular disorganization, together with substances found in the plasma under their influence, have a local action on the vascular endothelium; they are also transported by the blood to remote structures. There they

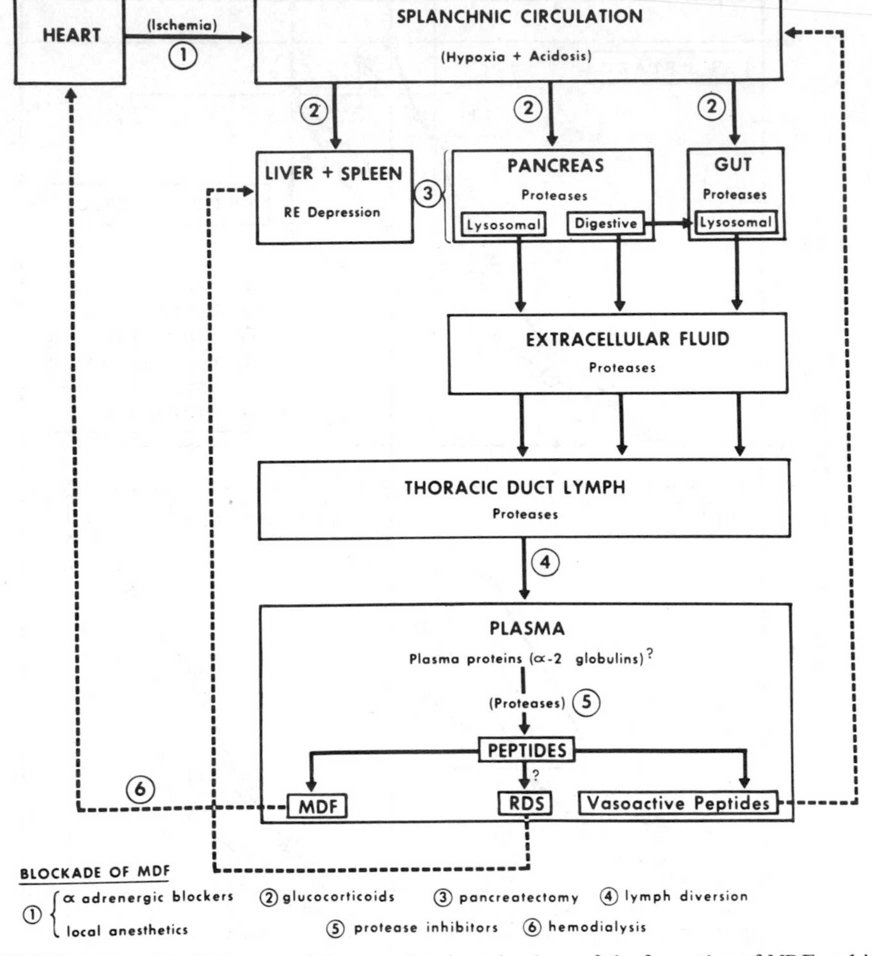

FIGURE 18C-4. Block diagram of the postulated mechanisms of the formation of MDF and its effect during hemorhagic shock. 1, ischemia of splanchnic circulation; 2, hypoxia of liver, spleen, and gut, leading to RE depression, and increased fragility and disruption of lysosomes in pancreas and intestine; 3, release of proteases into extracellular fluid and into thoracic lymph; 4, entry of proteases into plasma to act on substrates; 5, production of peptides, including MDF; 6, feedback effect of MDF on heart, RDS (reticuloendothelial depressing substance) on RES, and vasoactive peptides on splanchnic circulation. Below: points of scheme and method of blockade of MDF production. (From Lefer, 1970.)

have similar effects on the vascular endothelium and cause secondary injury to cells. The secondarily injured tissues may be regarded as a new source of pathogenic factors, resulting in a self-perpetuating process, and the final step in the causation of shock. Thus, a positive feedback mechanism is established which may account for irreversible shock. Since the peripheral resistance does not change enough in shock to explain the progressive fall in blood pressure, mechanisms such as the above must be seriously considered. Furthermore, the influences of humoral factors on the capacitance vessels and on the heart remain to be evaluated.

Another interesting ramification has been disclosed by Bounous, Hampson, and Gurd (1964), who have developed the following story: With the use of ^{32}P-labeled nucleotides, they observed profound depression of oxidative phosphorylation and nucleotide synthesis in the intestinal mucosa of the dog in hemorrhagic shock. These changes occurred in the irreversible stage before the appearance of detrimental alterations in hemodynamics of the gut in the whole animal. The change in the energy production capabilities of the gut apparently made the mucosal cells permeable to intraluminal proteolytic enzymes such as trypsin. The characteristic change noted was the *regional enteritis* so typical of the late stage of shock in the dog. This would lead to a breakdown of the barrier function of the gut and could be involved in the production and release of the toxic materials into the systemic circulation. It is of further interest that this trend was inhibited by a protease inhibitor, Trasylol, which when combined with lavage of the gut tended to prevent the mucosal enteritis. When the pancreatic duct was ligated and the same hemorrhagic shock stress was employed, no regional enteritis was seen, nor did the dogs die in circulatory crisis. The uptake of blood during the hypotensive phase was reduced and the tolerance to hypovolemia increased, according to these workers. Of great importance is the note that the nonsurvivors died some five days later, but with renal, myocardial, or pulmonary complications. The point was made that the typical picture of shock in the dog had been converted to one which more resembled the case in man and other primates, a conclusion weighted with important implications.

Figure 18C-5 summarizes the interrelationships in hemorrhagic shock. The complex interplay of peripheral circulatory factors and the pumping organ is evident. The damage created by a prolonged period of reduced effective circulating blood volume is only partly or temporarily corrected by restoration of blood volume. Continued loss of circulating blood volume, as discussed above, is an important feature of the downward trend seen in normovolemic shock. The myocardium may be weakened but undoubtedly pumps quite adequately if filling pressure is maintained. To this end "overtransfusion" has been resorted to, under which circumstance cardiac output is good, and peripheral circulation is well maintained. However, repeated transfusion is necessary in order to maintain an adequate circulatory status.

In summary, as Rothe has emphasized, the sites of failure are many. This fact points up the complexity of designing adequate treatment.

TREATMENT OF SHOCK

Comments on the treatment of shock are pertinent here because of the physiological implications. The obvious immediate treatment is transfusion of blood or a suitable substitute, depending on the nature of fluid loss. The value of this procedure depends upon the lapse of time since the fluid or blood loss occurred, the quantity transfused, and the character of the transfusate.

The transfusion must be given as soon as possible after the initial blood loss and the development of hypotension, in order to avert irreversible shock. With regard to the quantity infused, the general principle is to give blood until the blood pressure (arterial and venous) and other essential cardiovascular signs, e.g., heart rate, have returned to normal. One of the most reliable of indices which can be rather simply employed is to monitor the central venous pressure; a catheter in the superior or inferior vena cava is most useful for this purpose.

It was the practice in the recent treatment of battle casualties to transfuse quantities of blood that greatly exceeded the apparent immediate or obvious loss, and results were highly satisfactory. With a weakened myocardium, however, there

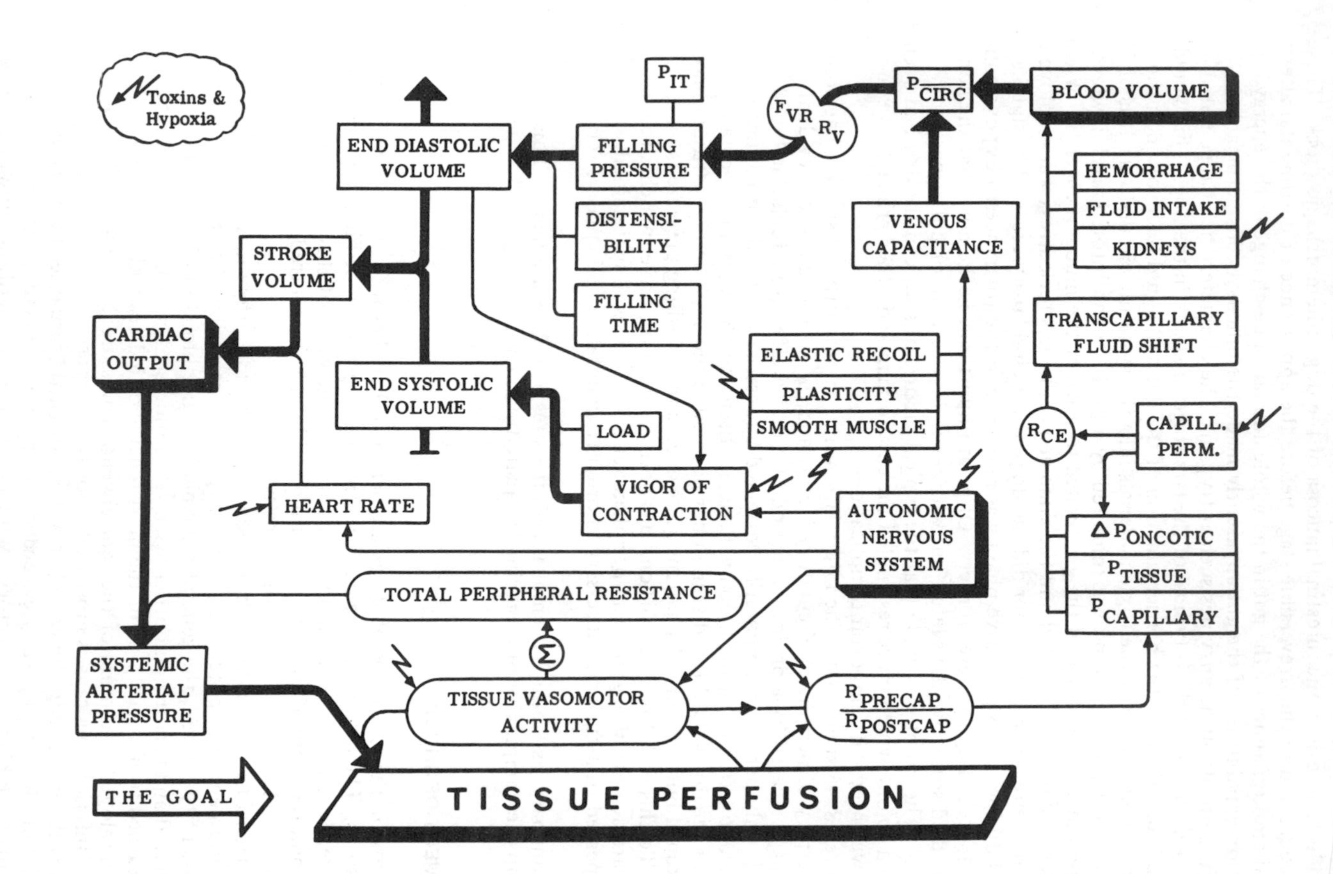

FIGURE 18C-5. Factors determining the cardiovascular response to hemorrhagic hypotension. Jagged arrows indicate probable sites of action of toxins, ischemia, and hypoxia. (From Rothe, 1970.)

is some danger of pulmonary congestion if pulmonary venous pressure is raised rapidly by excessive transfusion. It is well to remember that transfusion of 1 liter of blood does not mean that the circulating blood volume is necessarily increased by 1 liter. In the development of irreversible shock, dilated capillary channels might accumulate stagnant blood drawn out of the effective circulation, or rapidly leak out fluid.

The transfusate should conform to a number of requirements. It should be non-toxic and well retained by the vascular system. If a foreign substance, it should not deposit in the tissues and impair their function. It should not hemolyze or agglutinate cells. It should, of course, be isotonic and of proper pH. It should contain colloidal material to provide colloid osmotic pressure similar to that of the plasma protein. Hemorrhage is best treated with transfusion of fresh properly matched blood. If there has been loss of fluid and electrolytes by vomiting and diarrhea, saline infusion (Ringer's) is the fluid of choice; with burns, plasma, dextran, or PVP (polyvinyl pyrollidone). The latter are synthetic polysaccharide plasma expanders which have proved safer than those formerly used – gelatin, gum acacia, pectin, etc. If anemia supervenes as a sequel to severe burns, blood transfusion is necessary.

The use of vasopressor agents has often been considered in treatment of shock with hypotension. Norepinephrine (levarterenol, Levophed) and metaraminol (Aramine) have been utilized. Use of these drugs in the treatment of hypovolemic (oligemic) shock is of course not justified, in view of the high level of circulating catecholamines. However, they may be useful allies in the treatment of shock caused by myocardial infarction or bacteremic or hypersensitivity reactions, and they usually have a favorable effect against neurogenic shock.

Under the thesis that tissue and organ blood perfusion is impaired, it appears to be therapeutically advantageous to use vasodilator drugs, coupled with adequate restoration of blood volume, in the treatment of shock. Phenoxybenzamine (Dibenzyline) is the agent for which the largest clinical experience is available. Chlorpromazine, although a somewhat different pharmacological entity, produces equivalent and apparently equally effective peripheral vasodilation.

A limitation on the effectiveness of vasodilator drugs would appear to be the effective circulating blood volume. If the volume is severely depleted, blood perfusion of vital organs (e.g., the brain) might conceivably be further impaired by opening other peripheral vascular beds. Therefore, it cannot be overemphasized that for the most effective use of dilator drugs they must be coupled with adequate blood volume restoration.

Crowell (1970), being cognizant of the broad spectrum of disturbances leading to irreversible shock, has countered with a battery of therapeutic manipulations which have, in his hands, "reversed the irreversibility" of canine hemorrhagic shock. Hypotension and oxygen deficit were treated by blood transfusion coupled with ouabain to strengthen the heart and ensure an adequate cardiac output. Excessive secretion of fluid into the gut was minimized by atropine injection. Hypoxic cells show a lack of oxidative phosphorylation resulting in degradation of purine compounds, to adenosine or inosine, with subsequent loss into tissue fluids, then further degradation via hypoxanthine to uric acid. Since the change is reversible at the hypoxanthine stage, hypoxanthine was given to replace this loss. Dextrose or ketoglutaric acid was infused as a source of energy, and lastly, an antibiotic was given to minimize bacterial invasion. The net effect of the therapy was an 80 percent survival rate. The administration of hydrocortisone in pharmacological dosages has been found a useful empirical agent of benefit in shock therapy. Both experimentally and clinically the corticosteroids help fend off irreversibility if given in adequate dosage and as early as possible in the course of the acute disorder. They function by reducing the degree of cellular injury produced by noxious agents (both exogenous and endogenous), by cardiotonic effects, and, interestingly, in this dosage by reducing somewhat the peripheral arterial resistance.

Some investigators believe that constricted venous sphincters, e.g., at the outlet of the liver, are relaxed by corticosteroids, relieving splanchnic pooling and congestion and hence improving circulation. Lefer (1970) takes the view that glucocorticoids in massive doses achieve their beneficial effect by stabilizing the lysosomal enzyme activity and, in turn, minimizing release of proteases which in the final step of plasma interaction lead to production of vasoactive peptides, such as MDF (see Fig. 18C-4).

Adrenocortical hormones are thought by some workers to provide particular benefit in shock resulting from bacteremia or hypersensitivity reactions when an overwhelming response to inflammation threatens life. They must be used judiciously and with caution. One of them, cortisone, has several undesirable side-effects.

REFERENCES

Abel, F. L., J. A. Waldhausen, W. J. Daly, and W. L. Pearce. Pulmonary blood volume in hemorrhagic shock in the dog and primate. *Amer. J. Physiol.* 213:1072–1078, 1967.

Blattberg, B., and M. N. Levy. Mechanism of depression of the reticulo-endothelial system in shock. *Amer. J. Physiol.* 203:111–113, 1962.

Bock, K. D. (Ed.). *Shock* (Ciba Foundation Symposium). Berlin: Springer, 1962.

Bounous, G., L. G. Hampson, and F. N. Gurd. Cellular nucleotides in hemorrhagic shock: Relationship of intestinal metabolic changes to hemorrhagic enteritis and the barrier function of intestinal mucosa. *Ann. Surg.* 160:650–666, 1964.

Crowell, J. W. Oxygen transport in the hypotensive state. *Fed. Proc.* 29:1848–1853, 1970.

Gillett, D. J., and D. F. J. Halmagyi. Blood volume in reversible and irreversible posthemorrhagic shock in sheep. *J. Surg. Res.* 6:259–261, 1966.

Green, H. D. (Ed.). *Shock and Circulatory Homeostasis.* Transactions of Five Conferences. New York: Josiah Macy, Jr. Foundation, 1951–1955.

Hershey, S. G. (Ed.). *Shock.* Boston: Little, Brown, 1964.

Lefer, A. M. Role of myocardial depressant factor in the pathogenesis of hemorrhagic shock. *Fed. Proc.* 29:1836–1847, 1970.

Matthews, R. E., and G. J. Douglas. Sulphur-35 measurement of functional and total extracellular fluid in dogs in hemorrhagic shock. *Surg. Forum* 20:3–5, 1969.

Mills, L. C., and J. H. Moyer (Eds.). *Shock and Hypotension: Pathogenesis and Treatment.* New York: Grune & Stratton, 1965.

Rothe, C. F. Heart failure and fluid loss in hemorrhagic shock. *Fed. Proc.* 29:1854–1860, 1970.

Seeley, S. F., and J. R. Weisiger (Eds.). Recent progress and present problems in the field of shock. *Fed. Proc.* 20:Pt. II, 1961.

Selkurt, E. E. Primate kidney function in hemorrhagic shock. *Amer. J. Physiol.* 217:955–961, 1969.

Selkurt, E. E. Status of investigative aspects of circulatory shock. *Fed. Proc.* 29:1832–1835, 1970.

Shizgal, H. M., G. A. Lopez, and J. R. Gutelius. Extracellular fluid volume changes following hemorrhagic shock. *Surg. Forum* 18:35–36, 1967.

Slonim, N., and W. M. Stahl, Jr. Sodium and water content of connective versus cellular tissue following hemorrhage. *Surg. Forum* 19:53–54, 1968.

Wiggers, C. J. *Physiology of Shock.* New York: Commonwealth Fund, 1950.

19

The Dynamics of Respiratory Structures

Thomas C. Lloyd, Jr.

RESPIRATION, in the broadest sense, applies both to the processes whereby O_2 and CO_2 are exchanged with the environment and to the utilization of O_2 and production of CO_2 by individual cells. Cellular respiration, i.e., oxidative metabolism, is a subject more appropriately presented in biochemistry texts and will not be considered here. In birds and mammals the exchange of respiratory gases with the environment is a function essentially limited to the lung.

Lung function can be divided into four major divisions for convenience in presentation, but the reader should remember that this separation is arbitrary. Functionally, interactions among the several divisions are profound, and changes in one will result in alteration in some aspect of each of the others. The arbitrary divisions to be used in this text are as follows:

1. Airflow mechanics, which concern movements of the chest wall, of the lungs, and of air itself.
2. Blood flow mechanics, concerning the way blood is distributed to different parts of the lungs.
3. Diffusion and gas exchange, encompassing transfer of gases through the alveolar membrane, the interacting effects of blood and air flows on the concentrations of O_2, CO_2, and N_2 in blood and alveolar gas, and blood-tissue gas exchanges in other organs.
4. Regulation of respiration, concerned primarily with the control of rate and depth of breathing.

AIRFLOW MECHANICS

The lung, although it contains both muscle and nerve tissue which are in many ways important to the regulation of its function, acts passively as a gas exchanger. As its blood flow is determined by cardiac pumping, so lung airflow is caused by active motion of the chest wall. It is therefore useful to begin with an understanding of the thoracic wall.

MECHANICS OF THE THORAX

The chest wall is a closed container having several notable properties. Of the many thoracic muscles, only four groups (diaphragm, external intercostals, scaleni, and sternocleidomastoids) are important for generating the forces which normally enlarge the volume of the thorax and cause inspiration. Interestingly, none of the thoracic muscles plays a prominent role in expiration, although the internal intercostals may exert a minor expiratory force. In addition to muscles, which supply active forces, the ribs, by acting as rather springy supports, exert passive forces which influence the actions of the respiratory muscles and also help establish the state of inflation of the lungs even in the absence of active muscle tension.

Before proceeding to a more detailed description of the thoracic wall, it should be noted that a second set of active and passive forces involved with breathing arises from the abdomen. The importance of the abdomen lies in the fact that its upper border, the diaphragm, is the most easily displaced boundary of the thorax. Descent of the diaphragm presses the abdominal viscera against the elastic recoil of the abdominal wall, and conversely, contraction of the abdominal muscles will displace viscera cephalad and push the diaphragm upward. In fact, contraction of the abdominal wall is the major active expiratory force, though, as will be seen, it is not necessary to use these muscles during quiet breathing. Note also that in a standing man the abdominal viscera can be considered to be suspended from the diaphragm. Their weight causes a downward displacement of the diaphragm and outward movement of the lower abdomen, if the abdominal muscles are relaxed. This gravitational enlargement of the thorax can be overcome passively by lying down, or it can be negated or reversed by the high intra-abdominal pressures attending obesity, fluid accumulation, or pregnancy.

In normal quiet breathing, called *eupnea*, inspiration is caused by contraction of the diaphragm and external intercostal muscles essentially unassisted by other muscles, while expiration occurs passively. At the end of a eupneic expiration the relaxed diaphragm assumes a dome-like bulge into the thorax and away from the plane of its attachments to the lower ribs. On inspiration, the dome flattens as the muscles shorten, and the thorax is enlarged by this piston-like descent of its caudal boundary. Because the lung remains in contact with the thoracic wall, it tends to follow the volume expansion of the thorax and in so doing generates a pressure less than atmospheric within the alveolar spaces. The inrush of air through the open airways proceeds because of this pressure gradient. At the end of inspiration, the volume of air which has entered the lungs almost exactly equals the volume change of the thoracic cavity, the small difference being made up by an increase in thoracic blood volume. The lung, however, is an elastic structure which has been stretched by this volume increase. The diaphragm has not only done work moving air but worked against the elastic recoil of the lungs. At the end of inspiration the forces of elastic recoil just equal the muscle force necessary to hold this new volume. If the diaphragm then relaxes, elastic recoil expels air and pulls the diaphragm upward to its initial position. The energy for this passive expiration was stored in the elastic elements of the system during inspiration.

A second effect of diaphragmatic contraction is that it decreases the amount of overlap between lower rib cage and upper abdomen not only by displacing the abdominal viscera anteriorly but also by lifting the rib cage. That is, not only does the center of the dome descend, but its lateral borders ascend and take the rib cage with it. As the ribs move upward, rotations about their spinal attachments cause expansions to occur in both the lateral and anteroposterior directions (see Fig. 19-1).

Greater expiratory flow rates are achieved if expiration is assisted by contraction of the muscles of the abdominal wall. This "active" expiration is used when the demand for breathing exceeds about 40 liters of air per minute, or about 10 times the resting "minute volume." At the higher flow rates inspiration is assisted by contraction of accessory muscles, notably the scaleni and sternocleidomastoids.

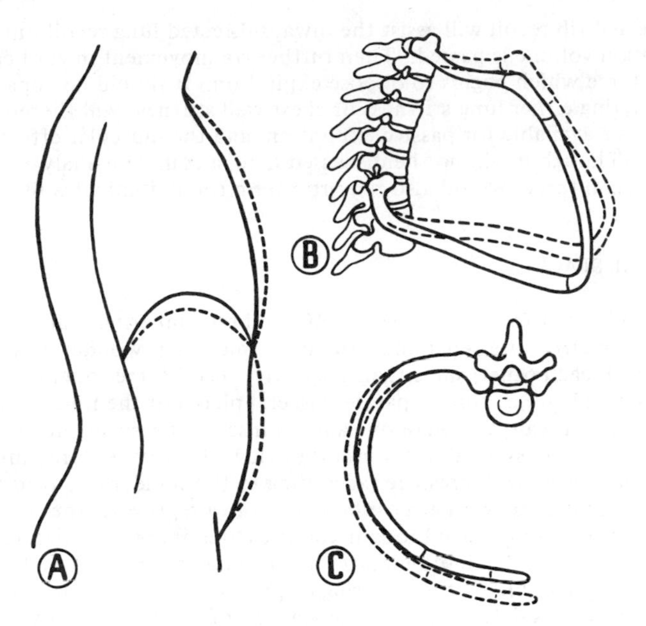

FIGURE 19-1. A. Changes in diameter of chest and abdomen during breathing. Positions in expiration are shown by solid lines; positions in inspiration, by dashed lines. B. Increase in thoracic diameter due to forward movement of the ribs in inspiration. C. Increase in lateral thoracic diameter in inspiration. (B and C from C. M. Goss [Ed.]. *Gray's Anatomy of the Human Body,* 28th ed. Philadelphia: Lea & Febiger, 1966. P. 317.)

These muscles, like the external intercostal muscle, bring about inspiration by raising the ribs. During very heavy breathing, muscles of the spine and shoulder girdle also assist in expanding the thorax.

At the end of a quiet expiration all muscle contraction has ceased, and any other forces which remain are balanced so that no motion occurs. The volume of the lungs at this point is named the functional residual capacity (FRC). If, however, the lungs were to be removed from the chest they would deflate to a much smaller volume by virtue of their elastic recoil property. Furthermore, the isolated rib cage, due to its own elastic recoil, would spring *outward* from the position of FRC to a position equivalent to about 60 percent of a maximal inspiration. Thus, the relaxation volumes of the rib cage and lungs taken separately differ in opposite directions from the relaxation volume of the combined system. At FRC, the force of inward recoil of the lung when in the chest is exactly offset by the outward recoil force of the rib cage. The action of the diaphragm in quiet breathing is to upset this balance, so that at the end of inspiration the inward-directed force of lung recoil has been increased by the increased stretch, and the outward rib cage recoil force has been decreased by moving toward its relaxation volume. It is this imbalance of forces that causes passive expiration.

With this information it can be understood that eupneic expiration does not simply proceed until the lungs are empty, but proceeds until the declining force of lung recoil is exactly offset by the increasing opposing force of rib cage recoil. Unlike eupnea, whenever there is a very large inspiration the thorax is drawn *above* its resting volume, and, during the first part of expiration from this large inspiration, an inward-directed rib recoil will assist the inward-directed lung recoil until the rib cage relaxation volume is passed. Then further rib movement inward causes a rib cage recoil force which begins to oppose expiration. It should be apparent that anything altering either lung stiffness or chest wall stiffness will thereby alter the FRC, the force available for passive expiration, and the muscular effort required to breathe. (Though it will not be discussed here, a complete analysis of the balance of recoil forces would also require a term for abdominal wall stiffness.)

THE PLEURAL SPACE

Returning to look at forces present at FRC, and remembering that the lungs and chest wall are really quite separable structures, one might wonder why they do not pull apart from each other, since their respective recoil forces operate to that end. Normally, a thin layer of fluid separates visceral pleura of the lung from parietal pleura of the chest wall, and were one able to insert a measuring device into this fluid-filled space a pressure about 4 mm Hg below atmospheric pressure would be recorded. This "negative" pressure (according to the standard convention of referring all pressures to local atmospheric pressure) represents the tendency for expansion of the pleural space brought about by the oppositely directed tissue recoil forces. Because of their high intermolecular attractive forces, liquids are essentially unexpandable by the lowering of pressure, so the distance separating the pleural surfaces is determined by the amount of fluid and not by the pleural pressure. However, if a hollow needle is inserted through the chest to connect this narrow space with the atmosphere, air will be drawn inward by the lower pressure and the lung will separate from the chest wall as both lung and wall recoil toward their relaxation volumes. Leaks from the atmosphere into the pleural space through tears either in the lungs or in the chest wall are not uncommon, and this condition is termed a *pneumothorax*. The presence of substantial volumes of gases or liquids in the pleural space can only result in a diminution of lung volume and is therefore potentially undesirable. Nonetheless, a thin layer of fluid can serve as both a mechanical coupling and a lubricant and thereby play a useful role in the to-and-fro motion of breathing.

The source of pleural fluid is the parietal pleura, which because of its blood supply from the systemic circulation has a capillary pressure high enough to cause a steady loss of fluid by transudation. The visceral pleura, on the other hand, has a larger capillary surface area and is perfused mostly by the lower pressures of the pulmonary circulation, so that fluid absorption is favored. Hence, pleural fluid is a transudate, the volume of which is determined by the balance between formation at the parietal pleura and removal in visceral pleura. The common medical finding of an increased pleural fluid volume (a pleural effusion) can arise from a disturbance in either absorption or transudation. Normally, absorptive forces exceed transudative, and one would expect to find no pleural fluid at all. What in fact happens is that absorption proceeds until the two surfaces contact at numerous points, to leave many intervening microscopic puddles of fluid. Complete absorption is prevented by pleural tissue stiffness, which limits the minimum size of these puddles.

Gases are kept from the pleural space, or are removed if they enter pathologically, by virtue of the fact that, even though the total pressure in the pleural space is slightly subatmospheric, the sum of the partial pressures of gases in pleural capillary blood is always even more subatmospheric. (For further details of gas partial pressures in blood, see Chapter 20.) The effect of this pressure gradient is to transfer gases from the pleural space to blood, preventing the existence of any gas space. Thus, even though lungs and rib cage may tend to separate because of their recoil forces, their surfaces remain near each other because absorption keeps the pleural space free of gases and nearly free of liquid.

STATIC MECHANICAL PROPERTIES OF LUNG

It has already been noted that lungs exhibit elastic recoil in much the same way as do stretched springs. In textbooks of elementary physics, it is usually stated that the lengthening of a spring is directly proportional to the applied force. That is, $F = k(\Delta l)$, where F represents the force exerted, Δl the change in length, and K is a constant which is determined by the elastic property of the material, the initial (unstretched) length, and the cross-sectional area of the material. If one remembers that pressure is a force and imagines the lungs to be three-dimensional structures made of many springs, it is apparent that a similar expression, $P = k(\Delta V)$, can be written, where P is the pressure causing a change in volume, ΔV, and k is again a coefficient of stiffness dependent on the material, the original volume, and the cross-sectional area of the stretched material. Note that k is not determined solely by the elastic property of the material itself. Nonetheless it defines the overall recoil property of the particular system at hand. This concept has long been used in pulmonary physiology where, instead of stiffness, its reciprocal, compliance, is measured and defined as $C = \Delta V/\Delta P$, where C is compliance and ΔV the change in lung volume produced by a change in pressure across the lung, ΔP.

To determine lung compliance it is necessary to measure the pressure gradient across the lung at times when the only determinant of that pressure is lung recoil – that is, at times of no airflow. In the intact individual that pressure gradient is given by the difference between alveolar space pressure (P_{alv}) and the pleural space pressure (P_{pl}) when respiration is temporarily arrested. If respiration is stopped with the glottis open, P_{alv} equals the local barometric pressure. At FRC, P_{pl} is about 4 mm Hg (5 cm H_2O) negative in respect to P_{alv}. P_{pl} could be measured through a needle inserted into the pleural space through the chest wall, but this is hazardous. The usual procedure is to measure pressure in the esophagus, which is essentially a flaccid tube in the pleural space whose walls are exposed to pleural pressure. The details of this measuring technique are unnecessary here; it is enough to note that when the esophagus is relaxed between peristaltic waves a pressure equivalent to P_{pl} can be obtained. Volume of air inhaled and exhaled is measured by means of any of several types of devices called spirometers.

If inhalation is done in steps, so that pressure can be determined at each volume plateau, one can obtain a graph of lung volume versus recoil pressure similar to that shown in Figure 19-2. The slope of the graph at any point is the compliance at that lung volume. Note from the figure that a single value for compliance applies to the lung over most states of inflation, but that near maximum inflation the compliance is less. This pressure-volume curve is characteristic not only of whole lungs but also of any subdivision. Consequently, the variations of compliance with extent of inflation will have important effects on the way an inspired breath is to be distributed throughout the lungs since the degrees of inflation of subsections of lung vary. Units which are most inflated will receive relatively less of a tidal inspiration because their lower compliances will allow them to undergo a smaller volume change for the same pressure change than would occur in less fully inflated subsections.

In an upright normal human, differences in degrees of inflation of subsections of the lungs always exist because there really is no single value of P_{pl} for all surfaces of the lung. Just as the pressure is greater at the bottom of a lake than at its top, so the pressure in the lower pleural space is more positive (less negative) than at the top. This is an effect of the weight of the lung, which though it is less dense than

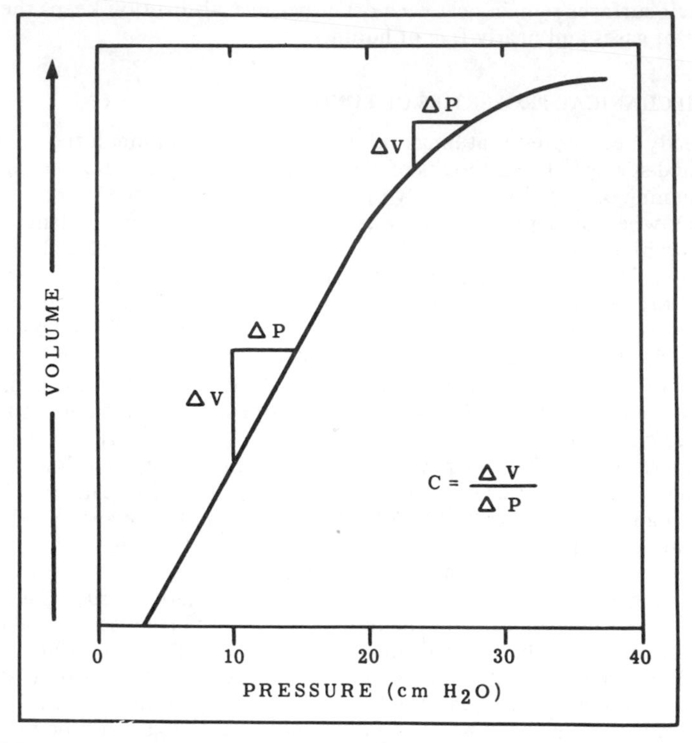

FIGURE 19-2. Relationship of static recoil pressure to lung volume between the limits of maximum expiratory and maximum inspiratory volumes.

water still is able to cause a pleural space pressure difference of about 7.5 cm H_2O between apex and base in an average adult man. Because the pressure gradient across upper units of lung is greater than across those at the bottom, the upper portions will tend to be more inflated than those lower down, and this situation will determine how the lungs expand with each breath. The same result is produced by the pull of lung tissue on itself, since the lung is effectively hanging from its apical surface. Each horizontal plane can be thought of as supporting the weight of all the lung below it, and in this way the downward pull on each level becomes progressively less at successively lower levels, so that lower units are least inflated.

THE ROLE OF ALVEOLAR SURFACE TENSION IN PULMONARY MECHANICS

While compliance is usually determined clinically by measuring V and P over only small volume ranges, the pressure-volume (P-V) relation of excised lungs covering the full volume range from complete lung collapse to full inflation can bring out several important lung characteristics. When lungs are removed from the chest they collapse to a smaller volume than they could achieve while in situ, but they do not collapse completely. Complete spontaneous collapse seems to be prevented because as the lung collapses the narrowest airways share in the collapse to the point where their moist walls touch and close off, thereby trapping air in the distal alveoli. Then, even suction applied to the trachea will not empty the lung.

However, it is possible to prepare completely degassed lungs by other means. When a degassed excised lung is inflated with air and deflated again by allowing the air to be returned by lung recoil, a P-V curve like that in Figure 19-3 is obtained.

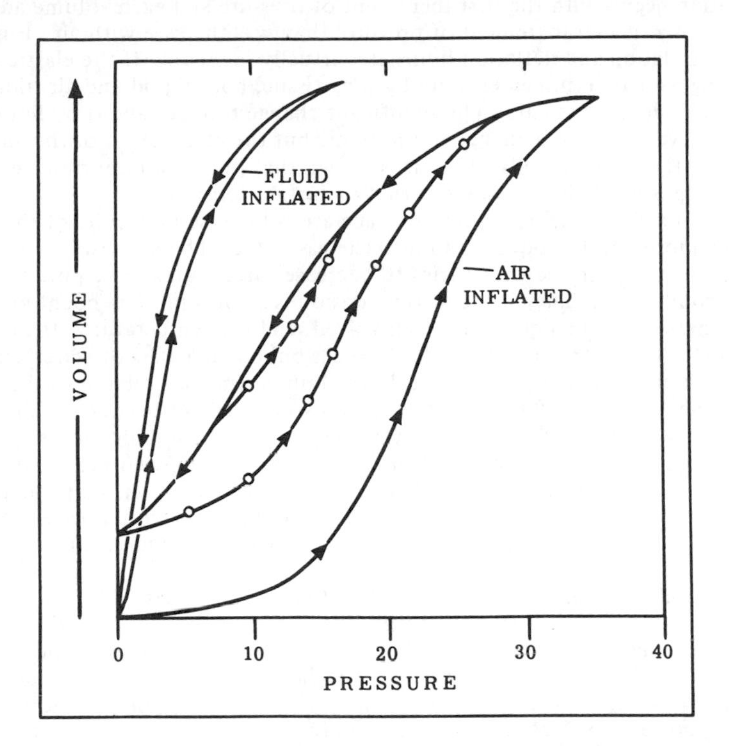

FIGURE 19-3. Relationship of static recoil pressure to lung volume when excised lungs are inflated from a degassed state with air and with 0.9 percent NaCl solution. Note that when air-inflated, the path of deflation differs from that of inflation, and that deflation to the completely degassed state is not achieved. The curves with open circles are subsequent inflations made in one case from the point of least volume after air inflation. In the other case, the path of a medium-deep inspiration from near mid-capacity is illustrated. All deflations when air-filled take place along the same curve.

Observe that initially there is very little increase in volume until the pressure reaches some critical value, but inflation beyond this is easily achieved. During deflation, pressure again changes more than volume in the first phase, but this gives way to a different slope for most of the expiration curve. The difference between the inflation and deflation curves is prominent, and most of it is in the slopes of the upper and lower quarters of total volume, while slopes of the middle halves are nearly equal. If one remembers that the slope of the curve is equivalent to compliance, it is immediately apparent that both the compliance and the recoil pressure depend on the state of inflation but are even more dependent on the direction of movement. If additional curves are obtained by beginning at lesser states of deflation than the degassed state, it is found that the inflation limbs lie in the area bounded by the loop already seen, whereas all the deflation limbs lie essentially on the same curve. Two inflation-deflation loops from intermediate volumes are included on the figure. Note that the openness of the loop is inversely related to the lung volume at the start of inflation. In striking contrast, if the same lungs are inflated from the degassed state with 0.9 percent NaCl solution instead of air, inflation begins with the first increment of pressure and each volume addition is made with a smaller increment of pressure than was the case with air. Furthermore, the inflation and deflation limbs are virtually identical. If the elastic properties of lung were determined entirely by lung tissue, both liquid and air filling should yield identical results. The significant change brought about by liquid filling, however, was not a change in the tissue but the elimination of the air-fluid interface at the alveolar walls. Thus it is shown that the most important recoil force of lung is not that of tissue but that of surface tension.

Surface tension at an air-liquid interface arises because molecules of the liquid are drawn more into the liquid than into the gas phase. The net result is equivalent to a tension at the surface which tries to decrease surface area. For pure liquids and true solutions the magnitude of this tension is a constant dependent upon the chemical natures of the liquid and gas involved, and the temperature. It can be shown that if the surface is spherical, like a gas bubble in a glass of water, surface tension produces a higher pressure inside the bubble which is related to the bubble size by the expression, $\Delta P = 2\gamma/r$. ΔP is the pressure gradient between the inside and outside of the bubble, γ is the surface tension, and r is the bubble radius. It is easy to see (though often hard to believe) that pressures are greater inside small bubbles than large ones. The important corollary is that if one were to attempt to add a small additional volume to a small bubble, a higher pressure would be required to bring about that addition than would be needed to add the same volume to a larger bubble.

This corollary helps explain the inflation limb of the degassed lung P-V curve. Had one looked at the lung surface during that inflation, he would have seen that the larger terminal respiratory units filled with air before the smaller ones, and that none began to fill until some pressure was reached at which the surface tension of the air interfaces at the collapsed airway ends could be overcome. A moment's reflection, however, will reveal that if alveoli really behave like bubbles, the smaller ones should never fill, since to add volume to a larger unit will take less pressure than to add to a smaller, and as the larger units begin to fill they should become even easier to fill! Nevertheless, alveolar expansion does not continue indefinitely, for at some point further expansion begins to stretch relatively uncompliant fibrous tissue elements which heretofore were slack. In this way the most-filled units become constrained and their pressures can be raised without further expansion, allowing smaller units to fill. The progressive opening of smaller and smaller units generates the inflation limb of the P-V curve beyond the sharp upward deflection where the first units open. When all units are air-filled, further inflation only stretches the constraining net of elastin and collagen. Since this net is relatively uncompliant, the pressure-volume curve will have a less steep slope at large volumes, and this change in slope is apparent in both the air- and fluid-filled lungs.

As deflation begins, another property of the alveolar surface becomes apparent, for if ordinary bubbles had been inflated and deflated, the inflation and deflation curves would have coincided. Furthermore, instead of small units deflating first, followed by large ones, as would be expected for soap bubbles, the whole lung has been found to deflate evenly. These and other findings have suggested that alveolar surface tension may be unusual, and many studies made of material obtained from lung bear out this suspicion. The alveolar surface tension is not that which would be found if the surface were covered with a pure liquid or a simple solution. Instead, the alveolar liquid layer appears to be covered by an insoluble material, perhaps di-palmitoyl-lecithin, which lowers the surface tension in an unusual way. The source of this material seems to be the alveolar lining cells themselves. This insoluble material, called a *surfactant* (for *surf*ace-*act*ive age*nt*) has two important properties. First, its presence lowers alveolar surface tension, with the result that less pressure is required to maintain any given alveolar volume. Second, this material causes the alveolar surface tension to vary as a function of alveolar area, and of the direction of change of area.

Since alveolar surface tension is a force related directly to transpulmonary pressure, and since surface area is a dimension immediately related to alveolar volume, a graph of surface tension of alveolar material versus area of the surface of that material can be used to examine the relation of alveolar surface tension to the state of inflation. Such a tension-area curve is shown in Figure 19-4. Note that

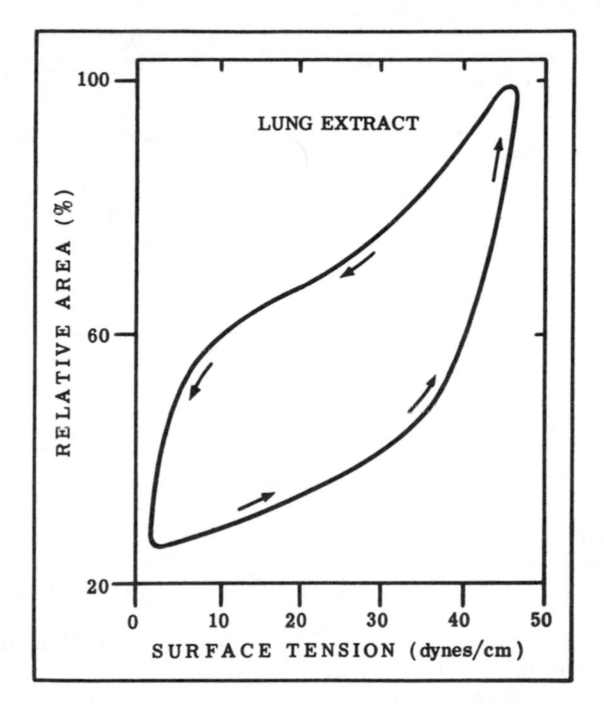

FIGURE 19-4. Surface tension of an aqueous preparation of lung extract as a function of area of surface. To generate such a curve, a sample of extract is placed on water in a pan which has a movable side wall. The surface tension is continuously measured while the surface area is varied by slowly moving the wall of the pan to and fro.

expansion and contraction of the surface also generates a looped curve, and that the surface tension for any given area is lower during "deflation" than during surface expansion. According to several elegant analyses, it is very likely that during lung deflation, collapse of small units is prevented because, as they begin to collapse, their areas, and hence their surface tensions, fall. This minimizes further collapse until the larger units with the still high surface tension have managed to "catch up" in the deflation process. As a separate test of this hypothesis it has been found that if lungs lack surfactant large pressures are needed to fully inflate them, and that during their deflation smaller terminal units collapse completely while large ones remain until last. Thus, the second significant role of surfactant is to allow coexistence of terminal respiratory units of different size, and to ensure that all units can inflate and deflate together.

AIRFLOW DYNAMICS

The passive pressures described in the preceding section were determined by the recoil properties of the lung, chest, and abdomen and therefore were dependent only on compliances and the lung volume. Pressure gradients due to elastic recoil are indifferent as to whether there is airflow or not. At times of no airflow the transpulmonary pressure (P_{tp}, the pressure gradient between mouth or nose and the pleural space) is determined solely by compliance, but during airflow an additional pressure must be added to achieve flow. That is, at *any* time, $P_{tp} = P_{el} + P_{res}$ where P_{el} is the pressure gradient due to elastic recoil and P_{res} is the pressure associated with flow resistance. In this section the determinants of P_{res} will be examined.

Air, like blood, is a fluid, and the concepts of fluid mechanics presented in Chapter 11 will be used again here. As before, resistance (R) can be *defined* as pressure gradient causing flow ÷ flow. Airflow ($\dot{V}$) is the rate of change of lung volume with respect to time and is usually given in liters per second. The units of airflow resistance are centimeters of water per liter per second. Since P_{tp}, V, C, and $\dot{V}$ can all be measured or calculated, it is possible to measure P_{tp} at several flow rates and lung volumes and calculate the associated P_{res} from the equation $P_{tp} = V/C + P_{res}$. When this is done, it is found that for airflow rates occurring during rest or mild exertion the relation of P_{res} to $\dot{V}$ is linear and the definition of resistance used above is also an adequate mathematical representation of behavior. With higher flow rates the pressure-flow relation deviates from linearity as flows become turbulent, and to describe adequately the P_{res}-V relation it becomes necessary to include another term related to the square of the flow rate, viz: $P_{res} = k_1\dot{V} + k_2\dot{V}^2$. R has been replaced by k_1 to indicate that k_1 is not entirely determined by the constants and parameters of Poiseuille's equation, which had been the case for lesser flow rates. Furthermore, unlike streamline flow, turbulent flow involves the important factor of fluid density. The constant k_2 contains terms for both density and viscosity, although the latter plays a minor role.

For the remainder of this discussion, flow rates will be considered to be low enough so that turbulent flow is not significant. The reader, however, must remember that at high airflows turbulence will occur and the pressures required may cause an inordinately high energy expenditure for breathing.

P_{res} is determined by viscous (i.e., frictional) losses from the movements of both air and lung tissue itself. The latter friction arises because lung tissue is deformed by the increase of its contained volume during breathing, much as syrup is deformed when a spilled drop spreads across a table top. This *tissue viscous resistance* amounts to 10–20 percent of the total pulmonary resistance. Note carefully that the energy *lost* in viscous deformation of lung is not the same energy accounted for in elastic recoil – the former depends on flow, while the latter depends on position; the former is lost as heat, while the latter is conserved and is available to cause passive expiration.

After tissue viscous resistance has been accounted for, the remainder of pulmonary resistance, related to airway sizes and gas viscosity, can be apportioned among several anatomical divisions. When one is breathing through the nose, airflow resistance of nose and nasopharynx accounts for two-thirds of the total. A switch to mouth breathing causes a substantial fall in total airway resistance. However, with mouth breathing, the resistance of upper airways – those included between mouth and the intrathoracic portion of the trachea – still accounts for a third of the new total. An average resistance with mouth breathing is 1.5 cm H_2O per liter per second. Surprisingly, the resistance of all airways distal to about the twelfth generation of branching (having diameters of about 1 mm and less) amounts to less than 10 percent of the total. The reason is that even though each individual airway is quite narrow, the large numbers which act in parallel at each order of branching cause the net resistance to be low. Thus the largest fraction of the airway resistance and the greatest pressure gradient occur between trachea and bronchi larger than 2 mm internal diameter. The gas volume contained by the airways in which most of the resistance occurs is less than 3 percent of the total thoracic volume.

It is usually difficult for students to visualize the interrelationships of P_{tp}, P_{res}, P_{el}, V, and $\dot{V}$ over the course of a breath. The following general statements about those relationships are always true, and it is useful to keep them in mind. (1) Alveolar pressure (P_{alv}) is *always* more positive than pleural pressure (P_{pl}). (2) At any time that $\dot{V} = 0$, $P_{tp} = P_{el}$. (3) During *expiration* P_{alv} is *always* more positive than pressure at the point where the airways open to the environment (P_{ao}), but during *inspiration* P_{ao} is always more positive than P_{alv}. (4) The absolute magnitude of P_{el} depends on V and is insensitive to $\dot{V}$. (5) The absolute magnitude of P_{res} depends on $\dot{V}$ and is insensitive to V. (6) P_{pl} may have *any* value (negative, zero, or positive), depending upon compliance, resistance, muscle strength, flow rate, and flow direction, but conditions 1–5 *always* hold.

Some of the above features can be verified in Figure 19-5. This shows an idealized curve, since breathing does not produce exact sine and cosine waves like those shown. However, any actual record can be mathematically handled as the sum of several sine curves, and all conclusions derived from the figure remain valid. Note how the algebraic addition of P_{res} (shown by the arrows) to P_{el} yields P_{tp}. Because P_{el} and P_{res} reach their maximum amplitudes at different times in the cycle, the point of maximum amplitude of P_{tp} corresponds in timing with neither of those maxima, nor does it correspond with the greatest displacements of V or $\dot{V}$. It can be shown that the point of maximum amplitude of P_{tp} lies nearer the time of maximum V if P_{el} is the major determinant of the work of breathing, but maximum P_{tp} lies closer to maximum $\dot{V}$ if P_{res} is the major load. In eupnea P_{tp} and V are nearly "in phase" with each other, indicating that most effort is used to overcome tissue elastic recoil.

Up to now it has been assumed that a single value for airway resistance pertains to all lung volumes and flow rates. This is quite untrue, and variations of resistance are of marked importance, particularly in relation to some lung diseases.

Airway resistance is inversely related to the extent of lung inflation because during expansion the intrapulmonary airways all participate in the volume increase. Though both lengthening and widening occur, changes in radii exert a more profound effect on resistance than do changes in length (see Poiseuille's equation). The graph of airway resistance upon lung volume is approximately hyperbolic, so the greatest changes occur at the smaller lung volumes. An example is shown in Figure 19-6.

Airway resistance can also be increased by contraction of tracheobronchial smooth muscle and by swelling of the mucosal layer. Reflex bronchoconstriction may arise from mechanical or chemical stimulation of a number of receptors within the airways, lung parenchyma, or nasopharynx, besides occurring as part of the reflex response to stimulation of the carotid chemoreceptors or pulmonary vascular

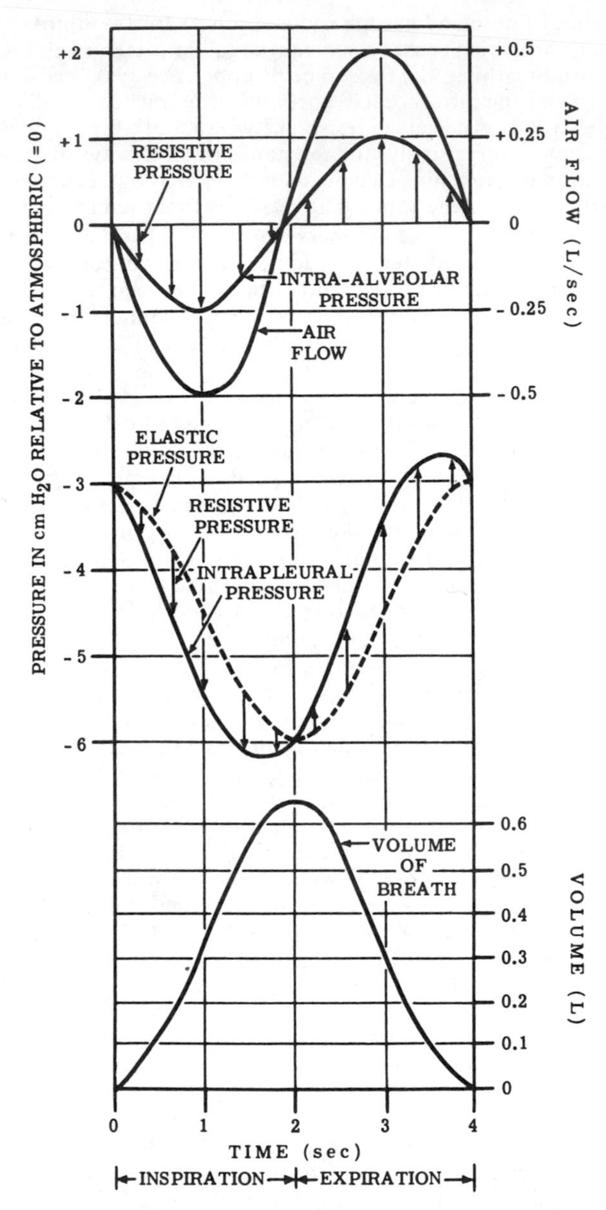

FIGURE 19-5. Idealized curves of the airflow, tidal volume, and intrapleural pressure which occur during the course of a breath. The two components of the intrapleural pressure, the resistive and elastic pressures, are also shown separately. (Reproduced from the article RESPIRATION in the *Encyclopaedia Britannica,* 1970.)

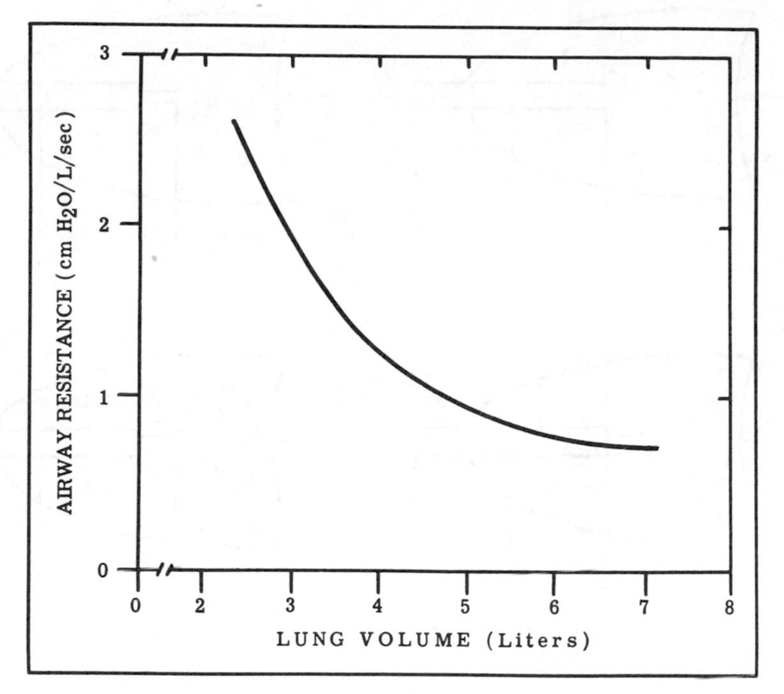

FIGURE 19-6. The effect of the state of lung inflation upon airway resistance in a normal human subject.

stretch receptors. Motor innervation is by the vagus nerve. Though bronchoconstriction from irritant vapors or particulate suspensions such as smoke is usually a reflex response to airway receptor stimulation, a number of chemicals directly alter bronchial muscle tone. Among the more important constrictors are acetylcholine and other drugs known to mimic or enhance the effects of parasympathetic nervous activity, and histamine. Epinephrine and other drugs which mimic the effects of sympathetic nervous activity cause relaxation of bronchial muscle. Bronchial muscle tension is also inversely related to the CO_2 concentration of the muscle environment – a characteristic common to most smooth muscle which may be important in homeostatic adjustments of airway resistance (see Chapter 20, Effects of Regional Variation in $\dot{V}/\dot{Q}$).

Contraction of airway muscle not only narrows the lumen but in cartilaginous airways also stiffens the walls. Bronchial cartilages, when pulled together by the muscles, interlock and overlap much like the plates of an armadillo. This can help prevent airway collapse caused by negative transmural pressures.

That airways can and do collapse is well known. In fact, airway collapse appears to be the mechanism whereby airflow rates become limited in both normal and diseased lungs. This is a "dynamic" collapse – it occurs during airflow. Because of its importance its mechanism will now be examined in some detail, by means of the technique of analogue modeling.

Imagine the chest to be only an air-filled cavity connected to the environment by a single airway which is collapsible (Figure 19-7A). Assume that in order for the cavity to empty at a particular rate, a pressure gradient of 5 units is required

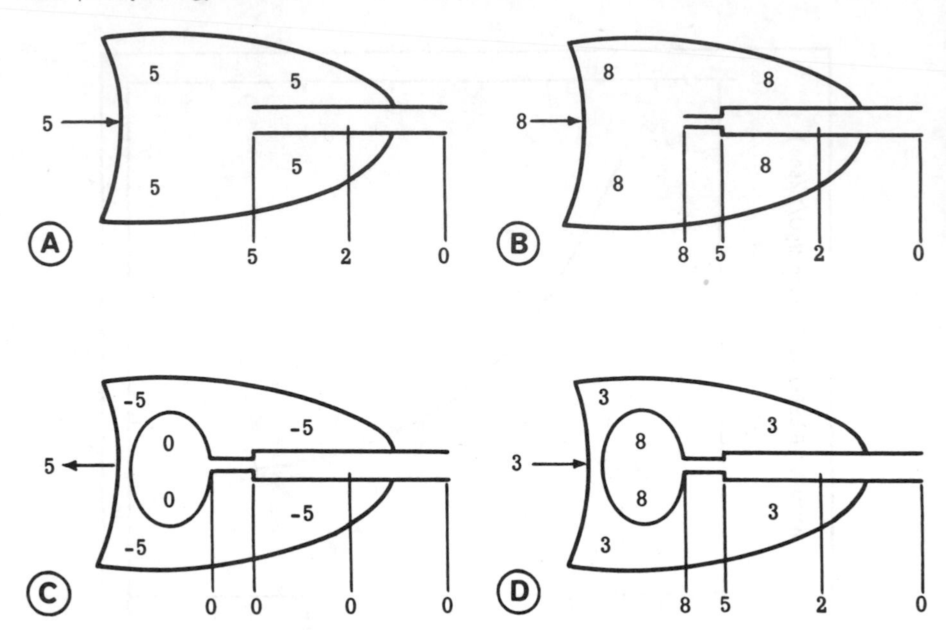

FIGURE 19-7. Models illustrating forces across airway walls. For details see text. Pressures within spaces are indicated by the numbers therein. Intra-airway pressures are indicated at several points by lines from those points to numbers below. The diaphragmatic pressure equivalent, and its direction, is indicated by arrow.

for the airway. The energy for this "expiratory" flow is supplied by a force at the diaphragm equivalent to a pressure in the amount and direction indicated by the arrow. The assumed conditions cause an intraluminal pressure (using the standard reference convention for zero) of +5 at the entrance of the airway and 0 at its exit. Since an airway's resistance is distributed along its entire length, there is a longitudinal pressure distribution such that at every point except the entrance the internal pressure is $<+5$. However, the pressure on the *outer* wall of the airway is +5 for all parts in the chest and 0 for all parts lying outside. This means that, except for the two ends, there is a *positive* (distending) transmural pressure across all the extrathoracic airway and a *negative* (collapsing) transmural pressure gradient across all the intrathoracic airway. Because of the progressive decline of internal pressure down the airway, the greatest negative transmural pressure occurs at the point just before the airway leaves the thorax. If this particular airway wall would always collapse at a transmural pressure of 6 units, no collapse would have occurred under the assumed conditions. If flow is doubled, however, all the pressures will be doubled and the intrathoracic portion will collapse at a point before it leaves the chest. Further analysis of the dynamics then becomes complicated, but it is sufficient to learn that the collapsed segment acts as a high resistance of variable magnitude which serves to limit the maximum flow. It does so because the harder one tries to force air outward, the greater becomes the resistance of the collapsed segment.

A change in behavior occurs if the model is modified by the addition of a length of higher-resistance airway at the internal end. This addition is shown in Figure 19-7B, where once again it is assumed that the expiratory flow rate is that which demands a 5-unit pressure gradient down the *original* airway. In order to meet the requirement, a greater pressure is needed in the thorax because the new section requires 3 units of pressure gradient for this flow. The necessary increase in thoracic pressure to 8 units results in a 6-unit negative transmural pressure across the airway just before it leaves the chest, and collapse will occur. Note that the same airway now collapses at a flow rate which did not cause collapse under the conditions of Figure 19-7A, and that collapse was caused by adding an *upstream* resistance.

Finally, the effect of a recoil element on dynamic airway collapse can be modeled by tying an elastic bag onto the airway as shown in Figure 19-7C and D. Assume that a pressure gradient of 5 units is necessary to hold the bag open to the smallest value it will be allowed to reach. Pressures in the system under the no-flow condition are given in Figure 19-7C. However, when one assumes air to be flowing out, the pressures shown in Figure 19-7D will pertain if the flow rate and airways are the same as they were for Figure 19-7B. Note that the requirement for 8 units of longitudinal pressure gradient is met by supplying 5 from elastic recoil and only 3 from an upward force on the diaphragm. Because of the elastic bag, the pressure outside the airway at the point of collapse in Figure 19-7B is now too low to cause collapse. The flow rate can be increased by a factor of 3 before transmural gradients of -6 appear at the airway wall near the thoracic outlet. It can also be shown that at volumes above the smallest ever allowed there is an even greater ability for recoil forces to *prevent* airway collapse.

If the assumed direction of airflow is reversed and a similar analysis made using the above models, it will be revealed that only *positive* transmural pressures occur during inspiration.

In summary, if lungs behave like the models, airway collapse during expiration should be promoted by high upstream resistances, by high lung compliance (little elastic recoil), and by a decrease in lung volume.

It is now only necessary to point out a few similarities to show that Figure 19-7D is a remarkably reliable lung model and to conclude this presentation of airway dynamics. The intrathoracic, extrapulmonary airways correspond well with the airway of the model between the elastic bag and the "chest" wall. The incompleteness of the tracheobronchial cartilaginous rings make considerable collapse possible, and it is immediately apparent that the pressure on the outer wall is P_{pl}. It is more difficult to visualize the force outside an intrapulmonary airway because this is not simply a pressure in a space but a force on the wall caused by the pull of tissue attachments. However, by an analysis of the mechanics it can be shown that for normal lungs intrapulmonary airways behave as if they were simply exposed to P_{pl} along their outer surfaces. Thus one can consider the pressure around all conducting airways to equal P_{pl} in conformance with the model.

In eupnea, the balance of forces is such that airway collapse does not normally occur. However, if a forced expiration is made, collapse of some part of the airway occurs at all lung volumes below about two-thirds of maximum inflation, and the collapsed segment becomes flow-limiting. There is, in fact, a characteristic direct relationship between maximum achievable flow and the state of inflation. This could have been predicted from the model and the information gained earlier about resistance and compliance. Recall that near maximum inflation, airway resistance is lowest and recoil forces are at their maximum, but these both change in a graded fashion as expiration proceeds toward the point of least achievable lung volume (termed the *residual volume*). During exhalation the increasing upstream resistance and decreasing P_{el} cause an upstream movement of that point in the airway where endobronchial and peribronchial pressures are equal. This progressively causes more of the airways to collapse. The result can be seen in Figure 19-8, where maximum expiratory flow rate is plotted as a function of lung volume from the point of

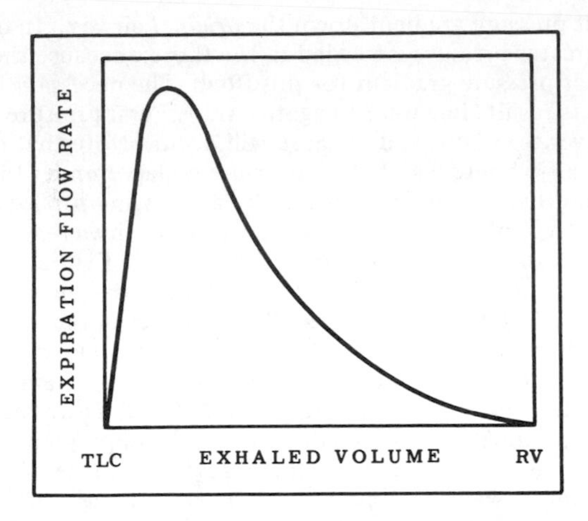

FIGURE 19-8. Airflow rate achieved by a normal subject during a maximally forced exhalation from total lung capacity to residual volume. The maximum airflow rate achievable at any volume decreases as the lungs become smaller. The initial upstroke deviates from vertical because the subject requires a finite time to get his chest wall moving with maximal force. In this subject the point of maximum flow occurred about 0.25 second after the start of exhalation.

maximum inflation (total lung capacity, TLC) to residual volume (RV, normally about 20 percent of TLC).

Normally, as long as one's lungs are inflated within the range of the upper two-thirds of the vital capacity, one achieves a maximum expiratory flow rate only by deliberate effort. Even during the heavy breathing of exercise, maximum flow rates are seldom reached. However, many pulmonary diseases result in increases of upstream resistances, or decreases in P_{el}, or both, and in such cases maximum flows may be reached with only modest effort. This typically makes it impossible for ventilation to supply the demands of exercise, and patients become "short of breath" during exertion.

PULMONARY FUNCTION TESTS

Measurements of total pulmonary resistance, airway resistance, lung compliance, or flow velocities achievable at graded levels of inflation can be used to assess pulmonary function in patients. While these more elaborate tests may be necessary to estimate pulmonary characteristics independently of some of their interactions, ventilatory ability is more often estimated from data contained in graphic recordings of single forced expirations. Many different derived values from such records are in use; the choice among them is based on empirical reasons. The ability of the tests to quantify pulmonary function arises from the dependence of airflow on intrinsic pulmonary factors such as resistance, compliance, and lung volume, rather than on muscle strength and subject volition. The interpretation of such tests in terms of specific tissue changes depends on understanding airflow dynamics and the static properties of the system. Some commonly used function tests are the following.

Vital Capacity (VC)

The largest amount of air which can be exhaled in a single breath after a maximum inhalation is known as the vital capacity. The vital capacity is determined by lung size, by subject size and age, and to a large extent by lung and chest wall compliances. It is also influenced by factors which alter thoracic mobility, such as muscle weakness, abdominal fluid, and chest pain. Results are expressed in liters. A graphic display of this and other commonly measured lung volumes is given in Figure 19-9.

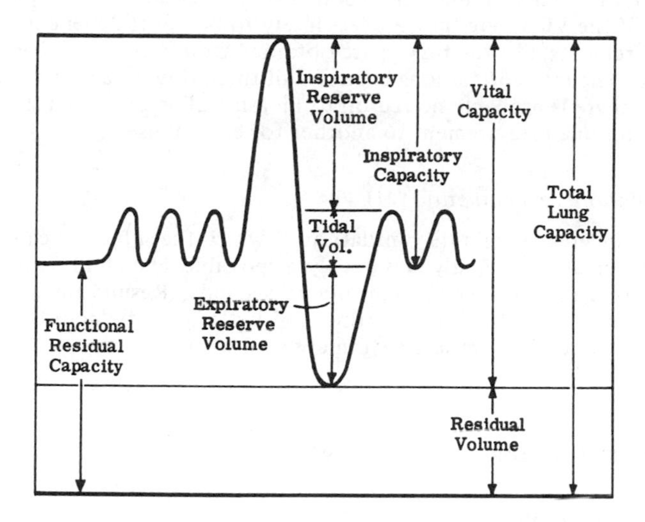

FIGURE 19-9. Schematic representation of the course of breathing versus time which includes both tidal breaths and a vital-capacity-sized breath. Note that the residual volume cannot be exhaled. Its size is determined by other methods.

Forced Expiratory Vital Capacity (FEV)

When exhalation of a vital-capacity-sized breath is done as rapidly as possible, several useful measures of expiratory flow rates can be obtained from a recording of volume delivered versus time. One of these is the amount exhaled in specific periods. Typically, a person should be able to exhale 70 percent of his VC in the first second (FEV_1) and empty more than 90 percent in three seconds (FEV_3). Diseases which decrease airflow as a result of either increase in resistance or decrease in tissue retractive force will often decrease the rate of delivery of a forced expiratory VC. This is a simple, useful clinical measurement, suitable for following the progress of many cardiopulmonary diseases and for recognizing patients with obstructive airway diseases. Results are expressed either as the percentage of the VC delivered in the chosen time interval or as the absolute volume of gas exhaled in that period. The FEV_3 is now rarely measured, but measurement of the FEV_1 is probably the commonest pulmonary function test.

Maximum Expiratory Flow Rate (MEFR)

Some instruments directly measure the greatest instantaneous expiratory flow rate achievable. If this measurement were to be obtained near the onset of a FEV, the result would be highly effort dependent. However, several varieties of instantaneous flow measurement equipment have been developed which can measure the maximum flow rate at any specified state of inflation. Note that these tests measure *instantaneous* flow rates. Results are expressed in liters per second.

Rather than use instantaneous flow rates, or the volume delivered in a specified time interval, some clinical physiologists measure the averaged rates at which specific segments of the FEV can be delivered. In common use are the mean flow rate at which the middle half of the forced expiratory VC was delivered (called the maximum mid-expiratory flow, MMF), and the mean flow rate present between 200 and 1200 ml of exhaled volume. The object of such tests is to measure flow rates over a part of the VC where they are less likely to be effort dependent. In practice, these measurements, like the FEV_1, are obtained from records of single forced VC exhalations. The intercorrelations between all these flow rate measurements are so high that there is probably no reason to measure all in any given subject, or to strongly prefer one measurement to another for clinical use.

Maximum Voluntary Ventilation (MVV)

To measure the greatest minute ventilation of which the subject is capable, he is instructed to breathe as rapidly and deeply as possible, and all his expired air is collected during a short interval, typically 15 seconds. Results are converted to liters per minute. In spite of being influenced by fatigability, coordination, and cooperation, the results correlate extremely well with the flow rate measurements described earlier.

MECHANICS OF THE LUNG CIRCULATION

BRONCHIAL CIRCULATION

The lung, like the liver, has two reasonably separate vascular beds. The smaller of the two is the bronchial vascular system. Since this is a division of the systemic circulation, arterial pressures are four to five times greater than pressures in the pulmonary circulation. The bronchial circulation is the principal blood supply of bronchi and bronchioles, whereas the respiratory bronchioles and more distal lung tissue are nourished by the pulmonary circulation. Bronchial arteries form the vasa vasorum of pulmonary arteries and provide the blood supply of the pulmonary nerves. The bronchial circulation joins the pulmonary circulation in several places, and under some pathological conditions these interconnections become important. The easiest intermingling to confirm experimentally is in the venous bed. Bronchial veins drain from the first one or two divisions of bronchi into the right atrium, but venous drainage from more peripheral airways goes into the left atrium through anastamoses with pulmonary veins. About 1 percent of pulmonary venous flow comes from bronchial veins. There are arterioarterial communications through both capillary and noncapillary networks. The volume of this arterial shunt flow is normally very small, but in congenital absence of part of the pulmonary arterial tree, or following pulmonary arterial occlusions, flow of inadequately oxygenated systemic arterial blood through these anastamoses may become great enough to be of value in gas exchange. Marked enlargement of the bronchial circulation also occurs in bronchi which are chronically infected. Under such conditions bronchial arterial flows as large as 20 percent of the cardiac output have been reported, whereas this circulation normally receives only about 2 percent of the cardiac output.

PULMONARY CIRCULATION

Turning now to the pulmonary circulation it should be apparent that the most efficient pulmonary gas exchange requires the optimum presentation of CO_2-rich and O_2-poor blood to CO_2-poor and O_2-rich inspired air. An understanding of the way lung blood flow becomes differentially distributed is as important as an understanding of airflow mechanisms, since the two are inexorably linked in gas exchange. Typically the pulmonary circulation has been described as a passive, low-pressure system capable of little regulation. Because of these characteristics pulmonary blood flow distribution can easily be profoundly altered; therefore an understanding of the mechanisms of the alterations is obligatory if one is to comprehend regional variations of gas exchange.

The pulmonary vessels on the arterial side are much more distensible than their systemic counterparts. Their high compliance allows these vessels to have their radii easily changed by variations in transmural forces. In the systemic circulation, except for the heart and other exercising muscles, the intravascular pressure alone is a good approximation of the entire distending force. This is not the case for the lungs, because the low intravascular pressure allows perivascular pressure and the radial pull of extravascular tissue to become significant parts of the total force. Perivascular pressure can be as high as P_{alv} or as low as P_{pl}. Variations of perivascular pressure from main pulmonary arteries to capillaries may be as large as the variation in intravascular pressure, so they play a prominent but often unquantifiable role in establishing the radii of these vessels.

A more readily computed variation in vascular transmural force occurs as a result of the weight of blood. That is, the pressure in a fluid column is higher at the bottom than at the top by an amount $P = \rho gh$, where ρ = blood density, g = gravitational acceleration, and h = column height. If pressure in the main pulmonary artery (with respect to atmospheric pressure) is 20 cm H_2O in an upright individual, the pressure 19 cm further cephalad in an arterial branch of the apical segment will be only 1 cm H_2O. (The densities of blood and water are essentially equal.) Conversely, at any lower level, intravascular pressure is above that at the midlevel. This phenomenon causes a gradation of pulmonary intravascular pressure which tends to distend vessels at the base and collapse vessels at the apex. As would be expected, distribution of blood flow becomes prominently affected by these variations in transmural forces, for they help define the resistance to flow at each horizontal level. Blood flow distributes differently when the person is lying down, because then the gravitational gradient is directed from front to back and acts over a smaller distance. If pulmonary arterial pressure were as high as systemic arterial pressure (about 100 cm H_2O), a difference in height of 10 cm would vary pressures by only 10 percent, whereas with arterial pressures of only 20 cm H_2O, the same difference in height causes a 50 percent change in intravascular pressure. All other things being equal, the proportionally larger stress would produce a proportionally large radial change, so it can be seen that low pulmonary vascular pressures act to augment vascular dimensional instability.

The *perfusion pressure* is the difference in pressure between two longitudinally separated points in a vascular bed. It is usually expressed as the difference between arterial and venous pressures. This gradient is a function of frictional loss between the two measurement sites and also of any differences in kinetic energy between the two sites (see Bernoulli's equation, Chapter 11). The latter effect is usually disregarded, though it is not always correct to disregard it. In vessels having transmural pressures everywhere sufficient to prevent collapse, the downward pull of gravity on arterial blood would be exactly offset by the pull on venous blood. If that were the way the lung vessels behaved, the perfusion gradient for all horizontal sections of the lungs would be given by the difference between pulmonary arterial pressure (P_A) and pulmonary venous pressure (P_V), where both are measured at any single horizontal level. However, in a vascular bed in which collapse can and does occur, any variation of the pressure downstream with respect to the collapse point could not make itself felt upstream beyond the collapse unless the downstream pressure were raised enough to open the collapsed zone. This makes the flow independent of the downstream pressure as long as the collapsed segment is present.

Such a vascular situation is analogous to the flow over a waterfall. The height of a river below a waterfall has no effect on the flow over the fall, or on the height of the river above the fall, unless the level of the river below the fall becomes as high as the fall itself. The flow of the river is determined by the slope of the river bed above the fall. For this reason the phenomenon of a collapsed vascular segment which determines the blood flow and dissociates the flow from P_V has been called the "vascular waterfall phenomenon." It is a phenomenon familiar to everyone who has attempted to drink through a collapsed straw. In the presence of a collapsed segment the perfusion gradient cannot be P_A minus P_V, but the correct replacement for P_V can easily be chosen after further examination of the collapse mechanism. When collapse occurs, flow stops, and in the absence of flow the pressure everywhere in that vessel upstream of the collapse rises to equal P_A. If P_A is above that necessary to open the collapse, the collapsed zone will open; flow will then resume and the intravascular pressure will fall because of the viscous loss. This reestablishes the collapse, and the cycle begins again. Actually the "fluttering" of the wall is more apparent than real, and a steady stream seems to be passing through the narrowed portion. In this collapsible system the arterial pressure is working against the pressure which causes collapse, and not against P_V, so the perfusion gradient is the difference between P_A and the pressure necessary to open the vessel. Viscous losses in the segments downstream from the collapse point would not be included in this perfusion gradient.

In lung, capillaries are the vessels one would most expect to collapse because they are least supported by tissue "guy wires." Furthermore, their perivascular pressures, being essentially P_{alv}, are the highest anywhere in the system. It has been found that the pulmonary capillaries collapse when their internal pressures are slightly below P_{alv}. The relevant perfusion gradient for the part of the lung in which collapse occurs is usually taken as P_A minus P_{alv}. It turns out that in a normal standing man there is sometimes an uppermost region of lung where $P_A < P_{alv}$, and where no perfusion occurs. (This region is known as zone I.) Immediately below zone I there is a region where $P_A > P_{alv}$ but $P_{alv} > P_V$, so "waterfall" flow occurs and the perfusion gradient is $P_A - P_{alv}$. This is zone II. Below this (in zone III) $P_A > P_{alv}$ and $P_{alv} < P_V$, so collapse does not occur and the perfusion gradient is $P_A - P_V$. Changes in position, P_{alv}, P_V, or P_A will all cause a redistribution of transmural pressures. Since all the above modulating pressures are large in comparison with pressures in the normal pulmonary artery, one would expect them to bring about significant dimensional alterations. A schema of these three zones is shown in Figure 19-10, which also includes a hypothetical plot of flow through each horizontal level as a function of its position.

Increases in P_V cause more of the lung to be in the zone III state and also cause a back pressure which raises P_A and the transmural pressures of all vessels. This increased transmural pressure distends some vessels and opens others which had had enough muscle tone to be closed at lower pressures. Dilation, and "recruitment" of additional parallel paths, lowers the net perfusion gradient, raises P_A somewhat, and increases the number of perfused capillaries.

Increases in P_{alv} cause more of the lung to function in the zone I and zone II conditions. If cardiac output remains constant, the expansion of zone I diverts

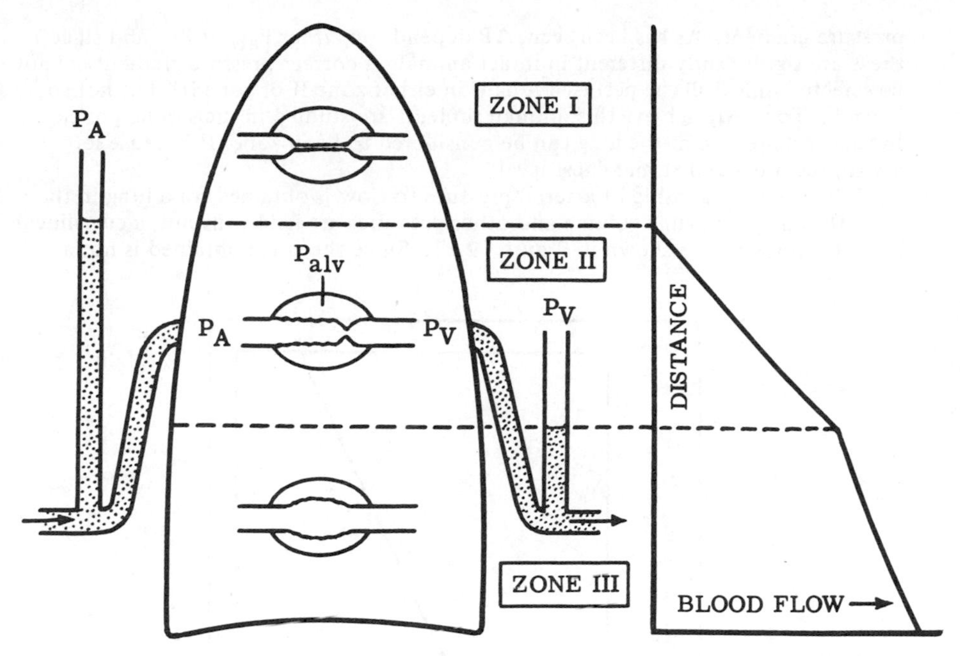

FIGURE 19-10. Schematic illustration of the three different conditions under which blood vessels exist in the lungs. For details see text. Note that no flow occurs through zone I, and that the way in which flow increases progressively at each lower level in zones II and III is different for each zone. The changes in flow through each zone are caused by gravitationally dependent transmural pressures which alter the length of the collapsed segment in zone II and the radii of distensible vessels in both zones. (From J. B. West, C. T. Dollery, and A. Naimark. *J. Appl. Physiol.* 19:713–724, 1964.)

blood elsewhere, causing concomitant elevation of P_A. This increase in their transmural pressure distends and recruits vessels in zones II and III.

Like airway caliber, vessel caliber depends upon lung volumes because vessel walls are tethered to the parenchyma. Unlike airways, however, progressive lung inflation does not cause a progressive decrease in perfusion resistance. Most studies have shown that the relationship of perfusion gradient to volume is U-shaped, and that the point of minimum pressure requirement lies, perhaps fortuitously, at the FRC. The detailed causes of the particular shape for this curve are too complex to be given here but include such things as the changing amount of radial change relative to length change that occurs with each increment of volume; the reapportionment of major sites of pressure drop among arterial, capillary, and venous beds; and the interaction of the (nonlinear) tissue length-tension relation with that characterizing the vessel wall. The effects of volume change are not small; a collapsed lung may require four times the pressure required at FRC to bring about the same blood flow. This becomes very important when lungs collapse pathologically and also in the way blood is distributed to the fetus, whose lungs are not yet expanded.

Up to this point I have carefully avoided defining pulmonary vascular resistance as pressure gradient ÷ flow ($R = \Delta P/\dot{Q}$). While such a number can tell something about the work of the right ventricle, it is for several reasons a bad way of assessing what the pulmonary vessels are doing. Part of the difficulty lies in defining the pressure gradient. As has been seen, ΔP depends on *either* P_{alv} *or* P_V, and since these are significantly different in intact animals, a correct pressure gradient cannot be selected unless all the perfused lung is in either zone II or zone III, but not some in each. To nearly achieve this, human subjects are studied in the supine position. In that instance, all of the lung can be considered to be in zone III because left atrial pressure is >0 at mid-chest level.

When the relationship of arterial pressure to flow is obtained for a lung in the zone III state while lung volume and other pressures are held constant, a curvilinear record is obtained, as shown in Figure 19-11. Since the curve obtained is not a

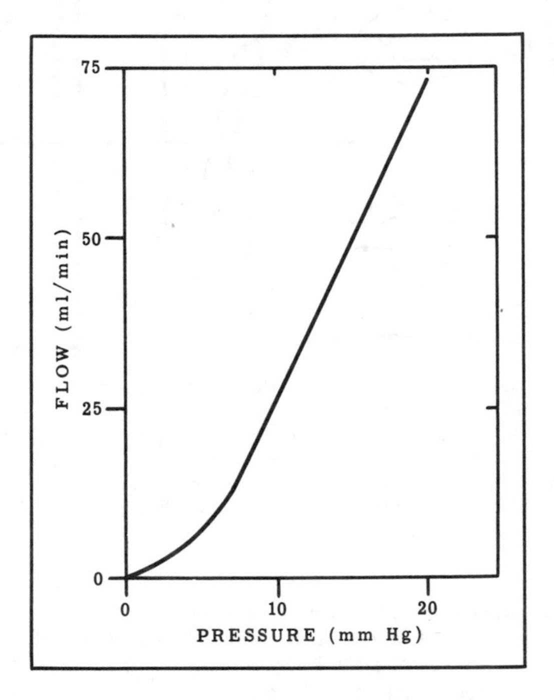

FIGURE 19-11. Perfusion pressure versus flow obtained with an excised dog lung lobe under zone III conditions using a newtonian fluid as perfusate.

solution set of the equation $P = R\dot{Q}$, a value for pulmonary vascular resistance determined as $P/\dot{Q}$ cannot be used to predict the flow consequent to any pressure gradient except the one at which R was first obtained. The P-$\dot{Q}$ curve deviates from linearity because as arterial pressure is raised it not only causes more flow but also widens vessels and opens channels previously closed. This change in net vascular bed cross-sectional area causes the increase in flow for each increase of pressure to be greater than expected. Widening and recruitment continue only up to a point. Beyond some pressure, all vessels are open, and further dilation is limited by the substantial decrease in wall compliance that occurs when previously slack fibrous tissue begins to be stretched. (In fact, pulmonary vessels are so distensible at low pressures that most vessels which are open at a normal zone III pressure are probably quite close to the greatest diameter that could be achieved even by a several-times increase in transmural pressure.) Once the vessels are fully open and distended, the P-$\dot{Q}$ relation becomes linear. Since vascular dimensions appear to remain constant above that pressure, it would be convenient for any calculated R to be constant also. In addition to having no predictive value, the P/$\dot{Q}$ quotient does not become constant when the P-$\dot{Q}$ curve becomes linear. For this reason some physiologists have defined resistance as the slope of the P-$\dot{Q}$ curve. The choice depends on the use to be made of it. Typically one wants to know whether something done experimentally or therapeutically has altered vessel size. Rather than use either of the above definitions for R, it is better to describe the pulmonary circulation with a P-$\dot{Q}$ curve having at least several points, or by collecting all data when flow is held constant and pressure is the only dependent variable (or vice versa).

Failure to recognize the limitations inherent in assigning a vascular resistance on the basis of the P/$\dot{Q}$ quotient, particularly when blood flow and ventilation both change, has for many years caused profound disagreements about the responses of the pulmonary circulation. Even after recognition of the difficulties it is usually very hard to obtain data unquestionably undisturbed by secondary influences. With this in mind, the presentation of factors altering vascular resistance can continue, but resistance will have to have an intuitive definition.

The part of the pulmonary vascular bed which has the largest share of the resistance is still unresolved. This dilemma arises, to some extent, because the major resistance may vary depending on extravascular conditions. Thus, in zone II the major site is probably the capillaries, but in zone III it is probably the small arteries. The veins are likely always to make up less of the resistance than either of the other beds. A distribution of 45 percent arterial bed, 35 percent capillary bed, and 20 percent venous bed could be considered typical.

The pulmonary circulation differs importantly from the systemic circulation in the amount of vascular smooth muscle present. Small pulmonary arteries, in the range of 100 μ to 1000 μ in diameter, are the vessels most likely to be capable of contracting effectively. The arterioles have too little muscle.

It is known that a number of things can cause vasoconstriction prominent enough to raise pulmonary arterial pressure under conditions in which cardiac output remains constant. Usually the same factors that constrict the systemic circulation will also cause pulmonary vasoconstriction. These vasopressor stimuli include epinephrine and norepinephrine, angiotensin, and increased sympathetic nervous activity. In some pathological conditions several other vasoactive materials may be the causes of pulmonary vasoconstriction. Serotonin, histamine, and various polypeptides may be released either in the general circulation or locally within discrete parts of the pulmonary circulation, and all these bring about vasoconstriction. The absolute changes in pressure caused by maximal pulmonary vasoconstriction under conditions of constant cardiac output are typically in the range of 10 to 30 mm Hg. While these changes may not seem great, the percentage changes in pressure are of about the same magnitude as are caused by intense vasoconstriction in the systemic circulation.

The effects of pulmonary vasoconstriction should be considered from the standpoint of whether vasoconstriction has been generalized or whether it has occurred only in some regions. Generalized vasoconstriction increases right ventricular work load. It also tends to make the distribution of pulmonary blood flow more even. The explanation for the latter is that when the arterial resistance everywhere increases, it becomes the largest component of the total resistance. In that event regional differences among arteriolar, capillary, and venous resistances are of less importance because they are then only small fractions of the total resistance in any region.

Regional vasoconstriction will divert blood flow to other parts of the lungs, and if the region of vasoconstriction is not too large (say, less than half the total circulation), there will be little or no increase in main pulmonary arterial pressure because the diverted blood opens additional paths. Regional vasoconstriction makes distribution uneven and typically does not result in increased right ventricular work.

One of the best-known stimuli for regional pulmonary vasoconstriction is alveolar hypoxia – that is, a lower than normal O_2 concentration in alveolar gas. In other organs, regional hypoxia causes vasodilation, but in the lungs vasoconstriction occurs. Its exact cause is unknown, but it is not a reflex, nor is it brought about by the usual neuromediators. It is known that the important factor is the O_2 concentration in the alveolar gas, not the amount of O_2 present in either pulmonary arterial or venous blood. Since a common cause of regional alveolar hypoxia is poor gas exchange resulting from regional impairment of airflow, local vasoconstriction consequent to alveolar hypoxia tends to divert pulmonary blood flow to regions which are best exchanging gas with the outside air. In this way the vascular response to hypoxia tends to enhance the overall efficiency of pulmonary gas exchange. The vasopressor effect of hypoxia is potentiated by acidemia. This potentiation also acts to aid in the diversion of blood to better-ventilated regions, because, as will be discussed in Chapter 20, poor ventilation results not only in low alveolar O_2 concentrations but also in high alveolar CO_2 concentrations. CO_2, acting as an acid anhydride, lowers regional pH, which in turn not only potentiates hypoxic vasoconstriction but may in itself cause vascular contraction.

If alveolar hypoxia is generalized, or severe enough to cause hypoxemia in the systemic arteries, part of the ensuing pulmonary vasoconstriction will be the result of generalized sympathetic nervous hyperactivity consequent to initiation of chemoreflexes by stimulation of carotid and aortic receptors. Other reflex effects in the pulmonary vessels are much less well established.

The importance of pulmonary vasomotion in adult animals, whether local or general, has not been conclusively decided. Probably the effects of vasomotion, like the passive effects already described, are of little hemodynamic consequence outside the lung. In the lung, however, all these factors act to determine regional gas exchange.

NEONATAL CARDIORESPIRATORY ADAPTATION

The ability of the fetus to convert from placental gas exchange to air breathing is a sine qua non of life. During gestation, respiratory exchange is achieved by circulating fetal systemic blood through the capillary bed immersed in the lakes of maternal blood of the placental sinusoids. The placenta does not provide nearly as complete an exchange or as much functional reserve as will the lungs, and this marginal ability makes any placental maldevelopment or derangement a potentially serious threat to fetal survival. "Fetal distress" from O_2 lack is a common obstetrical complication often requiring premature delivery in order to establish adequate fetal gas exchange.

In the fetus, oxygenated blood from the placenta returns to the heart through the inferior vena cava. Instead of mixing completely with superior caval blood, about two-thirds of inferior caval blood enters the left atrium through the foramen ovale. Any systemic venous blood which does enter the thick-walled fetal right ventricle is mostly ejected into the systemic circulation through the ductus arteriosus, and only about 20 percent of right ventricular output passes through the lungs. Pulmonary arterial pressure in utero slightly exceeds pressure in the aorta. The low pulmonary flow present at such high pressures is the result of a very high pulmonary vascular resistance, which is partly due to the small volume of the lung and partly due to constriction of a vascular smooth muscle layer several times thicker than that in adult lung.

In the latter part of pregnancy fetal lungs are not collapsed but are somewhat filled with fluid continuously elaborated within the lungs. This is not amniotic fluid, and there seems to be little admixture of amniotic and pulmonary fluids even though fetal respiratory movements sometimes occur. The high fluid viscosity (compared with air) militates against respiratory exchange.

In the last quarter of human gestation, pulmonary surfactant is normally present. This is a critical milestone to have passed, for without surfactant, stable air-filled lungs would be unattainable, and survival of infants born before the appearance of surfactant is unusual.

The conversion to pulmonary gas exchange at birth involves establishment of a gas-filled lung and dissociation of the systemic and pulmonary circulations. Compression of the thorax during delivery expels much of the fluid from conducting airways. The first inspiratory efforts suck the remaining fluid and outside air into alveoli and create the air-liquid interface. The alveolar fluid is apparently absorbed within the next few minutes. Considerable effort is necessary to bring about the first inflation, but once inflation has occurred, the pleural pressure changes required for breathing fall to essentially adult levels if surfactant is present. Most of the lung becomes gas inflated during the first few inspirations. This enlargement with air does two things: First, it mechanically lowers pulmonary vascular resistance; second, the increased alveolar O_2 removes a potent local stimulus for pulmonary vasoconstriction. The profound fall in pulmonary vascular resistance diverts the right ventricular output into the lungs, accompanied at least temporarily by some blood from the aorta which now traverses the ductus arteriosus in the opposite direction. The large pulmonary blood flow raises left atrial pressure above that in the right atrium and this closes the flaplike valve at the foramen ovale. Closure is aided by a fall in pressure on the right side of this valve consequent to termination of umbilical venous return. With the elimination of right-to-left shunts and appearance of pulmonary O_2 exchange, systemic arterial blood becomes well saturated with O_2. The ductus constricts sufficiently in the presence of the high O_2 content of its blood to prevent flow and complete the separation of the two circulations. Closure of the ductus does not occur instantaneously but may require several hours. Ductus constriction is apparently mediated entirely in response to the local O_2 level, though it may be aided by circulating epinephrine. The marked sensitivity to O_2 is unlike that of either pulmonary arteries or aorta, and its mechanism remains unexplained.

Over the next few weeks further changes take place. These include fibrotic obliteration of the ductus arteriosus and decreases in muscle thickness in the right ventricle and pulmonary vessels. Until the latter occurs, pulmonary arterial pressures are relatively high and responsiveness of this bed is greater than that of the adult. In fact, if neonates fail to breathe enough to oxygenate alveoli and systemic arterial blood, pulmonary vasoconstriction and ductus arteriosus dilation will occur and may reduce pulmonary perfusion while restoring the fetal right-to-left shunt. Occasionally the ductus fails to close, or atrial septal defects develop such that interatrial shunts remain. These generally divert blood from left to right and so increase pulmonary blood flow. In those instances regression of the pulmonary vascular muscle often does not occur, and persistence of the fetal type of vessels effects the ultimate physiological characteristics of children with congenital heart disease.

REFERENCES

Campbell, E. J. M. *The Respiratory Muscles and the Mechanics of Breathing.* Chicago: Year Book, 1958.

Dawes, G. S. *Foetal and Neonatal Physiology.* Chicago: Year Book, 1968.
Fenn, W. O., and H. Rahn (Eds.). *Handbook of Physiology.* Section 3: Respiration, Vol. I. Washington, D. C.: American Physiological Society, 1963.
Mead, J. Mechanical properties of lungs. *Physiol. Rev.* 41:281–330, 1961.
Pattle, R. E. Surface lining of lung alveoli. *Physiol. Rev.* 45:48–79, 1965.
West, J. B. *Ventilation/Blood Flow and Gas Exchange.* Oxford, Eng.: Blackwell, 1965.

20

Respiratory Gas Exchange and Transport

Thomas C. Lloyd, Jr.

GAS EXCHANGE in the body occurs by bulk flow of gases, by bulk flow of solutions of gases, and by diffusion of gases through tissues. The preceding chapter discussed some of the determinants of bulk flow but left the story of gas exchange far from complete.

To fully understand gas exchange processes in both lung and systemic tissues it is necessary next to examine some of the physical properties of gases and of solutions of gases in blood. This information can then be used to explore the process of diffusion, by which the internal exchanges occur. With that material in hand, it is possible to amalgamate it with what has been learned about airway and vascular mechanics from Chapter 19 to form a coherent story of pulmonary function. The symbols used to designate the various aspects of pulmonary physiology are listed in Table 20-1.

GASES, AND GASES IN SOLUTION

The composition of a gas mixture can be described by the percentage of each constituent, but composition gradients do not fully determine whether a specified component of a gas mixture will move by diffusion from one place to another. To quantify the tendency for movement, it is necessary to know the total gas pressure, as well as the percentage of the gas represented by any particular molecular type. These two variables are used to calculate an effective pressure for each component of the gas mixture, called the *partial pressure* of that component. The partial pressure (often called the *tension*) of any gas in a gas mixture is the pressure which that gas would exert if *it alone* were present. It is found by multiplying the total gas pressure by the fraction of the composition represented by the gas in question. For example, the partial pressure of O_2 (written P_{O_2}) in air at 1 atmosphere is 0.21×760 mm Hg = 159.6 mm Hg. Gas partial pressure is equivalent to voltage in electrical systems. Diffusive gas flow, like flow of electricity, occurs from points of high tension to points of lower tension.

As in meteorology, partial pressures are usually measured in millimeters of mercury. The unit of 1 millimeter of mercury is called a *torr*, in honor of Evangelista Torricelli, the seventeenth-century physicist who invented the barometer. Pressures may also be reported as fractions of a standard atmospheric pressure where 1 atmosphere = 760 torr. To help keep track of the various pressures in the

TABLE 20-1. Symbols Used in Pulmonary Physiology

FOR GASES

Primary Symbols	Secondary Symbols
V = gas volume	I = inspired gas
$\dot{V}$ = gas volume/unit time	E = expired gas
P = gas pressure	A = alveolar gas
$\bar{P}$ = mean gas pressure	T = tidal gas
f = respiratory frequency (breaths/unit time)	D = dead space gas
D = diffusing capacity	B = barometric
F = fractional concentration in dry gas	STPD = 0°C, 760 mm Hg, dry
R = respiratory exchange ratio	BTPS = body temperature and pressure saturated with water vapor
	ATPS = ambient temperature and pressure, saturated

FOR BLOOD

Primary Symbols	Secondary Symbols
Q = volume of blood	a = arterial blood
$\dot{Q}$ = volume flow of blood/unit time	c = capillary blood
S = % saturation of Hb with O_2	v = venous blood

Examples:

P_{O_2} – partial pressure of oxygen

$P_{I_{O_2}}$ – partial pressure of oxygen in inspired air

$F_{I_{O_2}}$ – fractional concentration of oxygen in inspired air (dry)

lung, *torr* will be used here for gas partial pressures, but *mm Hg* or *cm H_2O* will be used to quantitate vascular pressures and such pressures as P_{el}, P_{res}, or P_{pl}.

The concentration of a gas in a mixture may be given as the percentage or as the *mole fraction*, where percentage = 100 × mole fraction. The sum of the mole fractions of all the constituent gases of a mixture must equal 1, or, alternatively, the sum of all the partial pressures must equal the total gas pressure.

The concept of partial pressure also applies to gases dissolved in liquids, but gas partial pressure is not given simply by the product of hydraulic pressure times the amount of gas dissolved per unit volume. Rather, the partial pressure of a gas in a liquid is equal to that partial pressure in the gas phase *above* the liquid which would be found *under equilibrium conditions.* For instance, in a glass of water in equilibrium with room air at sea level, the P_{O_2} in both the air and the water is 159.6 torr. Even at the bottom of a lake where the hydraulic pressure might be equivalent to 2 or 3 atmospheres, the P_{O_2} would be 159.6 torr.

The *amount* of gas dissolved in a liquid at a given temperature is directly related to the gas tension by a coefficient of solubility peculiar to each gas-liquid combination. (Amount dissolved = constant × partial pressure.) Therefore, knowing the solubility coefficient for the system at hand, one can calculate the partial pressure in a solution by first determining the gas concentration in it. The partial pressure in the solution may be greater or less than the tension of that gas in the gas phase above the solution; this only means that an equilibrium does *not* exist, and more gas must either dissolve or leave solution in accordance with the partial pressure gradient. Temperature affects solubility in such a way that cooling increases the amount dissolvable.

One of the gases present above any solution is the vapor of the solvent itself. The *vapor pressure* of the solvent is determined by its own molecular properties and by temperature, but not by the local barometric pressure. At equilibrium, the partial pressure of solvent in the gas phase above a liquid is equal to the vapor pressure of the solvent. The vapor pressure of water at 37°C is 47 torr. Gas spaces like thoracic airways and alveoli which are fully equilibrated as far as water movement is concerned (i.e., have 100 percent relative humidity) have, therefore, a P_{H_2O} of 47 torr. Since this pressure must appear in the total gas pressure summation, the addition of water vapor to cooler and drier inspired air dilutes the inspired O_2 and N_2 somewhat. The water vapor pressure in ambient air depends on surface water temperature and also on the relative humidity. The relative humidity indicates the extent to which the liquid and air phases are in equilibrium. At a comfortable 68°F and 50 percent relative humidity, the P_{H_2O} is about 8 torr, but at an unpleasant 90°F and 100 percent humidity, the P_{H_2O} is 36 torr.

The solubility of O_2 in water and in biological fluids other than whole blood is low relative to the amount of O_2 contained in a similar volume of air. For instance, about 0.4 ml O_2 is dissolved in 100 ml of plasma equilibrated with room air, but 100 ml of air contains 50 times that quantity of O_2. Clearly, such liquids would be poor media to transport O_2 to tissues, and all animals more advanced than aquatic forms only a few cells thick have developed compounds which markedly increase the O_2-carrying capacity of their gas transport fluids. It is interesting to note that the type and amount of the specialized material (hemoglobin) in mammals multiplies the O_2-carrying capacity by a factor of about 50, i.e., the content ratio between air and water. The ways in which respiratory gases are carried in solution will next be presented in detail.

OXYGEN TRANSPORT IN BLOOD

When oxygen diffuses into the plasma from the alveolus, almost all of it finds its way into the red blood cell, where it combines with hemoglobin. Only a small portion of the oxygen remains in the plasma and is carried to the tissues in simple solution. Since the solubility coefficient for O_2 in aqueous solutions such as plasma is 2.44 ml O_2 per 100 ml plasma per atmosphere (760 torr) O_2 pressure, only 0.3 ml of oxygen will be dissolved in 100 ml of plasma with the usual P_{O_2} of 95 torr in arterial blood.

The erythrocyte carries almost all the oxygen (98.5 percent) in the blood stream by virtue of the capacity of hemoglobin to carry O_2. Hemoglobin is a protein with a molecular weight of 68,000, each molecule having the capacity to combine with 4 molecules of oxygen. When fully saturated, 1 gm of hemoglobin will hold 1.34 ml of oxygen. Since the normal hemoglobin content of the blood is 15 gm per 100 ml, the oxygen-carrying capacity is approximately 20 ml per 100 ml of blood. This is also expressed as 20 volumes percent.

Unlike the linear relationship between partial pressure and content of gases in true solution, the relationship of P_{O_2} to O_2 content is not a linear one (Fig. 20-1),

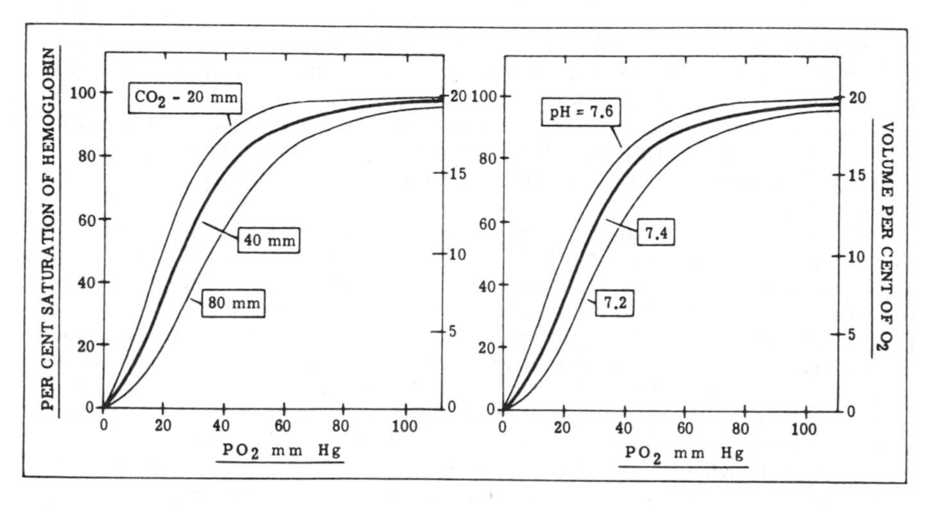

FIGURE 20-1. The normal oxyhemoglobin dissociation curve. Influences of partial pressure of CO_2 and pH on the curve.

a fact which has important physiological consequences. Thus there is no single constant usable to calculate blood O_2 content from the P_{O_2}. One must always use curves like Figure 20-1 to obtain such information. Close familiarity with this graph is necessary for an understanding of many aspects of gas exchange. The part of the O_2-hemoglobin dissociation curve of greatest importance is the "shoulder" in the P_{O_2} range of 40 to 100 torr. At the top of this shoulder, large changes in P_{O_2} have a small effect on the amount of oxygen carried by the hemoglobin; below that level, small changes in P_{O_2} have a large effect on the oxygen content. It should be noted that the "shoulder" covers the normal range of oxygen tension in the blood under resting conditions. At the upper end of the curve, a decrease in alveolar O_2 tension does not immediately jeopardize the adequacy of the arterial O_2 content. On the other hand, the steep part of the curve enables a lot of O_2 to be delivered from blood to tissues without causing a profound fall in P_{O_2}. This tends to keep the driving force which moves O_2 outward relatively constant along the whole length of a systemic capillary. Conversely, in the pulmonary capillaries a large gradient for O_2 movement into blood is maintained until most of the O_2 has been transferred.

The amount of oxygen combined with hemoglobin also depends upon other factors, such as the P_{CO_2} of the blood (Fig. 20-1). As the carbon dioxide tension increases, the oxygen and the hemoglobin tend to dissociate. This is an aid to normal function since, in the region of the systemic capillary, the P_{CO_2} is elevated.

The elevated carbon dioxide tension helps the blood to unload the oxygen as it passes through the capillary. In the lung, however, the loss of CO_2 by ventilation increases the affinity of hemoglobin for O_2.

The O_2-hemoglobin dissociation curve also depends upon the pH of the blood (Fig. 20-1). An increase in acidity tends to drive oxygen off the hemoglobin molecule. The reason for this will be considered later, along with the effect of carbon dioxide on the hemoglobin dissociation curve. This effect is also of physiological value since the pH in the systemic capillary region is lower than that of arterial blood. This also releases oxygen from hemoglobin to diffuse into the tissues.

The temperature of the blood has a significant influence on oxygen carriage, an elevated temperature causing dissociation of oxygen from hemoglobin. This effect may be of value in the muscles, where exercise increases temperature.

OXYGEN EXTRACTION IN SYSTEMIC CAPILLARIES

Under ordinary conditions the arterial blood P_{O_2} is 95 torr and the oxygen content is 20 volumes per 100 ml of blood. When the arterial blood reaches the capillaries, the oxygen in simple solution in the plasma begins to diffuse into the tissue because of the lower P_{O_2} there (< 30 torr). This in turn causes some of the oxygen combined with the hemoglobin to dissociate from it and diffuse first into the plasma and then to the tissues. By the time the blood leaves the capillaries the P_{O_2} has fallen to 40 torr and the oxygen content to 14 volumes per 100 ml.

The amount of oxygen which leaves the blood depends upon local needs, and conditions in the tissue being perfused. As metabolism increases, the increased O_2 utilization is not matched by an equal increase in capillary blood flow. The amount of oxygen extracted from each milliliter of blood will therefore increase, and it may be expressed in terms of the *oxygen utilization coefficient,* which is the ratio $\frac{\text{arteriovenous } O_2 \text{ difference}}{\text{arterial } O_2 \text{ content}}$. Normally this value is $\frac{20 - 14}{20} = 0.3$. In exercise this coefficient may approach 1.0 for the blood flowing through the active muscle and 0.7 to 0.8 for the mixed blood entering the right heart. At the same time, blood flow through the exercising muscle increases greatly, increasing the oxygen available to the tissues in this manner as well.

CARBON DIOXIDE TRANSPORT IN BLOOD

Carbon dioxide is carried by the blood in three forms: (1) as bicarbonate (HCO_3^-), (2) in combination with protein (carbamino), and (3) in simple solution. Among the three forms there is a total of approximately 48 volumes of CO_2 per 100 ml in arterial blood, which is divided as follows: bicarbonate, 43 volumes; carbamino hemoglobin, 2 volumes; carbamino plasma protein, 1 volume; and dissolved carbon dioxide, 2.4 volumes. The CO_2 in normal human arterial blood exerts a partial pressure of 40 torr. The relation between partial pressure and total CO_2 content for arterial blood is shown in Figure 20-2. As P_{CO_2} increases, the total CO_2 content rises in a fashion which is nonlinear. The shape of the curve favors greater CO_2 exchange at lower O_2 tensions.

The total CO_2-blood dissociation curve is made up of several separate curves representing the three forms of CO_2 carriage, as shown in Figure 20-2. The amount of CO_2 in simple solution is a linear function of the CO_2 tension. A small amount of this gas is actually in the form of carbonic acid (H_2CO_3).

The amount carried in the carbamino form (combined with hemoglobin and plasma protein) does not bear a linear relation to CO_2 tension (Fig. 20-2). Also,

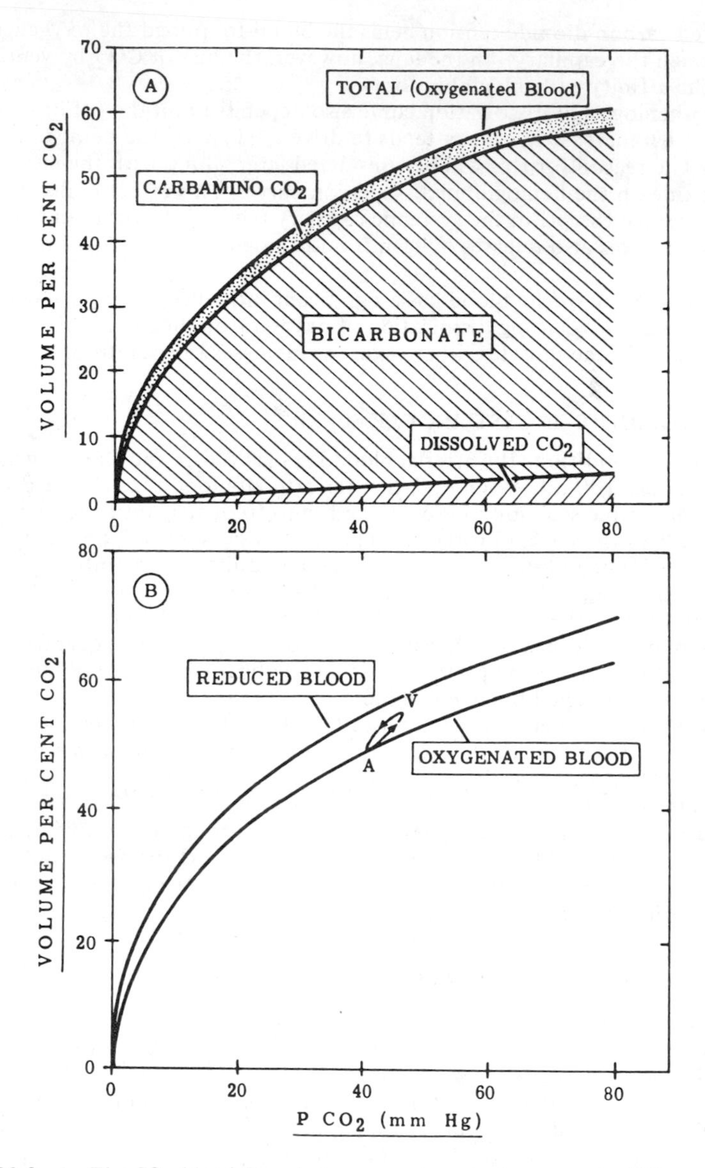

FIGURE 20-2. A. The CO_2-blood dissociation curve showing the amount of CO_2 present in each of the three forms. B. The dissociation curve of oxygenated and reduced blood. The loop indicates the changes which occur when arterial blood (A) becomes venous blood (V) and when it is again arterialized.

the O_2 partial pressure plays an important role here, since saturation of hemoglobin with oxygen drives off carbon dioxide from the molecule. In fact, if it were not for the reduction of the amount of oxyhemoglobin as the blood passes through the capillary, much less CO_2 would be taken up in the carbamino form.

The remainder of the CO_2 is carried in the form of bicarbonate. A solution of bicarbonate will take up appreciable amounts of CO_2 only at CO_2 tensions below 10 torr. However, in the company of other substances, such as hemoglobin, its behavior changes and it readily takes up CO_2 at higher partial pressures. Thus the curve for bicarbonate (Fig. 20-2) is characteristic only for bicarbonate in the blood. The amount of CO_2 which finds its way into the bicarbonate form is dependent upon the O_2 content of the hemoglobin also, as will be seen shortly. The difference in CO_2 dissociation curves of arterial and venous blood reflects the effect of oxygen on bicarbonate and carbamino formation.

The CO_2 tension in the tissues is over 50 torr, while in the arterial blood it is 40 torr, so carbon dioxide will diffuse into the plasma and red cells as the blood passes through the systemic capillaries. As noted earlier, the uptake is aided by the simultaneous outward movement of oxygen from the blood. When the blood leaves the capillary, its CO_2 content has increased from 48 volumes per 100 ml to 54 volumes per 100 ml and the P_{CO_2} from 40 to 46 torr. Since the oxyhemoglobin level has decreased at the same time, the pathway shifts to a new dissociation curve, as shown in Figure 20-2.

The CO_2 entering the blood in the capillary goes into all the three forms, with most (65 percent) becoming bicarbonate, 27 percent forming carbamino compounds, and 8 percent going into solution. The mechanisms by which these reactions occur are outlined in Figure 20-3.

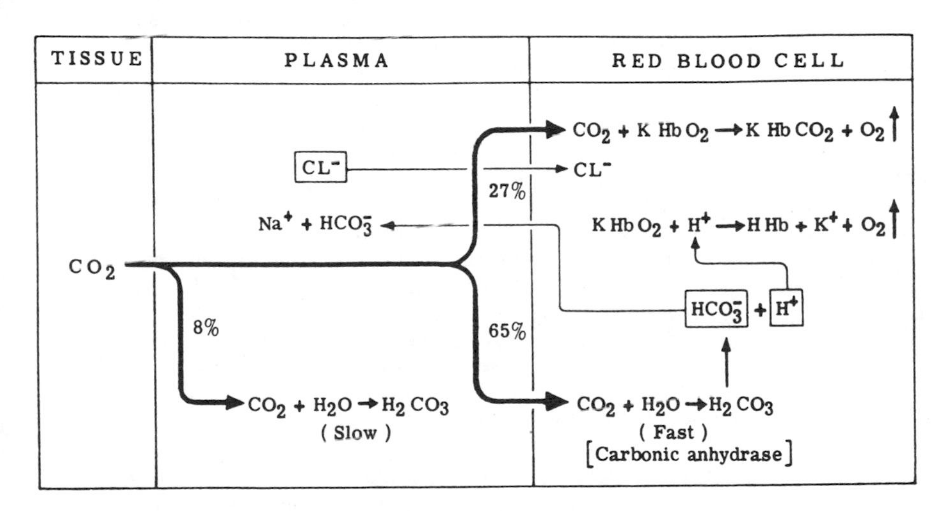

FIGURE 20-3. Chemical reactions involved in the three principal forms of CO_2 carriage by the blood. A small amount (8 percent) is carried as dissolved CO_2 by the plasma and red blood cell. A larger amount (27 percent) is carried by formation of carbamino compounds (primarily carbamino hemoglobin). The largest fraction (65 percent) is carried by formation of bicarbonate, which occurs within the erythrocyte. The CO_2 entering the cell is hydrolyzed to carbonic acid, which is neutralized by hemoglobin, with the bicarbonate ion diffusing out into the plasma. Simultaneously chloride moves into the cell to preserve electrical neutrality.

The CO_2 which goes into simple solution in the plasma is slowly converted to carbonic acid, although at equilibrium only a small amount is in acid form. Most of the CO_2 goes into the erythrocyte, where rapid hydration occurs through the catalyzing action of carbonic anhydrase. As the acid is formed, it dissociates to form bicarbonate and hydrogen ions. The bicarbonate concentration rapidly increases in the erythrocyte, and most of it diffuses out into the plasma. To preserve electrical neutrality, chloride ions diffuse into the cell (this is called the *chloride shift*). The hydrogen ions freed by the dissociation of carbonic acid combine with hemoglobin, freeing potassium, which balances the chloride that has diffused into the cell. Another portion of the CO_2 entering the cell combines with the imidazole group of histidine on the globin portion of the hemoglobin molecule. Although this site is different from that of the oxyhemoglobin combination, O_2 and CO_2 do not readily coexist on the hemoglobin molecule. It should be noted that hemoglobin plays an essential role in the bicarbonate and carbamino formation and is therefore indirectly responsible for most of the carriage of CO_2.

In the lung, these processes are reversed; as CO_2 diffuses out of the plasma, bicarbonate is drawn back into the red cell, where it is rapidly broken down to CO_2 through the action of carbonic anhydrase and again released to the plasma to diffuse into the alveoli. At the same time, carbamino CO_2 leaves the hemoglobin molecule, and O_2 joins it.

It might be supposed that movement of CO_2 in and out of the blood would entail considerable alteration in the blood pH. This would be true were it not for the fact that oxyhemoglobin is more acid than reduced hemoglobin. As the CO_2 enters the blood at the capillary and decreases pH, the hemoglobin at the same time shifts in the alkaline direction with the loss of O_2. The loss of one mole of O_2 causes a loss of hydrogen ion concentration of 0.7 mole. Thus, 0.7 mole of CO_2 could be added to the blood for every mole of O_2 lost without change in pH. The change in pH then will depend upon the relative contents of O_2 and CO_2 exchanged at the capillary. Release of acids from the tissues into the blood stream will also influence blood pH at this point.

The ratio of carbon dioxide produced to oxygen consumed in the tissues under resting, steady-state conditions is termed the *respiratory quotient* (RQ). The ratio varies from 0.7 to 1.0, depending upon diet, and is 0.82 in a fasted human. The change in acidity of the blood as it discharges O_2 and picks up CO_2 in the capillaries will depend upon this ratio: If it is 0.7, there will be no pH change; if 1.0, there will be a considerable change. Under ordinary conditions RQ is 0.82, arterial pH is 7.40, and venous pH is 7.36. It will be shown subsequently (Chap. 24) that these reactions also play an important role in the acid-base balance of the body.

OXYGEN AND CARBON DIOXIDE RECIPROCITY

The reciprocal effects of oxygen and carbon dioxide on gaseous transport seen in Figures 20-1 and 20-2 may be explained on the basis of the chemical reactions described above. For example, increasing the CO_2 level of the blood will increase the hydrogen ion concentration in the red cell and drive O_2 off the hemoglobin molecule. In addition, O_2 and CO_2 compete for hemoglobin directly as oxyhemoglobin and carbamino hemoglobin. The effect of decreased pH on oxygen carriage may also be readily deduced from the chemical reactions which occur in the blood.

Reciprocally, when the P_{O_2} of the blood is increased, CO_2 will be displaced from the hemoglobin and oxyhemoglobin will be formed. The more acidic oxyhemoglobin will provide hydrogen ions to combine with bicarbonate ions, forming carbonic acid, which is then dehydrated to free CO_2. As a result, at any fixed P_{CO_2} level the total amount of CO_2 carried by the blood will decrease as the P_{O_2} is increased.

DIFFUSION

GENERAL CONCEPTS

Diffusion is a process by which a difference in concentration (more correctly, a difference in chemical activity) of a substance between two points is evened out by a migration of molecules toward the region of lower activity. Several factors determine the time required for equilibration of activity to occur. Transport by diffusion is an important part of gas exchange, and certain aspects of the diffusion process must be detailed before diffusion as a physiological phenomenon can be discussed. Many of the concepts have already been presented in Chapter 1.

The determinants of transport by diffusion are intuitively obvious. Thus, the rate of movement of gas molecules through a single phase (either liquid or gas) depends on the differences in partial pressure of the gas between any two points and on the distance separating the points. (For gases, partial pressure gradients are equivalent to activity gradients, whereas concentration gradients are not.) In other words, diffusion depends on the partial pressure gradient per unit length of the diffusion path.

Since diffusion is caused by thermal motion of molecules, it will occur at a greater rate at higher temperatures, and at a greater rate for smaller molecules, since thermal agitation depends on molecular weight. The relation between rate of gas movement by diffusion and pressure gradient, path length, and temperature is linear, and can be expressed as

$$\dot{V} \propto \frac{\Delta PT}{L}$$

where $\dot{V}$ is volume rate of gas flow, ΔP is partial pressure gradient, T is absolute temperature, and L is path length.

The relation to molecular weight is not linear, but is to thc square root of the molecular weight because the energy of motion of a molecule at a particular temperature is given by kMc^2, where k is a constant, M is molecular weight, and c is the mean velocity of the molecule as it bounces about in its environment. Since diffusion rate depends on this velocity,

$$\dot{V} \propto c \propto \frac{1}{\sqrt{M}}$$

If one imagines diffusion to be occurring across a specified area perpendicular to the direction of movement, as in diffusion through a piece of cellophane separating two chambers, the net transport obviously depends on the area across which it can occur. Combining all the above factors,

$$\dot{V} \propto \frac{\Delta PTA}{L\sqrt{M}}$$

where the new term, A, represents the cross-sectional area.

At the alveolar interface, diffusion occurs from gas to liquid, and the above expression does not fully describe the situation. An additional term must be included for the solubility of the gas in the liquid. As might be expected, diffusivity is directly related to solubility. In the case of gas to liquid diffusion, then:

$$\dot{V} \propto \frac{\Delta PTA\alpha}{L\sqrt{M}}$$

where the new term, a, is solubility. For gases of physiological interest, a lies between 1.0 and 0.01; therefore diffusion rates into liquids are less than diffusion rates in the gas phase alone.

In physiological systems some of the above factors are more important than others. Thus, for the three gases O_2, N_2, and CO_2, differences in molecular weight make very little difference in relative rates of diffusion. Similarly, temperature is essentially constant since even during severe febrile responses the difference in absolute temperature is only about 1 part in 100. On the other hand, solubility and available capillary area play prominent roles in diffusion into blood, making this many times slower than diffusion in the gas phase. In consequence, diffusion from alveolar duct to alveolar interface is an insignificant impediment to gas transport even though the distance of migration may be 10^3 greater than the thickness of the pulmonary membrane. The major impediment is through the tissue phase.

In the lungs one cannot quantitate surface area or path length during life, so a *diffusion coefficient* for lung lacks the meaning of a rigorously derived physical constant. The closest one can approach a diffusion constant for lung is to obtain a *transport factor* or *diffusing capacity* by measuring the rate of uptake or removal of a particular gas and the average partial pressure gradient existing between alveolus and capillary blood which caused that exchange. The diffusing capacity of the lung, D_L, is defined by the relation $D_L = \dot{V}/\Delta P$, and the term D_L contains all the unmeasurable terms and physical constants noted in the proportionality expression given earlier. The dimensions of D_L are milliliters per minute per torr. Note that $1/D_L$ has the dimensions of resistance. After discarding the negligible term for diffusion within the gas phase, the total diffusion resistance to O_2 or CO_2 transfer in the lung is the sum of two major components: (1) diffusion through alveolar and capillary membranes and (2) diffusion within blood, across the red cell membrane, and uptake by hemoglobin. It has been estimated that for O_2, each of these two components of D_L accounts for about half the total, so the rate of uptake of O_2 by blood cannot be considered to be infinitely rapid.

CO_2 is about 20 times as soluble (hence as diffusible) as O_2; therefore it is doubtful that impairment of CO_2 diffusion would ever be critical to survival, since O_2 transfer would be so profoundly altered at that point that life would be unlikely. Actually, even the uptake of O_2 is limited by diffusion only under very unusual circumstances, as will be seen farther on.

DIFFUSIVE TRANSPORT AT THE PULMONARY CAPILLARY

Now that the principles of diffusion have been explained, the process of diffusion gas transfer as a function of transit time of blood through the pulmonary capillary will be examined. The normal adult human lung capillaries contain only about 90 ml of blood, but this is spread across a surface of about 70 square meters. The thickness of the liquid barrier between red cell and alveolar gas is on the order of 1 μ. While this would seem to provide more than enough surface for diffusion, the critical feature is that with a normal resting cardiac output each red cell spends only about 0.75 second in a pulmonary capillary.

Whether an equilibrium is established between gas in the alveolus and the capillary blood depends on the diffusing capacity, but it also depends on the amount of gas which needs to be transferred to reach equilibrium. A gas like N_2, which is less diffusible than O_2, comes into equilibrium very quickly in a unit of blood entering the capillary because it is so (relatively) insoluble that not much must cross the border to establish the same P_{N_2} on both sides. On the other hand, since O_2 and CO_2 have great solubilities in blood, many molecules must be transferred to achieve equality of tensions. It turns out that by the time blood reaches mid-capillary, the P_{O_2} of blood and the P_{O_2} of the alveolus are essentially identical. CO_2, because of its greater diffusivity, reaches equilibrium in a fraction of the time required for O_2, even though about the same number of molecules of CO_2 are transferred. It therefore follows that the blood flow rate could double and O_2 would still reach equilibrium, but at any higher flow with other things unchanged, insufficient time for complete O_2 equilibration may limit oxygenation while not influencing CO_2 removal.

If all pulmonary capillaries were open while the individual was at rest, there could be only a doubling of cardiac output if the condition of reaching equilibrium is to be met. However, the recruitment of additional parallel capillary paths caused by increasing the right ventricular output means that instead of capillary flow velocity increasing pari passu with cardiac output, the velocity through any *single* capillary increases only a small amount until all of the available bed is open. Note that the opening of more parallel pathways not only keeps the increase in capillary flow velocity to a minimum but increases the surface area available for diffusion. This property of the pulmonary circulation helps in allowing O_2 uptake to reach about 15 times the resting value without the appearance of a disequilibrium between capillary blood and alveolar gas. Under these conditions, capillary flow velocity will be approximately twice the resting rate, while the diffusing capacity increases about threefold.

Under such high O_2 demands as mentioned above, the blood entering the pulmonary capillary has less O_2 than when at rest, and therefore a lower P_{O_2}. This increase in partial pressure gradient will increase the rate of diffusion in the first part of the capillary. Refer to the hemoglobin dissociation curve (Fig. 20-1) and note that until the "shoulder" around 80 percent saturation is reached, addition of O_2 raises O_2 content more than it raises P_{O_2}. A large partial pressure gradient can be maintained until most of the O_2 has been transferred. The increase in the mean P_{O_2} gradient, as well as the increases in capillary area and flow velocity, is apparently what allows the adequate oxygenation of severe exertion.

O_2 uptake can be limited by diffusion if the capillary bed is reduced by disease or by surgical removal of lung. Diffusion limitations may also follow decreases in the P_{O_2} gradient caused by any mechanism which lowers alveolar P_{O_2}. Less often, diffusivity is compromised by thickening of the alveolar membrane by disease. As one would predict, diffusion impairment influences O_2 uptake during exercise long before interfering with oxygenation during rest.

PULMONARY GAS EXCHANGE

RESPIRATORY EXCHANGE RATIO

In most mammals the alveolar gas composition at 1 atmosphere ambient pressure and a body temperature of 37°C is P_{H_2O} – 47 torr, P_{CO_2} – 40 torr, P_{O_2} – 100 torr, and P_{N_2} – 570 torr. These particular values occur as a consequence of the balance between simultaneous CO_2 delivery and O_2 removal by blood, and CO_2 removal and O_2 delivery by breathing. N_2 is neither used nor produced in the body, so it is a passive filler. The alveolar P_{N_2} can be used as a mirror of the relation between O_2 uptake and CO_2 removal in the following way. If O_2 uptake (in milliliters per minute) is in excess of CO_2 outflow, alveolar N_2 will be more concentrated than is N_2 in air. Conversely, if O_2 uptake is less than the rate of CO_2 removal from pulmonary capillary blood, N_2 will be diluted by the excess of CO_2. Normally the number of moles of O_2 consumed is slightly greater than the amount of CO_2 produced, so the sum of P_{CO_2} and P_{O_2} is slightly less than the P_{O_2} of humidified room air at the same temperature. Since total alveolar pressure equals barometric pressure (the variations consequent to airflow being negligible), the P_{N_2} of average alveolar gas normally exceeds that of the wet warm air by about 10 torr. The relation between CO_2 output and O_2 uptake is expressed by the *respiratory exchange ratio,* abbreviated R. At a steady state, R for the lungs taken as a whole equals the respiratory quotient, the latter being a statement of the metabolic state of the whole animal. For short periods, such as with breath holding or panting, RQ and R may be unequal. R may be appraised either by measurements of the CO_2 and O_2 volumes exchanged or by comparison of alveolar P_{N_2} with P_{N_2} of ambient air. When $R < 1$, alveolar $P_{N_2} >$ ambient P_{N_2}; when $R > 1$, the reverse pertains.

When $R < 1$, exhaled volume is proportionally less than inhaled volume because the volume of O_2 removed from tidal air exceeds the volume of CO_2 added. The converse occurs when $R > 1$.

It can be shown that over wide ranges the rate of removal of CO_2 from a group of alveoli is most dependent on ventilation and is less sensitive to variations of pulmonary blood flow, whereas O_2 uptake is dependent on blood flow and is little altered by variations in ventilation. This allows R or alveolar P_{N_2} to be used as approximate measurers of the ventilation to blood flow ratio (called the $\dot{V}/\dot{Q}$ ratio), since they are measurers of the ratio of exchanged CO_2 to exchanged O_2. While so far R, P_{N_2}, and $\dot{V}/\dot{Q}$ have been used as expressions for *overall* gas exchange, it should be evident that these concepts could also be used to characterize *local* gas exchange in different parts of the lung. The material in Chapter 19 indicated that regional differences in both ventilation and perfusion must normally occur by a number of different mechanisms. Since it is possible to consider these differences, whatever their cause, in terms of R or the $\dot{V}/\dot{Q}$ ratio, this is a useful application that will be discussed at several points in the remainder of this chapter.

CO_2 EXCHANGE

CO_2 can be removed by breathing only because the act of inspiration continually dilutes alveolar gas, lowers its P_{CO_2}, and thereby establishes a diffusion gradient for movement from the capillary blood. Exhalations can then expel CO_2-laden gas. The balance is such that the alveolar P_{CO_2} at end expiration is normally about 41 torr, and at end inspiration about 38 torr, while the P_{CO_2} of blood entering the capillary has a P_{CO_2} of about 46 torr. If respiratory minute volume is decreased, while CO_2 production remains constant, the decrease in CO_2 removal will eventually raise the P_{CO_2} everywhere in the system as CO_2 “backs up”. For reasonably large decreases in minute volume, a new equilibrium can be established because the higher concentration of CO_2 in alveolar air will allow more CO_2 to be removed per unit volume of alveolar gas expelled. Thus, if effective ventilation were halved, a new equilibrium could be achieved by doubling the alveolar P_{CO_2}, since then each half-normal-sized breath would expel the normal number of moles of CO_2. The price paid is an increase in CO_2 stores. This readjustment is often seen in persons with chronic lung disease, but, as will be seen in Chapter 21, to accomplish it means that readjustments in the respiratory control mechanisms must also occur, since these act to hold blood P_{CO_2} quite constant.

If all of each tidal inspiration were equilibrated with capillary blood, the concentration of CO_2 in expired gas would be the same as in alveolar gas. However, part of the inspirate stops within conducting airways far removed from alveoli, and this volume takes no part in gas exchange. The volume of these airways in a normal adult man is between 100 and 200 ml. At end expiration, conducting airways are filled with gas pushed upward from below which has the composition of mixed alveolar gas. On inspiration, this volume is drawn downward, but because it already has alveolar gas composition, it does not alter alveolar gas concentrations. Only the volume inhaled which *exceeds* the volume of the conducting airways becomes a diluent for alveolar gas. The gas in conducting airways is simply shuttled back and forth. The internal volume of these airways is called the *anatomical dead space. Dead space ventilation* refers to the movement of gas in and out of some part of the respiratory system in which no actual transmembrane exchange occurs. The presence of a dead space minimizes the effectiveness of the tidal volume.

At the start of expiration, the anatomical dead space contains room air, and this volume is expelled before gas of alveolar composition appears. The composition of expired gas is, therefore, determined by the relative contributions of the alveolar and dead space compartments. Since the number of moles of CO_2 leaving the alveoli must equal the number collected in the expirate, one could calculate the sizes of the dead space and the effective alveolar exchanging volume if he knew the alveolar and mixed-expirate CO_2 concentrations, and the expired volume. A sample calculation brings out these features.

If all the expired air collected from an individual for one minute had a volume of 6.9 liters and a CO_2 content of 4 percent, the subject has expelled 275 ml CO_2 per minute (4% × 6.9 L/min). If his alveolar CO_2 content is determined to be 5.5 percent, the exhaled volume of CO_2 would be contained in 5 liters of alveolar gas (0.275 L/0.055). In this case the subject has had an *alveolar minute ventilation* of 5 liters. The difference between the actual volume exhaled (6.9 liters) and 5 liters expresses dead space ventilation. Since this 1.9 liters was distributed in equal parts to each breath, the dead space volume would be 147 ml if his respiratory rate had been 13 breaths per minute.

In the preceding example it was assumed that all alveoli had a CO_2 content of 5.5 percent. Assume for a moment that only half the alveoli receive any pulmonary blood flow. The remaining half will contain no CO_2 because of the absence of perfusion. If the perfused half has an alveolar CO_2 concentration of 5.5 percent, the average alveolar concentration for the lungs as a whole would be only 2.75 percent and the mixed expirate would be even less than this because of the anatomical dead space gas. In reality there is no way to sample alveolar gas directly. What is done is to use the P_{CO_2} of systemic arterial blood as the average alveolar value. In normal individuals the error is probably less than 2 torr. However, when alveoli receive no perfusion, those alveoli will not be represented in the systemic arterial blood. This means that the alveolar P_{CO_2} assumed by using the P_{CO_2} of arterial blood is weighted in the direction of the alveoli receiving the most blood flow. If alveoli are not perfused, they are physiologically just as dead (from the standpoint of gas exchange) as are conducting airways. When dead space is calculated by assuming the average alveolar P_{CO_2} to be equal to systemic arterial P_{CO_2}, the space so calculated is called the physiological dead space. Carefully note that because of the assumptions for its calculation, the physiological dead space is a *virtual* space and not a real one. Normally, the physiological dead space and anatomical dead space are essentially equal, but in diseases which occlude parts of the lung circulation physiological dead space exceeds anatomical dead space. The air entering and leaving unperfused alveoli represents work done fruitlessly.

So far it has been supposed that there are only two populations of alveoli: those containing CO_2 at a single concentration and those containing no CO_2 at all. But in truth the CO_2 contents of alveoli differ widely in different parts of the lung. As far as any one lobule is concerned, its alveolar P_{CO_2} is determined not by either ventilation or blood flow alone but by the ratio of one to the other (the $\dot{V}/\dot{Q}$ ratio). This is not in contradiction to the earlier statement that over reasonably wide limits the *volumes* of O_2 and CO_2 exchanged per minute were largely dependent on only one factor.

To illustrate: Imagine two situations where $\dot{V}/\dot{Q}$ is doubled but where the gas composition of pulmonary arterial blood remains constant. In the first, assume perfusion to remain normal while ventilation is doubled. This will approximately double the amount of CO_2 removed from blood (and the alveolar space), and since blood flow is normal, the end-capillary P_{CO_2} and mean P_{CO_2} within those alveoli will approximately halve. On the other hand, had $\dot{V}/\dot{Q}$ doubled from a halving of blood flow, the CO_2 lost *per unit of blood flow* would again double, because CO_2 loss rate remains nearly unchanged in the absence of a change in ventilation. In both cases, however, the alveolar P_{CO_2} would have decreased upon doubling the $\dot{V}/\dot{Q}$ (in these instances from about 39 to 28 torr) since in both cases the end-capillary blood P_{CO_2} was halved. The same phenomenon with regard to CO_2 loss rate and P_{CO_2} will be found if $\dot{V}/\dot{Q}$ is halved by changing either flow or ventilation. Thus it can be seen that, although alteration of $\dot{Q}$ has little effect on CO_2 loss rate, it significantly influences alveolar and end-capillary P_{CO_2}. O_2 is nearly immune to the vagaries of ventilation because blood O_2 content is rather insensitive to variations of P_{O_2} above about 50 torr (see Fig. 20-1). Hence, O_2 uptake depends predominantly on the rate of delivery of carrier, i.e., on perfusion, and is not too affected by variation in P_{O_2} caused by changes in ventilation. This would become less and less the case as alveolar P_{O_2} fell below 50 torr, but that represents a rather severe perturbation.

Alveolar P_{CO_2} could conceivably vary between the limits of 0 (no blood flow) and that of mixed systemic venous blood (no airflow). In Chapter 19 it was mentioned that differences in both ventilation and perfusion occur in the lung as a function of weights of blood and lung tissue. Though in general the bases of the lungs receive both more ventilation and more perfusion than the apices, the relative proportions are such that the $\dot{V}/\dot{Q}$ ratio is about 3.5 at the apices and 0.65 at the bases. The equivalent alveolar CO_2 tensions in these regions are 27 and 41 torr, respectively. Intermediate values occur at intermediate locations. The relations of $\dot{V}/\dot{Q}$ to alveolar gas tensions and R are shown in Figure 20-4. These are not simple linear relationships because they include combined effects of the hemoglobin dissociation curve, CO_2 blood solubility curve, and (for the gas tensions) influences of R. Note that the range of $\dot{V}/\dot{Q}$ in a normal human lung covers the steepest section of the $P_{CO_2}-\dot{V}/\dot{Q}$ curve. In other words, the normal range of $\dot{V}/\dot{Q}$ is the same range that most influences gas tensions. The overall effective $\dot{V}/\dot{Q}$ is about 0.8.

In view of the foregoing, how can it ever be said that there is *an* alveolar P_{CO_2}? This is a difficult problem in pulmonary physiology, for one must usually be content to use an average number and assume that the lungs consist of only one alveolus having a single airway and capillary. The average alveolar P_{CO_2} is most often assumed to be that of mixed pulmonary capillary blood. Systemic arterial blood is considered to have the same P_{CO_2} because there is very little CO_2 added to pulmonary capillary blood (a bit of bronchial and myocardial venous drainage) before it arrives in the systemic arteries. The amounts of CO_2 added by the small quantity of bronchial and cardiac venous blood is too small to be detected. Normally, systemic arterial P_{CO_2} is about 40 torr. Note that this is not the average of the extremes of alveolar P_{CO_2} known to be present in a normal lung; the alveoli having the lowest P_{CO_2} contribute the least blood, so their influence is minimized. This effect also makes the average $\dot{V}/\dot{Q}$ different from the mean of the extreme values.

Another measure of mean alveolar P_{CO_2} can also be made by assuming it to be equal to P_{CO_2} of the exhaled air which appears after the anatomical dead space has been flushed. This would not yield a value equal to systemic arterial P_{CO_2} if there were a large regional variation in $\dot{V}/\dot{Q}$ ratio, but in normal individuals the two are about equal. Expired air measurements have the advantage of not requiring needle puncture of an artery and thus may be more useful in some situations.

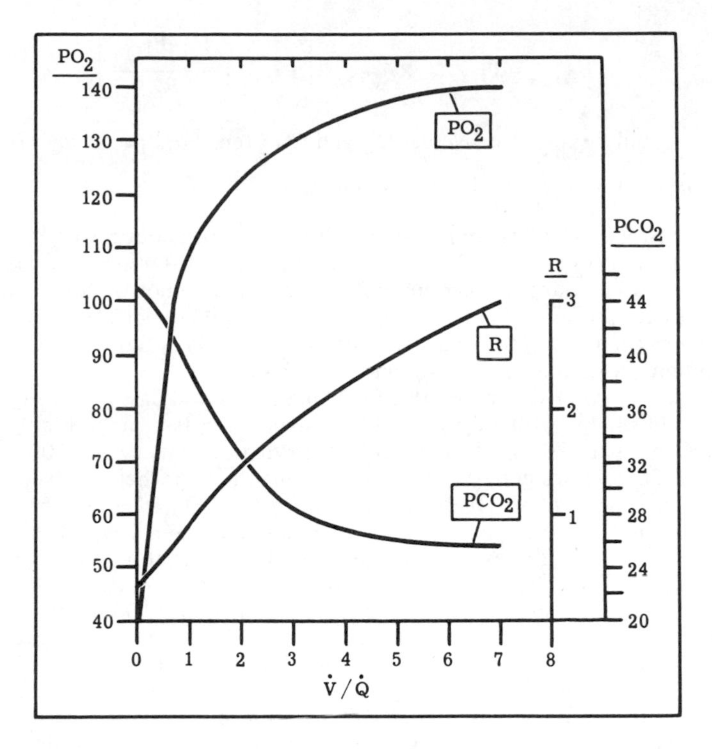

FIGURE 20-4. The relation of alveolar P_{O_2}, P_{CO_2}, and R to $\dot{V}/\dot{Q}$ ratio when inspired air has a normal sea-level composition and mixed systemic venous blood has a P_{O_2} of 40 torr and P_{CO_2} of 45 torr. Partial pressures are in torr. (From data of West, 1965. P. 110.)

O_2 EXCHANGE

It should be apparent that alveolar P_{O_2} also must vary throughout the lung as a function of $\dot{V}/\dot{Q}$, and this is borne out by the data shown in Figure 20-4. Once again, the steepest part of the P_{O_2}–$\dot{V}/\dot{Q}$ curve is that part contained by the normal range of $\dot{V}/\dot{Q}$ ratio. Note that the effect of $\dot{V}/\dot{Q}$ on P_{O_2}, in absolute terms, is much greater than the effect on P_{CO_2}. The range of P_{O_2} is from near 132 torr in the apices to about 90 torr at the bases. (See from Figure 20-1 that there will be little difference in O_2 content of blood coming from these two regions.)

For reasons soon to be presented, arterial P_{O_2} is a poor measure of mixed alveolar P_{O_2}. Because of this fact, and the questionable applicability of expired air analysis, the average alveolar P_{O_2} is usually calculated from the systemic arterial P_{CO_2} and R. This *alveolar air equation* is easily derived, but the derivation will be omitted here. One can intuitively understand it by remembering that if $R = 1$, the sum of P_{CO_2} and P_{O_2} must equal the P_{O_2} of humidified body-temperature room air. The equation simply does this sum after correcting for the concentration or diluting effects when $R \neq 1$. The calculation of alveolar P_{O_2} is given by

$$P_{A_{O_2}} = P_{I_{O_2}} - P_{A_{CO_2}} \left[F_{I_{O_2}} + \frac{1 - F_{I_{O_2}}}{R} \right]$$

Where $P_{A_{O_2}}$ and $P_{A_{CO_2}}$ are alveolar O_2 and CO_2 tension, $P_{I_{O_2}}$ is P_{O_2} of inspired air, and $F_{I_{O_2}}$ is the fraction of O_2 in ambient air.

In contrast with CO_2, there is usually a difference of about 5 to 10 torr between systemic arterial P_{O_2} and calculated alveolar P_{O_2}. The difference is caused by the addition of small amounts of incompletely oxygenated hemoglobin to pulmonary capillary blood. Because of the shape of the hemoglobin dissociation curve, addition of partially oxygenated blood causes a profound fall in P_{O_2} but only a small change in O_2 content. A sample calculation will bear this out.

Assume a P_{O_2} of 100 torr to exist in all alveoli. Blood passing through those capillaries will emerge with a P_{O_2} of 100 torr, and with its hemoglobin 95 percent saturated (see Figure 20-1). If the total O_2-carrying capacity, when 100 percent saturated, is 20 volumes per 100 ml, blood leaving the hypothetical alveoli will contain 19 volumes per 100 ml. If it is next assumed that 10 percent of the cardiac output will bypass the lungs and reenter the pulmonary veins beyond the capillaries, this "shunt" flow will add less-well-oxygenated blood to that coming from pulmonary capillaries. If the shunt blood has a normal systemic venous P_{O_2} of 40 torr, it is 75 percent saturated and therefore contains 15 ml O_2 per 100 ml blood. Since it was initially assumed that the relative flows were 10:1, each liter of capillary blood will be mixed with 100 ml of shunt blood, resulting in a total volume of 1100 ml which contains 205 ml of O_2 (190 from capillary blood, 15 from shunt flow). This represents 18.7 ml O_2 per 100 ml blood, or 90 percent saturation. Reference again to the dissociation curve shows that 90 percent saturation corresponds to a P_{O_2} of 63 torr. Thus, the addition of the 10 percent shunt of systemic venous blood has reduced the P_{O_2} of blood entering the left atrium from 100 to 63 torr, but has changed O_2 content by only 0.3 volume per 100 ml. This effect is particularly apparent if alveolar P_{O_2} is above the "shoulder" in the dissociation curve at about 50 torr. Addition of unsaturated hemoglobin has much less effect on blood P_{O_2} if alveolar P_{O_2} is lower than 50 torr, because over that part of the dissociation curve, changes in O_2 content have less effect on P_{O_2}.

EFFECTS OF REGIONAL VARIATION IN $\dot{V}/\dot{Q}$

While there is normally about a 2 percent shunt of venous blood into pulmonary veins, it is not necessary that unoxygenated hemoglobin arrive only in this way. It is more likely, particularly in disease states, for blood of less than average alveolar P_{O_2}, yet more than systemic venous P_{O_2}, to come from areas of lung where there is a low $\dot{V}/\dot{Q}$ ratio. Note from Figure 20-4 that if $\dot{V}/\dot{Q} < 1$, prominent falls in alveolar P_{O_2} will occur. These changes in $\dot{V}/\dot{Q}$ may result in less than 100 percent saturation of end-capillary blood from such regions. In these cases one can always define a *virtual* lung where all pulmonary end-capillary blood has the average alveolar P_{O_2} calculated by the *alveolar air equation* and where all the gradient between that P_{O_2} and the P_{O_2} of systemic arterial blood is ascribed to a *virtual* "shunt" of mixed systemic venous blood. This "physiological shunt" is akin to "physiological dead space," where all the difference between arterial blood and expired gas P_{CO_2} was attributed to the effects of a completely nonexchanging space. It is important to remember that one is calculating a *virtual* shunt or dead space, and unless real anatomical equivalents are known to exist, one should presume that inequalities in $\dot{V}/\dot{Q}$ ratio are really determining the partial pressure gradients. That is, decreases in regional perfusion or ventilation are more likely to occur than are complete absences of one or the other, though it is often convenient to express these abnormalities in terms equivalent to complete absences.

A true shunt is essentially a region where $\dot{V}/\dot{Q} = 0$: Anatomical dead space is a region where $\dot{V}/\dot{Q}$ = infinity. If some part of the lung has a $\dot{V}/\dot{Q}$ greater than normal, its effect approaches that of a dead space – it increases the P_{CO_2} difference between arterial blood and expired gas but has little effect on the P_{O_2} difference. Similarly, a region of low $\dot{V}/\dot{Q}$ approaches in its effect that of an anatomical shunt: It increases P_{O_2} differences between systemic arterial blood and calculated alveolar P_{O_2}; but because the difference between pulmonary arterial and pulmonary venous P_{CO_2} is so small, such shunts cause essentially no alveolar-arterial P_{CO_2} gradients.

The concept of regional $\dot{V}/\dot{Q}$ differences and their effects on systemic arterial blood as being a continuum of effects ranging from the result of a shunt to that of a dead space is a very important one. The normal lung is not a physiologically homogeneous structure. It seems to function like one as far as blood gas contents are concerned because *in the normal ranges of* $\dot{V}/\dot{Q}$, blood gas *contents* (not tensions) are rather insensitive to gas tension variations within alveoli. However, pulmonary pathology is especially prone to profoundly distort $\dot{V}/\dot{Q}$ ratios in an uneven way throughout the lungs, and the commonest cause of systemic hypoxia of pulmonary origin is as a result of alterations in $\dot{V}/\dot{Q}$.

It might seem that there could be an optimum $\dot{V}/\dot{Q}$ ratio for gas transport, and the supposition has been borne out by analysis. Gas exchange, in terms of airflow and blood flow requirements, occurs most efficiently at a $\dot{V}/\dot{Q}$ slightly greater than 1.0, which is quite close to the typical ratio of normal lungs taken as a whole. Beyond the limits for $\dot{V}/\dot{Q}$ of 0.4 and 4.0, gas exchange becomes much less efficiently accomplished.

As alluded to in Chapter 19, active variations of airway and vascular smooth muscle in response to local P_{O_2} and P_{CO_2} tend to restore toward normal any marked deviations of $\dot{V}/\dot{Q}$ away from unity. A fall in alveolar P_{O_2} below about 50 torr, such as would be produced by a decrease in regional ventilation, causes local pulmonary vasoconstriction which directs flow away from that region and so reduces perfusion to bring the $\dot{V}/\dot{Q}$ toward normal. In underperfused regions, such as would be caused by vascular changes, the relative hyperventilation leads to a low regional P_{CO_2} which brings about local bronchoconstriction. This increase in local airway resistance (and incidentally a fall in compliance from constriction of alveolar duct muscle) directs airflow elsewhere and returns the $\dot{V}/\dot{Q}$ toward normal. Because of the shapes of the O_2 and CO_2 dissociation curves, regulation of $\dot{V}/\dot{Q}$ is probably more often important for determining systemic arterial gas *tensions* and the efficiency of exchange than for determining arterial blood gas *contents*. However, when the range of $\dot{V}/\dot{Q}$ to be adjusted is expanded to include values found in disease conditions, these regulatory mechanisms significantly decrease systemic hypoxemia and the size of the ventilation wasted in dead space.

EFFECTS OF ABNORMAL CONCENTRATIONS OF RESPIRATORY GASES

Alterations in contents and tensions of the respiratory gases arise in a number of ways. Changes of pulmonary origin, or related to variations in the ambient environment, are always manifested by changes in both peripheral tissues and systemic arterial blood. This is not true of disturbances such as tissue hypoxia resulting from a fall in peripheral blood flow or from ingestion of agents which interfere with tissue oxidative metabolism, like potassium cyanide or dinitrophenol. In those cases, unless respiration also is changed, systemic arterial blood will have normal O_2 and CO_2 contents.

Unlike hypoxia, peripheral hypercapnia is almost always of pulmonary origin, and therefore systemic arterial blood and alveolar P_{CO_2} will be above normal. By far the most important cause of hypercapnia is generalized hypoventilation, i.e., a low effective alveolar ventilation in relation to metabolic demands. This reduction in the overall $\dot{V}/\dot{Q}$ ratio will always cause both hypercapnia and hypoxemia. Carbon dioxide tensions as high as 90 torr are not infrequently seen in people who continue to survive. The associated P_{O_2} may be in the forties. Were such disturbances to develop very suddenly, collapse or death would usually result, but when they come about gradually, some tolerance and adaptive changes seem to occur.

Hypoxia of systemic arterial blood in the presence of a normal or reduced P_{CO_2} usually is the result of disease which disturbs regional $\dot{V}/\dot{Q}$ ratios, but it also can appear from abnormal right-to-left shunts and it is seen at high altitudes in normal subjects. In the case of altitude, the cause is clear. If shunts are the cause, administration of 100 percent O_2 will not cause much increase in arterial blood O_2 content, since all the blood exposed to alveoli is already nearly fully saturated. If subjects who have regional hypoventilation as the cause of hypoxemia breathe 100 percent O_2, the hypoxemia is usually corrected; the poorly ventilated alveoli ventilate enough to wash out their N_2 and replace it with O_2 so that the P_{O_2} becomes much higher than when N_2 was present. Consequently, even underventilated units, which still retain their somewhat elevated P_{CO_2}, will now contribute blood with 100 percent O_2 saturation to the systemic circulation.

Rarely, a decreased diffusing capacity causes systemic hypoxemia. This is usually seen only upon exertion, since that puts a greater demand on diffusibility, as noted earlier. In such cases, breathing 100 percent O_2 restores the arterial O_2 content to normal because the gradient for O_2 transport is increased more than tenfold. Unfortunately, most of the conditions which increase the diffusion membrane thickness or decrease the capillary area also alter airways and other blood vessels, so that it is quite difficult to separate the effects of the altered $\dot{V}/\dot{Q}$ ratio from the decreased diffusivity. By and large, the $\dot{V}/\dot{Q}$ disturbance predominates in these diseases.

HYPOXIA

The effects of hypoxia are insidious. Acute lowering of alveolar P_{O_2} leads progressively to giddiness, lack of judgment, stimulation of respiration and cardiac and vasomotor activity, and finally unconsciousness and death. The first stages are a real threat to aviators and those scuba divers who use rebreathing-type units, for errors in judgment under those conditions can be fatal. Another acute cause of hypoxia is transit to high altitudes on the earth's surface. An acute "altitude sickness" consisting of sleeplessness, nausea, and headache in mild cases, or cerebral and pulmonary edema and retinal hemorrhages in severe cases, is not uncommonly seen when mountaineers quickly ascend to high-altitude base camps. Climbers often progress in stages to allow time for altitude adaptation when making ascents of the highest peaks. Prolonging the ascent time to several days or weeks almost always prevents symptoms, whereas abrupt transport usually causes trouble for new arrivals.

Residents at high altitudes, whether lowlanders who have remained for a time or natives, compensate in part for the decreased O_2 availability by an increase in red cell mass. Natives of Morococha, Peru (a town at an altitude of 15,000 feet) typically have hematocrit readings of 60 percent. Man and most domestic animals seem to develop pulmonary vasoconstriction of varying degree at high altitude.

In Morococha Indians, pulmonary arterial pressures were about two to three times higher than were found for the same flows at sea level. The effect of this is to even out perfusion distribution in a vertical lung, which may increase D_L and improve $\dot{V}/\dot{Q}$ distributions to cause more efficient gas exchange. High-altitude natives and adapted visitors have a decreased ventilatory response to O_2 lack. A high incidence of patent ductus arteriosus has been reported to exist among high-altitude natives, which lends credence to the theory that the ductus normally closes because of the large increase in blood P_{O_2} that occurs at birth.

HYPEROXIA

In recent years it has become clear that one can be exposed to too much O_2. O_2 toxicity is a function of P_{O_2} and time of exposure. Above 3500 torr, symptoms of central nervous system toxicity develop in man within a few minutes. At these high pressures, muscular twitching progresses rapidly to fully developed convulsions. Even at 1 atmosphere (100 percent O_2 breathing), within 24 hours there are signs of tissue damage in the lung, where exposure to the highest P_{O_2} is occurring. The changes include edema and hemorrhages of the airway mucosa, and decrease of surfactant, and they have been the cause of complications in patients treated for long periods with 100 percent O_2. It is to be hoped that such treatment is no longer done. Inspired O_2 at 70 percent concentration or less can be tolerated indefinitely, and 70 percent O_2 is adequate to overcome hypoxemia of any conditions that could be significantly aided by 100 percent O_2. The chemical changes of O_2 toxicity are incompletely understood, but interference with enzyme sulfhydryl groups has been consistently reported.

A further complication at pressures above 2100 torr is that the high P_{O_2} enables plasma to carry sufficient O_2 in solution to satisfy resting metabolic needs without recourse to hemoglobin transport. Because hemoglobin does not lose its O_2 in passage through tissue, CO_2 transport is hindered and tissue CO_2 retention occurs. The attendant acidosis may play a role in the toxic changes.

HYPERCAPNIA AND HYPOCAPNIA

Increased alveolar (and thus systemic arterial) P_{CO_2} most commonly occurs as a result of generalized hypoventilation. More insidious exposures are due to defective CO_2 absorbent in closed external breathing systems like scuba units, anesthesia machines, submarines, and space capsules. Acute exposure to very high ambient P_{CO_2} is a threat to explorers who venture into wet limestone caves – acid ground-water seepage may have liberated CO_2 into a confined space. Industrial exposures also occur among dry-ice workers, miners, and firemen.

The most prominent effect of acute hypercapnia is stimulation of respiration. If hypercapnia results from generalized hypoventilation, this tends to initiate a control mechanism which limits the severity of the underventilation (to be considered further in Chap. 21). In otherwise normal man, respiratory stimulation by CO_2 continues to increase for each increase in P_{CO_2} as far as it has been tested (about 30 percent CO_2 in inspired gas). Respiratory stimulation is probably caused by not one but several effects, both in the central nervous system and from peripheral chemoreflexes. The predominant mechanism depends on the CO_2 level. Inhalation of CO_2 concentrations above about 15 percent causes generalized convulsions and unconsciousness. Direct effects of CO_2 on vascular smooth muscle tend to produce dilation, but overwhelming effects from central nervous system and reflex stimulation result in hypertension from CO_2 excess, rather than the opposite.

The causes of responses to hypercapnia are not fully understood. Molecular

CO_2 itself is apparently responsible for some effects in several ways. In addition, CO_2 hydration leads to an acidosis which elicits some of the response and may become the life-limiting factor.

Hypocapnia occurs as a result of increased ventilation in excess of CO_2 production. This may be the result of simple anxiety or of stimulation by reflexes (including those from O_2 lack). The predominant effects are an increase in blood and tissue alkalinity and a decrease in cerebral blood flow, the latter arising as the typical direct vascular response to P_{CO_2} variation. These changes manifest themselves in dizziness, visual blurring, and neuromuscular hyperexcitability.

NITROGEN HAZARDS

These are problems peculiar to high ambient pressures, or the abrupt change from high to low ambient pressure. Only two aspects will be described, and those briefly.

First, N_2 is soluble enough so that at high ambient pressure ($>$ 5 atmospheres) it acts as a general anesthetic. The change is not abrupt but progressively follows elevation of pressure. This effect is characteristic of many substances; indeed, it is what allows general inhalation anesthesia at $<$ 1 atmosphere. Narcosis by helium or hydrogen requires greater pressures than is required for N_2, so those gases are used for the O_2 diluent in deep diving, particularly for dives of long duration where there is more time for saturation of body tissues with the inert gas to occur.

The second way in which N_2 becomes important is in decompression. When divers return from depth, or when astronauts ascend to orbit, the N_2 which has been dissolved at the higher partial pressure must be lost. If ascent is done rapidly, the sudden lowering of pressure causes bubble formation in body fluids, which in turn gives rise to several symptoms collectively called "the bends." Small vascular bubbles of N_2 occlude flow to parts of the central nervous system, particularly the spinal cord, and may cause permanent disability unless recompression is quickly accomplished to redissolve them. Bubbles in joint fluid cause local pain, and those in venous blood entering the pulmonary circulation give rise to labored breathing, a condition called "the chokes." These problems are obviated by allowing only gradual return from depth so that N_2 loss can be accomplished without supersaturation and bubble formation. The rate of ascent of astronauts cannot be slowed down economically, so they are depleted of N_2 prior to lift-off by breathing 100 percent O_2 for a short period. This technique could not be used in divers, because O_2 toxicity would supervene before much N_2 was eliminated.

REFERENCES

Dejours, P. *Respiration.* New York: Oxford University Press, 1966.

Fenn, W. O., and H. Rahn (Eds.). *Handbook of Physiology.* Section 3: Respiration, Vol. I. Washington, D. C.: American Physiological Society, 1964.

Forster, R. E. Exchange of gases between alveolar air and pulmonary capillary blood: Pulmonary diffusing capacity. *Physiol. Rev.* 37:391–452, 1957.

Rahn, H. A concept of mean alveolar air and the ventilation-bloodflow relationship during pulmonary gas exchange. *Amer. J. Physiol.* 158:21–30, 1949.

Riley, R. L., and A. Cournand. "Ideal" alveolar air and the analysis of ventilation-perfusion relationships in the lungs. *J. Appl. Physiol.* 1:825–847, 1949.

West, J. B. *Ventilation/Blood Flow and Gas Exchange.* Oxford, Eng.: Blackwell, 1965.

21

Nervous and Chemical Control of Respiration

Paul C. Johnson

THE RATE and depth of breathing are under the immediate control of the respiratory centers of the central nervous system. These centers integrate information from mechanical receptors and chemoreceptors and provide an output proportional to the summated demands for ventilation as seen by the sensors. The respiratory system therefore constitutes a classic type of feedback network in which the system is driven to maintain the tissue environment at the desired level or set point of the control network. The sensory input to the control centers includes such diverse data as the degree of lung inflation, body temperature, P_{O_2}, pH, and P_{CO_2}. The respiratory centers are also influenced by higher centers, which exert voluntary control over the system behavior.

In addition to integrating this information, the respiratory center generates a rhythmic discharge pattern which produces the inspiratory and expiratory cycles. The origin of this periodicity is a matter of considerable interest in neurophysiology.

THE RESPIRATORY CENTER

The neurons which control respiratory activity are located in the brain stem. The importance of this region was first shown by Legallois (1812), who experimentally ablated portions of the brain stem and abolished respiratory movements. Further evidence for the primacy of the brain stem was provided by the observation that decerebration did not abolish respiratory movements. More precise localization of the respiratory neurons has proved to be a formidable problem. Pitts and co-workers (1939) stimulated various regions of the medulla electrically to localize the area of respiratory control. They found that inspiratory responses were elicited by stimulation of the ventral reticular formation in a region overlying the cephalic portion of the inferior olive. Expiratory responses were also induced by stimulation of the ventral reticular formation, but they were limited to an area slightly cephalad to the "inspiratory center" and cupped over it. While the experiments suggest the existence of separate inspiratory and expiratory centers, there are no morphologically identifiable nuclei corresponding to these areas. More important is the fact that such stimulation would excite nerve fibers rather than cell bodies. The experiments therefore indicate the existence of a preponderance of fibers having expiratory connections in a particular area. This may be regarded as indicative but not conclusive evidence for a preponderance of expiratory neurons in the same area.

More refined studies in which single-unit neuronal activity was recorded suggest that there is a considerable degree of overlap and intermixing of the two neuronal types. Salmoiraghi and Burns (1960), for example, found an intermixing of inspiratory and expiratory neurons throughout the medullary respiratory center. The extent of the center as defined by their technique is shown in Figure 21-1. The

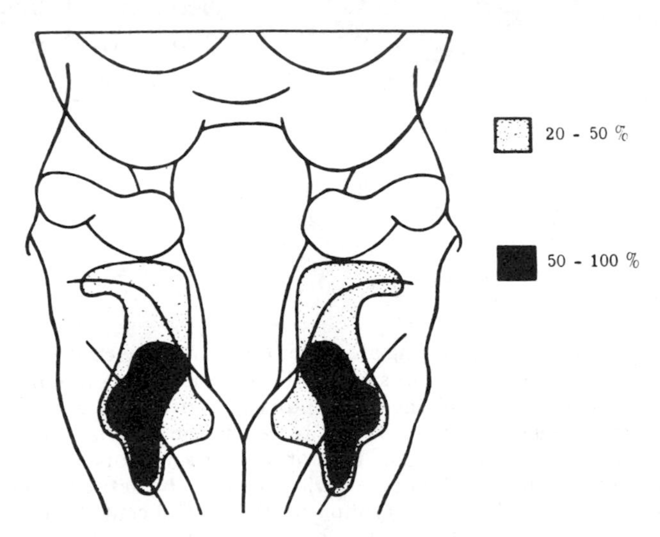

FIGURE 21-1. Localization of respiratory neurons in the brain stem of the cat. Percentages are for density of respiratory neurons per stab. (From Salmoiraghi and Burns, 1960.)

respiratory neurons were localized in the reticular formation, mostly from 2 to 4 mm below the floor of the fourth ventricle. Inspiratory and expiratory potentials were commonly found within 100 μ of one another throughout the area. Thus, it appears that present evidence favors the concept of a mixed population of inspiratory and expiratory neurons scattered throughout the respiratory center in the medulla.

The respiratory center in the medulla may be considered as a motor nucleus giving rise to fiber tracts which pass down the spinal cord. The fibers synapse at various levels with spinal motoneurons whose axons form the innervation of the respiratory muscles. The pattern of discharge in this efferent pathway has been studied to determine the behavior of the respiratory center.

The rate of discharge of single motor fibers to the diaphragm and external intercostal muscles increases during inspiration. In addition, quiescent muscle fibers are recruited as the inspiratory act proceeds. As a result of these two factors the force available to move gas into the lung increases, becoming maximal in the middle of the inspiratory cycle. At the end of inspiration, discharge to the inspiratory muscles diminishes, and expiration takes place by passive recoil of the lung tissue. During heavy or labored breathing, expiration is an active process, and comparable patterns are seen in the motor fibers to the expiratory muscles.

RESPIRATORY RHYTHMICITY

The discharge patterns recorded from the motor fibers give evidence of graded, rhythmic activity of the inspiratory and expiratory neurons in the medullary respiratory center. The means by which this activity, particularly the rhythmic alteration of inspiratory and expiratory neurons, is achieved has been the subject of much research and speculation.

The medulla seems to be the dominant region in the brain stem for respiratory control. There are, however, other regions which have important influences on the respiratory pattern. In fact, it was at one time believed that respiratory rhythmicity required the interaction of centers in the pons with those in the medulla. Lumsden (1923), for example, found that transection at the level of the upper pons in vagotomized animals abruptly altered the normal respiratory rhythm to a series of prolonged inspiratory spasms. This was not due to cerebral influences, for a similar transection at the level of the colliculi had little effect on the respiratory pattern. Lumsden suggested that the pons contained two centers vital to respiratory function: an *apneustic* center in the lower pons which tonically excited inspiratory effort and a *pneumotaxic* center in the midpontine region which interrupted the apneustic center periodically to allow expiration. Studies by Pitts, Magoun, and Ranson (see Pitts, 1946) led to a further development of this theory, which initially found wide favor. These workers found that section at the midpontine level in the vagotomized animal led to a prolonged inspiratory spasm, as found by Lumsden. If the vagus nerves were not cut, the respiratory rhythm was not altered by the brain stem transection. On the basis of these findings Pitts and co-workers suggested that respiratory automaticity was brought about by negative feedback networks, as shown in Figure 21-2. They proposed that the inspiratory center in the medulla was perpetually active unless actively inhibited by feedback from one of the networks. As inspiration began, the inspiratory center would send

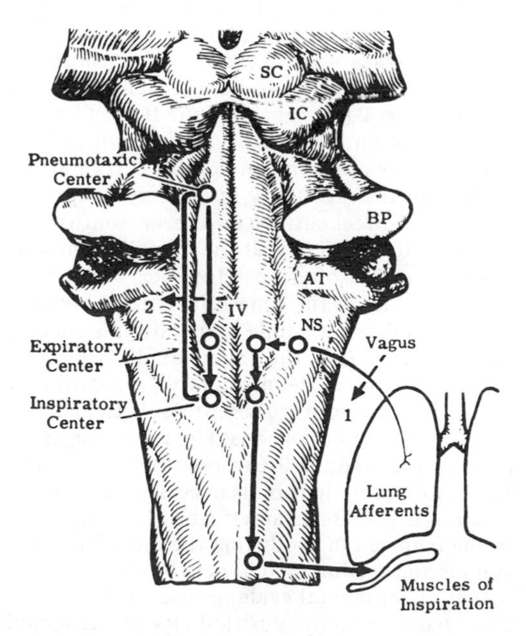

FIGURE 21-2. Diagram of respiratory control center according to the Pitts theory. Section of vagus nerves and brain stem (at arrows 1 and 2) should permit unchecked inspiratory activity (apneusis). (From Hoff and Breckenridge. In J. F. Fulton [Ed.], *Textbook of Physiology*, 17th ed. Philadelphia: Saunders, 1955. P. 852.)

impulses to the pneumotaxic center in the pons, which in turn stimulated the expiratory center. The latter would be expected to have an inhibitory influence on the inspiratory center and, when the feedback was sufficiently intense, would break through the inspiratory phase to cause expiration. A section of the brain stem at level 2 in Figure 21-2 would interrupt this feedback network. However, rhythmicity would be maintained through a second feedback system, involving the vagus nerve and stretch receptors in the lung. It is well known that inflation of the lung by a puff of air has an inhibitory effect on respiration which lasts for many seconds. Expansion of the lung fires the stretch receptors, and the impulse volley passes up to the respiratory center by way of the vagus nerve. This was also believed to constitute a normal feedback network for respiratory control since inflation of the lung during the normal inspiratory process would stretch the lung and presumably deliver inhibitory information to the inspiratory center. Thus, if the vagus were also sectioned, the inspiratory center would then be without negative feedback and continue to fire unabated.

Objections to this theory were raised by Hoff and Breckenridge (1955), who found that if sections were made in the lower region of the pons of vagotomized animals the sustained inspirations disappeared and roughly normal respiratory patterns appeared, although not as uniform as the normal breathing pattern – and of much greater amplitude. The effect on respiration of sectioning the brain stem at several different levels is shown in Figure 21-3.

Thus it appears that the medulla itself is capable of generating a respiratory rhythm. The difference between the medullary pattern and the normal pattern apparently represents the contribution of the pons. According to Hoff and Breckenridge, the pons contains facilitatory and inhibitory areas which normally play into the medullary respiratory center. The suppressor area is located in the region of the inferior colliculi and may be destroyed by discrete bilateral lesions. Normally it restrains the activity of the facilitatory areas located below it. The suppressor and facilitatory areas appear to be sufficiently separated so that a section in the upper pons (Fig. 21-3) separates the two and releases the facilitatory area to exert its full influence on the medulla. The result is the apneustic breathing pattern. This pattern was previously interpreted by Pitts and co-workers as evidence for upper pons inhibition of the medullary inspiratory neurons breath by breath.

The findings of Hoff and Breckenridge have been confirmed by other laboratories. Thus it seems clear at this time that the medullary neurons are capable of generating the basic rhythm of respiration. However, it is not evident to what extent the pons may participate in the genesis of normal breathing patterns. As shown in Figure 21-3, the isolated medulla preparation is characterized by an all-or-none breathing pattern; that is, the animal exhibits a gasping type of respiration. Since blood pressure is low in the medullary preparation, it is difficult to determine whether the ataxic pattern truly represents the natural behavior of the medulla or is secondary to the hypotension. Those who regard ataxic breathing as the normal behavior of the medulla look upon the pons as an integrative and regulative mechanism whose participation is necessary for eupneic breathing. Others look upon the ataxic pattern as an effect of hypotension and the necessarily large amount of trauma involved in producing the medullary preparation. The relative roles of the pons and medulla in eupneic breathing cannot be definitely established until further experimental evidence is available.

Localization of the basic respiratory periodicity to the medulla raises the possibility that the respiratory center is driven by a spontaneous "pacemaker" activity of its cells. This alternative was presented in its most complete form by Gesell (1940). He suggested that cellular metabolism caused a gradual increase in negativity of the dendritic end of the cell, and that this buildup occurred during

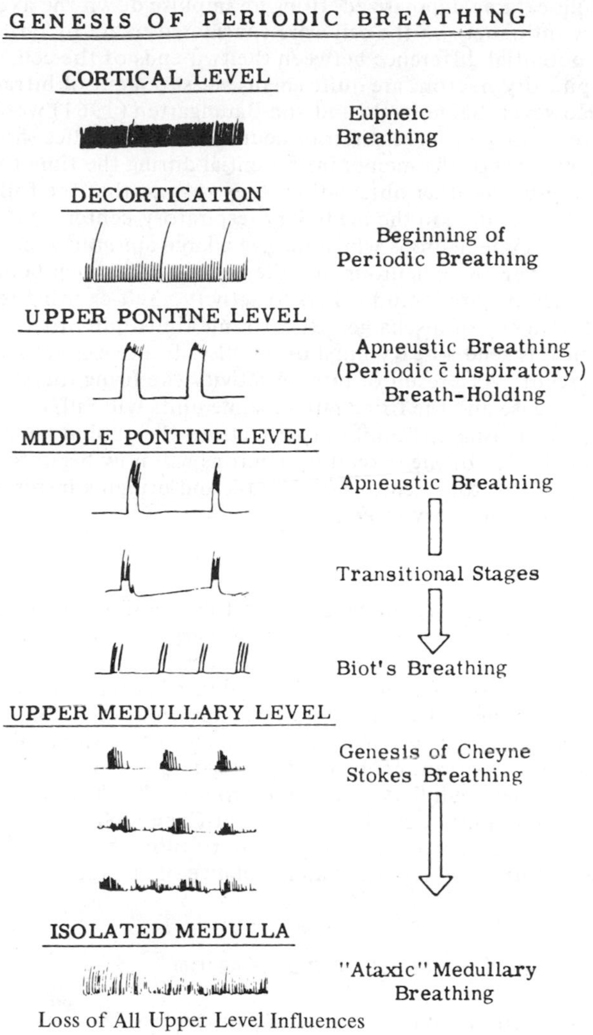

GENESIS OF PERIODIC BREATHING

CORTICAL LEVEL

Eupneic Breathing

DECORTICATION

Beginning of Periodic Breathing

UPPER PONTINE LEVEL

Apneustic Breathing (Periodic c̄ inspiratory) Breath-Holding

MIDDLE PONTINE LEVEL

Apneustic Breathing

Transitional Stages

Biot's Breathing

UPPER MEDULLARY LEVEL

Genesis of Cheyne Stokes Breathing

ISOLATED MEDULLA

"Ataxic" Medullary Breathing

Loss of All Upper Level Influences

FIGURE 21-3. Effect of transections at various CNS levels on respiratory rhythmicity after vagal section. Note the apneustic pattern (prolonged inspiration) with upper and midpontine sections and the emergence of periodic breathing at lower levels. (From Hoff and Breckenridge. In J. F. Fulton [Ed.], *Textbook of Physiology*, 17th ed. Philadelphia: Saunders, 1955. P. 862.)

a period in which the cell was quiescent – for example, during the expiratory pause for the inspiratory cells. As the negativity increased, a potential difference between the two ends of the cell would develop, giving rise to a current flow. The current was postulated to flow in an intracellular and extracellular circuit, passing out through the cell wall at the axon hillock and back in again at the dendritic end. As the potential difference increased, the current flow also increased until the membrane in the axon hillock area depolarized, firing an impulse down the axon. All inhibitory and excitatory influences on the cell were considered to act by either raising or lowering the potential difference between the two ends of the cell.

Since respiratory neurons are quite small, measurement of intracellular potential is difficult. However, Salmoiraghi and von Baumgarten (1961) were able to obtain intracellular recordings from respiratory neurons. Their studies show a gradual decrease in negativity of the membrane potential during the time the neuron is firing. From these and other observations they suggest that the following pattern of activity may take place in the medullary respiratory center. Within the inspiratory network are some neurons which receive a large amount of excitatory influence from without. When these neurons fire, they excite other members of the inspiratory population, which in turn recruit others to activity. Self-exciting feedback networks increase the frequency of discharge. At the same time, inhibitory collaterals to the expiratory neurons send a heavy burst of impulses to the expiratory neurons, holding them quiescent. As a result of intense activity the firing threshold of inspiratory neurons will increase and the firing rate of some units will fall, decreasing the self-excitation and decreasing the inhibitory effects on the expiratory neurons. The most highly facilitated of the expiratory neurons will now begin to discharge, recruiting other expiratory neurons to the task and bringing inspiration to a close through reciprocal inhibitory effects.

PERIODIC BREATHING

While the pons may not play the main role in the genesis of the respiratory rhythm, it is apparently important in the production of *periodic breathing patterns.* Two of the most important are *Biot's breathing* and *Cheyne-Stokes respiration* (Fig. 21-3). Biot's breathing consists of a series of maximal inspirations followed by a prolonged expiratory pause *(apnea).* Cheyne-Stokes respiration is a rhythmic waxing and waning of the amplitude of respiration. Cheyne-Stokes respiration is seen frequently in the clinic as an accompaniment to respiratory and circulatory insufficiencies. Sections at the upper medullary level often produce the rhythmic waxing and waning of respiratory amplitude so characteristic of Cheyne-Stokes respiration, whereas sections in the midpontine level often give rise to Biot's breathing. The slow rhythm seen in these patterns may represent an imbalance of the facilitatory and suppressor systems.

CENTRAL AND REFLEX EFFECTS ON RESPIRATION

The respiratory pattern produced by the medullary respiratory center is modified not only by the pons but by other influences, both higher and lower in the central nervous system. These influences consist of changes in the rate or amplitude of breathing, or both.

Certain terms are used to describe each of the commonly observed variations in the breathing pattern. *Eupnea* refers to a normal breathing pattern; *apnea* to an absence of respiratory movements. Apnea is generally considered to consist of a *maintained expiration.* It is thereby distinguished from *apneusis* – a *maintained inspiration* (as shown in Fig. 21-3). Other common terms are *hyperpnea* (an increase in ventilation by increasing either the rate or the amplitude of breathing) and *polypnea* (an increase in the frequency of breathing). *Tachypnea* is also commonly used for the latter condition. *Dyspnea* is a clinical term ordinarily used to indicate labored or difficult breathing and its accompanying distress.

The respiratory center is affected by structures in the midbrain and cerebral cortex. Anesthesia or coagulation of tuberal and mammillary areas of the hypothalamus leads to a reduction in respiratory activity. Tonic discharge from the hypothalamus apparently has an excitatory effect on the inspiratory neurons of the medulla. Certain areas of the cortex have a profound effect on respiration. The most important are concerned with speech, smell, chewing, and swallowing. This is to be expected, since respiration is greatly modified during these activities. During speech, for example, respiration consists of a rapid inspiration followed by a prolonged expiration. During swallowing the respiratory structures are fixed in the inspiratory position (apneusis).

The sensory input to the respiratory center constitutes the reflex control of respiration. The Hering-Breuer reflex has been mentioned in connection with its role in early theories of respiratory periodicity. It is also known as the *inhibito-inspiratory reflex.* As was explained, it is initiated by stretch receptors in the lung and carried by vagal afferents to the brain stem. Stimulation of this pathway causes a prolonged expiration. This is the reflex effect seen most frequently when the central end of the cut vagus nerve is stimulated electrically. When the vagus nerves are cut, inspirations become longer and deeper because of loss of this reflex. The pathway in the brain stem by which vagal impulses reach the inspiratory neurons is not known. It has usually been suggested that the impulses are routed through the expiratory center, but another proposal is that the reflex feeds into the bulbar suppressor system and thus acts quite indirectly in inhibiting inspiration.

A second reflex sometimes referred to as a Hering-Breuer reflex is the *excitato-inspiratory reflex.* It is elicited by collapse of the lung and brings forth inspiration. This reflex is not as strong as the inhibito-inspiratory reflex, nor is it as easily observed. It can be produced in the laboratory by suddenly sucking air out of the lungs during the expiratory pause. A premature inspiration then often occurs. This reflex appears to be of some importance in normal breathing, although it is not as active as the inhibito-inspiratory reflex. It is of greater consequence in exercise, and clinically it may reinforce respiration during a pneumothorax. When the central end of the cut vagus nerve is stimulated, an excitatory effect on inspiration is occasionally observed which is due to excitation of the excitato-inspiratory reflex pathway. This is particularly the case at high stimulus strengths.

The vagus contains other fibers, such as the aortic body chemoreceptor efferents – whose effects may be elicited by central stimulation.

A third reflex arising from *rapidly adapting stretch receptors* in the lung has been described by Gesell. It is active only when lung volume is large and produces further inspiration when the lung is stretched; hence the name *paradoxical reflex.* Its afferent limb is also the vagus nerve. The reflex is thought to be of some value in exercise.

These three reflexes are believed to arise from two basic types of receptor. The first, a *slowly adapting receptor,* is said to be excited by small increases in lung volume and is purely inhibitory to inspiration. This is the basis of the inhibito-inspiratory reflex. The second, a *rapidly adapting receptor,* is said to be excited by lung deflation or excessive inflation of the lung. It is believed to be responsible for the two reflexes excitatory to inspiration. It is excited only by large changes in lung volume, and its activity persists for less than one second.

The respiratory center is also considerably influenced by the carotid sinus reflex, so important in cardiovascular control. An increase in pressure in the sinus immediately decreases the rate and depth of breathing. This is comparable to its effect on the cardiovascular control center. It is generally true that the vasomotor and respiratory centers respond in the same manner to nervous stimuli. As will be noted later, this generalization holds also for chemical stimuli.

Stimulation of pain fibers has a strongly excitatory effect on the respiratory center, causing both hyperpnea and polypnea. This effect apparently involves inspiratory and expiratory neurons alike, since the depth of both inspiration and expiration is increased.

Other sensory modalities, such as temperature and touch, may affect the respiratory center. Also, there are receptors in the upper area of the respiratory tract which are stimulated by local irritation or mechanical movement. They initiate organized responses such as coughing and sneezing, in which the respiratory apparatus plays an important role. The major reflex inputs to the respiratory center are listed in Table 21-1.

TABLE 21-1. Reflex Effects on Respiration

Type of Reflex	Receptor	Afferent Pathway	Effect on Inspiratory Neurons	Effect on Expiratory Neurons
Hering-Breuer (inhibito-inspiratory)	Slowly adapting stretch receptor in lung	Vagus	Inhibitory	Excitatory*
Hering-Breuer (excitato-inspiratory)	Rapidly adapting stretch receptor in lung	Vagus	Excitatory	Inhibitory*
Paradoxical (overdistention of lung)	Rapidly adapting stretch receptor in lung	Vagus	Excitatory	Inhibitory*
Carotid sinus	Baroreceptor	Glossopharyngeal	Inhibitory	Inhibitory
Aortic arch	Baroreceptor	Vagus	Inhibitory	Inhibitory
Nociceptive	Pain	Visceral and somatic afferents	Excitatory	Excitatory
Cough†	Receptors in larynx and tracheobronchial system sensitive to mechanical or chemical irritation	Vagus, sympathetics, and possibly glossopharyngeal	Excitatory then inhibitory	Inhibitory then excitatory
Sneeze†	Same as for cough but located in nasal passages	Trigeminal and olfactory nerve	Excitatory then inhibitory	Inhibitory then excitatory

*These effects would be expected on the basis of the reciprocal innervation of the inspiratory and expiratory neurons.

†The cough and the sneeze reflexes are biphasic and consist of a slow, prolonged inspiration, followed by a sudden expiration.

CHEMICAL CONTROL OF RESPIRATION

The ultimate regulator of respiration, as stated previously, must be the demand for oxygen and the need for eliminating carbon dioxide. The oxygen requirement is in turn determined by the rate of metabolism of the body. This link between respiration and metabolism has been suspected, if not known, for many years. Mayow (1673) suggested that a combustive process occurred in the body to provide energy for the muscles in the same manner that gunpowder burned in air. He recognized the role of the lungs and the blood stream in providing a necessary component of the air for tissue combustion. Lavoisier (1789) noted that oxygen consumption increased with muscular exercise although the fundamental relation between these two factors was not appreciated until 50 years later.

As a consequence of this relation between respiratory activity and metabolic activity, the ultimate control of respiration may be said to reside not in the medulla but in the tissues. The summated metabolic activity of the individual cells is the ultimate stimulus and governor of respiratory activity. If the ventilation is insufficient to meet the metabolic needs of the tissues, the stimulus to respiration increases. This may be readily appreciated by holding the breath for a short period. On the other hand, if ventilation is in excess of body needs, the respiratory drive is reduced to less than normal levels. This can be appreciated by voluntarily hyperventilating for a brief time – after which it will be noted that normal respiratory movements are absent or reduced.

EFFECTS OF LOW OXYGEN

Rosenthal (1862) suggested that respiratory activity was regulated solely by the oxygen content of the blood stream. This is consistent with the foregoing observations, since hypoventilation reduces blood oxygen content, which should increase respiratory drive. Conversely, hyperventilation increases blood oxygen content somewhat, which was postulated to decrease respiratory drive. Also, breathing gas mixtures deficient in oxygen (less than 12 percent) causes respiration to increase.

The excitatory effect of breathing oxygen-deficient gas mixtures was for many years conjectured to be a direct effect on the respiratory center. However, in 1930 Heymans showed that there is a chemical receptor zone near the bifurcation of the carotid artery which is sensitive to the oxygen tension of the blood. The carotid bodies also respond to changes of 10 to 20 mm Hg P_{CO_2} and 0.1 pH unit, but this effect is probably of secondary importance. The chemoreceptor cells become increasingly active as the oxygen tension of the arterial blood falls (Fig. 21-4). Present evidence seems to indicate that some chemoreceptor cells are active at the normal P_{O_2} of arterial blood and can be inactivated by inhalation of high concentrations of oxygen. Most of the chemoreceptors have a lower oxygen threshold and are not stimulated maximally until the P_{O_2} has fallen to 40 or 50 mm Hg. This corresponds to the inhalation of 8 to 10 percent oxygen. Located around the arch of the aorta are similar chemoreceptor cell groups known as the *aortic bodies* which respond to changes in arterial P_{O_2}. Both chemoreceptor areas respond to a decrease in local blood flow. Physiological stimulation of either chemoreceptive area results in hyperpnea with an increase in tidal volume, frequency, and minute volume of breathing. Stimulation of carotid chemoreceptors in the dog causes bradycardia, while stimulation of aortic chemoreceptors causes tachycardia and hypertension, thus showing a qualitative difference between carotid and aortic body response. It should be emphasized that there is a wide variation in the respiratory response to anoxia. For example, when normal men were given a gas mixture containing 10 percent oxygen to breathe, arterial blood

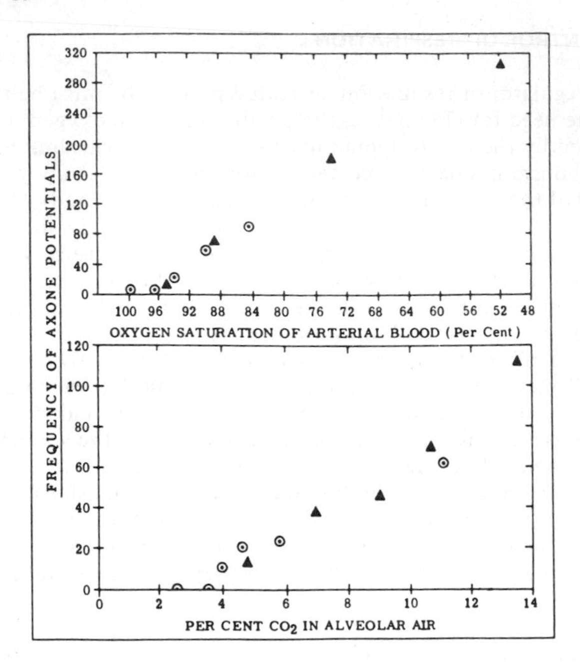

FIGURE 21-4. (Upper) Frequency of chemoreceptor firing during administration of oxygen mixtures to an anesthetized cat. (Lower) Frequency of firing during CO_2 administration. A relatively greater sensitivity to low oxygen is evident. (From U. S. von Euler et al. *Skand. Arch. Physiol.* 83:132, 1940.)

oxygen saturations ranging from 85 percent down to 50 percent were recorded. While low oxygen has an excitatory effect on the peripheral chemoreceptors, it has at the same time a depressant effect centrally. When P_{O_2} levels in the arterial blood fall below 30 mm Hg, respiration is depressed through this central effect.

EFFECTS OF CARBON DIOXIDE

Haldane and Priestley (1905) did extensive studies on the composition of alveolar gas. They found that under a variety of circumstances, such as decompression and compression to supernormal pressure, the partial pressure of carbon dioxide in the alveoli remained very nearly constant. They also noted that the addition of carbon dioxide to the inspired air greatly increased ventilation. Their conclusion from these observations was that respiratory activity was geared to maintain a constant level of carbon dioxide in the blood stream. An attractive feature of this theory was that it did provide an explanation for many of the commonly observed respiratory phenomena. The apnea after hyperventilation, for example, could be explained as simply due to blowing off carbon dioxide from the blood. The normal breathing pattern would resume when enough carbon dioxide accumulated in the blood to approximate normal levels. Low oxygen was still considered to have a stimulating effect on respiration.

In some instances when breathing resumes after hyperventilation, and in some clinical states, Cheyne-Stokes respiration is seen. Haldane correlated this periodic breathing pattern with changes in the oxygen and carbon dioxide levels of the blood. In the period of apnea after hyperventilation the carbon dioxide level is below normal and the oxygen level is higher than normal. Consequently, there is little chemical stimulus to respiration. As the apneic period continues, the carbon dioxide accumulates and the oxygen is depleted. Both of these factors, but particularly the carbon dioxide level, cause resumption of respiratory activity. As breathing begins, the oxygen content increases while the carbon dioxide content falls, reducing respiratory drive and again causing respiratory activity to cease. During the period of apnea the oxygen and carbon dioxide levels are again changing in the direction of increasing respiratory drive. Thus the cycle repeats when the drive is strong enough to initiate the respiratory act.

Carbon dioxide is an extremely potent stimulus to respiration, a 10-fold increase in ventilation occurring when 10 percent of CO_2 is present in inspired air (Fig. 21-5).

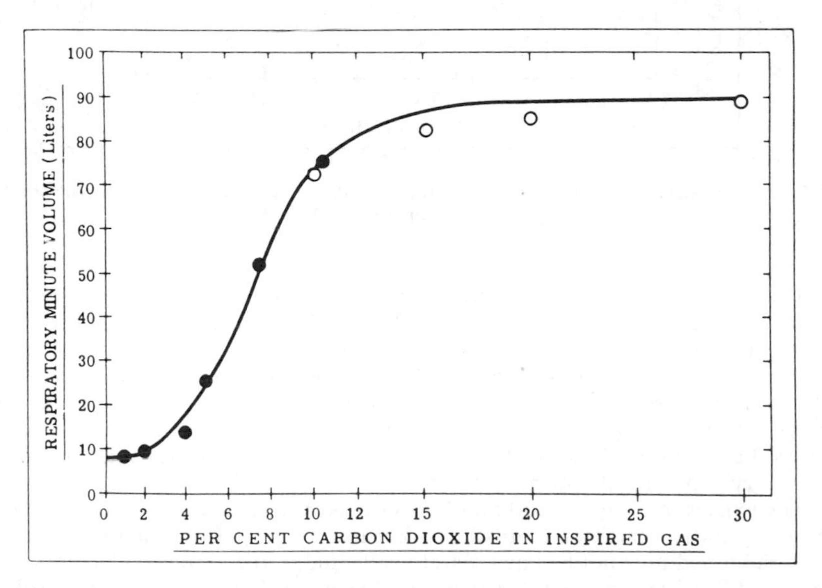

FIGURE 21-5. Effect of inhaled carbon dioxide on respiratory minute volume in conscious human subjects. (From C. J. Lambertsen. Therapeutic Gases, Oxygen, Carbon Dioxide, and Helium. In J. R. DiPalma [Ed.], *Drill's Pharmacology in Medicine,* 3d ed. New York: McGraw-Hill, 1965.)

When the inspired gas contains more than 30 percent carbon dioxide, *respiratory depression* occurs, and beyond 40 percent a combination of excitatory and depressant effects on the central nervous system can be fatal. Carbon dioxide exerts its effect primarily through centers in the medulla oblongata, although high arterial blood concentrations do stimulate the peripheral chemoreceptors.

EFFECT OF pH

Ventilation is directly affected by alterations in arterial pH. Ordinarily, blood pH is held very close to 7.4 by chemical and physiological buffer mechanisms, one of which is the respiratory system itself. When the pH of the blood decreases, the ventilation increases and carbon dioxide is blown off. The relation between arterial pH and ventilation rate is shown in Figure 21-6 for unanesthetized human subjects.

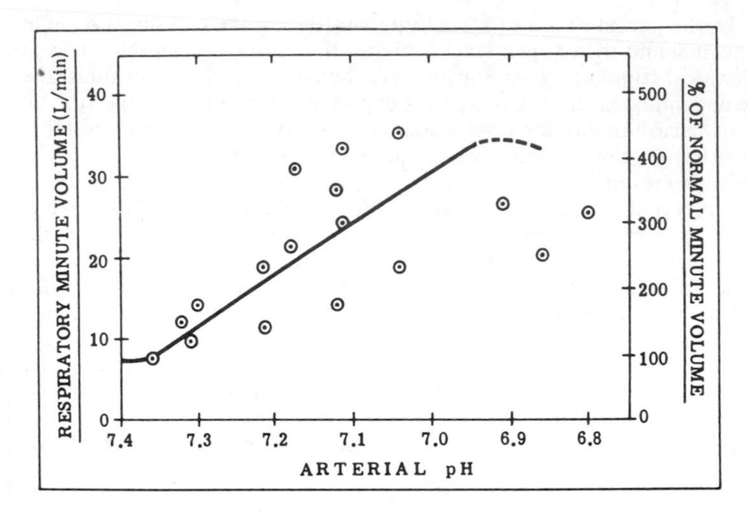

FIGURE 21-6. Effect of arterial pH on respiratory minute volume. Studies in unanesthetized human subjects. (From S. S. Kety et al. *J. Clin. Invest.* 27:500, 1948.)

The mechanism by which arterial pH influences respiratory activity appears to involve receptors in the area of the medulla. These receptors respond to changes in the pH of the interstitial fluid, which in turn is closely related to that of the cerebrospinal fluid (CSF). The same receptors are rapidly excited when blood CO_2 levels increase. However, the CO_2 molecule is believed not to be directly capable of stimulating these receptors. Rather, CO_2 alters CSF pH, which in turn excites the receptors. Some of the evidence for hydrogen ion as a mediator of respiratory control will now be considered.

The existence of the blood-brain barrier makes the study of the chemical control of respiration particularly difficult. Because of this barrier the pH changes in arterial blood may not be immediately reflected in the extravascular fluids. Gesell and Brassfield showed, for example, that immediately upon infusion of $NaHCO_3$ arterial blood becomes more alkaline, while at the same time the CSF becomes more acid. Recent studies indicate that carbon dioxide is free to diffuse into the CSF while bicarbonate ion enters slowly. As a result, administration of $NaHCO_3$ causes first an increase in carbon dioxide concentration in the CSF, which then causes the pH to decrease. Sodium bicarbonate administration has an immediate stimulatory effect on respiration, which heretofore has been difficult to reconcile with its alkaline properties. If sufficient time is allowed for equilibration after bicarbonate administration, the CSF does become alkaline and respiratory activity decreases correspondingly.

The pH of CSF therefore appears to change in a manner which may explain the effect of arterial CO_2 and pH on the respiratory centers. This requires the assumption that appropriate structures in the medulla are responsive to CSF pH. Von Euler and Soderberg have developed evidence that specific cells in the medulla, separate from the inspiratory and expiratory neurons, appear to mediate the response to carbon dioxide. They found that chloralose, an anesthetic agent, acts on the medulla to block the action of carbon dioxide while not altering the reflex activity of the respiratory center itself. They suggested that certain neurons in the medulla function as chemoreceptors. Further direct support for this concept was provided by Mitchell and his co-workers (1963), who found an area just below the surface of the medulla in the ventrolateral region which apparently mediates the chemical response. When the area was perfused with CSF adjusted to a low pH, respiratory activity increased. Conversely, if the CSF was cooled or a local anesthetic was added, the respiratory activity diminished, or in some instances ceased entirely. During this period of hypoactivity diminished responses to CO_2 administration were also apparent. The effect can be likened to the respiratory behavior seen after prolonged hyperventilation.

These findings appear to fit well with a unified concept of chemical respiratory control based upon a central chemoreceptor area which mediates the response to altered blood chemistry as it affects CSF pH and the peripheral chemoreceptors responding primarily to lowered P_{O_2} in arterial blood. A detailed schema of this concept of respiratory control is shown in Figure 21-7. As mentioned earlier, CO_2 readily passes the blood-brain barrier while H^+ and HCO_3^- do not. The CO_2 forms carbonic acid, which rapidly alters the H^+ concentration of extracellular fluid since there are no buffer systems here comparable to those existing in blood. The chemoreceptors on the medullary surface are stimulated by the increased H^+ concentration and cause ventilation to increase. Thus, a rapid and proportionate change in ventilation would be expected with elevated blood CO_2 levels. When blood pH is altered, the effect would be less rapid since H^+ does not diffuse rapidly across the barrier. Moreover, in the long term, the H^+ which does come across may be buffered to a degree by the blood-CSF barrier ion pumps. The latter tend to control CSF pH by altering the HCO_3^- of the CSF. This regulatory mechanism would also bring the CSF pH back to normal during prolonged alteration of blood CO_2. In clinical conditions such as emphysema there appears to be some adaptation to elevated CO_2 levels. By contrast, there is no adaptation of the carotid bodies to lowered blood P_{O_2}.

INTERACTION OF CHEMICAL STIMULI

While CO_2 and pH of arterial blood both act on the pH of CSF to alter respiration, the two factors often must be considered as separate entities. For example, increased carbon dioxide levels in arterial blood will reduce arterial pH and thereby act by this mechanism as well as the one cited above to exert its effect on the pH of the CSF. Thus it is found that the stimulatory effect of CO_2 on respiration is somewhat less if blood pH is held constant relative to the circumstance where pH is allowed to fall.

If the pH of arterial blood is lowered by adding fixed acids to the blood stream, blood P_{CO_2} is not directly affected. However, the increased ventilation resulting from the reduced arterial pH would influence arterial CO_2 levels. This effect is illustrated in Figure 21-8. If blood P_{CO_2} is held constant by artificial means, the addition of fixed acids to the blood is a rather potent stimulus to respiration. During normal breathing CO_2 is blown off during hyperventilation, and its contribution to the overall respiratory drive is lessened. As a result, the actual ventilation is somewhat less than expected from the pure pH change. Conversely, if the blood is made alkaline, respiration will diminish drastically if blood CO_2 can be held at normal levels. In the natural course of events the blood CO_2 will rise in the alkalotic animal, and this very potent stimulus will maintain respiration at near normal levels.

The relation between ventilation, blood CO_2, and blood pH has been examined by Gray (1949). He found that it was possible to express the interplay of these factors in a quantitative manner as follows:

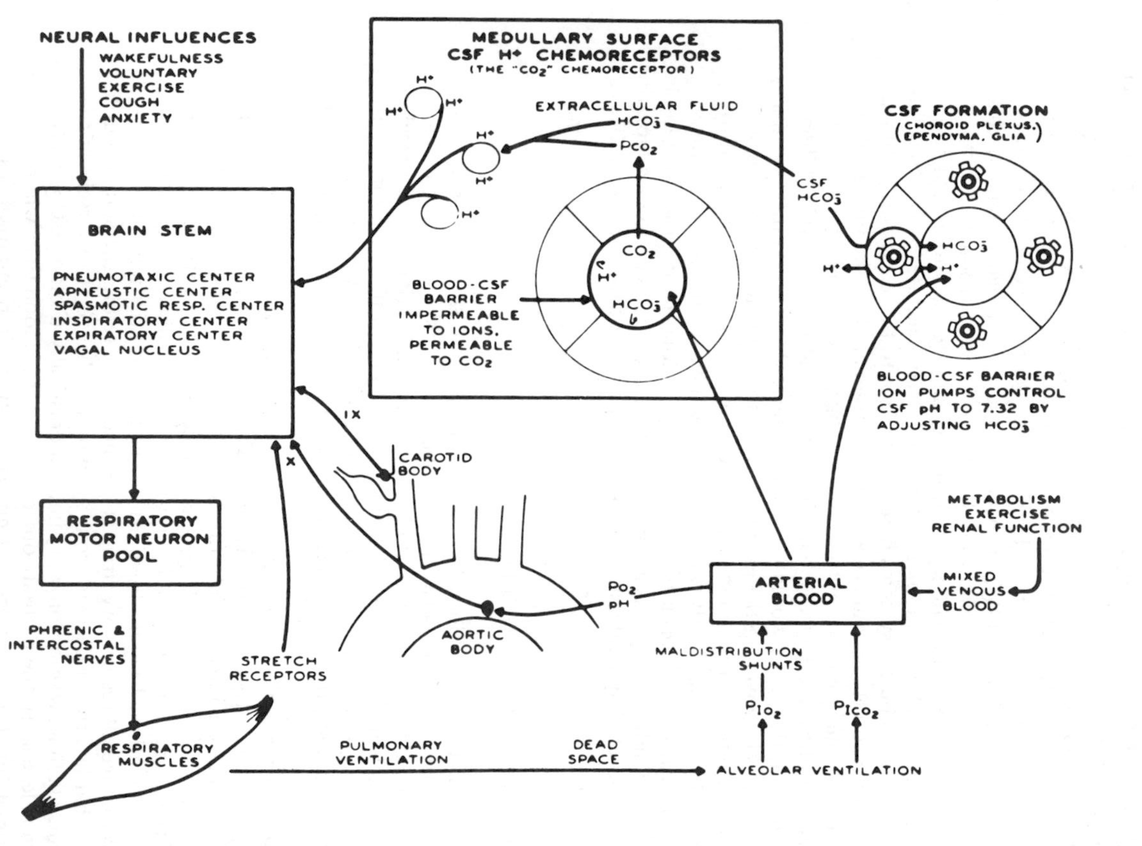

FIGURE 21-7. A schema of the chemical control of respiration. The medullary chemoreceptors are believed to respond to the pH of the extracellular fluid, which generally follows that of cerebrospinal fluid. The blood-brain barrier allows H^+ and HCO_3^- to diffuse only slowly while CO_2 passes rapidly. The respiratory center in the brain stem integrates this and other stimuli from the cortex and periphery and adjusts ventilation in accordance with the summated stimuli. An active transport mechanism in the formation of CSF adjusts the bicarbonate concentration to maintain the pH of the fluid at 7.32. (From J. W. Severinghaus and C. P. Larson, Jr. Respiration in Anesthesia. In *Handbook of Physiology*. Section 3, Vol. II. Washington, D.C.: American Physiological Society, 1965. Chap. 49, p. 1223.

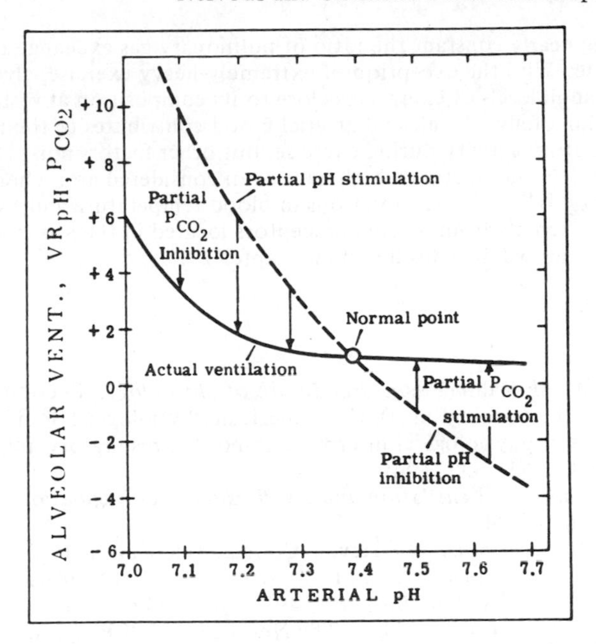

FIGURE 21-8. Effect of arterial pH on minute volume. Hyperventilation from lowered pH causes decrease of alveolar P_{CO_2} which attenuates the ventilatory stimulation. (From J. F. Gray, *Pulmonary Ventilation and Its Physiological Regulation,* 1949. Courtesy of Charles C Thomas, Publisher, Springfield, Illinois.)

$$\text{Ventilation rate} = 0.22\ H + 0.262\ P_{CO_2} - 18.0$$

where H is the hydrogen ion concentration of arterial blood in 10^{-9} moles per liter, P_{CO_2} is in mm Hg and ventilation rate is expressed as a fraction or multiple of the normal value in the resting subject. In light of current concepts of respiratory control, this equation probably expresses the effect of these two factors on the pH of CSF, to which the ventilation ultimately responds.

The interaction between stimuli is also found when low O_2 is administered. When the carotid bodies excite the respiratory center to increase ventilation, blood CO_2 levels are reduced in proportion to the increased ventilation, as discussed in Chapter 20. This diminishes the drive to the central chemoreceptor system, and the resultant ventilation is only moderately elevated because of the loss of CO_2 as a stimulus.

REGULATION OF RESPIRATION DURING EXERCISE

Vigorous exercise is the most potent stimulus to respiration known. Ventilation levels of 100 to 120 liters per minute may be attained as contrasted to 60 to 80 liters per minute produced by CO_2 inhalation. Careful study has shown that oxygen consumption increases as a direct function of the work rate, and that respiratory minute volume increases in direct proportion to the oxygen consumption. The mechanism by which these increases in ventilation occur is not well understood. The selection of a proper respiratory pattern during exercise aims at maintaining nearly constant the ratio of pulmonary gas exchange to alveolar minute volume. With the exception of extremely heavy exercise, alveolar air composition at all levels of exercise is close to its composition at rest. Certainly, the composition of alveolar air and arterial blood contributes to the fine adjustment of respiratory activity during exercise, but other factors must also be important. Among the other factors which have been considered are: chemical mediators other than CO_2, H^+, and O_2; alterations in blood temperature and hydrostatic blood pressure; and afferents from mechanoreceptors located in the joints and limbs. These mechanisms will be considered in Chapter 29.

REFERENCES

Fenn, W. O., and H. Rahn (Eds.). *Handbook of Physiology.* Section 3: Respiration, Vol. I. Washington, D. C.: American Physiological Society, 1964.

Gesell, R. A neurophysiological interpretation of the respiratory act. *Ergebn. Physiol.* 43:481–639, 1940.

Gray, J. F. *Pulmonary Ventilation and Its Physiological Regulation.* Springfield, Ill.: Thomas, 1949.

Haldane, J. S. *Respiration.* New Haven, Conn.: Yale University Press, 1922.

Heymans, S. C., J. J. Bouckaert, and L. Dautrebande. Sinus carotidien et reflexes respiratoires. *C. R. Soc. Biol.* 103:498–500, 1930.

Hoff, H. E., and G. C. Breckenridge. The Neurogenesis of Respiration. In J. F. Fulton (Ed.), *Textbook of Physiology,* 17th ed. Philadelphia: Saunders, 1955. Chap. 42.

Lambertsen, C. J. Chemical Factors in Respiratory Control. In P. Bard (Ed.), *Medical Physiology,* 11th ed. St. Louis: Mosby, 1961. Chap. 39.

Mitchell, R. A., H. H. Loeschke, W. H. Massion, and J. W. Severinghaus. Respiratory responses mediated through superficial chemosensitive areas on the medulla. *J. Appl. Physiol.* 18:523–533, 1963.

Pitts, R. F. Organization of the respiratory center. *Physiol. Rev.* 26:609–630, 1946.

Salmoiraghi, G. C., and B. D. Burns. Localization and patterns of discharge of respiratory neurones in brain-stem of cat. *J. Neurophysiol.* 23:2–13, 1960.

Salmoiraghi, G. C., and R. von Baumgarten. Intracellular potentials from respiratory neurons in brainstem of cat and mechanism of rhythmic respiration. *J. Neurophysiol.* 24:208–218, 1961.

22

Renal Function

Ewald E. Selkurt

THE KIDNEY plays a dominant role in the regulation of the constancy of the internal environment. It does so in the first instance by the regulation of the water content of the body with the aid of the antidiuretic hormone (ADH) released from the hypothalamo-neurohypophysial system. This regulation is closely integrated with the maintenance of proper salt balance in the body, i.e., the proper proportion of sodium, potassium, calcium, chloride, phosphate, bicarbonate, and sulfate. In the control of electrolyte balance, other endocrine regulations are important (Chap. 35).

An important area of regulation of proper electrolyte balance is the renal adjustment of the acid-base balance. Since the problem is primarily one of ridding the body of the acids created by metabolism, the kidney favors excretion of acid radicals through an ion exchange mechanism whereby the hydrogen ion is secreted in exchange for sodium, making the urine acidic. Second, the kidney forms ammonia, which is secreted by the tubular cells and replaces valuable base. Thus, the renal mechanism for acid-base regulation is essentially a base conservation mechanism. The mechanism will be described in greater detail in Chapter 24.

The kidney has the further responsibility of the excretion of waste, such as urea, uric acid, creatinine, and creatine. It must perform this function and others while conserving valuable foodstuffs, such as glucose and amino acids. These must be reabsorbed selectively, while the undesirable waste products are eliminated. Certain additional functions include detoxification, e.g., the conjugation of glycocol and benzoic acid to form the more innocuous hippuric acid, which is rapidly secreted by the tubular cells into the urine. Other metabolic functions, such as amino acid oxidation and deamination, also occur in the tubular cells.

In summary, the problem of excretion by the kidney involves filtration at the glomeruli of all but cellular constituents and the plasma proteins; further, it is the function of the tubules to reabsorb selectively the necessary valuable substances and to reject waste products and undesirable excesses of anything taken into the body. In addition, certain substances may be added to the urine by tubular secretion.

FUNCTIONAL ANATOMY OF THE KIDNEY

The functional unit of the kidney is the nephron, and there are approximately 1¼ million nephrons per kidney in the human. The nephron is diagrammed in

Figure 22-1 with the associated blood supply. It consists of the malpighian corpuscle and the attached tubular system. The malpighian corpuscle includes Bowman's capsule, which consists of an inner *visceral* layer and an outer *parietal* layer. The filtering area of the capsule has been estimated to be about 0.8 square millimeter, which would represent 2 square meters of filtering surface for both kidneys. The visceral

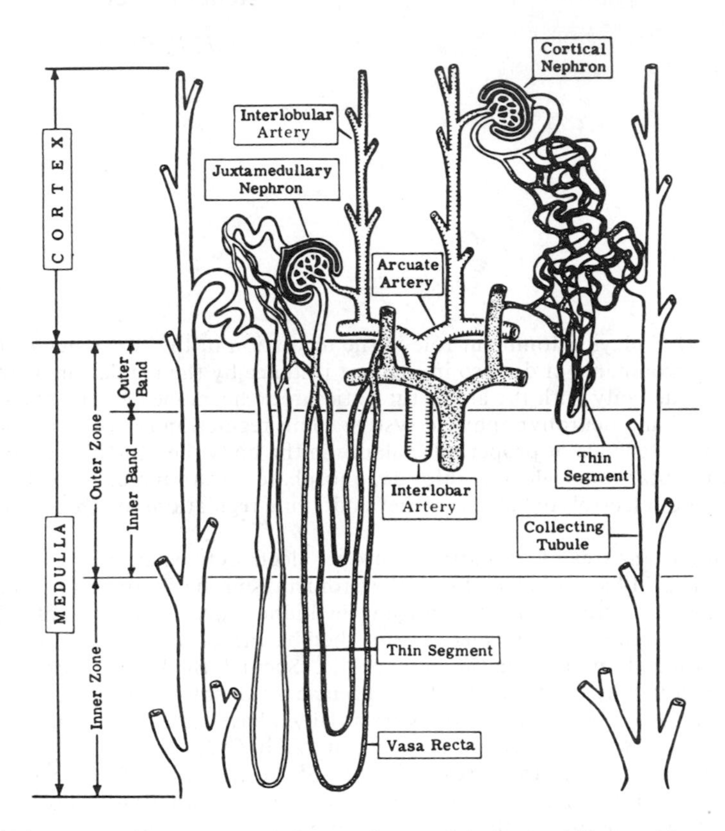

FIGURE 22-1. Cortical and juxtamedullary nephrons of the human kidney, with representative blood supply. (From Smith, 1951. P. 12.)

layer of Bowman's capsule is in intimate contact with the capillary loops important in filtration; the unit is called the *glomerulus.* Bowman's capsule continues into the *proximal convoluted tubule.* The proximal segment has the widest diameter of any portion of the nephron. It is made of truncated pyramidal cells characterized by a *brush border* on the inner or luminal aspect. At their basal margins they have striations perpendicular to the basement membrane. These striations are related to the distribution of the mitochondria in the cell. The proximal convoluted tubule turns toward the medulla to become the *loop of Henle.* The descending limb is at first of a thickness comparable to the proximal convoluted tubule but is then replaced by highly attenuated cells of the thin segment, which are characterized by flattened epithelium. This makes a hairpin turn in the medulla, then turns up toward the cortex into the ascending thick limb of the loop. The thin segment is found only in mammals and in a small percentage of the nephrons of birds. The average length varies considerably in different mammals and, indeed, varies considerably within the human kidney, depending on the location of the nephron (Fig. 22-1).

The ascending limb of the loop of Henle has cells which are at first cuboidal but become more columnar as they approach the cortex. This becomes the *distal convoluted tubule,* the cells of which lack brush borders but do exhibit basal striations. This segment joins the treelike system of *collecting ducts.*

RENAL CIRCULATION

ANATOMICAL ASPECTS

The renal artery divides into *interlobar arteries,* which subdivide into primary, secondary, and tertiary *arcuate arteries,* from which spring *interlobular arteries.* The *afferent arterioles* arise from these; in the dog, the afferent arterioles usually supply one glomerulus but rarely may branch to supply two to four glomeruli, with a total of 200,000 per kidney. This number compares with estimates ranging from 600,000 to 1,700,000 in each human kidney.

The glomerular capillaries converge in the efferent arterioles, then in a second capillary bed around the convoluted tubules, which drain into the interlobular veins. These drain into the *arcuate veins* and *interlobar veins.*

Arterioarterial and *venovenous* anastomoses have been found in the kidneys of man and dog. Although arteriovenous anastomoses probably exist, their occurrence is infrequent.

JUXTAGLOMERULAR APPARATUS

In the normal human kidney the afferent arteriole, as it approaches the glomerulus, shows a significant increase in the number of cells in the media, thickening the arteriolar wall to form an asymmetrical cap *(polkissen)* on one side of the glomerulus. The polkissen is composed of small, spindle-shaped, afibrillar cells. They may have a particular role in regulation of renal blood flow. In rats they show changes in granularity with variations in systemic blood pressure. Some investigators feel that renin, which plays a part in production of angiotensin, the pressor principle in renal hypertension, is formed in these cells.

A portion of the distal convoluted tubule which lies opposite the juxtaglomerular apparatus (JGA) shows a distinct modification, characterized by condensation of nuclei on one side of the tubule in an epithelial plaque, the *macula densa,* and together with the JGA is called the *juxtaglomerular complex.*

It has been suggested that the JGA and macula densa form a regulatory system responding to changes in the composition of the distal tubular period, particularly the sodium concentration. This in turn regulates renin production (Nash et al., 1968).

A recent hypothesis supported by Thurau (1968) is that a local control system involving renin release and production of angiotensin II within the JGA cells regulates the caliber of the afferent arterioles in response to sodium concentration in the tubular fluid.

BLOOD SUPPLY TO THE MEDULLARY ZONE

The glomeruli and associated tubules which lie deep in the cortex adjacent to the medulla *(juxtamedullary glomeruli)* have distinctive modification of their circulation compared with the cortical glomeruli and nephrons. They are typified by long loops of Henle dipping into the medullary zone and papillary portions of the kidney, and have associated modified vascular structures. The efferent arterioles from these glomeruli not only break up into the typical capillary supply to the convoluted portions but also form long hairpin loops of thin-walled blood vessels, the *vasa recta*, which accompany the loops of Henle. Although thin-walled, their diameter is several times greater than that of the typical peritubular capillary. In man the juxtamedullary glomeruli comprise about 20 percent of the total number of glomeruli in each kidney.

MEASUREMENT OF RENAL BLOOD FLOW

Both direct and indirect methods have been employed to measure renal blood flow. Direct methods employ flowmeters of various types, which record either arterial inflow or venous outflow. The indirect methods are an application of the *renal clearance principle*. This is based on the clearance ratio, $C = UV/P$, in which U is the urinary concentration in milligrams per milliliter, V is the minute urine volume, and P is the plasma concentration (usually systemic vein) in milligrams per milliliter.

Fick Principle Method

The requisite for the clearance of a substance to measure plasma flow is that it be entirely removed (or nearly so) from the plasma in one transit through the kidney. This is verified by examination of the concentration in the renal vein (V_c) and application of the *extraction ratio*

$$E = \frac{A_c - V_c}{A_c}$$

Hence, if V_c equals 0, E will equal 1.0. E is less than unity to the extent that the material is not removed by urinary excretion. It is clear that if the renal clearance UV/P is divided by the extraction ratio

$$\frac{UV/P}{(A_c - V_c)/A_c} \text{ or } \frac{C}{E}$$

the resultant quotient will yield the total plasma flow. The latter is an expression of the Fick principle. In order to obtain total renal blood flow (RBF), the hematocrit measurement of the blood is introduced:

$$RBF = \frac{C}{E} \times \frac{1}{1 - \text{hemat.}}$$

Several substances are so efficiently removed by combined processes of glomerular filtration and active tubular (proximal) transport (secretion) at low plasma concentrations that the renal vein concentration is very low (i.e., extraction nearly complete, and E close to unity). These include Diodrast (D) – E_D is 0.74 – and p-aminohippurate (PAH) – E_{PAH} is 0.80 to 0.85 in the dog and 0.90 in man. Then C_D or C_{PAH} is nearly equivalent to plasma flow. The fact that extraction is not complete is interpreted as indicating that a small fraction of blood does not perfuse excretory tissue: this would include capsule and inert supportive tissue, medullary tissue (loops of Henle, collecting ducts), calycine mucosa, and pelvis. On this basis, Smith has referred to the measured clearance as the "effective" plasma or blood flow.

The Fick principle can be employed with any substance cleared by the kidney which shows a measurable arteriovenous (A-V) difference. Obviously, the smaller the A-V difference, the more prone to error the calculation will be. Thus, phenol red, urea, mannitol, and inulin have been employed, but they have considerably smaller A-V differences than Diodrast and PAH.

Other Indirect Methods

The nitrous oxide method employed for the measurement of cerebral blood flow (Chap. 6) is also based on the Fick principle, adapted for measurement of renal blood flow. This involves inhalation of the gas by the subject and measurement of uptake from the blood by kidney tissue. The method yields an average of 3.2 ml per minute per gram of kidney weight in anesthetized dogs and in man. An obvious advantage is that the nitrous oxide method can be employed during conditions of anuria. A similar application, involving uptake, then washout of radioactive gases krypton (^{85}Kr) and xenon (^{133}Xe) injected into the renal artery, has been used in dog and man. The washout curve can be broken down into different components, signifying differential blood flow rates through different compartments. Following this procedure, computations of cortical blood flow and medullary blood flow have been made (Thorburn et al., 1963; Ladefoged, 1966). Representative values for the human kidney are: cortex, 3.6 to 6.3 ml per minute per gram of cortical tissue; outer medulla, 1.25 ml per minute per gram of tissue. (See also below for details of regional renal circulation.)

RENAL BLOOD FLOW VALUES

Total renal blood flow averages about 350 ml per minute in the dog (average weight, 15 kg) and about 1200 ml per minute in man, or about one-fifth of the cardiac output. Renal blood flow averages about 3.5 to 4.0 ml per minute per gram of kidney weight.

OXYGEN UTILIZATION

The renal venous blood contains considerably more O_2 than that draining other tissues. The resulting small A-V O_2 difference (1.7 volumes per 100 ml in man and about 3.0 volumes per 100 ml in the dog and cat) remains constant over a wide range of renal blood flows, although it may increase at very low rates of flow. Thus, O_2 consumption (normally 0.08 to 0.10 ml per gram per minute in the dog) is related to flow, so that when flow is reduced, as in shock, the organ ordinarily does not increase its extraction but apparently suffers curtailment of oxidative metabolism.

Warburg kidney tissue–slice studies have been done in the guinea pig, dog, and cat. The average Q_{O_2} values are as follows: in cortex, 21.3 mm^3 per milligram dry tissue per hour; outer medulla, 15.1; inner medulla, 6.2. These findings lead to the conclusion that the structures of the inner medulla (loops of Henle and collecting ducts) do not have important energy-requiring functions, or, alternatively, operate in part anaerobically. This zone is important in the countercurrent urinary concentration process. The blood flow to the medulla has been calculated to be small, perhaps less than 10 percent of the total, 0.7 to 1.0 ml per minute per gram in the outer medulla and only about 0.2 ml per minute per gram in the inner medulla compared with 4.0 in the cortex (Fig. 22-2). The cortex, on the other hand, contains the segments of the nephron involved in active sodium transfer, and it is the sodium-reabsorptive mechanism which largely accounts for the renal O_2 utilization.

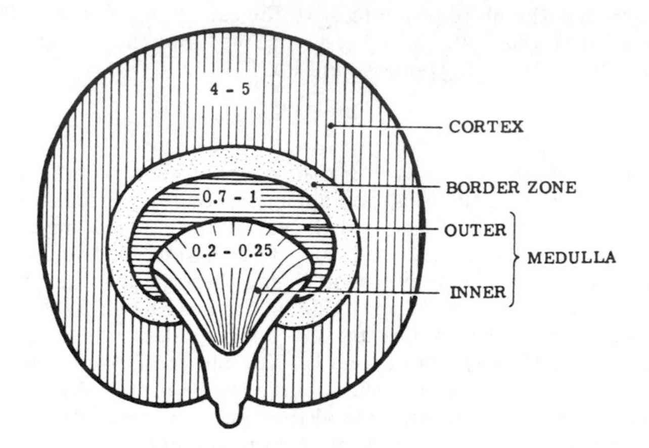

FIGURE 22-2. Regional blood flow in the kidney. (From K. Thurau. In C. H. Best and N. B. Taylor [Eds.]. *The Physiological Basis of Medical Practice,* 8th ed. Baltimore: Williams & Wilkins, 1966. P. 370. Copyright © 1966, The Williams & Wilkins Company, Baltimore, Md. 21202, U.S.A.)

AUTONOMY OF THE RENAL CIRCULATION

Rein in 1931, in his study of regional blood flow in dogs during changes in systemic arterial pressure, was struck by the relative constancy of the renal blood flow as compared with that in other tissues. This observation has been confirmed many times since. An example of the constancy of renal blood flow with variation in arterial perfusion pressure appears in Figure 22-3. The relative constancy of flow

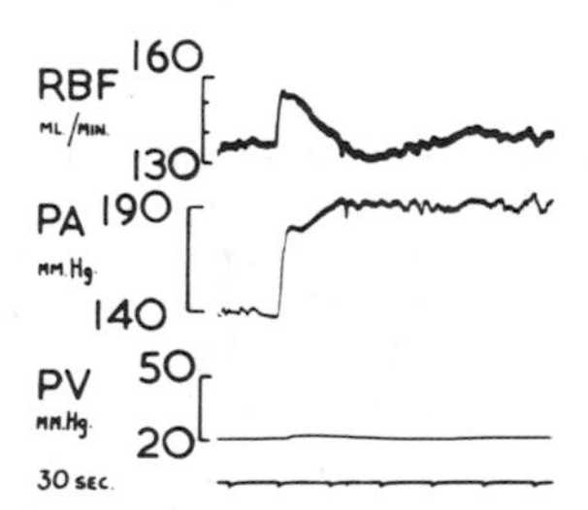

FIGURE 22-3. Immediate and stabilized relationship of renal blood flow (RBF) to perfusing pressure in the dog kidney. PA = renal arterial blood pressure; PV = renal vein pressure. (From S. J. G. Semple and H. E. DeWardener. *Circ. Res.* 7:643, 1959.)

as revealed by this figure is an example of *autoregulation* of the circulation. The mechanism is intrinsic, for it is manifested by the denervated kidney. The constancy of flow is maintained through a range of about 80 to 200 mm Hg of arterial perfusion pressure.

Several hypotheses have been advanced to explain renal autonomy. Of those considered in Chapter 7, the myogenic has been most popular, although the tissue pressure and metabolic theories have received some support. (For a detailed account, see Selkurt, 1963.) Evidence for the myogenic type of response appears to be offered in Figure 22-3, but participation of metabolic factors cannot be ruled out in this type of experiment.

When arterial pressure is suddenly increased, renal blood flow increases, but in 30 to 60 seconds internal adjustments occur, probably in the afferent arterioles, which bring the flow down to the original level despite maintenance of elevated pressure (Fig. 22-3). When pressure is suddenly dropped, an upward adjustment can be observed.

The dynamic reactivity implied in these fairly rapid adjustments corresponds to the type of reactivity anticipated from smooth muscle of the vasculature; hence the term *myogenic*. Isolated arterial segments have been shown to manifest similar responses. Moreover, that a vital phenomenon is involved is supported by the action of a number of agents known to impair smooth muscle activity, which eliminate autoregulation: papaverine, KCN, procaine, ethanol alcohol, and the anesthetic chloral hydrate. Cooling removes autoregulation, and it is depressed by hemorrhage and anoxia.

In summary, it seems likely that any single theory thus far advanced would oversimplify the complex adjustments that the kidney undergoes in its efforts to maintain the constancy of its blood flow. The possibility has been considered that several of the above mechanisms operate together in a composite mechanism.

FACTORS WHICH MODIFY THE RENAL CIRCULATION

Despite the inherent autonomy of its circulation, a number of factors, neurogenic and humoral, are able to decrease or increase kidney blood flow.

Innervation and Neurogenic Regulation

It is generally agreed that the major nerve supply to the kidney has its origin largely from the levels thoracic (T) XII through lumbar (L) II of the sympathetic nervous system in man, and in the dog from T IV to L II, but most abundantly from T X to T XII. In relation to its size, the kidney receives a more profuse and widespread supply than almost any other viscus. The thoracolumbar sympathetic supply to the kidney is a rich source of vasoconstrictor fibers for the kidneys, but presumptive evidence of dilator fibers is at hand (see below). The vagus nerve sends fibers to the kidney, but no evidence exists for vasodilator fibers in this circuit. Hence, the extrinsic vasomotor status of the kidney is maintained largely by variations in vasoconstrictor tone.

Evidence favors the notion that extrinsic regulation is low or absent in the basal state, to be invoked only in emergency states of heightened sympathetic nervous system activity. The principal evidence is that when one kidney is denervated, or denervated and transplanted, function is equal in the experimental and control kidneys in dog and man. This includes concordance of glomerular filtration rate (inulin or creatinine clearance), plasma flow (Diodrast or PAH clearance), diuretic activity, and electrolyte excretion. These findings are in accord with the concept of autonomy of the renal circulation.

Histochemical techniques have recently revealed the presence of both adrenergic and cholinergic fibers in the sympathetic nerve supply to the dog kidney. The cholinergic nonmedullated nerve fibers originate in ganglion cells in the hilus of the kidney and therefore persist after renal denervation (McKenna and Angelakos,

1968). Adrenergic innervation dominates the cortex; some norepinephrine is found in the outer medulla, but little, if any, in the inner medulla. Adrenergic nerve fibers travel along interlobar, arcuate, and interlobular arteries and along afferent arterioles, presumably innervating the smooth muscle cells to provide a vasoconstrictor action. None was seen in association with the glomeruli, efferent arterioles, or tubules.

The cholinergic fibers, on the other hand, appear to innervate predominantly efferent arterioles, most prominently those of the juxtamedullary nephrons, and the sphincters of the vasa recta (Moffat, 1967). A role in the regulation of the medullary circulation is strongly indicated for these vasodilator fibers. The cholinergic fibers may function to maintain blood flow during reduction of arterial pressure and may participate in alleged renal vasodilator reflexes. In concert with adrenergic constrictor fibers to the afferent arterioles, dilator fibers to efferent arterioles may provide a more versatile and effective control of glomerular filtration pressure, especially in juxtamedullary glomeruli.

Humoral and Pharmacological Factors

The catecholamines L-epinephrine and arterenol (levarterenol, norepinephrine) are active vasoconstrictors of the renal vasculature. Norepinephrine infused intra-arterially causes rather selective vasoconstriction in the cortex, implying that all the adrenergic terminals are here (Hollenberg et al., 1968). Angiotensin is another important vasoconstrictor. Several well-known and physiologically important humoral agents which cause renal vasodilation are acetylcholine (ACh), bradykinin, serotonin (in low dosage), and prostaglandins (found in high concentration in the renal medulla). When ACh is infused intra-arterially, RBF increases in both cortex and medulla, but mostly in the medulla, implying that the greater proportions of the ACh-sensitive receptors are here (Pinter et al., 1964).

For the pharmacology of a number of drugs, including anesthetic agents, that exert vasomotor influences on the renal vasculature, see Selkurt (1963).

RESPONSE OF RENAL BLOOD FLOW TO PHYSIOLOGICAL STRESS

Exercise

Renal blood flow is decreased in proportion to the severity of exercise. Both the amount of work involved and the duration of the exercise influence the results. As an example, subjects who had run the 440-yard dash at full speed had reductions of 18 to 54 percent below control, which remained for 10 to 40 minutes after exercise. It has been estimated that a saving of 0.5 to 1.0 liter of blood is made available to active tissue by renal vasoconstriction. Figure 22-4 illustrates how renal blood flow and hepatic blood flow are reduced in exercise, to supplement an increase in cardiac output, so that peripheral blood flow (largely to muscles) is enhanced substantially.

Posture and Orthostatic Hypotension

In normal young males, C_{PAH} in the sitting position is 91 percent of that in the supine; in the erect position, it is 85 percent. Motionless standing, or tilting of the subject, leads to progressive venous stagnation, reduced cardiac output, and neurogenic vasoconstriction, until the cerebral circulation becomes inadequate, when syncope occurs. Kidney flow is reduced to about one-half during the compensatory phase.

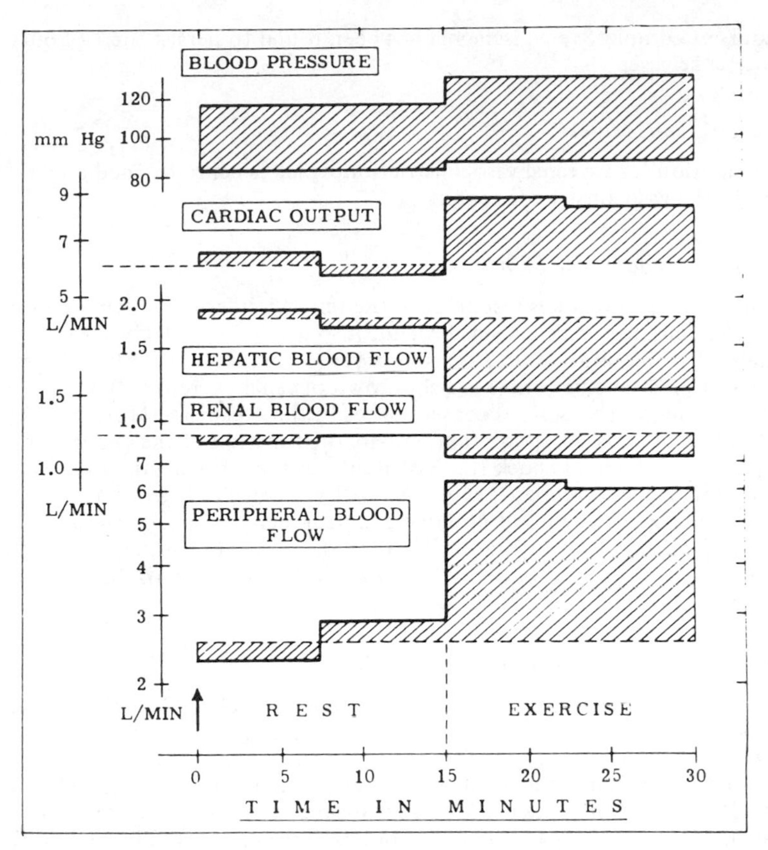

FIGURE 22-4. Effect of exercise on estimated hepatic blood flow, renal blood flow, peripheral blood flow, and arterial pressure in man. Exercise (bicycling against weights in recumbency) was started at 15 minutes. Hepatic blood flow by Bromsulphalein method; renal blood flow, PAH clearance; cardiac output, direct Fick. (From S. E. Bradley. *New Eng. J. Med.* 240:457, 1949.)

Renal Hypoxia and Ischemia

Hypoxia due to low O_2 intake or simulated altitude induces at first a slight increase in renal blood flow, but with more severe hypoxia this decreases. Apparently mild hypoxic states, unaccompanied by significant reflex vascular alterations, manifest a slight renal hyperemia, but more severe hypoxic states or asphyxia trigger vasoconstrictor reflexes in which the kidneys participate.

Acute Renal Ischemia

Very brief periods (1 to 5 minutes) result in a flow overshoot *(reactive hyperemia)* after release. Longer periods of ischemia in dogs (20 minutes) produce slight, inconsistent results on renal blood flow. Thirty minutes to two hours of ischemia result in marked, persistent reduction in renal blood flow. Since there is tubular damage, one must employ caution in interpreting clearance methods (e.g., C_{PAH}), although employment of the Fick principle (C_{PAH}/E_{PAH}) serves to correct for effects of tubular damage in the blood flow estimates. Reduction in renal blood flow after two hours of complete renal ischemia has been found to persist for 24 hours to two weeks after release.

Hypercapnia and Acidosis

These conditions cause renal vasoconstriction which is centrally mediated (enhanced activity of the vasoconstrictor centers).

Hemorrhagic Hypotensive Shock

Acute hemorrhage provokes responses in the renal circulation which are typical of general compensatory mechanisms brought into play, viz., reflex vasoconstriction, and shunting of blood to other tissues in order to compensate for blood loss. If blood loss is great enough, there is a shutdown of renal excretory function, which if prolonged might have serious consequences to the organism. Moreover, a prolonged period of anoxic hypotension impairs the function of the tubular epithelium, adding to the problem of shock the probability of renal failure and uremia.

In *tourniquet* and *traumatic shock*, as well as in *hemorrhagic shock*, blood flow is markedly reduced, as a result of enhanced neurogenic vasoconstrictor activity and of increase in renal vasoconstrictor substances such as epinephrine. Evidently, under these circumstances the renal circulation is subordinate to the welfare of the body as a whole.

Blood Flow in Renal Disease

Relevant information on renal blood flow in acute and chronic nephritis, nephrosis, and hypertensive kidneys has accrued from studies employing C_D and C_{PAH}. Varying degrees of tubular damage, reflected in decrease in *tubular excretory maxima* of Diodrast and PAH (Tm_D and Tm_{PAH}) and extraction (E_D and E_{PAH}) have complicated interpretation. When allowance is made for the influence of impaired tubular function, it is apparent that renal blood flow may actually be elevated in early nephrosis and the acute, inflammatory phase of nephritis. Later, as the disease becomes chronic, blood flow is undeniably reduced to varying degrees and may be very low in chronic nephritis accompanying the resultant disorganization of the kidney vascular pattern. The accompanying renal hypertension seen in nephritis is a manifestation of reduced renal blood flow.

RENAL LYMPHATIC SYSTEM

The lymphatic circulation is prominent in the cortex, where the lymphatic capillaries begin blindly. Lymphatics within the cortex follow the course of the interlobular vessels and drain in a centripetal fashion toward the arcuate vessels. Some evidence has been presented that lymphatics occur in the medulla of the human kidney, but their presence or absence in other species, including the dog, is controversial. The medullary lymphatics, when they occur, drain the vasa recta system, join the cortical branches at the arcuate level, then pass out with the interlobar vessels toward the renal pelvis. After converging at the hilus of the kidney, the lymphatic vessels course as perivascular channels to the cisterna chyli and the thoracic duct. Lymph flow can be increased by renal venous pressure increment and ureteral blockade.

Slight differences in electrolyte and organic content of renal lymph, as compared to plasma filtrate and thoracic lymph, have suggested a more complex origin than systemic interstitial fluid. Renal lymph involves the tubular reabsorbate and is complicated by the operation of the countercurrent system. The renal lymph protein concentration averages 2.9 gm per 100 ml. Evidently, the medullary renal lymphatics, when present, subscribe an important function for operation of the countercurrent mechanism by draining off excessive protein filtered by the vasa recta, which might otherwise accumulate in the interstitial spaces of the medulla. Removal of such protein could act to maintain a more favorable gradient of movement of the tubular reabsorbate into the vasa recta, attracted by the relatively higher colloid osmotic pressure there.

THE GLOMERULUS AND ITS FUNCTION

The glomerular capillaries are not simple loops but form a freely branching anastomotic network (Fig. 22-5). More specifically, larger *through channels* exist with an

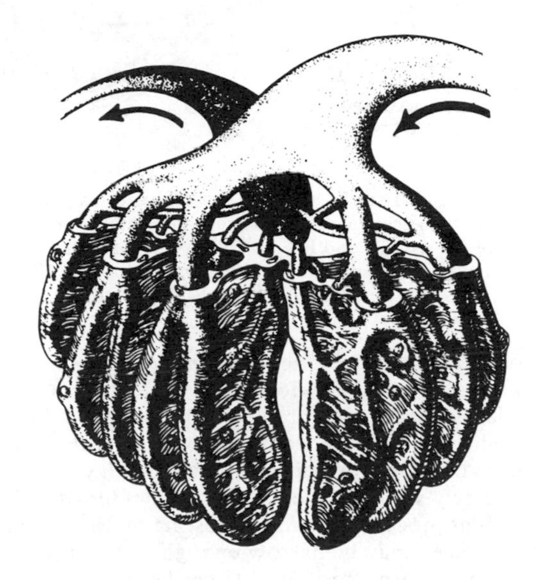

FIGURE 22-5. Pattern of distribution of *through channels* and *anastomotic channels* in the glomerulus. (From H. Elias et al. *J. Urol.* 83:795, 1960. Copyright © 1960, The Williams & Wilkins Company, Baltimore, Md. 21202, U.S.A.)

associated capillary network of smaller anastomotic channels. This arrangement may afford a structural basis for the skimming of plasma relatively freed of cells into the network of small capillaries, where the actual filtration proceeds, whereas the greater mass of blood cells directly and rapidly flows through the glomerular lobules to the efferent arterioles as an axial stream. This arrangement facilitates filtration processes by slowing the flow and reducing turbulence, and by possibly exposing more of the plasma which is to be filtered to the effective filtration surface.

The details of the filtering structures have been clarified by the use of the electron microscope (Fig. 22-6). This has revealed that the capillary endothelium, called the *lamina attenuata* or *lamina fenestrae* (0.05 μ thick), has pores 400 to 900 A in size. The pores are too large to restrain the plasma constituents. Instead,

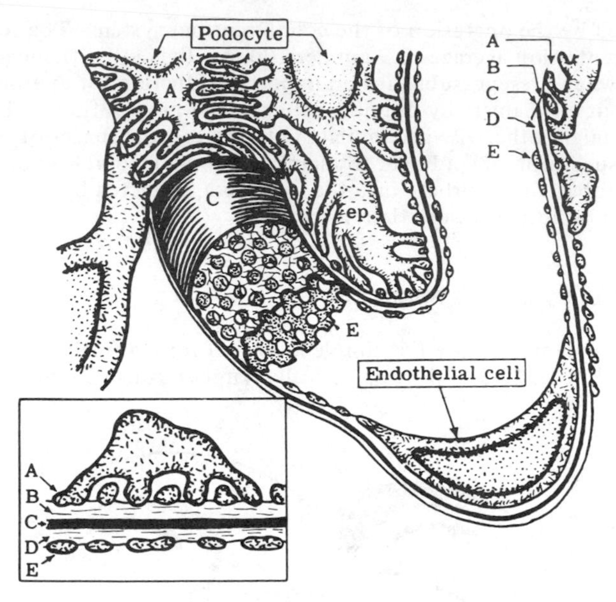

FIGURE 22-6. Glomerular capillary and associated portions of Bowman's capsule. Epithelial cells (ep.), the *podicytes* of the visceral layer of the capsule, form layer A. Layer C is the basement membrane and with layers B and D (cement layers) forms the *lamina densa.* Layer E is the capillary endothelium *(lamina fenestrae)*. (From D. C. Pease. *J. Histochem. Cytochem.* 3:297, 1955. Copyright ©1955, The Williams & Wilkins Co., Baltimore, Md. 21202, U.S.A.)

they expose the ultrafiltration membrane, the *lamina densa* (0.1 μ thick) to the free flow of plasma by removing the endothelial cytoplasmic barrier. The lamina densa is the glomerular basement membrane. Although it exhibits differences in stratification, it is probably a homogeneous layer and appears to be the limiting membrane for restraint of plasma proteins. The visceral layer of Bowman's capsule is organized into extremely specialized cells, the *podocytes* (foot cells), which possess thousands of foot processes *(pedicels)* resting on the lamina densa. The spacing between the pedicels may be narrow enough (100 A) to be a limiting dimension in restriction of plasma proteins. It has been designated by some authorities as a *slit-pore.*

DYNAMICS OF GLOMERULAR FILTRATION

The forces involved in glomerular filtration are the same as those which were discussed in Chapter 12 under the consideration of transcapillary molecular exchange. Glomerular filtration rate (GFR) is proportional to what is designated as *net filtration pressure.* This is expressed as:

$$\mathrm{GFR} = K\,(P'' - P' - \pi'' + \pi')$$

$$80 \quad 15 \quad 30 \quad 0^{*} \ (\mathrm{mm\ Hg}) \qquad (1)$$

(*assumes no filtration of plasma protein)

K is a proportionality coefficient embodying the volume of fluid filtered per minute per unit area per mm Hg pressure. Net filtration pressure (P_f) equals

35 mm Hg in equation 1. The double primes refer to the blood side of the capillary wall, and the single primes refer to the glomerular capsule fluid side. P denotes hydraulic pressure (P'', glomerular capillary pressure; P', the capsular pressure). π'' denotes colloid osmotic pressure of plasma proteins. The forces shown in mm Hg are values for the human kidney estimated from direct measurements made in other animals and done by puncture of single superficial nephrons (Gertz et al., 1969).

As was explained in Chapter 12, the free diffusion of molecules in solution is described by Fick's equation (Chap. 1). The concept of diffusion can be regarded as basic to the consideration of passage of water and solutes across the glomerular capillaries (Chinard, 1952). Diffusion occurs because of chemical potentials or electrochemical potentials of individual substances. The rate of passage will be determined by the diffusion coefficient of the substances (small molecules, including inulin, are close to H_2O, while large substances, like proteins, have diffusion coefficients so small as to be negligible). The gradients are determined by (1) difference in pressure, (2) chemical composition, and (3) temperature. (Electrochemical potentials involve the partial molal free energy of ions, plus the membrane potential.)

The nature of the filtering membranes is important. Chinard (1952), as a staunch proponent of the concept, believes that glomerular fluid is formed by diffusion rather than by bulk filtration through pores and feels that most of the capillary surface is involved in the passage of water and most dissolved substances.* Another concept, favored by Pappenheimer (1955), assumes that pores exist in the glomerular membranes as straight-bore cylindrical channels about 400 to 600 A (0.04 to 0.06 μ) in length. Average pore diameter was taken as 75 to 100 A, a pore size which would permit only minute quantities of albumin to be filtered. The total pore area was calculated as about 5 percent of the capillary surface. The membrane resistivity to movement of fluid could thus be described by Poiseuille's equation and characterized as bulk filtration, with fluid and solutes driven by hydrostatic pressure.

Consideration of the filtering membranes suggests that the large pores ("fenestrae") (up to 900 A in diameter) of the capillary endothelium in fact freely expose the plasma to the basement membrane. The basement membrane can be viewed as a hydrated gel, the supporting structures being protein micelles manifesting a thixotropic behavior. Since the membrane is relatively homogeneous

---

*The thermodynamic expression for GFR favored by Chinard is as follows:

$$\mathrm{GFR} = \sum^{J} \bar{V}_J \cdot \frac{dS_J}{dt} = \sum^{J} K_J \cdot \frac{A_J}{\Delta X} \cdot [J]'' \cdot \bar{V}_J \left[\bar{V}_J(P'' - P') + RT \ln \frac{N''_J f''_J}{N'_J f'_J} \right] \tag{2}$$

$\bar{V}_J$ is the partial molar volume of substance J, $\frac{dS_J}{dt}$ is the amount of J crossing per unit time, K_J is a proportionality coefficient, A_J is the area across which passage of J takes place, and ΔX is the distance across the capillary wall.

The gas constant terms, $RT \ln \frac{N''_J f''_J}{N'_J f'_J}$, can be replaced by the terms derived from osmometry: $-\bar{V}_{H_2O}(\pi'' - \pi')$.

Equation 2 can be simplified further for passage of H_2O:

$$\mathrm{GFR} = \frac{I}{[H_2O]''} \cdot \frac{dS_{H_2O}}{dt} = K'_{H_2O}[P'' - P' - \pi'' + \pi'] \tag{3}$$

in which K'_{H_2O} has been set equal to $K_{H_2O} \cdot A_{H_2O} \cdot \bar{V}_{H_2O}$.

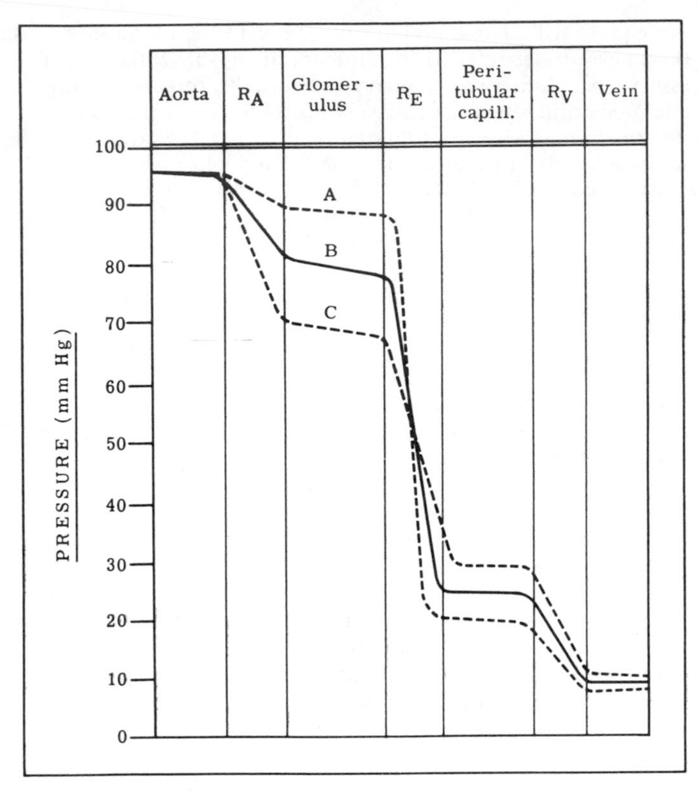

FIGURE 22-7. The pressure gradients in the renal arterioles and glomerular circuit. R_A = zone of resistance in afferent arterioles; R_E = efferent arteriolar resistance; R_V = venular resistance. Curve A is the relationship anticipated with afferent arteriolar dilation and efferent arteriolar constriction; curve B is the baseline relationship; curve C shows increased sympathetic activity (implying afferent arterial constriction and increased efferent arteriolar dilation), adjusted to values anticipated in the human kidney.

and pore-free as viewed by electron microscopy, the concept of bulk filtration through "pores" seems less appropriate than a hypothesis based upon physical–chemical laws of diffusion.

The pressure gradients across the glomerular circuit are shown in Figure 22-7. The capillary hydrostatic pressure is considerably higher than is typical of systemic capillaries, partly because the afferent arteriole, which leads directly into the glomerular capillary, is short and broad. Furthermore, the resistance to outflow of glomerular capillaries is higher than that of systemic capillaries because of the presence of the narrower efferent arteriole.

In summary, it can be stated that the filtration mechanism of the kidney is almost ideal since it provides a highly permeable membrane to diffusion and also provides relatively large forces for ultrafiltration, which are capable of a high degree of regulation by extrinsic and intrinsic (autoregulating) mechanisms.

Glomerular filtration pressure regulation is probably dominated by the influence of adrenergic constrictor activity to the afferent arterioles. If cholinergic (dilator) fibers simultaneously operate on the efferent arterioles, as may well be the case for the juxtamedullary glomeruli, a more efficient regulatory mechanism is provided (see Fig. 22-7). Curve C would represent the situation with increased afferent arteriolar constriction and increased efferent dilation (increased renal sympathetic nerve activity). Curve A represents the situation with dilation of afferent arterioles and efferent arteriolar constriction. B is the normal relationship.

Micropuncture studies of the capsular space have revealed that the qualifications for filtration are met; although the membranes are largely impermeable to protein, water and solutes pass freely, and the latter are found in approximately equal concentration on both sides of the glomerular membrane, allowing for slight differences in electrolyte concentration due to the Gibbs-Donnan effect (Chap. 1). Thus the classic investigations of Richards (1935) in amphibia and of Walker et al. (1941) in mammals (see Pitts, 1968) have shown that glucose, inorganic phosphate, creatinine, urea, uric acid, sodium and chloride, total osmotic pressure, and pH are approximately the same in the capsular fluid as in the plasma water. When the substance inulin was given, the clearance of which is taken to be a measure of glomerular filtration rate, it was found in equal concentration on both sides of the membrane.

GLOMERULAR FILTRATION RATE

Direct methods are not applicable to the measurement of filtration rate in the mammalian kidney, and indirect methods have been devised. These involve an understanding of the *renal plasma clearance concept.* The rate at which a substance (X) is excreted in the urine is the product of its urinary concentration (U_X, mg/ml) and the volume of urine per minute (V). The rate of excretion ($U_X \cdot V$) per minute depends upon the concentration of X in the plasma (P_X, mg/ml). It is therefore reasonable to relate $U_X V$ to P_X, and this relation is called the *clearance ratio:*

$$\frac{U_X \cdot V}{P_X} \tag{4}$$

or more generally UV/P. The calculated value has the dimensions of volume and is in reality the *smallest volume of plasma from which the kidneys can obtain the amount of X excreted per minute.* It must be understood that the kidneys do not usually clear the plasma completely of a substance in transit through the kidneys, but clear a larger volume incompletely. The clearance therefore is not a real but a *virtual* volume. When substances are being cleared simultaneously, each has its own clearance rate, depending on the amount reabsorbed from the glomerular filtrate or added to it by tubular secretion. The former has the lower clearance, the latter the higher. Those cleared only by glomerular filtration are intermediate, and their clearance, in effect, measures the rate of glomerular filtration in milliliters per minute.

The best-known substance which can be infused into the blood to provide a clearance equal to glomerular filtration rate is *inulin,* a polymer of fructose containing 32 hexose molecules (molecular weight 5200). Strong evidence exists that it is neither reabsorbed nor secreted by the tubular cells, that it is nonmetabolized in the body, that it has no physiological influences, and that it is freely filterable at the glomerular membranes. Several lines of evidence exist that the clearance of inulin is at the level of *glomerular filtration,* i.e., that it is an index of the total volume of fluid filtered by the glomerular membranes per unit of time. When simultaneous clearances of inulin and other sugars such as glucose, xylose, and fructose are performed, the clearances of the sugars are less than the clearance of inulin. The differences in clearance between these and inulin depend upon their particular reabsorptive mechanisms. However, when the transfer mechanism for sugar is inhibited by the glucoside *phlorizin,* the reabsorption of sugars is blocked and their clearances become identical with that of inulin; i.e., under these circumstances, the sugars are cleared only by the physical process of filtration. Evidence that inulin is not secreted by tubular cells is somewhat more indirect. One factor against this possibility is that inulin is not found in the urine of aglomerular kidneys of certain marine forms even when tremendously high plasma concentrations are attained by intravenous injection of inulin.

The clearance rate of inulin (C_{in}) in man is 120 to 130 ml per minute. This is taken to be the glomerular filtration rate (also C_F [clearance of filtrate]), or the amount of plasma water which is filtered through the glomeruli per minute. Substances other than inulin have been used to provide an estimate of GFR, the most common of which is creatinine. It has been shown that exogenous (true) creatinine clearance is approximately equal to that of inulin in the dog, rabbit, sheep, seal, frog, and turtle and thus provides a measure of glomerular filtration rate in these animals. However, in addition to being filtered through the glomeruli, creatinine is secreted in part by the tubular cells in man and other primates; hence, exogenous creatinine clearance cannot be used as an estimate of filtration rate.* Other substances which appear to have a clearance close to that of inulin in the dog and other animals are thiosulfate, thiocyanate, and mannitol.

A knowledge of the glomerular filtration rate permits quantitation of the amount of any substance freely filtered (C_F, ml/min $\times$ P_X, mg/ml). When dealing with a substance which is reabsorbed, one may subtract one minute's excretion, U_XV, from the filtered load to find the amount reabsorbed by the tubules in milligrams per minute:

$$T_R = C_F \cdot P_X - U_X \cdot V \tag{5}$$

Furthermore, under the same principle, the knowledge of glomerular filtration rate permits the calculation of tubular secretory activity, as follows:

$$T_S = U_X \cdot V - C_F \cdot P_X \tag{6}$$

A number of the substances studied for tubular secretion are bound to plasma protein in varying degrees and cannot be freely filtered. The filtered moiety therefore requires correction for the binding. This introduces the factor f, which represents the fraction of the substance not bound by the plasma proteins and therefore free to be filtered. The factor f may vary with the species, e.g., it is 0.92 for p-aminohippurate in the dog and 0.83 in man. Also, it differs for various substances. Thus the f value is 0.40 for phenol red and 0.72 for Diodrast. The estimates are made by dialysis experiments. The completed equation is

$$T_S = U_X \cdot V - C_F \cdot P_X \cdot f \tag{7}$$

It should be added that plasma binding does not affect the fraction of clearance which involves tubular secretion. Such mechanisms operate by action of transfer systems which rapidly remove the substances from the plasma water, to establish a gradient whereby the plasma binding is readily broken down, so that the substances continue to be rapidly removed by tubular secretion. However, the diffusion process across the glomerular membranes does not affect the protein binding, and it is the moiety which is free in plasma water that is filtered.

In summary, clearances less than that of inulin represent mechanisms involving filtration and reabsorption; clearances higher than that of inulin represent mechanisms involving glomerular filtration plus some degree of tubular secretion with allowance for possible plasma binding.

---

*A creatinine-like chromagen is found in human plasma which gives the same chemical reaction for colorimetric analysis (Jaffe reaction) as does true creatinine, but whose calculated clearance approximates that of inulin. Its clearance is widely used in patients as an estimate of GFR.

TUBULAR REABSORPTION OF ORGANIC SUBSTANCES

GLUCOSE

A classic example of tubular reabsorption is the glucose mechanism. Usually none or only a trace of this hexose appears in the urine. When plasma glucose is elevated in man to about 180 to 200 mg per 100 ml, the so-called *threshold,* glucose appears in the urine. As the concentration is further raised by continued infusion of glucose, the reabsorptive mechanism becomes progressively saturated until the rate of reabsorption becomes constant and maximal. This indication of saturation of the transport system is referred to as the *tubular maximum* or Tm – in this instance Tm_G. In the human, Tm_G has a value of 340 mg per minute; it is about 200 mg per minute in the dog.

Figure 22-8 illustrates the principle of saturation of the reabsorptive system by a progressive elevation of plasma glucose concentration experimentally. It will be noted that U_GV/P or C_G increases progressively. This approaches asymptotically the clearance of inulin (C_F) as the plasma levels increase further beyond the point of saturation of the tubular mechanism. To put it another way, the clearance becomes more and more preponderantly the consequence of glomerular filtration,

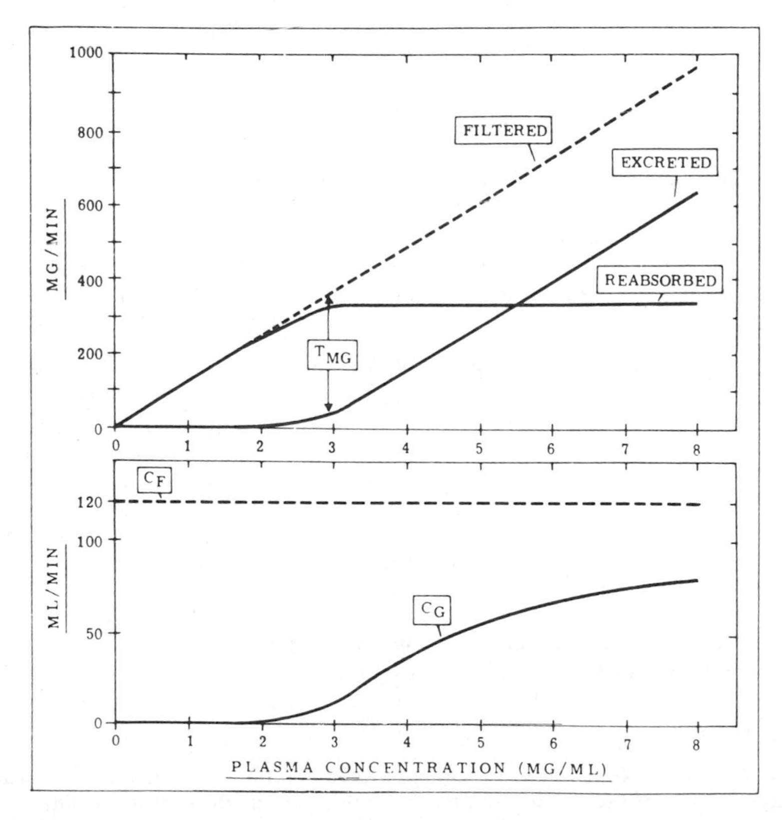

FIGURE 22-8. Glucose tubular reabsorption and clearance of glucose as related to increase in plasma concentration. C_F = filtration rate; C_G = glucose clearance. (From Smith, 1956. P. 39.)

and the reabsorbed moiety becomes a fractionally smaller part of the total excretory component.

A description of the kinetics of the transport system is as follows (Pitts, 1968, p. 74): A carrier substance, C, is present in the luminal membrane of proximal tubular cells in fixed and limited amount. The carrier combines reversibly with glucose, G, from the tubular fluid, to form the complex G · C within the membrane. This complex moves to the cytoplasmic side of the membrane, where it is split; the glucose is delivered to the cytoplasm, and the membrane carrier returns to the luminal surface to accept another glucose molecule.

$$\begin{array}{ccc} \text{Glucose} & \text{Membrane} & \text{Cytoplasm} \\ G & + C \rightleftharpoons G \cdot C \rightarrow & C + G \end{array} \tag{8}$$

From the cytoplasm, G moves out of the basal end of the cell into the interstitium by another step, apparently not rate-limiting.

Glucose reabsorption involves an active (enzymatic) process and the expenditure of energy, which is typical of all mechanisms manifesting a Tm. A widely accepted theory has been that the transfer system for glucose involves phosphorylation at the luminal border of the tubular cell at the expense of adenosine triphosphate (ATP) (hexokinase reaction), then uncoupling of the phosphate bond at the interstitial border, where the glucose molecule is picked up by the peritubular capillaries. In apparent support of the theory of phosphorylation has been the finding that phlorizin inhibits the oxidative reactions of the citric acid cycle and the generation of ATP which accompanies the formation of hexose phosphates. In addition, the ability of phlorizin to inhibit the phosphatases that hydrolyze hexose phosphates is well known. Taggart (1958) pointed out that phlorizin does not necessarily act primarily on the cellular transport system but may act at the cell membrane to limit penetration of sugars. It now seems clear that phlorizin competitively inhibits glucose reabsorption by binding with great affinity at the membrane site which binds or orients glucose (Diedrich, 1963). Phlorizin structurally is glucose with the C_1 hydroxyl replaced with phloretin; hence, competitive binding is not surprising. Other hexoses (galactose, fructose, and, slightly, xylose) compete with glucose for the transport site. The disaccharide sucrose shows very little reabsorption in the in vivo system.

Kleinzeller et al. (1967), using rabbit cortical slices and isolated tubules, have confirmed the notion that the active transport process is localized in the brush border at the luminal face of the cells, analogous to the intestinal glucose transport mechanism. Importantly, as in the gut, a Na^+ dependency for D-glucose transport has been confirmed, as well as phlorizin sensitivity. This mechanism causes accumulation of D-glucose within the tubular cells before its passive outflow, probably by facilitated diffusion, at the basal membrane. D-Galactose behaved very much the same as D-glucose in their system, and D-fructose and α-methyl-D-glucoside were also readily accumulated against a concentration gradient by the phlorizin-sensitive Na^+-dependent active transport system. Considerable specificity was indicated by the fact that other monosaccharides were transported poorly if at all (e.g., D-xylose, D- and L-arabinose, 6-deoxy-D-glucose, and 6-deoxy-D-galactose).

AMINO ACIDS

Amino acid clearance is low in man (1 to 8 ml per minute). Much of the earlier characterization of the mechanism has been done in the dog. More recently, rat kidney studies have supplied additional information, utilizing isolated nephrons, kidney slices, and micropuncture techniques. Several amino acids (glycine, arginine, lysine, proline, and hydroxyproline) demonstrate relatively poor reabsorption with small Tm; others, like histidine, methionine, leucine, isoleucine, tryptophan, valine, threonine, and phenylalanine, are so effectively reabsorbed that saturation is not achieved by plasma concentrations which do not cause severe nausea or other physiological disturbances.

There are probably no less than three renal tubular mechanisms for reabsorption of amino acids (Pitts, 1968). One transports lysine, arginine, ornithine, cystine, and possibly histidine; a second handles aspartic and glutamic acid; a third has been shown to involve proline, hydroxyproline, and glycine (Scriver and Goldman, 1966).

The amino acid transport mechanisms display saturability and substrate specificity, and, in vitro, Michaelis-Menten kinetics. An interesting series of "inborn errors of membrane transport" is manifested by some of the amino acids – cystinuria, hyperprolinemia, and hydroxyprolinemia, among others. The third is associated with mental deficiency.

ASCORBIC ACID

Ascorbic acid has a Tm of 2.0 mg per minute in man (about 1.5 mg per 100 ml of filtrate). In the dog, which is able to synthesize ascorbic acid, reabsorption is 0.5 mg per 100 ml of filtrate. The Tm represents the net activity of a three-component system (filtration, proximal tubular reabsorption, and distal tubular secretion). Distal secretion is promoted by pretreatment with adrenal steroid (deoxycorticosterone acetate), which stimulates distal sodium-potassium exchange. Additional stimulation of distal secretion results from increasing the filtered load of sodium, by alkalinization of the urine with $NaHCO_3$, and administration of acetazolamide. Evidently, a linkage between the ascorbic acid secretory site and the distal sodium-potassium exchange mechanism exists. This may be related to the vitamin's acidic properties.

UREA

Urea, a major product of protein metabolism, is filtered and reabsorbed to varying degrees (40 to 70 percent) throughout the nephron. Its reabsorption is inversely related to the rate of urine production because reabsorption is largely a process of back diffusion in man, dog, rabbit, and chicken. In such a reabsorptive mechanism, the return of urea from the tubular urine, where it is in relatively high concentration on account of water uptake by the nephron, proceeds by the process of diffusion across the tubular epithelium into the peritubular capillaries, where it is in relatively low concentration. With rapid urine flow, the time for diffusion is limited. However, active reabsorptive mechanisms for urea (involving a transfer system and expenditure of energy) operate in the kidneys of Elasmobranchii. Protein-depleted rats on a low-protein, high-salt diet show urea/inulin clearance ratios as low as 0.01, with papillary tissue urea concentration about three times that of the final urine, suggesting an adaptive development of an active reabsorption mechanism for urea, necessary to maintain the high papillary solute concentration required for the countercurrent mechanism (Truniger and Schmidt-Nielsen, 1964). On the other hand, in the amphibian (anuran) kidneys the tubules *secrete* urea by an active process.

The importance of the urea clearance stems from the fact that it has been used in the past as a clinical indication of glomerular filtration, which it reflects reasonably well at high rates of urine flow, when its reabsorption is minimal. The relationship of the clearance of urea to the rate of urine flow is shown in Figure 22-9.

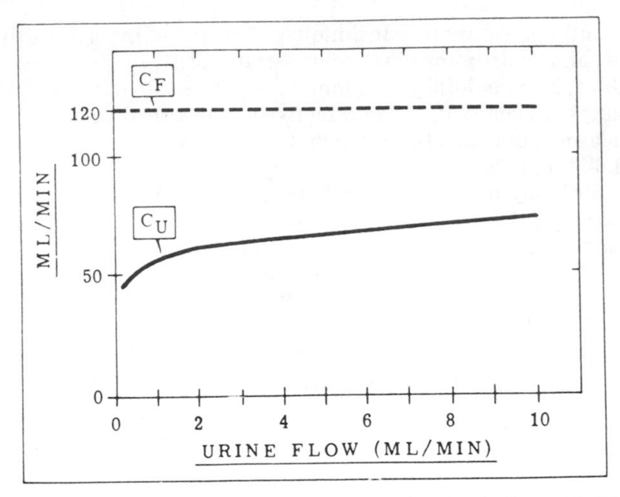

FIGURE 22-9. Relationship of urea clearance (C_U) to urine flow. (From Smith, 1956. P. 79).

URIC ACID

In mammals, uric acid appears as a consequence of metabolism of purine bases. In most mammals it is oxidized to allantoic acid, but not in primates or the Dalmatian coach dog. It has been generally assumed that a Tm characterizes its reabsorption, averaging 15 mg per minute in man, a value so high compared with the amounts normally existing in the plasma that saturation should not occur under normal circumstances. But under conditions of injections of a uricosuric drug (sulfinpyrazone) which blocks tubular reabsorption of uric acid, combined with vigorous mannitol diuresis, the ratio of excreted urate over filtered urate exceeds unity and may go up to 1.23 in man, demonstrating tubular secretion. Thus, a three-component system of filtration, reabsorption, and secretion is suggested, so that the amount excreted is normally the *net effect* of these operations. Definite evidence of tubular secretion has also been found in birds and reptiles and in the guinea pig and Dalmatian coach dog. It should be added that other uricosuric agents exist, all of which have been used in the treatment of gout, a disease characterized by deposition of uric acid crystals in the joints. These include cinchophen, salicylate, acetylsalicylic acid, the mercurial Salyrgan, Carinamide, and Benemid.

CREATINE

Creatine is a product of muscle metabolism which disappears from the urine of humans after adolescence. It is filtered and reabsorbed in concentrations below 0.5 mg per 100 ml. At higher concentrations, reabsorption is incomplete and excretion is enhanced. No Tm has been demonstrated. At higher plasma levels, the creatine/C_F ratio becomes constant at 0.8.

OTHER SUBSTANCES

Other substances involved in tubular reabsorption have been demonstrated in dog and man. These include acetoacetic acid, β-hydroxybutyric acid, lactic acid, and guanidoacetic acid. Maximal rates of transport have been demonstrated for β-hydroxybutyric and lactic acids.

In summary, tubular reabsorption of various substances is limited by a maximal rate, and this varies markedly from one substance to another. In some instances, such as the sugars and certain amino acids, a transport mechanism may be shared in common and involve the reabsorption of several related substances, but there appear to be many transport systems which operate wholly independently of one another. Besides glucose and the amino acids, these include lactic acid, uric acid, acetoacetic acid, and the vitamin mechanisms.

There is considerable evidence that most of these substances are reabsorbed either in entirety or in part in the proximal convoluted tubule. Information on localization has been obtained by micropuncture of the tubules and by the stop-flow technique. The technique is as follows: The ureter of the experimental animal is clamped during vigorous diuresis, bringing the movement of urine in the nephrons to a halt. Stoppage of the tubular movement of urine for a brief interval (about eight minutes) permits the tubular reabsorptive and secretory processes to proceed to greater completion. Fractional collection of the urine which issues upon release of the ureter permits allocation of observed changes in tubular urine concentration to various segments of the nephron. This is aided by the injection of a marker substance (such as inulin or creatinine) during the period of ureteral blockade to tag the entry of new glomerular fluid or, in effect, to mark the end of the stagnant urine column at the approximate level of the glomerulus. The knowledge of enzymatic and cellular mechanisms involved with these transport systems is rather limited at the present time. Further investigative effort is necessary to clarify the multiplicity of mechanisms involved.

TUBULAR SECRETION

Evidence has accumulated from techniques such as the above that the proximal convoluted tubule is the site of active secretion of some physiologically occurring substances as well as certain foreign substances when injected into the circulation.

TUBULAR SECRETION OF FOREIGN SUBSTANCES

p-Aminohippuric Acid (PAH)

At low concentrations PAH is almost completely cleared from the plasma by a combination of glomerular filtration and efficient tubular secretion. Hence, C_{PAH} measures approximately 90 percent of the total plasma flow through the kidneys. In man, the clearance is 600 to 700 ml per minute and corrected for hematocrit gives the *effective renal blood flow.* The proportion of the total plasma passing through the kidneys which is filtered at the glomeruli is called the *filtration fraction.* It can be estimated by the ratio of C_{in}/C_{PAH}, recognizing that C_{PAH} represents the "effective" plasma flow, not total plasma flow. It has a value of 0.20 in man and 0.32 in the dog.

When plasma levels are elevated to the range of 30 to 50 mg per 100 ml, the secretory mechanism becomes saturated, and the Tm can be discerned. Tm for PAH in man is about 77 mg per minute per 1.73 square meters of surface area (SA). In the dog it is about 33 mg per minute per 1.73 m^2 SA. The relationship of the mechanism of urinary excretion as it relates to plasma concentration appears in Figure 22-10. As the tubular secretory mechanism becomes saturated and is exceeded by progressive increments in plasma PAH, C_{PAH} is observed to decrease. This happens because, with saturation of the tubular transfer system, the total UV and the calculated UV/P become progressively more a function of glomerular

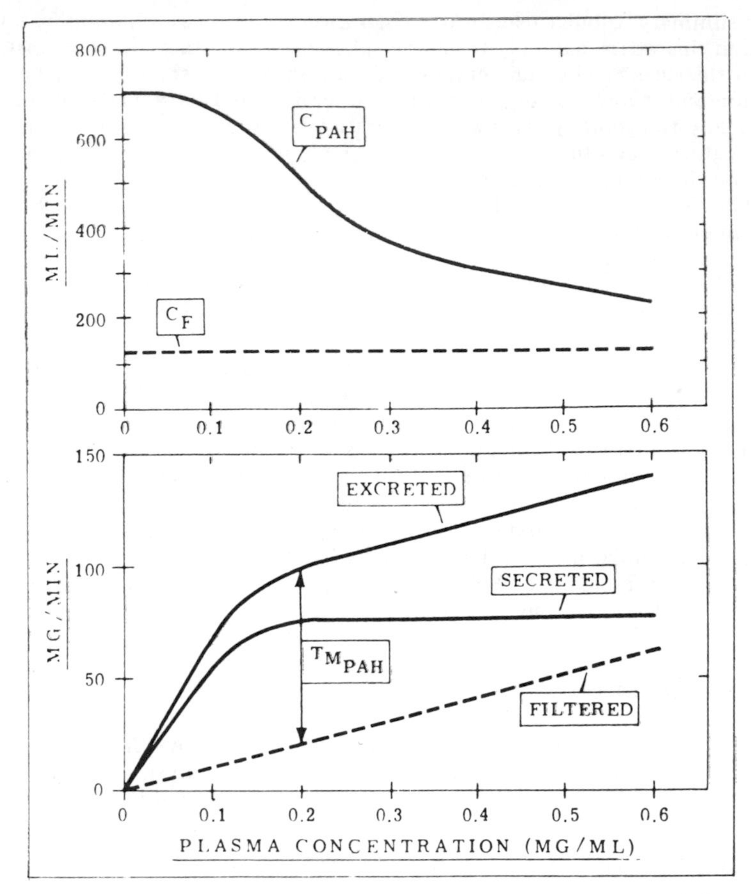

FIGURE 22-10. The relationship of p-aminohippurate clearance (C_{PAH}) and tubular secretion to increase in plasma concentration. (From Smith, 1956.)

filtration; hence UV/P approaches that function asymptotically. The portion of UV contributed by tubular secretion becomes a constant but fractionally smaller part of the total UV.

The mechanism of transfer of PAH and other substances secreted by the tubular cells has been approached by studies involving isolated nephrons and kidney slices. This approach offers the opportunity to study the effect of substances which accelerate or depress accumulation of PAH in the slices as observed in the Warburg apparatus. Taggart (1958) has shown that the accumulation of PAH in tissue slices is dependent on aerobic metabolism by the fact that it is suppressed by lack of oxygen, by chilling, or by any metabolic poisons which inhibit respiration. Uptake is also suppressed by 2,4-dinitrophenol and by related compounds, which do not inhibit the uptake of oxygen but uncouple phosphorylation. Accumulation was increased greatly by the addition of acetate to the medium, and to a lesser extent by addition of lactate and pyruvate. Acetate and lactate increase Tm_{PAH} markedly when administered intravenously. On the basis of such experiments, it was suggested that a reaction which involves acetyl coenzyme A is a metabolic pathway common to the processes by which the tubular epithelium secretes PAH and other substances competing with it.

The site of secretion of the organic acid PAH is the proximal convolution, and its active transport depends on generation of high-energy phosphate compounds (e.g., ATP) by oxidative metabolism. It is not understood how the energy is coupled to the transport process. Uptake of PAH from the blood side of the tubules probably involves an active pump at the peritubular cell membrane. Foulkes (1963) concluded, from a kinetic analysis in the rabbit kidney, that intracellular PAH is in a free, not bound, form. Transfer from cell to lumen probably involves a facilitated diffusion mechanism, with no energy-requiring step at the luminal membrane. In the mammal, the large volume of glomerular filtrate renders unnecessary an active concentration step at the luminal membrane.

Other Substances

Additional foreign substances secreted by the tubules are *Diodrast, phenol red,* and *penicillin.* Each has its characteristic Tm; e.g., Diodrast Tm averages 50 mg per minute in man, and Tm for phenol red is 36 mg per minute. The transfer systems for the different substances so far described appear to be mutually interrelated, because competitive depression of Tm is noted when they are simultaneously cleared at high plasma levels. Such blocking agents as Carinamide and Benemid inhibit the secretion of the above-mentioned substances.

A group of *organic bases* is actively secreted by the proximal tubule: tetraethylammonium (TEA), tetramethylammonium (TMA), tetrabutylammonium (TBA), Darstine (mepiperphenidol), Priscoline (tolazoline), and hexamethonium. The Tm's are small (e.g., Tm_{TEA} is 1.0 to 1.4 mg/min/m^2 SA in dogs). The mechanism is distinct from that of strong organic acids such as PAH.

TUBULAR SECRETION OF PHYSIOLOGICAL SUBSTANCES

Creatinine

Creatinine is derived from creatine in muscles. Exogenous creatinine is cleared by glomerular filtration plus tubular secretion in man and other primates. The Tm is small and about 16 mg per minute. It is also secreted by tubules of certain fish, the alligator, chicken, goat, rat, cat, and guinea pig. Animals from which it is cleared only by glomerular filtration have been cited above.

It has been disclosed that creatinine transport is shared by both the organic acid and organic base transport mechanisms in the monkey and guinea pig (Selkurt et al., 1968; Arendshorst and Selkurt, 1970). This finding may be ascribable to the amphoteric nature of creatinine.

Naturally Occurring Organic Bases

Naturally occurring organic bases include N-methylnicotinamide, guanidine, piperidine, thiamine, histamine, and choline. N-methylnicotamide, a metabolic derivative of nicotinic acid, has a clearance up to three times GFR, suggesting tubular secretion.

EXCRETION OF ELECTROLYTES

Virtually all segments of the nephron are concerned with the electrolyte economy of the body, to help preserve the proper balance of cations and anions. Reabsorptive mechanisms involve active pumps and energy expenditure.

Electric Potentials

A knowledge of electrical potentials across the tubular membrane is necessary in order to decide which ions are actively transported. The origin of such potentials was discussed in Chapter 1 and is based upon application of the Nernst equation. This has been examined experimentally in the Necturus kidney by Giebisch (1961). The Necturus kidney can be doubly perfused (renal portal system and aorta), so that the composition of fluid bathing the peritubular membrane (outside) and luminal membrane (inside) of the tubular cells can be altered independently.

If one increases progressively the K^+ concentration (as K·Cl) of the perfusion fluids outside the cell, the potential measured across the peritubular membrane decreases from a normal value of 72 mv (at 4 mEq per liter of K^+) to zero (Fig. 22-11). The slope of the function is –55 mv per 10-fold change in concentration over the linear portion. If perfectly K^+ permselective, the slope would be

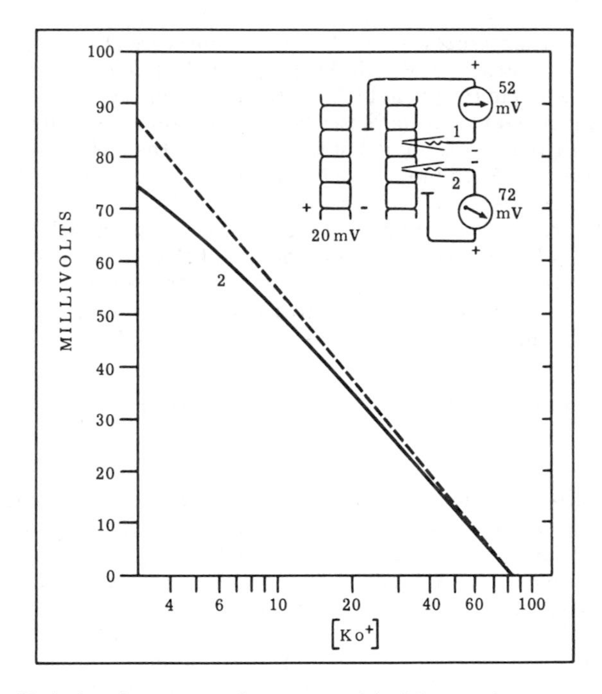

FIGURE 22-11. Variations in transmembrane potential with potassium concentration of the bathing medium across the luminal membrane (1) and across the peritubular membrane (2). (Adapted from Giebisch, 1961.)

58 mv at room temperature. The relationship of K^+ on both sides of the membrane to the potential is given by the Nernst equation:

$$E(\text{mv}) = 58 \log \frac{K_i}{K_o} \tag{9}$$

(The slight deviation from the theoretical K^+ potential might be due to electrical shunting between lumen and peritubular fluids via extracellular routes.)

The above brief review of transtubular potentials is helpful in the understanding of the nature of tubular ion transport. Thus, strong evidence for active transfer of Na^+ is that it moves out of the proximal tubule against a gradient of electrochemical potential.

CATIONS

Sodium

About 99 percent of filtered Na^+ is reabsorbed by active transport mechanisms (Table 22-1). Estimates vary from 65 to 80 percent reabsorption in the proximal tubules and the remainder in the ascending portion of the loop of Henle, the distal convoluted tubules, and the collecting duct. Reabsorption in the proximal tubule is isosmotic. Reabsorption of NaCl and $NaHCO_3$ here is primary, by an active process; water follows passively and is not influenced by ADH. Another site of active reabsorption occurs in the ascending thick portion of the loop of Henle, significant because this is relatively water-impermeable, and as a consequence the urine becomes hypotonic to plasma at this site.

TABLE 22-1. Reabsorption of Sodium and Related Anions

Plasma Conc. (mEq/L)	Donnan Factor	GFR	Filtered (mEq/min)	Excreted (mEq/min)	Reabs. (mEq/min)	Percent Reabs.
Na 140	0.95	125	16.6	0.135	16.465	99.3
Cl 103	1.05	125	13.5	0.135	13.37	99.1
HCO_3 27	1.05	125	3.55	<0.002	3.54	99.9

Evidence of the active nature of sodium reabsorption has been cited in terms of its being pumped against an electrochemical gradient. Other evidence is that the TF/P (tubular fluid/plasma concentration) of Na^+ has been observed to go below 1.0 in the face of brisk osmotic diuresis.

No Tm for sodium has been discerned. In man, about 200 μEq per minute is excreted. Increased loading of the kidney tubules by Na^+ is followed by increased total reabsorption, but with decreased efficiency, since the total load is not quite so effectively reabsorbed. As a result, urinary excretion of Na^+ increases. The hormone aldosterone, secreted by the glomerulosa zone of the adrenal cortex, is necessary for efficient reabsorption. Stop-flow experiments suggest that the distal convoluted tubule is the most important side of aldosterone activity, although some evidence suggests a proximal tubular action also.

Sodium Transport Mechanisms. Figure 22-12 represents a model of a proximal tubular cell. K^+ content of the cell is high (ca 140 mEq per liter), while Na^+ and Cl^- content is low (10 mEq per liter). The cell is electrically negative with respect to lumen and peritubular fluid. Na^+ can diffuse along an electrochemical gradient from lumen to cell; its concentration within the cell is kept low by a pump which extrudes it into the peritubular fluid. Diffusion of K^+ and Cl^- is the major determinant of the potential across the peritubular membrane. The peritubular membrane pump functions also to return K^+ that has diffused out.

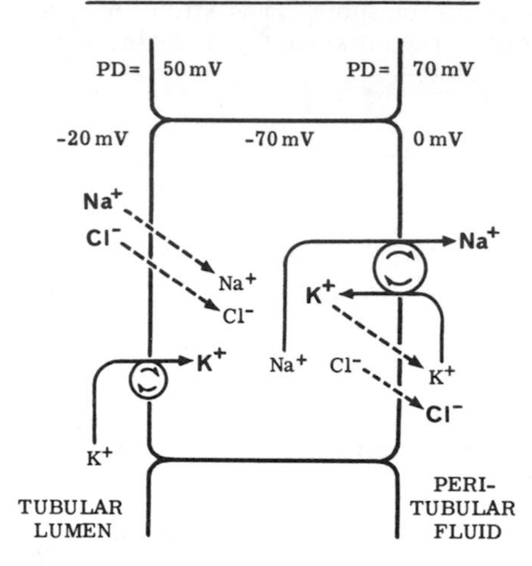

FIGURE 22-12. Model of a proximal tubular cell showing mechanisms for Na^+, K^+, and Cl^- transport. (From R. F. Pitts. A comparison of modes of action of certain diuretic agents. *Progr. Cardiovasc. Dis.* 3:537, 1961. By permission.)

The potential across the luminal cell membrane is produced by a current × resistance drop as current flows from lumen into cell. The electromotive force (EMF) is created by the diffusion potential of Na^+ and the sodium pump at the peritubular membrane. The circuit is closed by the existence of intercellular channels in parallel with the cellular pathways.

Chloride ions appear to be passively reabsorbed by the proximal tubular cells, accompanying sodium, but the exact nature of their entry is not well understood.

Essentially similar mechanisms to those operating in the proximal tubular cell occur in the distal convoluted tubule (Fig. 22-13). The potential across the peritubular membrane is greater, as is the transtubular potential. Na^+ still diffuses into the cell from the lumen, but along a less steep gradient, since the luminal fluid content of sodium has been considerably reduced to this point (to ca 10 percent of the filtered sodium at the beginning of the distal tubule, and to ca 2 percent at the end). Finally, the collecting duct reabsorbs almost all the remaining sodium. Na^+ reabsorbed in the distal tubule and collecting duct is exchanged with H^+, K^+, and NH_3, which are secreted in these segments. Operation of this ion exchange mechanism will be considered in greater detail in Chapter 24.

Energy Supply of Active Sodium Transport. Extrusion of sodium from the cell against an electrochemical gradient requires expenditure of metabolic energy. A linear relationship has been found between the rate of oxygen consumption and tubular sodium reabsorption by several groups of investigators. If blood pressure is reduced, and with it GFR, less sodium is filtered and reabsorbable, and O_2 consumption decreases proportionally. It has been calculated that 1 micromole (μM) of oxygen is used per 30 mEq of sodium reabsorbed.

Potassium

Potassium clearance is usually well below that of inulin, suggesting fairly complete reabsorption. Under certain circumstances, however, entailed with giving large

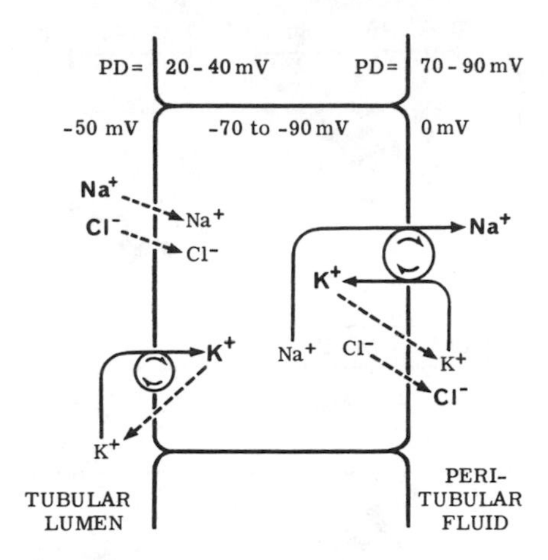

FIGURE 22-13. Model of distal tubular cell transport mechanisms for Na^+, K^+, and Cl^-. Note that K^+ "pump" in luminal membrane has potential of being bidirectional. (From *Physiology of the Kidney and Body Fluids,* 2nd. Edition, by Robert F. Pitts. Copyright © 1968 Year Book Medical Publishers, Inc. Used by permission.)

amounts of potassium salts of foreign anions, more potassium is excreted than is filtered. Thus, it appears that a three-component system operates with proximal tubular reabsorption and distal secretion. This conclusion is supported by micropuncture analysis. Figure 22-12 indicates that K^+ is actively reabsorbed by a pump in the tubular membrane of the proximal tubular cell. Figure 22-13 shows that the theoretical K^+ pump can operate either way in the distal tubule, reabsorbing K^+ under conditions of low intake or secreting K^+ during potassium loading.

The micropuncture studies of Giebisch et al. (1964) show this more clearly in Figure 22-14. Progressive reabsorption of filtered K^+ is indicated in the proximal tubule. In the distal, the K^+ concentration drops to very low levels with a low-K diet (part A), suggesting continued reabsorption through the collecting duct. In B (normal diet), the rising concentration along the distal segment clearly demonstrates a secretory process. This is magnified under the influence of a high-K diet (part C), where the TF/P ratio of K definitely exceeds the plasma concentration.

An important consideration in the distal tubular secretion is the fact that K^+ is exchanged for Na^+, so that the degree of secretion is limited by the amount of Na^+ available in the distal convoluted tubules and collecting duct. Secretion of K^+ decreases if there is little Na^+ available and increases if there are larger amounts of Na^+ available in the distal nephron.

Calcium

Excretion of calcium is complicated by the fact that a significant part of its plasma content is combined with plasma proteins. However, urinary excretion is nominally low (about 8.5 μEq per minute), suggesting efficient tubular reabsorption of the moiety which is filtered. Parathormone (PTH) of the parathyroid gland promotes tubular reabsorption of calcium (Chap. 33).

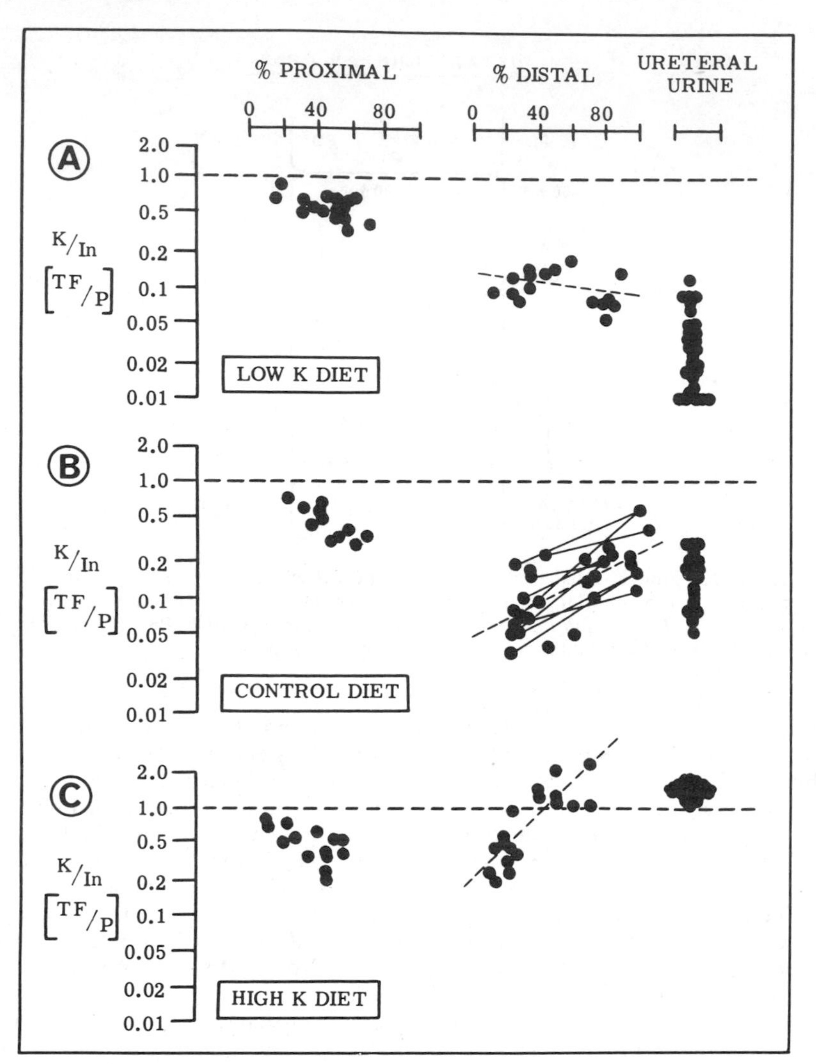

FIGURE 22-14. Potassium excretion along the nephron of rats on control diet (B), on low-K diet (A), and on high-K, low-Na diet (C). Potassium/inulin TF/P concentration ratios given as a function of tubular length. The K/in [TF/P] ratio corrects for the influence of tubular water reabsorption. (From G. Malnic, R. M. Klose, and G. Giebisch. *Amer. J. Physiol.* 206:674, 1964; and from G. Giebisch and E. E. Windhager. *Amer. J. Med.* 36:634, 1964.)

Magnesium

Magnesium not bound by plasma proteins is filtered and reabsorbed. Excretion of about 5 to 6 μEq per minute occurs in man.

Other Electrolytes

In this category can be considered NH_3 and H^+. Ammonium is synthesized in the tubular epithelium from glutamine and other amino acids and is secreted into the distal tubular urine. It exchanges in this process with Na^+ as a mechanism in acid-base regulation and Na^+ conservation. Finally, the distal tubule is able to generate and secrete H^+ as a means of acidifying the urine by the aid of the enzyme carbonic anhydrase. (Further details in Chap. 24.)

ANIONS

Chloride

Because chloride is the chief indifferent anion that accompanies Na^+ through the kidney tubular epithelium, this renal mechanism is conditioned by sodium reabsorption. Thus the Cl^- is pulled by its electrical charge as a result of the active transfer of Na^+. However, there is evidence (Chap. 23) that Cl^- may behave independently of Na^+.

Bicarbonate

Data on bicarbonate reabsorption are expressed in milliequivalents per 100 ml of glomerular filtrate with an apparent physiological relationship between HCO_3^- reabsorption and filtration rate. As plasma HCO_3^- increases, the quantity filtered increases in direct proportion. Essentially all of the filtered HCO_3^- is reabsorbed until the load exceeds 2.5 mEq per 100 ml of filtrate, when frank excretion of HCO_3^- begins. The quantity reabsorbed remains constant at this value despite further increases in HCO_3^- plasma concentration, the excess being excreted. Thus, there is a grossly maximal reabsorptive capacity which is reached at a filtered load of about 2.5 mEq per 100 ml of glomerular filtrate per minute. The critical value of plasma HCO_3^- required to cause frank excretion is about 27 mEq per liter in man.

Phosphate

In the plasma, about 80 percent of phosphate occurs as $HPO_4^=$ with about 20 percent as $H_2PO_4^-$. Both are usually combined with Na^+. The ultimate ratio of $H_2PO_4^-/HPO_4^=$ in the urine is determined by the pH. In reabsorption of phosphate there is manifested a Tm which has a magnitude of about 0.13 mM per minute. Reabsorption is in the proximal convoluted tubule. Excretion in man is from 7 to 20 μM per minute. Excretion is modified by PTH action (Chap. 33), which tends to block reabsorption.

Sulfate

Sulfate is actively reabsorbed in both dog and man and shows a well-defined, although small, Tm. It is 0.05 mM per minute in the dog and 0.110 mM per minute in man.

EXCRETION OF WATER: URINARY CONCENTRATION AND DILUTION MECHANISMS

The mechanisms which concern tubular reabsorption of water are intimately related to the handling of osmotic constituents, primarily NaCl and urea. Another factor is the action of the ADH which is elaborated by the supraoptic and paraventricular nuclei of the hypothalamus. Finally, the composite mechanisms are integrated in the light of a *countercurrent diffusion multiplier system,* operating particularly in the nephrons which project long loops of Henle and vasa recta into the papillary zones of the renal medulla. This anatomical arrangement is especially exemplified by the kidneys of the golden hamster and kangaroo rat but is applicable also to some of the nephrons of the white rat, the dog, and the human kidney.

The basis for this theory is the finding that the osmotic constituents of the kidney are arranged so that they are isosmotic with plasma in the cortex, with the concentration rising three to four times and even more in certain species, at the tip of the papilla. Using a cryoscopic method which depends upon a polarizing microscope to detect tiny crystals of ice formed during the cooling of sections of tissue of 30 μ thickness, Wirz et al. (1951) found that points of equal osmotic pressure form shells concentric to the tip of the papilla and parallel to the interzonal boundary. Uniform distribution of osmotic constituents among the loops of Henle, vasa recta, and collecting tubules has been proved by micropuncture in several animal species. This means that as the osmotic concentration progressively increases from the cortex to the papillary tip, at any given level, it is approximately equal in the loop of Henle, the vasa recta, and the collecting tubule.

The arrangement of the loop of Henle system and vasa recta, so that the currents of the fluids in the two limbs of this hairpin flow in opposite directions, has given a foundation for the theory of the *hairpin countercurrent system.* Basically, this operates because of the principle of countercurrent exchange, which has several applications in thermodynamics. The principle of countercurrent exchange has been used by heating engineers in furnace exhaust and air-intake ducts. Thus, if the ducts lie side by side, warm air moving from the inside to the outside passes along the duct which brings the cold air from the outside. The heat diffuses across to the incoming cold air, raising its temperature, while the air being exhausted to the outside is cooled. Thus, overall heat conservation results (Fig. 22-15A). This principle is utilized for conservation of heat in the living organism, based upon the countercurrent arrangements of the blood vessels in the limbs (see Chap. 28 for further discussion). When the two systems are connected at one end, an opportunity for a *countercurrent multiplier system* is created. This is illustrated by the use of a physical model in Figure 22-15B. Suppose that two parallel circuits are separated by a semipermeable membrane. A solution is driven through the circuit by the hydrostatic pressure, P, of an elevated reservoir. Because the membrane is permeable to water, but not to solute, water is driven across by the hydrostatic force, indicated by the vertical arrows. As fluid leaves the upper tube, the solute (C_1) becomes progressively concentrated to the end. On making the turn, the highly concentrated solution (C_2) becomes rediluted as it picks up the water flowing through the membrane.

The principle applies to the kidney, but the forces are different, involving changes in osmotic pressure based upon the movement of ions, rather than hydrostatic pressure forces. The analysis of several models corresponding to the loop of Henle and the collecting duct in Figure 22-16 serves to clarify the mechanism. In A, if it is considered that the membrane is unidirectionally permeable, that is, that Na^+ can move from the ascending limb across to the descending limb, the opportunity for countercurrent multiplication exists. Thus, the Na^+ from the ascending limb added to the isotonic fluid of the descending limb increases the tonicity of the fluid approaching the hairpin turn. This process continues as the hypertonic fluid rounds the turn. The process is multiplied over and over again, and the Na^+ is, in effect, trapped in relatively high concentration at the tip. Thus, the fluid would be isotonic entering the descending limb, hypertonic at the bend, and hypotonic issuing at the upper end of the ascending limb. This system assumes complete impermeability to water.

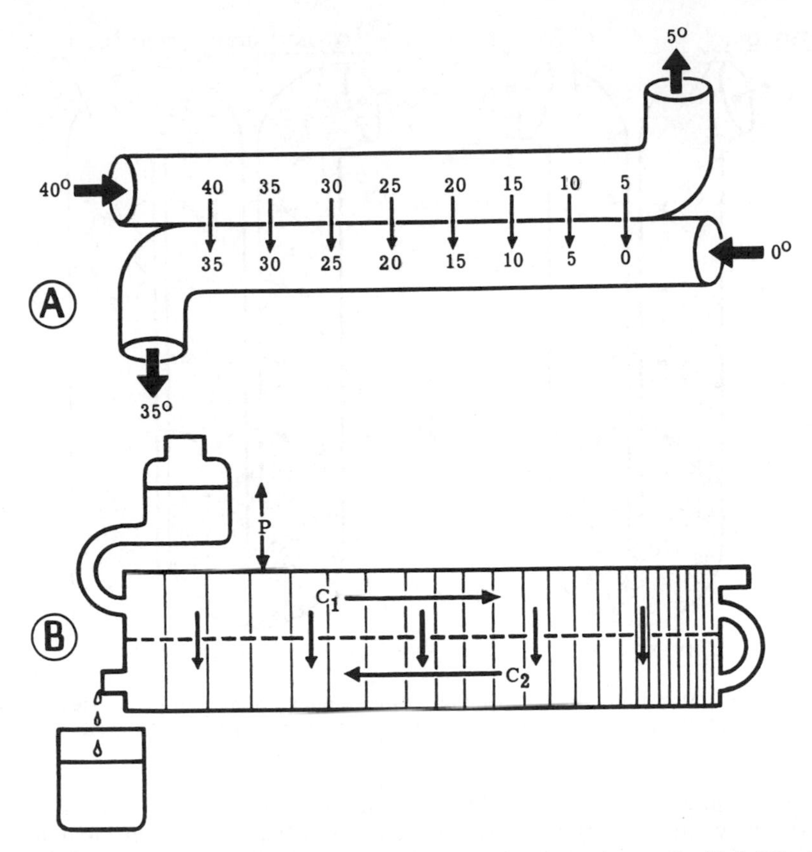

FIGURE 22-15. A. Diagram of the principle of countercurrent exchange. B. Model illustrating the principle of the countercurrent multiplier. (A from Wonderful net; with biographical sketch, by P. Scholander. *Sci. Amer.* 196:96, 1957. Copyright © 1957 by Scientific American, Inc. All rights reserved. B from B. Hargitay and W. Kuhn. *Z. Elektrochem.* 55:539, 1951.)

The mechanism for final concentration of the urine becomes more apparent if a tube representing the collecting tubule of the nephron is added to this system, as in B. If the upper loop to the right, analogous to the distal convoluted tubule, and the descending tube, analogous to the collecting tubule, are made water-permeable, then urinary concentration becomes possible. It will be seen later that this occurs because of the action of ADH.

Removal of water to the interstitium of the kidney (from the upper right-hand loop analogous to the distal convoluted tubule) establishes isotonicity in this region, but now water is drawn from the collecting tubule to the zone of hyperosmolality established by the countercurrent multiplier system, so that the fluid issuing from the descending right-hand limb, analogous to the collecting tubule, becomes hypertonic.

Although these models illustrate the basic principles involved in the countercurrent multiplier as it is utilized in the concentration of the urine, some immediate technical difficulties become apparent from the use of the simplified model. One is that the volume of the ascending limb of the loop of Henle and the distal convoluted

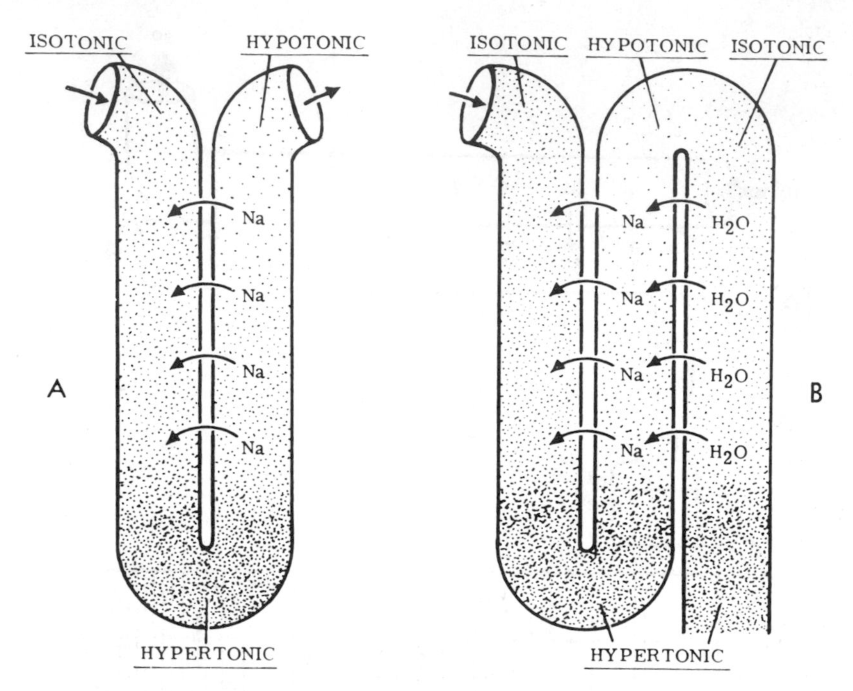

FIGURE 22-16. Models illustrating operation of the countercurrent multiplier system of the nephron. A. The loop of Henle system. B. Adjoined operation of the collecting duct.

tubular system would be increased by the return of water from the collecting duct. An extremely important consideration is that interstitial fluid is interposed between the loops of the tubule, and that the vasa recta lie in juxtaposition to the loop of Henle system and the surrounding interstitial fluid (Fig. 22-17A). Another important facet is the finding that water leaves the descending limb of the loop of Henle, as indicated by micropuncture studies which show that the inulin U/P ratio increases to the tip of the papilla in the hamster kidney, and probably also in the rat, proving that water has left the descending limb of the loop. This would contribute to the operation of the countercurrent multiplier system by producing an increment of osmotic pressure in the descending limb.

Last, consideration should be given to the distribution of osmolality in the vasa recta system. Since their course parallels the loop of Henle system, and they are freely permeable to water and solutes, the vasa recta should manifest the same stratification of osmotic pressure. This then becomes the final important link in the mechanism, for it permits the water taken out of the collecting tubule to be picked up by the plasma moving through the vasa recta system and be carried away into the venous channels of the kidney. The force is the colloid osmotic pressure of the plasma protein. The active reabsorption of sodium by the collecting tubular cells may contribute to hyperosmolality of the vasa recta system.

The vasa recta operate also as *countercurrent diffusion exchangers* in terms of preserving the osmotic gradient in the renal medulla. This is illustrated in Figure 22-17B. Blood enters the loop with a concentration of 300 mOsm per liter. As it dips down, water diffuses out, and osmotically active particles (NaCl, urea, ammonia) diffuse in from the medullary interstitium, increasing the concentration

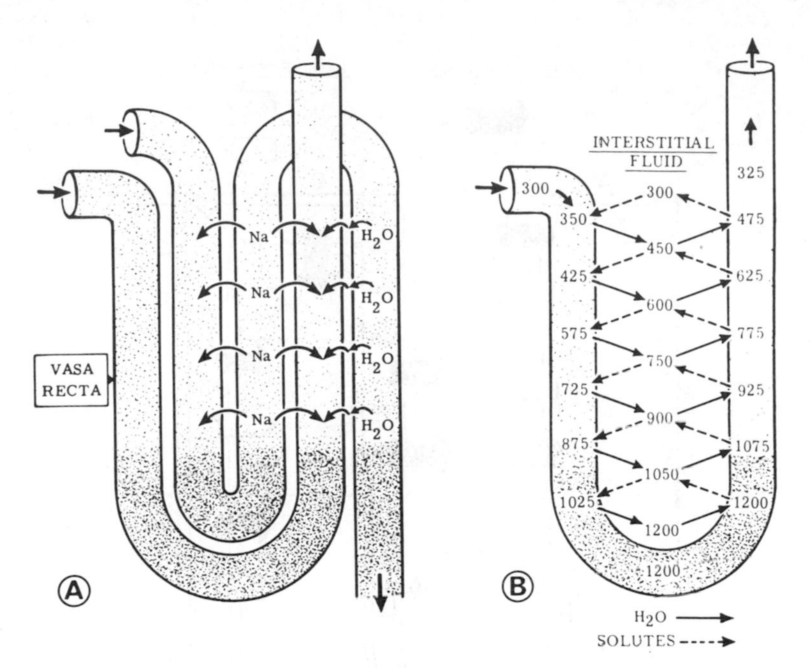

FIGURE 22-17. A. Relationship of vasa recta to the loop of Henle and the collecting duct, to illustrate role in removal of sodium and water. B. Operation of the countercurrent vascular loops in preserving the osmotic gradient in the renal medulla.

in the descending portion of the loop as this process is repeated. Then, as blood traverses the loop and ascends, osmotically active particles diffuse out, and water enters, thus reducing the gradient. The net effect is to trap solutes in the tip of the vascular loop, in equilibrium with the loop of Henle, in a concentration about four times that of the entering blood. Because of the short-circuiting effect of water across the tops of the vascular loops, red cells and plasma proteins are also concentrated, the latter aiding in picking up water reabsorbed from the collecting duct, facilitated by ADH action.

The situation as it exists in the nephron system is illustrated in Figure 22-18, based on tubular puncture techniques. Osmolality values have been arbitrarily assigned as they apply to the human kidney and become the basis for a more extensive recapitulation of the mechanism. The greatest proportion of the filtered sodium and attendant anions is reabsorbed by an active process in the proximal convoluted tubule (A). Water follows passively, not requiring mediation of ADH, so that the proximal tubular fluid remains essentially isosmotic. Micropuncture studies in the hamster and rat have revealed that 67 percent of the filtrate is absorbed in the proximal convolution (Lassiter et al., 1961). The descending limb of the thin segment of the loop of Henle does not transport salt actively. Another site of active Na^+ reabsorption is the ascending limb of the loop of Henle, especially the thick portion, which is relatively water-impermeable. Sodium, by an active mechanism, and Cl^- as a result of the electrochemical gradient established, are transported into the interstitium of the medulla until a gradient of perhaps 200 mOsm per kilogram of water has been established between the fluid of the ascending limb and the interstitium. This single effect is multiplied as the fluid

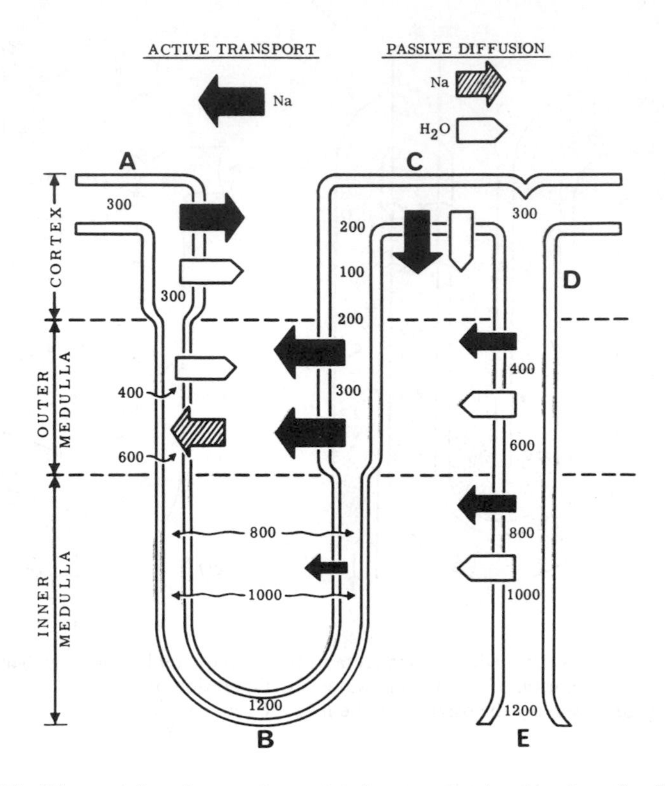

FIGURE 22-18. Diagram of nephron and associated vasa recta, showing sites of sodium and water reabsorption. Figures represent the distribution of osmolality values. (From H. W. Smith. The fate of sodium and water in the renal tubules. *Bull. N.Y. Acad. Med.* 35:293–316, 1959; and from Gottschalk and Mylle, 1959.)

in the thin descending limb comes into osmotic equilibrium with the interstitial fluid by the diffusion of water out of and probably the diffusion of some NaCl into the descending loop, thus raising the osmolality of the fluid making the hairpin turn to be presented to the ascending limb. In this fashion an increasing osmotic gradient is established in the direction of the tip of the papilla (B) by operation of the countercurrent multiplier, and yet at no level is there a large osmotic difference between the luminal and interstitial fluid. An additional 10 to 15 percent of the filtered water is reabsorbed in the pars recta and descending limb of the loop of Henle (passive efflux), and 15 percent in the distal convolution, leaving about 5 percent of the original volume to enter the collecting ducts.

In contrast to the ascending limb, the epithelium of the collecting ducts (D) in the presence of ADH is believed to be water-permeable. This results in diffusion of water out of the collecting ducts into the hyperosmotic medullary interstitium. The volume of fluid remaining in the collecting ducts becomes correspondingly concentrated (E). The degree to which distal tubular fluid is restored to isotonicity (300 mOsm per liter) appears to vary with the species. This is achieved in the rat and hamster kidney before the distal convolution is left, but not in the dog and monkey kidney, where restoration to isotonicity is delayed until the collecting duct.

For the urine to be significantly concentrated, the flow through the loops of Henle must exceed considerably the flow through the collecting ducts. This is accomplished by the action of ADH facilitating diffusion of water out of the distal convolution (C) into the interstitium of the cortex, thereby reducing the volume and increasing the osmolality to the isosmotic level before the fluid is presented to the collecting ducts. The rapid flow of blood through the cortex facilitates this removal.

Although the sodium salts are most important, urea probably also functions in this system, from present evidence. It has been proposed that the urea diffuses out of the distal tubule, and, more importantly, the collecting duct, to be concentrated in the loop of Henle by countercurrent exchange, adding to the gradients established in the loop. Ammonia produced in the proximal tubule also becomes concentrated in the papilla by operation of the medullary countercurrent exchange system (see p. 565).

To work most effectively, the medullary blood flow should be low, and experimental evidence at the present time indicates that it is. Probably the osmotic equilibration of the plasma in the vasa recta with the medullary interstitial fluid is due not only to the diffusion of solute into the descending and out of the ascending limbs but also largely to the diffusion of water in the opposite direction. This short-circuiting across the tops of the vascular loops may be in part the cause of the apparently rich content of cells and plasma protein which has been found in the vasa recta at the tip of the papilla.

The importance of the vasa recta is that the blood entering the medulla picks up not only water that diffuses out from the thin descending limb of the loop of Henle but also that from the collecting ducts. The water movement to the vasa recta system is the result of the favorable gradient established by the electrochemical potential created by the colloid osmotic pressure of the plasma proteins. If indeed, as seems to be the case, the plasma proteins are concentrated in this zone, the greater effectiveness of this mechanism becomes apparent.

The next question to consider is the mechanism which operates in the diluting kidney. This is the kidney which operates in absence of ADH, as occurs following water ingestion. In rats in which the final urine was quite dilute, Wirz (1956) found that the fluid remained dilute throughout the distal convolution and the collecting ducts. Thus an important role of ADH is that it renders the epithelium of the collecting ducts and perhaps the distal convolution freely permeable to water, so that the osmotic gradients established can operate in the presence of the hormone and are reduced or eliminated in the absence of the hormone. Studies with tritiated water (HTO) show that ADH acts by increasing the permeability of the luminal surface of the cellular membranes of the distal convolution and/or collecting ducts to water diffusion, by enlarging the mean functional diameter of so-called pores within the luminal portion of the membrane of the epithelial cells.

ADH may have an additional role in regulating medullary blood flow. It is well known that ADH is a vasoconstrictor. Reduction in medullary blood flow should favor further increase in osmolality of the papillary tip and thus favor further concentration of the urine. On the contrary, increased blood flow through the medullary circuit would tend to "wash out" the hypertonic zone, thus blunting the capacity for water uptake from the collecting tubule as outlined above. Figure 22-19 contrasts the mechanisms in the concentrating and diluting kidneys.

THE CONCEPT OF OSMOLAR CLEARANCE AND FREE WATER CLEARANCE

The highest concentration of urine achieved by the human kidney is about 1200 to 1400 mOsm per liter, compared with other mammalian kidneys that can

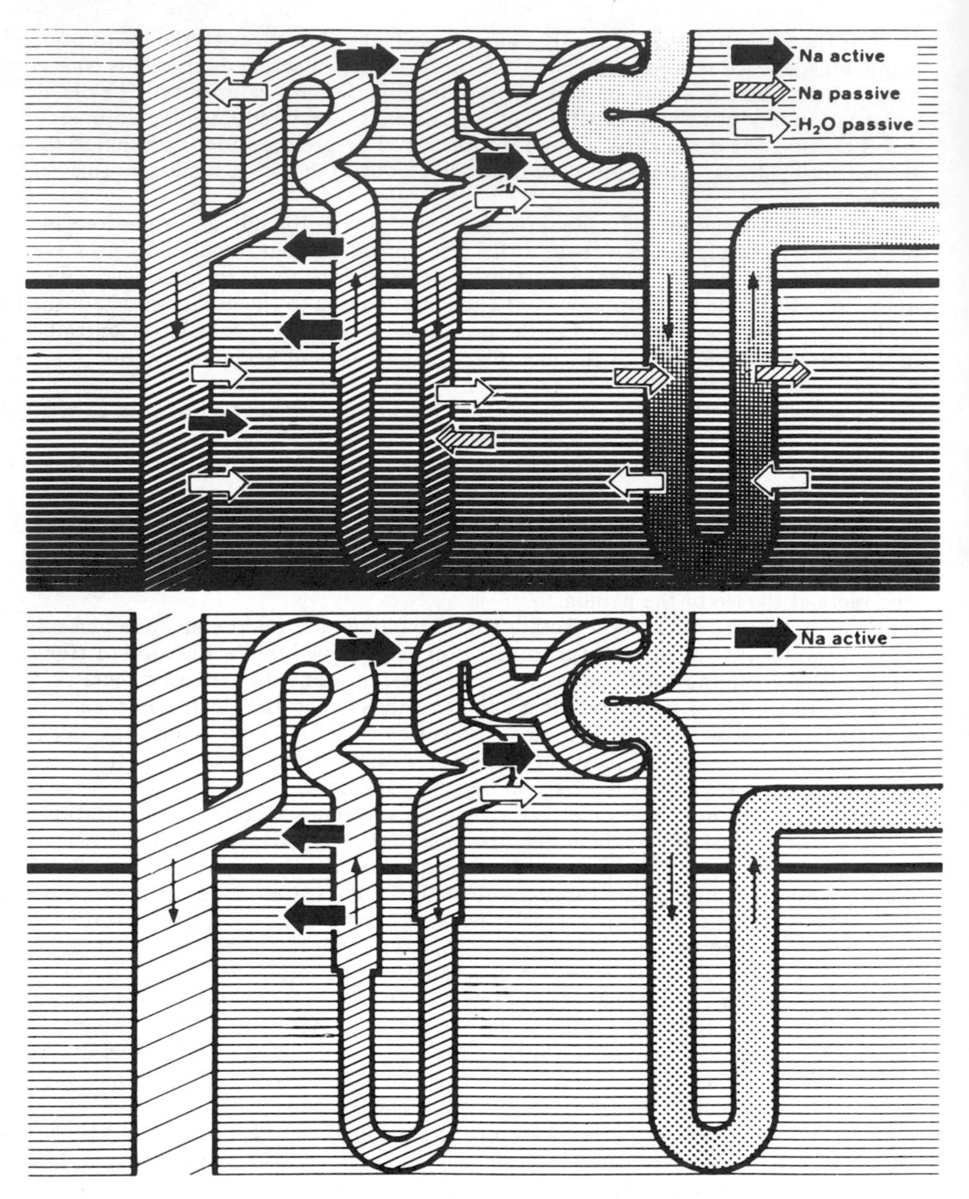

FIGURE 22-19. Comparison of the concentrating nephron and the diluting nephron. (From H. Wirz. *Mod. Probl. Paediat.* 6:86–98, 1960. [S. Karger, Basel/New York].)

concentrate to higher maxima, e.g., dog, 2500; white rat, 3000; and kangaroo rat, 5000. The lowest concentration observed is about 40 mOsm per liter. In terms of the more commonly employed clinical measurement, the *specific gravity,* this concentration involves a range between about 1.035 and 1.002 for human urine as compared with 1.010 for plasma. Understanding of the mechanism of change in urinary concentration requires the introduction of several concepts which involve the expressions for *free water clearance* and *osmolar clearance.* The volume of water required to contain all the urine solutes in an isosmotic solution with contemporaneous plasma is designated as the osmolar clearance (C_{Osm}). It is given by the equation

$$C_{Osm} = \frac{U_{Osm}V}{P_{Osm}} \tag{10}$$

Free water clearance (C_{H_2O}) is the difference between an osmolar clearance and the rate of urine flow:

$$C_{H_2O} = V - C_{Osm} \tag{11}$$

When C_{H_2O} is positive, as in water diuresis, the urine is more dilute than plasma; when C_{H_2O} is negative, as in dehydration, the urine is more concentrated than the plasma. Negative C_{H_2O} is more conveniently expressed by the designation $T_{C_{H_2O}}$.

The mechanism of formation of osmotically concentrated urine in hydropenic subjects has been studied by giving additional ADH by Pitressin infusion. The subjects were then given an infusion of hypertonic mannitol solution to produce a progressively increasing osmotic diuresis. Figure 22-20 illustrates the result of

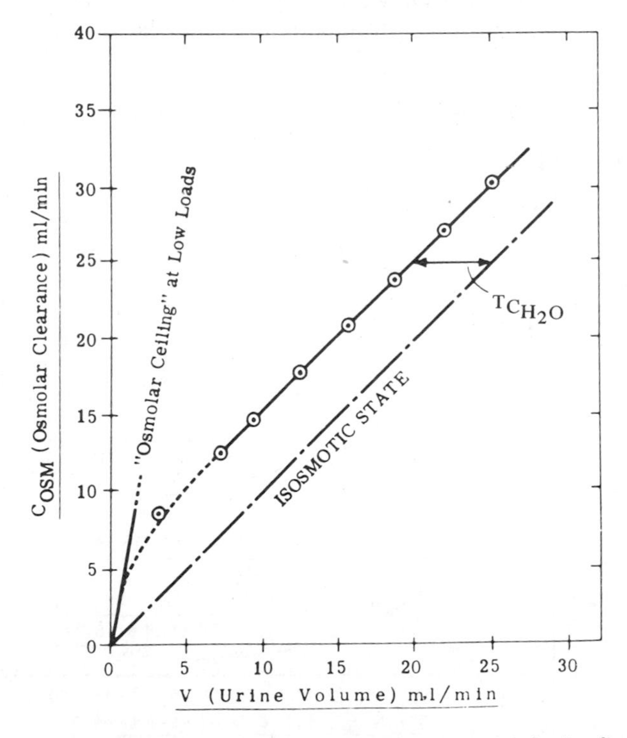

FIGURE 22-20. Relationship of osmolar clearance (C_{Osm}) to rate of urine flow during osmotic diuresis. (From Smith, 1956. P. 121.)

plotting of osmolar clearance against urine flow. The dashed line in the figure represents the isosmotic parameter, i.e., the line that would be generated if the osmolar clearance equaled the urine volume at all values, in which case the urine would always be isosmotic with the plasma. The actual line is shifted to the left in the concentrating kidney, indicating that the urine is hypertonic. The horizontal distance between these lines, labeled Tc_{H_2O}, represents the amount of pure water which, if added to the concentrated urine, would restore it to isosmolarity, or conversely, that which must have been removed from the isosmotic tubular fluid to result in the concentrated urine. In this case, the value for Tc_{H_2O} is 5 ml per minute at the higher rates of flow and clearance. The removal of the same volume of pure water from varying volumes of originally isosmotic fluid would lead to a progressive fall in urinary concentration as solute load and flow increase during osmotic diuresis. Thus, removal of 5 ml of pure water from 10 ml of isosmotic tubular fluid would give a final U/P ratio equal to 2, whereas removal of 5 ml from 30 ml of tubular fluid would concentrate the urine to only 1.2, because the amount of solute concentrated in 5 ml of tubular urine would now be dissolved in 25 ml (hence, U is one-fifth of the original concentration instead of twice, and U/P is 1.2 instead of 2.0). Any substance which will institute an osmotic diuresis will produce the same effect. Thus, increasing osmotic diuretic activity would result in increased amounts of isosmotic fluid passing from the proximal system to the loop of Henle and distal system. With removal of a fixed amount of pure water, the U/P ratio would decline (Fig. 22-21).

At very low solute loads and low urine flows, the value for Tc_{H_2O} decreases, the limiting factor being an osmolar ceiling above which the urine cannot be

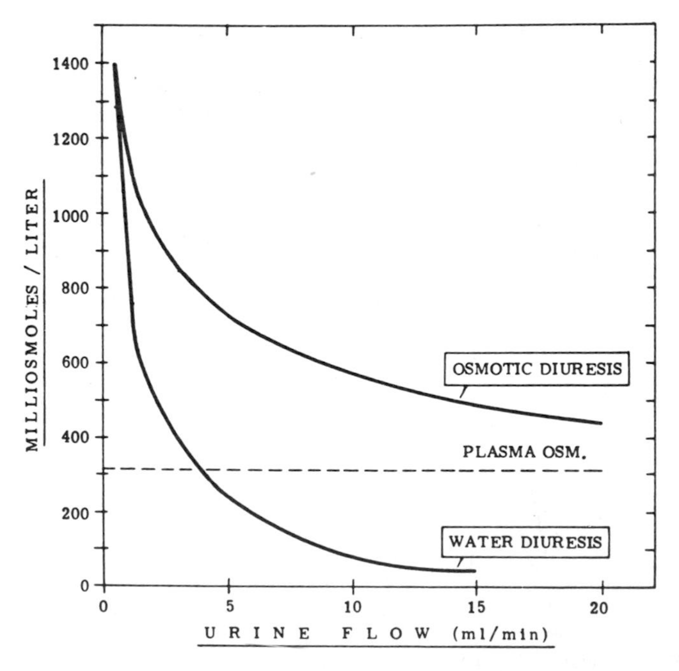

FIGURE 22-21. Urinary osmolar concentration related to urine flow during osmotic diuresis and water diuresis.

concentrated (at about 1400 mOsm per liter), representing an osmotic U/P ratio of approximately 4.2 (Figure 22-22).

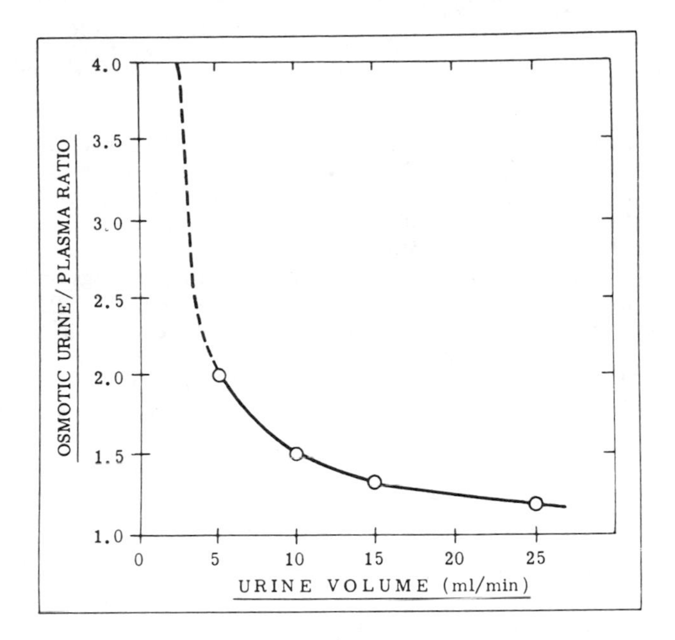

FIGURE 22-22. Osmolar U/P related to urine volume during osmotic diuresis. (From Smith, 1956. P. 121.)

When it is dilute, the urine flow exceeds the osmolar clearance, and the difference becomes the *free water clearance,* C_{H_2O}, which represents the *volume of pure water which originally contained the salt actively reabsorbed in the ascending limb of Henle, the distal convoluted tubule, and the collecting duct.* The capacity of these segments to reabsorb sodium salts, therefore, sets the limit on the amount of water which can be freed and excreted *above that containing the remaining urinary solutes in isotonic solution.* The rather limited capacity to excrete water freed of the salts accounts for the increasing concentration of dilute urine as solute excretion increases, because the isotonic portion of the total flow, C_{Osm}, constitutes an even larger fraction of total flow, V (see Fig. 22-20).

The role of ADH is primarily to convert the dilute urine of the distal system to isotonicity and then to hypertonicity. The relative quantitative importance of the dilution and concentration processes is illustrated by considering the excretion of 2 liters of isotonic urine in 24 hours. If the subject were to produce maximally diluted urine (absence of ADH), he would excrete 20 liters of urine per 24 hours. Therefore, if ADH were to effect a rise in urinary concentration only to isotonicity, 18 liters of water would be conserved. If the solute were to be excreted in maximally concentrated urine (maximal ADH effect), the urine volume would be reduced to 500 ml, with a further saving of only 1.5 liters. Viewed in this light, it is clear that water conserved by not putting out dilute urine is quantitatively much more important than that conserved in forming concentrated urine.

Discussion of the renal role in body water balance and the control systems related thereto will be deferred until the latter part of Chapter 23, after the facts concerning body water and electrolyte distribution have been considered.

REFERENCES

Arendshorst, W. J., and E. E. Selkurt. Renal tubular mechanisms for creatinine secretion in the guinea pig. *Amer. J. Physiol.* 218:1661–1670, 1970.

Chinard, F. P. Derivation of an expression for the rate of formation of glomerular fluid (GFR). Applicability of certain physical and physico-chemical concepts. *Amer. J. Physiol.* 171:578–586, 1952.

Diedrich, D. F. The comparative effects of some phlorizin analogs on the renal reabsorption of glucose. *Biochim. Biophys. Acta* 71:688–700, 1963.

Foulkes, E. C. Kinetics of p-aminohippurate secretion in the rabbit. *Amer. J. Physiol.* 205:1019–1024, 1963.

Gertz, K. H., M. Brandis, G. Braun-Schubert, and W. J. Boylan. The effect of saline infusion and hemorrhage on glomerular filtration pressure and single nephron filtration rate. *Pflügers. Arch. Ges. Physiol.* 310:193–205, 1969.

Giebisch, G. Measurements of electrical potential differences on single nephrons of the perfused Necturus kidneys. *J. Gen. Physiol.* 44:659–678, 1961.

Gottschalk, C. W., and M. Mylle. Micropuncture study of the mammalian urinary concentrating mechanism: Evidence for the countercurrent hypothesis. *Amer. J. Physiol.* 196:927–936, 1959.

Hollenberg, N. K., M. Epstein, S. M. Rosen, R. I. Basch, D. E. Oken, and J. P. Merrill. Acute oliguric renal failure in man: Evidence for preferential renal cortical ischemia. *Medicine* 47:455–474, 1968.

Kleinzeller, A., J. Kolinská, and I. Beneš. Transport of glucose and galactose (and other monosaccharides) in kidney-cortex cells. *Biochem. J.* 104:843–860, 1967.

Ladefoged, J. Renal cortical blood flow and split function test in patients with hypertension and renal artery stenosis. *Acta Med. Scand.* 179:641–651, 1966.

Lassiter, W. E., C. W. Gottschalk, and M. Mylle. Micropuncture study of net transtubular movement of water and urea in nondiuretic mammalian kidney. *Amer. J. Physiol.* 200:1139–1147, 1961.

Lotspeich, W. D. *Metabolic Aspects of Renal Function.* Springfield, Ill.: Thomas, 1959.

McKenna, O., and E. T. Angelakos. Adrenergic innervation of the canine kidney. *Circ. Res.* 22:345–354, 1968; Acetylcholinesterase-containing nerve fibers in the canine kidney. *Circ. Res.* 23:645–651, 1968.

Moffat, D. B. The fine structure of the blood vessels of the renal medulla with particular reference to the control of the medullary circulation. *J. Ultrastruct. Res.* 9:532–545, 1967.

Morel, F., M. Mylle, and C. W. Gottschalk. Tracer microinjection studies of effect of ADH on renal tubular diffusion of water. *Amer. J. Physiol.* 209:179–187, 1965.

Nash, F. D., H. H. Rostorfer, M. D. Bailie, R. L. Wathen, and E. G. Schneider. Renin release: Relation to renal sodium load and dissociation from hemodynamic changes. *Circ. Res.* 22:473–487, 1968.

Pappenheimer, J. R. Über die Permeabilität der Glomerulummembrane in der Niere. *Klin. Wschr.* 33:362–365, 1955.

Pinter, G. C., C. C. C. O'Morchoe, and R. S. Sikand. Effect of acetylcholine on urinary electrolyte excretion. *Amer. J. Physiol.* 207:979–982, 1964.

Pitts, R. F. *Physiology of the Kidney and Body Fluids.* Chicago: Year Book, 1968.

Scriver, C. R., and H. Goldman. Renal tubular transport of proline, hydroxyproline, and glycine: II. Hydroxy-L-proline as substrate and as inhibitor *in vitro. J. Clin. Invest.* 45:1357–1363, 1966.

Selkurt, E. E. Sodium excretion by the mammalian kidney. *Physiol. Rev.* 34:287–333, 1954.

Selkurt, E. E. Renal Circulation. In W. F. Hamilton and P. E. Dow (Eds.), *Handbook of Physiology.* Section 2: Circulation, Vol. II. Washington, D. C.: American Physiological Society, 1963. Chap. 43, pp. 1485–1489.

Selkurt, E. E., R. L. Wathen, and J. Santos-Martinez. Creatinine excretion in the squirrel monkey. *Amer. J. Physiol.* 214:1363–1369, 1968.

Smith, H. W. *The Kidney: Structure and Function in Health and Disease.* New York: Oxford University Press, 1951.

Smith, H. W. *Principles of Renal Physiology.* New York: Oxford University Press, 1956.

Taggart, J. V. Mechanisms of renal tubular transport. *Amer. J. Med.* 24:774–784, 1958.

Thorburn, G. D., H. H. Kopald, J. A. Herd, M. Hollenberg, C. C. C. O'Morchoe, and A. C. Barger. Intrarenal distribution of nutrient blood flow determined with Krypton^{85} in the unanesthetized dog. *Circ. Res.* 13:290–307, 1963.

Thurau, K., H. Valtin, and J. Schnermann. Kidney. *Physiol. Rev.* 30:441–524, 1968.

Truniger, B., and B. Schmidt-Nielsen. Intrarenal distribution of urea and related compounds: Effects of nitrogen intake. *Amer. J. Physiol.* 207:971–978, 1964.

Wirz, H. Der osmotische Druck in den corticalen Tubuli der Ratteniere. *Helv. Physiol. Pharmacol. Acta* 14:353–362, 1956.

Wirz, H., B. Hargitay, and W. Kuhn. Lokalisation des Konzentrierungsprozesses in der Niere durch direkte Kryoskopie. *Helv. Physiol. Pharmacol. Acta* 9:196–700, 1951.

Body Water and Electrolyte Composition and Their Regulation

Ewald E. Selkurt

FLUID VOLUMES

The principle of measurement of fluid volumes was introduced in Chapter 10, where the technique was explained in terms of its application to the determination of plasma volume. The same principle is employed for the measurement of total body fluid and extracellular volumes, based upon the use of a substance which distributes itself throughout the compartment to be measured. The equation is modified from that used for plasma volume, $V = \frac{A}{C}$, as follows:

$$V = \frac{A - E}{C}$$

where V is the volume in liters, A is the amount of the substance administered, E is the amount excreted in the urine at the time that C, the concentration per liter, is determined. The introduction of E, a subtraction for the amount excreted in the urine, is necessary because substances commonly used for these calculations are removed from the body by the kidney.

TOTAL BODY WATER

Total body water (water in the *cellular* plus *intracellular* compartments) is measured as the volume of distribution in the body of an appropriate indicator after a single intravenous injection. The requirement is that the substance must distribute itself uniformly in all the fluid spaces. Apparently, the most reliable and commonly used indicators today are antipyrine and the heavy isotopes, deuterium oxide and tritium oxide. Confirmatory evidence has been supplied by desiccation of human cadaver material to constant weight at a temperature of approximately 105°C. On the basis of determinations with the indicators, total body water ranges from 500 to 600 ml per kilogram of total body weight (50 to 60 percent) for the average adult human subject. Desiccation studies reveal a figure of 640 ml per kilogram. The total body water volume for a 70-kg subject is about 42 liters. Some authorities feel it is more desirable to express the total body water in terms of the lean body mass, since adipose tissue is relatively deficient in water. On this basis, the generally accepted figure is 70 percent of the lean body mass.

The *extracellular* fluid compartment can be directly determined by the use of a substance which distributes itself throughout this particular compartment. Although there is some controversy as to the exact space such a substance measures, at the present time mannitol appears to be the best for measuring this space, and the distribution of sucrose, thiosulfate, and inulin supplies confirmatory evidence. One of the technical problems involved is whether or not the fluid of the connective tissue and bone should be considered part of the extracellular fluid. The tracer substances, mannitol, inulin, sucrose, and others, do not penetrate connective tissue fluid and bone water readily, nor is it likely that they enter the cerebrospinal fluid freely; hence, the space measured may be actually smaller than the true physiological extracellular space. The average figure of 180 ml per kilogram includes the plasma compartment, the interstitial-lymph space, and transcellular fluids. If dense connective tissue and cartilage, and inaccessible bone water, are added, the total becomes 270 ml per kilogram (Table 23-1). *Transcellular*

TABLE 23-1. Body Water Distribution in a Healthy Young Adult Male*

Compartment	Percent of Body Weight	Percent of Total Body Water	Liters
Plasma	4.5	7.5	3
Interstitial lymph	12	20	8.5
Dense connective tissue and cartilage	4.5	7.5	3
Inaccessible bone water	4.5	7.5	3
Transcellular	1.5	2.5	1
Total extracellular	27	45	19
Functional extracellular	21	35	14.5
Total body water	60	100	42
Total intracellular	33	55	23

*All figures rounded to nearest 0.5.
Source: From Edelman and Leibman, 1959.

water is that concerned with the transport activity of specialized cells (salivary, pancreas, liver, mucous membrane of respiratory and gastrointestinal tract, cerebrospinal fluid, and ocular fluid). The *intracellular* water can be estimated as the difference between total body water and the total extracellular water. This would yield a volume of 23 liters.

As indicated in Chapter 10, the plasma volume can be measured by the distribution of suitable indicators such as T-1824 or iodinated albumin, giving a plasma volume of about 4.5 percent of body weight, or 3.2 liters for the average-size adult. The breakdown of the various volumes as a percentage of total body water appears in Figure 23-1.

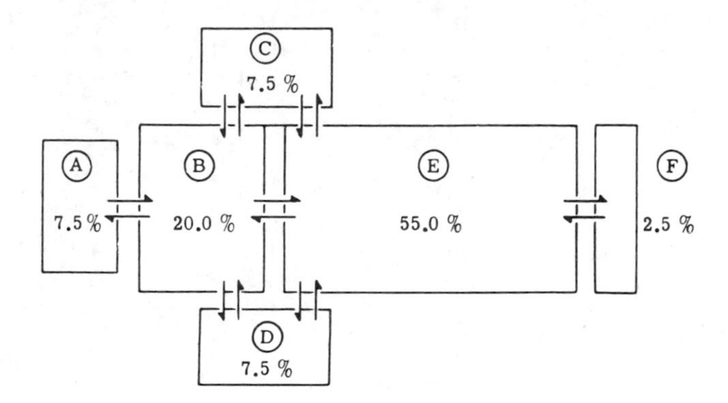

FIGURE 23-1. Volumes of distribution of water in the body fluid compartments. A. Plasma water – 7.5% of body water. B. Interstitial – lymph water – 20.0%. C. Dense connective tissue and cartilage water – 7.5%. D. Bone water – 7.5%. E. Intracellular water – 55.5%. F. Transcellular water – 2.5%. (From Edelman and Leibman, 1959.)

COMPOSITION OF THE EXTRACELLULAR FLUID

The principal constituents of extracellular fluid are shown in Table 23-2. Note that the concentrations of the interstitial fluid are corrected for the Gibbs-Donnan effect (Chap. 1). Thus, the essential differences between the serum water and interstitial fluid concentrations are the small inequalities resulting from the Gibbs-Donnan equilibrium effect. The exception is the much lower amount of protein which is found in the interstitial fluid. The important cation is Na^+, and the important anions are Cl^- and HCO_3^-. The last column supplies, for comparative purposes, the approximate values found in the intracellular fluid.

Specialized Extracellular Fluids: Aqueous Humor and Cerebrospinal Fluid

The aqueous humor of the eyeball and the cerebrospinal fluid (CSF) are examples of specialized interstitial fluids that cannot be described as simple filtrates from the plasma. Not only is each fluid different in composition from plasma dialysate, but in some respects they differ from each other (see Table 23-3). In general, CSF has higher Na^+ and Cl^- concentrations than plasma dialysate, while the concentration of K^+ is significantly lower in this fluid. The aqueous humor and CSF have similarities (lower urea and glucose concentrations than plasma) but differ from each other in concentrations of Cl^- and K^+.

Because the solutes are not distributed as one might expect in an ultrafiltrate, the suggestion has been made that these fluids are secretions. Reasons for not viewing the differential concentrations as the result of simple filtration through a highly selective membrane have been advanced (Davson, 1956). Thus, the osmotic pressure required to maintain the difference in concentration of potassium between plasma and CSF can be calculated to be 54 mm Hg. The pressure required to maintain the difference in concentration of protein is 30 mm Hg. Consequently, the total pressure available must be 84 mm Hg, a figure well beyond capillary

TABLE 23-2. Composition of the Body Fluids

Substance	Serum (mEq/L)	Serum Water* (mEq/L)	Interstitial Fluid† (mEq/L)	Intracellular Fluid (mEq/L)
Na^+	138	148	141	10
K^+	4	4.3	4.1	150
Ca^{++}	4	4.3	4.1	—
Mg^{++}	3	3.2	3	40‡
Cl^-	102	109	115	15
HCO_3^-	26	28	29	10
$PO_4^{\equiv}$	2	2.1	2	100
$SO_4^=$	1	1.1	1.1	20
Organic acids	3	3.2	3.4	—
Protein	15	16	1	60

*Correction to concentration per liter of serum water is based on the value 93 percent.
†Values derived by use of Gibbs-Donnan factor of 0.95 for cations and 1.05 for anions in serum water.
‡The figure 40 mEQ/L here represents largely Mg.

pressure. In other words, the supply of energy necessary to maintain this difference would be inadequate if the hydrostatic pressure in the capillaries were the sole force involved. Moreover, the concentration of Na^+ and Cl^- is higher in the CSF than in plasma, as are HCO_3^- and ascorbic acid in the aqueous humor. These facts appear to deny simple filtration. The conclusion that active transport is involved appears warranted. It has been suggested that the mechanism involves a sodium pump and probably diffusional transport of water as a result of osmotic forces.

Formation and Drainage: Aqueous Humor. The ciliary body of the eye is considered to be the source of the aqueous humor. The highly vascular ciliary processes have been specifically assigned the role of production of fluid. The principal drainage route is the *canal of Schlemm,* which connects via very fine ducts *(collectors)* with the intrascleral venous plexus.

CSF. The CSF is elaborated by the choroid plexuses which project with many folds and villi into the roofs of the third and fourth ventricles, and into the sides of the lateral ventricles. The fluid passes from the lateral ventricles to the third ventricle through the foramen of Monro, then to the fourth ventricle via the aqueduct of Sylvius. It leaves the fourth ventricle through the foramina of Magendie and Luschka, to reach the subarachnoid spaces, here expanding to form the *cisterna magna.* From this region it passes into the dural sinus. Differing from the eye, wherein the canal of Schlemm enters directly via fine ducts into the venous system, the CSF is separated from the blood by the mesothelial lining which covers the

TABLE 23-3. Distribution of Various Ions and Nonelectrolytes Between Aqueous Humor and Plasma (R_{Aq}), Cerebrospinal Fluid and Plasma (R_{CSF}), and Plasma-Dialysate and Plasma (R_{Dial})*

Substance	R_{Aq}	R_{CSF}	R_{Dial}
Na^+	0.96	1.03	0.945
K^+	0.955	0.52	0.96
Mg^{++}	0.78	0.80	0.80
Ca^{++}	0.58	0.33	0.65
Cl^-	1.015	1.21	1.04
HCO_3^-	1.26	0.97	—
H_2CO_3	1.29	1.61	—
Br^-	0.98	0.715	0.96
I^-	0.32	0.004	0.85
CNS^-	0.46	0.06	0.59
Phosphate	0.58	0.34	—
Glucose	0.86	0.64	0.97
Urea	0.87	0.81	—
Ascorbic acid	18.5	1.55	—
pH	7.48	7.27	7.46 (Plasma)

*R represents the ratio: concentration in fluid-water/concentration in plasma-water.
Source: From Davson, 1956, P. 229.

arachnoid villi, so that outward passage into the blood involves filtration-diffusion processes.

COMPOSITION OF INTRACELLULAR FLUID

The predominant cations of the intracellular fluid are K^+ and Mg^{++}. The predominant anions are organic $PO_4^{\equiv}$, $SO_4^{=}$, and proteins (see Table 23-2). It should be stated that these are only approximations of the composition of cell fluid and are based on the findings in muscle. Inadequate data are available to form conclusions concerning the complex forms in which the materials exist, the valence of the organic anions, and the degree of dissociation of the compounds, the extent to which the cations are bound, etc. In any event, the osmolar balance between the intracellular and extracellular fluids is operationally equivalent. A bar graph summary of the distribution of cations and anions for the fluids of the body is presented in Figure 23-2.

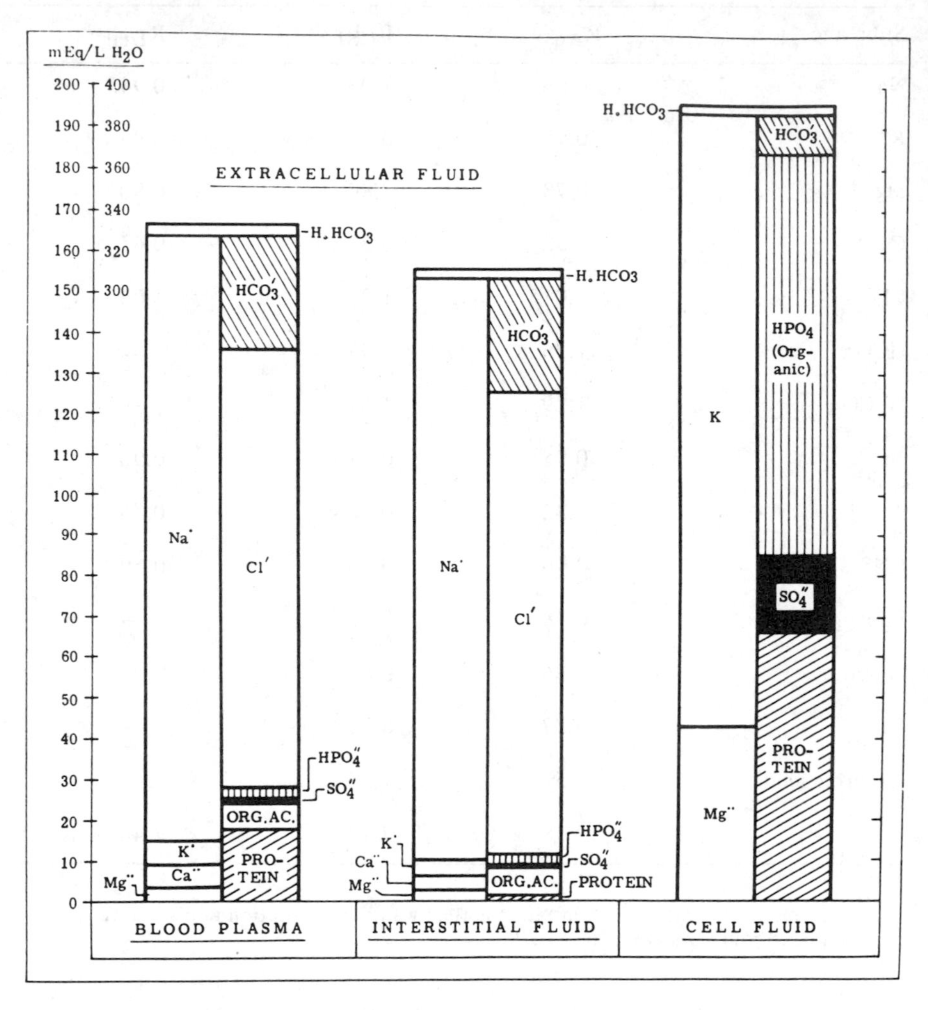

FIGURE 23-2. Ion distribution in the fluid compartments of the body. (From J. L. Gamble. *Chemical Anatomy, Physiology and Pathology of Extracellular Fluid,* 6th ed. Cambridge, Mass.: Harvard University Press, 1958.

FLUID EXCHANGE

EXCHANGE BETWEEN EXTRACELLULAR AND INTRACELLULAR COMPARTMENTS

Most cell membranes are apparently completely permeable to water, and the total exchange is enormous. The *net exchange* of water is governed by the osmotic pressure changes in the two compartments. When osmotic pressure changes, water moves across the cellular membrane to maintain an isosmotic state. The situation is diagrammed in Figure 23-3. Part A shows the normal situation, i.e., concentrations of 300 mOsm per liter both within the cell (C) and in the extracellular com-

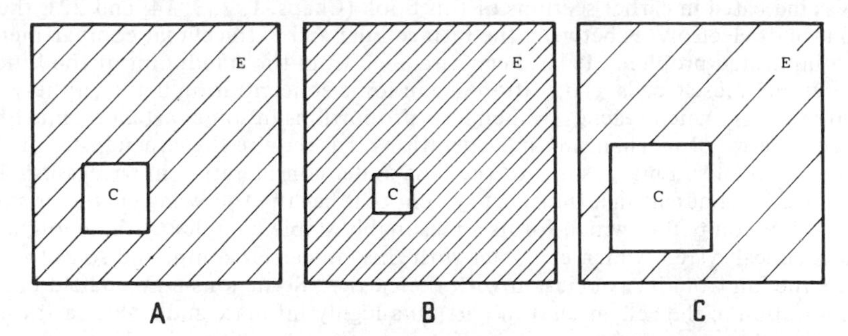

FIGURE 23-3. Regulation of water exchange between the extracellular and intracellular compartments. The representative volume of extracellular fluid, E, is taken as 1 liter. Intracellular volume is C. A. Normal (300 mOsm per liter both in E and C). B. Effect of addition of 300 mOsm to E (initially 300 mOsm per liter). C. Osmolarity of E is reduced to 150 mOsm per liter; C is 300 mOsm per liter. The result is net water gain by the cell, as fluid moves into the relatively hyperosmolar zone, C.

partment (E). Part B shows the result of adding 300 mOsm to compartment E. Water moves out of the cell (space C) to dilute the concentration of the materials in E, so the final concentration in each compartment will be somewhat less than 600 mOsm per liter. The net effect will be cellular dehydration. Part C illustrates what will happen if the osmolarity of the extracellular compartment is reduced to 150 mOsm per liter. The cellular fluid is hyperosmolar, and water moves in. Both compartments will become isosmotic, and a value greater than 150 and less than 300 mOsm per liter will be attained, because of the resulting movement of water from compartment E into C. The cell swells.

EXCHANGE BETWEEN VASCULAR AND INTERSTITIAL COMPARTMENTS

When fluid moves between the vascular and interstitial compartments, the total exchange of water and salts in solution is enormous because the capillaries are highly permeable to water and contained solutes, and the area of exchange is very great. The total surface area of the capillary bed (muscle) in man has been estimated at 6000 square meters. The limiting factor to diffusion is, in fact, the rate of circulation of the blood through the tissue. However, the net exchange is small because of the normal maintenance of volume of the interstitial and plasma compartments within relatively fixed limits. The protein concentration of the plasma becomes very important in the net exchange of fluid. It will be recalled that the osmotic pressure exerted by the proteins opposes the hydrostatic filtering forces that exist in the capillaries, so that, according to the Starling hypothesis (Chap. 12), the filtration of water is opposed by the retaining action of the plasma protein. Stated simply, the fluid that has been filtered outward at the arteriolar end of the capillaries is returned by the effective gradient of plasma protein at the venular end. The importance of the role of the lymphatics in the overall net water exchange between the capillaries and the interstitial space should be emphasized. The lymphatic system is important in draining off excesses of interstitial fluid.

ELECTROLYTE EXCHANGE

EXCHANGE BETWEEN EXTRACELLULAR AND INTRACELLULAR COMPARTMENTS

As was indicated in earlier sections of this book (Chaps. 1, 2, 3, 14, and 22), the exchange of electrolytes between the intracellular and extracellular compartments is a complicated problem. It has been discussed from the standpoint of the function of nerve and muscle cells. The difference in ionic concentration between these two compartments, which reaches tremendous proportions in some instances, must be viewed in terms of mechanisms and operations that require the expenditure of energy. In brief summary, some of the ions under consideration move against the concentration gradient in a transport system that has the following characteristics: (1) a carrier compound which forms a reasonably stable complex with an ion, and (2) a chemical system which can react with this ion carrier compound so as to release the ion from its association with the carrier. Some ions may be held in high concentration in the cell, at least in part, in a highly intimate and stable relationship with other large complex molecules – for example, the relationship between potassium and myosin in muscle cells.

In explaining mechanisms of transport through cellular membranes, considerable stress is placed upon the generation of electrical currents by the active ion transport. Some authorities believe that active transport occurs when there is a net transfer against an electrochemical potential gradient, or a net transfer in the absence of a gradient against resistance offered by the membrane. With active transport of an ion there is a net transfer of charge from one side to the other of the membrane. The electrical current so generated can be measured by short-circuiting the two faces. Such techniques have been employed by Ussing (1952) and Huf (1951) in studying the transport of ion across the frog skin, by Hogben (1951) for chloride transport in the gastric mucosa, and by Giebisch (1961) and others for electrolyte movement across the renal tubular epithelium.

The thorough treatment of this topic in the above-cited chapters obviates the need for greater development at this time. The reader is referred to these areas for detailed presentation of the facts.

EXCHANGE BETWEEN VASCULAR AND INTERSTITIAL COMPARTMENTS

The movement of electrolytes and other solutes between the vascular compartment and the interstitial compartment is limited only by the movement of the water in which they are in solution. Superimposed upon this is the slight influence exerted by the Gibbs-Donnan effect.

WATER BALANCE

Water balance is achieved between the forces responsible for *intake* of water as manifested by thirst, and the metabolic water and contained water of ingested foodstuffs, and *water loss* through various organs of the body, namely, the skin, the lungs, the gastrointestinal tract, and the kidneys.

WATER INTAKE

Thirst is a subjective sensory impression that results in ingestion of water. Physiological factors that seem to be involved are primarily concerned with increased osmolarity of the extracellular fluid brought about either by salt excess or by water deficit. A decrease in circulating vascular volume may be involved. In addition, influences which are related to habits of salt and water intake have an effect. Basic to stimulation is the sensation of dryness of the mucous membranes of the mouth and pharynx. The central nervous system areas involved are not known precisely, although it is believed that they lie in or near the ventromedial nuclei of the hypothalamus.

WATER LOSS

Average daily losses are as follows: insensible loss (vaporization through lungs and skin), 800 to 1200 ml; urine, 1500 ml; stool, 100 to 200 ml. In addition, there may be variable loss of water through sensible perspiration (sweat), depending upon environmental conditions and degree of physical activity.

Insensible and Sensible Perspiration

Insensible water loss is modified by the effective osmotic pressure of the body fluids. Thus, there are data suggesting that the rate of insensible loss is inversely related to the concentration of sodium in the extracellular fluids. It follows that the rate of loss of water from the skin and lungs by this mechanism may be related to vapor pressure of the body fluids. Insensible perspiration goes on at a reasonably constant rate when body temperatures remain constant. About half is lost by the skin, half by the lungs.

Sensible perspiration may vary from 0 to 2 liters per hour and more. The output of sensible perspiration is greatly increased in elevated temperatures and with exercise (Chap. 28). Centers regulating sweating are located in the anterolateral portions of the hypothalamus. These respond to elevation in blood temperature or reflex afferent nerve stimulation, e.g., gustatory nerve stimulation. Sensible perspiration is a hypotonic fluid. Average values for the more important solutes appear in Table 23-4.

TABLE 23-4. Average Composition of Sweat

Solute	mOsm/liter
Na^+	47.9
K^+	5.9
Cl^-	40.4
NH_3	3.5
Urea	8.6

Gastrointestinal Tract Contribution

Gamble (1958) has estimated that the volume of digestive secretions provided by an average adult in one day is about 8 liters, of which gastric and intestinal mucosal secretions contribute about 5.5 liters. The electrolytic composition of these secretions differs considerably in the various regions of the gastrointestinal tract (Chap. 26). However, the average solute concentration in the various secretions is reasonably close to that of the extracellular fluid except for saliva, which is distinctly hypotonic. Although the amount of fluid lost from the gastrointestinal tract is normally small, persistent and excessive vomiting, diarrhea, or drainage from an intestinal fistula very quickly causes serious reduction in volume of the extracellular compartment. In addition, serious disturbance in electrolyte balance is produced. This results from both dehydration and differential electrolyte loss. For example, because potassium is found in these fluids in somewhat higher concentration than in extracellular fluid, potassium deficit would develop. Since disproportionate concentrations of HCO_3^-, Cl^-, and H^+ occur, it is to be expected that disturbances in acid-base balance result from excessive vomiting or loss of intestinal secretions.

Role of the Kidney

The kidney serves as the buffering organ for adjustment of the fluid balance in the sense that it is able to retain water under circumstances of dehydration, or alternatively it can lose large excesses of water when it is necessary to maintain isotonicity and normal volumes in the fluid compartments of the body. Extremes of volumes have been registered between 500 ml for 24 hours to 20 liters or more, the latter in the absence of regulation by antidiuretic hormone (ADH) in diabetes insipidus.

The output of urine is the net result of kidney operation on the original 170 liters of fluid filtered per 24-hour period by the glomeruli. The factors which govern the release of ADH are discussed in greater detail in Chapter 31. In brief, the important controls involve the osmolarity of the extracellular fluid and the volume of the body fluids. Increases in osmolar concentration of the plasma extracellular fluid supplying the zones in the hypothalamus where the osmolar receptors are located cause increased output of ADH and ultimate water conservation by the kidney, according to the findings of Verney (1947). Excessive intake of water causes a reduction in the osmolarity of the extracellular fluid and inhibits the output of ADH by the hypothalamo-neurohypophysial system and leads to water diuresis. Second, changes in volume of the extracellular fluids are perceived by volume receptors and baroreceptors which cause the requisite correction in total fluid volume, by intermediation of ADH and the kidney.

A related important limiting factor upon the renal regulation of water content of the body is the solute load which must be excreted. The solute load is composed of metabolites, such as urea, creatinine, uric acid, and excess electrolytes. The maximal osmolar concentration of the urine of man, 1400 mOsm per liter, obtains only when the solute load for excretion and obligatory urine volume are both low, the latter being about 0.5 ml per minute. With increasing solute loads, in spite of presumed maximal ADH activity, urine volume increases, and urine osmolarity actually declines progressively. The reason is that the ADH mechanism has attained, in effect, a "ceiling" of operation, so that water is progressively lost with increasing solute loss (see Chap. 22).

SUMMARY OF FACTORS THAT REGULATE URINE VOLUME

MODIFICATION OF GLOMERULAR FILTRATION RATE

It is well known that changes in glomerular filtration rate (GFR) may produce parallel changes in urine volume. Changes in GFR can be brought about by changes in systemic arterial blood pressure, which in turn produce corresponding changes in the glomerular filtration pressure. An obvious example is the oliguria or anuria following hemorrhage, under which circumstance arterial pressure and, in turn, glomerular capillary pressure are significantly reduced. Besides the hydrostatic filtering pressure, another important factor is the colloid osmotic pressure which opposes filtration. Hence, if plasma protein concentration is altered, changes in glomerular filtration rate and subsequently changes in urine volume are effected. For example, rapid infusion of isotonic salt solution decreases the plasma protein concentration by dilution and increases the effective filtration pressure.

Certain drugs affect glomerular pressure by acting on the afferent or efferent arterioles. As an example, the xanthines (caffeine, theobromine, and theophylline) dilate the afferent arterioles and increase glomerular filtration rate. They also secondarily impair tubular reabsorption of Na^+.

The changes in filtration rate involve changes in solute load to the tubules, which must be considered in explaining the final effect on urine volume. Thus, the extent of solute excretion governs the amount of water that will follow because of osmotic obligation (see below).

FACTORS AFFECTING TUBULAR REABSORPTION OF WATER

The role of ADH in regulation of tubular water handling has been discussed. Control mechanisms are detailed in Chapter 31. Mechanisms which inhibit ADH production and release include reduction of plasma osmotic pressure – for example, by the oral ingestion of large quantities of water or infusion of hypotonic NaCl or isotonic glucose solutions. (When glucose is metabolized, the water of the solution is freed.) Figure 23-4 shows typical water diuresis resulting from ingestion of 1 liter of pure water. Specific gravity of the urine rapidly changes from 1.030 to 1.002 at the peak of diuresis.

Effect of ingestion of 1 liter of saline is also shown. The modest diuresis following saline ingestion has been explained by a small increment in GFR. Specific gravity of the urine shows little if any change, for body fluid osmolarity is not altered, and ADH output should not change. In well-hydrated subjects, saline ingestion produces a greater diuresis, with decrease in specific gravity of the urine. It is probable that ADH output is decreased in this circumstance by activation of volume receptors located in the left atrium, or by altering of the carotid sinus baroreceptor influence on ADH release (Chap. 31).

Among the factors stimulating ADH secretion and thereby induce antidiuresis are water deprivation and dehydration, which increase the plasma osmotic pressure, in turn, acting on the osmolar receptors in the hypothalamus. Also, as the result of reducing plasma volume and, in consequence, stimulation of left atrial volume receptors and carotid sinus baroreceptor activity, ADH output is further enhanced. Pain and emotional stress likewise cause an increase in ADH output. Various drugs such as nicotine, acetylcholine, morphine, and the anesthetic agents barbiturates, ether, and chloroform all act on the hypothalamo-neurohypophysial system to increase output of ADH and lead to antidiuresis.

Another means of tubular regulation of urine volume concerns the role of solutes delivered to the tubular cells, as discussed in the previous section. *Osmotic diuresis* results if the filtered load of a given solute exceeds the reabsorptive capacity for it by the cells at the site of reabsorption in the nephron. The unreabsorbed solutes osmotically retain some of the water filtered at the glomeruli and carry it on through the rest of the system. Common osmotic diuretics include certain organic substances such as mannitol, urea, sucrose, and glucose. Hypertonic NaCl and Na_2SO_4 are others. The acid-forming salts ammonium chloride, ammonium nitrate, and calcium chloride have been used as diuretic agents, alone or in conjunction with others (e.g., mercurials). The ammonium salts act in part by increasing urea formation, which acts as an osmotic diuretic. For example, NH_4Cl is converted to urea

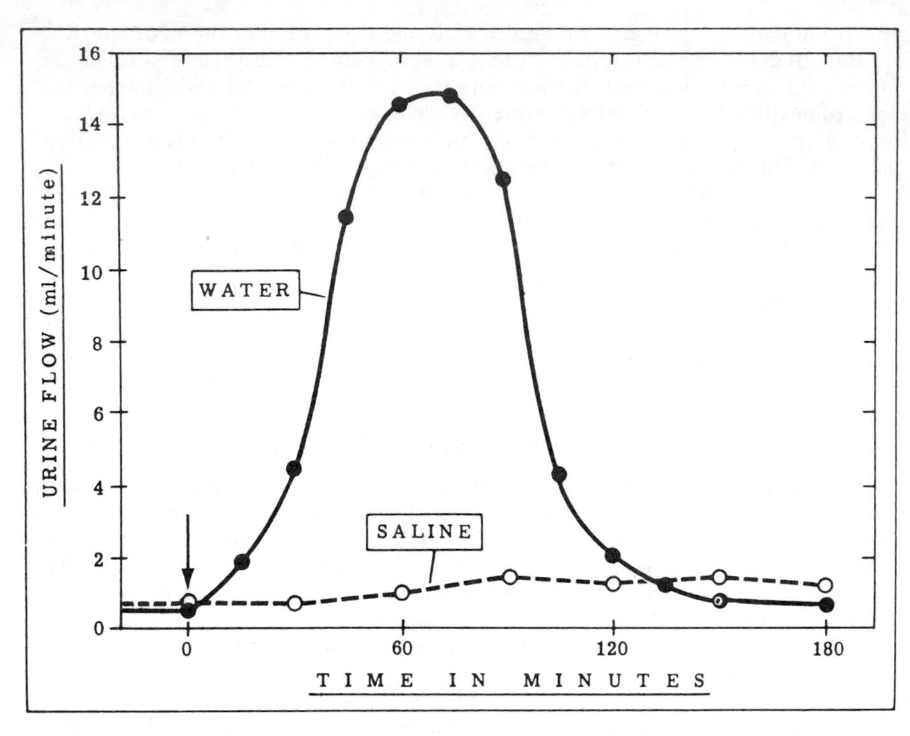

FIGURE 23-4. Comparison of water diuresis and saline diuresis after ingestion of 1 liter. (From H. W. Smith. *Principles of Renal Physiology*. New York: Oxford University Press, 1956. P. 113.)

and hydrochloric acid and the latter is buffered as follows:

$$HCl + NaHCO_3 \rightarrow NaCl + H_2CO_3$$

H_2CO_3 dissociates to carbon dioxide and water. The resulting acidosis is compensated in part by the release of carbon dioxide by the lungs. There is, however, an accelerated excretion of the residual Cl^- accompanied by Na^+ into the urine, and the loss of NaCl is accompanied by an equivalent loss of water. Oral ingestion of $CaCl_2$ acts by giving up Cl^- to be lost with Na^+; the poorly absorbed Ca^{++} tends to stay in the intestine.

OTHER DIURETIC AGENTS

Carbonic Anhydrase Inhibitors

Carbonic anhydrase inhibitors are heterocyclic sulfanilamide derivatives of which acetazolamide (Diamox) is the one commonly employed. These compounds are effective inhibitors of carbonic anhydrase in vitro, and the mechanism involved may be related to the mechanism for producing diuresis. It is probable that the drugs reduce the availability of hydrogen ions to both proximal and distal tubular ion exchange mechanisms. The agents promote the renal loss of NaCl and $NaHCO_3$.

More indirectly, increased loss of Na^+ favors enhanced K^+ secretion by supplying greater amounts of the ion needed for the exchange mechanism to operate in the distal nephron. Loss of all these ions promotes loss of water by their osmotic diuretic action.

Benzothiadiazine Diuretics

Sulfonamyl benzothiadiazines such as chlorothiazide (Diuril) have a weak carbonic-anhydrase-inhibiting action and act chiefly by blocking tubular reabsorption of sodium and chloride ions; potassium loss is also favored. Since they do not interfere with the production of maximally concentrated urine during hydropenia, it has been conjectured that they exert their action predominantly in the distal convoluted tubule.

Aldosterone Antagonists

The antialdosterone substances are spirolactones, the most effective of which is spironolactone (Aldactone). They are structurally similar to aldosterone and act by competitive inhibition, thus blocking aldosterone action and favoring Na^+ loss.

Mercurials

Numerous organic mercurial diuretics exist which will not be discussed in individual detail. Diuresis results because of Na^+ and Cl^- loss, being caused by impaired tubular reabsorption. K^+ excretion is variably influenced; if secretion of K^+ is initially low, excretion will be increased, and vice versa.

The discrete mechanism of action of the mercurials has not been settled. Earlier views that the action was due to the affinity of Hg for sulfhydryl groups has been challenged on the basis that there is little good evidence that sulfhydryl-combining enzymes play a significant role in electrolyte transport.

Xanthines

Xanthines also act in part by blocking Na^+ reabsorption.

MICTURITION

Urine collected in the pelvis of the kidney passes through the ureters to the bladder, not only by virtue of forces of gravity in the erect position but by contractions of muscle layers of the ureter. The muscular contractions are necessary in order to develop sufficient pressure to overcome the gradually increasing tension in the bladder as urine accumulates. Peristaltic waves observed are probably of myogenic origin but are modified by the action of the splanchnic nerves via the renal plexus to the upper portions of the ureters, and via the hypogastric plexus to the lower portions. These are excitatory fibers; inhibitory fibers have been described in the inferior mesenteric plexus. Afferent fibers are present, as is evidenced by the painful sensation which results from the passage of calculi through the ureter.

The ureters enter the base of the bladder obliquely, thus forming a valvular flap which passively prevents reflux of urine, although some authorities believe that there may be a sphincter at this point. The bladder musculature is capable of changing its capacity tremendously by means of changes in the tonus in its smooth muscle fibers, which are arranged in three layers. The ureteral orifice is guarded by an internal smooth muscle sphincter, and the urethra is surrounded by striated muscle which functions as an external sphincter and is under voluntary control. Both sphincters are normally in a state of tonic contraction. A considerable amount of fluid, about 400 ml, can be accommodated by stretching the bladder without a significant rise in pressure. As the volume increases further, the tension begins to rise, in part because of the beginning of tonic contractions of the bladder musculature. When vesicular pressure reaches a threshold level of about 18 cm H_2O, stretch and tension receptors are excited which create afferent impulses responsible for the sensations of distention and the desire to urinate. They also cause the act of micturition when spinal reflexes are released from cerebral control. Voluntary control can be exerted until the vesicular pressure increases to about 100 cm H_2O, at which point involuntary micturition begins.

As with the ureter, extrinsic nerves of the bladder are important in regulating the normal evacuation of the bladder, to modify changes of tonus, to enhance or depress the vigor of rhythmic contractions, and, reciprocally to these, to alter the tonus of the external and internal sphincters. This innervation is by way of the parasympathetic *pelvic* nerves, the sympathetic *hypogastric* nerves, and the somatic *pudendal* (pudic) nerves. The pelvic nerves function to maintain tonus of the bladder, and the pudic nerves mediate impulses which contract the striated musculature known as the external sphincter. The role of the sympathetic innervation is not so well understood.

Micturition is normally a voluntary act. Corticospinal impulses sent to the lumbosacral region of the cord cause intensive contraction of the bladder, opening of the sphincters, relaxation of the perineum, and thus lead to voiding. Afferent nerves are chiefly in the pelvic nerves, although afferent fibers are found also in the hypogastric and pudendal nerves. They enter the spinal cord at sacral levels III and IV, then go on up to the hypothalamus and ultimately the cortex, where voluntary control resides.

ELECTROLYTE BALANCE

ELECTROLYTE INTAKE

The average adult diet contains 10 to 12 gm of NaCl per day, either in the food (higher in meat diets than in vegetable diets) or added to it. Examples of excessive ingestion of salt or salt craving have been cited in the literature. This is usually associated with deficiency of adrenal cortical secretion, as in Addison's disease. As an extreme example, Strauss (1957) cites the case of a 34-year-old addisonian patient who put approximately one-eighth-inch layer of salt on his steak, used nearly one-half glass of salt for his tomato juice, put salt on oranges and grapefruit, and even made lemonade with salt. In fact, salt craving is often an early manifestation of Addison's disease that is of considerable diagnostic significance.

The urine always contains potassium, and if none is administered or ingested in the diet, a deficit of this cation will develop. Ingestion of about 3 gm of KCl a day appears to be adequate under normal circumstances. Other necessary cations are Ca^{++} and Mg^{+}.

ELECTROLYTE LOSS

Sodium

Total body sodium is 58 mEq per kilogram of body weight, 25 mEq per kilogram as bone sodium, of which about half is exchangeable (Edelman and Leibman, 1959). The intake of Na^{+} is normally quite variable, as is loss by sensible perspiration, so that it falls upon the kidney to regulate the homeostasis of its content in the body. Excesses of Na^+ are excreted by the kidney. Conversely, if the dietary intake is reduced, the urine will become virtually sodium free in several days in an attempt to maintain sodium balance.

It is known that the urinary excretion of Na^+, the principal extracellular electrolyte, must somehow be related to changes in volume of the extracellular fluid, most probably to some intravascular portion of that fluid phase. A number of investigators have provided evidence that changes in intravascular fluid volume and hemodynamics, when induced in the head, the thorax, the abdominal cavity, or the extremities, will lead to changes in rate of excretion of Na^+ and water. For example, hemorrhage, upright posture, abdominal compression, and cuffing of the proximal portion of the limbs lead to increased sodium reabsorption, and hence decreased excretion. Contrarily, supine posture, compression of the neck, compression of the limbs by elastic bandages, and infusion of hypotonic or isotonic saline lead to sodium loss. These effects have been observed in the absence of measurable changes in glomerular filtration rate. Elkinton (1960) has made the following summary statement: "In view of the variety of conditions that lead to changes in sodium excretion, it is most probable that the degrees of filling of the vascular tree in both the venous and arterial sides are stimuli in multiple receptor sites. From these receptor sites afferent neutral pathways must lead to an integrating center in the CNS, again location unknown."

With regard to mechanisms on the efferent arc of the control mechanism for aldosterone, a humoral factor may be involved. Davis (1961, 1962) believes that increased renin production and angiotensin II formation is a mechanism for aldosterone regulation; in his view, angiotensin stimulates the glomerulosa cells. Factors which reduce renal blood flow would be trigger mechanisms. It is well to keep in mind that not only aldosterone but other steroids of the adrenal cortex, such as hydrocortisone, are concerned with Na^+ turnover; they will be discussed in Chapter 35. ADH has been shown to have a Na^+-losing action in some species.

Osmotic diuretic agents, such as mannitol and glucose, have an influence on Na^+ excretion. The rapid transit of the osmotic agent and obligated water through the nephron washes out some of the Na^+ with it in a nonspecific manner. Finally, there are conditions of excess acid load wherein the load of the acid ions is so great as to exceed the capacity of the kidney to secrete H^+ and NH_3 in exchange for Na^+. Under these conditions the acid ions will be lost in the urine with fixed base.

Chloride

In general, Cl^- behaves similarly to Na^+ in regard to renal handling and sweat loss. However, there are exceptions. During acid-base alterations, proportions of Cl^- and HCO_3^- associated with Na^+ may be noted in the plasma and in the urine differing from the normal relationship. It has also been observed that mercurial and thiazide diuretics cause greater loss of Cl^- than of Na^+ in the urine.

Although chloride movement is ordinarily viewed as a passive mechanism secondary to that of sodium, some important exceptions occur. The active transport of chloride by the gastric mucosa is an outstanding example. Keynes (1963) has disclosed that the axoplasm of the giant squid axon actively takes up labeled chloride. The uphill inward transport was inhibited by dinitrophenol, and the lack of effect of ouabain indicated that the influx was not linked to transport of cations by the sodium pump. Evidence for an active chloride pump in amphibian cornea was supplied by Zadunaisky (1965), who showed that the short-circuiting current could account for the active transport of Cl^- from the aqueous to the tear side of the cornea. This mechanism appears to have a function in maintaining corneal transparency (by keeping the intracellular ion concentration low and thus preventing swelling).

Potassium

The major cation for maintaining proper pH of the intracellular fluid is potassium. All cellular activities involving electrical phenomena, such as skeletal and cardiac muscle contraction and nerve impulse conduction, are dependent on the gradients of K^+ and Na^+ across cell membranes. In considering the shifts of K^+ between the extracellular and intracellular compartments, it is well to keep in mind the quantitative disproportion that exists. Thus a total of 350 mEq in the extracellular fluid compartment contrasts with 3500 mEq in the intracellular fluid (Edelman and Leibman, 1959). Therefore, small losses out of the body cells or small uptakes can cause relatively significant changes in the concentration of potassium in the extracellular fluid.

The potassium and sodium shifts that occur at the cell border during nerve impulse transmission and skeletal muscular contraction were discussed in an earlier section. Tissue repair and growth require K^+; conversely, protein breakdown releases K^+ from the cells into the extracellular compartment, from which it is lost in the urine. The transfer of glucose from the extracellular to the intracellular phase requires K^+ by certain cells. In alkalosis, K^+ is replaced in the cells by Na^+ and H^+. In severe metabolic acidosis, as in diabetic acidosis, K^+ leaves the cells to enter the extracellular fluid compartment and is excreted in the urine. Contributing to cellular loss of potassium is cellular protein catabolism associated with increased gluconeogenesis.

About 10 percent of K^+ loss occurs in the stool, the remainder in the kidney. Renal regulation involves the adrenal cortical hormones. Potassium loss may be excessive in certain types of chronic renal diseases associated with polyuria and in the diuretic phase of recovery from acute lower nephron diseases. In a condition called renal tubular acidosis, abnormal amounts of HCO_3^- are excreted in the urine along with accompanying fixed cations, including K^+. In the oliguria of acute renal shutdown or chronic renal disease, K^+ retention results, which may cause an elevation in extracellular and plasma concentration to the level that fatal hyperkalemia eventuates with stoppage of the heart by K^+ inhibition.

Calcium and Magnesium

Because of their low concentration, the role of calcium and magnesium ions is small in the control of volume and osmolarity of the body fluids, and in acid-base equilibrium. Calcium, a major constituent of bone, is also important because of its influence on cell membrane permeability, neuromuscular excitability, transmission of nerve impulses, blood coagulation, and activation of certain enzyme systems. It is both protein-bound and ionized in the plasma; the extent of ionization is determined by the acid-base equilibrium, being increased in acidosis and decreased in alkalosis. The action of parathyroid hormone in renal regulation of Ca^{++} balance is also related to $PO_4^{\equiv}$ handling. The hormone favors increased tubular Ca^{++} reabsorption, while apparently depressing tubular reabsorption of $PO_4^{\equiv}$, with increased urinary loss. This leads to a secondary rise in serum Ca^{++}. Continued mobilization from bone leads to further hypercalcemia and ultimate renal loss by overloading the reabsorptive mechanism (Chap. 33).

Like calcium, magnesium is in either a bound or an ionized phase in the plasma. Its physiological action is noted mostly in connection with the functioning of the neuromuscular and cardiovascular systems. Thus, magnesium excess results in depression of the CNS with loss of the tendon reflexes, drowsiness, and finally coma. An excess may cause bradycardia and depression of conduction through the conducting tissue and myocardium.

DISTURBANCES OF FLUID AND ELECTROLYTE BALANCE

DEHYDRATION

When water loss exceeds intake, the interstitial fluid yields water as long as possible, but eventually desiccation of cells occurs *(dehydration).* Water deficits are incurred in normal individuals under two common conditions: (1) excessive loss of sweat; (2) prolonged deprivation of water.

During preliminary stages of negative water balance, skin and muscle give up fluid first, so that vital organs are protected. Following depletion of interstitial fluid, the plasma water is increasingly removed. The blood becomes more concentrated *(anhydremia).* With prolonged loss or deprivation, or both, intracellular water is also lost.

Characteristic symptoms result from dehydration. Water deficit gives rise to a shrunken appearance of the face and body. The skin loses its elasticity and becomes hard and leathery. There is rapid loss of body weight. When the deficiency reaches such a degree that the water is no longer sufficient for removal of heat of metabolism, high fevers may occur. As the condition worsens, circulatory failure develops. Anuria results, and acid products are retained, leading to acidosis. Cerebral disturbances, excitement, delirium, and coma terminate the episode.

Clinical dehydration may be a consequence of (1) failure of absorption from the alimentary tract (as in pyloric stenosis or high intestinal obstruction); (2) excessive loss from copious sweating, prolonged vomiting, diarrhea, and excessive diuresis; (3) drainage from wounds or burns.

Water loss under these conditions involves also loss of electrolytes, predominantly NaCl. If fluid balance is therefore restored with only pure water, a hypotonic state results in the extracellular compartment. As a consequence, water moves in abnormal amounts into the cells now relatively hyperosmotic, producing symptoms referable to *water intoxication,* among them the well-known heat cramps. Obviously, in treatment, electrolyte (chiefly NaCl) must be given as well as water.

EXCESS WATER LOADS

A condition of cellular overhydration may result also from attempts to produce diuresis by forcing hypotonic fluids, particularly in the presence of renal impairment. On the other hand, prolonged and excessive diuresis may result in excessive loss of extracellular solutes, and water will migrate from a zone of relative hypotonicity into the cells. The symptoms of water intoxication which result are particularly referable to the CNS: salivation, nausea, and vomiting; restlessness, asthenia, muscle tremors, ataxia, and tonic and clonic convulsions that may eventuate in stupor and death.

EDEMA

The fundamental bases for formation of edema fluid have been discussed in Chapter 12 in connection with capillary fluid dynamics. Since edema formation represents an abnormal accumulation of extracellular fluid, particularly that of the interstitial compartment, added facets of the problem of edema are presented here.

CARDIAC EDEMA

This important type of edema was discussed in detail in Chapter 18.

CIRRHOTIC EDEMA

The edema of cirrhosis is localized in the abdominal cavity (ascites). Degeneration of the parenchymal cells of the liver leads to impairment of circulation through the liver, interfering particularly with the low-pressure portal vein system. The resulting portal hypertension favors the exudation and accumulation of fluid in the abdomen, both from the capillaries within the liver and from the capillaries in the intestine. However, portal hypertension is not the only factor in this type of edema. Impaired liver function results in hypoalbuminemia, contributing to the effective outward gradient for filtration of fluid. The increased titer of ADH in the urine of cirrhotics is evidence of greater antidiuretic activity. This is presumed to be due to failure of the liver to inactivate this hormone. There is also evidence that aldosterone activity is enhanced under these circumstances, acting to favor salt and water retention. Other circumstances characterized by obstructive conditions in the liver give a similar picture, e.g., sclerosis of the hepatic vessels or compression of the portal vein by tumors, aneurysms, and the acute icteric phase of infectious hepatitis.

RENAL EDEMA

The basic causes of renal edema are (1) decrease in oncotic pressure of the plasma because of loss of albumin via the kidney; (2) coexisting increase in systemic capillary permeability with loss of protein to the interstitial spaces; (3) coexisting congestive heart failure of the hypertensive type when associated with chronic nephritis; and (4) Na^+ retention. In chronic nephrosis, in the nephrotic stage of glomerulonephritis, and in the amyloid kidney, the kidneys excrete 10 to 20 gm of protein a day into the urine because of glomerular damage. Plasma albumin decreases from a normal of about 5 percent to about 2 percent, resulting in a marked reduction of oncotic pressure. The protein content of the edema fluid is low, about 0.1 percent. With acute glomerulonephritis there is widespread capillary injury in all parts of the body; hence, the protein content of the edema fluid may be high (over 1 percent). The edema is a combination of decreased oncotic pressure of the plasma and increased oncotic pressure of the interstitial fluid. Hypertensive heart disease accompanies chronic diffuse glomerulonephritis, pyelonephritis, and polycystic renal disease. With this, the ultimate cardiac decompensation contributes factors which are characteristic of congestive heart failure, including the elevated venous pressure and sodium retention by mechanisms previously described.

NUTRITIONAL EDEMA

Nutritional edema can be caused by faulty metabolism or nutritional deficiency. Although vitamin lack – e.g., lack of vitamin B in beriberi and of vitamin C in scurvy – is contributory, perhaps most important is protein lack in the diet. Inadequate protein intake leads to hypoproteinemia and reduction of the oncotic pressure of the plasma proteins, favoring outward movement of fluid from capillaries into the interstitial spaces. The decrease in tissue pressure which results from the wasting of tissue may favor this outward movement of fluid, and a contributing factor may be the changes in capillary permeability caused by the attendant vitamin deficiency. Some authorities feel that the general reduction in cardiac activity and blood flow impairs kidney function and leads to secondary Na^+ retention as a final contributory factor.

REFERENCES

Christensen, H. N. *Body Fluids and Their Neutrality.* New York: Oxford University Press, 1963.

Davis, J. O. Control of aldosterone secretion. *Physiologist* 5:65–86, 1962.

Davis, J. O., C. R. Ayers, and C. C. J. Carpenter. Renal origin of an aldosterone stimulating hormone in dogs with thoracic caval constriction and in sodium depleted dogs. *J. Clin. Invest.* 40:1466–1474, 1961.

Davson, H. *Physiology of the Ocular and Cerebrospinal Fluids.* Boston: Little, Brown, 1956.

Deane, N. *Kidney and Electrolytes: Foundations of Clinical Diagnosis and Physiologic Therapy.* Englewood Cliffs, N.J.: Prentice-Hall, 1966.

Dittmer, D. S. (Ed.). *Blood and Other Body Fluids.* Washington, D.C.: Federation of American Societies for Experimental Biology, 1961.

Edelman, I. S., and J. Leibman. Anatomy of body water and electrolytes. *Amer. J. Med.* 27:256–277, 1959.

Elkinton, J. R. Regulation of water and electrolytes. *Circulation* 21:1184–1192, 1960.

Fishman, A. P. (Ed.). *Symposium on Salt and Water Metabolism.* New York: New York Heart Association, 1959.

Gamble, J. L. *Chemical Anatomy, Physiology and Pathology of Extracellular Fluid,* 6th ed. Cambridge, Mass.: Harvard University Press, 1958.

Hogben, C. A. M. The chloride transport system of the gastric mucosa. *Proc. Nat. Acad. Sci.* 37:393–395, 1951.

Huf, E. G., J. Parrish, and C. Weatherford. Active salt and water uptake by isolated frog skin. *Amer. J. Physiol.* 164:137–142, 1951.

Keynes, R. D. Chloride in the squid giant axon. *J. Physiol.* 69:690–705, 1963.

Maxwell, M. H., and C. R. Kleeman. *Clinical Disorders of Fluid and Electrolyte Metabolism.* New York: McGraw-Hill, 1962.

Pitts, R. F. *The Physiological Basis of Diuretic Therapy.* Springfield, Ill.: Thomas, 1959.

Shanes, A. H. (Ed.). *Electrolytes in Biological Systems.* Washington, D.C.: American Physiological Society, 1955.

Sodeman, W. A. *Pathologic Physiology.* Philadelphia: Saunders, 1961.

Strauss, M. B. *Body Water in Man.* Boston: Little, Brown, 1957.

Ussing, H. H. Ion Transport Across Living Membranes. In *Renal Function: Transactions of the Fourth Conference.* New York: Josiah Macy, Jr. Foundation, 1952. Pp. 88–122.

Verney, E. B. Antidiuretic hormone and the factors which determine its release. *Proc. Roy. Soc. [Biol.] (Lond.)* 135:25–106, 1947.

Zadunaisky, J. A. Chloride active transport in the isolated frog cornea. Abstracts of the 9th Annual Meeting of the Biophysics Society, San Francisco, February, 1965. P. 9.

24

Respiratory and Renal Regulation of Acid-Base Balance

Ewald E. Selkurt and
Paul C. Johnson

THE CHIEF acid product of metabolism is $CO_2(H_2CO_3)$, of which an adult produces about 288 liters per day or 26 Eq, which is equal to 2.6 liters of concentrated HCl. Adults subsisting on a mixed diet produce in addition to CO_2 a substantial quantity of nonvolatile (fixed) acids. For example, 100 gm of protein produces approximately 60 mEq of sulfate by the oxidation of proteins and a quantity of phosphate, requiring 50 mEq of base for its neutralization at pH 7.4. An additional 50 mEq of base is required to neutralize the phosphate from 100 gm of fat (phospholipids). The acids produced are mainly sulfuric and phosphoric but include some hydrochloric, lactic, uric, and β-hydroxybutyric. Approximately one-half of these metabolically produced acids are neutralized by base in the diet, but the remainder must in some fashion be neutralized by the buffer systems of the body.

BUFFER SYSTEMS OF THE BODY

Several buffer systems prevent the organism from being overwhelmed by its own acid products. The one which reacts most readily is the blood buffer system. This depends only upon chemical processes which occur in the blood to attenuate the change in hydrogen ion concentration. Ultimately the buffering capacity of the tissues is brought into play also, since blood pH influences interstitial and intracellular pH as well.

For long-term stability of pH it is necessary to have an effective means of excreting the acids formed in the metabolic processes. This consists of the responses of the respiratory control center and renal tubules to changes in pH and carbon dioxide of the arterial blood. Moreover, these acids must be excreted as acids or as ammonium salts by the kidney since, if they were excreted as neutral salts, the buffers of the body would be rapidly depleted. This becomes increasingly apparent when it is recognized that the total cation available in the blood as buffer salts does not exceed 150 mEq. In the body as a whole, including the protein salts and phosphates of the tissues, the total cation available is 1000 mEq.

BLOOD BUFFERS

The Brønsted-Lowry concept of acids and bases finds wide acceptance in the consideration of acid-base balance (see Muntwyler, 1968). According to this view, an acid is a proton (H^+) donor, and a base, a proton acceptor. On this basis, the reaction indicating the acidity of an acid, A, takes the form

$$A \rightleftharpoons B + H^+$$

where B is a base since it can accept a proton to form the acid, A, the proton donor. Some examples are given in Table 24-1.

TABLE 24-1. Proton Donors and Acceptors

Acid		Proton	Base
HCl	$\rightleftharpoons$	H^+	Cl^-
H_2SO_4	$\rightleftharpoons$	H^+	HSO_4^-
NH_4^+	$\rightleftharpoons$	H^+	NH_3

Buffers are substances that tend to stabilize the pH of a solution. They are partially ionized salts formed by the combination of a strong acid and a weak base, or vice versa (see Table 24-2). A buffer is effective only when there are appreciable quantities of the alkali salt and either the acid or the base from which

TABLE 24-2. Most Important Buffer Acids at Physiological pH

Proton Donor		Proton Acceptor		Proton
H_2CO_3	$\rightleftharpoons$	HCO_3^-	+	H^+
$H_2PO_4^-$	$\rightleftharpoons$	$HPO_4^=$	+	H^+
H protein	$\rightleftharpoons$	Proteinate$^-$	+	H^+
$HHbO_2$	$\rightleftharpoons$	HbO_2^-	+	H^+
HHb	$\rightleftharpoons$	Hb^-	+	H^+

Source: From E. Muntwyler. *Water and Electrolyte Metabolism and Acid-Base Balance.* St. Louis: Mosby, 1968.

it derives. It is most effective when the salt and the acid or the base are present in equal quantities. For this reason, buffers are usually thought of in terms of buffer pairs. Common buffer pairs in the blood are $NaHCO_3/H_2CO_3$ and Na_2HPO_4/NaH_2PO_4. The proteins of the blood, hemoglobin, oxyhemoglobin, albumin, and globulin, are extremely important buffers. Because of the amphoteric nature of proteins, each species represents a buffer pair. The major protein buffer in the blood is hemoglobin because of its high concentration. It is also considerably more important than the bicarbonate and phosphate buffers. The bicarbonate, however, has proved to be an accurate and easily determined index of acid-base balance.

MECHANISM OF BUFFER ACTION

The mechanism of buffer action may be understood from consideration of the bicarbonate system outlined in equation 1.

$$CO_2 + H_2O \rightleftharpoons H_2CO_3 \rightleftharpoons H^+ + HCO_3^- \quad (1)$$

If hydrogen ions are added to the system, some will combine with bicarbonate ions and drive the reaction to the left. Thus, not all the hydrogen ions added will stay in solution in the ionic form. Conversely, if base is added, some of the H^+ will be removed, shifting the reaction to the right and causing the H_2CO_3 to dissociate into H^+ and HCO_3^-. Thus, there is an inverse relationship between H^+ and HCO_3^- concentration when acid or base is added. This relation may be plotted as shown in Figure 24-1. The pH-bicarbonate diagram used here and in other portions of this chapter was introduced by Davenport (1958). It has proved extremely useful in describing the underlying causes of acid-base disturbances from a limited amount of data. This figure represents the buffer properties of plasma, of which bicarbonate is the principal buffer. The initial point A represents the normal condition of the plasma in arterial blood, namely, a pH of 7.4 and a HCO_3^- concentration of 24 mM per liter. When an acid is added to plasma, the pH decreases and, as a result of the reactions described above, the HCO_3^- concentration falls also. When base is added to plasma, both the pH and the HCO_3^- concentration increase. The effects of adding acid and base to blood or to plasma are as shown in Figure 24-1A only when the CO_2 tension is maintained constant. The above effects are seen when nonvolatile acids or bases are added to the blood. These are called *fixed* acids and bases, since they are not removed by respiratory activity.

The pH of the blood can also be influenced greatly by the CO_2 level. However, the effects of CO_2 on the pH and HCO_3^- concentration are quite different from those of the fixed acids and bases, as may be seen by referring to equation 1. If CO_2 is added or removed, the concentration of H^+ and HCO_3^- changes in the same direction. The behavior of plasma alone in vitro is shown in Figure 24-1B. Thus, the effect of CO_2 can be readily distinguished from that of fixed acid or base simply by noting the shift produced on the graph.

BUFFERING EFFECT OF WHOLE BLOOD

The changes produced in *separated plasma* studied in vitro as described above will differ in some important respects from those seen in plasma which is part of whole blood *(true plasma)*. When fixed acid or base is added to whole blood, all the buffers participate in the process of attenuating the pH shift. As a result, the directional changes in the pH-bicarbonate diagram will be the same as in separated plasma, but the magnitude of changes will be decreased. This effect may be seen from equation 2.

$$2H \text{ lactate} + NaHCO_3 + NaPr \rightleftharpoons H_2CO_3 + HPr + 2Na \text{ lactate} \quad (2)$$

When a *fixed acid* is added to whole blood, the buffering power of the red blood cell, which is due to the hemoglobin, tends to attenuate the change in pH. If the amount of acid added is plotted against the pH of the solution, the slope would be much different for whole blood as compared with separated plasma. However, the slope of the fixed acid line on the pH-bicarbonate diagram remains the same (Fig. 24-1C). This is to be expected since those hydrogen ions which combine with protein will not influence the bicarbonate concentration.

When CO_2 is added to whole blood, there is a lesser effect on pH than would be the case with separated plasma. In contrast to the effects of fixed acid, the slope of the buffer line (Fig. 24-1C) changes because H^+ and HCO_3^- are formed in equal quantities but some of the H^+ is neutralized by other buffer systems while all the HCO_3^- remains. As a result, in whole blood more bicarbonate ions are formed

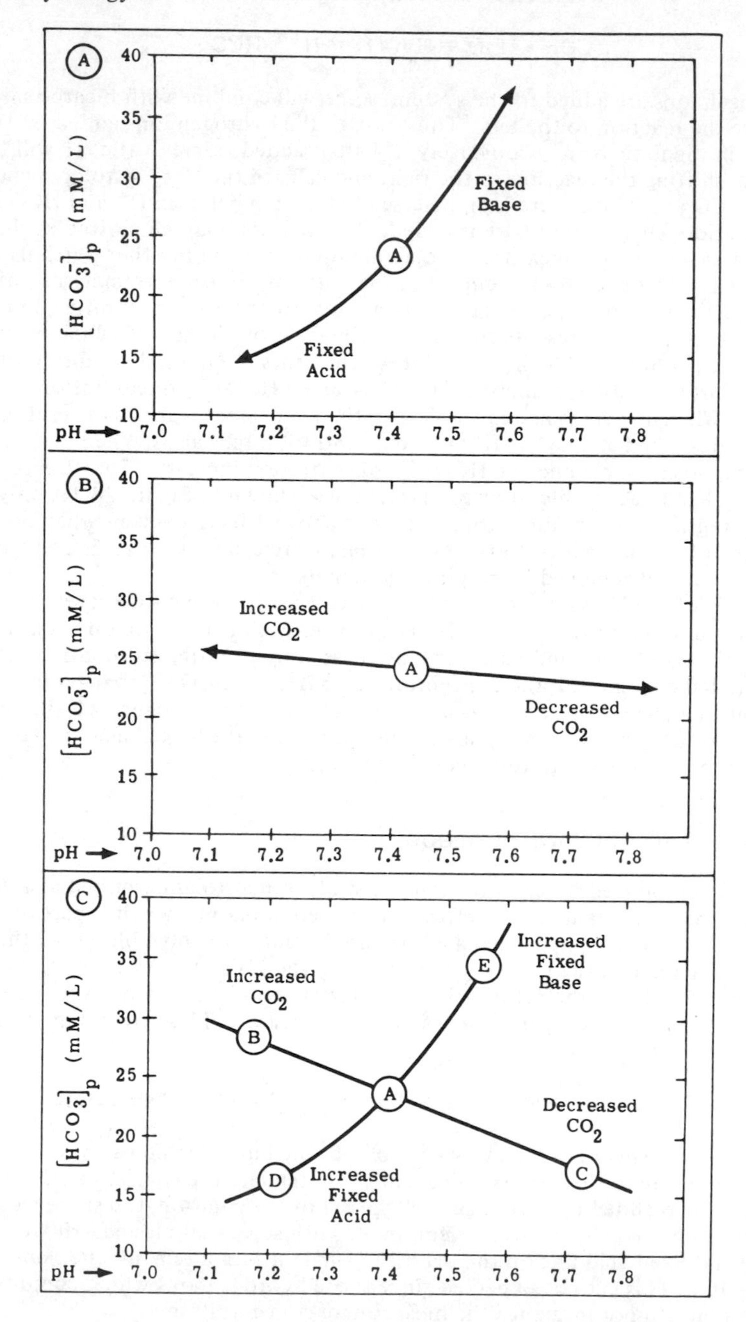

FIGURE 24-1. A. Effect of fixed acid and fixed base on pH and HCO_3^- content of separated plasma. B. Effect of changes in CO_2 on pH and HCO_3^-. C. Combination of the effects shown in A and B on true plasma. Note increased slope of buffer line with true plasma. (Reprinted from *The ABC of Acid-Base Chemistry,* 3d ed., by H. D. Davenport, by permission of The University of Chicago Press. Copyright 1947, 1949, 1950 and 1958 by The University of Chicago.)

per unit change in pH than in separated plasma. It should be borne in mind that the concentration of HCO_3^- being considered is that of the plasma portion of the blood only.

The concentration of H_2CO_3 and dissolved CO_2 in the plasma is much smaller than that of $NaHCO_3$. The ratio between these is approximately 1:20. Ordinarily, this would mean that such a buffer would be rather ineffective, particularly when base is added to the blood. However, since CO_2 is continuously manufactured and to some degree stored in the tissues, an additional supply is readily available. This makes it unnecessary for a large quantity of CO_2 to be present to permit the buffer system to operate effectively.

The curves of Figure 24-1C represent the response of true plasma to fixed acid or base and changes in CO_2 level. The four variations from the normal buffer point represent the four types of acid-base disturbance seen clinically: *respiratory acidosis* (CO_2 excess), *respiratory alkalosis* (CO_2 deficit), *metabolic acidosis* (fixed acid excess), and *metabolic alkalosis* (fixed base excess). By plotting the pH and HCO_3^- levels of a sample of arterial blood on the graph, it is possible to determine whether the subject is normal or in one of these states of acid-base disturbance.

To carry these considerations farther, it is necessary to analyze their quantitative aspects. The pH-bicarbonate diagram is actually a graphic representation of the Henderson-Hasselbalch equation:

$$pH = pK + \log \frac{[HCO_3^-]}{[H_2CO_3]} \tag{3}$$

The pK of the bicarbonate system is known and the pH and HCO_3^- concentration can be measured, but the H_2CO_3 concentration, which is extremely low, cannot be directly measured. This limits the utility of the equation as it now stands. However, reference to equation 1 reveals that H_2CO_3 concentration is directly dependent upon the concentration of dissolved CO_2, which is in turn dependent upon its partial pressure and solubility coefficient of CO_2 in plasma. When these factors are substituted into the equation, the following is obtained:

$$pH = pK + \log \frac{[HCO_3^-]}{a\,P_{CO_2}} \tag{4}$$

where the solubility constant, a, = 0.0301 and pK = 6.10.

Substituting, equation 4 becomes:

$$pH = 6.10 + \log \frac{[HCO_3^-]}{0.0301\,P_{CO_2}} \tag{5}$$

The equation now relates three measurable quantities; if any two are known, the third may be calculated.

The interrelation between the three factors may also be presented in the form of a logarithmic pH–P_{CO_2} plot (Fig. 24-2). This permits solution of the equation in a rapid graphic manner. For example, if the CO_2 tension and the pH of the blood are known, the HCO_3^- concentration can be determined simply by finding the point on the pH–P_{CO_2} and interpolating between the adjacent HCO_3^- lines.

This plot could be used to determine quantitatively the acid-base status of the individual. However, for assessment of the underlying clinical condition, the pH-bicarbonate diagram possesses certain advantages. For this reason, its construction and use will subsequently be considered in detail.

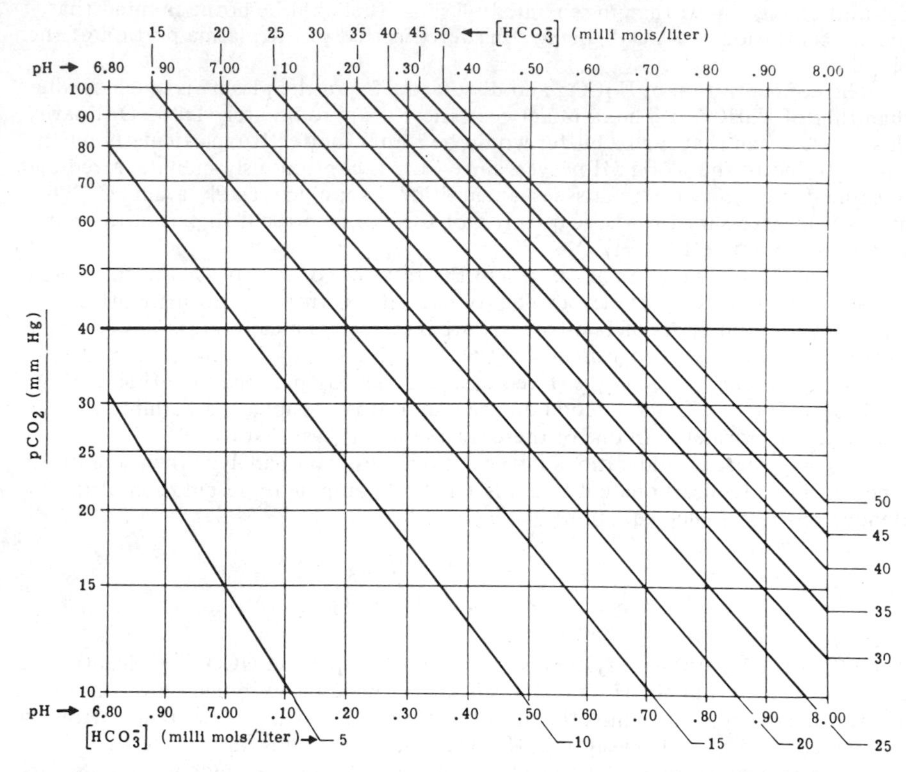

FIGURE 24-2. Nomogram of relation between pH, HCO_3^-, and P_{CO_2} of plasma. (Modified from Diagram P, Radiometer Co., Copenhagen, Denmark. Courtesy of Poul Astrup, M.D.)

Since the pH-bicarbonate diagram is a representation of the Henderson-Hasselbalch equation, the curves on it can be obtained simply by substituting appropriate values into the equation. For example, the curves relating pH and HCO_3^- concentration at a constant P_{CO_2} can be determined. Such a curve is called a P_{CO_2} *isobar.* In Figure 24-3, the values of HCO_3^- concentration are indicated at selected values of pH when P_{CO_2} is 40 mm Hg. Similar isobars are drawn for other P_{CO_2} levels. Such curves are obtained when fixed acid or base is added to the blood.

Calculation of the *buffer lines* is not so readily performed, since none of the three values is constant along the P_{CO_2} isobar. However, the slope of the line relating pH and HCO_3^- is constant and has been experimentally determined. It is 21.6 mM per liter per pH unit. This line may be drawn through the normal point, C, in Figure 24-3 (normal buffer line) and is shown drawn through point A in Figure 24-4; or it can be drawn through any other point on the graph. The direction which purely respiratory changes would take from any point would then be known, since changes along this line are due to alterations in CO_2 level.

Respiratory Acidosis

Since the directional changes on the graph of any form of acid-base disturbance are known, it is possible to determine the type and degree of any abnormality

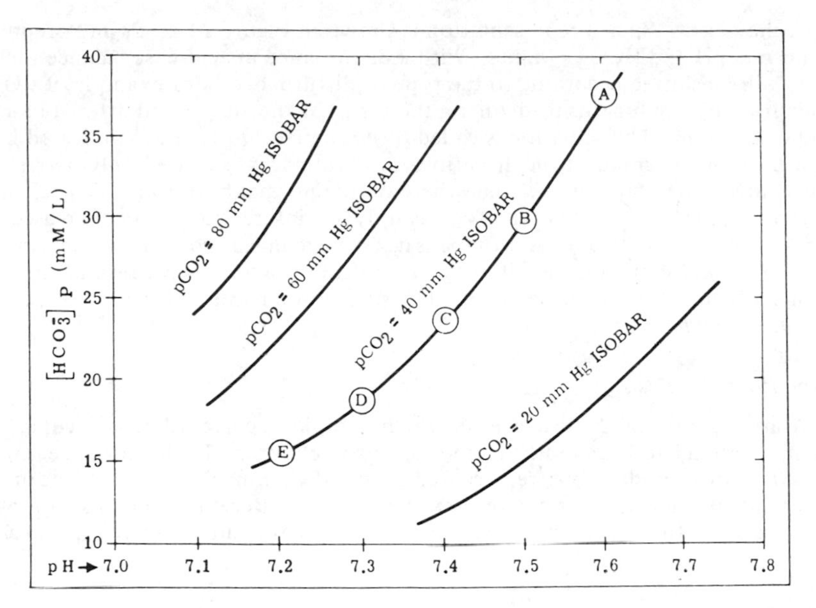

FIGURE 24-3. Calculation of P_{CO_2} isobars from Henderson-Hasselbalch equation. (Reprinted from *The ABC of Acid-Base Chemistry,* 3d ed., by H. D. Davenport, by permission of The University of Chicago Press. Copyright 1947, 1949, 1950 and 1958 by The University of Chicago.)

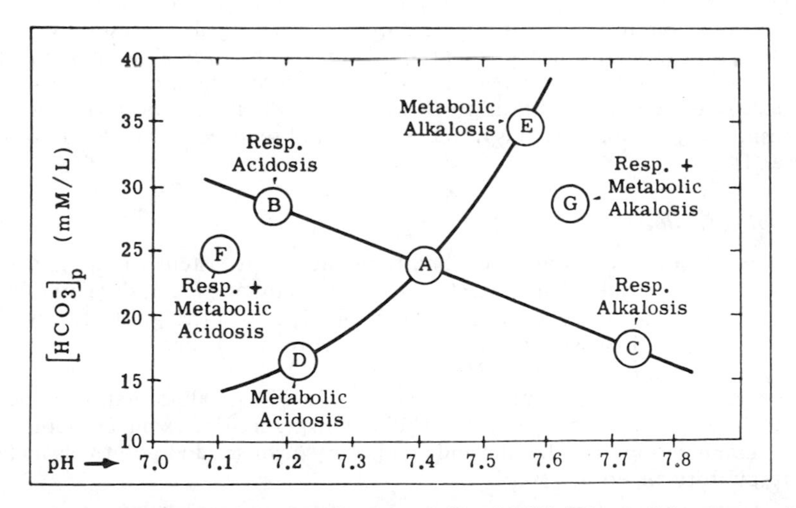

FIGURE 24-4. Effect of acid-base disturbances on pH and HCO_3^- level of the plasma, plotted on pH-bicarbonate diagram. Note that each type of disturbance occupies a unique place on the diagram. (Reprinted from *The ABC of Acid-Base Chemistry,* 3d ed., by H. D. Davenport, by permission of The University of Chicago Press. Copyright 1947, 1949, 1950 and 1958 by The University of Chicago.)

which may exist. Such a representation is shown in Figure 24-4. Point A represents the normal pH and HCO_3^- values. With a disturbance in acid-base balance these values will be shifted according to the type of disturbance. For example, if CO_2 elimination by the lungs is inadequate, the P_{CO_2} of the alveoli and arterial blood will be increased. This situation is called *respiratory acidosis* and is indicated by point B on the diagram. If the deviation from normality is caused only by respiratory acidosis, the point will lie somewhere on the normal buffer line. The change in pH and HCO_3^- produced represents simply the interaction of the increased CO_2 and the blood buffers. As compensatory mechanisms come into play (in this instance the kidney), point B will move away from the normal buffer line in a manner which will be described later. The situation illustrated is *uncompensated respiratory acidosis.*

Respiratory Alkalosis

There are also clinical disturbances in which chronic hyperventilation occurs. Examples are hysteria and salicylate poisoning in children. In this instance *respiratory alkalosis* is produced, as represented by point C. Since it lies on the normal buffer line, it is *uncompensated respiratory alkalosis.* Because compensation by the kidney is slow, this uncompensated state will be occasionally observed in the clinic.

Metabolic Acidosis

Metabolic acidosis is produced by conditions such as ketosis induced by diabetes mellitus or excessive loss of alkaline fluids from the lower digestive tract, as occurs in diarrhea. When such a condition occurs, the pH of the blood decreases and the HCO_3^- concentration decreases also (point D). In this manner it can be differentiated from respiratory acidosis. If the P_{CO_2} is not changed as a result of the acidosis, point D will lie on the same P_{CO_2} isobar as the normal point and is therefore termed *uncompensated metabolic acidosis.* However, reduction of pH has a stimulatory effect on respiration which will reduce the P_{CO_2} levels. The effect of this respiratory compensation is to increase pH and reduce HCO_3^- levels, as will be seen later.

Metabolic Alkalosis

Metabolic alkalosis is produced by conditions such as persistent vomiting, which causes loss of a considerable amount of acid. In metabolic alkalosis (point E) the pH and HCO_3^- increase together, which enables it to be distinguished from respiratory alkalosis. If there is no respiratory response to the change in pH, the condition is *uncompensated metabolic alkalosis.*

In some instances there may be a mixture of acid-base abnormalities which can also be detected by use of the pH-bicarbonate diagram. For example, point F represents a combination of metabolic and respiratory acidosis, and point G is metabolic and respiratory alkalosis.

Summary of Changes in Plasma Bicarbonate–Carbonic Acid Buffer System in Disturbances of Neutrality

Figure 24-5 attempts to summarize the primary and secondary changes in the HCO_3^-/H_2CO_3 buffer system with various disturbances in acid-base balance.

In *respiratory acidosis,* the retention of CO_2 and rise in H_2CO_3 are the primary cause. The kidney is the important compensatory organ, secreting H^+ and conserving bicarbonate.

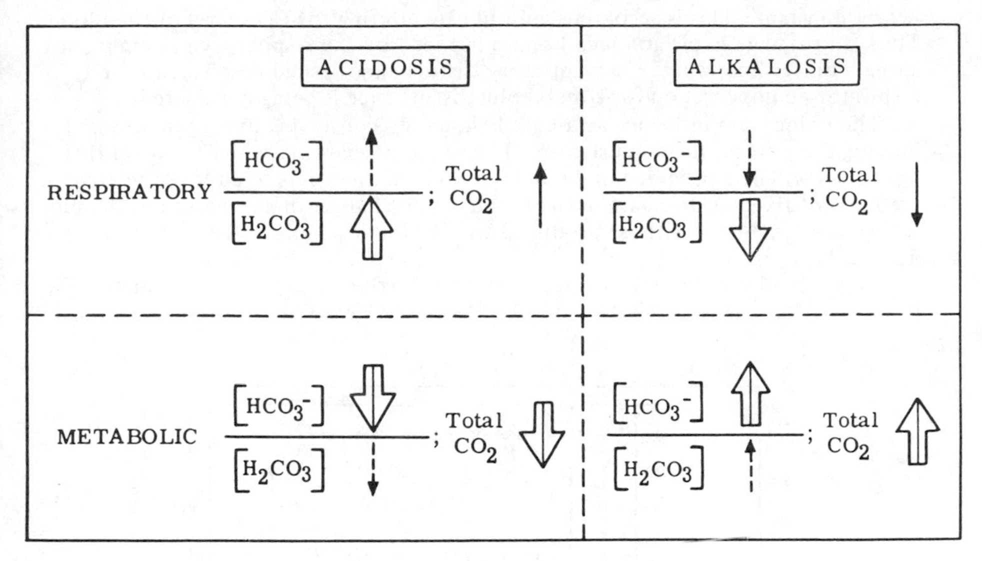

FIGURE 24-5. Primary and secondary changes in the bicarbonate–carbonic acid buffer system descriptive of the various types of disturbances in acid-base balance. Note that the total CO_2 does not always fall in acidosis or rise in alkalosis. (From H. N. Christensen. *Body Fluids and Their Neutrality*. New York: Oxford University Press, 1963. P. 122.)

In *respiratory alkalosis* the net loss of carbonic acid by overbreathing is primary, and any tendency to compensate would decrease the bicarbonate–ion concentration secondarily. The decrease is achieved by the kidney.

In *metabolic acidosis* the entering hydrogen ion reacts with the bicarbonate ion to produce a primary decrease in its concentration (shown by the heavy arrow). Respiratory activity causes a smaller, secondary decrease in the carbonic acid level (shown by the dotted arrow). Because the numerator is decreased more than the denominator, the pH must be lowered. In this instance the total CO_2 must be distinctly decreased. The total CO_2 content is a commonly employed laboratory determination, in which the serum is acidified and the CO_2 all brought into the gaseous form and measured. The compensatory mechanisms are discussed in greater detail below.

Likewise in *metabolic alkalosis,* the bicarbonate ion is primarily increased, the carbonic acid being secondarily elevated and to a lesser degree. The total effect on the CO_2 content is an elevation. Renal compensation involves loss of HCO_3^-.

Note that the overall direction in the change of the CO_2 content is downward in both metabolic acidosis and respiratory alkalosis and upward in metabolic alkalosis and respiratory acidosis. The reliability of any association between low CO_2 content and respiratory acidosis is vitiated. Reliance upon the serum CO_2 content (or capacity) alone as an assay of the acid-base balance obviously may lead to mistaken conclusions regarding acid-base balance disturbances of both respiratory and metabolic types.

RESPIRATORY AND RENAL COMPENSATION IN pH-BICARBONATE DIAGRAM

It has already been indicated that the respiratory system acts as a buffer system by responding to changes in blood pH in such a manner as to maintain the pH very nearly constant. This is accomplished only by altering the CO_2 level of the blood. Thus, any change in pH and bicarbonate induced by the respiratory system occurs along a buffer line, whether a respiratory disturbance of acid-base balance or a respiratory compensation for a metabolic disturbance is being considered.

The kidney can influence acid-base balance by conserving hydrogen ions or removing them from the blood stream. The renal mechanisms which perform this operation will be considered in detail later. The kidney acts to alter the relative amounts of fixed acid or base in the blood. Any changes of this nature in the blood follow the same path as those produced by metabolic acids or bases – that is, up or down a P_{CO_2} isobar.

Figure 24-6 visualizes the response of an animal to an acid injection in a stepwise fashion to produce a metabolic acidosis. As in the Henderson-Hasselbalch equation,

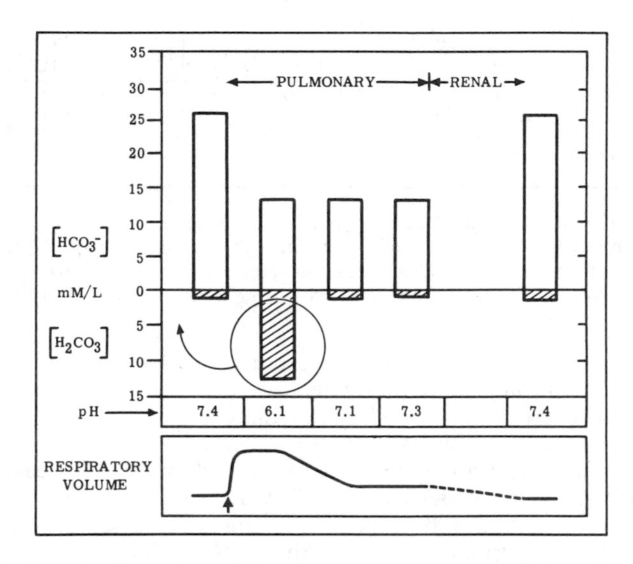

FIGURE 24-6. Stepwise illustration of the pulmonary response to acid invasion (not carbonic acid). Enough acid is administered at the arrow to cause almost one-half the HCO_3^- to be converted to H_2CO_3. (At the same time other buffer anions also accept H^+.) An explosive increase in respiration results, so that the new-formed H_2CO_3 is swept out as CO_2, and the hypothetical pH of 6.1 (second stage) is never reached. The third stage shows the transient picture when the P_{CO_2} has returned to a normal value. (From H. N. Christensen. *Body Fluids and Their Neutrality*. New York: Oxford University Press, 1963. P. 118.)

the serum bicarbonate-ion concentration is shown above the zero line; the carbonic acid, below. An amount of acid was injected to change the normal ratio of perhaps 26:1.3 millimoles to one of say 13.7:13.7. By the Henderson-Hasselbalch equation the pH should be 6.1. The lower curve illustrates diagrammatically that the incoming acid strongly stimulates the respiratory rate and volume, so that the new CO_2 is quickly swept out and the pH never actually falls as low as 6.1.

At the third stage the accelerated respiration has brought the carbonic acid concentration to normal. Since $[HCO_3^-]/[H_2CO_3]$ is now about 10, the pH is 7.1. But the respiratory system cannot bring the bicarbonate concentration back to normal. The injection of hydrogen ion has decreased the bicarbonate concentration to 12.5 mM per liter, and the injected chloride has taken its place. Although respiration has minimized the pH change, the compensation is not complete. The final stage of compensation is turned over to the kidney, which accomplishes it by excreting H^+ in exchange for Na^+, as $NaHCO_3$, and excreting Cl^-.

These compensatory mechanisms may now be considered in terms of their effect on the pH-bicarbonate diagram. The pathways followed by the compensatory mechanisms are shown in Figure 24-7A. When *respiratory acidosis* occurs, for

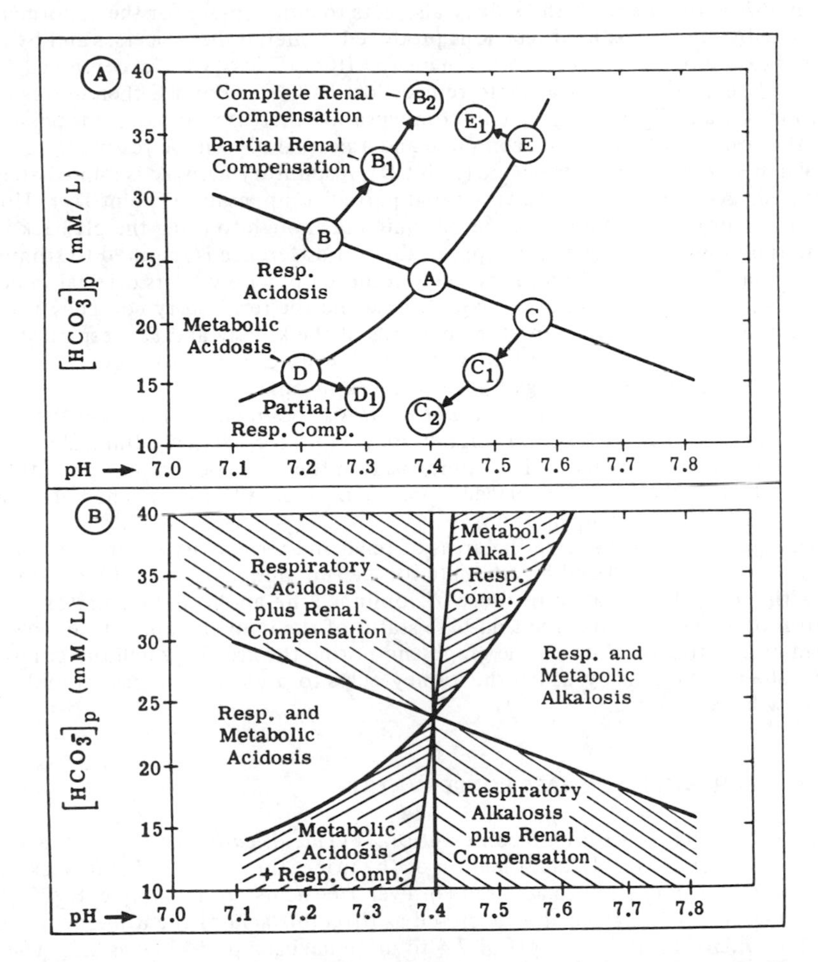

FIGURE 24-7. A. Effects of compensation on the pH and bicarbonate content. B. Zones of acid-base disturbance and compensation. (Reprinted from *The ABC of Acid-Base Chemistry*, 3d ed., by H. D. Davenport, by permission of The University of Chicago Press. Copyright 1947, 1949, 1950 and 1958 by The University of Chicago.)

example (point B), renal compensation follows, in the form of an abnormally large excretion of acid by the kidney. With this, HCO_3^- tubular reabsorption is increased. Plasma Cl^- is lost to the extent that HCO_3^- concentration is raised. This acid loss causes movement of the point up the P_{CO_2} isobar toward a normal pH. At point B_1

the pH has been only partially restored to its normal value; at point B_2 it has been fully restored – hence the designations *partially compensated* and *fully compensated respiratory acidosis.* The kidney is able under some circumstances to bring the pH back to 7.4 as indicated by the fully compensated state. This shows a highly effective mechanism of the kidney for hydrogen ion disposal. The response is, however, quite slow. Several days may be required for it to compensate for a chronic condition.

In *respiratory alkalosis* the kidney also acts to compensate for the abnormality, this time by conserving hydrogen ions produced in metabolic processes and by increased excretions of HCO_3^-. With increased HCO_3^- loss, Cl^- is conserved, raising plasma Cl^- concentration to replace lost HCO_3^-. Point C then moves to positions C_1 and C_2 as the degree of compensation increases. Again, it is possible that the compensation may be complete and a normal pH will be restored.

When *metabolic acidosis* occurs (point D), respiratory activity is immediately increased. As a result there is a very rapid partial compensation (point D_1). However, respiratory compensation is never complete enough to bring the pH back to the normal value. Apparently an appreciable pH difference is required to stimulate respiration. It is apparent that if pH were returned precisely to its original value the stimulus for increased respiration would cease and the respiratory compensation would disappear. The compensatory patterns of the kidney and the respiratory system therefore differ in the respect that respiratory compensation is never complete. The kidney may later again restore pH to normal.

A similar pattern is seen in *metabolic alkalosis* (points E, E_1). Under the influence of the high pH of the arterial blood, respiratory activity diminishes, allowing CO_2 to accumulate. The compensation therefore occurs along a buffer line which is parallel to but displaced from the normal buffer line. The kidney is again needed for full compensation.

In Figure 24-7B are shown the areas of renal and respiratory compensation. Any set of values of pH and bicarbonate for a given sample of arterial blood must fall within one of these areas or those of combined metabolic and respiratory acidosis or alkalosis. Thus, the acid-base status of the individual may be determined by plotting the data on the graph and noting the area in which the point falls. The mechanisms by which the kidney helps to preserve acid-base equilibrium will now be considered.

RENAL MECHANISMS FOR BASE CONSERVATION

Two mechanisms exist for base conservation: (1) *acidification of the urine* and (2) *ammonia synthesis.* In acidification of the urine, the neutral salts of weak buffer acids are converted to acid salts or even free acids. For example, H_3PO_4, NaH_2PO_4, uric acid, etc., can be excreted as titratable acid of the urine, reducing the glomerular filtrate from a pH of 7.4 to urine having a pH as low as 4.5. The general equation for this process is

$$Na_2HPO_4 + HHCO_3 \longrightarrow \underset{\text{(excr.)}}{NaH_2PO_4} + \underset{\text{(reabs.)}}{NaHCO_3} \qquad (6)$$

The other main mechanism is ammonia synthesis. Where there are strong acids (H_2SO_4, HCl), they are finally excreted as ammonium salts.

$$Na_2SO_4 + 2H^+ + 2NH_3 \longrightarrow \underset{\text{(excr.)}}{(NH_4)_2SO_4} + \underset{\text{(reabs.)}}{2Na^+} \qquad (7)$$

Replacement by hydrogen ions means (as illustrated in equation 6) that the acid radicals are excreted as free acids. There is a limit to the amount of acid which can be excreted in this form, however, because the kidneys cannot excrete urine which is more acid than about pH 4.5. Excretion of strong acids by this means is accompanied by their full equivalent of base. Weaker acids can be excreted in the free form to an extent which is determined by their pK values. This becomes clear from application of the Henderson-Hasselbalch equation:

$$pH = pK + \log \frac{\text{concentration of salt}}{\text{concentration of acid}}$$

Recall also that pK is the pH at which there are equal quantities of acid and the accompanying salt. If pK is the same as the pH of the urine, one-half of the acid will be present in the free form. In the case of a weak acid whose pK is 1 pH unit above the pH of the urine, nine-tenths of the acid will be free and one-tenth combined with base. However, with a strong acid whose pK is 1 pH unit below the pH of the urine, only one-tenth of the total amount of acid can be excreted in the free form, so that nine-tenths of its equivalent of base will have to accompany it. Stronger acids with pK values of more than 1 unit below the minimal urinary pH must always be excreted with more than nine-tenths of the equivalent base.

Those weaker acids which can be excreted to a considerable extent in the free form in the urine within the physiological range of pH are known collectively as the *buffer acids* of the urine. The most important of these is H_3PO_4, but amino acids and creatinine (pK 4.97) are also in this category. The pK for the dissociation of the second hydrogen ion of phosphoric acid is 6.8; therefore, each equivalent of phosphate in the urine of pH 4.8 requires one equivalent of base to cover it, whereas in the plasma at pH 7.4 it requires nearly two equivalents. Hence, by excreting urine of maximal acidity the kidneys can salvage one equivalent of fixed base for each equivalent of acid phosphate in the urine, as well as additional amounts corresponding to the other buffer acids which are present. The process of acidification of the kidney is reversed when the urine is titrated with alkali back to the pH of the plasma, 7.4, to determine the "titratable acidity." This reflects the amount of base the kidneys have conserved by acidifying the urine. Such base conservation may range from about 30 mEq to as high as 150 mEq per day during diabetic acidosis, when the urine contains large quantities of β-hydroxybutyric acid (pK 4.7), and acetoacetic acid.

MECHANISM OF ACIDIFICATION OF THE URINE

Pitts et al. (1945) examined older theories of phosphate and bicarbonate reabsorption with simultaneous measurement of glomerular filtration rate and found that both fell far short of accounting for the possible titratable acidity of the urine, in each case only 20 percent or less, the limitation being the amount of acid in the glomerular filtrate (see Pitts, 1968). To explain the much greater availability of the acid, it was postulated that the tubular cells are able to exchange ions between the tubular urine and the blood stream. Secretion of H^+ as a basis for an ion exchange mechanism was postulated, and this is now accepted.

Ion Exchange Mechanism

According to the ion exchange mechanism, H^+ is secreted by the tubular cells into the lumen in exchange mainly for Na^+ from the glomerular filtrate. The source of H^+ is the CO_2 of the metabolic activity of the tubular cells, or that brought in by the blood. The mechanism is facilitated by the enzyme *carbonic anhydrase* (CA). The evidence for this is that certain sulfonamide derivatives, which have the ability to inhibit CA enzyme, block H^+ secretion and produce excess excretion of sodium in the urine. The acetazolamide Diamox has been found most effective.

The overall rate of secretion of H^+ is determined by the following factors: (1) the degree of acidosis; (2) the quantity of buffer acid excreted; and (3) the strength of the buffer, i.e., how strongly it resists giving up Na^+ for H^+ (e.g., phosphate with a pK of 6.8 permits greater secretion of H^+ than creatinine with a pK of 4.97). The ion exchange theory is diagrammed in Figure 24-8.

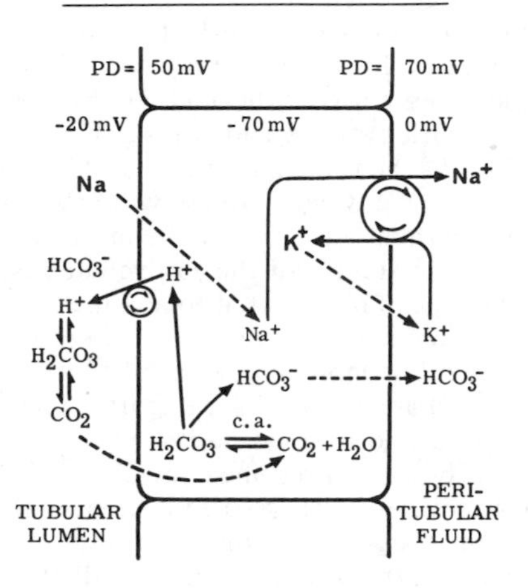

FIGURE 24-8. Ion exchange mechanisms in a representative proximal tubular cell. (PD = potential difference.) (From R. F. Pitts. A comparison of modes of action of certain diuretic agents. *Progr. Cardiovasc. Dis.* 3:537, 1961. By permission.)

Role of Potassium

A second component of the ion exchange mechanisms involved in acid-base regulation of the urine by the kidney concerns potassium ion, which has been shown to play a reciprocal role with hydrogen. It will be recalled that K^+ is filtered at the glomeruli and at least to some extent reabsorbed in the proximal convoluted tubular cells, or possibly some more distal segment (beginning of distal convoluted tubule?). This mechanism also involves secretion of K^+. Evidence from use of the micropuncture technique suggests that this occurs in the more distal portions of the distal convoluted tubule (Chap. 22). Recent microcatheterization techniques suggest that it may also be secreted in the collecting ducts (Fig. 24-9A).

There is little information concerning the actual mechanisms by which the K^+ - Na^+ exchange is accomplished in the distal tubular zone of urinary acidification. It appears that the K^+ - Na^+ exchange underlies the sodium extrusion mechanism which characterizes virtually all animal cells. However, the reciprocal action of H^+ with K^+ in the exchange with Na^+ indicates that no simple model will explain the exchange mechanism. In any event, in acidosis, H^+ is predominantly exchanged with Na^+; thus, the urine contains a large amount of HA and very little KA (A representing the acid anion). The same situation would prevail in potassium deficiency, and the urine would be highly acid. During conditions of alkalosis,

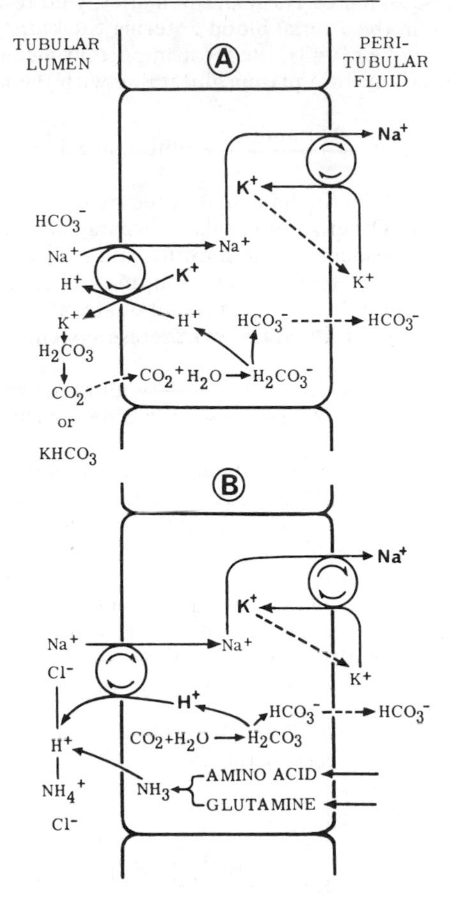

FIGURE 24-9. A. Collecting duct cell ion exchange mechanisms, emphasizing interrelationship between H^+ and K^+ secretion (the CA mechanism operates here as in Fig. 24-8). B. Mechanism of production of NH_3 and base ($NaHCO_3$) conservation. (From *Physiology of the Kidney and Body Fluids,* Second Edition, by R. F. Pitts. Copyright ©1968, Year Book Medical Publishers, Inc. Used by permission.)

predominantly greater amounts of K^+ are secreted in exchange for Na^+ and relatively little H^+, and the urine contains little or no HA but large amounts of KA. The same situation would obtain in conditions of K^+ excess. These interdependencies between H^+ and K^+ extraction do not stand in a one-to-one relation, however, because H^+ has a much greater affinity for the transport mechanisms.

Ammonia Production

It has been stated that the ability to excrete strong acid in free titratable form is limited because the urine pH cannot go below about 4.5. Another mechanism for conservation of base is created by secretion of NH_3 into the urine, which binds hydrogen ion and forms the ammonium ion NH_4^+. The ability of the kidney to form ammonia was discovered by Nash and Benedict, who found more NH_4^+ in the venous blood than in the arterial blood entering the kidney.

According to Van Slyke (1943), the most important mechanism for formation of ammonia is its formation from plasma glutamine with the aid of glutaminase:

$$\text{glutamine} \xrightarrow{\text{(glutaminase)}} \text{glutamic acid} + NH_3$$

Thus, glutamine appears to be the immediate precursor of ammonia normally produced by the kidney. Other amino acids (L-asparagine, DL-alanine, L-histidine, L-aspartic acid, glycine, L-leucine, L-methionine, and L-cysteine) when infused into acidotic dogs also increase the secretion of ammonia. These acids are transferred to α-ketoglutarate by transaminases to form glutamate, and this in turn is converted to additional glutamine by the glutamine synthetase system:

$$\alpha\text{-ketoglutarate} + \alpha\text{-amino acid} \longrightarrow \text{glutamate} + \alpha\text{-keto acid}$$
$$\text{glutamate} + ATP + NH_3 \longrightarrow \text{glutamine} + ADP + PO_4$$

Additional ammonia can theoretically come from glutamate directly (by undergoing deamination within the tubular cells containing a specific glutamic dehydrogenase), or from the glutamine formed, as above. In summary, glutamine and amino acids contribute amino and amide groups to a nitrogen pool from which ammonia is formed and diffuses into the urine, where it is trapped as ammonia ion (Fig. 24-9B). The mechanism of action is as follows: each molecule of NH_3 (a proton acceptor) secreted binds one hydrogen ion and permits one sodium ion to be reabsorbed. Thus, it is to be expected that the pH of the urine will determine the rate of diffusion of NH_3 into the tubular lumen. The mechanism is understood better by the following series of equations:

$$NH_3 + H^+ \rightleftharpoons NH_4^+ \tag{8}$$

The ionization constant, K_A, is:

$$K_A = \frac{[H^+]\ [NH_3]}{[NH_4^+]} \tag{9}$$

Or

$$pH = pK_A + \log \frac{[NH_3]}{[NH_4^+]} \qquad (pK_A = 8.9) \tag{10}$$

It is accepted on good evidence that the cells generally are permeable to the molecular species NH_3 but not the ionic species NH_4^+, perhaps because NH_3 is lipid-soluble, whereas NH_4^+ is water-soluble. Since it may be considered that NH_3 is in the same concentration on both sides of the cell membrane because of its easy diffusibility, the numerator of the last term of the equation can be considered to be a constant; or, to put it simply, the log of NH_4^+ will vary inversely with the pH. Then, as the urine becomes increasingly acid, the mass law operates to increase the fraction of the total ammonia produced in the distal segment that is captured in the urine as nondiffusible NH_4^+ and excreted as such. The capture of NH_4^+ in the urine serves to neutralize free acid and thus permits a larger quantity of acid radical to be excreted as NH_4^+ salt at a given urine hydrogen ion concentration than would otherwise be possible. Stated another way, the excretion of NH_4^+ salt reduces the quantity of Na^+ which would otherwise accompany the acid radical into the urine. The tubular excretion of NH_4^+ thus serves to conserve an equivalent quantity of Na^+ for the body.

Site of Production of NH_3

Micropuncture data in the rat reveal that the proximal tubule is an important potential source of final urine ammonia, and that increased proximal ammonia addition occurs in response to chronic ammonium chloride acidosis (which leads to metabolic acidosis and increased renal glutaminase activity). Distal tubular ammonia addition is also important in the final ion exchange mechanism (Fig. 24-10). The collecting duct was found to be a significant source of final urine ammonia in the hamster kidney but not in the rat kidney.

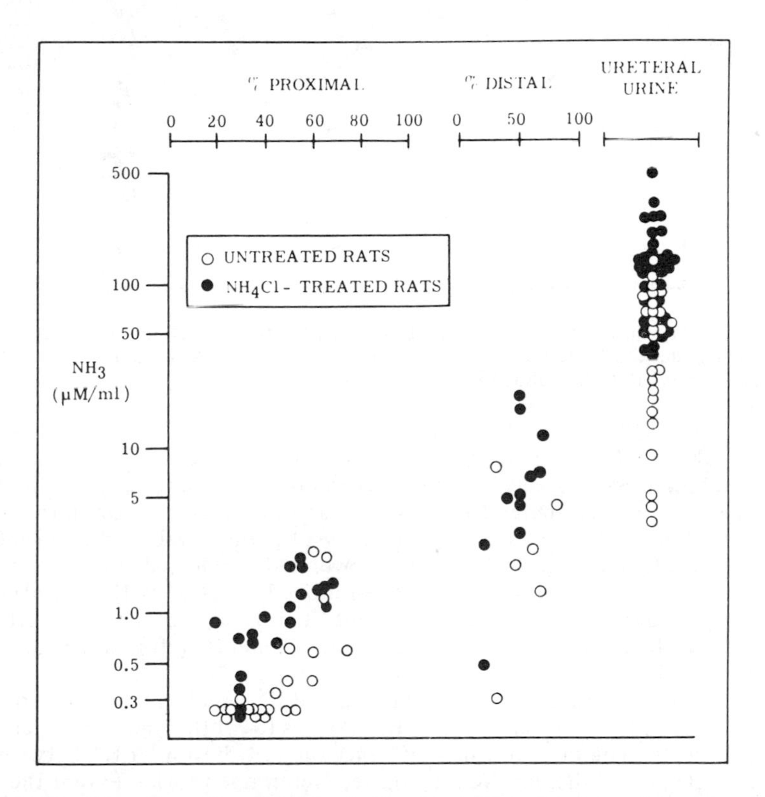

FIGURE 24-10. Sites of secretion of ammonia in untreated rats as compared to rats made acidotic by NH_4Cl ingestion. (From S. Glabman, R. M. Klose, and G. Giebisch. *Amer. J. Physiol.* 205:127–132, 1963.)

SUMMARY OF BASE CONSERVATION MECHANISMS

The micropuncture techniques of Gottschalk et al. (1960) revealed that in the nondiuretic rat the pH of the proximal tubular urine decreased below that of the plasma (Fig. 24-11). This is in keeping with the finding, also by micropuncture techniques, that considerable absorption of bicarbonate goes on in the proximal con-

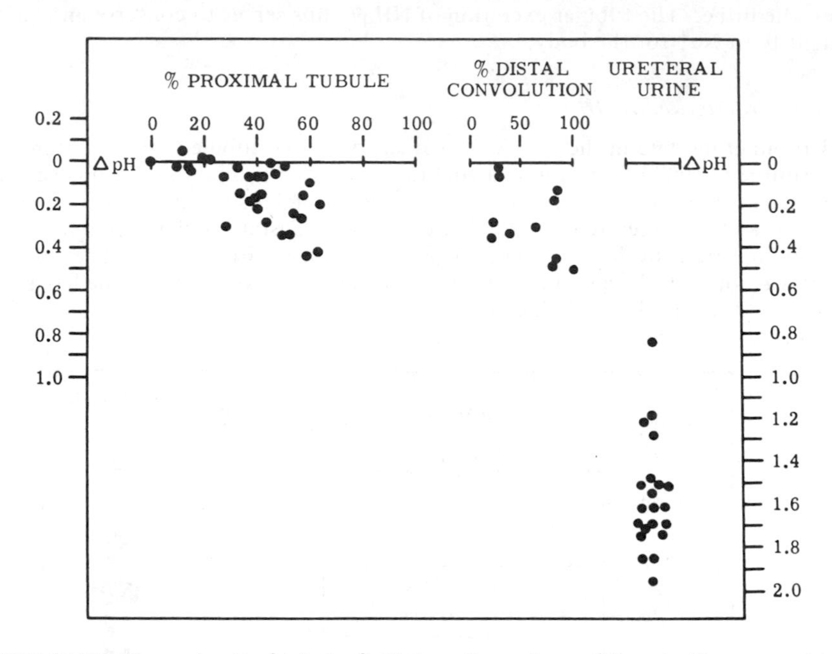

FIGURE 24-11. Change in pH of tubular fluid along the nephron of the rat. Measurements were made under equilibrium conditions. (From C. W. Gottschalk, W. E. Lassiter, and M. Mylle. *Amer. J. Physiol.* 198:581–585, 1960.)

voluted tubules. The drop of pH in the proximal system is not as great as in the distal nephron (less than one-half pH unit) and is explained in terms of the bulk exchange that proceeds here. In terms of bicarbonate reabsorption in the proximal tubule, the amount of H^+ exchanged here must be considerably greater (3500 mEq per day) than that secreted in the distal convoluted tubule and collecting duct (100 mEq per day) but proceeds on a one-for-one basis ($Na^+ \leftrightharpoons H^+$), so pH does not change significantly in the proximal segment. This is because the H^+ which reacts with filtered HCO_3^- is recycled back into the cell and therefore does not ordinarily contribute to the net elimination of acid.

Only that small fraction of the total secreted H^+ which is bound to urinary buffer, either as titratable acid or as NH_4^+, serves to rid the body of metabolically produced acids. The factors which influence the distribution of H^+ between filtered HCO_3^- and the non-HCO_3^- buffers determine to what extent the potentially available H^+ will be utilized to conserve filtered HCO_3^- and to what extent to generate new HCO_3^-. If, for example, metabolic acid is administered, $NaHCO_3$ will be decomposed, resulting in a fall in the [HCO_3^-] in glomerular filtrate. Consequently, less H^+ will be utilized in the reabsorption of filtered HCO_3^- and more available for excretion as titratable acid and urinary ammonia. If there is a plentiful supply of filtered buffers (e.g., phosphate, creatinine), titratable acid will be produced. If, on the other hand, there is a paucity of filtered buffers, the removal of H^+ from the renal cell will be restricted and an intracellular acidosis will supervene. The intracellular acidosis will stimulate the formation of new buffer in the form of ammonia, which is then added to the tubular fluid, thus permitting H^+ secretion to return toward its normal rate.

The particular importance of the distal nephron mechanism stems from the fact that the amount of H^+ that can be pumped into the urine is greater than the excretion of base, thus aiding in base conservation. This is abetted by capitalizing on the NH_3 production mechanism and the K exchange system. However, in view of the fact that ammonia is synthesized in the proximal convolution and because pH under certain conditions decreases along this segment, some overlap of function between proximal and distal convolutions can be anticipated.

EXCRETION OF ALKALINE URINE

Like the excretion of acid urine which helps to conserve the body stores of fixed base, the excretion of alkaline urine, which cannot take place without the loss of fixed base from the body, is desirable under certain circumstances of disturbed acid-base balance. There is an upper limit to the alkalinity of the urine at about pH 8.0, and excess cations in alkaline urine are balanced by HCO_3^-. Bicarbonate ordinarily appears in the urine when its concentration in the plasma exceeds the normal value of about 27 mEq per liter (Fig. 24-12). When the concentration in

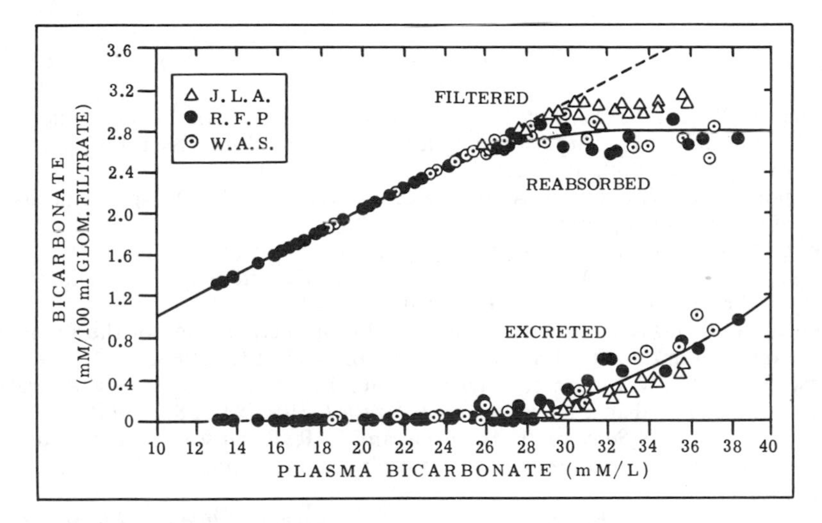

FIGURE 24-12. Filtration, reabsorption, and excretion of bicarbonate as functions of plasma concentration in normal man. (From R. F. Pitts, J. L. Ayer, and W. A. Schiess. *J. Clin. Invest.* 28:35–49, 1949.)

the plasma falls below this, only traces of bicarbonate are found in the urine. However, it is not correct to call this a "bicarbonate threshold," for the statement that bicarbonate is practically absent from the urine at plasma concentrations below 27 mEq per liter is true only so long as the respiratory center is free to regulate pulmonary ventilation in response to the prevailing level of plasma CO_2. During periods of overventilation the urine may become alkaline and contain considerable amounts of bicarbonate although the plasma bicarbonate may be well below the so-called normal "threshold" concentration. This situation appears less paradoxical if the kidneys are regarded as regulating not the level of bicarbonate in the plasma but the concentration of fixed base in the urine. This happens because, when CO_2 is blown off by the lungs, the cations are balanced in the plasma by the anionic groups of plasma proteins. Since the proteins do not pass into the glomerular filtrate in appreciable amounts, the cations, particularly Na^+, pick up HCO_3^- in the tubular fluid. This appears as bicarbonate in the urine to make it alkaline, even though the concentration of bicarbonate in the plasma is relatively low. Thus, the bicarbonate comes from the kidney, where it is synthesized in the epithelial cells and captured by the alkaline fluid passing them.

In summary of the regulatory mechanism: When plasma bicarbonate ($BHCO_3$) is reduced, all of the filtered HCO_3^- is converted by the substitution of H^+ for Na^+ to H_2CO_3. The CO_2 thus formed escapes from the tubular urine by diffusion. Reabsorbed Na^+ combines with intracellular CO_2 to re-form $NaHCO_3$, which is then returned to the blood. When the plasma $BHCO_3$ is elevated, the resulting decrease in activity of the hydrogen exchange mechanism permits excess $BHCO_3$ to be excreted in the urine. Thus, urine buffers are converted to the acidic form in acidosis, increasing the excretion of titratable acid, or to the alkaline form in alkalosis, increasing the titratable alkali. In acidosis, the acidity of the urine favors the capture of NH_3 and permits the excretion of additional acid as the NH_4 salt, conserving an equivalent quantity of Na^+.

REFERENCES

Berliner, R. W. Ion Exchange Mechanisms in the Nephron. In A. P. Fishman (Ed.), *Symposium on Salt and Water Metabolism.* New York: New York Heart Association, 1959. P. 892.

Davenport, H. D. *The ABC of Acid-Base Chemistry,* 4th ed. Chicago: University of Chicago Press, 1958.

Dick, D. A. T. *Cell Water.* Washington, D.C.: Butterworth, 1966.

Gamble, J. L. *Chemical Anatomy, Physiology and Pathology of Extracellular Fluid.* Cambridge, Mass.: Harvard University Press, 1954.

Glabman, S., R. M. Klose, and G. Giebisch. Micropuncture study of the ammonia excretion in the rat. *Amer. J. Physiol.* 205:127–132, 1963.

Gottschalk, C. W., W. E. Lassiter, and M. Mylle. Localization of urine acidification in the mammalian kidney. *Amer. J. Physiol.* 198:581–585, 1960.

Hayes, C. P., Jr., J. S. Mayson, E. E. Owen, and R. R. Robinson. A micropuncture evaluation of renal ammonia excretion in the rat. *Amer. J. Physiol.* 207:77–83, 1964.

Maxwell, M. H., and C. R. Kleeman. *Clinical Disorders of Fluid and Electrolyte Metabolism.* New York: McGraw-Hill, 1962.

Muntwyler, E. *Water and Electrolyte Metabolism and Acid-Base Balance.* St. Louis: Mosby, 1968.

Pitts, R. F. *Physiology of the Kidney and Body Fluids,* 2d ed. Chicago: Year Book, 1968.

Smith, H. W. *Principles of Renal Physiology.* New York: Oxford University Press, 1956.

Van Slyke, D. D., R. A. Phillips, P. B. Hamilton, R. M. Archibald, P. H. Futcher, and A. Hiller. Glutamine as source material of urinary ammonia. *J. Biol. Chem.* 150:481–482, 1943.

Welt, L. G. *Clinical Disorders of Hydration and Acid-Base Equilibrium,* 2d ed. Boston: Little, Brown, 1959.

25

Digestive Tract

A. Gastrointestinal Circulation

Paul C. Johnson

THE ACT of chewing and swallowing is carried out by the voluntary muscles of the upper gastrointestinal (GI) tract. The blood flow through such tissues is closely related to the metabolic requirements of the muscle, as discussed in Chapter 29. When the food reaches the stomach, the digestive process begins; this ultimately involves a number of organs – stomach, small and large intestine, liver, and pancreas – which together comprise what is known as the splanchnic region. While having no digestive function, the spleen is also included in this domain. Almost all the available information on blood flow to the GI tract relates specifically to the splanchnic region, which will be the focal point of this discussion.

SPLANCHNIC BLOOD SUPPLY

ANATOMICAL FEATURES

The blood supply to the stomach, liver, pancreas, and spleen is provided by the celiac artery. The superior mesenteric artery supplies the entire small intestine except for the superior part of the duodenum, which is perfused from the gastroduodenal branch of the celiac. The inferior pancreatoduodenal artery arises from the superior mesenteric and supplies the head of the pancreas as well as the duodenum. The large intestine is supplied in part from the superior mesenteric by way of the ileocolic, right colic, and middle colic arteries. The inferior mesenteric artery supplies a portion of the transverse colon, as well as the ileac and descending colon, the sigmoid colon, and much of the rectum.

The venous drainage of the splanchnic area is unique in that all the flood from the digestive organs is channeled into the portal system which empties into the liver. Only the spleen, pancreas, and lower part of the rectum bypass the liver. It is important to note that there are some collateral connections between the gastric veins and the esophageal veins which empty into the azygos veins, and between the inferior mesenteric veins and hemorrhoidal veins which empty into the hypogastric veins. When portal inflow to the liver is impeded, as in cirrhosis, collateral channels may constitute an important route of drainage. A substantial increase in portal pressure is involved, however.

The microanatomy of the splanchnic circulation shows diverse features in different organs and even within different tissues in the same organ. This is to be expected since the function of the muscular coat of the digestive tract is far different from that of the mucosa for example. Studies by Mall showed a basket-like network of capillaries in the intestinal villi. On the other hand, the capillaries of the intestinal muscle form a rectilinear network. In the villus the central artery and vein are in close proximity, allowing for the possibility of countercurrent exchange between vessels. The diameter of the villus capillaries is 5 to 7 μ while in the muscle it is 4 to 5 μ.

The circulation of the liver is unique in that it receives a dual supply from the hepatic artery and the portal vein. The portal vessels break up into small branches which ultimately join the hepatic sinusoids. The hepatic artery gives rise first to arterioles, which empty into the hepatic sinusoids. Some of the arterioles may initially divide into capillaries, which in turn empty into the sinusoids.

The capillary wall varies considerably in form and structure in different parts of the splanchnic region. Generally the wall structure is less formidable than it is in striated muscle, for example. The capillaries in the mucosa of the intestine possess an endothelium which is fenestrated and a basement membrane which is indistinct. The liver sinusoids have an endothelial lining which is perforated with large openings and a basement membrane which is sparse and incomplete.

SPLANCHNIC BLOOD FLOW

The blood flow differs greatly between these organs. As shown in Table 25A-1, flow varies from 18 ml per minute in the pancreas to 180 ml per minute in the

TABLE 25A-1. Blood Flow of the Gastrointestinal Organs in an "Average" 15-kg Dog

Organ	Weight (gm)	Fraction of Cardiac Output (%)	Blood Flow (ml)	Perfusion Rate (ml/min X 100 gm)
Stomach	100	1.9	50	50
Intestine	270	6.5	180	70
Colon	50	1.6	40	80
Pancreas	30	0.7	18	60
Gallbladder	2	0.04	1	40

Source: After Delaney and Custer, 1965.

small intestine. The perfusion rate on an organ weight basis is, however, roughly the same in all organs. Liver blood flow is the sum total of portal flow and hepatic artery flow. In man this is about 500 ml per minute, of which the artery contributes 150 ml.

Blood flow through the splanchnic circulation is regulated primarily by the arterioles and precapillary sphincters. The region of greatest pressure drop is in the arterial vessels smaller than 50 μ in diameter. By the time the blood reaches the arterial end of the capillaries, the pressure has fallen to approximately 35 mm Hg. In the capillary network the pressure falls an additional 20 mm Hg, and thus at the capillary-venous junction it is approximately 15 mm Hg. These figures were obtained on mesentery but are probably applicable to the small intestine and may be appropriate for stomach and colon as well. The pressure in the portal vein is usually about 10 mm Hg, so the venous pressure gradient is only 5 mm Hg.

The pressure drop in the hepatic artery circuit is believed to be similar in most respects to that in the small intestine. However, an important difference is that the liver sinusoids have a very low hydrostatic pressure (3 to 4 mm Hg). Thus it appears that approximately 95 percent of the driving pressure is dissipated in the arterial vessels of the liver. It should be noted that this low pressure is necessary to allow a positive pressure gradient to exist from portal vein to the hepatic sinusoids, which even so is only 6 to 7 mm Hg. The hepatic vein gradient is 3 to 4 mm Hg.

TRANSCAPILLARY EXCHANGE

Rapid absorption of fluid and foodstuffs from the GI tract obviously requires a capillary network with structural and physiological properties which maximize the exchange process. Among these are large surface area, high H_2O filtration constant, and adequate blood perfusion. The mucosal capillaries of the intestine form a basket-like network around the villus, presenting a large and readily accessible area for exchange. The perfusion rate of the mucosal area is high, 30 to 70 ml per minute $\times$ 100 gm tissue. The absorption of foodstuffs involves the blood capillaries, but the wall structure is readily penetrated by molecules of the size of sugars and amino acids without hindrance. Molecules as large as those of inulin (molecular weight 5200) produce no osmotic effect across the intestinal capillaries. The structure is apparently fenestrated to a sufficient degree to allow molecules of this size to pass readily. However, albumin and globulin do not pass the capillary wall easily and the leakage rate is quite small.

The modified wall structure of the intestinal capillary is also reflected in the high rate of water exchange. Intestinal capillaries filter water at a rate of 0.4 ml per minute $\times$ 100 gm tissue when a hydrostatic head of 1 mm Hg is applied. This is approximately 20 times higher than in capillary beds of skeletal muscle and skin.

The intestine provides an intesting example of the adaptability of the Starling forces governing capillary exchange. The pressure in the arterial capillaries is about 35 mm Hg and that in the venous capillaries is about 15. But the filtration constant of the venous capillary is probably twice as great as that of the arterial capillary. In addition, the venous capillary is typically somewhat larger, so the surface area is about twice as great. The fluid exchange across the venous capillary is four times as fast, therefore, and the Starling forces are balanced at some point in the venous capillary rather than at the midpoint of the capillary. This can be seen in studies of the mean capillary pressure of the intestine done by the isogravimetric technique, where the mean hydrostatic pressure in the capillaries required to maintain equilibrium is 16 to 17 mm Hg rather than 25. Thus the balance point of the capillaries is placed well into the venous portion of the network. Moreover, a restatement of the balance of forces across the capillary membrane is required, achieved in the main by a readjustment of tissue colloid osmotic pressure. Intestinal lymph has a protein concentration of approximately 4 percent, which if present in interstitial fluid would provide a tissue colloid osmotic pressure of about 10 mm Hg. This acts in the same direction as the capillary hydrostatic pressure to produce a combined force balancing the colloid osmotic pressure of the plasma.

Portal pressure may vary under some circumstances, which would cause capillary hydrostatic pressure to change also. If portal pressure rises, mean capillary pressure increases by 60 percent of the increment in portal pressure. A rise in capillary pressure causes an outward filtration of fluid, diluting the protein in the capillary spaces. This decreases the tissue colloid osmotic pressure and tends to restore equilibrium of the Starling forces. In moderate portal hypertension the mechanism would produce a somewhat elevated tissue fluid volume, but a balance would be maintained. However, if portal pressure is much greater than 20 mm Hg, capillary pressure will exceed plasma colloid osmotic pressure and the forces will not be in equilibrium even if tissue protein concentration falls to zero. In this circumstance lymph flow is greatly elevated and its protein content is very low, for the reason cited above. The high filtration constant of the intestinal capillary now becomes a handicap rather than an asset. Fortunately, the rate of formation is modified by closure of precapillary sphincters, which reduces the available capillary surface. This factor alone reduces capillary filtration coefficient of the intestine by a factor of 4.

Transcapillary exchange in the liver is very different from that described above. The wall of the sinusoids is so thin and tenuous as to pose a rather limited barrier to fluid exchange and molecular diffusion. The protein concentration of lymph draining the liver is between 80 and 95 percent of the plasma protein concentration. Moreover, dextran molecules as large as 400,000 molecular weight reach a concentration in liver lymph which is 70 to 80 percent of that in plasma. The filtration constant of the sinusoids is so high as to be undeterminable by present techniques.

REGULATION OF FLOW AND PRESSURE

Local blood flow is often considered to be determined by the immediate metabolic needs of the tissues. It is readily apparent, however, that blood flowing through the gastrointestinal tract performs other functions as well. Between 4 and 6 liters of fluid is secreted into the tract each day by the salivary glands, stomach, pancreas, liver, and intestine. This fluid is ultimately supplied by the local circulation of the secretory cells. Foodstuffs and water absorbed from the intestinal lumen are carried away by the portal circulation. Gastrointestinal hormones such as secretin and enterocrinin are carried to their target sites by the blood stream. In this context it should also be noted that blood flow to the intestine is approximately three times that required to maintain the oxygen supply.

In meeting these diverse needs a basic constraint is placed upon the system since it is necessary to maintain a reasonably constant capillary pressure. This is true to some degree of all vascular beds, but, as noted earlier, the filtration coefficient of intestinal and liver capillaries is quite high; thus small changes in capillary pressure could produce rapid filtration and edema formation. Effective regulation of the peripheral circulation requires mechanisms which provide adequate blood flow to meet the needs while maintaining a nearly constant capillary pressure.

Splanchnic blood flow is substantial, even in the resting, postabsorptive state. As shown in Table 25A-1, the perfusion rate to the splanchnic area is between 50 and 80 ml per minute × 100 gm. This is approximately the same perfusion rate (on a tissue weight basis) as found in the myocardium of a resting individual. The blood flow is not uniformly distributed among the various tissues within the organs. Approximately two-thirds of the intestinal blood flow goes to the mucosa and submucosa and the remainder to the muscularis. The perfusion rate of the smooth muscle is approximately 10 to 20 ml per minute × 100 gm. This corresponds also to the flow rate in gastric smooth muscle (approximately 25 ml per minute × 100 gm) and esophagus (20 ml per minute × 100 gm). By contrast, resting skeletal muscle requires only 5 ml per minute × 100 gm tissue. Perfusion of the mucosa-submucosa region is 30 to 70 ml per minute × 100 gm.

BLOOD FLOW DURING DIGESTION

During digestion, blood flow to the splanchnic region increases. However, the pattern of response is not simple. Fronek and Stahlgren (1968) showed that during anticipation of food ingestion cardiac output in the dog increases from 157 to 187 ml per minute times body weight in kilograms, while blood pressure rises from 95 to 125 mm Hg. The increased cardiac output actually goes to areas such as those served by the brachiocephalic and external iliac arteries. Superior mesenteric artery blood flow does not increase. Since arterial pressure rises almost 30 mm Hg in the period of anticipation, the constancy of mesenteric flow is achieved by increased vascular resistance in this area. During food ingestion there is a further rise of cardiac output and blood pressure. Resistance to flow in the mesentery rises further, but despite this, flow to the region increases by 10 to 20 percent. Flow to other regions is variable. External iliac artery flow falls to control levels despite the increased arterial pressure, while brachiocephalic flow is doubled. After feeding, blood pressure and cardiac output return to control levels, as do iliac and brachiocephalic flows. However, superior mesenteric artery flow is increased by about 30 percent for a period of hours, as shown in Figure 25A-1. The mechanism of this

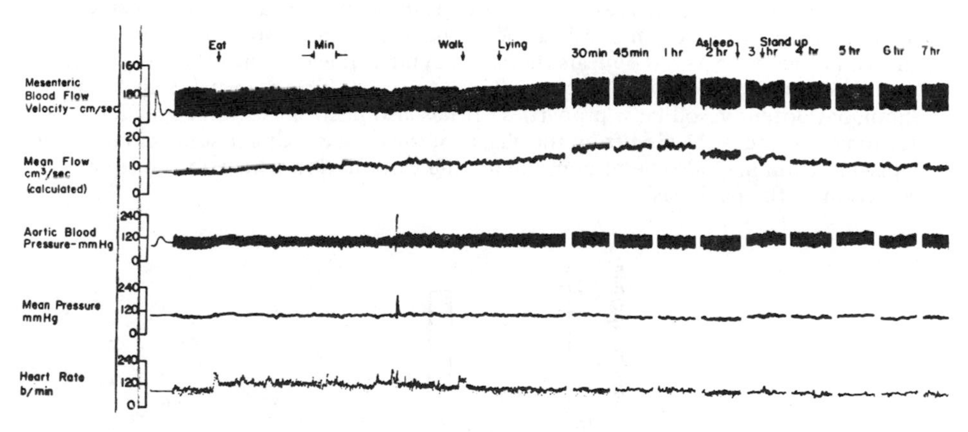

FIGURE 25A-1. Mesenteric blood flow during eating and digestion. (From Vatner et al., 1970.)

increase in flow is not clear. Placing foodstuffs directly into the intestine of anesthetized animals does not produce comparable changes.

The fact that blood flow to the stomach increases during digestion is assumed to be linked to the secretory function of the stomach in producing HCl. Studies using Heidenhain pouches have shown that flow to the mucosa specifically increases during increased secretory activity (Fig. 25A-2). In this instance total flow does not increase, suggesting that muscularis flow is probably changing in a direction opposite to that in the mucosa. The close relation between mucosal flow and secretion is presumably related to increased metabolic requirements during secretion. However, the direct effect of secretagogues such as histamine on arteriolar smooth muscle cannot be ruled out.

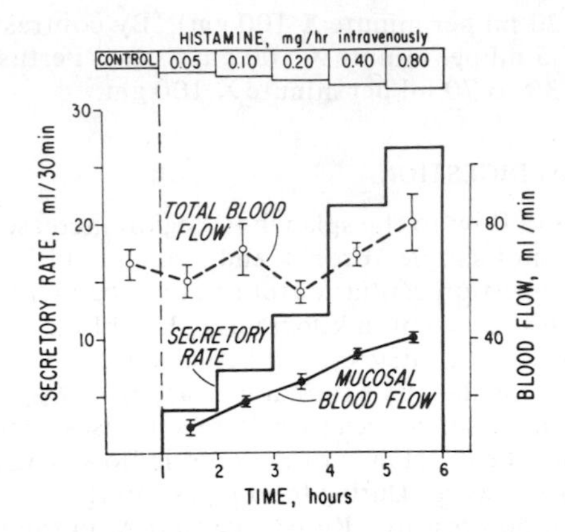

FIGURE 25A-2. Blood flow and secretory rate in a Heidenhain pouch during intravenous infusion of histamine. Note the close relation between secretion and mucosal blood flow. (From E. D. Jacobson et al. *Gastroenterology* 52:414, 1967. Copyright © 1967, The Williams & Wilkins Co., Baltimore, Md. 21202, U.S.A.)

The relation between secretion and blood flow has been investigated in the salivary glands and pancreas as well. Stimulation of the chorda tympani nerve produces a profound secretion of saliva and an associated increase in blood flow. As shown in Figure 25A-3, it appears that an enzyme is released into the tissue spaces during secretion which causes a plasma kinin to form. This material, called bradykinin, has potent vasodilator properties. It has also been shown that such a kinin-forming enzyme is released from the glands of the tongue. Under some circumstances a plasma kinin may also be responsible for the vasodilation which accompanies secretion of the pancreas.

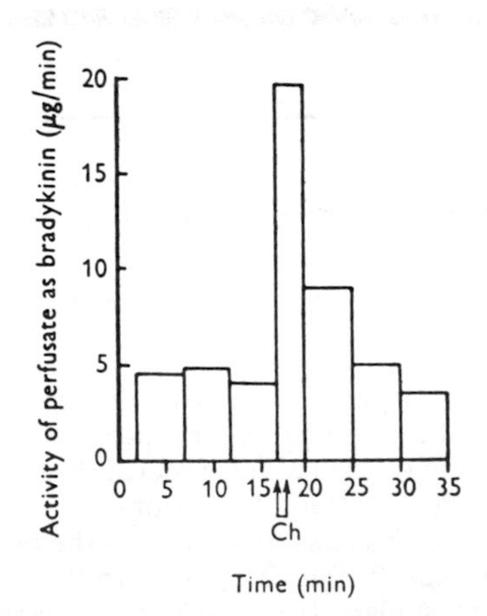

FIGURE 25A-3. Bradykinin production by the salivary gland during stimulation of the chorda tympani nerve. (Modified from G. P. Lewis. *Gastroenterology* 52:406, 1967. Copyright © 1967, The Williams & Wilkins Co., Baltimore, Md. 21202, U.S.A.)

While the period of digestion is a specific case of blood flow adjustment, other instances can be cited as well. The digestive tract possesses the property of autoregulation of blood flow, which is seen in many vascular beds of the body. When arterial pressure is reduced, intestinal and liver blood flow decreases, but it tends to return to control levels even while the pressure remains low. The phenomenon is not as pronounced in the intestine and liver as it is in the kidney and brain, for example, where flow autoregulates almost perfectly in the pressure range 60 to 160 mm Hg. Intestinal vascular resistance falls about 25 percent when arterial pressure is reduced from 100 to 40 mm Hg. The liver seems to have the same degree of self-regulatory capability. The large intestine has a distinctly less prominent capability to autoregulate. Studies of the stomach have not revealed any autoregulatory capacity in that vascular bed.

The mechanism of autoregulation is believed to be twofold. First, a reduction of arterial pressure reduces the internal pressure in the arterioles and precapillary sphincters. This diminishes the stimulus for myogenic activity of these vessels, especially of the sphincters, where the myogenic response is most pronounced. Second, the reduced flow which accompanies the pressure reduction would cause metabolites to accumulate and produce a general vascular relaxation. This relaxation has been found to occur in the arterioles (Fig. 25A-4, top) and in the precapillary sphincters. As a consequence of these effects, blood flow in some capillaries may actually increase when the arterial pressure is reduced. As shown in Figure 25A-4, bottom), capillary red cell velocity is phasic at normal arterial pressure. The periods of reduced flow represent the contraction phase of the precapillary sphincters. When arterial pressure is reduced, this periodicity disappears and average flow is in fact greater at one stage although arterial pressure is diminished. Relaxation of the sphincter and of the arterioles upstream to it so reduced resistance that the driving head of pressure to the capillary was actually increased. Not all capillaries autoregulate to such a degree, however.

The pressure sensitivity of the precapillary sphincters is also apparent when venous pressure is elevated. Small increases in venous pressure (3 to 4 mm Hg) can reduce flow through some capillaries by 50 percent or more. An increase of 10 mm Hg will totally stop flow in other capillaries. The effect of such a response of the sphincters is to reduce the effective capillary surface available for exchange. This is reflected in measurements of capillary filtration coefficient, which falls from 0.4 ml per minute to 0.1 ml per minute when venous pressure is raised from 0 mm Hg to 20 mm Hg. Total blood flow falls greatly when venous pressure is elevated. Some flow reduction would be expected since the arteriovenous pressure gradient is diminished. But increasing venous pressure from 0 to 20 mm Hg reduces blood flow by 50 percent while reducing the pressure gradient across the bed by 20 percent. The flow reduction is probably due primarily to precapillary sphincter closure.

Local regulation of blood flow is manifested in the liver in somewhat the same manner as described above for the intestine. Autoregulation and a marked resistance increase with venous pressure are both present in the liver. However, because of its dual blood supply, the liver does possess the unique feature of reciprocity between portal and hepatic arterial inflow. If portal inflow is reduced, hepatic artery inflow increases. This maintains a more nearly constant blood supply to the liver tissues. Apparently, a reduction of portal inflow reduces pressure in the terminal portion of the hepatic arteriolar bed, causing these vessels to dilate. A myogenic mechanism is probably responsible. While this phenomenon is usually termed reciprocity of flow, it is not truly reciprocal since a reduction of hepatic arterial inflow does not alter portal flow, which is determined by the resistance vessels of the organs of the digestive tract.

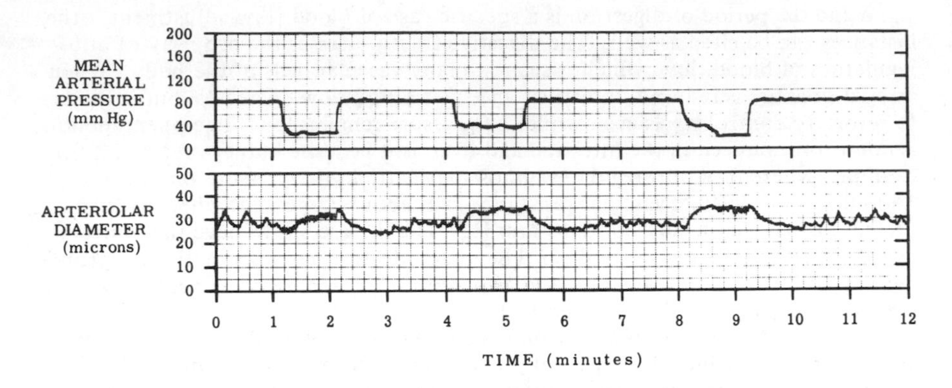

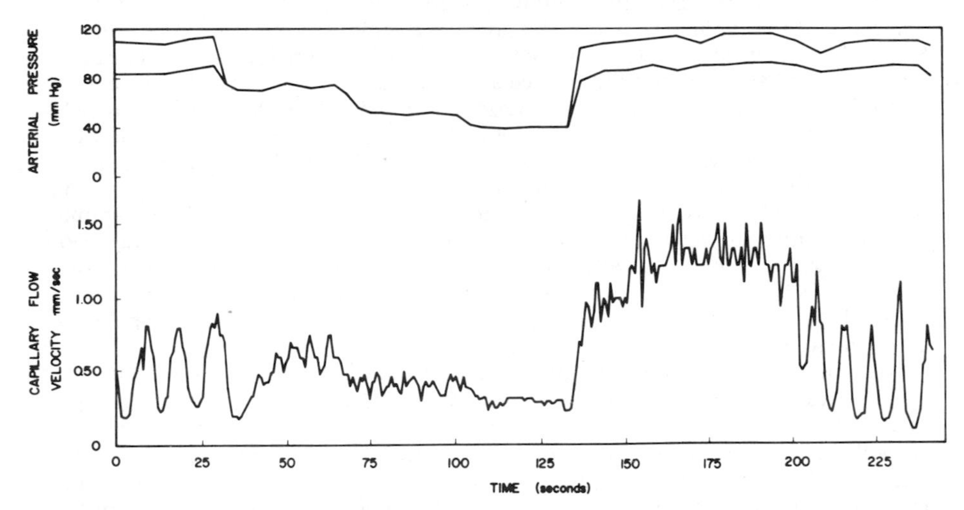

FIGURE 25A-4. (Top) Change in diameter of mesenteric arteriole during reduction of large artery pressure. (From P. C. Johnson. *Circ. Res.* 22:199, 1968.) (Bottom) Red cell velocity in a mesenteric capillary at normal and at reduced arterial pressure. (From P. C. Johnson. *Gastroenterology* 52:435, 1967. Copyright © 1967, The Williams & Wilkins Co., Baltimore, Md. 21202, U.S.A.)

SPLANCHNIC BLOOD VOLUME

Estimates in the dog employing tagged red cells (^{32}P) and labeled albumin (^{131}I) techniques have been variable but roughly approximate 30 percent of the total blood volume. Of this, about one-half is in the intestine, and the remainder is equally divided between the liver and spleen.

SPLEEN

ANATOMICAL ASPECTS

Splenic artery branches pass along the *trabeculae* of the spleen, through the *white pulp,* and break up into arterioles with husklike sheaths *(penicilli),* which constitute the arteriolar stopcocks. These branch into arterial capillaries, which supply the sinusoids (venous sinuses) in the *red pulp* of the spleen. The collecting *venules* converge ultimately into the splenic vein, which drains into the portal vein. Sphincters have been described at the sinusoid outlets, which close during storage phases. The sinusoids are composed of rodlike "barrel stave" endothelial cells, longitudinally arranged but not contiguous, and surrounded by rings or spirals of reticular fibers, reminiscent of barrel hoops, creating a lattice-like structure.

A commonly accepted view is that the system is a closed one but that the latticework is large enough to permit occasional erythrocytes to slip through, as shown by observations on the transilluminated mouse spleen. Thus, the sinusoids represent a mechanism creating a "separatory" circulation, filtering off plasma and storing red cells in high concentration in the dilated sinusoid, and operating in a cycling manner. The separated plasma may return through smaller veins and lymphatics to the general circulation. In a spleen exhibiting a high degree of storage, constant, rapid flow is seen primarily in the capillary shunts which directly connect arterioles with venules. When the splenic capsule contracts under physiological demand, the concentrated cells are discharged directly into the portal venous system. Filtered red cells reenter the venous sinuses by diapedesis, which leads to destruction of fragile erythrocytes. After rupture, the released hemoglobin and stoma are ingested by the reticuloendothelial system of the spleen, or elsewhere in the body.

Others have taken the position that the splenic circulation is an "open" one (whole blood comes into direct contact with splenic cells). Thus, in the mouse, the terminal arterioles are said to end in flared ampullae, the last structure to show signs of endothelial walls. Blood issues forth into the pulp spaces through the ampullae and reenters the vascular system via the open ends of the venous vessels and perforations in the venous walls. However, evidence appears good that the system is closed in man and the rat.

NEUROGENIC FACTORS

Nervous regulation is achieved by sympathetic outflow in splanchnic nerve branches accompanying the artery into the spleen, presumably mediated by norepinephrine. Regulation occurs by way of the smooth muscle of the capsule and trabeculae, and also by direct vasomotor action on the arterioles and sphincters. Pressure as high as 100 mm Hg can be developed by splenic contraction (if the vein is occluded). This contraction, combined with constriction of arterioles, forces the blood out. Simultaneous relaxation of the efferent sphincters aids discharge. Discharge is evidently initiated by reflex mechanisms, e.g., via chemoreceptor stimulation during hypoxia, or baroreceptor stimulation during hemorrhage.

Since dilator fibers to the spleen have not been demonstrated, filling is taken to be a passive process, achieved by relaxation of arterioles by decrease in splanchnic vasomotor tone, and/or by constriction of efferent sphincters and possibly the splenic veins by action of venomotor fibers.

RESERVOIR FUNCTION

In addition to hemorrhage and anoxia, discharge of stored blood is stimulated by exercise, raised external temperature, certain types of anesthesia (e.g., chloroform), certain humoral agents and drugs, and subacutely in estrus, pregnancy, and lactation. Under pentobarbital anesthesia, which apparently promotes greater than normal storage, severe hypoxia caused the spleen of the dog to discharge 16 to 20 percent of the normal circulating blood volume. Because of the high hematocrit value, the

O_2 capacity of the splenic vein blood can be as much as 80 percent higher than arterial blood. During hypoxia, this may serve as a small emergency store of O_2, for about 30 to 60 ml of O_2 is released with the 100 to 200 ml of blood stored in the canine spleen.

The volume of blood per gram of tissue is high in the spleen compared with other organs: 0.420 ml per gram in dogs as compared with 0.200 in the liver and 0.060 in the bowel.

REFERENCES

Delaney, J. P., and J. Custer. Gastrointestinal blood flow in the dog. *Circ. Res.* 17:394–402, 1965.

Fronek, K., and L. H. Stahlgren. Systemic and regional hemodynamic changes during food intake and digestion in nonanesthetized dogs. *Circ. Res.* 23:687–692, 1968.

Hanson, K. M., and P. C. Johnson. Local control of hepatic arterial and portal venous flow in the dog. *Amer. J. Physiol.* 211:712–719, 1966.

Jacobson, E. D. (Ed.). Symposium on the gastrointestinal circulation. *Gastroenterology* 52:332–471, 1967.

Johnson, P. C. (Ed.). Autoregulation of blood flow. *Circ. Res.* 15 (Suppl. 1): 1964.

Johnson, P. C., and K. M. Hanson. Capillary filtration in the small intestine of the dog. *Circ. Res.* 19:766–772, 1966.

Vatner, S. F., D. Franklin, and R. L. Van Citters. Mesenteric vasoactivity associated with eating and digestion in the conscious dog. *Amer. J. Physiol.* 219:170–174, 1970.

25

Digestive Tract

B. Movements of the Digestive Tract

Leon K. Knoebel

THE DIGESTIVE TRACT receives food which is moved through its structurally different parts. A number of digestive juices are secreted at various points along the route, and the enzymes contained in these secretions catalyze the conversion of complex foodstuffs to simple molecules which can be readily absorbed into the blood and the lymph. Unabsorbed food residues and a variety of waste products are propelled to the end of the tract and eliminated from the body.

It follows that the musculature of the digestive tract must possess the mechanical ability to propel these materials from one end of the tract to the other. Movements which mix the materials with the digestive juices and thus facilitate both digestion and absorption are also a feature of certain regions of this organ system. Furthermore, changes occur in the state of the tonus of the musculature of all parts of the digestive tract. These types of muscular activity will be discussed in the following sections in relation to each of the organs of the alimentary canal.

CHEWING

When solid food enters the mouth, chewing occurs. This process is important from a number of standpoints. As food is moved about in the mouth, the taste buds are stimulated, and the odors which are released stimulate the olfactory epithelia. These events are significant because much of the satisfaction of eating is derived from these stimuli. Reflex secretion of saliva also occurs during the chewing of food. The food is mixed with saliva, which softens and lubricates the food mass and thereby facilitates swallowing. In addition, chewing reduces the food to a particle size convenient for swallowing.

A crushing force of 100 to 160 pounds can be exerted by the molars and 30 to 80 pounds by the incisors of man. Since a force of 115 to 173 pounds is sufficient to crack hazelnuts, the maximum biting force which can be generated by the muscles of mastication is far greater than that required for the chewing of ordinary food. The force of biting is not the major factor in determining the efficiency of the chewing process. The occlusive contact area between the molars and premolars is much more important in this respect.

Although the act of chewing is under voluntary control, it is also partly reflex in nature. That reflexes can be involved is shown by the fact that an animal decerebrated above the mesencephalon will chew reflexly when food is put in the mouth.

The process of mastication is carried out by the combined action of the muscles of the jaws, lips, cheeks, and tongue. The movements of these skeletal muscles are coordinated by impulses over the V, VII, IX, X, XI, and XII cranial nerves. Once chewing is accomplished to the satisfaction of a particular individual, the food mass or bolus is ready for swallowing.

SWALLOWING

THE STAGES OF SWALLOWING

The act of swallowing has been divided into three stages on the basis of the regions through which a bolus passes on its way to the stomach. These regions are the mouth, the pharynx, and the esophagus. Many of the events which occur during swallowing can be visualized by means of x-ray motion pictures of a human subject swallowing a radiopaque suspension of barium sulfate. Furthermore, the pressures at various points along the route of transit which are developed during a swallow can be measured by small pressure transducers.

Oropharyngeal Stages

During the *first stage,* the bolus is passed from the mouth through the isthmus of the fauces into the pharynx. The food mass, which can be either liquid or solid, is rolled toward the back of the tongue, and the front of the tongue is pushed up against the hard palate. At the same time, the mylohyoid muscles contract rapidly and force the bolus into the pharynx.

In the *second stage,* the bolus is passed through the pharynx into the esophagus. This requires about one-fifth of a second. X-ray motion pictures show that a number of events occur simultaneously. The continued contraction of the mylohyoid muscles and the position of the tongue prevent the reentrance of food into the oral cavity. Respiration is briefly inhibited. The soft palate is elevated and shuts off the posterior nares. Food is prevented from entering the larynx by the elevation of the larynx and by the approximation of the vocal cords, both of which serve to close the glottis. The epiglottis may or may not be pressed down over the laryngeal orifice, but even if it is, it probably acts only as an auxiliary mechanism to keep food from entering the respiratory passages, since the epiglottis can be removed without having abnormalities in swallowing result. As these openings close, the pharyngeal constrictors contract and force the bolus into the esophagus. As a result of the various muscular activities which occur during this stage, pressure in the pharynx rises from ambient to about 100 cm H_2O (Fig. 25B-1).

Esophageal Stage

The bolus passes down the esophagus and into the stomach during the *third stage* of swallowing. When the esophagus is in the resting state, the upper part is closed over a distance of 1 to 3 cm by the tonic contraction of a band of striated muscle, the cricopharyngeal muscle. This structure is the *upper esophageal sphincter.* Since the sphincter is normally closed during rest, large volumes of air are prevented from entering the stomach during normal respiration. This sphincter relaxes following a swallow, allows the bolus to pass into the esophagus, and then contracts again, thereby preventing reflux from esophagus to pharynx. The pressures which exist in the pharyngoesophageal junction at rest and during a swallow are shown in Figure 25B-1.

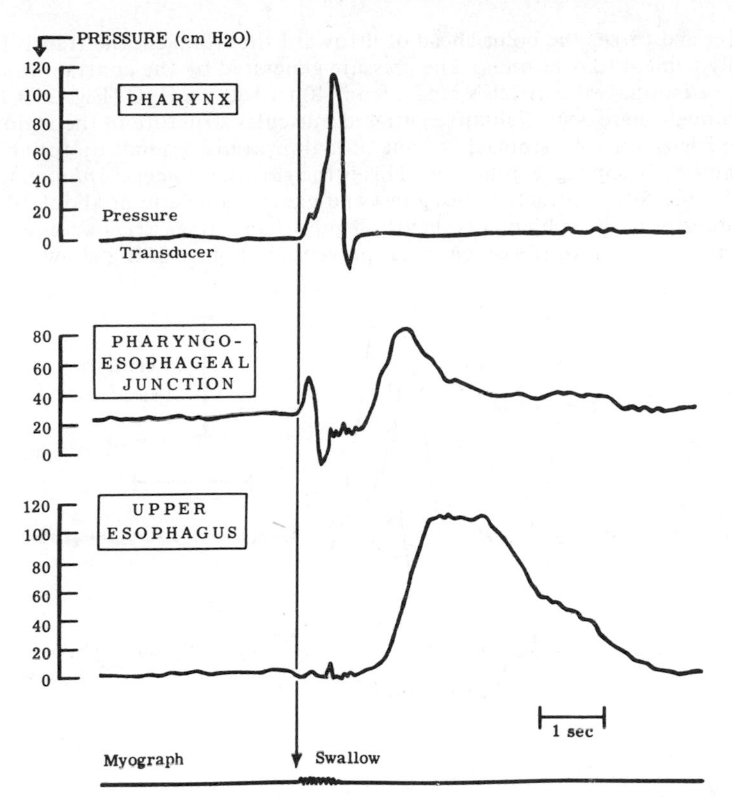

FIGURE 25B-1. Temporal sequence of pressure changes from rest with swallowing in the pharynx, pharyngoesophageal junction, and upper esophagus. Pressures were measured in the three sites during three different swallows by means of swallowed miniature pressure transducers. The beginning of a swallow is signaled by a myograph, which records action potentials from the jaw muscles. At rest, pressure in the pharyngoesophageal junction is about 40 cm H_2O higher than in the pharynx and upper esophagus. When pressure in the pharynx is high during a swallow, that in the sphincter falls to ambient pressure. Then, when pressure in the pharynx falls to rest value, sphincter pressure rises to about twice its resting level and remains high for about one second before falling to resting pressure once again. Peak pressure in the upper esophagus is attained at the time sphincter pressure is decreasing to resting level. (From E. F. Fyke and C. F. Code. *Gastroenterology* 29:29, 1955. Copyright © 1955, The Williams & Wilkins Company, Baltimore, Md. 21202, U.S.A.)

If the bolus is liquid, it is shot through the esophagus by the initial force of swallowing and travels by gravity to the stomach in about one second. If semisolid, the bolus is propelled down the esophagus by a type of muscular movement known as *peristalsis.* The main feature of esophageal peristalsis is the occurrence of a contraction (4 to 8 cm in length) of the muscular coat of the proximal esophagus, which then passes as a wave of contraction toward the stomach at the rate of 2 to 4 cm per second. The wave of contraction is thought by some to be preceded by a wave of inhibition. However, since the esophagus is normally relaxed at rest, it is difficult to detect any further esophageal relaxation. On the other hand, since the resulting tone of the upper sphincter is high, this structure does relax as the wave of inhibition passes over it. The wave of contraction which follows closes the upper sphincter and forces the bolus ahead of it toward the stomach, the transit time generally being 4 to 6 seconds. The pressure generated by the contractile component of esophageal peristalsis ranges from 40 to 160 cm H_2O (Fig. 25B-1).

Although there is no well-differentiated muscular structure in the region where the esophagus joins the stomach, a zone of high pressure extends up for about 7 cm from the gastroesophageal junction. This is the *gastroesophageal sphincter.* This region is normally contracted during rest and exerts a pressure of about 10 cm H_2O. Since pressure in the sphincter is slightly higher than intragastric pressure, the reflux of gastric contents into the esophagus is prevented. Figure 25B-2 shows that, almost

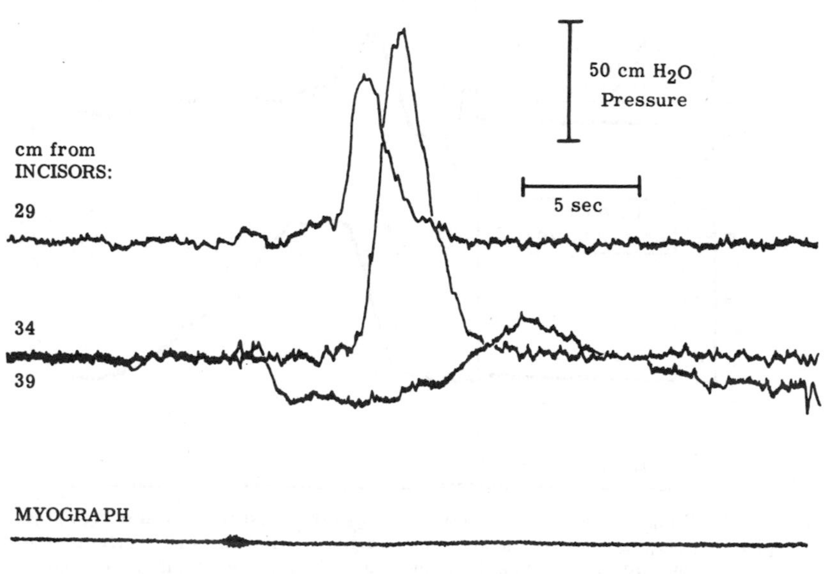

FIGURE 25B-2. Temporal sequence of pressure changes from rest with swallowing in the lower esophagus and gastroesophageal junction. Pressures were measured simultaneously in the three sites by means of swallowed miniature pressure transducers, the positions of which are given as distances from the incisors. The beginning of a swallow is signaled by a myograph, which records action potentials from the jaw muscles. (From E. F. Fyke, C. F. Code, and J. F. Schlegel. *Gastroenterologia* 86:146, 1956.)

immediately following the initiation of a swallow, pressure at the gastroesophageal junction drops and remains low during the time a peristaltic wave is traversing the lower esophagus. Presumably, the wave of inhibition causes relaxation of the gastroesophageal sphincter so that when a solid bolus is propelled down the esophagus by the wave of contraction, it is easily able to enter the stomach. Occasionally, a liquid bolus can be seen to accumulate momentarily above the sphincter, because a liquid bolus travels so rapidly that it sometimes arrives at the sphincter before relaxation is sufficient to allow entrance of the bolus. Once pressure in the lower esophagus falls to resting level, the pressure in the gastroesophageal junction rises and remains elevated for about 10 seconds before falling to resting level once again.

THE SWALLOWING REFLEX

The coordination of swallowing depends on nervous mechanisms. The first stage is initiated voluntarily, with the cerebral cortex playing the dominant role. However, there must be some stimulus to the mucous membranes of the mouth to enable one to initiate a swallowing movement. At least a moderate amount of fluid must be present in the mouth, as is shown by the fact that it is virtually impossible to swallow when the mouth is dry.

The remainder of the swallowing response and all other movements of the gastrointestinal tract with the exception of defecation are independent of the will. It is widely held that there is a swallowing center, located in the medulla, which is activated by stimulation of receptors in the mouth and pharynx. This center sends impulses to the muscles involved in the swallowing response, and the complete act of swallowing occurs in the proper sequence.

It can be shown by a simple experiment that the peristaltic wave is not just a conduction of contraction along the muscular wall of the esophagus. If the esophagus is transected, the lower end will still contract at the proper time after a swallow, providing the extrinsic innervation of the esophagus remains intact. Since total esophageal paralysis results following bilateral vagotomy in the neck, the extrinsic nerves which are important in coordinating the orderly progress of the peristaltic wave must be the vagi. A biphasic complex of inhibition and then excitation is probably mediated by the vagi after a swallow. This would result in the passage of a wave of relaxation followed by a wave of contraction over the upper sphincter, esophagus, and gastroesophageal sphincter.

The paralysis resulting from bilateral vagotomy is permanent in the upper esophagus of man, which consists predominately of striated muscle. Weak peristaltic activity does return in the extrinsically denervated distal esophagus, which contains mostly smooth muscle. The myenteric plexus of Auerbach, which constitutes the parasympathetic postganglionic nerve supply of the esophagus, lies between the longitudinal and circular smooth muscle layers, and it is probable that this cholinergic plexus is largely responsible for the coordination of any peristaltic activity occurring in the distal esophagus both under normal conditions and after vagotomy.

Peristalsis which follows a conscious effort of swallowing is known as *primary peristalsis.* Peristalsis can also be elicited by local stimulation of the esophageal mucosa and without being preceded by a swallowing movement. This is *secondary peristalsis.* The latter type, which like primary peristalsis is reflex and depends on the integrity of the vagi, is important when food remains in the esophagus following the passage of a primary peristaltic wave. Secondary peristalsis facilitates the removal of such residues from the esophagus.

ABNORMALITIES OF SWALLOWING

A condition known as *dysphagia* arises when there are abnormalities of any of the three stages of swallowing. Abnormalities may result from a variety of disorders associated with the nerves and muscles involved in these processes. There are abnormal motility patterns of the esophagus and gastroesophageal junction in *achalasia.* The esophageal musculature may show uncoordinated and spastic contractions, and the gastroesophageal sphincter often fails to relax following a swallow. Food has difficulty entering the stomach and tends to accumulate above the sphincter, causing dilation of the esophagus. This abnormality is apparently due to a lack of or degeneration of the nerve cells making up Auerbach's plexus.

GASTRIC MOTILITY

The functions of the stomach include the storage and the liquefaction of swallowed food. Furthermore, this organ is able to discharge its contents *(chyme)* into the small intestine at a rate suitable for optimal digestion and absorption. To understand fully how the stomach performs these functions, it is necessary to review the physiological anatomy of the stomach, which is illustrated in Figure 25B-3.

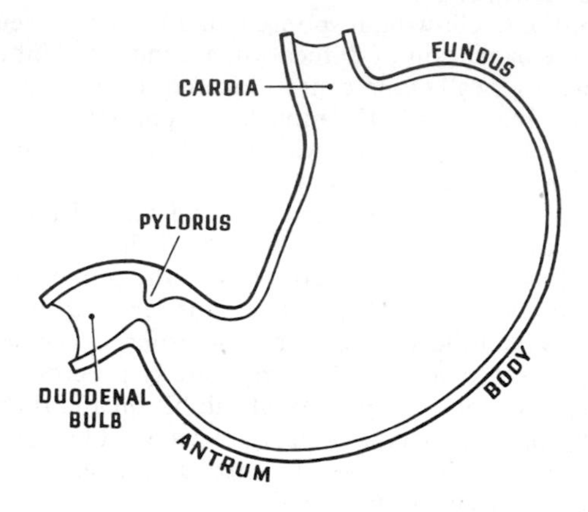

FIGURE 25B-3. Physiological anatomy of the stomach.

THE EMPTY STOMACH

During periods of fasting, the volume of contents present in the stomach of man is only about 50 ml or less. When the stomach is devoid of food, the sensation of hunger is often perceived. As part of the overall sensation of hunger, some people feel a sense of emptiness accompanied by occasional sharp pangs referred to the abdominal region. At one time it was felt that so-called hunger pangs were directly associated with strong contractions of the stomach and that the sensation of hunger might be due, at least in part, to these gastric contractions. However, improved techniques have shown that although the stomach does exhibit small contractions part of the time, rarely are vigorous contractions observed. In addition, a number of workers have not been able to demonstrate a relationship between contractions of the empty stomach and the perception of the hunger sensation. Thus, present evidence minimizes the possible contribution of gastric movements to the production of hunger pangs. The control of food intake as related to the sensations of hunger and satiety is discussed in Chapter 8.

GASTRIC FILLING

One function of the stomach is the storage of variable volumes of contents, and it is subserved by the smooth muscle of the fundus and body of the stomach. These regions possess the ability to adapt to the volume of contents they contain in that relatively large volumes can be introduced into them with little change in intragastric pressure (Fig. 25B-4). The volume adaptation phenomenon has been called *receptive* or *stress relaxation.* In order to maintain a fairly constant intragastric pressure as the stomach is progressively distended, the smooth muscle fibers of the gastric wall must lengthen without increasing tension. This is at least partially the result of the plastic properties of smooth muscle (Chap. 3B). When stretched, smooth muscle fibers rearrange internally by the sliding of muscle fibers past one another in such a manner that tension development is minimized. In addition, there is associated

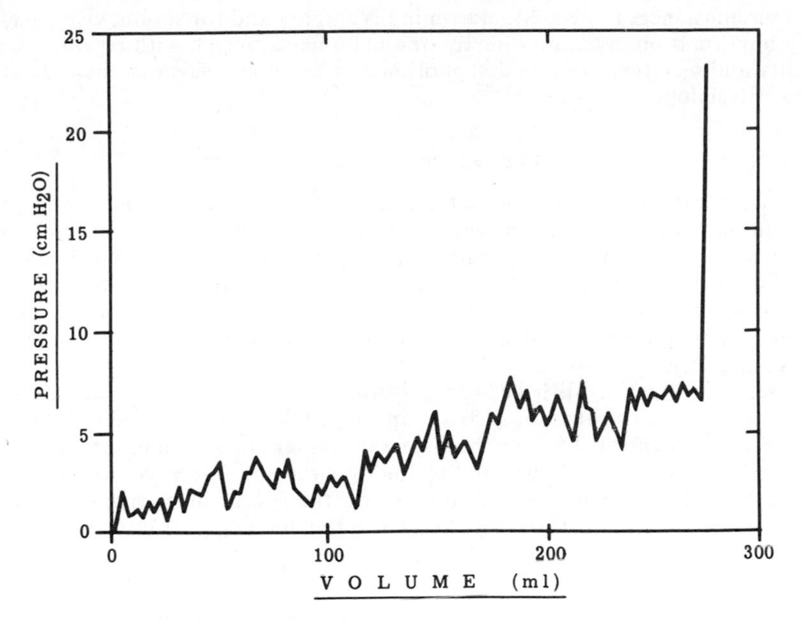

FIGURE 25B-4. Intragastric pressure during the addition of 2.5 ml of fluid every 2.5 minutes through a pyloric cannula into the cardia-ligated stomach of the rabbit. (From E. G. Grey. *Amer. J. Physiol.* 45:276, 1918.)

with the act of swallowing a reflex inhibition of the smooth muscle of the body and fundus. That this neural mechanism is mediated via the vagus nerves is shown by the fact that there is relaxation of the vagally innervated body and fundus following a swallow, but not if the vagi to these regions are sectioned.

MOVEMENTS OF THE FULL STOMACH

A pressure of about 6 to 10 cm H_2O is always maintained on the gastric contents by virtue of a tonic contraction of the musculature of the fundus and body. It might be said that these portions of the stomach have at all times "a grip on their contents." The tonic contraction continually presses on the food mass and aids its delivery to the pyloric antrum.

Repetitive peristaltic contractions of the stomach of man occur at the rate of three contractions every minute. These waves of contraction become visible in the body of the stomach, and because the antral musculature is much thicker than that of the rest of the stomach, peristalsis becomes very vigorous when the pyloric antrum is reached. Thus, the peristaltic contractions of the pyloric antrum are largely responsible for mixing ingested food with gastric secretions and for generating the forces which are required for gastric emptying.

Gastric movements have been observed by means of x-ray films, through windows in the abdominal wall of animals, and through fistulas in the stomach of man. The most famous human gastric fistula was that of Alexis St. Martin, a Canadian adventurer, who acquired a gastric fistula in his left side as the result of a gunshot wound. Dr. William Beaumont, a United States Army surgeon, capitalized on this set of circumstances to keep St. Martin in his employ and for several years made many important observations directly on the human stomach with regard to both motility and secretion. His results, published in 1833, are classic in the field of gastric physiology.

MECHANICS OF GASTRIC EMPTYING

For years it was thought that the state of contraction of the pylorus was the primary factor in the regulation of gastric emptying. However, the currently accepted view is considerably different and minimizes the importance of an anatomical pyloric sphincter. It has been shown that the gastric emptying times of a wide variety of substances are not changed from normal following the surgical removal of the pylorus. Consequently, this structure cannot be of primary importance in the control of gastric evacuation.

A series of experiments have been performed which have greatly enriched our knowledge of the mechanics of gastric emptying. Gastric motility, activity of the pylorus, motility of the duodenal bulb, and movement of a barium meal through these regions have been followed in the dog by x-ray and fluoroscopic observation. When a peristaltic wave is observed to pass over the pyloric antrum, contents can be seen to move through the relaxed pylorus into the duodenum (Fig. 25B-5). When

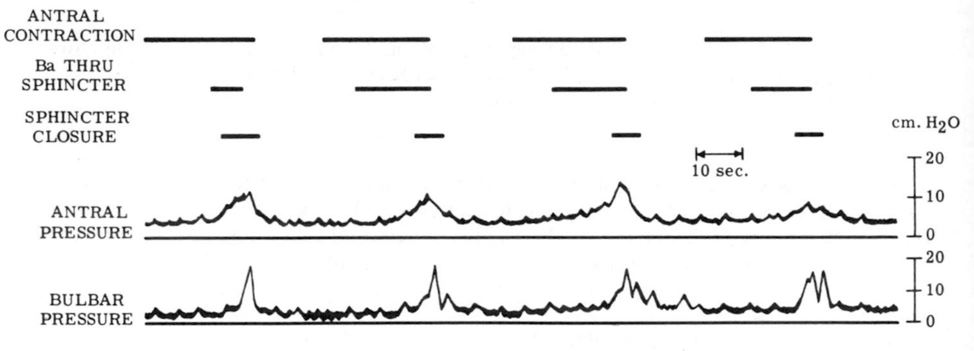

FIGURE 25B-5. Gastric evacuation-pressure cycle. (From J. M. Werle, D. A. Brody, E. W. Ligon, M. R. Read, and J. P. Quigley. *Amer. J. Physiol.* 131:608, 1941.)

the contraction wave reaches the pylorus, this structure contracts; and after this, the duodenal bulb contracts. The antrum, pylorus, and bulb all react in the same manner to the passage of a peristaltic wave; the pylorus is no different from the adjacent structures in this respect. Contents continue to escape from the stomach as the gradually contracting pylorus closes, until the increased resistance offered by this musculature prevents further evacuation. The pylorus remains contracted somewhat longer than does the duodenal bulb. This prevents to some extent regurgitation of duodenal contents into the stomach. The contraction of the duodenal bulb helps to propel contents down the intestine. All these structures finally relax again, and the cycle of emptying is repeated with the coming of the next peristaltic wave.

In addition to these observations, measurements of the pressures developed immediately on each side of the pylorus were made, using small pressure transducers placed through antral and duodenal fistulas (Fig. 25B-5). Although absolute pressures may rise 15 to 30 cm H_2O in the antrum and bulb during a cycle of gastric emptying, a pressure gradient of 3 to 4 cm H_2O is maintained between these regions throughout most of the cycle. However, at about the time the pyloric sphincter begins to contract, the antral-duodenal pressure gradient may increase to 20 to 30 cm H_2O. It was concluded that the driving force behind the emptying process is the pressure differential between the antrum and duodenum. If the pressure is higher in the antrum, and high enough to overcome the resistance offered at the pylorus, chyme will leave the stomach. Since the muscular activity of the stomach determines to a large extent the magnitude of this pressure differential, gastric emptying is regulated by mechanisms which control the movements of this organ.

NEURAL CONTROL OF GASTRIC MOTILITY

Extrinsic Innervation

The extrinsic innervation of the stomach is by the vagus (parasympathetic) and splanchnic (sympathetic) nerves. These autonomic nerves, in addition to providing afferent fibers from the stomach, serve as the efferent limbs of a variety of reflex arcs which can be initiated from many visceral and somatic receptors throughout the body and can cause either excitation or inhibition of gastric movements. Thus, the major function of the extrinsic nerves of the stomach is to correlate activities between the stomach and other parts of the gastrointestinal tract as well as other regions of the body.

The neural control of gastric motility is not completely understood. The results obtained by stimulating the parasympathetic and sympathetic nerve supplies to the stomach have not been uniform. No matter which set of nerves is stimulated, the musculature usually relaxes if it is in a state of contraction, and contracts if the tonus is low. In general, stimulation of the parasympathetic nerves usually produces increased muscular activity, whereas stimulation of sympathetic nerves most often results in inhibition. These generalities can be applied not only to the stomach but also to most of the smooth muscle of the digestive tract.

As will be pointed out below, the vagus nerves play the dominant role in the regulation of gastric motility. Although the sympathetic nerves do participate in the reflex regulation of gastric motor activity, they play only a minor role in the regulation which occurs during the normal course of digestion and absorption.

Extrinsic Denervation

When the stomach is denervated by having its extrinsic nerves cut, motility is at first abolished. However, after a period of recovery the stomach does evacuate its contents, although gastric emptying is often considerably delayed. In this circumstance the sequence of contractile events is unaltered, but gastric contractions are weaker.

The stomach and those parts of the digestive tract having a musculature of the smooth type seem to possess considerable autonomy, as shown by the fact that movements are not permanently abolished in the absence of extrinsic innervation. The autonomy probably depends on two factors: (1) the presence of the nerve plexuses of Auerbach and Meissner, which are in close association with the smooth muscle of the tract and are primarily cholinergic, and (2) the intrinsic property of smooth muscle to contract in the absence of any innervation. That the nerve plexuses are of primary importance for gastric peristalsis has been demonstrated by anesthetization of the gastric wall to eliminate all neural influences. In this circumstance the gastric wall contracts if stimulated properly, but the normal pattern of muscular activity cannot be elicited; i.e., propagated peristaltic contractions are absent. Thus, it would appear that the coordination of gastric movements is a function of the local nerve nets, since the pattern of gastric muscle activity is normal only when these are functional.

The nerve plexuses contain all components necessary for a local control of gastric movements. A local reflex arc system is subserved by receptors in the gastric wall which are capable of responding to a variety of intraluminar stimuli, the result being transmission of impulses over afferent fibers through one or more synapses to the effector smooth muscle cells. It might also be visualized that parasympathetic preganglionic fibers synapse on cell bodies somewhere in the nerve net, perhaps on those of the final efferent fibers of the local reflex arc, which then could be considered to represent the parasympathetic postganglionic nerve supply of the muscle fibers. Thus, it would be at this point where the vagus nerve manifests its regulatory influences. Apparently the function of the extrinsic nerves is to modify activity which can be initiated and maintained in the gastric wall itself, and this modification can be in the direction of either augmentation or suppression of activity, depending on which extrinsic nerves are involved.

ELECTRICAL ACTIVITY OF GASTRIC MUSCLE

Studies of the electrical activity of the stomach have shown that two types of potential changes can be recorded from gastric smooth muscle (Chap. 3B). One wave shape consists of single or multiple short-lived oscillations which represent typical action potentials. The action potential is the electrical event directly associated with contractile activity. Contraction occurs when action potentials are present, but not when they are absent.

The second potential change is distinct from the action potential in that it is of lesser amplitude and much longer duration. This slow wave represents a relatively small and slow oscillation in "resting" membrane potential. It originates in a pacemaker located in the cardiac region of the stomach and in man is propagated down to the pylorus through the longitudinal muscle layer at a rate of three per minute. Since this potential change shows cyclic activity, it has been called the *basic electrical rhythm* (BER) of the stomach.

The BER is not necessarily associated with contractile activity because, unlike the action potential, it is continually recorded both in the absence and in the presence of contractions. However, there is a relationship between the BER and contraction in that action potentials, when these are recorded, occur at the peak of the slow wave. It would be expected that spiking activity would be most likely at this point because the excitability of the muscle would be highest at this time – i.e., closest to threshold. It is here that any prevailing excitatory influences, such as muscle stretch or release of acetylcholine by nerve endings, would probably cause threshold to be reached. Since the slow wave propagates as a circumferential ring and encompasses a certain population of cells at any one moment, the muscle cells in that ring have an increased probability of spike discharge at about the same time. Thus, a function of the BER is to synchronize spiking activity in a specific region of the stomach, the result being a coordinated and efficient mechanical effort.

Since the BER and gastric peristalsis occur at precisely the same frequency and are propagated at similar rates, it seems apparent that the slow wave is the electrical event which sets the pace for the propagated mechanical event. The frequency of the BER, and thus the rate of peristalsis, is not altered by a variety of excitatory and inhibitory stimuli. On the other hand, these stimuli, by virtue of their ability to enhance or inhibit spiking activity, are of primary importance in determining the strength with which the gastric muscle responds to the exciting signals represented by the BER. For example, if the vagus is stimulated, the vigor of peristaltic contractions is increased, but the rate is unchanged. It might be anticipated that the vagal stimulation induced by feeding increases the strength of gastric contractions, so that these become manifested as typical strong peristaltic contractions.

REGULATION OF GASTRIC EMPTYING

When food is ingested, gastric peristalsis is stimulated in response to distention of the walls of the stomach. Receptors in the gastric wall are stimulated and send impulses over afferent fibers to the brain, which in turn relays impulses back to the stomach by way of the vagi to stimulate contraction. This is the *gastrogastric reflex*, and it, along with any myogenic contractile activity which might be elicited directly in response to muscle stretch, is a major mechanism involved in providing the excitation required to empty the stomach of its contents.

There is a basic pattern of gastric emptying. It has been found that the rate of gastric evacuation of liquid meals decreases with time after ingestion of a meal and that the initial rate of emptying varies directly with the size of the meal. These effects are probably related to the degree of distention of the gastric wall. A linear relationship is obtained when the square root of the volumes of a meal remaining in the stomach is plotted versus time (Fig. 25B-6). That there may be a physical basis

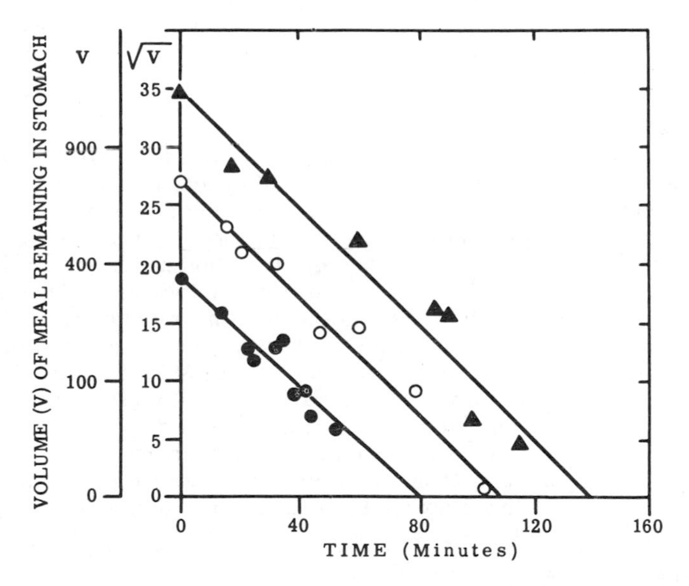

FIGURE 25B-6. The patterns of gastric emptying of meals of 330, 750, and 1250 ml according to the square root hypothesis (From A. Hopkins. *J. Physiol.* 182:144–149, 1966.)

for the square root pattern of gastric emptying is suggested by the fact that the radius of a cylinder varies with the square root of the volume and that circumferential tension is proportional to the radius (law of Laplace).

A number of mechanisms participate in the control of gastric emptying, many of them initiated in the duodenum. Chyme must come into contact with the duodenal mucosa in order to maintain the normally slow gastric evacuation. If the duodenum is transected close to the pylorus, the stomach empties almost as rapidly as ingested food enters it. It will be obvious from the discussion below that practically any stimulation of the duodenum tends to check gastric emptying. These inhibitory mechanisms are probably utilized to prevent the absorptive powers of the intestinal mucosa from being overwhelmed by a flood of ingested materials and to prevent undue chemical, mechanical, and osmotic irritation of the duodenum.

Both the chemical and the physical properties of the chyme which enters the duodenum have a profound influence on gastric emptying. With regard to the major foods, the presence in the duodenum of the digestion products of carbohydrate, protein, and especially fat impedes gastric emptying. Solutions of pH 3.5 or less greatly retard gastric evacuation. The osmotic pressure of the gastric contents is also an important factor in gastric emptying. Water is evacuated only half as rapidly as equal volumes of isotonic saline. The introduction of hypotonic and hypertonic solutions into the duodenum causes inhibition of gastric motility. Although such results suggest that the osmotic environment most favorable to the duodenum is one of isotonicity (310 mOsm per liter), solutions empty fastest if their osmotic pressure is about 200 mOsm per liter. Distention of the duodenum by increasing intraluminal pressure 10 to 15 mm Hg inhibits gastric emptying. Another factor which has a profound influence on gastric emptying is the consistency of the gastric contents. Dogs fed large chunks of meat take a much longer time to empty the stomach than dogs fed finely ground meat. Liquids are generally evacuated much faster than solids. These observations indicate that the contents of the stomach must be in a finely divided and liquid form prior to evacuation.

Enterogastric Inhibitory Reflex

The gastric inhibitory responses to the presence in the duodenum of the products of protein digestion, sugars, fats, mineral acids, and hypotonic and hypertonic solutions depend to varying degrees on a neural mechanism called the *enterogastric inhibitory reflex*. The vagus is an important component of this reflex arc, since the inhibitory effect is either partially or completely abolished if both vagi are severed. Section of sympathetic nerves has no effect on the reflex. Thus, the enterogastric inhibitory reflex is mediated from intestine to brain and back to the stomach with both afferents and efferents in the vagi. The manner in which vagal inhibition is mediated is unknown, but this may be the result of either inhibition of the vagal nuclei in the medulla by impulses over afferent vagal fibers or the existence of vagal efferent inhibitory fibers. It is also possible that there are a variety of intestinal receptors, such as osmoreceptors, hydrogen ion receptors, various chemoreceptors, and mechanoreceptors, which respond to specific intraluminar stimuli to initiate the reflex inhibition of gastric motor activity.

Enterogastrone

The inhibition in response to fat and acid solutions in the duodenum is not completely removed following vagotomy. The inhibitory effect of fat has been investigated using a denervated gastric pouch. A small section of fundus with intact blood and nerve supplies was removed from the stomach, shaped into the form of a pouch, and transplanted to the subcutaneous tissue of the mammary gland of the same dog. Since the pouch was placed so that it opened onto the surface of the body wall, the motility of the pouch could easily be observed or recorded. After a collateral circulation had been established two to four weeks later, the pedicle containing the original blood supply and all nerves to the pouch was cut. When fat was introduced into the duodenum of such a preparation, not only was the motility of the main stomach inhibited, but so was that of the pouch. Since the pouch was denervated, the only way inhibition could have resulted was from the passage of an inhibitory substance through the blood to the pouch. The inhibition was not due to the absorbed digestion products of fat, since the intravenous injection of these substances produced little or no inhibition. When acid extracts of duodenal mucosa were prepared and injected intravenously, a diminution of gastric motility was observed. On the basis of these results it has been postulated that when fat comes into contact with the duodenal mucosa, it causes the release of a hormone from the mucosa. This hormone, which has been named *enterogastrone,* travels through the blood to the stomach and is largely responsible for the inhibition of gastric peristalsis which occurs in response to fat in the duodenum. The nerve plexuses of the duodenal wall may play a role in enterogastrone release, because the application of a local anesthetic to the duodenum prevents release of the hormone. Enterogastrone has not yet been chemically identified, and the suggestion has recently been made that enterogastrone and secretin (Chap. 26) are the same hormone.

Other Gastric Reflexes

Although the regulation of gastric emptying is controlled from the duodenum to a large extent, gastric motility may be influenced reflexly from any sensory region of the body. Gastric emptying is delayed when the ileum is full (ileogastric reflex) and when the anus is mechanically distended (anogastric reflex). Stimulation of visceral and somatic pain receptors results in inhibition of gastric movements. It is also known that various emotional states produce changes in gastric motility. Anxiety and resentment increase the activity of the stomach, whereas fear and anger inhibit movements.

VOMITING

The act of vomiting accomplishes the purpose of rapidly emptying the stomach of its contents. Vomiting is generally preceded by a feeling of nausea and an active secretion of saliva. Forced inspirations are made, and the glottis and nasal passages are usually closed by the contraction of the appropriate muscles. The fundus, the cardia, and the esophagus become flaccid, and the pyloric region of the stomach contracts. The diaphragm descends on the stomach and there is a forcible contraction of the abdominal muscles. The major force for vomiting is supplied by the contraction of the abdominal muscles, rather than by contraction of the gastric musculature. If an animal is given curare, an agent which paralyzes the striated muscle of the abdominal wall, vomiting cannot occur.

Vomiting is an extremely complex reflex act and is coordinated by a center located in the medulla. Afferent impulses arrive at this center from many regions of the body. The most potent stimuli arise from the sensory nerve endings of the fauces and the pharynx. Other prominent receptor areas include almost any part of the digestive tract, other abdominal viscera, and the labyrinths during motion sickness.

MOTILITY OF THE SMALL INTESTINE

Ingested food, which is liquefied and partially digested in the stomach, passes into the small intestine, where the major part of digestion and absorption occurs in the duodenum and jejunum. Waste products and food residues are moved into the ileum and then into the colon. Different types of movements which accomplish these functions can be observed and recorded in the small intestine.

MOVEMENTS OF THE SMALL INTESTINE

The sequence of events in the type of movement most frequently seen in the small intestine and known as *segmenting contractions* is illustrated diagrammatically in Figure 25B-7. Although segmenting probably does help move chyme down the

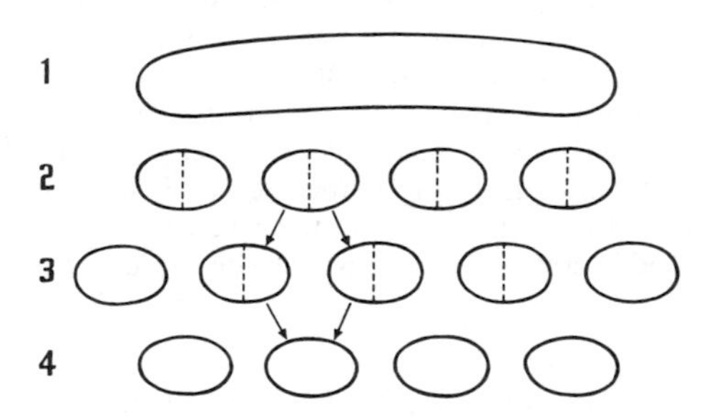

FIGURE 25B-7. Diagram of segmentation movements of the small intestine. An animal is fed a radiopaque meal, and the intestine is observed by means of x-rays. (1) A string of barium-impregnated meal can be seen when the gut is quiet. (2) The gut then contracts at a number of places, dividing it into a series of segments. (3) Each segment can be seen to divide, with adjoining halves coming together to form new segments. (4) The original pattern is established once again.

small intestine, this type of activity serves primarily to mix the chyme with the digestive juices and to facilitate absorption by bringing fresh portions of chyme into contact with the absorbing epithelium. This process is repeated about 10 times a minute in the upper intestine of man and may continue for as long as 30 minutes in one section of gut. A peristaltic wave may then be initiated and move chyme down the gut for a short distance, where segmenting movements are repeated.

Another type of movement observed in the small intestine is *peristalsis,* the function of which is to propel chyme along the gut. Under normal circumstances the wave moves slowly, at a rate of 1 to 2 cm per second, and travels only 4 to 5 cm. Occasionally peristalsis occurs as a swift movement which sweeps along the entire intestine. This rapidly propagated contraction is known as a *peristaltic rush* and occurs when the excitability of the intestine is high.

PROPULSION OF INTESTINAL CONTENTS

Evidence suggests that there is a one-way transmission mechanism built into the wall of the gut. If a segment of intestine is removed and reversed, the continuity of the gut being maintained, material will accumulate and distend the intestine at this point. The propulsive movements of the reversed segment work in the direction opposite to normal and impede the movement of chyme.

One idea that has been advanced to explain this polarity of the intestine is based on the fact that mechanical stimulation of the intestine results in contraction above and relaxation below the point of stimulation. This response, termed the *law of the intestine,* has suggested that peristalsis of the small intestine consists of two components, a wave of inhibition followed by a wave of contraction, the natural consequence of which would be to move contents in a downward direction. However, there is considerable uncertainty regarding the extent to which the law of the intestine contributes to normal intestinal motility. Moreover, although the contraction wave is always seen, there has been much controversy regarding the existence of a wave of inhibition, since not everyone has been able to record it.

Alvarez (1948) advanced the *gradient theory* in explanation of the polarization of the intestine. This theory is based on studies which demonstrate that rhythmic contractions, tonus, excitability, and metabolism of the intestine show the greatest activity in the duodenum, but become progressively less active toward the end of the small intestine. These activity gradients presumably account for the normally unidirectional movement of contents through the intestine.

CONTROL OF INTESTINAL MOTILITY

Extrinsic Reflex Control

Stimulation of the vagi generally causes increased intestinal motility, whereas sympathetic (splanchnic) stimulation usually produces inhibition. The importance of the extrinsic nerves in modifying intestinal movements is shown by the fact that intestinal motility can be altered reflexly by stimulation of many sensory areas.

Distention of any part of the intestine results in inhibition of the whole intestine. This effect, known as the *intestinointestinal inhibitory reflex,* does not occur after section of the splanchnic nerves. Intestinal motility is also inhibited when the anorectal region is distended *(anointestinal inhibitory reflex).* Trauma to organs outside the digestive tract, such as irritation of the peritoneum or urinary tract, causes intestinal inhibition. The atonic or flaccid gut, a condition known as *paralytic ileus,* which often follows surgery in these areas is a result of reflex inhibition.

Feeding results in increased motility of an innervated, isolated loop of intestine, and this effect is abolished by cutting the extrinsic nerves to the intestinal segment. This reflex is probably initiated when food enters either the stomach or intestine and would be termed a *gastrointestinal or intestinointestinal excitatory reflex.* It decreases propulsive activity, the reduction in the rate of downward movement of contents presumably being effected by narrowing of the intestinal lumen as a result of increased contraction of the intestinal musculature. This helps allow adequate time for optimal digestion and absorption.

Local Control

Since the small intestine retains normal movements following extrinsic denervation, regulation of intestinal motility must be mainly on a local basis; i.e., the mechanisms which generate and coordinate intestinal movements reside in the wall of the intestine. However, although movements are maintained in the absence of extrinsic innervation, as pointed out above, intestinal motility is influenced by impulses over these nerves.

The segmenting contractions of an extrinsically denervated intestine are not affected by the application of cocaine. Since cocaine paralyzes all nerve structures which are present in the intestinal wall, this type of contraction is initiated in the smooth muscle itself. Peristalsis is eliminated, showing the dependence of this type of muscular activity on the integrity of the intrinsic nerve plexuses of Auerbach and Meissner. It would appear that propagation of contraction depends on a local reflex between the intestinal smooth muscle layers and their intrinsic nerve plexuses.

Local mechanical and chemical stimulation by intestinal contents is probably largely responsible for the initiation and continuance of the movements of the small intestine. Factors which have been shown to increase motility of extrinsically denervated intestine are increased intraluminal pressure; hypotonic, hypertonic, and acid solutions; and products of digestion. These effects are abolished by anesthetization of the mucosa, thus demonstrating the participation of neural elements. The stimuli intensify the slight activity of the resting intestine by way of local reflexes between mucosa and muscle and probably by directly stretching the muscle itself.

The Peristaltic Reflex

Peristalsis first involves contraction of intestinal longitudinal muscle, which shortens the segment of intestine involved, and this is followed shortly by contraction of the circular muscle layer. When hexamethonium, an agent which blocks transmission in nerve ganglia, is applied to the extrinsically denervated gut, longitudinal muscle contraction is not affected, but there is no longer contraction of circular muscle. These results suggest that two neuronal pathways are required for the complete peristaltic reflex. It would appear that one set of neurons without intervening ganglia responds to intraluminal stimuli and initiates the reflex as manifested by contraction of the longitudinal muscle. In addition, another neuronal path with ganglia would subserve the propagated wave over circular muscle.

When the lumen of a segment of intestine exhibiting peristalsis is perfused, the perfusion fluid is found to contain serotonin (5-hydroxytryptamine), and the amount of serotonin released is increased with more vigorous peristalsis. Furthermore, peristaltic activity is increased by adding precursors of serotonin to the perfusion fluid and by the intravenous infusion of serotonin. These observations, plus the fact that the enterochromaffin cells of the intestinal mucosa contain a relatively high concentration of serotonin, have generated the idea that this substance plays a role in the control of peristalsis. In this regard, it has been shown that it takes less stretch to elicit peristalsis when serotonin is applied to the mucosa, presumably because the threshold of receptors responsive to stretch is reduced. However, the exact role, if any, of serotonin in the peristaltic reflex remains uncertain.

Electrical Activity of Intestinal Muscle

The electrical activity of the small intestine is similar in a number of respects to that of the stomach. The segmenting movements are associated with an intestinal basic electrical rhythm. These slow waves originate in a pacemaker region close to the entrance of the bile duct and move down the intestine through the longitudinal muscle as a circumferential ring. If a stimulus is present in a certain region of the gut and is sufficient to bring excitability to threshold from the peak of a slow wave as it traverses the region, spikes are fired, and a segmenting contraction results. In man, the duodenal BER is 11 per minute. When on the relatively rare occasions every slow wave is accompanied by action potentials, the maximal rate of segmenting contractions is attained and a very rhythmic contraction pattern is observed. There is a gradient from proximal to distal intestine in the frequency of the BER. In the dog, the BER is 17 to 18 per minute in the duodenum, 15 to 16 per minute in the jejunum, and 12 to 14 per minute in the ileum. Thus, it is possible that different regions of the intestine have their own pacemakers.

The frequency of the intestinal BER is not altered by stimulation of extrinsic nerves, fasting, feeding, or a variety of other manipulations. However, the vigor of any contraction which might accompany a slow wave can vary over a considerable range. Vagal stimulation augments segmentation, whereas sympathetic stimulation either depresses or abolishes this movement. These alterations in the strength of the segmenting movements are due to the specific influences of the mediators, acetylcholine and norepinephrine, on spiking activity and once again demonstrate the importance of extrinsic nerves in modifying activity initiated in smooth muscle itself.

ILEOCECAL SPHINCTER

The terminal ileum receives unabsorbed food residues about three and a half hours after the beginning of gastric evacuation. Contents arrive at a region immediately proximal to the cecum, where the last 2 to 3 cm of the muscular coat is thicker than the rest of the ileum. This is the *ileocecal sphincter,* which is normally closed. Activity here seems to be much like that of the pylorus. As a peristaltic wave travels over the ileum, the sphincter relaxes and allows chyme to enter the colon. It then contracts and prevents regurgitation of contents into the ileum. Peristalsis is generally not very active in the ileum, but it is increased when food enters the stomach *(gastroileal reflex).*

THE BILIARY SYSTEM

Bile is manufactured and continuously secreted by the liver. This secretion passes from the liver by way of the hepatic duct into the common bile duct and is stored in the gallbladder during periods when the upper intestine is devoid of food. In addition to acting as a storage organ, the gallbladder provides a safety factor for regulating pressures in the biliary system. However, this organ can be removed without interfering with normal digestion and absorption.

FILLING AND EVACUATION OF THE GALLBLADDER

Gallbladder function can be evaluated by watching this organ fill or evacuate its contents. This can be accomplished by administering a radiopaque substance, such as tetraiodophenolphthalein, which is secreted by the liver into the bile. The outline of the gallbladder and bile ducts can then be visualized by means of x-rays. The gallbladder can be seen to fill during periods of fasting. Since relatively little bile enters the duodenum under this circumstance, there must be a mechanism available to prevent the passage of bile into the small intestine and at the same time divert it into the gallbladder. It is widely accepted that there is a sphincter in the region where the common bile duct enters the lumen of the small intestine. If a catheter is inserted into the common bile duct at this point, the gallbladder will not fill because the sphincter is unable to exert its restraining influence. The resistance offered to the flow of bile by this *sphincter of Oddi* has been investigated by measuring the pressure required to force bile through the region. This pressure is a measure of the sphincteric resistance, and in fasting, unanesthetized dogs it can be as high as 30 cm H_2O. It is probably this resistance which prevents bile from entering the intestine during fasting.

When food is taken, the gallbladder can be seen to empty by means of a series of rather sluggish contractions. The intrabladder pressure rises to about 20 to 30 cm H_2O, and under most circumstances the pressure exerted by the contracting gallbladder is enough to overcome the sphincteric resistance. Further, it is believed that the gallbladder and sphincter act as a functional unit; i.e., when the gallbladder contracts, the sphincter relaxes. The entrance of bile into the gut is also influenced by the state of contraction of the duodenal musculature. During active contraction of the duodenum, the flow of bile is decreased or completely blocked by compression of the duct, but during the relaxation phase, resistance to flow is reduced. Therefore, as the result of duodenal movements, bile may be observed to enter the intestine in squirts.

REGULATION OF GALLBLADDER EVACUATION

There is a division of opinion regarding the extent of participation of nerves in the control of evacuation of the gallbladder. Many workers feel that contraction of the gallbladder is mediated reflexly, with the efferent limb of the reflex arc being in the vagus. Other investigators believe that evacuation of the gallbladder is only slightly influenced by nervous factors.

That a regulatory mechanism other than nervous is involved in contraction of the gallbladder is shown by the fact that normal function is maintained when all extrinsic nerves to this organ are cut. It has been shown, using dogs in cross-circulation experiments, that the control of gallbladder motility is at least partly hormonal in nature. When acid was introduced into the duodenum of a donor dog, the gallbladder contracted and 6 to 10 minutes later the gallbladder of the recipient dog contracted. Since the only connection between the two dogs was through the circulatory system, a substance must have been transported from the donor through the blood to cause contraction in the recipient. The gallbladder also contracts when extracts of the upper intestinal mucosa are injected intravenously. On the basis of these results it has been postulated that when chyme comes into contact with the intestinal mucosa, a hormone is released from the mucosa into the blood. This hormone, *cholecystokinin,* travels to the gallbladder and stimulates it to contract. Since cholecystokinin has never been separated from pancreozymin (Chap. 26), these two hormones may be the same. A variety of substances release cholecystokinin when placed in the duodenum. The most notable are fat and its digestive products. Proteins are effective, but not carbohydrates. Agents which promote emptying of the gallbladder are called *cholecystagogues.*

MOTILITY OF THE COLON; DEFECATION

The material which is discharged from the ileum into the proximal colon resembles diluted feces. The absorption of ingested carbohydrates, proteins, lipids, and other nutrients is practically complete at this time. The large intestine absorbs water and electrolytes from these contents and acts as a temporary storage organ for such waste materials as cellular debris, connective tissue, and cellulose. The colon also contains a variety of microorganisms, some of which contribute to the welfare of the host by synthesizing certain nutritional factors.

COLONIC MOVEMENTS

The large intestine of man is inactive for a large proportion of the time. However, when material is present in the proximal colon, segmenting movements (haustral churnings) occur and produce a limited back-and-forth movement of the contents. This is the major type of movement of the colon and it serves to expose the contents to the mucosa and facilitate the absorption of water and electrolytes. It is in this region that a firm fecal mass is formed by the absorption of water. Although the colon does absorb water, it is not the water-conserving organ of the digestive tract. About 8000 ml of water is absorbed every day by the small intestine and only about 300 to 400 ml by the colon. It is usually a misfunction of the small intestine that causes diarrhea. In this instance, large quantities of fluid enter the colon from the small intestine and overwhelm the colonic absorptive mechanism, which is not nearly as efficient as that of the small intestine.

At infrequent intervals of three to four times a day in man, a massive contraction of the proximal colon drives contents into the distal colon, where material accumulates distal to the pelvirectal flexure. These propulsive *mass movements* represent a simultaneous contraction over a relatively long length of colon. That they are rather strong contractions is shown by the fact that the pressure in a segment undergoing such a contraction may reach a peak of 100 mm Hg.

It generally takes about 18 hours for contents to reach the distal colon after leaving the small intestine, and they are stored here for varying lengths of time until defecation occurs. This may be 24 hours or longer following the ingestion of food. Although the rectum is normally empty, contents are occasionally shifted into the rectum after one of the mass contractions, and the resultant distention of the region elicits the desire to defecate. The act of defecation is partly voluntary and partly involuntary. The involuntary movements are concerned with smooth muscle; i.e., the distal colon contracts and the internal anal sphincter relaxes. Relaxation of the external anal sphincter, which consists of striated muscle, is voluntary. Other voluntary movements, which can supply one-half of the force involved in evacuation of the rectum, are a contraction of the abdominal muscles and forcible expiration with the glottis closed (straining movements).

REGULATION OF COLONIC MOTILITY

The proximal colon possesses a large degree of autonomy and functions in a relatively normal manner in the absence of its extrinsic motor innervation, which is derived from the vagus nerve. Movements here are probably largely initiated by distention of the colonic walls, but they can also be initiated reflexly, since when food enters the stomach or duodenum a mass contraction often occurs in the proximal colon. These *gastrocolic* and *duodenocolic reflexes* are usually most evident after the first meal of the day and are often followed by the desire to defecate.

The distal colon is somewhat more dependent on its extrinsic nerve supply, and movements in this region, including the act of defecation, disappear after transection of these nerves. However, weak movements do return after a period of time and there is a semblance of the act of defecation. Defecation as it normally occurs is under voluntary control. There are subsidiary centers, since a certain region of the medulla can be stimulated to cause defecation. If the spinal cord is transected in the thoracic region, defecation can still occur without voluntary control after spinal shock has passed off. On the other hand, if the sacral cord is destroyed, defecation becomes very imperfect. This portion of the cord presumably serves as a local reflex center for the act of defecation. Distention of the rectum causes impulses to be sent over afferent fibers in the pelvic nerve to the sacral cord, which relays impulses over parasympathetic nerve fibers to the distal colon and anal sphincters. If defecation is inhibited by higher centers, the rectum relaxes, the stimulus of distention disappears, and defecation is postponed. The efferent fibers to the distal colon and internal anal sphincter are in the pelvic nerve, while those to the external anal sphincter are in the pudendal nerve. During periods of nonactivity, the internal and external anal sphincters are maintained tonically contracted by impulses over the lumbar sympathetics and pudendal nerve, respectively.

REFERENCES

Alvarez, W. C. *An Introduction to Gastroenterology*, 4th ed. New York: Hoeber Med. Div., Harper & Row, 1948.

Bass, P. In Vivo Electrical Activity of the Small Bowel. In C. F. Code (Ed.), *Handbook of Physiology.* Section 6: Alimentary Canal, Vol. IV. Washington, D.C.: American Physiological Society, 1968. Chap. 100, pp. 2051–2074.

Beaumont, W. *Experiments and Observations on the Gastric Juice and the Physiology of Digestion.* Plattsburgh, N.Y.: Allen, 1833.

Code, C. F., N. C. Hightower, and C. G. Morlock. Motility of the alimentary canal in man. *Amer. J. Med.* 13:328–351, 1952.

Hightower, N. C., Jr. Motor Action of the Small Bowel. In C. F. Code (Ed.), *Handbook of Physiology.* Section 6: Alimentary Canal, Vol. IV. Washington, D.C.: American Physiological Society, 1968. Chap. 98, pp. 2001–2024.

Hunt, J. N., and M. T. Knox. Regulation of Gastric Emptying. In C. F. Code (Ed.), *Handbook of Physiology.* Section 6: Alimentary Canal, Vol. IV. Washington, D.C.: American Physiological Society, 1968. Chap. 94, pp. 1917–1935.

Ingelfinger, F. Esophageal motility. *Physiol. Rev.* 38:533–584, 1958.

Ivy, A. C. The physiology of the gall bladder. *Physiol. Rev.* 14:1–102, 1934.

Thomas, J. E., and M. J. Baldwin. Pathways and Mechanisms of Regulation of Gastric Motility. In C. F. Code (Ed.), *Handbook of Physiology.* Section 6: Alimentary Canal, Vol. IV. Washington, D.C.: American Physiological Society, 1968. Chap. 95, pp. 1937–1968.

Truelove, S. C. Movements of the large intestine. *Physiol. Rev.* 46:457–512, 1966.

Youmans, W. B. *Nervous and Neurohumoral Regulation of Intestinal Motility.* New York: Interscience, 1949.

26

Secretion and Action of Digestive Juices; Absorption

Leon K. Knoebel

SUBSTANCES contained in ingested food which are important in the nutrition of the body include carbohydrates, lipids, and proteins. These nutrients are structurally complex and are not readily absorbed from the digestive tract in their natural states. However, a variety of enzymes are secreted into the lumen of the alimentary canal, and these help catalyze the conversion of complex organic foodstuffs to simple molecules, which are absorbed into the circulation. The juices which contain these enzymes are elaborated and secreted by glands located in different regions of the digestive tract. This chapter discusses the secretory mechanisms of the digestive glands, describes the actions of the digestive enzymes, and considers the manner in which the end products of digestion are absorbed into the circulation.

MECHANISM OF SECRETION

Plasma is the ultimate source of any secretion in the sense that it supplies the constituents necessary for the elaboration of a secretion. The function of secretory cells is to modify plasma in some way in order to produce a secretion. The secretion of digestive juices consists in a variety of processes: (1) the transfer of water and electrolytes from plasma through the glandular epithelium into the lumen of the secretory duct system; (2) the liberation of organic compounds, which are synthesized by and stored in the secretory cells, into this fluid as it passes through the cells; and (3) the production and secretion of certain ions by the secretory cells as the result of the metabolic activity of these cells. In addition, during passage through secretory ducts, the composition of the primary cellular secretion may be modified as a result of movement of substances to or from the blood through the duct epithelium.

A number of lines of evidence show that secretion is an active, energy-requiring process and not simply the result of passive movement of water and dissolved materials from blood through glandular epithelium into the lumen of the digestive tract. For example, the classic experiment of Ludwig showed that the secretion of saliva is not simple filtration of fluid from the blood. The duct of the submaxillary gland of the dog was blocked by cannulating the duct and connecting the cannula with a mercury manometer. When the gland was stimulated to secrete, the secretion of saliva continued in spite of the fact that the pressure in the duct rose considerably above arterial pressure. Since filtration is the movement of fluid from a region of high pressure to one of low pressure, the secretion of saliva cannot be produced by this mechanism. This principle can probably be applied to the secretion of the other digestive juices. Most digestive juices are isosmotic to plasma, but a few are not. For example, parotid saliva is decidely hypotonic. Furthermore, the concentrations of individual electrolytes in secretions are often quite different from those of plasma. The gastric parietal cell secretes a fluid which is thought to be isosmotic to plasma, but which contains hydrogen as its major cation rather than sodium. The acinar cells of the pancreas secrete a juice which contains four to five times as much bicarbonate as does plasma. These facts suggest that secretory cells must do considerable work in order to elaborate from plasma a secretion which can have not only a different osmotic activity from that of plasma but a different electrolyte pattern as well. Consequently, it is to be expected that the energy expenditure of an actively secreting gland should be greater than that of a resting gland. That this is so is indicated by the fact that the oxygen consumption and utilization of carbohydrate are increased in a number of digestive glands during secretion.

Although the mechanisms involved in the formation of a secretion are not well understood, it is known that the various types of secretory epithelia are stimulated to secrete by either nervous or humoral mechanisms, and in most instances by both. The exact manner by which nervous and humoral factors act on secretory cells to cause secretion is unknown. Stimulation of a secretory cell could generate an osmotic gradient between the cell and its exterior or change the permeability of the cell membrane. These events, singly or in combination, could cause movement of fluid and dissolved materials through and out of the cell in the form of a secretion.

SALIVARY SECRETION

Man secretes 1 to 1.5 liters of saliva every day. By virtue of its lubricating properties, saliva facilitates chewing, swallowing, and speech. It also has a cleansing action by diluting noxious substances and flushing them from the oral cavity. Furthermore, the amylase which is contained in saliva initiates in the mouth the digestion of certain polysaccharides.

METHODS OF STUDY

Saliva secreted in response to various stimuli can be collected by a variety of methods. A cannula can be inserted into a salivary duct of an anesthetized animal in an acute experiment. This approach has the advantage that specific nerves can be stimulated. A chronic preparation can be used in which a permanent fistula of a salivary duct is made with the aid of surgery. In this instance many observations can be made on the same animal over long periods of time and in the absence of anesthetics.

COMPOSITION OF SALIVA

Both the volume and the composition of saliva secreted in response to different stimuli are quite variable and depend on the strength and nature of the stimulus. In general, saliva contains the usual electrolytes of the body fluids, the principal ions being sodium, potassium, chloride, and bicarbonate. As secretory rate increases, there are rises in the concentrations of sodium, chloride, and bicarbonate, and potassium concentration declines to a small extent (Fig. 26-1). At maximal secretory rates, potassium and bicarbonate concentrations are higher than those of plasma, whereas the concentrations of sodium and chloride are considerably less

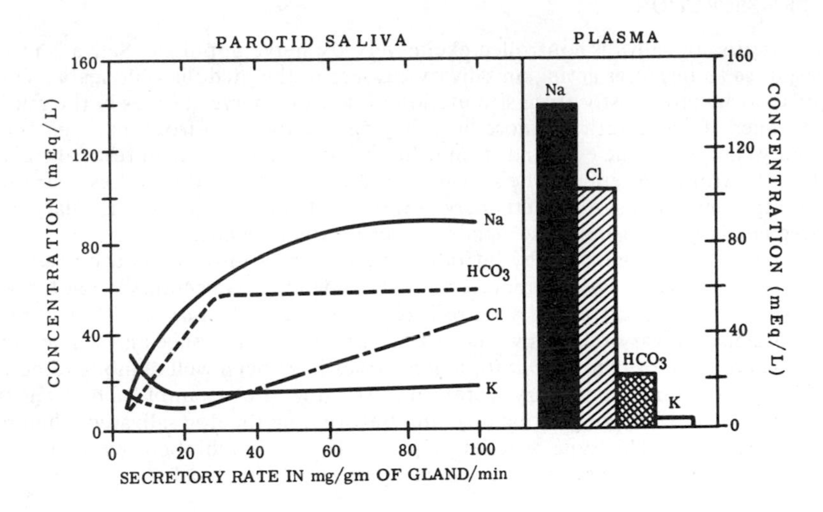

FIGURE 26-1. Electrolyte composition of human parotid saliva as a function of secretory rate. (From F. Bro-Rasmussen, S. Killmann, and J. H. Thaysen. *Acta Physiol. Scand.* 37:97, 1956.)

than the plasma levels of these ions. The overall result is that at high rates of secretion the osmotic pressure of saliva is only about two-thirds that of plasma. The pH of saliva rises from 6.2 to 7.4 with increasing rates of secretion as the result of the concomitant increase in bicarbonate concentration.

One theory in explanation of these results is that the primary secretion of the secretory cells may be isosmotic to plasma and contains potassium, chloride, bicarbonate, and very little sodium. During passage through the duct system, the primary secretion is modified by the movement of potassium from duct lumen to blood and by the movement of sodium in the opposite direction. Since the movement of sodium is slower than that of potassium and since the duct cells are relatively impermeable to water, the final secretory product is hypotonic. Furthermore, since the permeability of the duct cells increases with increased rates of secretion, the tendency is for more complete equilibration of ions and water across the tubular cells with the result that the concentrations of sodium and chloride in saliva approach those of plasma.

The content of organic matter, the most notable constituents being mucus and amylase, is quite variable and seems to depend on the type of stimulus applied. For example, dogs with permanent fistulas of the submaxillary gland have been stimulated to secrete reflexly by introducing either meat powder or dilute hydrochloric acid into the mouth. It was found that even though the volumes of secretion, electrolyte concentrations, and blood flows through the gland were the same for both stimuli, the concentration of organic matter in the saliva was considerably greater for meat powder as compared with acid.

It is interesting that a response is often well adapted to the function it must perform. To cite one example, if a dog is given fresh meat, a viscous saliva containing much mucus is secreted which lubricates the bolus, thereby facilitating its passage into the stomach. If the same dog is given dry meat powder, a large volume of a watery secretion is produced and washes the powder out of the mouth.

REFLEX SECRETION

The secretion of saliva is controlled exclusively by nerve impulses. Saliva is produced in response to impulses acting on salivary centers in the medulla oblongata. These impulses originate mostly from stimulation of sensory nerve endings in the mucous membranes of the mouth and nose, but they can be initiated from many parts of the body, i.e., from the eyes and from other regions of the gastrointestinal tract as part of the vomiting reflex. The salivary centers, in turn, send impulses over the parasympathetic and sympathetic nerve supplies of the parotid, submaxillary, and sublingual glands, causing these glands to secrete their specific juices.

Stimulation of taste buds by introducing substances into the mouth results in the secretion of saliva by decerebrate animals. Higher nerve centers are not required for this secretion of saliva, and such reflexes have been called *unconditioned reflexes.* There are also salivary reflexes which require previous experience and which involve higher nerve centers. These *conditioned reflexes* have been well demonstrated in Pavlov's experiments on salivary secretion in the dog. For example, a bell was rung every time food was brought to a dog, and before long the dog salivated when only the bell was rung. This type of reflex requires that the cerebral cortex be intact, and comes into play, for example, upon thinking of appetizing food.

Role of Autonomic Nerves

There has been difference of opinion over the years regarding the precise role of the autonomic nerves in salivary secretion. When the chorda tympani nerve (parasympathetic) is stimulated, the submaxillary and sublingual glands secrete a large volume of watery saliva which has a low concentration of organic matter. On the other hand, stimulation of the sympathetic nerves to these glands results in the formation of a scanty secretion which is thick and viscous and which has a high concentration of organic matter. Histologists have demonstrated two cell types in the submaxillary and sublingual glands: (1) demilune cells, which are serous and responsible for the secretion of water, salts, and organic matter; and (2) mucous cells, which secrete mucin in addition to those substances mentioned for the demilunes. These are a few of the observations which, in addition to the fact that it is mucin which gives the viscous character to certain types of saliva, have suggested to some workers that each type of secretory cell is innervated by only one division of the autonomic nervous system; i.e., the parasympathetic nerves would innervate only the demilune cells, and the sympathetic nerves, only the mucous cells.

Other observations are in conflict with this idea. For example, although the parotid gland contains only serous cells, it responds to both parasympathetic and sympathetic stimulation, thereby implying a dual innervation of the cells of this gland. Electrophysiological evidence suggests that the same situation may exist for the cells of the mixed salivary glands. When sympathetic stimulation was superimposed on chorda tympani stimulation, the potential difference measured between these glands and the rest of the body was not changed from that of chorda stimulation alone, as might be anticipated if sympathetic stimulation resulted in the stimulation of an additional cell type. Moreover, when secretory potentials were recorded from the interior of a single cell, potential changes were observed during stimulation of either the chorda tympani or sympathetic nerves.

Although the precise innervation of the salivary secretory cells remains to be defined, the parasympathetic nerves are probably the important secretory nerves of the salivary glands, because parasympathetic denervation, either by sectioning or by administration of atropine, results in abolition of normal reflex secretion. The role of the sympathetic nerves remains unclear at the present time.

SALIVARY DIGESTION

The one digestive enzyme present in saliva is *ptyalin* or *salivary amylase.* The action of this enzyme is to degrade complex polysaccharides, such as starch and glycogen, through a variety of intermediate stages (dextrins) to maltose. Small quantities of glucose are also formed. Since salivary amylase does not function in an acid medium such as is present in the stomach, and since food remains in the mouth for such a short time, it might be thought that the action of this enzyme is extremely limited. However, the bolus of food does not disintegrate immediately upon entry into the stomach, and digestion can continue inside the bolus for as long as half an hour.

GASTRIC SECRETION

TYPES OF CELLS

Man secretes 2 to 3 liters of gastric juice every day. The gastric mucosa contains a variety of secretory cells, and there are a number of mechanisms which regulate the activities of these cells. Gastric juice is most commonly regarded as the juice secreted by the body and the fundus, and at least three types of cell are present in the glands located in these regions of the stomach. All these cells secrete water and salts, but other special components are present in their secretions. The *parietal cell* is responsible for the secretion of hydrochloric acid, whereas the *chief cell* is the source of pepsin. Cells located in the neck of these glands and in the glands in the cardiac and pyloric regions of the stomach secrete an alkaline fluid which contains a soluble mucus. In addition to the glandular cells, the surface epithelial cells of the stomach secrete an alkaline juice containing an insoluble mucus.

It is impossible to make a single statement concerning the composition of gastric juice, since so many types of secretory epithelia are available for its production. The composition of gastric juice is determined by the relative activities of the different types of secretory cells. These, in turn, depend on the mechanism or combination of mechanisms which stimulate the various cells to secrete their specific juices.

METHODS OF STUDY

Surgical procedures have been devised in attempts to obtain pure gastric juice. In 1878 Heidenhain removed a small portion of the greater curvature of the stomach and formed it into a pouch. The secretions of the pouch could be removed through a fistula made by bringing the opening of the pouch through a stab wound in the belly wall. A pouch such as this may be considered to represent a miniature stomach which mirrors the secretory events occurring in the main stomach. Furthermore, the secretions obtained from the pouch are not contaminated with food, saliva, and material regurgitated from the small intestine. The vagal connections to the Heidenhain pouch are completely severed, so results obtained with this preparation may not be representative of secretion as it occurs in the main stomach. However, the pouch is useful when gastric secretion is to be studied in the absence of vagal innervation. In 1902 Pavlov made a pouch to which many vagal connections were maintained intact. Continuity between the pouch and the main stomach was maintained by a bridge of tissue through which vagal nerve fibers traveled to the pouch. This pouch secretes reflexly under the proper circumstances. In addition to the pouches of the body and fundus of the stomach just described, both innervated and denervated pouches of the pyloric antrum find wide usage in the study of gastric secretion.

Another preparation is an animal with an esophagostomy. Since ingested material can be drained from the upper esophageal fistula, food can be administered orally without coming into contact with the more distal regions of the digestive tract. On the other hand, food can be introduced through the lower esophageal fistula in order to eliminate stimulation of the oral cavity. A gastric pouch is most often used in conjunction with esophagostomy.

CONTROL OF SECRETION

Both neural and humoral mechanisms are involved in the regulation of gastric secretion. It is well established that there are two substances in the body which act directly to stimulate secretion by the gastric parietal and chief cells. One is acetylcholine, which is liberated by parasympathetic postganglionic nerve fibers ending on these cells; thus, the vagus nerve is the important secretory nerve to the stomach. The other is the hormone gastrin, which manifests its excitatory influences following release into the blood from the pyloric antrum of the stomach.

Proof of the existence of gastrin has been provided by a variety of experiments. For example, when certain substances are placed in a transplanted denervated pouch of the antrum, there is secretion by a denervated (Heidenhain) pouch of the body of the stomach. Since no nerve connections exist between the two pouches, there must be a humoral link between them. The stimulating agents do not act by being absorbed and transported by the blood, because little secretion occurs when these are injected intravenously. Furthermore, distention of a denervated antral pouch produces a secretory response of a denervated fundic pouch, and no chemical substances are available for absorption under this circumstance. The present concept is that certain stimuli cause the release of gastrin from the mucosa of the pyloric antrum, and this hormone then travels by the blood to excite secretion by the body and fundus of the stomach. Pyloric antral mucosal extracts devoid of histamine (see p. 609) have been prepared and when injected intravenously result in the secretion of gastric juice. More recently, two gastrins have been isolated, their structures determined, and the molecules synthesized. These are heptadecapeptides, the terminal tetrapeptide of which is the active fragment of the total molecule. Even though the potency of the terminal tetrapeptide is only one-fifth that of the whole molecule, it displays all the physiological properties of natural gastrin.

Excitation of gastric secretion by either acetylcholine or gastrin has been divided into three phases as based on the region in which a stimulus acts to ultimately cause secretion of gastric juice: the *cephalic, gastric,* and *intestinal* phases. Each phase will be discussed in the order in which it occurs following the ingestion of food.

Cephalic Phase

Pavlov, using dogs with both an esophageal fistula and a vagally innervated gastric pouch, described the mechanism involved in the cephalic phase. When these animals are fed, food does not reach the stomach but is returned to the outside through the esophageal fistula *(sham feeding)*. As long as the dogs chewed and swallowed food, the Pavlov pouch continued to secrete gastric juice. However, if the vagi were transected just above the stomach, no secretion could be elicited from the pouch under the same experimental conditions – strong evidence that this secretion is produced by reflex action. The reflex is initiated by the contact of food with the mucous membranes of the mouth, and the vagus nerve is the efferent component of the reflex arc. Stimulation of the taste buds involves unconditioned reflexes, but there are also conditioned reflexes. For example, Pavlov's dogs secreted gastric juice when they heard the attendant approach with food or when they heard food being prepared.

Since the mere introduction of solid material into the mouth and the chewing of inert substances do not cause secretion, mechanical stimulation of the oral cavity is not an important factor. Furthermore, the chewing of different foods is not always sufficient to produce a secretion. The secretion produced during the cephalic phase usually fails if the food is not taken with appetite. Consequently, this secretion has been referred to as *appetite juice.* The fact that food must be agreeable to elicit the secretion is supported by the observation that there is little gastric secretion during certain types of emotional upsets, presumably because no food is palatable under these conditions.

The cephalic secretory response is due both to the release of acetylcholine at the site of secretion and to the vagally mediated release of gastrin from the pyloric antrum. The former pathway was demonstrated using dogs equipped with a Pavlov pouch and with the pyloric antrum removed. When the dogs were sham-fed, the vagally innervated pouch secreted acid. Since antrectomy obviated any possible stimulation of gastric secretion by gastrin, this result can be accounted for only on the basis of direct vagal excitation of the secretory cells. On the other hand, the vagal stimulation which results during the cephalic phase does exert part of its effect by liberating gastrin. This was shown using dogs with an innervated antral pouch and a Heidenhain pouch. When the animals were sham-fed, there was copious secretion of the vagally denervated pouch. Furthermore, denervation of the antral pouch abolished this response.

The juice secreted during the cephalic phase contains much hydrochloric acid and is rich in pepsin. The same type of juice is obtained when the vagus nerve is stimulated. Since high acidity and high pepsin levels, singly or together, are thought to contribute to ulcer formation, bilateral vagotomy has been used in the treatment of ulcer in an attempt to reduce acid and pepsin secretion in such patients. One way to check whether all vagal fibers to the stomach have been severed is to inject insulin. Insulin reduces the blood glucose level, causing excitation of the vagal nerve centers, which in turn stimulate the stomach to secrete by way of the vagi. If the gastric glands do not secrete following insulin injection, the vagi have been cut.

Gastric Phase

The gastric phase of gastric secretion begins when food enters the stomach and continues for as long as food remains there, which may be three to four hours. In man, a highly acid juice with a high concentration of pepsin is secreted during this phase. As is the case for the cephalic phase, the stimulating actions of acetylcholine and gastrin are responsible for exciting secretion during the gastric phase.

Although gastrin is secreted in response to impulses over the vagus nerve as a part of the cephalic phase, this hormone is also released as the result of local stimulation of the mucosa of the pyloric antrum. There are substances which possess the ability to excite the secretion of a variety of digestive juices. They have been called *secretagogues* and either are present in food or result from the digestion of food. With respect to excitation of gastric secretion by means of gastrin release, the most potent secretagogues are protein digestion products, meat extracts, dilute ethanol, and caffeine. Distention of the antrum also results in enhanced gastric secretion. On the other hand, either undigested proteins or fats and carbohydrates in any form are not effective.

Stimulation of gastric secretion by the contact of secretagogues with or without distention of the pyloric antrum persists following extrinsic denervation of the antrum. However, local neural elements are involved in the release of gastrin because, if the antral mucosa of an extrinsically denervated antrum is treated with an anesthetic in order to inactivate all nerves in the antral wall, gastrin is no longer released in response to intraluminar stimuli. Furthermore, that there is at least one synapse in the local neural chain is shown by the fact that the ganglionic blocker, nicotine, abolishes the secretory response. Secretion is also prevented when atropine is used, and this shows that it is acetylcholine which is released at and stimulates the gastrin-secreting cell. Based on this evidence (Fig. 26-2), the idea has evolved that nerve

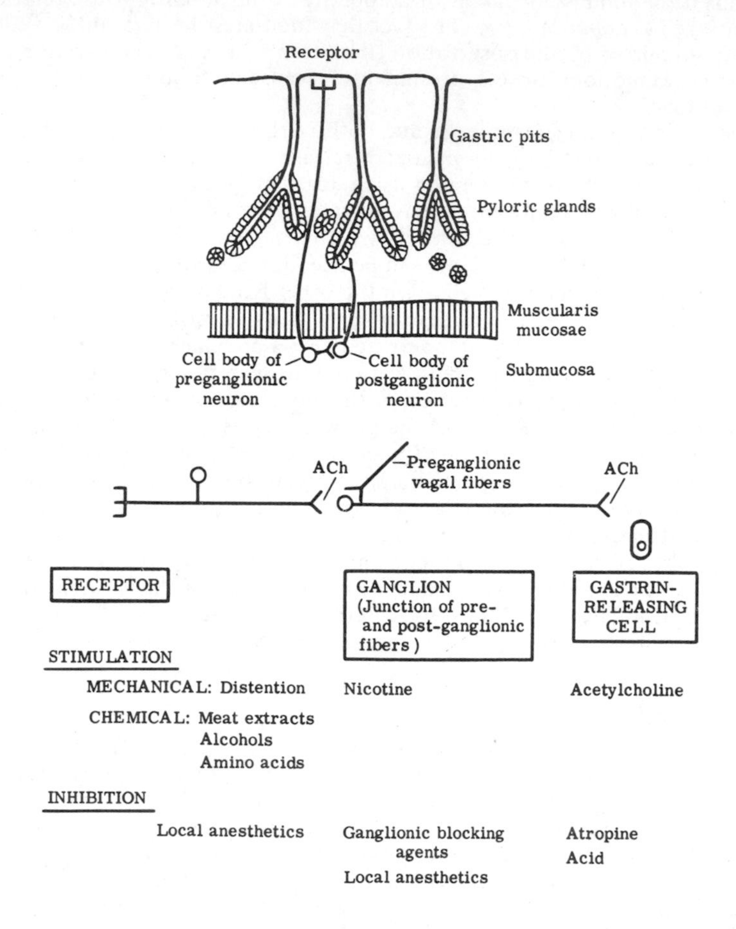

FIGURE 26-2. Hypothetical mechanism of regulation of release of gastrin and factors that influence it. (From M. I. Grossman. *Handbook of Physiology*. Section 6: Alimentary Canal, Vol. II. Washington, D.C.: American Physiological Society, 1967. P. 844.)

receptors in the antral mucosa are stimulated by secretagogues and distention. Impulses are sent over nerve fibers in the antral wall with at least one intervening synapse, and acetylcholine is ultimately released in the vicinity of the gastrin-producing cells, stimulating the release of the hormone into the blood. It might be further visualized that the vagal preganglionic fibers synapse on the final efferent cholinergic fiber, which could then be considered to represent the parasympathetic postganglionic nerve supply of the gastrin-releasing cells.

Part of the gastric juice which is secreted during the gastric phase is the result of stimulation of the body and fundus of the stomach. For example, if the body of the stomach is distended in an animal with the antrum removed in order to negate stimulation of secretion by gastrin, secretion of acid and pepsin occurs. This response is very much reduced if the vagi are cut, showing that a long reflex arc over the vagus nerve is primarily responsible for this effect. However, even under the circumstance of vagotomy, a small response remains. Since this is abolished by atropine, there must be, in addition to the vagovagal reflex, a local cholinergic reflex over the nerve nets in the wall of the stomach. It remains to be determined whether chemical stimuli will elicit these long and short reflexes.

Intestinal Phase

The secretion of gastric juice which is initiated during the cephalic phase is maintained during the gastric phase. However, gastric secretion is prolonged by an additional mechanism, the intestinal phase. This phase is demonstrated by the fact that a denervated gastric pouch secretes either when food is introduced directly into the duodenum through a fistula or when the small intestine is distended. It is apparent that stimulation of the intestinal mucosa causes secretion of gastric juice. The mechanism involved appears to be similar to one of those postulated for the gastric phase; i.e., there is an intestinal gastrin. In this regard, the gastric secretion which occurs in response to intestinal stimuli is prevented by anesthetization of the intestinal mucosa. Thus, as in the case of antral gastrin, nerves seem to be involved in the release of intestinal gastrin.

The secretion which results during this phase is low in acid. That the intestinal phase is relatively weak is shown by the fact that when both cephalic and gastric phases are excluded, gastric secretion is reduced by 90 percent.

Inhibition of Secretion

When food is ingested, gastric secretion is stimulated by a variety of mechanisms. Nevertheless, there are also inhibitory mechanisms, the function of which probably is to prevent excessive secretion.

If liver solution at pH 7 is introduced into an antral pouch, a denervated pouch of the body responds to the resultant release of gastrin by secreting acid. On the other hand, if the pH of the liver solution is 2, no response is obtained. Evidently, acid in contact with the antral mucosa inhibits secretion of acid by the body of the stomach. This demonstrates an autoregulatory mechanism for the secretion of hydrochloric acid and, since the pH of the contents of the antrum is often low during gastric evacuation, this inhibitory mechanism must play an important role in regulating gastric secretion. Although it has been suggested that acid inhibition is the result of the release of an inhibitory hormone, the bulk of the evidence is in favor of the idea that acid acts by suppressing the release of gastrin. For example, the gastric secretory responses that are inhibited by bathing the antral mucosa with acid are those known to be elicited specifically by stimuli which release gastrin. On the other hand, there is no inhibition for stimuli which operate by means other than gastrin release. The mechanism by which acid suppresses gastrin release is unknown. However, there is no involvement of nerves in the acid inhibition, because this is not blocked by the application of anesthetics to the antral mucosa. Acid must, therefore, manifest its inhibitory effect distal to the site of acetylcholine release; i.e., the gastrin-releasing cell may be sensitive to hydrogen ions.

A variety of substances are known to inhibit gastric secretion when they contact the duodenal mucosa. Examples are fat digestion products, acid, and hypertonic solutions. That a humoral mechanism is involved is demonstrated by the fact that secretion of a denervated pouch in response to gastrin release is inhibited when these substances are introduced into the duodenum. Uncertainty exists regarding the nature of the agent or agents mediating the inhibition. There is evidence for the involvement of enterogastrone, the duodenal hormone which is known to inhibit gastric motility (Chap. 25B). Another duodenal hormone, secretin (see p. 615), has been implicated as possibly playing a role in the gastric secretory inhibition which is mediated from the small intestine.

Figure 26-3 summarizes diagrammatically most of the mechanisms involved in the control of gastric secretion.

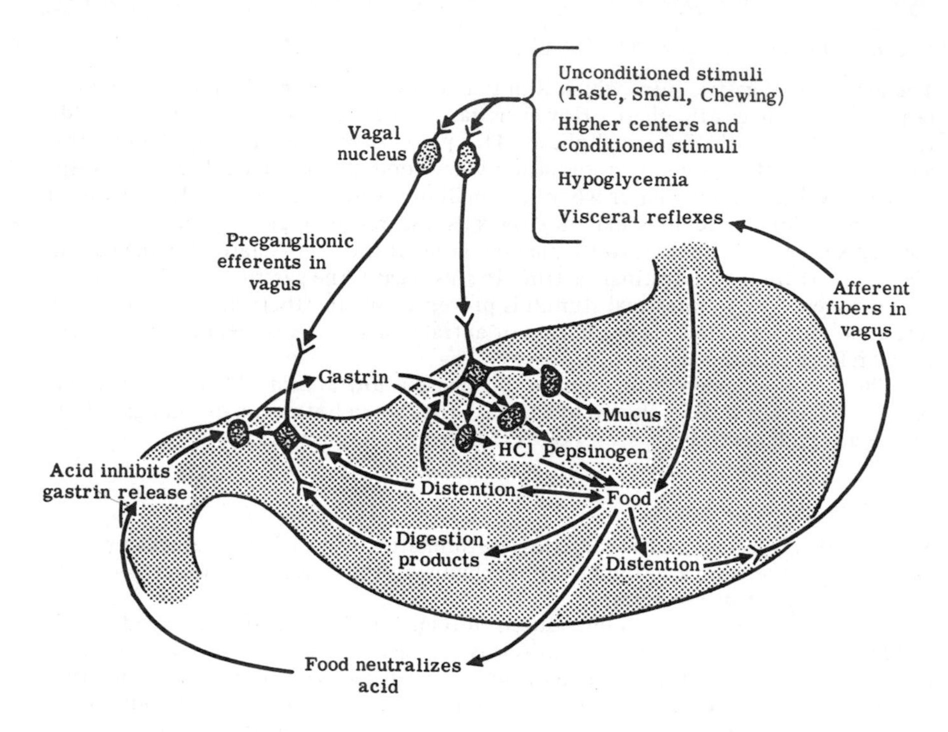

FIGURE 26-3. The control of gastric secretion. (From *Physiology of the Digestive Tract*, Second Edition, by Horace W. Davenport. Copyright © 1966 Year Book Medical Publishers, Inc. Used by permission.)

SECRETION OF THE PARIETAL CELL

Gastric juice is unique in that it contains a relatively high concentration of a free mineral acid, hydrochloric acid. Evidence that has accumulated suggests strongly that this acid is elaborated and secreted by the parietal cell.

Gastric Acidity

In the study of acid secretion, most work has been done using gastric pouches in attempts to obtain the secretion of the parietal cell in its purest form. It has been concluded that the parietal cell secretes acid at a constant concentration regardless of the rate of secretion. Since parietal cell secretion and plasma are thought to have identical osmotic pressures, the theoretical maximum hydrogen ion concentration obtainable in this secretion would be about 170 mEq per liter. Since the hydrogen ion concentration of parietal cell secretion has been estimated to fall in the range of 150 to 170 mEq per liter, practically all of the cation of this juice is in the form of hydrogen ion.

The concentration of acid found in the secretions of the intact stomach is always less than that secreted by the parietal cell. The secretions of the nonparietal cells, such as the chief and mucous cells, are slightly alkaline; when present these secretions dilute and neutralize the acid secretion of the parietal cells. Consequently, gastric acidity can vary over a considerable pH range, and the actual pH observed depends on the relative amounts of the acid and alkaline secretions present in the mixture. The acidity of gastric contents is even less following the ingestion of food. Parietal cell secretion obviously is further diluted and neutralized not only by food but by saliva and duodenal contents.

The concentration of acid in gastric contents is the result of the interplay of a number of factors. Since gastric acid is secreted at constant concentration, variations in gastric acidity cannot be accounted for on the basis of either hypoacidity or hyperacidity of parietal cell secretion. It is the rate of secretion that is important, and this depends on the balance between the activities of the excitatory and inhibitory mechanisms which influence the parietal cell. The other major factor governing the final acidity of gastric contents is the adequacy of the diluting and neutralizing mechanisms relative to parietal cell secretory rate. For example, a normal rate of acid secretion accompanied by inadequate dilution and neutralization causes hyperacidity. Hyperacidity also occurs when dilution and neutralization mechanisms are normal, but either excessive activity of the excitatory mechanisms or subnormal activity of the inhibitory mechanisms results in hypersecretion by the parietal cell.

Role of Histamine

A powerful endogenous stimulus of parietal cell secretion other than acetylcholine and gastrin is histamine, and at one time this substance was considered the most potent stimulus of acid secretion. It is now recognized that on a molar basis of comparison gastrin is about 500 times more effective than histamine. However, when it is desirable to assess the acid-secreting capacity of the human stomach, frequently histamine or one of its analogues is administered. When histamine is injected subcutaneously or intramuscularly, and if a subject is capable of secreting acid, a juice is obtained which is very high in acid but low in pepsin.

It has been postulated that histamine is the final link in the secretion of hydrochloric acid; i.e., all neural and humoral stimuli may act by causing the release of histamine, and this would be the substance stimulating the parietal cell to secrete. Evidence in favor of this idea includes the following: (1) There are large quantities of histamine in the region of the parietal cells relative to adjacent regions. (2) Vagal stimuli and gastrin reduce the amount of histamine in the gastric mucosa. (3) Histamine appears in gastric juice which is secreted in response to a variety of stimuli. (4) Agents that inhibit the activity of histidine decarboxylase, an enzyme which catalyzes the conversion of histidine to histamine, also inhibit parietal cell secretion in response to diverse stimuli. (5) When diamine oxidase, a histamine-destroying enzyme, is injected into animals which are then vagally stimulated by sham feeding, acid secretion is considerably reduced from control levels. However, since these results are subject to alternative interpretations, not all of them supporting a physiological role of histamine in acid secretion, and since there is a considerable body of evidence in opposition to this idea, the concent that histamine is an obligatory intermediate in the secretion of acid cannot at present be accepted as fact.

Mechanism of Hydrochloric Acid Production

The mechanism of hydrochloric acid production must be extremely efficient because the hydrogen ion concentration of parietal cell secretion is about 4 million times that of plasma. An active transport process such as this requires a considerable expenditure of energy, and this is provided in the form of high-energy phosphate bonds which are derived from oxidative metabolism and aerobic glycolysis of the mucosal cells.

Quantitative data have been obtained in studies in vitro using closed tubes of gastric mucosa of the frog. The tubes were incubated in a saline-bicarbonate solution through which was bubbled a gas mixture consisting of 95 percent oxygen and 5 percent carbon dioxide. When histamine was added to this medium, hydrochloric acid appeared inside the tubes (mucosal surface) and the same number of bicarbonate ions appeared outside the tubes (serosal surface). Furthermore, for every bicarbonate ion and every molecule of acid formed, a molecule of carbon dioxide disappeared from the medium. Although many theories have been proposed in attempts to explain the mechanism of gastric acid secretion, these experiments can be used as the basis for offering a simple working hypothesis in explanation of the formation of hydrochloric acid. The hypothesis is illustrated diagrammatically in Figure 26-4.

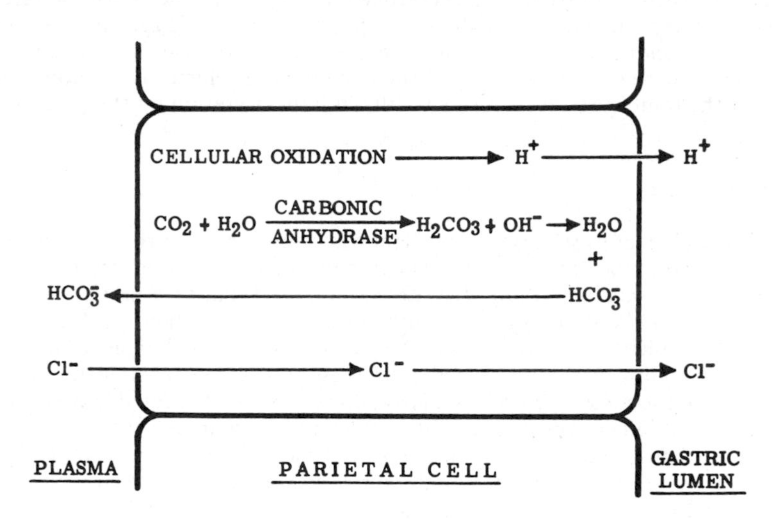

FIGURE 26-4. A possible mechanism for the secretion of gastric hydrochloric acid.

Hydrogen ions originate as a result of certain oxidative processes which occur in the parietal cell, and these are actively secreted into the lumen of the stomach. The alkali which is necessarily produced during the secretion of hydrogen ions is represented as hydroxyl ions, and they must be buffered in order to maintain cellular pH constant. Carbon dioxide, which is formed by the metabolic processes of the cell, is hydrated to form carbonic acid. The parietal cell is known to contain relatively large amounts of carbonic anhydrase, and this enzyme greatly increases the rate at which this reaction takes place. The hydroxyl ions are buffered in the parietal cell by reacting with carbonic acid with the resultant formation of water and bicarbonate ions. This scheme would explain the production of a bicarbonate ion and the loss of a molecule of carbon dioxide for every hydrogen ion that is produced. In addition, chloride ions are removed from the blood and secreted with hydrogen ions to form hydrochloric acid. Chloride ions do not just move passively along with hydrogen ions, because the surface of the mucosa is negative relative to the serosa. Thus, chloride ions move across the gastric mucosa against a potential difference. In addition, since chloride ions move from a concentration in plasma of about 108 mM per liter to a concentration of about 170 mM per liter in gastric juice, they are transported against a chemical as well as an electrical gradient; the transport of this ion is therefore an active, energy-requiring process. In order to maintain electrical neutrality, for all chloride ions which move from plasma into gastric juice, an equivalent number of bicarbonate ions move from the parietal cell into plasma. As a result of the movement of bicarbonate ions, both the blood leaving the stomach and the urine become more alkaline following a meal *(alkaline tide)*.

GASTRIC DIGESTION

Pepsinogens and Pepsins

There are at least three pepsins, and they are secreted in the form of the inactive precursors, pepsinogens I, II, and III. Pepsinogen I can be recovered from all regions of the stomach and the proximal duodenum, whereas pepsinogens II and III occur only in the oxyntic gland area. Pepsinogen II is synthesized by and stored as granules in the chief cells, but the cellular origin of the other pepsinogens is unknown. An extensively studied pepsinogen is one recovered from hog fundic mucosa; it is a protein having a molecular weight of 42,000. Since the corresponding pepsin has a molecular weight of 34,500, the inactive pepsinogen must be split to form active pepsin. The conversion in vitro of all pepsinogens occurs in the presence of either hydrochloric acid or small amounts of pepsin. Thus, it seems likely that hydrochloric acid initiates the breakdown of pepsinogen in vivo and that the small amounts of pepsin so liberated then carry the process on autocatalytically. Human pepsins split linkages in both proteins and polypeptides, the pH for optimal activity of pepsin I being 3.0 to 3.2 and that of pepsin II being 1.5 to 2.0. As the result of peptic digestion, mostly polypeptides along with some amino acids are found in the stomach following ingestion of a protein meal.

Since the most effective stimulus for the secretion of pepsinogen is acetylcholine, the vagal reflexes which occur during the cephalic and gastric phases and result in the release of acetylcholine in the vicinity of the pepsinogen-secreting cells are undoubtedly of considerable importance in determining pepsin levels in gastric contents. Stimulation of pepsinogen secretion by gastrin and histamine is weak compared to that of acetylcholine. However, there is good correlation between acid and pepsin outputs in man in that most stimuli which increase the secretion of acid also increase secretion of pepsinogen.

Gastric Lipase

A lipase is also present in gastric juice but is of little importance because, in order to act, this enzyme requires a pH close to neutrality. Only small amounts of free fatty acids and partial glycerides are recovered from the contents of the stomach after feeding triglycerides containing long-chain fatty acids.

SECRETION OF MUCUS

All regions of the stomach possess cells which secrete an alkaline fluid containing mucus. Prime stimuli for the secretion of this juice are chemical, mechanical, and thermal irritation of the gastric mucosa. Vagal impulses also stimulate the secretion of mucus.

A layer of mucus, 1.0 to 1.5 mm thick, adheres to the gastric wall and serves as a protective barrier against various forms of irritation. It provides protection against mechanical injury by serving as a lubricant, and against chemical injury by virtue of its neutralizing properties. Mucus holds the alkaline fluid within its gel-like structure, and when acid diffuses into the gel, it can be neutralized before coming into direct contact with the epithelium; 100 ml of mucus neutralizes 40 ml of 0.1N HCl. When pepsin diffuses into the mucus barrier, this enzyme is inactivated in the medium of high pH, so that the chance of attack on the protein structure of the underlying epithelium is minimized. However, the ultimate barrier is the mucosal cell membrane, the permeability characteristics of which restrict the entrance of molecules to the interior of the cell. If membrane permeability is increased by toxic agents or mechanical trauma, hydrochloric acid and pepsin enter the cells and cause considerable damage.

INTRINSIC FACTOR

An important substance which is associated with the parietal cell secretion of man is the *intrinsic factor*. The intrinsic factor is necessary for optimal absorption of vitamin B_{12}, which is required for the formation of normal red blood cells. When the gastric mucosa is deficient or lacking in intrinsic factor, vitamin B_{12} is poorly absorbed and pernicious anemia develops.

PANCREATIC SECRETION

The digestion of foodstuffs, which is initiated in the mouth and continued in the stomach, is carried to completion in the small intestine. Since the digestive enzymes secreted into the lumen of the small intestine require a pH close to neutrality for optimal activity, the acidity of the chyme which enters the small intestine from the stomach must be reduced. This may be partly accomplished by the transfer of hydrogen ions across the intestinal membrane, or the exchange of these ions with extraluminar sodium and potassium, or both. In addition, the alkaline juices of the pancreas, liver, and small intestine are of considerable importance in the neutralization of the acid gastric contents.

COMPOSITION OF PANCREATIC JUICE

As is the case for other digestive secretions, the composition of pancreatic juice depends on the type of stimulus applied. One reason is that pancreatic juice consists of two distinct components: (1) an aqueous juice containing electrolytes and with very little enzymatic activity which constitutes the bulk of the volume secreted by the pancreas; and (2) a juice which contains the digestive enzymes of the pancreas. Thus, the total secretion of the pancreas represents a mixture of these two components in varying proportions.

It is undecided which cells are responsible for the secretion of the 200 to 800 ml of aqueous juice produced every day. These may be either the acinar cells or the cells lining the intercalated ducts, or both. Regardless of the origin of this juice, its composition has been well characterized (Fig. 26-5). The principal cations are sodium, potassium, and calcium, and they are present in the same concentrations as in plasma at all rates of secretion. The outstanding feature of this juice is that, relative to plasma, it has a high bicarbonate content. The bicarbonate concentration

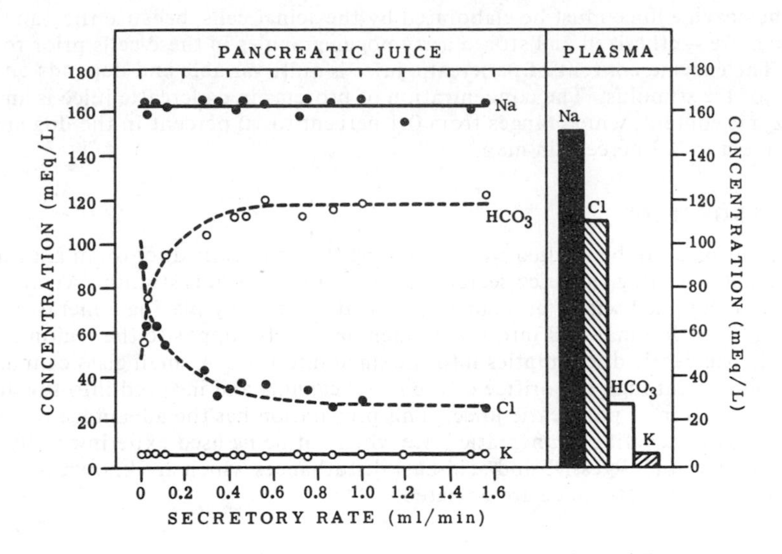

FIGURE 26-5. Relation between rate of secretion and concentrations of sodium, potassium, chloride, and bicarbonate in the pancreatic juice of the dog (after secretin injection). (From F. Bro-Rasmussen, S. Killmann, and J. H. Thaysen. *Acta Physiol. Scand.* 37:97, 1956.)

increases with increased rates of secretion and ranges from 66 mEq per liter at low to 140 mEq per liter at high secretory rates. The other anion is chloride, the content of which is low relative to that of plasma, but the sum of the chloride and bicarbonate concentrations at any rate of secretion is the same as the sum of these in plasma. This secretion has the same osmotic pressure as that of plasma and, because of its high bicarbonate content, has a pH from 7.6 to 8.2.

The bicarbonate of the aqueous juice is derived partly from plasma and partly from carbon dioxide produced as the result of pancreatic cell metabolism. Carbonic anhydrase plays an important role in the formation of that bicarbonate which results from cellular metabolism. For example, the intercalated duct cells contain a high concentration of this enzyme compared to other pancreatic cells. Furthermore, inhibitors of carbonic anhydrase markedly depress pancreatic bicarbonate secretion.

The variations in bicarbonate and chloride concentrations which occur with alterations in rate of secretion are the basis for a number of theories regarding the mechanism of secretion of the aqueous juice. One of these suggests that the primary cellular secretion is an isosmotic solution of bicarbonate and that bicarbonate is actively transported into the duct lumen. In order to maintain electrical neutrality, sodium and potassium passively follow bicarbonate in the same proportion in which these cations are present in plasma, and water moves passively with the electrolytes in order to satisfy osmotic dictates. As the bicarbonate solution moves through the duct system, the secreted bicarbonate exchanges with chloride of the interstitial fluid, the extent to which the exchange occurs being determined by the rate of secretion. The greatest exchange would take place at lower flow rates, because more time would be available for it. In this instance, the concentration of bicarbonate in pancreatic juice as it enters the intestine would be relatively low and that of chloride relatively high.

The enzyme juice must be elaborated by the acinar cells, because the pancreatic enzymes are synthesized and stored as zymogen granules in these cells prior to secretion. The enzyme content of pancreatic juice is quite variable and depends on the nature of the stimulus. The concentration of proteins in pancreatic juice is an index of enzyme content, which ranges from 0.1 percent to 10 percent in the dog and 0.1 percent to 0.3 percent in man.

METHODS OF STUDY

Pancreatic juice can be studied by cannulating the pancreatic duct of an anesthetized animal and collecting the juice secreted in response to various stimuli. An animal can also be supplied with a permanent pancreatic fistula by placing a metal cannula through the abdominal wall into the duodenum directly opposite the region where the main pancreatic duct empties into the small intestine. A small glass cannula can be introduced through the orifice of the metal cannula and inserted into the duct for the collection of pancreatic juice. This preparation has the advantage that the animal has the benefit of pancreatic juice when not being used experimentally. Thus the profound digestive and acid-base disturbances which are known to accompany loss of pancreatic juice are obviated.

CONTROL OF SECRETION

The secretion of pancreatic juice is regulated by both nervous and hormonal mechanisms.

Reflex Regulation

If a dog is sham-fed, the pancreas secretes a small volume of juice which is very rich in enzymes. It has also been claimed that the sight or smell of food results in this same type of secretion. When the vagi are cut or atropine is administered, the juice produced by these stimuli is not obtained. Furthermore, the secretion can be elicited when either the vagi are stimulated or parasympathomimetic drugs, such as acetylcholine, are injected intravenously. These observations are proof for a cephalic phase of pancreatic secretion and show that stimulation of the pancreas occurs reflexly by way of the vagus nerves.

A gastric phase of pancreatic secretion can be demonstrated by distending the body of the stomach. The response of the pancreas is then primarily one of increased enzyme output, but there is also a small increase in volume flow. This effect is abolished by vagotomy. The reflex involved is called the *gastropancreatic reflex,* and both afferent and efferent pathways are in the vagus nerve.

Hormonal Regulation

Since there is little impairment of the digestive function of the pancreas when all extrinsic nerves to this organ are cut, there must be other mechanisms involved in the regulation of the secretion of pancreatic juice. In 1901 Bayliss and Starling performed one of the classic experiments in physiology. Their investigations led to the discovery of the first of many hormones now known to be highly important in the regulation of body function. They denervated a segment of jejunum by grossly dissecting the nerves to this section of gut and discovered that a profuse flow of pancreatic juice resulted when dilute hydrochloric acid was introduced into the lumen of the denervated segment. The response obtained was not due to the absorption of hydrochloric acid, since secretion did not occur when this acid was injected intravenously. They concluded that acid in contact with the intestinal mucosa caused the liberation of a substance into the blood and that this substance was transported to the pancreas, stimulating it to secrete. They also found that the same effect could be obtained by injecting intravenously extracts of the intestinal mucosa. The substance which was thought to be present in the mucosal extracts was named *secretin.*

Bayliss and Starling were sharply criticized at the time for suggesting such a drastically different regulatory concept, but time has proved them correct in their conclusions. Unequivocal proof has been supplied using dogs with denervated pancreatic and denervated jejunal transplants. A profuse secretion of a denervated pancreatic transplant follows the introduction into a jejunal transplant of acid and various foods, such as peptones, soaps, and amino acids.

Secretin has been isolated in pure form, is a basic linear polypeptide containing 27 amino acids, and has a minimal molecular weight of 3200 to 3500. When highly purified secretin is injected intravenously, a relatively large volume of alkaline juice with little enzyme activity is secreted. It has been theorized on the basis of this result that pancreatic juice, which is secreted in response to the secretin mechanism, may have its function in the neutralization of acid gastric contents. This argument is supported by the observation that the secretin mechanism responds only to a pH lower than 4.5, and it may be that the mechanism simply guards against high acidities in the gut.

In addition to the neural controls involved in the regulation of the secretion of enzymes and the intestinal hormonal control for the secretion of water and electrolytes, there is an intestinal hormonal mechanism for the secretion of enzymes. Since enzymes are secreted in the absence of pancreatic innervation and since pure secretin preparations produce a juice with low enzymatic activity, another factor must be responsible for stimulating the secretion of enzymes. That an intestinal hormone is involved was suggested by the fact that various foodstuffs elicit a pancreatic secretion of relatively high enzyme content when placed directly into the intestine of dogs with denervated pancreatic transplants. Products of protein digestion, such as proteoses, peptones, and amino acids, are potent in this respect, as are free fatty acids and soaps. Acid is a moderate stimulus, but carbohydrates are not particularly effective. Furthermore, the older and relatively impure secretin preparations stimulate enzyme secretion. The implication is that, in addition to secretin, these impure preparations contain another hormone which is responsible for the secretion of enzymes, and it has been called *pancreozymin.* Pancreozymin has been isolated in fairly pure form from the standpoint that good secretory responses can be obtained using minute quantities of it. However, the structure of pancreozymin has not yet been determined, and it has never been separated from cholecystokinin.

The excitatory effects of secretin and pancreozymin are manifested during the intestinal phase of pancreatic secretion. For many years, little emphasis was placed on a gastric phase of pancreatic secretion. However, first the gastropancreatic reflex was demonstrated, and, more recently, evidence has been accumulating for the involvement of gastrin as a stimulator of pancreatic secretion. It is now known that there is a weak stimulation of pancreatic flow and a strong enzyme response under conditions which result in the release of gastrin from the pyloric antrum. A similar result is obtained when pure gastrin is administered intravenously. Thus, secretion by the pancreas is apparently stimulated by the gastrin which is released as the result of impulses over the vagi during the cephalic phase and direct stimulation of the antrum during the gastric phase. Since the flow rate which can be attained with gastrin is about one-third that of secretin and the enzyme response is about three-fourths that of pancreozymin, the gastrin mechanism undoubtedly plays a quantitatively important role in the regulation of pancreatic secretion.

Figure 26-6 summarizes the neural and hormonal mechanisms which control secretion by the pancreas.

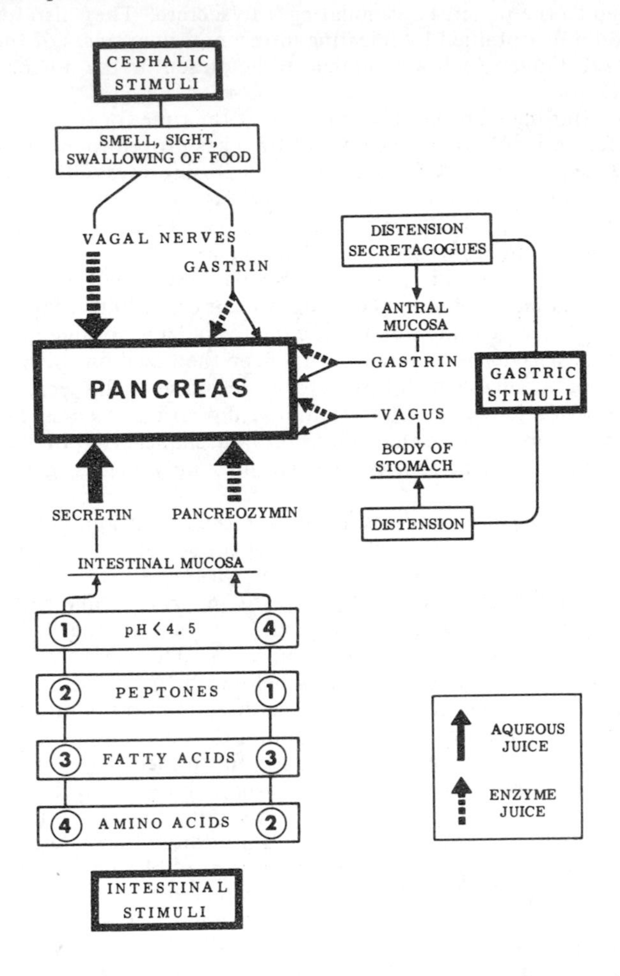

FIGURE 26-6. Neural and hormonal mechanisms for the secretion of pancreatic juice. The intestinal intraluminal factors listed at the bottom of the diagram are numbered to indicate the relative magnitude of the effect of each of these agents.

PANCREATIC DIGESTION

Pancreatic juice is the most versatile and active of the digestive secretions. Its enzymes are capable of almost completing the digestion of all foodstuffs in the absence of the other digestive secretions.

Proteases

Trypsin, a proteolytic enzyme, is secreted as an inactive precursor, *trypsinogen.* Trypsinogen is converted to active trypsin by *enterokinase,* an enzyme which is present in intestinal juice. Small amounts of trypsin are also able to catalyze this conversion.

Trypsin hydrolyzes proteins and the polypeptides which are supplied by peptic digestion of protein. The end products of tryptic digestion are peptides of various sizes and amino acids. The optimal pH for tryptic activity in vitro is 7.8, which is higher than that usually found in the small intestine. Other enzymes secreted into the lumen of the small intestine also show optimal activities at a pH higher than that of intestinal contents. However, this apparently is not a limiting factor, since ingested foods are absorbed efficiently in spite of an environment which is suboptimal for digestion.

There is a trypsin inhibitor present in the pancreas, and this combines with trypsin in the ratio of one molecule of inhibitor to one molecule of trypsin. Since the end product of the combination is enzymatically inactive, the inhibitor prevents the pancreas from digesting itself during those times when small amounts of trypsin are activated in the pancreas. However, if large amounts of trypsin are activated, the inhibitor is overcome and the pancreas may be damaged or destroyed completely.

Another proteolytic enzyme of pancreatic juice is *chymotrypsin.* It is secreted as inactive *chymotrypsinogen* and converted to the active form by trypsin. Except for a few differences, chymotrypsin acts in a manner similar to that of trypsin. Both trypsin and chymotrypsin are endopeptidases and split linkages in the interior of the protein molecule.

The combined activities of pepsin, trypsin, and chymotrypsin result in the degradation of native proteins to polypeptides and amino acids. There are also a number of peptidases present in pancreatic juice. One peptidase which has been isolated is *carboxypeptidase.* This enzyme is secreted as *procarboxypeptidase* and is converted to the active form by either trypsin or enterokinase. Carboxypeptidase is an exopeptidase and splits amino acids from peptides with free carboxyl groups, but not from peptides with free amino groups.

Lipase

The pancreas supplies a lipolytic enzyme which is important in the digestion and absorption of fats. If pancreatic juice is drained away through a fistula so that it cannot enter the intestine, excessive amounts of fats are excreted in the feces. The action of *pancreatic lipase* is to split the ester linkages between fatty acids and glycerol with the production of free fatty acids, partial glycerides, and glycerol. It is doubtful that intraluminar hydrolysis of triglycerides is complete. This point will be considered in more detail in the discussion of the absorption of fats. The optimum pH for activity of pancreatic lipase is about 8.

Amylase

The pancreas also secretes an *amylase,* the action of which is similar to that of salivary amylase. Starch and glycogen are degraded to maltose, the optimal activity occurring at pH 7. Some pancreatic amylase escapes into the blood during pancreatic secretion, because stimulation of the pancreas results in a rise in plasma amylase. Plasma amylase levels have been used to evaluate the functional status of the pancreas. For example, trauma to this organ elevates plasma amylase levels, presumably by increasing the escape of pancreatic amylase into the blood. During autolysis of the pancreas by tryptic digestion, both plasma and urinary amylase show marked rises.

Other Pancreatic Enzymes

Additional enzymes secreted by the pancreas include ribonuclease, deoxyribonuclease, elastase, cholesterol esterase, and lecithinase.

BILE

The result of the secretory activity of the liver is the production of bile. This secretory process is not simply ultrafiltration from blood, as demonstrated by the fact that secretory pressure can be higher than blood pressure. In addition, a number of substances, including the bile salts and bile pigments, are transported from blood into bile against large concentration differences. These facts show that the secretion of bile by the polygonal cells of the liver is an active, energy-expending process.

COMPOSITION OF BILE

The amount of bile secreted by the liver every day in man is 500 to 1000 ml, far greater than the capacity of the gallbladder, which is about 50 ml. However, this inequality of volumes is compensated for by the considerable absorptive capacity of the epithelium of the gallbladder. A comparison of the compositions of bile obtained from the liver and gallbladder as shown in Table 26-1 strongly suggests that water and salts are absorbed by the gallbladder.

TABLE 26-1. Composition of Human Bile

Constituent	Liver Bile (percent)	Gallbladder Bile (percent)
Water	97.48	83.98
Mucin and pigments	0.53	4.44
Bile salts	0.93	8.70
Fatty acids	0.12	0.85
Cholesterol	0.06	0.87
Lecithin	0.02	0.14
Mineral salts	0.83	1.02

The water content of gallbladder bile is much less than that of liver bile, whereas the dissolved materials of gallbladder bile, with the exception of the mineral salts, are much more concentrated than those of the liver. Thus, water and salts are absorbed by the gallbladder, and the other constituents, which are not appreciably absorbed, become concentrated in the process. Normally, the gallbladder concentrates liver bile about 10-fold, but if the epithelium is damaged or irritated, the ability to concentrate is decreased or lost completely.

Both liver bile and gallbladder bile have the same osmotic pressure as plasma. Since the osmotic activity of gallbladder bile is not altered by absorption, there must be an isosmotic absorption of water and electrolytes. The evidence indicates that sodium is actively transported by the gallbladder epithelium and that water follows sodium passively. Sodium transport must be active, because this ion moves against both an electrical and a chemical gradient. That water movement is passive is pointed out by the observations that the flow of water is directly proportional to the transport of sodium and that water does not move in the absence of sodium. As the result of the absorption of electrolytes, the reaction of gallbladder bile (pH 5 to 6) is more acid than that of liver bile (pH 7.4).

METHODS OF STUDY

Bile, which is secreted in response to various stimuli, can be collected from a cannula inserted in the common bile duct of an anesthetized animal. If only secretion by the liver is to be studied, the cystic duct must be ligated in order to remove influences from the gallbladder. Bile can also be collected from a surgically prepared bile fistula in a conscious dog. In this instance, provision is made to return bile to the small intestine when the experiment requires this.

CONTROL OF SECRETION

The liver secretes bile continuously, and it has been shown that fasting, unanesthetized dogs secrete 6 to 10 ml of bile per kilogram per day. When these dogs were fed without allowing bile to enter the small intestine, the amount of bile secreted increased to 13 to 18 ml per kilogram per day. When bile was returned to the intestine after feeding, the amount of bile secreted was 24 to 27 ml per kilogram per day. Thus, the presence of bile in the intestine during digestion increases the secretion of bile over that of feeding alone. This is due specifically to the stimulatory effects of the bile salts on the polygonal cells of the liver following absorption of these substances from the intestine and their subsequent transport by the blood to the liver.

Certain studies have quantitated the influence of the bile salts on the rate of flow of bile. These studies involved the use of trained, unanesthetized dogs with duodenal cannulas installed in such a manner that a catheter could be inserted into the common bile duct for the collection of bile. The bile salt, sodium taurocholate, was infused at different rates into a vein in order to vary the plasma levels of this bile salt. As the plasma concentration of taurocholate was increased with increased rates of infusion, the rate of secretion of taurocholate also increased and was about the same as the rate of infusion. Figure 26-7 shows that bile flow increased in direct proportion to the rate of taurocholate secretion. A similar relationship was obtained for the output of chloride and bicarbonate ions. These data suggest that both the flow of bile (water output) and electrolyte output are dependent on the rate at which bile salts are secreted. One idea that attempts to explain these observations is that the active transport of bile salts provides a primary osmotic force for the secretion of bile. It is visualized that, when bile salts are actively transported into the canaliculi, a certain amount of water follows by virtue of osmotic drag. In addition, since there will be a diffusion gradient for ions, these move into the canaliculi accompanied by more water, the final result being the formation of a secretion isosmotic to plasma. When plasma bile salt concentration is elevated, the extent to which these compounds are actively transported will be increased, with the result that greater quantities of water and electrolytes follow passively. That there is a specialized system for the transport of bile salts is indicated by the fact that these substances can be concentrated more than 100 times in passage from blood to bile. Furthermore, there is an upper limit to transport capacity at high plasma bile salt concentrations; i.e., the transport system becomes saturated.

It is apparent from the data presented above that there is some stimulation of bile secretion when an animal is fed and bile salts are diverted from the small intestine so that these compounds are not able to manifest their excitatory influences. This effect is thought to be due at least partly to the release of secretin from the

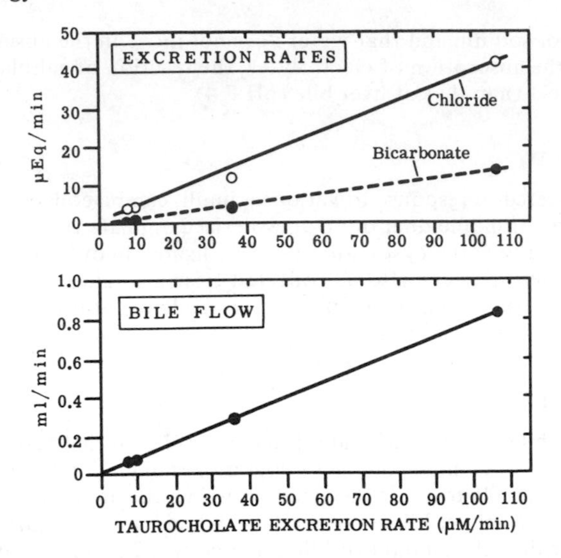

FIGURE 26-7. Relation between secretion rate of bile salt and the output of chloride, bicarbonate, and water (bile flow) in the dog. Sodium taurocholate was infused intravenously at rates of 8, 40, 118, and 8 μM per minute (in that order) to obtain the four points. Output of water appeared to be directly proportional to taurocholate secretion rate, as did the output of chloride and bicarbonate ions. (From H. O. Wheeler. *Handbook of Physiology.* Section 6: Alimentary Canal, Vol. V. Washington, D.C.: American Physiological Society, 1968. P. 2417.)

intestinal mucosa, because any stimulus which is known to release secretin and cause the pancreas to secrete also stimulates the secretion of bile. Furthermore, intravenous administration of pure secretin increases the flow of bile. However, the role of secretin in the stimulation of secretion by the liver is secondary to that of the bile salts, since secretin produces an increase in bile flow only about one-tenth that of the bile salts. It is of interest that the response of the liver to secretin consists in an increased output of water and electrolytes without any increase in output of bile salts and pigments. This is analogous to secretin stimulation to the pancreas, which has little effect on enzyme output but a decided excitatory influence on the water-electrolyte component of pancreatic secretion. There is also some evidence that the site of action of secretin is the duct cells of the biliary system and that, as the secretory rate is increased in response to secretin, bicarbonate concentration rises and chloride concentration falls.

The response to feeding may be partly accounted for by vagal stimulation, because this produces an increased liver bile flow. However, it remains to be proved that vagal stimulation has a direct stimulatory effect on the secretory cells of the liver. Since gastrin produces a small secretory response by the liver, vagal stimulation may partly manifest its excitatory effect through the release of gastrin.

Any agent which produces an increased secretion of bile by the liver is a *choleretic,* and the process by which this substance produces this effect is known as *choleresis.* Bile salts are the most important single stimulus for choleresis.

THE BILE SALTS

The bile salts are synthesized from cholesterol in the liver and secreted mostly as glycocholic acid and taurocholic acid. Glycocholic acid is a conjugate of the amino acid glycine with the steroid cholic acid. Taurine is derived from the amino acid cystine and is conjugated with cholic acid to form taurocholic acid. There are species differences in the type of bile salt secreted. The dog secretes mostly taurocholic acid, whereas the ratio of glycocholic acid to taurocholic acid in man is 2:1.

If bile salts are fed, 90 percent of the ingested bile salts is absorbed from the intestine and secreted in the bile during the next 6 to 12 hours. The remaining 10 percent is either lost in the feces or metabolized. Once bile salts enter the intestine from the biliary system, they are absorbed from the terminal ileum by an active transport process, enter into the portal circulation, and are transported to the liver, where they stimulate the secretion of more bile. They are then resecreted into the intestine, where the cycle is repeated, with only relatively small amounts of bile salts being lost in the process. This is the *enterohepatic circulation of bile salts.* During fasting, the process is interrupted when the bile salts are stored and concentrated in the gallbladder.

The major function of the bile salts is in the digestion and absorption of fats. If bile is excluded from the intestine, large quantities of fats appear in the feces. Since bile salts possess both polar and nonpolar groups, they have an affinity for both oil and water. This hydrotropic or detergent property enables the bile salts to reduce the interfacial tension between the oil and water phases in the lumen of the intestine following ingestion of a meal containing fats. The result is the formation of a fine emulsion of oil in water, which according to most workers is a necessary prerequisite for optimal digestion and absorption of fats. There is also evidence that, in addition to having an emulsifying action, the bile salts are able to solubilize certain water-insoluble fats which are present in the lumen of the intestine. This is accomplished by forming with these fats either water-soluble complexes of molecular dimensions or aggregates of micellar dimensions. The participation of the bile salts in the absorption of fats will be discussed in more detail later in this chapter.

OTHER COMPONENTS OF BILE

The bile pigments are excretory products of the liver and are derived from the metabolism of hemoglobin. The color of bile depends on the type and concentration of pigment that it contains. If the pigment is bilirubin, as in carnivorous animals, the bile is golden yellow. If the pigment is biliverdin, an oxidation product of bilirubin, the color varies from green to black; this is the color of bile secreted by herbivorous animals. The bile salts and the bile pigments may be handled by the same transport system, because these compounds compete with one another for secretion.

An important constituent of bile from a clinical standpoint is cholesterol, the chief constituent of gallstones. Absorption of fluid by the gallbladder occurs to a much greater extent than normal when there is stasis of bile in the biliary system. Gallbladder bile becomes very concentrated as a result, and cholesterol crystallizes out of solution, the crystals so formed gradually growing into large gallstones. Since bile salts maintain cholesterol in solution, a deficiency of these substances often results in the formation of gallstones. Many individuals who show a tendency to form gallstones are fed bile salts. This therapy not only helps maintain cholesterol in solution but also stimulates liver bile flow, thereby reducing the possibility that the cholesterol concentration will rise to a value great enough to form gallstones

SECRETION OF INTESTINAL JUICE

Relatively little is known about the mechanisms involved in the secretion of intestinal juice. Furthermore, experiments dealing with intestinal secretion are always complicated by the fact that absorption proceeds at the same time as secretion. Consequently, the juice obtained is the net result of both processes.

TYPES OF CELLS

The secretion of intestinal juice is the result of the secretory activities of a variety of cell types. Two types of intestinal gland contribute to this secretion: Brunner's glands, which are present in the duodenum, and the crypts of Lieberkuhn, which are scattered through the entire length of the small intestine. Brunner's glands contain only mucous cells. Columnar epithelial cells and mucous cells line the surface of the villi and extend into the crypts, which also contain argentophil cells and Paneth cells.

The combined secretions of these cells are composed of water and the usual electrolytes of plasma and contain a small amount of organic matter consisting of mucin and a host of digestive enzymes. The pH of intestinal juice ranges from 6.5 to 7.6 and increases from proximal to distal intestine.

METHOD OF STUDY

Samples of intestinal juice can be collected from an intestinal transplant. A segment of intestine is removed and transplanted to the mammary region of the abdominal wall. One end of the segment is closed, and the other is placed so that it opens onto the surface of the abdominal wall. The secretions of the transplant are collected from the fistulous opening. If a preparation devoid of extrinsic innervation is desired, the pedicle containing the original blood and nerve supplies can be cut after a new circulation has grown to the transplant.

DUODENAL SECRETION

The major contribution to the secretion of the upper duodenum is probably by the glands of Brunner. Since the volume of juice secreted by this region is low during fasting and increases only to a small extent following feeding, such secretion plays a negligible role in the neutralization of gastric contents. However, it is known that this region possesses a greater resistance to ulceration as compared to the remainder of the small intestine, probably because the mucus secreted by the upper duodenum is thick and forms a coating over the underlying epithelium. This serves as a barrier against contact of acid and pepsin with the duodenal mucosa.

CONTROL OF SECRETION

Mechanical Stimulation

The mechanisms involved in the secretion of intestinal juice are not well elucidated. Mechanical stimulation of any region of the intestine seems to be a potent stimulus for intestinal secretion. When the intestinal mucosa is rubbed or when a balloon is inflated in the gut, there is a considerable increase in the volume of secretion formed, and it might be concluded that under physiological circumstances local mechanical stimulation by chyme is important in eliciting the secretion of intestinal juice. Since the same effect can be produced in a segment of gut which does not possess extrinsic innervation, the response to mechanical stimulation is not necessarily mediated through extrinsic nerves. The secretion produced by mechanical stimulation has clinical importance, since in intestinal obstruction the secretion of intestinal juice is stimulated by the obstruction itself. Fluid accumulates above the obstruction, distends the intestine, and stimulates additional secretion. A vicious cycle develops, since, as more fluid collects, the stimulus for secretion is increased.

Neural Influences

The role of extrinsic nerves in intestinal secretion is uncertain. It has been reported that vagal stimulation increases the rate of secretion of both duodenal juice and that obtained from the rest of the intestine. Extrinsic denervation of an intestinal segment results in a copious secretion, which subsides in a day or two. This *paralytic secretion* of the intestine is also obtained when only the sympathetic nerve supply is cut. This result suggests that the sympathetic nerves send tonic inhibitory impulses to the secretory cells of the intestine. When the sympathetic nerves are severed, the inhibition is removed and the intestinal cells secrete unchecked until some readjustment of secretory activity occurs.

Humoral Influences

There is evidence that a hormone is involved in regulating the secretion of intestinal juice. When dogs with denervated jejunal transplants are fed, the transplanted segment of intestine responds with an increased secretion of fluid and enzymes. An extract of the mucosa of small intestine has been prepared which produces the same type of secretion when injected intravenously. The substance contained in this extract is called *enterocrinin,* and it is specific for the secretion of intestinal juice; i.e., it does not affect the secretion of pancreatic juice or bile. A hormonal mechanism (duocrinin) has also been postulated for stimulation of secretion of duodenal juice. This idea is based on the fact that a denervated transplant of the Brunner's gland area responds with an increased rate of secretion following feeding.

INTESTINAL DIGESTION

A variety of enzymes has been reported to be present in intestinal juice. However, there is uncertainty as to whether all these enzymes are actually secreted. It is probable that most are really endocellular enzymes and that they are present in the lumen of the intestine as the result of desquamation of the intestinal epithelium. That desquamation is a quantitatively important source of enzymes is shown by the fact that there is a complete turnover of the intestinal epithelium every one to three days. In addition, the quantity of enzymes present in juice collected from the intestine is usually related to the content of cellular debris. In any event, regardless of the source, some digestive activity is derived from the small intestine.

Intestinal juice does not contain an enzyme which is capable of digesting native protein, but it does contain a family of enzymes which completes the work initiated by pepsin, trypsin, and chymotrypsin. Included in this group is *aminopolypeptidase,* which hydrolyzes the terminal amino acid from polypeptides with terminal amino groups. There is also *dipeptidase,* which splits certain dipeptides. *Prolinase* and *prolidase* liberate proline from the end of peptide chains. Enterokinase has been mentioned in connection with the activation of trypsin, and there are probably other proteolytic enzymes not yet demonstrated, which hydrolyze specific linkages in the molecules of protein digestion products.

Intestinal juice also possesses weak lipolytic activity. The presence of an *intestinal lipase* is indicated by the fact that small quantities of free fatty acids and monoglycerides are present in the contents of the small intestine of depancreatized dogs fed triglycerides.

Intestinal juice contains *intestinal amylase,* which is similar in action to salivary and pancreatic amylases. There is also *maltase,* which splits maltose, and *lactase,* which hydrolyzes lactose. In addition, *sucrase,* sometimes called *invertase,* hydrolyzes sucrose.

Man possesses no enzyme capable of splitting certain complex polysaccharides, such as lignin and cellulose, and these substances are excreted in the feces. However, it is obvious that there are many enzymes in intestinal juice and that these aid in the digestion of food initiated by the other digestive secretions.

COLONIC SECRETION

The colonic mucosa has numerous tubular glands containing goblet cells which secrete mucus. The fluid secreted by these glands is very viscous and alkaline (pH 8.0 to 8.4). Stimulation of the parasympathetic fibers of the pelvic nerves, chemical irritation, and mechanical irritation increase the volume of colonic secretion. The function of the secretion is to lubricate and facilitate the passage of feces and to protect the mucosa of the colon from mechanical and chemical trauma. The alkaline secretion serves to neutralize irritating acids formed by bacterial action, a process which occurs to a considerable extent in the colon. When there is excessive secretion of mucus by the colon, this substance can constitute the major portion of a bowel movement. This condition is called *mucous colitis* and is thought to be caused by excessive parasympathetic activity, most often attributable to emotional tension.

INTESTINAL ABSORPTION

Absorption has been defined as the passage of substances into the circulation. The absorption of foodstuffs occurs from the gastrointestinal tract, and the process consists of the transfer of various digestion products through the barrier imposed by the semipermeable intestinal membrane into the blood and the lymph. One driving force involved in the transport of molecules across this membrane is the difference in concentration of a substance on the two sides of the membrane. The rate of transfer is dependent not only on this diffusion gradient but also on the molecular size and lipid solubility of the substance being transported. Certain chemical reactions which occur in the mucosal cells also supply a driving force for the transport of many substances across the intestinal epithelium. Energy is required for this type of transport (active), and transfer is impeded when oxidative metabolism of the mucosal cells is inhibited. Chapter 1 supplies a more detailed description of the mechanisms involved in the movement of substances across biological membranes.

Absorption from the mouth, esophagus, and stomach is not important from a nutritional standpoint. The principal site of absorption is the small intestine. Anatomically, the small intestine is well suited to this task, since the surface area available for absorption is greatly increased, not only by the presence of the villi, but also by the microvilli. That the small intestine is a very efficient absorbing organ is shown by the fact that chyme recovered from the terminal ileum contains no digestible carbohydrate, very little lipid, and only 15 to 17 percent nitrogen-containing substances. Most of this material can be accounted for as bacteria, desquamated epithelial cells, the remains of various digestive secretions, and undigested and unabsorbed residues of food, such as cellulose and connective tissue. Although the intestinal epithelium does act as a barrier to many materials, it also permits, and in many instances facilitates, the absorption of a large variety of substances. The following discussion will deal with the absorption of carbohydrates, proteins, fats, water, and electrolytes.

ABSORPTION OF CARBOHYDRATES

The major dietary carbohydrates are polysaccharides, such as starch and glycogen, and a variety of disaccharides, including maltose, lactose, and sucrose. However, it is widely accepted that practically all ingested carbohydrates are absorbed in the form of monosaccharides. This idea is supported by the observation that, following ingestion of a wide variety of carbohydrates, only monosaccharides appear in the portal blood system, which is the route of absorption for carbohydrates. The reason polysaccharides are not absorbed as such is probably twofold: (1) The intestinal epithelium is relatively impermeable to carbohydrates of high molecular weight, and (2) hydrolysis of polysaccharides by amylase proceeds so rapidly that, for the most part, only sugars of lower molecular weight are available for absorption. The major end products of amylase activity are disaccharides, and the activities of the enzymes which hydrolyze disaccharides are very low in intestinal contents. Nevertheless, it has been shown that the brush border of the mucosal cell contains all the disaccharidases that are present in these cells. They include maltase, lactase, and sucrase. Prevailing opinion is that disaccharides, such as maltose, lactose, and sucrose, are hydrolyzed to their constituent monosaccharides in the brush border of the epithelial cells during the absorptive process.

Evidence accumulated over the years demonstrates that among the naturally occurring monosaccharides, glucose and galactose are absorbed by an active, energy-requiring process, whereas most of the others are absorbed by passive processes not requiring the expenditure of metabolic energy. In 1925 Cori made observations which showed that the absorption of certain monosaccharides cannot be completely explained on the basis of diffusion. Cori measured the rates of absorption of different monosaccharides in rats, and the results are summarized in Table 26-2. Rate of

TABLE 26-2. Absorption of Monosaccharides by Rats*

Sugar	Absorption Rate
Galactose	110
Glucose	100
Fructose	43
Mannose	19
Xylose	15
Arabinose	9

*Values are expressed in relative terms with the absorption of glucose equal to 100.
Source: From C. F. Cori. *J. Biol. Chem.* 66:691, 1925.

diffusion is inversely related to molecular size, and if absorption takes place by diffusion only, the pentoses, xylose and arabinose, should be absorbed more rapidly than the other four monosaccharides, which are hexoses. However, the pentoses are absorbed least rapidly. In addition, even though all four hexoses diffuse at the same rate, there is considerable variation in their rates of absorption. It was concluded from these results that there is a special mechanism in the cells of the intestinal epithelium which increases the rate of absorption of glucose, galactose, and perhaps fructose over the absorption rate of mannose and the pentoses.

More recent work supports and extends the conclusions drawn from the older experiments. Wilson and Wiseman in 1954 used the everted intestinal sac method to study the movement of various sugars across the epithelium of isolated segments of hamster intestine. This method involves the use of small segments of intestine turned inside out, filled with fluid, and tied at both ends. The absorptive mucosal surface is exposed to solutions of known composition. Following a period of incubation, the absorption of specific substances can be measured by determining their concentrations in the mucosal and serosal solutions. Figure 26-8 shows the result

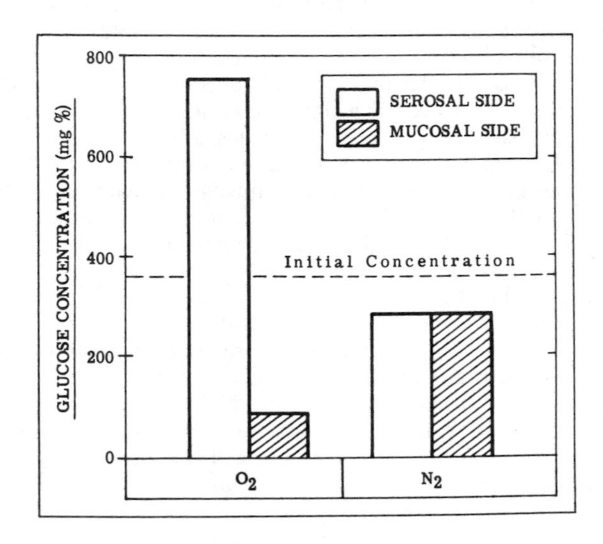

FIGURE 26-8. The transport of glucose by everted sacs of hamster jejunum under aerobic and anaerobic conditions. (From T. H. Wilson et al. *Fed. Proc.* 19:870, 1960.)

of one such experiment in which the mucosal and serosal sides of an everted sac of jejunum were simultaneously exposed to glucose solutions of the same concentration. This approach obviates any possibility of a diffusion gradient for glucose. After a period of incubation it was found that glucose moved from the mucosal to the serosal side of the sac against a large concentration difference. The same was true when galactose was used. Since under the conditions of the experiment diffusion cannot be the mechanism involved in the transport of these sugars across the intestinal membrane, some special mechanism must be available. This idea is supported by the fact that the transfer of both glucose and galactose was abolished under anaerobic conditions (Fig. 26-8), thereby demonstrating the importance of oxidative metabolism for the normal functioning of this mechanism. It can be reasoned that certain oxidative metabolic reactions supply energy to facilitate the absorption of glucose and galactose.

There was no net movement of a variety of other monosaccharides in the everted sac studies, showing that no special mechanism is available for transporting these sugars against a concentration difference. The evidence indicates that they are absorbed by passive diffusion. For example, the rates of absorption of these sugars increase in linear relation to intraluminal concentration. This is typical of a diffusion process; i.e., the greater the concentration difference between lumen and blood, the greater the rate of diffusion, and in direct proportion to the difference. On the other hand, an intestinal transport maximum can be demonstrated for glucose absorption at high intraluminar glucose concentrations. A transport saturation phenomenon such as this is typical of specialized transport systems.

The relative absorption rate of fructose (Table 26-2) is intermediate between the rates of monosaccharides which are actively absorbed and those of sugars which are absorbed by diffusion. There is an enzyme system in the intestinal mucosa which transforms fructose to glucose. In addition, in certain species the mucosal cells convert fructose to lactate. Since fructose is not transported against a concentration gradient, it is probably absorbed passively. The reason for the rapid rate of absorption of fructose relative to other passively absorbed sugars may be the maintenance between intestinal lumen and mucosal cell of a relatively high diffusion gradient as the result of partial conversion in the cell to glucose and lactic acid.

The process involved in the absorption of glucose and galactose is a clear example of what has been termed an active process; i.e., these sugars move against a concentration difference, the movement depending on certain energy-yielding reactions. The nature of the reactions and the manner in which they participate in the absorption of certain sugars are unknown. It has been suggested that glucose and galactose are phosphorylated in the intestinal epithelial cell and that this is the primary reaction in the active transport of these sugars. However, all attempts to show an alteration in the structure of these sugars during absorption have failed. Present thinking is that a membrane carrier is located on the luminal border of the epithelial cells which transports sugars across the membrane unchanged. The evidence indicates that a common pathway is involved in active sugar transport, because the presence of one actively absorbed sugar inhibits the transport of another actively absorbed sugar; there is competition for the same transport system.

A recent idea is that there is a coupling of the sodium and sugar transport systems of the intestine. It has been shown in studies in vitro that active transport of glucose ceases when the medium is devoid of sodium, and that as the sodium content is increased, absorption of glucose increases. Furthermore, cardiac glycosides which are known to inhibit sodium transport also inhibit glucose absorption. One line of thought is that sodium increases the affinity of glucose for its carrier and facilitates the subsequent transport of the sugar-carrier complex into the cell (Fig. 26-9).

An assumption is that there is a mobile carrier in the brush border of the epithelial cell and that this carrier has a binding site which has substrate specificity for actively absorbed sugars. It is of interest in this regard that all actively absorbed sugars have certain structural features in common, some of which are lacking in the passively absorbed sugars. These structural characteristics are at least six carbon atoms, a D-pyranose ring structure, and a hydroxyl group at carbon 2. It might be visualized that such a structure is required to "fit" into the binding site. In addition to a binding site for sugar, it is theorized that there is a specific binding site on the same carrier for sodium, and that the carrier affinity for sugar is greatest when sodium is bound to the carrier (Fig. 26-9B). Moreover, when the carrier is loaded with both sugar and sodium, it travels across the membrane, the driving force being the concentration difference which exists for sodium from luminal to interior surface of the brush border membrane (Fig. 26-9A). At this point the carrier arrives in an environment which is favorable for the release of sugar into the cell interior – an environment of high potassium concentration. Potassium is known to compete with sodium for the sodium binding sites, and besides, when these binding sites are occupied by potassium, the sugar binding site possesses much less affinity for sugar (Fig. 26-9B). Since potassium concentration is high inside the cell, whereas that of

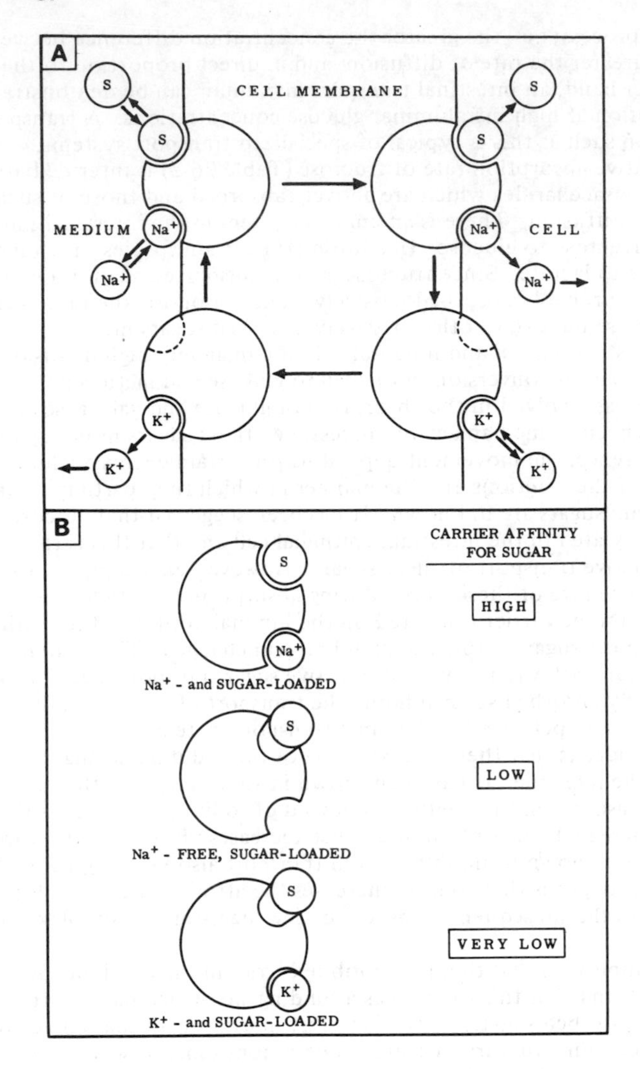

FIGURE 26-9. A. Model of mobile carrier with two sites, one specific for sugar and one specific for sodium. The influence of intracellular potassium on carrier movement is illustrated. B. Some possible affinities of the carrier for sugar. (From R. K. Crane. *Handbook of Physiology*. Section 6: Alimentary Canal, Vol. III. Washington, D.C.: American Physiological Society, 1968. Pp. 1335–1336.)

sodium is low, potassium should be able to compete successfully for the binding sites, thereby displacing sodium and causing sugar to unbind. Once potassium attaches to the sodium binding site, the carrier is driven back to the luminal surface according to the concentration difference for potassium (Fig. 26-9A). Here potassium is replaced by sodium and more sugar is transported into the cell.

However, in order to ensure continued translocation of the sodium-loaded and sugar-loaded carrier across the brush border membrane and maintain the release of sugar in the cell interior as the result of potassium binding, it is essential that a low intracellular sodium concentration be preserved. This is accomplished by an energy-dependent pump which moves sodium from the inside to the exterior of the cell. Since the sodium pump requires energy derived from metabolism for normal function, it would be here that the energy requirement for active sugar transport would be manifested. If sodium pump activity were abolished by anaerobic conditions or metabolic inhibitors, sugar could not accumulate in the cell, because the increased cellular sodium content would not only reduce the sodium gradient across the brush border membrane, thereby decreasing the driving force for movement of the carrier, but also reduce the effectiveness of potassium for competing with sodium for the binding sites.

ABSORPTION OF PROTEINS

The intestine of many newborn mammals can absorb by pinocytosis considerable quantities of intact protein molecules. Although the capacity to absorb proteins is lost shortly after birth, this process is important in a number of species because absorption of antibodies in the colostrum confers passive immunity against infection. From a nutritional standpoint, absorption of native proteins in the adult is insignificant. However, minute amounts of proteins are occasionally absorbed, as shown by the fact that some adults have allergic reactions to various food proteins. When a protein is absorbed unchanged, it sensitizes the individual to future doses of the same protein.

The gastrointestinal tract possesses a large variety of proteolytic enzymes, the combined action of which reduces the large protein molecules to their constituent amino acids. Proteins are absorbed into the portal blood system mostly as amino acids. In one study when different proteins were fed to dogs, there was a large increase in many free amino acids in the portal vein during absorption. On the other hand, higher products of protein digestion, such as peptides, were not present. It would appear that peptides, even though water-soluble and diffusible, are poorly absorbed. This is substantiated by a study in vitro reporting that when simple dipeptides such as glycylglycine and leucylglycine were put on the mucosal side of rat intestine only small amounts of these dipeptides appeared on the serosal side. Since large amounts of free glycine were present on both sides of the intestine, the inference is that the intestine is far more capable of transporting amino acids than peptides. On the other hand, there are enzymes in the brush border of the mucosal cells which hydrolyze peptides. Thus, it is possible that different peptides do enter the epithelial cells and are hydrolyzed there by specific peptidases. The amino acids formed as the result of intracellular digestion would then be transported into the portal blood.

As is the case for certain monosaccharides, many amino acids are subject to selective absorption. The absorption of amino acids has been compared with the absorption of certain compounds, such as polyhydric alcohols and acid amides, which are not normally present in the diet. Although all compounds fed were of approximately the same molecular size, the amino acids were absorbed much more rapidly than were the nonphysiological compounds. The absorption of amino acids cannot be explained solely on the basis of diffusion, since all substances fed had about the same molecular weight and therefore about the same rate of diffusion. These results suggest that the intestinal mucosa is equipped with special mechanisms which facilitate the transport of amino acids across the intestinal membrane. It also appears that the mechanism involved is selective with respect to the amino acid absorbed. Outright proof of selective transport was provided by experiments in which the absorption of mixtures of the L- and D-forms of individual amino acids was studied in loops of rat intestine. In all instances the naturally occurring L-amino acid was absorbed much more rapidly than the corresponding D-form, although the only difference between the two is their stereoconfiguration. Absolute proof for active transport of certain amino acids has been provided by the observation that they are transported across the intestinal epithelium against a concentration difference, and that this movement is prohibited by anaerobic conditions.

There is probably more than one transport system for amino acids. For example, neutral amino acids compete with one another for absorption, but their absorption is not inhibited by basic amino acids. Furthermore, basic amino acids compete with one another for absorption, and although a few neutral amino acids do inhibit the absorption of some basic amino acids, these results have suggested that there is one transport system for neutral amino acids and another having a primary affinity for basic amino acids. It is probable that there are other transport systems for other groups of amino acids, such as the dicarboxylic amino acids and the imino acids.

Although the nature of the systems which participate in the transport of amino acids is unknown, normal function of the mechanisms involved is dependent on oxidative metabolism because transport is inhibited by anaerobic conditions, cyanide, and dinitrophenol. As is the case for sugars, it has been suggested that a membrane carrier is involved in the transfer of amino acids across the intestinal epithelial cell, and there is evidence that the carrier mechanism is sodium-dependent. The amino acid molecular configuration which is required for optimal active transport includes the L-form of an amino acid, an unsubstituted carboxyl group attached to the alpha-carbon atom, an alpha-amino group, and an alpha-hydrogen atom. Although knowledge of the intestinal transport of amino acids is growing rapidly, the more intimate features of the mechanisms involved remain to be clarified.

ABSORPTION OF FATS

Triglycerides constitute the major portion of the lipids of the diet. It has been difficult to elucidate the mechanism by which triglycerides are absorbed, particularly since uncertainty has existed regarding the form in which triglyceride fat enters the epithelial cells from the intestinal lumen. In the case of carbohydrate and protein absorption, only monosaccharides and amino acids can be detected in the portal blood during absorption, and it can be deduced that these are the forms in which carbohydrates and proteins are transported across the intestinal membrane. However, not all products of fat digestion leave the intestinal epithelium unchanged. It is known that considerable quantities of fatty acids with 12 carbon atoms or less are transferred into the portal blood when either fed in the free form or incorporated into triglycerides. On the other hand, the lymph is the major route of absorption for fatty acids possessing 14 or more carbon atoms. Because the triglycerides of the diet are composed chiefly of 16 and 18 carbon atom fatty acids, most of the absorbed fat is transported into the lymph. Since analysis of lymphatic fat of dietary origin shows it to be almost completely in the form of triglycerides, and since pancreatic lipase hydrolyzes ingested triglycerides to varying degrees in the lumen of the intestine, the absorbed digestion products must be resynthesized to triglycerides in the cells of the intestinal epithelium. The resynthesis makes difficult any attempt to determine by analysis of lymphatic fat the form in which triglycerides leave the intestinal lumen to enter the absorptive epithelium.

One problem has been to determine the extent to which triglycerides are hydrolyzed in the intestine prior to absorption. Lipid recovered from intestinal contents of animals fed triglycerides contains diglycerides, monoglycerides, free fatty acids, and unhydrolyzed triglycerides. The material absorbed could be one of these four lipids, all of them, or some particular combination of the four.

Studies in vitro of the pancreatic lipase hydrolysis of triglycerides have shown that this enzyme preferentially catalyzes splitting of the 1- and 3-ester bonds of triglycerides with the result that there is the consecutive formation of 1,2-diglyceride and 2-monoglyceride along with the liberation of fatty acids. Since lipase attacks the 2-ester linkage with difficulty, 2-monoglyceride tends to accumulate in the system. If these results can be extended to the intact animal, the inference is that triglycerides may not be completely hydrolyzed to glycerol and free fatty acids during the relatively short time required for fats to traverse the length of the small intestine. Since practically all ingested fats are absorbed during transit through the small intestine, a portion would have to be absorbed in the form of glyceride as well as fatty acid. Since hydrolysis to the monoglyceride stage occurs with relative ease, it might be anticipated that a large part of the absorbed fat is in the form of free fatty acid and monoglyceride.

If it is accepted that both glycerides and free fatty acids are absorbed, the problem remains regarding the mechanism of entrance of these substances into the epithelial cells. Long-chain fatty acids are the most common type of dietary fatty acid, and these, as well as the triglycerides, diglycerides, and monoglycerides which contain them, are water-insoluble. Without the availability of some special mechanism, fat would exist as large globules in the aqueous medium which constitutes intestinal chyme, and optimal absorption of fat could not occur. It is known that lipid-soluble substances, by virtue of their solubility in the lipid portion of cell membranes, are able to cross the membranes. But these substances must be present in some form which assures ready entrance into the cell; i.e., in order to gain access to the surface of a cell, lipids must be in the form, for example, of a molecular dispersion, micelles, or a very fine emulsion.

Two theories are in vogue regarding the mechanism of entrance of fats into the epithelial cells. The *particulate theory* states that fats are absorbed in the form of a finely dispersed emulsion. This requires that particles be able to pass through the intestinal membrane, and there is evidence that the intestinal epithelium is capable of absorbing very small particles. In support of this theory are studies in vitro showing that the addition of relatively small amounts of free fatty acids, monoglycerides, and bile salts to triglycerides contained in a medium which resembles that in the intestine results in the formation of an emulsion having characteristics similar to those found in the intestine following the ingestion of triglycerides. This triple combination of free fatty acids, monoglycerides, and bile salts lowers the surface tension between the oil-water interface to such an extent that an extremely stable emulsion consisting of particles about 5000 A in diameter is formed. This result suggests that the emulsion of intestinal chyme may be formed in the same manner. The mechanism by which particles pass through the intestinal membrane is not clear. One suggestion which has some experimental basis is that the epithelial cells engulf the particles by the process of pinocytosis. Electron microscope studies have demonstrated the occurrence of membrane invaginations and intracellular vesicles during fat absorption. Presumably a fat particle on contacting the membrane is engulfed in one of these pinocytotic vesicles, which then moves rapidly into the supranuclear region of the cell. It has not yet been possible to quantitatively determine the extent of participation of this mechanism in the absorption of fats.

A more recent and widely accepted idea, the *micellar theory* proposed by Borgstrom, is that fats are absorbed as monoglyceride and free fatty acid, these substances being prepared for absorption by forming micelles with bile salts. A micelle is an aggregate of molecules, the dimensions of which (50 A) more closely approach the state of molecular dispersion than does a fine emulsion. Thus the micellar theory allows the penetration of fat into the epithelial cell to be explained on the basis of diffusion rather than by resorting to a rather specialized process such as pinocytosis. Evidence in support of this concept is that monoglycerides and free fatty acids do form micelles in vitro when mixed with bile salts. Furthermore, such a micellar phase is present in intestinal contents. If intestinal contents of a human fed triglycerides are centrifuged at high speed, an oily top phase and a clear bottom phase are obtained. The oil phase contains mostly triglycerides and diglycerides with some monoglycerides and free fatty acids, whereas the clear micellar phase contains mostly monoglycerides and free fatty acids.

The physical state of intestinal intraluminar fat is represented diagrammatically in Figure 26-10. When certain aspects of the particulate theory are combined with

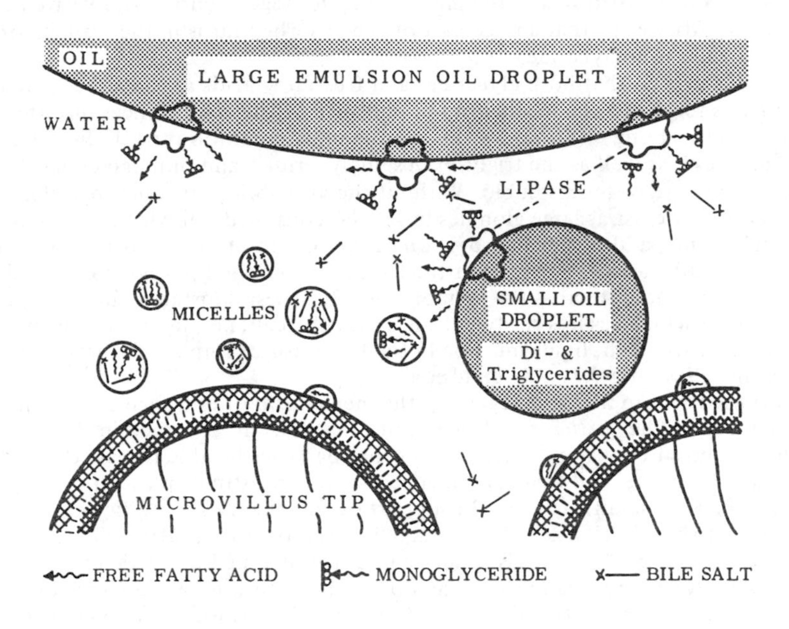

FIGURE 26-10. The physical state of intestinal intraluminal lipids during fat absorption. (From Senior, 1964.)

the micellar theory, a reasonable working hypothesis can be constructed for the mechanism of transport of fats from the lumen of the intestine into the epithelial cell. The situation that might be visualized is that when triglycerides enter the duodenum, pancreatic lipolysis is initiated, free fatty acids and monoglycerides are formed, and the presence of these digestion products along with the bile salts results in the formation of a particulate emulsion. However, since free fatty acids, monoglycerides, and bile salts possess the capacity to form micelles, a micellar phase also results. Since the micellar form of lipid more closely approaches molecular dimensions than does the emulsified fat, the bulk of the intraluminal fat would be absorbed as micelles. The driving force for movement of micelles could be a concentration difference between lumen and epithelial cell, and free fatty acids and monoglycerides would diffuse through the lipid portion of the cell membrane. Furthermore, there would be an equilibrium between the emulsion and micellar phases in that as the micellar phase is depleted by absorption, there would be continual replenishment of the micellar phase because lipolysis of the triglycerides and diglycerides of the particulate fat continues with the resultant production of more free fatty acids and monoglyceride. It is of interest that the bile salts, which are required for micelle formation, are not absorbed until they reach the terminal ileum. This implies that they perform their fat-dispersing function quite efficiently throughout the entire length of the small intestine in that a minimal amount of bile salts is necessary to prepare for absorption a relatively large quantity of fat.

Two pathways have been defined for the intracellular synthesis of triglycerides from absorbed digestion products. One involves the conversion of fatty acid to triglyceride and the other, monoglyceride to triglyceride. This is in accord with the concept that triglycerides are absorbed as micelles of free fatty acids and monoglycerides. The triglycerides delivered into the lymph are aggregated into droplets called *chylomicrons.* Chylomicrons have diameters ranging from 0.05 to 1.0 μ and are stabilized by an enclosure in a layer of phospholipid and protein.

Triglycerides containing fatty acids with chain lengths less than 12 carbon atoms are both hydrolyzed and absorbed more rapidly than the longer-chain triglycerides. Thus, there is less interference in the absorption of the shorter-chain triglycerides under conditions of decreased pancreatic lipase and bile salt levels in the intestinal lumen, and it has been considered to be of value to feed medium-chain triglycerides (8 to 10 carbon atoms in length) to patients with either pancreatic insufficiency or biliary obstruction. Fatty acids with less than 12 carbon atoms are absorbed into the portal blood and without being synthesized to triglycerides. This situation may be due to a low activity of the esterifying enzymes of the intestinal mucosa toward these fatty acids, which in turn may be related to the fact that the shorter-chain fatty acids are water-soluble.

ABSORPTION OF WATER AND ELECTROLYTES

Water and electrolytes, which are present in the digestive tract as the result of ingestion and secretion of these substances, are absorbed primarily from the small intestine and to a much lesser extent from the colon. About 1.5 liters of fluid is ingested every day by man and another 5 to 10 liters is secreted by the various regions of the gastrointestinal tract. Obviously any malfunction of the water and electrolyte absorptive mechanisms of the intestine can quickly result in serious depletion of body water and salt.

Although it has been suggested that water can be actively absorbed from the small intestine, most workers feel that the absorption of water occurs by the simple physical process of osmosis. When solutions of different tonicities are introduced into the lumen of the small intestine, there is a tendency to adjust the osmotic activity of these solutions to isotonicity; i.e., water is rapidly absorbed from hypotonic solutions, and hypertonic solutions are diluted by the entrance of water into the gut. These results suggest that the movement of water across the intestinal epithelium depends on the solute concentration gradient. Chyme probably becomes hypotonic during the absorption of dissolved materials, and water is then driven across the intestinal membrane by the concentration gradient of water which results from the absorption of solute.

Monovalent ions, such as sodium, potassium, chloride, and bicarbonate, are much more readily absorbed than are the polyvalent ions, calcium, magnesium, and sulfate. Studies employing the short-circuit technique (Chap. 1) have shown that sodium can be absorbed against an electrochemical gradient, and the present concept is that the intestinal epithelium possesses a mechanism for the active transport of sodium. Chloride also may be actively absorbed, but it is possible that the electrical potential created by the movement of sodium across the intestinal membrane is responsible for the simultaneous movement of chloride ions.

That the intestine possesses special mechanisms for the transport of certain divalent cations is demonstrated by the observations that both calcium and iron are transferred from mucosa to serosa against a concentration difference, and that such movement is abolished by the procedures which interfere with oxidative metabolism. Proteins and lactose increase the absorption of calcium, whereas carbonate and phosphate ions, especially in alkaline solution, tend to form insoluble salts with calcium and inhibit the absorption of this ion. Insoluble calcium soaps are formed in the presence of large quantities of fatty acids and are excreted in the feces, thereby decreasing the absorption of calcium. A deficiency of dietary vitamin D leads to impaired calcium absorption, and the unabsorbed calcium forms insoluble salts with phosphorus. Since these salts are not readily absorbed, the absorption of phosphorus is also decreased under these conditions.

Phosphorus can be rapidly absorbed under normal circumstances, as shown by the fact that radioactive inorganic phosphorus appears in the circulation within five minutes after introduction into the small intestine.

Iron is absorbed only when the body becomes deficient in this substance. Ferrous iron of the mucosal cells is in equilibrium with circulating iron and with ferritin, a protein-iron complex in the mucosal cells. When the iron of the blood is decreased below normal values, this element is liberated from the mucosal stores of ferritin-iron, and more iron is absorbed from the gut in order to replenish these stores.

REFERENCES

Burgen, A. S. V., and N. G. Emmelin. *Physiology of the Salivary Glands.* Baltimore: Williams & Wilkins, 1961.

Code, C. F. (Ed.). *Handbook of Physiology.* Section 6: Alimentary Canal, Vol. II, Secretion. Washington, D.C.: American Physiological Society, 1967.

Code, C. F. (Ed.). *Handbook of Physiology.* Section 6: Alimentary Canal, Vol. III, Intestinal Absorption. Washington, D.C.: American Physiological Society, 1968.

Crane, R. K. Intestinal absorption of sugars. *Physiol. Rev.* 40:789–825, 1960.

Davenport, H. W. *Physiology of the Digestive Tract.* 2d ed. Chicago: Year Book, 1966.

Florey, H. W., R. D. Wright, and M. A. Jennings. The secretions of the intestine. *Physiol. Rev.* 21:36–69, 1941.

Gregory, R. A. *Secretory Mechanisms of the Gastro-intestinal Tract.* London: Arnold, 1962.

Grossman, M. I. Gastrointestinal hormones. *Physiol. Rev.* 30:33–82, 1950.

Grossman, M. I. (Ed.). *Gastrin.* Berkeley: University of California Press, 1966.

Hunt, J. N. Gastric emptying and secretion in man. *Physiol. Rev.* 39:491–530, 1959.

James, A. H. *The Physiology of Gastric Digestion.* Monographs of the Physiological Society. London: Arnold, 1957.

Senior, J. R. Intestinal absorption of fats. *J. Lipid Res.* 5:495–521, 1964.

Taylor, W. (Ed.). *The Biliary System; A Symposium of the NATO Advanced Study Institute.* Oxford, Eng.: Blackwell, 1965.

Thomas, J. E. *The External Secretion of the Pancreas.* Springfield, Ill.: Thomas, 1950.

Wilson, T. H. *Intestinal Absorption.* Philadelphia: Saunders, 1962.

27

Energy Metabolism

Leon K. Knoebel

METABOLISM is the sum of all transformations of both matter and energy which occur in biological systems. By virtue of metabolism, cells are endowed with the capacity to grow, reproduce, contract, conduct, secrete, and absorb. Thus, metabolism is the basis for all physiological phenomena that one can observe or measure. Transformations of matter concern the chemical reactions which occur in the body, and these fall mainly in the province of the biochemist. However, transformations of matter are accompanied by transformations of energy. The following discussion will be confined to this phase of metabolism, energy metabolism.

TRANSFORMATIONS OF ENERGY

A diagrammatic scheme of biological transformations of energy is given in Figure 27-1, and it is suggested that the reader refer to this illustration during the subsequent discussion. Four major forms of energy are encountered in the living organism. These are chemical, mechanical, electrical, and thermal energies. The cells of the body are able to use energy from one source only: the chemical energy which is liberated by chemical reactions. Chemical energy of the body can be transformed into mechanical, electrical, and thermal energies, but these transformations are irreversible.

The source of chemical energy used by cells to effect transformations of energy is the *metabolic pool,* which can be thought of as existing in the fluids of the body. A variety of substances are present in the pool, and they can be readily used by the cells as a source of energy. The pool, in turn, is supplied with energy from two sources. A direct source of chemical energy for the metabolic pool is chemical energy stored in tissue cells. There is a free interchange (dynamic state) between the cells of the body and the metabolic pool of certain energy-containing substances derived from carbohydrates, lipids, and proteins. The energy stores of certain cells can be mobilized and utilized as a source of energy by other cells of the body. However, if the energy status of an animal is to be maintained at normal levels, the chemical energy of the cells must be replenished. Cellular energy is replaced by chemical energy of ingested food. Since the absorption products of ingested foods are similar to the materials in the metabolic pool, the body can make free use of these substances, either for replenishing tissue energy reserves or for transformations to other forms of energy.

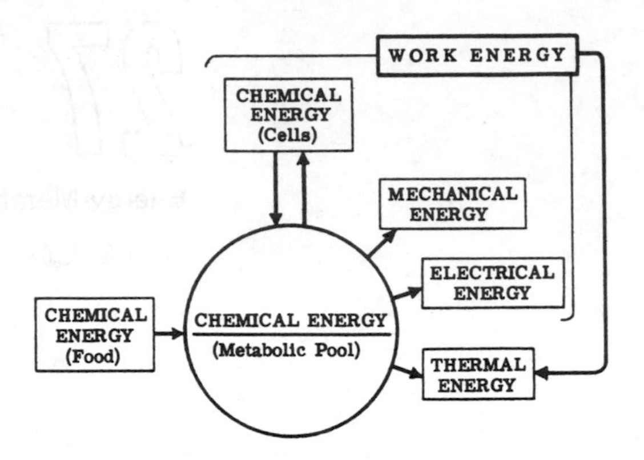

FIGURE 27-1. Diagram of biological transformations of energy.

Chemical energy is utilized for the purpose of doing work. Energy supplied in the form of work energy enables various types of cells to maintain the processes of life. In the category of work energy are three forms of energy mentioned previously: mechanical, electrical, and chemical energies. The conversion of chemical energy to work in the form of mechanical energy is exemplified by the muscle which shortens and lifts a load. The transformation of chemical energy to work in the form of electrical energy occurs during the transmission of the nerve impulse and the electrical activity of the various types of muscle. Chemical energy can also be used to do chemical work by supplying the energy necessary for synthetic reactions. When complex molecules are synthesized from simple molecules, energy is stored and chemical work is done. However, in order to store energy in a large molecule, chemical energy must be supplied by the breakdown of other molecules; i.e., chemical work is done at the expense of chemical energy. Two terms have been widely used in the discussion of this aspect of metabolism. These are *anabolism* and *catabolism*. Anabolism means the synthesis of complex molecules from simple molecules, a process which is accompanied by the storage of chemical energy. Catabolism designates those reactions which involve the breakdown of complex molecules to simple molecules, with the release of chemical energy. It is difficult to dissociate the two processes, since anabolic reactions are accompanied by catabolic reactions.

The efficiency of the body in converting chemical energy to work energy is in the neighborhood of only 20 percent. A large part of the chemical energy which is expended, therefore, must be converted to a form of energy other than work, and it appears as thermal energy or heat. In addition to the thermal energy which is directly liberated by chemical reactions, work energy is also ultimately converted to heat. For example, the heart does mechanical work in pumping blood. However, the work energy of the heart is converted to heat in overcoming friction as blood passes through the circulatory system. Other examples can be cited for other functions. Thus the chemical energy of the body ultimately appears as heat, generated either directly from chemical reactions or indirectly from work energy.

Thermal energy cannot be used for doing work, because cells have no mechanism available for this purpose. Heat derived from metabolic processes may be used to maintain the body temperature at a level which is optimal for the enzymatically regulated reactions occurring in the body. However, much of the time more heat is generated than can be used for this purpose, and the body has the problem of getting rid of excess heat (Chap. 28).

ENERGY BALANCE

According to the first law of thermodynamics, energy can be neither created nor destroyed. This law can be applied to living systems in that studies can be made of energy balance as well as of energy transformations. The total amount of energy which is taken in by the body must be accounted for by the energy put out by the body. The relationship between the factors involved in the balance between input and output of energy is given in the following equation:

$$\text{ENERGY INPUT} = \text{ENERGY OUTPUT} \quad (1)$$
$$\text{Chemical Energy of Food} = \text{Heat Energy} + \text{Work Energy} \pm \text{Stored Chemical Energy}$$

All energy of the body is ultimately derived from one source, the chemical energy of food. The major output of the body is in the form of heat, and this is produced under all circumstances. The extent of heat production varies with the conditions prevailing at the time of measurement. For example, heat production depends on whether an individual is resting or working, is in the postabsorptive state or is ingesting food; it is also influenced by the type of food ingested and the environmental temperature. In addition to releasing energy as heat, the body puts energy out in the form of work. When the amount of energy ingested as food is sufficient to balance the amount of energy put out in the form of heat plus work, the chemical energy of the body remains constant. However, this situation is the exception rather than the rule. For example, during growth the chemical energy of the body increases markedly over a period of years. The same is true when an adult becomes obese. On the other hand, the chemical energy of the body decreases under conditions of nutritional insufficiency, the most marked effect occurring in starvation. Furthermore, there are daily fluctuations in the chemical energy stores of normal individuals, and these are reflected by changes in body weight over the period of a day.

It is obvious from these considerations that another factor, stored chemical energy, must be included in the energy balance equation in order to provide for changes in the chemical energy of the body. If intake of food energy is greater than the energy put out as heat and work, the body stores of energy increase and storage energy is positive in this equation. If more energy is released in the form of heat and work than is ingested, the body stores of energy are depleted and storage energy is negative in order to balance the equation.

To study the energy balance of a person under the conditions normally encountered in life, three of the variables present in the energy balance equation must be measured so that the fourth can be determined by difference. Although all four variables can be measured, it is not always convenient to measure them. This is especially the case in the clinic, where the facilities are usually inadequate and time precludes such an approach. However, a measurement of energy balance can be simplified by eliminating some of the variables. For example, the energy balance of a subject in the postabsorptive state can be determined, thereby excluding chemical energy derived from food. Moreover, when voluntary movement is restricted, energy in the form of work can be disregarded. With these two variables eliminated, the equation representing energy balance is simplified:

$$-\text{Stored Chemical Energy} = \text{Heat Energy} \quad (2)$$

This equation simply states that a person in the resting, fasting condition uses a certain amount of stored chemical energy (as shown by the minus sign) with the resultant production of a certain amount of heat. Under these conditions, the chemical energy which is utilized from the metabolic pool is completely converted to heat. Furthermore, since the subject is at rest and postabsorptive *(in the basal state),* the chemical energy is used solely for the purpose of maintaining the vital activities of the body; i.e., for the maintenance of heart action, respiration, etc. Only two variables are concerned in the energy balance of the body under these circumstances. If one can be determined, the other is known, since they are equal. Both the heat production and the loss of stored chemical energy can be measured, and a few of the methods used for these determinations will be considered in the following sections.

ENERGY UNITS

The unit of energy which has been most commonly used in the study of energy metabolism and which will be employed in the following discussion is the kilocalorie (Kcal). In terms of heat, 1 kilocalorie is the amount of energy required to raise the temperature of 1 kg of water 1°C. One kilocalorie is equal to 1000 calories (cal). It has recently been suggested that the joule (J) be used as the measure of energy, and for purposes of conversion, one kilocalorie is equal to 4187 joules. Rate of energy conversion (power) has been expressed as kilocalories per hour, but this can also be stated in terms of watts (joules per second) by multiplying kilocalories per hour by 1.16.

DIRECT CALORIMETRY

The heat production of man and experimental animals can be determined by *direct calorimetry.* The subject is placed in an insulated chamber (calorimeter) through which cool water is circulated. The rate at which the water flows through the chamber is adjusted so that the temperature of the calorimeter is maintained constant. In addition, a meter measures the volume of water which flows through the chamber in a given time, and the temperature of the water entering and leaving the chamber is determined. Thus, a knowledge of the volume of water passing through the chamber and the rise in temperature of the water enables one to calculate in terms of kilocalories the amount of heat transmitted from the subject to the water.

Although this accounts under most circumstances for most of the heat produced by the subject, another avenue of heat loss from the body is represented by the water which leaves the body from the skin and the respiratory membranes in the form of water vapor. For every gram of water vaporized, 0.58 Kcal is lost from the body as heat (heat of vaporization). Water vapor produced by the subject is collected in a suitable chemical absorbent and determined by weight. The amount of heat dissipated by evaporation is calculated by multiplying the weight of water vapor by the heat of vaporization. The rate at which heat is lost by evaporation plus the amount of heat absorbed per unit time by the circulating water represents the rate at which the subject produces heat. This method for the assessment of energy exchange is tedious and difficult to perform; although it was used extensively in the past, simpler techniques have now replaced it.

INDIRECT CALORIMETRY

The chemical energy utilized from the body stores is measured by *indirect calorimetry.* By measuring oxygen consumption, carbon dioxide production, and urinary nitrogen excretion, and by knowing certain experimentally predetermined factors, it is possible to determine accurately the type and amount of substances utilized in the metabolic pool. Furthermore, as a measure of metabolic rate, the total amount of heat produced per unit time by the oxidation of these substances can be calculated; i.e., an indirect measurement of heat production is made.

CALORIC VALUE OF FOODS

Both carbohydrates and fats are completely oxidized to carbon dioxide and water in the body, and the same is true when these substances are combusted in vitro. Since the initial reactants and the final products are the same in both instances, the amounts of energy released as heat in vivo and in vitro are identical. Consequently, it is a relatively simple matter to determine the caloric value of carbohydrates and fats in the body by measuring the heat of combustion of these materials in a bomb calorimeter. A bomb calorimeter is a metal chamber in which is placed a weighed amount of substance. Contact is made between the substance and an iron wire, the chamber is sealed, and oxygen is introduced into the chamber under high pressure. The calorimeter is immersed in a water bath of known volume and temperature, and when an electric current is sent through the wire the material is ignited and complete combustion occurs. Depending on the substance combusted, a certain amount of heat is liberated. The heat of combustion of the substance (kilocalories per gram) can be calculated from the volume of water in the bath and the rise in temperature of the water.

The heat of combustion of various carbohydrates differs. That of glucose is 3.7 Kcal per gram, whereas that of starch is 4.2 Kcal per gram. However, an average value for carbohydrate is generally taken as 4.1 Kcal per gram. The energy content of fat, which is more than double that of carbohydrate, varies, depending on the constituent fatty acids, but an average value of 9.3 Kcal per gram is most often used.

Protein differs from carbohydrate and fat in that it is not completely oxidized in the body. Certain end products of protein metabolism appear in the excreta, for the most part in the urine as urea. Urea contains chemical energy, and if it were ignited in a bomb calorimeter, heat would be liberated. This means that the amount of heat which is produced by the oxidation in vivo of 1 gm of protein is less than that which is produced when the same amount of protein is combusted in a bomb calorimeter (5.6 Kcal per gram). By taking into account the energy lost in the excreta, it has been calculated that the caloric value of protein in vivo is 4.3 Kcal per gram.

CALORIC VALUE OF OXYGEN

When indirect calorimetry is used, it is necessary to relate the caloric values of carbohydrates, fats, and proteins to the volumes of oxygen consumed during the oxidation of each of these substances in the body – i.e., to determine how many kilocalories are produced when 1 liter of oxygen is used to oxidize each type of substance. Carbohydrate and fat react with definite amounts of oxygen to produce certain quantities of carbon dioxide and water. Using glucose (molecular weight 180) as representative of the oxidation of carbohydrate, the following equation can be written:

$$C_6H_{12}O_6 + 6\,O_2 \longrightarrow 6\,CO_2 + 6\,H_2O \qquad (3)$$

In this instance, 180 gm of glucose reacts with 6 M of oxygen or 134.4 liters of oxygen (6 M oxygen × 22.4 liters oxygen/M oxygen). The same volume (134.4

liters) of carbon dioxide is formed. One gram of glucose reacts with 0.75 liter of oxygen (134.4 liters oxygen ÷ 180 gm glucose) to produce 0.75 liter of carbon dioxide. Since 3.7 Kcal is liberated when 1 gm of glucose is oxidized, it follows that the caloric value of oxygen is 5 Kcal per liter of oxygen (3.7 Kcal/gm ÷ 0.75 liter oxygen/gm).

A similar calculation can be made for a typical fat. The following equation represents the oxidation of a triglyceride (molecular weight 860), the fatty acids of which are oleic, palmitic, and stearic acids:

$$C_{55}H_{104}O_6 + 78\ O_2 \longrightarrow 55\ CO_2 + 52\ H_2O \qquad (4)$$

In this case, 860 gm of fat combines with 1747 liters of oxygen to form 1232 liters of carbon dioxide. It can be calculated that 1 gm of fat combines with 2.03 liters of oxygen to form 1.43 liters of carbon dioxide. Since 9.3 Kcal is liberated by the oxidation of 1 gm of fat, the caloric value of oxygen is 4.7 Kcal per liter of oxygen (9.3 Kcal/gm ÷ 2.03 liters oxygen/gm).

Determining the caloric value of oxygen is more difficult for proteins, since these substances are structurally complex and are incompletely oxidized in the body. However, by using the empirical formula for a typical protein and by taking into account the amount of urea that would be formed by the oxidation of this protein, the volumes of oxygen consumed and carbon dioxide produced during the oxidation of 1 gm of protein have been calculated. These values are 0.97 liter of oxygen per gram and 0.78 liter of carbon dioxide per gram. Since 4.3 Kcal is released when 1 gm of protein is oxidized, 1 liter of oxygen is equivalent to 4.5 Kcal (4.3 Kcal/gm ÷ 0.97 liter oxygen/gm). These data are summarized in Table 27-1.

TABLE 27-1. Metabolic Values for Carbohydrates, Fats, and Proteins

Unit of Measurement	Carbohydrates	Fats	Proteins
Kilocalories per gram	4.1	9.3	4.3
Liters of CO_2 per gram	0.75	1.43	0.78
Liters of O_2 per gram	0.75	2.03	0.97
Respiratory quotient	1.00	0.70	0.80
Kilocalories per liter of O_2	5.0	4.7	4.5

RESPIRATORY QUOTIENT

It is obvious that there are variations in the volumes of oxygen consumed and carbon dioxide produced by the individual oxidations of 1 gram of carbohydrates, fats, or proteins. Furthermore, the ratio of the volume of carbon dioxide produced to the volume of oxygen consumed during the oxidation of each type of food also varies (Table 27-1). This ratio (vol. CO_2 ÷ vol. O_2) is called the *respiratory quotient* (RQ). A calculation of the RQ from the rate of oxygen consumption and carbon dioxide production of a subject furnishes qualitative information regarding the substances utilized in the metabolic pool. If only carbohydrate is oxidized, an RQ of 1.00 is obtained, whereas the specific oxidation of fat results in an RQ of 0.70. Values which fall between these extremes represent the oxidation of various mixtures of carbohydrates, fats, and proteins. This will be discussed in more detail in the following section. The RQ of a subject on an ordinary mixed diet is about 0.85, and that of a fasting subject is about 0.82.

The respiratory quotient is based solely on the rate of exchange of the respiratory gases, since it is calculated from measurements of oxygen consumption and carbon dioxide production. Consequently, this ratio can be affected by factors other than oxidative metabolism. For example, much carbon dioxide is excreted during hyperventilation. This tends to increase the RQ, and values as high as 1.5 to 1.7 are obtained during severe exercise. On the other hand, carbon dioxide is retained during hypoventilation, and the RQ may fall below 0.70. An RQ obtained under these circumstances simply reflects the state of the respiration and supplies no information about the composition of the metabolic mixture being oxidized. Large amounts of carbon dioxide are also excreted in conditions of acidosis, and the RQ is high, whereas the opposite effect is obtained during alkalosis. In addition, when carbohydrates are transformed to fats in the body, as in the fattening of a farm animal, the RQ can reach 1.4. Carbohydrates are oxygen-rich substances, and fats are relatively deficient in oxygen. Oxygen is liberated from carbohydrates during the conversion to fats, and less oxygen is required from the external environment for oxidative metabolism. The body uses more oxygen than can be accounted for by a measurement of oxygen consumption, and the RQ so obtained is not a true index of oxidative metabolism. Conversely, when fats are converted to carbohydrates, as in hibernating animals, the RQ can fall below 0.70. Enough oxygen must be consumed not only to carry out oxidative processes but also to facilitate the conversion of oxygen-poor fats to oxygen-rich carbohydrates.

It is apparent that an RQ determined under these circumstances is not a reliable measure of oxidative metabolism. A better term to apply to this ratio is *respiratory exchange ratio* (Chap. 29). This has the advantage that the ratio need not always be considered to represent an index of oxidative metabolism; i.e., it is a ratio based simply on the rate of exchange of the respiratory gases. However, if the measurements on which the ratio is based are made under well-controlled conditions, the respiratory exchange ratio can be used to evaluate oxidative processes.

CALCULATION OF METABOLIC RATE

It is possible to determine quantitatively the type and amount of substances oxidized in the metabolic pool when oxygen consumption, carbon dioxide production, and urinary nitrogen excretion are measured and when the data given in Table 27-1 are available. Furthermore, the energy liberated as heat by the individual oxidations of carbohydrates, fats, and proteins can be calculated, and from this the total amount of heat produced per unit time can be used as a measure of metabolic rate. The principles of indirect calorimetry can be exemplified by considering a resting human subject in the postabsorptive state on whom the following measurements have been made: (1) urinary nitrogen excretion, 0.5 gm per hour; (2) oxygen consumption, 16.0 liters per hour; and (3) carbon dioxide production, 13.5 liters per hour.

Practically all nitrogen which is derived from the oxidation of protein is excreted in the urine. Since the amount of nitrogen contained in a typical protein molecule represents, on the average, 16 percent of the total weight of a protein molecule, the weight of protein oxidized per unit time is determined by multiplying the urinary nitrogen excretion by a factor of 6.25:

$$0.5 \text{ gm N/hour} \times 6.25 \text{ gm protein/gm N} = 3.1 \text{ gm protein/hour} \quad (5)$$

Since the oxidation of 1 gm of protein produced 4.3 Kcal,

$$3.1 \text{ gm protein} \times 4.3 \text{ Kcal/gm protein} = 13.4 \text{ Kcal/hour} \quad (6)$$

The quantity of heat produced by the oxidation of carbohydrate and fat is determined from the oxygen consumption and carbon dioxide production. The total volumes of oxygen consumed and carbon dioxide produced during the oxidation of all three foods are measured. However, it is necessary to determine the amounts of these gases which are involved only in the metabolism of carbohydrates and fats. This necessitates knowing the volumes of oxygen and carbon dioxide involved in the oxidation of protein, and they are calculated in the following manner:

$$3.1 \text{ gm protein/hour} \times 0.97 \text{ liter } O_2\text{/gm protein} = 3.0 \text{ liter } O_2\text{/hour} \quad (7a)$$
$$3.1 \text{ gm protein/hour} \times 0.78 \text{ liter } CO_2\text{/gm protein} = 2.4 \text{ liter } CO_2\text{/hour} \quad (7b)$$

The total respiratory exchange is then corrected for the amounts of oxygen consumed and carbon dioxide produced by the oxidation of protein:

$$16.0 \text{ liter } O_2\text{/hour} - 3.0 \text{ liter } O_2\text{/hour} = 13.0 \text{ liter } O_2\text{/hour} \quad (8a)$$
$$13.5 \text{ liter } CO_2\text{/hour} - 2.4 \text{ liter } CO_2\text{/hour} = 11.1 \text{ liter } CO_2\text{/hour} \quad (8b)$$

The ratio of the volume of carbon dioxide produced to the volume of oxygen consumed during the oxidation of mixtures of carbohydrates and fats is called the *nonprotein RQ.* This is calculated:

$$11.1 \text{ liter } CO_2\text{/hour} \div 13.0 \text{ liter } O_2\text{/hour} = 0.85 \quad (9)$$

Recall that when pure carbohydrate is oxidized in the body the RQ is 1.00, and when only fat is oxidized the RQ is 0.70. The nonprotein RQ determined in this experiment shows that a mixture of carbohydrate and fat is oxidized. The caloric value of oxygen for carbohydrate is 5.0 Kcal per liter of oxygen and for fat is 4.7 Kcal per liter of oxygen. Every mixture of carbohydrate and fat between the extremes of pure carbohydrate and pure fat has a specific nonprotein RQ and caloric value of oxygen. These relationships have been determined and are given in Table 27-2. At the nonprotein RQ (0.85) determined for the subject under consideration, the consumption of 1 liter of oxygen results in the production of 4.862 Kcal. Consequently, the heat produced by the oxidation of this particular mixture of carbohydrate and fat is

$$13.0 \text{ liter } O_2\text{/hour} \times 4.862 \text{ Kcal/liter } O_2 = 63.2 \text{ Kcal/hour} \quad (10)$$

The amount of heat liberated by the oxidation either of carbohydrate or of fat can also be calculated from data supplied in Table 27-2. At a nonprotein RQ of 0.85, carbohydrate supplies 51 percent and fat 49 percent of the heat produced during the oxidation of this mixture of carbohydrate and fat.

$$0.51 \times 63.2 \text{ Kcal/hour} = 32.2 \text{ Kcal/hour (from carbohydrates)} \quad (11a)$$
$$0.49 \times 63.2 \text{ Kcal/hour} = 31.0 \text{ Kcal/hour (from fats)} \quad (11b)$$

The amounts of carbohydrates and fats that are oxidized can also be calculated from these data:

$$32.2 \text{ Kcal/hour} \div 4.1 \text{ Kcal/gm carbohydrate} = 7.9 \text{ gm carbohydrate/hour} \quad (12a)$$
$$31.0 \text{ Kcal/hour} \div 9.3 \text{ Kcal/gm fat} = 3.3 \text{ gm fat/hour} \quad (12b)$$

It is apparent that less than half as much fat as carbohydrate need be oxidized to supply approximately the same amount of heat.

TABLE 27-2. Caloric Value of a Liter of Oxygen at Various Nonprotein Respiratory Quotients

Nonprotein Respiratory Quotient	Kilocalories per Liter Oxygen	Kilocalories Derived from	
		Carbohydrate (percent)	Fat (percent)
0.70	4.686	0	100.0
0.71	4.690	1.10	98.9
0.72	4.702	4.76	95.2
0.73	4.714	8.40	91.6
0.74	4.727	12.0	88.0
0.75	4.739	15.6	84.4
0.76	4.751	19.2	80.8
0.77	4.764	22.8	77.2
0.78	4.776	26.3	73.7
0.79	4.788	29.9	70.1
0.80	4.801	33.4	66.6
0.81	4.813	36.9	63.1
0.82	4.825	40.3	59.7
0.83	4.838	43.8	56.2
0.84	4.850	47.2	52.8
0.85	4.862	50.7	49.3
0.86	4.875	54.1	45.9
0.87	4.887	57.5	42.5
0.88	4.899	60.8	39.2
0.89	4.911	64.2	35.8
0.90	4.924	67.5	32.5
0.91	4.936	70.8	29.2
0.92	4.948	74.1	25.9
0.93	4.961	77.4	22.6
0.94	4.973	80.7	19.3
0.95	4.985	84.0	16.0
0.96	4.998	87.2	12.8
0.97	5.010	90.4	9.58
0.98	5.022	93.6	6.37
0.99	5.035	96.8	3.18
1.00	5.047	100.0	0

Source: After N. Zuntz and H. Schumberg. In G. Lusk, *The Elements of the Science of Nutrition,* 4th ed. Philadelphia: Saunders, 1928. P. 65.

The total heat production is the heat liberated by the oxidation of proteins plus that liberated by the oxidation of carbohydrates and fats:

$$13.4 \text{ Kcal/hour} + 63.2 \text{ Kcal/hour} = 76.6 \text{ Kcal/hour} \quad (13)$$

It can be further calculated that, of the total heat production, 42 percent is derived from carbohydrates, 41 percent from fats, and 17 percent from proteins. It is evident that all three major foods are used to supply energy for the maintenance of the vital activities of the body. The total heat production is a measure of metabolic rate; i.e., it is equivalent to the rate at which the chemical energy of the body stores is metabolized.

DETERMINATION OF METABOLIC RATE FROM OXYGEN CONSUMPTION

The method of indirect calorimetry described in the preceding sections requires a moderate amount of equipment and considerable time. However, since the determination of metabolic rate is a clinical tool of some importance, it has been necessary to devise simpler methods. One means of assessing metabolic rate in the clinic is based on a single measurement, that of oxygen consumption. It was found, using the more complicated techniques of indirect calorimetry, that under standard conditions of resting and fasting the average RQ of a large group of normal subjects was 0.82. On the basis of these determinations, the assumption is now made that the RQ of a subject in the basal state is 0.82. At a nonprotein RQ of 0.82, the caloric value of oxygen is 4.825 Kcal. Heat production is determined by measuring oxygen consumption over a short interval of time, usually six minutes, converting this value to an hourly basis, and multiplying it by 4.825 Kcal per liter of oxygen. This approach considerably simplifies the determination of metabolic rate.

One type of apparatus which is commonly used in the clinic to measure oxygen consumption is shown in Figure 27-2. This apparatus consists of a spirometer bell which is arranged to write on a moving kymograph by means of a pulley system. The bell is filled with oxygen, and during a determination a motor-blower draws oxygen through the inspiratory tube and returns the expired gas mixture through

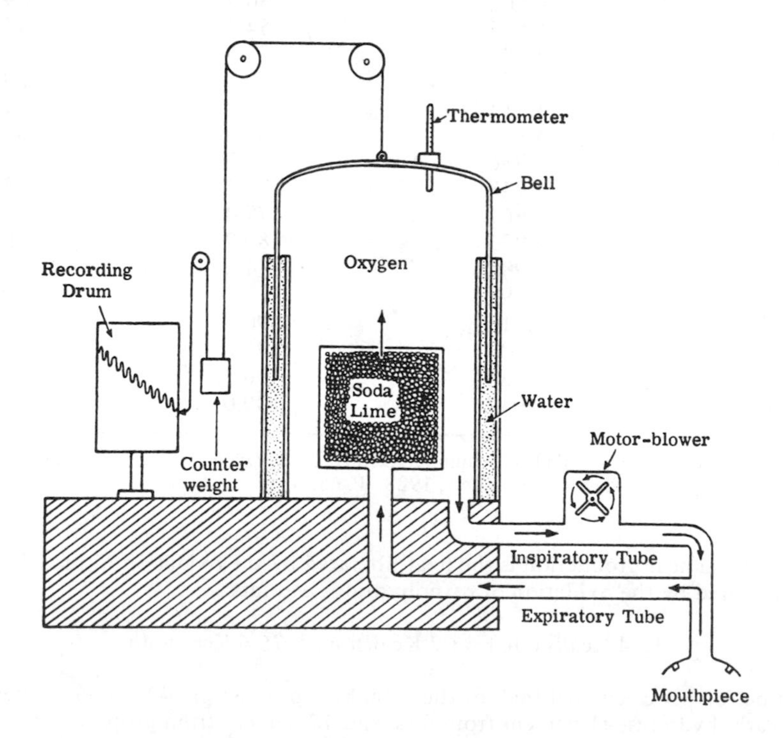

FIGURE 27-2. Diagram of the Sanborn apparatus for measuring human oxygen consumption. (From G. Lusk. *J. Biol. Chem.* 59:41–42, 1924.)

the expiratory tube. The expiratory tube is connected to a canister of soda lime in order to remove all carbon dioxide present in the expired air. As the subject consumes oxygen and as carbon dioxide is removed from the system, individual respiratory excursions and the drop of the bell are recorded. The difference between the height of the bell at the beginning and the end of the experiment is a measure of the volume of oxygen consumed during this time. To determine the distance the bell drops, a line is drawn which best follows either the peaks or the troughs of respiration as recorded, and the vertical distance between the top and the bottom of this sloping line is measured. Since the spirometer is calibrated in terms of volume per unit distance fall of the bell, the total volume of oxygen consumed during the experiment is calculated by multiplying this conversion factor by the distance the bell falls. The oxygen consumption is corrected for water vapor tension, reduced to standard conditions of temperature and pressure, and converted to liters of oxygen consumed per hour. Some subjects find it hard to breathe normally through a mouthpiece and valves, and one serious criticism of this method is that the accuracy of the determination depends on the ability of a subject to breathe regularly. If the rate and/or the amplitude of respiration are irregular, it is difficult to decide precisely where to draw the sloping line.

It is recognized that a determination of metabolic rate by this method introduces certain errors into the calculations. Since a nonprotein RQ is used to determine the caloric value of oxygen, protein metabolism is neglected. However, the RQ (0.80) and the caloric value of oxygen (4.5 Kcal per liter of oxygen) for the oxidation of protein are only slightly less than the average values used. Furthermore, protein accounts for only a relatively small part of the total energy exchange. Consequently, the error introduced by neglecting protein oxidation is small and has been calculated to be only about 1 percent. In addition, the assumption is made that the RQ of a subject is normal; but if the RQ is not 0.82, it is not correct to use 4.825 as the caloric value of oxygen. However, this introduces only a small error into the calculations, because the caloric value of oxygen varies so slightly between the extreme respiratory quotients of 1.00 and 0.70. By using an RQ of 0.82, which falls midway between the extremes, there can be at most an error of only 3.5 percent.

The errors inherent in the short method are small and not significant enough to outweigh the advantages gained. Moreover, since a large number of studies have been carried out using this method, good standards are available for comparison. One drawback to the method is that it does not provide information regarding the type and amount of substances oxidized in the body stores. The only information that can be obtained by measuring oxygen consumption is the total heat production of a subject. However, this value can be extremely useful as an aid to diagnosing certain disorders.

BASAL METABOLIC RATE

The methods and calculations of indirect calorimetry described have dealt with the energy exchange of a fasting and resting subject. The term applied to the exchanges of energy which occur under these conditions is *basal metabolism* or *basal metabolic rate* (BMR). Clinically, BMR is determined 12 to 14 hours after the last meal, usually in the morning after at least 8 hours of sleep. There should be no voluntary muscular movement during the test and no muscular exertion within half an hour to an hour prior to the test. In addition, there should be no stress due to extremes of environmental temperature, and the subject should be both mentally and physically at rest. Many factors which produce variable influences on metabolic rate are eliminated by carrying out the determination under these circumstances. The heat production so determined is a measure of the energy exchange required to maintain the vital activities of the body. The basal metabolic state does not represent the minimal functional activity of the body, since energy exchange is about 10 percent lower during sleep. This has been attributed to the more complete muscular relaxation achieved during sleep.

Once the basal heat production of a subject has been determined in terms of kilocalories per hour, it is necessary to ascertain whether this value is normal. Consequently, normal values must be available for comparison. If a comparison is to be valid, the heat production of a subject should be compared with the average heat production of a large population of normal subjects having physical and biological characteristics similar to those of the subject. Among the more important factors to consider when making such comparisons are the body size, sex, and age of a subject.

BODY SIZE

The total energy exchange of a large animal is greater than that of a small animal. For example, the heat production of an elephant is many times that of a mouse. Likewise, a 21-year-old man weighing 140 kg has a greater total energy exchange than a man of the same age weighing 70 kg. Consequently, when comparisons of metabolic rate are made between individuals of the same or different species, it is essential that there be some basis for comparison which takes body size into consideration. Various criteria of body size which have been considered include body weight, body surface area, and lean body mass.

Metabolic rate is not directly related to body weight, since the caloric output per kilogram of body weight of a small animal is greater than that of a large animal. However, energy exchange is thought to be proportional to certain power functions of body weight ($Kcal/kg^{0.67}/hour$ or $Kcal/kg^{0.73}/hour$). When heat production is expressed in this manner, large and small animals have approximately the same metabolic rates.

The trend commonly followed in the clinic today is to express metabolic rate in terms of body surface area ($Kcal/m^2/hour$). This concept is based on the idea that heat is lost at the surface of the body and that an amount of heat equivalent to what is lost must be produced in order to maintain a constant body temperature; i.e., an animal presumably produces heat in proportion to its surface area. Some relationships between heat production and body surface area are given in Table 27-3.

TABLE 27-3. Relationship Between Metabolic Rate and Body Surface Area

Dogs		Various Animals		
Body Weight (Kg)	$Kcal/m^2/day$	Animal	Body Weight (Kg)	$Kcal/m^2/day$
31.20	1036	Hog	128.00	1074
24.00	1112	Man	64.00	1042
19.80	1207	Dog	15.00	1039
18.20	1097	Guinea pig	0.50	1246
9.61	1183	Mouse	0.018	1185
6.50	1153			
3.19	1212			

Source: From M. Rubner. *Z. Biol.* 19:535, 1883, and 30:73, 1894.

Individuals of the same species (dogs) show considerable variation in body weight, but metabolic rate, when expressed in terms of surface area, is approximately the same for all. The same relationship is evident when different species are compared. Although a hog weighs about 7000 times more than a mouse, the caloric outputs are similar when based on body surface area. The same is true for other species, such as man, dog, and guinea pig. Surface area obviously cannot be measured every time metabolic rate is determined. However, the surface areas of people of many different sizes and shapes have been measured by determining the total area of pieces of paper required to cover each body surface. It has been possible from these measurements to derive an equation which can be used to calculate body surface area from the weight and height of a subject, both of which are easily measured variables. Furthermore, surface area can be conveniently determined from nomograms which have been constructed from this equation.

There has been some opposition to the use of surface area for relating energy exchanges of individuals of different sizes, since heat loss is affected by factors other than total surface area. Posture and the insulating effects of hair and clothing influence heat loss to varying degrees by modifying the actual surface exposed to the environment. It has been suggested that a better standard might be the weight of the active tissue of the body, i.e., those tissues which are actively engaged in metabolic processes and primarily responsible for the production of heat. Since adipose tissue is relatively inactive with respect to energy exchange, lean body mass might serve as a better reference for comparing metabolic rates of individuals of different sizes. Methods for estimating the weight of the body tissues exclusive of adipose tissue have been proposed and are now being evaluated.

SEX

Since the BMR of women is 6 to 10 percent lower than that of men of the same size and age, it is necessary to have standard values for both males and females.

AGE

It is well known that BMR (Kcal/m^2/hour) is at its peak during the early years and gradually declines throughout the remainder of life (Table 27-4). On the basis of size, the growing animal has a higher metabolic rate than an adult, because the rate of turnover of the body tissues is greater in a growing animal and this is reflected by an increased production of heat. The size of the body increases during growth, and much energy is stored by the syntheses of carbohydrates, fats, and proteins. Since relatively more synthetic activity occurs in a growing animal as compared with an adult, relatively more energy is liberated as heat by the catabolic reactions which necessarily accompany the synthetic reactions. As a result, standards for energy exchange must be based on age as well as on sex.

STANDARD VALUES

Standard normal values based on age, sex, and body surface area have been established for basal metabolic rate. One such set of standard values is given in Table 27-4. When a value in terms of kilocalories per square meter per hour is obtained for a subject, it is then compared with the normal value obtained from individuals of the same age and sex. The usual clinical method of reporting BMR is in terms of the percentage deviation from the standard value. For example, if the BMR of a male

TABLE 27-4. The Mayo Foundation Normal Standards of Basal Metabolic Rate (Kilocalories per Square Meter per Hour)*

Males		Females	
Age	BMR	Age	BMR
6	53.0	6	50.6
7	52.5	6 1/2	50.2
8	51.8	7	49.1
8 1/2	51.2	7 1/2	47.8
9	50.5	8	47.0
9 1/2	49.4	8 1/2	46.5
10	48.5	9-10	45.9
10 1/2	47.7	11	45.3
11	47.2	11 1/2	44.8
12	46.7	12	44.3
13-15	46.3	12 1/2	43.6
16	45.7	13	42.9
16 1/2	45.3	13 1/2	42.1
17	44.8	14	41.5
17 1/2	44.0	14 1/2	40.7
18	43.3	15	40.1
18 1/2	42.7	15 1/2	39.4
19	42.3	16	38.9
19 1/2	42.0	16 1/2	38.3
20-21	41.4	17	37.8
22-23	40.8	17 1/2	37.4
24-27	40.2	18-19	36.7
28-29	39.8	20-24	36.2
30-34	39.3	25-44	35.7
35-39	38.7	45-49	34.9
40-44	38.0	50-54	34.0
45-49	37.4	55-59	33.2
50-54	36.7	60-64	32.6
55-59	36.1	65-69	32.3
60-64	35.5		
65-69	34.8		

*The normal limits are usually taken as ± 10 percent, and divergence beyond these limits indicates an abnormal BMR for a subject of this age and sex.

Source: From Boothby, W. M., J. Berkson, and H. L. Dunn. *Amer. J. Physiol.* 116:468–484, 1936.

subject, age 28, is determined to be 38.3 Kcal per square meter per hour and compared with the standard value of 39.8 Kcal per square meter per hour:

$$\% \text{ Deviation} = \frac{38.3 - 39.8}{39.8} \times 100 = -3.7\% \qquad (14)$$

FACTORS THAT AFFECT METABOLIC RATE

EFFECT OF FOOD

When a resting subject ingests food, heat production increases over the basal level. The increased production of heat which results from eating is called the *specific dynamic action* (SDA) of foods. Specific dynamic action varies with the type of food ingested. If protein is eaten, the heat production over a period of hours rises above the basal level by 25 to 30 percent of the caloric value of the protein ingested. Heat production also increases when either carbohydrates or fats are ingested, but to a lesser extent. For example, suppose that the basal heat production of a subject is 75 Kcal per hour. The total heat production of this subject over a period of 4 hours should be 300 Kcal. If the subject is fed an amount of protein equivalent to 300 Kcal, enough energy should be available in the food to balance that lost as heat over the 4-hour period. However, a measurement of metabolic rate over the 4 hours following the ingestion of the protein would show that the heat production of the subject is 380 Kcal or, on the average, 95 Kcal per hour. It can be calculated that 80 more kilocalories were produced during the 4-hour interval than can be accounted for on the basis of the basal heat production. In terms of the caloric value of the protein fed (80 Kcal ÷ 300 Kcal × 100), 27 percent of the caloric value of the ingested protein appears as extra heat.

These results show that there cannot be kilocalorie-for-kilocalorie substitution of food energy for energy stored in the body reserves. In the example cited, proteins spare some of the body reserves of energy, but it is obvious that some of the reserve energy is needed to utilize the ingested proteins. Consequently, in order to spare the body stores of energy more completely, the total caloric intake of protein and of other foods should be somewhat greater than the basal energy exchange.

It is not certain why heat production increases following the ingestion of food. The consensus is that the excess heat is the result of intermediary metabolism of absorbed foods. If amino acids are injected intravenously into dogs, heat production increases. However, if the animal is first hepatectomized, heat production increases only slightly. These results suggest that the liver is the major site of the extra heat production, and the importance of the liver with respect to the processes of intermediary metabolism of amino acids and other foods is well established.

MISCELLANEOUS FACTORS

Many factors influence the total energy exchange of an individual. Among these are environmental temperature and fever (Chap. 28) and muscular exercise (Chap. 29). Thyroid hormone exerts considerable influence on the rate at which cellular oxidations occur, and excesses and deficits of circulating thyroid hormone modify metabolic rate. Basal metabolic rate can be 25 to 40 percent lower than normal in hypothyroidism, and 40 to 80 percent higher in hyperthyroidism. The administration of male sex hormone and growth hormone and the release of epinephrine and norepinephrine which results from sympathetic stimulation increase metabolic rate by virtue of the stimulatory effects of these substances on cellular activity. The nature of the diet seems to have little influence on basal metabolic rate, but prolonged malnutrition may cause a decrease of 20 to 30 percent, presumably because of a deficiency of cellular nutrients. Certain psychic influences, such as intense mental work, do not modify metabolic rate. However, the increased muscular tone, increased respiratory activity, and tachycardia which accompany some emotional states can produce increases of 5 to 20 percent.

REFERENCES

Brody, S. *Bioenergetics and Growth.* New York: Reinhold, 1945.

Clark, W. M. *Topics in Physical Chemistry,* 2d ed. Baltimore: Williams & Wilkins, 1952.

DuBois, E. F. *Basal Metabolism in Health and Disease,* 3d ed. Philadelphia: Lea & Febiger, 1936.

DuBois, E. F. Energy metabolism. *Ann. Rev. Physiol.* 16:125–134, 1954.

Garry, R. C. (Ed.). Energy Expenditure in Man (Symposium). *Proc. Nutr. Soc.* 15:72–99, 1956.

Kleiber, M. Body size and metabolic rate. *Physiol. Rev.* 27:511-541, 1947.

Richardson, H. B. The respiratory quotient. *Physiol. Rev.* 9:61–125, 1929.

Swift, R. W., and C. E. French. *Energy Metabolism and Nutrition.* New Brunswick, N. J.: Scarecrow, 1954.

Wilhelmj, C. M. The specific dynamic action of food. *Physiol. Rev.* 15:202–220, 1935.

28

Temperature Regulation

Robert W. Bullard

MAMMALS AND BIRDS *are homeotherms.* That is, they are capable of maintaining a near-constant internal body temperature. The advantages of this constancy are great because so many physiological processes are the result of chemical reactions, and slight temperature changes bring about great changes in reaction rates. For example, a 10°C increase in temperature will double or triple the rate of many biological processes. If body temperature were not held close to a constant, but varied with environment, as in cold-blooded (poikilothermic) animals, the more subtle neuroendocrine regulatory processes would be rendered ineffective.

The term *homeothermy* must be qualified. In man, considered to be a precise temperature regulator, the deep body temperature may fluctuate 1°C in daily activity cycles. It is usually lowest in the early morning and reaches a peak in late evening. These temperature changes are probably of a controlled nature; the depression at night appears to be effected through the use of physiological cooling mechanisms and is not passively dependent on metabolic depression. A considerable variation occurs in the so-called normal internal temperature of different individuals. In one study, rectal temperatures in a group of healthy individuals varied from 36.2°C to 37.6°C, with a mean of 36.9°C.

Opposite man on the homeothermic scale are the hibernators. These birds and mammals, to save energy, lower their body temperatures as much as 35°C or to just a few degrees above environmental temperature. Hibernators remain dormant for varying periods and eventually rewarm spontaneously in order to enter a typical regulated homeothermic existence. Between man and the hibernators are many homeothermic species showing differing degrees of daily or seasonal temperature fluctuations, but capable of maintaining a body temperature independent of environmental temperature.

The processes utilized by man and mammals in regulating the internal temperature in spite of environmental temperature changes are the subject of this chapter. The study of temperature regulation may range from pure physical to psychological concepts and is important, aside from its pure biological interest, for a rational approach to (1) problems of the febrile state in disease, (2) methods of inducing and facilitating recovery from hypothermia as an adjunct to surgery, (3) therapy for excess heat or cold exposure, and (4) problems of military, exploratory, and industrial operations in climatic extremes.

PRINCIPLES OF PHYSIOLOGICAL HEAT TRANSFER

CONCEPT OF CORE TEMPERATURE

The body may be considered as possessing an inner core at a temperature near 37°C and an outer shell of variable temperature. The body strives for rather precise regulation of core temperature, but often, as we shall see later, this is achieved at the expense of the shell temperature. Physiologists usually consider that about two-thirds of the body mass is at the core temperature, represented by the rectal temperature (T_r) or some other internally measured temperature, while one-third of the body mass is at the shell or mean skin temperature ($\overline{T}_s$). The mean body temperature ($\overline{T}_b$), therefore, may be expressed by the equation:

$$\overline{T}_b = 0.33\ \overline{T}_s + 0.67\ T_r \qquad (1)$$

The mean skin temperature is obtained by temperature measurements at several sites on the body surface. Although an error may occur because of relative shifts in shell and core sizes, an important use of the relationship between the core and skin temperatures is that an estimate may be made of the actual heat loss or heat gain of the body.

To characterize temperature regulation quantitatively, a measurement of the changes in body heat stores must be utilized. This measurement is the heat capacity or specific heat defined as the ratio of heat supplied (or removed) to the corresponding temperature rise (or decrease) – or:

$$\text{Specific heat} = \frac{\Delta \text{ kilocalories/kg}}{\Delta T} \qquad (2)$$

When 1 Kcal is added to 1 kg of water, the temperature is raised 1°C. Thus, the specific heat of water is 1.0 and the kilocalorie as an energy unit is defined.

The body is mostly water but contains proteins and lipids of lower specific heat, and its average specific heat is generally considered to be about 0.83. Thus for every 0.83 Kcal of heat energy added or subtracted per kilogram, the mean temperature changes 1°C. For the physiologists the change in heat content rather than the overall total content is the important factor and can be estimated from the following equation:

$$S \text{ (in Kcal/hours)} = \frac{0.83\ (\overline{T}_{b_1} - \overline{T}_{b_2}) \times \text{body weight}}{\text{Time in hours}} \qquad (3)$$

where $\overline{T}_{b_1}$ and $\overline{T}_{b_2}$ are the mean body temperatures at the beginning and end of the time period.

Recently, terminology in thermal physiology has been standardized and now watts rather than kilocalories per hour are used. To convert kilocalories per hour to watts, one simply multiplies by 1.16. Both *watts* and *kilocalories* will be used in this chapter.

HEAT SOURCES AND SINKS

Metabolism

The metabolic activities of the organism constantly supply heat and may always be considered as a heat source. Metabolism or heat production in man may range from a basal level of 80 watts (70 Kcal per hour) up to possibly 20 times this level in heavy muscular activity. In cold exposure, man can increase skeletal muscle activity by shivering, which may bring about a fivefold increase in metabolic heat production. It should be remembered that in exposure to heat, basal metabolic rate cannot be effectively lowered as part of the temperature regulation process.

Evaporation

Water evaporation is considered to be a heat sink, as the body, under most conditions, loses some heat by this route. Even though sweating is not occurring there is a continuous diffusion of water molecules through the skin, because the integument of man is not waterproof. This is the so-called *insensible perspiration* which was first observed by Sanctorius in 1616 during careful body weight studies in man. Such water diffusion accounts for a loss of about 11 watts, and 8 watts more is lost by water evaporation in the respiratory tract. A fairly constant 20 to 25 percent of the basal heat production is lost by the evaporative routes. Evaporative heat loss can be increased tremendously in man by sweating and in animals by panting and increased salivation. This function of sweating, as a mechanism of heat dissipation, was first explained by Benjamin Franklin and decisively demonstrated in 1775 by Blagden, who placed men in dry ovens at 250°F without harmful effects.

The heat lost by evaporation (E) is generally calculated from water loss, which may be determined from body weight changes after correction for metabolic and other weight losses. Thus,

$$\text{E (in watts)} = \frac{0.7 \times \text{grams of water evaporated}}{\text{Time in hours}} \qquad (4)$$

where 0.7 is the latent heat of vaporization (i.e., evaporation of 1 gm of water per hour requires 0.7 watts or 0.6 Kcal).

Radiation, Convection, and Conduction

If the environment is warmer than the body surface, the body gains heat; if cooler, it loses heat. Most of this exchange takes place by *radiation* to or from the surroundings. Heat transferred by radiation is in the form of electromagnetic waves, which can pass through a vacuum and travel at the speed of light. The first warmth felt after the clouds blocking the sun pass is an example of radiation or radiant heat.

Conduction refers to heat transfer from atom to atom or molecule to molecule by successive transfers of kinetic energy. Little body heat exchange takes place via this route, except in water immersion.

Convection indicates heat transfer by movement from molecules of a fluid (either gas or liquid). In cold, for example, the fluid (air) next to the body gains heat by conduction or radiation, becomes less dense, rises, and is replaced by cooler air in a continuous process. Ordinarily this is not an important avenue of heat exchange. However, wind, which may be considered *forced convection,* is quite effective as a heat exchange route, and in water immersion the major route of heat transfer is by convection.

Because radiation, convection, and conduction are separated only with difficulty, they will be grouped together under the term *H.*

$$\text{H (in watts)} = \text{h}\,(\overline{\text{T}}_\text{s} - \text{T}_\text{ambient}) \qquad (5)$$

where h is the combined heat transfer coefficient dependent on exposed area, surface geometry, and properties of the surrounding air. It is obvious that changes in heat transfer will occur when there are changes in the gradient or difference between mean skin temperature and the temperature of the surroundings (ambient temperature). Physiologically, this may be accomplished by changing skin blood flow. H is a heat sink when $\overline{T}_S$ is higher than $T_{ambient}$ and a heat source when $T_{ambient}$ is higher than $\overline{T}_S$.

PARTITIONAL HEAT EXCHANGE

A useful approach in temperature regulation studies is to draw a balance statement for heat sources and heat sinks in any given environment. When man is in a thermally steady state or thermal equilibrium, all the heat produced by metabolism equals all the heat dissipated, and there is no change in mean body temperature or heat stores:

$$\begin{gathered} M = H + E \\ \text{Sources} = \text{Sinks} \end{gathered} \tag{6}$$

Usually, an unclothed man in the cold is not in thermal equilibrium with the environment, and heat is lost from body heat stores (S):

$$M + S = H + E \tag{7}$$

in which M and S are heat sources and H and E are heat sinks. In such a relationship, all source heat must equal all heat going into the sinks. In cold, E is low and almost constant and cannot be greatly altered to prevent a large loss of heat stores (S); therefore, metabolism (M) must be increased or H must be decreased in any effective efforts toward core temperature maintenance.

In heat, man will not reach thermal equilibrium immediately and the relationship becomes

$$M + H = S + E \tag{8}$$

Here, because environmental temperature is greater than surface temperature, H is a source of heat as well as M. The body heat stores are now a sink since all the heat not lost by evaporation is increasing the mean body temperature or increasing S. The partitional heat exchange diagram in Figure 28-1 presents a summary of these avenues of heat loss and gain. The values shown are only examples and will be different when either relative humidity, exposure time, clothing, posture, or air velocity is varied.

It should be noted that algebraic addition of the sinks below the zero line and the sources above the line equals zero at any specific ambient temperature. The zone of ambient temperature between 33° and 25°C may be considered the zone of vasomotor regulation. Here, H may be altered by changing the gradient ($\overline{T}_S - T_{ambient}$) through skin and limb blood flow changes, and thermal equilibrium may be maintained with neither sweating on the right side nor increasing metabolism on the left side. At cool ambient temperatures, metabolism (M) must be increased to decrease the heat being given up from the body heat stores (S). On the right side of the diagram, sweating must occur to increase E because of the high heat gain from radiation and convection (H). The small change in S at the higher temperatures illustrates the effectiveness of evaporation as a heat loss mechanism.

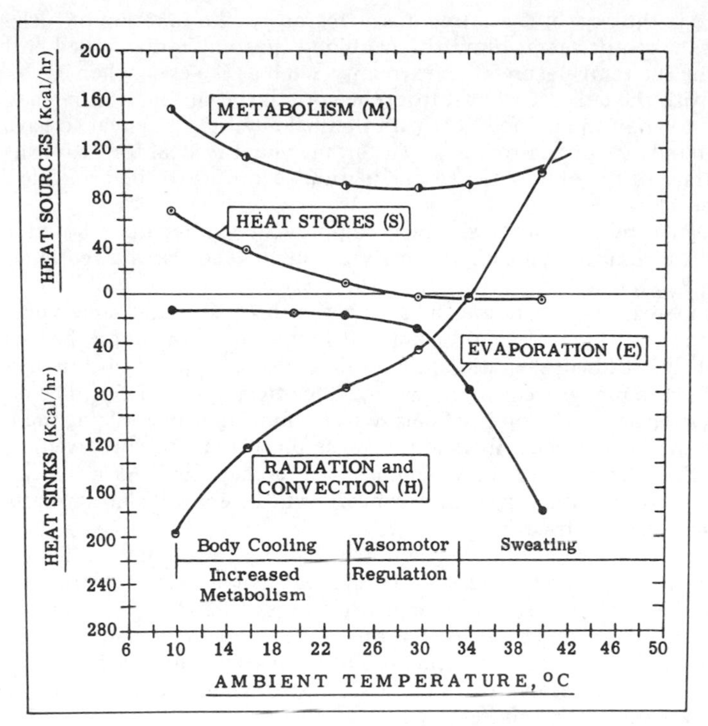

FIGURE 28-1. Partitional heat exchange of a human at different ambient temperatures. All heat sources are above the zero line; all heat sinks are below the line.

EFFECTOR ACTIVITIES IN TEMPERATURE REGULATION

This section will describe the functional mechanisms by which physiological heat transfer can be adjusted in an effort to regulate body temperature.

SKIN BLOOD FLOW

Because the skin temperature is so important in heat transfer to the environment and because delivery of heat to the surface is blood flow–dependent, the physiology of skin circulation is considered in this chapter.

Anatomically, human skin is supplied with an extensive system of capillary loops which arise in the corium and return to enter subpapillary venous plexuses. The amount of blood in the plexuses, which usually contain the greatest portion of skin blood volume, determines the intensity or depth of skin color not dependent on skin pigmentation. This latter factor is determinable by simply applying pressure to squeeze out the blood and observing the resulting color. The degree of oxygenation of hemoglobin of the contained blood usually determines the hue or shade, which, of course, would also be influenced by presence of methemoglobin or carboxyhemoglobin.

The second outstanding functional feature of the skin circulation is the large number of arteriovenous anastomoses. These units are well endowed with smooth muscle and nerve endings indicating functional control and are present in greatest abundance in the skin of the extremities. Obviously, the function of such structures is beyond the requirements of nutritional blood flow and is concerned both with maintaining the temperature of an extremity, such as the ears, when exposed to cold and with the delivery of heat from the core to the surface. Skin areas showing the greatest variation in blood flow with thermal alterations appear to have the greatest abundance of anastomoses. The arrangement is ideal for delivering heat to the surface of the extremities for dissipation according to the principle described by Newton's law.

The regulation of skin blood flow is complex, depending upon both local factors and impulses from the sympathetic nervous system. The classic experiment of Claude Bernard (1852), wherein the division of the cervical sympathetic chain of the rabbit caused an increased temperature and flushing in the ear on that same side, established the presence of a vasoconstrictor neural outflow. Depending upon the particular skin area being studied, the ratio of blood flow between the condition of severed nerve and maximum vasoconstriction may range from 1:4 to 1:50. The neural outflow for vasoconstriction is, of course, very important in reducing the flow of heat from core to periphery in conditions when heat must be conserved. The mediator substance released at the nerve endings is norepinephrine. Skin vessels are extremely sensitive to either injected norepinephrine or epinephrine and respond with intense vasoconstriction.

In several experiments it has been noted that the degree of vasodilation obtainable in certain procedures decreases after sympathetic nerves have been severed, suggesting that such nerves carry vasodilator fibers. Cholinergic vasodilator nerves have been described for specific organs of species other than man, but whether man's skin possesses this type of innervation is still an unsettled question.

The local influences upon skin blood flow are well documented. The examples given here are vascular responses that still occur following denervation of the test skin area, but, as one might expect, these local responses can be greatly altered by arriving vasomotor neural impulses.

A good example of a local response is reactive hyperemia, wherein blood flow following a period of ischemia produced by circulatory occlusion is increased in such fashion that repayment or total increase roughly approximates the incurred debt. This correspondence of debt and repayment, and the fact that repayment is less if the arm is to be cooled, suggests that a chemical substance produced by ischemia is being released to cause the vasodilation. Several substances have been implicated but none is established as a causative agent. The list of possibilities includes histamine, various Krebs cycle intermediates, various electrolytes, adenosine, adenine nucleotide, possibly vasoactive polypeptides, and carbon dioxide, and oxygen lack itself. These agents could be effective singly or in some synergistic combination.

Physical factors alone may play a major role. Observations are that lowered pressure in resistance vessels, the arterioles, will relax the smooth muscle. At present, the most likely explanation of reactive hyperemia is that several factors combine to produce the response.

Local temperature can have a marked effect on the status of resistance vessels of the skin. In general, with warming of an extremity (for example, the hand) blood flow will increase. This increase is greater if the subject's central temperature is higher, indicating that in cooler subjects the greater vasoconstrictor neural outflow to the skin cannot be completely overcome by local mechanisms.

Although the general trend is for vasoconstriction by skin cooling acting directly to contract vascular smooth muscle, there is one interesting exception. When an extremity such as a finger, a toe, or an ear is cooled by immersing in water at 0° to 12°C, a maximum constriction of blood vessels occurs, followed by a vasodilation in 10 to 30 minutes. The vasodilation, often recorded as an increased local skin temperature, is then reversed by vasoconstriction, and a cyclic course of skin temperature and blood flow increases followed by reversals ensues in what has been called a "hunting reaction." This response is of local origin, and the most plausible explanation is that the vascular smooth muscle loses its functional ability at low temperature. The vasodilation, which is probably not maximal, rewarms the tissue and thus it can contract again and thereby give rise to the oscillatory behavior. In support of this is the observation that cooled strips of vascular smooth muscle lose their reactivity to norepinephrine. With the hunting reaction or cold-induced vasodilation, body heat loss in water immersion is increased, although no data are available on the magnitude of this extra loss.

Role of Skin Blood Flow in Temperature Regulation

The overall human skin blood flow has been variously measured and ranges from 150 to 200 ml per minute in a cool environment up to 2000 ml per minute for conditions of heat stress. In some extremities the range of variation may be 100-fold. This wide range greatly surpasses any variation in metabolic requirements of the skin, but it is necessary for adjusting the flow of heat from core to periphery.

An expression of the delivery of heat from core to surface is the so-called tissue conductance, or whole body conductance, calculated by a relation analogous to Ohm's law:

$$\text{Conductance} = \frac{\text{Total heat flow from core to surface}}{T_r - \overline{T}_s} \tag{9}$$

The numerator would thus include heat produced by metabolism and that lost from body heat stores and have subtracted from it that amount of heat lost through transfer to respired air. The lower the gradient between the rectal temperature and mean skin temperature, the greater must be the flow of blood to the surface. If blood flow to the periphery were reduced to a minimum, heat would still be transferred by simple conduction through the tissues. This value for conductance would be about 5 to 10 Kcal per hour per C°, depending on the thickness of fat, which is the best insulator of the body tissues. With heat exposure and maximum blood flows, the conductance may reach 150 Kcal per hour per C°. Thus there is, on the average, about a 20-fold variation in conductance through the tissues. As a general approximation, the maximum vasoconstriction is equivalent to adding one suit of clothes as far as insulation is concerned.

The fact that heat flow does not always parallel blood flow is of interest physiologically. It has been shown that in cool ambients, as blood proceeds to the periphery, heat can be transferred from the arterial side to the venous side and returned or kept within the core. The reason is that the routes of venous return from the limbs may be shifted to set up a *countercurrent heat exchanger.* Venous blood is warmed as it returns to the deep core, and arterial blood is cooled as it proceeds to the extremities. Prime examples of this exchanger are found in the legs of wading birds that stand in freezing water, and in the extremities of aquatic mammals. Bazett (see Newburgh, 1968) suggested that the mechanism is effective in human subjects, and some of his data are shown in Figure 28-2. He found that the temperature of arterial blood near the chest was about 37°C, but by the time this blood had reached the wrist it could be as low as 25°. The important finding is that venous blood as it nears the chest reaches temperatures of almost 37°C; heat is thus short-circuited back to the core.

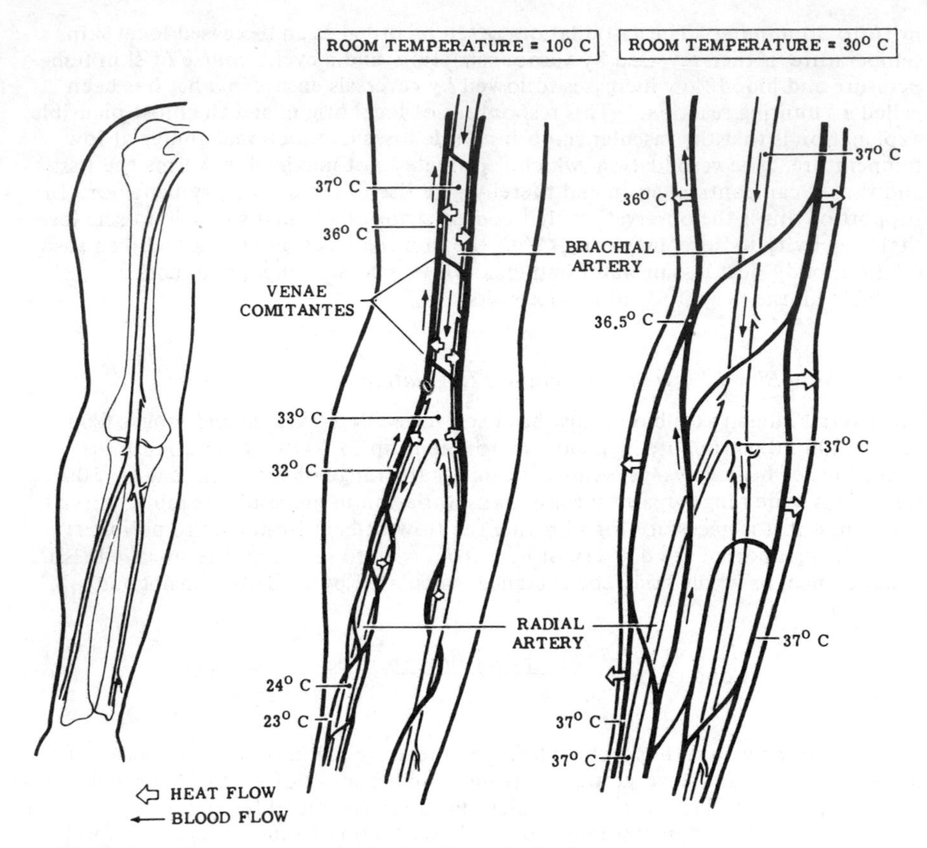

FIGURE 28-2. The countercurrent heat exchange in the human arm.

As a response to cold, venous flow shifts from the surface veins, so prominent in heat exposure, to the deep venous plexuses or so-called *venae comitantes,* where heat is gained from the arterial blood, as is shown in Figure 28-2. As the cool venous blood moves toward the core and passes the warmer arterial blood in countercurrent fashion, heat passes through the vessel walls at a rate proportional to the arteriovenous temperature gradient and the contact surface area. The mechanism can be demonstrated by a simple experiment: When the hand is placed in ice water, the arterial blood at the elbow will cool because of this countercurrent heat exchange. Thus, blood flow in the limbs can be maintained while heat loss is diminished, since heat is not carried from the core by this flow. In a warm environment, venous return can be simply shifted to the surface to eliminate countercurrent exchange, but with increased transfer of heat to the environment.

The vasoconstriction and blood flow shifts are so effective that upon exposure to cold a rise in rectal temperature of as much as 0.5°C may ensue. The vascular response efficiency is due to a decrease in the amount of heat carried to the surface – or, in effect, to increased body insulation. This necessarily brings about considerable shell cooling and perhaps increases the shell size with a rather large loss of heat stores. Therefore, a decrease in heat stores and shell temperature appears essential for economical regulation of the more vital core temperature, and will eventually allow the individual to reach thermal equilibrium.

CHEMICAL REGULATION

At temperatures below 25°C the peripheral circulation of the nude man is maximally vasoconstricted, and he cannot maintain heat balance by further vasomotor activity. With sufficient cold stimulus, an increase in metabolism will occur by either overt skeletal muscle activity or shivering. Shivering consists in muscle contractions which continuously wax and wane. Synchronous contractions occur in small groups of motor units or whole muscles out of phase with others and coinciding with antagonists, so that gross limb movements do not occur. Because of the antagonistic contraction, shivering is more effective in increasing metabolism than are other types of tremor. From the viewpoint of energetics, shivering is effective in that all of the chemical energy released in the contractile process is converted to heat and no energy goes into useful work.

Less heat is lost in shivering than in contraction involving limb movement with its associated forced convection. A seminude man exposed at 5°C for 60 minutes will double his heat production by shivering. Eightfold increase in metabolism by shivering appears to be near the upper limit. This increase can be obtained in the nude man immersed in water at 6°C. However, even at this high metabolic rate, body heat stores are still being lost, and at these severe exposure levels thermal equilibrium cannot be attained.

SWEATING

At temperatures of 32 to 34°C circulatory adjustments alone are not adequate for dissipating heat by radiation because of the reduced gradient or negative gradient between the skin and surroundings ($\bar{T}_s - T_{ambient}$). When ambient temperature is greater than skin temperature and a negative gradient results, the body is gaining heat by radiation. Increased evaporative heat loss by sweating must be utilized at these temperatures. If metabolism is increased by work, sweating must be called forth at lower temperatures, as the gradients for heat transfer are not great enough to dissipate the extra heat load.

Sweat may be considered a true secretion and not a filtrate and can be produced even when arterial pressure is markedly reduced by occlusion procedures. It is a water solution of an osmolar concentration well below that of plasma (see p. 537). The principal constituent is sodium chloride, and there are traces of potassium, urea, and lactate, but no protein or glucose, as are found in plasma. Contractile elements in the gland duct function by periodic pumping of droplets of the secreted fluid from the lumen onto the skin surface. The ejection occurs simultaneously in most skin areas and thus must be under neural control. The rate of this pumping or cyclic activity ranges from 1 to 2 per minute at the lowest sweating rates up to 16 to 20 at the highest rates. In addition, cycles lasting several minutes are seen in the sweating rate, probably representing oscillations in, or adjustments by, the controlling mechanisms.

Sweating is a particularly effective cooling mechanism. It has been estimated that man has about 2.5 million sweat glands spread over the skin. The total area of all the duct openings on the skin is 90 cm^2. To simply fill the ducts of all the sweat glands 40 ml of fluid must be secreted. Thus when all the glands are hypothetically combined one has an organ of some magnitude. Sweating rates for commonly encountered activities and temperatures are shown in Figure 28-3. Men working in severe heat and high relative humidity have sweated at rates as high as 4 liters per

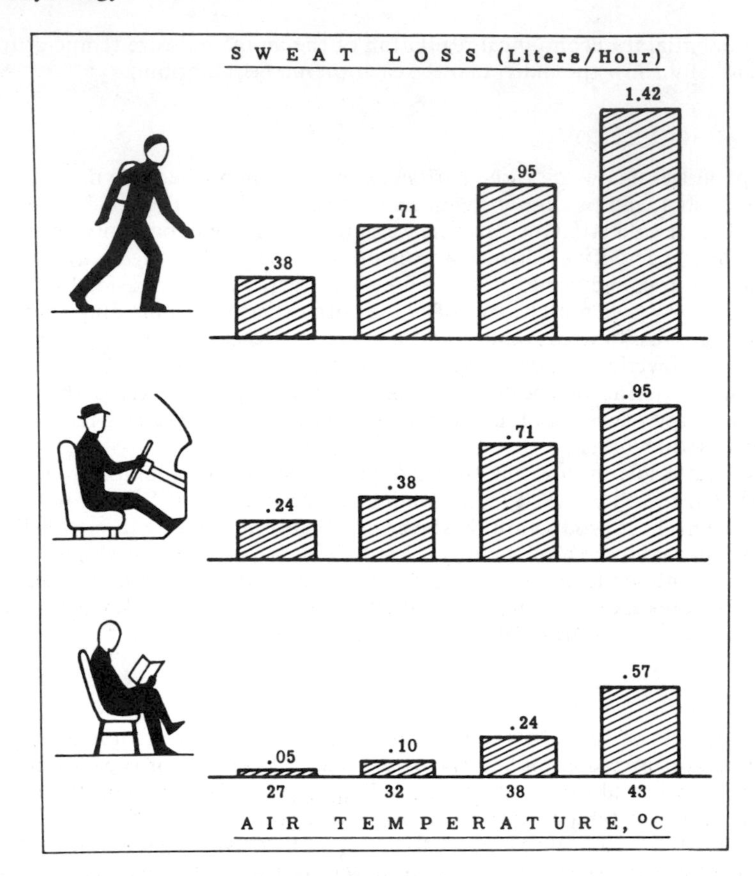

FIGURE 28-3. Sweating rates for a clothed man in various activities and at various ambient temperatures. (From Adolph et al., 1947.)

hour for short periods. Working humans can maintain a sweat output of 2 liters per hour for more than 5 hours. In cooling power this is equal to 2000 × 0.7 or 1400 watts.

To be effective, sweat must evaporate. The driving force for evaporation is the difference between the vapor tension of water at the skin surface and in the air. Thus, when air temperature is warm and relative humidity is high, the air-water vapor tension is high, and sweating is not effective. This accounts for the well-known discomfort experienced on warm, humid days.

NEURAL CONTROL OF TEMPERATURE REGULATION

The striking observation that over a very wide range of thermal stress, from extreme heat to extreme cold, the core temperature may vary only a fraction of a degree (C) dramatically emphasizes the precision of the controlling centers in calling forth proper effector mechanisms. The mass of experimental evidence indicates that thermal responses are controlled from centers above the spinal level. The chronic spinal dog or man shivers only in muscles or parts of muscles innervated from above the lesion. Sherrington (1924) pointed out that the precise segmental level of the lesion could be determined by tracing the nonappearance of shivering, which occurs only in muscles innervated from below the lesion. Vasomotor responses are absent, for the most part, in areas deprived of sympathetic innervation, although some local responses of blood vessels to heating and cooling may complicate the picture.

It has also been noted that decorticated animals show only slight impairment of temperature-regulating ability. Pathological sections and experimental procedures ranging from crude extirpation to production of precise electrolytic lesions, localized electrical stimulation, and localized heating and cooling have shown universally that the primary control centers for temperature regulation lie in the hypothalamus, located in the ventral diencephalon. The entire diencephalon except for a small portion of the thalamus and hypothalamus may be removed with only minor impairment of the mammal's ability to regulate body temperature.

All the experimental evidence points to the hypothalamus and its rostral preoptic area as the center of thermoregulatory control. In animal experiments with implanted thermodes (small devices for heating or cooling selected brain regions) the preoptic area is the most responsive to local temperature alteration. With heating, the animal (dog, cat, or monkey) will respond with the "antirise" responses of vasodilation and panting to increase evaporative heat loss. With cooling through the thermode, the animal will call forth the "antidrop" actions, vasoconstriction and shivering. With the same procedure in behavioral experiments the trained animal will utilize bar pressing to call for heating when the preoptic area is cooled, and will call for cold air with preoptic heating.

Probing with minute electrodes has shown that in the preoptic area there are many neurons that fire spontaneously and have a firing rate which increases when the local temperature is raised. A smaller number of neurons show a higher rate of firing with cooling, and these would be regarded as cold sensors.

It is considered that these neurons make up the primary elements of the regulatory actions. Although there is a tendency to think of thermoregulatory centers, the sensitive elements are diffusely located and not arranged in anatomically discrete nuclei. The output of the primary elements courses back through the lateral, ventral, and posterior parts of the hypothalamus, but not in discrete bundles.

The output from the preoptic region and hypothalamus is greatly modified by thermal information from the skin and possibly from receptors located deep in the body core. Although histological studies have suggested that cold is sensed on the skin by the end-bulbs of Krause, and heat through the end-organs of Ruffini, the active investigators in this field agree that most peripheral thermal detectors are naked nerve endings. Deep body thermal detection has been suggested by physiological experiments, but the receptors have not been identified.

It appears that the thermostatic system of the body utilizes both peripheral and central information, and the relative importance of these two inputs is an area of active research. Local hypothalamic cooling by 1°C or more appears necessary to evoke a strong cold-protective response when the skin is warm. This amount of cooling is far greater than that occurring in the usual exposure to cold; therefore, the peripheral receptors must play an important role in calling forth mechanisms of protection. Cold receptors respond to a sudden thermal decrease with a marked increase of firing rate, then show accommodation or a decreased rate if the new temperature is maintained. Shivering also shows this "rate" response, as subjects shiver intensely in the early exposure to severe cold when skin temperature is changing rapidly.

Experimental evidence has been presented suggesting that the primary regulation of sweating depends mainly on a proportional control system. That is, the rate of sweating is proportional to the difference between the actual hypothalamic temperature and a "set point" or threshold temperature, resembling systems used in many engineering applications. The physiological set point is probably the temperature at which activities of various antirise and antidrop neurons are equivalent and opposite in effects. As the hypothalamus warms, antirise neurons rapidly increase their firing rate, and sweating and vasodilation ensue. Further experiments have suggested that the set point is adjustable and changes with bodily activity.

It must also be noted that alteration of skin temperature without central temperature changes can markedly influence sweating rate. Experiments with heating or cooling of small skin areas demonstrate a marked thermal influence on local sweat production. Here the rate at which neural impulses driving sweating leave the central nervous system stays constant, but the effect of these impulses at the sweat gland is modulated by local temperature. When much larger skin areas peripheral to an arterial occlusion are heated or cooled, sweating in nonoccluded regions can be strongly affected. Because the thermally altered blood cannot reach the central nervous system, the mechanism must be of a neural nature, possibly acting through thermal receptors and the hypothalamus. The apparent effect of this type of stimulation is either to change the preoptic-hypothalamic set point or threshold or to alter its sensitivity or gain. The regulation of shivering is similarly dependent on skin temperature.

HYPOTHALAMIC INTEGRATION WITH OTHER NEURAL MECHANISMS

The neurophysiology of the integration of hypothalamic efferent outflow is a subject of great complexity and will be considered here from a functional rather than a neuroanatomical viewpoint. The hypothalamus, in functioning as a dual thermostat, or reacting to prevent both heating and cooling, can act only through other autonomic and somatic effector systems, as shown in Figure 28-4. The sympathetic nervous system plays an important role in temperature regulation. Antidrop outflow for vasoconstriction and piloerection is mediated through sympathetic centers. The part of the antirise efferent outflow responsible for a vasodilation acts by inhibition of sympathetic vasomotor tone and possibly by activation of sympathetic vasodilation outflow. Antirise sweating outflow is also mediated by the sympathetic system. In mammals, antirise panting outflow acts via the respiratory centers and must be closely associated with increased parasympathetic stimulation of salivation and salivary gland blood flow. This is the only clear-cut parasympathetic contribution to temperature regulation.

The shivering efferent pathways are not completely demarcated at the present time. For the most part, the outflow from the brain travels downward not in the pyramidal tracts but in the lateral white matter. However, the final pathway to the muscle is the lower motoneuron. It appears that impulses leave the brain in a continuous fashion, and experimental attempts to record action potentials in the cord with a rhythm similar to that recorded at the muscle have been unsuccessful. The actual shivering rhythm may be due to the proprioceptive impulses which go back to the cord and inhibit efferent outflow. When the proprioceptive impulses from the inactive muscle stop, shivering returns. Such an oscillating system can give rhythmic tremors occurring at the observed rate of 10 to 20 per second. It has been demonstrated by decortication and local electrical stimulation that shivering may be suppressed from the cerebral cortex. The necessity of a suppressor mechanism is obvious when one considers the two functions of skeletal muscle contraction: movement and postural functions, and shivering. An animal or a man exposed to

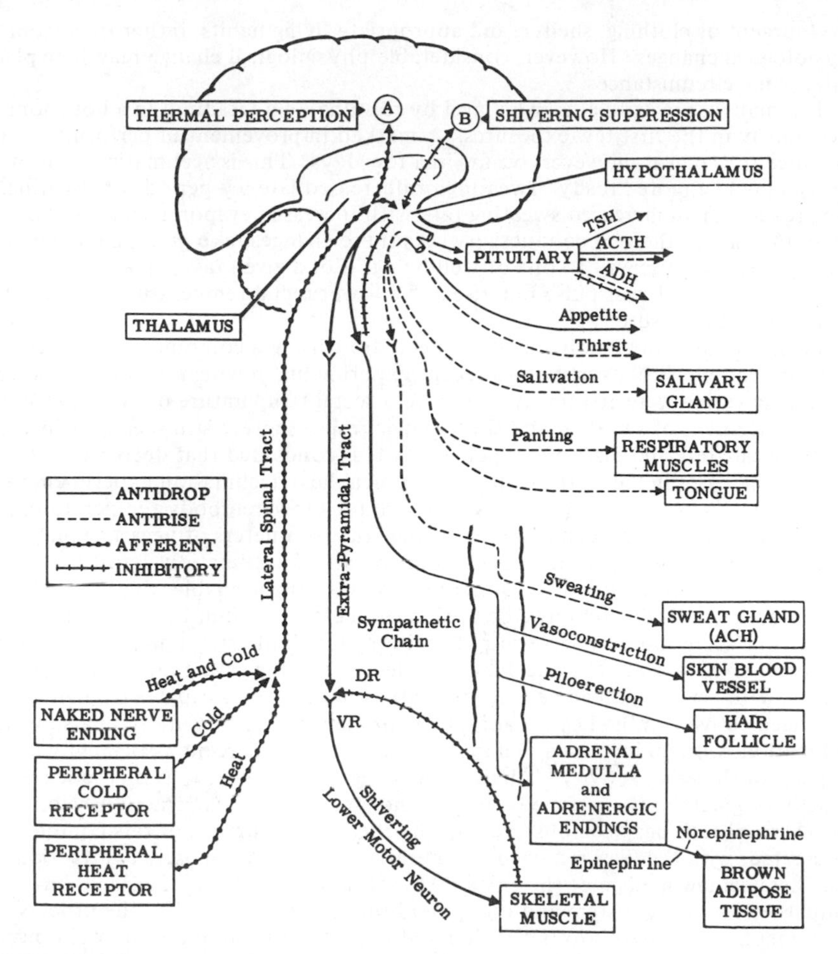

FIGURE 28-4. Integration of temperature regulation mechanisms. (From Burton and Edholm, 1955.)

cold is still capable of controlled muscular action because the shivering may be readily halted in favor of voluntary motor activity, perhaps through this suppressor mechanism.

Physiological responses to cold and heat also involve food and water intake and probably many hormonal changes. In fact, almost all tissues or organ systems are involved in, or influenced by, the thermal environment or the responses to it.

ACCLIMATION TO HEAT AND COLD

Acclimation refers to functional changes occurring in an organism upon continued or repeated exposure to a hot or cold environment. This definition is limited here to include only functional changes which enhance performance or survival in the environment. Most of man's ability to combat climatic extremes has come from development of clothing, shelter, and appropriate living habits, rather than from physiological changes. However, considerable physiological change may take place under some circumstances.

If a man works daily at a given level by running on a treadmill in a hot room, he does poorly in the first few exposures. A marked improvement in performance and tolerance to the heat, however, occurs in a few days. This is acclimation to heat and consists in (1) a more "ready" sweating or decreased latency period for sweating occurrence, (2) an increased sweating rate with increased evaporation heat loss, and (3) an increase in thermal conductivity. As these changes take place, a feeling of heat oppression gives way to one of well-being. For a given task, an acclimated individual shows a lower pulse rate rise and a lower rectal temperature rise than he did upon first exposure.

Most acclimation experiments have been done using a combination of heat exposure and physical exertion. In recent experiments in which men at rest wearing a plastic covering were artificially heated to a rectal temperature of 40°C every day, the enhancement of sweating and the vascular response were still seen but the augmented heart rate and discomfort persisted. It is concluded that decreases in heart rate and increases in comfort of the more typical heat acclimation experiments are consequences of the increased ability to maintain a lowered body temperature, and sweating and circulatory adjustments are the true parameters of heat acclimation.

Decreased sodium chloride loss in sweat is part of acclimation, but it has been found that sweat salt content is primarily related to the available supply to the sweat glands. If the individual enters a condition of negative sodium balance, increased aldosterone secretion may serve to produce a rather dilute sweat and greatly reduce the salt loss. Man cannot, by acclimation, decrease his sweating rate and his water intake. With acclimation and sufficient water intake, man is capable of high rates of prolonged activity in the heat. Laboratory animals kept at 5°C definitely show an acclimation, with an increased metabolism of 50 to 100 percent associated with a decrease of shivering activity. This *nonshivering thermogenesis*, as it is called, may actually originate in the muscle tissue, but not from muscle contraction. Nonshivering thermogenesis has been observed during the limited survival of rats completely eviscerated, with the liver removed and curarized to prevent muscle contraction. The chemical mechanism of the extra metabolism has not been described, but the sympathetic nervous system and norepinephrine appear essential to the process. Norepinephrine infused into the acclimated rat may bring about a threefold increase in metabolism. The thyroid hormone is also involved and appears to have a synergistic action with norepinephrine. The completely thyroidectomized rat does not show the metabolic increase or "calorigenic" action with injected norepinephrine but may still shiver.

Whether or not a similar metabolic acclimation to cold occurs in man is an unsettled question which requires further study. One of the few remaining groups of humans that are consistently exposed to cold are the Australian aborigines, who sleep almost naked in near freezing temperatures. They do not show a metabolic increase but allow considerable body cooling at night with repayment of the heat deficit with morning activity. Energy is conserved at the expense of body temperature. The diving women of Korea, accustomed to cold water immersion, demonstrate great decreases in tissue conductance in cold exposure. This group shows in the winter what appears to be nonshivering thermogenesis.

FEVER

Fever presents a most interesting aspect of temperature regulation and is extremely important as a reliable and common sign of disease. The mechanism of action of the agent, such as a bacterial pyrogen (e.g., endotoxin from gram-negative bacilli) or an endogenous pyrogen (e.g., leukocyte extracts), is not known, but experiments with highly purified pyrogen suggest a direct effect on firing rates of neurons in the preoptic area. The induction process of most fevers appears similar to response seen upon cold exposure. The first sign in many fevers is vasoconstriction, which may be blocked or reduced by sympathectomy and adrenergic blocking agents. From 10 to 30 minutes later, shivering may occur, with a resultant increase in heat production. In one study of malarial "chills," it was found that metabolism (M) increased to 290 watts, vasoconstriction reduced heat loss by radiation (H), and a large increase in heat stores (S) occurred, with a rapid rise of core temperature to 40°C. Eventually the temperature reached 41.5°C.

The evidence for a fever mechanism has been summarized by DuBois (1948): "In fever it appears as if the thermoregulatory center were suddenly set at a higher level just as one might change the regulator in a thermostat or in a house with automatic heating." The limited studies that have been made indicate that fever patients still possess precise temperature regulation ability and prevent overheating in a warmer room or with increased activity by increased sweating. Interestingly enough, the temperature in fever rarely rises above 41.5°C. When the agent responsible for the febrile state is no longer effective, the physiology of the patient is similar to that seen as a response to a hot environment, i.e., vasodilation and sweating. The so-called crisis, for example in the termination of pneumonia, consists in this vasodilation and sweating, which rapidly brings the body temperature back to normal. Just why the body reacts to certain agents by increasing body temperature and whether or not this increase of temperature is beneficial are questions which cannot be answered at present.

HYPERTHERMIA AND HEAT DEATH

In prolonged or severe heat stress man's performance may fail through several pathophysiological mechanisms. The simplest are heat syncope and asthenia with vague manifestations of fatigue, headache, and mental and physical inefficiency resulting from the vasodilation induced by a warm environment. Second in severity are the disorders of water and electrolyte metabolism resulting from sweating and dehydration, including heat cramps and heat exhaustion. Also related to this category would be the tetany observed in resting subjects exposed to hot humid environments which can induce hyperventilation.

Heat stroke, a third category, is characterized by *severe disturbances of consciousness and brain function,* and by *total absence of sweating,* which results in self-perpetuating hyperthermia. As body temperature rises, heat production of the tissues also rises, being forced upward by thermally forced increases of chemical reaction rates. Thus, a vicious cycle is started in which temperature is continually pushed upward to the lethal level unless rapid cooling is utilized. For man such a level appears to be 45° or 46°C, although serious impairment occurs in some cases near 40°C. The actual mechanism of heat stroke or heat pyrexia is not very well understood. It appears to be related to cell damage, particularly in the nervous system. Liquefaction of lipid components in the cell membrane has been suggested, but there is evidence that cell protein is being destroyed by denaturation faster than it can be rebuilt.

HYPOTHERMIA AND COLD DEATH

When mammals are exposed to severe cold, such as ice water immersion, the circulatory and metabolic capacities for body temperature maintenance are rapidly exhausted and body temperature falls, resulting in *hypothermia.* With the great loss of body heat stores the rate of body heat production is reduced, and cooling of the central nervous system leads to a suppression of hypothalamic control centers. As temperature decreases, there is progressive depression of nervous system excitability, slowing of muscular movement, and respiratory failure followed by circulatory failure. For many years it was thought that mammals other than hibernators had distinct lower limits of core temperature compatible with survival; temperatures once considered lethal are 17°C for the dog and 15°C for the rat. Experiments in Nazi prison camps in World War II, although of questionable validity, established a lethal temperature for man of about 25°C. However, there are authenticated cases in which individuals, usually in a drunken stupor, have survived core temperatures as low as 16°C. Circulatory failure and resulting anoxia, either general or localized, have been suggested as the cause of death.

A revolution has occurred in cold death theory as the result of some experiments by Andjus, in Belgrade, in 1951. Rats were sealed in 2-liter jars and placed in a refrigerator. Because of the decrease in oxygen and buildup of carbon dioxide, thermal regulation was depressed and rapid cooling took place. With localized rewarming of the chest it was found that rats could survive after cooling to 0 to 2°C. In the usual cooling methods, survival was impossible after cooling to below 15°C. Further experiments showed that hamsters could be cooled to well below freezing, with ice formation in 40 percent of the body tissues, and then be rewarmed with a high percentage of survival. Dogs may also survive lowering of body temperature to near 0°C when rapid methods of cooling are utilized. A patient has been cooled to core temperature of 9°C, with stoppage of the heart for one hour, and yet survived the cooling. The results of these studies have led to a reevaluation of cold death theories, since lethal temperature seems dependent upon methods of cooling and rewarming. In deep hypothermia, tissue metabolism is so low that blood circulation is hardly necessary. However, at present only a limited duration of life is possible at core temperatures near 0°C. The ultimate cause of death is unknown. In studies employing the oxygen electrode it was found that death was more frequent in animals having lower tissue oxygen tension.

Clinically hypothermia is a useful adjunct to procedures such as cardiac surgery and correction of various aneurysms. The rationale is simple: The reduced body temperature decreases metabolism and the tissue oxygen demand. Thus, blood flow to vital areas may be halted for variable time periods without anoxic damage. A particular hazard of hypothermia is the high incidence of ventricular fibrillation (Chap. 14). Induced hypothermia may be used more extensively in clinical work as more of its physiology is understood.

REFERENCES

Adolph, E. F., et al. *Physiology of Man in the Desert.* New York: Interscience, 1947.

Burton, A. C., and O. G. Edholm. *Man in a Cold Environment.* London: Arnold, 1955.

Dill, D. B. (Ed.). *Handbook of Physiology.* Section 4: Adaptation to the Environment. Washington, D.C.: American Physiological Society, 1964.

DuBois, E. F. *Fever and the Regulation of Body Temperature.* Springfield, Ill.: Thomas, 1948.

Greenfield, A. D. M. The Circulation Through the Skin. In W. F. Hamilton and P. Dow (Eds.), *Handbook of Physiology.* Section 2: Circulation, Vol. II. Washington, D.C.: American Physiological Society, 1963. Pp. 1325–1353.

Hammel, H. T. Regulation of internal body temperature. *Ann. Rev. Physiol.* 30:641–710, 1968.

Hardy, J. D. (Ed.). *Temperature, Its Measurement and Control in Science and Industry,* Vol. 3, Pt. 3: Biology and Medicine. American Institute of Physics. New York: Reinhold, 1963.

Hardy, J. D., A. P. Gagge, and J. A. J. Stolwijk. *Physiological and Behavioral Temperature Regulation.* Springfield, Ill.: Thomas, 1970.

Kuno, Y. *Human Perspiration.* Springfield, Ill.: Thomas, 1956.

Newburgh, L. H. (Ed.). *Physiology of Heat Regulation and the Science of Clothing.* New York: Hafner, 1968.

Strom, G. Central Nervous Regulation of Body Temperature. In J. Field (Ed.), *Handbook of Physiology.* Section 1: Neurophysiology, Vol. II. Washington, D.C.: American Physiological Society, 1960. Pp. 1173–1196.

29

Physiology of Exercise

Robert W. Bullard

THE primary physiological event in exercise is contraction of skeletal muscle. This extremely complex event requires a score of enzymes acting in series to promote the conversion of chemical energy into mechanical energy. The basic biochemical alterations at the muscle cell level are accompanied by increase of integrated activity of many systems, beginning with impulses from the central nervous system which initiate coordinated skeletal muscle activity. Extra oxygen must be carried to active cells, and carbon dioxide away from them, at rates dependent upon the severity and duration of the exercise. The greatest activity of circulatory and respiratory functions may be required in exertion. Contracting muscle cells may increase total heat production 10 to 20 times and thus place extra demands on thermal regulatory mechanisms. These complex interrelationships of metabolic, circulatory, respiratory, and neural activities present a most challenging field to the biologist. Only a thorough understanding of exercise physiology will permit the physician to give reasoned answers to those vitally important questions concerning physical activity of the young, the middle-aged, the aged, and the patient with ischemic heart disease.

METABOLIC ASPECTS OF EXERCISE

ENERGY REQUIREMENTS IN EXERTION

As explained in Chapter 27, the biological oxidation of fats, proteins, and carbohydrates produces energy which may eventually be dissipated from the body as heat or work. The physiologist, utilizing physical theories, measures the total work output and energy turnover of the exercising individual. In many activities, such as most contact sports, this measurement would be exceedingly difficult. Therefore, the most thorough physiological studies have been done in situations in which work can be measured. Common methods are running on a treadmill on a belt moving at a known rate and set on a known grade, and pedaling a bicycle ergometer against either a known resistance or a generator from which energy output may be measured.

A simplified example of an energetics calculation is presented as a review of the basic principles involved:

When a 70-kg individual climbs a flight of stairs 100 meters in height, he has then done 70 kg times 100 meters or 7000 kg-meters of work. Because 427 kg-meters is equivalent to 1 kilocalorie, the individual has put 7000/427 or 16.4 Kcal into useful work. Performance of this work would require about 3.5 gm of glycogen or glucose (1 gm = 4.7 Kcal) and 3 liters of oxygen (caloric equivalent = 5 Kcal per liter). However, at the level of efficiency of human work of this type, only about one-fifth of the total energy turnover can go into useful work; thus, the activity described would require 5 times more energy or about 17.5 gm of glucose and 15 liters of oxygen.

In most exercise experiments it would be difficult to measure the loss of body glucose or glycogen content or other chemical energy stores. However, the amount of oxygen consumed by a working individual may be readily measured by standard techniques. The volume of oxygen consumed and the volume of carbon dioxide produced for any time period may be determined at various work levels by collecting the expired air in a Douglas bag or large spirometer, to measure its volume and then to analyze for carbon dioxide and oxygen content (Chaps. 19 and 20). Assumptions can be made about the proportion of carbohydrates, fats, and proteins being utilized as metabolic fuel from the ratio of CO_2 to O_2. From this ratio the caloric equivalent of 1 liter of oxygen consumed may be approximated from standard charts (Chap. 27) with small error. For an individual utilizing an average mixture of foodstuffs, 4.825 Kcal are produced for each liter of oxygen consumed.

OXYGEN REQUIREMENT DURING WORK

The average individual in the basal state consumes about 0.250 liter of O_2 per minute (or $\dot{V}_{O_2} = 0.250$). The abbreviation $\dot{V}_{O_2}$ will be used throughout this chapter for O_2 consumption in liters per minute. Light or moderate industrial tasks may involve a threefold increase in $\dot{V}_{O_2}$. Heavy work may cause an eightfold increase in O_2 consumed, but usually this work rate is not maintained for long periods. Athletic activity such as running, cycling, swimming, cross-country skiing, and wrestling may raise the $\dot{V}_{O_2}$ well above an eightfold increase.

When a bout of heavy exercise is performed which exhausts an individual in 3 to 10 minutes, a plateau, or the so-called *maximum limit* of $\dot{V}_{O_2}$, is attained in 1 to 2 minutes (Fig. 29-1). The maximum intake, or $\dot{V}_{O_2}$ max, of an untrained individual is 8 to 12 times as large as the basal intake. On the other hand, the trained athlete may show a $\dot{V}_{O_2}$ increase of 20 times with all-out exertion. Indeed, the upper limit of $\dot{V}_{O_2}$ is one of the best indices of physical fitness.

Many well-trained athletes have $\dot{V}_{O_2}$ max values between 4 and 5 liters per minute. However, values as high as 5.88 in a world class cross-country skier, 6.0 in a cross-country runner, and 6.20 in a world champion cyclist have been recorded. It is preferable to express $\dot{V}_{O_2}$ max as oxygen consumed per kilogram of body weight. When this is done, values between 50 and 60 ml per kilogram per minute are obtained for individuals in "good" condition, and 85 ml per kilogram per minute in an endurance athlete appears to be the highest value recorded. There is not a perfect correlation between $\dot{V}_{O_2}$ max and performance; often an athlete can have a high $\dot{V}_{O_2}$ max and be a mediocre performer, indicating the important influences of other factors less objectively measured.

OXYGEN DEBT

A scheme of the sources of contraction energy is presented in Figure 29-2. The ultimate energy source is oxidation of glycogen and glucose or lipid and protein fragments to CO_2 and H_2O. Such oxidation results in the buildup of energy in the form of "packets" or high-energy phosphate bonds. The breaking or hydrolysis of these bonds supplies the energy for the sliding of actin filaments by the myosin

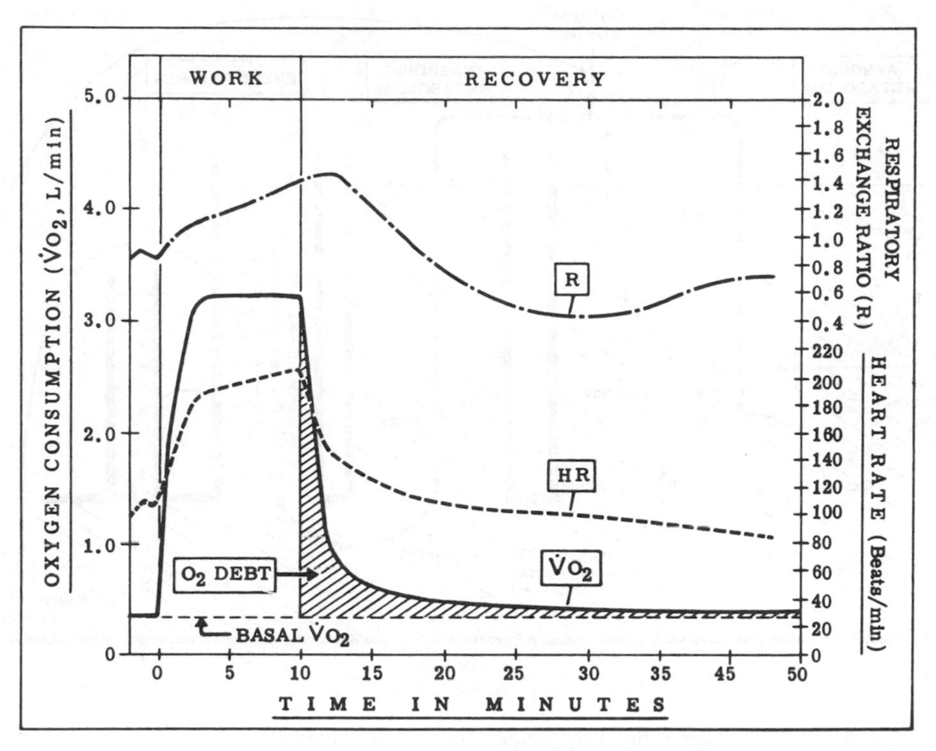

FIGURE 29-1. Changes in oxygen consumption, respiratory exchange ratio, and heart rate during and following heavy work. The shaded area represents the oxygen debt.

filaments in the contractile process of skeletal muscle. The actual contraction does not require the presence of oxygen but proceeds in its absence as long as the source of high-energy bonds remains.

From Figure 29-2, it can be seen that the primary energy source, adenosine triphosphate (ATP), with hydrolysis yields the energy for contraction, loses one phosphate, and forms adenosine diphosphate (ADP). From creatine phosphate (CP or phosphoryl creatine), which also contains the high-energy phosphate bond, ADP is converted to ATP. Biopsy studies reveal that 100 gm (dry weight) of human quadriceps muscle contains about 8 mg of CP and 2 mg of ATP. With exercise, there is little fall in the ATP reservoir, but CP falls to 2 to 4 mg in moderate exercise and to near-zero concentration with heavy exercise. Thus, CP breakdown must rather quickly serve to replenish ATP.

Energy for replenishment of the high-energy bonds of CP may come from glycolysis, which consists in the phosphorylation and breakdown of glycogen to pyruvate or lactate under anaerobic conditions. A net yield of two high-energy bonds results from such breakdown of each 6-carbon unit. When oxygen is available and oxidation by means of the Krebs citric acid cycle and electron transport is complete to CO_2 and water, 38 bonds are generated, as shown in Figure 29-2.

As exertion is increased in intensity and exhaustion approaches, the cardiopulmonary system, delivering oxygen to the active tissue, falls behind the oxygen demand. Following work, the $\dot{V}_{O_2}$ will remain elevated from several minutes up

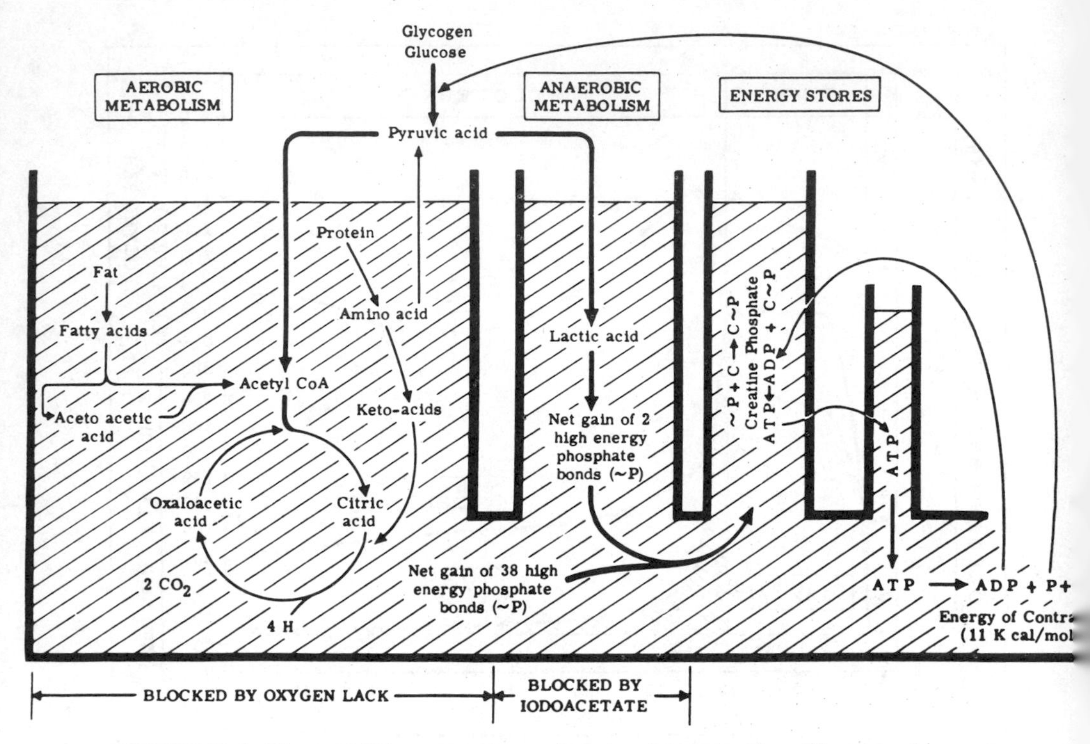

FIGURE 29-2. Diagram of the energy sources for muscular contraction. The adenosine triphosphate and creatine phosphate may be considered storage depots for high-energy phosphate bonds. The anerobic glycolysis and aerobic metabolism function to restore these storage depots.

to one or two hours. All postexercise $\dot{V}_{O_2}$ above the basal oxygen consumption level is defined as the oxygen debt. Characteristically, this debt is repaid rapidly at first and then at a slower rate, as shown in Figure 29-1.

The buildup of oxygen debt to a critical level appears to be an important but poorly understood factor limiting the duration of heavy exertion and represents one form of exhaustion. The untrained individual will stop activity when he has surpassed a debt of about 10 liters, but with endurance training the debt ceiling may be increased to 17 to 18 liters. The determinant of the ceiling, whether muscle glycogen, CP depletion, or other factors, has not been fully evaluated. However, it is obvious that if the effort is of long duration, the athlete cannot allow the continuous buildup of the debt but must work aerobically or at a rate low enough for oxygen delivery to parallel oxygen cost or demand. This pay-as-you-go exertion is known as *steady state* work, and its upper limit is a $\dot{V}_{O_2}$ of 3 to 4 liters per minute or about 70 percent of the maximum $\dot{V}_{O_2}$ of the trained athlete.

Although the oxygen debt is usually attributed to the cost of oxidation and reconversion of lactic acid, the metabolic end product of anaerobic breakdown of carbohydrate, no very firm correlation of the debt volume and lactic acid content has been found. At fairly low levels of work ($\dot{V}_{O_2}$ less than 2.5 liters per minute) an oxygen debt occurs without any increased blood lactate concentration. This is known as the *alactic* oxygen debt. When the duration of work is increased, the debt stays constant, is paid off in about two minutes, and is equal to or only slightly greater than the initial lag in the oxygen consumption rise at the start of work.

The alactic debt is generally attributed to a replenishment of body oxygen stores including (1) restoration of oxygen saturation of muscle myoglobin and blood hemoglobin, which had been reduced because of the initial circulatory lag in O_2 supply; (2) oxidation of reduced H^+ acceptors and metabolites including lactate; (3) and possibly the O_2 stores of tissues, which may change slightly. Recently it has been shown that the alactic O_2 debt corresponds quite closely to the amount of oxygen required to regenerate the creatine phosphate store to its original level.

With more severe exertion, in the range of 70 to 100 percent of the $\dot{V}_{O_2}$ max, the oxygen debt will increase with duration, and it can become many times greater than the lag phase in $\dot{V}_{O_2}$ at the start of exertion. If all the oxygen required to replenish all stores, including venous blood, tissues, myoglobin, and arterial blood, and the oxygen amount required to regenerate all CP and ATP that could be used, is computed, the total is only 3.4 liters. Apparently the repayment, which is actually what is measured, greatly exceeds the accountable debt. Harris (1969) discounts the involvement of lactic acid oxidation or conversion in the postexercise "recovery volume" or debt, but attributes it to correction of chemical disturbances resulting from contraction. Another viewpoint is that the rapid part of the recovery is due to replenishment of O_2 stores and creatine phosphate, a process requiring only a few minutes, and that the slower phase (Fig. 29-2) is due to an increased or disturbed resting metabolism. This is an area requiring advanced biochemical approaches.

TOTAL OXYGEN CONSUMPTION AND OXYGEN COST

Both aerobic ($\dot{V}_{O_2}$) and anaerobic (oxygen debt) processes are utilized to various degrees in different athletic events. The oxygen cost of any exertion is the sum of oxygen used during the work and the recovery volume or O_2 debt. This is conveniently expressed on a per minute basis, abbreviated as $\dot{V}_{O_2}$ total, and computed as follows:

$$\dot{V}_{O_2}\text{ total} = \frac{\text{Total } O_2 \text{ consumed during work} + O_2 \text{ debt}}{\text{Duration of work in minutes}} \qquad (1)$$

With this equation, the cost of a work task can be compared to summed totals of aerobic and anaerobic power as reflected in oxygen utilized to determine whether an athlete can complete the task. Figure 29-3 is a plot of linear speed of walking and running versus oxygen cost per minute. From the plot, consider a hypothetical individual attempting a 4-minute mile which requires a pace of 7.3 yards per second costing 8.7 liters of O_2 per minute. Assume that this individual can maintain an average $\dot{V}_{O_2}$ of 4 liters per minute for the 4 minutes and can tolerate an oxygen debt of 16 liters. His $\dot{V}_{O_2}$ total then equals

$$\frac{4\text{ min} \times 4.0\text{ liters/min} + 16}{4\text{ min}} = 8.0\text{ liters/min} \qquad (2)$$

Thus, for this pace and duration, the average oxygen used including the recovery is 8.0 liters per minute, which falls short of the cost of 8.7 liters per minute. The 4-minute mile is beyond this individual's physiological capabilities. For the half-mile run in two minutes:

$$\frac{2.0 \times 4.0 + 16}{2} = 12.0\text{ liters/min} \qquad (3)$$

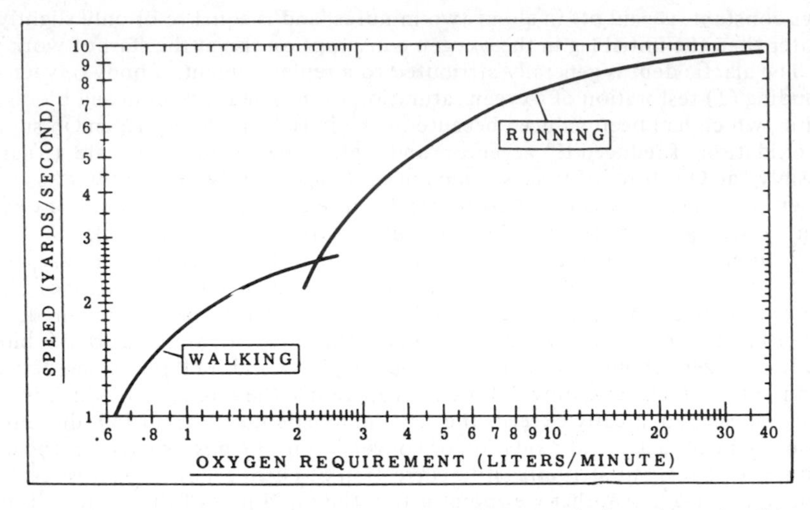

FIGURE 29-3. Oxygen cost (liters per minute) versus linear speed of running and walking. (From F. Henry. *Res. Quart.* 24:169, 1963.)

As 12.0 greatly exceeds the cost of 8.7 liters per minute, he can certainly maintain that pace for two minutes. The number of athletes running the half-mile in two minutes far surpasses the number who can run the mile in four minutes.

These calculations are more instructive than practical; the athlete can be tested far more simply on the track. They do, however, emphasize the relative importance of aerobic and anaerobic metabolism at varying work intensities and durations. In the mile, half the O_2 consumed was in the recovery. In the sprints, most of the O_2 of effort was consumed in recovery. It has been reported that the trained sprinter can go 100 yards while holding his breath. At best, the O_2 consumed in a 10-second sprint would be only 0.5 liter, but as such, a run requires about 6 liters; a 5.5-liter O_2 debt is incurred. Here 90 percent of the cost is paid for in recovery.

At the other extreme is the marathoner running at a pace of 5.3 yards per second for 144 minutes. Figure 29-3 shows that this pace costs 4.4 liters of O_2 per minute. Assume our runner can maintain a $\dot{V}_{O_2}$ of 4.2 liters per minute and can tolerate a 16-liter debt:

$$\dot{V}_{O_2} \text{ total} = \frac{4.2 \times 144 + 16}{144} = 4.4 \tag{4}$$

This man can meet the cost of the run. Only 2 percent of the total oxygen cost comes from the oxygen debt, and thus for long-duration exertion, steady state or pay-as-you-go work rates must be utilized.

MECHANICAL EFFICIENCY

Obviously two contributing factors in high-level performance during severe exertion are (1) ability to reach a high $\dot{V}_{O_2}$ and (2) ability to tolerate a high oxygen debt. Another extremely important factor is the ability to perform a given task with the lowest possible oxygen cost or energy requirement, or with the greatest *mechanical efficiency* (ME). Efficiency is usually calculated in percentage, by the following relationship:

$$\text{Percent ME} = \frac{\text{External work in kilocalories} \times 100}{\text{Total kilocalories required for work}} \tag{5}$$

The exercise must be such that the useful work may be measured, a difficult task in most athletic activities. Total kilocalories required is determined usually from oxygen consumption data. The sum of oxygen consumed during work and oxygen debt is multiplied by the energy equivalent.

Physiologists have utilized two distinct methods for calculating efficiency. The first and simplest is *gross efficiency,* the method described above, in which the total energy output is divided into the total measurable work. The second, *net efficiency,* assumes, perhaps incorrectly, that basal processes continue unchanged throughout the work period. Therefore, basal energy turnover, which is determined by the resting oxygen consumption, may be subtracted from total energy.

$$ME_{net} = \frac{\text{Kilocalories of work} \times 100}{\text{Total kilocalories} - \text{basal kilocalories}} \tag{6}$$

Net efficiency calculated in this way yields a value higher than gross efficiency. It is often difficult to measure the actual amount of work done when a man is walking or running on a level treadmill. Methods of calculating the work done have consisted of careful analysis of movies of the action, with consideration of limb weights, distances they are moved, and forces involved in their acceleration and deceleration. To eliminate this time-consuming process, the physiologist may determine the energy requirement for walking or running at a given pace on a level treadmill. The subject then goes at the same pace on a treadmill set at a grade. If the grade is 10 percent and the 70-kg individual goes 1000 meters against the moving belt, he has done $70 \times 1000 \times 10$ percent or 7000 kg-meters of work above that required for linear motion only. This work in kilocalories is 7000/427 or 16.4 Kcal. Only this work is considered. The energy involved in moving on the level treadmill is eliminated. By this method the so-called absolute efficiency is calculated:

$$ME_{absolute} = \frac{16.4 \times 100}{\text{Total Kcal, graded run} - \text{total Kcal, level run}} \tag{7}$$

Thus, the problems of calculating work for a run on the level are avoided. This third method may yield net efficiencies of over 30 percent.

Which method of mechanical efficiency calculation is the best is a moot point. Provided the same method is used when comparisons are being made, all methods are equally valid. By any method of determination, the efficiency of the organism compares favorably with that of most machines. It has been pointed out that the organism is perhaps highly efficient because it supplies its own food and automatically replaces or repairs worn parts. However, if it takes, for example, one day to recover from a bout of heavy exertion, then true mechanical efficiency is quite low.

Efficiency may be altered to a certain extent by training and practice. The beginning swimmer, thrashing wildly to move through the water, is far less efficient than the smooth-stroking champion. Economy of effort may be developed with training. Studies have shown that the trained swimmer may possess an efficiency at least six times that of the untrained swimmer.

Mechanical efficiency is also related to rate of work or velocity of contraction. Very slow movements as well as extremely rapid movements are inefficient. Climbing stairs at a rate of 50 steps per minute is more efficient for the average young person than a slower or faster rate. A slow rate of static contraction or "holding back" without production of useful work reduces efficiency. The break in the curve in Figure 29-3 is due simply to the fact that extremely fast walking is less efficient than running at that linear velocity. In running, oxygen cost does not rise linearly with velocity but exponentially. The exponent of the oxygen cost rise with velocity increase usually ranges from 2 to 3. At faster rates of movement the athlete may be reducing efficiency because greater forces are required for acceleration and decleration of limbs, and because he is fighting, more and more, the intrinsic or viscous resistance of the contractile system.

RESPIRATORY EXCHANGE RATIO AND THE FUEL OF EXERCISE

It will be recalled from Chapter 27 that the respiratory exchange ratio (R) is equal to the volume of CO_2 expired/volume of O_2 consumed. When this is determined under resting, steady state conditions, it is known as the respiratory quotient or RQ and reflects the metabolic foodstuff being oxidized. In the unsteady state of heavy exercise this ratio (R) is not dependent upon the metabolic fuel but on the immediate alterations of respiratory exchange. Just prior to exertion some individuals show an anticipatory increase in pulmonary ventilation. This results in a temporary pumping out of CO_2 stores and may temporarily increase R. Further increases in R may be seen in exercise as lactic acid begins to diffuse from cells into body fluids. Lactic acid acts as a fixed acid, and reactions are as follows:

$$\begin{aligned} &\text{Lactic acid} + NaHCO_3 \rightleftharpoons \text{Na lactate} + H_2CO_3 \\ &H_2CO_3 \rightleftharpoons H_2O + CO_2 \end{aligned} \tag{8}$$

The CO_2 is then blown off, a process that compensates for metabolic acidosis. The extra blowing off then may raise R to quite high values, perhaps 1.6, as shown in Figure 29-1. Actually, R may keep rising for a short time following exertion while O_2 consumed (the denominator) drops rapidly and lactic acid is still diffusing from cells and forcing CO_2 formation. As recovery progresses and lactic acid is removed from body fluids, these reactions are reversed. Metabolically produced CO_2 will be retained in bicarbonate to replace the lactate. With the CO_2 retention, the R value will reach extremely low values, possibly 0.5. Complete recovery is indicated by an increase to normal values.

The respiratory exchange ratio (R) at any instant gives little information about the chemical fuel being utilized in exertion. However, some workers feel that the integrated or average R for the entire exertion period and recovery period will be the same as the RQ and accurately reflect the fuel oxidized. The assumption necessary for this approach is that CO_2 stores of the body are completely replenished by the end of the recovery period.

Thus considerable caution is in order when using R with heavy or intermittent exercise. With steady state work, the R value appears to be a fairly accurate reflection of the respiratory quotient or the chemical fuel utilized, equaling 1.0 with all carbohydrate to 0.7 with lipid utilization. It is generally believed that protein metabolism is not greatly increased in exercise and is, therefore, not an important factor. Recent studies have employed techniques such as measuring arteriovenous differences in blood lipid and glucose content, and advances have also been made using radioactively labeled glucose and fatty acids. With these techniques, the available supply of the substrate can be assessed, and the amount of radioactive carbon dioxide produced is a measure of the utilization of the labeled substrate.

The most important technical advance has been the development of muscle biopsy techniques used on the exercising individuals as mentioned above. With these techniques, the utilization is directly measurable and can be compared to the recorded R values and total oxygen consumption.

Both lipid and carbohydrate fuels are utilized in exercise in proportions varying with work intensity, duration, and availability of stored fuels. In heavy work above 70 percent of the $\dot{V}_{O_2}$ max, the fuel is carbohydrate deriving primarily from glycogen stored in the active muscle. In one typical experiment, subjects maintained an average $\dot{V}_{O_2}$ of 3.3 liters per minute for 90 minutes. The R value leveled off between 0.9 and 0.95, suggesting that 65 to 80 percent of the utilized fuel was carbohydrate. By biopsy measurement, glycogen in active muscles decreased from 1.6 gm per 100 gm of muscle (wet weight) to 0.8 gm at 40 minutes and to 0.01 gm at 90 minutes. Toward the end of this work period, when muscle glycogen content was low, the R value had decreased only slightly. The decrease, of course, suggests a greater lipid utilization. However, the fact that the R value decreased slightly suggests that glucose uptake from the blood increased. Follow-up studies, in which the glucose concentration differences across the liver circulation and the liver blood flow were measured, demonstrated that the liver added about 25 gm of glucose to the circulation. As blood glucose concentration remained relatively constant, it is assumed that this glucose was utilized in exertion. Yet the total carbohydrate used, as estimated by changes in glycogen, and indirectly from the R value and total $\dot{V}_{O_2}$, was over 150 gm; thus with heavy exertion, most of the fuel is the glycogen from the active muscle itself. Attempts have been made to maintain glycogen levels by increasing blood glucose concentration. It appears that the conversion of glucose to muscle glycogen or direct utilization of blood glucose is too slow to prevent or greatly slow the fall in muscle glycogen.

The greatest drop in muscle glycogen appears early in work. This is also the time at which the greatest production of lactic acid occurs. This, of course, is an indication of anaerobic breakdown of glycogen, where metabolic demand is exceeding oxygen supply. Under anaerobic conditions, carbohydrate is the only source of contractile energy as it is broken down to lactic acid. Lipid requires a supply of oxygen to provide a contribution to contraction. It appears that physical training enhances the ability to replace muscle glycogen rapidly.

With lower rates of work, lipid supplies a larger portion of energy. For the human, with light to moderate exercise, 50 percent of the fuel derives from lipids, primarily free fatty acids. In careful experiments with dogs, this ranges up to 71 percent of total fuel requirements. It has been shown by studying arteriovenous blood lipid differences across the muscle that about half of the free fatty acids utilized derive from the blood and the other half from local stores of lipids.

There exist physiological mechanisms for mobilizing free fatty acids from fat deposits utilizing adrenergic activation. It should be remembered that in such a conversion glycerol also is released which is readily utilizable as a carbohydrate fuel contraction. Injection of radioactive glycerol into the exercising human forearm indicates that 25 to 30 percent of the produced CO_2 is derived from this source.

It appears that the body can substitute carbohydrate for lipid, but lipids cannot be substituted for carbohydrates, particularly at a fairly high rate of work. Muscle fatigue at such rates occurs, probably from glycogen exhaustion prior to the point at which the R value is greatly reduced. However, the individual may be able to go to a lower rate of work and continue with greater lipid utilization and a lower respiratory quotient. This is also the case when a direct low in carbohydrate is utilized. It is not clear what factors in lipid oxidation are limiting its usefulness as a substrate in heavy exertion.

BODY TEMPERATURE IN EXERCISE

Prolonged severe exertion markedly increases body temperature. Heat production is greatly augmented because only about one-fifth of the energy released in the

10- to 20-fold metabolic increase can go into external work. The remaining energy appears as heat and must be dissipated or added to body heat stores. Increasing body temperature up to 39° to 40°C has been observed frequently in exertion even when the environment was cool. Rectal temperatures (representing internal body temperature) of 41.1°C (106°F) were recorded by Robinson (1949) after a 3-mile race between champion athletes Don Lash and Greg Rice. Strangely enough, many record-breaking performances have been made by athletes with postexercise body temperatures in the high fever range.

One important question is, How does an athlete tolerate this body temperature increase? If one attempts to raise the rectal temperature of a resting individual to 40°C, a vigorous defense effort by the sweating mechanism is encountered. Thus, dehydration and perhaps heat exhaustion may occur before this temperature is exceeded. The internal body temperature rise with exercise is independent of environmental temperature except at very extreme ranges, but dependent on the metabolic load. In one series of experiments, athletes performed a standard work test on a bicycle ergometer at three environmental temperatures: 22°C, 16°C, 11°C. In all environments the rectal temperature leveled off at 39.3°C. One subject's starting rectal temperature had been reduced to 36.4°C, but it also rose to this same level. There is no evidence that the temperature rise is due to a deficiency in thermoregulatory mechanisms (except under extremely warm conditions) because approximate thermal equilibrium seems to be attained at the higher temperature level. Internal body temperature then will stay almost constant even though heat production remains elevated.

All these observations suggest that in exertion, as in fever, a resetting to a higher temperature of the thermostatic (antirise) control center takes place. However, this issue is far from settled. Precise thermoregulatory experiments on dogs in which hypothalamic temperatures were measured suggest that thermoregulatory responses for a certain central temperature are greater in work than at rest. Human experiments have shown that exertion calls forth sweating responses prior to body temperature increases. A lowering of a thermostatic set point seems to be implied, yet the fact is not disputed that body temperature shows a marked increase in physical exertion.

A second question, Does the increase in body temperature enhance physical performance? It has not been answered because no one has designed the critical experiment. There is some evidence that skeletal muscle works more efficiently at temperatures higher than 37°C. It has been observed that the sources of increased heat production, namely, contracting muscles, may be 1°C higher than core temperature. This local heat may be beneficial in that it shifts the hemoglobin and myoglobin dissociation curves to the right to promote unloading of oxygen in the tissues (Chap. 20). A higher temperature may also slightly reduce the resistance to blood flow in the muscles by lowering blood viscosity. Diffusion rates of respiratory gases and metabolites to and from the tissues are likewise enhanced by temperature increases.

One advantage of allowing body temperature to rise is that less strain is put upon the thermoregulatory mechanisms. This may be important in regard to blood flow distribution. Maintenance of a constant body temperature would necessitate a considerable flow of blood to the skin and reduce the essential skeletal muscle blood flow. Sweating rates would need to be greatly increased also, thus resulting in possible dehydration earlier in prolonged work effort. Allowing a rise in body temperatures would increase the gradient with the ambient temperature (if cooler). By Newton's law (Chap. 28), a greater amount of heat could leave the body by convection and radiation, thus sparing some sweat loss.

These considerations do not mean that heat is always a beneficent friend of the athlete; it may be his worst enemy. The fate of many athletic contests has been decided by the differences in heat tolerance of the participants. A combination of high environmental temperature and greatly elevated heat production may rapidly bring on deterioration of performance. A marked increase in sweating rate promotes dehydration during work, and if fluids are not replaced, eventual circulatory collapse can be precipitated. The augmented skin blood flow necessary to facilitate heat loss can seriously impair muscle blood flow, especially if combined with the decreased blood volume of dehydration. To compete successfully in heat, an athlete must be acclimated (Chap. 28) and be fully aware of the hazards involved.

RESPIRATION IN EXERCISE

VENTILATION AND CONTROL OF RESPIRATION

The increased aerobic metabolism of skeletal muscle in exercise is supported by an increase in functions responsible for the supply of oxygen to the tissue. The augmented oxygen consumption of the working individual is paralleled quite precisely by increased pulmonary ventilation. This relationship is expressed as a ventilatory efficiency or the ventilation coefficient, or $\dot{V}_E/\dot{V}_{O_2}$ (in which $\dot{V}_E$ = pulmonary ventilation in liters per minute, E = expired, and $\dot{V}_{O_2}$ = oxygen consumed per minute). It is about 20 during rest, but reaches higher than 20 in heavy exertion. Some workers have reported no change in the ventilation coefficient over a wide spectrum of activity. Because $\dot{V}_E$ is the product of tidal volume (Chap. 19) and respiratory rate, it can be increased by augmentation of either factor. In moderate work the greater portion of increase is in tidal volume. In hyperpnea of heavy work, both respiratory rate and tidal volume must be greatly increased. The $\dot{V}_E$ of the trained athlete may surpass 100 liters per minute with a tidal volume of 3 liters and respiratory rate of over 33 breaths per minute.

The precise attunement of pulmonary ventilation to oxygen demand is one of the finest examples of physiological regulation; yet no clear-cut explanation of the controlling mechanisms has emerged which is universally accepted by physiologists in this field. This section briefly considers some of the factors that have been studied as drives for the exercise hyperpnea.

HUMORAL STIMULI

Carbon Dioxide

The parallelism between ventilation rate and metabolic rate logically suggests that the ventilatory drive may be from a metabolic by-product or chemical constituent that is altered at rates proportional to metabolism. As CO_2 is considered the prepotent respiratory stimulus, it was reasonable to explain the hyperpnea as due to an increase in alveolar or arterial P_{CO_2}. Actual measurements of arterial P_{CO_2} indicate that changes with exercise are slight. A small increase is often reported for moderate work and no change or a slight decrease for heavy work. A large P_{CO_2} increase would have to occur, in order to approach the ventilation rates of heavy exertion, if CO_2 were the stimulus. For example, inhalation of 8 percent CO_2 would markedly increase arterial P_{CO_2} and increase ventilation to 100 liters per minute, yet the trained athlete may reach 120 liters per minute without any detectable arterial P_{CO_2} change. The sensitivity of centers to increased arterial P_{CO_2} does not change with exercise. In the dog an increased P_{CO_2} of 10 mm Hg will raise the ventilation rate 50 per minute whether the dog is resting or exercising.

Lactic Acid and pH Change

Because CO_2 failed to meet the tests as a stimulus for exercise hyperpnea, lactic acid was considered. However, it has been shown that marathon runners may attain tremendously high ventilation rates with only a slight increase in the blood lactic acid level. After exercise, lactic acid levels may increase slightly while ventilation rate rapidly falls. Injection of large quantities of lactic acid does not increase ventilation to even a small fraction of that seen with exercise. Decreases in blood pH may augment ventilation, but pH changes in exercise are too slight, and sometimes not even in the proper direction, for driving the hyperpnea.

Arterial Hypoxemia

Several laboratories have observed that inhalation of air mixtures containing 33 to 100 percent oxygen reduces ventilation during moderately heavy work. However, explanations of this effect differ. In heavy exercise, arterial hypoxemia is not usually seen; if it is, many workers feel that the oxygen levels are not low enough to force chemoreceptor activity unless sensitivity is greatly increased. When the inspired oxygen percentage is increased from 33 to 100 percent, a further ventilatory decrease occurs. It is felt that the arterial P_{O_2} changes undergone with this change in inspired O_2 percent do not influence the chemoreceptor activity.

Catecholamines

The role of epinephrine and norepinephrine in the hyperpneic drive has not been fully evaluated but needs further consideration. The concentrations in blood and urine of the catecholamines increase with the intensity of work. The injection of catecholamines produces a rather marked increase in $\dot{V}_E$ in resting subjects. The mechanism of this response is poorly understood at present but may involve threshold or sensitivity change of regulatory centers.

Not one of the presented humoral factors can explain the hyperpnea of exercise. However, these factors singly or in combination, working with neurogenic stimuli, possibly chemoreceptors, and altered characteristics of the centers and receptors, may play a prominent role.

NERVOUS STIMULI

Peripheral Receptors

There are several neural mechanisms that may be involved in the hyperpnea of exercise, but it is difficult to understand how purely nervous activity can be so finely tuned to metabolic activity. Many investigators have observed that passive limb movement evokes a ventilatory response. This may be a proprioceptor response, as the ventilation increase is roughly proportional to the number of joints involved in the movement, and thus muscle spindles and articular receptors have been implicated. Nervous mediation is indicated by the fact that such responses are still seen after the leg blood flow is halted with a tourniquet. The magnitude of the ventilatory response is much lower than in actual exercise. However, some workers have suggested that muscle contraction alters the physicochemical environment of the receptors and augments their sensitivity.

Kao (1963) has done some promising experiments with direct electrical stimulation of the hind legs of anesthetized dogs. The induced muscular work increased both $\dot{V}_{O_2}$ and $\dot{V}_E$ to values equal to the value seen during actual exercise. It was demonstrated that the augmented $\dot{V}_E$ was a neural response by using pairs of dogs in cross-circulation experiments designed to separate the drives. The head of dog B received all of its blood supply from dog A. When the hind limb of A was stimulated to exercise, A had increased ventilation but B did not. When the hind limb of B was stimulated, A showed no response, but ventilation of B was markedly increased. Since the neural connections were intact in both dogs, the hyperpneic drive could be neural. In the same experimental setup Kao has found that if dog A is allowed to breathe air enriched with carbon dioxide, B will exhibit increased ventilation also, thus showing that a humoral influence, if present, would be carried from dog A to affect the respiratory center of dog B.

It must be emphasized that the receptors have not been characterized for any of these mechanisms.

Irradiation of Impulses

It has been proposed that as motor impulses leave the brain cortical area and pass through the reticular formation, some impulses irradiate to the respiratory center to evoke a ventilatory increase. Thus, ventilation would be proportional to the total activation of the muscle mass. Proof of such a mechanism is extremely difficult to obtain and awaits the design of a critical experiment. The experiments of Kao (see above) tend to discount the central mechanisms, but others have suggested the involvement of the gamma loop in increasing the traffic at neural impulses. It must also be kept in mind that the ventilatory response may of a volitional nature and learned by experience. Tests with hypnotized subjects who exercised without full awareness of work intensity have resulted in a lower ventilatory response than would be predicted from metabolic rates.

Afferent Impulses from Lung Stretch Receptors

Once tidal volume is increased, augmentation of respiratory rate could be promoted by feedback from lung stretch receptors (Chap. 22). This probably occurs but is not considered a prime stimulus of hyperpnea since voluntary ventilation is not self-perpetuating. It should be noted, however, that dogs with denervation of lungs show a greatly reduced work performance.

Increased Body Temperature

It has been noted that hyperthermia increases ventilation. However, the ventilatory increase in exercise is of far greater magnitude and is seen before any rise in rectal temperature is seen. Perhaps the temperature of the respiratory center does show an earlier rise, but sufficient study has not been made. In subjects made slightly hypothermic, a normal ventilatory increase in exercise is seen. It has been suggested too that increased temperature augments the responsiveness of the above-described joint and muscle receptors.

Increased Sensitivity of Respiratory Centers

Increased respiratory center sensitivity has been suggested but it is hard to approach experimentally. Some workers have found that when breath-holding time is measured during moderate work and correlated to the alveolar CO_2 tension, the breaking point occurs at a lower alveolar CO_2 in exercise than in rest. It is difficult to ascribe this effect purely to respiratory center sensitivity because the neural effects described above may be operating to increase the activity of the respiratory center.

RELATIVE IMPORTANCE OF HUMORAL AND NEURAL STIMULI

Obviously a satisfactory explanation awaits more intensive experimental evidence. Dejours (1964) has presented a most reasonable approach. As shown in Figure 29-4,

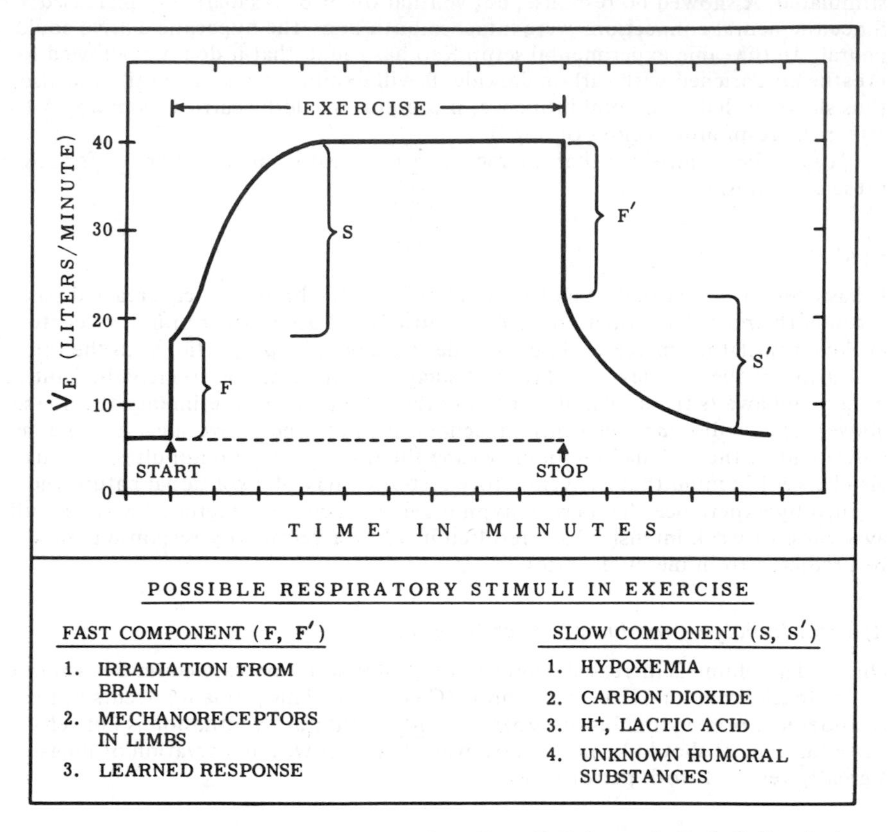

FIGURE 29-4. Ventilation responses in exercise. (From P. DeJours. In D. J. C. Cunningham and B. B. Lloyd [Eds.], *The Regulation of Human Respiration.* Oxford, Eng.: Blackwell, 1963.)

there is a double component of the $\dot{V}_E$ response to exercise. A very sharp rise at the beginning, proportional in magnitude to the intensity of work, is followed by a slower rise, a sharp decrease at the end of exercise, and then a slower decrease to resting levels. The steep rise and fall must be attributed to neurogenic components, for they occur prior to any possible metabolite change.

The slower components, particularly after exercise is stopped to eliminate neurogenic factors, are attributed to humoral factors. The precise nature of both the neural and humoral ventilatory drives has not been determined, but they may represent "potentiated" combinations of the factors presented above working in a summative or multiplicative fashion.

CIRCULATION IN EXERCISE

A coordinated response of the heart, resistance vessels, and capacitance vessels occurs with exercise in an effort to fulfill the oxygen demand of contracting muscles. This response is controlled by both humoral and neural factors. In recent years, studies utilizing direct and indirect Fick methods, dye dilution procedures, and more unusual techniques such as cineradiography of silver-tantalum markers sutured to ventricular walls have yielded enough data so that the cardiac response of the exercising human is fairly well characterized. The range of response levels is fairly wide among different individuals, and much data from the trained athlete will be presented here as exemplification. One should be aware, however, that the overall spread of response values for the rest of the population will range between those obtained from the athlete in vigorous exercise and the usual resting values.

Cardiac output does not increase to the same extent with heavy exercise as do pulmonary ventilation and metabolic rate. In one study, for example, on a trained distance runner in a maximum exercise effort, the $\dot{V}_{O_2}$ reached 6 liters per minute and the cardiac output attained 40 liters per minute. This is an approximate 24-fold increase in metabolic rate and an eight-fold increase in cardiac output. Many studies have shown that the cardiac output increase is roughly one-third of the metabolic rate increase and its parallel ventilatory response. However, oxygen delivery to the tissue is adjusted by increasing the tissue O_2 uptake from each unit volume of blood, or increasing the arterial-venous oxygen difference (see Fig. 29-5B). In the above example, this difference increased threefold. The increase in cardiac output is brought about by increases in both heart rate and stroke volume.

HEART RATE

Because heart rate is readily measurable by electrocardiographic, pulse recording, and various pressure recording techniques during rest and exercise, it has been studied extensively. This single factor quite accurately reflects the cardiovascular adjustment required for the intensity of work. The heart rate of a normal subject may rise to 100 beats per minute in very light work, to 130 beats per minute in moderate work, and to almost 200 beats per minute in very heavy work. The heart rate of 180 has been considered an upper limit, but thorough studies have shown much faster heart rates in young people and trained athletes. For example, among a group of Scandinavian athletes, heart rates between 195 and 205 beats per minute were often attained with maximum effort. As a general rule, the heart rate of the trained endurance athlete is lower at rest and can increase about fourfold while that of the untrained individual increases twofold to threefold.

STROKE VOLUME

The question of stroke volume changes with exercise has been the subject of rather lively controversy in the last few years. This arguing has been stilled with the development of appropriate measuring techniques, and with the use of a wider range of exertion levels. The supine subject has a greater stroke volume at rest and shows a smaller increase with exercise, whereas the erect subject may have a resting stroke some 40 percent less than in the supine position. The usual increase in stroke volume is about twofold. Thus with the athletes, a fourfold increase in heart rate will yield an eightfold cardiac output increase. In a study of nine top athletes in heavy work ($\dot{V}_{O_2} > 4$ liters per minute), the stroke volume in 19 determinations averaged 190 ml. This value can be compared to the averaged stroke volume of

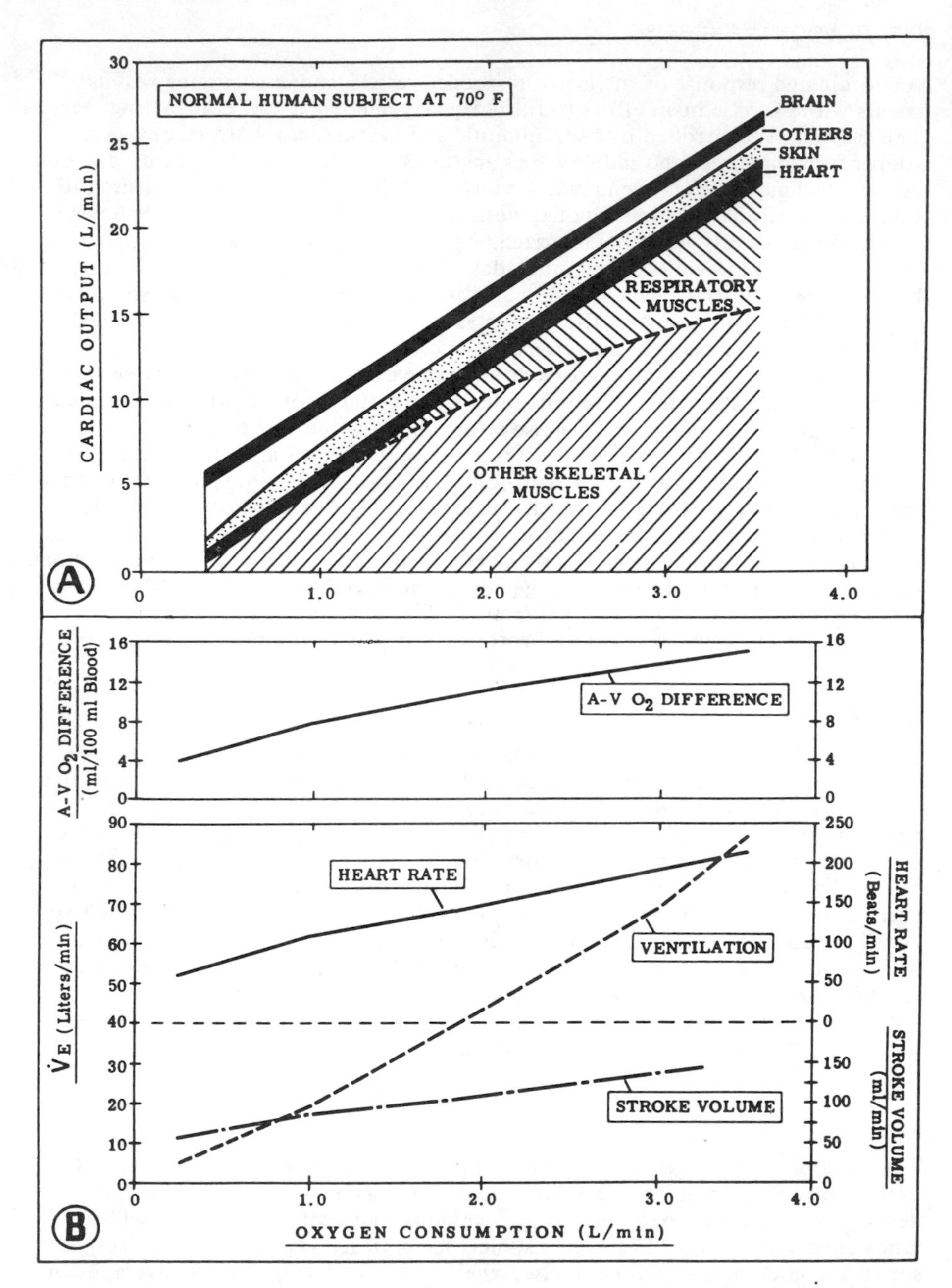

FIGURE 29-5. Circulatory and respiratory functions versus oxygen consumption ($\dot{V}_{O_2}$) and work load (kilogram-meters per minute). (A from *Science and Exercise of Medicine and Sports,* edited by Warren R. Johnson: Brouha, L., and Radford, E. P., Jr., Fig. 10.4, p. 190 [Harper & Row, 1960]. The data for arteriovenous oxygen differences and stroke volume are from Y. Wang, R. J. Marshall, and J. P. Shephard. *J. Clin. Invest.* 39:1051, 1961.)

152 ml obtained in a group of trained but less proficient athletes at equivalent rates of work.

BLOOD PRESSURE

During exertion, systolic pressure increases approximately in proportion to the increase in work load, while diastolic pressure decreases somewhat or remains almost constant. The systolic pressure seldom rises over 180 mm Hg in the normotensive subject, and mean blood pressure may increase only 10 to 20 mm Hg.

An arterial pressure tracing during exertion gives some indication of changes in peripheral resistance. The time required to reach the systolic peak is shorter and the actual systolic level is higher because of increased vigor of ventricular ejection. Pulse pressure may more than double, while cardiac cycle time is decreased more than 50 percent. Thus the arterial runoff curve is much steeper, suggesting a greatly reduced total peripheral resistance (TPR). When enough data are obtained so that peripheral resistance may be actually calculated, a decreased resistance is apparent:

$$\frac{\text{Mean blood pressure}}{\text{Cardiac output}} = \text{TPR mm Hg/liter} \qquad (9)$$

$$\text{Control TPR} = \frac{90}{5} \text{ or 18 mm Hg/liter} \qquad \text{Exercise TPR} = \frac{100}{25} \text{ or 4 mm Hg/liter}$$

Thus, in heavy exertion, total peripheral resistance may be reduced to one-fourth that of the resting control. This great reduction in resistance is largely due to an opening of skeletal muscle vascular beds (Fig. 29-5).

CHANGE IN LOCAL BLOOD FLOWS

With exercise, it is evident that active muscles, including those of respiration, receive a bigger share of the increased cardiac output. With a cardiac output of 25 liters per minute, over 20 liters goes to muscle, as shown in Figure 29-5. Brain blood flow is constant, while kidney, liver, and intestinal blood flow may actually decrease while cardiac output is increasing (Chap. 22). More studies are required on coronary circulation in exercise of even moderate severity. In one study in which cardiac output was increased by 60 percent coronary blood flow increased 40 percent.

Studies on skin blood flow have shown a dual effect of exercise. At the start of exercise there is a greatly reduced cutaneous blood flow, which supposedly facilitates augmented muscle blood flow. However, as exercise continues, skin blood flow increases to greater than control levels to allow dissipation of heat.

MUSCLE BLOOD FLOW

It is evident in Figure 29-5A that the greatest portion of the increased cardiac output is going to the active muscles to support the increased oxygen demand. Because oxygen extraction from the blood increases, the muscle blood flow and oxygen consumption do not increase in a one-to-one relationship. Regulation of muscle blood flow is a well-studied area and, like many other physiological regulatory systems, shows redundancy, or several mechanisms apparently of similar action. These mechanisms are summarized in Figure 29-6, which presents several experimental procedures. A study of each curve and the relationships between the curves will yield an understanding of the involved mechanisms.

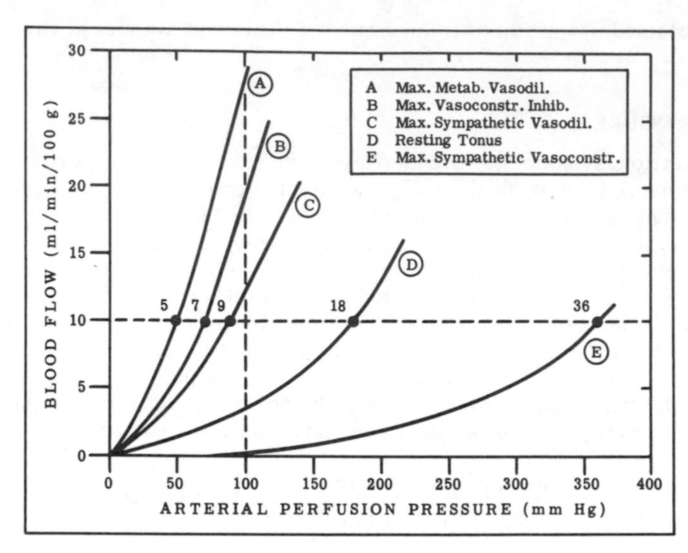

FIGURE 29-6. A plot of blood flow versus perfusion pressure for the dog gracilis muscle and the cat gastrocnemius-plantar muscle. The numbers indicate the peripheral resistance at a flow of 10 ml per minute for 100 gm of tissue. See text for further explanation. (From Renkin and Rosell, 1962.)

In resting conditions, skeletal muscle possesses a high degree of vascular tone. Curve D in Figure 29-6 represents the resting condition at an arterial perfusion pressure of 100 mm Hg. The flow rate is approximately 3 ml per minute per 100 gm tissue. With maximum vasoconstriction and nerve activity, flow at 100 mm Hg decreases to about 0.5 ml per minute (curve E). Physiologically, the ability to restrict muscle blood flow by sympathetic vasoconstriction is important in making those adjustments necessary to maintain flow to the brain, for example, with postural changes.

The effect of the stimulation of sympathetic vasodilator fibers is shown in curve C. In the dog and cat, the vasodilator tract appears to originate in the motor cortex and has relay stations in the hypothalamus and collicular region. The efferent outflows appear to be via the brain stem and ventrolateral part of the medulla oblongata to the lateral spinal horns with final distribution from the sympathetic ganglia to the vessels. Stimulation along this tract produces vasodilation primarily in skeletal muscles. This tract has not been found in the primates that have been studied, but there is evidence that in man active vasodilation may occur. One interesting aspect of active vasodilation in muscle is that nutritional blood flow in capillaries is not increased in rest upon stimulation of the appropriate fibers. It is assumed that the decrease in resistance occurs in A-V capillaries and thoroughfare channels (see Fig. 12-2). With more blood now available in the muscle, concomitant metabolic influences with exercise open up the capillaries, and thus the nutritional flow itself can be greatly augmented.

The inhibition of sympathetic vasoconstrictor neural outflow to the muscle can enhance flow considerably, as curve B indicates. A sixfold increase occurs at a perfusion pressure of 100 mm Hg, when vasoconstrictor activity is halted. While vasoconstrictor activity of the contracting muscle is halted, it may be enhanced in resting muscle and other organs.

The most effective single vasodilator mechanism is that of local metabolites. Many factors or agents released or altered by muscle contraction have been implicated, but none can be singled out as the primary vasodilator. Locally increased carbon dioxide may have an effect, but it is probably a weak one, while the concomitant oxygen decrease is perhaps stronger in its vasodilating action. Some of the metabolites of intermediary metabolism, including those of glycolysis and the Krebs cycle, have weak vasodilator activity. Lactic acid and hydrogen ion concentration changes have been suggested, but the evidence is not very compelling. Potassium appears to be a more likely candidate, and it is released by muscle activity. Histamine content of muscle decreases in contraction, and this is a potent vasodilator. It has been suggested that ATP "leaks" from muscle in contraction. This is a substance described as having "great vasodilator power." Epinephrine, when infused, causes a transient vasodilation in muscle vascular beds. It is followed, however, by vasoconstriction, and epinephrine blood levels do not seem to correlate well with vascular resistance in exercise.

It seems reasonable that several substances working together may account for local vasodilation. The local aspect of this vasodilation must be emphasized because no evidence exists that metabolites released from contracting muscle circulate to other resting muscles or other organs to produce vasodilation.

The regulation of muscle blood flow thus involves both central and local factors. This redundancy ensures a rapid response, as resistance in the muscle vasculature has been reported to decrease within one second after exercise initiation. It also guarantees that the regulation will be maintained and be in the appropriate direction. Even in the sympathectomized muscle of patients, vasodilation will still occur.

Muscle blood flow possesses a unique feature not shared by other vascular beds. When a muscle contracts, a mechanical compression of all the associated vessels occurs. From the veins, either valved or unvalved, blood is squeezed toward the heart, while in arteries and capillaries flow may be stopped altogether. In rhythmic contraction, this squeezing and refilling may actually serve as a pump to enhance muscle blood flow, providing the duration of the contraction phase is not too long. In sustained or continuous contraction, blood flow may be completely arrested. When one stands on tiptoe, blood flow through the gastrocnemius soleus practically reaches zero, but this is followed by a hyperemia upon relaxation. However, with sustained contraction of grip or forearm muscles, vessels do open up and flow increases during the contraction period.

Figure 29-7 shows the effect of intermittent contraction on leg blood flow. As can be seen, rather sharp decreases in flow are followed by equally sharp increases. The overall integrated or averaged blood flow during contraction would obviously be much greater than the resting blood flow. How much of this increase is due to mechanical squeezing and how much to the factors described above as vasodilators is not easily ascertained.

CAPACITANCE VESSELS

In addition to the muscle pump aiding the venous return of blood to the heart, there are supporting alterations in the postcapillary vessels or veins which contain the greatest portion of the blood volume. These alterations consist in contractions of the smooth muscle and are centrally mediated by the sympathetic system and there is no evidence that the local influences described above for enhancing muscle blood flow have anything to do with the response. A simple demonstration involves recording a pressure increase through a needle in an arm vein isolated by both venous and arterial occlusion when the legs start to exercise.

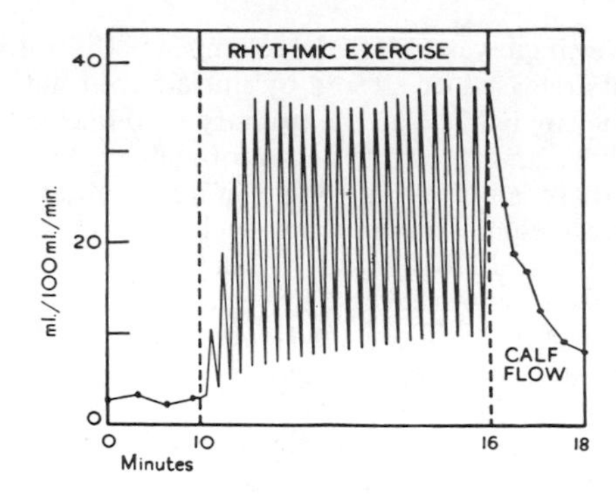

FIGURE 29-7. Effect of rhythmic muscle contraction on human muscle (calf) blood flow. (From H. Barcroft. *Sympathetic Control of Human Blood Vessels.* London: Arnold, 1953.)

This "tightening" of the venous or capacitance system appears essential for guaranteeing rapid return of blood to the heart. It is easy to conceive of a muscle pump simply forcing more blood into capacitance vessels where it would be sequestered. However, the increased venous tone prevents this, and it must be considered an important part of the total integrated cardiovascular response.

INTEGRATED CONTROL OF THE CARDIOVASCULAR SYSTEM IN WORK

With exercise, there are several mechanisms which aid the return of venous blood to the heart. These consist of the factors discussed above, including the decreased peripheral resistance in muscle which permits blood to flow through a more "open channel." The increase in mean arterial pressure, although small, may be effective as an aid. The muscle pump and the "tightening" of the capacitance vessels are also important in this regard. It is extremely difficult to sort these factors out with respect to the quantitative contributions as they occur in an integrated relationship.

In addition to the above, the respiratory pump or abdominothoracic pump plays a supporting role. Here the increased negative intrathoracic pressure necessary for the exercise hyperpnea increases, thus increasing the pressure gradient for the return of blood to the thorax. The intrathoracic pressure changes necessary for greater expiration also tend to facilitate the pumping.

It should be noted that all these changes tend to improve the distribution of pulmonary blood flow which ordinarily is higher in the lower regions of the lung. Upon the first contraction of a large muscle mass, the return of blood to the heart is increased, and one might suspect increased filling would now raise the stroke volume by the action described by Starling's "law of the heart." However, there is evidence that the end diastolic volume of the heart is not increased with exercise. The greater vigor of contraction may pump out more blood, resulting in a lower end systolic volume and higher stroke volume. Functional comparisons of normal dogs with dogs possessing chronic excision of the cardiac extrinsic nerves reveal that neural influences markedly alter the characteristics of the heart in exercise. After cardiac denervation, there is little impairment in work capacity and in the ability to increase cardiac output, but the heart rate of the denervated preparation increases in two minutes of severe work from 100 to 140 beats per minute by some unexplained mechanism. In the intact dog, the heart rate sometimes surpassed 240 beats per minute. Stroke volume of the denervated hearts almost doubled, while only small changes occurred in the hearts with an intact nerve supply. The product of the stroke volume and heart rate, or cardiac output, was the same in both types of dogs in exercise.

An important lesson here is that in the denervated dog the cardiac output increase depended solely on the Frank-Starling mechanism and required much more time to reach its maximum level. The neural response in exercises increased the rate of beating and the vigor of contraction and decreased the time delay in which a high cardiac output was reached.

The sympathetic nervous system, in adjusting vascular resistance and capacitance and altering the cardiac function by enhancing rate and, in effect, shifting the ventricular function curve to produce a more vigorous contraction, plays a major part in adjusting circulatory performance to metabolic requirements. Attempts to mimic precisely the exercise changes by infusing norepinephrine or epinephrine have met with failure. Rushmer and co-workers have found an area in the brain of the dog which produced, upon stimulation, cardiac responses similar to those occurring with exercise. The effective area is located in the diencephalon or subthalamus (H_2 fields of Forel). Properly placed diencephalic lesions will abolish the cardiac responses to exertion. The tract for active vasodilation, a response of unknown importance in the human, has been described above. Sympathetic centers controlling vasoconstriction are differentially stimulated or inhibited as smooth muscle of the capacitance vessels, and the arterioles in the kidney, gastrointestinal tract, and resting muscles may show increased contraction, while the resistance in active muscles is always decreased. The means by which these centers are activated is not understood. Somehow signals from the motor cortex must activate the centers, but direct pathways have not been located.

Considerable compensation for exercise could take place without any sympathetic influence. Local vasodilators would increase muscle blood flow, the muscle pump and respiratory pumps would return blood to the heart, and the mechanism described in Starling's law would enhance the cardiac output. However, there would be no increased resistance in inactive beds and no tightening of the capacitance vessels, and all changes would require more time. One might look at the neural influence first as reducing the time constants for any compensations and second as a sort of overall integrating mechanism needed in the organism's periods of maximum energy turnover.

RELATION OF CIRCULATION, RESPIRATION, AND METABOLISM

The most obvious and immediate effect of enhanced circulatory and respiratory activity is to deliver oxygen to the active tissues at a greater rate. Several important factors play a role in this process.

The increased ventilation is rendered more effective by an increase in the diffusing capacity of oxygen (D_{O_2}). This is primarily accomplished by increasing the size of the pulmonary capillary bed and increasing the alveolar area for diffusion. Thus, the oxygen going into the blood for each mm Hg of oxygen pressure gradient is greater (Chap. 30).

Because cardiac output does not increase to the extent that pulmonary ventilation and oxygen consumption do, many investigators consider that the circulation is the limiting factor. To a certain extent, oxygen delivery to active cells may be increased by simply removing a greater amount of oxygen from the blood.

Total oxygen consumption equals cardiac output times mean arteriovenous (A-V) oxygen difference. Thus, if cardiac output lags, oxygen consumed may be kept up by increasing the mean arteriovenous oxygen difference. Figure 29-5 shows the change in arteriovenous oxygen difference at different work levels.

$$\text{Coefficient of oxygen utilization} = \frac{\text{A-V } O_2 \text{ difference}}{\text{Arterial } O_2} \quad (10)$$

$$\frac{5 \text{ vol./100 ml}}{20 \text{ vol./100 ml}} = 25\% \text{ (resting)} \quad (11)$$

$$\frac{14 \text{ vol./100 ml}}{20 \text{ vol./100 ml}} = 70\% \text{ (severe exertion)} \quad (12)$$

In exercise the overall coefficient of oxygen utilization may increase to 60 or 70 percent because several factors act to promote the movement of oxygen into the tissues. (1) The increase in oxidation lowers the tissue P_{O_2}, resulting in a greater gradient for oxygen diffusion. (2) Lactic acid diffuses from the cells, and decreasing pH, increased carbon dioxide, and increased temperature all act to shift the hemoglobin dissociation curve to the right, which promotes the unloading of oxygen in the tissues. (3) The increased number of open capillaries enlarges the surface area available for diffusion of oxygen and also decreases the distance required for diffusion

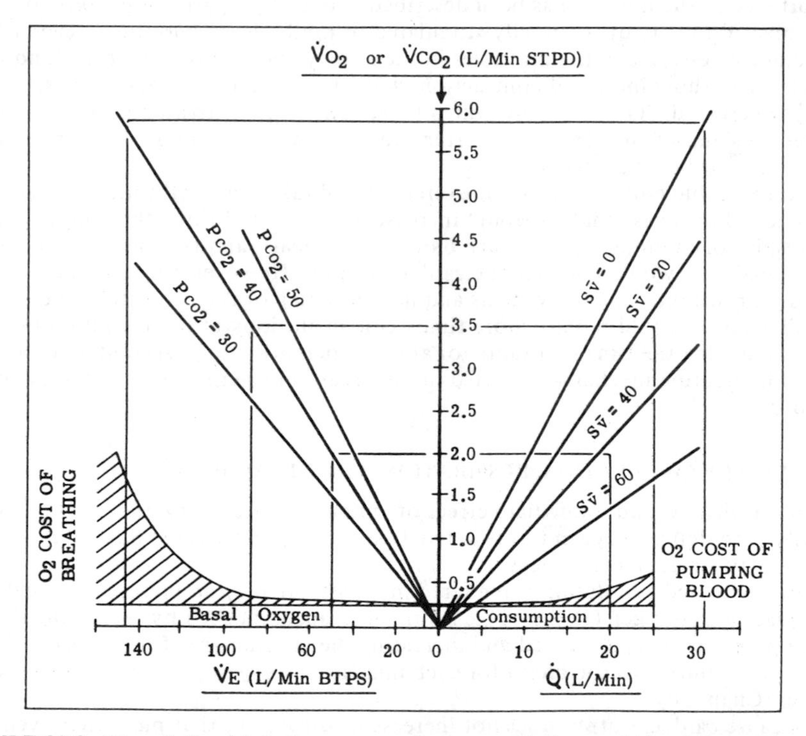

FIGURE 29-8. The relationship between gas exchange, ventilation, and blood flow. The cross-hatched areas represent oxygen cost of breathing (left) and pumping blood (right). The ordinate is $\dot{V}_{O_2}$ or $\dot{V}_{CO_2}$ in liters per minute. The right abscissa is cardiac output ($\dot{Q}$) in liters per minute; left abscissa is ventilation ($\dot{V}_E$) in liters per minute. At any oxygen consumption level the relationship of $\dot{V}_E$ to arterial or alveolar P_{CO_2} (left isopleths) and the relationship of $\dot{Q}$ to percent oxygen saturation of mixed venous blood ($S_{\bar{v}}$, right isopleths) may be approximated. Basal oxygen consumption is at a $\dot{V}_{O_2}$ of .250 liter per minute. (From *Science and Medicine of Exercise and Sports,* edited by Warren R. Johnson: Riley, R. L., Fig. 9.1, p. 193 [Harper & Row, 1960].)

to all active cells. In no circumstance has venous oxygen content been observed to approach zero. Most studies have indicated a low limit of about 5 volumes per 100 ml.

The increased oxygen demand of the tissues is met by an increased ventilation, increased blood flow or cardiac output, and increased removal of oxygen from the blood flowing through the tissues. Figure 29-8 is a convenient diagram for expressing the approximate quantitative aspects of these relationships. A $\dot{V}_E$ of about 6 liters will maintain a normal alveolar P_{CO_2}, and a $\dot{Q}$ of 5 liters per minute will maintain a normal venous oxygen saturation. In hard work at a $\dot{V}_{O_2}$ of 3.5 liters, a $\dot{V}_E$ of 88 liters per minute would be required to maintain a P_{CO_2} of 40 mm Hg; and with a cardiac output of 25 liters per minute, venous oxygen saturation would fall to 25 percent. If cardiac output could go to only 18 liters per minute, venous oxygen saturation would drop to zero percent, which would be intolerable.

The graph represents only an approximation for normal humans. However, it is a good way to express the load on the systems responsible for respiratory gas transport during various work levels.

In summary, the metabolic increase in exercise is supported by increased activity of the systems responsible for oxygen and metabolite transport. Respiratory activity quite closely parallels metabolic activity. Cardiac output appears to lag behind metabolism, but several factors effect a more efficient oxygen delivery into the tissues.

REFERENCES

Asmussen, E., and M. Nielsen. Pulmonary ventilation and effect of oxygen breathing in heavy exercise. *Acta Physiol. Scand.* 43:365–378, 1958.

Astrand, P. O., and K. Rodahl. *Textbook of Work Physiology.* New York: McGraw-Hill, 1970.

Barcroft, H. Circulation in Skeletal Muscle. In W. F. Hamilton (Ed.), *Handbook of Physiology.* Section 2: Circulation, Vol. II. Washington, D.C.: American Physiological Society, 1963. Pp. 1353–1368.

Beregard, B. S., and J. T. Shepherd. Regulation of circulation during exercise in man. *Physiol. Rev.* 47:178–209, 1967.

Chapman, C. B. (Ed.). Physiology of muscular exercise. *Circ. Res.* 20 (Suppl.), 1967.

Dejours, P. Control of Respiration in Muscular Exercise. In W. O. Fenn and H. Rann (Eds.), *Handbook of Physiology.* Section 3: Respiration, Vol. I. Washington, D.C.: American Physiological Society, 1964. Pp. 631–648.

Ekelund, L. G. Exercise. *Ann. Rev. Physiol.* 31:85–116, 1969.

Falls, H. B. (Ed.). *Exercise Physiology.* New York: Academic, 1968.

Harris, P. Lactic acid and the phlogiston debt. *Cardiov. Res.* 3:381–390, 1969.

Johnson, W. R. (Ed.). *Science and Medicine of Exercise and Sports.* New York: Harper, 1960.

Kao, F. F. An Experimental Study of the Pathways Involved in Exercise Hyperpnea Employing Cross Circulation Techniques. In J. C. Cunningham and B. B. Lloyd (Eds.), *The Regulation of Human Respiration.* Oxford, Eng.: Blackwell, 1963.

Karpovich, P. V. *Physiology of Muscular Exercise,* 6th ed. Philadelphia: Saunders, 1965.

Renkin, E. M., and S. Rosell. Effects of different types of vasodilator mechanisms on vascular tonus and transcapillary exchange of diffusible materials in skeletal muscles. *Acta Physiol. Scand.* 54:241–251, 1962.

Ricci, B. *Physiological Basis of Human Performance.* Philadelphia: Lea & Febiger, 1967.
Rushmer, R. F. *Cardiovascular Dynamics,* 2d ed. Philadelphia: Saunders, 1961. Chap. 8.
Uvnas, B. Central Cardiovascular Control. In J. Field (Ed.), *Handbook of Physiology.* Section 1: Neurophysiology, Vol. II. Washington, D.C.: American Physiological Society, 1960. Chap. 44, pp. 1131–1162.

30

Physiological Problems of Unusual Environments: Space and Undersea Activities

Robert W. Bullard

MAN'S PROCLIVITY to invade those environments for which he is physiologically unsuited is well documented by both historical and recent attempts to penetrate, survive, and perform useful functions in outer space or at great depths beneath the sea. Such activities may be accompanied by extremes of pressures, temperatures, gravitational forces, radiation, and psychophysiological stresses. Engineering may do its share in alleviation of these problems, but it is the biomedical investigators and practitioners who are challenged to provide adequate tolerance level information for engineering standards. Biomedicine must determine the ultimate tolerance limits of the encountered combined stresses in which men can survive or perform in a useful fashion as well as investigate the possible means by which tolerance to stress can be augmented.

Man's experience in diving precedes written history. However, experiments with compressed-air diving, to extend the depth duration, were not feasibly attempted until the middle of the seventeenth century. The modern scientific approach was really initiated in the early twentieth century with the classic studies of Haldane and his co-workers. Since then a multitude of scientific observations have been assimilated, and the maximum depths attainable have been greatly increased. Undersea activity by necessity exposes man to a more severe physiological stress than does space exploration, but it is easier to simulate diving conditions in the laboratory environment. Man's history of space exploration is of course much more recent. At this writing the United States has accumulated a little more than 6000 man-hours of space activities in the Mercury, Gemini, and Apollo programs. It is against this background that the present chapter is written.

THE PRESSURE ENVIRONMENT

In both ascent into space and descent below the surface of the sea, serious pressure problems are encountered. Unprotected man, even when supplied with oxygen, has a ceiling of perhaps 45,000 feet above which pressure protection must be supplied. On the other hand, the diver can tolerate the markedly increasing pressure at least to as far as 36 atmospheres, or roughly 1100 feet. However, for the diver other problems occur having to do with the increased solubility of gases with pressure and the direct effect of high partial pressures of these gases on certain tissue functions.

SPACE ACTIVITY

Until quite recently, man's conquest of the vertical frontier has been slow. The first limited encroachments were made just prior to and in the nineteenth century, with the early balloon flights. Even in these pioneer efforts man's first and perhaps major problem of aviation physiology was encountered, that of hypoxia. Human dependence upon a fairly high partial pressure of oxygen is illustrated by the fact that the highest permanent habitations, in the mining camps of the Andes, are under 18,000 feet (P_{O_2} = 75 mm Hg). Between the altitudes of 20,000 feet and 25,000 feet the unacclimated individual encounters serious difficulty, to the point of losing consciousness. The highest mountain peak attained without extra oxygen is 26,811 feet and it was climbed only by individuals with a high degree of altitude acclimation. For the unacclimated even at altitudes as low as 10,000 feet moderate impairment may be encountered. To avoid any loss of cerebral function, oxygen must be supplied in routine aviation operations above 10,000 feet. A major accomplishment of aviation physiology was the development of oxygen supply systems to maintain pilot and crew in a state of effectiveness. Needless to say, the space flight ambience is completely lacking in oxygen, but the oxygen supply systems which were developed for aviation uses have been improved to meet the demands of space travel.

At the barometric pressure of 87 mm Hg encountered at approximately 50,000 feet the sum of the alveolar carbon dioxide tension (40 mm Hg) and the alveolar water vapor tension (47 mm Hg) is equal to the ambient barometric pressure of 87 mm; hence, breathing 100 percent oxygen at normal ventilation rates would be ineffective. It is obvious that supplying both oxygen and pressurization becomes necessary at this altitude.

Ebullism

An absolute altitude limit for unprotected man could be set at 63,000 feet, the altitude at which barometric pressure equals 47 mm Hg. The vapor tension of water at body temperature is also 47 mm Hg. Boiling occurs when vapor tension of a liquid equals atmospheric pressure, and at this altitude the body fluids would therefore boil.

When various mammals were suddenly exposed experimentally to this ambient pressure, the first signs were a deepening of respiration and abdominal distention. The animals immediately collapsed with mild convulsions and became quiescent except for respiratory gas changes which ceased after 30 seconds. Lacrimation, salivation, urination, vomiting, and defecation were observed. All fluids were bubbling upon emission. After 30 to 40 seconds a secondary swelling was seen in the hind limbs and abdomen. All the effects that occurred can be attributed to either anoxia or the reduced pressure resulting in vaporization of tissue fluids and expansion of body gases, a condition sometimes called ebullism, or the ebullism syndrome. The respiratory passages become filled with water vapor, a condition known as vapothorax, which renders breathing completely ineffective. The syndrome has not been described for man, and resuscitative or therapeutic measures have not been developed. As far as the pressure environment is concerned, altitudes above 63,000 feet are equivalent to outer space, and obviously the survival of unprotected man would be indeed brief. However, the pressurized cabin eliminates this hazard and the full pressure suit permits extravehicular activity either during flight or on the lunar surface.

Cabin Pressure

The factors of oxygen supply and pressurization must be considered in designing and constructing a cabin that is livable in the vacuum of outer space. Opinions vary somewhat as to the ideal atmospheric constituency of such a cabin. Although it has not been thoroughly studied from a physiological viewpoint, an atmosphere having a total pressure of approximately 200 to 250 mm Hg consisting of 100 percent oxygen has been used almost entirely in the United States' manned space flights. Total cabin pressure about one-third or less of the sea level ambient pressure, but with the partial pressure of oxygen greater than at sea level, provides several advantages: (1) The lower pressure gradient across the cabin wall diminishes any tendencies to leak and to push the cabin wall outward into the vacuum of space, thus reducing the strain on the vehicle hull. (2) The atmosphere of 100 percent oxygen denitrogenates the body. In the event of a pressure loss, the tendency to a decompression sickness or the bends is reduced. This partial pressure of oxygen is lower than levels known to produce oxygen toxicity. (3) A one-gas system is easier to control than a two-gas system, particularly with reabsorption of the metabolically produced carbon dioxide. If the artificial atmosphere contained inert gas such as nitrogen, extra control would be required to maintain the partial pressure or the percentage of both gases. However, it should be noted that the atmosphere in the cabin of the recent Apollo flights contained 36 percent nitrogen at launch. This nitrogen is slowly purged by bleeding oxygen into the cabin and permitting a small leakage to the outside through a valve. While only oxygen is being replaced, the nitrogen diminishes until it levels off at a mean of about 7 percent.

There are alternatives concerning the space cabin atmosphere. In the Vostok Sputnik which carried Gagarin on the first successful orbital flight and in subsequent Russian flights, the artificial ambient pressure ranged between 750 and 770 mm Hg.

Dysbarism

Other than the general problem of hypoxia and the dramatic sequelae of ebullism, two conditions related to altitude or space flight deserve consideration. The first problem is that of *dysbarism,* a general term which includes a wide variety of effects within the body, exclusive of hypoxia, caused by changes in ambient barometric pressure; it also includes ebullism. As shown in Figure 30-1, ambient pressure decreases markedly as altitude increases. If an individual in a sealed cabin at sea level ambient pressure were suddenly exposed to the prevailing pressure at 35,000 to 40,000 feet, he would probably experience decompression sickness. The manifestations result from expanding gas evolved from solution and produce the so-called bends, characterized by pain in the bones, joints, and muscles, or choking, characterized by breathing difficulty.

The physiological reactions to pain may then produce other symptomatology. The individual who is denitrogenated after prolonged oxygen breathing will not have as great a tendency toward the bends as the person whose tissues and blood stream are saturated with nitrogen. With vertical flight, a considerable height is required before decompression sickness or the bends can occur. This is generally a decompression to one-fourth or at least one-third of the initial ambient pressure.

Erythrocyte Loss

A second problem which is attributed to the cabin atmosphere of pure oxygen is the reduction in erythrocyte mass. During the eight-day flight of Cooper and Conrad in the Gemini V, the erythrocyte mass of both astronauts decreased about 20 percent. Red cell survival time dropped from an expected value of 82 or 83 days to 69 days in one case and 70 days in the other. The decreased postflight reticulocyte count suggested also that the rate of red cell production decreased. However, the data obtained in the longer flights in the Gemini and Apollo programs

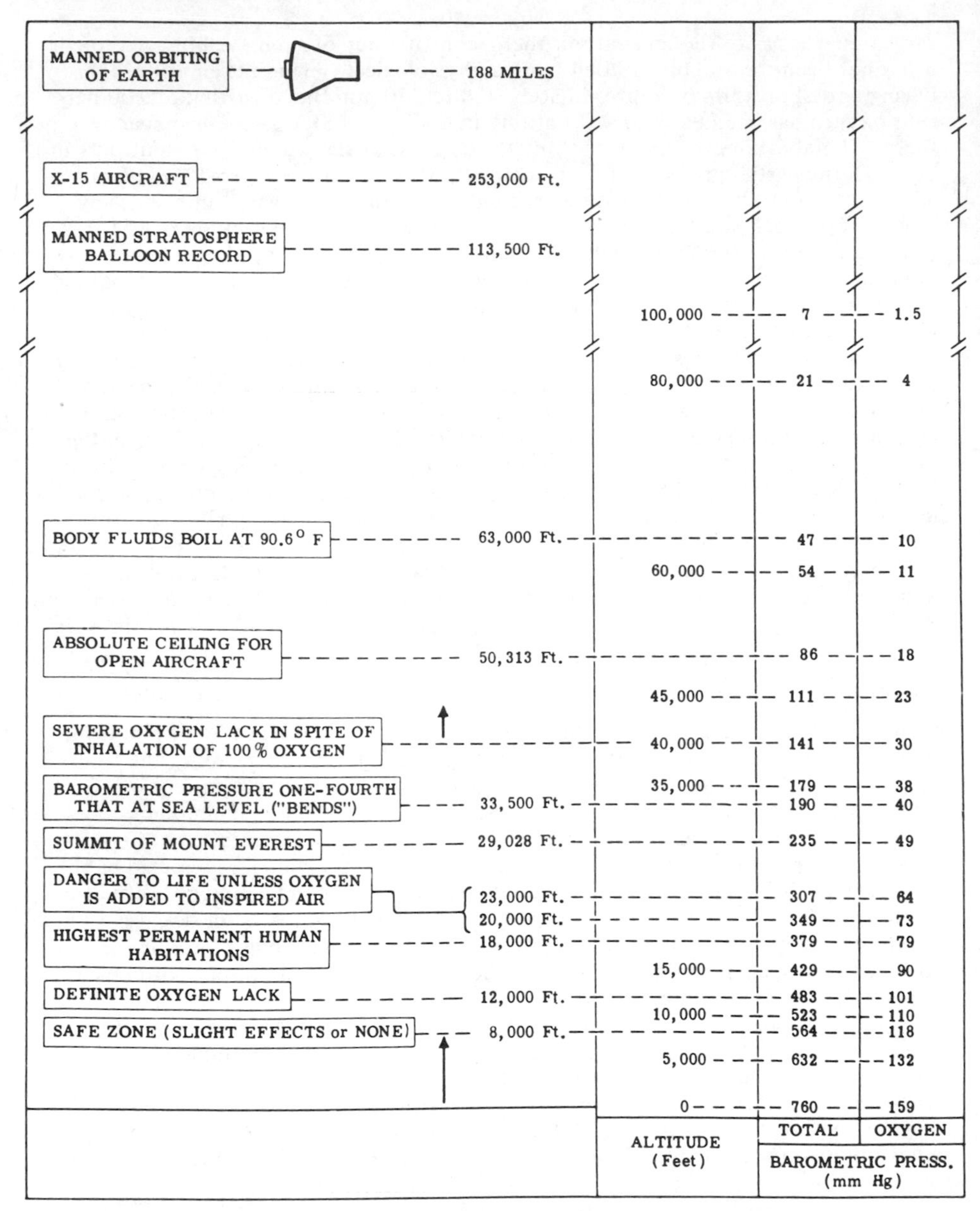

FIGURE 30-1. Ambient pressure and physiological changes with altitude. (From C. J. Lambertsen. Anoxia, Altitude, and Acclimatization. In P. Bard [Ed.], *Medical Physiology,* ed. 11. St. Louis: Mosby, 1961. P. 700.)

suggest that following these initial changes red cell production and destruction equilibrate, and the decrease in red cell mass is not progressive. In laboratory efforts to duplicate this process, it has been found that the oxygen environment may tend to destroy the oldest cells in the population; once this is completed, the red blood cell mass stays about the same.

In the recent Apollo flights (9, 10, and 11) of 8 to 10 days' duration, there was essentially no reduction in the red cell volume. It is thought that possibly nitrogen remaining in the cabin atmosphere even at the 7 percent level serves to prevent this decrease. The Russian cosmonauts, who travel with nitrogen in the cabin, have not shown a decreased red cell mass. It must be borne in mind that erythrocyte loss may not simply be a problem of the oxygen environment but could also be a response to the weightless environment, decreased food intake and physical activity, changes in fluid balance, or some other factor.

THE UNDERSEA PRESSURE ENVIRONMENT

Pressure on the Body

From ground level to the outer reaches of space the entire pressure change is only 1 atmosphere. However, the diver starts with this 1 atmosphere at the surface, and for every 33 feet of descent 1 atmosphere of pressure is added. As water is quite incompressible, the relationship holds even to great depths. As the diver descends, several factors with serious physiological consequences are encountered. Boyle's law states that the product of pressure and volume is a constant and thus, if the pressure doubles, the volume must be halved. Consider an individual starting out with perhaps 4 liters of gas in the lungs at one atmosphere. If he goes to 33 feet below the surface, this volume will be reduced to 2 liters, and by the time he reaches 100 feet, the gas will be compressed to 1 liter. The resulting condition, dramatically named the "squeeze," tends to collapse the chest and brings severe pain and physical damage. The depth of 100 feet thus appears to exceed the limit for safe breath-hold diving.

Depths slightly greater than this are repeatedly attained by the divers of the Tuamotu Archipelago of the South Pacific, who after hyperventilation descend rapidly on an anchored line. The very high incidence of a multitude of pathophysiological consequences grouped under the title of Taravana, or diving syndrome, which includes severe neurological and mental signs, such as unconsciousness, vertigo, mental anguish, paralysis, and quite often death, emphasizes the extreme hazards of free diving to such depths.

Carbon Dioxide Narcosis

A prerequisite for safe undersea activity at reasonable depth, then, is the maintenance of pressure in the airways and lungs closely equivalent to the external hydrostatic pressure. Historically, carbon dioxide, although this was not substantiated with precise measurement, appeared to be a factor producing impairment as depths were extended. Carbon dioxide at, for example, a level of 5 percent in the inspired air is uncomfortable but is primarily hyperventilatory in its effects. However, 5 percent CO_2 at 4 atmospheres or 100 percent would have a partial pressure of 160 mm Hg. This is a concentration in the narcotic range and would produce a rather serious performance detriment. Thus it is essential that for whatever respiratory apparatus is used the ventilation be high enough to remove CO_2 before it accumulates.

Oxygen Toxicity

Thus pressure, as well as adequate gas flow or effective techniques of carbon dioxide absorption, must be utilized to maintain performance levels below 100 feet. Use of compressed air, however, can be limited even with high flow rates, because of oxygen toxicity (see Chap. 20). Air with 21 percent oxygen content will have a P_{O_2} of 1 atmosphere at a depth of 132 feet. With long exposure, this may result in moderate pulmonary irritation, not of a very serious or irreversible nature. But a P_{O_2} above this level should be avoided. For example, a P_{O_2} of 3.15 atmospheres, which would be encountered in breathing air at 462 feet, would incur central nervous system effects between one and two hours. These effects may include convulsions and result in complete impairment of the diver. One can well imagine the serious consequences upon an unattached, freely swimming individual at this depth.

Obviously, for diving at great depths, the percentage of oxygen in the respiratory gas mixture must be reduced. Pure oxygen has some advantage in that it reduces the tendency toward decompression sickness, but it should never be used at depths lower than 33 feet because the P_{O_2} would be above 2 atmospheres and produce oxygen toxicity effects on the central nervous system.

Nitrogen Narcosis

Solutions to the problems of diving have come stepwise in the history of its hazardousness. That oxygen pressure and carbon dioxide can be controlled was found early in these endeavors, yet divers were still experiencing difficulties at depths from 150 to 300 feet or at 6.0 to 11.6 atmospheres of pressure. Serious accidents were occurring as a result of semi-loss of consciousness accompanied by subjective signs of euphoria, hyperexcitability, and reduction of intellectual and perceptive facilities. Various names have been given to this such as "rapture of the deep," but *inert gas narcosis* or *nitrogen narcosis* is more appropriate (see Chap. 20). It is suspected that most gaseous substances can produce some type of narcosis if they are presented to the tissues at high enough pressure. For nitrogen approximately 5 to 10 atmospheres will produce this narcotic effect which has been a cause of fatality among divers.

The Use of Helium

Helium is definitely safer than nitrogen at higher pressures. Mammalian experiments have shown that 163 atmospheres of helium is required to produce central nervous system effects. Solubility of helium in lipid is about one-third that of nitrogen, and the narcotic tendencies of the molecule supposedly are related to this solubility, or perhaps to the van der Waals forces of the molecules. Helium is thus the gas of preference for deep sea operations. No instances of helium narcosis in man have been observed at the highest pressure tested. Helium provides another advantage in that its solubility in body tissue is less than nitrogen's, which reduces the tendency to form bubbles or produce disabling decompression sickness upon ascent. A major problem of helium in diving operations is expense, since at the extremely high pressures in which CO_2 is detrimental, ventilation must be maintained at fairly high rates.

Limiting Depths

With helium replacing nitrogen in the respired gas and with the oxygen percentage reduced so that reasonable oxygen partial pressures are maintained, gaseous problems are at least partially solved. A depth of 600 feet has proved to be a practical level for diving operations. Long-term simulated dives have been undertaken in which the pressure in a chamber reached 31 to 36 atmospheres for an 80-hour period without ill effects on human subjects. Compression was rather rapid and consequently produced symptoms of vertigo, nausea, and tremor, at present of unexplained origin. Such dives are called saturation-excursion dives. That is, the divers were at this pressure long enough so that they became completely equilibrated with the gaseous environment; in this case, with helium as the primary atmospheric constituent equilibration required about 24 hours. The excursion aspect of the title means that from this saturation level the exploratory or work excursions are taken at such depths that desaturation cannot occur. In the particular experiment in question, the oxygen content in the atmosphere supplied to the divers was 1 percent. However, at the atmospheric pressure utilized, this would give a P_{O_2} of only 200 mm Hg.

Decompression Problems

Ascent from sea depth is more likely to result in decompression sickness than ascent into higher altitudes simply because of the greater pressure changes. There is no simple formula for determining the safe rate of ascent because the respiratory gas mixture, the depth, and the duration of the dive are all determinants. Generally, above 35 feet ascent can be made as rapidly as possible no matter what the duration of the dive has been. A diver ascending from 130 feet can do so immediately if the dive has been less than 10 minutes in duration because there is a time lag for saturation of the blood and tissues of the body with the gas at the higher pressures. For this depth of dive with a duration longer than 10 minutes, a slower decompression regime must be utilized. In the long saturation-excursion dives described above with 31 to 36 atmospheres of pressure, 88 hours was utilized for the simulated ascent and one out of the three subjects had moderate joint pains requiring recompression. In a similar experiment 285 hours was utilized for the ascent or decompression.

THE THERMAL ENVIRONMENT

Activities both in outer space and under the surface of the sea pose thermal problems quite different from those encountered on the surface of the earth.

SPACE ACTIVITY

Physical heat transfer in the airless reaches of space is very much different from that in the highly absorbent atmosphere of earth. In a certain sense the space capsule is in a condition somewhat similar to the inner chamber of a vacuum bottle. Although the actual temperature, which is really a measure of molecular agitation, is low in space, approximately −172°C, this temperature has little effect because of the lack of conduction and convection with their molecular requirements. Heat transfer is solely by radiation, the solar side of the capsule receiving, at the distance the earth and moon are from the sun, about 2 cal per second per square centimeter. This is an irradiance that will produce pain sensation on unprotected skin. The dark side of the capsule will radiate heat outward to space, and simple rotation of the capsule will almost ensure heat balance, especially if its surface has proper optical properties.

Heat problems can occur and must be anticipated for adequate protective planning. The radiant heat load at the distance Venus is located from the sun is twice what it is at earth's distance, and the surface of Venus has been estimated to approach temperatures of 800°C. At the distance of Mercury from the sun the irradiance is seven times what it is above our own atmosphere. Thus interplanetary expeditions toward the sun require advanced engineering concepts. One idea is to utilize a heat shield or "solar parasol" always oriented between the craft and the sun. Reentry of any spacecraft into the earth's atmosphere at high speeds will produce aerodynamic heating by friction of the moving vehicle with gaseous particles. The amount of this heat reaching the interior of the capsule is determined by the cabin engineering, and the amount of heat produced depends upon the reentry pattern and velocity. Temperatures as high as 3000°F have been attained in reentry of some of the U.S. space capsules. The very effective heat shield on the leading surface has alleviated any thermal problems for the astronaut except transient heat exposure of short duration.

As lunar surface temperatures range from close to 200°C at the subsolar point directly below the sun to −200°C on the dark side, the early exploratory expeditions have taken place away from the subsolar point, at areas where the shadows are long and the temperatures manageable. No difficulties have been encountered in the maintenance of the lunar module cabin temperature, which has ranged around 25°C. The astronauts, now utilizing a liquid-cooled suit for extravehicular activity, have not had any trouble because the suit is of very high insulative properties and can readily handle the peak work loads of 300 cal per hour. The early extravehicular activities in the Gemini program produced some heat stress as energy costs were higher than originally anticipated. However, this served as motivation for development of the effective liquid-cooled garment.

UNDERSEA ACTIVITY

The physical properties of water are quite different from those of outer space in regard to heat transfer. Since radiation from the body is immediately absorbed, this is not an effective heat transfer mechanism. Convective heat loss is the primary mode of transfer, with conductive loss accounting for 1 or 2 percent. Any movement of the diver increases conversion markedly, and heat loss by this means, which is velocity dependent, can range from 5 to 30 times above that in air at the same temperature. Most of the ocean water is below 16°C, especially a few meters below the surface. This is a temperature at which unprotected man can survive only five hours. In fact, the neutral temperature of water (temperature at which heat balance is maintained without increased metabolism) is 33°C. Thus most diving operations are in water where heat loss is a problem. Formed neoprene wet suits can provide considerable protection, but it has proved extremely difficult to protect the extremities in cold water while at the same time maintaining any manipulative ability.

The diver appears always to face the problem of cold exposure, and body cooling is a complaint in most undersea operations. It has been noted that in underwater operations a considerable drop in body temperature can take place without the diver's being aware of the change. This can markedly depress work performance, and if body temperature reaches 34°C dangerous and disabling loss of consciousness may occur.

The high-pressure atmosphere of saturation-excursion diving has led to another interesting thermal problem even while the aquanaut is out of the water in a chamber. Whereas a lightly dressed man may be quite comfortable at 23°C for a long period of time, a helium-oxygen environment of 31 atmospheres requires a temperature of 29°C to 31°C for the same degree of comfort. The reason is that helium, even at 1 atmosphere, has a higher specific heat than air and will carry more heat away from the body by convection. The pressurized gases with their increased density are even more effective heat sinks. It has been estimated that in operations below 600 feet the entire metabolic heat production may be lost from the respiratory tract; thus the diver faces a serious heat drain which may lead to brain temperature depressions conceivably exaggerating any narcotic effects of inert gases and hastening impairment. It is highly likely that techniques of warming the inspired gas and the maintenance of a warmer level of chamber temperature will be utilized to prevent body heat loss.

THE RADIATION ENVIRONMENT

Life as we know it has evolved in an environment essentially free of ionizing radiation. The human organism possesses no sensory receptors for detection of radiation, no known regulatory mechanisms for self-protection, no ability to adapt to ionizing radiation. No methods are available, nor will they be for a number of years, for selecting astronauts of high radiation resistance. An underestimation of the radiation hazard may leave the astronaut vulnerable, and an overestimation may impose a serious weight penalty because of excessive shielding. Two possible sources of radiation may be encountered in space flight: (1) radiations intrinsic to space itself and (2) man-made contamination from either nuclear weapons or by-products of nuclear-powered space vehicles. At the present time, the biologically important radiations are those intrinsic to space. Radiation is not a hazard of undersea activity.

Some of the known effects of radiation have been garnered from atomic bomb explosions, an accident that occurred after an atomic bomb test in the Marshall Islands, and isolated accidents in various radiation laboratories, as well as radiation therapy programs. The classic symptoms of radiation sickness include nausea, vomiting, diarrhea, a drop in the white blood cell count, alopecia, bone marrow destruction resulting in anemia, hemorrhage, ulceration in the gastrointestinal tract, sterility, and finally death. A dose level of 25 to 50 rad may decrease the lymphocyte count. This response is considered the most sensitive indicator of acute radiation.

The geomagnetic field surrounding most of the earth is both a protection and a hazard, in that it traps radiation that otherwise would be harmful but to earthbound residents provides a danger zone that must be crossed in lunar or interplanetary flight. Van Allen first characterized the great radiation belts in 1958, and since then additional data on intensities have been gathered by both United States and Russian probes. The inner zone, 3600 km above the earth, consists of a concentration of protons with great penetrating power. The astronaut in a lunar mission passing through the Van Allen belt is shielded by capsule walls and traverses the belt in a very short time. The actual measured radiation dosage on the skin of the astronauts has been measured at 10 mrad, an exposure that has no noticeable effect.

The primary danger to the lunar explorer would be a solar flare of large magnitude, or cosmic radiation consisting of high-energy protons and alpha particles. In this event with minimum protection the skin dose would reach 691 rad. This is a dose which causes rather intense skin irritation, and possibly desquamation ulceration, subcutaneous edema, and fibrosis of a serious enough nature to impair performance. The chances of such a flare's occurring are less than one in 5000 missions, and most of the flares would be of an intensity yielding a skin dose of 237 rad, which has a minimum pathological effect. Other possibilities exist which would be damaging. These include prolonged orbiting flights within the Van Allen belt or exploration of larger planets that may have greater intensities of trapped radiation within their geomagnetic fields.

METABOLIC PROBLEMS OF SEALED CABIN EXISTENCE

Much valuable experience in the engineering of habitable sealed cabins has been gained in the development of the nuclear submarines. The cabin toxicology problems are quite similar to those of spacecraft. Paints completely free of noxious vapors, and efficient air purification systems have been developed. Tolerance standards for contaminants such as carbon dioxide have been carefully determined. Although the problems are somewhat akin, the weight limitations of the spacecraft complicate existence in the sealed cabin.

Two general types of system can be employed for both space flight and prolonged saturation diving, depending on the duration of the flight or the dive: the open supply system and the closed ecological system. In short-duration orbital or lunar flights the open supply system is utilized. Food, water, and oxygen are supplied and carbon dioxide and waste materials are absorbed and stored or ejected from the cabin. In order to have an effective supply and absorption system without a weight penalty, precise and complete quantitative knowledge of man's metabolic turnover is required. Figure 30-2A gives some estimates of this daily turnover. The required food, oxygen, and water must be stored. Three forms of oxygen supply have been developed: the compressed gas, liquid oxygen, or oxygen-generating chemicals.

Equipment must be utilized for carbon dioxide absorption and excretory product storage or removal. Chemicals such as lithium hydroxide are effective carbon dioxide absorbents and have been used in the Mercury and Gemini flight programs. As carbon dioxide is absorbed, oxygen is added to the environment. Water vapor has been condensed from the cabin atmosphere, with the dual benefit of humidity control and possible water reusage. Excretory material has been saved to permit detailed analyses after recovery of the capsule. In the Apollo flights some waste material was ejected from the cabin. One idea is to fire the packaged waste material toward the earth's atmosphere with a rocket. The frictional heating as earth is approached will completely burn the material, it is hoped.

The closed ecological system or recycling system is illustrated in Figure 30-2B. The best example of such a closed system is the entire earth. However, under space flight conditions, or in undersea activity, only a partial or limited ecological system is possible, since the great variety of organisms required for assimilation of all man's waste materials and production of all his dietary requirements, or the chemical required, could probably not be carried aloft or below the sea surface.

Because of weight limitations, space flight requires careful consideration of the astronaut's metabolic needs. In the manned Apollo flights the food intake averaged only 1680 Kcal per day. This is lower than the expected 2500 or 3000 Kcal of the average man's diet. The anorexia may be related to inactivity of confinement and weightlessness, to difficulty in eating, or to questionable palatability of the prepared foods. Concerning the latter point, however, the astronauts have reported they are quite satisfied. Gastrointestinal function is not gravity-dependent because peristalsis is quite effective. Gastric distress problems have occurred but were transient in nature and probably of viral etiology. Part of the problem of diet may be moderate motion sickness, which has consistently occurred. In Apollo flights 7 through 11 ranging in duration from 147 hours to 260 hours the average weight loss was 6.2 pounds, half of which has been attributed to body water loss. In longer flights it is anticipated that the rate of weight loss will level off and nutritional balance will be maintained with a food intake of about 2300 Kcal per day.

ACCELERATION FORCES

For ballistic global flight, orbital flight, or interplanetary travel, man must start from rest and accelerate rapidly to a velocity of perhaps 25,000 miles per hour and then

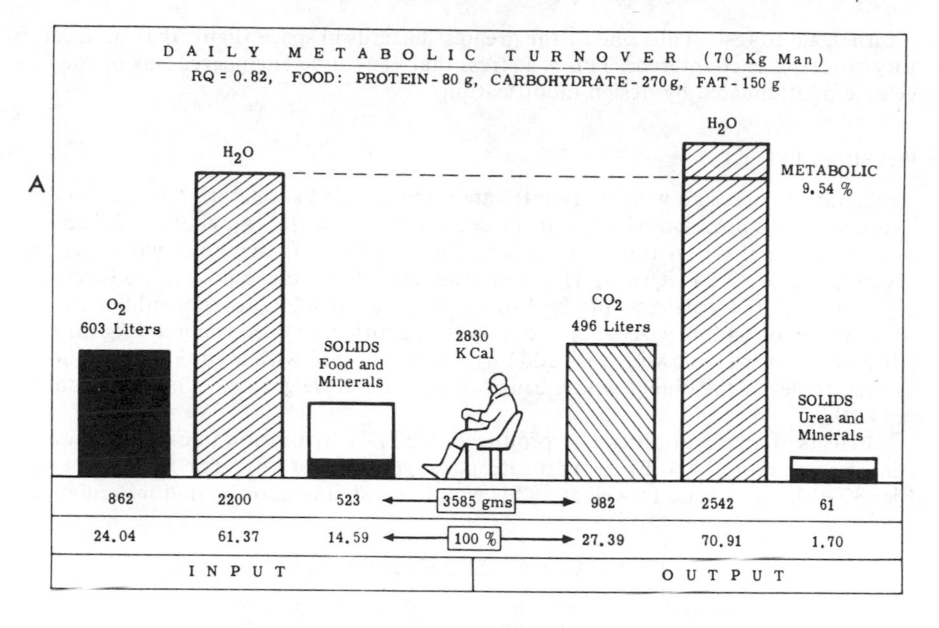

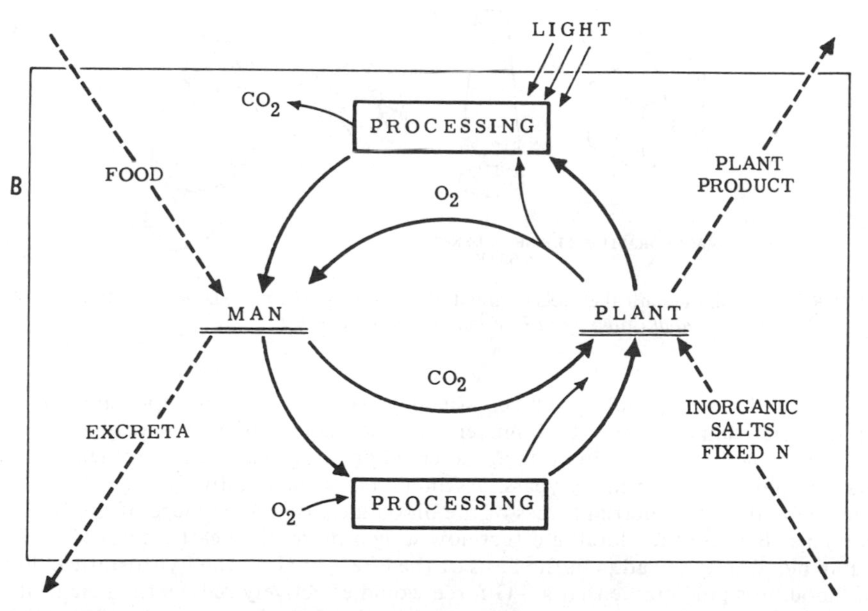

FIGURE 30-2. A. Daily metabolic turnover. B. Simplified scheme of a partial ecological system. (A reprinted from "The Engineered Environment of the Space Vehicle" by Dr. Hans G. Clamann, *Air University Quarterly Review* [Summer 1958, Vol. X, No. 2]. B from J. Myers. In O. Benson and O. Strughold [Eds.], *Physics and Medicine of the Upper Atmosphere and Space.* New York: Wiley, 1960. P. 389.)

return again to rest. Thus one of the greatest hazards in space flight, that of accelera-tory forces, is encountered. This is a stress that cannot be engineered out of the vehicle or attenuated by design modifications.

PHYSICAL EFFECTS

Acceleration force and weight are intricately related, and acceleration forces are expressed in comparison with the force exerted by the earth's gravitational field (abbreviated as 1 G). A force of 1 G acting upon a freely falling body will cause an acceleration of 32.2 feet/sec^2. If a man weighing 70 kg were exposed to a force of 2 G, his measured weight would be 140 kg. In free fall, a 2-G force would cause an acceleration of 64.3 feet/sec^2. To look at it in another way, consider a 1-kg object sitting on a spring scale which is suddenly moved upward with a 10-G force. The inertial forces of the object would cause an increase in weight, and the scale would read 10 kg.

Effects of acceleration can be produced in a body by motion in one direction (linear acceleration) and by centrifugation (radial acceleration). The first effect (see Fig. 30-3) is an increase in weight. This would impair any coordinated muscular

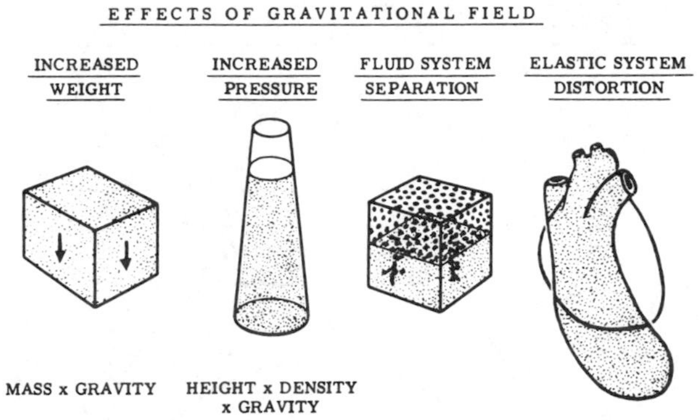

EFFECTS OF GRAVITATIONAL FIELD

INCREASED WEIGHT | INCREASED PRESSURE | FLUID SYSTEM SEPARATION | ELASTIC SYSTEM DISTORTION

MASS x GRAVITY | HEIGHT x DENSITY x GRAVITY

FIGURE 30-3. Mathematical effects of gravitational forces. (From J. D. Hardy. In *Proceedings of the XXI International Congress of Physiology*, 1959. P. 6.)

activity, and if great, could produce tissue damage. Thus, if it is important for a pilot to maintain control of limb movements and manipulative ability, high G force must be avoided. The second effect, increased pressure, has been a problem of aviation physiology for many years. As an aircraft such as a dive bomber starts the rise after a dive, the inertial forces of the blood increase the pressure of the fluid columns. Because the blood in effect now weighs more, the heart cannot effectively pump blood to the head. On the basis of the effect of G on the hydrostatic column of blood, it is predictable that a 3-G force would effectively reduce the height of the hydrostatic column maintained by the heart so that blood would no longer reach the cerebral arteries. Actually, even though reflex compensation initiated by the carotid sinus is of some aid, the tolerance to acceleration in this direction is only slightly above 3 G. The increased pressure in the lower extremities and abdomen would result in vessel distention. Blood pooling would occur, decreasing venous return to the heart and thereby decreasing cardiac output to contribute to the cere-bral hypoxia. The visual system, including the retina and the optic nerve and cerebral visual centers, is extremely sensitive to hypoxia. Loss of vision usually precedes loss of consciousness. This is the well-known "blackout."

Headward acceleration along the long axis of the body is conventionally known as *positive G*. The pilot in maneuvers, such as an outside loop, may impose *negative G* forces upon himself. Because of the rigid cranium, pooling of blood in the head does not occur to a great extent. However, the eyes and face, unprotected by the cranium, are subject to hyperemia. A temporary blindness or "redout" may occur. Suggested explanations of this phenomenon include intraocular hemorrhage, forcing of the lower lid over the cornea, and bleeding into the lacrimal fluid.

Distention of elastic systems also is due to a weight change and is augmented by the pressure increases if the system is liquid-filled. The empty heart exposed to angular acceleration could show the change depicted, but with contained blood the distention would be greater. Accelerations applied to the body transversely have resulted in a transposition of the heart of almost 1 inch. Finally, separation of fluid systems is illustrated by the action of the common laboratory centrifugation. The acceleratory forces met in space flight are probably not great enough to produce this effect.

Fortunately, the stresses of the linear acceleration forces of space flight may be simulated almost exactly on the human centrifuge. This has proved to be a useful tool for tolerance testing and also for training the prospective astronaut. Figure 30-4 depicts the various positions of the pilot and the corresponding acceleration tolerance. As may be seen, G tolerance is related to the intensity, the duration, and the direction of the stress applied. Man's tolerance to acceleration perpendicular to the long axis of the body is more than twice as much as his tolerance to headward acceleration.

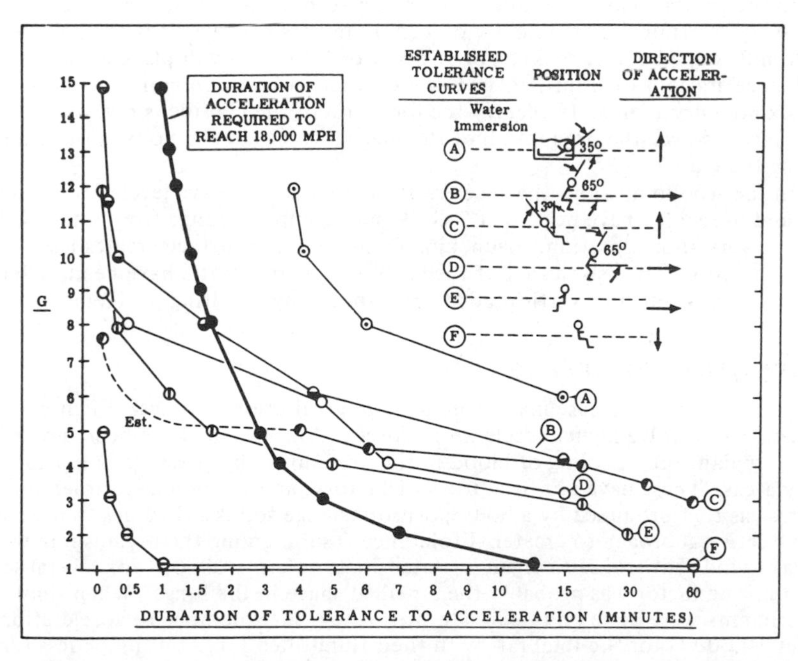

FIGURE 30-4. Position versus acceleration tolerance. The heavy line represents the total G force necessary for attainment of orbital flight. Note the improved tolerance with water immersion. (From S. O. Bondurant et al. *U. S. Armed Forces Medical Journal* 9:1095, 1958.)

The main reason is that the acceleration is perpendicular to the fluid columns of blood. Pressure increments upon the circulation result in less impairment, and there is less excitation of reflex activity of the circulatory system.

PHYSIOLOGICAL EFFECTS

With the pilot exposed primarily to transverse acceleration, the problems of blackout and redout are largely eliminated. Even in transverse acceleration, however, cardiovascular effects still occur, particularly in the pulmonary circuit, where the air-containing lungs would not hydrostatically balance a pressure increase in blood vessels. Circulatory studies on man undergoing 10 minutes of 5-G transverse acceleration have shown a mean increase in heart rate of 59 percent, and a mean increase in aortic pressure of 3.9 percent. The most striking finding was the increase in right atrial pressure, which averaged 12 mm Hg. Presumably this was the result of increased blood volume in the thorax. The other changes may be reflex responses, although interpretation is complicated by the psychological state of the subject.

Heart rates as high as 162 beats per minute have been recorded in U.S. astronauts upon launch and as high as 184 during reentry. In these complex situations it is impossible to separate acceleration effects from vibrational, heat, and emotional influences.

An efficient rocket launch involves high acceleration. When a rocket rises slowly, a considerable amount of the expended fuel energy is used only to hold the rocket off the ground. The acceleratory forces required to reach the velocity for attaining an orbit or escaping from the earth are calculable and can be expressed as the product of the G force times seconds of exposure. A total of 828 G-seconds is required to attain the orbital velocity of 8000 meters per second. The total acceleratory force necessary to attain escape velocity of 11,600 meters per second is 1152 G-seconds. With multistage booster rockets, brief peaks of 7 to 12 G will place a craft in orbit. Peak accelerations of from 7 to 10 G were experienced upon launch and reentry of the six Mercury flights. In reentry the direction of the capsule is reversed, and as the capsule slows, the orientation of the inertial forces to the body is the same as during the launch.

In the Apollo missions the accelerations at launching have leveled off at about 4 G and in reentry have peaked at 7 G. Landing impacts range from 6 to 8 G but are of a very short duration. Launching from the lunar surface creates less G force, as gravity is only one-sixth that of the earth's. Lunar lift-offs have been undertaken with the astronauts in a vertical position without any resulting problems.

AUGMENTED PROTECTION

Are there ways to increase man's tolerance to gravitational forces? Flights to other planets may involve higher acceleratory forces. Anti-G suits have been devised to counterbalance the pooling of blood in the extremities by pressured inflation over body areas. These have not been too satisfactory, and the increased tolerance is not as great as that produced by a body position change to take G forces transversely. An effective approach to greater G tolerance is submerging the organism in water. It was found that rats submerged in water could tolerate 1000 G for several seconds. The limiting factor was probably the air-filled space in the lungs. When similar experiments were done on rats in late pregnancy, the tolerance to acceleration was about 10,000 G for the fetal rats with their fluid-filled lungs but much less for the mother. The principles of water protection have been known for some time and were actually experimented with in Germany prior to World War II. In Figure 30-4 it may be seen that the tolerance of submerged individuals is higher. The subjects in these tests breathed by an underwater breathing valve in which the pressure supplied was increased as the pressure of water increased in acceleration. Thus breathing could be maintained. In experiments done at the U.S. Naval Acceleration Laboratory, in Johnsville, Pennsylvania, the tolerance reached was 16 G with subjects submerged to eye level. When these experiments were extended so that submersion of the subject was complete, a considerable increase in protection was seen. Even though no provision was made for breathing and the subject had to hold his breath, a record tolerance of 31 G for five seconds has been established by R. F. Gray (Hardy, 1964).

The benefits of water submersion are twofold. First, no displacement of the body fluids in an elastic system and no elastic tissue distortion occur under water because these tendencies are almost balanced out by the equal buildup of pressure in the surrounding water. Second, water submersion can alleviate some of the immobilizing and tissue-damaging effects of weight increases produced by high G forces. For example, the human arm may weigh 10 pounds in air, but in water, because of buoyancy effects, it will weigh 0.1 pound. Under 10 G acceleration the arm in air would be immobilized, weighing 100 pounds, whereas the submerged arm would weigh 1 pound. Thus the astronaut could maintain a high degree of mobility while under water and continue manual control. Although the buoyancy may also be effective in attenuating any structural or tissue damage due to the increased weight with acceleration, the heart and mediastinum are not immersed in water but are surrounded by air-filled lungs. Water immersion would offer no protection for these structures, and they may ultimately prove to be the limiting factors in high G stress. Although water is heavy, it is a valuable substance in space flight, and with more powerful vehicles this means of protection may be feasible.

Acceleration tolerance is related to the exposure time and to the gravitational force involved. Rather high G force may be tolerated for extremely brief periods. Experiments at the Holloman Air Force Base have produced high G forces of short duration by abruptly stopping a fast-moving rocket sled. Although this is deceleration rather than acceleration, the mathematical treatment and the physical and physiological effects are exactly the same, in that high G forces are developed. Colonel Stapp, riding the sled in a forward-facing position, has lived through a total exposure of 40 G for 0.2 second. This deceleration and resulting G force is equal to that obtained when a car moving at 120 mph is brought to a complete stop in 19 feet. The subject was hospitalized in a state of shock for several days. However, no permanent injury occurred. Captain Beeding, also at Holloman, during a backward-facing ride was exposed to 80 G for about 0.4 second, which may represent a record of sorts. Beeding fared much better than Stapp and did not sustain any permanent injury. In comparison, however, it appears that the water immersion experiments resulted in much less physical impairment with high G forces.

THE WEIGHTLESS STATE

In manned space flights, brief periods of increased gravitational forces are encountered upon launch and reentry, but the intervening period is characterized by reduced gravitational force, or the so-called *weightless state*. Because weightlessness could not be simulated in the laboratory, it had given rise, prior to actual flight, to much speculation on its physiological consequences. The list of possible effects was lengthy and included anorexia, nausea, disorientation, sleeplessness, fatigue, restlessness, sleepiness, euphoria, hallucinations, gastrointestinal disturbances, urinary retention, diuresis, muscular incoordination, muscle atrophy, and bone demineralization. Because man has evolved in structure and function in the 1-G state for millions of years and depends on gravity for many sensory cues, particularly cutaneous touch and pressure, otolith stimulation, and kinesthetic sensing, it was assumed that lack of such information would lead to many of the above conditions. Now that data are accumulating on weightlessness almost faster than it can be assimilated, much of the above list is no longer tenable.

The weightless state or subgravity condition may occur in a constant-velocity vehicle which is far enough out into space so that earth's gravitational influence is greatly diminished. The gravitational field encountered in lunar exploration is one-sixth of that on earth because of the moon's smaller mass. In orbital flights the weightless state is attained by exact balancing of centrifugal or inertial forces with the earth's gravitational forces at a speed of 18,000 to 25,000 mph.

ALTERED SPATIAL ORIENTATION

Because much of the sensory input involved in human spatial orientation is gravity-dependent, it was suspected that disorientation could be a serious problem. Involved in spatial orientation are the gravity-activated otolith organs of the labyrinthine system of the inner ear; the kinesthetic system with sensory receptors located in muscles, skin, and viscera; and the nongravity-oriented visual system.

In the early weightless experiences it was found that after some minor sensations of disorientation visual function could effectively replace the otolithic and kinesthetic affectors. However, in the larger Apollo cabin where more movement is possible a rather high incidence of motion sickness symptoms have occurred in the first few days of flight. Five out of the six astronauts on the Apollo 8 and 9 missions reported such symptoms. In what was considered a program of adaptation, crewmen had been instructed to undergo a head movement exercise for several minutes. This, however, produced stomach awareness, and the above symptoms are assumed to be related to augmented activity of the otolith organs in the weightless environment. In Apollo 11 the astronauts were briefed on utilizing more cautious movement, and symptoms did not occur. The limited duration of symptoms when they have occurred suggests that some form of adaptation can take place. This, however, appears to be a significant problem which bears continuous investigation.

BONE METABOLISM

The human skeleton is far from being an inert structure. Wolff pointed out in 1868 that any change in its function is followed by changes in internal structure. Therefore, removal of those forces normally exerted in weight bearing, which will occur with the weightless state or in prolonged bed rest, causes the internal architecture to be changed, principally through reabsorption of calcium or osteoporosis. This can lead to two serious consequences: (1) increasing the proneness to fracture and (2) increasing the tendency toward formation of urinary calculus. In Gemini flights 5 and 6, of four and eight days' duration, bone density measured radiologically decreased. Urinary excretion of calcium and phosphate increased, and losses compared to intake definitely indicated a negative calcium balance. These changes were also observed in the Russian Vostok flights 3 and 4. However, there has been no evidence of urinary calculi production.

In bed rest studies of four weeks' duration the total loss of calcium was only 0.5 percent of the total body pool. If this can be extended to space flight, it may not be serious. However, the evidence is that the loss is of a preferential nature, occurring principally from those bones with the greatest alteration of function, the weight-bearing leg bones, the vertebrae, and the os calcis. Specific densitometric determinations on the os calcis have shown a greater density loss than would have been predicted from overall calcium excretion data. All these changes are readily reversible within a few postflight days. There is evidence from the Gemini 7 flight of 14 days that a diet supplemented with additive calcium and performance of especially designed isometric or isotonic exercises may prevent the bone demineralization. Prolonged weightless exposure of from 60 days to 6 months' duration will provide the full assessment as to the seriousness of this process.

CARDIOVASCULAR DECONDITIONING

Although the human cardiovascular system is designed to compensate for rather marked changes in pressure or heights of hydrostatic columns, it was correctly predicted that this system would be altered by prolonged exposure to the weightless state. Diminished orthostatic tolerance occurred following both the fairly brief Mercury flights and the longer Gemini flights. Cooper, in a tilt test 2.5 hours after landing in the Gemini V flight of 191 hours, showed a heart rate increase to 147 beats per minute and blood pressure of 98/82. This, compared to preflight tilt tests in which the heart rate went to 82 and the blood pressure equaled 100/86, is strong evidence for cardiovascular deconditioning. The increased heart rate represents a compensation for lack of suitable vascular adjustments as the height of hydrostatic column is altered. Responses return to normal in about two days. The quarantine period required for the Apollo 11 astronauts prevented similar postflight testing after a lunar mission, but several of the Apollo astronauts have undergone postflight tests in which the lower portion of the body was exposed to a stepwise decrease in pressure until –50 mm Hg was attained. All astronauts tested showed a significant postflight increase in the heart rate response (range 13 to +68) over preflight responses (range 15 to +9). Again, a return to normal occurred in about two days.

Physiologically this deconditioning has probably occurred because of prolonged lack of stimuli from the 1-G forces usually acting on the system, and it resembles quite closely the orthostatic hypotension found in patients. Several other mechanisms have been suggested, such as alteration of smooth muscle responsiveness, alteration of viscoelastic characteristics of the vessels, particularly capacity vessels, and decreased activity or production of vasoactive hormones. Because the average body water loss of all the Apollo astronauts is about 3 liters, some shift of body fluid compartments may have contributed to the response. After 14 days of weightlessness, deconditioning has not produced any serious impairment even during the increased G forces during reentry. Whether or not longer exploration will produce debilitating cardiovascular effects remains to be tested.

DECREASED EXERCISE CAPACITY

The weightless environment requires only a minumum of muscular contraction for movement and performance of many work tasks. It was suspected, therefore, that work capacity might decrease. A series of tests were designed for the Apollo astronauts in which they pedaled a bicycle ergometer at a rate to yield given heart rates for three minutes of 120, 140, 160, and 180 beats per minute. Average oxygen consumption was 26.2 percent lower at equivalent heart rates than in preflight tests. If this state were progressive and continued throughout the duration of the flight, future explorations of, for example, Jupiter – with its higher gravity – might markedly impair the performance of the explorer.

PROTECTION IN WEIGHTLESSNESS

Although weightlessness does alter the crewman's physiology, such alteration has not reached levels where impaired performance occurs. Most astronauts and cosmonauts have adapted quite readily and appear to enjoy this condition. With a little practice, movement is quite easy and relaxing. The skills for handling food and performing work tasks are rapidly learned. Various exercise regimes and pressure suit deflation or inflation to keep the cardiovascular system in practice may be required if some of the changes leading to impairment continually progress in prolonged operations. Rotation of a space station at a rate dependent upon its radius can establish an artificial gravity of 1 G through centrifugal force. Thus, it does not appear that lack of gravity need be a constraint for prolonged space activity.

ENGINEERING OF THE HUMAN

The prevailing concept at the present time in human space technology is known as human engineering. That is, an attempt is made to engineer the hardware of the spacecraft so that man will carry with him a small chunk of his terrestrial environment. To the biologist a more exciting concept, and one which may offer real promise, is to fit the man to the problem, or to engineer the human rather than the hardware. What is meant by this will be discussed in several aspects.

DRUG USES

The pharmacological approach to space flight problems offers advantages in several areas. The vibratory and rotatory motions of a space vehicle during launch and active flight may result in motion sickness, a specific syndrome consisting of vertigo, nausea, drowsiness, cold sweating, salivation, pallor, and headache. It is thought that at least some of the symptoms originate from overstimulation of the labyrinthine system. Rather than engineer out of the craft those influences that produce motion sickness, perhaps a more economical method in terms of both money and weight load would be to administer anti-motion-sickness drugs. Many effective drugs are available, such as Dramamine, Benadryl, Bonamine, and Lergigan, which probably act by an anticholinergic action and some cerebral depression. These drugs can no doubt even be improved and could play a useful role in space flight.

Metabolic depression by drugs is another possibility. Man's metabolic rate may easily range from 1800 to 3600 Kcal per day. If a crew of men had to be carried on a long voyage, a considerable economy could be attained by keeping the metabolic rate near the 1800 Kcal level. Tranquilizers such as chlorpromazine may be of benefit in keeping metabolism near this basal level when high-level performance is not a requirement.

The environment of space has been described as a "hypodynamic environment." Because of sensory deprivation or the loss of afferent information from receptors responding to gravity, to noise, possibly to light, and possibly to temperature, it has been supposed that psychic disorientation may occur in flights far beyond the moon. The usual clues for pacing normal biological rhythms are greatly altered. If this becomes a problem, one possible approach would be administration of drugs, such as amphetamines, which evoke definite stimulation of the sensory cortex. Similarly, excitants could be used perhaps to maintain muscle tone, and vasoactive agents could be used to maintain vascular tone, if these factors become problems. Toward the end of the longer U.S. flights, dextroamphetamine sulfate (Dexedrine) was taken by tired astronauts to increase alertness in the critical reentry phase of flight. Other drug usage has not been made as yet.

Of direct benefit to the space man as well as the terrestrial man would be the development of an antiradiation drug or a method of alleviating radiation damage.

Of the procedures tried, the most effective appears to be replacement after radiation exposure of bone marrow cells with "banked cells." If the problem in space flight is of sufficient magnitude, astronauts could bank part of their bone marrow for future use.

HYPOTHERMIA AND HIBERNATION

A concept that could fit under the classification of engineering of the human would be the use of hypothermia or artificial hibernation in prolonged space travel. This approach has long been a favorite of the science fiction writer. Physiological studies of man in the hypothermic state have indicated that a body temperature reduction of 8° to 9°C reduces metabolism to less than 50 percent. Small mammals have been cooled to near 0°C and even to the point where considerable body mass is frozen. The metabolism here is less than 1 percent of that of the normothermic animal. Such an approach could, of course, result in a considerable economy of oxygen and food supplies. At the present stage of physiology the problems of keeping a hypothermic individual alive, and rewarming him at the desired time, are great. The weight of equipment necessary to produce and maintain a prolonged hypothermia in man by present techniques would be prohibitive for space travel.

Prolonged hypothermia would also be useful in flights lasting several years. This advantage is well illustrated in nature. The bat, which spends a large portion of its life in hibernation, has a much longer life-span than the shrew, which is of similar body size. Hibernation appears to be a state in which life is maintained at the minimum level possible for continued existence. The life-span of the bat may reach 20 years, while the shrew, which continuously maintains a high metabolic rate, lives only about 20 months. However, the total caloric turnover per lifetime is approximately the same for both mammals. The bat takes considerably longer for the turnover. The potential application of these facts to long space journeying is obvious. An army of men could be carried for many years with only a few months of aging and the minimum metabolic requirements.

Hibernation of the astronaut must be considered quite remote at present; much more must be learned of the physiology of both the human and the hibernating species. Hibernators possess genetic adaptations which range in influence from the central nervous system to the simplest somatic cell. All these adaptations could not be simulated in the human to obtain true hibernation. However, a prolonged, profound hypothermia which could be terminated with external rewarming aid may be possible. A body temperature depression of less than 10°C would possibly produce a 50 percent decrease in aging rates and in the weight requirements to support metabolism.

PHYSIOLOGICAL ADAPTATION

Current training programs for astronauts and cosmonauts do not emphasize adaptation programs. Preconditioning to gravitational, thermal, and isolation stresses is done, but full investigation of the usefulness of biological adaptation may prove valuable. Man may show considerable adaptations to hypoxia and heat. For example, the period of useful consciousness of the natives who live at 14,000-foot altitude in the Andes is much greater than that of sea level residents when both are suddenly exposed to an altitude of over 30,000 feet. Certainly this would be a useful adaptation to incur prior to any space program in which there is a possibility of sudden hypoxic exposure. Adaptation to heat could decrease cabin weight requirements by permitting less emphasis on a cabin cooling system which may be of use only during the heat loads of high-velocity reentry.

From undersea activity there is evidence that the diver does adapt to cold water. Part of the adaptation is simply becoming used to the cold. However, in the diving women of Korea, an occupational group regularly exposed to cold water, both increased metabolism and decreased tissue conductance have occurred as adaptive changes.

Psychic adaptation may also be utilized. For example, in isolation experiments of long duration, considerable anxiety and hostility were manifested by the subjects. Training or adaptation to isolation is a possible approach. The six months' isolation of Admiral Byrd in the Antarctic and prolonged isolations of other properly motivated individuals on the ocean have been experienced with some mental depression but without disability. Studies of adaptation are also important in determining those stresses to which man cannot adapt. For example, it is known that man may not adapt to decreasing his water intake below a certain level, and proper cabin engineering must take this into account.

Space flight and undersea existence present an entirely new dimension to the biologist. The study of responses of organisms to new stresses never before encountered, and of means to combat such stresses effectively, is an exciting challenge. Whether the effort should be to keep the astronaut or aquanaut in a state of maximum physiological and psychological efficiency or in a psychophysiologically depressed state depend upon the duration and the objectives of the mission. Physiological and technical problems of lunar missions have been solved. The next step, to Mars or Venus or a nearby asteroid requiring five to six months, will be a major one requiring additional physiological experience gained from earth-orbiting laboratories. Within a few short years diving operations will be extended to 1500 feet. The intense scientific effort and the resulting knowledge gained should be of direct benefit to all mankind.

REFERENCES

Armstrong, H. G. *Aerospace Medicine.* Baltimore: Williams & Wilkins, 1961.

Bennett, P., and D. N. Elliott. *The Physiology and Medicine of Diving and Compressed Air Work.* Baltimore: Williams & Wilkins, 1969.

Berry, Charles A. Summary of medical experience in the Apollo 7 through 11 manned space flights. *Aerospace Med.* 41:500–519, 1970.

Branley, F. M. *Exploration of the Moon.* Garden City, N.Y.: Natural History Press, 1964. P. 129.

Brown, J. H. V. (Ed.). *Physiology of Man in Space.* New York: Academic, 1963.

Burns, N. M., R. M. Chambers, and E. Hendler (Eds.). *Unusual Environments and Human Behavior.* New York: Free Press of Glencoe, 1963.

Busby, D. E. *Space Clinical Medicine.* Dordrecht, Holland: O. Reidel, 1968.

Campbell, P. A. (Ed.). *Medical and Biological Aspects of the Energies of Space.* New York: Columbia University Press, 1961.

Gauer, O. H., and G. D. Zuidema. *Gravitational Stress in Aerospace Medicine.* Boston: Little, Brown, 1961.

Hardy, J. D. (Ed.). *Physiological Problems in Space Exploration.* Springfield, Ill.: Thomas, 1964.

Hock, R. The potential application of hibernation to space travel. *Aerospace Med.* 31:485–489, 1960.

Lansberg, M. P. *A Primer of Space Medicine.* New York: American Elsevier, 1960.
National Aeronautics and Space Administration. NASA SP, Washington.

Pogrund, R. S. Human engineering or engineering of the human being – which? *Aerospace Med.* 32:300–315, 1961.

White, W. J. *A History of the Centrifuge in Aerospace Medicine.* Santa Monica, Calif.: Douglas Aircraft Co., 1964. P. 90.

Wunder, C. C. *Life into Space: An Introduction to Space Biology.* Philadelphia: Davis, 1966.

31

The Hypophysis: Neuroendocrine and Endocrine Mechanisms

Ward W. Moore

INTRODUCTION

The degree of complexity of functions observed in man and other vertebrates has been attained in part by the evolution of the two primary integrating systems, the nervous system and the endocrine system. Each participates in the regulation and coordination of the activities of the organism.

The endocrine system is a regulatory system. It functions to maintain the internal environment (the body fluids) at a relatively constant level with respect to volume and concentration, and within the limits of life in the face of changes in the activity of the body. It transmits information by means of chemical messengers, the hormones. These are dispersed without direction throughout the body via the circulatory system, from which they move to act upon genetically conditioned and differentiated cells of the body, the target cells. The effects of the hormones are much more diffuse and relatively slower than are those of the nervous system, which transmits its information rapidly with point-to-point precision via the neurons. The fundamental microscopic anatomy of the glandular elements of the endocrine glands is similar to that of the exocrine glands, with but two exceptions. The endocrine gland does not possess a duct system, and each glandular cell of the endocrine system has a surface which abuts against a venous sinusoid or a capillary. Thus, the morphology of the system is such that the secretions can be released directly into the circulation.

Emphasis will be placed in this and succeeding chapters on the mechanisms which regulate the rate of secretion of hormones and the role of the hormones in homeostasis. The hormone-secreting elements to be considered are the secretory or glandular cells of the hypothalamus, the hypophysis or pituitary gland, the pancreas, the parathyroid glands and derivatives of the ultimobranchial body, the thyroid gland, the adrenal cortex, the gonads, and the placenta.

HISTORICAL BACKGROUND

Berthold's studies in 1849 were probably the first experimental demonstration of an internal secretion, although they were overlooked as such for nearly half a century. Berthold showed that the grafting of testicular tissue would prevent atrophy of the comb of the capon. Following these studies, nothing approaching the modern concepts of endocrinology was recognized for over 50 years. However, various clinical observations had correlated certain glands with specific disease states and provided the eventual stimulus for the study of the endocrine glands and their functions. In 1902 Bayliss and Starling showed the existence and physiological role of secretin, and the term *hormone* was first used by Starling in 1905. It was Starling who suggested that the constancy of the internal environment, first postulated by Claude Bernard, might be maintained in part through hormonal mechanisms, and he thus linked the function of the endocrine system with Bernard's concept.

The field of endocrinology was advanced on a firm experimental foundation during the first third of this century. During the 1940's the study of endocrinology reached a yet more advanced stage of development. Marked progress in steroid chemistry was stimulated by the clinical studies of Hench and co-workers. New experimental tools and methods such as radioactive isotopes, the phase microscope and electron microscope, tissue culture and perfusion techniques, chromatographic techniques, and electronic recorders were put to use in the study of the hormones, and as a result notable advances in the field have been made during the last 15 years. Certain steroids (cortisone and progesterone) were first produced on a mass basis. Studies utilizing isolated adrenal glands and perfusion of the system with ^{14}C-labeled acetate and cholesterol brought major advances in knowledge of the biosynthesis of the hormonally active steroids. The active substance of the amorphous fraction of the adrenal cortex was finally determined in 1953 by Simpson and Tait to be the steroid aldosterone. Using chromatographic and isotopic techniques, two independent groups isolated a second thyroid hormone. In 1954 Sanger and his associates established the amino acid sequence and full chemical structure of insulin. Du Vigneaud and his associates determined the structure of the active principles of the neurohypophysis, vasopressin and oxytocin, and succeeded in synthesizing the two polypeptides. The development of techniques utilizing in vitro systems led to improvements concerned with the effects of many hormones on chemical reactions and enzyme systems in many cells. Modifications of the original techniques of Hess (chronic electrode implantation) were utilized to great advantage by C. H. Sawyer and G. W. Harris and their co-workers to expand knowledge of the relationships between the central nervous system (CNS) and the anterior pituitary. Recently the radioimmunological techniques introduced by Berson and Yalow for the assay of polypeptide and protein hormones in the plasma have added much to an understanding of endocrine functions.

METHODS OF STUDY

The evidence required to show that an organ functions as an endocrine gland involves the demonstration of specific effects in the absence of the organ, and specific physiological restorative responses following exogenous administration of the hormone by transplantation of the glandular tissue or injection of suitable extracts of the gland. A further step needed to verify the endocrine activity of a gland requires the administration of extracts of the gland to intact animals with the production of exaggerated effects of the hormone. Other requirements are concerned with evaluation of the chemical and physical characteristics of the hormone, the characterization of effects of the hormone, and factors which regulate the rate of secretion of the hormone. The endocrinologist also concerns himself with the problem of synthesizing the hormone and determining the mode of action of the hormone. The known hormones of the thyroid, the gonads, the adrenal glands, and the neurohypophysis have been synthesized, and this has been accomplished recently with the larger polypeptide and protein hormones. Little is understood of the fundamental action of hormones, including the primary molecular actions.

FUNCTIONS OF HORMONES

The effects of hormones fall into three general groups. First, they influence reactions which aid in the maintenance of a constant internal environment. This includes the regulation of the rates at which carbohydrates, fats, proteins, electrolytes, and water are deposited in or removed from the tissues of the body. For example, insulin participates in regulating the chemical and/or physical factors which ensure an adequate supply of glucose to most extrahepatic tissues by increasing the permeability of the cell membrane to glucose. Second, the hormones have a morphogenic action; this includes the effects on metamorphosis, growth, maturation, trophic actions, and aging processes. Examples of the morphogenic actions are the effects of the ovarian or testicular hormones (under the influence of the pituitary gland) during the growth and development of the accessory sex organs and secondary sex characteristics at puberty. Finally, the hormones regulate autonomic activity, as well as certain CNS activities and behavioral patterns. For instance, maternal behavioral patterns are linked to the presence of various hormones. Also, the sensitivity of effector cells to epinephrine and norepinephrine is altered by the circulating levels of some hormones.

It must be emphasized that the action of all the hormones is *not* to initiate chemical reactions but to alter the rates of preexisting reactions without contributing significant amounts of either matter or energy to the process. The hormones can function at peak efficiency only when the concentrations of substrate, cofactors, hydrogen ion, and the temperature are at optimal levels. It follows that, in experimental tests utilizing hormones, the subjects should be in nutritional, temperature, and fluid balance, and as free as possible from psychological stimuli, if representative results are to be expected.

The mechanisms by which hormones exert their control are varied, and recently much information has become available concerning the mechanism of action of hormones. In general, some hormones may be said to act upon genes and alter the rate of protein (enzyme) synthesis, some hormones may act to release small molecules or ions which alter the rate of protein (enzyme) synthesis or activity and/or membrane permeability, and some hormones may act upon membranes and alter their permeability characteristics. In each case the hormones appear to alter the activity of rate-limiting steps in a metabolic pathway.

On the basis of experimental studies many hormones have been assumed to act as "gene activators," because they alter the rate of deoxyribonucleic acid (DNA)–directed ribonucleic acid (RNA) synthesis and the rate of protein (enzyme) synthesis in target cells. Many of the effects of the gonadal hormones, the thyroid hormones, insulin, growth hormone, and adrenocorticotropic hormone on their target cells are probably mediated via this mechanism. A more specific illustration of this type concerns the effect of an adrenal steroid, aldosterone, on sodium reabsorption in the toad bladder. It has been shown that aldosterone becomes concentrated in the nuclei of the cells concerned with sodium reabsorption and promotes messenger RNA (mRNA) synthesis. The latter then serves as a template for the ribosomal synthesis of a "permease," which increases the permeability of the mucosal surface of the cell to sodium, thus increasing the amount of sodium available for active transport into the extracellular fluid.

The work of Sutherland et al. with 3′, 5′-cyclic adenosine monophosphate (cyclic AMP) has added greatly to the understanding of hormone action. Sutherland refers to cyclic AMP as "the 2nd messenger," whereas the hormone is termed the "1st messenger." In this concept many hormones activate an adenyl cyclase at the target cell which produces from adenosine triphosphate (ATP) many "messengers" of cyclic AMP. The cyclic AMP, in turn, may then act to alter the permeability of a membrane, or promote the synthesis or increase the activity of an enzyme, which would alter the rate of physiological response. In many respects cyclic AMP could be considered to amplify the hormonal message. Experimental work has indicated that many hormones, such as vasopressin, glucagon, adrenocorticotropin, luteinizing hormone, thyrotropin, and epinephrine, exert some of their effects by increasing the accumulation of cyclic AMP in their respective target cells.

A hormone may also produce an effect by increasing the permeability of a cell to substrate. Insulin has an effect which facilitates the entry of glucose into muscle cells, but the mechanism or mode of action underlying this increase in permeability to glucose is not known. It has been hypothesized that insulin removes a barrier which acts to prevent glucose entry into the cell.

The actions of a hormone upon an end-organ may vary considerably, depending upon the nature of the effector cell. Thus, thyroid-stimulating hormone stimulates the thyroid gland but has little, if any, effect upon the adrenal cortex. Also, the estrogens stimulate growth of the female reproductive tract but have little effect upon skeletal muscle. A hormone through its primary actions may produce a stimulating effect on one type of cell and an inhibitory effect on another type. For example, the hormones of the thyroid gland cause an increase in the oxygen consumption of hepatic tissue but depress the oxygen consumption of the pituitary.

PRINCIPLES OF HORMONE ASSAY

The endocrine glands contain only minute amounts of the hormones which they secrete, and the blood contains amounts in even smaller concentration. The physiological activity of the hormones is, in most cases, the only guide to their presence, and these effects form the basis for bioassay procedures. In such cases, the functional effect of a hormone is used as the basis for the quantitative estimation of the amount of hormone present. The concentration of the steroid and amino acid hormones can be estimated by using chemical techniques. Suitable assay procedures are now available for most of the hormones.

The level of activity of many of the endocrine glands can be estimated by observing the morphology of the glandular cells themselves. Such criteria as size and shape of cells, as well as quantity and characteristics of intracellular granules, supply indices of the level of activity. Similarly, the morphology of cells and functional activity of tissues upon which the hormones act are used to estimate the activity of an endocrine gland.

The hormonal content of each of the endocrine glands can be estimated by using biological and/or physiochemical techniques. In addition, the level at which a gland is functioning can be assessed by analyzing the body fluids for the secretion of the gland involved. However, only very small amounts of hormone are present in the blood, and small amounts can be recovered in the urine. The extent to which the urinary concentration of hormones and their metabolites reflects the amount in the blood is unknown in many cases. It has become much too convenient to assume that urinary concentration of hormones and their metabolites truly reflects the blood levels. It is the body fluids, exclusive of the urine, from which the hormones must eventually exert their action. Nevertheless, in the absence of adequate techniques for the assay of hormones in the blood, studies on the urinary concentration of hormones have yielded valuable information.

Recently radioimmunological methods have been applied in quantitative assays of protein and polypeptide hormones in the plasma. These methods are sensitive, precise, specific, and accurate. They can be applied to very small quantities of plasma and depend upon highly specific reactions between a protein or polypeptide hormone, the antigen, and its antibody. The assay is based upon competition between unlabeled hormone and radioactively labeled hormone for specific sites on the antibody. Initially radioactively labeled ("hot") hormone is complexed with antibody, and then known amounts of unlabeled ("cold") hormone are added and serve as standards. Some of the "cold" hormone displaces some of the "hot" hormone from the antibody. The free or unbound hormone is then separated from the complexed or antibody-bound hormone, and the amount of radioactivity in both samples is determined. The lower the concentration of "cold" hormone, the more labeled or "hot" hormone will remain bound to the antibody. A standard curve can be prepared by using different concentrations of unlabeled hormone, and unknowns, or plasma samples, can be compared with it. Assays of this type are now available for the determination of plasma levels of insulin, glucagon, parathyroid hormone, calcitonin, growth hormone, adrenocorticotropic hormone, thyroid-stimulating hormone, luteinizing hormone, and follicle-stimulating hormone, utilizing very small quantities of plasma.

REGULATION OF ENDOCRINE ACTIVITY

A variety of stimuli are capable of altering the activity of many of the endocrine glands. Changes in both the external environment and the internal environment precipitate factors which lead to changes in the output of hormones in an attempt to maintain the internal environment within the limits compatible with life. The stimuli which influence the output of a hormone may be transmitted solely via neural channels, solely through humoral influences, or through a combination of both, i.e., neurohumoral channels.

Neural Control of Endocrine Activity

Either directly or indirectly, a normally functioning system of endocrine glands is dependent upon a normal CNS and its many processes. The release of the secretions of two of the endocrine glands, both arising from embryonic neural tissue, is regulated directly by their innervation; i.e., their efferent nerves are secretomotor. These two structures are the adrenal medulla and the neurohypophysis, and both undergo atrophy following denervation (Fig. 31-1).

The anterior pituitary, which has been considered the master gland, is in fact the target of various stimuli, both neural and humoral, which stimulate or inhibit its activity. It now appears certain that the supreme control of the endocrine system resides within the CNS, but this regulation is not expressed via secretomotor fibers to these glands. It is expressed rather through the secretion of mediator substances released by nerves of the hypothalamus which stimulate the release of the respective trophic hormones from the pituitary. The CNS also regulates the function of all the endocrine glands indirectly by exerting its effects via the vasomotor innervation of the various glands. However, it should not be inferred that the rate of hormone secretion by a gland is necessarily related to the rate of blood flow through the gland, for it has been demonstrated that the rate of blood flow through the thyroid gland does not reflect the level of thyroidal activity. It has been shown, however, that the sympathetic mediators, epinephrine and norepinephrine, increase the responsiveness of the thyroid gland to a constant dose of thyroid-stimulating hormone. Thus, release of the mediators at the vasomotor endings may act at the level of the glandular cells to increase their sensitivity to trophic hormones.

Humoral Control of Endocrine Activity

The endocrine glands should not be considered individually but should be viewed as an integrated system, because a given hormone seldom acts independently and

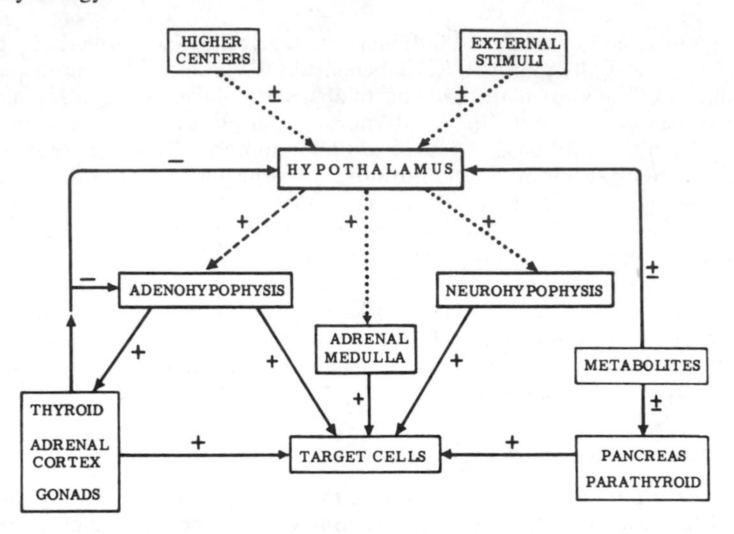

FIGURE 31-1. Relationship between the endocrine glands and the central nervous system. + stimulation; – inhibition; ·····→ neural pathway; -----→ neurohumoral pathway; ——→ humoral pathway.

generally affects other endocrine glands and alters their rate of secretion. In this action, the hormone may assist or oppose the effect of another hormone. Such interrelationships are illustrated by the following observations. (1) The administration of estrogens inhibits the production of the pituitary gonadotropins, follicle-stimulating hormone, and luteinizing hormone. The administration of either of the two gonadotropins increases the weight of the ovaries, but the ovaries do not respond by increasing their output of estrogen. However, the simultaneous administration of the two gonadotropins produces marked ovarian growth and increased estrogen secretion. (2) The two pancreatic hormones, insulin and glucagon, induce opposite effects upon the blood sugar; the former lowers it and the latter elevates it. In addition, the functional state of an endocrine gland may be altered by a hormone secreted by a gland other than the pituitary; e.g., the administration of thyroid hormone may increase the functional activity of the adrenal cortex. Such interrelations between hormones of various glands can account for many of the paradoxical effects observed in experimental and clinical studies.

The pituitary secretes several hormones whose effects may vary considerably. Several endocrine glands are directly affected by the pituitary hormones; these – the gonads, the thyroid, the adrenal cortex – are referred to as the target glands. These glands can secrete small amounts of their hormones in the absence of the pituitary, but normal trophic hormone secretion by the pituitary is necessary for normal target gland function. An increase in the rate of secretion of a trophic hormone results in an increase in the rate of secretion from the respective target gland. The secretions of the target gland, in turn, tend to produce countereffects which oppose the secretion of its particular trophic hormone by the pituitary and thus inhibit the initial source of stimulus. For example, when there is an increase in the rate of secretion of thyroid-stimulating hormone by the pituitary, the rate of secretion of thyroxine from the target gland, the thyroid, increases. The increasing secretion of thyroxine then inhibits the secretion of thyroid-stimulating hormone. The end result is a balance of the forces involved, and the level of thyroid function and of heat production by the cells of the body is maintained within relatively narrow limits. This serves to illustrate the concept of the "negative feedback system" of Figure 31-1.

In addition to the trophic hormones, which exert their primary effects upon specific glands, the pituitary gland also secretes a hormone, growth hormone, which acts upon all cells of the body. The hormones of the thyroid gland also appear to act upon all cells of the body.

The rate of secretion of some hormones is relatively independent of the pituitary, being regulated by the concentration of nonhormonal metabolites in the blood. An example of this type of regulation concerns secretion of the hormone of the parathyroid gland: A low serum calcium concentration stimulates the secretion of parathyroid hormone, and, conversely, hypercalcemia induces a decrease in the rate of secretion. In each case the calcium acts directly on the gland to alter its secretion rate.

In addition to the effects of the trophic hormones and the reciprocal relations between the target glands and the pituitary, the hormones of the target glands may affect one another. For instance, for normal ovarian function the level of thyroidal activity must be optimal.

THE HYPOPHYSIS (PITUITARY GLAND)

The hypophysis secretes at least nine hormones, each being of polypeptide or protein nature. Some of the pituitary hormones regulate the functional capacities of other endocrine glands, and these are the trophic hormones: adrenocorticotropic hormone (ACTH), thyrotropic hormone (TSH), follicle-stimulating hormone (FSH), luteinizing hormone or interstitial cell–stimulating hormone (LH or ICSH), and luteotropic hormone (LTH, prolactin, or lactogenic hormone). Somatotropic hormone (STH or growth hormone) appears to act upon all cell types, whereas the actions of vasopressin (ADH), oxytocin, and melanocyte-stimulating hormone (MSH) are restricted to particular cell types.

The hypophysis is a compound gland of ectodermal origin, arising from two different sources. One part, the *neurohypophysis,* arises from the ventral floor of the diencephalon and remains connected to the hypothalamus throughout life by means of its stalk. The glandular portion of the hypophysis, the *adenohypophysis,* stems from oral ectoderm as Rathke's pouch, which is an outgrowth from the roof of the mouth. This outgrowth meets the embryonic neural portion of the gland and then loses its connection with the oral epithelium.

The adenohypophysis is described as consisting of three parts: the *pars tuberalis,* the *pars intermedia,* and the *pars distalis* (often referred to as the anterior pituitary or AP). The neurohypophysis is also divided into three parts: the *median eminence of the tuber cinereum,* the *infundibular stem,* and the *infundibular process* (neural or posterior lobe). The median eminence and infundibular stem are collectively referred to as the infundibular or *neural stalk,* whereas the *hypophysial stalk* includes the neural stalk plus the sheath portions of the adenohypophysis, the pars tuberalis (Fig. 31-2).

The well-protected hypophysis is one of the most inaccessible organs of the body, being located in the sella turcica, a depression in the sphenoid bone. The gland is encapsulated by the dura mater, but the hypophysial stalk penetrates the dura through the diaphragma sellae. The neurohypophysis is characterized by its rich innervation of hypothalamic origin, whereas the adenohypophysis is characterized by rich vascularization.

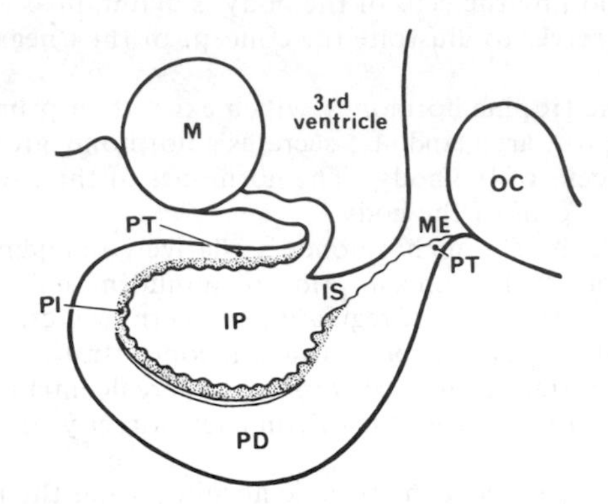

FIGURE 31-2. Midsagittal section through the hypophysis. Parts of the adenohypophysis are the pars distalis (PD), the pars intermedia (PI), and the pars tuberalis (PT). The neurohypophysis is composed of the median eminence of the tuber cinereum (ME), the infundibular stem (IS), and the infundibular process (IP). The mammillary body (M) and the optic chiasm (OC) are also depicted.

THE NEUROHYPOPHYSIS

Strictly speaking, the neurohypophysis is not an endocrine organ, because it lacks glandular elements and serves merely as a storage place for certain secretions of the hypothalamus. However, it does play an essential role in the release of the stored hormones.

Microscopically, the neurohypophysis is composed of four basic structures: unmyelinated nerve fibers, glial cells, and capillaries surrounded by argyrophilic connective tissue. The nerve fibers have their origin in the cells of the supraoptic and paraventricular nuclei, and according to an early hypothesis these fibers composed the secretomotor innervation of cells called *pituicytes*. The latter were held to be sites of formation of the posterior lobe hormones. However, nerve fibers do not end on these cells and they have been reclassified as glial cells. Furthermore, the morphological behavior in vitro of the pituicytes indicates that they are not secretory cells.

It has been observed that the cells of the supraoptic and paraventricular nuclei possess stainable cytoplasmic granules which are regarded as neurosecretory products. These neurosecretory granules have been demonstrated throughout the extent of the supraopticohypophysial tract, from the nuclear region to where the fibers end within the infundibular process. One can identify, by using a variety of staining and histochemical procedures, the neurosecretory granules along the tracts, and they appear as swelling on the fibers. The neurosecretory granules also appear in the interstitial spaces of the infundibular process. The *Herring bodies* are now considered to represent the swellings or beads on the neurosecretory fibers and therefore the Herring bodies do not lie free in the posterior lobe. The manner in which the neurosecretory material is released into the blood stream is not known. The chemical composition of the neurosecretory material also is not known. However, it is protein in nature, and substances which possess all the known biological activities of the posterior lobe hormones have been obtained from extracts of this material.

The above observations, plus the facts that posterior lobe hormones can be extracted from hypothalamic tissues, that the posterior lobe can be depleted of stainable neurosecretory material by procedures which deplete the organ of hormonal activity, and that neurosecretory material piles up at the proximal end of a transected infundibular stem, suggest the following concept: The hormones vasopressin and oxytocin are formed in the cell bodies of the supraoptic and paraventricular nuclei and are transported down the respective nerve tracts to the infundibular process, where they are stored. The terminations of the neurons are regarded as the site of hormone release. The release of the hormone is apparently triggered by the nerve action potential passing down the hypothalamo-hypophysial tracts. In view of these considerations, it appears proper to refer to this system as the hypothalamo-neurohypophysial system (HNS) and regard it as an organ of internal secretion.

Four different activities have been attributed to the hormones of the HNS. First, in 1895 Oliver and Schafer observed that the intravenous administration of extracts of the whole pituitary resulted in a marked rise in systemic blood pressure (pressor effect). The effect was purely one of vasoconstriction, and Howell showed later that this effect was obtainable if only neural lobe extracts were administered. Second, Sir Henry Dale demonstrated that extracts which caused contraction of the uterus (oxytocic effect) could be prepared from this organ. Third, it was demonstrated that neural lobe extracts were helpful in ameliorating the symptoms of diabetes insipidus (antidiuretic effect). Finally, extracts of the HNS have been shown to induce contractions of the myoepithelial cells of the alveoli of the lactating mammary gland and cause evacuation of milk from the alveoli (milk letdown or draught effect).

The brilliant research of Du Vigneaud and his associates has resulted in the isolation of two distinct polypeptides from the neural lobe. Later work led to the synthesis of the two compounds. Each has a molecular weight of about 1000 and contains eight amino acids, six of which are common to both hormones. One compound, vasopressin, possesses a high degree of antidiuretic activity and no oxytocic activity. The other compound, oxytocin, shows little antidiuretic activity and high oxytocic activity.

It should not be at all surprising, considering the close structural similarity, that the two compounds have overlapping activities. A protein complex (molecular weight about 30,000) has also been isolated from the neural lobe, and it contains all four activities of the organ. It is not known, however, whether this protein is an artifact of the extraction procedure or is the neurosecretion in some stage of formation.

Vasopressin (ADH)

The most obvious consequence of neurohypophysectomy is the induction of polyuria. This is characterized by the excretion of a high-volume, low-osmolar-concentration urine. As a result of the large volume of water lost, there is an intense thirst and a high water intake (polydipsia). ADH plays an important role in the maintenance of water balance because it regulates the excretion of free water from the kidney. The site of action of ADH on the nephron is discussed in Chapter 22.

Control of ADH Secretion

The rate at which the secretion and release of ADH occur is under nervous control and is subject to changes in the effective osmotic concentration of the extracellular fluid (ECF), the volume of the ECF or plasma compartments, exteroceptive stimuli, and psychic stimuli (Fig. 31-3).

Various drugs such as nicotine, acetylcholine, morphine, barbiturates, ether, and chloroform act on the HNS to increase the output of ADH. On the other hand, alcohol inhibits the secretion of ADH.

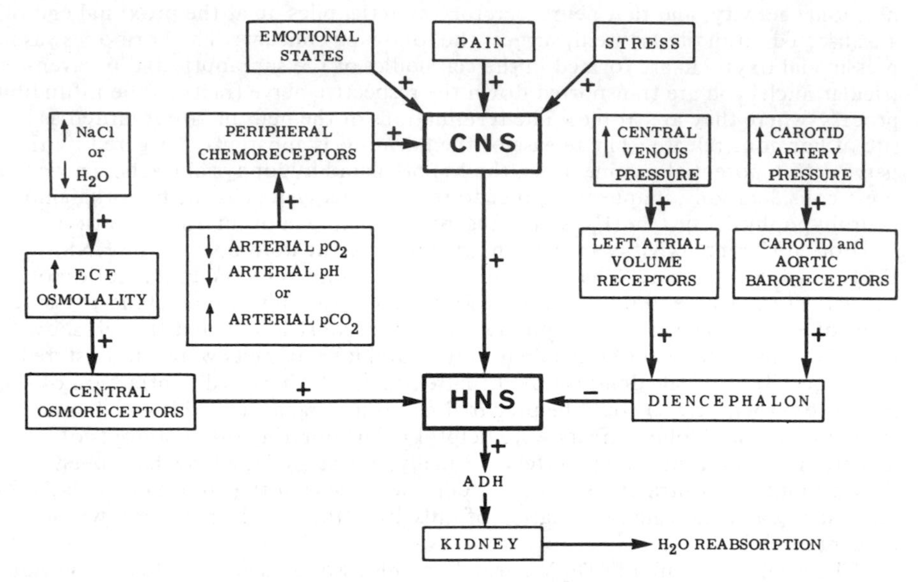

FIGURE 31-3. Regulation of antidiuretic hormone (ADH) secretion by the hypothalamo-neurohypophysial system (HNS).

Verney has shown that the injection of hypertonic solutions of NaCl, glucose, or sucrose into the carotid artery results in a diminution of water diuresis. Injection of urea (which is ineffective osmotically at the cell) is not an effective inhibitor of water diuresis. Subsequent work has shown that a substained (about 30 minutes) increase in the effective osmotic pressure of only 2 percent causes an increase in ADH release and a decrease in free water clearance. Similarly, the ingestion of a large volume of water lowers the osmotic concentration, and the secretion of ADH is depressed. It has been suggested that receptors exist which are sensitive to differences in the osmotic concentrations between the ECF and the intracellular fluid (ICF). These are probably located in the anterior hypothalamus near or within the supraoptic nuclei and therefore transmit their stimuli to the neural lobe over this tract. Recently, Harris stimulated the area around the supraoptic nucleus, causing inhibition of water diuresis and increased chloride excretion. Similarly, it has been shown that the injection of hypertonic saline increases the electrical activity of single neurons in and adjacent to the supraoptic nuclei.

Superimposed upon the basic mechanism whereby changes in the effective osmotic pressure of ECF alter ADH secretion and release by the HNS are alterations induced by nervous reflexes. These alterations may be due to excitation of the HNS by emotional stress or pain.

Recent studies pioneered by Strauss and by Henry and Gauer and their co-workers have firmly established that "volume" receptors and pressure receptors play a prominent role in the control of vasopressin secretion during isotonic contraction or expansion of the ECF volume. The best available evidence shows that the monitoring system for such reflexes has an intravascular origin.

It has been shown that distention of the left atrium or the pulmonary vein within the pericardium is invariably followed by a diuresis. The diuresis occurs in spite of the fact that the manipulations markedly diminish the cardiac output.

Distention of the pulmonary artery, the right atrium, or the pulmonary vein outside of the pericardium does not lead to either a diuresis or an antidiuresis. Procedures such as negative pressure breathing, the infusion of iso-oncotic or hyperoncotic albumin solutions, isotonic saline infusion, or changing from the sitting to the supine position, each of which results in an increased left atrial volume, lead to diuresis. These procedures are in effect only when the vagus is functional, and associated with each is an increased neuronal activity of afferent fibers running in the vagus trunk. That the volume and not pressure is the stimulus is shown by the fact that bursts of activity in the nerve are minimal during the a wave of the atrial pressure cycle, which is a time when the pressure is highest. Conversely, procedures such as venous occlusion of the legs, hemorrhage, orthostasis, positive pressure breathing, and changing from the supine to the sitting position, all of which lead to a decreased left atrial volume, result in an antidiuresis.

The above evidence suggests that activation of the left atrial stretch receptors inhibits the release of ADH from the HNS (Fig. 31-3). Share and his co-workers have shown that the circulating level of ADH increases following isotonic contraction of the blood volume, following bilateral vagotomy, or following carotid occlusion. However, the blood ADH concentration decreases during left atrial distention. They have also shown that carotid sinus baroreceptor activity and carotid body chemoreceptor activity influence vasopressin secretion. Activity of the baroreceptors appears to be inversely related to vasopressin secretion, whereas chemoreceptor activity (induced by hypoxia, hypercapnia, or a decrease in pH) appears to be directly related to vasopressin secretion. It has also been shown that the ADH concentration in the blood is elevated in the human in the standing position when compared to that of the individual in the recumbent position. Similarly, the individual acutely exposed to a cold environment shows a decrease in blood ADH, whereas an increase in blood ADH follows acute exposure to a hot environment. Apparently, therefore, the rate of receptor discharge from the left atrial receptor is a function of the central blood volume.

Action of ADH

The role of ADH in the regulation of renal tubular water handling and homeostasis of the extracellular fluid have been discussed in Chapter 22. Recent evidence suggests that the action of ADH on the distal renal tubules is mediated through cyclic AMP. However, the nature of the change in permeability to water remains obscure.

Thus it appears that a type of negative feedback mechanism affords a reciprocity between the kidney, gastrointestinal tract, lungs, skin, and sweat glands and the HNS and ensures a relatively constant extracellular fluid with respect to its effective volume and osmolality. This system is directly dependent upon its connections with the CNS.

The regulation of the secretion of oxytocin, its role in sperm transport, and its effects upon uterine motility and milk ejection are discussed in Chapter 36.

THE ADENOHYPOPHYSIS

The glandular portion of the pituitary is made up of irregular masses and columns of epithelial cells which are supported by a delicate framework of connective tissue. The groups and columns of cells are separated by sinusoids. Two main types of cell are evident. One type does not show any conspicuous stainable cytoplasmic granules, and these cells are referred to as *chromophobes.* The other contains cytoplasmic granules which take up stains readily, and the cells are called *chromophils.* The latter are considered to be secretory in nature. Many cytologists believe them to be daughter cells of the chromophobes, but it is also possible that the chromophobes represent resting cells. The chromophils may be further divided, according to the stainability of their granules, into *alpha cells* (acidophils) and *beta cells* (basophils). The chromophils may also be divided on the basis of their reaction with the periodic acid–Schiff method (PAS stain). This procedure is used in the identification of mucopolysaccharides and mucoprotein, and a positive PAS reaction may be interpreted as the staining of a carbohydrate-containing substance. Additional procedures have shown that there are at least six epithelial cell types in the anterior pituitary.

There is general agreement on the cellular origin of five of the six anterior pituitary hormones. This agreement has been achieved on the basis of the stainability and cytological appearance of cell types following target gland removal or following excessive (endogenous or exogenous) secretion of target gland hormones and functional changes which occur in pituitary tumors. This will be considered later when each hormone is discussed.

Hormones of the Adenohypophysis

Six hormones have been isolated in a partially purified state from the AP, and all are polypeptide or protein in nature. They are TSH, FSH, LH, LTH, ACTH, and STH. The existence of several other fractions has been postulated, but in all probability the effects of each of the postulated hormones can be accounted for through the effects of the six established hormones.

Partially purified TSH preparations have been obtained, and such preparations have the properties of a relatively low-molecular-weight (10,000) glycoprotein. TSH and other protein hormones are ineffective when administered orally because they are inactivated in the gastrointestinal tract. The cells which secrete TSH are PAS-positive beta cells and are referred to as *thyrotrophs* or *delta cells.* The cells undergo degranulation and become hyperplastic and hypertrophic following thyroidectomy. Following hypophysectomy, the thyroids become atrophic, and all criteria of thyroid function decline rapidly. Any procedure which leads to a marked decrease in the output of thyroid hormone is followed by a marked increase in the secretion and/or release of TSH from the AP and morphological changes in the thyrotrophs. The morphological and functional changes can be prevented by the administration of thyroid hormone. Thus, all available evidence indicates that TSH secretion and/or release is regulated by the level of circulating thyroid hormones. A negative feedback system would then operate between the thyroid and AP.

Two of the gonadotropic hormones, FSH and LH, are also carbohydrate-containing proteins, and both are secreted by PAS-positive beta cells. The cells which secrete FSH and LH undergo hypertrophy and degranulate following castration. Large cytoplasmic vacuoles also form, and the nucleus is compressed against the cell membrane to form the "signet ring" or castration cell. The cells which secrete FSH are located near the periphery of the AP, whereas those which secrete LH are located centrally. FSH induces ovarian follicular growth and in conjunction with LH leads to estrogen secretion and ovulation in the female. In the male, FSH leads to proliferation of the seminiferous tubules, whereas LH (ICSH) stimulates the Leydig cells of the testes to secrete androgen. A negative feedback regulation also exists between the AP and the gonads.

LTH, another gonadotropin, is a protein hormone with a molecular weight of about 26,000. It is secreted by PAS-negative alpha cells. These cells have a high affinity for azocarmine. The number of carmine-staining cells increases after stimulation in species which ovulate and form corpora lutea only following coitus, and in most species the number of carmine cells increases during pregnancy and lactation.

LTH aids in maintaining the functional and morphological integrity of the corpus luteum and plays a fundamental role in the development and secretion of the mammary gland.

The pituitary cells which secrete ACTH are probably beta cells. This conclusion is based upon marked changes (hyalinization) in the beta cells which follow spontaneous hyperadrenal states or which follow the chronic administration of high doses of adrenal steroids. It has also been noted that fluorescent ACTH-induced antibodies combine with ACTH (specific antigen) and localize in a specific type of beta cell. ACTH is a polypeptide; it contains 39 amino acids, and the full sequence of amino acids is known. This hormone stimulates adrenal cortical growth and hormonogenesis. The relationship between the AP and adrenal cortex is a reciprocal one in that hormones of the adrenal cortex inhibit the output of ACTH and thus tend to maintain adrenal cortical hormonogenesis at an optimal level.

There is general agreement that the PAS-negative alpha cells secrete STH. Pituitary dwarfism and gigantism and acromegaly correlate well with deficiency and hyperplasia, respectively, of alpha cells. There are practically no alpha cells in the pituitaries of genetic-dwarf mice, and the pituitaries contain only small amounts of STH. The acromegalic has a pituitary alpha cell tumor rich in STH.

The trophic hormones of the AP induce growth and various metabolic changes in the respective target organs, and these changes are not necessarily different from those which take place in tissues in general. Nevertheless, under physiological conditions, the primary effect of the trophic hormones is on the target gland (adrenal cortex, thyroid, gonads) because these tissues possess an unusual sensitivity to the respective trophic hormones. If the trophic hormones are circulating at high levels, such as those after removal of the target gland or after excessive exogenous administration, they may be able to affect other tissues. Thus, the pituitary trophic hormones might be capable of extending their direct effects to tissues other than their specific target organs. For example, ACTH stimulates oxygen consumption and ^{32}P uptake in the adrenal cortex and also increases oxygen consumption in the adrenalectomized animal. The functions and the regulation of the secretion of each of the trophic hormones of adenohypophysial origin will be considered later in the appropriate sections.

The Hypophysial Portal Vessels

For many years it was quite evident that nervous mechanisms could affect the actions of the AP. There was a considerable amount of anatomical evidence against direct neural and vascular connections between the CNS and the AP. Evidence was available which supported the concept of vasomotor control of the AP circulation, and in 1930 Popa and Fielding described what was to be later called the hypophysial portal system. In most species, including man, the hypophysial portal system is arranged as follows: Small branches of the internal carotid and posterior communicating arteries form a rich vascular network within the pars tuberalis and median eminence. A large number of capillary tufts or loops arise from this network and penetrate into the median eminence and form the primary plexus. This plexus comes into intimate contact with neural elements of the hypothalamus. The primary plexus then forms large portal trunks which lie on the anterior or ventral surface of the hypophysial stalk. These vessels then break up and distribute the blood into the sinusoids of the AP. Direct observation has revealed that the direction of blood flow in this system is from the median eminence to the AP. The hypophysial portal vessels have a marked capacity to regenerate after transection of the hypophysial stalk, because capillary outgrowths can bridge the gap of a divided stalk within 24 hours. Such regeneration can occur following transplantation of the AP in the subarachnoid space below the median eminence. The functional significance of the hypophysial portal vessels is apparent if one considers that the AP does not possess a direct secretomotor innervation, yet the CNS shows a strong regulatory action on the AP. The present concept states that nerves end in and around the primary plexus (within the median eminence) and release specific substances which are transported via the hypophysial portal vessels to the AP. The substance (mediator or neurohumor) then serves as a stimulator (or inhibitor?) of a specific AP activity.

It should be emphasized that the anterior pituitary is as much under hypothalamic control as is the posterior pituitary, but control of the secretions of the AP is not mediated via direct neural connections as are those of the neural lobe. The control over AP secretions is exerted through the release of specific mediators into the capillary plexus of the portal vessels in the median eminence by hypothalamic neurons (see Fig. 31-4). The aspects of the neurohumoral regulation of the trophic hormones of the AP are considered in the appropriate chapters.

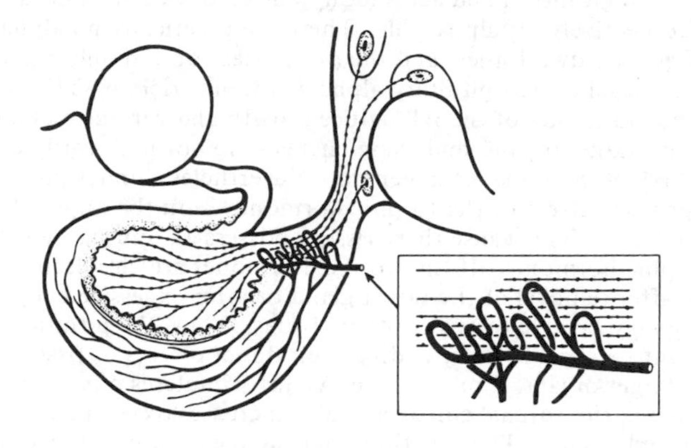

FIGURE 31-4. Blood supply to the anterior pituitary and the innervation of the neural lobe of the pituitary. Detail of the primary plexus (capillary loops) of the hypophysial portal vessels at the right. Fibers which contribute to the innervation of the neural lobe are from the paraventricular and supraoptic nuclei. Note that various axons of hypothalamic origin terminate in and/or traverse the capillary loops.

Effects of Hypophysectomy

Several notable morphological and functional alterations result from total hypophysectomy in the young animal: (1) failure of the gonads to mature, with resultant infantile sexual development and sterility; (2) atrophy of the thyroid gland and the characteristics of thyroid insufficiency; (3) atrophy of the adrenal cortex and signs of hypoadrenalism without salt loss; (4) cessation of growth and failure to attain an adult stature; (5) a decided tendency toward hypoglycemia, hypersensitivity to insulin, and a loss of body nitrogen accompanied by diminished fat catabolism. However, if optimal environmental conditions are maintained, removal of the pituitary is not incompatible with life. All parameters of functions of the gonads, thyroid, and adrenal indicate a sharp decline following hypophysectomy in the adult human. Similarly, metabolic derangements coincident with a lack of STH are evident.

Following hypophysectomy or during the development of hypopituitarism the earliest and most frequent deficiencies observed are failure of growth and gonadotropic hormone secretion, and resultant gonadal failure. Evidences of thyroidal insufficiency next appear and are followed by signs of adrenal cortical insufficiency. This sequence of malfunction generally follows regardless of the species. In several species it has been shown that ablation of increasing amounts of AP tissue results in gonadal, thyroidal, and adrenal cortical failure in that order. The removal of up to 80 percent of AP is compatible with normal gonadal, thyroidal, and adrenal cortical function. Thus, it appears that a large safety factor is present, and the same is true with respect to all the other endocrine glands.

The properly treated hypophysectomized human retains a normal appearance, is able to gain weight and attain a positive nitrogen balance, can repair bone, and can be maintained in good health indefinitely. It follows that the administration of extracts of the AP to intact individuals results in marked development of the gonads and accessory sex glands, adrenal and thyroidal hypertrophy with signs of hyperadrenalism and hyperthyroidism, increase in growth rate with the laying down of nitrogen and increase in long bone length, and a tendency toward diabetes with hyperglycemia and glucosuria.

Functional Role of the Adenohypophysis

The adenohypophysis occupies a crucial position in the body economy because it directly influences the output of hormones from the adrenal cortex, thyroid, and gonads through the elaboration of its trophic hormones. It acts directly on all body structures through STH, and it also indirectly affects the output of the hormones from the pancreas and parathyroids as a result of its action, both direct and indirect, on various metabolic pathways.

The effects of an excess or deficiency of the trophic hormones of the AP are thus mediated through the various target glands. The trophic hormones act only on preexisting chemical reactions and do not initiate the reactions. However, the trophic hormones are essential for complete morphological and physiological development of their target glands. In general, functional changes in target glands are induced prior to anatomical changes following the administration of hormones.

According to the negative feedback concept of control systems, the trophic hormones are released in amounts varying with the functional state of the individual and not at a constant rate. In all probability, at least two types of negative feedback system operate between the trophic hormone secretion of the AP and secretion of hormones by the target gland. Initially, it was believed that the hormones of the target glands acted directly upon the AP to inhibit the trophic hormone output. In this instance, when target hormone secretion rate was high, trophic hormone secretion would be inhibited, and when target hormone concentration was low, inhibition would be removed and trophic hormone output would be increased, etc. However, it has also been shown that various target gland hormones might act upon CNS elements and, depending on the circulating concentrations, induce inhibition or stimulation of trophic hormone output. Moreover, other changes in the internal environment and changes in the external environment can affect AP hormone secretion by regulation of the hormone output through a neurosecretory pathway via the hypothalamus and the hypophysial portal system. This aspect is discussed in Chapters 34, 35, and 36.

Growth Hormone or Somatotropic Hormone (STH)

As the name implies, growth hormone plays a central role in the growth of the organism. The hormone is protein in nature, but its chemical composition varies with the species; e.g., human growth hormone contains 241 amino acids, whereas that of the bovine contains 396 amino acids. The human responds to growth hormone of human or monkey origin but is unresponsive to all other available types. Growth hormone exerts many biological effects, and it has been only recently that its effects could be studied in man. Early observation concerning the role of growth hormone in human physiology was limited to clinical observations in acromegaly, gigantism, and dwarfism. The hormone affects several parameters of protein, fat, and carbohydrate metabolism, and in contrast to the other adenohypophysial hormones it does not require a target gland as an intermediary to produce its widespread effects, but acts upon the effector or target cell directly.

Probably the most striking gross effect of STH is that on the skeleton. Hypersecretion of STH or the exogenous administration of STH can lead to gigantism if instituted prior to closure of the epiphysial plates of the long bones. However, the presence of high levels of STH in the adult (after closure of the epiphysial plates) results in acromegaly. The condition is characterized by enlargement of the skeleton, especially the skull, hands and feet, the skin, the subcutaneous tissue, and the viscera; it is generally the result of an acidophilic adenoma. The disease is accompanied by sellar enlargement, gonadal atrophy, changes in the visual fields, and diabetes mellitus.

Growth hormone plays a prominent role in protein metabolism and the regulation of growth. It promotes protein synthesis and nitrogen retention. This is accomplished by accelerating the rate of transfer of amino acids from the extracellular to the intracellular compartment and incorporating the transferred amino acids into cell proteins. Some evidence indicates that STH promotes protein synthesis via gene activation, because it stimulates the synthesis of messenger ribonucleic acid (RNA), ribosomal RNA, and transfer RNA in liver. This activity is blocked by actinomycin, a substance which forms a complex with deoxyribonucleic acid (DNA), the genetic material which participates in the formation of RNA. The DNA and RNA are required for continued protein synthesis.

STH also leads to a reduction in the rate of lipid synthesis, promotes the mobilization of fatty acids from adipose tissue, and increases fatty acid oxidation. The first indication that STH played a role in carbohydrate metabolism was the observation that hypophysectomy, when performed on the depancreatized animal, led to the amelioration of the symptoms of diabetes mellitus. Later work showed that the chronic administration of STH could lead to the development of permanent diabetes mellitus (Chap. 32).

Recent evidence indicates that the CNS plays a role in the regulation of STH secretion. It has been shown that hypothalamic lesions depress the rate of body growth and cause a marked decrease in the STH content of the pituitary. Either crude or purified extracts of hypothalamic tissue stimulate the release of STH from anterior pituitary tissue both in vivo and in vitro. The mediator which is contained in these extracts has been called *somatotropin-releasing factor* (SRF).

A substance, also of hypothalamic origin, has been isolated which inhibits STH secretion. Evidence is available indicating that STH may inhibit its own secretion by acting on the hypothalamus and depressing SRF release. This type of control has been referred to as the "short" feedback mechanism. Secretion of each of the other adenohypophysial hormones is controlled, in part, via the "short" feedback loop as well as via the "long" feedback loop.

Using a radioimmunoassay technique which is able to detect 0.25 mμg (10^{-9} grams) of human STH, it has been shown that hypoglycemia induces a prompt elevation in the STH concentration in plasma. On the other hand, the administration of glucose (the induction of hyperglycemia) induces a fall in STH concentration. The inference is, therefore, that the release of STH from the anterior pituitary is governed by the blood glucose concentration.

Several types of stressful stimuli have been shown to increase STH secretion, and several factors such as exercise, fasting, and sleep stimulate STH secretion.

The administration of arginine, increasing the plasma amino nitrogen level, or decreasing the plasma free fatty acid concentration leads to an increase in STH secretion.

Pars Intermedia (Intermediate Lobe)

The pars intermedia, although of the same embryonic origin as the pars distalis, becomes associated anatomically with the neurohypophysis. It secretes the melanocyte-stimulating hormone (MSH), sometimes called intermedin. The hormone β-MSH has a molecular weight of 2177 and is a polypeptide composed of 18 amino acids. This hormone is structurally similar to ACTH in that it shares a common sequence of seven amino acids with ACTH. This similarity accounts for the overlapping actions of the two substances in man.

In lower vertebrates, such as the amphibia, MSH causes dispersion of the pigment granules in the chromatophores. The granule dispersion causes the skin to be darkened and thus affords protection to the animal when placed upon a dark background. Conversely, adaptation to a light background is effected by concentration of the pigment granules around the nucleus of the chromatophores following a decrease in MSH secretion. The stimulus for MSH release, the intensity of light as reflected by the environment, is monitored by the retina, and the greater the intensity, the less the MSH released. Thus, impulses transmitted via the optic tract exert an inhibitory effect on MSH release.

In man, MSH stimulates the synthesis of melanin by the melanocytes and thus leads to darkening of the skin. In view of the structural similarity of MSH and ACTH, it is not altogether surprising that ACTH also leads to increasing pigmentation.

REFERENCES

Antoniades, H. N. *Hormones in Human Plasma.* Boston: Little, Brown, 1960.

Berde, B. (Ed.). Neurohypophysial Hormones and Similar Polypeptides. In *Handbook of Experimental Pharmacology,* Vol. 23. Berlin: Springer–Verlag, 1968.

Ganong, W. F., and L. Martini. *Frontiers in Neuroendocrinology.* New York: Oxford University Press, 1969.

Gauer, O. H., J. P. Henry, and C. Behn. The regulation of extracellular fluid volume. *Ann. Rev. Physiol.* 32:547–584, 1970.

Gray, C. H., and A. L. Bacharach. *Hormones in Blood,* Vols. 1 and 2. New York: Academic, 1967.

Harris, G. W. *Neural Control of the Pituitary Gland.* London: Arnold, 1955.

Harris, G. W., and B. T. Donovan. *The Pituitary Gland,* Vols. 1, 2, and 3. Berkeley: University of California Press, 1966.

Haymaker, W., E. Anderson, and W. J. H. Nauta (Eds.). *The Hypothalamus.* Springfield, Ill.: Thomas, 1969.

Knobil, E. The pituitary growth hormone: An adventure in physiology. *Physiologist* 9:25–44, 1966.

McCann, S. M., and J. C. Porter. Hypothalamic pituitary stimulating and inhibiting hormones. *Physiol. Rev.* 49:240–284, 1969.

Martini, L., and W. F. Ganong. *Neuroendocrinology,* Vols. 1 and 2. New York: Academic, 1967.

Migley, A. R., G. D. Niswender, and R. W. Rebar. Principles for the assessment of the reliability of radioimmunoassay methods. *Acta Endocr.* (Kobenhavn) 142 (Suppl.):163–184, 1969.

Moore, W. W. Antidiuretic hormone levels in normal subjects. *Fed. Proc.* 30:1387, 1971.

Orloff, J., and J. Handler. The role of adenosine 3', 5'-phosphate in the action of antidiuretic hormone. *Amer. J. Med.* 42:757–768, 1967.

Sawin, C. T. *The Hormones: Endocrine Physiology.* Boston: Little, Brown, 1969.

Segar, W. E., and W. W. Moore. Regulation of antidiuretic hormone release in man. *J. Clin. Invest.* 47:2143–2151, 1968.

Share, L. Vasopressin, its bioassay and the physiological control of its release. *Amer. J. Med.* 42:701–712, 1967.

Tepperman, J. *Metabolic and Endocrine Physiology,* 2d ed. Chicago: Year Book, 1968.

Turner, C. D. *General Endocrinology,* 4th ed. Philadelphia: Saunders, 1968.

Williams, R. H. (Ed.). *Textbook of Endocrinology,* 4th ed. Philadelphia: Saunders, 1968.

Zarrow, M. X., T. M. Jochim, J. L. McCarthy, and R. C. Sanborn. *Experimental Endocrinology: A Sourcebook of Basic Techniques.* New York: Academic, 1964.

32

Endocrine Functions of the Pancreas

Ward W. Moore

THE RELATIONSHIP between the pancreas and altered carbohydrate metabolism was first shown by the classic experimental work of Von Mehring and Minkowski in 1889. They observed that removal of the pancreas of the dog resulted in the production of all the features of diabetes mellitus. This disease state has been known for centuries, and it is the most frequent, and probably the most studied, metabolic disorder. The endocrine component of the pancreas secretes two hormones, insulin and glucagon, and each functions in the regulation of carbohydrate metabolism. The most prominent pancreatic hormone, *insulin,* was named and its major biological properties were elucidated before Banting and Best obtained the first stable insulin preparation in 1922. In 1926 Murlin reported that some of the extracts of pancreatic tissue contained a hyperglycemic principle, in addition to the hypoglycemic material insulin, and proposed the name *glucagon* for this factor. The effects of insulin have probably been studied more extensively than those of any other hormone.

ANATOMY

The pancreas is derived from gut endoderm and remains connected to the small intestine by means of two ducts (duct of Wirsung and duct of Santorini). Glandular tissue appears during the third month of intrauterine life and develops as side buds from the ducts and duct branches. The acinar tissue remains connected to the duct system and becomes the exocrine portion of the pancreas. Small islets of cells proliferate from the ducts, differentiate, and lose their connection with the duct system and become the endocrine portion of the pancreas. The blood supply of the pancreas is via the left gastric, splenic, and hepatic arteries and the inferior pancreatoduodenal artery. This organ is drained by the splenic and superior mesenteric veins and celiac lymph nodes. It receives a sympathetic innervation via the celiac plexus, and parasympathetic fibers from the vagus. The latter are the only contribution to the endocrine portion of the pancreas.

Langerhans first described small nests or islets of highly vascularized cells which are independent of the duct system of the pancreas. These islets maintain their functional integrity following ligation of the pancreatic ducts, whereas the acinar tissue becomes atrophic. The total volume of the islet tissue makes up about 1 to 3 percent of the entire pancreas. Three cell types have been shown to be present in the islets of Langerhans, and they have been differentiated on the basis of their solubility and affinity for certain stains. The alpha cells contain fine granules which are insoluble in alcohol and which stain bright red with Mallory or Masson staining methods. They are relatively few in number and are the source of glucagon. The beta cells contain small alcohol-soluble granules which stain orange-brown with the above methods. They are the source of insulin and make up about 75 percent of the islet tissue. The delta cells do not contain granules and their functional significance is unknown, but they may act as mother cells of the alpha and beta cells.

CHEMISTRY OF INSULIN AND GLUCAGON

The fundamental work on the extraction of insulin from pancreatic tissue was done by Banting and Best in 1922. In 1926 Abel and his associates were the first to prepare crystalline insulin, and in 1934 Scott described the crystallization of insulin in the presence of small amounts of metallic ions, such as zinc. Insulin is a simple protein and has a molecular weight of about 6000. In 1954 Sanger and associates established the number and sequence of amino acids composing the two dissimilar polypeptide chains which form the insulin molecule. One chain (chain A) contains 21 amino acids, another (chain B) contains 30 amino acids, and the chains are linked by disulfide (-S-S-) bridges. The biological activity appears to be a function of the entire molecule rather than of specific groupings within the molecule, because potency is lost following slight alterations in the molecular structure. Slight variations in amino acid sequence within the disulfide ring of the A chains do exist between species. Nevertheless, all the insulins do cross species barriers with respect to their physiological activities. Insulin is readily inactivated either by acid hydrolysis or by proteolytic enzymes and hence is rendered ineffective in the gastrointestinal tract and must be administered parenterally.

Insulin is degraded rapidly in the body; less than 0.1 percent may be recovered in the urine following the administration of large doses. The factor responsible for the hepatic inactivation and degradation of insulin is probably a series of proteolytic enzymes which has been termed the insulinase system. The hormone also becomes firmly bound to target tissue such as muscle, adipose tissue, and mammary gland.

Glucagon is composed of one long chain consisting of 29 amino acid residues and has a molecular weight of 3485. It has been synthesized. This molecule, like insulin, is cleared by proteolytic enzymes, and the integrity of most of the molecule appears to be required for activity, because none of its degradation products possess hyperglycemic activity.

Radioimmunoassays are now available for both insulin and glucagon. With these techniques insulin has been shown to have a half-life of about 30 minutes, whereas that of glucagon is about 10 minutes. Normal fasting plasma levels in man have been determined for both insulin and glucagon and they vary widely, depending upon the laboratory and the technique used. However, they are very low – on the order of 0.5 to 10 mμg per milliliter of plasma.

METABOLIC INTERRELATIONSHIPS

The pancreas, via its hormones, plays a prominent role in the regulation of carbohydrate metabolism. A disturbed utilization and regulation of carbohydrate metabolism results from a relative or absolute deficiency in insulin. Any essential fault in carbohydrate metabolism necessarily involves the metabolism of protein and fat as well, because the metabolic pathways through which the organism derives energy from food sources are known not to be separate and distinct, but are intermingled. Carbohydrate is the active fuel of the body and is ordinarily the primary source of energy of the cell, but fatty acids can be utilized by resting muscle. Carbohydrate also contributes part of its substance to fatty acid and amino acid formation. Consequently, the proper utilization and formation of protein and fat is dependent upon carbohydrate metabolism. The digestion and absorption of each of these foods has been discussed in Chapter 26.

CARBOHYDRATE METABOLISM

Following their absorption from the gastrointestinal tract, the nutrient sugars are transported directly to the liver. This organ converts the sugars to a phosphorylated hexose and acts as a central clearing house for their disposition. The hexose phosphate which results from various transformations may be converted to glycogen (glycogenesis) and stored in the liver; it may be broken down (glycolysis) to intermediates, which may be used either as sources of energy or in synthetic processes; or glucose may be released into the general circulation.

Most extrahepatic tissues (brain, muscle, adipose tissue, etc.) utilize glucose for their supply of energy. In order that such tissue may utilize glucose, it must permeate the cell membrane and be trapped by the cell. Immediately after entering a cell, glucose is phosphorylated to glucose-6-phosphate (G-6-P), and thus trapped by the cell, because the reaction is irreversible and the cell membrane is impermeable to phosphate esters. G-6-P may be converted to uridine diphosphate glucose (UDP glucose) and either converted to glycogen and stored, or utilized in mucopolysaccharide synthesis. It may be oxidized via the hexose monophosphate shunt (HMP) and yield ribose phosphate, which may be either utilized in the synthesis of ribonucleic acid (RNA) or converted to pyruvate. An important by-product of this pathway is reduced nicotinamide-adenine dinucleotide phosphate (NADPH). G-6-P may also undergo anaerobic glycolysis to form lactic acid via the Embden-Myerhof pathway (EMP). Under aerobic conditions, either pyruvate formed via this route undergoes oxidative decarboxylation and irreversibly reacts with coenzyme A (CoA) to form acetyl CoA, or it may be carboxylated to form malate. Therefore, pyruvate may serve as precursor to both ingredients necessary for the tricarboxylic acid (TCA) or Krebs cycle (Fig. 32-1). The TCA cycle is a major source of CO_2 and contributes to the production of high-energy phosphate bonds via oxidative phosphorylation.

The energy for biological work is made available from the degradation and transformation of various substrates. Compounds such as adenosine triphosphate (ATP) which contain high-energy bonds are produced in the process of metabolism of energy-yielding substrates (exergonic processes). The cleavage or hydrolysis of the high-energy bonds is then coupled with energy-requiring processes (endergonic) for biological work, such as synthetic and secretory processes, muscle contraction, nerve conduction, etc. Compounds such as ATP thus act as the currency which maintains the metabolic economy of the cell. High-energy phosphate bonds may be generated in glycolysis, in the hexose monophosphate shunt, in the formation of acetyl CoA, and in the TCA cycle. Various monographs and biochemistry textbooks contain detailed maps of enzymatic function and biological oxidations.

Acetyl CoA plays an important pivotal role in intermediary metabolism because it is the compound which is common to carbohydrate, protein, and fat metabolism. Furthermore, through this compound the metabolic products of each enter the common pathways which lead to oxidative processes furnishing the organism with energy. When acetyl CoA is oxidized to CO_2 and H_2O, phosphorylation is coupled with oxidation, and 12 moles of ATP are formed per revolution of the TCA cycle

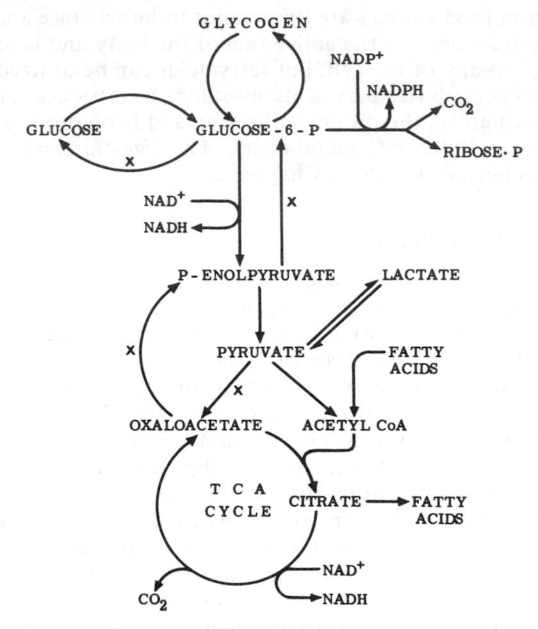

FIGURE 32-1. Pathways of carbohydrate metabolism.

(per mole of acetate utilized). Thirty-eight moles of ATP are formed by complete oxidation of 1 mole of glucose by way of the glycolytic and TCA pathways. This amounts to an energy storage of a considerable magnitude, because 1 mole of ATP stores about 11,000 calories of free energy.

FAT METABOLISM

A small amount of fat is synthetized in the liver, but most of it is synthesized in adipose tissue. Because 1 gm of fat yields about 9 Kcal as compared with 4 Kcal per gram of protein or carbohydrate, the fat depots of the body constitute a concentrated store of energy. Formerly, the adipose depots were considered relatively inert, static forms of energy, but newer techniques have shown them to be exceedingly active metabolically. They furnish about 40 percent of the energy content of a normal individual and represent the major source of stored energy.

The fatty acids synthesized in cells derive their carbon from acetyl CoA and malonyl CoA. The synthesis of long-chain fatty acids (lipogenesis) results from carboxylation of acetyl CoA with the formation of malonyl CoA. Repeated condensations of malonyl CoA plus decarboxylation and subsequent reductions in the presence of NADPH lead to the formation of fatty acids. The free fatty acids then can combine with glycerol and form triglyceride. The degradation of fatty acids (lipolysis) proceeds stepwise by the removal of two carbons at a time and results in the formation of acetyl CoA. Thus the fatty acid is broken down to acetyl CoA units, and these are eventually channeled into the TCA cycle. It is quite evident that any excess of caloric intake, as either carbohydrate or fat, can be channeled via acetyl CoA to result in increased lipogenesis, and that the bulk of storage fat depots must increase if the rate of the lipogenesis exceeds that of lipolysis. Acetyl CoA may also be converted to ketones instead of entering the TCA cycle. The circulating ketone bodies are formed in the liver, and production of them increases as one increases the supply of fatty acids to the liver. Oxidation of the ketone bodies occurs in the extrahepatic tissue, and the circulating level increases when their production exceeds their utilization (Fig. 32-2).

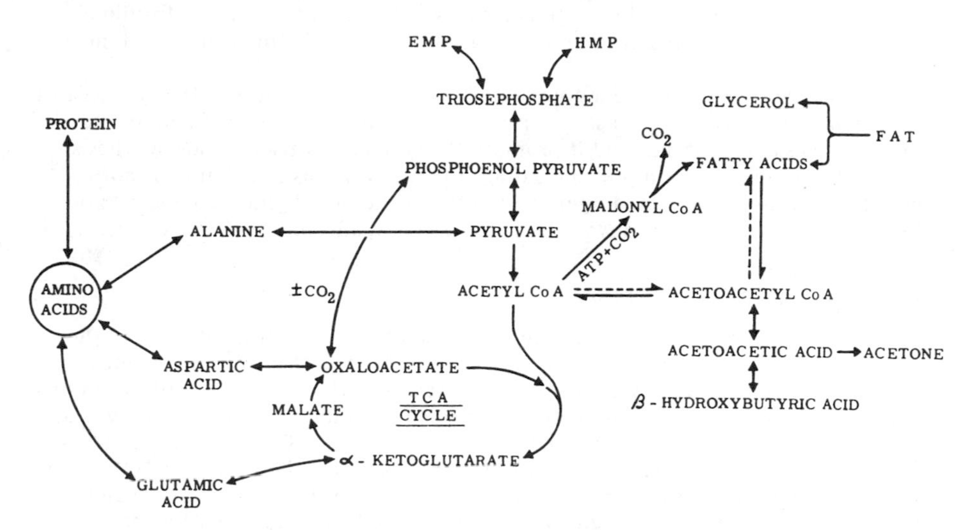

FIGURE 32-2. Relationship of certain phases of lipid and protein metabolism with carbohydrate metabolism.

PROTEIN METABOLISM

The amino acids absorbed from the gastrointestinal tract either are used by the metabolic mill for the synthesis of protein or contribute their carbon chains for the synthesis of fatty acids, for glucose and glycogen formation (gluconeogenesis), or as an energy source. Eight amino acids must be supplied in the diet of man, because none of these can be synthesized. They are isoleucine, leucine, lysine, methionine, phenylalanine, threonine, tryptophan, and valine. Other amino acids such as alanine, glycine, glutamic acid, and aspartic acid may be synthesized from specific intermediates of the glycolytic pathway or TCA cycle.

All or a portion of the carbon chain of an amino acid may be converted to glycogen or to ketone bodies. Those amino acids which yield pyruvate are channeled into carbohydrate metabolism and classified as glycogenic; those which form acetoacetyl CoA and contribute primarily to reactions of fat metabolism are ketogenic. The great majority of the amino acids are glycogenic.

Man's need for utilizable dietary protein during growth is great, and he has a continuing constant need for protein in order to maintain and repair cells. As indicated in Figures 32-1 and 32-2, liver and muscle glycogen stores may be replaced with glucose derived from protein, and fat may be synthesized and stored in adipose tissue as a result of ingesting protein calories in excess of needs.

ROLE OF INSULIN IN METABOLISM

Numerous effects on the body have been attributed to the administration of insulin. However, many of these alterations in body functions are probably subsidiary to the effect of insulin on glucose transfer from the extracellular fluid into certain cell types. Levine and co-workers (1955) have shown that insulin facilitates the transfer of glucose across the cell membrane. Tissues such as muscle and adipose tissue are very sensitive to this action of insulin, but the permeability of cells of the liver, brain, kidney, and gastrointestinal tract to glucose is not altered by insulin. Insulin also increases the rate of transfer of amino acids and fatty acids through the cell membrane.

Insulin can be considered a potent anabolic agent. It promotes the synthesis of glycogen in both liver and muscle, it promotes the incorporation of amino acids into peptides and proteins, and it promotes fatty acid and triglyceride synthesis. Thus, overall, insulin promotes the storage of energy and is necessary for normal growth. It is also an effective agent in regulating the blood glucose concentration, being the only hormone to lower the blood glucose; all others tend to elevate it.

EFFECTS OF PANCREATECTOMY

The many and varied effects of insulin on metabolism are best illustrated by the derangements in metabolism which occur in the absence of insulin in diabetes mellitus. Insulin deficiency brings about decreases in the peripheral utilization of glucose and glycogenesis. These effects, coupled with increased glycogenolysis and gluconeogenesis, lead to an increased blood glucose level. The oxidation of glucose via the HMP is decreased in relation to that oxidized over the EMP and TCA cycle in liver and adipose tissue. However, to gain entry into the cells where it can be utilized, the concentration of glucose in the extracellular fluid (ECF) must be significantly elevated. When the degree of hyperglycemia reaches a high level, the amount of glucose filtered at the renal glomerulus exceeds the tubular maximum (Tm) for glucose and glucosuria results. The excessive filtered load of glucose acts as an osmotic diuretic, and the volume of urine excreted increases markedly. Large amounts of Na^+ and Cl^- are also lost as a result of the high urine flow. The ECF volume tends to be depleted, and dehydration, thirst, and polydipsia occur as a consequence of the large amounts of water and salt lost in the urine. The decrease in carbohydrate utilization results in an essential carbohydrate starvation, and, by an unknown mechanism, centers within the hypothalamus are affected so as to augment the food drive (Chap. 8) and polyphagia results in experimental animals.

The rate of lipogenesis is decreased to about 5 percent of normal. This decline in fatty acid synthesis may be the result of a deficiency in NADPH. However, the decreased lipogenesis may be related to a deficiency in end products of carbohydrate metabolism which are necessary for the formation of fatty acids. The rate of catabolism of fats is increased greatly in insulin deficiency. As the fat stores are mobilized, the fat content of the blood increases. In the face of decreased carbohydrate utilization, most of the energy source of the diabetic must be derived from adipose tissue fat. In this state, lipolysis is increased to such an extent that acetyl CoA is produced in excess of the amount which can be handled by the TCA cycle, and acetoacetyl CoA tends to accumulate in the liver. This excess is channeled at an increasing rate to ketone body (acetoacetic acid, β-hydroxybutyric acid and acetone) production. When the rate of production of ketone bodies exceeds the capacity of extrahepatic cells to oxidize them, ketonemia and ketonuria result.

The ketone bodies are acids and are buffered by the bicarbonate buffer system. Therefore, ketonemia decreases the alkali reserve. Ketones are excreted mainly as sodium and potassium salts, so that the loss of base contributes further to acidosis. Compensatory reactions, such as hyperventilation and the resultant increase in loss of carbon dioxide via the lungs, are triggered by stimulation of the respiratory centers by the lowered pH of the blood. Thus, the HCO_3^-/H_2CO_3 ratio returns toward normal, and the pH is elevated. The respiratory alkalosis may compensate for nearly all the base lost in mild diabetes acidosis, and, along with renal compensations (increase in loss of fixed anions by excreting them with H^+ or NH_4^+), complete compensation may be attained and a normal pH restored. However, in severe insulin insufficiency, uncompensated metabolic acidosis occurs. The presence of an excess of ketone bodies depresses the oxygen consumption of the CNS and results in depression of CNS activity. If the depression becomes marked, mental disorientation and coma follow.

Protein metabolism is also affected by the lack of insulin. There is an increase in the rate of protein catabolism, and this, coupled with impaired protein synthesis, leads to negative nitrogen balance, impaired growth, delayed healing, and increased susceptibility to infections. The main source of glucose (other than exogenous) excreted in the diabetic is probably protein. In order that protein may be converted to glucose, it must be hydrolyzed to amino acids, which are deaminated and converted to keto acids. The latter may be converted to pyruvate or to ingredients of the TCA cycle. The glycolytic pathway must then be traversed in a reverse direction to form glucose. Because of the increase in protein catabolism and gluconeogenesis, there occur increases in urinary nitrogen and glucose excretion, respectively. Both of the latter contribute further to the osmotic diuresis and the tendency to dehydration.

The effects observed in insulin deficiency appear to result from a diminution in the utilization of glucose for oxidative purposes, lipogenesis, and glycogenesis. The concentration of glucose in the ECF becomes elevated because the rate of transport of the hexose across the cell membrane of most tissues (muscle, adipose tissue) is reduced. A loss of large amounts of available glucose follows because the renal Tm for glucose is exceeded. The relative carbohydrate deficit precipitates lipolysis and proteolysis, which lead to metabolic events whereby the carbon chains of fatty acids and amino acids are utilized as sources of energy. Parts of the substances also may be lost in the urine as ketones, glucose, and nitrogen. The body economy of the diabetic resembles that produced by chronic starvation, and even though the diabetic may avoid acidotic crises, he may succumb eventually to depletion of the body tissues.

EFFECTS OF INSULIN

Insulin increases the rate of glucose uptake by muscle cells and adipose tissue and thus induces a hypoglycemia and makes more glucose available for utilization. The mechanisms whereby insulin increases glucose transport are poorly understood. Insulin does decrease G-6-Pase activity and promotes the synthesis of glucokinase in the liver, thereby permitting more G-6-P to enter the glycogenic and glycolytic pathways and allowing less free glucose transfer from the liver to the ECF. In addition to increasing the synthesis of glucokinase, insulin increases the rate of synthesis of phosphofructokinase and pyruvate kinase. It increases HMP activity, and more glucose is oxidized via this pathway.

Insulin probably exerts its effects by directly stimulating ribosomal protein synthesis and does not require new mRNA synthesis. It may also suppress the accumulation or antagonize the action of cyclic AMP. The latter appears to be the mediator of the actions of several catabolic hormones. Because of the initial actions, the glucose concentration of the ECF is depressed, the amount of glucose filtered at the renal glomerulus becomes less than the Tm for glucose, glucosuria ceases, and the diuretic effect of glucose is lost. As a consequence, less glucose, water, and salt are lost from the ECF.

The effect of insulin on glucose assimilation in muscle is much more rapid than are its effects on the two hepatic enzymes, glucokinase and G-6-Pase. However, the slower response of these two enzymes to insulin may serve as an important buffer against abrupt changes in blood glucose concentration. Severe hypoglycemia might be expected to be a consequence of an elevated insulin secretion which would follow the ingestion of a high carbohydrate meal in the normal individual if the activity of the two enzymes paralleled glucose assimilation by extrahepatic tissue.

The administration of insulin to the diabetic returns the rates of lipogenesis and lipolysis to normal levels. The rate of fatty acid synthesis is increased, because the greater assimilation of glucose increases the availability of the intermediates of carbohydrate metabolism which are utilized in fat synthesis. The rate of lipolysis is decreased, thus reducing ketone body production to a level at which extrahepatic tissue oxidation of these substances is equal to or less than their production. Therefore, ketonemia and the resultant ketonuria are inhibited. As the circulating ketone bodies return to normal, the alkali reserve returns to normal, less base is lost via the urine, and electrolyte metabolism returns to normal.

The rates of protein synthesis and gluconeogenesis are returned to normal levels following administration of insulin to the diabetic. The stimulating effect of insulin on protein synthesis is dependent upon adequate glucose metabolism, which supplies the intermediates and cofactors necessary for the process. Insulin also stimulates the transport of amino acids into cells and thus makes them available to enter into synthetic reactions. Decreased gluconeogenesis results in a decrease of urinary nitrogen excretion, and the contribution of protein to the formation of blood glucose declines.

All the metabolic alterations of the diabetic are readily reversed by insulin. These effects of insulin increase the utilization of glucose by peripheral tissue, promote the formation of hepatic and muscle glycogen, stimulate the synthesis of fat and protein, and decrease lipolysis and gluconeogenesis. The administration of excessive amounts of insulin or hypersecretion of insulin causes hypoglycemia. Because the CNS is largely dependent upon blood glucose for its nutrition, this tissue reacts as if it were deprived of oxygen (Chap. 6).

Nonendocrine Regulation

Many agents regulate the rate of secretion of insulin, and most of them are humoral, but not necessarily hormonal, in nature. The transplanted pancreas is capable of maintaining the blood glucose level and insulin level with a high degree of precision. The major physiological stimulus for insulin secretion is glucose. The amount of insulin secreted from an isolated pancreas can be increased or decreased by the perfusion of the preparation with blood containing high or low glucose concentrations, respectively. Several other sugars, as well as amino acids and fatty acids, stimulate the secretion of insulin.

Neurogenic mechanisms may also play a role in the regulation of insulin secretion, because stimulation of vagal fibers to the pancreas results in increased insulin secretion. Similarly, acetylcholine and several cholinergic drugs stimulate insulin secretion. The adrenergic agents, norepinephrine and epinephrine, directly inhibit insulin secretion.

Agents such as theophylline and cyclic AMP enhance or promote the release of insulin from the pancreatic beta cells. The compound *alloxan* exerts a selective destruction of the beta cells of the islets; hence its administration leads to diabetes mellitus. The severity of the diabetic state may be of varying degrees. Following pancreatectomy of an alloxanized animal, the insulin requirement is decreased, presumably as a result of the removal of a source of glucagon and pancreatic enzymes which participate in the digestion and absorption of sugars. This method of production of diabetes has advantages over pancreatectomy in that the experimental animal does not lack the exocrine function of the gland. However, alloxan may damage the liver and kidneys if too high a dose is used.

Endocrine Regulation

The endocrine portion of the pancreas is independent of direct control by the pituitary gland. However, the endocrine functions of the pancreas can be either directly or indirectly altered by several hormones.

The intestinal hormones, secretin and pancreozymin, each stimulate insulin secretion directly. Both pancreatic and intestinal glucagon also stimulate insulin secretion directly. Many agents influence the release and/or action of insulin. Some are capable of producing temporary or permanent diabetes by virtue of their ability to antagonize the action of insulin or to destroy the beta cells, the source of insulin.

The insulin level of the blood may also play a part in the regulation of insulin secretion because, if insulin is perfused prior to and during glucose infusion, degranulation of the beta cells is less than that observed following the infusion of glucose alone, in spite of the presence of marked hyperglycemia in both cases.

The anterior pituitary gland has indirect influences on insulin secretion. Carbohydrate metabolism is affected by growth hormone, ACTH, and TSH. The effects of the latter two hormones on carbohydrate metabolism are indirect because their influence is mediated via the hormones of their respective target organs.

Houssay demonstrated that hypophysectomy reduces the effects of pancreatectomy in the dog. Furthermore, the hypophysectomized animal is extremely sensitive to the action of insulin, and growth hormone causes a decided aggravation of diabetes in the hypophysectomized diabetic. In acromegaly and gigantism, there is a strong tendency toward hyperglycemia and the diabetic state. It has been demonstrated that the administration of growth hormone over prolonged periods produces degeneration of the beta cells and results in the production of diabetes. The effect of growth hormone on insulin secretion is indirect and stems from the ability of the hormone to elevate the blood glucose. Growth hormone decreases the peripheral utilization of glucose, increases the release of glucose from the liver, increases the release of a pancreatic hyperglycemic factor, decreases the fixation of insulin to tissues, and thus leads to the diabetic state.

Adrenalectomy ameliorates the symptoms of diabetes whereas the administration of cortisol-like compounds increases the severity of the symptoms. The diabetogenic effect of these hormones results from a marked increase in gluconeogenesis, lipolysis, and ketogenesis and a decrease in the utilization of glucose. Therefore, their administration promotes the secretion of insulin because they produce hyperglycemia.

Diabetes may be produced by the administration of thyroid hormones in the partially depancreatized animal. It results from beta cell degeneration and subsequent decline in insulin secretion. Diabetes increases in severity in hyperthyroidism and decreases in hypothyroidism. The effects of thyroid hormones on carbohydrate metabolism include an accelerative rate of gluconeogenesis and increased glucose oxidation in tissue, both of which contribute to a decrease in the rate of insulin secretion.

Epinephrine and norepinephrine are hyperglycemic agents by virtue of their ability to increase the rate of hepatic and muscle glycogenolysis. The lactic acid produced as a result of muscle glycogenolysis is liberated into the circulation and used in the hepatic synthesis of glucose. Hypoglycemia is a potent stimulus to the sympathetic nervous system with increased catecholamine release. This might afford an important buffer mechanism against acute episodes of hypoglycemia. The catecholamines also increase the release of free fatty acids from adipose tissue. This ability is shared with thyroxine and cortisol, whereas insulin decreases the release of free fatty acids from adipose tissue.

ROLE OF GLUCAGON IN METABOLISM

Glucagon, in contrast to insulin, can be considered a catabolic agent. It acts to decrease energy stores and to increase the supply of glucose and fatty acids for oxidation.

Because glucagon has a rapid hyperglycemic effect, it has been regarded as being primarily involved with the regulation of glucose metabolism. It is a powerful hepatic glycogenolytic agent and increases the rate of gluconeogenesis. These actions are mediated by cyclic AMP. Glucagon also has a lipolytic action in that it stimulates lipase activity in both liver and adipose tissue. Thus, it stimulates free fatty acid release from adipose tissue and increases fatty acid uptake and oxidation in liver and muscle. Being a protein-catabolic agent, glucagon increases the rate of uptake and deamination of amino acids by the liver. However, it depresses the uptake of amino acid by muscle cells. The net outcome of the lipolytic and proteolytic effects of the hormone is to supply more fatty acids and amino acids for the hepatic gluconeogenic pathway.

Recently glucagon has been shown to have a stimulating effect upon the adrenal medulla and on cardiac function. It increases the rate of secretion of catecholamines from chromaffin tissue of the adrenal. It also has two important effects upon the heart; they are direct and not mediated through catecholamine release. Glucagon acts directly upon the sinoatrial node and causes an increase in heart rate. In addition to its chronotropic effect, the hormone also exerts a positive inotropic effect upon the myocardium and papillary muscle.

Control of Glucagon Secretion

The plasma glucose concentration appears to be the primary regulator of glucagon secretion. A rise in plasma glucose concentration inhibits the secretion of glucagon, and a fall stimulates secretion. It appears that a reciprocal relationship exists between the rates of secretion of glucagon and insulin, and that both are regulated by the concentration of glucose perfusing the endocrine pancreas.

REFERENCES

Cahill, G. F., J. Ashmore, A. F. Renold, and A. B. Hastings. Blood glucose and liver. *Amer. J. Med.* 26:264–282, 1959.

Frohman, L. A. The endocrine function of the pancreas. *Ann. Rev. Physiol.* 31:353–382, 1969.

Lawrence, A. M. Glucagon. *Ann. Rev. Med.* 20:207–222, 1969.

Levine, R., and M. S. Goldstein. On the mechanisms of action of insulin. *Recent Progr. Hormone Res.* 11:343–380, 1955.

Lynen, F. Biosynthesis of saturated fatty acids. *Fed. Proc.* 20:941–951, 1961.
Mahler, H. R., and E. H. Cordes. *Biological Chemistry.* New York: Harper & Row, 1966.
Mayhew, D. A., P. H. Wright, and J. Ashmore. Regulation of insulin secretion. *Pharmacol. Rev.* 21:183–212, 1969.
Sutherland, E. W., and G. A. Robinson. The role of cyclic AMP in the control of carbohydrate metabolism. *Diabetes* 18:797–819, 1969.
West, E. S., W. R. Todd, H. S. Mason, and J. T. Van Bruggen. *Textbook of Biochemistry,* 4th ed. New York: Macmillan, 1966.
White, A., P. Handler, and E. L. Smith. *Principles of Biochemistry,* 3d ed. New York: McGraw-Hill, 1964.
Wool, I. G., W. S. Stirewalt, K. Kurihara, R. B. Low, P. Bailey, and D. Oyer. Mode of action of insulin in the regulation of protein synthesis in muscle. *Recent Progr. Hormone Res.* 24:139–213, 1968.

33

Endocrine Control of Calcium Metabolism

Ward W. Moore

Several endocrine glands and their hormones are involved in the regulation of calcium and phosphorus metabolism. Two of them, however, both of bronchial origin, secrete hormones which have marked effects on the calcium activity of the ECF and on bone. The parathyroid glands secrete parathormone or parathyroid hormone in response to a low serum calcium activity or hypocalcemia. The ultimobranchial glands secrete calcitonin or thyrocalcitonin in response to an elevated serum calcium activity or hypercalcemia. Together, the two hormones exert a precise homeostatic control over the serum calcium.

The functional significance of the parathyroid glands was not realized until about 1909, when MacCullum and Voegtlin demonstrated a lowering of blood calcium in parathyroidectomized animals. Before then, effects of their removal were generally attributed to the removal of thyroidal tissue. The relationship of the parathyroids to calcium and phosphorus metabolism had been realized for several years before their endocrine nature was fully established in 1925 by Collip.

The functional significance of the ultimobranchial glands or their derivative cells has been realized only a short time. Initial observations indicating that thyroparathyroidectomy impaired the control of hypercalcemia were not appreciated. In 1961 Copp and co-workers suggested that the parathyroid glands secreted a rapidly acting hypocalcemic agent and called it calcitonin. However, it was subsequently shown that this agent could be extracted from mammalian thyroid glands but not from parathyroid glands. The name *thyrocalcitonin* was introduced. More recently comparative studies on a wide variety of vertebrates indicate that the hypocalcemic agent is secreted by cells of the ultimobranchial glands.

ANATOMY AND BIOCHEMISTRY

The parathyroid glands are small paired bodies found in the region of the thyroid gland. There are usually four glands, which develop from the third and fourth pairs of branchial pouches. The blood supply to the glands is through the inferior thyroid arteries, although the glands may receive some blood from the superior thyroid arteries. Each gland is surrounded by a connective tissue capsule, and septa divide the glands into lobules. Microscopically, the glands resemble hyperplastic thyroid tissue. Two types of epithelial cells are present: the *chief cells,* the source of the hormone of the gland, and the *oxyphil cells,* which do not appear until about the sixth year of life. The functional significance of the latter is obscure.

The ultimobranchial glands are derived from the terminal branchial pouch and in nonmammalian vertebrates appear as separate bodies. However, in mammals the cells become embedded in the thyroid along with parathyroid tissue. The source of calcitonin is the parafollicular, or "light," or "C" cells. These cells appear in both the thyroid and internal parathyroids.

The hormone of the parathyroid glands, *parathormone* (PTH), is a straight-chain polypeptide containing 83 amino acids and has a molecular weight of about 9500. The biologically active portion of the molecule has a sequence of 35 amino acids, and this portion also has full immunological activity. Earlier studies of the actions of the hormone were complicated as a result of relatively crude preparations, but recently essentially pure PTH has become available. It is known that the hormone acts at least at three sites, namely, bone, kidney, and gastrointestinal tract, affecting calcium and phosphorus handling by each tissue. The hormone of ultimobranchial origin, *calcitonin* (CT), contains 32 amino acid residues and has a disulfide bridge between amino acids 1 and 7. Its molecular weight is about 3580, and it has been synthesized. The entire molecule seems to be necessary for complete biological activity. Thus far CT has been shown to act at two sites, bone and kidney, and it acts on calcium and phosphorus handling by each. Radioimmunoassays are available for both PTH and CT.

FUNCTIONS OF CALCIUM AND PHOSPHORUS

Calcium is an indispensable mineral constituent of all body tissues and participates in several physiological processes, among which are (1) coagulation of blood, (2) maintenance of cardiac rhythmicity, (3) membrane permeability, (4) neuromuscular excitability, (5) production of milk, and (6) formation of bone and teeth. Primary attention here is focused on this last function, because the other roles of calcium are discussed in preceding chapters.

Calcium is absorbed in the upper intestinal tract, and its absorption is facilitated by the presence of HCl, vitamin D, and protein. Alkali and excessive fat, excessive quantities of phosphate or oxalate, and the presence of chelating agents interfere with the absorption of calcium from the intestinal tract. Most of the calcium which is excreted is lost in the feces, and only a small part is excreted by the kidneys.

Following absorption, the greater part of the calcium is stored in the skeleton and only a small fraction of the total body calcium is present in other tissue, including the circulating body fluids. The normal serum calcium level is from 9 to 11 mg per 100 ml (5 m Eq per liter); about 60 percent of this is ultrafiltrable and diffusible, whereas the remainder is bound to protein. A small fraction of the diffusible calcium is nonionized, but most is ionized.

Phosphorus plays an important role in biological systems because it is involved in the transfer of energy in the intermediary metabolism of foodstuffs; it helps in maintenance of pH of body fluids and is an important constituent of bone. Phosphorus absorption from the gastrointestinal tract is facilitated by acids and an excess of fat and is decreased by high calcium and other cations, and by alkaline salts. About 80 percent of the total body phosphorus is in the skeleton, about 11 percent is in muscle, and the rest is present in the body fluids or distributed in other tissue as organic compounds. The main avenue of phosphate excretion is the kidney, but a small amount appears in the feces. The total phosphorus present in the serum is about 12 mg per 100 ml. This may be divided into three portions: lipid phosphorus (8 mg per 100 ml), ester phosphorus (1 mg per 100 ml), and inorganic phosphorus (3 mg per 100 ml or 2 mEq per liter).

BONE

The manner in which hormones maintain the calcium and phosphorus levels of the internal environment requires a brief discussion of the development and maintenance of bone. Bone has been generally considered an inert tissue. However, studies utilizing isotopes of calcium and phosphorus show that bone is constantly being torn down and remodeled. Bone, in addition to its ground substance, is composed of an organic matrix of connective tissue, collagen, onto which are deposited a complex salt of calcium and phosphate, hydroxyapatite [$3Ca_3(PO_4)_2 \cdot Ca(OH)_2$], and calcium carbonate. More than 98 percent of the calcium of the body and about 80 percent of the body phosphorus are contained in bone.

Three cell types in bone are involved with bone formation or accretion and bone resorption: *osteoblasts, osteoclasts,* and *osteocytes.* The role of the osteoblasts appears to be confined to the formation of calcifiable collagen fibers in a suitable ground substance and the alignment of the fibers to form the bone template. The osteoblasts contain a high concentration of the enzyme alkaline phosphatase, which hydrolyzes organic esters of phosphoric acid. An increase in the local activity of such an enzyme would lead to an increase in the local concentration of $PO_4^{\equiv}$ and result in the deposition of insoluble calcium phosphate on available protein matrix. Therefore, this enzyme might provide the phosphate ions necessary for bone calcification. When sufficient concentrations of Ca^{++} and $HPO_4^{=}$ are present in the extracellular fluid (ECF), each interacts with specific sites on the collagen fibers and forms the hydroxyapatite crystals. These grow and during the process of growing constitute the exchangeable mineral of bone, and an equilibrium is established between the Ca^{++} and $HPO_4^{=}$ of the ECF and bone. However, as crystal formation increases, more water is excluded, and because the crystals do not release their ions readily, part of the bone structure becomes a relatively inert, solid, nondiffusible mass. Nearly all (about 99 percent) of the mineral of bone is in this state, the so-called *nonexchangeable bone.* All bone would achieve this state if it were not constantly being remodeled.

The osteoclasts perform a role in processes involved in dissolving bone mineral and the destruction of bone collagen. It is believed that bone resorption results from increased acid production and increased acid phosphatase activity, which lead to the release of additional Ca^{++} and $HPO_4^{=}$ into the ECF. It has been estimated that about 500 mg of calcium from old bone are resorbed under the influence of the osteoclasts each day. Naturally, this old bone must be replaced by an equivalent amount of new bone formation, under osteoblast activity, if equilibrium is to be attained. Conceptually, one may regard the ions of the ECF as being in simple equilibrium with those of exchangeable bone. Thus, the exchangeable bone acts as a buffer in controlling the concentration of Ca^{++} and $HPO_4^{=}$ in the internal environment.

The osteocyte is required for the homeostatic regulation of bone metabolism, and it appears to be capable of promoting both bone formation and bone resorption.

Vitamin D is also concerned with the formation of bone, for this vitamin aids in maintaining Ca^{++} levels of the ECF. It reduces the excretion of Ca^{++} in the feces and that of $HPO_4^{=}$ in the urine. In the absence of vitamin D, the Ca^{++} and $HPO_4^{=}$ concentrations decline, calcification of collagen ceases, and bone growth stops. The bones readily bend and break and the condition *rickets* develops. Although vitamin D can promote the maximal retention of Ca^{++} and $HPO_4^{=}$ and maintain the product of the two concentrations above a level at which bone calcification can occur, it cannot do so unless the dietary Ca^{++} is maintained above a critical level.

EFFECTS OF PARATHORMONE

A considerable amount of controversy has existed concerning the primary effects of parathormone. One school contended that the effect of the hormone was primarily on bone; another proposed that the primary site of action was in the kidneys, and that all other effects followed as a consequence of its renal action; another suggested that there are two hormones secreted by the parathyroids, one acting on the kidney, the other on bone. Present data indicate that only one hormone, *parathormone,* is secreted by the parathyroid gland, and it has three primary sites of action, the aforementioned bone and kidney, and the gastrointestinal tract.

The actions of PTH help to maintain the concentration of calcium ion activity in the plasma and protect the organism from hypocalcemia. PTH is a potent hypercalcemic agent.

EFFECT ON BONE

Administration of parathormone or excessive release of its secretion in parathyroid hyperplasia causes softening of the skeletal system as a result of withdrawal of mineral from bone. One of the earliest responses to PTH administration is an elevation of the blood citrate level, and this is followed by an elevated serum phosphate and calcium. Recent evidence indicates that the elevated calcium level is secondary to and dependent upon citrate response. Diuresis with increased renal calcium and phosphate loss, the tendency toward increased coagulability of blood, and decreased excitability of nerve result from the elevated Ca^{++} level.

That a primary action of the hormone occurs with bone is shown by the following observations: (1) transplantation of parathyroid tissue to membranous bone of the skull leads to bone resorption at the site of contact with the transplant, while bone deposition occurs on the opposite surface; (2) the addition of parathyroid tissue or PTH to bone grown in tissue culture leads to increased osteoclast activity and bone resorption; and (3) a prompt fall in the plasma calcium level occurs in the nephrectomized animal following parathyroidectomy. The data indicate that PTH, through its effects on bone cells, increases the rate of bone resorption. This process includes the breakdown of collagen, ground substance, and hydroxyapatite crystals. The manner in which this is accomplished remains to be ascertained.

EFFECT ON THE KIDNEY

The outstanding metabolic changes which follow parathyroidectomy are *hypocalcemia, hyperphosphatemia, hypocalciuria,* and *hyperphosphaturia.* As a result of the fall of Ca^{++} activity in plasma, symptoms attributed to increased neural and muscular excitability are evident. However, the initial change observed following parathyroidectomy is an increase in renal calcium excretion; the subsequent decrease in renal excretion of calcium occurs only after a significant fall in plasma calcium. Similarly, the initial effect of administration of PTH is a fall in urinary calcium excretion, presumably a result of increased Tm for calcium, and the plasma Ca^{++} becomes elevated. Hypercalciuria may develop but only after hypercalcemia is evident. PTH administration also results in rapid sustained increase in urinary phosphate excretion in the absence of a change in glomerular filtration rate (GFR). Recently it has been unequivocally demonstrated that PTH has a direct effect upon the kidney. PTH infused directly into a renal artery produced unilateral phosphaturia on the infused side without changes in GFR, by blocking tubular reabsorption; no changes in the handling of phosphate or GFR were observed in the contralateral kidney.

EFFECT ON THE GASTROINTESTINAL TRACT

Several clinical observations indicate that PTH influences the intestinal absorption of calcium. Experimental studies have indicated that parathyroidectomy results in a 50 percent decrease in the rate of absorption of radiocalcium from the intestine within four hours after the operation. The effects of PTH are diagrammed in Figure 33-1.

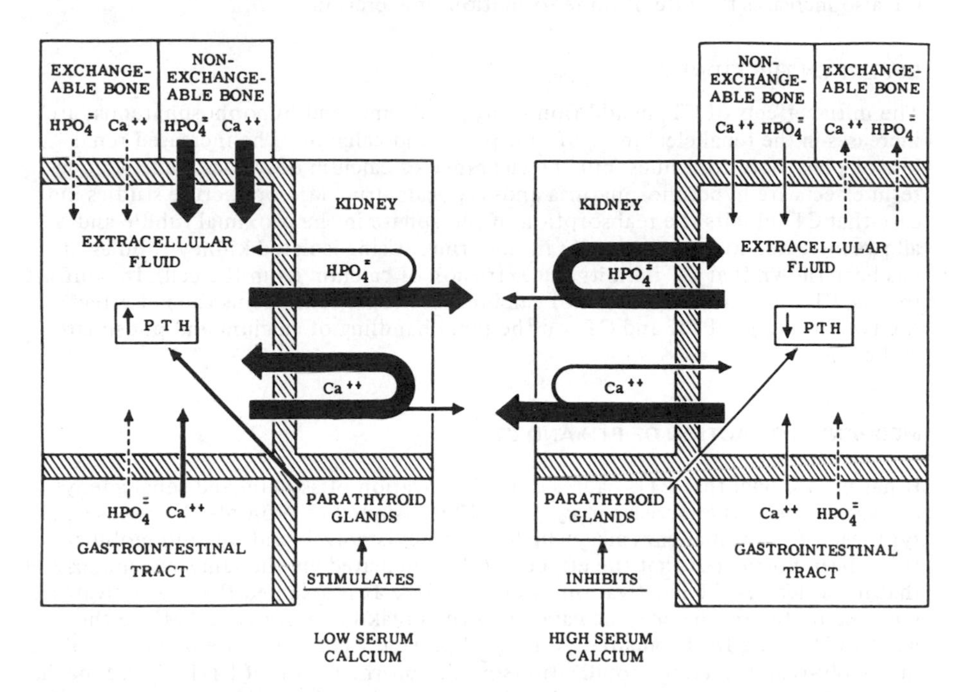

FIGURE 33-1. Actions and regulation of PTH secretion. The figure illustrates the factor regulating PTH secretion and depicts the *early* changes in Ca^{++} and $HPO_4^{=}$ metabolism over which PTH has a regulatory function. The dashed lines indicate a lack of PTH effect. The width of the arrows indicates the magnitude of response.

EFFECTS OF CALCITONIN

Calcitonin, in contrast to PTH, protects the organism from hypercalcemia, and this protection is exerted through the actions of CT on bone and on calcium and phosphate handling by the kidney. The net effect of CT is to lower the serum calcium level, and this is accomplished by hypophosphatemia.

EFFECT ON BONE

The administration of CT decreases the rate of resorption of bone and inhibits the effect of PTH on bone. In vitro studies have shown that this is a direct effect on bone. As noted above, PTH added to bone in tissue culture leads to a decrease in osteoblast activity and to increased osteocyte and osteoclast activity, hydroxyproline release, and demineralization. However, CT added to PTH-stimulated bone in vitro stimulates osteoblast activity and reduces osteocyte and osteoclast activity and causes a decrease in the release of calcium. In addition, CT inhibits citrate production. It also appears to decrease the rate of release of hydroxyproline from collagen. These studies, coupled with in vivo studies on the rate of removal of calcium from PTH-sensitive pools, the urinary excretion of hydroxyproline, and urinary excretion of ^{85}Sr and ^{45}Ca following CT administration, indicate that the primary effect of CT is to prevent bone resorption. Evidence is also accumulating to indicate that CT also increases the rate of bone formation or accretion.

EFFECT ON THE KIDNEY

The initial effects of CT, in addition to hypocalcemia and hypophosphatemia, are increases in the renal clearances of phosphate and calcium. The increased renal loss of phosphate continues, but the clearance of calcium decreases, and the long-term effects are hyperphosphaturia and hypocalciuria. Micropuncture studies indicate that CT inhibits the reabsorption of phosphate in the proximal tubule and allows increased loss of phosphate in the urine. Using isolated kidney tubules, it has been shown that CT inhibits the extrusion of calcium from the cell. In contrast to this, PTH increases the influx of calcium into renal cells. Thus the net effect of the two hormones, PTH and CT, on the renal handling of calcium and phosphates is the same.

MECHANISM OF ACTION OF PTH AND CT

It has been shown that PTH causes a rapid activation of skeletal and renal adenyl cyclase. In bone the increase in cyclic AMP is assumed to promote the release and synthesis of lysosomal enzymes, which promote osteolysis and calcium mobilization. It has been postulated that the effect of CT is mediated via the same system, except that it accelerates the inactivation of cyclic AMP. It is assumed that CT activates phosphodiesterase, the enzyme catalyzing the breakdown of cyclic AMP to the inactive 5'-AMP. In this manner, CT could prevent the release of enzymes involved in osteolysis and calcium mobilization and inhibit the action of PTH. These mechanisms, which suggest a role of the "2nd messenger," cyclic AMP, in bone resorption, are depicted in Figure 33-2. The ways in which PTH acts on the kidney and GI tract and CT acts upon the kidney to produce their effects are not known.

REGULATION OF SECRETION OF PTH AND CT

The rates of secretion of PTH and CT are independent of the adenohypophysis and the central nervous system. Their secretion rates are controlled by the Ca^{++} concentration in the plasma. Recent studies, involving perfusion of isolated glands, have shown that the rate of secretion of either hormone can be altered by either increasing or decreasing the Ca^{++} concentration of the perfusate. The secretion rates of PTH and CT appear to be reciprocally related.

Studies have shown that there is a simple linear inverse relationship between plasma Ca^{++} and plasma PTH. These studies were conducted over a range of plasma Ca^{++} of 4 to 12 mg per 100 ml (2 to 6 mEq per liter) and they lead to the conclusion that Ca^{++} regulates PTH secretion through a proportional control mechanism. When the concentration of Ca^{++} rises, the glands are inhibited and secrete less hormone, and when the Ca^{++} falls, the glands are stimulated to secrete more hormone. Similarly, high-calcium diets lead to parathyroid hypoplasia, and

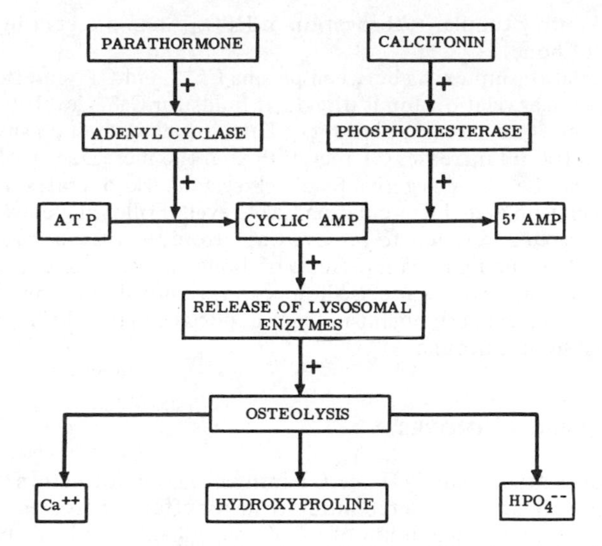

FIGURE 33-2. Theoretical mechanism of action of PTH and CT on bone. In this scheme PTH increases the concentration of cyclic AMP and CT prevents the accumulation of cyclic AMP. The + indicates a stimulating effect.

low-calcium diets result in hypertrophy and hyperplasia of parathyroid glands. Thus, a negative feedback system exists between the parathyroid glands and the circulating Ca^{++} level. Following a depression of Ca^{++} levels, PTH is secreted, and it acts on the intestinal cells to increase calcium absorption; it acts on the kidney tubules to increase reabsorption of Ca^{++}, and loss of $HPO_4^{=}$ from the glomerular filtrate; and it acts upon bone cells to increase osteoclast activity, which leads to a release of Ca^{++} from nonexchangeable bone. The hormone also increases the rate of phosphate released by bone, which is offset by increased urinary excretion of phosphate. The final result is an increased plasma Ca^{++} and decreased $HPO_4^{=}$. When the plasma Ca^{++} concentration reaches a critical level, the parathyroid gland is inhibited and the PTH secretion decreases and the changes are reversed.

PTH is responsible for the hour-to-hour regulation of Ca^{++} concentration of the blood. Studies utilizing isotopic calcium have shown that an amount of calcium equivalent to the total blood calcium is replaced every minute in the young animal. Two mechanisms are responsible for this rapid turnover. The more rapid mechanism is one of ion exchange between blood and exchangeable bone and is independent of hormonal action. As a result of this exchange, the serum calcium concentration rarely falls below 6 mg per 100 ml, even after parathyroidectomy. The second mechanism is a function of PTH and involves a calcium-mobilizing effect of the hormone on bone. This effect maintains the serum calcium level at about 10 mg per 100 ml. In addition, an increased rate of bone resorption leads to an increased release of $HPO_4^{=}$ into serum. If the effect of PTH on bone were the only means of regulating the Ca^{++} level of the serum, the feedback system would lead to wide oscillations in the Ca^{++} level of the serum, for this effect is relatively slow in onset. The effect of PTH on gastrointestinal tract absorption of calcium is more rapid than that on bone and would tend to reduce the oscillations in the serum Ca^{++} level. However, the kidney is capable of responding rapidly to PTH and tends to maintain the level of serum Ca^{++} within narrow limits. It also facilitates the excretion of

$HPO_4^=$ by inhibiting tubular reabsorption of $HPO_4^=$ and offsets the rise produced by resorption of bone.

A similar relationship exists between plasma Ca^{++} and CT secretion and plasma CT, except the linear relationship is direct. It holds for Ca^{++} levels between 8 and 20 mg per 100 ml (4 to 10 mEq per liter). Thus, an increase in plasma calcium above 8 mg per 100 ml increases CT secretion, and the increase is in direct proportion to the plasma Ca^{++}. A negative feedback system also operates, therefore, between CT secretion and the circulating Ca^{++} level. Following elevation of Ca^{++} levels, CT is secreted and it acts to remove Ca^{++} from the ECF or prevent the addition of Ca^{++} to the ECF via its effects on bone and the kidney.

The two hormones exert a precise homeostatic control over the plasma Ca^{++}. PTH prevents or protects the organism from hypocalcemia, and CT prevents or protects against hypercalcemia.

EFFECTS OF OTHER HORMONES

Several hormones other than PTH and CT influence bone formation and resorption and calcium and phosphorus metabolism. A major effect of growth hormone is to make available an increased amount of calcifiable collagen, and this brings about an increase in the rate of bone formation. It decreases the renal loss of phosphate and increases renal calcium loss. Because of the effects of growth hormone on the plasma Ca^{++} there is a tendency toward hypocalcemia, which leads to more secretion of PTH and thus to a higher rate of bone remodeling.

At puberty the estrogens and androgens promote bone formation and mineralization, and they are required for normal nitrogen, calcium, and phosphorus content in bone of the adult. The thyroid hormones are necessary for normal bone growth and maturation; they increase the rate of turnover of bone and tend to increase the Ca^{++} plasma level. Insulin also appears to function as a growth hormone in bone.

The hormones of the adrenal cortex, particularly cortisol and corticosterone, have a marked effect on bone. They increase the rate of protein breakdown in bone and decrease the rate of collagen synthesis. They thus decrease the total mass of calcified bone. They tend to lower the rate of intestinal absorption of Ca^{++} and increase the loss of renal Ca^{++}. These effects should lead to a rise in PTH secretion.

In summary, negative feedback mechanisms for the regulation of serum calcium and phosphate concentration involve the parathyroids, cells derived from the ultimobranchial body, bone, the gastrointestinal tract, and the kidney. The renal regulator is rapid in response but has limited capacity. The gastrointestinal regulator is slower to respond than is the kidney and also has limited capacity. On the other hand, the bone regulator is slow to respond to PTH but responds rapidly to calcitonin; it is relatively insensitive to PTH but is rather sensitive to the calcitonins; and it has an unlimited capacity for responding to both hormones. Other hormones also play a role in bone and calcium and phosphorus homeostasis.

REFERENCES

Borle, A. B. Effect of thyrocalcitonin on calcium transport in kidney cells. *Endocrinology* 85:194–199, 1969.

Care, A. D., C. W. Cooper, T. Duncan, and H. Orimo. A study of thyrocalcitonin secretion by direct measurement of in vivo secretion rates in pigs. *Endocrinology* 83:161–169, 1968.

Copp, D. H. Endocrine control of calcium homeostasis. *J. Endocr.* 43:137–161, 1969.
Copp, D. H. Endocrine regulation of calcium metabolism. *Ann. Rev. Physiol.* 32:61–86, 1970.
Gaillard, P. Parathyroids and bone in tissue culture. In R. O. Greep and R. V. Talmadge (Eds.), *Parathyroids.* Springfield, Ill.: Thomas, 1961. Pp. 20–48.
Hirsch, P. F., and P. L. Munson. Thyrocalcitonin. *Physiol. Rev.* 49:548–622, 1969.
Melson, G. L., L. R. Chase, and G. D. Aurbach. Parathyroid hormone-sensitive adenyl cyclase in isolated renal tubules. *Endocrinology* 86:511–518, 1970.
Rasmussen, H. Parathyroid hormone: Nature and mechanism of action. *Amer. J. Med.* 30:112–128, 1961.
Sammon, P. J., R. E. Stacy, and F. Bonner. Role of parathyroid hormone in calcium homeostasis and metabolism. *Amer. J. Physiol.* 218:479–485, 1970.
Sherwood, L. M., G. P. Mayer, C. F. Ramberg, D. S. Kronfeld, G. D. Aurbach, and J. T. Potts. Regulation of parathyroid hormone secretion: Proportional control by calcium, lack of effect of phosphate. *Endocrinology* 83:1043–1051, 1968.
Talmadge, R. V., and L. F. Belanger (Eds.). *Parathyroid Hormone and Thyrocalcitonin (Calcitonin).* Amsterdam: Excerpta Medical Foundation, 1968.

34

Thyroidal Physiology

Ward W. Moore

ANATOMY

Phylogenetically speaking, the thyroid gland is one of the oldest of endocrine glands. The human thyroid first appears in the 3-week-old embryo as a ventral diverticulum from the midventral floor of the pharynx between the first and second pharyngeal pouches. By the fifth week, a small hollow sac has developed which remains joined to the pharynx by the thyroglossal duct. The latter atrophies during the sixth week, and with this change the thyroid loses its central cavity and adopts a bilobed form. Discontinuous cavities become manifest within the gland by the eighth week. These represent the beginnings of the follicles of the adult gland; the follicles soon become filled with colloid substance.

The characteristic and dominant structural feature of the adult thyroid gland is the follicle, which is roughly spherical, lined by epithelium, and filled with colloid. The colloid is a viscid homogeneous substance, rich in iodine, and clear in appearance in the fresh state. The lining epithelium is cuboidal and rests directly upon the vascular connective tissue without the intervention of a basement membrane. The normal epithelium varies in appearance with the state of activity of the gland; generally speaking the cells are low when the gland is resting and high when it is active. The low resting epithelium is usually associated with a large follicular lumen and an increase in colloid content. A columnar epithelium, on the other hand, is usually observed in the active gland, with a small amount of colloid and a consequent folding of the epithelium upon itself. The cells lining the follicle have the unique capacity both to release their secretions into the follicular lumen and to permit the passage of the active principles of the cells from colloid to blood stream. The mitochondria of the thyroid cells increase in number when the gland is active in the phase of formation of colloid, rather than with the release of hormone into the blood. On the other hand, hypertrophy of the Golgi apparatus is associated with an increase in the rate of release of hormone into the circulation.

The thyroid gland receives its blood supply from paired superior and inferior thyroid arteries which arise from the external carotids and subclavians, respectively. The rate of blood flow through the gland is high (5 to 7 ml per gram per minute), more per gram of tissue even than through the kidney. The gland is innervated by laryngeal and pharyngeal branches of the vagus and by cervical sympathetic ganglia. These fibers are probably vasomotor and possibly secretomotor, but the transplanted or denervated gland can function normally. However, various neurohumors can alter the sensitivity of the thyroidal cells to stimuli.

ROLE OF THE THYROID GLAND

The thyroid gland plays an important role in a homeostatic control system which aids in maintaining an optimal level of oxidative metabolism and heat production of the organism. Other parts of this regulatory mechanism inchude the central nervous system (CNS), the adenohypophysis (AP), the general circulation and certain plasma proteins, and the metabolic machinery of all cells of the body. The thyroid gland also plays a vital role in the organization and maturation processes of several systems. In its role, the thyroid gland excels in three characteristic functions: (1) trapping of iodide, (2) synthesis of organic iodine, and (3) storage and secretion of hormones which are iodothyronines.

The isolation of the hormone of the thyroid gland, *thyroxine,* by Kendall in 1915, its identification as 3,5,3',5'-tetraiodothyronine by Harington in 1926, and its synthesis by Harington in 1927 gave great impetus to the study of thyroidal physiology and biochemistry. In 1952, two independent groups, Gross and Pitt-Rivers, and Roche, Michel, and Lissitzky, demonstrated the presence of another iodothyronine in the thyroid gland and blood. This compound has three to four times the biological activity of thyroxine and has been identified as 3,5,3'-triiodothyronine (Fig. 34-1). The extensive use of one of the radioactive isotopes of iodine, ^{131}I, in conjunction with chromatographic techniques has led to important progress in every aspect of thyroid biology.

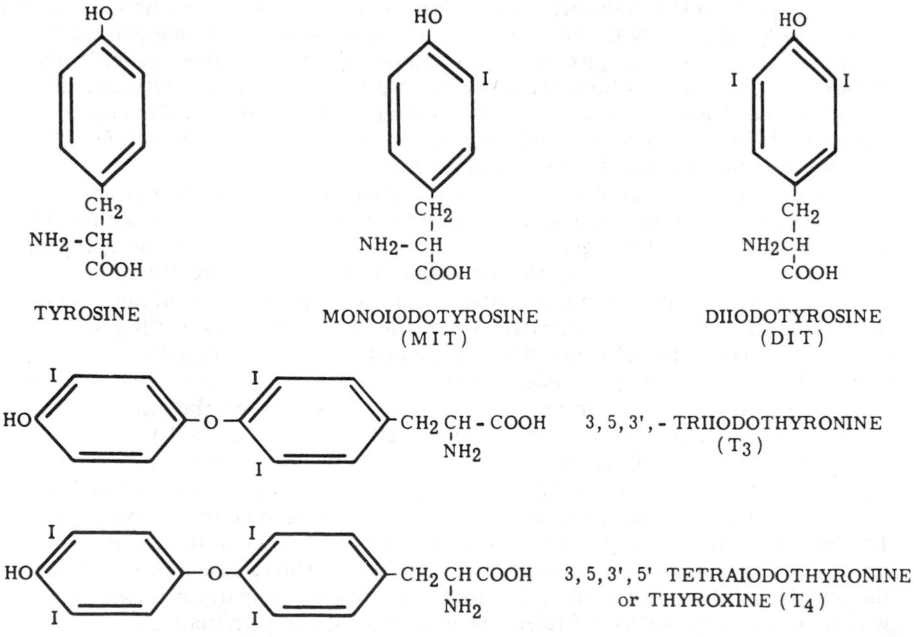

TYROSINE

MONOIODOTYROSINE (MIT)

DIIODOTYROSINE (DIT)

3, 5, 3', - TRIIODOTHYRONINE (T_3)

3, 5, 3', 5' TETRAIODOTHYRONINE or THYROXINE (T_4)

FIGURE 34-1. Compounds involved in thyroid hormone production.

FORMATION AND RELEASE OF THYROID HORMONES

It has been shown that the biosynthetic mechanisms involved in the formation of hormones are autonomous. Nevertheless, the rate of formation of the hormones is regulated by extrathyroidal factors. The most important factors are adenohypophysial thyroid-stimulating hormone (TSH) and the availability of iodide. However, intrathyroidal mechanisms also regulate hormonogenesis. Several key steps are involved in the synthesis, storage, and secretion of thyroid hormones (see Fig. 34-2).

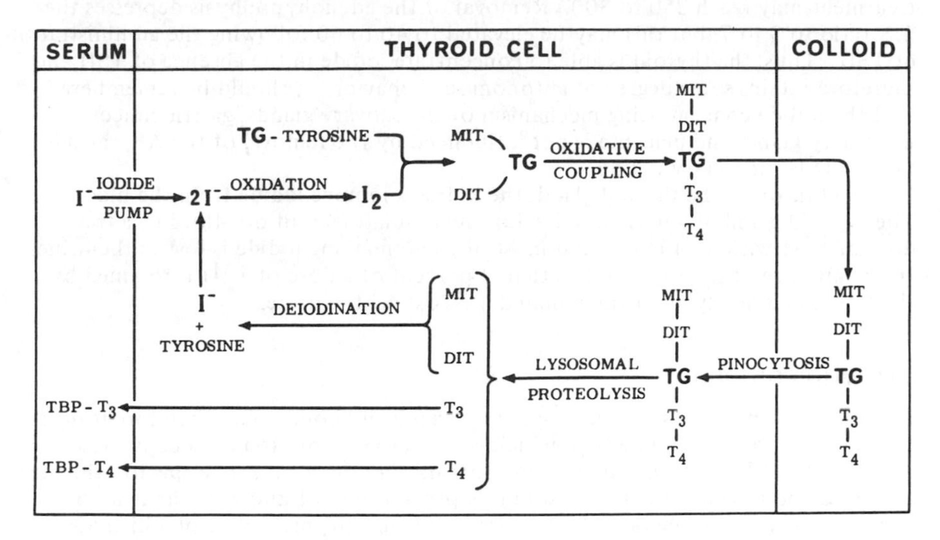

FIGURE 34-2. Pathways of thyroid hormone formation and secretion. TG=thyroglobulin; TBP=thyroxine-binding protein.

CONCENTRATION OF IODIDE

Normally, a concentration gradient for iodide (I^-) of about 25 to 1 is maintained between the thyroid and the serum. The mechanism by which the thyroid accumulates iodide from the serum (with an iodide concentration of about 0.1 to 0.5 μg per 100 ml) is not understood. Iodide is transported into the thyroid cell from the interstitial fluid against an electrochemical gradient (a negative cell potential with respect to the extracellular fluid [ECF] and lumen). The transfer is energy-dependent, and therefore iodide is actively transported. The iodide-concentrating mechanism is referred to as the iodide pump. The thyroid gland removes iodide from the blood and clears the plasma of iodide at a rate of about 20 ml per minute.

That the iodide pump is of physiological importance is best illustrated following the administration of thiocyanate, perchlorate, and other similar anions. These anions inhibit the iodide pump, with the result that the thyroid iodide concentration cannot exceed the serum concentration. The mechanism by which the anions interfere with the capacity of the thyroid to concentrate iodide is not known, but it has been postulated that the action of thiocyanate is competitive; i.e., it competes with iodide at the site of the concentrating mechanism. It should be pointed out that other tissues, particularly the salivary glands, gastric mucosa, mammary gland, and placenta, also possess iodide-concentrating mechanisms. Inhibition of the iodide pump in these organs has been observed following treatment with anions of the perchlorate group.

The best-known regulator of the iodide pump is the thyroid-stimulating hormone (TSH). TSH increases the membrane potential of thyroid cells, whereas hypophysectomy decreases it. The performance of the iodide pump may be assessed by determining the thyroid/serum iodide concentration ratio (T/S) at equilibrium. In a standardized test, the organic binding of iodide is inhibited by a single dose of propylthiouracil (PTU) (see p. 763) followed by a tracer dose of radioiodide. The T/S of normal rats following a single dose of PTU is 25 but following chronic PTU treatment may reach 250 to 300. Removal of the adenohypophysis depresses the T/S ratio to 3 to 7, but this may be elevated to 40 to 60 following the administration of TSH. Thus, the thyroid is able to concentrate iodide in the absence of TSH and therefore retains some degree of autonomous behavior. It should be remembered that the iodide-concentrating mechanism of the salivary glands, gastric mucosa, mammary gland, and placenta is not influenced by the removal of the AP, the addition of TSH, or dietary iodide.

In addition to the thyroid gland, the kidneys remove iodide from the blood. The thyroid gland of a normal individual accumulates about one-third of a tracer dose of ^{131}I within 24 hours, and most of the remaining iodide is lost in the urine. In the absence of the thyroid, less than 2 percent of a dose of ^{131}I is retained by the body, and nearly all of the remainder is lost via the urine.

OXIDATION OF IODIDE

Over 90 percent of the thyroidal iodine is organically bound, and therefore it must have appeared as free iodine (I_2), which is a reactive form. Iodide is converted to iodine ($2\,I^- \rightarrow I_2$), which can enter into organic combination. The specific chemical reactions and the enzymes involved in this process are not known. The reaction ($2\,I^- \rightarrow I_2$) can be accelerated by various oxidizing enzymes, none of which has been isolated and characterized as specific. Peroxidase activity has been demonstrated histochemically in the thyroid cell, and this activity is strongly depressed by thiouracil and its derivatives. According to some endocrinologists, the presence of a specific oxidase for this reaction is physiologically not necessary.

Agents such as propylthiouracil depress the amount of iodine available for iodination of tyrosine. Following the use of extremely high doses of propylthiouracil, only traces of iodotyrosines may still be formed, but significant amounts of iodotyrosines and iodothyronines are still formed after small doses of propylthiouracil.

IODINATION OF TYROSINE AND IODOTHYRONINE FORMATION

Following the formation of elemental iodine, iodination of tyrosine is accomplished first at the third position and then at the fifth position (Fig. 34-1). The tyrosine which is iodinated exists as a tyrosyl radical, part of the thyroglobulin molecule. Thus, the thyroglobulin acts as a substrate for a series of reactions leading to the formation of the thyroid hormones. The steps involved are the formation of monoiodotyrosine (MIT) and diiodotyrosine (DIT), followed by the oxidative coupling of MIT and DIT with a loss of alanine within the thyroglobulin molecule. These reactions result in the formation of triiodothyronine (T_3) and tetraiodothyronine (T_4) or thyroxine. The above-mentioned steps in hormone formation occur in the cells of the follicle, and thence the thyroglobulin is secreted into the lumen of the follicle to become part of the colloid. In general, it can be stated that TSH increases the rate of organification of iodine and iodothyronine formation. However, the mechanism or mechanisms by which TSH accelerates organification of iodide are not known. TSH administration decreases the MIT/DIT and T_3/T_4 ratios and increases the absolute amount of T_3 and T_4 in thyroglobulin. It follows that TSH might play a role in stimulating the oxidative coupling of MIT with DIT and DIT with DIT in the thyroglobulin molecule.

It has been indicated that there is an intrathyroidal regulation of the synthesis of iodotyrosines and iodothyronines. This is seen most clearly when an animal is placed upon an iodine-deficient diet. In such a situation, the MIT/DIT and T_3/T_4 ratios within the thyroglobulin molecule are markedly increased, but the concentration of each compound is decreased. If the elevated T_3/T_4 ratio in the gland should be maintained in the serum, this would demonstrate an important physiological adjustment in cases of iodine deficiency. However, it has not been determined whether or not the more potent mixture of T_3 and T_4 is carried over to the serum under such circumstances.

PROTEOLYSIS OF THYROGLOBULINS

The manner in which the thyroid hormones are stored is unique among the endocrine glands. Thyroxine and triiodothyronine are contained, via peptide linkage, within the large thyroglobulin molecule (molecular weight about 650,000). This is stored in an extracellular site, the follicular colloid, and the rate of its formation and removal is dependent, in part, upon TSH. Recent work, combining electron microscope, biochemical, differential centrifugation, and isotopic techniques, has shed light upon the manner by which T_4 and T_3 are removed from their storage sites and secreted into the plasma. Following TSH administration droplets or particles of colloid appear in the apex of the follicular cells after their engulfment by pinocytosis. These droplets migrate toward the base of the cell and fuse with lysosomes, which have also migrated from the base of the cell. The lysosomes contain enzymes which can hydrolyze thyroglobulin (TG). T_4 and T_3 release occurs by TG lysis within the fused droplet and lysosome or "derived lysosome." MIT and DIT are also released into the thyroid cell by the hydrolysis of TG.

RELEASE OF THYROID HORMONES

The iodotyrosines, MIT and DIT, are not normally secreted into the blood stream because they are deiodinated rapidly by dehalogenases in the thyroid cells. Thus the iodide may be reclaimed immediately by the gland. However, the iodothyronines, T_3 and T_4, are resistant to the dehalogenases and diffuse from the gland as a result of gradients which may be as high as 100 to 1 between the gland and the blood. The diffusion gradient of thyroxine is also aided as a result of competitive binding favoring certain plasma proteins over that of the thyroidal proteins. It should be remembered that thyroglobulin does not appear in the blood except under abnormal conditions.

TSH leads to a rapid increase in thyroidal cyclic AMP. The effects of this hormone on thyroid hormone synthesis and secretion are probably mediated via the "2nd messenger." TSH accelerates the rate of proteolysis of thyroglobulin and the rate of release of T_3 and T_4, and following diffusion into the blood stream each is bound by a specific plasma protein.

Several observations indicate that an intrathyroidal factor participates in the regulation of the release of organic iodine from the gland. When individuals are placed on chronic PTU ingestion, the daily output of hormonal iodine falls in proportion with the store of hormonal iodine in the gland. However, an increase in circulating TSH (which must occur in this instance) should cause an increase in proteolysis of thyroglobulin, with a resultant increase in release of thyroid hormones from the gland. Under this circumstance, one would expect an increase in thyroid hormone output instead of a decline, and it must follow that some mechanism within the thyroid induces a conservation of thyroidal iodine. Apparently the three intrathyroidal factors which regulate thyroid function lead to physiological adjustments which aid in the maintenance of an adequate level of thyroidal iodine stores, yet at the same time lead to the secretion of a mixture of T_3 and T_4 of increased potency when challenges which tend to decrease hormonogenesis are put to the thyroid-pituitary axis.

TRANSPORT OF THYROID HORMONES IN THE CIRCULATION

Three plasma proteins readily bind the iodothyronines. A serum alpha globulin, called the thyroxine-binding globulin (TBG), binds approximately 200 μg of thyroxine per liter of serum. The concentration of TBG is elevated by estrogen therapy, during pregnancy, during the neonatal period, and in certain hepatic diseases, but it is depressed by androgen administration. The binding capacity of TBG is increased in hypothyroidism. It appears that TBG accounts for most of the transport of thyroxine at physiological concentrations. One of the prealbumins, the thyroxine-binding prealbumin (TBPA), apparently contributes to thyronine binding. Its saturation capacity for thyroxine is greater than that of TBG, but its binding power is less. A serum albumin also reversibly binds thyroxine, but avidity and capacity for thyroxine are much less than those of TBG and TBPA. T_3 is bound by TBG, but not by TBPA or the albumin. It has been shown, by using in vitro systems, that the rate of passage of the iodothyronines from the medium into the tissue cells is inversely related to the concentration of the binding proteins in the medium. This might indicate that the concentration of TBG and the other binding proteins controls the rate at which the thyroid hormones are taken up by the cells in vivo. An indication of the concentration of the circulating thyroid hormones is the protein-bound iodine (PBI) or the butanol-extractable iodine (BEI) of the serum. These chemical assays entail the determination of the iodine of thyroxine which is combined with TBG. A PBI level of 3.5 to 8.5 μg per 100 ml of serum is said to be compatible with normal thyroidal activity, whereas values below this range are found in hypothyroidism, and those above are associated with hyperthyroid states. However, as noted above, factors other than the thyroid gland can influence the PBI.

It is obvious that there are many points in the synthesis and release processes at which a failure could cause a decrease in thyroid activity, i.e., a deficiency of hormone secreted by the gland and/or delivered to extrathyroidal tissues. Thus, the exogenous supply of iodine may be inadequate, or a deficiency of one of the enzymes needed for trapping, oxidation, coupling, or hydrolysis may result in reduced hormone production. A deficiency of TSH which exerts its effects at several stages in hormonogenesis would result in a deficiency of thyroid hormone. Similarly, an elevated TBG might lead to a decrease in thyroxine uptake by the cells, or an altered extrathyroidal dehalogenase activity might lead to a relative thyroid hormone deficiency (Fig. 34-2).

The primary source of iodine is dietary. However, the iodine of T_3 and T_4 is not completely lost when they are metabolized. In some tissues, notably liver, kidney, heart, brain, and spleen, deiodination of the iodothyronines occurs. The ether linkage within the compounds is broken, with the liberation of MIT and DIT, and further deiodination is accomplished. The iodide released can then be reclaimed by the thyroid. The circulating iodide is about 3 μg per liter of serum in the normal individual with a normal iodine intake. The renal tubules reabsorb about 95 percent of the filtered iodide load. The resulting renal plasma clearance of iodide is about 33 ml per minute, and therefore about 150 μg of iodide is lost in the urine per day. Most of the iodine lost from the body fluids is in the form of iodide, but a small part is lost as thyroxine and T_3 in the urine, feces, and sweat. Thyroxine is present in the bile as a glucuronide, and as it passes down the small intestine, part of it is reabsorbed and the remainder appears in the feces. Most of the fecal iodine is in the form of organic iodine.

REGULATION OF THYROID FUNCTION

Many factors alter the rate of thyroid activity, but the common denominator in states of altered thyroid activity is TSH activity. In all instances, the only agent which stimulates thyroid growth and thyroid hormonogenesis is TSH.

Every type of abnormal growth of the thyroid and/or change in hormone output by the gland of a normal individual must be accompanied by a change, either relative or absolute, in TSH output.

The actions of TSH are many. As already stated, it increases the thyroid-serum (T/S) iodide ratio, the rate at which iodothyronines are formed, and the rate at which thyroglobulin is hydrolyzed. In addition, TSH causes an increase in the rate of thyroidal uptake of ^{131}I. It causes an increase in the rate at which organic ^{131}I disappears from the gland, an increase in the rate at which labeled iodothyronines appear in the serum, and an increase in the serum concentration of iodothyronines. It decreases the volume of colloid, causes hypertrophy and hyperplasia of the cells of the thyroidal follicles, and increases the uptake of ^{32}P by the thyroid. Furthermore, TSH induces hypertrophy of retro-orbital tissues, which leads to the condition known as *exophthalmos.* The concentration of TSH in the normal human is about 0.2 milliunit per milliliter of blood, and any maneuver which leads to a decrease in the level of circulating thyroid hormones results in an increased TSH output. Conversely, any procedure which elevates the circulating thyroid hormone concentration leads to a decreased TSH output. Thus, in the normal individual a balance is established by a negative feedback system which tends to maintain, within certain limits, an optimal level of thyroid activity in the face of altered external or internal environments. Exceptions to these generalizations occur in certain disease states.

From the foregoing it follows that thyroidal enlargement (goiter) will result from any condition which leads to a marked and sustained imbalance between TSH output and thyroid hormone output, if the imbalance favors TSH output (Fig. 34-3). The reverse is also true, because a chronically low level of thyroidal activity will induce an abnormal growth in the basophils of the adenohypophysis, the so-called *thyrotrophs,* which secrete TSH.

The output of thyroid hormone in "normal" individuals with an iodine-deficient diet is generally normal, but is maintained at normal levels only by an increase in TSH output. It is obvious that iodine plays a very important role in the synthesis of thyroxine, since in addition to being part of the thyroxine molecule it decreases the output of thyroid hormone either by inhibiting TSH secretion or by decreasing TSH activity. Therefore, at the outset of iodide deprivation the output of thyroxine is decreased, so a degree of inhibition for the release of TSH is removed. This means that TSH output would rise, increasing the avidity of the thyroid for iodide and making more iodide available for synthesis of thyroid hormone. Thus, a marked sustained iodine deficiency indirectly maintains an increased TSH secretion, which results in hypertrophy and hyperplasia of thyroid cells. The increase in the rate of iodide uptake from the low serum levels can then maintain a level of thyroid hormone production within normal limits, as judged by the observation that all criteria which assess the effects of thyroid hormones appear to be normal; nevertheless, simple goiter is produced. However, if the increase in TSH output cannot restore normal thyroid hormonogenesis over a prolonged period, the gland will eventually become exhausted. On the other hand, if the supply of iodide is increased slightly, the exhaustion is reversible. But if the iodide supply to the gland cannot be increased,

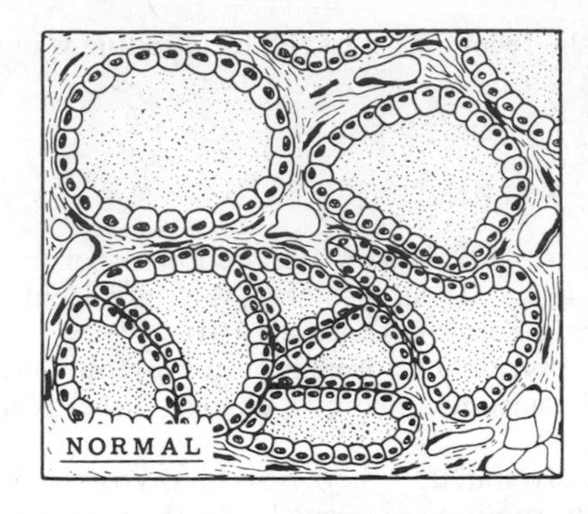

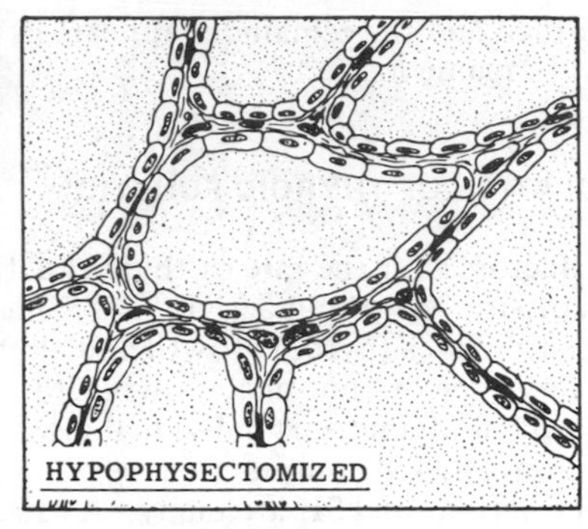

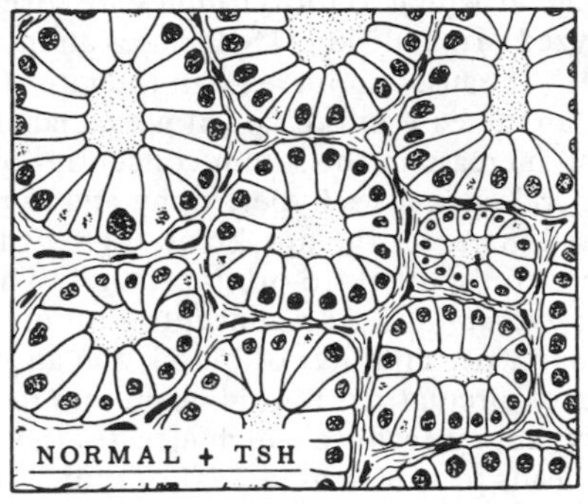

FIGURE 34-3. Effect of TSH on the morphology of the thyroid gland.

the exhaustion is irreversible. It must follow that TSH will lead to thyroid atrophy if the amount of iodide available to the thyroid is inadequate.

The ingestion of large amounts of iodide can also strongly influence thyroidal hormonogenesis. It acts to decrease the T/S ratio through the intrathyroidal mechanism, and by decreasing TSH output or by inhibiting TSH action. It decreases the rate of organification of iodine in the thyroid and may lead to the development of hypothyroidism and goiter. It has no effect upon the rate of release of hormone in the normal individual but decreases the rate of hormone release in the hyperthyroid individual or a normal individual treated with TSH. The manner in which high levels of circulating iodide influence thyroidal physiology is not known.

EFFECTS OF ANTITHYROID COMPOUNDS

The antithyroid compounds or goiterogenic drugs are so named because they inhibit in some manner the synthesis of thyroid hormones and indirectly induce goiter production by facilitating an increase in TSH output. The first substance observed to interfere with thyroid function was thiocyanate. The administration of thiocyanate and related monovalent anions, such as perchlorate, chlorate, and periodate, inhibits the ability of the thyroid to concentrate iodide. It was shown that the goiterogenic activity of these chemicals could be prevented with simultaneous administration of thyroxine. The anion must therefore act upon some stage of hormonogenesis. It has been demonstrated that the goiter produced as a result of thiocyanate can be reversed by treatment with iodide. The anions inhibit the ability of the thyroid to concentrate iodide. None of the anions completely inhibits movement of iodide into the thyroid, but the iodide which enters by simple diffusion can enter into the synthesis of thyroxine and its intermediates. Therefore (with a maximal thiocyanate effect), one would expect to observe a T/S ratio of 1.0.

Another agent, thiouracil, was shown to produce hyperplastic changes in the thyroid which were accompanied by signs of hypothyroidism. The administration of thyroxine will prevent the goiterogenic effects of thiouracil also, but the administration of iodide will not. Following thiouracil treatment, the T/S ratio increases markedly, and most of the iodine present in the gland is present as iodide. Thiouracil acts on the thyroid by keeping the iodine in the reduced state (I^-), probably by inhibiting peroxidase activity in the thyroid gland. Much work has led to the production of substituted thiouracil and thiocarbamide derivatives with a high degree of therapeutic effect and minimal toxicity for use in hyperthyroid states. The actions of the goiterogenic agents of the thiocyanate and thiouracil type have been used extensively for studying the concentration of iodide by the thyroid and its incorporation into thyroglobulin.

REGULATION OF TSH SECRETION

It has been stated that the thyroid and the adenohypophysis control each other's functions by a feedback mechanism which ensures in the normal individual an adequate, but not excessive, amount of thyroid hormone for the regulation of metabolic processes. Two primary factors participate in the regulation of the rate of TSH secreted by the AP: the hypothalamus, and a direct negative feedback action of the circulating thyroid hormones on the AP.

Several types of experimental procedures have shown that the hypothalamus is intimately concerned with TSH secretion. First, transplantation of the AP to extrasellar sites leads to a reduction in TSH secretion, and thyroid function is decreased, approaching, but not reaching, that observed in the hypophysectomized animal. Second, effective hypophysial stalk section results in a decreased rate of TSH secretion. Third, lesions of the anterior hypothalamus, particularly in the supraoptic region, depress thyroid function which is a reflection of decreased TSH secretion. Finally, electrical stimulation of the anterior hypothalamus or rostral portion of the median eminence results in an increase in circulating TSH.

Both hypophysial stalk section and lesions of the anterior hypothalamus lead to a reduction in total TSH content of the AP and a decrease in serum TSH, but the pituitary TSH concentration and the appearance of the thyrotrophs remain normal. Alterations which follow each procedure are: decreases in the T/S iodide concentration ratio, the rate of release of organic ^{131}I, the serum PBI, and thyroid weight; and thyroid function comparable to that observed in the hypophysectomized animal. Another feature observed in these preparations is the failure to respond to stimuli which in the intact animal lead to either an increase or a decrease in TSH secretion. The goiterogenic effects of thiocyanate and PTU are prevented, and the reflex alteration of TSH secretion by such stimuli as cold, trauma, restraint, and emotional stress is absent. Recent studies have shown that stimulation of the hypothalamus by local cooling causes an increase in TSH secretion and a marked rise in thyroid function. Extracts of hypothalamic tissue have been prepared which contain TSH-releasing activity when used on pituitaries either in vivo or in vitro. The active principle of these extracts has been called the thyrotropin-releasing factor (TRF). TRF is a polypeptide and is quite distinct from the other releasing factors, i.e., SRF, CRF, LRF, and PIF (Chaps. 31, 35, and 36).

The inhibition of TSH release by thyroxine is not lost after effective stalk section or after placement of anterior hypothalamic lesions. It has been shown that local infusions of thyroxine into the AP inhibit TSH secretion, and that the transplantation of minute amounts of thyroidal tissue directly into the AP prevents the development of thyroidectomy cells after removal of the thyroid.

The thyrotrope cells of the AP appear to be affected by two chemical stimuli: TRF, which stimulates TSH secretion; and the thyroid hormones, which inhibit it. Thus present evidence indicates that the principal feedback mechanism of pituitary TSH secretion is a direct one by the thyroid hormones, but for full stimulatory or inhibitory adaptations to external stimuli, the functional integrity of the hypothalamo-hypophysial system must be maintained. Very little is known concerning the influence of fibers connecting the hypothalamus with the remainder of the CNS upon the release of TSH, but it is apparent that the afferent input to the anterior hypothalamic area must in some way influence this system.

Experimentation has shown that stressful stimuli powerfully inhibit the uptake of ^{131}I and the release of thyroid hormones. The alterations in thyroid function occur prior to the withdrawal of TSH, and the changes in thyroid function result from the increased circulating adrenal cortical steroids. These hormones enhance urinary iodide loss and slow thyroid circulation, thus contributing to the reduced thyroidal iodide uptake. Chronic stress produces depressed thyroid function, yet an excess of T_4 decreases the survival to a variety of stresses. Emotional stress has long been thought to precipitate a marked increase in the rate of TSH release and lead to hyperthyroidism in predisposed individuals (see Fig. 34-4).

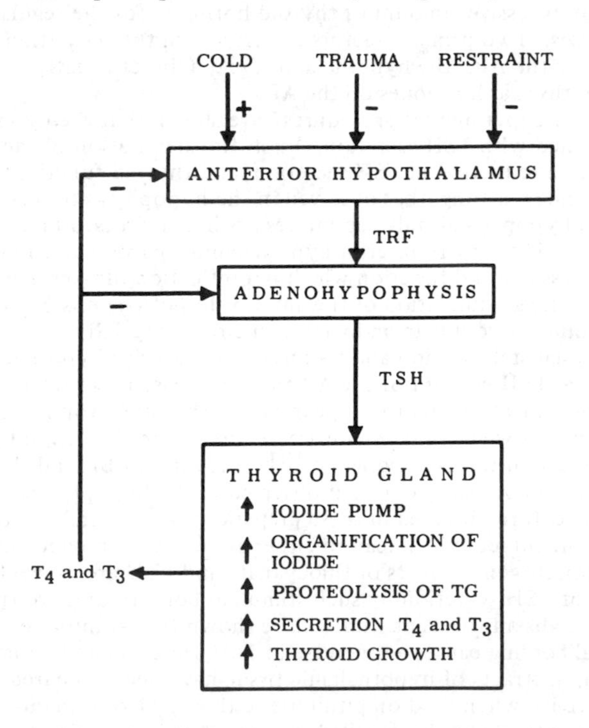

FIGURE 34-4. Regulation of the secretion of TSH and thyroid hormones. + stimulation; – inhibition.

PHYSIOLOGICAL EFFECTS OF THYROID HORMONES

The physiological effects of the thyroid hormones have been extensively studied by observing structural and functional changes which occur in spontaneous and experimental hypothyroidism and hyperthyroidism. The thyroid gland is not indispensable to life, but its presence is necessary for normal growth, maturation, heat production, and well-being of the individual.

Probably the most prominent effect of the thyroid hormones, first noted by Magnus-Levy in 1895, is their effect upon respiratory exchange. His classic discovery that respiratory exchange is decreased in Gull's disease (hypothyroidism) and increased in Graves' disease (hyperthyroidism) has led to the publication of many works concerned with the effects, both pharmacological and physiological, of the thyroid hormones.

The increased oxygen consumption and heat production (calorigenic effect) which follow the administration of the thyroid hormones to animals are marked. The calorigenic effect can also be demonstrated in certain excised tissues obtained from animals with experimentally induced hyperthyroidism. However, when thyroid hormones are added to excised tissues taken from a normal animal, no increase in oxygen consumption results. In contrast to other tissues, the respiration of the adenohypophysis is not affected by thyroid insufficiency or excess. Similar observations have been made concerning the brain and spleen. The observation that heat production is less in the hypophysectomized individual than in the athyroid individual indicates that the thyroid hormones are not the only hormones involved in the regulation of heat production. Rather it is regulated by a balance of hormonal factors, which includes growth hormones and the adrenal cortical and medullary hormones, all of which are calorigenic.

It has been estimated that 300 μg of thyroxine daily is required to maintain man in a state of normal metabolic activity under ordinary circumstances. Disturbances of the heat-regulating mechanism are readily observed in the thyroidectomized individual. The normal individual responds to a cool environment by increasing the output of thyroid hormones, as judged by an increase in the basal metabolic rate (BMR), and it has been shown that the thyroxine requirement is inversely related to the environmental temperature. However, the twofold to threefold increase in heat production over the resting level which follows sudden exposure to a cool environmental temperature is not due solely to increased thyroidal activity, for shivering and epinephrine release account for much of the rise. It should be borne in mind that many mechanisms other than hormonal are involved in thermoregulation, and reference should be made to earlier sections (Chaps. 27 and 28) on energy metabolism and temperature regulation for a discussion of factors which participate in thermoregulation.

The manner in which the thyroid hormones act at the cellular level is not known. Many observations have been made on alterations in the activity and/or concentration of enzyme systems which play a vital role in the reactions concerned with the respiratory activity of cells, and on the release and utilization of energy which might be related to the action of the thyroid hormones. Most of the theories presented for the mode of action of the thyroid hormones have their basis at this level. Since an excess of thyroid hormone uncouples oxidative phosphorylation, cellular respiration and energy are less effectively controlled. These hormones also increase mitochondrial permeability and size. It might be recalled that the mitochondria are regarded as the principal "power plant" of the cell. The thyroid hormones also increase the rate of transfer of transfer-RNA-bound amino acids to microsomal protein, and this property may form the basis on which they function to stimulate metamorphosis, development, and anabolic processes. Because an enhanced turnover of nuclear and cytoplasmic RNA precedes the increase in protein anabolic activity and oxygen consumption following the acute administration of thyroxine, it may well act primarily via gene activation.

The effects observed following the administration of thyroid hormones or following thyroidectomy have a considerable latent period. When one considers heat production, for instance, after thyroidectomy, the BMR falls steadily until it reaches a minimum of about −35 to −50 percent in about 40 days. On the other hand, a change in the BMR is not detectable until about 36 hours after a large dose of thyroxine is given to a normal individual, and a maximal effect is not observed until five to seven days later.

Many effects observed in hyperthyroid and hypothyroid states are in all probability results of both direct calorigenic effects and reflex responses to them. Most of these effects change in proportion to the BMR and therefore are assumed to stem from the calorigenic effect of the hormones. Some changes observed in hyperthyroid states are: an increase in fasting nitrogen excretion, increases in the rates at which carbohydrate and fatty acids are absorbed from the gastrointestinal tract, and a diminished serum cholesterol which results from an increased catabolism of the substance. Cardiac output and plasma volume increase, and pulmonary ventilation goes up. Increases in glomerular filtration rate, tubular maxima for Diodrast and glucose, and urine volume occur. Many adaptations take place, where the increase in thyroid hormone output leads to an increase in oxygen consumption, whereby the organism provides adequate means of satisfying the increased demand for oxygen.

On the other hand, in hypothyroid states varying degrees of anemia and decreased bone marrow activity commonly occur. Obesity is observed, but only if the appetite and caloric intake are not decreased in proportion to the decline in BMR. Sensitivity to external stimuli decreases, and tendon reflex time increases. An accumulation of considerable amounts of water, salts, and nitrogenous material occurs in the extravascular spaces. Certain other changes, opposite to those noted in hyperthyroidism, are also observed (e.g., decreased cardiac output and renal function). Finally, thyroid hormones are essential for normal reproduction and lactation.

The thyroid hormones must also participate in some manner in the organization of cells, for when the thyroid is absent, certain cellular functions become disorganized. Two typical examples of this phenomenon are the epiphysial dysgenesis and the myxedema which are seen in the absence of the hormone. The epiphysial dysgenesis is manifested by spotty and irregular calcification of the developing skeletal epiphyses. The abnormal cartilaginous ossification pattern in the epiphyses is not seen, however, if ossification has occurred prior to the onset of thyroid deficiency. This abnormal calcification pattern is commonly found at the head of the long bones, and the administration of thyroid hormones to hypothyroid individuals causes normal patterns of calcification to follow. The myxedema of hypothyroidism is characterized by the accumulation of an abnormal protein (mucoprotein) in the interstitial spaces. It persists in the hypothyroid individual even when the metabolic rate is elevated above normal by agents such as dinitrophenol, and therefore is probably not a result of a diminished rate of cellular metabolism.

It was observed initially that metamorphosis does not occur in thyroidectomized amphibia, and that premature metamorphosis could be induced with thyroid hormone. The thyroid hormones have a profound influence on the rate of body growth and maturation. With normal thyroid hormone concentrations the tendency for cells to grow and mature is maximal, and the presence of normal thyroid function is essential for normal growth and maturation. The absence of thyroid hormone also causes a retardation in the rate of skeletal maturation, and prevents full body growth to adult dimensions.

Thyroid deficiency during human fetal life, or postnatal life prior to puberty, produces a marked diminution in the rate of growth and maturation of the individual. The earlier the onset of the deficiency, the more marked are the alterations from normal, particularly when one considers the nervous and skeletal systems.

Probably no tissue suffers more than does nervous tissue from lack of thyroid hormone during fetal life. The deficiency results in *cretinism.* In the cretin there is a decided loss of intellect. However, in adult hypothyroidism the mental capacity alteration is manifested as reduced mental alertness. Recent studies concerned with the thyroid hormone and central nervous system structure and function have shown that in the absence of fetal thyroid hormones there are a delayed appearance and decreased content of myelin in fiber tracts and a decrease in size and number of cortical neurons, but all changes may be reversed if thyroid hormone is given within the third week of postnatal life. It has also been observed that brain succinic dehydrogenase decreases in concentration following thyroidectomy of rats at birth. Such activity can be restored to normal with thyroid hormone if treatment is instituted within 10 days. All these observations might provide an explanation for the irreversible retardation of mental development in cretins. On the other hand, the presence of thyroid hormone in excess does not cause growth and development to be accelerated, and may even retard development. The mechanism of growth stimulation is unknown. However, the observation that thyroid hormone has a mitogenic action in thyroidectomized rats only in the presence of growth hormone might suggest a vital role for thyroid hormones in cell division and differentiation. It also strengthens the postulate that the combined effects of growth hormone and thyroid hormone are essential for normal development.

REFERENCES

Alexander, W. D., R. M. Harden, and J. Shimmins. Studies of the thyroid iodide "trap" in man. *Recent Progr. Hormone Res.* 25:423–446, 1969.

Blizzard, R. M. Inherited defects of thyroid synthesis and metabolism. *Metabolism* 9:232–247, 1960.

Deiss, W. P. Transport of thyroid hormones. *Fed. Proc.* 21:630–634, 1962.

Deiss, W. P., and R. L. Peake. The mechanism of thyroid hormone secretion. *Ann. Intern. Med.* 69:881–890, 1968.

Halasz, B., W. H. Florsheim, N. L. Corcoran, and R. A. Gorski. Thyrotrophic hormone secretion in rats after partial or total interruption of neural afferents to the medial basal hypothalamus. *Endocrinology* 80:1075–1082, 1967.

Harris, G. W. *A Summary of Some Recent Research on Brain-Thyroid Relationships.* (Ciba Foundation Study Group No. 18) Boston: Little, Brown, 1964. Pp. 3–16.

Tata, J. R. Biological Action of Thyroid Hormones at the Cellular and Molecular Levels. In G. Litwack and D. Kritchevsky (Eds.), *Actions of Hormones on Metabolic Processes.* New York: Wiley, 1964. Pp. 58–131.

Werner, S. C., and J. A. Nauman. The thyroid. *Ann. Rev. Physiol.* 30:213–244, 1968.

Wolff, J. Transport of iodide and other anions in the thyroid gland. *Physiol. Rev.* 44:45–90, 1964.

35

The Adrenal Cortex

Ward W. Moore

ANATOMY

The adrenal gland consists of two distinct organs, the adrenal cortex and the adrenal medulla. Embryologically, the cortical tissue is first to appear. It becomes evident at about the fifth week of intrauterine life in the human and arises from mesoderm in the urogenital zone. During the seventh week, the mass of presumptive cortical cells is invaded by cells migrating from neural crest material. These cells are surrounded by the mesodermal cells and form the adrenal medulla. The adrenal medulla is regulated directly through its nerve supply, which is part of the thoracolumbar division of the autonomic nervous system. The functional significance of this organ is discussed in Chapter 7. The direct regulation of the secretion of the adrenal cortex is mediated solely by humoral means.

The adrenal cortex consists of epithelioid cells arranged in continuous cords or sheets of cells which are separated by capillaries. Structurally, the cortex may be divided into three zones. The *zona glomerulosa,* the outer zone next to the fibrous capsule, has cells arranged in irregular masses. In the middle zone, the *zona fasciculata,* the cords or sheets of cells are straight and radially disposed. In the inner cortical zone, the *zona reticularis,* the cords form an irregular network. All the cortical cells show evidence of accumulation, storage, and secretion of lipid. The chief lipid constituent is cholesterol, which is a precursor of all the adrenal cortical hormones. The adrenal cortex is also rich in ascorbic acid, and the role of this substance in cortical hormone synthesis remains to be ascertained.

ROLE OF THE ADRENAL CORTEX

The adrenal cortex is essential to life, and part of its functional integrity is dependent upon adenohypophysial adrenocorticotropic hormone (ACTH). More than 40 steroids have been isolated from the adrenal cortex. Many of these compounds are biologically inactive and may represent precursors or metabolites of the biologically active steroids. The latter may possess glycogenic, electrolytic, progestational, androgenic, or estrogenic activity depending upon their molecular structure.

BIOCHEMISTRY OF ADRENAL STEROIDS

The hormones of the adrenal cortex are closely related structurally to cholesterol. Information regarding the synthesis of the steroids has come from studies using isolated perfused cow and human adrenal glands, the collection of adrenal venous blood from several species, and slices or homogenates of adrenal cortical tissue. Although the results of such experiments (in vitro) do not necessarily reflect synthetic processes (in vivo), these methods afford a very good basis for the understanding of adrenal cortical hormone biosynthesis. Isotopic methods, in particular, have led to an understanding of the intermediates involved in the condensation of acetate molecules to form cholesterol.

Nine active steroids have been characterized. These substances are cortisol (hydrocortisone, Cpd F), corticosterone (Cpd B), aldosterone, cortisone (Cpd E), desoxycorticosterone, 11-dehydrocorticosterone (Cpd A), 17-hydroxydesoxycorticosterone (Cpd S), Δ^4-androstene-3,17-dione, and Δ^4-androstene-11-β-ol,-3,17-dione (Fig. 35-1).

The reaction sequence of cortical steroid synthesis (Fig. 35-2) from cholesterol is initiated when cholesterol is cleaved to yield a 21 carbon compound with a ketone group at the 20 position to form Δ^5-pregnenolone. The latter, under the influence of 3-β-hydroxydehydrogenase, is converted into progesterone. Progesterone occupies a key position in adrenal cortical hormone biosynthesis, and in the human the greater part of the progesterone is subjected to three successive hydroxylating steps catalyzed by the enzymes 17-α-hydroxylase, 21-hydroxylase, and 11-β-hydroxylase to form 17-α-hydroxyprogesterone, 17-α-hydroxydesoxycorticosterone (Cpd S), and cortisol (hydrocortisone or Cpd F), respectively. The last is the predominant circulating adrenal steroid in the human (10 to 15 μg per 100 ml of plasma). Each of the three cortical zones possesses the 11- and 21-hydroxylases, but the 17-α-hydroxylase is present in only the zona fasciculata and the zona reticularis. That amount of progesterone which escapes hydroxylation at the 17-α position undergoes hydroxylation at the 21 position to form desoxycorticosterone. This is hydroxylated at the 11 carbon to form corticosterone. The circulating level of corticosterone in the human is about 1.0 to 1.5 μg per 100 ml of plasma. Recent evidence indicates that corticosterone is a direct precursor of aldosterone. This occurs as the result of a substitution of an aldehyde group at the 18 carbon of corticosterone under the influence of "18-aldolase," an enzyme localized only in the zona glomerulosa. The concentration of aldosterone in human peripheral venous plasma is about 0.1 μg per 100 ml.

Knowledge concerning the biosynthesis of androgens and estrogens by the normal adrenal cortex is incomplete, although the manner in which the ovary and testes synthesize these substances is fairly clear. In the adrenal, as in the ovary and testes, 17-α-hydroxyprogesterone is presumed to play a pivotal role in the formation of androgens. Δ^4-androstene-3,17-dione is formed from 17-α-hydroxyprogesterone as a result of removal of carbons 20 and 21 and the substitution of ketone group at carbon 17. It is subsequently acted upon by an 11-β-hydroxylase to yield Δ^5-androstene, 11-β-ol-3,17-dione. It is also probable that the adrenal may convert some Δ^5-pregnenolone to dehydroepiandrosterone, which in turn is transformed to Δ^4-androstene-3,17-dione.

With reference to the biosynthesis of estrogens, Δ^4-androstene-3,17-dione appears to be the pivotal point. Using ovarian tissue under the stimulus of gonadotropin, Δ^4-androstene-3,17-dione is transformed into estradiol. This is accomplished by aromatization of ring A and the removal of carbon 19. The assumption is made that adrenal cortical tissue can form estradiol in a similar manner. It is certain, however, that the adrenal cortex is capable of secreting a potent estrogen.

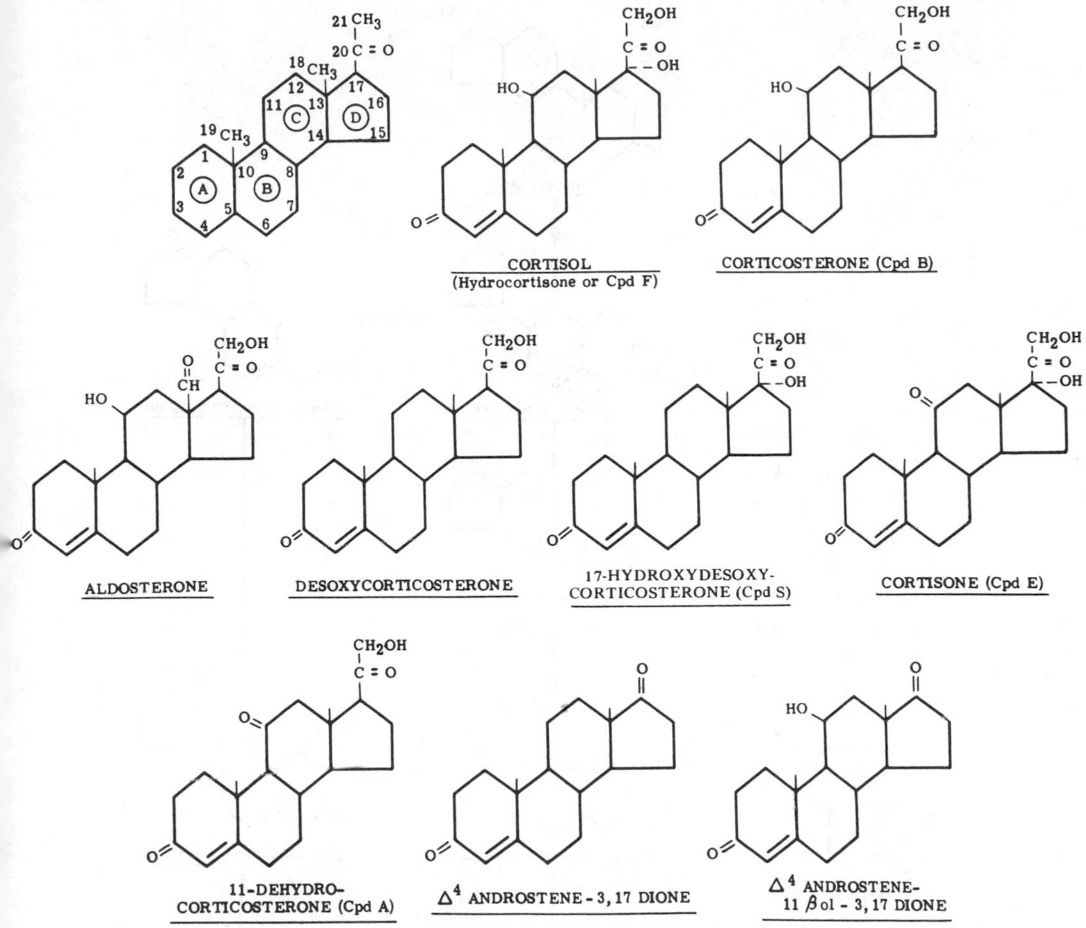

CORTISOL (Hydrocortisone or Cpd F)

CORTICOSTERONE (Cpd B)

ALDOSTERONE

DESOXYCORTICOSTERONE

17-HYDROXYDESOXY-CORTICOSTERONE (Cpd S)

CORTISONE (Cpd E)

11-DEHYDRO-CORTICOSTERONE (Cpd A)

Δ^4 ANDROSTENE-3,17 DIONE

Δ^4 ANDROSTENE-11 βol-3,17 DIONE

FIGURE 35-1. Structures of some biologically active adrenal cortical steroids and the numbering system of the steroid nucleus.

The steroids secreted by the adrenal cortex are excreted mainly as inactive forms, although a small amount of free active steroid may appear in the urine. Cortisol, for instance, is released by the adrenal cortex and is transported in the blood partly bound to an alpha globulin, transcortin, and partly in the free state. It is thought that only the free cortisol is metabolically active. The blood level of cortisol is constantly reduced as a result of degradation to an inactive form in the liver. In this organ and others, the adrenal hormones undergo further changes to form the inactive tetrahydro forms (Fig. 35-3). The active hormones are rendered inactive by saturation of ring A and substitution of hydroxyl groups at carbons 3 and/or 20. The metabolites are rendered water-soluble by conjugation with glucuronic acid in the liver prior to excretion by the kidney. The adrenal cortex of the normal human adult secretes about 20 mg of cortisol, 3 mg of corticosterone, and 0.15 mg of aldosterone per day.

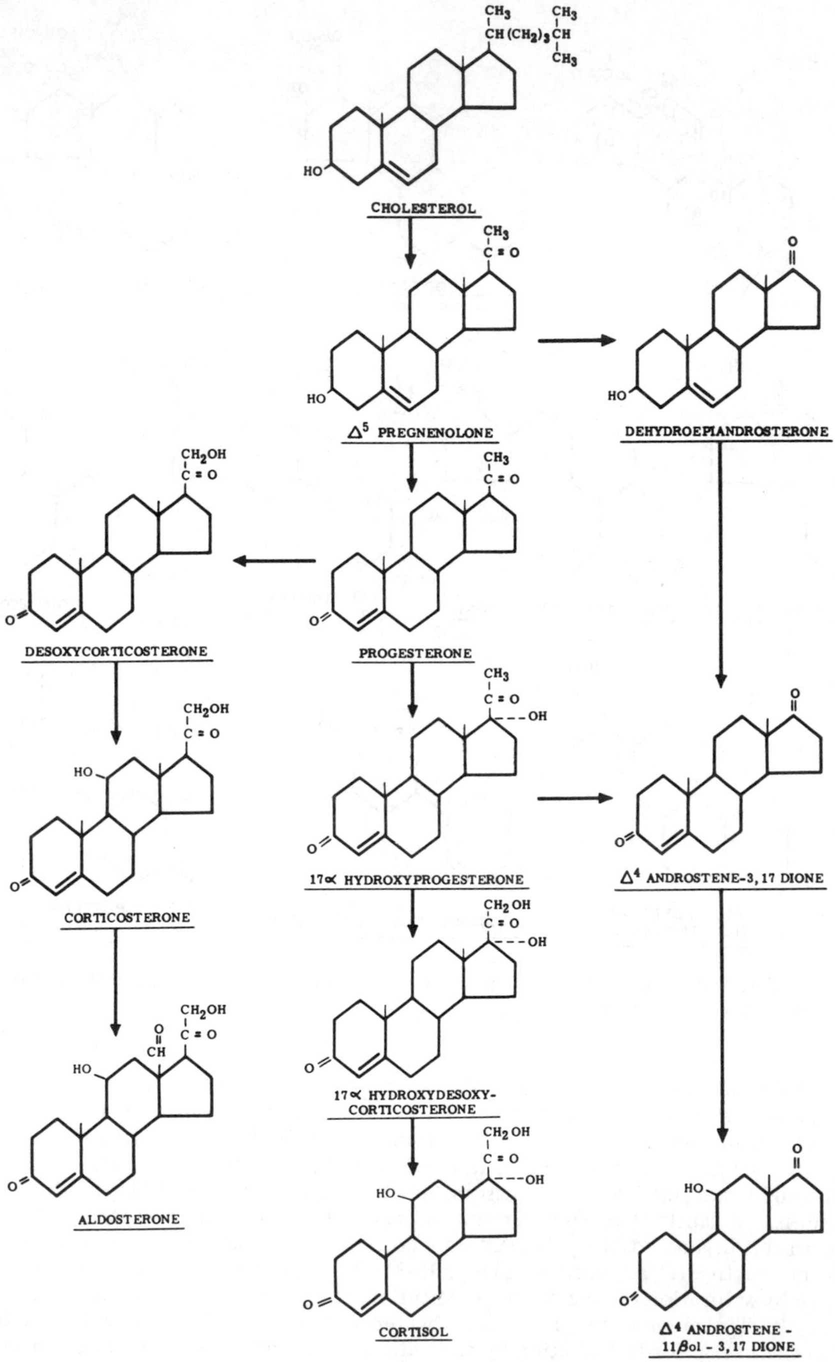

CHOLESTEROL

Δ^5 PREGNENOLONE

DEHYDROEPIANDROSTERONE

DESOXYCORTICOSTERONE

PROGESTERONE

CORTICOSTERONE

17∝ HYDROXYPROGESTERONE

Δ^4 ANDROSTENE-3, 17 DIONE

ALDOSTERONE

17∝ HYDROXYDESOXY-CORTICOSTERONE

CORTISOL

Δ^4 ANDROSTENE - 11βol - 3, 17 DIONE

FIGURE 35-2. Pathways of adrenal cortical hormone biosynthesis.

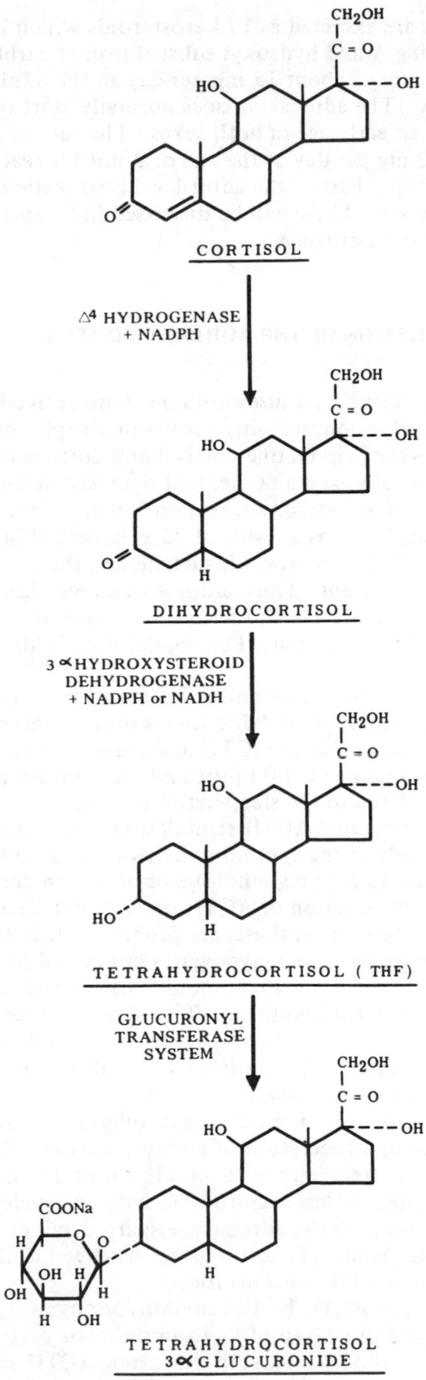

FIGURE 35-3. Pathway for the degradation and excretion of cortisol.

The adrenal androgens are excreted as 17-ketosteroids which have been rendered inactive by saturation of ring A and hydroxyl substitution at carbon 3. These are excreted as sulfates at the rate of about 15 mg per day in the adult male and 10 mg per day in the adult female. The adrenal cortices normally start to contribute to the urinary 17-ketosteroids at an early age in both sexes. The rate of excretion increases with age from about 1 to 2 mg per day at the age of 2 until it reaches the adult level at about 23 to 25 years of age. Part of the adult level is of testicular or ovarian origin, depending upon the sex. These will be discussed in Chapter 36, as will the secretion and excretion of the estrogens.

REGULATION OF THE SECRETION OF THE ADRENAL CORTEX

The removal of the anterior pituitary causes notable changes in adrenal cortical function and morphology. Hypophysectomy results in atrophy of the adrenal cortex, a decrease in the level of circulating cortisol and corticosterone to about 10 percent of that seen normally, and a 60 percent decrease in the rate of aldosterone secretion. The level of aldosterone secretion is maintained to such an extent that the animals do not usually die as a result of adrenocortical insufficiency. Although they show signs of adrenocortical insufficiency, the marked alterations in electrolyte metabolism are absent. Thus, aldosterone secretion rates can be maintained at levels which are compatible with life in the absence of ACTH, but ACTH is required for maximal aldosterone output. The regulation of aldosterone secretion is discussed later.

ACTH is a polypeptide containing 39 amino acids and has a molecular weight of about 4500. A radioimmunoassay is available for its plasma determination. There is a sharp diurnal variation in the plasma ACTH and plasma cortisol levels, and in humans they peak at about 0600 to 0700 hours and are minimal at 1600 to 1700 hours. These changes are related to the sleep-activity cycle.

Available evidence indicates that ACTH stimulates the secretion of cortisol by affecting enzymatic steps early in the synthetic pathway, possibly by accelerating the conversion of cholesterol to Δ^5-pregnenolone or of Δ^5-pregnenolone to progesterone; there appears to be little action of ACTH on the later steps. The early effects of ACTH on adrenal steroid synthesis are probably mediated through cyclic AMP. This activates phosphorylase, and glycogen is converted to glucose-1-phosphate and then to glucose-6-phosphate. Increased amounts of generated NADPH then serve as an energy source for steroid synthesis. The more long-term effects of ACTH on protein synthesis and growth of the gland appear to be dependent upon an increased synthesis of mRNA, increased total RNA content, an increased number of polysomes, and increased mitotic activity.

When ACTH is administered to a normal or hypophysectomized animal, there is a marked increase in the rate of secretion of cortisol, corticosterone, aldosterone, and the adrenal androgens. If treatment with ACTH continues, histological changes indicative of increased adrenal cortical secretory activity are evident. Other early effects of ACTH administration on the adrenal are sharp declines in cholesterol and ascorbic acid content of the gland. The latter effect has been utilized as a sensitive assay method for circulating ACTH concentration.

Changes in the secretion of ACTH by the adenohypophysis follow removal of the adrenal cortex or the administration of large amounts of cortisol. In the former case, an increase in ACTH secretion occurs; in the latter, ACTH secretion may be completely inhibited.

FUNCTIONS OF THE ADRENAL CORTEX

Cortisol, corticosterone, and aldosterone are the physiologically significant steroids secreted by the normal mammalian adrenal cortex. Cortisol exerts its effects on the metabolism of protein, carbohydrate, and fat, whereas aldosterone is concerned primarily with electrolyte metabolism. The effects of corticosterone are somewhat intermediate between those of cortisol and aldosterone. Those steroids which exert an androgenic effect alter the metabolism of protein in a manner opposite to that of cortisol.

The effects of the adrenal steroids are best illustrated by the changes which take place in the adrenalectomized animal. Following adrenalectomy, animals invariably die if no supportive means are instituted. The observation that adrenalectomized animals can be maintained indefinitely by a variety of adrenal steroids is taken as proof that the cortical portion of the gland, and not the medullary portion, is essential for life. Symptoms which appear in the adrenalectomized animal include loss of appetite, asthenia, gastrointestinal disturbances, hemoconcentration, reduced blood pressure and body temperature, and renal failure.

ELECTROLYTE METABOLISM

A primary abnormality resulting from adrenalectomy is the disturbance in electrolyte metabolism. This alteration depends in part upon the effect of renal handling of sodium, potassium, and water. In the absence of the steroids (aldosterone) a decreased tubular reabsorption of sodium, chloride, and water and an increased reabsorption of potassium occur. Therefore, adrenalectomy results in hyponatremia, hypochloremia, hyperkalemia, and extracellular dehydration. Each abnormality then contributes to hemoconcentration, acidosis, hypotension, decreased glomerular filtration rate, extrarenal uremia, and shock. The administration of aldosterone reverts these changes toward normal because it has an action on the distal and renal tubules. In the presence of excessive amounts of aldosterone there is an increase in renal reabsorption of Na^+, and along with it Cl^- and water; urinary loss of K^+ and H^+ also increases.

The effects of aldosterone are mediated through DNA-directed RNA synthesis and "permease" synthesis. The "permease" increases the permeability of the mucosal or tubular cell to sodium. A "sodium pump" then moves the sodium out of the cell on the serosal side into the extracellular fluid (ECF) and conserves sodium. The alterations induce an expansion of the ECF volume, a shift of sodium to the intracellular compartment, and a depletion of potassium in both the extracellular and intracellular compartments. An extracellular metabolic alkalosis which is characterized by increased serum pH and carbon dioxide combining power, mild hypernatremia, and hypochloremia results (Fig. 35-4).

In addition to the renal influences, a deficiency or excess of adrenal steroidal activity exerts a direct effect upon the passage of sodium, potassium, and hydrogen ions across cell membranes. Excessive amounts of aldosterone may lead to an expansion of the ECF volume which is greater than the amount of exogenous water retained. Thus, intracellular fluid (ICF) dehydration must occur. Adrenalectomy or hypoaldosteronism is accompanied by changes which lead to hyponatremia, hypochloremia, hyperkalemia, extracellular dehydration, and circulatory collapse. It should be pointed out that during continued administration of aldosterone to normal subjects an "escape" from the intense sodium-retaining effect is observed within 5 to 10 days even though the steroid administration is continued. Aldosterone also decreases the sodium concentration and increases the potassium concentration in sweat and the secretions of the salivary and intestinal glands.

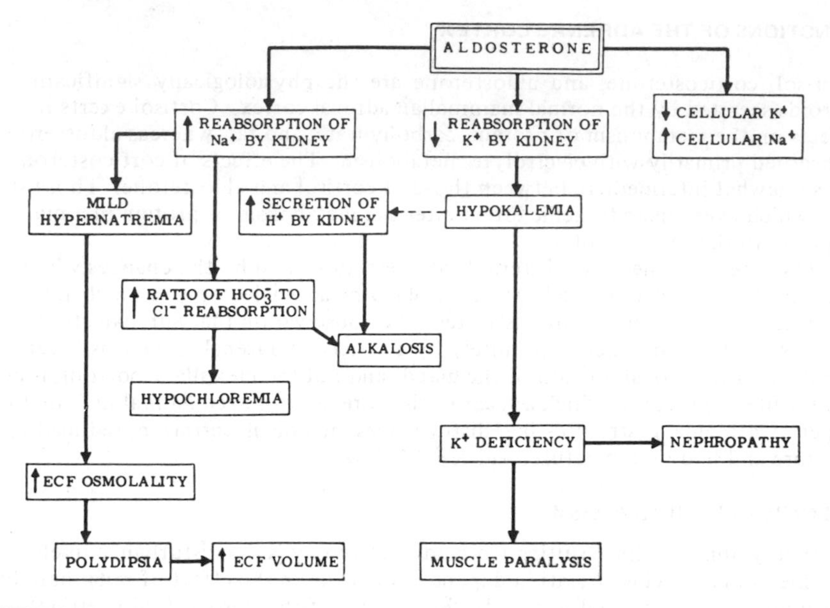

FIGURE 35-4. Actions of aldosterone on electrolyte balance.

Aldosterone is approximately 25 times more potent than desoxycorticosterone in its sodium-retaining effect, about 800 times more potent than cortisol, and 200 times more potent than corticosterone. The synthetic steroid 9-α-fluorocortisol is an extremely potent salt retainer (one-fourth the potency of aldosterone).

METABOLISM OF CARBOHYDRATES, PROTEINS, AND FATS

Other defects observed in adrenalectomized animals are those concerned with protein, carbohydrate, and fat metabolism. In the adrenalectomized animal, lacking cortisol or corticosterone, there is a tendency for the liver to lose glycogen. The ability of the animal to mobilize the precursors of glycogen from the peripheral tissues is reduced, the rate at which glucose is utilized by the peripheral tissue is elevated, and the deposition of body fat is reduced. Therefore the adrenalectomized animal is prone to develop hypoglycemia and is extremely sensitive to the action of insulin. Such an animal also exhibits an elevated blood nonprotein nitrogen (NPN) which results not from increased protein breakdown but from renal decompensation that develops as a result of circulatory failure.

Cortisol, on the other hand, is diabetogenic. This steroid can reverse all the changes in carbohydrate and fat metabolism observed in the adrenalectomized animal. Following the administration of cortisol to a fasting subject, there occurs within hours a marked increase in liver glycogen. This is accompanied by an increase in the blood sugar and a marked increase in the excretion of nonprotein nitrogen, but the muscle glycogen stores are increased slightly. The increase in total body carbohydrate in the "steroid diabetic," therefore, must be the result of decreased glucose oxidation and/or an accelerated liver glycogen formation from tissue protein (gluconeogenesis). Various studies have been interpreted as indicating a primary action of cortisol on protein catabolism in peripheral tissue, and on the ability of the liver to concentrate amino acids. It has been observed that cortisol has a protein catabolic effect with respect to muscle and lymphatic tissue, and it leads to an overall negative nitrogen balance. However, cortisol does exert a strong protein anabolic effect on hepatic tissue. This differential effect on the liver may explain the diabetogenic effect, when one takes into account the myocytolytic and lymphocytolytic effect plus the increased hepatic enzyme protein synthesis. One can induce hyperglycemia and glucosuria in intact animals fed a high-carbohydrate diet by the prolonged administration of cortisol-like compounds. Adrenalectomy will partially alleviate the symptoms of diabetes mellitus.

Cortisol-like compounds also mobilize fat from depots to the liver. Together with increased mobilization of fat, there is a redistribution of fat depots which produces "moon face" (rounding of the cheeks), "buffalo hump" (growth of supraclavicular and upper dorsal fat pads), increased axial fat, and a loss of fat on the extremities. As in the diabetic state, there is a decreased formation of fat from carbohydrate. In general, the cortisol-like steroids tend to act antagonistically to insulin with respect to its action on carbohydrate and fat metabolism. The androgenic adrenal steroids have a protein anabolic effect, which is opposite to that produced by cortisol-like compounds.

Cortisol is the most active naturally occurring steroid in the human with respect to its effects on liver glycogen formation. On a molar basis, cortisone is about 65 percent as effective, corticosterone about 35 percent as effective, whereas aldosterone and desoxycorticosterone show less than 1 percent of the activity of cortisol. The synthetic steroid 9-α-fluorocortisol possesses about 10 times the activity of cortisol. The effects of cortisol on protein, carbohydrate, and fat metabolism are illustrated in Figure 35-5. The mechanisms whereby cortisol-like compounds exert their effects are poorly understood.

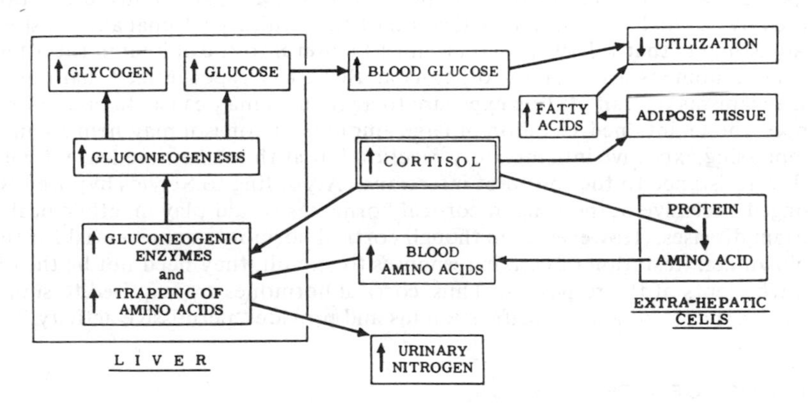

FIGURE 35-5. Effects of cortisol on carbohydrate, protein, and fat metabolism.

MISCELLANEOUS FUNCTIONS

The cortisol-like adrenal steroids cause a marked involution of thymus and lymph nodes and a decrease in the circulating lymphocytes and eosinophils. Such steroids also profoundly influence the inflammatory responses of the tissues. Local inflammatory reactions to irritating substances are greatly reduced or delayed, hypersensitivity reactions to antigens are suppressed, and healing of wounds is delayed when active steroids are given in excess over a period of hours or days. Therefore, the actions of excess steroids can be deleterious to the combating of infections and the healing of wounds, although alterations of these reactions may be of benefit in certain acute hypersensitivity states.

The adrenal steroids may also have many diverse effects, such as muscular weakness as a result of muscle wasting, increased rate of secretion of gastric HCl and pepsin leading to the development of gastric and duodenal ulcers, hypertension and increased capillary resistance and fragility, increased excitability of brain tissue with euphoria and restlessness, and stimulation of bone marrow. In physiological levels the steroids will prevent the deposition of melanin in the skin of the adrenal-insufficient individual by preventing the release of ACTH, and they are required for normal thyroid, gonadal, and pituitary function.

ROLE OF ADRENAL CORTEX IN STRESS

The adrenalectomized animal or adrenal-insufficient individual has little ability to tolerate changes in the internal and external environment, such as cold, heat, infections, trauma, or prolonged exercise. The resistance to environmental change in such subjects is very low, and as a result of the stimulus they will die, whereas the normal individual can tolerate a much greater stimulus. Selye (1956) has shown that when an animal is subjected to a severe but sublethal injury it reacts in a stereotyped manner. This reaction is biphasic and has been termed the *alarm reaction.* The initial phase is the *phase of shock,* characterized by changes which are similar to those seen in acute adrenal insufficiency. A *countershock phase* follows during which changes similar to those following cortisol administration occur. If the same stimulus is given to a hypophysectomized or adrenalectomized animal, the shock phase is extremely severe and the animals show no evidence of the countershock phase prior to death. However, survival can be attained if adrenal cortical extracts or anterior pituitary extracts are given to the hypophysectomized animals. Thus, protection is afforded by adrenal steroids, and activation of the pituitary-adrenal axis must ensue. The adjustments that follow a stimulus must be useful to the animal in the attempt to maintain homeostasis. On the other hand, Selye has hypothesized that the adaptive mechanisms operant during exposure to a stimulus may cause disease. For example, the maintained secretion of large amounts of cortisol may help an animal in suppressing excessive inflammatory reaction but at the same time leave it with a very low resistance to the spread of infections. According to Selye's hypothesis, the prolonged excessive outpouring of cortical hormones would play an etiological role in certain diseases. However, even though cortical hormones are essential for the full-blown manifestation of certain responses to stimuli, they need not be the direct causative agents of the responses. Thus, cortical hormones are required to support a normal response to a nonspecific stimulus and provide "permissive activity."

REGULATION OF ACTH SECRETION

An intricate homeostatic mechanism regulates the normal secretion of cortisol by the adrenal cortex. A lowering of the level of circulating free cortisol presumably influences some metabolic process in the hypothalamus. This leads to a release of a neurohumor by the hypothalamus into the hypophysial portal system and stimulates the secretion of ACTH from the adenohypophysis. The adrenal cortex synthesizes cortisol at an increased rate in response to the elevated ACTH secretion. Adrenal atrophy can be induced by the administration of cortisol and related compounds, because these inhibit ACTH secretion. Similarly, following the removal of the adrenals, the circulating level of ACTH increases markedly. The foregoing probably depicts the steady state relationship between the adenohypophysial ACTH and the secretion of cortisol by the adrenal cortex. It does not, however, explain the increase in ACTH secretion and the subsequent adrenal cortical responses which follow a variety of stimuli.

The discussion of physiological triggers which initiate ACTH secretion in the stimulated animal revolves around three theories, based on (1) the circulating level of epinephrine, (2) the circulating level of cortisol, and (3) the influence of afferent nerve impulses impinging on the diencephalon.

Epinephrine from the adrenal medulla was postulated by Long to be the agent responsible for activating the release of adenohypophysial ACTH. This was based upon the observation that nonspecific stimuli cause the secretion of large amounts of epinephrine and ACTH. Epinephrine may induce ACTH release in the intact animal. Pituitary transplants (to the anterior chamber of the eye) in hypophysectomized rats respond to the local application of epinephrine by increasing ACTH output, as judged by the eosinopenic response. Briefly, the pertinent observations regarding the validity of this theory are as follows: (1) Increased ACTH release follows most stresses and is normal after removal of the adrenal medulla, sympathectomy, or total adrenalectomy. (2) Adrenergic blocking agents prevent ACTH release following epinephrine administration, but not that resulting from other stressful agents. (3) The intravenous administration of histamine causes ACTH release within 10 seconds, whereas intravenous epinephrine is ineffective and subcutaneous epinephrine requires several minutes to release ACTH. (4) Certain hypothalamic lesions prevent the increase in cortisol secretion in response to stress and do not affect epinephrine secretion, and vice versa. (5) Stimulation of various parts of the hypothalamus may cause increased epinephrine secretion and no increase in cortisol secretion, and vice versa. Therefore, it appears that neither endogenous nor exogenous epinephrine is necessary for the increase in ACTH release following stress, but that the administration of large amounts of epinephrine serves as an effective stimulus and activates ACTH release.

The second theory involves the level of circulating cortisol. Administration of cortisol results in a decreased ACTH secretion and adrenal atrophy. Pretreatment of animals with cortisol prior to stress depresses the secretion of ACTH, and the adrenalectomized animal secretes large amounts of ACTH. It is known that a reciprocal relationship holds between the adrenal cortex and the adenohypophysis. These basic facts were used by Sayers in support of the view that ACTH secretion is directly regulated by the blood level of adrenal hormones. It was suggested that the peripheral tissues utilize adrenal cortical hormones more rapidly under conditions of stress. Thus, the inhibition placed upon the adenohypophysis by cortisol is decreased, and an increase in ACTH secretion results.

The level of circulating steroids must play some role in regulating ACTH secretion under resting conditions, but the following observations are not consistent with the view that the circulating level of cortisol has a regulating function of ACTH secretion in stress. The transplanted adenohypophysis will not maintain the adrenal cortex in a normal state, because the adrenals of such animals approach a functional level which is observed following hypophysectomy. Furthermore, no acute drop in blood cortisol has been demonstrated following stimulation in the intact animal, and actually the rate of removal of cortisol is decreased. The elevated blood level of ACTH of the adrenalectomized animal is further elevated following stressful stimuli. Thus, it seems improbable that a decreased level of cortisol following stress causes an increase in ACTH secretion. However, it is likely that this mechanism acts to set a baseline of secretion, with other factors adjusting adenohypophysial activity following stressful stimuli.

Finally, stimulation of adenohypophysial release of ACTH following stress appears to be the result of afferent impulses or humoral agents stimulating the hypothalamus and producing the release of a neurohumor into the hypophysial-portal circulation. It has been shown many times that if lesions are placed in the hypothalamus (posterior) or impinge upon the median eminence, the common responses of the pituitary-adrenal system to nonspecific stress are prevented. In such preparations the adrenal cortical tissue does not undergo atrophy but maintains a basal level of hormone output. It is not able to increase its hormone output because its source of stimulus, ACTH, is not increased. The adenohypophysis is also capable of secreting ACTH but does not increase its secretion rate because its source of stimulation in the hypothalamus has been removed. Likewise, an increase in ACTH output can be induced by electrical stimulation of appropriate areas in the CNS, through chronically implanted electrodes. Furthermore, various drugs, such as a combination of barbiturate and morphine, block hypothalamic connections and inhibit the pituitary response to stressful stimuli. Suitable extracts of hypothalamic tissue activate the release of ACTH in the presence of hypothalamic lesions and/or pharmacological blockade. Similar extracts also have the ability in vitro to stimulate the release of ACTH from adenohypophysial cells. A like substance has been isolated from hypophysial portal blood. This substance has been called corticotropin-releasing factor (CRF) and has been characterized as being similar to, but not identical with, vasopressin. Therefore, there is every indication that the increased secretion of ACTH following nonspecific stimuli is mediated through the hypothalamus. The latter releases a specific neurohumor into the portal system which acts on the adenohypophysis to stimulate the release of ACTH.

The problem arises as to how the CNS and its connections may influence the hypothalamo-pituitary-adrenal system. Psychic factors such as anxiety readily elevate the blood cortisol level. Various sedatives, tranquilizers, and anesthetics on initial administration increase ACTH output, but on prolonged administration may inhibit ACTH release. Each must act via the CNS. Stimulation of the amygdala results in elevated cortisol content of blood, whereas stimulation of the hippocampus produces the opposite effect. Lesions in the posterior midbrain cause an increased steroid output, whereas lesions in the rostral midbrain and posterior diencephalon depress adrenal steroid output. Similarly, the placement of minute amounts of cortisol in the latter two areas inhibits the secretion of ACTH, but placement of the steroid directly into the adenohypophysis does not inhibit ACTH secretion. Peripheral nerve stimulation induces an increased secretion of ACTH, but stimulation of the hind limbs following cervical transection produces no change in ACTH output. It seems probable that a tonic bombardment of the hypothalamus by stimuli from peripheral receptors helps maintain "basal" ACTH secretion, and this is accomplished via the release of CRF. A schematized representation of the regulation of ACTH output is shown in Figure 35-6.

CONTROL OF ALDOSTERONE SECRETION

That the zona glomerulosa does not atrophy following hypophysectomy (short term) was first observed by Houssay. Greep and Deane proposed that this zone secretes an electrolyte-active substance relatively independent of the adenohypophysis. Later, as had been predicted by Greep and Deane on the basis of their morphological work, Ayres demonstrated that aldosterone is preferentially synthesized by the zona glomerulosa.

The rate of secretion of aldosterone is not completely independent of ACTH, because following hypophysectomy it is reduced to one-half the normal rate. Also, varying degrees of atrophy of the zona glomerulosa follow hypophysectomy, and the degree of atrophy is said to be directly related to the time interval between hypophysectomy and autopsy.

The questions concerning the manner in which the secretion of aldosterone is controlled are not readily answered. The circumstances which alter aldosterone

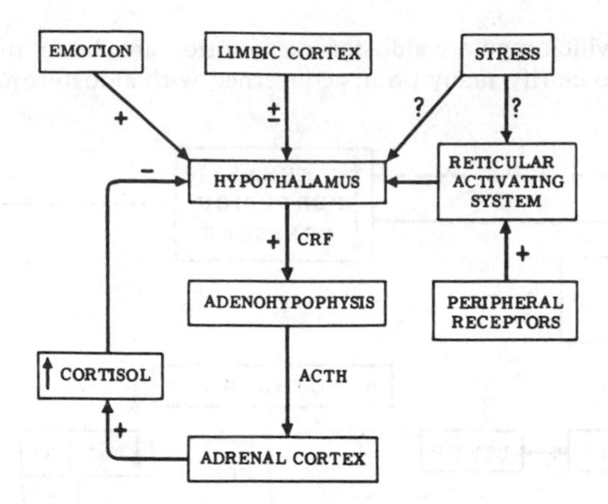

FIGURE 35-6. Regulation of ACTH secretion. + stimulation; – inhibition.

output are many, but the alterations in sodium balance and in the volume of the body fluids stand out as especially important. Increased output of aldosterone, without a simultaneous rise in cortisol output, has been shown to follow (1) restriction of dietary sodium, (2) high potassium intake, (3) water depletion, (4) thoracic postcaval constriction, and (5) aortic constriction. Hemorrhage, anxiety, surgical stress, and trauma also stimulate aldosterone secretion, but there is a concurrent elevation of cortisol secretion. This would indicate that the hypothalamo-hypophysial activation of ACTH release has occurred.

Davis and associates (1962) have presented evidence showing that the kidney secretes an aldosterone-stimulating hormone (ASH). Hyperaldosteronism resulting from chronic thoracic caval constriction, during chronic sodium depletion, acute blood loss, or acute aortic constriction is accompanied by a decrease in renal blood flow, an increase in renin content of the kidney, and hypergranulation (with occasional hyperplasia) of the juxtaglomerular cells of the kidney. Nephrectomy in the hypophysectomized animal consistently causes a marked fall in aldosterone output (0.033 to 0.006 μg per minute). The liver, adenohypophysis, brain, and pineal gland have been eliminated in these studies as sources of aldosterone-stimulating hormone. Present evidence would indicate that a decrease in renal perfusion pressure or a decrease in filtered sodium load acts upon the renal juxtaglomerular complex, and an increase in renin production follows. The renin then acts upon the circulating angiotensinogen to form the decapeptide angiotensin I. The latter, in the presence of chloride and a converting enzyme, is converted to the octapeptide angiotensin II. Angiotensin II acts directly on the zona glomerulosa to increase aldosterone secretion. The aldosterone then acts on the nephron and increases sodium reabsorption and, along with it, water reabsorption. The ECF is expanded, renal perfusion pressure is elevated, and the stimulus for renin production is removed. Angiotensin II can stimulate aldosterone secretion in the hypophysectomized, nephrectomized animal without increasing cortisol secretion, and it stimulates the in vitro conversion of cholesterol to aldosterone.

The circulating Na^+ and K^+ exert an effect upon aldosterone secretion. High K^+ and low Na^+ act directly on the zona glomerulosa to increase aldosterone secretion, because in the absence of the adenohypophysis and/or kidneys, aldosterone secretion remains high in either circumstance. Also, using the isolated perfused adrenal gland, lowering the Na^+ or increasing the K^+ concentration of the perfusate provokes a rise in aldosterone secretion.

The factors which regulate aldosterone secretion are shown in Figure 35-7. Much work is needed to clarify many points concerned with aldosterone secretion.

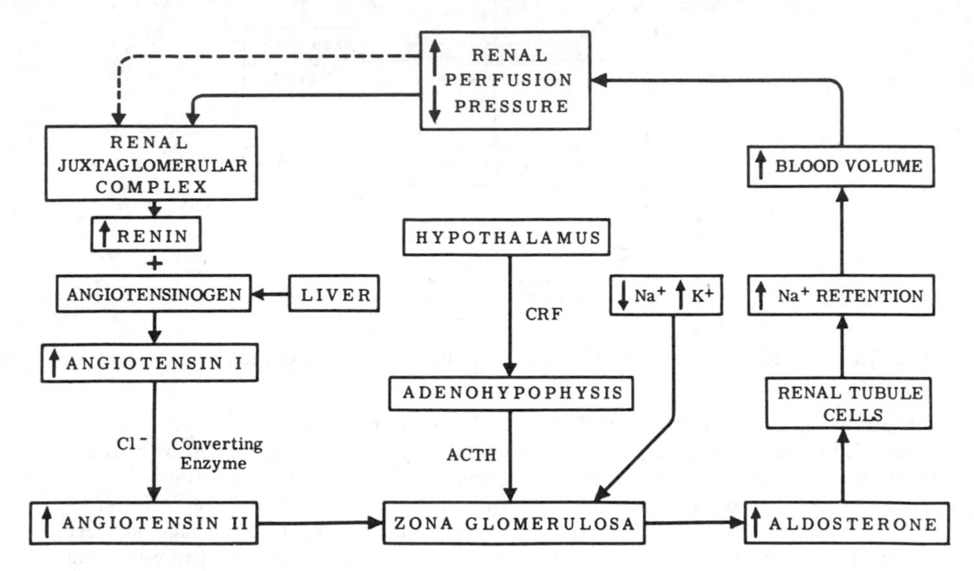

FIGURE 35-7. Regulation of aldosterone secretion. ——→ stimulation; - - - - -→ inhibition.

ADRENOCORTICAL DISEASES

Observation of the scheme of biosynthetic pathways of adrenal cortical hormone leads to the conclusion that many types of functional disorders may be produced as a result either of enzyme deficiency or excess, or of faulty pituitary function. The most generalized adrenal disorders are most evident in conditions in which there is either a total deficiency of ACTH or a marked excess of ACTH secretion. No disorder has been reported wherein the pituitary is deficient only in ACTH. Cushing's disease, a hyperadrenal state secondary to increased ACTH secretion, and Cushing's syndrome, a primary hyperadrenal state, are both characterized by an increased secretion of all the adrenocortical hormones. On the other hand, Addison's disease, a primary adrenocortical deficiency disease, is characterized by changes similar to those seen in the adrenalectomized animal. Exaggerated production of one or more of the hormones may be observed in hyperplastic or neoplastic changes in the adrenal cortex. Salt-retaining and salt-losing syndromes result from hyperaldosteronism and hypoaldosteronism respectively. Virilization may occur with or without salt loss, or with or without hypertension. A feminizing syndrome of adrenal origin has been described.

REFERENCES

Branson, E. D. Adrenal cortex. *Ann. Rev. Physiol.* 30:171–212, 1968.
Davis, J. O. The control of aldosterone secretion. *Physiologist* 5:65–86, 1962.
Denton, D. A. Evolutionary aspects of the emergence of aldosterone secretion and salt appetite. *Physiol. Rev.* 45:245–295, 1965.

Eberlein, W. R., and A. M. Bongiovanni. Pathophysiology of congenital adrenal hyperplasia. *Metabolism* 9:326–340, 1960.
Edelman, I. S., and G. M. Fimognari. On the biochemical mechanism of action of aldosterone. *Recent Progr. Hormone Res.* 24:1–44, 1968.
Eisenstein, A. B. (Ed.). *The Adrenal Cortex.* Boston: Little, Brown, 1967.
Feigelson, P., and M. Feigelson. Studies on the Mechanism of Cortisone Action. In G. Litwack and D. Kritchevsky (Eds.), *Actions of Hormones on Molecular Processes.* New York: Wiley, 1964. Pp. 218–233.
Ganong, W. F. The Central Nervous System and the Synthesis and Release of Adrenocorticotropic Hormone. In A. V. Nalbandov (Ed.), *Advances in Neuroendocrinology.* Urbana: University of Illinois Press, 1963. Pp. 92–149.
Renold, A. E., and J. Ashmore. Metabolic Effects of Adrenal Corticosteroids. In R. A. Williams (Ed.), *Diabetes.* Baltimore: Williams & Wilkins, 1960. Chap. 16.
Selye, H. *The Stress of Life.* New York: McGraw-Hill, 1956.
Urquhart, J. Blood-borne signals: The measuring and modelling of humoral communication and control. *Physiologist* 13:7–41, 1970.
Vander, A. J. Control of renin release. *Physiol. Rev.* 47:359–382, 1967.

36

Endocrinology of Reproduction

Ward W. Moore

GENETIC BASIS OF SEX

The sex of an individual is the outcome of two distinct processes: sex determination and sex differentiation. Sex determination is regulated by genetic phenomena which are fixed at the time of fertilization and which determine the genotype of an individual. Sex differentiation refers to the course of development of the gonads, genital ducts, external genitalia, and secondary sex characteristics; it results from two developmental processes: the direction of differentiation of the gonads toward ovary or testes, and the development of the primordial duct systems toward femaleness or maleness. These changes define the somatic sex (phenotypic sex) of the individual. It is generally held that the two processes are predetermined at the time of fertilization by the particular chromosomal pattern of the zygote. Forty-six chromosomes are present in all somatic and immature germ cells of man, and two of them are partly responsible for the determination of sex. These are designated X and Y chromosomes. The male cells possess an XY and the female an XX pair of sex chromosomes.

The first specific process of division (meiosis) which leads to the formation of the mature male germ cell involves a division of chromosomes, half of each pair to one daughter cell and half to another. Thus, the two daughter cells bear different sex chromosomes, one an X and the other a Y. In the female, the corresponding division produces two cells, each of which carries an X chromosome. When the ovum of the female and spermatozoon of the male unite, the chances are equal for the formation of an XX zygote (female genotype) or an XY zygote (male genotype). The genotypic or true sex of the zygote is regulated by the components contributed by the male germ cell and is determined at conception. However, the phenotypic or somatic sex may be the outcome of a balance between the influence of the sex hormones and that of an undefined number of genes carried on autosomal chromosomes.

It is now possible to determine the true sex of individuals, varying in age from very young embryos to very old adults, by the use of the so-called *sex chromatin.* This mass of chromatin is present in all somatic nuclei of most mammals, including man. The sex chromatin is usually found lying next to the inner surface of the nuclear membrane, is Feulgen-positive, is about 1 μ in diameter, and is found only in genetic females. It is thought to result from the fusing of heteropyknotic portions of the two X chromosomes. The determination, by this technique, of the true sex of an individual has widespread use in clinical medicine as an aid in diagnosis of a variety of conditions.

ROLE OF HORMONES IN THE DEVELOPMENT OF THE GENITAL SYSTEM

The reproductive system passes through a period of early embryonal development during which it is impossible, either grossly or microscopically, to tell the sexes apart. This period is referred to as the indifferent stage. Nalbandov (1964) has stated, "In this stage all *anlagen* for subsequent differentiation into complete male and female systems are present in rudimentary form; the blueprint, as it were, is finished. All the materials for the later elaboration of the fittings and furnishing of structures are present, but no attempt is made to arrange the internal furnishings permanently until final orders are received for the emphasis of either male or female aspects of the different structures."

A bilateral longitudinal ridge, the genital ridge, appears medial to the mesonephros during the fifth week of embryonic life in the human. The surface of this ridge is covered by germinal epithelium and contains the primordial germ cells which have migrated by ameboid movement from yolk-sac endoderm. Specific histological changes occur in this structure (the *indifferent gonad)* during the seventh week, and they indicate that the gonad is committed to either the male or female role. If the gonad is to develop into a testis, the cells of the germinal epithelium become organized within the medulla of the gonad, form the anlagen of the seminiferous tubules, and subsequently join the cords of the mesonephros to form a continuous network of tubules. Mesenchymal cells in the medulla develop into the cells of Leydig, which become apparent at the end of the eighth week. If the gonad is to become an ovary, no changes are apparent until the tenth week. At this time, the cortex of the gonad accumulates nests of cells, and these are differentiated into ovarian follicles, each containing an ovum.

If the ducts are to differentiate in accordance with the sex of the gonads, either the mesonephric system or the müllerian ducts must degenerate. The duct system and the gonads pass through an undifferentiated stage, because the fetus possesses the basic duct systems for both sexes. Only after continued development does one system gain ascendancy and become definite, whereas the other regresses and becomes vestigial. The müllerian ducts begin to degenerate in the male at about the twelfth week, and the mesonephric system develops. In the female the mesonephric system begins to disappear late in the thirteenth week and the müllerian system becomes dominant. The origins of various parts of the genital system in both sexes are given in Table 36-1.

Removal of the gonads from the early embryo of either sex prior to the development of the genital ducts leads to the development of the female-type internal and external genitalia, regardless of the genetic sex of the fetus. This observation is interpreted as meaning that the testes secrete a substance which leads to the development of the male system, and that in the absence of this substance the genital duct system develops in the direction of the female.

There is little doubt that certain aspects of the development of the reproductive system are effected or facilitated by the sex hormones, the androgens and the estrogens. Both nature and experimentation have shown that while the developing embryo or fetus is in an undifferentiated or differentiating state, the phenotypic sex of the individual may be altered or reversed. The classic papers of Lillie describe a naturally occurring case of sex reversal, the freemartin of cattle. The freemartin is a genetic female and a twin of a genetic male, and the twins possess a common fetal circulation. The genetic female is sterile, has inhibited ovaries and female

TABLE 36-1. Homologies of the Male and Female Reproductive Systems in the Human

Male	Indifferent Stage	Female
Testis	Indifferent gonad	Ovary
Rete testis		Rete ovarii*
Vas efferens Paradidymis* Vas aberrans	Mesonephric tubules	Epoophoron* Paroophoron*
Epididymis Vas deferens Ejaculatory duct Seminal vesicle	Mesonephric duct	Gartner's duct*
Prostatic utricle*	Müllerian duct	Hydatid* Oviduct Uterus Vagina
Urethra	Urogenital sinus	Urethra Vagina Vestibule Vestibular glands
Glans penis	Genital tubercle	Glans clitoris Corpus clitoris
Raphe penis Raphe scrotum	Urethral folds	Labia minora
Scrotum	Labioscrotal swellings	Labia majora

*Rudimentary.

genital ducts, while the accessory sex ducts and secondary sex characteristics approach full maleness. As noted before, the testes differentiate prior to the ovaries and are capable of secreting androgen at an early stage. Thus, androgen acts on the undifferentiated duct system of both twins and causes differentiation of the indifferent duct system in the male direction and regression of the primordial female system.

Varying degrees of masculinization are seen in the human female with congenital adrenal hyperplasia. This occurs as a result of a genetic defect in which simple mendelian recessive genes combine and lead to a deficiency in an enzyme system concerned with adrenal steroid synthesis. The result is excess production of adrenal androgen. The earlier in embryonic life the defect is manifested, the more marked are the aberrant changes observed in the newborn female genital system.

Under normal circumstances, the effect of the maternal hormones is insufficient to cause profound changes in the fetus. The maternal estrogens and progestin cause cervical enlargement, hypertrophy of the vaginal epithelium, and growth of the mammary gland of the fetus. However, certain synthetic estrogens and progestins, when administered early in pregnancy, may produce marked structural alterations (masculinization) in the female fetus. The placenta is permeable to most hormones, but at birth the infant normally shows little evidence of excessive stimulation by maternal hormones.

It should be noted that none of the fetal endocrine glands studied has been found to be indispensable for fetal survival; removal of the hypophysis, thyroid, parathyroid, gonads, or adrenals has been accomplished in the fetuses of laboratory animals, and these fetuses survive. Homeostasis of the fetus is generally maintained by way of maternal homeostatic mechanisms, but fetal hormones do play an indispensable role in the development of a normal newborn.

ENDOCRINOLOGY OF THE MALE

The testes of the male have a dual function, one of gamete formation and the other of hormone production. The gametogenesis is dependent to a considerable extent on androgen production by the testes, and both are dependent upon the adenohypophysis (AP). In turn, the AP is dependent upon normal hypothalamic function and intact hypophysial portal vessels. Each is indispensable if the male is to reach full reproductive capacity, but both are dispensable to life.

The essential histological picture of the testes is that of seminiferous tubules separated by interstitial connective tissue. This pattern is observed from prenatal stages throughout life with decided variations according to age. At birth, the seminiferous tubules have not yet formed lumina, are small, and contain only spermatogonia. Leydig cells are present at birth, but then regress, and are not again identifiable until the approach of puberty. At this time maturation changes such as thickening of the tunica propria, initiation of spermatogenesis, proliferation of the Leydig cells, and secretion of androgens occur. The changes are dependent on hypophysial gonadotropin production. Thus, from the third month of life, by which time maternal hormone influences have disappeared, until the eighth to the eleventh year, sexual development is largely in abeyance, and individuals of both sexes are in a state of physiological hypogonadism.

During the infantile period, the gonads are potentially able to function in an adult fashion, for they respond to gonadotropins. The secondary sex organs respond either to androgens or to estrogens. Therefore, gross insensitivity of the target organs cannot be responsible for the lack of sexual maturation in childhood. The gonads of immature animals when transplanted to mature animals function as mature gonads; but the gonads of mature animals, when transplanted to immature animals, become quiescent and atrophy.

The newborn adenohypophysis transplanted into the sella turcica of a hypophysectomized mature animal stimulates and repairs gonadal function more rapidly than the pituitary of a mature animal when transplanted into the sella of an infant animal. Thus, the fetal hypophysis contains gonadotropins, and the gland of the prepubertal individual is potentially capable of stimulating the gonads. These observations form the basis of the hypothesis that the secretion of gonadotropins at puberty is dependent upon a neural or neurohumoral phenomenon which is inhibited during childhood. The hypothesis is strengthened by the experimental observations that lesions in the amygdala, stria terminalis, or anterior hypothalamus result in precocious puberty. Similarly, many naturally occurring lesions in the CNS induce the production of gonadotropins at an early age. These lesions have one characteristic in common, namely, involvement of the hypothalamus. The manner in which the CNS-induced inhibition of gonadotropic secretion is removed at puberty is not known.

REGULATION OF TESTICULAR FUNCTIONS

The transition from boyhood to manhood normally begins between the eighth and eleventh years, although it is usually not apparent on gross examination until the tenth to thirteenth year of age. Each stage which occurs with age transition is initiated by the action of gonadotropins, the secretion of which is dependent upon altered CNS activity.

As a result of the increase in follicle-stimulating hormone (FSH) and luteinizing hormone (LH) secretion by the adenohypophysis, the seminiferous tubules are stimulated and the Leydig cells commence to elaborate androgenic substances *(testosterone).* Figure 36-1 depicts the relation between the number of Leydig

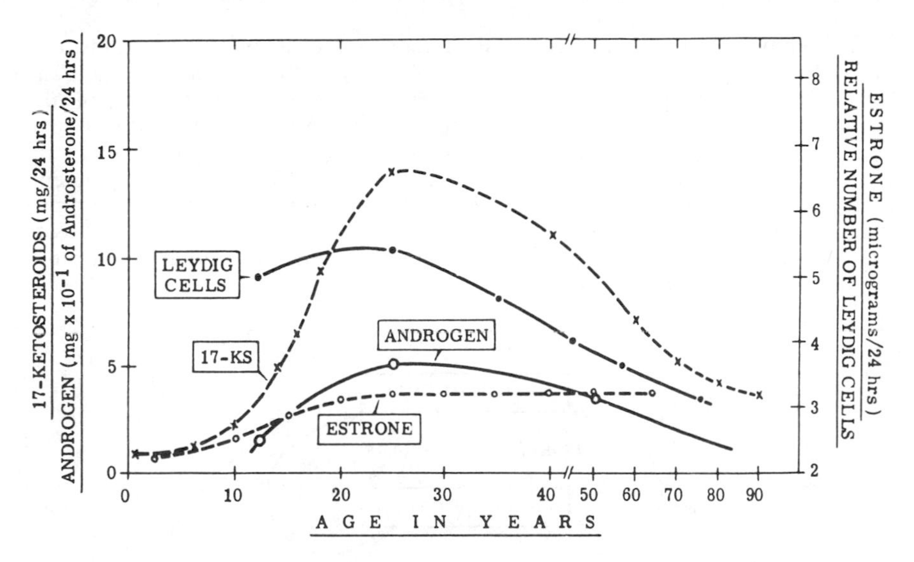

FIGURE 36-1. Urinary excretion of androgen, 17-ketosteroids, and estrogens throughout the life-span of the male. (After C. Hamburger, *Acta Endocr.* 1:19–37, 1948; J. B. Hamilton, *J. Clin. Endocr.* 14:452–471, 1954; K. G. Tillinger, *Acta Endocr.* Suppl. 30:5–192, 1957; and from *Reproductive Physiology,* Second Edition, by A. V. Nalbandov. W. H. Freeman and Company. Copyright © 1964.)

cells and the amounts of androgen and 17-ketosteroid (17-KS) excreted from birth to old age. The urinary 17-KS excretion is constant and equal in both boys and girls up to the age of 8 or 9 years, and is of adrenal origin. Of the total of approximately 14 mg of 17-KS excreted per day by the adult male, about 10 mg is of adrenal origin, because the eunuch excretes only 10 mg per day. The rate of excretion of urinary estrogen also rises markedly from low levels at puberty in the male; these changes also are depicted in Figure 36-1.

The mechanism of the initiation and maintenance of gametogenic and secretory activity of the testes is best illustrated in the hypophysectomized male. The seminiferous tubules and Leydig cells are atrophic in such an individual, gametogenesis is halted, and androgen production is minimal.

The administration of LH to the hypophysectomized animal stimulates the Leydig cells to respond by secretion of androgen. FSH has no demonstrable effect upon the Leydig cells but can cause a significant stimulation of the seminiferous tubules. FSH alone cannot support spermatogenesis and spermiogenesis. In order to maintain complete spermatogenesis, both FSH and LH or androgen must be administered. It is clear that, in the male, LH is the primary adenohypophysial hormone responsible for stimulation of the seminiferous tubule and Leydig cells. In the latter case, there is direct stimulation, while in the former an indirect action is mediated via the secretion of androgen by the cells of Leydig. When LH is administered immediately after hypophysectomy, it has the capacity for stimulating the Leydig cells and advancing spermatogenesis and spermiogenesis. That FSH has a physiological role in spermatogenesis is shown when administration is delayed in order to allow posthypophysectomy atrophy of the testes. In this instance, LH cannot act as a complete gonadotropin but must be supplemented with FSH in order to allow complete repair of testicular function.

The administration of large amounts of androgen to the intact male leads to the inhibition of gonadotropin secretion by the pituitary and results in testicular damage. The degree of testicular damage is directly related to the amount of androgen administered. Castration of the male removes the inhibitory effects of the endogenous androgens on the pituitary and results in an increase in gonadotropin secretion and urinary gonadotropin excretion. The secretion of both FSH and LH is dependent upon the release of the mediators, FSHRF and LRF, respectively, from the hypothalamus. FSH and LH also act back, via a "short loop," to inhibit the release of their mediator. The relations between the hypothalamus, the pituitary, and the testes are depicted in Figure 36-2.

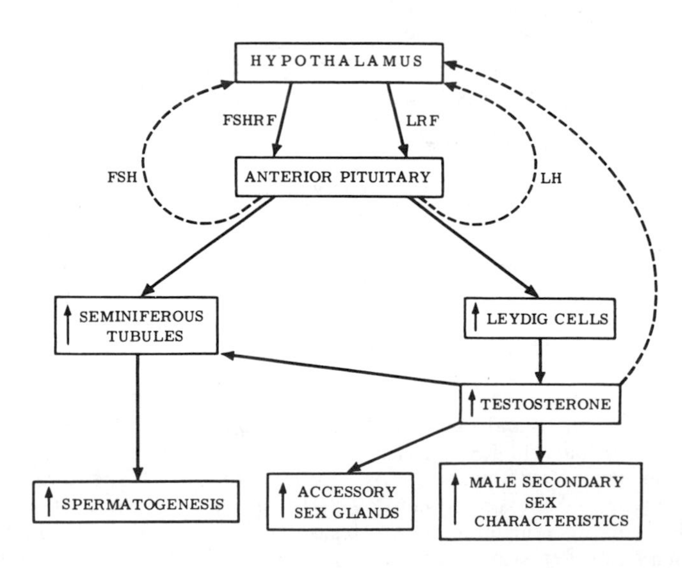

FIGURE 36-2. Regulation of reproduction in the male. Solid arrows = stimulation; dashed arrows = inhibition.

HORMONES OF THE TESTES

The androgens are the chief hormones of the testes and are defined as masculinizing compounds. Testosterone is the chief testicular androgen; it is secreted by the Leydig cells.

The testicular androgens are steroid in nature and contain 19 carbon atoms with methyl groups at carbons 10 and 13. In the testis, as in the adrenal cortex, the androgens arise from cholesterol. The synthetic pathways are presumed to be the same in both organs to the formation of 17-hydroxyprogesterone. This intermediate is converted to Δ^4-androstene-3,17-dione, which is acted on by a reductase to yield testosterone. These secretions are stimulated by LH (ICSH). Circulating testosterone is bound to proteins, is not stored in the body, and is either utilized or degraded to inactive forms which are excreted through the bile or urine. The major site of steroid catabolism is the liver, where the testicular androgens are converted to androsterone, epiandrosterone, and etiocholanolone (Fig. 36-3). These metabolites are conjugated with glucuronide or sulfate, are excreted in the urine as water-soluble salts, and are regarded as the 17-ketosteroids.

Estradiol has been isolated from the human testes, and the urine contains a significant amount of estrogens. The latter are diminished following bilateral orchidectomy and increased by the administration of gonadotropins.

EFFECTS OF THE ANDROGENS

Reproductive Tract

The entire male reproductive tract, including the accessory glands and external genitalia, is dependent on androgen for full morphological and functional development. Removal of the source of androgen results in marked reduction of the size of the seminal vesicles, prostate, and Cowper's gland, and in cessation of the secretory function of these tissues. On the other hand, the administration of androgen to a castrate restores the structure and function to normal, whereas the administration of androgen to an intact male results in a marked hypertrophy of the accessory glands. It has already been noted that androgens are necessary for spermatogenesis.

Androgens initiate the recession of the male hairline and dictate the distribution and growth of hair on the face, body, and pubes, and they stimulate hair growth in the axilla. The androgens cause thickening of the laryngeal mucosa and lengthening of the vocal cords, thus leading to the deep voice of the adult male. The androgens, particularly testosterone, are necessary for both the qualitative and quantitative development of the secondary sex characteristics which distinguish the man from the boy.

Metabolism

The most notable effect of testosterone on metabolism is nitrogen retention. This is accomplished by increasing the rate of protein synthesis. Following androgen administration, the rate of urinary nitrogen excretion decreases, and a positive nitrogen balance results. In many species, testosterone has been shown to increase the rate of protein anabolism and to decrease the rate of protein catabolism. It should be noted that the male accessory sex apparatus accumulates nitrogen in advance of other tissues. This effect, as well as the effect of androgens on the accessory sex glands and external genitalia, is probably mediated through increased DNA-directed RNA synthesis.

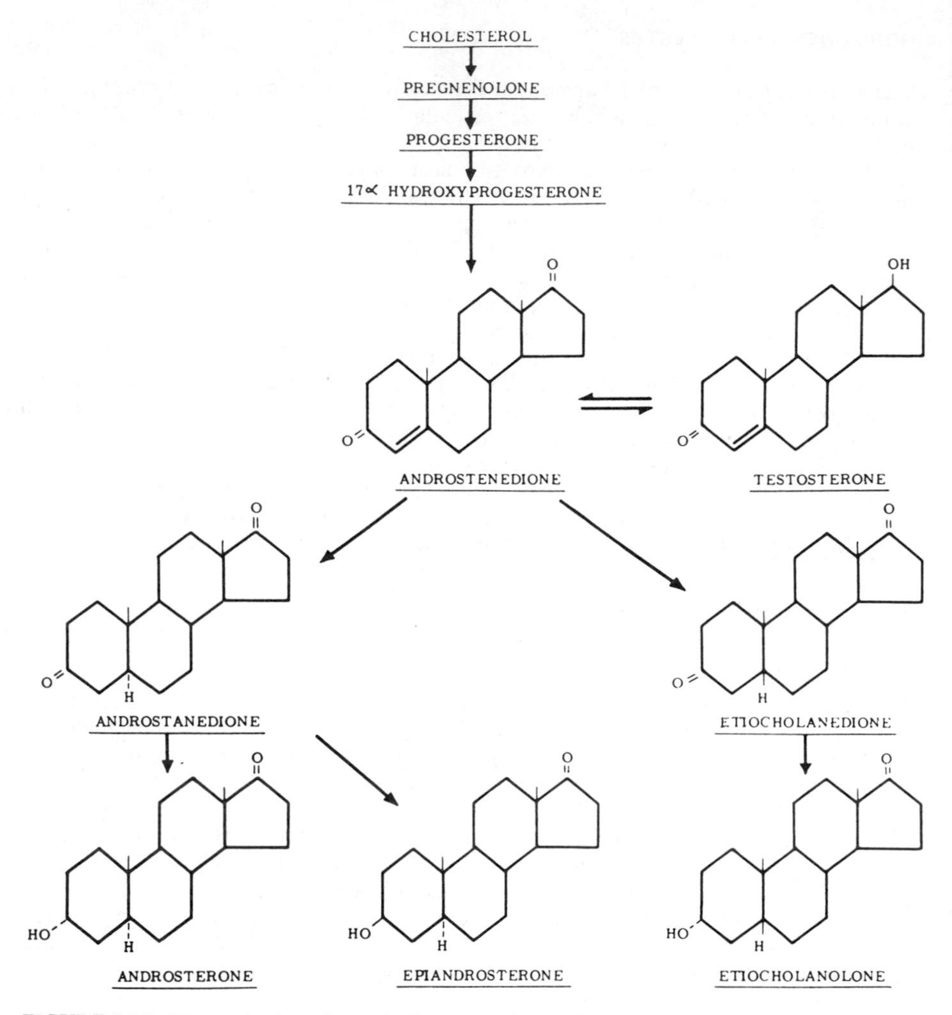

FIGURE 36-3. Biosynthesis and metabolic end products of testicular androgens.

Following castration, some skeletal muscles cease to accumulate nitrogen, some accumulate nitrogen at a reduced rate, and others are not affected. Androgens cause an increase in the working ability of muscles as a result of their stimulating action on the number, thickness, and tensile strength of muscle fibers. They are responsible for the greater muscle development in men and regulate in part the distribution of body fat and skeletal configuration of the adult male.

The androgens also stimulate the renal retention of sodium and potassium, and with these water, but they are considerably less effective in this respect than is aldosterone. The effects of an androgen, such as testosterone, on linear growth and bone metabolism are such that it stimulates growth of bone but limits the final size attained by long bones. It stimulates the deposition of protein matrix and causes the retention of calcium and phosphorus, yet at the same time promotes closure of the epiphyses. Testosterone is a potent growth stimulator, but its growth-promoting effect is observed only in the presence of an otherwise normal hormonal environment.

Following castration of the male, respiratory activity of the accessory glands decreases markedly. The levels of glycogen, phosphocreatine, and adenosine triphosphate (ATP) also decrease in the skeletal muscle but can be reversed by testosterone administration.

PHYSIOLOGY OF SPERMATOZOA

The gametogenic function of the testes is dependent upon testosterone production by the Leydig cells. Normal function of the accessory sex glands is also dependent upon testicular androgen production.

The seminiferous tubules constitute about 90 percent of the testicular mass. In adult testes, the spermatogonia give rise to primary spermatocytes. Both of these cell types are diploid cells; i.e., they contain XY sex chromosomes. The primary spermatocytes undergo meiotic division and form haploid secondary spermatocytes, which in turn divide to form spermatids. After a series of transformations, called spermiogenesis, the spermatids yield sperm cells or spermatozoa. The process of spermatogenesis is fairly slow but continues uninterrupted and may be maintained to extreme old age. It has been estimated that a man will produce 1×10^9 sperm for every ovum shed by a woman's ovaries.

The testes descend into the scrotum during late prenatal life. The main function of the scrotum in man is to provide a testicular environment that is about 5°F lower than the core temperature. Failure of testicular descent, *cryptorchism,* commonly occurs in man and may be the result of androgen deficiency or anatomical obstruction. In either naturally occurring or experimental cryptorchism, spermatogenesis eventually stops completely. If spermatogonia are still present in the testes, spermatogenesis can be reinstituted on removal of the testes from the body cavity. However, after prolonged exposure to elevated temperature, testicular damage becomes irreversible. Temporary or permanent sterility in man may result from prolonged fever, even though the testes have descended into the scrotum. It is not known why high temperatures are injurious to sperm.

The manner in which sperm are transported through the seminiferous tubules and epididymis is unknown. Various mechanisms have been proposed; they include a continuous *vis a tergo* provided by the release of new sperm, conveyance in seminiferous tubule secretions, ciliary action by cells of the efferent ducts, and contraction of smooth muscles of the epididymal tubules. However, sperm within the seminiferous tubules and epididymis are nonmotile and do not become motile until they come into contact with the seminal plasma. That sperm may survive a long time in the epididymis can be demonstrated by blocking the ducts between the epididymis and the testes. By using this technique and studying the ejaculate, it can be demonstrated that such epididymal sperm remain viable up to 60 days. Fertile sperm may be present in the ejaculate for as long as six weeks after castration. On the other hand, sperm in the vas deferens lose their fertilizing capacity rapidly. Sperm are not found in the seminal vesicles, prostate, or any of the other accessory glands.

The seminal plasma consists of secretions of the testes and epididymis (less than 5 percent), seminal vesicles (about 30 percent), prostate (about 60 percent), and Cowper's glands (less than 5 percent). These secretions have two main functions: They furnish the sperm with metabolizable substrate and function as a suspending and activating medium.

A major constituent of the secretion of the seminal vesicles is fructose, which serves as the chief source of energy for the sperm. Major constituents of the prostatic contribution to seminal plasma are citric acid, calcium, acid phosphatase, and proteolytic enzyme. The latter is primarily responsible for the liquefaction of coagulated semen. The functional significance of acid phosphatase in seminal fluid is not known. This enzyme has clinical significance because it enters the blood stream when malignant growth of the prostate occurs. The secretions of the prostate and seminal vesicles are not necessary for the production of viable sperm because both testicular sperm and sperm taken from the epididymis are viable. However, the viability of such sperm is decreased.

Erection is a vascular phenomenon which is dependent upon the structural pattern of the penis. It is the result of venous constriction and arterial dilation which allows blood to flow under high pressure into the erectile tissue. The erectile tissue is a spongelike system composed of vascular spaces which are relatively collapsed and contain little blood in the flaccid state. However, during erection these spaces become distended with blood, and the pressure approaches that in the carotid artery. Erection may result from either physical or psychic stimuli; the motor pathways are via the sacral component of the craniosacral system, the *nervi erigentes,* and a center for reflex erection exists in the sacral cord. The subsidence of erection can result from stimulation of the sympathetic innervation of the penis.

Ejaculation actually refers to two distinct actions, emission and ejaculation, both of which are reflex phenomena. The afferent arc of these reflexes originates in the sense organs of the glans penis, and the impulses are transmitted centrally through the internal pudendal nerves. The first action of ejaculation, which delivers sperm and seminal plasma into the urethra, is the result of contraction of smooth muscles of the genital tract and is called emission. This action includes peristaltic contractions of the testes, epididymis, and vas deferens which cause the expulsion of sperm into the internal urethra. It also includes the contractions of the seminal vesicles, prostate, and bulbourethral glands which result in the movement of their respective secretions into the urethra. Emission is evoked by stimulation of the hypogastric nerves. The second action, ejaculation, causes the expulsion of seminal fluid from the urethra to the exterior as a result of striated muscle contraction (bulbocavernous muscle). This results from increased neural activity transmitted peripherally via the internal pudendal nerves.

The volume of ejaculate is from 2 to 6 ml in man. The ejaculate contains more than 10^7 sperm per milliliter, has a slightly alkaline reaction (pH = 7.1 to 7.5), and has a specific gravity of about 1.028. Coagulation of normal semen occurs promptly after ejaculation. However, it liquefies within several minutes, and the sperm attain full motility.

The life-span of sperm, or the time during which they are able to fertilize ova, is relatively short, about 24 to 36 hours. However, sperm may be frozen to $-169°C$, stored, and thawed without impairing their fertilizing capacity. It has been claimed that sperm concentrations of less than 50×10^6 sperm per milliliter of ejaculate indicate sterility. Out of this number introduced into the female reproductive tract, less than 10^4 sperm reach the oviducts, less than 10^2 reach the vicinity of the ovum, 10^1 may penetrate the zona pellucida of the ovum, but only 1 sperm enters the ovum and accomplishes fertilization. Therefore, the sine qua non for fertility in the male is the number of sperm per ejaculum. Other criteria, such as percentage of abnormal sperm (with respect to morphology) and volume of ejaculate, are also used in assessing semen quality.

ENDOCRINOLOGY OF THE FEMALE

In addition to the production of gametes, the basic function of the ovaries is to secrete hormones which regulate the activity of the female reproductive tract and determine the secondary sex characteristics. The gametogenic and endocrine aspects of ovarian function fluctuate rhythmically during the active reproductive life of the female. The periodic changes in the functional activities of the ovaries are determined by the interrelationships between the activity of the adenohypophysis and the ovaries. The adenohypophysial gonadotropins, FSH and LH, are concerned with ovarian function as well as with testicular function. In addition, a third pituitary hormone, *prolactin* or luteotropic hormone (LTH), is concerned with cyclic ovarian function. The female experiences a shorter reproductive life than does the male. The active reproductive life of the female commences at puberty (10 to 15 years of age) and is maintained for approximately 30 years. Reproductive activity gradually decreases, the ovaries involute, and reproductive cycles cease during the fifth decade of life. This transition is termed the menopause and is followed by a period of hypogonadism for the remainder of the life-span.

Reproductive functions and behavior are markedly influenced by the gonadal hormones. That these are not the only factors involved is emphasized by the adverse effects on the functions of the pituitary-gonadal axis of both female and male of the lack of certain nutritional factors and caloric intake, nongonadal endocrine disorders, changes in the external environment, and CNS and psychological factors.

The reproductive cycle of the female is a complex series of coordinated events involving the CNS, adenohypophysis, ovaries, oviducts, uterus, cervix, and vagina. The most complete studies on the physiology of reproduction have utilized laboratory primates, rodents, and farm animals. Many of those studies have been extended to the human. Although certain quantitative differences exist, the fundamental factors which are known to regulate reproductive phenomena in the rat and bovine probably regulate the same phenomena in the human.

THE OVARY

The size and microscopical appearance of the paired ovaries in the adult vary with the period of the reproductive cycle. The most important functional components of the mature ovary (Fig. 36-4) are the *follicles* and *corpora lutea.* Three stages of growth of the follicles can be noted. The immature stage, i.e., the primary follicle, is observed from embryonic to late life of the ovary. This structure makes up the bulk of the population of follicles. In the primary follicle the ovum is surrounded by several layers of granulosa cells and has no vitelline membrane. The secondary follicle is formed as a result of the development of the *zona pellucida,* which is a membrane surrounding the ovum. The tertiary follicle is characterized by the presence of an *antrum,* which is a fluid-filled space surrounding the ovum. The ovum, enclosed in granulosa cells, is bathed by liquid, the *liquor folliculi.* As the follicle grows, it moves toward the cortex of the ovary, and the antrum enlarges to produce the mature follicle. The mature follicle, called the *graafian follicle,* extends throughout the thickness of the cortex and bulges as a blister on the free surface of the ovary (Fig. 36-4). In addition, the mature ovary contains follicles which exhibit varying degrees of degeneration; these are the *atretic follicles.*

The *corpus luteum* is a temporary endocrine organ which forms following ovulation as a result of luteinization of the *granulosa* and *theca interna* cells. Luteinization involves hypertrophy and hyperplasia of the granulosa and theca interna cells with the development of lipid inclusions within the cells. Immediately after ovulation, the cavity of the follicle becomes filled with blood and lymph. Gradually the space occupied by the fluid is filled by the luteal cells, and a well-developed blood supply is formed within the mass of lutein cells. Seven to eight days after ovulation, regressive changes are seen, such as loss of lipid material and invasion by connective tissue. The span of functional activity of the corpus luteum is approximately 10 to

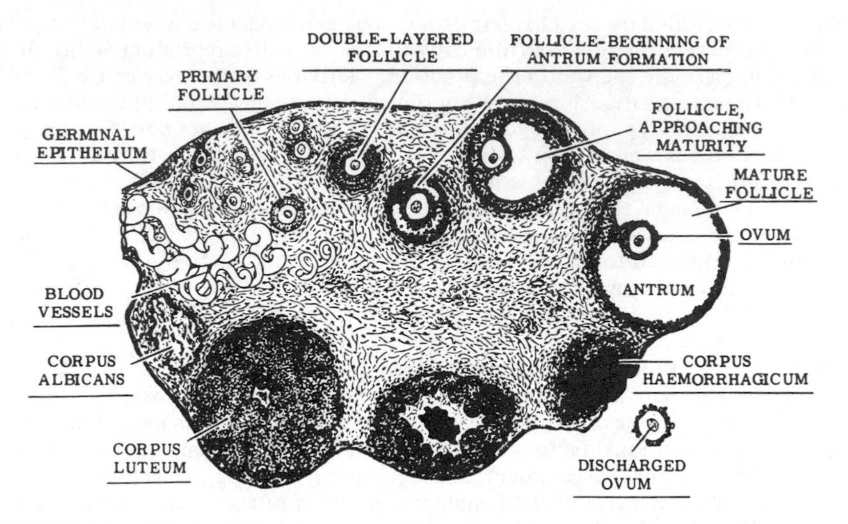

FIGURE 36-4. Schematic diagram of mammalian ovary showing sequence of events in origin, growth, and rupture of ovarian follicle, and formation and retrogression of a corpus luteum. (From *Embryology of the Pig* by Bradley M. Patten. Copyright 1948 by McGraw-Hill, Inc. Used with permission of McGraw-Hill Book Company.)

12 days in the human, after which the gland is nonfunctional and referred to as the *corpus albicans*. During the ensuing weeks, further degenerative changes occur, and the organ is replaced by connective tissue.

Ovarian Hormones

The ovaries secrete three steroid hormones (estrogens, progestins, and androgens) and a protein hormone, *relaxin*. The term *estrogen* refers to any substance which produces vaginal cornification, whereas the term *progestin* refers to any substance which produces secretory changes in the estrogen-primed uterus. These hormones are formed by the follicle and corpus luteum. Cells of the theca interna of the ovarian follicle and thecal lutein cells of the corpus luteum are the source of ovarian estrogens. Progestins are produced by luteinized granulosa cells. The cellular origin of the ovarian androgens and relaxin is uncertain, although some endocrinologists believe that androgens are produced by cells in the hilus of the ovary. The estrogens and progestins stimulate the growth and differentiation of the female reproductive tract and exert a variety of metabolic effects on other tissues.

The human ovary secretes the estrogens *estradiol* and *estrone;* a third compound with estrogenic activity, *estriol,* is a metabolic end product of estradiol and estrone. It is well established that all the steroid hormones, regardless of their site of origin, are derived from cholesterol. The compound *17-α-hydroxyprogesterone* plays a pivotal role in the ovarian synthesis of estradiol and estrone. This becomes apparent if one refers to the pathways of adrenal and testicular steroid biosynthesis. The pathways of ovarian estrogen and progestin biosynthesis are depicted in Figure 36-5. The ovarian estrogens are secreted into the blood stream, where they combine with plasma protein. However, approximately one-third of the circulating estrogen remains in a free form and is in equilibrium with protein-conjugated estrogens, estroprotein. That the liver inactivates estrogens can be demonstrated by transplanting

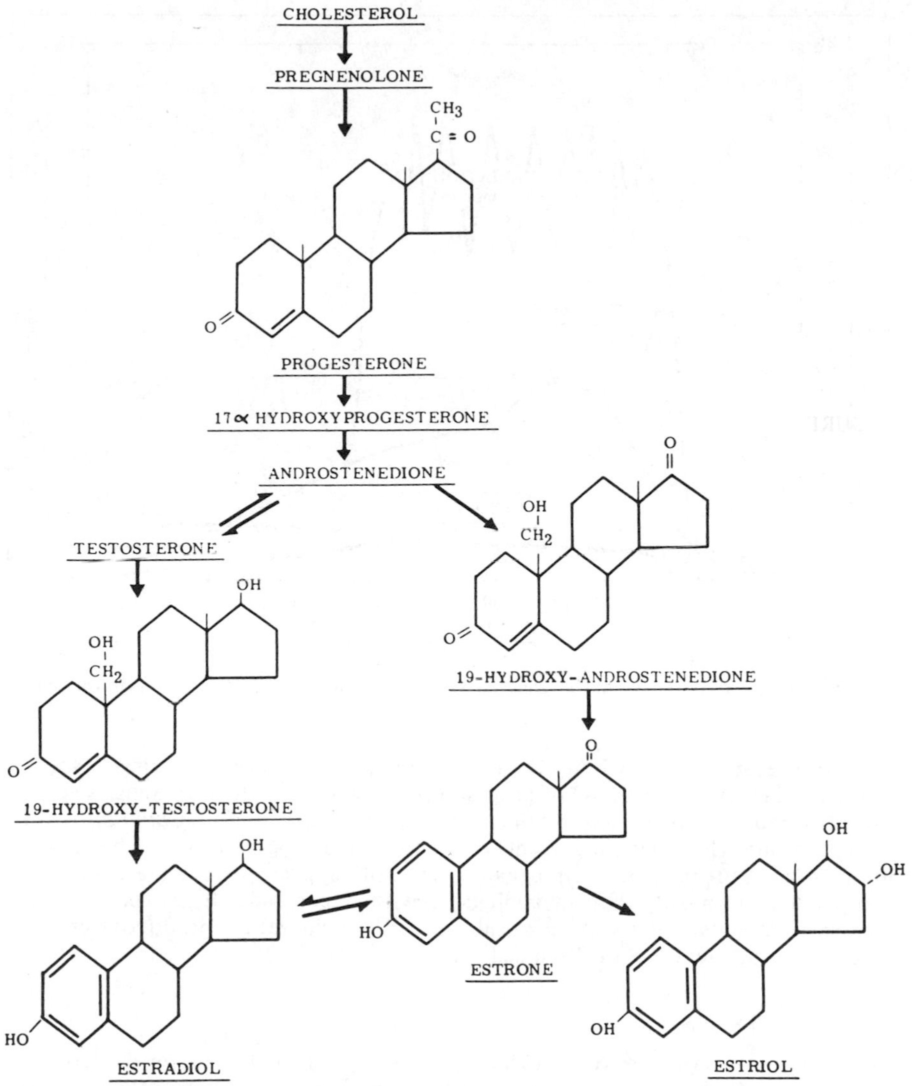

CHOLESTEROL

PREGNENOLONE

CH_3
C = O

O

PROGESTERONE

17α HYDROXYPROGESTERONE

ANDROSTENEDIONE

O

OH
CH_2

O

TESTOSTERONE

OH

OH
CH_2

O

19-HYDROXY-ANDROSTENEDIONE

O

19-HYDROXY-TESTOSTERONE

OH

OH

OH

HO

ESTRONE

OH

HO

ESTRADIOL

ESTRIOL

FIGURE 36-5. Biosynthesis of ovarian progesterone and estrogens.

the ovaries to the spleen, so that the estrogens pass first to the liver. In this instance the estrogens are immediately inactivated and a hypoestrogenism develops. The important role of the liver in estrogen inactivation is further emphasized by the fact that hyperestrogenism accompanies certain types of hepatic disorders. The hyperestrogenism occurs as a result of failure of hepatic removal of estrogens from the circulation. The estrogens undergo oxidation in the liver to form *estriol,* and the liver conjugates each of the estrogens with sulfate or glucuronide. Large amounts of conjugated estrogens are excreted in the bile, and small amounts appear in the urine. The urinary excretion of estrogens and androgens from birth through life is depicted in Figure 36-6.

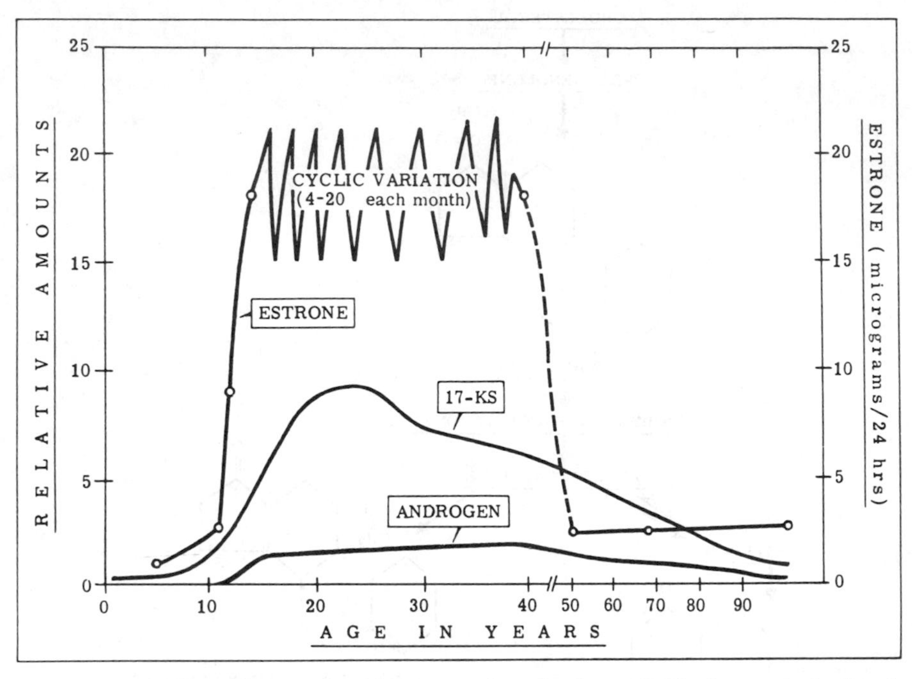

FIGURE 36-6. Effect of age on the urinary excretion of androgens and estrogens in the female.

Progesterone is secreted into the blood stream by luteal tissue during its short life-span. The circulating level of progesterone, like that of the estrogens, is very low. It is rapidly metabolized by the liver to *pregnanediol* and *pregnanolone.* Only a small amount (less than 10 percent) of administered progesterone can be recovered in the urine as pregnanediol or pregnanolone. Both appear in the urine as the free compounds, but mostly they are conjugated with glucuronide. The excretion of progesterone metabolites will be considered further when the reproductive cycle, pregnancy, and lactation are discussed.

EFFECTS OF ESTROGENS

The estrogens produce striking effects upon the cells of the female reproductive tract, and hardly a tissue in the body is not affected by them. Their effects on the genital tract of the female are most dramatically observed when estrogens are administered to the ovariectomized animal.

Vagina

The vaginal epithelium of either the immature or ovariectomized animal is very thin and shows marked atrophic changes. Estrogen administration results in a sharp increase in mitotic activity and stratification of the cells of the vaginal epithelium. Cornification of the epithelial cells results from a loss of blood supply to the superficial cells, which occurs because of the rapid growth (increase in thickness) of the vaginal epithelium. Estrogens increase the glycogen and mucopolysaccharide content of the vaginal mucosa, decrease the pH (about 4.5), increase the vascularity of the structure, and produce a slight degree of edema. The appearance of cornified cells in a vaginal smear or lavage is diagnostic of estrogen action.

Uterus

Estrogen administration results in proliferation of the uterine cervical mucosa and the secretion of alkaline mucus from the cervical glands. The estrogens also have a pronounced effect upon the endometrium, inducing endometrial mitoses (hyperplasia), increasing the height of the lining epithelium (hypertrophy), increasing the blood supply and capillary permeability of uterine vessels, and increasing the uterine water and electrolyte content. They also enhance uterine anaerobic and aerobic glycolysis, and increase the rate of uterine ribonucleic acid (RNA), and deoxyribonucleic acid (DNA), protein, and glycogen synthesis. Thus, the uterus increases in weight and size as a result of marked proliferative changes following estrogen administration. Contrarily, ovariectomy results in definite atrophic changes in all uterine tissue, particularly the endometrium and myometrium.

The smooth muscle is relatively inactive in an ovariectomized animal. Estrogen administration increases myometrial contractility and motility, and increases the uterine smooth muscle to oxytocin. These changes result from the ability of estrogens to increase the content of the myometrial contractile substance, actomyosin; thus, a greater isometric tension can be developed. Action potentials are rarely observed in the uterine muscle taken from an ovariectomized animal, but the frequency increases following estrogen administration and results in greater myometrial activity. It has also been shown that the membrane potential of uterine muscle is elevated by estrogen over that of the castrate. This effect of estrogen is dependent upon the presence of Ca^{++}. Apparently, estrogens set the membrane potential at a critical level so that the myometrial cells show an increased excitability (decreased threshold), increased spontaneous activity, and increased sensitivity (reactivity) to appropriate stimuli.

Estrogen administration also stimulates proliferation of the epithelium and smooth muscle of the uterine tubes or the oviducts; the effects here are similar to those on the body of the uterus. Estrogen administration can lead to failure of implantation of ovum. This may be due largely to its effects on the musculature of the oviduct. Estrogen stimulates the smooth muscle of the oviduct to increase the peristaltic action in the direction from the uterus to the ovary. This may be of benefit to sperm transport, but if the activity is too great, it prevents the passage of the ovum down the tubes to the uterus.

Metabolism

Estrogen administration results in renal retention of Na^+ and Cl^- and decreased urine volume, with subsequent increases in blood volume, extracellular fluid (ECF) volume, and body weight. The action of estrogen on renal Na^+ handling is independent of changes in renal hemodynamics and the secretions of the adrenal cortex.

The estrogens also promote protein anabolism and affect skeletal growth by increasing osteoblast activity and the rate of deposition of calcium and protein in bone. Bone growth (length) is limited by estrogens because they promote closure of the epiphysial plates, which are the growth centers of bone.

Secondary Sex Characteristics

The secondary sex characteristics of the female are dependent upon the estrogens. Estrogens act on the mammary gland to induce duct growth and proliferation of stromal tissue. The tendency in the female for adipose tissue to concentrate in the buttocks, hips, thighs, and mammary gland, plus the presence of a uniform layer of subcutaneous fat over the body, leads to a rounded contour with varying convexities and results in a form which is distinctly less angular than that of the male. The female skeleton is smaller than the male, with marked differences in the pelvis, which is adapted in the female to aid in the carriage and delivery of young. The hair of the body is fine, and scalp hair grows faster and is more permanent. Ovariectomy in experimental animals abolishes mating responses and sexual behavior patterns. Libido in the human is relatively unaffected following ovariectomy in the adult, but may be decreased if ovariectomy is performed prior to puberty.

Mode of Action

It would be logical to assume that estrogens influence the enzyme systems responsible for protein anabolism, because of their mitogenic and growth-promoting properties. That estrogens act upon energy-supplying systems and provide energy for systems is shown by their stimulating action on oxygen consumption, and oxidation of glucose and pyruvate in endometrial tissue. Studies have shown that the estrogens are involved in stimulating an enzyme which transfers hydrogen ion from reduced nicotinamide-adenine dinucleotide phosphate (NADPH) to nicotinamide-adenine dinucleotide (NAD). Such a system would accelerate the activity of the Krebs cycle and the monophosphate shunt, thereby increasing the amount of energy available for synthetic reactions of the cells. This may be mediated via cyclic AMP. Treatment with the estrogens results in activation at the gene level, as evidenced by the observations that they stimulate, in sequence, the rate of synthesis of messenger RNA, transfer RNA, and ribosomal RNA, and this is followed by an increase in the rate of protein synthesis.

Each of the effects of estrogen on the reproductive tract and mammary gland may be considered as indicative and diagnostic of puberty, because the estrogen secretion heralds the onset of ovarian endocrine activity. The primary point is that estrogens cause proliferative changes in the female reproductive system. The role of estrogen in cyclic reproductive behavior will be discussed later.

EFFECTS OF PROGESTINS

The effects of progesterone, per se, are never seen under physiological conditions and are not normally apparent until after the actions of estrogens have been operative. The effects of estrogens are generalized as being proliferative, whereas progesterone acts upon proliferated tissue to differentiate it into a secretory type of tissue. Progesterone acts on the reproductive tract and mammary gland in such a manner as to prepare the tract for implantation of the fertilized ovum and to maintain gestation and lactation.

Reproductive Tract

Progesterone transforms the cornified vaginal epithelium to a mucified condition and decreases cervical secretions. The progestational response of the endometrium, which is a result of priming with estrogen followed by progesterone stimulation, is characterized by further endometrial growth, including thickening of the epithelium and accumulation of water. The proliferated endometrium is transformed into the secretory endometrium under the influence of progesterone. These changes include enlargement of endometrial stromal cells and the formation of tortuous glands with marked glycogen deposition. Thus the highly secretory endometrium provides the environment which is necessary for implantation of the blastocyst. That expulsion of the implanted blastocyst does not occur is assured by the effect of progesterone on the myometrium. The membrane potential of uterine smooth muscle is further elevated and results in a decreased excitability of these elements. The hyperpolarized membrane of the progesterone-dominated smooth muscle cell of the uterus shows poor conduction, lowered spontaneous activity, and reduced sensitivity to specific stimuli. It follows that the progesterone-dominated uterus cannot participate in the organized contractions which result in expulsion of the young. Progesterone also results in decreased smooth muscle activity in the oviducts. It acts on the estrogen-primed mammary gland and stimulates lobulo-alveolar growth, which is the hallmark of the mature mammary gland.

Body Temperature

Variations in the awakening basal body temperature occur and have been correlated with cyclic ovarian activity. The phase of the cycle during which estrogen is dominant is characterized by a low basal body temperature. At ovulation time, a slight drop in temperature occurs, but when progesterone secretion rises, the body temperature is elevated over that observed during the early stage of the cycle (Fig. 36-7). The basal body temperature remains elevated so long as progesterone secretion is maintained. The mechanism of this action is not understood. It should be obvious, because of the small temperature changes, that even minor infections which elevate the body temperature obscure the cyclic variations in basal body temperature.

Miscellaneous

Progesterone, because of its key position in steroid biosynthetic pathways, might be expected to mimic the effects of a variety of steroids. It possesses adrenal corticoid activity in that it prolongs survival of the adrenalectomized animals and has slight protein catabolic activity. It has androgenic activity and aids in the maintenance of testicular weight in the hypophysectomized animals. It also tends to induce a slight hyperventilation, which results in a lowered P_{CO_2} of arterial blood.

RELAXIN

The hormone relaxin, which is released from the uterus and placenta by action of the corpus luteum hormone progesterone, is protein in nature and plays a role in parturition by inducing relaxation of the pubic symphysis. The effectiveness of relaxin is dependent upon prior stimulation by estrogen. Relaxin is found in human blood only during pregnancy. It induces "softening" of the cervix and inhibits spontaneous uterine contractions. The maximal activity of relaxin in the blood occurs immediately prior to parturition, and no activity is observed 24 hours after delivery. The physiological role of relaxin is obviously that of ensuring an uneventful parturition, but the mechanism of its control and action remains to be ascertained.

REGULATION OF CYCLIC OVARIAN ACTIVITY

The cyclic variations in reproductive activity occur as the result of humoral interactions between the ovaries, pituitary, and CNS. The normal cycle is characterized by maturation of ovarian follicles, ovulation, and preparation of the reproductive tract for the implantation of a fertile ovum. If conception and implantation do not occur, these processes are repeated. The gonadotropic hormones, FSH, LH, and

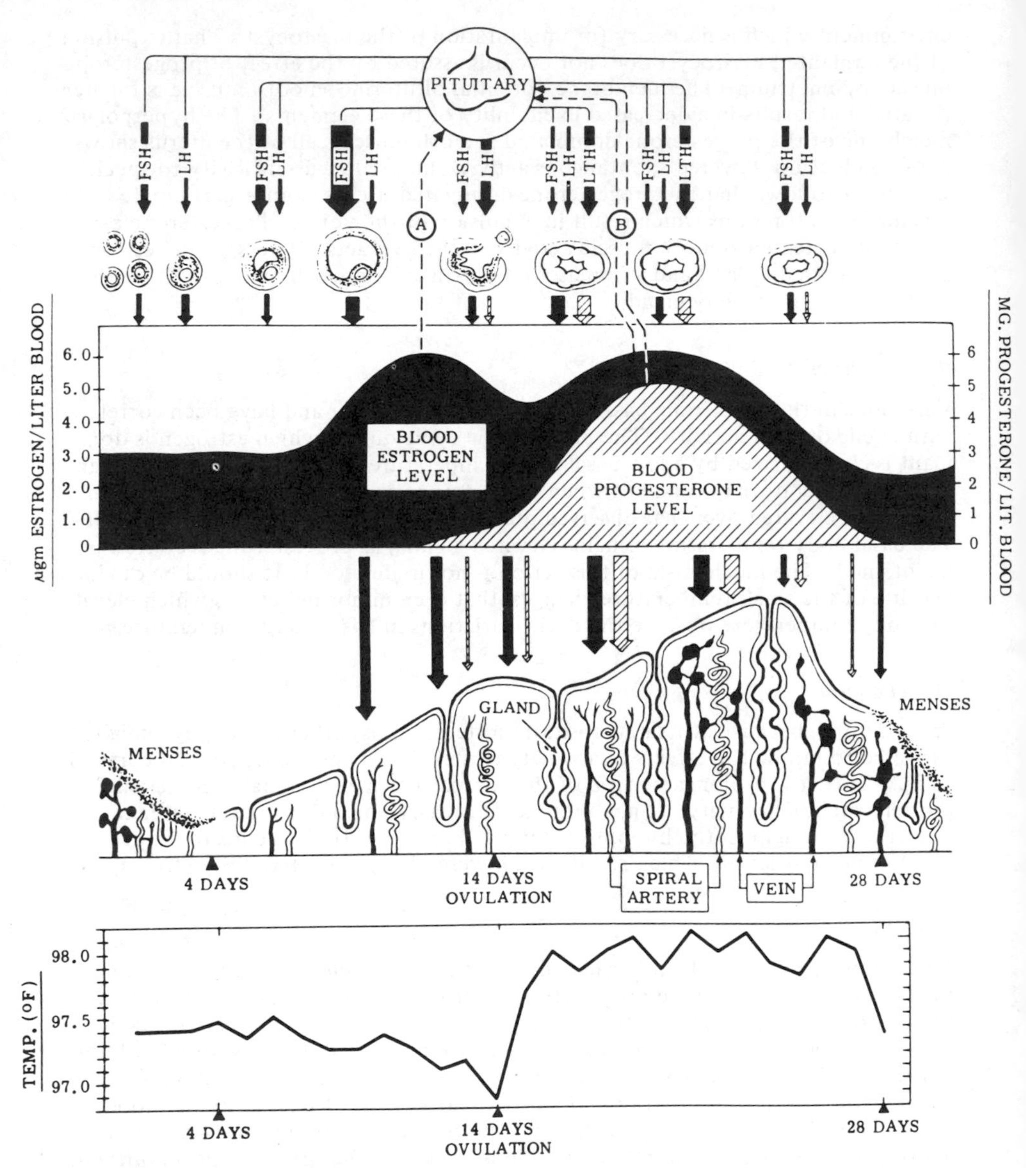

FIGURE 36-7. Variations in pituitary and ovarian hormone secretion, endometrial morphology, and basal body temperature during the menstrual cycle. The widths of the arrows indicate the relative circulating levels of the hormones. A indicates inhibition of FSH and stimulation of LH secretion by estrogens. B indicates the net inhibitory influence of estrogens and progesterone on LTH secretion. These influences are probably mediated via the CNS.

LTH, and the ovarian hormones, estrogens and progesterone, are involved in the regulation of these processes. The CNS is also involved in regulation of cyclic ovarian activity, for it is known that each of the five hormones, either directly or indirectly, affects or is affected by the CNS.

The observation that the ovaries of hypophysectomized animals remain immature, but that gonads of such animals can be restored or maintained with pituitary extracts, particularly the gonadotropins, is taken as proof that the pituitary regulates ovarian function. However, the pituitary is not autonomous, because the secretions of the ovary in turn affect pituitary gonadotropic hormone formation and release. Moreover, the pituitary which is removed from its normal site and transplanted to some extrasellar site is incapable of maintaining normal cyclic gonadal function.

FSH, as the name indicates, stimulates growth of the ovarian follicle, which includes hyperplasia of the granulosa cells, enlargement of the antrum, and a marked increase in ovarian weight. These changes occur when FSH is administered to immature hypophysectomized rats. Changes in the reproductive tract indicative of estrogen secretion do not occur, and therefore it is concluded that FSH alone is incapable of stimulating estrogen secretion. If the immature hypophysectomized animal is treated with LH, the ovarian interstitial cells are stimulated, but no estrogen secretion results. However, the combination of FSH and LH notably increases ovarian weight, follicle growth, and estrogen secretion. In the intact animal, the rate of follicular growth and estrogen secretion increases under the influence of FSH and LH until the point is reached where the increased estrogen titer inhibits FSH. Thus, a negative feedback mechanism operates between estrogen and FSH secretion, because marked increases in FSH secretion result when estrogen titers are lowered. Examples of this are seen in the ovariectomized or menopausal individual, each of whom is characterized by a decreased circulating estrogen and high urinary FSH. The increased estrogen titer also stimulates the release of LH from the pituitary; LH then acts on the "ripe" follicles and causes ovulation.

Luteinization of ovarian follicles is also a result of LH stimulation, and in some manner, probably as a result of increasing estrogen titers, LTH secretion increases. LTH produces further luteinization and corpus luteum formation with the resultant initiation and maintenance of progesterone and estrogen secretion by the luteal cells. The mechanisms which induce regression of the corpus luteum and the subsequent decline in progesterone and estrogen secretion are not known. It is very likely that the ovarian hormones indirectly lead to regression of the corpus luteum. Nevertheless, as the corpus luteum regresses, the amounts of ovarian hormones are reduced, with consequent removal of inhibition from the pituitary. The pituitary then produces more FSH to repeat the cycle. Studies concerned with the circulating and excretory levels of LTH are meager, although LTH has been detected in the urine throughout the cycle. The LTH levels, therefore, are theoretical (Fig. 36-7).

REGULATION OF GONADOTROPIC HORMONE FUNCTION BY THE CNS

As a result of observing the effects of emotional states, malnutrition, light, and temperature upon gonadal function, modulation of gonadal activity by the central nervous system has been obvious for many years. It has been well known that several species of animal ovulate only after appropriate stimulation, e.g., coitus. These species have been referred to as the induced or reflex ovulators. There was, however, considerable doubt concerning the direct role of the CNS on adenohypophysial activity because of the paucity of nerve fibers in this structure. Adequate explanation of the participation of the CNS in reproductive activity became available only after the presence of the hypophysial portal vessels was realized (Chap. 31).

Several alterations in ovarian function have been observed following destruction or stimulation of various parts of the brain. Lesions of the median eminence result in gonadal atrophy which approaches that observed in hypophysectomy. Animals with midline lesions in the anterior hypothalamus show persistent vaginal cornification but fail to ovulate, and it is assumed that constant amounts of LH that are too low to cause ovulation are secreted. The implantation of ovarian grafts or minute amounts of estrogen in this area decreases LH output. However, such transplants to the adenohypophysis or posterior hypothalamus have no effect on the output of this hormone. Therefore, LH secretion, as a result of the local estrogen effect in the CNS, has been inhibited, and a stimulus for ovarian estrogen secretion is withdrawn.

Precocious puberty may be produced by anterior hypothalamic lesions, amygdaloid lesions, or interruption of the stria terminalis. Electrical stimulation of the ventral anterior hypothalamus induces ovulation in animals in which spontaneous LH release has been blocked. Recently, it has been demonstrated that extracts of hypothalamic tissue will cause an increase in the blood LH of ovariectomized rats. Electrical stimulation of the amygdala also results in ovulation, but stimulation of the adenohypophysis does not cause ovulation. That the reticular formation might be involved in ovulation is shown by the observations that lesions anterior to the mesodiencephalic juncture inhibit ovulation, and certain substances which block the spontaneous release of LH also elevate the threshold of the reticular activating system. On the other hand, lesions of the midbrain tegmentum, which result in somnolence, do not interfere with ovulation.

Transplantation of the adenohypophysis to a heterotopic location results in gonadal atrophy. However, if corpora lutea are present, such a transplant may maintain the corpora lutea in a functional state indefinitely, indicating that the transplants are capable of secreting luteotropin. Furthermore, if the transplants are returned to the sella turcica and revascularization with the portal vessels occurs, cyclic ovarian activity is resumed. These observations indicate not only that a substance is released by the hypothalamus and transmitted via the portal vessels to the adenohypophysis, stimulating the release of LH and FSH, but also that another substance is released by the hypothalamus, resulting in the inhibition of LTH secretion. Furthermore, experimentation has indicated that the two substances are reciprocally related. Relatively purified preparations of hypothalamic tissue have been prepared, and they have been shown to exert LH-releasing, FSH-releasing, or LTH-inhibiting activities, respectively. Consequently, they have been termed *luteinizing hormone releasing factor* (LRF), *follicle-stimulating hormone releasing factor* (FSHRF), and *prolactin* (LTH) *inhibiting factor* (PIF).

Available evidence indicates that the hypothalamus acts as a center which integrates brain stem and spinal reflexes into patterns of behavior observed in mating in the female. The activity of the hypothalamus may be modified by influences from the neocortex, rhinencephalon, and brain stem reticular formation. This integrating mechanism may also be sensitized by estrogens, because the implantation of minute amounts of estrogen into the posterior hypothalamus may lead to mating responses in the absence of any stimulating effect by the estrogen on the reproductive tract. Therefore, when referring to the pituitary-gonad axis, one must interpose the CNS because the hormones of the gonads may feed back to the CNS, influence its activity, and indirectly alter the activity of the adenohypophysis, the organ through which direct stimulation of the gonads is mediated.

REGULATION OF THE FEMALE REPRODUCTIVE TRACT

The structural and functional changes in the reproductive tract are well known and are repeated without much variation throughout the year. Changes in uterine and vaginal function are directly regulated by the ovarian estrogens and progestins. The changes which occur during the preovulatory stage or early part of the cycle depend upon estrogen secretion, whereas the postovulatory stage is influenced by estrogen and progesterone. Thus, the early part of the cycle is characterized by follicle development and proliferation of the reproductive tract and has been termed the *proliferative* or *follicular* phase of the cycle. During this phase the vaginal epithelium undergoes rapid growth and the vaginal smear is eventually transformed into one dominated by cornified cells. The uterus begins to accumulate fluid, becomes highly contractile, and undergoes proliferative changes under the influences of estrogen until the time of ovulation. Following ovulation and corpus luteum formation, reproductive tract function is altered by progesterone.

The phase of the cycle which follows ovulation is called the *secretory* or *luteal* phase. During this phase the vagina is infiltrated with leukocytes and the vaginal smear is characterized by the presence of leukocytes, mucin, and cornified cells. The vaginal epithelium becomes thin, and leukocytes are abundant, so that the vaginal smear contains nucleated epithelial cells and leukocytes. Immediately after ovulation the uterus diminishes in vascularity and contractility. The increasing titer of progesterone transforms the proliferated endometrium into an organ which has highly coiled glands possessing the ability to secrete large amounts of glycogen. Thus, the uterus is prepared for nidation. Following regression of the corpus luteum a new set of follicles is stimulated by the rising levels of FSH, and the cycle is repeated. Little is known concerning the cyclic alterations in function of the oviducts and uterine cervix.

The reproductive cycle differs among species in several ways. In general, the nonprimate mammalian species exhibit a period of sexual receptivity during a particular phase of the cycle. This period coincides with the time of maximal estrogen secretion prior to ovulation. The period of sexual receptivity, *estrus,* or "heat," is characterized by behavioral changes designed to attract the male at the time when ova are most easily fertilized.

CONTROL OF FERTILITY

Currently a great deal of attention is being given to the control of fertility and family planning. The prevailing new attacks on these problems are via the use of contraceptives taken orally and intrauterine devices. The common oral contraceptives in use today are progestational agents whose actions are potentiated by the inclusion of small amounts of estrogenic substances. These agents inhibit the release of LH from the adenohypophysis and thus prevent ovulation. They may also make implantation of fertilized ova difficult through their action on the endometrium. Since primitive times, intrauterine devices have been used in an attempt to control conception. But their side-effects, i.e., infection, bleeding, etc., have made them undesirable. Recently new designs and materials have made the use of such devices practical. Their mode of action may be via a reflex inhibition of LH, or their physical presence may prevent implantation. Well-designed and unbiased studies have revealed that both the oral contraceptives and the intrauterine devices possess minimal side-effects and that both are effective contraceptive agents.

THE MENSTRUAL CYCLE

The reproductive cycle in the primates is called the menstrual cycle, and in the normal nonpregnant individual the interval which extends from the onset of a period of uterine bleeding to the onset of the next period of bleeding describes its duration.

The mean cycle length is 28 days, but the range is very large, from 20 to 35 days. The menstrual cycle has been divided into three stages, and these divisions are based on the histology of the endometrium.

The first stage, menstruation or *menses,* lasts about 4 to 6 days and is characterized by the occurrence of hemorrhage in the endometrial stroma. The endometrium degenerates and, together with interstitial blood, is sloughed into the uterine lumen and exits through the vagina as menstrual discharge. The second or *proliferative* stage is characterized by a period of repair and proliferation of the endometrium and proceeds under the influence of estrogens during the next 8 to 10 days. The endometrium undergoes marked changes which provide the appropriate environment for the implantation of the fertilized egg during the third or *secretory* stage (about 9 to 10 days' duration). This final stage is dependent upon ovulation and the secretion of progesterone by the corpus luteum. If a fertilized ovum is not available or if implantation does not occur, the corpus luteum and the endometrium degenerate and the cycle is repeated.

Menstruation should not be considered an actively induced process, because it occurs as a result of cessation of stimulation of the endometrium by the ovarian steroid hormones. It is the result of withdrawal of the ovarian steroids, as shown by the following observations. (1) Ovariectomy, if performed during the last two or three weeks of the cycle, precipitates uterine bleeding two to six days postoperatively. (2) The administration of estrogens to ovariectomized individuals results in a proliferative type of endometrium, and the withdrawal of estrogen treatment results in menstruation. (3) The simultaneous administration of estrogen and progesterone to an ovariectomized individual causes the development of a secretory endometrium, and the withdrawal of both steroids results in menstruation within two to three days. (4) The administration of large amounts of estrogen prevents menstruation for long periods of time. Figure 36-7 depicts the changes in the endometrium throughout the menstrual cycle.

Transplantation of the endometrium to the anterior chambers of the eye has enabled investigators to observe directly those factors which are associated with menstruation. The observations have shown that alterations in blood flow through the endometrial vessels are responsible for menstruation. Prior to menstruation (two to six days), the coiled arteries deep in the endometrium offer an increased resistance to blood flow, and because of the inadequate blood flow the endometrium regresses. Immediately prior to menstruation, the coiled arteries constrict further and reduce, to a greater degree, the flow to the endometrium. Following this at varying intervals, the arteries dilate and hemorrhage occurs. The arteries then again constrict, and hemorrhage ceases from the artery involved. It should be remembered that each artery bleeds only once during each cycle; thus each artery dilates, bleeds, then constricts, but not all coiled arteries do so simultaneously, and a small area may be sloughed and repaired before other areas are sloughed. Therefore, as a result of decreasing estrogen and progesterone secretion, the endometrium is deprived of essential materials, and it regresses. The spiral arteries undergo intermittent contraction and cause recurrent ischemia, so that small hemorrhages occur in the tissue; cellular necrosis occurs, and menstruation ensues.

PREGNANCY

Pregnancy may be attained only following a series of events which terminate in the oviducts, where zygote formation takes place as a result of union of an egg and a sperm. Human ova are probably not fertilizable for more than 12 to 16 hours after ovulation, whereas sperm retain their viability in the female reproductive tract for

24 to 36 hours. It has been estimated that in women of proved fertility only one in four conceives when exposed during the ovulation phase. Sperm deposited in the vagina must move or be transported up the reproductive tract to the oviducts. In many species, sperm appear in the oviducts within several minutes after having been deposited in the vagina during midcycle or estrus. The rapid transport of sperm through the uterus and into the oviducts is the result of mechanical propulsion provided by contraction of the uterine musculature. On the basis of work in experimental animals, it has been suggested that the increase in uterine muscle activity is the result of the reflex release of oxytocin from the hypothalamo-neurohypophysial system (HNS).

Union of the gametes occurs in the oviducts, and the resultant zygote, which is in the morula stage, reaches the uterus during the fourth day after ovulation. By the seventh day, the cell mass has become the blastocyst and made contact with the endometrium; and by the tenth day it has adhered to the uterus, penetrated the uterine epithelium, and passed through the epithelium. It soon becomes covered by the uterine epithelium. The functional life of the corpus luteum is prolonged, with a consequent increase in estrogen and progesterone secretion.

The mechanism by which the corpus luteum is caused to persist is unknown. Nevertheless, as a result of the maintained estrogen and progesterone secretion, the endometrium fails to undergo its usual menstrual regression, the menstrual period does not occur, and overt cyclic ovarian and uterine activity are absent throughout the pregnancy. Pregnancy can be established and maintained only if adequate amounts of progesterone are secreted prior to and throughout the course of that pregnancy.

The placenta serves as an organ of exchange between the maternal and fetal organisms. It also serves as an endocrine organ, secreting at least four types of hormones, including a gonadotropin, estrogens, progestins, and relaxin. The gonadotropin, human chorionic gonadotropin (HCG), is the first to appear, and its presence serves as the basis of pregnancy tests. Positive pregnancy tests have been obtained using 4 ml of serum obtained by the fifth day after the missed menstruation (19 days after ovulation). However, HCG can be detected in a 24-hour urine sample about 11 to 12 days after ovulation. The amount of HCG in the serum rises with extreme rapidity, reaches a peak during the sixth or seventh week of pregnancy, and then declines (Fig. 36-8). This hormone has been shown to be luteotropic in primates.

The pregnancy tests on which the presence of HCG is based are numerous. All are dependent upon the ability of HCG to stimulate gonadal function in a test animal. (1) The Aschheim-Zondek test utilizes the ability of the urinary extract of HCG to induce ovarian hyperemia in the immature mouse ovary. (2) In the Friedman test, urine is injected intravenously in a rabbit and the end point of the assay is the production of ovulation. (3) The Hogben test depends upon the ability of HCG to cause ovulation in the South African clawed toad (Xenopus laevis). (4) The Galli Mainini test depends upon the ability of HCG to cause the expulsion of spermatozoa by the male frog.

The endocrine function of the ovary can be replaced by the placenta early in the second month of human pregnancy, as shown by the fact that bilateral ovariectomy may be performed during the second month of gestation, yet pregnancy continues to a successful termination at the end of the fortieth week. The levels of urinary estrogens and pregnanediol increase gradually throughout the pregnancy and reach a maximum a few days prior to parturition (Fig. 36-8).

In addition to estradiol, estrone, estriol, and progesterone, several other steroids have been isolated from the human placenta. Three of the most active are hydrocortisone, cortisone, and aldosterone. Figure 36-8 also depicts the changes in

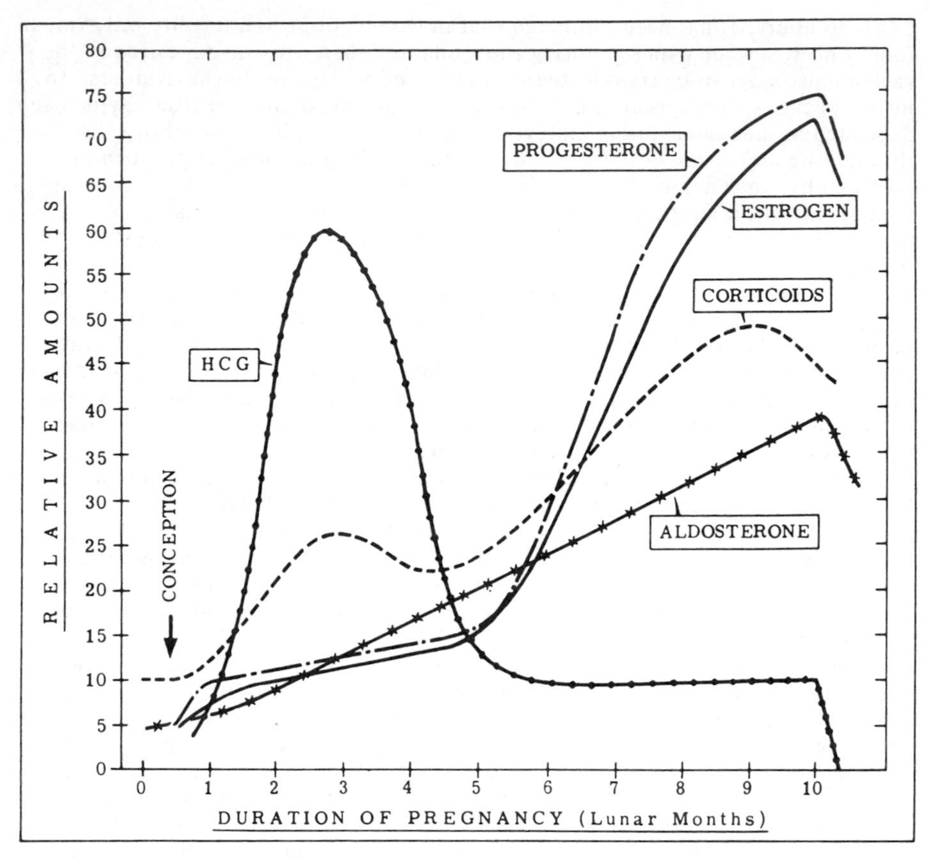

FIGURE 36-8. Excretion of hormones during pregnancy.

excretion of the adrenal steroids throughout pregnancy. The thyroxine-binding globulin also increases during pregnancy, resulting in an increase in the protein-bound iodine (PBI).

The uterus must serve several related functions during pregnancy. Initially, it must be prepared to receive the fertilized ovum and allow implantation. Once implantation has occurred, it must grow and adapt to the growing products of conception; and it must be prepared to assume delivery at term.

The pregnant state imposes widespread changes on the maternal organism. Nearly all of these changes represent adaptive responses which permit the mother to deliver a normal infant without harm to either the mother or child. In order to supply adequate amounts of oxygen and other nutrients to the fetus, the blood flow through the gravid uterus must increase. In an experimental animal such as the rabbit, the uterine blood flow increases from 5 ml per minute prior to gestation to about 30 ml per minute. Studies in the human indicate that uterine blood flow at term in a single pregnancy is about 500 ml per minute, whereas in a twin pregnancy it is over 1000 ml per minute. On the other hand, 24 to 48 hours after delivery the blood flow is decreased to about 205 ml per minute. No data are available on uterine blood flow throughout the reproductive cycle. Vasomotor fibers to the uterus are thoracolumbar in origin; their stimulation results in constriction of the vessels they supply. However, the dominant control over the uterus vessels is exercised by hormones. For instance, the administration of estrogen induces uterine hyperemia prior to any metabolic changes in the uterine tissue. In the face of the increased blood flow to the pregnant uterus, various adaptive changes in the cardiovascular system must occur, increased cardiac output, blood volume, heart rate, and pulse pressure. In any event, it appears that the uterine blood flow is maintained so that oxygen is supplied in excess of the oxygen requirement of the tissue.

PARTURITION

The cause of parturition, the termination of a normal pregnancy, is unknown. The duration of a normal pregnancy in the human is 10 lunar months. Several experimental observations show that progesterone plays a predominant role in the maintenance of pregnancy, and massive doses of progesterone may maintain pregnancy beyond term. Csapo has advanced the concept that a progesterone block of uterine smooth muscle maintains pregnancy, and that withdrawal of the block is responsible for the onset of parturition. This concept is based on the theory that the placenta gives up the bulk of its progesterone by direct diffusion to the myometrium. Under this circumstance, one would expect that the concentration gradient of progesterone would be greatest at implantation sites rather than at interplacental sites. The latter has been shown to be the case in experimental animals. This concept is further strengthened by observations in women who bear twins in a bicornuate uterus (septum divides uterus into two horns). The twins have separate placentas and can be born several weeks apart. Thus, in the same woman, conditions at the same moment in pregnancy can be appropriate for the maintenance as well as the termination of pregnancy, indicating that local effects of progesterone appear to predominate.

It is also well known that the rate of clearance in vivo of oxytocin decreases progressively throughout pregnancy, and that the sensitivity of the myometrium to a constant dose of oxytocin increases up to the thirty-sixth week of pregnancy. There is little doubt that the induction of labor can be accomplished by the administration of oxytocin after progesterone withdrawal and the ascendancy of an estrogen-dominated uterus. However, the manner in which the onset of labor is accomplished is unknown.

The hormone relaxin has been extracted from ovarian, placental, and uterine tissue and is apparently concerned with preparation for parturition. It causes softening of the pelvic ligaments and the cervix in a variety of experimental animals. Its functional role during pregnancy in the human remains to be ascertained. It has been shown to soften the cervix of pregnant women, and it increases the rate of success of oxytocin-induced labor.

LACTATION

Lactation consists of *mammogenesis* (development of the mammary glands), *lactogenesis* (milk secretion), *galactopoiesis* (maintenance of lactation), and *milk ejection.* The most extensive studies have been conducted using dairy animals and laboratory rodents.

Mammary glands are modified sweat glands and are made up of 15 to 25 lobes. The immature gland is comprised of short ducts which radiate from the nipple to each lobe. Each duct has many side branches extending from it to supply the lobules. The estrogens are primarily concerned with growth of the duct system, since, following the administration of estrogen, the duct system becomes extensively arborized. Progesterone acts in conjunction with estrogens to stimulate duct and alveolar growth. Physiological doses of estrogen and progesterone produce full mammary growth in either the castrate male or female. However, the administration of estrogen plus progesterone to the hypophysectomized animal fails to induce mammary growth. Similarly, the administration of the pituitary hormones, FSH and LH, which stimulate estrogen secretion by the ovary, fails to induce mammary development in the hypophysectomized animal. In order to obtain full mammary growth comparable to that observed in late pregnancy in the hypophysectomized castrate, estrogen, progesterone, prolactin, STH, and either ACTH or adrenal corticoids must be administered. Thus, five hormones are required to build up the mammary gland to full functional development. The hormonal environment at the end of pregnancy is such that the mammary gland is morphologically ready to begin lactogenesis (Fig. 36-9).

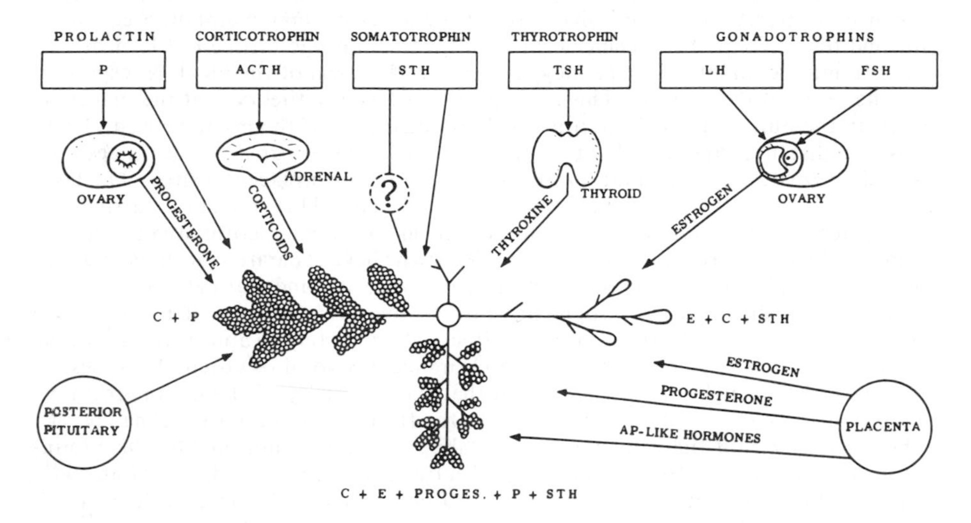

FIGURE 36-9. The action of hormones on mammary growth and lactation. In the diagram of the gland: Upper – rudimentary gland; right – prepuberal to puberal gland; lower – prolactational gland of pregnancy; left – lactating gland. (From W. R. Lyons et al. *Recent Progr. Hormone Res.* 14:246, 1958.)

Some synthesis and storage of milk may begin prior to parturition, but only at or shortly after parturition does a copious secretion and flow of milk occur. The integrity of the hypophysis is necessary for the initiation of milk secretion, because failure of lactation occurs following hypophysectomy, and the administration of extracts of the hypophysis provides a positive lactogenic stimulus. The increased secretion of milk at parturition is probably a result of diminished circulating titers of estrogen and progesterone when the placenta is delivered. It is known that high levels of estrogen tend to inhibit lactation, whereas low levels increase the secretion of prolactin. STH, ACTH, TSH, prolactin, and insulin certainly all participate in milk secretion and its maintenance, although their exact roles in this phenomenon remain obscure. Mechanical factors also play a prominent part in the maintenance of lactation, because as milk production begins, pressure in the glands rises, and if not relieved, the pressure rises to a point where milk secretion is retarded. As a result, the mammary glands involute when nursing terminates (Fig. 36-9).

The involution of the mammary gland which normally follows removal of the young can be delayed by the administration of oxytocin. It has been shown that the young of neurohypophysectomized lactating rats die of starvation, not because of lactational failure, but because of failure to remove milk from the mammary tissues. However, if the mother is administered oxytocin, milk can be obtained by suckling young. The stimulus of suckling, with or without the removal of milk, decreases the amount of prolactin in the pituitary and indicates that such a stimulus maintains prolactin secretion at a high level.

Spinal cord transection between the last thoracic and first lumbar vertebrae leads to denervation of the caudal nipples of the rat but leaves the more cranial nipples intact. If the cranial nipples are covered so as to force suckling from the denervated nipples, the milk cannot be removed from these glands, and the glands involute. However, if suckling is permitted from two intact nipples, full lactation and milk removal occurs from the denervated nipples. Thus, it appears that suckling induces the release of galactopoietic hormones (prolactin, STH?) via a neural pathway. Suckling also stimulates the release of oxytocin from the CNS. Both cases are clear examples of neuroendocrine reflexes.

The latter reflex, the *milk ejection* reflex, is composed, on the afferent side, of sensory stimuli associated with suckling acting on the receptors in the nipple of the mammary gland. These are conveyed centrally where they facilitate the release of oxytocin from the HNS (Chap. 31). Oxytocin forms the efferent limb of the reflex arc and is carried via the blood to the mammary gland, where it causes contraction of the myoepithelial cells which surround the alveoli of the mammary gland. The milk is forced out of the alveoli of the mammary gland into the duct system and cisterns, from which it can be removed by suckling. This reflex may be easily conditioned in either experimental animals or the woman. On the other hand, it may be inhibited either by strong emotions or by the administration of epinephrine. As a result of experiments in vitro, the action of epinephrine has been shown to be due to vasoconstriction. The effect of emotional stress on the reflex may act through epinephrine.

In summary, lactation is the result of the action of many hormones and neural elements. The mammary gland develops through the action of estrogen and progesterone, and full secretory function is attained through the action of prolactin, STH, TSH, ACTH, and probably insulin. Milk secretion is maintained only in the animal with intact innervation of the mammary glands and neurohypophysis, and it appears that active stimulation of the mammary gland through suckling participates in the milk letdown reflex by stimulating oxytocin release.

REFERENCES

Armstrong, D. T. Reproduction. *Ann. Rev. Physiol.* 32:439–470, 1970.

Assali, N. S. (Ed.). *Biology of Gestation,* Vols. I and II. New York: Academic, 1968.

Davidson, J. M., and G. J. Black. Neuroendocrine aspects of male reproduction. *Biol. Reprod.* Suppl. 1, pp. 67–92, 1969.

Diamond, M. (Ed.). *Perspectives in Reproduction and Sexual Behavior.* Bloomington: Indiana University Press, 1968.

Everett, J. W. Neuroendocrine aspects of mammalian reproduction. *Ann. Rev. Physiol.* 31:383–416, 1969.

Frieden, E. H. Sex Hormones and the Metabolism of Amino Acids and Proteins. In G. Litwack and D. Kritchevsky (Eds.), *Actions of Hormones on Molecular Processes.* New York: Wiley, 1964. Pp. 509–559.

Ganong, W. F., and L. Martin. *Frontiers of Neuroendocrinology.* New York: Oxford University Press, 1969.

Gier, H. T., and G. B. Marion. Development of mammalian testes and genital ducts. *Biol. Reprod.* Suppl. 1, pp. 1–23, 1969.

Gorski, J., D. Toft, G. Shyamala, D. Smith, and A. Notides. Hormone receptors: Studies on the interaction of estrogens with the uterus. *Recent Progr. Hormone Res.* 24:45–80, 1968.

Interdisciplinary Conference on Initiation of Labor, Princeton, N.J., 1963. Bethesda, Md.: National Institute of Child Health and Human Development (PHS Publication #1390), 1966.

Lloyd, C. W. (Ed.). *Human Reproduction and Sexual Behavior.* Philadelphia: Lea & Febiger, 1964.

Martin, L., and W. F. Ganong. *Neuroendocrinology,* Vols. I and II. New York: Academic, 1967.

Masters, W. H., and V. E. Johnson. *Human Sexual Response.* Boston: Little, Brown, 1966.

Nalbandov, A. V. *Reproductive Physiology,* 2d ed. San Francisco: Freeman, 1964.

Nalbandov, A. V., and B. Cook. Reproduction. *Ann. Rev. Physiol.* 30:245–278, 1968.

Rock. J., C. Garcia, and M. F. Menkin. A theory of menstruation. *Ann. N.Y. Acad. Sci.* 75:831–839, 1959.

Villee, D. B. Development of endocrine function in the human placenta and fetus. *New Eng. J. Med.* 281:473–484, 533–541, 1969.

Young, W. C. (Ed.). *Sex and Internal Secretions,* 3d ed., Vols. I and II. Baltimore: Williams & Wilkins, 1961.

Index